FREE ELECTRON LASERS 2000

FREE ELECTRON LASERS 2000

Proceedings of 22nd International Free Electron Laser Conference
and 7th FEL Users Workshop,
Free Electron Laser Laboratory, Duke University,
Durham, NC, USA, August 13–18, 2000

Editors

V. N. Litvinenko
*Free Electron Laser Laboratory, Physics Department, Duke University,
Durham, NC, USA*

Y. K. Wu
ALS, Lawrence Berkeley National Laboratory, Berkeley, CA, USA

2001

ELSEVIER

Amsterdam – London – New York – Oxford – Paris – Shannon – Tokyo

ELSEVIER SCIENCE B.V.
Sara Burgerhartstraat 25
P.O. Box 211, 1000 AE Amsterdam, The Netherlands

First edition 2001
ISBN: 0 444 50939 9

Part 1 is reprinted from:
NUCLEAR INSTRUMENTS AND METHODS IN PHYSICS RESEARCH
Section A: Accelerator, Spectrometers, Detectors and Associated Equipment, Vol. 475, Nos. 1–3.

The Manuscript of the Proceedings was received by the Publisher: March 2001

Printed and bound by Antony Rowe Ltd, Eastbourne

Preface

The 22nd International Free Electron Laser Conference and 7th FEL User Workshop were held August 13–18, 2000 at Washington Duke Inn & Golf Club in Durham, North Carolina, USA. The conference and the workshop were hosted by Duke University's Free Electron Laser (FEL) Laboratory. Following tradition, the FEL prize award was announced at the banquet. The year 2000 FEL prize was awarded to three scientists propelling the limits of high power FELs: Steven Benson, Eisuke Minehara and George Neil (see the special tribute).

The conference program was comprised of traditional oral sessions on First Lasing, FEL Theory, Storage Ring FELs, Linac and High Power FELs, Long Wavelength FELs, SASE FELs, Accelerator and FEL Physics and Technology, and New Developments and Proposals. Two sessions on Accelerator and FEL Physics and Technology reflected the emphasis on the high quality of accelerators and components for modern FELs. The breadth of the applications was presented in the workshop oral sessions on Materials Processing, Biomedical and Surgical Applications, Physics and Chemistry as well as on Instrumentation and Methods for FEL Applications. A special oral session was dedicated to FEL Center Status Reports for users to learn more about the opportunities with FELs. As usual, the oral sessions were supplemented by poster sessions with in-depth discussions and communications. The FEL physicists and FEL users had excellent opportunities to interact throughout the duration of the event, culminating in a Joint Session. The year 2000 was very successful being marked by lasing with two SASE and one storage ring short-wavelength FELs, and by the first human surgery with the use of FEL, to mention but a few. The International Program Committee and chairs of the sessions had the challenging and exciting problem of selecting invited and contributed talks for the conference and the workshop from the influx of abstracts mentioning new results and ideas. The success of the conference was determined by these contributions. Scientists from 15 countries gave 70 talks, presented 176 posters and submitted 146 papers, which are published in the present volume of proceedings. Thanks to the generous support of Air Force Office of Scientific Research and Duke University's Vice Provost for Academic Affairs, Vice Provost for Research and Dean of Natural Sciences, we were able to support 38 students and scholars from 11 countries. We deeply regret that five of these scientists were unable to attend on account of visa problems, in spite of the best efforts put forward by the Duke University International Office.

In addition to the intense scientific and technical program, about 200 participants had the opportunity to tour Duke Free Electron Laser Laboratory (DFELL) housing two FELs and the user facilities located in the new two-story Keck Life Science Building. Visitors saw the Mark-III infrared FEL, the 1 GeV Duke storage ring, and the OK-4 FEL generating a wide range of spontaneous radiation, coherent light from the visible to below 200 nm and intense semi-monochromatic γ-ray beams via Compton back-scattering. DFELL users showed their IR and UV FEL based research stations for medical, biological and material sciences. The nuclear physics vault for experiments with the beams from OK-4's High Intensity γ-ray Source

(HIγS) and gargantuan skeleton of the pion detector in its center were some of the tour's attractions.

We would like to thank the Local Organizing committee and physics graduate students for their hard and inspiring work, and most especially the conference secretary, Denise Gamble, and the chair of the local organizing committee/computer guru, Steve Hartman, for their highly professional and dedicated contributions to the success of the conference. Our special thanks belong to the chair of the workshop, Michelle Shin, who did an enormous job, to Donna Gilchrist, whose help and experience made the event run smoothly, and to the DEFLL Director, Glenn Edwards, for his continuous support and efforts.

We express our gratitude to the Duke Conference Services, especially to Steve Burrell, for providing trouble-free registration using modern technology. We were happy with the choice of Washington Duke Inn & Golf Club for the conference site, whose professional staff, led by Jim Redmond and Erin Riley, offered a sense of authentic Southern hospitality.

We hope this volume conveys to the reader the level of scientific achievements by FEL community and the level of excitement with FEL breaking into new wavelength regions, new levels of peak and average power, and new applications. We are most thankful to the authors, to the reviewers and to the editing team of Duke University physics students, especially to Timothy Burt and Jacob Foster, for making the reading of these proceedings a pleasure.

Vladimir N. Litvinenko
Ying K. Wu
(*Guest Editors*)

Conference Chairman:

V.N. Litvinenko (Duke University)

Applications Workshop Chairwoman:

M. Shinn (Jefferson Lab)

Program Committee Chairman:

Y.K. Wu (Lawrence Berkeley National Lab)

Local Organizing Committee:

S. Hartman (Chairman)
W.C. Fowler
J. Patterson

G. Edwards
D. Gamble (Secretary)
I. Pinayev

J. Faircloth
R. Nemanich
G. Swift

International Executive Committee:

I. Ben-Zvi (BNL)
A. Gover (Tel Aviv Univ.)
K.J. Kim (ANL)
V.N. Litvinenko (Duke Univ.)
K. Mima (Osaka Univ.)
J.M. Ortega (LURE)
A. Renieri (ENEA)
J. Rossbach (DESY)
P. Sprangle (Naval Res. Lab)
N.A. Vinokurov (BINP)

W.B. Colson (Naval Postgraduate School)
H. Hama (UVSOR)
Y. Li (IHEP)
J.M.J. Madey (Hawaii Univ.)
G.R. Neil (Jefferson Lab)
M.W. Poole (Daresbury Lab)
C.W. Roberson (Office of Naval Research)
T. Smith (Stanford Univ.)
M.J. van der Wiel (Eindhoven Univ.)

International Program Committee:

S. Benson (Jefferson Lab)
G. Dattoli (ENEA)
H. Hama (UVSOR)
A.N. Lebedev (FIAN)
S. Milton (ANL)
S. Okuda (Osaka)
Liu Shenggang (UESTC)
R.P. Walker (TRIESTE)
M. Xie (LBL)

M. Cornacchia (SLAC)
H.P. Freund (Sci. A.I. Corp)
E. Jerby (TAU)
B.C. Lee (KAERI)
E.J. Minehara (JAERI)
J. Pflueger (DESY)
T. Tomimasu (FELI)
J.E. Walsh (Dartmouth Univ.)
L. Yu (BNL)

M.E. Couprie (LURE)
N.S. Ginzburg (IAP)
S. Krinski (BNL)
A. Lumpkin (ANL)
D.C. Nguyen (LANL)
G. Ramian (UCSB)
P.J.M. Van der Slot (NCLR)
K. Wille (Dortmund Univ.)

Applications Workshop Committee:

W. Gabella (Vanderbilt Univ.)
L. van der Meer (FELIX)

G. Edwards (Duke Univ.)

E.D. Jansen (Vanderbilt Univ.)

PII: S0168-9002(01)01744-2

Free Electron Laser Conference 2000

13–18 August 2000
Duke FEL Laboratory
Duke University
Durham, North Carolina

Sponsors

2001

NORTH-HOLLAND

PII: S0168-9002(01)01745-4

FEL Prize

Benson, Neil, and Minehara share the 2000 International Free-Electron Laser Prize

The International Free-Electron Laser Prize is awarded each year to recognize individuals for pioneering contributions to the field of free-electron lasers. At the 22nd International Free-Electron Laser Conference in Durham, NC, the prize for 2000 was awarded jointly to Steven V. Benson and George R. Neil, of the Thomas Jefferson National Accelerator Facility, and Eisuke J. Minehara, of the Japan Atomic Energy Research Institute, in recognition of their achievement of high-power operation of free-electron lasers.

Since its conception three decades ago, the free-electron laser has held out the promise of high-power operation owing to its unique ability to discard the waste heat in the gain medium at nearly the speed of light. However, this promise has proved elusive due to the difficulty of producing high-brightness electron beams, and the limitations of high-power laser optics, not to mention a host of surprises lurking in the physics of free-electron lasers themselves! It is therefore a milestone for the FEL community that these three distinguished scientists and the groups that they have lead have finally placed free-electron lasers among the most powerful lasers in the world.

Both groups have exploited superconducting accelerator technology, which lends itself to long-macropulse and continuous operation. Dr. Minehara and his group at JAERI have focused their attention on achieving high power in long macropulses, and their best achievement has been an average power of 2.3 kW in 1-ms macropulses, repeated at 10 Hz. Dr. Benson and Dr. Neil have focused their attention on continuous operation at high average power, and their best achievement has been 1.7 kW cw. These results eclipse the previous record average power of 11 W, and they are already working on increasing the power by another order of magnitude!

The scientists receiving the award are as distinguished as the results they have achieved. Dr. Minehara has long been active in the development of superconducting linacs, and has been part of the JAERI FEL

PII: S0168-9002(01)01746-6

project for more than a decade. Dr. Neil headed the effort at TRW that demonstrated the first tapered wiggler before joining TJNAF. Dr. Benson began his FEL career on the Mark III at Stanford, and then spent several years at Duke before joining TJNAF. They have all been leaders in the FEL community, and their stature among their peers, as much as the worldwide respect for their outstanding achievements, is reflected in the overwhelming number of nominations they received for these awards. It is with great satisfaction that we award them the Free-Electron Prize for 2000.

Charles Brau; Chair
Giuseppe Dattoli
Kwang-Je Kim
Nikolai Vinokurov

Tribute to John E. Walsh

John Walsh, Professor of Physics at Dartmouth College, died quite unexpectedly on December 5th, 2000, of complications following surgery to repair a broken tibia. One of the most respected and beloved members of the Free-Electron Laser Community, John had not only a powerful intellect but a warmth and enthusiasm that infected everyone that he knew and met. We will miss him dearly.

John was born in New York in 1939, and grew up in Montauk, at the very tip of Long Island. He got his BS in electrical engineering from Nova Scotia Technical College in 1962 and joined the US Army Signal Research Laboratory at Fort Monmouth. His experience there was valuable for the rest of his career, but Fate had other things planned for him. His old rag-topped convertible, used to commute between Montauk and Fort Monmouth, broke down in front of the Columbia University Applied Physics Department. He went in to seek help and emerged several years later with the MS (1965) and Sc.D. (1968) degrees. His thesis, under the supervision of Perry Schlesinger and Tom Marshall, was in plasma physics. Although his degree was in electrical engineering, John immediately joined Dartmouth as an Assistant Professor in the Department of Physics.

At Dartmouth John continued his research in plasma physics, and became an outstanding teacher and mentor. But the focus of his research shifted following a sabbatical at Columbia in

0168-9002/01/$ - see front matter © 2001 Published by Elsevier Science B.V.
PII: S 0 1 6 8 - 9 0 0 2 (0 1) 0 1 9 7 9 - 9

1975–6. Working with Marshall, again, he investigated the stimulated Cherenkov effect. When an intense electron beam travels along a loaded waveguide, "slow waves" traveling at the velocity of the electrons grow exponentially. This results in MW-level radiation in the microwave and millimeter regions. On his return to Dartmouth John expanded this research to include a number of free-electron laser devices using electron-beam interactions with both dielectrics and gratings. His work was always characterized by originality and simplicity. He rarely traveled where others were going, but used his imagination to create ideas that fitted his small-scale university environment in a field dominated by huge national laboratories. An example of John's imaginative approach to research was his recent development of a far-infrared source based on the Smith–Purcell effect. The interaction between an electron beam and a grating causes an instability in the beam much like the stimulated Cherenkov effect, and produces useful radiation in the 200–1000 µm region, where there are few alternative sources. For his experiments he used a cast-off scanning electron microscope to provide an electron beam that could be passed very close to the grating, and succeeded in demonstrating operation above the instability threshold. For higher energy experiments John required electron beams not available at Dartmouth. His enthusiasm and expertise made him welcome everywhere, and he led successful experiments at the ENEA Center in Frascati, at Oxford University, and at Brookhaven on Cherenkov and Smith–Purcell devices. In recognition of the importance of his research he was awarded the International Free-Electron Laser Prize in 1998.

In addition to his research, John always found time for his students, even in those periods when he was head of the Physics Department or associate dean of the science faculty. He was always popular with his students, and mentored 40 Ph.D.s during his career. He was very close to his students, and continued to maintain contact with them over the years. He was also an effective administrator, undoubtedly due in part to his dislike of unnecessary meetings and paperwork that kept him from his beloved research. His legacy to the College as associate dean includes a splendid extension to the Wilder Physics Laboratory.

John was a big man in every sense of the word. Besides his enormous intellect, he had a warm heart, a great sense of humor and a fund of good stories. In his youth he worked on fishing boats out of Montauk, and it was probably there that he acquired the happy knack of treating all people equally. Pomposity was foreign to him. A considerable athlete in his younger days, he remained an excellent skier until his untimely death, and was a trustee of the Northeast Ski Slopes Association. His close-knit family, the Dartmouth community, and his countless friends and colleagues in many countries have suffered a grievous loss.

Charles A. Brau
J. Hayden Brownell
Maurice F. Kimmitt

FEL Prize Winners at FEL'2000 conference. Left to right: Eisuke Minehara, George R. Neil, Nikolay A. Vinokurov, Todd I. Smith, John E. Walsh, Kwang-Je Kim, William B. Colson, Stephen V. Benson, Charles A. Brau and Claudio Pellegrini.

PII: S0168-9002(01)01978-7

ELSEVIER

Nuclear Instruments and Methods in Physics Research A 475 (2001) xxi–xxix

NUCLEAR
INSTRUMENTS
& METHODS
IN PHYSICS
RESEARCH
Section A

www.elsevier.com/locate/nima

Contents

FEL 2000
Proceedings of the 22nd Free Electron Laser Conference
and 7th FEL Users Workshop
Duke University, Free Electron Laser Laboratory
Durham, NC, USA, August 13–28, 2000

Editors: V.N. Litvinenko, Y.K. Wu

Part I. Reviewed papers

Section I. FEL Prize Lecture and First Lasing

Section VII. New Developments and Proposals

Section VIII. Accelerator and FEL Physics and Technology

Section IX. Applications of FELs

Part II (Extended Abstracts: The text of these papers is included only in the book edition of the Proceedings, ISBN 0-444-50939-9)

ELSEVIER

Nuclear Instruments and Methods in Physics Research A 475 (2001) 1–12

NUCLEAR
INSTRUMENTS
& METHODS
IN PHYSICS
RESEARCH
Section A

www.elsevier.com/locate/nima

Design considerations for a SASE X-ray FEL ☆

Claudio Pellegrini*

Department of Physics, University of California at Los Angeles, 405 Hilgard Avenue, Los Angeles, CA 90095, USA

Abstract

The well developed theory of short wavelength SASE-FELs is now being used to design two X-ray lasers, LCLS and Tesla-FEL. However, the physics and technology of these projects present some unique challenges, related to the very high peak current of the electron beam, the very long undulator needed to reach saturation, and the importance of preserving the beam phase-space density even in the presence of large wake-field effects. In the first part of this paper, we review the basic elements of the theory, the scaling laws for an X-ray SASE-FEL, and the status of the experimental verification of the theory. We then discuss some of the most important issues for the design of these systems, including wake-field effects in the undulator, and the choice of undulator type and beam parameters. © 2001 Elsevier Science B.V. All rights reserved.

PACS: 41.60.Cr; 52.59.Rz; 42.55.Vc

Keywords: Free-electron laser; X-ray laser; Radiation from relativistic electrons

1. Introduction

Undulator radiation is particularly useful because of its small line width and high brightness, is being used in many, if not all, synchrotron radiation sources built around the world, and provides at present the brightest source of X-rays. For long undulators, the free-electron laser (FEL) collective instability gives the possibility of much larger X-ray intensity and brightness. The instability produces an exponential growth of the radiation intensity, together with a modulation of the electron density at the radiation wavelength. The radiation field necessary to start this instability is the spontaneous radiation field, or a combination of the spontaneous radiation field and an external field. In the first case, we call the FEL a Self Amplified Spontaneous Emission (SASE) FEL. If the external field is dominant we speak of an FEL amplifier.

The existence of an exponentially growing solution for the FEL has been studied in Refs. [1–13]. This work led to the first proposals [14–17] to operate a SASE-FEL at short wavelength, without using an optical cavity, difficult to build in the Soft X-ray or X-ray spectral region.

The analysis of a SASE-FEL in the one-dimensional (1-D) case has led to a simple theory of the free-electron laser collective instability, describing all of the free-electron laser physics with one single quantity, the FEL parameter ρ [6],

☆Work supported by grant DOE-DE-FG03-92ER409693.

*Corresponding author. Tel.: 310-206-1677; fax: 310-206-5251.

E-mail address: pellegrini@physics.ucla.edu (C. Pellegrini).

a function of the electron beam density and energy, and of the undulator period and magnetic field. The extension of the FEL theory to three dimension, including diffraction effects [18–22] has been another important step toward a full understanding of the physics of this system. From the collective instability theory, we can obtain a scaling law [23] of a SASE-FEL with wavelength, showing a weak dependence of the gain on wavelength, and pointing to the possibility of using this system to reach the X-ray spectral region. This analysis shows that to reach short wavelengths one needs a large electron beam six-dimensional phase-space density, a condition until recently difficult to satisfy.

The development of radio frequency photo-cathode electron guns [24], and the emittance compensation method [25,26] has changed this situation. At the same time, the work on linear colliders has opened the possibility to accelerate and time compress electron beams without spoiling their brightness [27–30]. At a Workshop on IV Generation Light Sources held at SSRL in 1992 it was shown [31] that these new developments make possible to build an X-ray SASE-FEL. This work led to further studies [32,33], and to two major proposals, one at SLAC [34], the other at DESY [35], for X-ray SASE-FELs in the 1 Å region, with peak power of the order of tens of GW, pulse length of about 100 fs or shorter, full transverse coherence, peak brightness about ten orders of magnitude larger than that of III generation synchrotron radiation sources.

While the theory of the SASE-FEL has been developed starting from the 1980s, a comparison with experimental data has become possible only during the last few years, initially in the infrared to visible region of the spectrum [36–46]. These experimental data agree with our theoretical model, in particular the predicted exponential growth, the dependence of the gain length on electron beam parameters, and the intensity fluctuations. A recent experiment at DESY [47] has demonstrated gain of about 1000 at the shortest wavelength ever reached by an FEL, 80 nm. These results give us confidence that we can use the present theory to design an X-ray SASE-FEL.

2. Free-electron laser physics

The physical process on which a FEL is based is the emission of radiation from one relativistic electron propagating through an undulator. We consider for simplicity only the case of a helical undulator, and refer the reader to other books or papers, for instance Refs. [48,49], for a more complete discussion.

Let us consider the emission of coherent radiation from N_e electrons, that is the radiation at the wavelength

$$\lambda = \frac{\lambda_u}{2\gamma^2}(1 + a_u^2) \tag{1}$$

within the coherent solid angle

$$\pi \theta_c^2 = \frac{2\pi\lambda}{\lambda_u N_u} \tag{2}$$

and line width

$$\frac{\Delta\omega}{\omega} = \frac{1}{N_u}. \tag{3}$$

In Eq. (1) $E_{beam} = \gamma mc^2$ is the beam energy, and $a_u = eB_u\lambda_u/2\pi mc^2$ is the undulator parameter.

When there is no correlation between the field generated by each electron, as in the case of spontaneous radiation, the total number of coherent photons emitted is $N_{ph} = \pi\alpha N_e a_u^2/(1 + a_u^2)$, where α is the fine structure constant. Hence, the number of coherent photons is about 1% of the number of electrons. If all electrons were within a radiation wavelength, the number of photons would increase by a factor N_e. Even when this is not the case, and the electron distribution on the scale of λ is initially random, the number of photons per electron can be increased by the FEL collective in-stability [6].

The instability produces an exponential growth of the field intensity and of the bunching parameter

$$B = \frac{1}{N_e}\sum_{k=1}^{N_e}\exp(2\pi i z_k/\lambda)$$

where z_k is the longitudinal position of electron k. The growth saturates when the bunching parameter is of the order of one. For a long undulator the intensity is approximately given by

$$I \approx \frac{I_0}{9} \exp(z/L_G) \qquad (4)$$

where L_G is the exponential growth rate, called the gain length, and I_0 is the spontaneous co-herent undulator radiation intensity for an undulator with a length L_G, and is proportional to the square of the initial value of the bunching factor, $|B_0|^2$.

The instability growth rate, or gain length, is given, in a simple 1-D model by

$$L_G \approx \frac{\lambda_u}{4\sqrt{3}\pi\rho} \qquad (5)$$

where ρ is the free-electron laser parameter [6],

$$\rho = \left(\frac{a_u}{4\gamma} \frac{\Omega_p}{\omega_u}\right)^{2/3} \qquad (6)$$

$\omega_u = 2\pi c/\lambda_u$ is the frequency associated to the undulator periodicity, and $\Omega_p = (4\pi_e c^2 n_e/\gamma)^{1/2}$, is the beam plasma frequency, n_e is the electron density, and r_e is the classical electron radius.

A similar exponential growth, with a different coefficient, occurs if there is an initial input field, and no noise in the beam (uniform beam, $B_0 = 0$), i.e. amplified stimulated emission. In the SASE case, saturation occurs after about 20 gain lengths, and the radiated energy at saturation is about $\rho N_e E_{beam}$. The number of photons per electron at saturation is then $N_{sat} = \rho E_{beam}/E_{ph}$. In a case of interest to us, an X-ray FEL with $E_{ph} \approx 10^4$ eV, $E \approx 15$ GeV, $\rho \approx 10^{-3}$, we obtain $N_{ph} \approx 10^3$, i.e. an increase of almost 5 orders of magnitude in the number of photons produced per electron.

The instability can develop only if the undulator length is larger than the gain length, and some other conditions are satisfied:

(a) Beam emittance smaller than the wavelength:

$$\varepsilon < \frac{\lambda}{4\pi}. \qquad (7)$$

(b) Beam energy spread smaller than the free-electron laser parameter:

$$\sigma_E < \rho. \qquad (8)$$

(c) Gain length shorter than the radiation Raleigh range:

$$L_G < L_R \qquad (9)$$

where the Raleigh range is defined in terms of the radiation beam radius, ω_0 by $\pi\omega_0^2 = \lambda L_R$.

Condition (a) says that for the instability to occur, the electron beam must match the transverse phase-space characteristics of the radiation. Condition (b) limits the beam energy spread. Condition (c) requires that more radiation is produced by the beam than what is lost through diffraction.

Conditions (a) and (c) depend on the beam radius and the radiation wavelength, and are not independent. If they are satisfied, we can use with good approximation the 1-D model. If they are not satisfied and the gain length deviates from the one-dimensional value (5)—as in the LCLS case where the emittance is about 3 times larger than $\lambda/4\pi$—it is convenient to introduce an effective FEL parameter, defined as

$$\rho_{eff} = \frac{\lambda_u}{4\sqrt{3}\pi L_{G3D}} \qquad (10)$$

where L_{G3D} is the three-dimensional gain length obtained from numerical simulations, including the effects of diffraction, energy spread, and emittance. This quantity is a measure of the three-dimensional effects present in the FEL, and can be used to obtain more realistic information on the system.

2.1. Scaling laws

Analyzing Eqs. (6)–(8), one obtains the scaling law for a SASE-FEL at a given wave-length. We assume that the beam is focused by the undulator and an additional focusing structure, to provide a focusing function β_F of the order of the gain length. We use the emittance $\varepsilon = \sigma_T^2/\beta_F$ and the longitudinal brightness $B_L = ecN_e/2\pi\sigma_L\gamma\sigma_\gamma$, where σ_L and $\gamma\sigma_\gamma$ are the bunch length and the bunch absolute energy spread, to describe the bunch density and energy spread. The result is that the FEL ρ parameter scales like the beam longitudinal brightness [23]

$$\rho \sim \sqrt{2\pi}\, \frac{a}{b} \frac{a_u^2}{1+a_u^2} \frac{B_L}{I_A} \qquad (11)$$

where we have assumed $\sigma_E = a\rho$, $\varepsilon = b\lambda/4\pi$, and $I_A = ec/r_e$ is the Alfven current. In the LCLS case, we have $a \simeq 1/4$, $b \simeq 4$, $K = 3.7$ requiring $B_L > 100$ A for ρ to be of the order of 0.001. Since this discussion does not consider explicitly gain losses due to diffraction, undulator errors and beam misalignment, we need in practice a larger value of B_L.

To obtain an emittance which satisfies condition (7) at about 0.1 nm, using a photo-cathode gun and a linac, we need a large beam energy, of the order of several GeVs, to reduce the emittance by adiabatic damping. The beam longitudinal brightness is determined by the electron source. Wakefields in the linac can, however, reduce it considerably. For LCLS the photo-cathode gun gives a slice emittance $\varepsilon = 6 \times 10^{-8}$ m rad, slice longitudinal brightness $B_L = 8000$ A at 10 MeV. Acceleration and compression in the SLAC linac then gives $\varepsilon \simeq 4 \times 10^{-11}$ m rad, $B_L \simeq 1500$ A at 15 GeV, good enough to produce lasing, even considering the gain losses due to diffraction, imperfections and misalignment.

2.2. Slippage, fluctuations and time structure

When propagating in vacuum, the radiation field is faster than the electron beam, and it moves forward, "slips", by one wavelength λ for each undulator period. The slippage in one gain length defines the "cooperation length" [55],

$$L_c = \frac{\lambda}{\lambda_u} L_G. \qquad (12)$$

For the SASE case the radiation field is proportional to $I(\omega)$, the Fourier component at $\omega = 2\pi c/\lambda$ of the initial bunching factor B_0, and the intensity to $|I(\omega)|^2$. If the bunch length, L_B is such that $L_B \gg \lambda$, and the beam is generated from a thermionic cathode or photo-cathode, the initial bunching and its Fourier component $I(\omega)$ are random quantities. The initial value of B_0 is different for each beam section of length λ, and has a random distribution. The average values are

$\langle I(\omega) \rangle \sim \langle B_0 \rangle = 0$, and $\langle |I(\omega)|^2 \rangle \sim \langle |B_0|^2 \rangle$ $\sim N_e$.

As the beam and the radiation propagate through the undulator, the FEL interaction introduces a correlation on the scale length of L_c, producing spikes in the radiation pulse, with a length of the order of L_c, and a random intensity distribution. The number of spikes is [55,56] $M = L_B/4\pi L_c$. The total intensity distribution is a Gamma distribution function

$$P(I) = M^M \frac{I^{M-1}}{\langle I \rangle^M \Gamma(M)} \exp(-MI/\langle I \rangle) \qquad (13)$$

where $\langle I \rangle$ is the average intensity. The standard deviation of this distribution is $1/\sqrt{M}$. The line width is approximately the same as for the spontaneous radiation, $\Delta\omega/\omega \approx 1/N_u$.

3. Experimental results on SASE-FELs

A SASE-FEL is characterized by L_G and the intensity fluctuations, the distribution of $|B_0|^2$. Very large gain in the SASE mode has so far been observed in the centimeter [36–38] to millimeter wavelength. Gain between about 1000% and 100% has been observed at Orsay [39] and UCLA [40] in the infrared, and at Brookhaven [43] in the visible. Larger gain in the infrared has also been observed at Los Alamos [41], and gain as large as 3×10^5 at 12 μm has been measured by a UCLA-LANL-RRIK collaboration [42]. The intensity distribution function has been previously measured for spontaneous undulator radiation [57], with no amplification, and long bunches, and more recently for amplified radiation, and a short bunch length [40,42].

A BNL group [44], has demonstrated high gain harmonic generation seeding the FEL with a 10.6 μm external laser and producing a 5.3 μm FEL output, with intensity 2×10^7 larger than spontaneous radiation. The VISA group is commissioning a 0.8–0.6 μm experiment, using a 4 m long undulator with distributed strong focusing quadrupoles. Initial results have shown a gain of about 100 [45]. An experiment at Argonne uses the APS injector, with an energy of 220–444 MeV, wavelength 500–20 nm, and an 18 m long undulator. SASE amplification as a function of undulator length has been demonstrated recently at 530 nm [46]. A DESY group is using the TESLA Test Facility superconducting linac. In Phase 1 the electron beam energy can reach up to 390 MeV, and a wavelength of 42 nm. Phase 2 will reach 1000 MeV, and 6 nm, with an undulator length of 30 m. Initial results have shown a gain of about 10^3 over a wavelength range from 180 to 80 nm, the shortest wavelength ever obtained in an FEL [47]. Similar experimental programs on SASE-FELs are being prepared also in Japan, at Spring 8 and other laboratories, and in China.

The main results of the UCLA-LANL-RRIK-SSRL experiment are shown in Figs. 1 and 2. Fig. 1 shows an increase in output intensity by more than 10^4, when changing the electron charge by a factor of seven. The bunch radius, energy spread, and length change with the charge, making impossible to have a simple analytical model to evaluate the intensity. The experimental data and the theory have been compared using the simulation code Ginger [58], and the measured values of all bunch parameters. The results are plotted in Fig. 1, and, within experimental errors, agree with the data. The intensity measured at a charge of 2.2 nC corresponds to a gain of 3×10^5, the largest measured until now in the infrared. The measured intensity fluctuations, shown in Fig. 2 are well

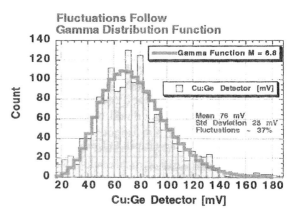

Fig. 2. Intensity distribution over many events for the same experiment. The experimental data are fitted with a Gamma function distribution [42].

described by a Gamma function with the M parameter evaluated from the experimental data, and is in agreement with the theory.

4. LCLS: an X-ray SASE-FEL

The first proposal for an X-ray SASE-FEL was made in 1992 [31], it was then developed by a study group until 1996 and by a design group that has prepared the LCLS design report [34]. The LCLS parameters are given in Table 1. The average brightness, $\langle B \rangle$, and peak brightness B_p, are measured in photons/s/mm^2/mrad^2/0.1% bandwidth.

The LCLS experimental setup is shown in Fig. 3. The electron beam is produced in a photo-cathode gun, developed by a BNL-SLAC-UCLA collaboration [50], and producing a bunch with a charge of 1 nC/bunch, normalized emittance, rms, 1 mm mrad; pulse length, rms, 3.3 ps [51]. The beam is then accelerated to 14.3 GeV and compressed to a peak current of 3400 A in the SLAC linac. During acceleration and compression the transverse and longitudinal phase-space densities are increased by space charge, longitudinal and transverse wake-fields, RF-curvature, coherent synchrotron radiation effects. The acceleration and compression system has been designed to minimize all these effects simultaneously, and it

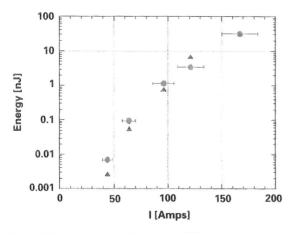

Fig. 1. Measured values of the mean FEL intensity vs beam current, compared with a Ginger simulation for the UCLA-LANL-RRIK-SSRL 12 μm SASE-FEL [42].

limits the transverse emittance dilution to about 10% or less.

The planar hybrid LCLS undulator has vanadium permendur poles, Nd–FeB magnets, and $K = 3.7$ [52]. It is built in sections about 3 m

Table 1
LCLS electron beam, undulator, and FEL parameters

LCLS Electron beam parameters	
Electron energy, GeV	14.3
Peak current, kA	3.4
Normalized emittance, mm mrad	1.5
Energy spread, %, at undulator entrance	0.006
Bunch length, fs	67
LCLS undulator parameters	
Undulator period, cm	3
Undulator length, m	100
Undulator field, T	1.32
Undulator K	3.7
Undulator gap, mm	6
LCLS FEL parameters	
Radiation wavelength, nm	0.15
FEL parameter, ρ	5×10^{-4}
Field gain length, m	11.7
Effective FEL parameter, ρ_{eff}	2.3×10^{-4}
Pulses/s	120
Peak coherent power, GW	9
Peak brightness	10^{33}
Average brightness	4×10^{22}
Cooperation length, nm	51
Intensity fluctuation, %	8
Linewidth	2×10^{-4}
Total synchrotron radiation energy loss, GW	90
Energy spread due to synchrotron radiation emission	2×10^{-4}

long, separated by 23.5 cm straight sections [53]. Since the natural undulator focusing is weak at the LCLS energy, additional focusing is provided by permanent magnet quadrupoles located in the straight sections. Optimum gain is obtained for a horizontal and vertical beta function of 18 m, giving a transverse beam radius of 30 μm, radiation Raleigh range of 20 m, twice the field gain length, making diffraction effects small. The FEL gain is sensitive to errors in the undulator magnetic field, and to deviation in the beam trajectory. Simulations of these effects, including beam position monitors and steering magnets along the undulator to correct the trajectory, show that the field error tolerance is 0.1%, and the beam trajectory error tolerance is about 2 μm [54].

LCLS generates coherent radiation at $\lambda \simeq$ 1.5 nm and its harmonics [59]. It also generates incoherent radiation, which, at 14.3 GeV, has a spectrum extending to about 500 keV, and a peak power density on axis of 10^{13} W/cm^2. The power density of the coherent first harmonic is about 2×10^{14} W/cm^2, and the peak electric field is about 4×10^{10} V/m. Filtering and focusing the radiation and transporting it to the experimental areas is a challenge. A normal incidence mirror at 100 m would see an energy flux of about 1 J/cm^2, about 1 eV/atom, large enough to damage exposed materials. The LCLS large power density will push the optical elements and instrumentation into a new strong field regime, but offers also new opportunities for scientific research.

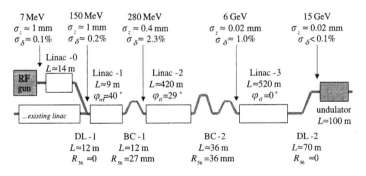

Fig. 3. LCLS experimental setup.

5. Effects of undulator wake-fields and spontaneous radiation

The emission of spontaneous radiation by the electrons in the undulator has two main effects, a decrease of the electron energy, W_{eR}, and an increase of energy spread, $\sigma_{\lambda,R}$. Both effects can reduce the gain if the conditions $W_{eR}/E \ll \rho_{eff}$, $\sigma_{\gamma,R} \ll \rho_{eff}$, are not satisfied. The two quantities W_{eR}, $\sigma_{\gamma,R}$, have been evaluated in Ref. [60]. For the LCLS case we have $W_e/E_{beam} \simeq 1.8 \times 10^{-3} > \rho_{eff}$, $\sigma_{E,R} \simeq 1.5 \times 10^{-4} \simeq \rho_{eff}$, and both effects have to be considered, even if the effect on the gain is not large. The average energy loss can be compensated by tapering the wiggler. The energy spread could be reduced using a shorter undulator.

For an X-ray FEL, with a large peak current, and a long undulator, the wake-fields in the undulator vacuum pipe can have an important effect on the lasing process, and can reduce the output power and change the temporal structure of the X-ray pulse. To evaluate these effects, we use a model which considers the effects of the vacuum pipe resistivity and roughness. The resistive longitudinal wake-field is [61]

$$W_z(t) = -\frac{4ce^2 Z_0}{\pi R^2}\left[\frac{1}{3}e^{t/\tau}\cos(\sqrt{3}t/\tau)\right.$$
$$\left.-\frac{\sqrt{2}}{\pi}\int_0^\infty \frac{x^2}{x^6+8}e^{x^2 t/\tau}\,\mathrm{d}x\right] \quad (14)$$

where t measures the longitudinal position of the test particle respect to the particle generating the field, Z_0 is the vacuum impedance, $\tau = (2R^2/Z_0\sigma)^{1/3}/c$, σ is the conductivity of the material, and R the pipe radius.

The effect of the pipe roughness has been evaluated by several authors. The first models of roughness impedance, based on a random distribution of surface bumps, were developed by Bane, Ng and Chao [62–64], and confirmed by Stupakov [66]. They give an inductive impedance proportional to $1/R$, depending on the ratio of bumps height to length. If this ratio is about one, and the field wavelength is larger than the bump height and width, then the rough surface can support the propagation of a wave synchronous

with the beam. In this case the wake-field is rather strong, and the tolerance for LCLS is a bump height of about 40 nm. For a bump length much larger than the height, a different model, due to Stupakov, applies and the effect is much weaker. Electron microscope observations of the surface of a metal similar to that a vacuum pipe, reported in this paper [65], support this case.

In another model [67,68] the roughness is considered equivalent to a thin dielectric layer on the surface of the pipe, and the pipe can support a wave synchronous with the beam, giving a wake-field

$$W_z(t) = -\frac{ce^2 Z_0}{\pi R^2}\cos(k_0 t) \quad (15)$$

where δ is the thickness of the layer, Z_0 the vacuum impedance, $k_0 = \sqrt{2\varepsilon/(R\delta(\varepsilon-1))}$, and it is assumed $\varepsilon \sim 2$.

To have no gain reduction from the wake-fields, we must satisfy the condition that the variation in energy that they induce be small compared to the gain bandwidth, $(\Delta E/E)_{wake} < \rho_{eff}$. In the LCLS case this gives the condition $W_z < 30$ KV/m.

6. Options for the choice of undulator and beam characteristics

The LCLS design shows the feasibility of an X-ray FEL. It is, however, possible to optimize the system by reducing the undulator saturation length; reducing in the ratio of total spontaneous synchrotron radiation to amplified coherent radiation; choosing electron beam parameters to reduce wake-field effects; controlling the X-ray pulse output power, pulse length, line-width.

The undulator saturation length is controlled by the FEL parameter ρ (6) (5), by the ratio ε/λ (7), and by the electron energy spread (8). A reduction of the beam charge and emittance, keeping their ratio constant, leaves the 1-D gain length unchanged, and reduces the ratio ε/λ. For systems, such as LCLS, where this ratio is larger than 1, this reduces the 3-D gain length. Reducing the charge can also reduce the FEL intensity, giving a way to control the output power [69], and reduces wake-field effects in the linac [70] and undulator.

Table 2
Parameters for helical undulator and low charge cases

	LCLS	A	B	C	D
Beam energy, GeV	14.3	14.7	14.7	12	14.7
Bunch charge, nC	1.0	1.0	1.0	1.0	0.2
Normalized emittance, mm mrad	1.1	1.1	1.1	1.1	0.3
Peak current, kA	3.4	3.4	3.4	3.4	1.17
Energy spread, rms, %	0.006	0.006	0.006	0.008	0.006
Undulator type	Planar	Helical	Helical	Helical	Helical
Undulator period, cm	3	3	3	4	3
Undulator parameter, K	3.7	2.7	2.7	1.8	2.7
Undulator gap, mm	6	8.5	8.5	7.5	8.5
Focusing beta function in undulator	18	17.7	73	20.5	5
Total synchrotron radiation loss, GW	90	50	50	11.6	10
Gain length, m	4.2	2.8	3.4	4.2	1.8

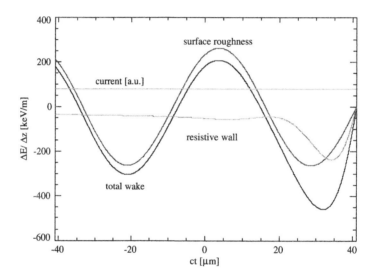

Fig. 4. Resistive and roughness wake-field along the electron bunch for LCLS [71].

An optimization of a SASE-FEL has been done in Ref. [71], where the five cases shown in Table 2 are discussed. One case is the LCLS. Cases A, B, D, use permanent magnet helical undulators with large gap and large field. For A, we use additional FODO focusing, while B uses only the natural undulator focusing; case D uses a lower beam charge, emittance and peak current. The beam parameter for this case have been obtained using the photo-cathode gun scaling laws [72], and discussed later in this sections. Case C uses a lower field helical undulator. The FEL power growth along the undulator for the 5 case has been evaluated using the numerical simulation code Genesis, and including the effect of synchrotron radiation emission, and of the resistive (14) and roughness wake-fields (15) in the undulator vacuum pipe. The total wake-field for the LCLS case is shown in Fig. 4. The wake-field violates the condition $W_z < 30\,\mathrm{KV/m}$ by almost a factor of ten. Even if we consider only the resistive wake-field this condition is violated in part of the bunch.

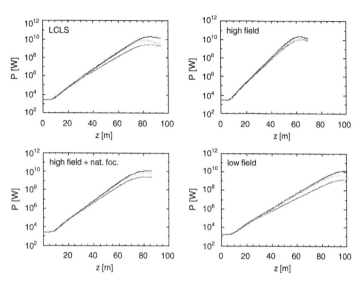

Fig. 5. SASE-FEL power, in W, versus undulator length, in m, for LCLS and cases A, B, D. The upper curve describes the ideal case of no wake-fields, the intermediate includes the resistive wall effect, and the lower curve includes resistivity and roughness [71].

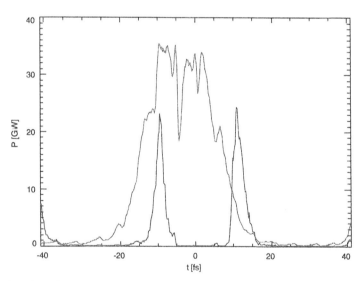

Fig. 6. Power distribution along the bunch length for the LCLS and case A, when wake-fields are included in Genesis [71].

The results from Genesis show that the undulator wake-fields produce an order of magnitude reduction in output power for the LCLS case, Fig. 5, and a smaller reduction in cases A, B, D. The reason for this reduction can be seen clearly in Fig. 6: only the electrons that have a small energy loss in traversing the undulator show gain, and this electrons are in a position in the bunch were the wake-field is near zero.

The power loss is less in case B, because of the larger undulator gap and of the smaller

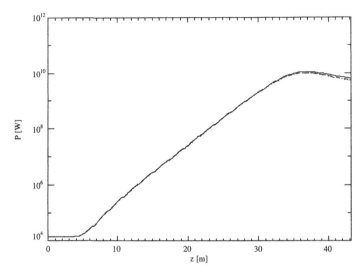

Fig. 7. Power as a function of undulator length for case C; full line, no wake-fields; dotted line, including wake-fields [71].

undulator length. Wake-field effects are negligible in case C, Fig. 7, with a small peak current and undulator length. Due to the larger value of ρ, we have in case C the same output power as in the standard LCLS case, about 10 GWatt, while the spontaneous synchrotron radiation power is reduced from 90 to about 10 Gwatt.

Two methods can be considered to reduce the charge and the emittance. One controls the laser intensity, spot size and phase on photocathode gun to minimize the emittance as a function of charge [72,73]. The scaling laws, neglecting the effect of thermal emittance, are

$$\varepsilon_N = 1.45\times(0.38Q^{4/3} + 0.095Q^{8/3})^{1/2}, \quad\text{mm mrad, } Q \text{ in nC,} \tag{16}$$

$$\sigma_L = 6.3\times10^{-4}Q^{1/3}, \text{ m, } Q \text{ in nC.} \tag{17}$$

Another approach [74] is to produce a 1 nC bunch and then reduce the emittance and charge by collimation. As an example, with a collimator to beam rms radius ratio of 1.5, one can reduce the charge by a factor of 2.5 and the emittance by a factor of 5. The effect of the collimator wake-field on the emittance has also been studied and found to be small.

7. Conclusions

The possibility of large amplification of the spontaneous undulator radiation has been demonstrated in the recent SASE-FELs experiments in the infrared, visible, and UV spectral regions. The experimental results on the gain length and the intensity fluctuation distribution are in good agreement with the FEL collective instability theory. Gain as large as 3×10^5 have been observed in the infrared, bringing us near the saturation level, and large gain has been measured at a wavelength of 80 nm. Experiments over a range of wavelengths will continue, to study saturation, and the spectral, temporal, and angular properties of the SASE radiation, and completely characterize the FEL. These results, and the continued progress in the production, acceleration, measurements, and wake-field control of high brightness electron beams, together with the construction of high quality planar and helical undulators, will lead to a successfully operation of X-ray SASE-FEL in the next few years.

Acknowledgements

I wish to thank all the members of the LCLS group, and I. Ben Zvi, M. Cornacchia, D. Nguyen, J. Rosenzweig, R. Sheffield, A. Varfolomeev, H. Winick, the UCLA students and Postdocs, who have made this work possible.

References

[1] N.M. Kroll, Phys. Quantum Electron. 5 (1978) 115.
[2] P. Sprangle, R.A. Smith, NRL Memo-Report 4033, 1979.
[3] D.B. McDermott, T.C. Marshall, Phys. Quantum Electron. 7 (1980) 509.
[4] A. Gover, P. Sprangle, IEEE J. Quantum Electron. QE-17 (1981) 1196.
[5] G. Dattoli, et al., IEEE J. Quantum Electron. QE-17 (1981) 1371.
[6] R. Bonifacio, C. Pellegrini, L. Narducci, Opt. Comm. 50 (1984) 373.
[7] J.B. Murphy, C. Pellegrini, R. Bonifacio, Opt. Comm. 53 (1985) 197.
[8] J. Gea-Banacloche, G.T. Moore, M. Scully, SPIE 453 (1984) 393.
[9] P. Sprangle, C.M. Tang, C.W. Roberson, Nucl. Instr. and Meth. A 239 (1985) 1.
[10] E. Jerby, A. Gover, IEEE J. Quantum Electron. QE-21 (1985) 1041.
[11] K.-J. Kim, Nucl. Instr. and Meth. A 250 (1986) 396.
[12] J.-M. Wang, L.-H. Yu, Nucl. Instr. and Meth. A 250 (1986) 484.
[13] R. Bonifacio, F. Casagrande, C. Pellegrini, Opt. Comm. 61 (1987) 55.
[14] J.B. Murphy, C. Pellegrini, Nucl. Instr. and Meth. A 237 (1985) 159.
[15] J.B. Murphy, C. Pellegrini, J. Opt. Soc. Am. B 2 (1985) 259.
[16] Ya.S. Derbenev, A.M. Kondratenko, E.L. Saldin, Nucl. Instr. and Meth. A 193 (1982) 415.
[17] K.J. Kim, et al., Nucl. Instr. and Meth. A 239 (1985) 54.
[18] G.T. Moore, Nucl. Instr. and Meth. A 239 (1985) 19.
[19] E.T. Scharlemann, A.M. Sessler, J.S. Wurtele, Phys. Rev. Lett. 54 (1985) 1925.
[20] M. Xie, D.A.G. Deacon, Nucl. Instr. and Meth. A 250 (1986) 426.
[21] K.-J. Kim, Phys. Rev. Lett. 57 (1986) 1871.
[22] L.-H. Yu, S. Krinsky, R. Gluckstern, Phys. Rev. Lett. 64 (1990) 3011.
[23] C. Pellegrini, Nucl. Instr. and Meth. A 272 (1988) 364.
[24] J.S. Fraser, R.L. Sheffield, E.R. Gray, Nucl. Instr. and Meth. 250 (1986) 71.
[25] B.E. Carlsten, Nucl. Instr. and Meth. Phys. Res. A 285 (1989) 313.
[26] X. Qiu, K. Batchelor, I. Ben-zvi, X-J. Wang, Phys. Rev. Lett. 76 (1996) 3723.
[27] K. Bane, Wakefield effects in a linear collider, AIP Conference Proceedings, Vol. 153, 1987, p. 971.
[28] J. Seeman, et al., Summary of emittance control in the SLC linac, US Particle Accelerator Conference, IEEE Conference Proceedings 91CH3038-7, 1991, p. 2064.
[29] J. Seeman, et al., Multibunch energy and spectrum control in the SLC high energy linac, US Particle Accelerator Conference, IEEE Conference Proceedings 91CH3038-7, 1991, p. 3210.
[30] T. Raubenheimer, The generation and acceleration of low emittance flat beams for future linear colliders, SLAC-Report 387, 1991.
[31] C. Pellegrini, A 4 to 0.1 nm FEL based on the SLAC Linac, in: M. Cornacchia, H. Winick (Eds.), Proceedings of the Workshop on IV Generation Light Sources, SSRL-SLAC report 92/02, (1992), p. 364.
[32] C. Pellegrini, et al., Nucl. Instr. and Meth. A 341 (1994) 326.
[33] G. Travish, et al., Nucl. Instr. and Meth. A 358 (1995) 60.
[34] LCLS Design Study Report, Report SLAC-R-521, 1998.
[35] J. Rossbach, et al., Nucl. Instr. and Meth. A 375 (1996) 269.
[36] T. Orzechowski, et al., Phys. Rev. Lett. 54 (1985) 889.
[37] D. Kinfrared, K. Patrick, Nucl. Instr. and Meth. A 285 (1989) 43.
[38] J. Gardelle, J. Labrouch, J.L. Rullier, Phys. Rev. Lett. 76 (1996) 4532.
[39] R. Prazeres, et al., Phys. Rev. Lett. 78 (1997) 2124.
[40] M. Hogan, et al., Phys. Rev. Lett. 80 (1998) 289.
[41] D.C. Nguyen, et al., Phys. Rev. Lett. 81 (1998) 810.
[42] M. Hogan, et al., Phys. Rev. Lett. 81 (1998) 4867.
[43] M. Babzien, et al., Phys. Rev. E 57 (1998) 6093.
[44] L.-H. Yu, et al., Nucl. Instr. and Meth. A 445 (1999) 301, and Proceedings of this Conference, p. II-41.
[45] A. Tremaine, et al., Initial gain measurements of a 800 nm SASE-FEL, VISA, Proceedings of this Conference, p. II-35.
[46] S.V. Milton, et al., Phys. Rev. Lett. 85 (2000) 988, Nucl. Intr. and Meth. A 475 (2001) 28, these proceedings.
[47] J. Rossbach, et al., Nucl. Instr. and Meth. A 475 (2001) 13, these proceedings.
[48] J.B. Murphy, C. Pellegrini, Introduction to the physics of free-electron lasers, in: W.B. Colson, C. Pellegrini, A. Renieri (Eds.), Laser Handbook, Vol. 6, North Holland, Amsterdam, 1990.
[49] R. Bonifacio, et al., La Rivista del Nuovo Cimento 13 (1990) 9.
[50] D.T. Palmer, et al., Emittance studies of the BNL-SLAC-UCLA 1.6 Cell photocathode RF Gun, Proceedings of IEEE 1997 Particle Accelerator Conference, 1998, pp. 2687–2690.
[51] R. Alley, et al., Nucl. Instr. and Meth. A 429 (1999) 324. M. Ferrario, et al., New design study and related experimental program for the LCLS RF photoinjector, LCLS-TN-00-9, SLAC, 2000.

[52] R. Carr, Design of an X-ray free electron laser undulator, Proceedings of US Synchrotron Radiation Instrumentation Conference, Ithaca, NY 1997, Rev. Sci. Instr. November 1997, SLAC pub 7651.

[53] E. Gluskin, et al., Nucl. Instr. and Meth. A 475 (2001) 323, these proceedings.

[54] P. Emma, Electron trajectory in an undulator with dipole field and BPM errors, LCLS-TN-99-4 SLAC, 1999.

[55] R. Bonifacio, et al., Phys. Rev. Lett. 73 (1994) 70.

[56] E.L. Saldin, E.A. Schneidmiller, M.V. Yurkov, Statistical properties of radiation from VUV and X-ray free electron laser, DESY rep. TESLA-free-electron laser 97-02, 1997.

[57] M.C. Teich, T. Tanabe, T.C. Marshall, J. Galayda, Phys. Rev. Lett. 65 (1990) 3393.

[58] W. Fawley, private communication.

[59] K.-J. Kim, private communication, and proceedings of this Conference, p. II-15.

[60] E.L. Saldin, et al., Nucl. Instr. and Meth. A 381 (1996) 545.

[61] K. Bane, SLAC report AP-87, 1991.

[62] C.-K. Ng, Phys. Rev. D 42 (1990) 1819.

[63] K. Bane, C.-K. Ng, A. Chao, Estimate of the impedance due to wall surface roughness, SLAC-PUB-7514, 1997.

[64] K.L. Bane, A.V. Novokhatskii, SLAC, Technical Report SLAC-AP-117, 1999.

[65] G.L. Stupakov, et al., Phys. Rev. ST 2 (1999) 60701/1.

[66] G. Stupakov, Surface impedance and Synchronous modes, SLAC, SLAC-PUB-8208, 1999.

[67] A.V. Burov, A.V. Novokhatskii, Budker Institute of Nuclear Physics, Technical Report BudkerINP-90-28, 1990.

[68] A.V. Novokhatskii, A. Mosnier, Proceedings of the 1997 Particle Accelerator Conference, IEEE, New York, 1997, pp. 1661–1663.

[69] C. Pellegrini, X. Ding, J. Rosenzweig, Output Power Control in an X-ray FEL, Proceedings of the IEEE Particle Accelerator Conference, New York, 1999, p. 2504.

[70] P. Emma, LCLS Accelerator Parameters and Tolerances for Low Change Operations, LCLS-TN-99-3, SLAC, 1999.

[71] C. Pellegrini, et al., Nucl. Instr. and Meth. A 475 (2001) 328, these proceedings.

[72] J.B. Rosenzweig, E. Colby, in AIP Conference Proceedings, Vol. 335, 1995, p. 724.

[73] J.B. Rosenzweig, et al., Optimal scaled photoinjector designs for FEL applications, Proceedings of IEEE Particle Accelerator Conference, 1999, p. 2045.

[74] C. Schroeder, C. Pellegrini, H.-D. Nuhn, Proceedings of this Conference.

ELSEVIER

Nuclear Instruments and Methods in Physics Research A 475 (2001) 13–19

NUCLEAR
INSTRUMENTS
& METHODS
IN PHYSICS
RESEARCH
Section A

www.elsevier.com/locate/nima

Observation of self-amplified spontaneous emission in the wavelength range from 80 to 180 nm at the TESLA test facility FEL at DESY

J. Rossbach*

Deutsches Elektronen-Synchrotron DESY, Notkestrasse 85, D-22603 Hamburg, Germany

For the TESLA FEL Group[1]

Abstract

The first observation of Self-Amplified Spontaneous Emission (SASE) in a free-electron laser (FEL) in the Vacuum Ultraviolet range between 80 and 180 nm wavelength is presented. The observed FEL gain (typically above 1000) and the radiation characteristics, such as dependency on bunch charge, angular distribution, spectral width and intensity fluctuations are discussed. Some accelerator issues are covered, and the future plans for the TESLA Test Facility (TTF) FEL are mentioned. © 2001 Elsevier Science B.V. All rights reserved.

PACS: 41.60.Cr; 29.17.+w; 29.27.−a; Fh

Keywords: Free-electron laser; Self-amplified spontaneous emission; VUV radiation; Bunch compression

*Tel.: +49-40-8998-3617; fax: +49-40-8994-4305.

E-mail address: joerg.rossbach@desy.de (J. Rossbach).

[1]V. Ayvazyan, N. Baboi, R. Brinkmann, M. Castellano, P. Castro, W. Decking, M. Dohlus, H.T. Edwards, B. Faatz, A. Fateev, J. Feldhaus, K. Flöttmann, A. Gamp, T. Garvey, C. Gerth, E. Gluskin, V. Gretchko, U. Hahn, K. Honkavaara, M. Hüning, R. Ischebeck, M. Jablonka, T. Kamps, M. Körfer, J. Krzywinski, J. Lewellen, M. Liepe, A. Liero, T. Limberg, T. Lokajczyk, C. Magne, G. Materlik, J. Menzel, P. Michelato, A. Mosnier, A. Novokhatski, C. Pagani, F. Peters, J. Pflüger, P. Piot, L. Plucinski, K. Rehlich, S. Reiche, I. Reyzl, J. Rossbach, S. Roth, E.L. Saldin, W. Sandner, H. Schlarb, G. Schmidt, P. Schmüser, J.R. Schneider, E.A. Schneidmiller, H.J. Schreiber, S. Schreiber, D. Sertore, S. Simrock, B. Sonntag, F. Stephan, K. Sytchev, M. Timm, M Tonutti, E. Trakhtenberg, R. Treusch, D. Trines, V. Verzilov, R. Wanzenberg, T. Weiland, H. Weise, M. M. White, I. Will, K. Wittenburg, M.V. Yurkov, K. Zapfe.

1. Introduction

X-ray lasers are expected to open up new and exciting areas of basic and applied research in biology, chemistry and physics. Due to recent progress in accelerator technology, the attainment of the long sought-after goal of wide-range tunable laser radiation in the Vacuum-Ultraviolet (VUV) and X-ray spectral regions is coming close to realization with the construction of Free-Electron Lasers (FEL) [1] based on the principle of Self-Amplified Spontaneous Emission (SASE) [8,9]. In a SASE, FEL lasing occurs in a single pass of a relativistic, high-quality electron bunch through a long undulator magnet structure.

The photon wavelength λ_{ph} of the first harmonic of FEL radiation is related to the period length λ_u

of a planar undulator by

$$\lambda_{ph} = \frac{\lambda_u}{2\gamma^2}\left(1 + \frac{K^2}{2}\right) \qquad (1)$$

where $\gamma = E/m_e c^2$ is the relativistic factor of the electrons, $K = eB_u\lambda_u/2\pi m_e c$ the "undulator parameter" and B_u the peak magnetic field in the undulator.

At very short wavelengths, the generation of an electron beam of extremely high quality in terms of emittance, peak current and energy spread, and a high-precision undulator of sufficient length are the challenge to be met in order to achieve high gain or even laser saturation within a single pass. Provided the spontaneous radiation from the first part of the undulator overlaps the electron beam, the electromagnetic radiation interacts with the electron bunch leading to a density modulation (micro-bunching) which enhances the power and coherence of radiation. In this "high gain mode" [2–5], the radiation power $P(z)$ grows exponentially with the distance z along the undulator

$$P(z) = P_0 A \exp(2z/L_g) \qquad (2)$$

where L_g is the field gain length, P_0 the effective input power (see below), and A the input coupling factor [4,5]. A is equal to $\frac{1}{9}$ in one-dimensional FEL theory with an ideal electron beam.

The R&D program for the TeV-Energy Superconducting Linear Accelerator (TESLA) FEL aims at wavelength far below the visible. Therefore, there is no laser tunable over a wide range to provide the input power P_0. Instead, the spontaneous undulator radiation from the first part

of the undulator is used as an input signal to the downstream part. FELs based on this Self-Amplified-Spontaneous-Emission (SASE) principle [6,7] are presently considered the most attractive candidates for delivering extremely brilliant, coherent light with wavelength in the Ångström regime [8–11]. Compared to state-of-the-art synchrotron radiation sources, one expects full transverse coherence, larger average brilliance and, in particular, up to eight or more orders of magnitude larger peak brilliance at pulse lengths of about 200 fs FWHM.

2. Experimental set-up

The experimental results presented in this paper have been achieved at the TESLA Test Facility (TTF) FEL [12] at the Deutsches Elektronen-Synchrotron DESY. The TESLA collaboration consists of 39 institutes from 9 countries and aims at the construction of a 500 GeV (center-of-mass) e^+/e^- linear collider with an integrated X-ray laser facility [10]. Major hardware contributions to TTF have come from Germany, France, Italy, and the USA. The goal of the TTF FEL is to demonstrate SASE FEL emission in the VUV and, in a second phase, to build a soft X-ray user facility [13,14]. The layout is shown in Fig. 1. The main parameters for FEL operation are compiled in Table 1.

The injector is based on a laser-driven $1\frac{1}{2}$ cell RF gun electron source operating at 1.3 GHz [16]. It uses a Cs_2Te cathode [17] and can generate bunch charges more than 10 nC at 1 MHz repetition rate.

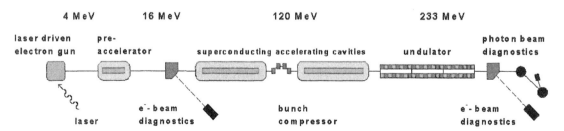

Fig. 1. Schematic layout of phase 1 of the SASE FEL at the TESLA Test Facility at DESY, Hamburg. The linac contains two 12.2 m long cryogenic modules each equipped with eight 9-cell superconducting accelerating cavities [14]. The total length is 100 m.

Table 1
Main parameters of the TESLA Test Facility for FEL experiments (TTF FEL, phase 1)

Parameter	Measured value for FEL experiment
Beam energy at undulator	181–272 MeV
Rms energy spread	0.3 ± 0.2 MeV
Rms transverse beam size	100 ± 30 µm
ε_n (normalized emittance) in the undulator	6 ± 3 π mrad mm
Electron bunch charge	1 nC
Peak electron current	400 ± 200 A
Bunch spacing	1 µs
Repetition rate	1 Hz
λu (undulator period)	27.3 mm
Undulator peak field	0.46 T
Effective undulator length	13.5 m
Typical betatron function horizontal/ vertical	1.1 m
λph (radiation wavelength)	80–181 nm
FEL gain	10^3–10^4
FEL radiation pulse length	0.4–1 ps

A loading system allows mounting and changing of cathodes while maintaining ultra-high vacuum conditions [17]. The cathode is illuminated by a train of UV laser pulses generated in a mode-locked solid-state laser system [18] synchronized with the RF. An energy of up to 50 µJ with a pulse-to-pulse variation of 2% (rms) is achieved [19]. The UV pulse length measured with a streak camera is $\sigma_t = 7.1 \pm 0.6$ ps . The RF gun is operated with a peak electric field of 37 MV/m on the photocathode. The RF pulse length was limited to 100 µs and the repetition rate to 1 Hz for machine protection reasons. The gun section is followed by a 9-cell superconducting cavity, boosting the energy to 16 MeV. The superconducting accelerator structure has been described elsewhere [15].

The undulator is a fixed 12 mm gap permanent magnet device using a combined function magnet design [20] with a period length of $\lambda_u = 27.3$ mm and a peak field of $B_u = 0.46$ T, resulting in an undulator parameter of $K = 1.17$. The beam pipe diameter in the undulator (9.5 mm) [21] is much larger than the beam diameter (300 µm). Integrated quadrupole structures produce a gradient

of 12 T/m superimposed on the periodic undulator field in order to focus the electron beam along the undulator. The undulator system is subdivided into three segments, each 4.5 m long and containing 10 quadrupole sections to build up 5 full focusing–defocusing (FODO) cells. The FODO lattice periodicity runs smoothly from segment to segment. There is a spacing of 0.3 m between adjacent segments for diagnostics [22]. The total length of the system is 14.1 m. The vacuum chamber incorporates 10 beam position monitors and 10 orbit correction magnets per segment, one for each quadrupole [21].

For optimum overlap between the electron and light beams, high precision on the magnetic fields and mechanical alignment are required. The undulator field was adjusted such that the expected rms deviations of the electron orbit should be smaller than 10 µm at 300 MeV [23]. The beam orbit straightness in the undulator is determined by the alignment precision of the superimposed permanent magnet quadrupole fields which is better than 50 µm in both vertical and horizontal direction. The relative alignment of the three segments is accomplished with a laser interferometer to better than 30 µm [24].

Different techniques have been used to measure the emittance of the electron beam [22,25]: Magnet optics scanning ("quadrupole scans"), tomographic reconstruction of the phase space including space charge effects, and the slit system method. All methods use optical transition radiation emitted from aluminum foils to measure the bunch profiles and yield values for the normalized emittance of (4 ± 1) π mrad mm for a bunch charge of 1 nC at the exit of the injector. The emittance in the undulator, as determined from quadrupole scans and from a system of wire scanners was typically between 6 and 10 π mrad mm (in both horizontal and vertical phase space). It should be noted that the measurement techniques applied determine the emittance integrated over the entire bunch length. However, for FEL physics, the emittance of bunch slices much shorter than the bunch length is the relevant parameter. It is likely that, due to spurious dispersion and wakefields, the bunch axis is tilted about a transverse axis such that the projected emittance is larger than the emittance of any slice. Based on these considerations, we estimate the normalized slice emittance in the undulator at (6 ± 3) π mrad mm.

A bunch compressor is inserted between the two accelerating modules, in order to increase the peak current of the bunch up to 500 A, corresponding to 0.25 mm bunch length (rms) for a 1 nC bunch with Gaussian density profile. Experimentally, it is routinely verified that a large fraction of the bunch charge is compressed to a length below 0.4 mm (rms) [26]. There are indications that the core is compressed even further. We estimate the peak current for the FEL experiment at 400 ± 200 A. Coherent synchrotron radiation in the magnetic bunch compressor may affect the emittance and the energy spread at such short bunch lengths [27,28]. Beam parameters like peak current and slice emittance determine the FEL gain length critically. Thus, we consider further improvements of beam diagnostics essential for any precise verification of FEL models at short wavelengths.

For radiation intensity measurements [29] we use a PtSi photodiode integrating over all wavelengths. The detector unit was placed 12 m down-stream the undulator exit. A 0.5 mm iris was placed in front of the photodiode in order to avoid saturation effects.

3. FEL measurements

A strong evidence for the FEL process is a large increase in the on-axis radiation intensity if the electron beam is injected such that it overlaps with the radiation during the entire passage through the undulator. Fig. 2 shows the intensity passing a 0.5 mm iris, located on axis 12 m downstream of the undulator, as a function of the horizontal beam position at the undulator entrance. The observed intensity inside a window of ± 200 μm around the optimum beam position is a factor of more than 100 higher than the intensity of spontaneous radiation. This intensity gain was first observed with the photodiode and later confirmed with the CCD camera of the spectrometer. The central wavelength for this first SASE demonstration at the TTF FEL was 108.5 nm [30].

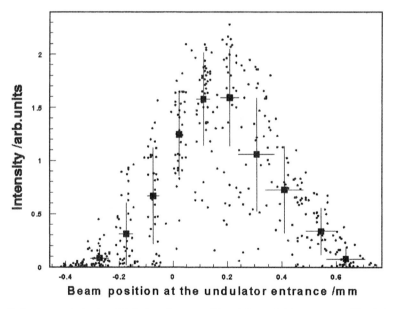

Fig. 2. Sensitivity of radiation power to horizontal electron beam position at the undulator entrance. The dots represent mean values of the radiation intensity for each beam position. The horizontal error bars denote the rms beam position instability while the vertical error bars indicate the standard deviation of intensity fluctuations, which are due to the statistical character of the SASE process, see Eq. (3).

SASE gain is expected to depend on the bunch charge in an extremely nonlinear way. An intensity enhancement by a factor of more than 100 was observed when increasing the bunch charge from 0.3 to 0.6 nC while keeping the beam orbit constant for optimum gain. The gain did not further increase when the bunch charge exceeded some 0.6 nC. This needs further study, but the most likely reason is that the beam emittance becomes larger for increasing Q thus, reducing the FEL gain.

The wavelength of 108.5 nm was consistent with the measured beam energy of 233 ± 5 MeV and the known undulator parameter $K = 1.17$, see Eq. (1). From the user point of view, a most important feature of FELs starting from noise is the arbitrary tunability of wavelength. At TTF FEL, this was demonstrated by tuning the electron beam energy between 272 and 181 MeV, corresponding to wavelengths from 80 to 181 nm. Within this range, SASE was achieved at any energy where the SASE search procedure was performed, see Fig. 3. Typically, a SASE gain (for definition, see below) above 1000 was observed. The spectral width was in most cases in agreement with theory. A possible source of spectral widening was energy jitter, since the spectra were taken by averaging over many bunches.

A characteristic feature of SASE FELs is the concentration of radiation power into a cone much narrower than that of wavelength integrated

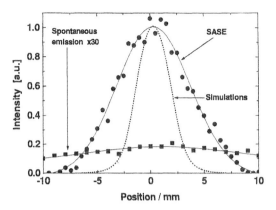

Fig. 4. Horizontal intensity profile of SASE FEL and spontaneous undulator radiation ($\times 30$), measured with a photodiode behind a 0.5 mm aperture in a distance of 12 m from the end of the undulator. The dotted line is the result of numerical simulation.

undulator radiation, whose opening angle is in the order of $1/\gamma$. Measurements done by moving the 0.5 mm iris horizontally together with the photodiode confirm this expectation, see Fig. 4. The spontaneous intensity is amplified by a factor of 30 to be visible on this scale.

In order to study which section of the undulator contributes most to the FEL gain, we applied closed orbit beam bumps to different sections of the undulator, thus disturbing the gain process at various locations along the undulator. It was seen that practically the entire undulator contributes, but with some variation in local gain. Some improvement in the over-all gain should be possible by optimizing the settings of the 30 orbit correction coils.

The energy flux was 2 nJ/mm^2 at the location of the detector and the on-axis flux per unit solid angle was about 0.3 J/sr. This value was used as a reference point for the numerical simulation of the SASE FEL at 108.5 nm with the code FAST [31]. The longitudinal profile of the bunch current was assumed to be Gaussian with an rms length of 0.25 mm. The transverse distribution of the beam current density was also taken to be Gaussian. Calculations have been performed for a Gaussian energy spread of 0.1%, and the normalized emittance was varied in the simulations between 2 and 10 π mrad mm. Our calculations show that in this range of parameters the value of the effective

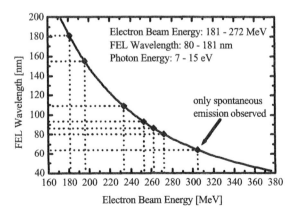

Fig. 3. Wavelength of the central radiation cone (collimation angle ± 0.2 mrad) as a function of electron beam energy. The FEL gain was typically > 1000. The bunch charge was 1 nC.

power of shot noise P_{in} and coupling factor $A \sim 0.1$ (see Eq. (2)) are nearly constant. A level of energy flux of 0.3 J/sr is obtained at five field gain lengths L_g. With these parameters, the FEL gain can be estimated at $G \approx 3 \times 10^3$ with a factor of 3 uncertainty which is mainly due to the imprecise knowledge of the longitudinal beam profile. If we assume that the entire undulator contributes to the FEL amplification process, we estimate the normalized emittance at 8π mrad mm in reasonable agreement with the measurements. However, as stated before, it is more likely that the normalized (slice) emittance is smaller and the electron orbit is not perfectly straight. This is supported by the observation that in a first systematic attempt of improving the orbit straightness in the undulator, the SASE gain at 109 nm was increased by another factor of 3. Thus, at 109 nm the maximum achieved gain was $G \approx 1 \times 10^4$. It should be noted that large SASE gain was achieved in a stable and reproducible way for several weeks.

It is essential to realize that the fluctuations seen in Fig. 2 are not primarily due to unstable operation of the accelerator but are inherent to the SASE process. Shot noise in the electron beam causes fluctuations of the beam density, which are random in time and space [32]. As a result, the radiation produced by such a beam has random amplitudes and phases in time and space and can be described in terms of statistical optics. In the linear regime of a SASE FEL, the radiation pulse energy measured in a narrow central cone (opening angle $\pm 20\,\mu$rad in our case) at maximum gain is expected to fluctuate according to a gamma distribution $p(E)$ [33]

$$p(E) = \frac{M^M}{\Gamma(M)}\left(\frac{E}{\langle E \rangle}\right)^{M-1}\frac{1}{\langle E \rangle}\exp\left(-M\frac{E}{\langle E \rangle}\right) \quad (3)$$

where $\langle E \rangle$ is the mean energy, $\Gamma(M)$ is the gamma function with argument M, and $M^{-1} = \langle (E - \langle E \rangle)^2 \rangle / \langle E \rangle^2$ is the normalized variance of E. M corresponds to the number of longitudinal optical modes. Note that the same kind of statistics applies for completely chaotic polarized light, in particular for spontaneous undulator radiation.

For these statistical measurements, the signals from 3000 radiation pulses have been recorded at

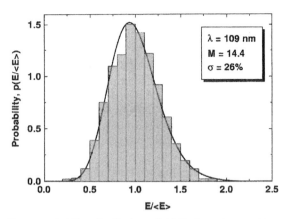

Fig. 5. Probability distribution of SASE intensity at 109 nm wavelength. The rms fluctuation yields a number of longitudinal modes $M = 14$. The solid curve is the gamma distribution for $M = 14.4$. The bunch charge is 1 nC.

109 nm wavelength, with the small iris (0.5 mm diameter) in front of the photodiode to guarantee that transversely coherent radiation pulses are selected. As one can see from Fig. 5, the distribution of the energy in the radiation pulses is quite close to the gamma distribution. The relative rms fluctuations are about 26% corresponding to $M = 14.4$. Similar measurements have been performed at other wavelengths. One should take into account that, these fluctuations arise not only from the shot noise in the electron beam, but the pulse-to-pulse variations of the beam parameters can also contribute to the fluctuations. Thus, the value $M \approx 14$ can be considered a lower limit for the number of longitudinal modes in the radiation pulse. Using the width of radiation spectrum, we calculate the coherence time [33] and find that the part of the electron bunch contributing to the SASE process is at least 100 μm long. From the quality of the fit with the gamma distribution we can also conclude that the statistical properties of the radiation are described with Gaussian statistics. In particular, this means that there are no FEL saturation effects.

4. Summary

High gain SASE in the VUV has become reality. Wavelength tuning between 80 and 181 nm as well as reliable operation was demonstrated at DESY over several weeks. To date, all observations are in agreement with the present SASE FEL models. More precise electron beam diagnostics is desirable for a more detailed verification of FEL models. Now, there is more than ever optimism justified in view of the feasibility of future X-ray user facilities. However, there is still a way to go: the SASE gain demonstrated so far is still some orders of magnitude below saturation. Also, stable operation with long pulse trains containing several thousand pulses and flexible timing pattern, as requested by users, remains a challenge for accelerator physicists.

Acknowledgements

The authors acknowledge that successful FEL operation has been achieved as a joint effort of all members of the TESLA collaboration. A list of participating institutes can be found in Ref. [30].

References

[1] J.M. Madey, J. Appl. Phys. 42 (1971) 1906.
[2] G. Dattoli, A. Marino, A. Renieri, F. Romanelli, IEEE J. Quantum Electron. QE-17 8, (1981) 1371.
[3] K.J. Kim, Phys. Rev. Lett. 57 (1986) 1871.
[4] S. Krinsky, L.H. Yu, Phys. Rev. A 35 (1987) 3406.
[5] E.L. Saldin, E.A. Schneidmiller, M.V. Yurkov, The Physics of Free Electron Lasers, Springer, Berlin, 1999 and references therein.
[6] A.M. Kondratenko, E.L. Saldin, Part. Accelerators 10 (1980) 207.
[7] R. Bonifacio, C. Pellegrini, L.M. Narducci, Opt. Commun. 50 (1984) 373.
[8] H. Winick, et al., Proceedings of the PAC Washington and SLAC-PUB-6185, 1993.
[9] R. Brinkmann, et al., Nucl. Instr. and Meth. A 393 (1997) 86.
[10] R. Brinkmann, G. Materlik, J. Rossbach, A. Wagner (Eds.), DESY 1997-048 and ECFA 1997-182, 1997.
[11] H.-D. Nuhn, J. Rossbach, Synchrotron Radiat. News 13 (1) (2000) 18.
[12] W. Brefeld, et al., Nucl. Instr. and Meth. A 393 (1997) 119.
[13] T. Åberg, et al., A VUV FEL at the TESLA Test Facility at DESY, Conceptual Design Report, DESY Print TESLA-FEL 95-03, 1995.
[14] J. Rossbach, Nucl. Instr. and Meth. A 375 (1996) 269.
[15] H. Weise, Proceedings of the 1998 Linac Conference Chicago, 1998, pp. 674–678.
[16] J.-P. Carneiro, et al., Proceedings of the Partical Accelerator Conference, 1999 New York, 1999, pp. 2027–2029.
[17] P. Michelato, et al., Nucl. Instr. and Meth. A 445 (2000) 422.
[18] I. Will, S. Schreiber, A. Liero, W. Sandner, Proceedings of the 1999 FEL Conference, Hamburg, Vol. II-99, 1999.
[19] S. Schreiber, et al., Nucl. Instr. and Meth. A 445 (2000) 427.
[20] Y.M. Nikitina, J. Pflüger, Nucl. Instr. and Methods A 375 (1996) 325.
[21] U. Hahn, et al., Nucl. Instr. and Meth. A 445 (2000) 442.
[22] G. Schmidt, et al., Nucl. Instr. and Meth. A 475 (2001) 545, these proceedings.
[23] J. Pflüger, P. Gippner, A. Swiderski, T. Vielitz, Proceedings of the 1999 FEL Conference, Hamburg, Vol. II-87, 1999.
[24] J. Pflüger, H. Lu, T. Teichmann, Nucl. Instr. and Meth. A 429 (1999) 386.
[25] H. Edwards, et al., Proceedings of the 1999 FEL Conference, Hamburg, Vol. II-75, 1999.
[26] M. Geitz, G. Schmidt, P. Schmüser, G.v. Walter, Nucl. Instr. and Meth. A 445 (2000) 343.
[27] M. Dohlus, A. Kabel, T. Limberg, Proceedings of the 1999 Partical Accelerator Coference, New York, 1999, pp. 1650–1652.
[28] P. Piot, et al., Nucl. Instr. and Meth., this conference.
[29] Ch. Gerth, et al., Nucl. Instr. and Meth. A 475 (2001) 481, these proceedings.
[30] J. Andruszkow, et al., Phys. Rev. Lett. 85 (2000) 3825.
[31] E.L. Saldin, E.A. Schneidmiller, M.V. Yurkov, Nucl. Instr. and Meth. A 429 (1999) 233.
[32] R. Bonifacio, et al., Phys. Rev. Lett. 73 (1994) 70.
[33] E.L. Saldin, E.A. Schneidmiller, M.V. Yurkov, Opt. Commun. 148 (1998) 383.

ELSEVIER

Nuclear Instruments and Methods in Physics Research A 475 (2001) 20–27

NUCLEAR
INSTRUMENTS
& METHODS
IN PHYSICS
RESEARCH
Section A

www.elsevier.com/locate/nima

First lasing and initial performance of the European UV/VUV storage ring FEL at ELETTRA ☆

R.P. Walker[a],*, J.A. Clarke[b], M.E. Couprie[c], G. Dattoli[d], M. Eriksson[e],
D. Garzella[c], L. Giannessi[d], M. Marsi[a], L. Nahon[c], D. Nölle[f,1], D. Nutarelli[c],
M.W. Poole[b], H. Quick[f], E. Renault[c], R. Roux[a], M. Trovò[a], S. Werin[e], K. Wille[f]

[a] Sincrotrone Trieste, Trieste 34012, Italy
[b] CLRC-Daresbury Laboratory, Warrington, UK
[c] CEA-SPAM and LURE, Orsay, France
[d] ENEA-Frascati, Frascati, Italy
[e] MAX-lab, Lund, Sweden
[f] University of Dortmund, Dortmund, Germany

Abstract

An overview of the European storage ring FEL project is presented, including a description of the main components, details of initial lasing at 350 and 220 nm and future prospects. © 2001 Elsevier Science B.V. All rights reserved.

PACS: 29.20.Dh; 41.60.Cr

Keywords: Storage ring; Free-electron laser

1. Introduction

The European storage ring FEL project at ELETTRA [1–3] began officially in May 1998 and is a collaboration between six laboratories interested in the development of storage ring free-electron lasers (FEL). The goal of the project is the development of the FEL as a future user facility in the UV/VUV, providing tuneable output between 350 nm and at least 190 nm. Important issues to be

addressed, other than the FEL itself, will be the compatibility with operation of other synchrotron radiation (SR) beamlines, and the carrying out of pilot user experiments. The main distinguishing features of the project are

- the integration of the FEL on a low emittance "third-generation" synchrotron radiation facility,
- the use of a helical undulator (optical klystron) to reduce power loading on the mirrors, and
- the use of sophisticated mirror chambers that allow in situ switching between different mirrors.

Hardware installation was completed in mid-February 2000 and FEL shifts started immediately afterwards. After only 30 h of beam time, first

☆ Partly funded under EC RTD Contract No. ERBFMGE-CT98-0102.

*Corresponding author. Tel.: +39-040-3758225; fax: +39-040-3758565.

E-mail address: richard.walker@elettra.trieste.it (R.P. Walker).

[1] Now at SD&M AG, Ratingen, Germany.

lasing occurred on February 29th at 351.6 nm. Lasing at 220 nm was subsequently obtained at the end of May. In this report, we describe the main FEL components, the initial lasing performance as well as future plans.

2. Main FEL components

2.1. Layout and optical cavity

The main parameters of the project were described in Ref. [1]. The location of the mirror chambers immediately outside the shielding wall leads to an optical cavity length of 32.4 m, synchronised to the four-bunch mode of operation. Since the mirror positions are not symmetric with respect to the undulator centre, mirror radii of 19 and 16 m were initially specified in order to locate the waist at the undulator centre, however, in order to provide greater flexibility for the selection of mirrors, at the cost of a small reduction in gain, this was later changed to 17.5 m for both mirrors, leading to a waist 1.5 m upstream of the undulator centre. The resulting Rayleigh length and stability parameter are 4.59 m and 0.72, respectively. The UHV mirror chambers designed and constructed at Daresbury Laboratory, containing three remotely interchangeable mirrors, were described in detail in a previous publication [4].

The location of the FEL allows radiation from the upstream end of the cavity to be directed into an existing diagnostic area, allowing simultaneous measurement of the electron and laser beam time structures. The downstream mirror chamber is followed by a standard switching mirror which allows the undulator radiation to be used during normal operation by an SR beamline, presently under construction. This also allows the possibility of using this beamline for experiments with the FEL radiation.

2.2. Electron beam

The operating energy of ELETTRA for the FEL mode is 1 GeV, corresponding to the usual injection energy. Stabilisation of longitudinal

coupled bunch motion in the four-bunch mode is achieved by means of a careful setting of RF cavity temperatures. Residual oscillation amplitudes are typically $<0.2°$ of RF phase, corresponding to 1 ps. Several series of measurements of the bunch length as a function of bunch current have been made using a Photonetics dual-sweep streak camera [3]. Measurements taken after installation of the new aluminium vacuum vessels required for the FEL in Aug./Sep. 1999 are consistent with the data taken previously, and are only about 7% larger than those measured in 1997 with an earlier system. The r.m.s. bunch length is well fitted by the following relationship:

$$\sigma_t(ps) = 3.39 + 10.67(I_{bunch}(mA))^{1/3}$$

from which the slope allows the broad-band impedance of $0.14\,\Omega$ to be calculated. At a typical current used presently for FEL work, 20 mA total, the 21.6 ps bunch length leads to a peak current of 80 A.

The installation of several insertion device vacuum chambers in the ring has, however, led to a progressive increase in vertical impedance [5] resulting in an increased sensitivity to the vertical orbit, affecting both the maximum single bunch current that can be injected, as well as the vertical stability. FEL work has so far been carried out using quite modest currents (up to 30 mA total) with respect to the maximum previous performance of 100 mA, which also eases problems associated with mirror heating and degassing. More attention will, however, be given to this topic in the future in order to produce higher laser power for experiments.

2.3. Optical klystron

The two permanent magnet undulator sections are based on the APPLE-2 design, each having 19 periods of 100 mm. The maximum wavelength available in the circular polarisation mode at 1 GeV and minimum gap (19 mm) is 450 nm, allowing some flexibility for an increase in operating energy in the future, and for greater compatibility with the operation of other SR beamlines. The undulator field quality was optimised to minimise both phase errors and field integrals [6].

The final r.m.s. phase errors are $<3°$, and the field integrals $<2\,G\,m$ and $<3\,G\,m^2$, at any gap and phase. Adjustment of the phase of the radiation emitted in the two undulators is carried out by a three-pole electromagnet. For FEL operation, this also allows an optical klystron mode with maximum phase delay (N_d) of 85 wavelengths (at 350 nm, 1 GeV).

The intrinsically poor field homogeneity of the APPLE design, combined with the relatively high magnetic field, long period length and low operating energy in the present case, gives rise to significant focusing effects in the ring [7]. At the settings for 350 nm operation (gap = 22.2 mm, phase = 30 mm), each undulator section gives a tune change of about $\Delta Q_x = -0.030$, $\Delta Q_y = +0.025$. To allow the undulators to be set without losing the beam, a temporary tune correction procedure was developed. Experiments have shown that a local tune correction (i.e. using two pairs of quadrupoles on either side of the undulator) gives a significantly better beam lifetime than a global correction (using quadrupoles in all straight sections). Studies are underway to explore other matching schemes and to prepare a program for dynamic tune correction during undulator closure [7]. The undulators also cause a variation in closed orbit, sufficiently small such that all the present FEL work could be carried out without the use of the correction coils, which will be implemented in the future.

Measurements of spontaneous radiation spectra were made using a visible/UV monochromator in order to verify the correct functioning of the undulator, to determine the radiation axis, and also to allow a determination of the electron beam energy spread. Fig. 1 shows an example of a spectrum taken under standard FEL operating conditions. The optical klystron spectrum can be written in the following form:

$$I_{OK} = 2I_{und}\left[1 + f\cos\left(2\pi(N + N_d)(\Delta\lambda/\lambda)\right)\right]$$

where I_{und} is the spectrum of one undulator and the modulation rate f is given by [8]:

$$f = f_0 \exp\left(-8\pi^2\left((N + N_d)\,\sigma_\gamma/\gamma\right)^2\right). \qquad (1)$$

Analysis of spectra at various N_d allows extraction of the energy spread σ_γ/γ and the residual

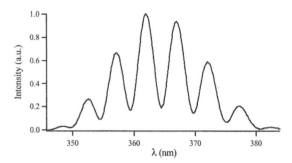

Fig. 1. Measured optical klystron spectrum with $N_d = 55$ and 2.8 mA/bunch.

modulation rate f_0 at zero energy spread. The high value of the latter, 0.91–0.99, indicates a small effect due to transverse emittance. The increase in energy spread with current follows the same trend as the bunch length data, but is consistently less than that of the bunch length [3]. Further measurements will be carried out to extend the current range and to determine if this is a real effect or a systematic measurement error.

3. Lasing at 350 nm

3.1. First lasing

An initial wavelength of 350 nm was chosen in order to profit from the experience of mirror characterisation and FEL operation at this wavelength at LURE. The first batch of silica and sapphire substrates prepared by Maris-Delfour (France) were subsequently coated by the Centre de Perfectionnement de Couches Minces (Lyon, France), with oxide (Ta_2O_5/SiO_2) multilayers for high reflectivity around 350 nm. The mirrors chosen for first lasing trials were those with the maximum transmission. The radii were 19.11 m (back) and 17.56 m (front) leading to a more stable cavity ($g_{12} = 0.59$) than the nominal one, with a waist 0.6 m upstream of the undulator centre. According to the measurements made at LURE, initial total loss (P) was 2.6% and total transmission (T) was 0.64% at 350 nm. Direct measurements of absorption (0.033% total at 350 nm) showed by comparison that the majority of the total losses resulted from scattered light due to the relatively large roughness of this first batch of substrates.

Transverse alignment of the mirrors was carried out using standard procedures, i.e. (i) using HeNe beams aligned to the spontaneous emission axis to align the mirrors by autocollimation, (ii) using a photomultiplier to observe multiple reflections with the storage ring operating in single bunch mode, and (iii) by visual observation of the image seen through the upstream mirror. The latter has so far proved to be a crucial element of the final alignment needed to establish lasing. The cavity length was adjusted so as to minimise the temporal width of the trapped radiation pulse using a streak camera. First lasing was observed at 351.6 nm with 17 mA total current in four bunches.

3.2. Detuning curve

Fig. 2 shows the measured average output power as a function of RF frequency. In the present case, 100 Hz corresponds to a cavity length change of 6.5 μm. The width of the detuning curves increases with beam current as expected. The variation of spectral linewidth and laser micropulse length as a function of detuning is shown in Fig. 3. Comparing with Fig. 2, it can be seen that minimum spectral and pulse widths are obtained at the detuning corresponding to maximum power.

Fig. 4 shows the minimum profiles with a relative spectral width of 4.2×10^{-4} (full-width at half-maximum—FWHM) and pulse length of 8.9 ps (FWHM), giving a product 7 times larger than that for a Fourier transform limited pulse (for which $c\, \Delta t_{[\text{FWHM}]}\, \Delta\lambda_{[\text{FWHM}]}/\lambda^2 = 0.44$).

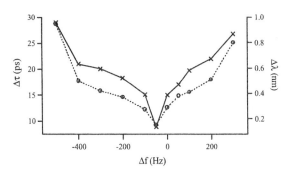

Fig. 3. Pulse length (upper, left scale) and spectral width (lower, right scale) as a function of cavity detuning, measured with 10.8 mA.

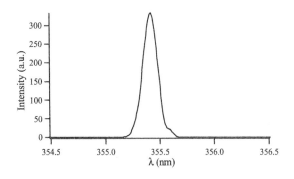

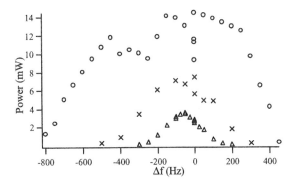

Fig. 2. Power output as a function of cavity length detuning at different (total) current levels: 19 mA (upper), 10.8 mA (middle), 6.6 mA (lower).

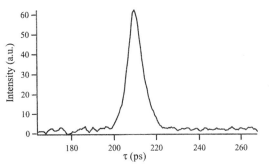

Fig. 4. Minimum linewidth (0.15 nm FWHM) and pulse length (8.9 ps FWHM) at 350 nm.

3.3. Gain estimations

Assuming that saturation is due only to energy spread and bunch length increase, the initial gain can be calculated from the following:

$$\frac{G}{P} = \frac{(\sigma_t)_{on}}{(\sigma_t)_{off}} \exp\left(8\pi^2(N+N_d)^2\left[\left(\frac{\sigma_\gamma}{\gamma}\right)^2_{on} - \left(\frac{\sigma_\gamma}{\gamma}\right)^2_{off}\right]\right)$$

$$= X \exp\left(8\pi^2(N+N_d)^2\left(\frac{\sigma_\gamma}{\gamma}\right)^2_{off}(X^2-1)\right). \quad (2)$$

Assuming the same factor increase in energy spread as in bunch length (X) thus allows the gain to be estimated. Table 1 shows the results of two sets of measurements of electron bunch length with the laser on and off and the corresponding values of gain, assuming the initial cavity losses, $P = 2.6\%$. It can be seen that the gain estimated in this way is in reasonable agreement with predictions.

Another measurement of gain has been made by the "maximum detuning method" [9], a technique used successfully at SuperACO [10]. The gain is calculated using the following expression:

$$G = \frac{P}{1-P}\sqrt{1+\left(\frac{\sigma_L}{\sigma_t}\right)^2}\exp\left(\frac{\tau_L^2}{2(\sigma_L^2+\sigma_t^2)}\right)$$

where τ_L is the position of the laser micropulse with respect to the centre of the electron bunch at the threshold of lasing at the edge of the detuning curve. A measurement with 10 mA total current gave $\sigma_t = 19$ ps, and $\sigma_L = 10.6$ ps and 14.9 ps on either side of the detuning curve with a movement of the laser by $2\tau_L = 70$ ps. The expression above then gives gain values (assuming $P = 2.6\%$) of

9.7–11.1%, somewhat higher than the predicted value of 8.4%.

Finally, another confirmation of the gain is given by the threshold for lasing. The predicted threshold when $G = P = 2.6\%$ occurs at 1.4 mA, and is in good agreement with the observed threshold at 1.3 mA (total).

Fig. 5 summarises the measured and predicted gain values, calculated using the standard expression:

$$G = 0.85g_0 + 0.19g_0^2 \quad (3)$$

with

$$g_0 = \frac{8\pi^2N^2(N+N_d)\lambda_0^2K^2}{\gamma^3\langle\Sigma\rangle}\frac{\hat{I}}{I_A}f(\sigma_\gamma/\gamma)$$

where f is the modulation rate of the spectrum (Eq. (1) with $f_0 = 1$), and $\langle\Sigma\rangle$ is the transverse overlap of the electron and photon beams:

$$\Sigma = 2\pi\left(\sigma_x^2 + w^2/4\right)^{1/2}\left(\sigma_y^2 + w^2/4\right)^{1/2}$$

averaged over the undulator length. In the present case, the electron beam sizes (σ_x, σ_y) are sufficiently small such that Σ is dominated by the size of the radiation mode, w. The calculation uses the measured bunch lengths (see Section 2.2) and assumes in the modulation rate that the energy spread scales with bunch length.

Table 1
Estimated gain from the measured increase in bunch length (X) compared to the theoretically predicted value

Total current (mA)	$(\sigma_t)_{off}$ (ps)	$(\sigma_t)_{on}$ (ps)	X	G (%) from bunch length increase, Eq. (2)	G (%) from Eq. (3)
4.0	13.4	17.8	1.33	4.3	5.2
28.0	22.5	32.7	1.45	12.0	12.0

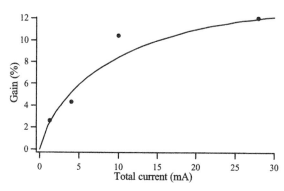

Fig. 5. Measured (points) and calculated gain (line) as a function of current at 350 nm.

3.4. Laser power

A maximum extracted power of 20 mW has been recorded, with 30 mA total beam current, in reasonable agreement with the predictions of the following model [11]:

$$P_L = 8\pi\frac{T}{P}(N + N_d)f\left[\left(\frac{\sigma_\gamma}{\gamma}\right)^2_{on} - \left(\frac{\sigma_\gamma}{\gamma}\right)^2_{off}\right]P_{SR}$$

$$= 8\pi\frac{T}{P}(N + N_d)f\left(\frac{\sigma_\gamma}{\gamma}\right)^2_{off}(X^2 - 1)P_{SR} \qquad (4)$$

where f is the laser-on modulation rate and where the ratio of laser-on to laser-off energy spread (X) is calculated from Eq. (2). In the present case, with 30 mA total current, the total synchrotron radiation power (P_{SR}) is 477 W. With $N_d = 56$, the measured electron bunch length at 30 mA (24.3 ps), energy spread (laser-off) scaled in proportion with bunch length (0.15%), the above predicts a total laser average power (P_L) of 65 mW, of which 28 mW is at the back mirror. A more sophisticated calculation based on parameterisations which take into account both bunch length and energy spread effects [12] gave a value of 35 mW for the power at the back mirror. Considering the fact that some losses occur because the power was measured after four reflections outside the cavity, and the fact that the model refers to the equilibrium state and not the pulsed state that has normally been observed so far, as well as the uncertainty in the energy spread behaviour, the agreement can be considered to be quite good.

3.5. Macrotemporal structure

The macrotemporal structure has been measured with both the streak camera and photomultiplier. Unlike many other storage ring FELs, close to perfect synchronism the laser is regularly pulsed (e.g. Fig. 6a). The frequency increases with beam current and in the range 10–30 mA lies typically in the range 230–330 Hz. Theory predicts a natural oscillation frequency, f_r, given by [13]:

$$f_r = \frac{1}{\pi\sqrt{2\tau_0\tau_s}}$$

where $\tau_0 = T_0/(G - P)$ is the laser rise-time and τ_s the synchrotron oscillation damping time. In the present case with $\tau_s = 65$ ms, T_0 (bunch spacing) = 216 ns, $P = 2.6\%$ the expression above predicts a frequency of 480–570 Hz. The discrepancy suggests a much slower laser rise time than expected from the simple expression above. Various other macrotemporal structures have been observed, but not yet studied in detail including c.w. (Fig. 6b), quasi-c.w. in which the laser intensity is modulated at 50 Hz (Fig. 6c), as well as randomly pulsed (Fig. 6d).

3.6. Polarisation

The polarisation of the spontaneous and FEL radiation has been investigated using a standard linear polarising filter. Measurements of spontaneous radiation spectra showed the expected result for circularly polarised radiation that the spectrum is independent of polariser angle. The laser radiation, however, showed a strong variation in intensity with polarisation angle, with a max/min ratio in the range 5–10. Similar effects have been observed at UVSOR and were found to be due to a variation in mirror reflectivity depending on the direction of polarisation [14]. Further studies will be carried out to determine if this is the cause in the present case.

4. Lasing at 220 nm

A number of silica substrates were coated with oxides (HfO_2–Al_2O_3–SiO_2) in a multilayer superstructure for 220 nm operation using a Plasma Ion Assisted Deposition (PIAD) technique by Fraunhofer Institut für Angewandte Optik und Feinmechanik (IOF) (Jena, Germany), in the framework of an R&D contract. The mirrors were installed on May 27th and lasing was obtained at 220.5 nm with a total current of 10 mA. The modulator current was set to provide the same $N_d = 56$ as used at 350 nm. Changing the undulator gap allowed lasing over the range 217.9–224.1 nm. Fig. 7 shows the narrowest line obtained at 223.6 nm. The width (FWHM) is 0.05 nm ($\Delta\lambda/\lambda = 2.2 \times 10^{-4}$).

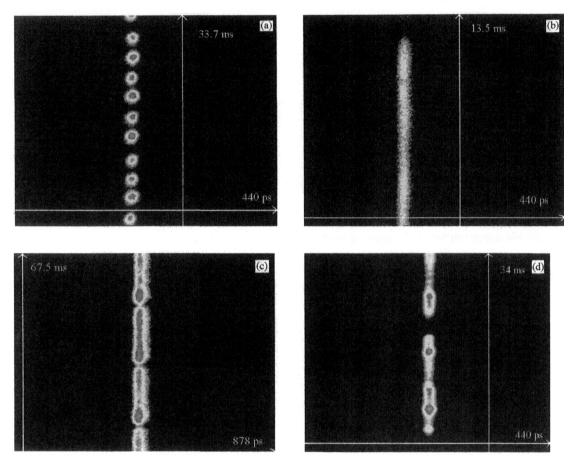

Fig. 6. Dual-sweep streak camera images showing the macrotemporal structure of the FEL: a—pulsed, near perfect synchronism; b—c.w.; c—c.w. modulated at 50 Hz; d—randomly pulsed.

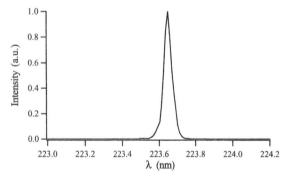

Fig. 7. Laser spectrum at 223.6 nm.

The detuning curve had a width of 1.1 kHz. A maximum outcoupled power of 10.5 mW was recorded at 27 mA, whereas Eq. (4) predicts 18 mW, assuming a mirror transmission of 0.3%

and total cavity losses of 4.2%. The lasing threshold was at about 3 mA (total current) in good agreement with the calculated gain at this current of 4.6%.

5. Conclusions and future prospects

The major features of the FEL on ELETTRA are in good agreement with expectations both at 350 and 220 nm. Due to the relatively short amount of time available since first lasing (8 days in total), many topics have yet to be studied in detail including the macrotemporal structure, wavelength tuning, as well as lasing stability, reproducibility and the long term effects of mirror degradation, etc.

One main future aim is to proceed to shorter wavelength, below 200 nm. Some initial tests have already been made around 190 nm using prototype fluoride coated mirrors produced by IOF (Jena, Germany). A parallel initiative is also underway, funded by an EC TMR Network, to develop resistant mirrors for FELs down to 200 nm [15].

Another main aim is to carry out pilot experiments using the beamline which shares the same undulator source. Sets of mirrors will be prepared to cover the 280–220 nm range for these experiments, with higher output coupling. Calculations show that a power of several hundred mW should be possible using mirrors with a few % transmittance and with total beam currents of 20–50 mA.

A further important aspect that has begun to be addressed is the operation of the FEL while other synchrotron radiation beamlines are in use, particularly those interested in exploiting the time structure of the radiation for time-resolved experiments.

Acknowledgements

It is a pleasure to thank the staff at Daresbury Laboratory responsible for mirror chamber design, construction and installation (N. Bliss, A.A. Chesworth, B. Fell, C. Hill, R. Marl, I.D. Mullacrane, R.J. Reid), staff at Sincrotrone Trieste responsible for optical klystron design and construction (R. Bracco, B. Diviacco, R. Visintini, D. Zangrando), controls (A. Abrami, L. Battistello, M. Lonza), layout, vacuum chambers, frontends and alignment (C. Fava, P. Furlan, A. Gambitta, R. Godnig, G. Loda, F. Mazzolini, G. Pangon, N. Pangos, M. Pasqualetto, F. Pradal), radiation safety (G. Tromba, A. Vascotto), electron beam studies and diagnostics (A. Fabris, M. Ferianis, F. Iazzourene, E. Karantzoulis, C. Pasotti, M. Svandrlik). The support of C. Rizzuto, President, Sincrotrone Trieste and D. Norman, as Director of Synchrotron Radiation, Daresbury Laboratory is also gratefully acknowledged. We are also grateful to A. Gatto, R. Thielsch and N. Kaiser (IOF) for an active collaboration in FEL mirror development. Support for R. Roux under EC TMR Network Contract ERBFMRX-C980245 is acknowledged.

References

[1] R.P. Walker, et al., Proceedings of the 20th International FEL Conference, Williamsburg, VA, August 1998, Nucl. Instr. and Meth. A 429 (1999) 179.

[2] R. Roux, et al., Proceedings of the 21st International FEL Conference, Hamburg, August 1999, Nucl. Instr. and Meth. A 445 (2000) 11–21.

[3] R.P. Walker, et al., Proceedings of the 7th European Particle Accelerator Conference, Vienna, June 2000, p. 93.

[4] M.W. Poole, et al., Nucl. Instr. and Meth. A 445 (2000) 448.

[5] E. Karantzoulis, Sincrotrone Trieste Internal Report ST/M-TN-00/8, August 2000.

[6] B. Diviacco, et al., Proceedings of the 7th European Particle Accelerator Conference, Vienna, June 2000, p. 2322.

[7] L. Tosi, et al., Proceedings of the 7th European Particle Accelerator Conference, Vienna, June 2000, p. 2349.

[8] P. Elleaume, J. Phys. 44 (1983) C1–333.

[9] R. Roux, Ph.D. Thesis, Université Paris VI, 28th January 1999.

[10] D. Garzella, et al., Nucl. Instr. and Meth. A 445 (2000) 143.

[11] M. Billardon, et al., IEEE J. Quantum Electron. QE-21 (1985) 805.

[12] G. Dattoli, et al., Nucl. Instr. and Meth. A393 (1997) 55; G. Dattoli, et al., ELETTRA FEL optical klystron optimisation, July 1998, unpublished.

[13] P. Elleaume, J. Phys. 45 (1984) 997.

[14] H. Hama, Private communication.

[15] A. Gatto, et al., Proc. SPIE 4102-37, Presented at the 45th Annual Symposium on Optical Science and Technology, San Diego, CA, 30 July–4 August 2000.

ELSEVIER

Nuclear Instruments and Methods in Physics Research A 475 (2001) 28–37

**NUCLEAR
INSTRUMENTS
& METHODS
IN PHYSICS
RESEARCH**
Section A

www.elsevier.com/locate/nima

Observation and analysis of self-amplified spontaneous emission at the APS low-energy undulator test line

N.D. Arnold[a], J. Attig[a], G. Banks[a], R. Bechtold[a], K. Beczek[a], C. Benson[a],
S. Berg[a], W. Berg[a], S.G. Biedron[a], J.A. Biggs[a], M. Borland[a], K. Boerste[a],
M. Bosek[a], W.R. Brzowski[a], J. Budz[a], J.A. Carwardine[a], P. Castro[b], Y.-C. Chae[a],
S. Christensen[a], C. Clark[a], M. Conde[c], E.A. Crosbie[a], G.A. Decker[a], R.J. Dejus[a],
H. DeLeon[a], P.K. Den Hartog[a], B.N. Deriy[a], D. Dohan[a], P. Dombrowski[a],
D. Donkers[a], C.L. Doose[a], R.J. Dortwegt[a], G.A. Edwards[a], Y. Eidelman[a],
M.J. Erdmann[a], J. Error[a], R. Ferry[a], R. Flood[a], J. Forrestal[a], H. Freund[d],
H. Friedsam[a], J. Gagliano[a], W. Gai[c], J.N. Galayda[a], R. Gerig[a], R.L. Gilmore[a],
E. Gluskin[a], G.A. Goeppner[a], J. Goetzen[a], C. Gold[a], A.J. Gorski[a], A.E. Grelick[a],
M.W. Hahne[a], S. Hanuska[a], K.C. Harkay[a], G. Harris[a], A.L. Hillman[a],
R. Hogrefe[a], J. Hoyt[a], Z. Huang[a], J.M. Jagger[a], W.G. Jansma[a], M. Jaski[a],
S.J. Jones[a], R.T. Keane[a], A.L. Kelly[a], C. Keyser[a], K.-J. Kim[a], S.H. Kim[a],
M. Kirshenbaum[a], J.H. Klick[a], K. Knoerzer[a], R.J. Koldenhoven[a], M. Knott[a],
S. Labuda[a], R. Laird[a], J. Lang[a], F. Lenkszus[a], E.S. Lessner[a], J.W. Lewellen[a],
Y. Li[a], R.M. Lill[a], A.H. Lumpkin[a], O.A. Makarov[a], G.M. Markovich[a],
M. McDowell[a], W.P. McDowell[a], P.E. McNamara[a], T. Meier[a], D. Meyer[a],
W. Michalek[a], S.V. Milton[c,*], H. Moe[a], E.R. Moog[a], L. Morrison[a], A. Nassiri[a],
J.R. Noonan[a], R. Otto[a], J. Pace[a], S.J. Pasky[a], J.M. Penicka[a], A.F. Pietryla[a],
G. Pile[a], C. Pitts[a], J. Power[c], T. Powers[a], C.C. Putnam[a], A.J. Puttkammer[a],
D. Reigle[a], L. Reigle[a], D. Ronzhin[a], E.R. Rotela[a], E.F. Russell[a], V. Sajaev[a],
S. Sarkar[a], J.C. Scapino[a], K. Schroeder[a], R.A. Seglem[a], N.S. Sereno[a],
S.K. Sharma[a], J.F. Sidarous[a], O. Singh[a], T.L. Smith[a], R. Soliday[a], G.A. Sprau[a],
S.J. Stein[a], B. Stejskal[a], V. Svirtun[a], L.C. Teng[a], E. Theres[a], K. Thompson[a],
B.J. Tieman[a], J.A. Torres[a], E.M. Trakhtenberg[a], G. Travish[a], G.F. Trento[a],
J. Vacca[a], I.B. Vasserman[a], N.A. Vinokurov[e], D.R. Walters[a], J. Wang[a],
X.J. Wang[f], J. Warren[a], S. Wesling[a], D.L. Weyer[a], G. Wiemerslage[a], K. Wilhelmi[a],
R. Wright[a], D. Wyncott[a], S. Xu[a], B.-X. Yang[a], W. Yoder[a], R.B. Zabel[a]

*Corresponding author. Advanced Photon Source, Argonne National Laboratory, Argonne, IL 60439, USA.
E-mail address: milton@aps.anl.gov (S.V. Milton).

[a] Advanced Photon Source, Argonne National Laboratory, Argonne, IL 60439, USA
[b] Deutsches Elektronen-Synchrotron DESY, Notkestrasse 85, Hamburg 22603, Germany
[c] Argonne Wakefield Accelerator, Argonne National Laboratory, Argonne, IL 60439, USA
[d] Science Applications International Corporation, 1710 Goodridge Drive, McLean, VA 22102, USA
[e] Budker Institute of Nuclear Physics, Novosibirsk 630090, Russian Federation
[f] Brookhaven National Laboratory, Upton, NY 11973, USA

Abstract

Exponential growth of self-amplified spontaneous emission at 530 nm was first experimentally observed at the Advanced Photon Source low-energy undulator test line in December 1999. Since then, further detailed measurements and analysis of the results have been made. Here, we present the measurements and compare these with calculations based on measured electron beam properties and theoretical expectations. © 2001 Elsevier Science B.V. All rights reserved.

PACS: 41.60.Cr; 52.75.Va; 41.75.Lx; 29.27.−a

Keywords: Free-electron laser; Self-amplified spontaneous emission

1. Introduction

Third-generation synchrotron-light sources such as the Advanced Photon Source (APS) at Argonne National Laboratory currently provide high-brightness X-ray beams to a wide range of users. These sources rely on the spontaneous emission of synchrotron radiation generated by electron bunches passing through undulator magnets. Due to the incoherent nature of the emission process between individual electrons, the intensity of the generated light is proportional to the total number of particles within the bunch. This situation can be improved dramatically by forcing the electrons to emit coherently. In such a case, the intensity is proportional to the square of the total number of coherently radiating electrons. This is the underlying essence of current thought when speaking of the "next" or "fourth-generation" of synchrotron radiation sources.

Present plans for fourth-generation synchrotron radiation facilities capitalize on the self-amplified spontaneous emission (SASE) process [1,2] to induce the required microbunching within the electron bunch needed for coherent emission. The advantage of this process is that it does not require mirrors or an input seed, but starts up naturally from spontaneous noise. The process is thus scalable to X-ray wavelengths.

Until recently, SASE had only been observed at 633 nm and longer wavelengths [3–6]. The requirements on the electron beam properties become increasingly stringent as the wavelength is reduced; however, significant progress has been made in electron beam production technology and beam control such that the attainment of SASE into the vacuum ultraviolet (VUV) and perhaps soft X-ray wavelength range is becoming feasible. Progress in the study of SASE at these wavelengths will be used to guide the design of X-ray facilities utilizing the SASE effect [7–9]. We report here on recent progress in the study of SASE at 530 nm using the Low-Energy Undulator Test Line (LEUTL) facility at the APS, and briefly describe our future plans with the system.

2. LEUTL description

2.1. Overview

The APS linac and LEUTL systems have been configured as a test bed for SASE research and development at wavelengths ranging from the visible into the VUV. Upgrades to the linac have included the addition of radiofrequency (RF) electron guns for high-brightness electron beam generation, improvements in the magnetic lattice structure, and performance upgrades of the RF modulators, low-level RF systems, and some power supplies [10]. The LEUTL is a 50-m tunnel in line with the linac and is capable of housing undulators or strings of undulators of over 30 m in length. The APS LEUTL system is designed to test concepts critical to the success of a linac-based fourth-generation light source, such as the proposed X-ray Linac Coherent Light Source (LCLS) at SLAC [9] and the TESLA-FEL at DESY [8], but at much longer wavelengths. Fig. 1 shows a schematic of the LEUTL SASE system and Table 1 lists the parameters for the first three phases of the operation.

2.2. Parameters

We have chosen to concentrate our initial efforts on investigation of SASE at 530 nm. Optics and optical diagnostics at this wavelength are readily available, the required electron beam performance at 217 MeV is readily achievable, and it is well within the operating envelope of the APS linac. As performance is improved, the energy of the electron beam will be increased to explore SASE at shorter wavelengths (see Table 1).

Table 1
LEUTL SASE design parameters for various phases of operation

	Phase 1	Phase 2	Phase 3
Wavelength (nm)	530	120	59
Electron energy (MeV)	217	457	650
Normalized rms emittance (μm)	5	3	3
Energy spread rms (%)	0.1	0.1	0.1
Peak current (A)	100	300	500
Undulator period (cm)	3.3	3.3	3.3
Magnetic field (T)	1.0	1.0	1.0
Undulator gap (mm)	9.3	9.3	9.3
Length of one undulator (m)	2.4	2.5	2.4
Power gain length (m)	0.81	0.72	0.77
Installed undulator length (m)	5 × 2.4 then 9 × 2.4	9 × 2.4	9 × 2.4

2.3. Linac

A schematic of the APS linac is shown in Fig. 2. It consists of 13 S-band, 3-m-long, constant gradient, travelling wave accelerating structures similar to the SLAC design. Three 35-MW klystrons each power four 3-m accelerating structures through a SLED. A photocathode gun (PC

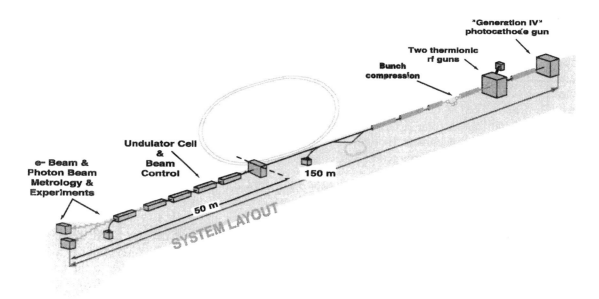

Fig. 1. Schematic of the APS LEUTL system.

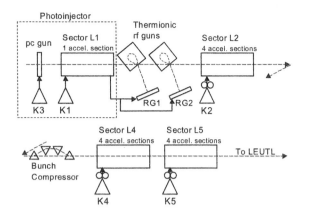

Fig. 2. Schematic of the APS linac system.

gun) is powered by a single 35-MW klystron, while either of two thermionic RF guns share power with a single accelerating structure that follows the photocathode RF gun. Power distribution to these RF guns and linac structure is through S-band RF switches. In the PC gun mode, the electron beam can be accelerated by all 13 structures. The maximum energy attainable in PC gun mode is 650 MeV. One accelerating structure, the first, is not used to accelerate beam in the thermionic RF gun configuration.

A bunch compression system has recently been installed and commissioned within the linac following the fifth accelerating structure [11,12]. It will provide higher peak currents for both photocathode and thermionic RF guns. It was, however, not installed during the measurements reported here.

2.4. Guns

High-brightness electron bunches are generated using either a photocathode RF gun system or a thermionic RF gun with alpha-magnet compression. The photocathode RFgun is a 1.6-cell Brookhaven S-band gun IV model [6,13] that employs a copper photocathode. A Nd : glass picosecond drive laser system is used to generate the electrons [14]. It is assembled from commercially available components and is timing stabilized to the RF within 1 ps. This system can generate a single electron bunch of roughly 1 nC at

a 6-Hz repetition rate. The thermionic RF gun is a 1.5 cell S-band gun with a tungsten dispenser cathode. An alpha magnet is used to both inject beam into the APS linac and to compress the bunch to very high peak currents. An 8-ns pulsed kicker magnet is used for safety purposes to limit the total charge delivered to the linac. The result is a bunch train of roughly 23 bunches, each with approximately 48 pC of charge. This thermionic RF gun system is extremely reliable and is used as the primary injector for the APS storage ring.

2.5. Undulators

The undulator system is built of identical cells. Each cell contains a fixed-gap 2.4-m-long undulator with a 3.3-cm period and an undulator parameter, K, of 3.1. There is a diagnostic station, a horizontal focusing quadrupole, and horizontal and vertical steering before the first undulator, between each of the currently installed five undulators, and after the final undulator. The longitudinal spacing between undulators, about 0.38 m, is set to insure proper phase matching of the optical fields and the electron beam at successive undulator sections. In its current configuration of five cells, the total installed undulator magnetic length is 12 m.

2.6. Diagnostic arrangement

Direct measurement of the exponential growth of the optical intensity as a function of length along the undulator is a hallmark of our diagnostics design. Fig. 3 is a representative diagram of the diagnostics station located between each undulator. Electron beam diagnostics include YAG and optical transition radiation (OTR) screens viewed by a CCD camera. Two sets of filter wheels afford both intensity and, to a coarse degree, wavelength selectivity. Completing the primary electron beam diagnostics at each station are capacitive pickup beam position monitors (BPMs) with single-shot resolution of less than 10 μm for bunch charges of 1 nC.

In-tunnel visible light detectors (VLDs) consist of three-position actuators with positions as

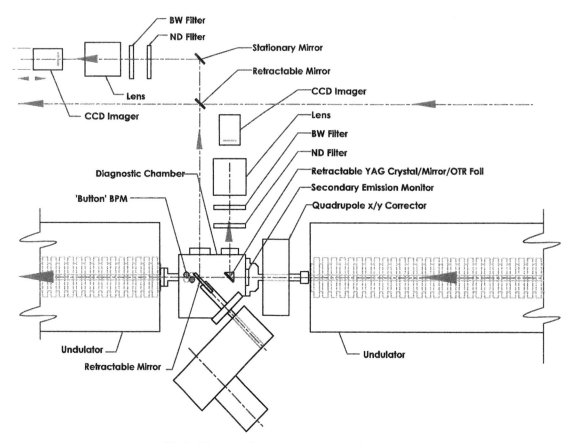

Fig. 3. Schematic of the undulator diagnostic stations.

follows: (1) out, (2) mirror, (3) mirror with 600-μm diameter hole. These are used to deflect the synchrotron light through a set of filters to a CCD camera. All five mirror hole centers have been aligned along the beamline to within 20 μm of the ideal electron beam trajectory. The cameras viewing these mirrors can be focused at the mirror, at infinity, or at any distance between the two. Focus at infinity is valuable as it allows us to measure the angular distribution of the optical radiation.

A green alignment laser is located at the entrance to the undulator string. It is used to insure that all mirrors and optical components are properly aligned. It also serves as an intensity and wavelength calibration source for the optical systems.

Further details of the LEUTL system, the undulators, and the diagnostics arrangements can be found in Refs. [15–19].

3. Gain analysis

3.1. Via fitted intensity growth

Measurements of the signal intensity as a function of length along the undulator can be used to extract the SASE gain length. In the SASE FEL, the Fourier harmonics of the field amplitude $E_\omega(s, z)$ at frequency ω grow exponentially with position along the undulator z:

$$E_\omega(s, z) \propto \exp(gz) \qquad (1)$$

where s is the position along the bunch. Near the optimal frequency ω_0 at the point s_0 within the bunch, where the longitudinal particle density is maximal, the factor g reaches a maximum, and therefore, can be represented in the form

$$g \approx \frac{1}{2L_G}\left[1 - a(\omega - \omega_0)^2 - b(s - s_0)^2\right] \qquad (2)$$

where L_G is the power gain length at the optimal frequency and peak current, and a and b are constants.

The radiated energy W is then proportional to

$$W \propto \int_{-\infty}^{\infty} \int_{-\infty}^{\infty} |E_\omega(s, z)|^2 \mathrm{d}s\,\mathrm{d}\omega$$

$$\approx \frac{L_G}{z} \exp\left(\frac{z}{L_G}\right) \frac{\pi}{\sqrt{ab}}. \qquad (3)$$

Thus, W can be written in the form

$$W(A, L_G, z) \approx A\frac{L_G}{z} \exp\left(\frac{z}{L_G}\right). \qquad (4)$$

The quantities L_G and A can be explicitly calculated in terms of the electron beam and undulator parameters [20–25].

3.2. Via opening angle

An alternative, albeit rough, way of determining the gain length is by observation of the opening angle of the optical radiation. An estimate of the gain length based on the FWHM angular divergence of the SASE radiation θ_{SASE} is

$$L_G \approx 2\lambda_r / \pi \theta_{SASE}^2 \qquad (5)$$

where λ_r is the observed resonant wavelength.

3.3. Intensity fluctuations

Intensity fluctuations of the optical signal are intrinsic to the SASE process [26,27]. This is due to the startup from noise developing into a number of longitudinal degrees of freedom within the resultant optical pulse. Roughly, the number of degrees of freedom is

$$M = \frac{\pi}{2} \frac{\sigma_\ell}{\lambda_r} \frac{\lambda_u}{z} \qquad (6)$$

where σ_ℓ is the rms electron bunch length, z is the length of the "exponential-growth" part of the undulator (the observation point along the undulator minus approximately two power gain lengths), and λ_u is the undulator period. Eq. (6) is valid for $M > 1$. The standard deviation of the shot-to-shot fluctuations is found to be equal to $1/\sqrt{M}$.

Although one clearly observes these fluctuations, their use for precise quantitative measurement of the SASE process is hindered by a lack of precise knowledge of the beam properties. This is further exacerbated by shot-to-shot fluctuation of the electron beam properties.

4. Gain measurements

4.1. Early data

4.1.1. Measured gain

First measurement of significant SASE using the PC gun was on December 22, 1999 [28]. Fig. 4 shows the optical intensity following the first four undulators. The signal is the integral over all measureable angles, within a 10-nm bandwidth, and over the duration of the electron bunch. At the time of the measurement, there was significant shot-to-shot fluctuation in the beam properties and trajectory. As a result, we plot the peak

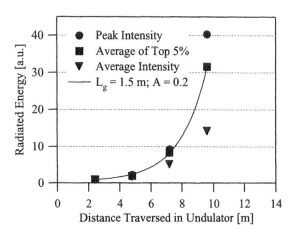

Fig. 4. Early data. Radiated energy vs. distance traversed in the undulator. The PC gun source was used for this experiment.

Table 2
Measured linac beam parameters at the end of the linac using beam from the PC gun on December 22nd

	Typical	Min–Max
Energy (MeV)	217	—
Normalized emittance (μm)	5	4–8
Charge/bunch (nC)	0.7	0.6–0.8
Bunch length FWHM (ps)	5	4–7
Energy spread rms %	0.1	<0.1–0.2

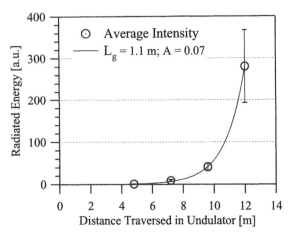

Fig. 5. Radiated energy vs. distance traversed in the undulator. Thermionic RF gun source.

measured intensity, the average of the top 5% measured intensities, and the average of all measurements at each diagnostic station. All measurements were taken with the electron beam properties shown in Table 2.

By fitting the average of the top 5% to the functional form given in Eq. (4), we found a gain length of 1.5 m. Similarly, the opening angle of the SASE signal following the fourth undulator implies, from Eq. (5), a gain length of 1.6 m.

The theoretical gain length, using the typical values listed in Table 2 and calculated by methods discussed in Refs. [20–25], is 0.7 m. The discrepancy between theory and the experimentally measured values was probably due to incomplete knowledge of the electron beam parameters.

4.2. Thermionic RF gun data

4.2.1. Beam properties

Fig. 5 shows SASE signal measurements as a function of length along the undulator system using beam from a high performance thermionic RF gun system. Electron bunches from the thermionic RF gun were optimally compressed by the alpha magnet by observing signals generated off a coherent transition radiation bunch length monitor [29]. Reconstruction of the longitudinal bunch profile was performed with the result tabulated in Table 3. Also shown in Table 3 are the measured electron beam properties for this experiment. Sequential measurements at each undulator diagnostics station were made for a number of electron bunches. Shown are the averages of all measurements. The intensity measured is an integration over all wavelengths, angles, and the duration of the electron bunch.

Table 3
Measured linac beam parameters at the end of the linac using beam from the thermionic RF gun

	Typical	Min–Max
Energy (MeV)	217	—
Normalized emittance (μm)	12	9–15
Charge/bunch (nC)	0.048	0.043–0.052
Bunch length FWHM (ps)	0.35	0.33–0.40
Energy spread rms %	0.1	<0.1–0.2

With this electron source the intensity signal is also an integral over the entire 23-bunch train generated during each linac macropulse. Error bars correspond to the rms fluctuation of the measured intensity signal. Fitting to the average value of the measured signal gives a gain length of 1.1 m.

As before, one can calculate the expected gain lengths. Using the typical values from Table 3, a gain length of 1.0 m is calculated. A more complete Monte Carlo analysis using normalized three-point estimates based on the beam parameters listed in the table predicts a most probable gain length of 1.1 m with a distribution width of ±0.2 m, which is in good agreement with the measured data.

4.2.2. Fluctuations

It is interesting to note that at 530 nm, the FWHM length of the thermionic gun electron bunch corresponds to roughly 200 optical periods and the undulator is 360 periods long. The bunch length is then less than the slippage length and Eq. (6) is not valid. M can, therefore, be assumed to be close to 1. Given that each measurement is an average over the 23 bunches in the bunch train, we would expect fluctuations in the intensity of $1/\sqrt{23} \approx 20\%$. Our data indicate fluctuations of nearly 30%, indicating that fluctuations in the electron beam properties were dominant.

4.3. Recent PC gun data

4.3.1. Measured gain

Fig. 6 shows our most recent data taken in July 2000 using the beam from the PC gun. Also included in the figure is the fit to Eq. (4). The data points are the average of all measurements, and the error bars indicate the rms spread in the measurements. Once again the intensity is an integration over all wavelengths, angles, and the duration of the electron bunch. All data were normalized to the intensity following the second undulator. Data following the first undulator were not used in order to insure that we were well into the exponential-growth regime. Data following the third undulator are missing due to problems

encountered with the diagnostics system at this location.

4.3.2. Comparison to theory

Table 4 lists the beam conditions for all three run conditions shown in Fig. 6. Also listed are the fitted and predicted gain lengths for the listed beam conditions. The agreement is very good. The total installed length of undulator at the time of these measurements was 12 m. For a gain length of 0.8 m, this would imply that 15 gain lengths have been traversed.

Further confirmation of the gain length comes from measurement of the opening angle. We measured an FWHM opening angle following the fourth undulator of approximately 600 μrad, in agreement with the angular divergence of the fundamental guided mode obtained by solving the eigenmode equation [23]. From Eq. (5), this implies a gain length of 0.9 m, again in rough agreement with the measurement of 0.8 m.

Error bars in Fig. 6 indicate the rms spread in the measured intensity. Measured spreads of the intensity signals at the last two diagnostics stations are in the order of 50–60%. Electron bunch lengths were 300 μm rms. For our parameters, M

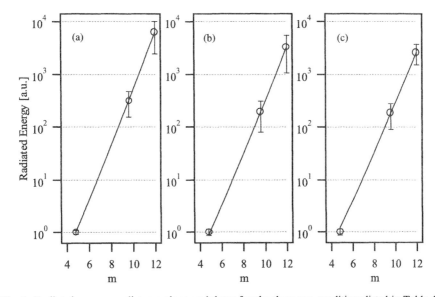

Fig. 6. Radiated energy vs. distance along undulator for the three run conditions listed in Table 4.

Table 4
Beam measurements, fitted gain length, and calculated gain lengths for the data displayed in Fig. 6[a]

	A	B	C
Peak current (A)	170	130	150
Charge (nC)	0.42	0.45	0.50
Normalized emittance (π mm-mrad)	9	7.5	7.5
Fitted gain length (m)	0.75	0.80	0.82
Calculated gain length (m)	0.77	0.83	0.76

[a] *Note*: rms energy spread $\sim 0.1\%$ for all cases.

is calculated to be 2.4, giving an expected fluctuation level of 64%.

4.4. Confirmation of beam bunching

As further confirmation of microbunching, we observed the OTR evolution to coherent transition radiation (CTR) as a function of distance along the undulator. The synchrotron radiation signal was blocked by a thin foil immediately preceding the 45° metal pickoff mirror [30]. The resultant CTR signal was then directed to the wall-mounted VLD camera system, and in the case of the fifth

undulator, it could also be transported to an optical spectrometer. A 530-nm filter was used for the VLD measurements to insure that we viewed the CTR signal at the expected bunching wavelength. Although the CTR signals as measured at the VLDs have not been fully analyzed, they do show indication of intensity gains well beyond linear. Perhaps the most significant signature of bunching, though, was via the spectrometer measurements that showed a CTR narrow-band signal at the SASE fundamental wavelength (Fig. 7).

5. Summary

Exponential growth of the optical signal as a function of length has been directly measured at 530 nm, providing clear evidence of the SASE process. Early measurements made using beam from the photocathode RF gun indicated a gain length of 1.5 m. This was later improved to 0.8 m in agreement with theory, giving a total of 15 power gain lengths within our undulator system. A thermionic RF gun with alpha-magnet compres-

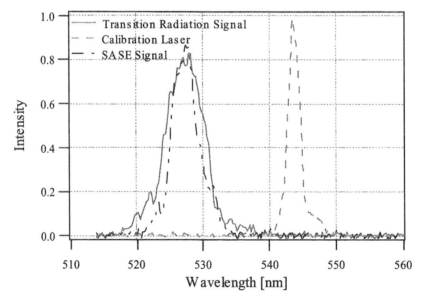

Fig. 7. CTR spectrum of the microbunched beam along with the SASE spectrum at the same location. The intensities were adjusted with neutral density filters to be roughly equal. (The SASE is ~ 200 times brighter.) The data were not taken simultaneously. The sharp line at 543.5 nm is the calibration laser.

sion has also been used and a gain length as low as 1.1 m at 530 nm was measured. Initial measurements of electron beam microbunching at the resonant wavelength also provide strong evidence of the SASE process. Measurements of the gain length now agree with the theoretical predictions, but we are still suffering from fluctuations in the electron beam properties.

We have recently installed a bunch compression system into the APS linac and are in the early stages of commissioning it. Four additional undulators are also being installed, which will bring the total length of installed undulator to 21.6 m. With this setup, we will explore SASE at and beyond saturation. We will also investigate nonlinear harmonic generation [31] and increase the beam energy of the linac to begin exploration of SASE at even shorter wavelengths.

Acknowledgements

This work is supported by the U.S. Department of Energy, Office of Basic Energy Sciences, under Contract No. W-31-109-ENG-38.

References

[1] Y.S. Derbenev, A.M. Kondratenko, E.L. Saldin, Nucl. Instr. and Meth. 193 (1982) 415.
[2] R. Bonifacio, C. Pellegrini, L.M. Narducci, Opt. Commun. 50 (1985) 6.
[3] T.J. Orzechowski, et al., Nucl. Instr. and Meth. A 250 (1986) 144.
[4] D.A. Kirkpatrick, et al., Phys. Fluids B 1 (7) (1989) 1511.
[5] M. Hogan, et al., Phys. Rev. Lett. 81 (1998) 4807.
[6] M. Babzien, et al., Phy. Rev. E 57 (1998) 6093.
[7] R. Tatchyn, et al., Nucl. Instr. and Meth. A 374 (1996) 274.
[8] R. Brinkmann, G. Materlik, J. Rossbach, A. Wagner (Eds.), Conceptual design of a 500 GeV e$^+$ e$^-$ linear collider with integrated X-ray laser facility, DESY Report No. DESY97-048, 1997.
[9] M. Cornacchia, et al., Linac Coherent Light Source (LCLS) Design Study Report, Stanford University, University of California Report No. SLAC-R-521 / UC-414, revised 1998.
[10] Travish, et al., High-brightness beams from a light source injector: the advanced photon source low-energy undulator test line linac, in: A. Chao (Ed.), Proceedings of the 20th International Linac Conference, SLAC-R-561 (CD available from the Stanford Linear Accelerator Center, Stanford, CA, 2000).
[11] M. Borland, Design and performance simulations of the bunch compressor for the APS LEUTL FEL, in: A. Chao (Ed.), Proceedings of the 20th International Linac Conference, SLAC-R-561 (CD available from the Stanford Linear Accelerator Center, Stanford, CA, 2000).
[12] M. Borland, J. Lewellen, S. Milton, A highly flexible bunch compressor for the APS LEUTL FEL, in: A. Chao (Ed.), Proceedings of the 20th International Linac Conference, SLAC-R-561 (CD available from the Stanford Linear Accelerator Center, Stanford, CA, 2000).
[13] S. Biedron, et al., in: Proceedings of the 1999 Particle Accelerator Conference, IEEE Press, New York, 1999, pp. 2024–2026.
[14] G. Travish, N. Arnold, R. Koldenhoven, in: J. Feldhaus, H. Weise (Eds.), Free Electron Lasers 1999, Vol. II, Elsevier, Amsterdam, 2000, p. 101.
[15] S.V. Milton, et al., Nucl. Instr. and Meth. A 407 (1998) 210.
[16] S.V. Milton, et al., Proc. SPIE Int. Soc. Opt. Eng. 3614 (1999) 86.
[17] I.B. Vasserman, et al., in: Proceedings of the 1999 Particle Accelerator Conference, IEEE Press, New York, 1999, pp. 2489–2491.
[18] I.B. Vasserman, N.A. Vinokurov, R.J. Dejus, Phasing of the insertion devices at APS FEL Project, Proceedings of the 11th National Conference on Synchrotron Radiation Instrumentation, Stanford, CA, 1999; AIP Conference Proceedings, Vol. 521, Melville, New York, 2000, pp. 368–371.
[19] E. Gluskin, et al., Nucl. Instr. and Meth. A 429 (1999) 358.
[20] L.-H. Yu, S. Krinsky, R.L. Gluckstern, Phys. Rev. Lett. 64 (1990) 3011.
[21] Y.H. Chin, K.-J. Kim, M. Xie, Phys. Rev. A 46 (1992) 6662.
[22] M. Xie, in: Proceedings of the 1995 Particle Accelerator Conference, IEEE, Dallas, 1995, pp. 183–185.
[23] M. Xie, Exact and variational solutions of 3D eigenmodes for high-gain FELs, Nucl. Instr. and Meth. A 445 (2000) 59.
[24] M. Xie, Nucl. Instr. and Meth. A 475 (2001) 51, these proceedings.
[25] Z. Huang, K.-J. Kim, Nucl. Instr. and Meth. A 475 (2001) 59, these proceedings.
[26] R. Bonifacio, et al., Phys. Rev. Lett. 73 (1994) 70.
[27] E.L. Saldin, E.A. Schneidmiller, M.V. Yurkov, DESY Report TESLA-FEL 97-02, 1997.
[28] S.V. Milton, et al., Phys. Rev. Lett. 85 (2000) 988.
[29] A. Lumpkin, et al., Nucl. Instr. and Meth. A 429 (1999) 336.
[30] A.H. Lumpkin, et al., Nucl. Instr. and Meth. A 475 (2001) 462, these proceedings.
[31] H.P. Freund, S.G. Biedron, S.V. Milton, Nucl. Instr. and Meth. A 445 (2000) 53.

ELSEVIER

Nuclear Instruments and Methods in Physics Research A 475 (2001) 38–42

**NUCLEAR
INSTRUMENTS
& METHODS
IN PHYSICS
RESEARCH**
Section A

www.elsevier.com/locate/nima

First lasing of KHI FEL device at the FEL-SUT

M. Yokoyama[a],*, F. Oda[a], K. Nomaru[a], H. Koike[a], M. Sobajima[a], H. Miura[a],
H. Hattori[a], M. Kawai[a], H. Kuroda[b]

[a] *Kanto Technical Institute Kawasaki Heavy Industries, Ltd., 118 Futatsuzuka, Noda, Chiba 278-8585, Japan*
[b] *IR FEL Research Center, Research Institute for Science and Technology, The Science University of Tokyo,
2641 Yamazaki, Noda, Chiba 278-8510, Japan*

Abstract

The first lasing of the KHI FEL device installed at the FEL-SUT was obtained at 9 μm on July 6, 2000. The FEL device consists of a 40 MeV accelerator, a beam transport system with two 25° bending magnets, an undulator of 40 periods and 32 mm period length, and a cavity mirror system. The output power was 10^5 times that of the spontaneous emission. The tuning range of the cavity length is about 12 μm. © 2001 Elsevier Science B.V. All rights reserved.

PACS: 41.60.C; 07.57.Hm

Keywords: FEL; Infrared; First lasing linac; RF-gun

1. Introduction

The IR FEL Research Center at the Science University of Tokyo (FEL-SUT) was established in April 1999 at the Noda Campus of the Science University of Tokyo. The main purpose of this research center is to develop new applications of the strong pico-second infrared pulse from the IR FEL in physics, chemistry, biology, materials science and so on [1]. Kawasaki Heavy Industries, Ltd. (KHI) has designed and assembled a compact FEL device to use as the IR FEL source for the center. It has a linac with a magnet, RF-gun, and 3 m accelerator tube similar to the Mark III. The

FEL device can provide a powerful and tunable (5–16 μm in the infrared region) light source [2–3]. The operation of the device was initiated in January 2000. Spontaneous emission was obtained in February 2000. The micro-bunch length was tuned and measured by using a streak camera. The beam energy spread was measured using a CCD camera. After completing the tuning of the micro-bunch, on July 6, the first lasing at 9 μm was achieved.

2. KHI FEL device

The FEL device consists of a 40 MeV accelerator, a beam transport system with two 25° bending magnets, an undulator of 40 periods, each 32 mm in length, and a cavity mirror system. The

---

*Corresponding author. Tel.: +81-471-24-0258; fax: +81-471-24-5917.

E-mail address: yokoyama_minoru@now.khi.co.jp (M. Yokoyama).

0168-9002/01/$ - see front matter © 2001 Elsevier Science B.V. All rights reserved.
PII: S 0 1 6 8 - 9 0 0 2 (0 1) 0 1 5 3 1 - 5

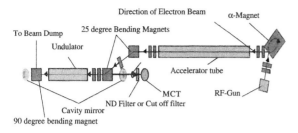

Fig. 1. Layout of the KHI-FEL device.

40 MeV accelerator is mainly composed of an OCS type RF-gun [4–6], an alpha magnet, and a 3 m accelerator tube. Fig. 1 shows the layout of the KHI FEL device and the set-up for optical experiments. The fundamental parameters of the accelerator components and the FEL light on the FEL device are shown in Table 1. The performance is shown in detail in Ref. [3].

A Fabry–Perot type cavity was adopted as the optical cavity, an important feature for the FEL experiment. The cavity length is 3359 mm, containing 32 RF periods. The distance from the end of the undulator to the downstream mirror is 1.217 m, while the distance from the entrance of the undulator to the upstream mirror is 0.862 m. Each mirror is held by a gimbal mount carried by a precision translation XYZ stage. Longitudinal 'Z' translation is controlled by a piezoelectric fine control in addition to a micrometer, in order to adjust the cavity length. Horizontal 'X' and vertical 'Y' tilts of these mirrors are also driven by micrometers. The cavity mirrors are made from copper coated with gold. The reflection of the mirrors is 99% in 5–16 μm. The curvatures of the upstream and the downstream mirrors are 2.0 and 1.6 m, respectively, to make the optical beam center coincide with the undulator center and to ensure the stability of the cavity. (G parameter is $0 < 0.746 < 1$.) The Rayleigh length is 0.45 m. The beam sizes at the waist point and the upstream mirror are 0.8–1.5 and 3.3–6.2 mm, respectively. The upstream mirror has a hole of $\varnothing$ 1 mm for the extraction of FEL light. The extraction loss is 0.78–0.14%. The loss in the 7 mm vertical undulator chamber is 0.02–2.3%. Hence the total cavity loss is 2.8–4.5%.

Table 1
The design values of the fundamental parameters of KHI FEL device and its main components

Accelerator	
Beam macropulse length (μs)	6
Beam energy in the undulator (MeV)	32
Micropulse peak current (A)	> 30
Energy spread(FWHM) (%)	~0.5
Normalized emittance (πmm rad)	11
Micro-pulse length (ps)	2
Repetition rate (Hz)	10
RF power supply	
Frequency (MHz)	2856
Power (MW)	45
Flat top of macro-pulse (μs)	7
Fluctuation of power(flat top) (%)	5
Fluctuation of phase(flat top) (°)	< 3
Accelerator tube	
Type	C.G
Length (mm)	3072
Number of cell	86
Shunt impedance (MΩ)	53
Filling time (μs)	0.83
Undulator	
Number of periods	40
Undulator period (mm)	32
Maximum magnetic field (T)	0.83
Minimum gap (mm)	8.6
Laser	
Wave length (μm)	5–16
Peak power (MW)	> 1
Micro pulse duration (ps)	~1
Macro pulse duration (μs)	~1

3. Experiments

3.1. Beam transport

A beam current of 200 mA with macro-pulse duration of 1.5 μs has been successfully passed through the undulator by tuning of the beam transport. The macro-pulse duration is restricted by the back-bombardment occurring at the RF-gun [6].

3.2. Measurement of the energy spread

The energy spread of the beam accelerated to 32 MeV was estimated by measuring the beam size at a position downstream of the 25° bending magnet. The lattice was designed to have a dispersion of 0.1 m and a horizontal betatron function of 0.1 m at the position. A polished aluminum slit was then inserted into the beam orbit. The OTR light from the slit was guided to a CCD camera. The intensity of the light obtained on the picture was analyzed. The relations between beam size and RF phase are shown in Fig. 2. In the case of setting the magnetic field gradient of the alpha magnet to 0.9–1.2 T/m, all minimum beam sizes were about 0.7 mm. The energy spread was estimated to be less than 0.7%.

3.3. Micro-bunch length measurement

The most important thing to do is to shorten the micro-pulse length to 2 ps, because shortening of the micro-pulse length contributes to the peak current term of FEL gain directly. The shortening is carried out by tuning the magnetic field gradient of the α-magnet [2]. A streak camera (HAMA-MATSU FESCA-200) was used for measurement of the micro bunch length. The OTR light from the same aluminum target used in the energy spread measurement was directed to the streak camera [6]. The marker triangles in Fig. 3 show the measured tuning curve of the micro-bunch length. At about 1.5 MeV beam energy from the RF-gun, the micro-bunch length reached the minimum magnetic field

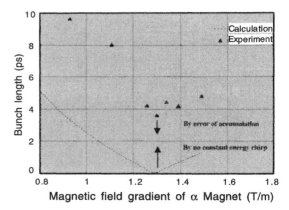

Fig. 3. Tuning curve of the micro-bunch length.

gradient of about 1.3 T/m. The minimum length of the micro-pulse was about 3.4 ps. However, in this measurement, 20 signals were accumulated to obtain 1 data point. This was because the signal obtained for one micro-pulse was small. Taking this effect into account, the minimum bunch length was estimated to be less than 2.5 ps. In Fig. 3, the dashed line is the calculated tuning curve when the negative energy chirp of the beam from the RF-gun is 10 keV/ps and constant in time domain. (The top energy and the energy cut by the slit are 1.5 MeV and less than 1.43 MeV, respectively.) The constant negative energy chirp of 10 keV/ps is estimated by calculation using EMSYS code [6]. The bunching point at 1.3 T/m of the experiment strongly agrees with the simple calculation. It is thought that the difference between the calculation bunch length and the measured one is due to the absence of a constant negative energy chirp in the time domain.

3.4. Lasing experiment

Spontaneous emission measurement was done as a first check of the cavity system. An MCT detector was set upstream of the upstream cavity mirror with its hole through a focusing lens. Instead of the spontaneous emission from the undulator, a probe laser light was reflected by a plane mirror near the downstream cavity mirror, passed through the hole of the upstream mirror and a ZeSe window, and directed to the sensitive

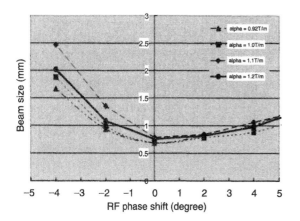

Fig. 2. Relation between the beam size behind the 25° magnet and the RF phase.

area of the MCT. The signal induced in the MCT by spontaneous emission was obtained by angular adjustments (horizontal and vertical) of the downstream mirror. Beam energy was fixed by two cut-off filters of 5 and 8.5 μm. They were set at the front of the MCT detector one at a time. The relation between the undulator gap and the wavelength of spontaneous emission was checked, as is shown in Fig. 4. The dashed curve is given by calculations using the measured magnetic field strength of the undulator. The two marker squares are the points measured by the two filters. These are in agreement with the estimated line.

First lasing was obtained at about 9 μm by tuning the cavity length and maintaining resonance between the two cavity mirrors. Fig. 5(a) shows the signals of the MCT detector and the beam current monitor at the first lasing. Effective macro-pulse duration of the beam current was 1.5 μs. At 0.7 μs, rapid increase of the signal occurred. From 0.8 μs, saturation of the detector occurred. Fig. 5(b) shows the signal of the lasing light passing through a ND filter of 0.001% transmittance. By analyzing the decrease of the signal, the cavity loss is estimated to be 3% and is in agreement with the estimated value. From the rising slope of the signal, the gain is estimated to be more than 20%. In order to reach the FEL saturation, a macro-pulse duration of more than 2.5 μs is required. The measured tuning curve of

(a) Without a ND filter

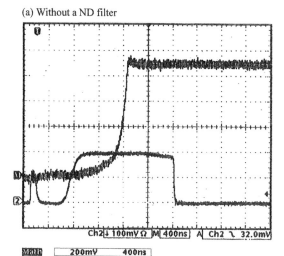

(b) With a ND filter of 0.001 % transmittance

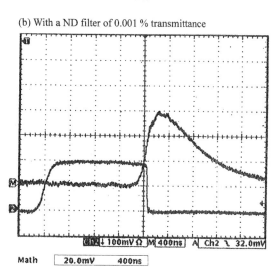

Fig. 5. Photographs of first lasing. Channel '2' and 'M' are the signals of the beam current monitor and the MCT signal, respectively. (a) the MCT signal of the lasing light without a ND filter, (b) the signal with a ND filter of 0.001% transmittance.

the cavity length is shown in Fig. 6. It has a width of approximately 12 μm.

4. Summary

First lasing of the KHI FEL device at the FEL-SUT was achieved at 9 μm. The output power was 10^5 times that of the spontaneous emission. The

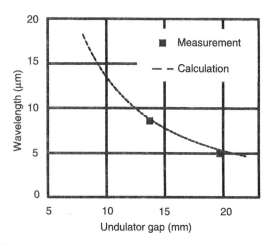

Fig. 4. Relation between the undulator gaps and the wavelength.

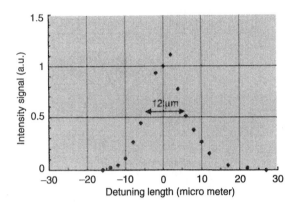

Fig. 6. Tuning curve of the cavity length.

obtained gain and the cavity loss were more than 20% and 3%, respectively. The tuning range is about 12 μm.

In a few months, the macro-pulse duration will be made longer by the improvement of the RF-gun cathode. Then saturation of the FEL lasing will be accomplished.

Acknowledgements

The authors would like to thank Dr. Tomimasu of the former FELI director, Prof. Brau and Dr. Gabella of Vanderbilt University for their helpful discussions and technical support.

References

[1] H. Kuroda, A. Iwata, M. Kawai, Proceedings of the 12th Russian Synchrotron Radiation Conference SR98, Novosibirsk, 1998.
[2] M. Yokoyama, F. Oda, A. Nakayama, K. Nomaru, M. Kawai, Nucl. Instr. and Meth. A 429 (1999) 269.
[3] M. Yokoyama, F. Oda, A. Nakayama, K. Nomaru, H. Koike, M. Kawai, H. Kuroda, FEL Conference Proceeding, 1999, 205pp.
[4] F. Oda, M. Yokoyama, M. Kawai, A. Nakayama, E. Tanabe, Nucl. Instr. and Meth. A 429 (1999) 332.
[5] F. Oda, M. Yokoyama, M. Kawai, A. Nakayama, E. Tanabe, Nucl. Instr. and Meth. A 445 (2000) 404.
[6] F. Oda, M. Yokoyama, M. Kawai, H. Koike, M. Sobajima, Nucl. Instr. and Meth. A 475 (2001) 583, these proceedings.

ELSEVIER

Nuclear Instruments and Methods in Physics Research A 475 (2001) 43–46

**NUCLEAR
INSTRUMENTS
& METHODS
IN PHYSICS
RESEARCH**
Section A

www.elsevier.com/locate/nima

Third-harmonic lasing at JAERI-FEL

R. Hajima*, R. Nagai, N. Nishimori, N. Kikuzawa, E.J. Minehara

Japan Atomic Energy Research Institute, 2–4 Shirakata-Shirane, Tokai-mura, Ibaraki 319-1195 Japan

Abstract

We have demonstrated FEL lasing at the third harmonic in JAERI-FEL driven by a superconducting linac. The lasing was achieved with a super enhanced gold-coated mirror tuned to 7 μm, which is a gold-coated copper mirror with Zn–Se multi-layer deposited on the surface. Lasing at the third harmonic is in the single-supermode regime with an average power of 15 W. The small-signal gain is estimated from the cavity-length detuning curve found to be 7.5%. © 2001 Elsevier Science B.V. All rights reserved.

PACS: 41.60.Cr; 42.65.Ky; 29.17.+w

Keywords: Free-electron laser; Superconducting linac; Harmonic lasing

1. Introduction

A high-power free-electron laser driven by a superconducting linac has been developed at Japan Atomic Energy Research Institute (JAERI). The FEL is operated at wavelengths between 16 and 23 μm and an average power of 1.7 kW in quasi-CW mode [1]. Operation of the FEL in a broader wavelength range is preferable for using the FEL in various applications. Lasing around 7 μm is especially useful, because there is a strong absorption of human tissue applicable to laser surgery. It is also known that toxic materials such as dioxin have absorption peaks around 7 μm, and decomposition of such materials can be realized with coherent radiation tuned to the absorption [2].

We made an experiment of the third-harmonic lasing around 7 μm at JAERI-FEL and observed the lasing successfully. In the present paper, the experimental procedure and the result of the third-harmonic lasing are described.

2. Experiment

For the lasing of higher-harmonics in FEL oscillators, we must suppress the lasing of the fundamental wavelength which has the largest gain among all the harmonics. The suppression of the fundamental lasing has been achieved in several ways: using intra-cavity dispersive material and separating round-trip time of the fundamental and the third harmonic [3], inserting a scraper to reduce the cavity quality-factor for the fundamental [4], using a dielectric multi-layer mirror having excellent reflectivity only at the higher-harmonics [5].

---

*Corresponding author. Tel: +81-29-282-6315; fax: +81-29-282-6057.

E-mail address: hajima@popsvr.tokai.jaeri.go.jp (R. Hajima).

0168-9002/01/$ - see front matter © 2001 Elsevier Science B.V. All rights reserved.
PII: S 0 1 6 8 - 9 0 0 2 (0 1) 0 1 5 3 2 - 7

The mirrors we used for the fundamental lasing are gold-coated copper mirrors with 120 mm in diameter. Because the mirrors have constant reflectivity of 99.4% for wavelengths longer than 4 μm, we cannot select the third harmonic only. We decided, therefore, to replace the cavity mirror by another one tuned to 7 μm for the third-harmonic lasing.

Two types of mirrors were prepared for the third-harmonic lasing around 7 μm: a Zn–Se multi-layer mirror of 80 mm in diameter and a super enhanced gold-coated mirror (SEG mirror) of 120 mm in diameter. The latter is a gold-coated copper mirror with several layers of Zn–Se deposited on the surface and has a reflectivity dependent on wavelength. The SEG mirror has a maximum reflectivity of 99.89% at 7.0 μm and still has good reflectivity over 99.2% between 5.5 and 9.1 μm. The former mirror, a Zn–Se multi-layer mirror, has reflectivity over 99.2% between 6.2 and 7.9 μm, which is narrower than the SEG mirror. The broadband reflectivity of a SEG mirror is useful for operating an FEL with varying wavelength. A SEG mirror is also easy to fabricate even for large diameter in comparison with a conventional dielectric multi-layer mirror. This is an advantage of a SEG mirror for JAERI-FEL as we see later.

Lasing at 7.13 μm was obtained by using the SEG mirror. The extraction of FEL power was made by a small scraper mirror and the total loss of the optical cavity was measured at 1.94% from the decay of FEL macro-pulses. Other experimental parameters are listed in Table 1. A cavity-

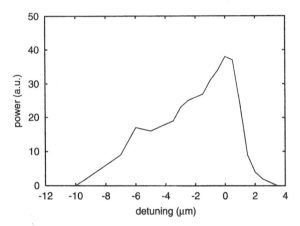

Fig. 1. Extracted FEL power as a function of cavity-length detuning, where zero-detuning is defined for the maximum power.

length detuning curve for the lasing is shown in Fig. 1. The maximum FEL power extracted is 15 W as an averaged value in a macro-pulse. The lasing spectrum was measured by a monochromator and a Hg–Cd–Te detector. Fig. 2 shows the lasing spectrum of the third harmonic at the maximum output power. This wavelength spectrum suggests that the lasing is in the single-supermode regime.

We made a similar experiment by using the Zn–Se mirror, but lasing never occurred. The reason why we could see lasing only with the SEG mirror is discussed in the next section.

3. Discussion

The small-signal gain in our experiment is estimated from the detuning curve together with an analytical formula [6] and found to be $G_0 = 7.5\%$, which is much smaller than the parameters obtained for the fundamental lasing. This difference of gain parameter explains lasing behavior observed for both the fundamental and the third harmonic. The lasing at the third harmonic is in the single-supermode regime, while lasing behavior at the fundamental shows characteristics of high-gain FEL oscillators: that is, spiking-mode and superradiance with broadening of the lasing spectrum [1, 7].

Table 1
Experimental parameters for the third-harmonic lasing

Electron beam	
Energy	16.5 MeV
Peak current	100 A
Bunch width	5 ps (FWHM)
Repetition	10.4 MHz
Undulator	
Undulator parameter (rms)	0.7
Undulator pitch	3.3 cm
Number of period	52

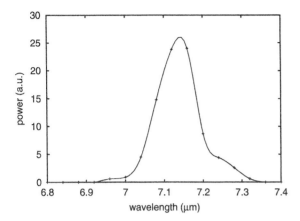

Fig. 2. Measured lasing spectrum at the maximum output power.

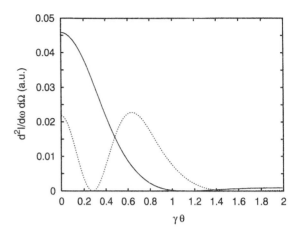

Fig. 3. Angular divergence of spontaneous radiation in the wiggling plane calculated for JAERI-FEL. Solid line is the fundamental and dotted line is the third harmonic.

The reduction of small-signal gain in the third harmonic is inevitable due to the smaller coupling between the electrons and the optical field for the higher-harmonics, which is represented by the Bessel function term in the formula of small-signal gain [8]. It is also known that the energy spread of electrons degrades the FEL gain. Inhomogeneous gain reduction caused by energy-spread in the single-supermode regime is estimated as $f_E = 1/(1 + 1.7(NN_w\sigma_E)^2)$, where N is the harmonic number and σ_E is rms energy-spread [9]. This formula shows that the gain reduction by energy spread becomes significant for the higher harmonic. The electron bunch in our experiments has relatively large energy-spread due to coherent synchrotron radiation force during a 180° arc before the undulator. The rms energy-spread at the undulator is estimated at 0.8% from the results of the fundamental lasing [7], and the reduction of the small-signal gain for the third harmonic is calculated as $f_E = 0.27$.

Another issue to discuss here is the reason why lasing did not occur with the Zn–Se mirror. We consider that the Zn–Se, which has smaller dimension than the SEG mirror, could not store enough spontaneous radiation to start up the lasing. The optical cavity of the JAERI-FEL is 14 m long and the undulator is installed at the center of the cavity. The angular divergence of the spontaneous radiation from an undulator is roughly expressed as $\gamma\theta \sim 1$, which corresponds to

the mirror diameter of 420 mm for the JAERI-FEL parameters.

For quantitative analyses, the angular distribution of the spontaneous radiation is calculated by using an analytical formula [10][1]. In the following, we show the angular distribution calculated only for the wiggling plane, because the undulator duct has flat shape with large dimension in the wiggling plane. Inhomogeneous broadening of the spontaneous radiation is ignored for simplicity.

The angular distribution of the fundamental and the third harmonic is plotted in Fig. 3. The third harmonic has large divergence, and the diameter of cavity mirror may be critical for storing the spontaneous radiation for the FEL seed light.

4. Summary

Lasing at third harmonic has been demonstrated in JAERI-FEL. The wavelength and the averaged-power extracted are 7.13 μm and 15 W, respectively. The small-signal gain is 7.5%, and the lasing is still in the single-supermode regime. We hope to increase the FEL power by optimizing the

---

[1] Disagreement between analytical and numerical results for a large undulator parameter seen in this paper can be completely fixed by correcting Eq. (I.4.27) : $(f'\xi) \rightarrow (f\xi)$, $(f'Z) \rightarrow (fZ)$.

electron beam and optical cavity and to explore possible applications in the wavelength.

References

[1] N. Nishimori, et al., Nucl. Instr. and Meth. A 475 (2001) 266, these proceedings.

[2] T. Yamauchi, et al., Nucl. Instr. and Meth., in these proceedings.

[3] S.V. Benson, M.J. Madey, Phys. Rev. A 39 (1989) 1579.

[4] K.W. Berryman, T.I. Smith, Nucl. Instr. and Meth. A 375 (1996) 6.

[5] S. Benson, M. Shinn, G.R. Neil, T. Siggins, Proceedings of the 21st International FEL Conference, Hamburg, Germany, 1999, Elsevier Science B. V., Amsterdam, 2000, p. II-1.

[6] G. Dattoli, et al., Lecture on the Free Electron Laser Theory and Related Topics, World Scientific, Singapore, 1993.

[7] R. Hajima, et al., Nucl. Instr. and Meth. A 475 (2001) 43, these proceedings.

[8] C.A. Brau, Free-Electron Lasers, Academic Press, New York, 1990.

[9] G. Dattoli, et al., IEEE Quantum Electron. 20 (1984) 637.

[10] R. Barbini, et al., Rev. Nuovo Cimento 13 (1990) 1.

ELSEVIER

Nuclear Instruments and Methods in Physics Research A 475 (2001) 47–50

NUCLEAR
INSTRUMENTS
& METHODS
IN PHYSICS
RESEARCH
Section A

www.elsevier.com/locate/nima

First lasing of the KAERI compact far-infrared free-electron laser driven by a magnetron-based microtron

Young Uk Jeong[a],*, Byung Cheol Lee[a], Sun Kook Kim[a], Sung Oh Cho[a],
Byung Heon Cha[a], Jongmin Lee[a], Grigori M. Kazakevitch[b], Pavel D. Vobly[b],
Nicolai G. Gavrilov[b], Vitaly V. Kubarev[b], Gennady N. Kulipanov[b]

[a] *Korea Atomic Energy Research Institute, P. O. Box 105, Yusong, Taejon 305-600, South Korea*
[b] *Budker Institute of Nuclear Physics, Lavrentyev ave. 11, 630090 Novosibirsk, Russia*

Abstract

The KAERI compact far-infrared (FIR) free-electron laser (FEL) has been operated successfully in the wavelength range of 97–150 μm. It is the first demonstration of FEL lasing by using a magnetron-based classical microtron. We developed a high precision undulator consisting of 80 periods, with each period being 25 mm. The field strength of the undulator can be changed from 4.5 to 6.8 kG with an amplitude deviation of only 0.05% in r.m.s. value. The kinetic energy of the electron beam is 6.5 MeV. The average current and pulse duration of the electron beam macropulses are 45 mA and 5.5 μs, respectively. The measured power of the FEL with the electron beam parameters was more than 50 W for a FIR macropulse having a duration of 4 μs. The spectral width of the FEL was measured to be 0.5% of the central wavelength. The FEL system, aside from the racks for the controlling units, is compact enough to be located inside an area of $3 \times 4 \, m^2$. © 2001 Elsevier Science B.V. All rights reserved.

PACS: 41.60.Cr; 52.75.Ms

Keywords: Lasing; Free-electron laser; Microtron; Far infrared; Compact FEL

1. Introduction

There is an increasing demand for tunable sources in the wavelength range of far infrared (FIR) for applications in solid-state physics, plasma physics, chemistry, bio-medical research, and surface science [1–3]. To make the free-electron laser (FEL) a potent source for such applications, its size and cost should be within the

*Corresponding author. Tel.:+82-42-868-8342; fax:+82-42-861-8292.

E-mail address: yujung@kaeri.re.kr (Y.U. Jeong).

means of an individual research laboratory. Progress has been made toward a compact FIR FEL using radio frequency (RF) guns as the electron beam source [4,5]. The compactness and beam quality of such an accelerator, having an energy range of several MeV, are crucial factors for FEL applications. A microtron using an RF generator for a magnetron has several noticeable advantages in cost, size, beam quality and operational convenience.

We have developed a compact FIR FEL driven by a magnetron-based microtron [6,7]. The first lasing in the wavelength range of 97–150 μm was achieved at the end of 1999, after successful results with the frequency stabilization of the RF, a high-performance undulator, and a complex optical resonator with a high coupling ratio between electron beam and radiation mode. The lasing results of the FEL are shown and discussed in this report, with a brief description of the facility.

2. Microtron and beam line

The schematic layout of the FEL system is shown in Fig. 1. The main parameters of the system are listed in Table 1. The electron energy from the microtron is 6–7 MeV depending on the magnetic strength of its main magnet. The macropulse duration of the electron beam is 5.5 μs and the pulse current is up to 70 mA. The microtron is composed of a magnetic-vacuum system, an RF system, and a modulator for the magnetron. The volume of the total system is only 2 m³. Details on the microtron are described in Ref. [8].

The RF system of the microtron was designed to stabilize the magnetron frequency. The stabilization of the magnetron due to the frequency pulling effect was realized by a short-length RF line and by allowing the reflected wave from an RF accelerating cavity to reach the magnetron. The length of the RF line from the magnetron to the

RF cavity is approximately 1 m and the isolation ratio for the reflected wave is <15 dB at the RF pulse power of 2 MW. The long-term fluctuation of the RF frequency is kept within the range of 40 kHz ($\Delta f/f_0 \sim 10^{-5}$) by keeping the temperature of the RF cavity constant to an accuracy of 1°C. The measured deviation of the RF frequency

Table 1
Main parameters of the KAERI FIR FEL

Electron beam	
Electron beam energy	6–7 MeV
Macropulse current	40–50 mA
Horizontal emittance	3.5 mm mrad
Vertical emittance	1.5 mm mrad
Energy spread	0.3–0.4%
Beam size at undulator entrance (FWHM)	2.1 mm × 0.6 mm
Macropulse duration	5.5 μs
Micropulse repetition frequency	2.8 GHz
Undulator (planar electromagnet)	
Period (Number of periods)	25 mm (80)
Peak magnetic Induction	4.5–6.8 kG
Gap	5.6 mm
Resonator	
Cavity length	2781 mm
Cylindrical mirror	R = 3000 mm
Waveguide gap	2.0 mm
Waveguide length	2778 mm
Radiation wavelength	97–150 μm

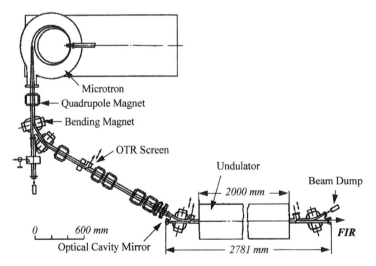

Fig. 1. Schematic layout of the KAERI compact FEL.

during the pulse is <150 kHz in the case of the optimized beam parameters for the laser. This value is much less than that measured for the magnetron itself. We have investigated the effect of the measured frequency fluctuations on the lasing process by simulation. The result shows sufficient gain of the FEL oscillation up to saturation power within the pulse width of 5 μs.

Electrons accelerated by the microtron are transported through the beam line with three bending magnets and six quadruples. The beam line is used for matching the input condition of the electron beam to the undulator for optimum laser operation. The beam line also includes a current transformer and an optical transition radiation (OTR) screen for electron beam diagnostics. An OTR screen with beam optics is being used to measure the Twiss parameters of the electron beam.

radiation mode. The cross-section of the waveguide is 2 mm × 30 mm. The 2780 mm long waveguide is centered to within 0.1 mm (height fluctuations). The diameter of the electron beam envelope in the vertical plane of the undulator is estimated to be <1 mm and the electron beam can be transported through the waveguide without loss. The out coupling mirror has a 0.7 mm diameter hole. The coupling ratio is approximately 2% of the intracavity power. The alignment accuracy and stability of the mirrors are better than 0.05 mrad. The electron beam is also aligned to the axis of the optical resonator with two OTR screens located at both ends of the undulator. The positioning accuracy and stability of the electron beam is approximately 0.1 mm. These accuracies satisfy the requirement of the confocal and waveguide mode resonator for lasing.

3. Undulator and optical cavity

The main issues concerning the design of a microtron-driven FIR FEL are the maximization of the FEL gain and the minimization of the gain reduction factors, including the resonator losses. To obtain enough gain for FEL oscillation, we have focused on developing a strong and high precision electromagnet undulator with an optimized length. The total length of the undulator is 2 m, containing 80 periods of length 25 mm. The configuration of the undulator is described in Ref. [9]. In this undulator, permanent magnets are used to reduce the saturation in iron poles. The magnetic field strength can be tuned from 4.5 to 6.8 kG by changing the current on the main coils. Deviation of the field amplitude from pole to pole all along the undulator was measured at the extremely low value of 0.05%. We have improved the accuracy of the undulator with a modular pole structure. This new structure makes fabrication and assembly much easier, as well as significantly reducing the cost and time for construction of the undulator.

The resonator of the FEL comprises a confocal scheme in the horizontal plane, with cylindrical mirrors and a parallel-plate waveguide with a gap of 2 mm in the vertical plane to increase the coupling between the electron beam and the

4. Lasing results

Two liquid helium cooled Ge : Ga detectors have been used to measure the power and the temporal evolution of the FIR radiation [10]. One of the detectors was calibrated at the wavelength of 118 μm by using a stable water-vapor laser.

Fig. 2 shows a typical waveform of the FEL oscillation. A temporal shape of the electron beam current is shown for comparison. The build-up time for the oscillation is <1.5 μs. The power of the FEL was stable and the fluctuation of the

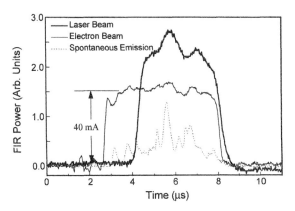

Fig. 2. Typical waveform of the FEL oscillation (solid line) and the spontaneous emission (dotted line) at a wavelength of 118 μm.

power is <1% in r.m.s value. The loss of the radiation in the resonator was measured to be 13% for a wavelength of 120 μm and is in good agreement with the calculated value. The calculated value of the net small-signal gain ranges from 15% to 25% depending on the radiation wavelength. The waveform of the spontaneous emission is shown in Fig. 2 as a dotted line. It should be noted that the power ratio between the laser and the spontaneous emission is more than 3000. The spontaneous emission power of the FEL is 100 times stronger than the calculated value. It is attributed to the coherent effect induced by the modulated structure of the electron beam micropulses [11,12]. The measured power of the spontaneous emission tends towards a quadratic dependence on the electron-beam current. The power enhancement of the spontaneous emission contributes to reducing the build-up time of the oscillation. The fluctuation of the laser pulse width is approximately 1 μs, which can be explained by the power fluctuation of the spontaneous emission.

Laser power dependence on the resonator length was measured for several wavelengths and it is shown in Fig. 3. The macropulse power of the laser was measured by using a pyroelectric detector. It was approximately 50 W for a wavelength of 120 μm. The calculated power from the simulation was approximately 80 W. The measured results of the power and dependence on the cavity detuning are in good agreement with those of the simulation.

Fig. 4 shows a spectrum of the laser beam measured using a Fabry–Perot spectrometer. The

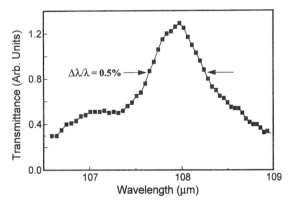

Fig. 4. Measured spectrum of the laser beam by using a Fabry–Perot spectrometer at a wavelength of 108 μm.

resolving power of the spectrometer is approximately 1000. The measured linewidth of the spectrum is $0.5\,\mathrm{cm}^{-1}$, $\Delta\lambda/\lambda = 0.5\%$, for a wavelength of 108 μm, and the Fourier transform limit associated with the laser micropulse is estimated to be $0.2\,\mathrm{cm}^{-1}$. The linewidth of the spectrum can be increased up to 1.2% by increasing the gain with a longer wavelength.

References

[1] S.J. Allen, Nucl. Instr. and Meth. B 144 (1998) 130.
[2] D. Dlott, M. Fayer, IEEE J. Quantum Electron. QE-27 (1991) 2697.
[3] E.V. Alieva, et al., Phys. Solid State 40 (1998) 213.
[4] I.S. Lehrman, et al., Nucl. Instr. and Meth. A 393 (1997) 178.
[5] J. Schmerge, et al., IEEE J. Quantum Electron. QE-31 (1995) 1166.
[6] R.R. Akberdin, et al., Nucl. Instr. and Meth. A 405 (1998) 195.
[7] Jongmin Lee, et al., Nucl. Instr. and Meth. A 407 (1998) 161.
[8] G. Kazakevich, et al., Compact Microtron Driven by a Magnetron as an Injector of the FIR Free Electron Laser, Presented at the 22nd International FEL Conference, Durham, USA, August 2000.
[9] Y.U. Jeong, et al., Nucl. Instr. and Meth. A 407 (1998) 396.
[10] V. Kubarev, et al., Proceedings on the Fourth Asian Symposium on FELs, Taejon, Korea, 1999, p. 369.
[11] Y.U. Jeong, et al., Phys. Rev. Lett. 68 (1992) 1140.
[12] G. Kazakevich, et al., Nucl. Instr. and Meth. A 475 (2001) 599, these proceedings.

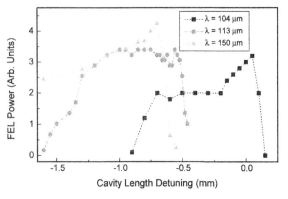

Fig. 3. Laser power dependence on the resonator length at wavelengths of 104, 113, and 150 μm.

ELSEVIER

Nuclear Instruments and Methods in Physics Research A 475 (2001) 51–58

NUCLEAR
INSTRUMENTS
& METHODS
IN PHYSICS
RESEARCH
Section A

www.elsevier.com/locate/nima

Grand initial value problem of high gain free electron lasers

Ming Xie*

Lawrence Berkeley National Laboratory, Berkeley, CA 94720, USA

Abstract

Initial value problem is one of the cornerstones in the framework of high gain free electron laser (FEL) theory. It determines the startup of FEL interaction from initial signal or noise in either laser field or electron beam. Yet, this problem was solved only for the cases without emittance and betatron oscillations. In this paper, we present the first solution to the initial value problem in a grand scale by expanding the startup theory into the six-dimensional phase space. In particular, with Gaussian beam distribution, explicit calculations of mode coupling from input field and noise power for self-amplified spontaneous emission (SASE) are presented. One of the major results is the discovery of excessively large noise power for SASE. © 2001 Elsevier Science B.V. All rights reserved.

PACS: 41.60.Cr; 05.40.−a; 04.20.Ex; 42.55.Vc

Keywords: 3D high gain FEL theory; SASE; Noise power; Initial value problem

1. Introduction

The most promising approach to generate intense coherent short wavelength radiation down to hard X-ray region is via the principle of single pass high gain free electron laser (FEL). A better understanding of high gain FEL requires more complete description and analysis of the physical reality, in which an electromagnetic field in three-dimensional (3D) configuration space interacts with a beam of electrons in six-dimensional (6D) phase space. For this purpose, the so called 3D high gain FEL theory has been developed persistently over the past two decades, becoming now the most active branch of FEL theory. Generally speaking, a high gain amplifier can be divided into three regions along the interaction distance:

initiation, growth, and saturation. Whereas the last region has to be handled by nonlinear theory and more often by simulation, the first two can be covered under the scope of linear theory. Furthermore, the linear theory can be classified into two major problems: eigenvalue problem (EVP) which is responsible for the power growth, and initial value problem (IVP) which determines the startup of FEL interaction from an initial condition. As FEL is pushed in the short wavelength frontier, it is becoming more important to take into account the reality associated with beam emittance and betatron oscillations in our theoretical framework. To this end, EVP has been solved to a level of great sophistication [1–5]. But for IVP, emittance and betatron oscillations have been neglected in all previous calculations [5–11]. As a result of lacking a solution to IVP in the presence of emittance and betatron oscillations, there has been no theory permitting a calculation of even the

*Tel.: + 510-486-5616; fax: + 510-486-6485.
E-mail address: mingxie@lbl.gov (M. Xie).

0168-9002/01/$ - see front matter © 2001 Elsevier Science B.V. All rights reserved.
PII: S 0 1 6 8 - 9 0 0 2 (0 1) 0 1 5 3 4 - 0

most rudimental quantities for the startup of an FEL amplifier, for instance, mode coupling from an input field and startup noise power for self-amplified spontaneous emission (SASE). In this paper, we present the first solution to IVP taking into account simultaneously the following effects: diffraction, optical guiding and finite bandwidth of laser field, energy spread, emittance and betatron oscillations of electron beam. In addition to deriving a general solution valid for arbitrary beam distribution, explicit calculations are performed for a Gaussian model.

2. Coupled Maxwell–Vlasov equations

Equations of motion for an electron in 6D phase space can be derived from the Hamiltonian $H_t(\mathbf{x}, \mathbf{P}_\perp, z, P_z; t)$ given the potentials: $\mathbf{A}_\perp = \mathbf{A}_w + \mathbf{A}_q + \mathbf{A}_r$, $A_z = 0$, $\Phi = 0$, where a planar wiggler field with natural focusing is specified by $\mathbf{A}_w = A_w [\text{ch}(k_{wx}x)\,\text{ch}(k_{wy}y)\sin(k_w z)\,\hat{\mathbf{x}} - (k_{wx}/k_{wy})\text{sh}(k_{wx}x)\,\text{sh}(k_{wy}y)\sin(k_w z)\hat{\mathbf{y}}]$, a sinusoidally alternating gradient quadrupole field by $\mathbf{A}_q = A_q[\text{sh}(k_{qx}x)\,\text{ch}(k_{qy}y)\sin(k_q z)\hat{\mathbf{x}} - (k_{qx}/k_{qy})\,\text{ch}(k_{qx}x)\,\text{sh}(k_{qy}y)\sin(k_q z)\hat{\mathbf{y}}]$, and a linearly polarized radiation field by $\mathbf{A}_r = A_r(\mathbf{x}, z, t)\sin[k_r z - \omega_r t + \phi_r(\mathbf{x}, z, t)]\hat{\mathbf{x}}$, with $k_w^2 = k_{wx}^2 + k_{wy}^2$ and $k_q^2 = k_{qx}^2 + k_{qy}^2$. In this paper, we limit our presentation to the case of equal focusing in both planes. Emphasizing short wavelength region, we also neglect space charge effect and consider only a long electron bunch. After some manipulations including taking wiggler-average over the scale $\lambda_w = 2\pi/k_w$, the Hamiltonian under paraxial approximation can be reduced to

$$H_z(\mathbf{x}, \mathbf{p}, \theta, \eta/k_r; z) = \frac{k_w}{k_r}\eta^2 + \frac{1}{1+\eta}\left[\frac{\mathbf{p}^2}{2} + \frac{K_1(z)\mathbf{x}^2}{2}\right.$$
$$\left. + K_2(z)(xp_x - yp_y) - \frac{f_B a_w |a_r|}{\gamma_0^2}\cos(\theta + \phi_r)\right] \quad (1)$$

where $\mathbf{p} = \mathbf{P}_\perp/\gamma_0 mc$, $\theta = (k_r + k_w)z - \omega_r t$, $\eta = (\gamma - \gamma_0)/\gamma_0$, γ_0 is the average value of γ related to the resonance condition by $k_r = 2\gamma_0^2 k_w/(1 + a_w^2)$, $a_w = eA_w/\sqrt{2}mc$, $a_q = eA_q/\sqrt{2}mc$, $A_w = B_w/k_w$, $A_q = \sqrt{2}B'_q/k_q^2$, B_w is the peak wiggler field, B'_q is the peak quadrupole field gradient,

$|a_r| = eA_r/\sqrt{2}mc$, $a_r = |a_r|\exp(i\phi_r)$, $f_B = J_0[a_w^2/2(1 + a_w^2)] - J_1[a_w^2/2(1 + a_w^2)]$, $K_1(z) = k_{\beta w}^2 + 2k_{\beta q}^2\sin^2(k_q z)$, $K_2(z) = \sqrt{2}k_{\beta q}\sin(k_q z)$, $k_{\beta w} = a_w k_w/\sqrt{2}\gamma_0$, $k_{\beta q} = a_q k_q/\sqrt{2}\gamma_0$. The Hamiltonian Eq. (1) can be used to study such effects as beam envelope modulations due to alternating gradient focusing, beam mismatching and synchro-betatron coupling. An extensive study of these subjects will be published elsewhere. In this paper we consider circumstances when betatron phase advance per cell is small in the periodic focusing channel. Under this condition, one may perform a quad-average of the Hamiltonian Eq. (1) over the scale $\lambda_q = 2\pi/k_q$, yielding

$$H_z = \frac{k_w}{k_r}\eta^2$$
$$+ \frac{1}{1+\eta}\left[\frac{1}{2}(\mathbf{p}^2 + k_\beta^2\mathbf{x}^2) - \frac{f_B a_w |a_r|}{\gamma_0^2}\cos(\theta + \phi_r)\right] \quad (2)$$

where $k_\beta^2 = k_{\beta w}^2 + k_{\beta q}^2$. Noting that with $|a_r| \ll a_w$ and $\eta \ll 1$, equations of motion to leading order can be derived from the Hamiltonian Eq. (2) as $d\mathbf{x}/dz = \mathbf{p}$, $d\mathbf{p}/dz = -k_\beta^2\mathbf{x}$, $d\theta/dz = 2k_w\eta - (k_r/2)(\mathbf{p}^2 + k_\beta^2\mathbf{x}^2)$, $d\eta/dz = -(k_r f_B a_w |a_r|/\gamma_0^2)\sin(\theta + \phi_r)$.

To properly treat beam fluctuations in SASE process, we use the Klimontovich distribution [12] to account for the discreteness of electrons in 6D phase space $\mathscr{F}(X; z) = (k_r/n_0)\sum_i^{N_e}\delta[X - X_i(z)]$, where $X = \{\mathbf{x}, \mathbf{p}, \theta, \eta\}$, n_0 is the peak volume density and N_e the total number of electrons. The distribution function can be further separated into two parts $\mathscr{F}(X; z) = F(Y) + f(X; z)$, where $Y = \{\mathbf{x}, \mathbf{p}, \eta\}$, F is the ensemble-averaged, unperturbed, smooth background distribution which is assumed to be uniform in θ and independent of z, and f contains both random fluctuations and coherent modulations. Upon introducing Fourier transform by $a_\nu = (1/\sqrt{2\pi})\int d\theta e^{-i\nu\theta}(e^{i\theta}a_r)$, $f_\nu = (1/\sqrt{2\pi})\int d\theta e^{-i\nu\theta}f$, where $\nu = \omega/\omega_r$, paraxial Maxwell equation and linearized Vlasov equation can be written as

$$\left[i\frac{\partial}{\partial z} + \frac{1}{2k_r}\frac{\partial^2}{\partial\mathbf{x}^2} - k_w\Delta\nu\right]a_\nu = -h_a\int d^2\mathbf{p}\,d\eta f_\nu \quad (3)$$

$$\left[i\frac{\partial}{\partial z} + i\left(\mathbf{p}\frac{\partial}{\partial\mathbf{x}} - k_\beta^2\mathbf{x}\frac{\partial}{\partial\mathbf{p}}\right) - \nu\xi\right]f_\nu = h_f\frac{\partial F}{\partial\eta}a_\nu \quad (4)$$

where $\Delta v = v - 1$, $\xi = 2k_w\eta - (k_r/2)(\mathbf{p}^2 + k_\beta^2\mathbf{x}^2)$, $h_a = 2\pi r_e a_w f_B n_0/\gamma_0 k_r$, $h_f = k_r a_w f_B/2\gamma_0^2$, $h_a h_f = h/16k_w L_{1d}^3$, $h = (2/\sqrt{3})^3$, L_{1d} is the 1D power gain length, and r_e is the classical radius of electron. Eqs. (3) and (4) are essentially those formulated by Kim [13], except the differences due to definition of the carrier wave and normalization of the distribution function. Our main objective is to solve IVP of Eqs. (3) and (4) completely.

3. General solution to the initial value problem

To obtain general solution to IVP, it is convenient to cast Eqs. (3) and (4) into a vectorized form of the Schrödinger type [10,14]

$$i\frac{\partial\Psi}{\partial z} = \mathcal{H}\Psi, \quad \Psi = \begin{bmatrix} a_v \\ f_v \end{bmatrix} \tag{5}$$

with the Hamiltonian operator expressed as

$$\mathcal{H} = \begin{bmatrix} (k_w\Delta v - \frac{1}{2k_r}\frac{\partial^2}{\partial\mathbf{x}^2}) & (-h_a\int \mathrm{d}^2\mathbf{p}\,\mathrm{d}\eta) \\ (h_f\frac{\partial F}{\partial\eta}) & (v\xi - i[\mathbf{p}\frac{\partial}{\partial\mathbf{x}} - k_\beta^2\mathbf{x}\frac{\partial}{\partial\mathbf{p}}]) \end{bmatrix}.$$

Since $\mathcal{H}$ is independent of z, Eq. (5) admits eigenvectors of the form $\Psi_n = V_n e^{-i\mu_n z}$. The determination of eigenvalues and eigenvectors $\{\mu_n, V_n\}$ constitutes an eigenvalue problem $\mathcal{H} V_n = \mu_n V_n$, where $V_n = [a_n(\mathbf{x}), f_n(\mathbf{x},\mathbf{p},\eta)]$. It is noted that the operator $\mathcal{H}$ is neither Hermitian nor self-adjoint, as such, its eigenvectors are not mutually orthogonal. However, one may construct an adjoint operator $\tilde{\mathcal{H}}$ and hence define an adjoint eigenvalue problem by $\tilde{\mathcal{H}}\tilde{V}_n = \tilde{\mu}_n\tilde{V}_n$, where $\tilde{V}_n = [\tilde{a}_n(\mathbf{x},\mathbf{p},\eta), \tilde{f}_n(\mathbf{x},\mathbf{p},\eta)]$, such that the adjoint eigenvectors $\{\tilde{V}_n\}$ are orthogonal to the original set $\{V_n\}$. This property is known as biorthogonality [15]. To carry out this procedure, we define the scalar product by $\langle V_n\tilde{V}_m\rangle_5 \equiv \langle a_n\tilde{a}_m + f_n\tilde{f}_m\rangle_5$, where $\langle\,\rangle_5 = \int \mathrm{d}^2\mathbf{x}\,\mathrm{d}^2\mathbf{p}\,\mathrm{d}\eta \equiv \int \mathrm{d}^5 Y$, and introduce two more shorthands for later use, $\langle\,\rangle_2 \equiv \int \mathrm{d}^2\mathbf{x}$ and $\langle\,\rangle_3 \equiv \int \mathrm{d}^2\mathbf{p}\,\mathrm{d}\eta$. Correspondingly, the adjoint operator is found to be

$$\tilde{\mathcal{H}} = \begin{bmatrix} (k_w\Delta v - \frac{1}{2k_r}\frac{\partial^2}{\partial\mathbf{x}^2}) & (h_f\frac{\partial F}{\partial\eta}) \\ (-h_a\int \mathrm{d}^2\mathbf{p}\,\mathrm{d}\eta) & (v\xi + i[\mathbf{p}\frac{\partial}{\partial\mathbf{x}} - k_\beta^2\mathbf{x}\frac{\partial}{\partial\mathbf{p}}]) \end{bmatrix}.$$

Indeed, it can be verified that biorthogonality holds, $\langle V_n\tilde{V}_m\rangle_5 = \delta_{nm}\mathcal{N}_n$, where $\mathcal{N}_n = \langle V_n\tilde{V}_n\rangle_5$, and the two sets of eigenvalues are identical, $\{\tilde{\mu}_n\} = \{\mu_n\}$. As a result, a formal solution to IVP can be expressed as $\Psi(z) = \sum C_n V_n e^{-i\mu_n z} + \cdots$, where $C_n = \langle \Psi(0)\tilde{V}_n\rangle_5/\langle V_n\tilde{V}_n\rangle_5$, and the four components $\{a_n, f_n, \tilde{a}_n, \tilde{f}_n\}$ in two eigenvectors $\{V_n, \tilde{V}_n\}$ can be determined through the following procedure. First, a_n is calculated by solving the eigenmode equation [4,5]

$$\left[\mu_n - k_w\Delta v + \frac{1}{2k_r}\frac{\partial^2}{\partial\mathbf{x}^2}\right]a_n(\mathbf{x}) = ih_a h_f$$

$$\times \int_{-\infty}^{\infty}\mathrm{d}^2\mathbf{p}\int_{-\infty}^{\infty}\mathrm{d}\eta\frac{\partial F}{\partial\eta}\int_{-\infty}^{0}\mathrm{d}s e^{-i(\mu_n-\xi)s}a_n(\mathbf{Q}^+) \tag{6}$$

then use $a_n = \langle \tilde{a}_n\rangle_3$, $f_n = -ih_f(\partial F/\partial\eta)\int_{-\infty}^{0}\mathrm{d}s \times \exp[-i(\mu_n-\xi)s]a_n(\mathbf{Q}^+)$, $\tilde{f}_n = ih_a\int_{-\infty}^{0}\mathrm{d}s\exp[-i(\mu_n-\xi)s]a_n(\mathbf{Q}^-)$, where $\mathbf{Q}^{\pm} = \mathbf{x}\cos(k_\beta s) \pm (\mathbf{p}/k_\beta)\sin(k_\beta s)$. To leading order we have set $v\xi = \xi$ since $|\Delta v| \ll 1$. The most complete solutions of Eq. (6), to be used later, were obtained for a Gaussian model [1,2]. Note the eigenvalue of Eq. (6) is related to that in Refs. [1,2] by $\mu = iq/2L_{1d} + k_w\Delta v$.

4. Applications of the solution

The general solution to IVP has opened the floodgate to solutions of many important problems that were simply not possible to solve before. Depending on how the initial condition $\Psi(0) = [a_v(\mathbf{x},0), f_v(\mathbf{x},\mathbf{p},\eta,0)]$ is specified, the solution can be applied, for example, to the following problems: mode coupling from an input field, startup noise power for SASE, dynamic connection between sections of a multi-segment wiggler, cascade harmonic generation, and transverse coherence of SASE when higher order modes are included in the analysis. We will investigate some of these problems in detail. First, let us express radiation power spectrum by a mode expansion $\mathrm{d}P(z)/\mathrm{d}\omega = \sum\sum \mathrm{d}P_{nm}(z)/\mathrm{d}\omega$, where the cross-terms are nonvanishing because the eigenmodes $\{a_n\}$ are in general not power-orthogonal.

For startup from an input field with initial condition $\Psi(0) = [a_v(\mathbf{x},0), 0]$, we have $\mathrm{d}P_{nm}(z)/\mathrm{d}\omega =$

$G_{nm} dP_{in}/d\omega \exp(-i\mu_n z + i\mu_m^* z)$, where $dP_{in}/d\omega = (mc^2/4\pi r_e l_r)\langle |a_v(0)|^2 \rangle_2$ is input power spectrum, l_r is input pulse length, and

$$G_{nm} = \frac{\langle a_v(0)a_n \rangle_2 \langle a_v(0)a_m \rangle_2^* \langle a_n a_m^* \rangle_2}{\mathcal{N}_n \mathcal{N}_m^* \langle |a_v(0)|^2 \rangle_2} \qquad (7)$$

is input coupling coefficient. If input bandwidth is significantly narrower than FEL gain bandwidth, one may use the form $a_v(\mathbf{x}, 0) = a_{in}(\mathbf{x})g_{in}(\nu)$ and express frequency integrated power as $P_{nm}(z) = G_{nm}P_{in} \exp(-i\mu_n z + i\mu_m^* z)$, where P_{in} is the total input power. For diagonal terms, the coefficient $G_n \equiv G_{nn}$ is a positive quantity which can be maximized by varying input field profile $a_{in}(\mathbf{x})$, yielding $a_{in}(\mathbf{x}) = a_n^*(\mathbf{x})$, where an asterisk indicates complex conjugate. This condition is known as conjugate input mode coupling [8], under which maximum power coupling to mode a_n is reached at $G_n = \langle |a_n|^2 \rangle_2^2/|\mathcal{N}_n|^2$. In 1D limit specified by $F(\mathbf{x}, \mathbf{p}, \eta) = \delta(\mathbf{p})\delta(\eta)$, where all transverse modes become degenerate, Eq. (7) gives the well-known result $G_{1d} = 1/9$ [11].

Initial condition for SASE can be specified by $\Psi(0) = [0, f_v(\mathbf{x}, \mathbf{p}, \eta, 0)]$. The term describing beam fluctuation, $f_v(\mathbf{x}, \mathbf{p}, \eta, 0)$, is governed by the statistics of shot noise that is completely uncorrelated in 6D phase space, defined by [12] $\langle \mathcal{F}(X; 0) \mathcal{F}(X'; 0) \rangle_{en} = (k_r/n_0)\delta(X - X')F(Y)$, where the angle bracket $\langle \rangle_{en}$ indicates ensemble average. SASE power spectrum after ensemble average is given by $dP_{nm}(z)/d\omega = P_{\omega nm} \exp(-i\mu_n z + i\mu_m^* z)$, where

$$P_{\omega nm} = \frac{mc^2 k_r^2}{8\pi^2 r_e n_0} \frac{\langle F \tilde{f}_n \tilde{f}_m^* \rangle_5 \langle a_n a_m^* \rangle_2}{\mathcal{N}_n \mathcal{N}_m^*}. \qquad (8)$$

It is conceptually useful to introduce an effective input power spectrum for startup from shot noise by expressing SASE power spectrum in the same form as for the case with an input field, hence, $P_{\omega nm} = G_{nm} dP_{sn}/d\omega$. Of course, unlike $dP_{in}/d\omega$, the effective noise power spectrum so defined, $dP_{sn}/d\omega$, depends on the modes indexed by n and m as well as on the choice of a fictitious input field profile. Consider the case when SASE is dominated before saturation by the fundamental mode, thus $dP(z)/d\omega = P_\omega \exp(2 \operatorname{Im}(\mu)z)$, where P_ω can be identified from Eq. (8) with $n = m = 0$. To obtain frequency integrated power, the factor with

dominant frequency dependence, $\exp[2 \operatorname{Im}(\mu)z]$, can be approximated by an expansion near the peak of the gain spectrum at ν_p, $2 \operatorname{Im}(\mu)z = z/L_g - \delta\nu^2/2\sigma_\nu^2 + \cdots$, where L_g is the power gain length corresponding to the peak gain, $\delta\nu = \nu - \nu_p$ and $\sigma_\nu = 1/\sqrt{\alpha_\omega(k_w L_{1d})(k_w z)}$. The factor α_ω can be determined by either a perturbation calculation or a fit to the gain curve near the peak. Defining effective SASE bandwidth by $\Sigma_\omega = \sqrt{2\pi\omega_r\sigma_\nu}$, one may express frequency integrated power as $P(z) = P_\omega \Sigma_\omega \exp(z/L_g)$, and hence by definition, $P_{sn} = P_\omega \Sigma_\omega/G_0$. In the 1D limit [5–7], we have $P_{\omega 1d} = (1/9)(\rho E_0/2\pi)$ and $\alpha_{\omega 1d} = 2/3$, where $\rho = 1/2\sqrt{3} k_w L_{1d}$ is the Pierce parameter and $E_0 = \gamma_0 mc^2$.

The first analysis of transverse coherence of SASE taking into account emittance and betatron oscillations was based on the solution to EVP [2,16]. There, a criteria on the transverse coherence was proposed. However, the criteria is approximate in the sense that assumptions were made on the quantities that cannot be calculated without a solution to IVP. Although reasonable, those assumptions are nevertheless heuristic. With the solution to IVP now at hand, the subject can be studied on a much more rigorous basis. A detailed treatment of this subject will be published elsewhere. Here, we provide some clarification of concept and definition of transverse coherence that are pertinent to SASE process. The theory of partial coherence can be formulated in terms of correlation function in either space-time domain $\{\mathbf{x}, \theta\}$, $\Gamma_c(\mathbf{x}_1, \mathbf{x}_2, \theta_1, \theta_2, z) = \langle a_r(\mathbf{x}_1, \theta_1, z)a_r^* (\mathbf{x}_2, \theta_2, z) \rangle_{en}$, or space-frequency domain $\{\mathbf{x}, \nu\}$, $W_c(\mathbf{x}_1, \mathbf{x}_2, \nu_1, \nu_2, z) = \langle a_v(\mathbf{x}_1, \nu_1, z)a_v^* (\mathbf{x}_2, \nu_2, z) \rangle_{en}$. Correspondingly, the coherence measures in long bunch limit are defined by [17]

$$\gamma_c(\mathbf{x}_1, \mathbf{x}_2, \Delta\theta, z) = \frac{\Gamma_c(\mathbf{x}_1, \mathbf{x}_2, \Delta\theta, z)}{\sqrt{\Gamma_c(\mathbf{x}_1, \mathbf{x}_1, 0, z)\Gamma_c(\mathbf{x}_2, \mathbf{x}_2, 0, z)}}$$

$$\mu_c(\mathbf{x}_1, \mathbf{x}_2, \nu, z) = \frac{W_c(\mathbf{x}_1, \mathbf{x}_2, \nu, z)}{\sqrt{W_c(\mathbf{x}_1, \mathbf{x}_1, \nu, z)W_c(\mathbf{x}_2, \mathbf{x}_2, \nu, z)}}$$

with $0 \leqslant |\gamma_c| \leqslant 1$ and $0 \leqslant |\mu_c| \leqslant 1$, where $\Delta\theta = \theta_2 - \theta_1$. When the fundamental mode is dominant, $a_v(\mathbf{x}, \nu, z) = C_0(\nu)a_0(\mathbf{x}, \nu)e^{-i\mu_0(\nu)z}$, we have $W_c(\mathbf{x}_1, \mathbf{x}_2, \nu, z) = e^{2 \operatorname{Im}[\mu_0(\nu)]z}Q_0(\nu)a_0(\mathbf{x}_1, \nu)a_0^*(\mathbf{x}_2, \nu)$, where $Q_0(\nu)$ is real. The factorized form with respect to

$\mathbf{x}_1$ and $\mathbf{x}_2$ gives $|\mu_c(\mathbf{x}_1, \mathbf{x}_2, \nu, z)| = 1$. Thus for SASE, the dominance of the fundamental mode inevitably implies full transverse coherence. It is important to note, however, that the coherence measures defined in different domains are not equivalent [18]. For the same single mode dominant case, full transverse coherence is not indicated in space-time domain. Using the relation $\Gamma_c(\mathbf{x}_1, \mathbf{x}_2, \Delta\theta, z) = (1/2\pi) \int d\nu \exp(-i\Delta\nu\Delta\theta) W_c(\mathbf{x}_1, \mathbf{x}_2, \nu, z)$, it can be shown that $\Gamma_c(\mathbf{x}_1, \mathbf{x}_2, \Delta\theta, z)$ cannot be expressed in a factorized form. As a result, it is true in general that $|\gamma_c(\mathbf{x}_1, \mathbf{x}_2, \Delta\theta, z)| < 1$. This is so because of the frequency dependence of the mode as in $a_0(\mathbf{x}, \nu)$. Only under circumstances when the mode variation is sufficiently small over the gain bandwidth determined by the term $\exp\{2\,\mathrm{Im}[\mu_0(\nu)]z\}$, the factorization can be recovered to give $|\gamma_c(\mathbf{x}_1, \mathbf{x}_2, \Delta\theta, z)| = \exp[-(\sigma_\nu\Delta\theta)^2/2]$, indicating that SASE in single mode has a longitudinal coherence length $l_c = 1/k_r\sigma_\nu$ and full transverse coherence since $|\gamma_c(\mathbf{x}_1, \mathbf{x}_2, 0, z)| = 1$. The frequency dependence of the fundamental mode is shown in the detuning curves [1] to be reasonably weak for the LCLS parameters [19].

5. Specific solution for Gaussian model

To perform specific calculations, the unperturbed distribution has to be given and correspondingly EVP must be solved at least for the fundamental mode. However, due to phase space mixing caused by emittance and betatron oscillations, the two integrals that have to be evaluated, $\langle f_n \tilde{f}_n \rangle_5$ and $\langle F \tilde{f}_n \tilde{f}_m^* \rangle_5$, become very complicated. As a result, the possibility of performing any specific calculation is ruled out practically for all solutions of the eigenmodes in numerical form. To overcome this complication, we make use of the analytical solutions obtained earlier by variational approximation for a Gaussian model [1,2], $F(\mathbf{x}, \mathbf{p}, \eta) = F_\perp(\mathbf{x}, \mathbf{p})F_\parallel(\eta)$, with $F_\perp(\mathbf{x}, \mathbf{p}) = (1/2\pi\sigma_x^2 k_\beta^2) \exp[-(k_\beta^2\mathbf{x}^2 + \mathbf{p}^2)/2k_\beta^2\sigma_x^2]$, $F_\parallel(\eta) = (1/\sqrt{2\pi}\sigma_\eta) \exp(-\eta^2/2\sigma_\eta^2)$, where σ_η is relative rms energy spread, $\sigma_x = \sqrt{\varepsilon/k_\beta}$ is rms beam size matched to the betatron focusing channel, and ε is rms emittance. These analytical solutions have been compared with the exact numerical ones [1,2] and

found highly accurate in the short wavelength region where diffraction is weaker. It is noted that another approximate solution, more suitable in the long wavelength region where diffraction is stronger, was also obtained for Gaussian model [3].

Next we present results for the fundamental mode with a profile given by $a_0(\mathbf{x}) = \exp[-\alpha(\mathbf{x}/\sigma_x)^2]$ [1], where α is a complex quantity known as the mode parameter. After lengthy manipulations, the final results are

$$G_0 = \frac{4\alpha\alpha^*}{(\alpha + \alpha^*)^2} \frac{1}{|1 + i4h\alpha I|^2} \qquad (9)$$

$$\frac{P_\omega}{P_{\omega 1d}^g} = \frac{72\alpha\alpha^*}{(\alpha + \alpha^*)} \frac{J}{|1 + i4h\alpha I|^2} \qquad (10)$$

where the coefficient G_0 is given under the condition of conjugate input coupling, P_ω is normalized by the Gaussian asymptotic 1D limit, $P_{\omega 1d}^g = (4/3)P_{\omega 1d}$, to be discussed later, I and J, originally each a seven-dimensional integral, are reduced to one-dimensional integrals as follows:

$$I = \int_0^\infty d\tau f_i(\tau), \quad J = \frac{1}{2\kappa_i}\left[\int_0^\infty d\tau f_j(\tau) + \text{c.c.}\right]$$

$$f_i(\tau) = \frac{\tau^2 e^{i\kappa\tau - 2\eta_\gamma^2\tau^2}}{(1 - i\eta_\varepsilon\tau)^2 + 4\alpha(1 - i\eta_\varepsilon\tau) + 4\alpha^2\sin^2(2\sqrt{\eta_d\eta_\varepsilon}\tau)}$$

$$f_j(\tau) = \frac{e^{i\kappa\tau - 2\eta_\gamma^2\tau^2}}{(1 - i\eta_\varepsilon\tau)^2 + 4\alpha_r(1 - i\eta_\varepsilon\tau) + 4|\alpha|^2\sin^2(2\sqrt{\eta_d\eta_\varepsilon}\tau)}$$

where $\kappa = 2L_{1d}\mu$, $\kappa_i = \mathrm{Im}(\kappa)$, $\alpha_r = \mathrm{Re}(\alpha)$, and the scaling parameters are defined by [1,20] $\eta_d = L_{1d}/2k_r\sigma_x^2$, $\eta_\varepsilon = 4\pi(L_{1d}/\lambda_\beta)k_r\varepsilon$, $\eta_\gamma = 4\pi(L_{1d}/\lambda_w)\sigma_\eta$, where $\lambda_\beta = 2\pi/k_\beta$. For LCLS nominal case [19], we have [1] $\eta_d = 0.0367$, $\eta_\varepsilon = 0.739$ and $\eta_\gamma = 0.248$. Another scaling parameter for frequency detuning, $\eta_\omega = 4\pi(L_{1d}/\lambda_w)\Delta\nu$, will be optimized for peak growth rate in all subsequent calculations. The input coupling coefficient for the fundamental mode, G_0 from Eq. (9), is plotted in Fig. 1, the normalized SASE spectral power, $P_\omega/P_{\omega 1d}^g$ from Eq. (10), is plotted in Fig. 2, all as functions of η_ε and η_γ with $\eta_d = 0.0367$.

It is shown in Fig. 1 that G_0 increases monotonically to magnitude even beyond unity for larger η_ε and η_γ. This result, suggesting that partial can be larger than the whole, seems to be in

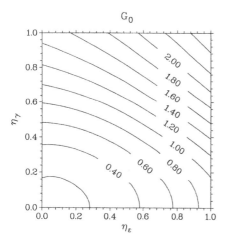

Fig. 1. Coutour plot of $G_0 \in \{\eta_\varepsilon, \eta_\gamma\}$ with $\eta_d = 0.0367$.

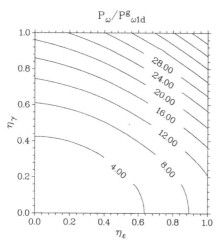

Fig. 2. Coutour plot of $P_\omega/P^g_{\omega 1d} \in \{\eta_\varepsilon, \eta_\gamma\}$ with $\eta_d = 0.0367$.

next to be responsible, at least partially, also for the excessively large noise power. It is noted that $G_0 > 1$ was first observed in a parallel beam case with energy spread [8].

One of the most surprising and important results of this paper is the discovery of excessively large noise power for SASE. It is shown in Fig. 2 that the noise power can be significantly larger than what previously known and the ratio increases monotonically without bound for larger η_ε and η_γ. Specific calculations of SASE noise power were performed previously only in two cases. In the case of 1D limit ($\eta_\varepsilon = \eta_d = 0$) specified by $F(\mathbf{x}, \mathbf{p}, \eta) = \delta(\mathbf{p})F_\|(\eta)$, where $F_\|(\eta)$ is a tophat profile, it was found that $P_\omega/P_{\omega 1d} \lesssim 1.7$ [6]. In the case of parallel beam limit ($\eta_\varepsilon = \eta_\gamma = 0$) specified by $F(\mathbf{x}, \mathbf{p}, \eta) = u_\perp(r)\delta(\mathbf{p})\delta(\eta)$, where $u_\perp(r)$ is again a tophat profile, it was found that $P_\omega/P_{\omega 1d} \lesssim 1$ [9]. Thus, one may take $P_{\omega 1d}$ as the reference magnitude for SASE noise power known from previous explorations of severely limited parameter space and models.

Although the normalized noise power increases for larger η_ε and η_γ, this effect is accompanied by a drop in normalized growth rate, κ_i. Nevertheless, the surprising effect may translate into real benefits for existing designs of SASE FELs. Let us define a factor of surprise by the ratio of frequency integrated noise power relative to the corresponding 1D value, $f_S = P(0)/P_{1d}(0) = (P_\omega/P_{\omega 1d})(\Sigma_\omega/\Sigma_{\omega 1d})$. Taking LCLS nominal case as an example, from Fig. 2 we have $P_\omega/P^g_{\omega 1d} = 7.2$, which gives $P_\omega/P_{\omega 1d} = 9.6$, and with $\Sigma_\omega/\Sigma_{\omega 1d} = 0.9$, the factor is $f_S = 8.6$. A reduction in saturation length is then given by $\Delta L_{sat} = -L_g \ln(f_S) = -13$ m, where $L_g = 6$ m [2]. Alternatively, the factor of surprise can be expressed as $f_S = (G_0/G_{1d})(P_{sn}/P_{sn1d})$. Thus, noting $G_0/G_{1d} = 6.7$ from Fig. 1, we see that most of the increase in noise power is due to the enhancement in input coupling for this case.

At first glance it is puzzling to see why should noise power, if it were just spontaneous undulator radiation from a collection of initially uncorrelated electrons, increase with emittance and energy spread. A preliminary remark is that it is the partial noise power coupled into the fundamental mode that concerns us directly. A

apparent violation with energy conservation. Since power in the fundamental mode at $z = 0$ is given by $P_0(0) = G_0 P_{in}$, thus $G_0 > 1$ would necessarily imply that initial power coupled into one mode out of many is larger than there is in the whole input field to start with. It turns out that this phenomenon is a direct consequence of the non-power-orthogonal nature of the eigenmodes. In this situation, conservation of energy is maintained by the nonvanishing cross-terms, of which some are negative. This intriguing paradox is shown

further observation is that the correlation between the partial power (Fig. 2) and the coupling coefficient for that mode (Fig. 1) is remarkable, though not complete. As explained before, the dramatic increase in the coupling coefficient, leading consequently to the paradox that partial can be larger than the whole, is a direct manifestation of non-power-orthogonal nature of the eigenmodes. In fact, we may conjecture from physical point of view that the non-power-orthogonal nature is a universal property of any high gain amplifier where radiation is coupled strongly with the source. Power-orthogonality, strictly speaking, is possible only in a "closed" system in which there is no exchange of energy between a laser and its environment. One such example is laser propagation in passive lossless fiber, where total power and power carried in each guided mode are all conserved quantities.

In principle, the noise power given by Eq. (10) can be calculated with 6D macroparticle simulations. To do that, statistics of startup noise has to be modeled effectively and a large number of required macroparticles have to be loaded efficiently. However, due to practical CPU limit and the lack of an appropriate multi-dimensional loading algorithm, no simulation code has been able to model the noise statistics in the full 6D phase space. For the well-known 6D code GINGER [21], startup noise is loaded only in 3D (θ and $\mathbf{x}$), whereas for another 6D code GENESIS [22], the noise is loaded only in 1D (θ). Noise loadings required for all other dimensions are neglected without proper justification. In fact, for all existing FEL simulation codes, benchmark on SASE startup has been done with only previously known theoretical results in 1D case [21] and parallel beam case [23]. With the first 6D calculation of SASE noise power presented here, we now have a unique benchmark, highly accurate in the short wavelength region, for further development in modeling and simulation of SASE process.

Finally, we discuss some newly discovered subtleties in 1D limit, which is an important reference point in the framework of 3D theory. Until recently, there has been only one 1D limit. Let us call it the fictitious 1D limit for it assumes an electron beam that is infinitely wide. In other

words, fictitious 1D limit is born 1D. Yet, it is generally believed that this limit can be approached asymptotically by increasing the transverse beam size from a finite distribution. This notion, however, needs further qualification because of a recent discovery of another 1D limit [1,2]. Only two asymptotic 1D limits have been carefully studied, tophat profile [8,9] and Gaussian profile [1,2,24]. For tophat profile, it is found that the asymptotic limit is same as the fictitious limit in all aspects, including eigenvalue, eigenmode (approaching a plane wave with flat phase front), input mode coupling coefficient and SASE noise power. Thus, the two limits can be considered identical. For Gaussian profile, however, agreement is found only in eigenvalue, not in eigenmode. As a result of the different asymptotic behavior in eigenmode, two more differences are discovered here. The two quantities from Gaussian asymptotic 1D limit, G_{1d}^{g} and $P_{\omega 1d}^{g}$, are related to that of the fictitious 1D limit by $G_{1d}^{g} = (4/3)G_{1d}$ and $P_{\omega 1d}^{g} = (4/3)P_{\omega 1d}$. For model dependent self-consistency, we have used the Gaussian asymptotic 1D limit as a normalization factor in Eq. (10).

6. Conclusions

We have presented the first solution to the grand initial value problem. It is the first time for the startup theory to reach 6D in climbing the mountain of dimensions. The solution has provided a rigorous foundation for the analysis of initiation and subsequent evolution of power and coherence in a high gain FEL amplifier before saturation. The specific solution developed for Gaussian model has been made so efficient that it is now possible to map out SASE noise power, input coupling and many other important quantities in the entire parameter space of interest to short wavelength FELs. In particular, the discovery of excessively large SASE noise power may have significant impact on current development of high gain FELs. Furthermore, the solution has provided a unique benchmark in noise power and other startup properties for modeling and simulation of SASE process in 6D phase space. Last but not least, this work has opened the floodgate to solutions of many more important pending problems.

Acknowledgements

This work was supported by the US DOE under Contract No. DE-AC03-76SF00098.

References

[1] M. Xie, Nucl. Instr. and Meth. A 445 (2000) 59.
[2] M. Xie, Nucl. Instr. and Meth. A 445 (2000) 67.
[3] Y. Chin, K.-J. Kim, M. Xie, Phys. Rev. A 46 (1992) 6662.
[4] L.-H. Yu, S. Krinsky, R. Gluckstern, Phys. Rev. Lett. 64 (1990) 3011.
[5] K.-J. Kim, Phys. Rev. Lett. 57 (1986) 1871.
[6] K.-J. Kim, Nucl. Instr. and Meth. A 250 (1986) 396.
[7] J.-M. Wang, L.-H. Yu, Nucl. Instr. and Meth. A 250 (1986) 484.
[8] G. Moore, Nucl. Instr. and Meth. A 250 (1986) 381.
[9] S. Krinsky, L.-H. Yu, Phys. Rev. A 35 (1987) 3406.
[10] M. Xie, D. Deacon, J. Madey, Nucl. Instr. and Meth. A 272 (1988) 528.
[11] A. Kondratenko, E. Saldin, Particle Accelerators 10 (1980) 207;
H. Haus, IEEE J. Quantum Electron QE17 (1981) 1427;
C. Shih, A. Yariv, IEEE J. Quantum Electron QE17 (1981) 1387.
[12] Y. Klimontovich, Statistical Physics, Harwood Academic, New York, 1986 (Chapter 8).
[13] K.-J. Kim, Nucl. Instr. and Meth. A 318 (1992) 489.
[14] K.-J. Kim, private communication, 1986.
[15] P. Morse, H. Feshbach, Methods of Theoretical Physics, McGraw-Hill, New York, 1953 (Chapter 7.5);
A. Siegman, Lasers, University Science Books, Mill Valley, 1986 (Chapter 21.7).
[16] M. Xie, Transverse coherence: from undulator radiation to self-amplified spontaneous emission, Invited Talk Presented at 1997 FEL Conference, also LBNL-40250, 1997, unpublished.
[17] L. Mandel, E. Wolf, J. Opt. Soc. Am. 66 (1976) 529.
[18] E. Wolf, G. Agarwal, J. Opt. Soc. Am. A 1 (1984) 541.
[19] LCLS Design Study Report, SLAC-R-521, 1998.
[20] M. Xie, IEEE Proceedings for PAC95, No.95CH3584, 1995, p. 183.
[21] W. Fawley, et al., IEEE Proceedings for PAC93, No.93CH3279-7, 1993, p. 1530.
[22] S. Reiche, Nucl. Instr. and Meth. A 429 (1999) 243.
[23] L.-H. Yu, Phys. Rev. E 58 (1998) 4991.
[24] M. Xie, D. Deacon, Nucl. Instr. and Meth. A 250 (1986) 426.

ELSEVIER

Nuclear Instruments and Methods in Physics Research A 475 (2001) 59–64

NUCLEAR
INSTRUMENTS
& METHODS
IN PHYSICS
RESEARCH
Section A

www.elsevier.com/locate/nima

Solution to the initial value problem for a high-gain FEL via Van Kampen's method ☆

Zhirong Huang*, Kwang-Je Kim

Advanced Photon Source, Argonne National Laboratory, 9700 S. Cass Avenue, Argonne, IL 60439, USA

Abstract

Using Van Kampen's normal mode expansion, we solve the initial value problem for a high-gain free-electron laser described by the three-dimensional Maxwell–Klimontovich equations. An expression of the radiation spectrum is given for the process of coherent amplification and self-amplified spontaneous emission. It is noted that the input coupling coefficient for either process increases with the initial beam energy spread. The effective start-up noise is identified as the coherent fraction of the spontaneous undulator radiation in one field gain length, and is larger with increasing energy spread and emittance mainly because of the increase in gain length. © 2001 Elsevier Science B.V. All rights reserved.

PACS: 41.60.Cr; 42.55.Vc

Keywords: Initial value problem; High-gain free-electron laser; Self-amplified spontaneous emission

1. Introduction

In a high-gain free-electron laser (FEL), a coherent external signal or the incoherent undulator radiation can initiate the FEL interaction for an exponentially growing coherent radiation. Such a radiation is a promising source for future-generation X-ray facilities. Thus, it is important to understand how the exponential process starts and how the incoherent radiation develops into a coherent source.

The FEL initial value problem was solved in one-dimensional (1-D) theory [1,2] using the La-

place transform technique. The three-dimensional (3-D) initial value problem for a parallel beam was studied by Van Kampen's method in Ref. [3] and by a Green's function technique in Refs. [4,5]. Extension of Van Kampen's method to include the emittance effect was made in Refs. [6,7]. Using an equivalent method, Xie [8] independently obtained the solution to the initial value problem including emittance and numerically found that the effective start-up noise in self-amplified spontaneous emission (SASE) becomes significantly larger with finite emittance and energy spread.

Inspired by the work of Xie, we explain the solution to the FEL initial value problem using Van Kampen's method applicable to the 3-D case including betatron focusing and emittance. We then attempt to provide an understanding of the dependence of the effective start-up noise on beam parameters. Two factors determining the start-up

---

☆Work supported by the US Department of Energy, Office of Basic Energy Sciences, under Contract No. W-31-109-ENG-38.

*Corresponding author. Tel.: +1-(630)252-6023; fax: +1-(630)252-5703.

E-mail address: zrh@aps.anl.gov (Z. Huang).

process are identified. The input coupling coefficient for both coherent amplification (CA) and SASE is found to increase with the initial energy spread. The effective start-up noise is shown to be the coherent fraction of the spontaneous undulator radiation in the first field gain length, generalizing the result of Ref. [5] for a beam with vanishing energy spread and emittance. The effective start-up noise appears to be larger with increasing energy spread and emittance mainly because of the increase of the gain length. Fluctuation in initial electron velocities (due to beam energy spread and angular spread) do not seem to contribute to any additional start-up noise.

2. The dynamic equations and the initial conditions

We follow closely to the notations of Ref. [7], which makes extensively use of the FEL parameter ρ [9] to scale all the dynamical variables. Assuming that the initial smooth electron distribution f_0 is matched to the undulator channel transversely and has a uniform longitudinal profile, the Maxwell–Klimontovich equations in the small signal regime can be written as [7]

$$\left(\frac{\partial}{\partial \bar{z}} - i\mathbf{M}\right)\Phi(\bar{z}) = 0 \qquad (1)$$

where the state vector is

$$\Phi(\bar{z}) = \begin{pmatrix} a_\nu(\bar{x}; \bar{z}) \\ f_\nu(\bar{\eta}, \bar{x}, \bar{p}; \bar{z}) \end{pmatrix} \qquad (2)$$

and the operator $\mathbf{M}$ is defined through

$$\mathbf{M}\Phi(\bar{z})$$

$$= \begin{pmatrix} (-\bar{v} + \frac{\bar{\nabla}_\perp^2}{2})a_\nu - i\int d^2\bar{p}\int d\bar{\eta} f_\nu \\ -ia_\nu \frac{\partial f_0}{\partial \bar{\eta}} + [-v(\bar{\eta} - \frac{1}{2}(\bar{p}^2 + \bar{k}^2\bar{x}^2)) \\ + i(\bar{p}\frac{\partial}{\partial \bar{x}} - \bar{k}_\beta^2 \bar{x}\frac{\partial}{\partial \bar{p}})]f_\nu \end{pmatrix}.$$

$$\qquad (3)$$

Here the scaled undulator distance $\bar{z}$ is the independent "time" variable, a_ν and f_ν are the νth fourier component of the scaled electric field

and the perturbed electron distribution function f_1, respectively, $(\theta, \bar{\eta}, \bar{x}, \bar{p})$ are the longitudinal and the transverse phase space variables, $\bar{\nabla}_\perp = \partial/(\partial \bar{x})$ is the scaled transverse Laplacian, $\bar{v} = (v - 1)/(2\rho)$, $\bar{k}$ is the scaled natural focusing strength, and $\bar{k}_\beta$ is the total scaled focusing strength (including the natural focusing and the average effects of the external focusing).

The evolution of the radiation field and the distribution function in the start-up and the exponential growth regimes is completely determined by Eq. (1) and the initial value $\Phi(0)$ of the state vector. The latter is specified by the external signal $a_\nu(0)$ and the shot noise $f_\nu(0) = \int (2\rho \, d\theta/2\pi)e^{-i\nu\theta}f_1(0)$. Although the ensemble average of $f_\nu(0)$ is zero, physically meaningful quantities such as intensity can be computed by using the relation [10] (see also Ichimaru [10]):

$$\langle f_\nu(\bar{\eta}, \bar{x}, \bar{p}; 0)f_\nu(\bar{\eta}', \bar{x}', \bar{p}'; 0)\rangle$$

$$= \frac{2k_1^2 k_u \rho^3 \theta_b}{\pi^2 n_0}\delta(\bar{\eta} - \bar{\eta}')\delta(\bar{x} - \bar{x}')\delta(\bar{p} - \bar{p}')f_0. \qquad (4)$$

Here $\lambda_1 = 2\pi/k_1$ is the resonant radiation wavelength, $\lambda_u = 2\pi/k_u$ is the undulator period, n_0 is the peak electron volume density, and θ_b is the bunch length in units of $2\pi/\lambda_1$.

3. Van Kampen's normal mode expansion

The initial value problem formulated in the previous section can be solved by expanding the solution in terms of the eigenvectors of Eq. (1). The coefficients of the expansion are determined from the initial conditions if the eigenvectors are mutually orthogonal under a suitably defined scalar product. The procedure is well-known in quantum mechanics in which all operators are Hermitian. Here $\mathbf{M}$ is not a Hermitian operator, and we employ the extension of the method developed by Van Kampen [11] in studying the 1-D plasma waves.

Let us first find the eigenvalues and the eigenvectors of Eq. (1), defined to be solutions

$$e^{-i\mu_n\bar{z}}\Psi_n = e^{-i\mu_n\bar{z}}\begin{pmatrix} A_n(\bar{x}) \\ F_n(\bar{\eta}, \bar{x}, \bar{p}) \end{pmatrix}. \qquad (5)$$

Solving the eigenvalue equation $(\mu_n + \mathbf{M})\Psi_n = 0$, we obtain the mode equation

$$\left(-i\mu_n + i\bar{v} + \frac{\bar{\mathbf{V}}_\perp^2}{2i}\right)A_n(\bar{\mathbf{x}})$$

$$-\int d^2\bar{p}\int d\bar{\eta}\int_{-\infty}^0 d\tau A_n(\bar{\mathbf{x}}_\beta(\tau))e^{i\phi_\beta(\tau)}\frac{\partial f_0}{\partial\bar{\eta}} = 0 \quad (6)$$

where

$$\bar{\mathbf{x}}_\beta(\tau) = \bar{\mathbf{x}}\cos\bar{k}_\beta\tau + \frac{\bar{\mathbf{p}}}{\bar{k}_\beta}\sin\bar{k}_\beta\tau$$

$$\phi_\beta(\tau) = [\bar{\eta} - \tfrac{1}{2}(\bar{p}^2 + \bar{k}_\beta^2\bar{x}^2) - \mu_n]\tau. \quad (7)$$

Eq. (6) is the dispersion relation derived in Refs. [3,12] for natural focusing only (i.e., $\bar{k}_\beta = \bar{k}$). For alternating-gradient focusing with $\bar{k}_\beta \gg \bar{k}$, it is shown to be also valid after averaging properly over many periods of the focusing structure [13,14]. It can be solved using a variational principle [12,15] and a matrix formalism [15]. In general, a discrete set of eigenvalues and eigenmodes exists.

Van Kampen orthogonality of these eigenvectors is constructed by introducing the adjoint eigenvalue equation $(\tilde{\mu}_n + \tilde{\mathbf{M}})\tilde{\Psi}_n = 0$, where $\tilde{\mu}_n$ and $\tilde{\Psi}_n$ are the adjoint eigenvalues and eigenvectors of the adjoint operator $\tilde{\mathbf{M}}$. The formal procedure can be found in Ref. [7]. In the high-gain regime where the fundamental mode $A_0(\bar{\mathbf{x}})$ dominates because its eigenvalue μ_0 has the largest imaginary part μ_I, we project the initial conditions to this mode à la Van Kampen and obtain the evolution of the electric field

$$a_v(\bar{\mathbf{x}};\bar{z}) = \frac{A_0(\bar{\mathbf{x}})e^{-i\mu_0\bar{z}}}{C_0}\left[\int d^2\bar{\mathbf{x}}'A_0(\bar{\mathbf{x}}')a_v(\bar{\mathbf{x}}';0)\right.$$

$$+ \int d^2\bar{\mathbf{x}}'\int d^2\bar{p}\int d\bar{\eta}f_v(\bar{\eta},\bar{\mathbf{x}}',\bar{\mathbf{p}};0)$$

$$\left.\times\int_{-\infty}^0 d\tau A_0(\bar{\mathbf{x}}_\beta(\tau))e^{i\phi_\beta(\tau)}\right] \quad (8)$$

where

$$C_0 = \int d^2\bar{x}A_0^2(\bar{\mathbf{x}}) + \int d^2\bar{x}\int d^2\bar{p}\int d\bar{\eta}\frac{\partial f_0}{\partial\bar{\eta}}$$

$$\times\left[\int_{-\infty}^0 d\tau A_0(\bar{\mathbf{x}}_\beta(\tau))e^{i\phi_\beta(\tau)}\right]^2. \quad (9)$$

These expressions have been obtained independently by Xie [8] using an equivalent method. The first term in the square bracket of Eq. (8) describes the process of coherent amplification, which starts from an external signal $a_v(0)$. The second term describes the process of self-amplified spontaneous emission, which starts from white noise. Eq. (8) for the parallel e-beam (with vanishing emittance) reduces to those of Refs. [3,4]. The ensemble averaged spectrum of the radiation intensity (power per unit area) can be computed with the help of Eq. (4):

$$\frac{1}{\rho I_{\text{beam}}}\frac{dI}{dv} = \frac{2\pi}{(2\rho)^2\theta_b}\langle|a_v(\bar{\mathbf{x}};z)|^2\rangle$$

$$= \frac{1}{|C_0|^2}|A_0(\bar{\mathbf{x}})|^2e^{2\mu_I\bar{z}}$$

$$\times\left[\frac{2\pi}{(2\rho)^2\theta_b}\left|\int d^2\bar{\mathbf{x}}'A_0(\bar{\mathbf{x}}')a_v(\bar{\mathbf{x}}';0)\right|^2\right.$$

$$+ \frac{2k_1^2k_u\rho}{2\pi n_0}\int d^2\bar{x}'\int d^2\bar{p}\int d\bar{\eta}f_0(\bar{\eta},\bar{\mathbf{x}}',\bar{\mathbf{p}})$$

$$\left.\times\left|\int_{-\infty}^0 d\tau A_0(\bar{\mathbf{x}}_\beta(\tau))e^{i\phi_\beta(\tau)}\right|^2\right] \quad (10)$$

where $I_{\text{beam}} = \gamma_0 mc^3 n_0$, and $\gamma_0 mc^2$ is the beam energy.

4. Effects of energy spread

To isolate the energy spread effects in the FEL start-up process, we look at the 1-D limit of the above results by setting $A_n(\bar{\mathbf{x}}) = 1$, $\int d^2\bar{x} = 2k_1k_u\rho\Sigma$ (Σ is the beam cross-section) and dropping $\int d^2\bar{p}$ and the transverse Laplacian. The mode Eq. (6) reduces to the 1-D dispersion relation [1]:

$$D(\mu) = \mu - \bar{v} - \int d\bar{\eta}\frac{dV/d\bar{\eta}}{\bar{\eta} - \mu} = 0 \quad (11)$$

where $f_0 = V(\bar{\eta})$ with $\int d\bar{\eta}V(\bar{\eta}) = 1$. For a monoenergetic beam (i.e., $V(\bar{\eta}) = \delta(\bar{\eta})$), this reduces to the cubic equation [9] with a growing, a decaying and an oscillatory solution. The intensity spectrum of Eq. (10) becomes the power spectrum of

Ref. [1]:

$$\frac{dP}{d\omega} = \frac{dI}{d\nu}\frac{\Sigma}{ck_1} = g_A e^{2\mu_I \bar{z}}\left[\left(\frac{dP}{d\omega}\right)_C + \left(\frac{dP}{d\omega}\right)_S\right], \quad (12)$$

where

$$g_A = \left[1 - 2\int d\bar{\eta}\frac{V(\bar{\eta})}{(\bar{\eta} - \mu)^3}\right]^{-2} = \left(\frac{dD}{d\mu}\right)^{-2}$$

$$\left(\frac{dP}{d\omega}\right)_C = \frac{\pi P_{\text{beam}}}{2\rho ck_1\theta_b}|a_\nu(0)|^2 \quad \text{(coherent source)}$$

$$\left(\frac{dP}{d\omega}\right)_S = g_S\frac{\rho\gamma_0 mc^2}{2\pi} \quad \text{(start-up noise)}$$

$$g_S = \int d\bar{\eta}\frac{V(\bar{\eta})}{|\bar{\eta} - \mu|^2} = \int d\bar{\eta}\frac{V(\bar{\eta})}{\mu_I^2 + (\bar{\eta} - \mu_R)^2}. \quad (13)$$

Here $P_{\text{beam}} = I_{\text{beam}}\Sigma$ is the beam power, and $\mu = \mu_R + i\mu_I$ is a function of the frequency detuning $\bar{\nu}$ through the dispersion relation. For CA, the amplification occurs at the frequency defined by the frequency of the coherent source. For SASE, the frequency dependence is determined by $\mu_I(\bar{\nu})$ in the exponent of Eq. (12). Thus, g_A and g_S, evaluated at the optimal detuning $\bar{\nu}_0$ where the growth rate μ_I reaches the maximum, determine the input coupling to the exponentially growing mode and the effective start-up noise in units of $\rho\gamma_0 mc^2/(2\pi)$, respectively.

In Ref. [1], $G = g_A g_S$ has been computed numerically for a flat-top energy distribution and has been found to increase initially with energy spread. For a Gaussian energy distribution

$$V(\bar{\eta}) = \frac{1}{\sqrt{2\pi}\bar{\sigma}_\eta}\exp\left(-\frac{\bar{\eta}^2}{2\bar{\sigma}_\eta^2}\right) \quad (14)$$

we compute μ_I, g_A and g_S as functions of the r.m.s energy spread $\bar{\sigma}_\eta = \sigma_\eta/(\gamma_0\rho)$ (see Fig. 1) and find that both g_A and g_S increase with $\bar{\sigma}_\eta$. For a monoenergetic beam, any initial signal (external or spontaneous) couples equally well to the three (growing, decaying and oscillatory) modes that have the same normalization factor, hence we have the well-known $g_A = \frac{1}{9}$. However, g_A is larger for a larger energy spread, approaching $\frac{1}{4}$ for the flat-top model and 1 for the Gaussian model. We note that the increase of the input coupling coefficient g_A is the same for both CA and

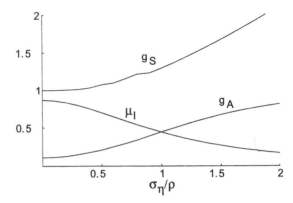

Fig. 1. The behavior of μ_I, g_A and g_S as functions of the r.m.s. energy spread $\bar{\sigma}_\eta = \sigma_\eta/\rho$ for a Gaussian energy distribution.

SASE, and a plausible explanation may be made [16] as follows: an electron beam with larger energy spread is less sensitive to the detuning effect due to the deceleration of the electrons caused by the FEL interaction. Thus, it can couple more effectively with the exponentially growing radiation.

The increase in the effective start-up noise through g_S may be interpreted in the following way. First of all, $g_S = 1$ for a monoenergetic beam, and the quantity $\rho\gamma_0 mc^2/(2\pi)$ is approximately the spontaneous undulator radiation in the first field gain length $L_{g0} = 1/(k_u\rho\sqrt{3})$ [5]. For a beam with a finite energy spread, the spontaneous radiation spectrum in the forward direction is the convolution of the beam energy spectrum and the undulator radiation spectrum with an intrinsic bandwidth $2\Delta\eta = \Delta\nu = \Delta\omega/\omega \sim 2\pi/(k_u z)$. After the first field gain length $z = L_g = (2k_u\rho\mu_I)^{-1}$, the spontaneous radiation spectrum becomes

$$\left(\frac{dP}{d\omega}\right)_{L_g}^{\text{spont}} = \frac{\rho\gamma_0 mc^2}{2\pi\mu_I^2}\int d\bar{\eta}V(\bar{\eta})S_u\left(\frac{\bar{\eta} - \bar{\nu}}{2\mu_I}\right) \quad (15)$$

where $S_u(x) = \sin^2(x)/x^2$ is the undulator spectral function [18]. Rewriting Eq. (13) as

$$\left(\frac{dP}{d\omega}\right)_S = \frac{\rho\gamma_0 mc^2}{2\pi\mu_I^2}\int d\bar{\eta}V(\bar{\eta})\left[\frac{1}{1 + (\bar{\eta} - \mu_R)^2/\mu_I^2}\right] \quad (16)$$

and comparing with Eq. (15), we may interpret the effective start-up noise as the fraction of the spontaneous undulator radiation in the first field gain length within the coherent gain bandwidth $\Delta\bar{\eta} \sim \Delta\bar{\nu} \sim \mu_I$ (much narrower than the intrinsic undulator bandwidth $2\pi\mu_I$). Using the Gaussian energy distribution in Eq. (14) and approximating the Lorentzian in the square bracket of Eq. (16) by another Gaussian, we can carry out the $\bar{\eta}$-integral to obtain

$$\left(\frac{\mathrm{d}P}{\mathrm{d}\omega}\right)_S \approx \frac{\rho\gamma_0 mc^2}{2\pi\mu_I^2}\exp\left(-\frac{\mu_R^2}{2\bar{\sigma}_\eta^2 + \mu_I^2}\right)\frac{1}{\sqrt{1 + 2\bar{\sigma}_\eta^2/\mu_I^2}}.$$

$$(17)$$

With increasing energy spread, the coherent fraction of the spontaneous radiation decreases, but the drop in the growth rate significantly increases the spontaneous radiation power in one field gain length, leading to the overall increase of the effective start-up noise through g_S (as seen in Fig. 1). In fact, for large values of the energy spread $\bar{\sigma}_\eta^2 \gg 1$, $\mu_R \approx \bar{\nu}_0 \approx -\bar{\sigma}_\eta$ and $\mu_I \approx 0.76/\bar{\sigma}_\eta^2$ [17], so that $(\mathrm{d}P/\mathrm{d}\omega)_S \propto \bar{\sigma}_\eta$ increases without bound because the noise required to start the SASE process is infinite.

5. Effects of emittance

We now return to the full 3-D Eq. (10) and consider SASE (the second term) only. Assuming that the betatron oscillations are slow on the scale of the gain length, we take $\bar{k}_\beta\tau \ll 1$, $\bar{x}_\beta(\tau) \approx \bar{x}'$ and integrate $\int \mathrm{d}x^2(\mathrm{d}I/\mathrm{d}\omega)$ to obtain the SASE power spectrum

$$\left(\frac{\mathrm{d}P}{\mathrm{d}\omega}\right)_{SASE} \approx g_A^{3D}\left(\frac{\mathrm{d}P}{\mathrm{d}\omega}\right)_S e^{2\mu_I\bar{z}}.$$

$$(18)$$

Here $g_A^{3D} = \int \mathrm{d}^2\bar{x}|A_0(\bar{x})|^2/|C_0|^2$ is the input coupling coefficient. The effective start-up noise is

$$\left(\frac{\mathrm{d}P}{\mathrm{d}\omega}\right)_S$$

$$= \frac{\rho\gamma_0 mc^2}{2\pi\mu_I^2}\int \mathrm{d}^2\bar{x}|A_0(\bar{x})|^2 \int \mathrm{d}^2\bar{p}\,U(\bar{p}^2 + \bar{k}_\beta^2\bar{x}^2)$$

$$\times \mathrm{d}\bar{\eta}\,V(\bar{\eta})\left[1 + \left(\frac{\bar{\eta} - (\bar{p}^2 + \bar{k}_\beta^2\bar{x}^2)/2 - \mu_R}{\mu_I}\right)^2\right]^{-1}$$

$$(19)$$

where $U(\bar{p}^2 + \bar{k}_\beta^2\bar{x}^2)$ is the electron transverse distribution function. Eq. (19) provides a similar phase-space convolution as the spontaneous undulator radiation when the effects of electron angular spread is taken into account [18], except that the spectral function is a Lorentzian instead of the undulator spectrum S_u at one field gain length. Identifying μ_I as the bandwidth of $\bar{\eta}$ (or $\bar{\nu}$) as in Section 4 and $\sqrt{\mu_I}$ as the angular spread of the fundamental mode (or $L_g = (2k_u\mu_I\rho)^{-1}$ as the Rayleigh length), we may interpret the effective start-up noise as the phase-space convolution of the spontaneous undulator radiation in the first field gain length with the coherent fundamental laser mode.

For numerical computation, we approximate $A_0(\bar{x}) = \exp(-w|\bar{x}|^2/\bar{\sigma}_x^2)$, where $w = w_R + w_I$ is a complex number characterizing the fundamental radiation mode, and $\bar{\sigma}_x$ is the scaled transverse e-beam size and is related to the r.m.s. emittance $\varepsilon = \bar{\sigma}_x^2\bar{k}_\beta/k_1$. Eq. (18) can be written as

$$\left(\frac{\mathrm{d}P}{\mathrm{d}\omega}\right)_{SASE} \approx g_A^{3D}g_S^{3D}\frac{\rho\gamma_0 mc^2}{2\pi}e^{2\mu_I\bar{z}}$$

$$(20)$$

where

$$g_A^{3D} \approx \left|1 - 4iw\int_{-\infty}^0 \mathrm{d}\tau_1\int_{-\infty}^0 \mathrm{d}\tau_2\right.$$

$$\left.\times\frac{(\tau_1 + \tau_2)\exp[-(\bar{\sigma}_\eta^2/2)(\tau_1 + \tau_2)^2 - i\mu_0(\tau_1 + \tau_2)]}{[1 + i\bar{k}_\beta^2\bar{\sigma}_x^2(\tau_1 + \tau_2)][1 + 4w + i\bar{k}_\beta^2\bar{\sigma}_x^2(\tau_1 + \tau_2)]}\right|^{-2}$$

$$g_S^{3D} \approx \frac{4|w|^2}{w_R}\int_{-\infty}^0 \mathrm{d}\tau_1\int_{-\infty}^0 \mathrm{d}\tau_2$$

$$\times\frac{\exp[-(\bar{\sigma}_\eta^2/2)(\tau_1 - \tau_2)^2 - i(\mu_0\tau_1 - \mu_0^*\tau_2)]}{[1 + i\bar{k}_\beta^2\bar{\sigma}_x^2(\tau_1 - \tau_2)][1 + 4w_R + i\bar{k}_\beta^2\bar{\sigma}_x^2(\tau_1 - \tau_2)]}.$$

$$(21)$$

For example, using the current design parameters of the Linac Coherent Light Source (LCLS) [19], we have $\bar{\sigma}_r = 2.8$, $\bar{\sigma}_\eta = 0.45$, and $\bar{k}_\beta = 0.29$. The fundamental guided mode has a complex growth rate $\mu_0 = -1.2 + 0.42i$ and a mode profile determined by $w = 0.64 - 0.50i$ at the optimal detuning $\bar{\nu}_0 = -1.0$ [7]. Hence we obtain $g_A^{3D} \approx 0.3$ and $g_S^{3D} \approx 2.6$, both larger than the values with vanishing energy spread and emittance by a factor of ~ 3 each.

References

[1] K.-J. Kim, Nucl. Instr. and Meth. A 250 (1986) 396.

[2] J.-M. Wang, L.-H. Yu, Nucl. Instr. and Meth. A 250 (1986) 484.

[3] K.-J. Kim, Phys. Rev. Lett. 57 (1986) 1871.

[4] S. Krinsky, L.-H. Yu, Phys. Rev. A 35 (1987) 3406.

[5] L.-H. Yu, S. Krinsky, Nucl. Instr. and Meth. A 285 (1989) 119.

[6] K.-J. Kim, unpublished notes.

[7] Z. Huang, K.-J. Kim, Phys. Rev. E 62 (2000) 7259.

[8] M. Xie, presented at the American Physical Society April Meeting, Long Beach, 2000; M. Xie, Nucl. Instr. and Meth. A 475 (2001) 51, These Proceedings.

[9] R. Bonifacio, C. Pellegrini, L.M. Narducci, Opt. Commun. 50 (1984) 373.

[10] Y.L. Klimontovich, Sov. Phys. JETP 6 (1958) 753; S. Ichimaru, Basic Principles of Plasma Physics, Benjamin, London, 1973.

[11] N.G. Van Kampen, Physica (Utrecht) 21 (1951) 949; K.M. Case, Ann. Phys. (NY) 7 (1959) 349.

[12] L.-H. Yu, S. Krinsky, R.L. Gluckstern, Phys. Rev. Lett. 64 (1990) 3011.

[13] S. Reiche, Nucl. Instr. and Meth. A 445 (2000) 90.

[14] E.L. Saldin, E.A. Schneidmiller, M.V. Yurkov, Nucl. Instr. and Meth. A 475 (2001) 86, These Proceedings and Ref. [20] therein.

[15] M. Xie, Nucl. Instr. and Meth. A 445 (2000) 59, 67.

[16] N.A. Vinokurov, private communication.

[17] E.L. Saldin, E.A. Schneidmiller, M.V. Yurkov, Nucl. Instr. and Meth. A 313 (1992) 555; E.L. Saldin, E.A. Schneidmiller, M.V. Yurkov, The Physics of Free Electron Lasers, Springer, Berlin, 1999.

[18] K.-J. Kim, in: M. Month, M. Dienes (Eds.), AIP Conference Proceedings 184, AIP, New York, 1989, p. 585.

[19] Linac Coherent Light Source Design Study Report, SLAC-R-521, 1998.

ELSEVIER

Nuclear Instruments and Methods in Physics Research A 475 (2001) 65–73

NUCLEAR
INSTRUMENTS
& METHODS
IN PHYSICS
RESEARCH
Section A

www.elsevier.com/locate/nima

Power limitations in the OK-4/Duke storage ring FEL ☆

V.N. Litvinenko*, S.H. Park, I.V. Pinayev, Y. Wu

FEL Laboratory, Department of Physics, Duke University, Box 90319, Durham, NC 27708-0319, USA

Abstract

In this paper, we present results of our experimental and theoretical studies of average power in the OK-4/Duke storage ring FEL. Our theoretical studies are based on the 3D FEL macro-particle model, which includes the local interactions, diffusion, radiation damping and spontaneous radiation. The OK-4/Duke storage ring FEL is operational since 1996 and demonstrated lasing in a wavelength range from 193.7 to 730 nm using electron beam energies from 220 to 800 MeV. It operated in both CW and giant pulse modes. During this period of time we collected substantial amounts of data regarding the FEL power and electron beam dynamics. We compare selected results on CW lasing with our theoretical predictions based on the rigorous numerical model. We also discuss a number of simplified scaling laws for the FEL gain and power as functions of electron beam energy and current, as well as, the cavity losses. © 2001 Elsevier Science B.V. All rights reserved.

PACS: 41.60.Cr; 29.20D; 78.66; 52.75

Keywords: Free electron laser; Storage ring; Optical klystron; Power limitations; Coherence

1. Introduction

All operational storage ring FELs (SR FELs) are based on the scheme of optical klystron (OK), suggested and theoretically studied by Vinokurov and Skrinsky [1] in 1977 (see Fig. 1). The use of a buncher provides control of the longitudinal dispersion and phasing between the two wigglers it separates. The optimized buncher setting, with $[N_D] \cong \max(0, [1/(4\pi\sigma_\varepsilon) - N_w])$ (see Table 1 for a list of symbols), enhances the gain and the power of SR FEL. The re-use of the same e-beam in a short wavelength SR

FEL causes a "soft" limitation of average lasing power. In 1977, Vinokurov and Skrinsky [2] derived this limit as

$$\bar{P}_{SR\ FEL} \cong \xi_\varepsilon P_{SR} \sigma_{\varepsilon induced}; \quad \xi_\varepsilon \sim 2 \tag{1}$$

using the balance between the energy diffusion caused by FEL interaction and radiation damping. Eq. (1), which is traditionally called the Renieri limit, was independently derived and published in 1979 [3]. Later papers developed the estimate in Eq. (1) slightly further using some specific assumption [4–6]. The exact solution of the self-consistent equations for SR FEL turned out to be a rather elaborate problem, which has very few, and still approximate, analytical solutions. In this paper, we attempt to present a self-consistent picture of SR FEL operation. We consider a case of a properly designed, optimized and aligned, low gain

☆ This work is supported by ONR Contract #N00014-94-1-0818.

*Corresponding author. Tel.: +1-919-660-2658; fax: +1-919-660-2671

E-mail address: vl@phy.duke.edu (V.N. Litvinenko).

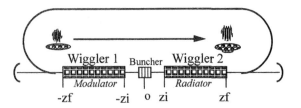

Fig. 1. Schematic layout of an OK installed onto a storage ring. The system, the optical and electron beams have bilateral symmetry with respect to the center.

short wavelength SR FEL with

$$\lambda \ll \lambda_c = C \max\{\alpha_c \sigma_\varepsilon, \langle \sigma_\theta^2 \rangle / 2\} \qquad (2)$$

installed in the dispersion free straight section and driven by an ultra-relativistic e-beam with $\gamma \gg 1$. A SR FEL based on the scheme of the OK is depicted in Fig. 1. The e-beam and short optical wave-packet (WP) co-propagate through the OK. In the first wiggler the e-beam interacts with the WP and obtains local energy modulation. After the buncher, more energetic electrons catch up with the lower energy ones, and the density modulation appears. The $\{N_D\} \cong 3/4$ [1] provides for optical pulse amplification in the second wiggler, where the e-beam has the net energy loss and the WP is amplified. Electrons leave the FEL with residual "λ-scale" energy modulation. One turn around the ring wipes out the phase correlation between electrons (see Eq. (2)) and causes the energy spread growth. This process repeats with the e-beam revolution frequency.

2. CW mode in SR FEL

In CW mode, the balance between the heating (diffusion) and the radiation damping establishes a steady-state, self-consistent distribution of the e-beam $(f(X), X = \{x, x', y, y', z, \varepsilon\})$ and the field of the FEL pulse $E = E(\vec{r}, z - ct)e^{ik(z-ct)}$. In this paper, we are not interested in polarization issues; hence, we use a scalar function for E. The "soft" power saturation provides for the linearity of SR FEL interaction. A typical change of individual electron energy in the FEL is $\Delta\varepsilon \propto \sigma_\varepsilon/\sqrt{N_\varepsilon}$. With typical $N_\varepsilon \sim 10^4$–10^6 and $\sigma_\varepsilon \sim 10^{-3}$, one gets

Table 1
List of symbols[a]

c	Speed of the light
e, m	Charge and mass of electron
$E_0, E_e = E_0(1 + \varepsilon) = \gamma m c^2$	Energy of electron
$\vec{v} = \vec{\beta}c$	Velocity of electron
C	Circumference of the ring
$f_0 = C/v, T_0 = 1/f_0$	Revolution frequency and time
$\bar{I} = eN_e f_0, \hat{I}$	Average and peak beam current
N_e, N_b	Number of electron and e-bunches
$f_{RF} = h_{RF} f_0, V_{RF} = V_0 \cos\phi_0$	RF frequency and voltage
ρ, α_c	Orbit radius and compaction factor
β_{xy}, η	Storage ring lattice functions
β_0	Rayleigh range[b]
$\nu_{x,y}, \varepsilon_{x,y}$	Betatron tunes and emittances
$\nu_s = \sqrt{eV_{RF} h_{RF} \alpha_c / 2\pi E_0}$	Synchrotron tune
$\Omega_s = 2\pi f_0 \nu_s, b = \alpha_c C/2\pi\nu_s$	
$\sigma_\varepsilon, \sigma_{z,\tau}$	Energy spread and bunch length
$\sigma_{x,y}, \sigma_{x',y',\theta}$	e-Beam sizes and angular spreads
$P_{SR} = \bar{I}U_0, U_0 = 4\gamma^4 \pi e^2 / 3\rho$	Power of synchrotron radiation
$\tau_{x,y,\varepsilon} = \mathcal{E}_0/J_{x,y,\varepsilon} \equiv f_0 \tau_{x,y,\varepsilon}$	Damping times, partition numbers
$\xi_{x,y,\varepsilon} \equiv 1/N_{x,y,\varepsilon} \equiv 1/f_0 \tau_{x,y,\varepsilon}$	
$D_{x,y} = \varepsilon_{x,y}/\tau_{x,y}, D_\varepsilon = \sigma_\varepsilon^2/\tau_\varepsilon$	Diffusion coefficients
Z_n/n	Longitudinal impedance [7]
N_w, a_w	Wiggler parameters
$\lambda_w \equiv 2\pi/k_w, L_w = N_w \lambda_w$	Wiggler period and length
$L_B = 2z_i$	Length of bunching section
$L_{OK} = 2L_w + L_B$	Length of optical klystron
$\lambda = \lambda_w(1 + a_w^2)/2\gamma^2$	FEL wavelength
$\Delta s, N_D = \Delta s/\lambda$	Slippage in buncher section
$N_t = N_w + N_D, \Xi = N_t \lambda$	Full slippage in FEL
$L, \delta = cT_0 - 2L$	Optical cavity length, detuning
$I_A = mc^3/e = 17.045\,\text{kA}$	Alfen current
$Z_0 = 4\pi/c = 376.73\,\Omega$	Impedance of vacuum
$\bar{P}_{FEL}, \hat{P}_{FEL}$	Average and peak FEL power
$x = [x] + \{x\}$	Integer and fractional parts of x

[a] For definitions of the accelerator parameters see [8]. Subscript "0" denotes natural values of parameters.
[b] β_0 is used to free symbol Z_0 for the impedance of vacuum.

$\Delta\varepsilon \sim 10^{-5}$–$10^{-6}$. This level of energy modulation causes $\Delta\varphi_{FEL} \approx 4\pi N_t \Delta\varepsilon \sim 10^{-2} - 10^{-4} \ll 1$.

2.1. Optical field and gain in SR FEL

The optical field in a SR FEL is paraxial and can be expanded in a complete set of Hermit-Gaussian functions. The evolution of the optical field in a linear FEL is described by a linear integral equation [9,10]. We will limit our discussions to a practical case for existing SR FELs with dominating TEM_{oo} mode and will use a single WP to represent its complex amplitude

$$E(\vec{r}, z - ct) = \mathrm{Re}\left\{ A(z - ct)e^{ik(z-ct)} \right.$$
$$\left. \times \frac{\beta_0}{\beta_0 - iz} \exp\left(-\frac{x^2 + y^2}{2(\beta_0 - iz)} \right) \right\}. \quad (3)$$

The expression for the complex, z-dependent gain (the kernel) for TEM_{oo} mode and e-beam with finite emittance and energy spread can be found elsewhere [6,11,12]. At Duke, we use a 3-D formula for the OK gain, which is tested and confirmed experimentally [11,13]

$$G_{peak} = 8\pi N_w \frac{\hat{I}}{I_A} \frac{g(a_w)}{\gamma\sigma_\varepsilon} \mathrm{FF}\left(\varsigma_{x,y,\varepsilon}, \varsigma_w, g_{0,x,y,i}\right) \quad (4)$$

$$g(a_w) = \frac{a_w^2}{1 + a_w^2} \left\{ \begin{array}{ll} 1 & \text{-helcial wiggler} \\ (J_0(\chi) - J_1(\chi))^2 & \text{-plane wiggler} \end{array} \right\},$$
$$\chi = a_w^2/(1 + a_w^2)/2$$

where $\varsigma_{x,y} = 4\pi\varepsilon_{x,y}/\lambda$, $\varsigma_\varepsilon = 4\pi N_t\sigma_\varepsilon$, $\varsigma_w = 4\pi N_w\sigma_\varepsilon$, $g_{0,x,y} = \beta_{0,x,y}/L_w$, $g_i = z_i/L_w$.[1] Details of the gain dependence on the g-factors are published elsewhere [6,13]. With optimal g-parameters, the FF-factor in Eq. (4) can be written as

$$FF \cong F_g F_x F_y F_\varepsilon;$$

$$F_{x,y} \cong \left(1 + \varsigma_{x,y} + \varsigma_{x,y}^2 \right)^{1/2};$$

$$F_g \cong g_0 \left(sh^{-1}\left(\frac{1 + g_i}{g_0} \right) - sh^{-1}\left(\frac{g_i}{g_0} \right) \right)^2. \quad (5)$$

The OK-4/Duke SR FEL with $g_i = 0.13$ and $g_0 = 0.985$ has $F_g = 0.71$. The energy spread term in Eq. (5), F_ε, is of main interest for this paper. It depends on the setting of the buncher, the energy spread and the form of distribution function. For a Gaussian distribution function it can be expressed as [6,13]

$$F_\varepsilon \cong \frac{\varsigma_\varepsilon}{\sqrt[4]{1 + \varsigma_w^4}} \exp\left(-\frac{\varsigma_\varepsilon^2}{2} \right). \quad (6)$$

The setting of the buncher can be optimized for maximum gain for any energy spread. For low energy spread $\varsigma_\varepsilon \cong 1$ provides maximum boost to the FEL gain. With high energy spreads when $\varsigma_w \geqslant 1$, the buncher works mostly for phasing ($0 < N_D < 1$) of the two wigglers and provides a modest gain enhancement of about 1.4. With the optimized buncher one has [6,13]

$$F_\varepsilon \cong e^{-0.5}/\sqrt[4]{1 + \varsigma_w^4}. \quad (7)$$

The ratio between the energy spread and bunch-length $\sigma_z = b\sigma_\varepsilon$ and the ratio between the peak and average current,

$$\hat{I} = \frac{\bar{I}}{N_b} \frac{C}{\sqrt{2\pi}\sigma_z} = \frac{\bar{I}}{N_b\sigma_\varepsilon\sqrt{\gamma}} \sqrt{\frac{4\pi V_{RF}h_{RF}}{\alpha_c I_A Z_0}} \quad (8)$$

reduce Eqs. (4)–(8) to a simple analytical dependence of the SR FEL gain on the energy spread in both OK and FEL modes:[2]

$$G_{peak} = \frac{G_0}{\sigma_\varepsilon^2 \sqrt[4]{1 + (4\pi N_w\sigma_\varepsilon)^4}};$$

$$G_0 = 8\pi e^{-0.5} \frac{N_w g(a_w)}{\gamma^{3/2}} \frac{\bar{I}}{I_A N_b} \sqrt{\frac{4\pi V_{RF}h_{RF}}{\alpha_c I_A Z_0}} F_g F_x F_y. \quad (9)$$

The known formula for the energy spread induced by self-saturating microwave (MW)

---

[1] we use a_w instead of k_w for uniformity of expressions. For planar wiggler $k_w = \sqrt{2}a_w$.

[2] We call it FEL mode when $N_D \sim 0$.

instability [7,13],

$$\sigma_{\varepsilon\,MW} = \sqrt{\sigma_{\varepsilon 0}^2 + \frac{4\pi V_{RF} h_{RF}}{\alpha_c \gamma Z_0 I_A} \left(\frac{\bar{I} Z_n/n}{2\pi N_b V_{RF} h_{RF}} \right)^{2/3}}$$

(10)

and Eq. (9) give a good estimate of the start-up gain. For well developed MW instability and $\varsigma_w < 1$, the scaling of SR FEL gain is[3]

$$G_{peak} \propto \left(\frac{\alpha_c}{\gamma} \right)^{1/2} \left(\frac{\bar{I}}{I_A N_b} \right)^{1/3} \left(\frac{Z_0}{Z_n/n} \right)^{2/3} \left(\frac{V_{RF} h_{RF}}{I_A Z_0} \right)^{1/6}.$$

(11)

We found that "easy-to-use" formulae (4–11) are in good agreement with the exact analytical 3-D theory [11], with our computer simulations [14], and with the direct gain measurement [13,15,16]. The relative accuracy of this gain formula is about 5% in the working range of the OK-4/Duke SR FEL.

In the CW mode [4] the FEL gain is very close to the round-trip loss of the optical cavity, i.e. to the lasing threshold gain G_{th}. We can use this fact for estimating the self-consistent RMS energy spread in CW mode:

$$G_{peak}(\sigma_\varepsilon) = G_0 / \sigma_\varepsilon^2 \sqrt[4]{1 + (4\pi N_w \sigma_\varepsilon)^4} \cong G_{th}.$$

(12)

The aproximate solution of Eq. (12) is

$$\sigma_\varepsilon \cong \sqrt{\frac{G_0}{G_{th}}} \left(1 + \left(4\pi N_w \frac{G_0}{G_{th}} 0.769 \right)^4 \right)^{-1/24}.$$

(13)

The self-consistent energy distribution in the CW lasing mode is slightly different from the Gaussian. Nevertheless, the above formula gives a reasonable estimation of the SR FEL lasing power [6]

$$P_{lasing} = \varsigma_e P_{SR} \frac{\sigma_\varepsilon^2 - \sigma_{\varepsilon 0}^2}{\sigma_\varepsilon} e^{-0.5} \kappa; \quad \kappa \approx 1$$

(14)

with a typical dependence on the e-beam energy for a given wavelength $P_{lasing} \sim \gamma^\alpha$, with $\alpha = 3.25$–3.5. Even though, estimation (14) is good within $\pm 25\%$, the exact solution of the problem is also possible.

---

[3] For low emittance, fixed geometry and fixed wavelength. Eq. (11) contrasts the widespread opinion of $G_{peak} \sim \gamma^{-3}$.

[4] With $|\delta| \ll G_{th} \sigma_z$, see Ref. [21].

2.2. Fokker–Plank equation for the e-beam

The self-consistent distribution function of the e-beam, $f(X)$, in a SR is described by the Fokker–Plank equation [18]

$$\dot{f} + \frac{\partial}{\partial X_\alpha}(\dot{X}_\alpha f) - \frac{\partial}{\partial X_\alpha} \frac{\partial}{\partial X_\beta} \left[\frac{1}{2} D_{\alpha\beta} f \right] = 0$$

(15)

where $D_{\alpha\beta}$ is the diffusion tensor. In low emittance storage rings used for driving FELs, the motion of electrons is linear and radiation damping is slow. These features allow X to be expanded in a set of eigen modes: $f(X) \to \prod_{n=1}^{3} f_i(I_n, \varphi_n)$, where (I_n, φ_n) are the canonical "angle-action" variables. The separation of the Fokker–Plank equations into three independent equations for transverse betatron $x-y$ oscillations and slow ($v_s \ll 1$) synchrotron oscillations simplifies analytical studies drastically. Transverse distributions in storage rings are described by Gaussian functions with a high degree of accuracy. It enables the use of analytical tools for taking the transverse emittances into account [11]. When FEL is installed in dispersion-free straight section ($\eta = 0$, $\eta' = 0$), the diffusion caused by FEL affects only the energy spread. The reduced e-beam density has indirect (positive) effect on the transverse emittances via reduced intra-beam scattering.[5] In the longitudinal phase space $\{I = \varepsilon^2/2 + l^2/2b^2, \varphi_s\}$ Eq. (15) becomes [6,12]

$$\frac{\partial f}{\partial t} - b\Omega_s \varepsilon \frac{\partial f}{\partial z} + \frac{\Omega_s}{b} z \frac{\partial f}{\partial \varepsilon} - \frac{\partial}{\partial \varepsilon}$$
$$\times \left[\frac{2}{\tau_\varepsilon} \varepsilon f + (D_0 + D_{MW} + D_{FEL}) \frac{\partial f}{\partial \varepsilon} \right] = 0.$$

(16)

In the CW mode the averaging over one synchrotron oscillation reduces Eq. (16) to [6,12]:

$$\frac{\partial f}{\partial t} - \frac{\partial}{\partial I} \left(\frac{2}{\tau_\varepsilon} I f \right)$$
$$- \frac{\partial}{\partial I} \left[I (D_0 + D_{MW} + \tilde{D}_{FEL}) \frac{\partial f}{\partial I} \right] = 0;$$

(17)

---

[5] The intra-beam scattering may substantially increase emittance of the e-beam, especially for low energy of e-beam. Details on the emittance growth by intra-beam scattering can be found elsewhere [7].

$$\tilde{D}_{FEL} \equiv \tilde{D}_{FEL}(I, t) = \frac{1}{\pi} \int_0^{2\pi} D_{FEL}(I, \varphi_s, t)$$
$$\times \cos^2 \varphi_s \, d\varphi_s$$

with an easy-to-find stationary solution as an integral [6]:

$$f(I) = \Phi_0$$

$$\times \exp\left[-\int_0^I \frac{2}{\tau_\varepsilon (D_0 + D_{MW} + \tilde{D}_{FEL}(I'))} dI' \right].$$

(18)

The diffusion is caused by quantum fluctuations of synchrotron radiation (D_{SR}), intra-beam scattering (D_{IBS}), broad-band micro-wave instability (D_{MW}), and, finally, the FEL interaction (D_{FEL}). The $D_0 = D_{SR} + D_{IBS}$ defines the natural energy spread without effects from the FEL and the MW instability. The FEL gain and the lasing start-up in SR are strongly affected by self-saturated MW instability. Nevertheless, when FEL interactions increase the energy spread of the e-beam to the level sufficient to suppress MW instability completely, one can use Eq. (18) to find a steady-state e-beam distribution for CW SR FEL with $D_{MW} \approx 0$. In OK-4/Duke SR FEL an increase of the energy spread by a factor of 2–4 is typical.

The diffusion caused by the FEL interaction depends on the energy, the position z (i.e. the amplitude of WP at $\Delta z = \Xi$), and the transverse motion of the electrons:

$$D_{FEL} = \frac{1}{2} f_0 \langle \delta \varepsilon_{FEL}^2 \rangle; \quad \delta \varepsilon_{FEL} = \frac{e}{\mathscr{E}_0} \int_{FEL} \vec{E} \, d\vec{r}$$

$$= -\frac{1}{4\pi\mathscr{E}_0} \left(\vec{E}_{FEL}(\vec{r}) \vec{E}_{SR}(\vec{r}) \right).$$

(19)

Our main theoretical tool-kit is based on two computer codes, #felTD and #vuvfel, written in FORTRAN. The codes are based on the exact 3D-theory of storage ring FELs [6,11] and include most of the known effects [6,14]. Both codes are using the same main input file specifying the FEL and the e-beam parameters. A typical example of an input file is given in Table 2. The #felTD code prepares nine arrays of 3D-integrals for 10,001 electron energies with a step $\delta\varepsilon = 4 \times 10^{-6}$. The integration is exact and includes both the TEM$_{oo}$

structure of the optical beam and the electron distribution in the transverse phase space [6,11].

The #vuvfel uses up to 30,000 macro particles (MPs). Each MP (with charge $q = -Me$) represents an ensemble of M electrons with the same energy of $E = E_0(1 + \varepsilon)$, with an uniform phase distribution on the scale of λ, and a Gaussian distribution in the 4D-phase phase space (x, x', y, y'). The #vuvfel code uses all the parameters in Table 2. When passing around the ring, MPs are treated as regular electrons with the radiation damping, quantum fluctuation and regular RF action:

$$\varepsilon_{N+1} = \varepsilon_N - \Delta\varepsilon_{SR}(1 + 2\varepsilon_N) + \frac{eV_{RF}}{E}$$
$$\times \cos\left(\frac{2\pi h_{RF} z_N}{C} + \phi_s \right) + \delta\varepsilon_{rand \, SR};$$

$$z_{N+1} = z_N + \alpha_c C \varepsilon_N.$$

In the FEL, the macro particles interact with the FEL WP according to their energy and their arrival time. The evolving TEM$_{oo}$ WP with complex amplitude $A(z - ct)$ is represented by a complex array with a maximum length of 30,000 bins. The bin length is equal to the full slippage in the FEL, Ξ, makes the #vuvfel code multi-frequency. Typical band-width $\delta k/k$ for OK-4 FEL is 0.5–1.5%. More information on the codes can be found elsewhere [14].

The #vuvfel start with zero field in the optical cavity. The initial distribution of the MPs in the (ε, z) phase space is generated using RMS energy spread from the input file. On each turn, MPs radiate spontaneous radiation. The radiation is bouncing back-and-forth in the optical cavity and, on each successive pass, the e-beam (MPs) amplifies it. The lethargy of induced radiation (i.e. the slippage) increases of the optical correlation lengths. Fig. 2 shows an example of self-consistent energy diffusion coefficient for MPs.

4. Selected predictions and experimental results

During past four years, we collected a very significant body of data with the OK-4/Duke SR FEL in the visible, UV and deep-UV ranges. With

Table 2
The main input parameters of the codes. All lengths are in cm[a]

Ring	E_e (GeV)	C	ε_x	ε_y	β_x	β_y
	0.60	10746.0	**1.1×10^{-6}**	**1.1×10^{-7}**	**330.0**	**400.0**
	h_{RF}	V_{RF} (GV)	α_c	σ_{enat}	$1/\zeta_\varepsilon$	N_e
	64	6.0×10^{-4}	8.601566×10^{-3}	0.348×10^{-3}	54,000	0.78×10^{10}
OK-4	Reflectrity ($\sqrt{R}$)	δ (bins)	β_0	z_i		
	0.98950	0.000	**330.0**	**44.0**		
	N_w	λ_w	λ	N_D	Energy de-tuning	
	33.5	**0.1**	**2.37×10^{-5}**	14.5425	0	
Other	# Particles	#Revolutions	Wave-Packet (bins)			
	20,000	200,000	30,000			

[a] Bold-faced are parameters which are used by both codes. The rest of parameters can be changed interactively during the run. Beam current for this run is 3.51 mA in a single bunch.

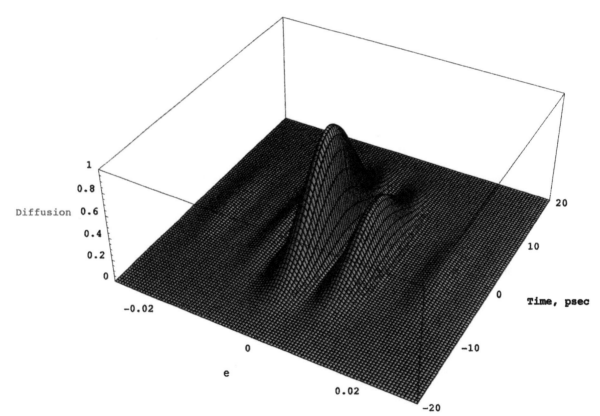

Fig. 2. Dependence of the MPs energy diffusion coefficient (vertical axis, normalized on the maximum value) on relative energy deviation, ε, (horizontal axis) and the arrival time $t = z/v$ into the OK-4 FEL. We used a self-consistent profile of the FEL pulse with RMS duration of 5.6 ps to generate this plot.

few exceptions, we found that the data on the average power, the e-beam and the optical beam time structure is in excellent agreement with numerical simulations. We present here one specific set of data with parameters listed in Table 2. The OK-4 FEL operates in so-called general lasing mode with $Y \approx 1$ (see [19]). By focusing on this typical, by no means the most impressive, case we present complete picture of the processes and parameters of the SR FELs. We used a dual-sweep Hamamatsu streak-camera for time-resolved studies of the electron and the FEL beams. Fig. 3 shows a typical profile of the 3.51 mA e-beam with and without lasing in the OK-4 FEL. Without lasing, the e-beam profile (i.e. $I(t)$) is practically an ideal Gaussian with an RMS duration of 32 ps. With the lasing, the e-beam profile is very different from the Gaussian and has a distinct almost-flat top ∼70 ps and 72 ps RMS duration. This behavior is expected and is well described by

Eq. (18) [6,12]. The FEL pulse has RMS duration of about 5.8 ± 0.4 ps when the resolution is taken into account. We run simulations for this case for 200,000 turns, i.e. for about four damping times, to establish CW mode of operation. Fig. 4 shows the predicted time average and instantaneous (at the turn 200,001) temporal profiles.

The time average profile has an RMS duration of 5.6 ps and fits the experimental data well. It is also consistent with the theory (see Figs. 3 and 4 in Ref. [19]). Simulations provide additional details which are hard to obtain experimentally. For example, we can see details of the WPs, the variation of the amplitude and phase, as well as the correlation function of the optical filed. We found that the correlation function has long oscillating tail with period of about 5 ps and correlation length of about 500 μm. It agrees well with the measured linewidth of 0.04% FWHM. Further, the #vuvFEL code provided all information on the e-beam distribution shown in Fig. 5. The predictions and the measurements compare remarkably well, including the average FEL power. The #vuvFEL calculated average intra-cavity power of 2.14 W for the measured cavity

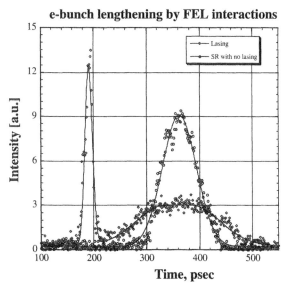

Fig. 3. Longitudinal e-beam profiles measured with Hamamatsu streak-camera without (○) and with lasing (◇). The FEL beam profile is also shown for the lasing mode (short pulse on the left). Continuous lines are fits. The synchrotron radiation from the bending magnet, carrying the information on the e-beam profile, is combined with the FEL pulse via a set of beam splitters and filters. The e-beam profile is intentionally delayed for clear separation from FEL pulse profile. In reality, the centers of FEL pulse and e-bunch coincided within few ps.

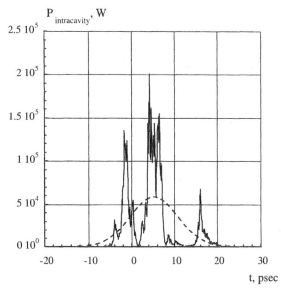

Fig. 4. The calculated instant (solid line) and time-averaged (dashed line) intra-cavity optical power after 200,000 turns in the OK-4 FEL. Parameters are given in Table 2.

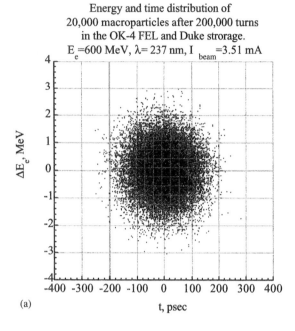

(a)

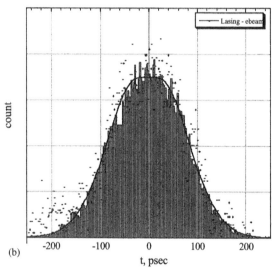

(b)

Fig. 5. (a) Pointcare (ε, z)-space plot of the 20,000 MPs after 200,000 turns in the OK-4/Duke SR FEL. (b) Shows the time histogram (columns) of the distribution with 1 psec bin size. The vertical scale is 370 MP/per bin. The points ($\blacklozenge$) show the measured e-beam profile from Fig. 3. The solid line is the fit with the flat-top ~ 70 ps and Gaussian wings. Fit: $I(t) = I_{peak} \exp\left(-1/2\sigma_\tau^2 \sqrt{\tau_1^4 + t^4}\right)$, $\tau_1 = 49.01$ ps, $\sigma_\tau = 70.43$ ps.

loss of 2.3% per pass. It gives average lasing power of 49 mW. The measured transparencies of two cavity mirrors at 237 nm were 0.13% and 0.22%.

The expected average power through the mirrors would be 3.2 and 4.5 mW correspondingly. The FEL beam passed the CaF_2 window and was focused by a quartz lens into the widow of Molectron power meter. The transparency of the window and the lens are 92% and 91%, respectively. Their resulting delivering efficiency is 83.7%. The average powers measured on both sides of the optical cavity were 2.2 ± 0.1 and 3.8 ± 0.1 mW agreeing well with the expectations.

Another, more typical, example is the OK-4 operating with two 800 MeV, 8.4 mA e-bunches at $\lambda = 223.8$ nm to generate 50.9 MeV γ-rays via intra-cavity Compton back-scattering [20]. At this wavelength the cavity loss is 3.98% per pass, 0.138% and 1.942% of which are due to transparencies of the up-stream (USM) and up-stream mirrors (DSM). The down-shift for 11.3 nm in transparency curve of the down-stream mirror was caused by its annealing by strong spontaneous radiation from the OK-4 wigglers. The simple formula (14) gives the lasing power of 473 mW, while the #vuvFEL code predicted 577 mW. The measured powers of 15.6 at USM and 236 mW at DSM agree better with the #vuvFEL code when the measurement system efficiency (83.7%) is taken into account. The average intra-cavity power of 14.5 W was indirectly confirmed by measured γ-ray flux of 10^8/s.

During our studies, we found some discrepancies between the measurements and predictions. For example, we expected twice larger FEL power than we had measured at 198 nm [17]. We believe that this serious discrepancy has two plausible explanations. The first is that the induced energy spread was insufficient to suppress the MW instability. This scenario is explored further in Ref. [17]. The second, and more probable, explanation is the degradation of the out-coupling down-stream CaF_2 window. After completion of the 200 nm-lasing run, this window was removed. It had dark-violet color and a transparency of about 10%. We have reason to believe that the damaged window absorbed $\sim 88\%$ of 19 mW out-coupled through down-stream mirror, whose transparency at 198 nm increased to about 0.6% after the exposure to the OK-4 spontaneous radiation. These down-shifts were observed for all of down-stream mirrors exposed to the high power radiation from the OK-4 wigglers [21].

5. Conclusions

Results of rigorous 3-D theory and computer simulations are in good agreement with experimental data from the OK-4/Duke SR FEL. We did not find any indication substantiating claims by other SR FEL groups of FEL power being well below Renieri limit.

We are improving #vuvFEL code for better description of threshold lasing in SR FELs, which has mostly academic interest, by adding the MW interaction routine.

Overall, we found that the OK-4/Duke SR FEL power and beam dynamics predicted by our simulations are consistent with our measurements. Presently, the average lasing power in the OK-4/Duke SR FEL is limited by the un-cooled down-stream mirror. We had destroyed two mirrors by the OK-4 FEL spontaneous radiation. The mirror problems will be less critical for the OK-5 FEL with helical wigglers. Instability of the e-beam limits single bunch current to about 20 mA. We plan to improve beam stability using a feed-back system. We also plan to build a full energy 1.2 GeV booster-injector in 2004, which will provide for an increased injected e-beam current.

The diversity of the effects in giant pulse mode operation is too large to fit in this paper and they are published elsewhere [22].

Acknowledgements

The authors are grateful to the Office of Naval Research MFEL program for financial support. The authors would like to thank the staff, the engineers of the Duke FEL laboratory, and especially the accelerator operation group led by Dr. Ping Wang, for their help in the experiments.

References

[1] N.A. Vinokurov, A.N. Skrinsky, preprint 77–59 (1977) BINP, Novosibirsk, Russia.N.A. Vinokurov, A.N. Skrinsky, Theory of Optical Klystron, MW Relativ. Electronics, Gorky, 1981, p. 204.

[2] N.A. Vinokurov, A.N. Skrinsky, preprint 77–67 (1977) BINP, Novosibirsk, Russia.

[3] A. Renieri, Nuovo Cimento 53B (1979) 160.

[4] G. Dattoli, A. Renieri, Nuovo Cimento 59B (1980) 1.

[5] P. Elleaume, Nucl. Instr. and Meth. A 237 (1985) 28.

[6] V.N. Litvinenko, Thesis, Novosibirsk, 1989.

[7] D. Boussard, CERN LABII/rf/INT/75-2, 1975;
J. LeDuff, Proceedings of CERN Accelerator School, Rhodes, Greece, 1993, CERN 95-06, Vol. II, 1995, p. 573.

[8] J. Murphy, Synchrotron Light Source Data Book, BNL 42333, Version 4, Brookhaven, New York, 1996.

[9] Y.H. Chin, K.J. Kim, M. Xie, Nucl. Instr. and Meth. A 331 (1993) 429.

[10] N.A. Vinokurov, ANL/APS/TB-27, Argonne, IL, 1996.

[11] V.N. Litvinenko, Nucl. Instr. and Meth. A 359 (1995) 50.

[12] S.H. Park, Thesis, Duke University, 2000.

[13] V.N. Litvinenko, Gain of a storage ring FEL, to be published.

[14] V.N. Litvinenko, et al., SPIE 2571 (1995) 78.
V.N. Litvinenko, B. Burnham, J.M.J. Madey, Y. Wu, Nucl. Instr. and Meth. A 358 (1995) 334, 369.
V.N. Litvinenko, The #felTD and #vuvfel codes for storage ring FELs, DFEL materials, 1994.

[15] V.N. Litvinenko, et al., Nucl. Instr. and Meth. A 407 (1998) 8.

[16] I.V. Pinayev, et al., Nucl. Instr. and Meth. A 475 (2001) 222, these proceedings.

[17] V.N. Litvinenko, S.H. Park, I.V. Pinayev, Y. Wu, Nucl. Instr. and Meth. A 475 (2001) 195, these proceedings.

[18] Yu.B. Rumer, M.Sh. Ryvkin, Thermodynamics, Statistical Physics and Kinetics, Nauka, Moscow, 1977, p. 241.

[19] V.N. Litvinenko, S.H. Park, I.V. Pinayev, Y. Wu, Nucl. Instr. and Meth. A 475 (2001) 240, these proceedings.

[20] V.N. Litvinenko, et al., Phys. Rev. Lett. 78 (1997) 4569.
V.N. Litvinenko, J.M.J. Madey, SPIE 2521 (1995) 55.

[21] V.N. Litvinenko, et al., Nucl. Instr. and Meth. A 470 (2001) 66.

[22] V.N. Litvinenko, et al., Giant and Super-Pulses in the OK-4/Duke storage ring FEL—predictions and experimental results, Phys. Rev. Lett., submitted for publication.

ELSEVIER

Nuclear Instruments and Methods in Physics Research A 475 (2001) 74–78

NUCLEAR
INSTRUMENTS
& METHODS
IN PHYSICS
RESEARCH
Section A

www.elsevier.com/locate/nima

Quasilinear theory of high-gain FEL saturation

N.A. Vinokurov[a,*], Z. Huang[b], O.A. Shevchenko[a], K.-J. Kim[b]

[a] *Budker Institute of Nuclear Physics, 11 Acad. Lavrentyev Prosp., 630090 Novosibirsk, Russia*
[b] *Advanced Photon Source, Argonne National Laboratory, 9700 S. Cass Avenue, Argonne, IL 60439, USA*

Abstract

Understanding of saturation behavior is important to assess the performance of a high-gain free-electron laser. In this paper, we study the saturation mechanism using a quasilinear approximation to the coupled Maxwell–Vlasov equations. It is found that the quasilinear theory correctly describes the evolution of the radiation field from the small signal regime to reaching saturation. © 2001 Elsevier Science B.V. All rights reserved.

PACS: 41.60.C

Keywords: Free-electron laser

1. Introduction

An electron beam traveling through a long undulator is unstable with respect to the longitudinal bunching at the frequency of the spontaneous undulator radiation in the forward direction. For small bunching the system is linear, and the bunching grows exponentially along the undulator. However when the value of bunching becomes not so small, the exponential growth stops. This phenomenon is referred to as "saturation". The reason for saturation is the obvious fact that the value of bunching cannot exceed 1, as the distribution function of particles is positive. In other words, the amplitude of the AC component of the electron current cannot exceed its DC component by more than twice. From the point of view of the linear small-signal theory, the growth of a particular harmonic of the distribution

function is limited by the nonlinear interaction with other harmonics. Sometimes the interaction with the zero harmonics (the average) dominates. The beam energy spread increases and therefore the bunching growth rate decreases. This mechanism is well known in plasma physics as the quasilinear relaxation (see, for example [1]).

In this paper, we derive the quasilinear equations for 1-D free-electron laser (FEL) theory and solve them numerically. It is shown that for large enough initial energy spread the maximum bunching is much less than one, and so the quasilinear approximation is applicable until the saturation length.

2. Derivation of quasilinear equations

The FEL equations may be obtained from the Maxwell equations and continuity equation in the phase space (Vlasov equation). To simplify the consideration we assume the electromagnetic wave to be quasimonochromatic, i.e., having a narrow spectrum near frequency ω. It is exactly true if we

---

*Corresponding author. Tel.: +7-3832-394003; fax: +7-3832-342163.

E-mail address: vinokurov@inp.nsk.su (N.A. Vinokurov).

0168-9002/01/$ - see front matter © 2001 Elsevier Science B.V. All rights reserved.
PII: S 0 1 6 8 - 9 0 0 2 (0 1) 0 1 5 2 5 - X

neglect radiation at higher harmonics of the fundamental frequency (for weak undulator field, for example). Using the standard approach [2] one can easily write down the following equations for the distribution function $f(\psi, \eta, \xi)$ and the dimensionless complex field amplitude $A(\psi, \xi)$:

$$\frac{\partial f}{\partial \xi} + \eta \frac{\partial f}{\partial \psi} - 2 \operatorname{Re}(A e^{i\psi}) \frac{\partial f}{\partial \eta} = 0$$

$$\frac{\mathrm{d}A}{\mathrm{d}\xi} = \frac{1}{2\pi} \int_0^{2\pi} \int_{-\infty}^{\infty} e^{-i\psi} f \, \mathrm{d}\eta \, \mathrm{d}\psi. \qquad (1)$$

Here $\xi = 2\rho k_w z$ is the dimensionless longitudinal coordinate, k_w is the undulator wavenumber, ρ is the Pierce parameter, $\eta = \delta\gamma/(\gamma/\rho)$ is the relative deviation of particle energy from the synchronous one $\gamma = (1 + K^2)\gamma_\parallel = (1 + K^2)\sqrt{\omega/2ck_w}$, K is the undulator deflection parameter, and $\psi = \omega [z/c(1 + 1/2\gamma_\parallel^2) - t]$ is the slow time variable.

In this work, we intend to show the applicability of the quasilinear approximation near the saturation point. For further simplification we restrict our consideration to the case of a monochromatic electromagnetic field. It is true for the case of amplification of the monochromatic seed signal at frequency ω. For SASE near saturation the relative spectral width is significantly less than ρ (or, more generally, than the "one-gainlength amplification bandwidth"), therefore the results

for the monochromatic signal, presented below, are probably applicable also. Then the field amplitude A does not depend on time ψ, and the distribution function f is assumed to be periodic in ψ: $f(\psi, \eta, \xi) = \sum_{n=-\infty}^{\infty} f_n(\eta, \xi) e^{in\psi}$ and normalized as $\int_{-\infty}^{\infty} f_0(\eta, \xi) \, \mathrm{d}\eta = 1$. This assumption allowed us to write down the chain of equations for the harmonics of the distribution function:

$$\frac{\partial f_n}{\partial \xi} + in\eta f_n = A \frac{\partial}{\partial \eta} f_{n-1} + A^* \frac{\partial}{\partial \eta} f_{n+1}. \qquad (2)$$

Truncation of Eq. (2) for $n \geqslant 2$ gives the closed system of equations that may be solved numerically:

$$\frac{\partial f_1}{\partial \xi} + in f_1 = A \frac{\partial f_0}{\partial \eta}$$

$$\frac{\partial f_0}{\partial \xi} = 2 \operatorname{Re}\left(A^* \frac{\partial}{\partial \eta} f_1\right)$$

$$\frac{\mathrm{d}A}{\mathrm{d}\xi} = \int f_1 \, \mathrm{d}\eta. \qquad (3)$$

3. Numerical solution

The numerical solution is based on the usage of the explicit centered difference scheme. Some results for centered Gaussian initial distribution with standard deviation 1 are presented in Figs. 1

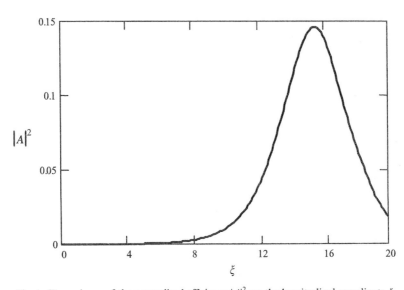

Fig. 1. Dependence of the normalized efficiency $|A|^2$ on the longitudinal coordinate ξ.

and 2. Fig. 1 shows the dependence of the field amplitude on the longitudinal coordinate. Note that the efficiency is $-\rho\langle\eta\rangle = \rho|A|^2$. One can see that at some value of ξ the exponential growth stops and the field amplitude reaches its first maximum. The cause of this becomes clear after considering the behavior of the average distribution function at different ξ (Fig. 2). The increase of the first harmonic in our approximation leads to the growth of the energy spread, which causes the "saturation". Fig. 3 shows the square of bunching

$|\int f_1\,\mathrm{d}\eta|^2$. It may be seen that its value remains significantly less than one until the first maximum of the field amplitude. This fact confirms the statement that the quasilinear mechanism of saturation dominates at the first stage.

4. Values of higher harmonics

To verify the self-consistency of the proposed quasilinear approximation we need to estimate the

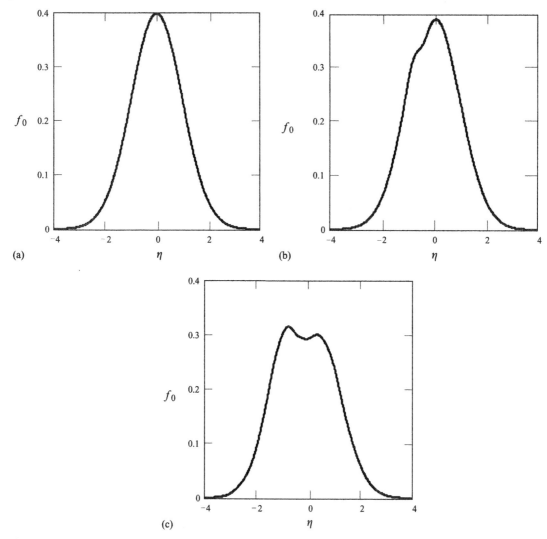

Fig. 2. Modification of the distribution function at different stages of the "saturation" process: (a) initial distribution ($\xi = 0$); (b) distribution at half of the distance before the first maximum ($\xi = 10$); (c) distribution at the first maximum of the field amplitude ($\xi = 15.3$).

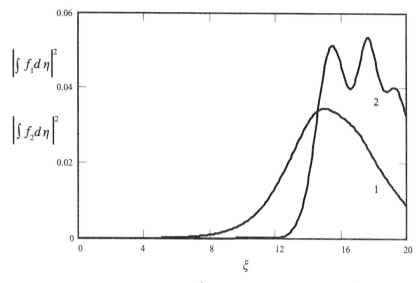

Fig. 3. (1) Square module of bunching $|\int f_1 d\eta|^2$ and (2) second harmonic $|\int f_2\,d\eta|^2$ versus distance.

values of higher current harmonics. The approximate solution for the second harmonic of the distribution function may be found from Eq. (2):

$$f_2 = \int_0^\xi e^{-2i\eta(\xi-\xi')} A(\xi') \frac{\partial}{\partial\eta} f_1(\xi',\eta)\,d\xi'. \tag{4}$$

Then the "improved" quasilinear system can be written as

$$\frac{\partial f_1}{\partial \xi} + i\eta f_1 = A\frac{\partial f_0}{\partial \eta} + A^*\frac{\partial}{\partial \eta}\int_0^\xi e^{-2i\eta(\xi-\xi')}$$
$$\times A(\xi')\frac{\partial}{\partial \eta} f_1(\eta,\xi')\,d\xi'$$
$$\frac{\partial f_0}{\partial \xi} = 2\,\mathrm{Re}\left(A^*\frac{\partial}{\partial\eta}f_1\right)$$
$$\frac{dA}{d\xi} = \int f_1\,d\eta. \tag{3'}$$

The first term on the right side of the first equation in Eq. (3′) is proportional to A, but the second term is proportional to A^3. Therefore, the contribution of the last term to the value of f_1 is small at least at the first stage of saturation (before the first field maximum). The second harmonic bunching is plotted in Fig. 3.

5. Conclusion

The quasilinear equations describe the saturation phenomena consistently. The calculated normalized efficiencies $-\langle\eta\rangle$ for different values of the initial normalized relative energy spread $\sigma = \sqrt{\langle\eta^2\rangle}$ are shown in Fig. 4. Further investigation of the applicability region of these equations and its comparisons with other models (for example [3]) is planned.

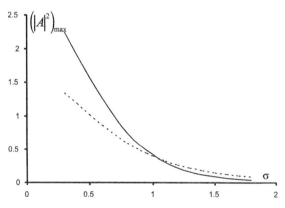

Fig. 4. Dependence of the maximum normalized efficiency $(|A|^2)_{max}$ on the normalized relative energy spread σ (solid line). Dashed line was obtained using fitting formula [4].

This work is partly supported by the US Department of Energy, Office of Basic Energy Sciences, under Contract No. W-31-109-ENG-38.

References

[1] N.A. Krall, A.W. Trivelpiece, Principles of Plasma Physics, McGraw-Hill, New York, 1973.

[2] K.-J. Kim, Phys. Rev. Lett. 57 (1986) 1871.

[3] R.L. Gluckstern, S. Krinsky, H. Okamoto, Phys. Rev. E 47 (6) (1993) 4412.

[4] M. Xie, IEEE Proceedings of the 1995 Particle Accelerator Conference, 183 (IEEE, 1996).

ELSEVIER

Nuclear Instruments and Methods in Physics Research A 475 (2001) 79–85

**NUCLEAR
INSTRUMENTS
& METHODS
IN PHYSICS
RESEARCH**
Section A

www.elsevier.com/locate/nima

Eigenmodes and mode competition in a high-gain free-electron laser including alternating-gradient focusing

Juhao Wu[a,b,*], Li Hua Yu[a]

[a] *National Synchrotron Light Source, Brookhaven National Laboratory, ATF, Bldg. 725 C, Box 5000, Upton, NY 11973-5000, USA*
[b] *Department of Physics and Astronomy, C.N. Yang Institute for Theoretical Physics, State University of New York at Stony Brook, Stony Brook, NY 11794-3840, USA*

Abstract

We solve the eigenvalue problem for a high gain free-electron laser in the "water-bag" model including alternating-gradient focusing by a variational-solution-based (VSB) expansion method. Such VSB expansion method is very efficient for finding the eigenvalue. The results agree with those obtained by numerical simulation quite well. We further discuss the mode degeneracy and mode competition. © 2001 Elsevier Science B.V. All rights reserved.

PACS: 41.60.Cr; 52.75.Ms

Keywords: Eigenmodes; Modes competition; High gain FEL; Alternating-gradient focusing

1. Introduction

The rapid progress in the Free-Electron Laser (FEL) research has now made it possible to go to the deep VUV region or even the X-ray region. The most important physical quantity that we need in practical design is the power exponential-folding length (L_G) of the guiding eigenmode in the exponential growth region. There have been lots of efforts in finding the eigenmode both analytically [1–7] and numerically [8–10]. Initially, this problem was attacked by a variational method within a "water-bag" electron distribution model [1–3] and a Gaussian electron distribution model [4]. The accuracy of the variational result could

achieve a few percent compared with the simulation results [8]. Later on, the accuracy was further improved by a precise solution [6,7] for the Gaussian model. All the above workers did not consider the longitudinal velocity oscillation due to the external alternating-gradient quadrupole focusing in the real experiment case. So, a variational method taking this effect into consideration was developed in Ref. [5]. The study of this work showed that such a longitudinal velocity oscillation effect itself could generate 10%–50% error, so without considering this effect, the results could be questionable. The solution given by this variational method [5] agrees with the simulation [8] within a few percent for most of the practical cases. In some cases, the discrepancy could be larger. So in this paper, we will develop a method to further improve the accuracy and to find the

---

*Corresponding author.

E-mail address: jhwu@bnl.gov (J. Wu).

0168-9002/01/$ - see front matter © 2001 Elsevier Science B.V. All rights reserved.
PII: S 0 1 6 8 - 9 0 0 2 (0 1) 0 1 5 2 6 - 1

reason why the discrepancy exists. Our preliminary result shows that when there is mode degeneracy there could be disagreement between the variational method result and the simulation.

The paper is organized as follows. In Section 2, we briefly review our previous work [5] and then present the analytical derivation. The preliminary results are presented in Section 3. Section 4 is our conclusion.

2. Analytical derivation

In the earlier work [3], we presented an analytical calculation of the FEL gain valid in the regime of exponential growth before saturation, based upon a dispersion relation derived from the Vlasov–Maxwell equations. This treatment included the effects of the energy spread, emittance, and the natural focusing of the electron beam, as well as the diffraction and guiding of the radiation field. Later [5], the treatment was extended to include the alternating-gradient quadrupole focusing. The dispersion relation was solved using a variational method. The results for the exponential-folding length of the power, i.e., L_G, of the fundamental guided mode were expressed in a scaled form. Here, we will introduce the variational-solution-based (VSB) expansion method to improve the variational approximation with all the above mentioned effects taken into consideration.

In our calculation, the electron beam's energy distribution $h(\gamma)$ is Gaussian, with an average energy $\gamma_0 mc^2$, and rms spread $\gamma_0\sigma$. Initially, we assume that the electron beam has a uniform longitudinal density and a uniform "water-bag" distribution inside a four-dimensional sphere in the four-dimensional transverse phase space. We assume that the electron beam focusing is due to both the natural focusing and the external alternating-gradient quadrupole focusing. We model the external quadrupole focusing by assuming it as resulting from a magnetic field

$$\mathbf{B}_Q = (-gy\cos k_Q z, -gx\cos k_Q z, 0) \qquad (1)$$

where $k_Q = 2\pi/\lambda_Q$ and λ_Q is the period of the external quadrupole field. In interesting practical

cases, $\lambda_w \ll \lambda_Q \ll \lambda_q \ll \lambda_\beta \ll \lambda_n$ can be satisfied. Under such a condition, the effective betatron wave number reads,

$$k_\beta = \sqrt{k_n^2 + k_q^4/2k_Q^2}, \qquad (2)$$

where $k_n = Kk_w/\gamma\sqrt{2}$ is due to natural focusing only, and $k_q^2 = eg/\gamma mc$ corresponds to the quadrupole focusing. Eq. (2) explicitly shows the enhancement of focusing due to the quadrupole.

2.1. Dispersion relation

We consider the linear region before saturation, and we write the electric field of the fundamental guided laser mode of frequency ω in the form

$$E(\mathbf{r}, \omega)\, e^{-i\mu k_w z}\, e^{-i\omega(t-z/c)}\hat{\mathbf{e}} + \text{c.c.} \qquad (3)$$

where $\hat{\mathbf{e}}$ is the polarization vector. The function $E(\mathbf{r}, \omega)$ describes the transverse-mode profile in terms of the dimensionless coordinates $\mathbf{r} = \sqrt{2k_r k_w}\, \mathbf{R}$. Assuming that $\gamma_0 \gg 1$, so that space charge effects are negligible, Vlasov–Maxwell equations have been used to derive a dispersion relation determining μ and $E(\mathbf{r})$:

$$(\nabla_\perp^2 + \mu)E(\mathbf{r}) = \frac{i}{2}(2\rho\gamma_0)^3 \int \frac{d\gamma}{\gamma^2} h'(\gamma)$$

$$\times \int d^2 p\, u(p^2 + k^2 r^2) \int_{-\infty}^{0} ds\, e^{-i\alpha s}$$

$$\times \exp\left\{ -i\frac{b}{2k}\left[\left(r^2 + \frac{p^2}{k^2} \right)2ks + \left(r^2 - \frac{p^2}{k^2} \right)\sin(2ks) \right. \right.$$

$$\left. \left. +\frac{\mathbf{p}\cdot\mathbf{r}}{k}2[1 - \cos(2ks)] \right] \right\} E\left(\mathbf{r}\cos ks + \frac{\mathbf{p}}{k}\sin ks \right) \qquad (4)$$

where $k = k_\beta/k_w$ characterizes the total focusing strength (Eq. (2)), $\alpha = \mu + (\omega - \omega_r)/\omega_r - 2 \times (\gamma - \gamma_0)/\gamma_0 + (p^2 + k^2 r^2)/4$, $b = -k_q^4/(16k_w^2 k_Q^2)$, $u(p^2 + k^2 r^2) = (1/\pi k^2 a^2)\theta(k^2 a^2 - k^2 r^2 - p^2)$, with $\theta(x) = 1$ for $x > 0$, and $\theta(x) = 0$ for $x < 0$. $(2\rho\gamma_0)^3 = e^2 Z_0 n_0 K^2 [JJ]^2/2mck_w^2$, with $[JJ] = 1$ for helical wiggler, and $[JJ] = J_0(K^2/2(1 + K^2)) - J_1(K^2/2(1 + K^2))$ for a linear wiggler. $a = \sqrt{2k_r k_w}\, R_0$, where R_0 is the electron beam radius, and $Z_0 = 377\,\Omega$ is the impedance of free space. The parameter b in the dispersion relation of Eq. (4) indicates the different focusing schemes. For natural focusing, we will have $b = 0$, hence the dispersion relation reduces to that of Ref. [3]; and if the alternating-gradient quadrupole focusing is used, then the effect is represented by the terms containing b as what was discussed in Ref. [5].

2.2. Variational method

To maintain a consistent notation, here we briefly review the procedure in our previous papers [1–3,5]. We previously made the following observation: the dispersion relation of Eq. (4) corresponds to the stationary solution of the variational form

$$\int d^2r\, E(\mathbf{r})(\nabla_\perp^2 + \mu)E(\mathbf{r})$$

$$= \int d^2r \int d^2r'\, E(\mathbf{r})\mathcal{K}(\mathbf{r}, \mathbf{r}')E(\mathbf{r}') \tag{5}$$

with the kernel $\mathcal{K}(\mathbf{r}, \mathbf{r}')$ given by

$$\mathcal{K}(\mathbf{r}, \mathbf{r}') = (2\rho)^3 \int d\gamma\, h(\gamma) \int_{-\infty}^{0} ds \frac{k^2 s}{\sin^2 ks} u\left[\frac{k^2}{\sin^2 ks}(r^2 + r'^2 - 2\mathbf{r}\cdot\mathbf{r}'\cos ks)\right]$$

$$\times e^{-is\{\mu + (\omega - \omega_r)/\omega_r - 2[(\gamma - \gamma_0)/\gamma_0] + 1/4(k^2/\sin^2 ks)(r^2 + r'^2 - 2\mathbf{r}\cdot\mathbf{r}'\cos ks)\}}$$

$$\times e^{-ib/2k\{(2ks/\sin^2 ks)(r^2 + r'^2 - 2\mathbf{r}\cdot\mathbf{r}'\cos ks) - (2/\sin ks)[(r^2 + r'^2)\cos ks - 2\mathbf{r}\cdot\mathbf{r}']\}} . \tag{6}$$

Then, we chose the following trial function:

$$E(r) = \begin{cases} e^{-\chi r^2/2a^2}, & r \leqslant a \\ AH_0^{(1)}(r\sqrt{\mu}), & r \geqslant a \end{cases} \tag{7}$$

and the problem was solved completely [5].

2.3. Variational-solution-based (VSB) expansion method

The success of the above mentioned concise variational method has been proven by its agreement with direct dynamical simulation [8] when the radiation frequency is up to UV region. When the radiation frequency is even higher, the discrepancy between the variational method and the simulation becomes visible. So an expansion method is developed in this paper in trying to solve this problem.

Eq. (4) already indicates that this is an eigenvalue problem. If we regard $\vec{r}$ as a continuous index,

then Eq. (4) is in a matrix form, i.e., it could be written as

$$\int d^2\vec{r}'\, \mathcal{M}(\vec{r}, \vec{r}', \mu)E(\vec{r}') = \mu E(\vec{r}) \tag{8}$$

where μ is the eigenvalue. Then if we discretize $E(\vec{r})$ numerically, Eq. (8) will be discretized into a matrix form. One could then find the eigenvalue and the corresponding eigenmode. In the following, we will discretize $E(\vec{r})$ by expanding it in some basis, so that the problem is reformulated into finding the eigenvalue of the expanding coefficient matrix, i.e., assume that

$$E(\vec{r}) = \sum_n a_n \vec{e}_n(\vec{r}) \tag{9}$$

where $\vec{e}_n(\vec{r})$ is a complete orthonormal basis, then Eq. (4) is rewritten as

$$\vec{M}(\mu) \cdot \vec{a} = \mu\vec{a} \tag{10}$$

where $\vec{a}$ is the expansion coefficient column matrix, and μ is the eigenvalue. Once μ is found, the corresponding eigenvector, i.e., the column vector $\vec{a}$ is found. According to Eq. (9), the corresponding eigenmode is obtained. The reason why we need to expand $E(\vec{r})$ is because we require to make full use of the variational solution, which will be regarded as the zeroth-order solution in our expansion method. The reason why we need to make full use of the variational method solution is because of the numerical difficulty in dealing with equations like Eq. (8) or Eq. (10).

We adopt Laguerre polynomials as the basis for expansion, i.e.,

$$E(r) = \sum_{n=0}^{\infty} a_n e^{-\zeta r^2/2} L_n(\Re(\zeta) r^2)$$

$$= \sum_{n=0}^{\infty} a_n e^{-\Re(\zeta) r^2/2 - i\Im(\zeta) r^2/2} L_n(\Re(\zeta) r^2), \tag{11}$$

where ζ is a complex number related to scale, and is arbitrary. $\Re(\zeta)$ is the real part of ζ, and $\Im(\zeta)$ is the imaginary part of $\zeta \cdot L_n(x)$ is the Laguerre polynomial. Now put Eq. (11) into Eq. (4), after several steps of straightforward derivation, we finally reach an equation as Eq. (10) with

$$M_{mn} = 2\zeta\delta_{mn} + 2\Re(\zeta)$$

$$\times \int_0^\infty dr \, r e^{-\Re(\zeta)r^2/2} \, e^{i\Im(\zeta)r^2/2} L_m(\Re(\zeta)r^2)$$

$$\times \int d^2r' \, \mathcal{K}(\mathbf{r},\mathbf{r}') \, e^{-\Re(\zeta)r'^2/2}$$

$$\times e^{-i\Im(\zeta)\,r'^2/2} L_n(\Re(\zeta)r'^2)$$

$$-\int_0^\infty dr \, e^{-\Re(\zeta)r^2} L_m(\Re(\zeta)r^2)$$

$$\times \{2\Re(\zeta)\zeta^2 r^3 L_n(\Re(\zeta)r^2)$$

$$- 8\Re(\zeta)^2 \, r L_{n-1}^1(\Re(\zeta)r^2)$$

$$+ 8\Re(\zeta)^2\zeta r^3 L_{n-1}^1(\Re(\zeta)r^2)$$

$$+ 8\Re(\zeta)^3 r^3 L_{n-2}^2(\Re(\zeta)r^2)\}. \tag{12}$$

So, the problem is cast into an eigenvalue problem. However the coefficient matrix depends non-linearly on the eigenvalue μ, which makes the problem non-trivial. Here, we will make full use of the solution obtained from the variational method. From the following discussion, we will show why it is called VSB expansion method. Let us make it explicit.

(1) As we mentioned above, the complex number ζ is arbitrary, and it is related to the scale.

(2) By comparing the variational trial solution (7) with the formal solution (11) of the expansion method, we realize, (a). In order to have correct asymptotic behavior for small r limit, we need put $\chi/a^2 = \zeta$, (b). $L_0(x) = 1.0$, (c). Since we are looking for the fundamental mode, we could always normalize the solution by setting $a_0 = 1.0$. Then, the variational solution (7) is just the lowest order solution (11) of this expansion method inside the electron beam. So by doing this, i.e. picking up (a), we fix the arbitrariness of ζ, as mentioned in (1).

Now let us write down the steps of our VSB expansion method.

(1) We first find the variational solution.
(2) From the variational solution, we get the value of χ/a^2, so we fix ζ.

(3) We use the variational solution as the initial try for the eigenvalue. In practical cases, it is not too different from the real solution. Hence, it converges to the real solution within just a few steps.

All the above discussion show that our expansion method is VSB, while a naive expansion is normally not a good way to solve a non-linear problem.

3. Preliminary results

Here, we report some preliminary results. First, we need to know normally how many terms we need to get a convergent result, and what is the accuracy of the VSB expansion method. Secondly, we will investigate the mode competition in the high-gain regime. In the literature, there is an extensive discussion about the mode competition and mode locking in low-gain oscillators, since in that case the cavity could be resonant to more than one mode. In the high-gain regime, it is normally believed that one mode will survive. In fact, since there are four or five parameters in the gain length formula, it is still fairly possible that there is mode degeneracy, i.e., there are possibly many modes having similar power e-folding gain length, hence they are all amplified. If this is the case, then the situation will be fairly complex. Besides the modification in the radiation spectrum, the laser will be degraded.

3.1. Convergence of the VSB expansion method

Typical results are plotted in Fig. 1, which represents FEL at 1000 Å. The results show that we need only up to five or six terms in expansion. The solution converges very quickly, which makes it feasible to adopt a VSB expansion method for parameter optimization in real experimental design. Further increase in the expansion terms does not improve the accuracy because it is limited by the machine precision. In Fig. 1, the corresponding result from the variational method is also plotted. It is clear that the result of the VSB expansion method agrees better with the simulation than the

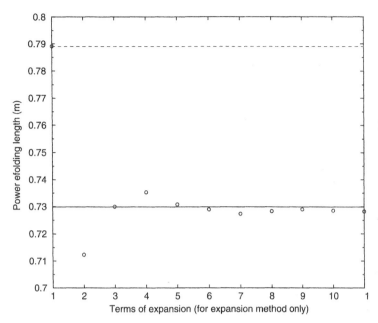

Fig. 1. Results of three different methods for a UV FEL. Solid line: numerical simulation; Dashed line: variational result; Circle: expansion method. Horizontal axis is the number of terms used in the expansion method.

variational solution. We worked out another example of an FEL at 72 Å. A similar situation is obtained. The simulation gives $L_G = 0.3884$ m, and the variational result is $L_G = 0.3815$ m. Five expansion terms in VSB method already gave a very good result $L_G = 0.3907$ m. Increasing expansion terms does not improve the agreement, obviously.

3.2. Mode competition

Once we have confidence in our VSB expansion method, we should now investigate mode competition in the high-gain regime. For some set of parameters, we will observe mode competition even in the high-gain regime. This is of practical application and theoretical interest as well. From a practical point of view, if there is more than one mode excited, then it will affect the radiation spectrum and the output power as well. From a theoretical point of view, it means that the single guiding mode assumption may break down. If there is degeneracy, i.e., if there are two modes

having similar power e-folding length with different frequency, then both of them survive and get amplified.

Shown in Fig. 2 is an example of an FEL at 1.5 Å with a set of parameters different from that of LCLS project. There are two modes excited at slightly different wavelengths. The discrepancy between the variational result and the simulation result is obvious. According to the simulation, these two modes have almost the same power e-folding length, $L_G = 7.8$ m at $\lambda = 1.505$ Å, and $L_G = 7.4$ m at $\lambda = 1.507$ Å. In the output, these two modes are expected to be both there. However, from the variational result we have $L_G = 9.2$ m at $\lambda = 1.505$ Å and $L_G = 6.9$ m at $\lambda = 1.507$ Å. There is still a large difference between the two modes so as to make only one mode survive. We resolved this puzzle by using the VSB expansion method. The VSB method found $L_G = 7.9$ m at $\lambda = 1.505$ Å and $L_G = 7.4$ m at $\lambda = 1.507$ Å. This agrees with the simulation quite well.

Now let us investigate why the variational method gives a much higher result at $\lambda =$

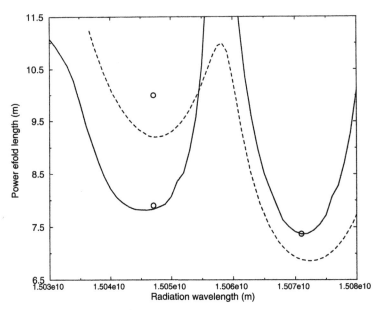

Fig. 2. Results of three different methods for a 1.5 Å FEL. Solid line: numerical simulation; Dashed line: variational result; Circle: expansion method.

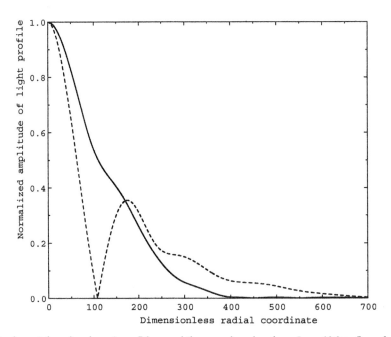

Fig. 3. Profile of the fundamental mode, whose $L_G = 7.9$ m, and the second mode, whose $L_G = 10.0$ m. In such a dimensionless radial coordinate, the edge of the electron beam is at $a = 211.4$.

1.505 Å. In Fig. 3, we plotted the lowest two modes at $\lambda = 1.505$ Å found by the VSB method. The fundamental mode, whose $L_G = 7.9$ m, is actually the one found by the simulation, while the second mode, whose $L_G = 10.0$ m, is fairly closer to the one found by the variational method. So, firstly, the accuracy of the variational method is not good enough in such a case. Secondly, there will be mode degeneracy for such a set of parameters.

4. Conclusion

In conclusion, we developed an expansion method making full use of the success of our previous variational method, and we also find a way out in case the previous variational method does not give a good result for some set of parameters. The VSB expansion method is very efficient for finding the solution, and it gives results agreeing with numerical simulation. As we have demonstrated, there is mode competition in some region and we should avoid such regions in real design. In the case where more than one mode gets excited, any calculation based on the single guiding mode assumption will fail to give reasonable results. The development of an analytical tool which could deal with multifrequency situation is underway.

Acknowledgements

J.W. wishes to thank Prof. C.N. Yang for his invaluable advice. The authors also wish to thank Dr. S. Krinsky for many suggestions and comments. This work has been performed under the auspices of the U.S. Department of Energy.

References

[1] S. Krinsky, L.H. Yu, Phys. Rev. A 35 (1987) 3406.
[2] L.-H. Yu, S. Krinsky, Phys. Lett. A 129 (1988) 463.
[3] L.H. Yu, S. Krinsky, R.L. Gluckstern, Phys. Rev. Lett. 64 (1990) 3011.
[4] Y.H. Chin, K.J. Kim, M. Xie, Phys. Rev. A 46 (1992) 6662.
[5] L.H. Yu, C.M. Hung, D. Li, S. Krinsky, Phys. Rev. E 51 (1995) 813.
[6] M. Xie, IEEE Proceedings of PAC95, No. 95CH3584, 1996, p. 183.
[7] M. Xie, Nucl. Instr. and Meth. A 445 (2000) 59.
[8] T.M. Tran, J.S. Wurtele, Comput. Phys. Commun. 54 (1989) 263.
[9] Roger J. Dejus, Oleg A. Shevchenko, Nikolai A. Vinokurov, Nucl. Instr. and Meth. A 445 (2000) 19.
[10] S.G. Biedron, Y.C. Chae, R.J. Dejus, B. Faatz, H.P. Freund, S.V. Milton, H.-D. Nuhn, S. Reiche, Nucl. Instr. and Meth. A 445 (2000) 110.

ELSEVIER

Nuclear Instruments and Methods in Physics Research A 475 (2001) 86–91

NUCLEAR
INSTRUMENTS
& METHODS
IN PHYSICS
RESEARCH
Section A

www.elsevier.com/locate/nima

The general solution of the eigenvalue problem for a high-gain FEL

E.L. Saldin[a], E.A. Schneidmiller[a],*, M.V. Yurkov[b]

[a] *Deutsches Elektronen-Synchrotron, DESY, Notkestrasse 85, 22607 Hamburg, Germany*
[b] *Joint Institute for Nuclear Research, Dubna, 141980 Moscow Region, Russia*

Abstract

The exact solution of the eigenvalue equation for a high-gain FEL derived in Xie (Nucl. Instr. and Meth. A 445 (2000) 59) is generalized in order to include the space charge effects. This solution is valid not only for natural undulator focusing, but also for alternating-gradient focusing under some condition that is presented. At such, the obtained solution includes all the important effects in the system of axially homogeneous electron beam and undulator: diffraction, betatron motion, energy spread, space charge and frequency detuning. It is valid for ground TEM_{00} mode as well as for high-order modes and can be used for calculation of high-gain FEL amplifiers operating in the wavelength regions from far infrared down to X-ray. In addition, a computationally efficient approximate solution for TEM_{00} mode is derived providing high accuracy (better than 1% in the whole range of parameters). It can be used for quick optimization of FEL amplifiers. © 2001 Elsevier Science B.V. All rights reserved.

PACS: 41.60.Cr; 52.75.M

Keywords: Free electron laser; High-gain FEL; Eigenvalue problem

1. Introduction

The solution of an eigenvalue problem is the first and very important step in the design and optimization of a high-gain FEL. The first solution of the eigenvalue problem was obtained in Ref. [2] in the frame of one-dimensional model taking into account space charge, energy spread, and frequency detuning. Diffraction effects for a parallel electron beam (no betatron oscillations) were considered in Refs. [3,4] neglecting space charge

and energy spread. In Ref. [3] the asymptotic solutions for weak and strong diffraction regimes were obtained, and in Ref. [4] the exact solution was derived for the first time for a stepped transverse profile of the axisymmetric electron beam. In the proceeding papers [5,6] the eigenvalue problem for the model of a parallel electron beam, including diffraction of radiation, was studied extensively in order to include different transverse profiles of the electron beam, high-order modes of the radiation, energy spread and space charge effects.

A more difficult task was the eigenvalue problem that includes diffraction of radiation and a transverse motion of electrons (betatron oscillations). The integro-differential equation,

---

*Corresponding author. Tel.: +49-40-8998-2676; fax: +49-40-8998-4475.

E-mail address: schneidm@mail.desy.de
(E.A. Schneidmiller).

0168-9002/01/$ - see front matter © 2001 Elsevier Science B.V. All rights reserved.
PII: S 0 1 6 8 - 9 0 0 2 (0 1) 0 1 5 4 9 - 2

taking into account both these effects as well as frequency detuning and energy spread, was derived in Ref. [7,8]. Different methods to find an approximate solution of the eigenvalue problem were used in Refs. [1,9–11] for different phase space distributions of the electron beam. The exact solution (in a sense that it can be evaluated numerically with any desirable accuracy) for all the eigenmodes was obtained in Ref. [1].

In this paper, we generalize the exact solution derived in Ref. [1] in order to include the space charge effect. Although this effect is negligible in X-ray wavelength range [12–14], it can be important for infrared [15–17], visible and ultraviolet [18,19] high-gain FELs. Thus, our solution is a universal tool for calculation and optimization of high-gain FELs. Indeed, it includes all the important effects in a system of axially homogeneous electron beam and undulator: diffraction, betatron motion, energy spread, space charge and frequency detuning. In addition, we derive here an approximate solution for the ground TEM_{00} mode. The numerical algorithm for finding this solution is very fast and accurate, so it can be used for quick optimization of high-gain FELs.

It is worth noticing that the authors of [1,8–11] assumed natural undulator focusing. Actually, the validity region of the obtained results (including the results of this paper) is wider: they can also be used in the case of a superposition of the natural undulator focusing and an alternating-gradient external focusing. In this case the following condition should be satisfied:

$$\frac{L_f}{2\pi\beta} \ll \min\left(1, \frac{\lambda}{2\pi\epsilon}\right)$$

where L_f is a period of the external focusing structure, β is an average beta-function, ϵ is rms emittance of an electron beam, and λ is a radiation wavelength. This condition is met in many practical situations.

2. Basic equation

Let us have at the undulator entrance a continuous electron beam with the current I_0, with a Gaussian distribution in energy

$$F(\varepsilon - \varepsilon_0) = \left(2\pi\langle(\Delta\varepsilon)^2\rangle\right)^{-1/2}\exp\left(-\frac{(\varepsilon - \varepsilon_0)^2}{2\langle(\Delta\varepsilon)^2\rangle}\right) \quad (1)$$

and in a transverse phase plane

$$f(x, x') = (2\pi\sigma^2 k_\beta)^{-1}\exp\left[-\frac{x^2 + (x')^2/k_\beta^2}{2\sigma^2}\right] \quad (2)$$

the same in y phase plane. Here $k_\beta = 1/\beta$ is the wavenumber of betatron oscillations and $\sigma = \sqrt{\epsilon\beta}$.

Using cylindrical coordinates, in the high-gain limit we seek the solution for a slowly varying complex amplitude of the electric field of the electromagnetic wave in the form [20]:

$$\tilde{E}(z, r, \varphi) = \Phi_n(r)\exp(\Lambda z)\begin{pmatrix}\sin(n\varphi)\\\cos(n\varphi)\end{pmatrix} \quad (3)$$

where n is an integer, $n \geqslant 0$. For each n there are many radial eigenmodes that differ by eigenvalue Λ and eigenfunction $\Phi_n(r)$. The integro-differential equation for radiation field eigenmodes [1,8] taking into account the space charge effect [20] can be written in the following normalized form:

$$\left[\frac{d^2}{d\hat{r}^2} + \frac{1}{\hat{r}}\frac{d}{d\hat{r}} - \frac{n^2}{\hat{r}^2} + 2iB\hat{\Lambda}\right]\Phi_n(\hat{r})$$

$$= -4\int_0^\infty d\hat{r}'\hat{r}'\{\Phi_n(\hat{r}') \quad (4)$$

$$+ \frac{\hat{\Lambda}_p^2}{2}\left[\frac{d^2}{d\hat{r}'^2} + \frac{1}{\hat{r}'}\frac{d}{d\hat{r}'} - \frac{n^2}{\hat{r}'^2} + 2iB\hat{\Lambda}\right]\Phi_n(\hat{r}')\}$$

$$\times \int_0^\infty d\zeta\frac{\zeta}{\sin^2(\hat{k}_\beta\zeta)}\exp\left[-\frac{\hat{\Lambda}_T^2\zeta^2}{2} - (\hat{\Lambda} + i\hat{C})\zeta\right]$$

$$\times \exp\left[-\frac{(1 - iB\hat{k}_\beta^2/2)(\hat{r}^2 + \hat{r}'^2)}{\sin^2(\hat{k}_\beta\zeta)}\right]$$

$$\times I_n\left[\frac{2(1 - iB\hat{k}_\beta^2\zeta/2)\hat{r}\hat{r}'\cos(\hat{k}_\beta\zeta)}{\sin^2(\hat{k}_\beta\zeta)}\right]$$

where I_n is the modified Bessel function of the first kind. The following notations are used here: $\hat{r} = r/(\sigma\sqrt{2})$, $B = 2\sigma^2\Gamma\omega/c$ is the diffraction parameter, $\hat{k}_\beta = k_\beta/\Gamma$ is the betatron motion parameter, $\hat{\Lambda}_p^2 = 2c^2(A_{JJ}\theta_s\sigma\omega)^{-2}$ is the space charge parameter, $\hat{\Lambda}_T^2 = \langle(\Delta\varepsilon)^2\rangle/(\rho^2\varepsilon^2)$ is the energy spread parameter, $\hat{C} = [k_w - \omega/(2c\gamma_z^2)]/\Gamma$ is the detuning parameter, $\Gamma = [A_{JJ}^2 I_0 \omega^2 \theta_s^2 (I_A c^2 \gamma_z^2 \gamma)^{-1}]^{1/2}$ is the gain parameter, $\rho = c\gamma_z^2\Gamma/\omega$ is the efficiency parameter, ω is the frequency of the electromagnetic wave, $\theta_s = K_{rms}/\gamma$, K_{rms} is the rms undulator parameter, γ is the relativistic factor, $\gamma_z^{-2} = \gamma^{-2} + \theta_s^2$, k_w is the undulator wavenumber, $I_A = 17\,\text{kA}$ is the Alfven current, $A_{JJ} = 1$ for helical undulator and $A_{JJ} = J_0(K_{rms}^2/2(1 + K_{rms}^2)) - J_1(K_{rms}^2/2(1 + K_{rms}^2))$ for planar undulator. Here J_0 and J_1 are the Bessel functions of the first kind. The space charge effect is included into (4) under the condition $\sigma^2 \gg c^2\gamma_z^2/\omega^2$.

3. Exact solution

As suggested in Ref. [1] we apply to (4) the Hankel transformation defined by the following transform pair:

$$\bar{\Phi}_n(p) = \int_0^\infty d\hat{r}\,\hat{r}\,J_n(p\hat{r})\Phi_n(\hat{r}),$$

$$\Phi_n(\hat{r}) = \int_0^\infty dp\,p\,J_n(p\hat{r})\bar{\Phi}_n(p).$$

Then we obtain the integral equation for the Hankel transform $\bar{\Phi}_n(p)$:

$$\bar{\Phi}_n(p) = -\frac{1}{2iB\hat{\Lambda} - p^2}\int_0^\infty dp'\,p'\,\bar{\Phi}_n(p')$$

$$\times \left[1 + \frac{\hat{\Lambda}_p^2(2iB\hat{\Lambda} - p'^2)}{2}\right]$$

$$\times \int_0^\infty d\zeta \frac{\zeta}{(1 - iB\hat{k}_\beta^2\zeta/2)^2}\exp\left[-\frac{\hat{\Lambda}_T^2\zeta^2}{2} - (\hat{\Lambda} + i\hat{C})\zeta\right]$$

$$\times \exp\left[-\frac{p^2 + p'^2}{4(1 - iB\hat{k}_\beta^2\zeta/2)}\right]I_n\left[\frac{pp'\cos(\hat{k}_\beta\zeta)}{2(1 - iB\hat{k}_\beta^2\zeta/2)}\right]. \quad (5)$$

When the space charge field is negligible, $\hat{\Lambda}_p^2 \to 0$, this equation is reduced to one obtained in Ref. [1].

To solve (5) we discretize it:

$$p_i = \Delta(i - 1/2), \quad i = 1, 2, \ldots, K,$$

$$p_j' = \Delta(j - 1/2), \quad j = 1, 2, \ldots, K$$

where Δ and K should be chosen in such a way that the required accuracy is provided. Then we obtain

a matrix equation

$$\bar{\Phi}_n(i) = M_n(i, j)\bar{\Phi}_n(j)$$

or $[M_n - I]\bar{\Phi}_n = 0$, where I is a unit matrix. Matrix M_n depends on an eigenvalue $\hat{\Lambda}$ as well as on the problem parameters: B, $\hat{k}_\beta$, $\hat{\Lambda}_T^2$, $\hat{\Lambda}_p^2$, and $\hat{C}$. The eigenvalues of all radial modes for a given azimuthal index n can be found by solving the equation $|M_n - I| = 0$. Then the calculation of the eigenmodes is straightforward.

This algorithm allows one to find with any desirable accuracy the eigenvalues and eigenfunctions of a high-gain FEL including all the important effects: diffraction, betatron motion, energy spread, space charge, and frequency detuning (see Fig. 1 as an example). Therefore, it can be considered as a universal tool for calculation and optimization of FEL amplifiers of wavelength range from infrared down to X-ray. The only disadvantage is that the algorithm becomes time-consuming when one needs a reasonable accuracy of calculations.

4. Approximate solution for TEM$_{00}$ mode

For quick optimization of high-gain FELs one needs a fast and, at the same time, pretty accurate procedure for the determination of eigenvalues. Usually the eigenvalue of ground TEM$_{00}$ mode is the subject of optimization because FELs are

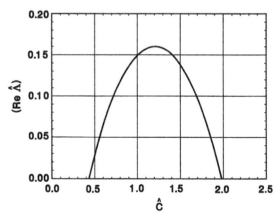

Fig. 1. Reduced growth rate of TEM$_{00}$ mode versus detuning parameter. Here $B = 1$, $\hat{k}_\beta = 1$, $\hat{\Lambda}_T^2 = 0.5$, and $\hat{\Lambda}_p^2 = 0.2$.

designed in such a way that only this mode survives in the end of amplification process. Thus, we will look for an approximate solution only for the ground mode. Different approaches for obtaining approximate solutions have been developed [1,9,10]. We consider here the one described in Ref. [1] and then we introduce the new method allowing us to improve the accuracy of the solution.

As it has been done in Ref. [1], we construct a variational functional from (4), use a trial function in the form

$$\Phi_0(\hat{r}) = \exp(-a\hat{r}^2) \qquad (6)$$

and apply the variational condition, $\delta\hat{\Lambda}/\delta a = 0$. As a result, we obtain two equations for two unknown quantities, $\hat{\Lambda}$ and a:

$$1 - \frac{iB\hat{\Lambda}}{a} - 2\int_0^\infty d\zeta\zeta\frac{\exp[-\hat{\Lambda}_T^2\zeta^2/2 - (\hat{\Lambda} + i\hat{C})\zeta]}{f_2(\zeta)} = 0 \qquad (7)$$

$$\frac{iB\hat{\Lambda}}{a^2} + 4\int_0^\infty d\zeta\zeta\frac{f_1(\zeta)\exp[-\hat{\Lambda}_T^2\zeta^2/2 - (\hat{\Lambda} + i\hat{C})\zeta]}{f_2^2(\zeta)} = 0 \qquad (8)$$

where

$$f_1(\zeta) = 1 - iB\hat{k}_\beta^2\zeta/2 + a\sin^2(\hat{k}_\beta\zeta)$$

$$f_2(\zeta) = (1 - iB\hat{k}_\beta^2\zeta/2)^2 + 2a(1 - iB\hat{k}_\beta^2\zeta/2)$$
$$+ a^2\sin^2(\hat{k}_\beta\zeta).$$

These equations are obtained under the condition $\hat{\Lambda}_p^2 \to 0$. The solution of the system of Eqs. (7) and (8) is accurate at large values of the diffraction parameter, $B \gg 1$, when the radiation field is concentrated inside the electron beam and is well described by function (6). On the other hand, it becomes highly inaccurate at small values of B. For instance, in the limit of parallel beam, $\hat{k}_\beta = 0$, negligible energy spread, $\hat{\Lambda}_T = 0$, and small diffraction parameter, $B \to 0$, we get $\max(\text{Re}\,\hat{\Lambda}) \to \sqrt{2}$ (at the optimal detuning $\hat{C}$) instead of the well-known logarithmic asymptote [3,4]. In this limit the mode size is much larger than the electron beam size and the actual behaviour of the field at large $\hat{r}$ is $\Phi_0(\hat{r}) \propto \exp(-g\hat{r})$ [20] rather than (6).

In order to improve the accuracy of the solution in the whole range of parameters we propose to find the second approximation to the eigenvalue and the eigenfunction in the following way. We modify the r.h.s. of (4) assuming the function $\Phi_0(\hat{r})$ to be of the form (6) with the definite value of the parameter a that is already found by solving the system of Eqs. (7) and (8). As a result, we approximate the integro-differential equation (4) by a differential equation (the space charge effect is again neglected here):

$$\left[\frac{d^2}{d\hat{r}^2} + \frac{1}{\hat{r}}\frac{d}{d\hat{r}} + \mu^2(\hat{r})\right]\Phi_0(\hat{r}) = 0 \qquad (9)$$

where

$$\mu^2(\hat{r}) = -g^2 + 2\exp(-\hat{r}^2)\int_0^\infty d\zeta\zeta f_1^{-1}(\zeta)$$

$$\times\exp\left[-\frac{\hat{\Lambda}_T^2\zeta^2}{2} - (\hat{\Lambda} + i\hat{C})\zeta\right]$$

$$\times\exp\left[\frac{i(1 - iB\hat{k}_\beta^2\zeta/2)B\hat{k}_\beta^2\zeta/2 + a(1 + a)\sin^2(\hat{k}_\beta\zeta)}{f_1(\zeta)}\hat{r}^2\right] \qquad (10)$$

and $g^2 = -2iB\hat{\Lambda}$. In the limit of parallel beam, $\hat{k}_\beta \to 0$, Eq. (9) is reduced to the well-known one [20]. New approximations to the eigenvalue $\hat{\Lambda}$ and the eigenfunction $\Phi_0(\hat{r})$ should be determined from (9). To do this we use the multilayer approximation method described in Refs. [6,20].

The accuracy of this algorithm was tested by comparison with the exact solution described above. For any set of parameters chosen the accuracy was better than 1%. This can be explained as follows. As we have already mentioned, at large values of the diffraction parameter B the eigenfunction is correctly described by (6) and the variational method gives accurate result. The second approximation (solution of the differential Eq. (9)) almost does not differ from the variational solution. In the opposite limit, $B \to 0$, the variational method gives the wrong shape for the eigenfunction. Nevertheless, this does not matter for our solution because in this limit the field is much wider than the electron beam and, therefore, is almost constant within the beam. In other words, in this limit the source term (r.h.s. of

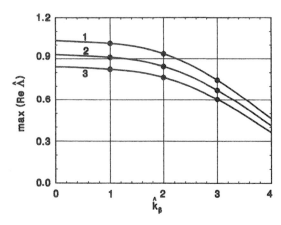

Fig. 2. Reduced growth rate of TEM_{00} mode at the optimal detuning versus betatron motion parameter. Here $B = 0.1$ and $\hat{\Lambda}_p^2 \to 0$. Curve (1): $\hat{\Lambda}_T^2 = 0$, Curve (2): $\hat{\Lambda}_T^2 = 0.1$, Curve (3): $\hat{\Lambda}_T^2 = 0.2$. Curves are the results of approximate solution and circles are the results of exact solution.

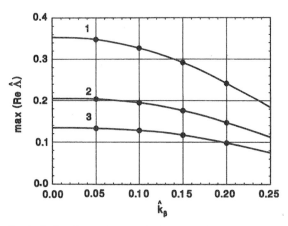

Fig. 4. Reduced growth rate of TEM_{00} mode at the optimal detuning versus betatron motion parameter. Here $B = 10$ and $\hat{\Lambda}_p^2 \to 0$. Curve (1): $\hat{\Lambda}_T^2 = 0$, Curve (2): $\hat{\Lambda}_T^2 = 0.1$, Curve (3): $\hat{\Lambda}_T^2 = 0.2$. Curves are the results of approximate solution and circles are the results of exact solution.

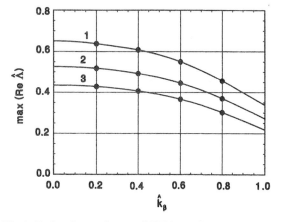

Fig. 3. Reduced growth rate of TEM_{00} mode at the optimal detuning versus betatron motion parameter. Here $B = 1$ and $\hat{\Lambda}_p^2 \to 0$. Curve (1): $\hat{\Lambda}_T^2 = 0$, Curve (2): $\hat{\Lambda}_T^2 = 0.1$, Curve (3): $\hat{\Lambda}_T^2 = 0.2$. Curves are the results of approximate solution and circles are the results of exact solution.

the numerical algorithm to find the approximate solution is much faster than in case of the exact solution. In Figs. 2–4 we present some results of the eigenvalue calculation obtained with both methods.

Eq. (4)) is independent of the first approximation to $\Phi_0(\hat{r})$ obtained from variational method. Then, solving (9) one gets accurate eigenfunction and eigenvalue. Thus, our solution has correct behaviour in both limits of weak and strong diffraction. In addition, it is always correct in the limit of parallel beam. In fact, the approximate solution presented here fits very well the exact solution in the whole region of parameters. At the same time,

References

[1] M. Xie, Nucl. Instr. and Meth. A 445 (2000) 59.
[2] N.M. Kroll, W.A. McMullin, Phys. Rev. A 17 (1978) 300.
[3] A.M. Kondratenko, E.L. Saldin, Part. Acc. 10 (1980) 207.
[4] G.T. Moore, Opt. Commun. 52 (1984) 46.
[5] M. Xie, D. Deacon, Nucl. Instr. and Meth. A 250 (1986) 426.
[6] E.L. Saldin, E.A. Schneidmiller, M.V. Yurkov, Opt. Commun. 97 (1993) 272.
[7] K.J. Kim, Phys. Rev. Lett. 57 (1986) 1871.
[8] L.H. Yu, S. Krinsky, Phys. Lett. A 129 (1988) 463.
[9] L.H. Yu, S. Krinsky, R.L. Gluckstern, Phys. Rev. Lett. 64 (1990) 3011.
[10] Y.H. Chin, K.J. Kim, M. Xie, Nucl. Instr. and Meth. A 318 (1992) 481.
[11] M. Xie, Nucl. Instr. and Meth. A 445 (2000) 67.
[12] J. Rossbach, Nucl. Instr. and Meth. A 375 (1996) 269.
[13] R. Brinkmann, G. Materlik, J. Rossbach, A. Wagner (Eds.), Conceptual Design of a 500 GeV e+e− Linear Collider with Integrated X-ray Laser Facility, DESY 97-048, Hamburg, 1997.
[14] Linac Coherent Light Source Design Study Report, SLAC-R-521, Stanford, 1998.

[15] M. Hogan, et al., Phys. Rev. Lett. 81 (1998) 4867.

[16] E.L. Saldin, E.A. Schneidmiller, M.V. Yurkov, Nucl. Instr. and Meth. A 429 (1999) 197.

[17] E.L. Saldin, et al., Photon Linear Colliders of TeV Energy Range, preprint JINR E-9-94-74, Dubna, 1994.

[18] B. Faatz, et al., preprint DESY-00-095, Hamburg, 2000.

[19] C. Pagani, et al., Nucl. Instr. and Meth. A 423 (1999) 190.

[20] E.L. Saldin, E.A. Schneidmiller, M.V. Yurkov, The Physics of Free Electron Lasers, Springer, Berlin, 1999.

ELSEVIER

Nuclear Instruments and Methods in Physics Research A 475 (2001) 92–96

NUCLEAR INSTRUMENTS & METHODS IN PHYSICS RESEARCH
Section A

www.elsevier.com/locate/nima

Limitations of the transverse coherence in the self-amplified spontaneous emission FEL

E.L. Saldin[a,*], E.A. Schneidmiller[a], M.V. Yurkov[b]

[a] *Deutches Elektronen Synchrotron (DESY), Notkestrasse 85, 22607 Hamburg, Germany*
[b] *Joint Institute for Nuclear Research, Dubna, 141980 Moscow Region, Russia*

Abstract

In this paper we analyze the process of the formation of transverse coherence of radiation from a self-amplified spontaneous emission (SASE) FEL. It is shown that in the high-gain linear regime the degree of transverse coherence approaches unity asymptotically as z^{-1}, but not exponentially, as one would expect from a simple physical assumption that the transverse coherence is established due to the transverse mode selection. It has been found that even after finishing the transverse mode selection process the degree of transverse coherence of the radiation from SASE FEL visibly differs from unity. This is a consequence of the interdependence of the longitudinal and transverse coherence. The SASE FEL has poor longitudinal coherence which develops slowly with the undulator length thus preventing a full transverse coherence. © 2001 Elsevier Science B.V. All rights reserved.

PACS: 41.60.Cr; 52.75.M

Keywords: Free electron laser; High-gain FEL; Initial-value problem

1. Introduction

A description of the properties of the output radiation from the SASE FELs is of great practical importance in view of the X-ray SASE FEL projects under development [1–4]. In paper [5] we developed a technique for an analytical description of the start-up from shot noise. The solution of the initial-value problem is based on an analytical approach originally developed by Krinsky and Yu [6,7] and a multilayer approximation method for

---
*Corresponding author. Tel.: +49-40-8998-2676; fax: +49-40-8998-4475.

E-mail address: saldin@vxdesy.de (E.L. Saldin).

calculations of the eigenfunctions developed in Ref. [8]. This systematic approach allows us to calculate the average radiation power, radiation spectrum envelope, the angular distribution of the radiation intensity in the far zone, and the degree of transverse coherence. Analytical results serve as a primary standard for testing the three-dimensional, time-dependent simulation code FAST [9]. Comparison with analytical results shows that in the high-gain linear limit there is a good agreement between the numerical and analytical results.

In this paper we analyze the process of the formation of the transverse coherence in the SASE FEL. Both analytical and numerical results show that, in the high-gain linear regime, the degree of transverse coherence approaches unity asymptotically as z^{-1}, but not exponentially, as one would expect from the assumption that transverse coherence is established due to the transverse mode selection. Even after finishing the transverse mode selection process the degree of transverse coherence of the radiation from SASE FEL visibly differs from unity. Since the gain in a SASE FEL always has a finite value, this effect imposes a limit on the maximum value of the degree of transverse coherence.

2. Analytical description of start-up from shot noise

We consider an FEL amplifier with a helical undulator and an axisymmetric electron beam having a transverse current density $j_0(r) = I_0 S \times (r/r_0)/[2\pi \int_0^\infty r S(r/r_0)\,\mathrm{d}r]$, where r_0 is the beam profile parameter (typical transverse size of the beam) and I_0 is the beam current. It is assumed that the electron beam has a Gaussian energy spread, $F(\mathscr{E} - \mathscr{E}_0) = (2\pi\langle(\Delta\mathscr{E})^2\rangle)^{-1/2}\exp[-(\mathscr{E} - \mathscr{E}_0)^2/(2\langle(\Delta\mathscr{E})^2\rangle)]$, where $\langle(\Delta\mathscr{E})^2\rangle$ is r.m.s. energy spread. We neglect the transverse variation of the undulator field and assume that electrons move along constrained helical trajectories in parallel with the z-axis. The electron rotation angle is considered to be small and the longitudinal electron velocity v_z is close to the velocity of light c.

Initial-value problem for start-up from shot noise is solved for the case of long rectangular electron bunch [5]. At a sufficiently large undulator length, the spectrum of the SASE radiation is concentrated within the narrow band near the resonance frequency ω_0 (see Fig. 1). Therefore, the electric field of the wave can be presented as

$$E_x + \mathrm{i}E_y = \tilde{E}(\vec{r}_\perp, z, t)\exp[\mathrm{i}\omega_0(z/c - t)] + \text{C.C.}$$

where $\tilde{E}$ is the slowly varying complex amplitude. Taking into account Parseval's theorem, and using the notation $\vec{\rho} = \vec{r}_\perp - \vec{r}'_\perp$ and $\vec{R} = (\vec{r}_\perp + \vec{r}'_\perp)/2$, we can write the average angular spectrum of the radiation in the far zone in the following form:

$$h(\vec{k}_\perp, z) = \frac{1}{(2\pi)^2}\int \gamma^{\mathrm{eff}}(\vec{\rho}, z)\exp(-\mathrm{i}\vec{k}_\perp \vec{\rho})\,\mathrm{d}\vec{\rho}$$

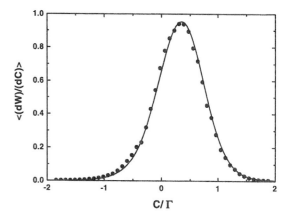

Fig. 1. Averaged spectrum of the radiation from the FEL amplifier starting from shot noise at the undulator length $\hat{z} = 15$. Here $B = 1$, $\Lambda_\mathrm{p}^2 \to 0$, $\Lambda_\mathrm{T}^2 = 0$, and $N_\mathrm{c} = 7 \times 10^7$. Solid curve represents analytical results calculated for nine beam radiation modes ($m, n = 0, 1, 2$). The circles are the results obtained with linear simulation code FAST.

where $\gamma^{\mathrm{eff}}(\vec{\rho}, z)$ is effective transverse correlation function:

$$\gamma^{\mathrm{eff}}(\vec{\rho}, z) = \frac{\int \langle \tilde{E}(\vec{R} + \vec{\rho}/2, z, t)\tilde{E}^*(\vec{R} - \vec{\rho}/2, z, t)\rangle\,\mathrm{d}\vec{R}}{\int \langle |\tilde{E}(\vec{R}, z, t)|^2\rangle\,\mathrm{d}\vec{R}}.$$

Here, the averaging symbol $\langle \ldots \rangle$ means the ensemble average over bunches. In the axisymmetric case the angular spectrum can be written in the following dimensionless form [5]:

$$h(\hat{\theta}, \hat{z}) = \left[\sum_{n,k,j}\int_{-\infty}^{\infty}\mathrm{d}\hat{C}\left(\Omega_{kj}^{(n)}\int_0^{\infty}\Phi_{nk}(\hat{r})J_n(\hat{\theta}\hat{r})\hat{r}\,\mathrm{d}\hat{r}\right.\right.$$
$$\left.\times \int_0^{\infty}\Phi_{nj}^*(\hat{r}')J_n(\hat{\theta}\hat{r}')\hat{r}'\,\mathrm{d}\hat{r}'\right)\right]$$
$$\times \left[2\pi\sum_{n,k,j}\int_{-\infty}^{\infty}\mathrm{d}\hat{C}\left(\Omega_{kj}^{(n)}\right.\right.$$
$$\left.\left.\times \int_0^{\infty}\Phi_{nk}(\hat{r})\Phi_{nj}^*(\hat{r})\hat{r}\,\mathrm{d}\hat{r}\right)\right]^{-1} \quad (1)$$

where

$$\Omega_{kj}^{(n)} = u_k^{(n)}(u_j^{(n)})^*\exp\left\{[\lambda_k^{(n)} + (\lambda_j^{(n)})^*]\hat{z}\right\}$$
$$\times \int_0^{\infty}\Phi_{nk}(\hat{r})\Phi_{nj}^*(\hat{r})S(\hat{r})\hat{r}\,\mathrm{d}\hat{r}$$

$$u_j^{(n)} = \hat{D}_0(\lambda_j^{(n)}) \left[B \int_0^\infty \Phi_{nj}^2(\hat{r})\hat{r}\,\mathrm{d}\hat{r} - \left(\frac{\mathrm{d}\hat{D}}{\mathrm{d}p} \right)_{p=\lambda_j^{(n)}} \right.$$

$$\left. \times \int_0^\infty \Phi_{nj}^2(\hat{r}) S(\hat{r})\hat{r}\,\mathrm{d}\hat{r} \right]^{-1}$$

$$\begin{pmatrix} \hat{D} \\ \hat{D}_0 \end{pmatrix} = \int_0^\infty \begin{pmatrix} i\xi \\ 1 \end{pmatrix} \exp[-\hat{\Lambda}_T^2 \xi^2/2$$

$$- (p + i\hat{C})\xi]\,\mathrm{d}\xi$$

with $\hat{C} = C/\Gamma = [k_w - \omega/(2c\gamma_z^2)]/\Gamma$ being the detuning parameter, $B = r_0^2\Gamma\omega/c$ the diffraction parameter, $\hat{\Lambda}_T^2 = \langle(\Delta\mathscr{E})^2\rangle/(\rho^2\mathscr{E}_0^2)$ the energy spread parameter, $\rho = c\gamma_z^2\Gamma/\omega$ the saturation parameter, $\Gamma = [I_0\omega^2\theta_s^2(2I_Ac^2\gamma_z^2\gamma\int_0^\infty \zeta S(\zeta)\,\mathrm{d}\zeta)^{-1}]^{1/2}$, $\hat{\theta} = \theta\omega_0 r_0/c$, ω is frequency, $\theta_s = K/\gamma$, $K = eH_w/(\mathscr{E}_0 k_w)$ is undulator parameter, H_w and k_w are the field and wavenumber of the undulator, respectively, $\gamma_z^{-2} = \gamma^{-2} + \theta_s^2$, $\gamma = \mathscr{E}_0/(m_e c^2)$, $\hat{z} = \Gamma z$, $\hat{r} = r/r_0$, and $I_A \simeq 17$ kA is the Alfven current. The eigenvalues and eigenfunctions of the beam radiation modes, $\lambda_k^{(n)}$ and $\Phi_{nk}(\hat{r})$, are calculated using multilayer approximation method [8].

3. Transverse coherence

We illustrate the process of formation of transverse coherence in the SASE FEL for specific example of diffraction parameter $B = 1$ and number of electrons in the volume of coherence $N_c = 7 \times 10^7$ which is typical for a VUV FEL [10]. Transverse distribution of the beam current density is Gaussian. Fig. 2 shows the evolution along the undulator of the partial contributions of different beam radiation modes into the total power. It is seen that analytical and simulation results agree well at $\hat{z} \gtrsim 7$. We also see that at large undulator length, $\hat{z} \sim 15$, contribution of the fundamental Φ_{00} mode to the total power is close to 99.9%. Analytical predictions for the averaged angular distribution of the radiation power are given by Eq. (1). Similar characteristic has also been calculated with numerical simulation code. It is seen from Fig. 3 that both approaches agree well in the high-gain linear regime. It is important to stress that even after finishing the transverse mode

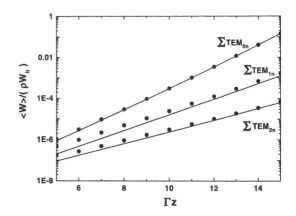

Fig. 2. Partial contributions to the total power of three azimuthal modes with $m = 0$, 1, and 2. Here $B = 1$, $\Lambda_p^2 \to 0$, $\Lambda_T^2 = 0$, and $N_c = 7 \times 10^7$. Solid curves represent analytical results for sum of three radial modes ($n = 0, 1, 2$). The circles are the results obtained with linear simulation code FAST.

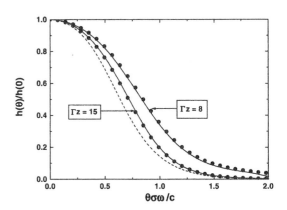

Fig. 3. Averaged angular distribution of the radiation intensity in the far zone for the FEL amplifier starting from shot noise. Here $B = 1$, $\Lambda_p^2 \to 0$, and $\Lambda_T^2 = 0$. Solid curves are the results of analytical calculations with (1), and the circles are the results obtained with linear simulation code FAST. Dashed line represents angular distribution of the fundamental Φ_{00} mode for maximum growth rate calculated in the framework of the steady-state theory.

selection process (which takes place after $\hat{z} \gtrsim 10$ for the considered numerical example) the intensity distribution in the far zone differs visibly from the angular distribution of the fundamental Φ_{00} mode for maximum growth rate calculated in the framework of the steady-state theory (dotted line in Fig. 3).

In the case of axisymmetric electron beam the radiation field is statistically isotropic. For such a field the effective correlation function depends only on the modulus $|\vec{\rho}|$ and the angular spectrum depends on the modulus $|\vec{k}_\perp|$. It is natural to define the area of coherence in this case as

$$\pi \hat{r}_{\text{coh}}^2 = 2\pi \int_0^\infty |\gamma_1^{\text{eff}}(\hat{\rho}, \hat{z})|^2 \hat{\rho} \, d\hat{\rho},$$

$$\gamma_1^{\text{eff}}(\hat{\rho}, \hat{z}) = 2\pi \int_0^\infty J_0(\hat{\rho}\hat{\theta}) h(\hat{\theta}) \hat{\theta} \, d\hat{\theta} \qquad (2)$$

where $\hat{\rho} = |\vec{\rho}|/r_0$. The degree of coherence, ζ, may be defined as

$$\zeta = \hat{r}_{\text{coh}}^2 / \hat{r}_{\text{max}}^2 \qquad (3)$$

where $\hat{r}_{\text{max}}$ is the radius of coherence $\hat{r}_{\text{max}}$ for the fully coherent radiation which is represented by the fundamental $\Phi_{00}(\hat{r})$ mode for maximum growth rate. Using angular distributions of the radiation field in the far diffraction zone we can trace the dependence of the degree of transverse coherence versus undulator length. Solid line in Fig. 4 is the result of analytical calculations, and the circles are the results obtained with numerical simulation code. It is clearly seen that the degree of coherence differs visibly from the unity in the high-gain linear regime, $\zeta \simeq 0.9$ at $\hat{z} = 15$.

Another possible way to define the degree of coherence is based on the statistical analysis of fluctuations of the instantaneous power [10,11]. Since in the linear regime we deal with a Gaussian random process, the power density at fixed point in space fluctuates in accordance with the negative exponential distribution [10]. If there is full transverse coherence, then the same refers to the instantaneous power W equal to the power density integrated over cross-section of the radiation pulse. If the radiation is partially coherent, then we have a more general law for instantaneous power fluctuations, namely the gamma distribution [10,11]:

$$p(W) = \frac{M^M}{\Gamma(M)} \left(\frac{W}{\langle W \rangle} \right)^{M-1}$$

$$\times \frac{1}{\langle W \rangle} \exp\left(-M \frac{W}{\langle W \rangle} \right) \qquad (4)$$

where $\Gamma(M)$ is the gamma function and $M = \langle W \rangle^2 / \langle (W - \langle W \rangle)^2 \rangle$. The parameter M of this distribution can be considered as the number of transverse modes. Then, the degree of coherence in the linear regime, may be defined as $\zeta = 1/M$. The value of M should be calculated with numerical simulation code producing time-dependent results for the radiation power. In Fig. 5 we

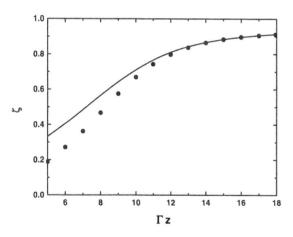

Fig. 4. Degree of transverse coherence of the radiation from the FEL amplifier versus the undulator length. Solid curve represents analytical results, and the circles are the results obtained with linear simulation code FAST. Here $B = 1$, $\Lambda_p^2 \to 0$, $\Lambda_T^2 = 0$, and $N_c = 7 \times 10^7$.

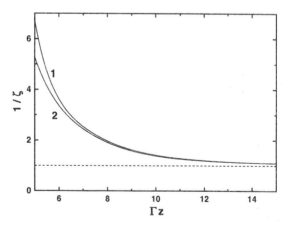

Fig. 5. Inverse value of transverse coherence versus undulator length. Here $B = 1$, $\Lambda_p^2 \to 0$, $\Lambda_T^2 = 0$, and $N_c = 7 \times 10^7$. Calculations have been performed with linear simulation code FAST. Curve 1 is calculated using instantaneous fluctuations of the radiation power. Curve 2 is calculated using angular distribution of the radiation power in far zone.

present the dependence of the number of transverse modes on the undulator length for the specific value of the diffraction parameter $B = 1$. It is seen that both definitions for the degree for the transverse coherence are consistent in the high-gain linear regime.

Let us discuss asymptotical behaviour of the degree of transverse coherence. At a large value of the undulator length it approaches unity asymptotically as $(1 - \zeta) \propto 1/z$, but not exponentially, as one can expect from simple physical assumption that transverse coherence establishes due to the transverse mode selection. Due to the latter effect the degree of coherence grows quickly at an early stage of amplification only. Starting from some undulator length the contribution to the total power of the fundamental mode starts to be dominant (see Fig. 2). However, one should take into account that the spectrum width has always finite value (see Fig. 1). Actually, this means that in the high-gain linear regime the radiation of the SASE FEL is formed by many fundamental Φ_{00} modes with different frequencies. The transverse distribution of the radiation field of the mode is also different for different frequencies. As a result of interference of these modes we do not have full transverse coherence. Taking into account this consideration, we can simply explain asymptotical behaviour of the degree of transverse coherence — this is reflection of the slow evolution of the width of the radiation spectrum as $z^{-1/2}$ with the undulator length.

All the results presented above have been obtained in the framework of the linear theory. Simulations with nonlinear code shows that for the considered numerical example the saturation occurs at $\hat{z} \simeq 18$. Using the plot presented in Fig. 4 we find that the value of the transverse coherence is < 0.9 in the end of the linear regime. A typical range of the values of $N_c = I_0/(e\omega_0\rho)$ is 10^6–10^9 for the SASE FEL of wavelength range from X-ray up to infrared. The numerical example presented in this paper is calculated for $N_c = 7 \times 10^7$ which is typical for a VUV FEL. It is worth to mention that the dependence of the saturation length of the SASE FEL on the value of N_c is rather weak, in fact logarithmic. Therefore, we can state that obtained effect limiting the value of transverse coherence might be important for practical SASE FELs.

References

[1] A VUV Free Electron Laser at the TESLA Test Facility: Conceptual Design Report, DESY Print TESLA-FEL 95-03, Hamburg, DESY, 1995.
[2] J. Rossbach, Nucl. Instr. and Meth. A 375 (1996) 269.
[3] R. Brinkmann, et al. (Ed.), Conceptual Design of 500 GeV e$^+$e$^-$ Linear Collider with Integrated X-ray Facility, DESY 1997-048, ECFA 1997-182, Hamburg, May 1997.
[4] Linac Coherent Light Source (LCLS) Design Study Report, The LCLS Design Study Group, Stanford Linear Accelerator Center (SLAC) Report No. SLAC-R-521, 1998.
[5] E.L. Saldin, E.A. Schneidmiller, M.V. Yurkov, Opt. Commun. 186 (2000) 185.
[6] K.J. Kim, Phys. Rev. Lett. 57 (1986) 1871.
[7] S. Krinsky, L.H. Yu, Phys. Rev. A 35 (1987) 3406.
[8] E.L. Saldin, E.A. Schneidmiller, M.V. Yurkov, Opt. Commun. 97 (1993) 272.
[9] E.L. Saldin, E.A. Schneidmiller, M.V. Yurkov, Nucl. Instr. and Meth. A 429 (1999) 233.
[10] E.L. Saldin, E.A. Schneidmiller, M.V. Yurkov, The Physics of Free Electron Laser, Springer, Berlin, Heidelberg, New York, 1999.
[11] E.L. Saldin, E.A. Schneidmiller, M.V. Yurkov, Nucl. Instr. Meth. A 429 (1999) 229.

ELSEVIER

Nuclear Instruments and Methods in Physics Research A 475 (2001) 97–103

NUCLEAR
INSTRUMENTS
& METHODS
IN PHYSICS
RESEARCH
Section A

www.elsevier.com/locate/nima

Predictions and expected performance for the VUV OK-5/Duke storage ring FEL with variable polarization ✩

V.N. Litvinenko[a,*], O.A. Shevchenko[b], N.A. Vinokurov[b]

[a] Free Electron Laser Laboratory (FEL), Department of Physics, Duke University, P.O. Box 90319, Durham, NC 27708-0319, USA
[b] Budker Institute of Nuclear Physics, Novosibirsk, Russia

Abstract

The OK-5 FEL is the first distributed optical klystron (DOK) comprised of four electromagnetic wigglers and three bunchers. In this paper we focus on the expected performance of the OK-5/Duke storage ring FEL which will operate in the UV and the VUV ranges of spectrum and will generate intense beam of Compton backscattered γ-rays. We present the calculation of the main parameters of the OK-5 VUV FEL such as its tuning range, gain and lasing power. Based on these calculations, we present the predictions for the OK-5 Compton γ-ray source. We extend our prediction to the OK-5/Duke storage ring FEL with the low-emittance lattice and low impedance vacuum chambers. We also discuss briefly the mode in which one of the wigglers is tuned on a harmonic of the OK-5 FEL to provide optimal conditions for harmonic generation in VUV and soft-X-ray ranges. © 2001 Elsevier Science B.V. All rights reserved.

PACS: 41.60.C; 29.20.D; 78.66; 78.40.D

Keywords: Free electron laser; Vacuum ultraviolet; Storage ring

1. Introduction

The 24.2-m long magnetic system of the OK-5 comprises four 4-m long electromagnetic wigglers and three matching midsections between them. The details on the OK-5 FEL and its lattice design are published elsewhere [1–3].

The polarization of the OK-5 FEL is controllable. The choice of the polarization includes but is not limited to the left and the right circular polarization, and the horizontal and the vertical linear polarization. The main mode of the OK-5 FEL operation will be based on the circular polarization. We will use the opposite helicity of the adjacent elliptically polarized wigglers to tune the OK-5 FEL onto linear polarization.

The lattice of the South straight section of the Duke storage ring will be modified to accommodate the OK-5 FEL and to provide three collision points for the generating semi-monochromatic γ-rays via Compton back-scattering [4]. The dispersion-free lattice of the 34-m long straight section has two end-matching sections and three triplets between the undulators. It is very flexible providing the tunability of the β-functions in center of

✩ This work is supported by ONR Contract #N00014-94-1-0818 and the Dean of Natural Sciences, Duke University.

*Corresponding author. Tel.: +1-919-660-2568; fax: +1-919-660-2671.

E-mail address: vl@phy.duke.edu (V.N. Litvinenko).

each wiggler from 4 to 10 m in both directions. The studies of the Duke storage ring with the OK-5 FEL showed the acceptable value of the dynamic aperture. The details on the lattice and the dynamic aperture are published separately [2].

This paper is focused on the predictions for the OK-5/Duke storage ring FEL. In Section 2 we present the results for the OK-5/DSR FEL gain, FEL power and γ-ray beam parameters. In Section 3 we discuss possible impact on the OK-5 FEL gain by a proposed low-emittance lattice for the Duke storage ring. We also briefly discuss configuration for harmonic generation in the OK-5 FEL.

2. Performance of the OK-5/Duke storage ring FEL

The description of the current performance of the 1 GeV Duke storage ring can be found in our recent publications [5]. In 2001 the Duke storage ring will undergo a reliability up-grade of the arcs and will operate in the energy range from 0.27 to 1.2 GeV. We plan to install the OK-5 FEL onto the Duke storage ring in 2002.

The RF system of the Duke storage ring operates on 64th harmonic and can support up to 64 electron bunches. The main mode of operation with the OK-5 FEL will use eight electron bunches with current of 20 mA per bunch. The single bunch and two bunch operations with 20 mA per bunch were already demonstrated with 270 MeV injection energy. We expect to build and to commission a full energy 1.2 GeV booster-injector in 2004 [14]. Higher injection energy will allow us to store and operate higher average current and higher current per bunch.

A development program is under way for establishing a stable multi-bunch operation. The developments involve a set of strip-lines for transverse feedback system, a high precision temperature control of the RF cavity, and a longitudinal feedback. In addition, we plan to replace the existing RF cavity with a new single-mode RF cavity in 2004. These developments will ultimately provide for stable multi-bunch operations with 20 mA per bunch in the near future. The following predictions are based on the assumption that broad-band impedance of the Duke storage ring vacuum chamber will remain at the present level of $Z/n = 1.6\,\Omega$ [10].

The structure of the OK-5 FEL is rather sophisticated. It is well suited for FEL 3D-code RON [6], which can handle multi-segmented systems with wigglers, bunchers and quadrupoles. The RON is benchmarked by experimental results at APS SASE FEL [7]. In addition, we benchmarked the RON with the 3D-FEL codes #felTD and #uvfel [8] for the OK-4/Duke storage ring FEL. The gains calculated by RON coincided with those calculated by #felTD/uvfel and those experimentally measured [9] with accuracy of ~5%. In contrast with the #felTD/uvfel codes, the RON can be used for both low and high FEL gain regimes. The wavelength of the OK-5 FEL wigglers can be tuned from 11 nm to 2.3 μm using e-beam energy from 0.3 to 1.2 GeV. The OK-5 tuning range is shown in Fig. 1. The lasing range will be limited by the OK-5 FEL gain and the reflectivity of the available mirrors. We expect to extend the lasing with OK-5 FEL installed on the existing Duke storage ring down to 100 nm or even lower.

With a given e-beam parameters and wavelength, the gain of the OK-5 FEL strongly depends on a large number of variables. These variables include the choice of the lattice, the Rayleigh length, the settings of three individually controlled bunches, and the detuning from the resonant wavelength δk. The non-zero detuning is required to compensate for the Guoy phase advance of the TEM$_{00}$ Gaussian mode. The large number of variables provides a flexibility, but also makes optimization quite a painstaking process. We limited the range of parameters by assuming the symmetry with respect to the center of the OK-5 FEL. The same settings of the first and the third bunchers and the bilaterally symmetric envelopes of the electron and the optical beam were used for the results presented here.

The real part of the gain is a fast *sin*-like oscillating function of the buncher strengths. To facilitate the calculations, we developed a semi-

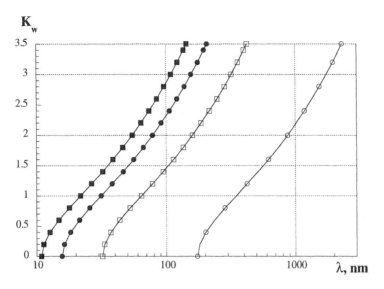

Fig. 1. Tuning range of the spontaneous radiation of the OK-5 helical wigglers. (○)$E = 0.3$ GeV; (□)$E = 0.7$ GeV, (●)$E = 1$ GeV, and (■)$E = 1.2$ GeV.

analytical program using Mathematica [11][1]. This code allowed us to optimize the choice of parameters with reasonable speed and accuracy. We initially optimized the absolute value of the complex gain G, for finding its maximum. At the next step, we optimized the $Re(G)$ by searching in the reduced range of parameters. Fig. 2 depicts a typical dependence of $|G|$ on the value of Rayleigh length in the OK-5 FEL, β_o, and on the detuning parameter δk. The optimal β_o for the OK-5 FEL is about 4.5 m. The choice of the focusing optics for the e-beam lattice was less thorough and there is still room for future optimization.

We used RON to calculate the dependencies of the OK-5 FEL gain on the e-beam energy, emittances and energy spread. We focused on the short wavelengths to explore the limits of the OK-5 FEL performance. The OK-5 has larger gains in the visible and the near-IR ranges.

Two typical dependencies of the OK-5 FEL gain (at 100 and 200 nm) on the beam current are shown in Fig. 3. In addition to the gain calculations, the RON calculates β_x and β_y, and the

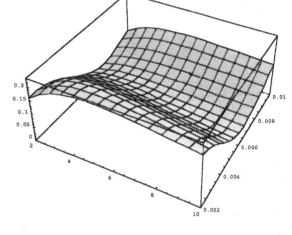

Fig. 2. The dependence of the absolute value of the OK-5 FEL gain $|G|$ on the Rayleigh length, β_o, and the detuning parameter δk. $E = 0.7$ GeV, $I_{peak} = 10$ A ($I = 2.5$ mA/bunch), $\lambda = 100$ nm, and the energy spread is 0.05%. The bunchers are set on $ND_{Edge} = 70.9823$, $ND_{Center} = 72.0423$. The vertical scale is $|G|$. The β_o varies from 2 to 10 m, and the $\delta k/k$ varies from 0.002 to 0.01. The absolute value of the gain peaks at $\beta_o \sim 4.5$ m and $\delta k/k \sim 0.0025$.

---

[1] The FEL gain can be defined as complex number. Imagine part of the gain describes the coherent radiation with $\pi/2$ phase shift with respect to the incoming wave.

degree of the bunching in the electron beam at the FEL wavelength. Typical output from RON is shown in Fig. 4. The degree of the bunching plays

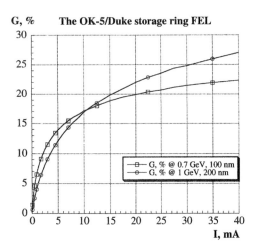

Fig. 3. The calculated OK-5 gain as a function of the bunch current. $V_{RF} = 700\,kV$, peak current is $I_{peak} = 40.3\,A$ for $I = 20\,mA/bunch$: (○)$\lambda = 200\,nm$, $E = 1\,GeV$, $\varepsilon_x = 19\,nm\,rad$, $\varepsilon_y = 1.9\,nm\,rad$; (□)$\lambda = 100\,nm$, $E = 0.7\,GeV$, $\varepsilon_x = 10\,nm\,rad$, $\varepsilon_y = 1.0\,nm\,rad$.

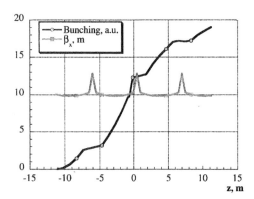

Fig. 4. The horizontal β-function and the bunching of the electron beam with period of $\lambda = 200\,nm$ calculated by RON. The OK-5 FEL operates with $E = 0.7\,GeV$, $I = 10\,mA/bunch$, $\varepsilon_x = 10\,nm\,rad$ and $\varepsilon_y = 1.0\,nm\,rad$ e-beam. $V_{RF} = 700\,kV$.

a role of the absolute value of the local complex gain. It was used for initial optimization followed by a fine tuning of the bunchers to maximize the value of the $Re(G)$.

It is worth mentioning that the UV/VUV OK-5 FEL has maximum gain with electron beam energy of 0.6–0.7 GeV [10]. For example, the RON predicts the OK-5 gain to be 44% per pass at 200 nm with a 0.7 GeV, 20 mA electron bunch. Reasonably, high gain will allow us to use

Al-broad-band mirror coating for tuning the OK-5 FEL wavelength in the UV and the VUV ranges in addition to multi-layer narrow-band dielectric mirror coatings. The use of the helical wigglers will hamper the mirror degradation problems. We predict that the OK-5/Duke storage ring FEL will lase in the UV and VUV range where mirrors with reflectivity above 85–90% are available. Coatings with this level of reflectivity are manufactured by a number of optical companies for wavelengths > 150 nm.

We plan to modify the self-consistent set of #felTD/uvfel codes to handle DOK of the OK-5 complexity. This modification will provide us with means to calculate exactly the OK-5 FEL lasing power, as we did for the OK-4 FEL [12]. Presently, we base our predictions on the approximate formula for the lasing power of an OK installed on the storage ring [12]. This formula is in a good agreement with the experimental results obtained for the OK-4/Duke storage ring FEL operated in both the OK and the conventional FEL configurations [12]. The optimized average power of storage ring FEL is

$$P_{lasing} \cong \xi_E P_{SR} \frac{\sigma_{E\,lasing}^2 - \sigma_{Eo}^2}{E \cdot \sigma_{E\,lasing}} e^{-1/2} \qquad (1)$$

where P_{SR} is the power of the synchrotron radiation, $\sigma_{E\,lasing}$ is the steady-state electron beam energy spread in the CW lasing mode. The rest of parameters are specified in Ref. [12] and Table 1.

Table 1
Example of the lasing power and γ-ray flux for OK-5 FEL

Wavelength (nm)	200–240	
Beam energy (GeV)	0.9–1.0	
Beam current in 8 bunches (mA)	160	
P_{SR} (kW)	4.4–6.7	
Energy partition number, ξ_s	2.07	
Natural energy spread, σ_{Eo}(%)	0.052–0.058	
Mode of operation	FEL users	γ-rays
Total optical cavity loss (%)	2.1	1.1
Mirrors transparency (%)	1.1	0.1
Lasing energy spread, $\sigma_{E\,lasing}$ (%)	0.288	0.374
Average lasing power (W)	15–23	20–31
Power delivered to users (W)	8–12	—
Average intracavity power (kW)	—	1.8–2.8
Peak energy of γ-rays (MeV)	—	60–87
Average flux, γ-rays per second	—	$2.8–4.2 \times 10^{10}$

Table 1 provides specific example of the lasing power calculated for the OK-5 FEL. The OK-5 FEL will induce $\sigma_{E\,lasing} \sim 0.3\%$, which is sufficient for complete suppression of the microwave instability (MWI). With the lasing, the peak e-beam current will be ~ 5–10 times lower than the threshold of MWI, which makes formula (1) valid [12].

To estimate the OK-5 lasing power, we used RON for calculating the OK-5 FEL gain dependence on the e-beam energy spread. From these dependencies we found $\sigma_{E\,lasing}$ by setting the OK-5 FEL gain to be equal the round trip loss in the optical cavity, i.e. a steady-state lasing. Two typical dependencies for 20-mA e-bunch are shown in Fig. 5 with arrows pointing at two selected points with 1.1% and 2.1% gains. Table 1 summarizes a specific case for the deep-UV 0.9–1.0 GeV operation of the OK-5 FEL.

In this range, dielectric coatings with losses for absorption plus scattering $\sim 0.5\%$ per mirror are available. For FEL users we plan to install mirrors with $\sim 1.1\%$ transparency and the total round trip losses of $\sim 2.1\%$. An average power ~ 10 W will be delivered to the user stations in this case. For γ-ray generation, we will optimize the intracavity power by using less transparent mirrors with the total round trip losses of $\sim 1.1\%$. The average intracavity power ~ 2 kW and three collision points will provide for average total flux $\sim 3 \times 10^{10}$ of γ-rays with 60–90 MeV peak energy. The γ-rays from the OK-5 will be collimated to attain a desirable energy resolution. This configuration gives 1.5% of the total flux for each 1% of FWHM energy resolution [13]. The γ-ray flux from the OK-5 FEL will be ~ 200 times higher compared to that attainable with the HIγS source using OK-4 FEL [14]. The factor 20 in this improvement is due to the use of eight bunches and three collision points, compared with the presently used two bunches and one collision point.

Naturally, the OK-5 will deliver lower average power (at level of a few watts), when operated at 150 nm and below. We do not have reliable data on the mirror reflectivity for the range of 100–150 nm for predicting the OK-5 FEL lasing parameters. Because of the high losses in

substrates, we plan to use VUV mirrors with an outcoupling hole of 0.5 mm in diameter. A prototype mirror will be tested with the OK-4 FEL in 2002 at 240 nm.

The OK-5 FEL will also operate in the giant-pulse mode using the existing gain modulator [9]. We expect a substantial improvements in the performance compared with the OK-4 FEL. With the parameters specified in Table 1, the OK-5 FEL will generate 170 mJ in a 35-μs deep-UV giant pulse. Eighty millijoules will be lost in the mirror substrates. The remaining 90 mJ will be delivered to users in the form of FEL pulses with peak power of 10 MW and rep-rate of 100 Hz. Energy per micropulse with duration of few psec will be ~ 0.1 mJ.

The peak intracavity power will reach 1 GW level and will be sufficient for effective generation of coherent harmonics. One of the OK-5 wigglers will be tuned on a harmonic, with harmonic number being from 2 to 7, for this mode of operation.

It is worth mentioning, that when the OK-5 FEL is reconfigured to lase with linear polarization, its gain will be 1/2 of that for circular polarization at the same wavelength.

3. Future options for the OK-5 FEL and the Duke storage ring

The present Duke storage ring has rather large horizontal emittance, 18 nm rad at the energy of 1 GeV, and high value of the impedance, $Z/n \sim 1.6\,\Omega$. These parameters strongly affect the performance of the OK-5 FEL, especially at wavelengths below 100 nm. For example, the RON predicts that the gain of the present OK-5/ Duke storage ring FEL will be $< 10\%$ at 50 nm. It makes impossible to consider a lasing in this wavelength range with the existing system seriously.

We are developing the basis for the future upgrades of the Duke storage ring to match its performance with the advanced OK-5 FEL system. First, we are considering the design and installation of the modern low-impedance vacuum chambers, similar to that used at ELETTRA

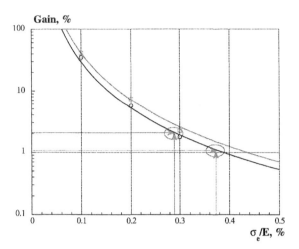

Fig. 5. The OK-5 gain for 20 mA electron bunch as a function of its energy spread. The arrows show the points used for calculations of the lasing power for Table 1: ($\bigcirc$)$\lambda = 200$ nm, $E = 1$ GeV, $\varepsilon_x = 19$ nm rad, $\varepsilon_y = 1.9$ nm rad, $V_{RF} = 700$ kV; ($\square$)$\lambda = 100$ nm, $E = 0.7$ GeV, $\varepsilon_x = 10$ nm rad, $\varepsilon_y = 1.0$ nm rad, $V_{RF} = 700$ kV.

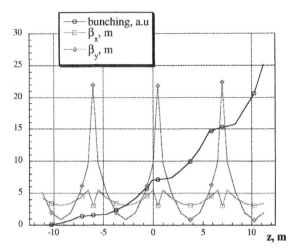

Fig. 6. The β-functions and the bunching of the electron beam calculated by RON for the OK-5/Duke storage ring FEL operating at $\lambda = 100$ nm. $E = 1$ GeV with $I_{peak} = 200$ A, $\varepsilon_x = 1.4$ nm rad, $\varepsilon_y = 0.14$ nm rad, $V_{RF} = 700$ kV.

storage ring with $Z/n \sim 0.14 \Omega$ [15]. The reduction of the impedance to $Z/n \sim 0.16 \Omega$ alone will bring the OK-5 FEL gain to 65% per pass at 100 nm. Second, we developed new lattices for the Duke storage ring with 1.4 and 1 nm rad horizontal emittance at the energy of 1 GeV [16,17]. The 1.4-nm rad "low-cost" proposal is based on the re-use of the existing magnets and power supplies, while the 1-nm rad one requires more radical changes. Both lattices will not modify the footprint of the Duke storage ring and will not impact the existing straight sections and beam-lines.

With these [16] and previously mentioned modifications, the Duke storage ring will be capable to operate electron bunches with peak current of 200 A with rather low-energy spread of 0.15% at 1 GeV. The RON predicts that the OK-5 FEL will have the gain of 10.1 (i.e. net gain of 910%) per pass at 100 nm with this e-beam parameters. The lattice and the bunching of the e-beam for this case are shown in Fig. 6. This level of gain will be sufficient to operate the OK-5 FEL at Duke in oscillator mode below 100 nm, where mirrors with reflectivity $\sim 50\%$ are available.

High gain at level of 300% is critical for efficient generation of coherent harmonics in a storage ring FELs [18]. The OK-5 lattice was designed with the capability of tuning one or two of its wigglers onto a harmonic of the FEL light. The wavelength of harmonics can be as short as 15 nm. The out-coupling of the coherent harmonic radiation will be arranged via the use of a hole in the down-stream mirror substrate or via the e-beam out-coupling scheme [19].

4. Conclusions

The manufacturing of the OK-5 FEL is close to completion in the Budker Institute of Nuclear Physics, Novosibirsk. We plan to install the OK-5 FEL onto the Duke storage ring in 2002. We expect it to be operational by summer of 2002.

The OK-5/Duke storage ring will provide sufficient gain for reliable lasing below 200 nm with few watts of average lasing power. Nevertheless, the best performance of the OK-5 FEL will be attainable with further improvements of the Duke storage ring.

Acknowledgements

The author would like to acknowledge support of this project by the Office of Naval Research and the Dean of Natural Sciences, Duke University. The authors are thankful to Dr. Stepan Mikhailov, Duke University, and Dr. Y. Wu, Lawrence Berkeley Laboratory, for their contributions and help with this paper.

References

[1] V.N. Litvinenko, S.F. Mikhailov, N.A. Vinokurov, N.G. Gavrilov, G.N. Kulipanov, O.A. Shevchenko, P.D. Vobly, Nucl. Instr. and Meth. A 475 (2001) 247, These Proceedings.

[2] Y. Wu, V.N. Litvinenko, S.F. Mikhailov, N.A. Vinokurov, N.G. Gavrilov, O.A. Shevchenko, T.V. Shaftan, D.A. Kairan, Nucl. Instr. and Meth. A 475 (2001) 253, These Proceedings.

[3] V.N. Litvinenko, S.F. Mikhailov, N.A. Vinokurov, N.G. Gavrilov, D.A. Kairan, G.N. Kulipanov, O.A. Shevchenko, T.V. Shaftan, P.D. Vobly, Y. Wu, Nucl. Instr. and Meth. A 475 (2001) 407, These Proceedings.

[4] S.H. Park, V.N. Litvinenko, W. Tornow, C. Montgomery, Nucl. Instr. and Meth. A 475 (2001) 425, These Proceedings;
V.N. Litvinenko, et al., Phys. Rev. Lett. 78 (1997) 4569;
S.H. Park, Thesis, Duke University, Durham, NC, USA, 2000.

[5] V.N. Litvinenko, S.H. Park, I.V. Pinayev, Y. Wu, Nucl. Instr. and Meth. A 475 (2001) 195, These Proceedings;
V.N. Litvinenko, S.H. Park, I.V. Pinayev, Y. Wu, Performance of the OK-4/Duke storage ring FEL, Nucl. Instr. and Meth A 470 (2001) 66.

[6] R.J. Dejus, O.A. Shevchenko, N.A. Vinokurov, Nucl. Instr. and Meth A 445 (2000) 19;

[7] S.V. Milton, et al., Nucl. Instr. and Meth. A 475 (2001) 28, These Proceedings.

[8] V.N. Litvinenko, B. Burnham, J.M.J. Madey, Y. Wu, SPIE 2521 (1995) 79;
V.N. Litvinenko, B. Burnham, J.M.J. Madey, Y. Wu, Nucl. Instr. and Meth. A 358 (1995) 334, p. 369.

[9] I.V. Pinayev, V.N. Litvinenko, S.H. Park, Y. Wu, M. Emamian, N. Hower, J. Patterson, G. Swift, Nucl. Instr. and Meth. A 475 (2001) 222, These Proceedings.

[10] V.N. Litvinenko, I.V. Pinayev, S.H. Park, Y. Wu, Nucl. Instr. and Meth. A 475 (2001) 195, These Proceedings.

[11] Mathematica 4, © Copyright 1988–1999 Wolfram Research, Inc.

[12] V.N. Litvinenko, S.H. Park, I.V. Pinayev, Y. Wu, Average and peak power limitations and beam dynamics in the OK-4/Duke storage ring FEL, Nucl. Instr. and Meth., These Proceedings;
V.N. Litvinenko, Thesis, Novosibirsk, 1989.

[13] V.N. Litvinenko, J.M.J. Madey, SPIE 2521 (1995) 55;
S.H. Park, Thesis, Duke University, Durham, NC, USA, 2000.

[14] V.N. Litvinenko, S.F. Mikhailov, HIγS γ-ray source, Technical Description, proposal to US Department of Energy, 2000.

[15] R.P. Walker, et al., Nucl. Instr. and Meth. A 475 (2001) 20, These Proceedings.

[16] S.F. Mikhailov, V.N. Litvinenko, Y. Wu, Nucl. Instr. and Meth. A 475 (2001) 417, These Proceedings.

[17] S.F. Mikhailov, V.N. Litvinenko, Y. Wu, Lattice for the Duke storage ring with 1-nm horizontal emittance, in preparation.

[18] V.N. Litvinenko, X-ray storage ring FELs: new concepts and new directions, Proceedings of fourth Generation Light Sources ICFA Workshop, Grenoble, France, January 1996.

[19] N.A. Vinokurov, Concept of the e-beam outcoupling, BINP, 1985.

N.A. Vinokurov, The integral equation for a high gain FEL, ANL/APS/TB-27, Argonne National Laboratory, 1996.

ELSEVIER

Nuclear Instruments and Methods in Physics Research A 475 (2001) 104–111

**NUCLEAR
INSTRUMENTS
& METHODS
IN PHYSICS
RESEARCH**
Section A

www.elsevier.com/locate/nima

Coherent hard X-ray production by cascading stages of High Gain Harmonic Generation

Juhao Wu[a,b,*], Li Hua Yu[a]

[a] *National Synchrotron Light Source, Brookhaven National Laboratory, Upton, NY 11973-5000, USA*
[b] *Department of Physics and Astronomy, C.N. Yang Institute for Theoretical Physics, State University of New York at Stony Brook, Stony Brook, NY 11794-3840, USA*

Abstract

We study a new approach to produce coherent hard X-rays by cascading several stages of High Gain Harmonic Generation (HGHG). Our calculation shows that such a scheme is feasible within the present technology. Compared with the Self-Amplified Spontaneous Emission (SASE) scheme for the Linac Coherent Light Source (LCLS), the HGHG scheme can provide radiation with full longitudinal coherence, narrower bandwidth, more stable central wavelength, more controllable pulse length, and much higher performance stability. The electron bunch parameters used in our calculation are those of the DESY TTF and SLAC LCLS projects. To achieve similar output power, i.e. about 15 GW, the total length of the HGHG scheme will be 88 m, while that of SASE scheme will be around 100 m. © 2001 Elsevier Science B.V. All rights reserved.

PACS: 41.60.Cr; 42.25.Kb; 42.55.Vc; 42.65.Ky

Keywords: High Gain Harmonic Generation FEL; Coherent hard X-ray; Cascading; Stability

0. Introduction

Short wavelength Free-Electron Lasers (FELs) are perceived as the next generation of synchrotron light sources. In the past decade, significant advances have been made in the theory and technology of high brightness electron beams and single pass FELs. These developments facilitate the construction of practical VUV FELs and make X-ray FELs possible. Self-Amplified Spontaneous Emission (SASE) [1–8][1] (see also Ref. [10][2]) and High Gain Harmonic Generation (HGHG) [11,12] are the two leading candidates for VUV to X-ray FELs. The results of the first HGHG proof-of-principle experiment [13,14] agree with the theoretical predictions. Compared with the SASE FEL, the following advantages of HGHG FEL were confirmed: (1) Full longitudinal coherence; (2) Much narrower bandwidth; and (3) More stable central wavelength.

---

*Corresponding author. National Synchrotron Light Source, Brookhaven National Laboratory, Upton, NY 11973-5000, USA.

E-mail address: jhwu@sun2.bnl.gov (J. Wu).

[1] SASE gain of 10^5 at 12 μm was reported in Ref. [9].
[2] SASE gain at 530 nm was recently observed at the LEUTL facility at APS/ANL in S.V. Milton, et al., and at 109 nm at TTF/DESY in J. Andruszkow et al.

In this paper, we first briefly describe the principle of HGHG. Then we give details about how to produce hard X-rays by the HGHG scheme and discuss the stability of the HGHG process. Finally, we give additional discussion and present our conclusion.

1. The HGHG principle

Harmonic generation using a seed laser is well known and has been verified experimentally and analyzed [15–17]. However, HGHG is different from such pure harmonic generation in the sense that it is harmonic generation followed by an exponential growth. In the HGHG scheme, there are three components: one wiggler used as the modulator, one dispersion section, and a second wiggler used as the radiator. A seed laser, together with an e-beam, is introduced into the modulator, where the seed laser interacts with the e-beam to modulate its energy. Next the energy-modulated e-beam passes through the dispersion section (a three-dipole chicane), where the energy modulation is converted into spatial modulation, producing abundant harmonics in the e-beam density distribution. When the spatially modulated e-beam enters the radiator, which is designed to be resonant to one of the harmonics of the seed laser, rapid coherent emission at this resonant harmonic is produced. This harmonic is further amplified exponentially until saturation. In the first HGHG experiment [13,14], the input CO_2 seed laser power was 0.7 MW at 10.6 μm, and the output HGHG FEL power was about 35 MW at 5.3 μm, an order of magnitude larger than the input.

2. The HGHG scheme to produce hard X-ray

We now investigate whether it is possible to produce hard X-rays using an HGHG scheme. To achieve 1 Å hard X-rays from commercially available lasers having wavelengths at thousands of Angstroms by one step of HGHG would require extremely high harmonics, of the order of several thousands. Previous analysis [18] showed that one needs very high input seed laser power in

order to generate high harmonics. Beyond 60th harmonic, this becomes difficult. Also, when the harmonic is too high, the stability of the output is not good. A better approach is to cascade several stages of HGHG. During each stage, a moderately high harmonic of the seed laser will be produced at the end of the radiator. This harmonic will be subsequently used as the seed for the next stage. Thus after several stages, we generate hard X-ray radiation.

To cascade several stages of HGHG, we need some extra components. Each stage consists of one modulator, a dispersion section, and one radiator. So the physical process in each stage will be the same as in the recent experiment [13,14]. During the process, the output radiation disturbs a part of e-beam, which interacts with the seed. In order to achieve the best efficiency for carrying out the next stage of HGHG, we must use a fresh part of the e-beam. There are two methods. The first is to shift the laser (i.e., the output radiation from the previous HGHG stage) toward the front part of the same e-beam, so that the laser will interact with a "fresh" part of the same e-beam. The second is to introduce a new electron bunch for each stage, so that again the laser will interact with a "fresh" bunch. This is the "fresh bunch technique" [19]. For the first case, we use a "shifter" to "shift" the laser to the "fresh" part of the same e-beam. For the purpose of comparison, we restrict ourselves to the parameters from DESY TTF [10], and SLAC LCLS project [21]. Without this restriction, further optimization should be possible.

Let us now present the details. As illustrated in Fig. 1, we consider a laser with a wavelength of 2250 Å and a peak power of $P_{in} = 180$ MW. The corresponding start-up shot-noise power [8] is only about $P_{noise} \approx 60$ W. So, the input seed laser power dominates the shot-noise power. This remains true for all seeds into the five stages and the last amplifier. After five stages, we get 1.5 Å radiation, which is then amplified to saturation with a peak power around 15 GW by traversing the last wiggler called the amplifier. The parameters for the electron beams, the wigglers, and the dispersion sections are given in the table of Fig. 1. Let us first explain the meaning of each parameter in Fig. 1. The numbers on the first row above the

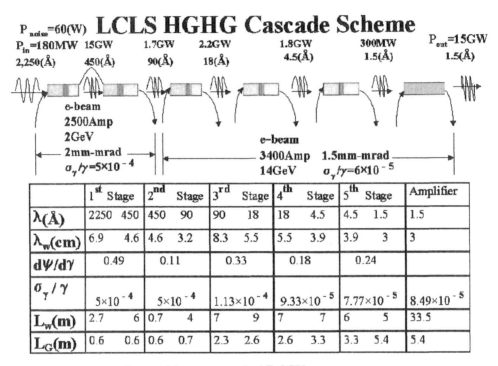

Fig. 1. The schematic device to produce hard X-ray with an HGHG-based approach.

schematic system stand for the output power of each stage. The output power of one stage is the input power of the next stage. The second row stands for the corresponding wavelength of the radiation. The e-beam parameters are printed just below the schematic device. The local energy spread given in the plot is the initial energy spread before the e-beam goes into the first modulator. The energy spread is increased because of spontaneous radiation. We take this into account [20], and we give its value at the end of the modulators in the table. For the first two stages, where the wavelength of the radiation is relatively long, we use a lower energy e-beam. The parameters are those of DESY TTF, except the energy is $E = 2$ GeV. The e-beam has a peak current I_{pk} of 2500 A, normalized emittance $\varepsilon_n = 2\pi$ mm-mrad, and initial local energy spread $\sigma_\gamma/\gamma = 5 \times 10^{-4}$. For the following stages, a higher energy e-beam is used. It has the parameters of those of SLAC LCLS project, i.e. $I_{pk} = 3400$ A, $E = 14$ GeV, $\varepsilon_n = 1.5\pi$ mm-mrad, and $\sigma_\gamma/\gamma = 6 \times 10^{-5}$.

For the table, the first row gives the radiation wavelength λ. The second row gives the wiggler period λ_w. The third gives the dispersion strength $d\psi/d\gamma$. The fourth gives the relative local energy spread σ_γ/γ at the end of the modulator in each stage. The fifth gives the length of the wigglers L_w (modulators, radiators, and the amplifier). Finally, the sixth row stands for the power e-folding length L_G in each wiggler. The table has six boxes: the first five boxes stand for the five stages, while the last one stands for the amplifier. In each of these five boxes, the left column gives the parameters for the modulator, while the right column gives those for the radiator. The numbers in the middle stand for $d\psi/d\gamma$ and σ_γ/γ at the end of the modulator. Here, $\psi = (k_s + k_w)z - \omega_s t$ is the pondermotive phase in the radiators. For example, the second box stands for the second stage. The left column in the second box stands for the modulator of the second stage. The table shows that in the modulator the resonant radiation is $\lambda = 450$ Å, the modulator has $\lambda_w = 4.6$ cm,

$L_w = 0.7$ m, and the corresponding $L_G = 0.6$ m. The right column shows that the radiation in the radiator is $\lambda = 90$ Å, the radiator has $\lambda_w = 3.2$ cm, $L_w = 4$ m, and the corresponding $L_G = 0.7$ m. The numbers in the middle stand for $d\psi/d\gamma = 0.11$ and $\sigma_\gamma/\gamma = 5 \times 10^{-4}$ at the end of modulator. The other boxes are similar (except for the sixth, which stands for the amplifier), so there is no $d\psi/d\gamma$. Also, σ_γ/γ given is the average value along the amplifier. The effect of the global energy spread (or correlated energy spread, in the terminology of certain other workers in this field) is addressed in the following discussion of sensitivity to parameter variation, for its effect is essentially an issue of detuning.

Let us now explore the physics of the process. Shown in Fig. 1, the 2250 Å laser, with a peak power of 180 MW, together with the 2 GeV e-beam, are introduced into the modulator of the first stage. The modulator and the radiator are resonant to 2250 and 450 Å, respectively. The first stage generates 450 Å output according to the HGHG principle. To go to next stage, we need a shifter, where the e-beam is magnetically delayed. Hence, the 450 Å radiation is effectively shifted to the front part of the same e-beam, where the e-beam is still "fresh". In our example, we assume the output pulse is Gaussian with rms width of 10 fs and the e-beam pulse is nearly a longitudinally flat pulse 250 fs long. The 450 Å radiation serves as the seed laser in the second stage, where the 450 Å radiation input generates a 90 Å output with 1.7 GW. Up to this stage, we are using the lower energy e-beam and a relatively long wavelength. In order to achieve high-power, hard X-rays, the 2 GeV e-beam is dumped after this stage, and a 14 GeV e-beam is introduced for the next stage. Now, the 90 Å radiation is the seed laser for the next stage to be converted to 18 Å. In our example, we use one bunch for each stage. This process is repeated at the fourth, fifth, and amplifier stages, except that there is no HGHG process in the amplifier and the radiation is amplified exponentially until saturation. Finally, with a total wiggler length of about 88 m for the whole device, we obtain about 15 GW radiation at 1.5 Å, well into saturation. We emphasize that in the radiator of the third, fourth and fifth stages,

there is no exponential growth of the harmonic, but rather, after the coherent emission is finished, the harmonic is introduced to the next stage directly.

3. Stability and the sensitivity to parameter variation

We need to check the stability of the performance. For each stage, the fluctuation in any parameter of the e-beam or the seed leads to output power fluctuation of the harmonic at the end of the radiator. However, this output harmonic is the input seed for the next stage. Hence, the output power fluctuation of the harmonic in one stage is just the input power fluctuation of the seed for the next stage. Therefore, the stability consideration could be simplified to check whether each stage of HGHG could reduce the fluctuation.

In Fig. 2, we plot the relation between the output power and the input power for the fifth stage. The output power fluctuation is only about 20% when the input power changes from 1 to 3 GW. This is an attractive feature of the HGHG. For the first, second, third and fourth stages, a variation of a factor of 3 in the input power

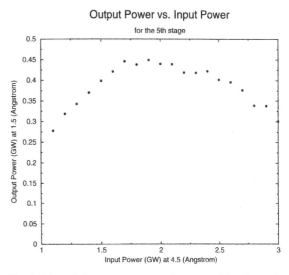

Fig. 2. Plot of the output power change resulting from the input power change in the fifth stage.

generates an output fluctuation of 10%, 15%, 45% and 30%, respectively. This result arises from a trade off between better stability and total wiggler length, i.e., if we use lower harmonic number and one more stage, the stability will be further improved. Analytical study shows that this attractive feature holds as long as the harmonic number is not too high. So, in our scheme, we use harmonic numbers, 3, 4 and 5. Since each stage reduces the fluctuation, the radiation fluctuation caused by the fluctuation in the parameters of the previous stage will be stabilized in the following stage. Therefore, not much fluctuation is expected after the full five stages. The stability of the whole system is determined mostly by the last amplifier, which is only 33.5 m long. Hence the system is very stable.

The calculation is carried out by the modified version [12] of the TDA code [22] and compared with an analytical estimate (for the sake of space, it will be published elsewhere in more details). For each stage, the analytical estimate agrees with the simulation to within a factor of 2. In the calculation, we also consider the diffraction effect of the laser when it traverses the "shifter" and the connection regions between stages. In our calculation, the connection region is designed to be $L_c = 1.55$ m long, and the "shifter" $L_s = 25$ cm long. For the 14 GeV e-beam, the β-function in wigglers is chosen to be 18 m, the same as in LCLS. For the 2 GeV e-beam, the β-function is proportionally reduced to be 2.5 m. We assume to use hybrid wigglers of Nd–Fe–B type, so according to Halbach's formula, the wiggler period λ_w and the wiggler parameter K satisfy $K = 3.44 \times 93.4\lambda_w \exp[-5.08g/\lambda_w + 1.54(g/\lambda_w)^2]$, where g is the wiggler gap. For all the wigglers, g is restricted not to be smaller than 6 mm.

To test the stability, we check the sensitivity of the output to the initial local energy spread, the global energy detuning, the peak current, and the normalized emittance of the e-beam. For the first stage, we vary each of these four parameters independently to obtain an output power fluctuation of the harmonic (the 450 Å radiation). This is just the input power fluctuation of the seed laser (the 450 Å radiation) for the second stage. Such input power fluctuation leads to a smaller output

power fluctuation of the harmonic at 90 Å, similar to what is shown in the Fig. 2. But the parameters of the e-beam in the second stage fluctuate also, so we vary each of these four parameters independently again to get the output power fluctuation of the harmonic (the 90 Å radiation). So, the total output power fluctuation for each stage is produced by the fluctuation in each of the parameters of the e-beam as well as the input power of its seed laser. Repeating the same procedure, we finally obtain the fluctuation of the 1.5 Å radiation.

Each stage the output power fluctuation leads to an energy modulation fluctuation at the end of the modulator of the next stage. The dispersion strength is adjustable. If we want to optimize the system, then we should tune the dispersion strength, so that the largest energy modulation will be favored to produce the highest coherent emission. However, in order to have good stability, we will overtune the dispersion strength, so that the largest energy modulation will produce over-bunching in the e-beam, while the medium energy modulation will produce the best spatial bunching. By doing such "overtuning" for the dispersion strength in the smallest energy modulation case, the e-beam is underbunched, and radiation rises slowly in the beginning, but continues growing till the end of the radiator; on the other hand, the e-beam is overbunched in the largest energy modulation case and radiation grows fast at first but drops later. As a result, the smallest energy modulation will produce amounts of coherent radiation similar to those produced by the largest energy modulation. Once the dispersion strength is fixed, the fluctuation range of the radiation is a function of the radiator length, for example, as shown in Fig. 3 for the second stage. We then determine the length of the radiator to be where the output power fluctuation is the minimum, so the radiator length for this second stage should be around 4 m long. In Fig. 3, the solid line stands for the case when the input power is P_{in}^{solid}, the dotted line: $P_{in}^{dotted} = e^{-0.25} \times P_{in}^{solid}$; the dashed line: $P_{in}^{dashed} = e^{-0.5} \times P_{in}^{solid}$; and the dot-dashed line: $P_{in}^{dot-dashed} = e^{-1} \times P_{in}^{solid}$. In so doing, the large input power fluctuation will result in a small output power fluctuation, as shown in Fig. 2. So,

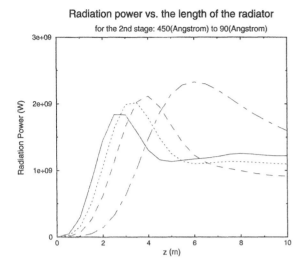

Fig. 3. The radiation power along the second radiator for a range of input powers, showing that at around 4 m, the output fluctuation is the minimum.

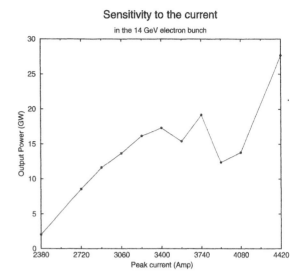

Fig. 4. The sensitivity of the output power to current fluctuations. $I_{pk} = 3400$ A is the nominal value.

we trade off the best performance point for improved stability.

The sensitivity to the e-beam parameters is investigated for the initial local energy spread σ_γ/γ, the global energy detuning $\Delta\gamma/\gamma$, the peak current, and normalized emittance ε_n, respectively. For the sake of space, here we only plot the sensitivity to the peak current in Fig. 4. For the initial local energy spread, when σ_γ/γ increases to about 9.5×10^{-5}, the final output power decreases monotonically to about half of that produced by an e-beam with the nominal value $\sigma_\gamma/\gamma = 6 \times 10^{-5}$. For the global energy spread (its effect corresponds to a detuning), when $\Delta\gamma/\gamma$ changes from $-\rho$ to 0.5ρ, where $\rho = 4 \times 10^{-4}$, is the Pierce parameter [2] in the 1.5 Å amplifier, the output power varies between 11.4 GW and the nominal value 17.3 GW. For the normalized emittance, when ε_n increases from 1.5π mm-mrad to about 1.8π mm-mrad, the output power decreases from the nominal value 17.3 GW to about 9.4 GW monotonically. As shown in Fig. 4, the output power fluctuates between 11.6 and 19.2 GW, when the peak current changes from 3000 to 4000 A. Tolerance for the 2 GeV part of the system is much more relaxed. These data clearly show that

the scheme provides the stability of the system based on "overtuning".

4. Pulse length, pulse shape, and spectrum

In Fig. 5, we plot the final output pulse shape. During the calculation, considering that the full radiation pulse length of 50 fs only occupies a small part of the whole bunch, which is 250 fs, we assume that the e-beam parameters remain constant within the radiation pulse length range, except that there is an energy linear chirp of 1.5%/mm for the 14 GeV e-beam as part of the global energy spread. For a comparison, we include the input Gaussian pulse in the plot. Both pulses are normalized. We find such an HGHG device could produce a more or less "square" shape pulse 40 fs long. Once again, this is due to our "overtuning scheme", because, in every stage of HGHG, the variation of the output power is reduced. The spectrum of this pulse is calculated. The central peak has a FWHM bandwidth of about 1.8×10^{-5}. This is to be compared with the LCLS bandwidth of about 7×10^{-4}.

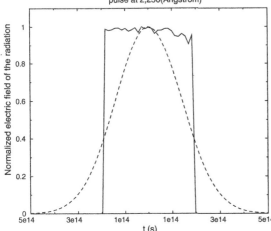

Fig. 5. The pulse shape and pulse length of the final output pulse, assuming a Gaussian input pulse. Both the output pulse and the input pulse are normalized for comparison.

In conclusion, the cascading HGHG scheme is an attractive scheme for generating coherent hard X-rays. It will have the advantages mentioned in the beginning of the paper. Among them, the most attractive feature is that the HGHG FEL will achieve full longitudinal coherence. One other feature is that the device could provide excellent stability. Compared with the SASE scheme, HGHG will need more electron bunches, using appropriate RF kicker, this can be achieved [23]. In this paper, we use a five-bunches scheme to illustrate the principle. An alternative scheme using only two-bunches has also been studied, where the HGHG process from 2250 to 18 Å uses a 4.5 GeV bunch, while from 18 to 1.5 Å a 14 GeV bunch is used (for both energies, a set of LCLS parameters is used). The total wiggler length is about 98 m to achieve similar or better performance. For example, the initial local energy spread σ_γ/γ can be as large as 10.5×10^{-5} before the final output power drops to half of that produced by an e-beam having the nominal value $\sigma_\gamma/\gamma = 6.0 \times 10^{-5}$. Our study indicates that for larger initial local energy spread, we can achieve similar stability by lowering the harmonic number and

increasing the number of stages, trading off improved stability with an increase of the total wiggler length. The increase of the total wiggler length is not very sensitive to the initial local energy spread increase.

Acknowledgements

J.W. wishes to thank Prof. C.N. Yang for his invaluable advice. The authors wish to thank Dr. S. Krinsky for many suggestion and comments. The authors also wish to thank Prof. C. Pellegrini and Dr. H.-D. Nuhn for pointing out the quantum diffusion effect [20]. This work is performed under the auspices of the U.S. Department of Energy.

References

[1] Y.S. Derbenev, A.M. Kondratenko, E.L. Saldin, Nucl. Instr. and Meth. A 193 (1982) 415.
[2] R. Bonifacio, C. Pellegrini, L.M. Narducci, Opt. Commun. 50 (1984) 373.
[3] J.M. Wang, L.H. Yu, Nucl. Instr. and Meth. A 250 (1986) 484.
[4] K.J. Kim, Nucl. Instr. and Meth. A 250 (1986) 396.
[5] K.J. Kim, Phys. Rev. Lett. 57 (1986) 1871.
[6] S. Krinsky, L.H. Yu, Phys. Rev. A 35 (1987) 3406.
[7] L.H. Yu, S. Krinsky, R.L. Gluckstern, Phys. Rev. Lett. 64 (1990) 3011.
[8] Li Hua Yu, Phys. Rev. E 58 (1998) 4991.
[9] M.J. Hogan, et al., Phys. Rev. Lett. 82 (1998) 4867.
[10] S.V. Milton, et al., Phys. Rev. Lett. 85 (2000) 988. J. Andruszkow, et al., Phys. Rev. Lett. 85 (2000) 3825.
[11] I. Ben-Zvi, L.F. Di Mauro, S. Krinsky, M.G. White, L.H. Yu, Nucl. Instr. and Meth. A 304 (1991) 151.
[12] L.H. Yu, Phys. Rev. A 44 (1991) 5178.
[13] L.-H. Yu, M. Babzien, I. Ben-Zvi, L.F. DiMauro, A. Doyuran, W. Graves, E. Johnson, S. Krinsky, R. Malone, I. Pogorelsky, J. Skaritka, G. Rakowsky, L. Solomon, X.J. Wang, M. Woodle, V. Yakimenko, S.G. Biedron, J.N. Galayda, E. Gluskin, J. Jagger, V. Sajaev, I. Vasserman, Nucl. Instr. and Meth. A 445 (2000) 301.
[14] L.-H. Yu, M. Babzien, I. Ben-Zvi, L.F. DiMauro, A. Doyuran, W. Graves, E. Johnson, S. Krinsky, R. Malone, I. Pogorelsky, J. Skaritka, G. Rakowsky, L. Solomon, X.J. Wang, M. Woodle, V. Yakimenko, S.G. Biedron, J.N. Galayda, E. Gluskin, J. Jagger, V. Sajaev, I. Vasserman, Science 289 (5481) (2000) 932.

[15] I. Boscolo, V. Stagno, Nucl. Instr. and Meth. A 188 (1982) 483.

[16] R. Bonifacio, L. De Salvo Souza, P. Pierini, E.T. Scharlemann, Nucl. Instr. and Meth. A 296 (1990) 787.

[17] R. Prazeres, P. Guyotsionnest, J.M. Ortega, D. Jaroszynski, M. Billardon, M.E. Couprie, M. Velghe, Y. Petroff, Nucl. Instr. and Meth. A 304 (1991) 72.

[18] L.H. Yu, Ilan Ben-Zvi, Nucl. Instr. and Meth. A 393 (1997) 96.

[19] I. Ben-Zvi, K.M. Yang, L.H. Yu, Nucl. Instr. and Meth. A 318 (1992) 726.

[20] E.L. Saldin, E.A. Schneidmiller, M.V. Yurkov, Nucl. Instr. and Meth. A 381 (1996) 545.

[21] The LCLS Design Study Group, Linac Coherent Light Source (LCLS) Design Study Report, Stanford Linear Accelerator Center (SLAC) Report No. SLAC-R-521, 1998.

[22] T.M. Tran, J.S. Wurtele, Comput. Phys. Commun. 54 (1989) 263.

[23] Paul Emma, private communication.

ELSEVIER

Nuclear Instruments and Methods in Physics Research A 475 (2001) 112–117

NUCLEAR
INSTRUMENTS
& METHODS
IN PHYSICS
RESEARCH
Section A

www.elsevier.com/locate/nima

Nonlinear harmonic generation of coherent amplification and self-amplified spontaneous emission

Zhirong Huang*, Kwang-Je Kim

Advanced Photon Source, Argonne National Laboratory, 9700 S. Cass Avenue, Argonne, IL 60439, USA

Abstract

Nonlinear harmonic generation in a uniform planar undulator is analyzed using the three-dimensional Maxwell–Klimontovich equations that include both even and odd harmonic emissions. After a certain stage of exponential growth, the dominant nonlinear harmonic interaction is caused by strong bunching at the fundamental. As a result, gain length, transverse profile, and temporal structure of these harmonic radiations are eventually determined by those of the fundamental. Transversely coherent third-harmonic radiation power is found to approach one percent of the fundamental power level for current high-gain FEL projects, while the power of the second-harmonic radiation is less but still significant for relatively low-energy FEL experiments. © 2001 Elsevier Science B.V. All rights reserved.

PACS: 41.60.Cr; 42.55.Vc; 42.65.Ky

Keywords: Nonlinear harmonic generation; Coherent amplification; Self-amplified spontaneous emission

1. Introduction

In a high-gain free-electron laser (FEL) employing a planar undulator, a one-dimensional (1-D) model [1] and a three-dimensional (3-D) simulation [2] indicate that strong bunching at the fundamental wavelength can drive substantial harmonic bunching and sizable power levels at the harmonic frequencies. A 3-D analysis of harmonic generation in a uniform planar undulator has been given in Ref. [3], for the process of coherent amplification (CA) and self-amplified spontaneous emission (SASE). In this paper, we extend the formalism of Ref. [3] to include the generation of even harmonics and present analytic

formulas for computing the second and the third nonlinear harmonic powers as well as the bunching parameters. Explicit calculations based on current high-gain FEL projects show that the power of the transversely coherent third-harmonic radiation can approach one percent of the fundamental power level, while the power of the second-harmonic radiation is less but still significant for relatively low-energy FEL experiments.

2. Nonlinear harmonic generation

For an electron in a planar undulator (with the undulator parameter K), the transverse wiggling motion in the x plane is accompanied by a longitudinal oscillation (at twice the transverse frequency ck_u) about the average longitudinal

---

*Corresponding author. Tel.: +1-630-252-6023; fax: +1-630-252-5703.

E-mail address: zrh@aps.anl.gov (Z. Huang).

0168-9002/01/$ - see front matter © 2001 Elsevier Science B.V. All rights reserved.
PII: S 0 1 6 8 - 9 0 0 2 (0 1) 0 1 5 5 3 - 4

position ct^*. This figure-eight motion (in the comoving frame) can give rise to harmonic emissions. Let us represent the electric field in the form

$$\hat{x}\int_{-\infty}^{\infty}\frac{\mathrm{d}v}{2}E(v,\mathbf{x};z)\mathrm{e}^{\mathrm{i}vk_1(z-ct)} \qquad (1)$$

where $\mathbf{x}=(x,y)$ represents the transverse coordinates, $ck_1=2\gamma_0^2 ck_u/(1+K^2/2)$ is the fundamental resonant frequency, and $|E(v)|$ is the field amplitude at frequency $\omega=vk_1c$.

It is convenient to treat z, the distance from the undulator entrance, as the independent variable, and change the dependent coordinate from t to θ by $\theta(z)=(k_u+k_1)z-ck_1t^*=(k_u+k_1)z-ck_1t+\xi\sin(2k_uz)$, where $\xi=K^2/(4+2K^2)$. The Maxwell equation under the paraxial approximation becomes

$$\left(\frac{\partial}{\partial z}+\frac{\nabla_\perp^2}{2\mathrm{i}vk_1}\right)E(v,\mathbf{x};z)$$

$$=-\frac{eK}{\varepsilon_0\gamma_0}\int_{-\infty}^{\infty}\frac{k_1\,\mathrm{d}\theta}{2\pi}\mathrm{e}^{-\mathrm{i}v\theta}\mathrm{e}^{\mathrm{i}vk_uz+\mathrm{i}v\xi\sin(2k_uz)}$$

$$\times\cos(k_uz)\sum_{j=1}^{N_e}\delta\left(x-\frac{K}{\gamma_0k_u}\sin(k_uz)-x_j^\beta\right)$$

$$\times\delta(y-y_j^\beta)\delta(\theta-\theta_j) \qquad (2)$$

where N_e is the total number of electrons, and x_j^β and y_j^β describe the transverse betatron oscillations. Because the transverse wiggling amplitude is normally smaller than the transverse dimension of the electron beam, we approximate

$$\delta\left(x-\frac{K}{\gamma_0k_u}\sin(k_uz)-x_j^\beta\right)$$

$$\approx\delta(x-x_j^\beta)-\frac{K}{\gamma_0k_u}\sin(k_uz)\delta'(x-x_j^\beta) \qquad (3)$$

where $\delta'=\mathrm{d}\delta/(\mathrm{d}x)$. Since the FEL interaction and the betatron oscillation occur on a scale much longer than the fast wiggling motion, we average Eq. (2) over the undulator period λ_u with the help of the Bessel function expansion

$$\mathrm{e}^{\mathrm{i}v\xi\sin(2k_uz)}=\sum_{p=-\infty}^{+\infty}J_p(v\xi)\mathrm{e}^{\mathrm{i}2pk_uz}. \qquad (4)$$

Inserting the first term of Eq. (3) into Eq. (2), we find that the wiggling average is nonzero only

when v is close to an odd integer $h=-(2p\pm1)$ [4] and obtain the equation for odd harmonics [3]:

$$\left(\frac{\partial}{\partial z}+\frac{\nabla_\perp^2}{2\mathrm{i}hk_1}\right)E_h(\Delta v_h,\mathbf{x};z)=-\frac{eK_h}{2\varepsilon_0\gamma_0}\mathrm{e}^{\mathrm{i}\Delta v_hk_uz}$$

$$\times\int\frac{k_1\,\mathrm{d}\theta}{2\pi}\mathrm{e}^{-\mathrm{i}v\theta}\sum_{j=1}^{N_e}\delta(\mathbf{x}-\mathbf{x}_j^\beta)\delta(\theta-\theta_j) \qquad (5)$$

where $\Delta v_h=v-h\ll1$ is the frequency detuning and the effective coupling strength is

$$K_h=K(-1)^{(h-1)/2}[J_{(h-1)/2}(h\xi)-J_{(h+1)/2}(h\xi)],$$

$$h=1,3,5,\ldots\ . \qquad (6)$$

Inserting the second term of Eq. (3) into Eq. (2), we find that the wiggling average is nonzero only when v is close to an even integer $h=-(2p\pm2)$ [5] and obtain the equation for even harmonics:

$$\left(\frac{\partial}{\partial z}+\frac{\nabla_\perp^2}{2\mathrm{i}hk_1}\right)E_h(\Delta v_h,\mathbf{x};z)$$

$$=-\mathrm{i}\frac{eK_h}{2\varepsilon_0\gamma_0}\mathrm{e}^{\mathrm{i}\Delta v_hk_uz}\int\frac{k_1\,\mathrm{d}\theta}{2\pi}\mathrm{e}^{-\mathrm{i}v\theta}$$

$$\times\frac{K}{\gamma_0k_u}\sum_{j=1}^{N_e}\delta'(x-x_j^\beta)\delta(y-y_j^\beta)\delta(\theta-\theta_j) \qquad (7)$$

where the effective coupling strength is

$$K_h=K(-1)^{(h-2)/2}J'_{h/2}(h\xi),\quad h=2,4,\ldots\ . \qquad (8)$$

Hence, in the forward z direction of a perfectly aligned undulator trajectory, even harmonic emissions are present due to the transverse gradient of the electron current in the wiggling plane.

The electron distribution in phase space is described by the Klimontovich distribution function $f(\theta,\eta,\mathbf{x},\mathbf{p};z)$, where $\eta=(\gamma-\gamma_0)/\gamma_0$, and $\mathbf{p}=\mathrm{d}\mathbf{x}/\mathrm{d}z$ are the conjugate variables to θ and $\mathbf{x}$. Using the Pierce parameter ρ [6], we introduce the following scaled variables:

$$\bar{z}=2\rho k_uz,\qquad\bar{\eta}=\frac{\eta}{\rho},\qquad\bar{v}_h=\frac{\Delta v_h}{2\rho}$$

$$\bar{\mathbf{x}}=\mathbf{x}\sqrt{2k_1k_u\rho},\qquad\bar{\mathbf{p}}=\mathbf{p}\sqrt{\frac{k_1}{2k_u\rho}}$$

$$a_h(\bar{v}_h,\bar{\mathbf{x}};\bar{z})=\frac{-eK_h}{4\gamma_0^2mc^2k_u\rho}\mathrm{e}^{-\mathrm{i}\Delta v_hk_uz}E_h(\Delta v_h,\mathbf{x},z). \qquad (9)$$

Eqs. (5) and (7) can be written as

$$\left(\frac{\partial}{\partial \bar{z}} + i\bar{v}_h + \frac{\bar{\nabla}_\perp^2}{2ih}\right)a_h = \left(\frac{K_h}{K_1}\right)^2 \int d^2\bar{p}$$

$$\times \int d\bar{\eta} \int \frac{2\rho\, d\theta}{2\pi} e^{-iv\theta} \begin{cases} f(\bar{z}) & \text{odd } h \\ \frac{iK}{\gamma_0 k_u}\frac{\partial f}{\partial x} & \text{even } h. \end{cases} \quad (10)$$

The evolution of the distribution function is governed by the Klimontovich equation integrated along the unperturbed trajectory [3]:

$$f(z) = f(0) + \int_0^{\bar{z}} d\bar{s} \sum_{\text{odd } h} \int d(\bar{v}_h)e^{iv\theta^{(0)}}$$

$$\times a_h(\bar{v}_h, \bar{\mathbf{x}}^{(0)}, \bar{s})\frac{\partial}{\partial\bar{\eta}} f(\theta^{(0)}, \bar{\eta}, \bar{\mathbf{x}}^{(0)}, \bar{\mathbf{p}}^{(0)}; \bar{s})$$

$$+ \int_0^{\bar{z}} d\bar{s} \sum_{\text{even } h} \int d(\bar{v}_h)e^{iv\theta^{(0)}} a_h(\bar{v}_h, \bar{\mathbf{x}}^{(0)}, \bar{s})$$

$$\times \frac{iK}{\gamma_0 k_u}\frac{\partial^2 f}{\partial x \partial\bar{\eta}} + \text{c.c.} \quad (11)$$

Here the summation of h is extended to include the interactions with the even harmonics. The unperturbed trajectory is described by

$$\theta^{(0)}(s) = \theta + \phi(\bar{s} - \bar{z}) \quad \text{with}$$

$$\phi = \bar{\eta} - (\bar{\mathbf{p}}^2 + \bar{k}_\beta^2\bar{\mathbf{x}}^2)/2$$

$$\bar{\mathbf{x}}^{(0)}(s) = \bar{\mathbf{x}}\cos(\bar{k}_\beta(\bar{s} - \bar{z})) + \frac{\bar{\mathbf{p}}}{\bar{k}_\beta}\sin(\bar{k}_\beta(\bar{s} - \bar{z}))$$

$$\bar{\mathbf{p}}^{(0)}(s) = -\bar{k}_\beta\bar{\mathbf{x}}\sin(\bar{k}_\beta(\bar{s} - \bar{z})) + \bar{\mathbf{p}}\cos(\bar{k}_\beta(\bar{s} - \bar{z}))$$

$$(12)$$

where $\bar{k}_\beta = k_\beta/(2k_u\rho)$ is the scaled betatron focusing strength. $f(0) = f_0 + \delta f_0$ contains the initial fluctuation δf_0 as well as the initial smooth distribution f_0, which is assumed to be

$$f_0(\bar{\eta}, \bar{\mathbf{p}}^2 + \bar{k}_\beta^2\bar{\mathbf{x}}^2) = \frac{1}{2\pi\bar{\sigma}_x^2\bar{k}_\beta^2}$$

$$\times \exp\left[-\frac{(\bar{\mathbf{p}}^2 + \bar{k}_\beta^2\bar{\mathbf{x}}^2)}{2\bar{\sigma}_x^2\bar{k}_\beta^2}\right]\frac{e^{-\bar{\eta}^2/(2\bar{\sigma}_\eta^2)}}{\sqrt{2\pi}\bar{\sigma}_\eta} \quad (13)$$

where $\bar{\sigma}_x = \sigma_x\sqrt{2\rho k_u k_1}$ and $\bar{\sigma}_\eta = \sigma_\eta/\rho$ are the scaled beam size and scaled energy spread, respectively.

Coherent harmonic radiation is generated through nonlinear harmonic interactions. After a certain stage of exponential growth, the dominant nonlinear term has been shown to be predominantly driven by the fundamental field [3]. Thus, we consider the nonlinear harmonic bunching determined by the fundamental field only. In the small signal regime, we keep the a_1 term only in Eq. (11) and solve it by iteration

$$f(z) \approx f_0 + \delta f_0 + \sum_{\text{all } h} f_h(z) + \text{c.c.} \quad (14)$$

where

$$f_h(z) = \left[\prod_{m=1}^h\left(\int_0^{\bar{s}_{m-1}} d\bar{s}_m \int d(\bar{v}_m)e^{i(1+2\rho\bar{v}_m)\theta^{(0)}(s_m)}\right.\right.$$

$$\left.\left.\times a_1(\bar{v}_m, \bar{\mathbf{x}}^{(0)}(s_m); \bar{s}_m)\frac{\partial}{\partial\bar{\eta}}\right)\right]f_0 \quad (15)$$

for $h \geqslant 1$ and $s_0 = \bar{z}$. We note that Eq. (14) with Eq. (15) is the approximate solution of Eq. (11) when the nonlinear harmonic generation dominates over the linear harmonic generation or the spontaneous harmonic emission [3]. It becomes an increasingly good approximation as the fundamental field is significantly amplified.

The evolution of the fundamental field is obtained by solving Eq. (10) with $h = 1$ and f replaced by $\delta f_0 + f_1(z)$ [3,7]:

$$a_1(\bar{v}, \bar{\mathbf{x}}; \bar{z}) = e^{-i\mu_1\bar{z}}A_1(\bar{\mathbf{x}})\left[\int d^2\bar{x}' A_1(\bar{\mathbf{x}}')a_1(\bar{v}, \bar{\mathbf{x}}'; 0)\right.$$

$$+ \int d^2\bar{x}' \int d^2\bar{p} \int d\bar{\eta}\delta f_0(\bar{v}, \bar{\mathbf{x}}', \bar{\mathbf{p}}, \bar{\eta})$$

$$\left.\times \int_{-\infty}^0 d\tau A_1(\bar{\mathbf{x}}^{(0)})e^{i(\phi - \mu_1)\tau}\right] \quad (16)$$

where μ_1 is the complex growth rate of the fundamental mode $A_1(\bar{\mathbf{x}})$ with the largest imaginary part of μ_1. The first term of Eq. (16) describes the process of coherent amplification from the initial coherent signal $a_1(\bar{v}, \bar{\mathbf{x}}; 0)$, and the second term of Eq. (16) describes the process of self-amplified spontaneous emission from the initial shot noise δf_0. Inserting Eq. (14) into Eq. (10), we find that a_h ($h > 1$) is determined by f_h with a complex growth rate $h\mu_1$, and that the characteristics of the nonlinear harmonic generation are all determined by the fundamental field. While the transverse profile of the odd harmonics is azimuthally symmetric just as the fundamental mode, the transverse profile of the even harmonics possesses the odd symmetry in the wiggling plane (the x plane) due to the transverse gradient effect in Eq. (10).

3. Third-harmonic radiation

The most significant nonlinear harmonic generation occurs at the third harmonic, given by Eq. (10) for $h = 3$ and f replaced by f_3 of Eq. (15). For a seeded FEL, we assume that the external signal matches optimal detuning $\bar{v}_0$ for the fundamental field (with a complex growth rate μ_0 that has the maximum imaginary part). We can set $\bar{v}_1 = \bar{v}_0$ and $\bar{v}_3 = 3\bar{v}_0$ and drop the frequency dependence of a_1 and a_3 in Eq. (10). In view of Eq. (16), we write $a_1(\bar{\mathbf{x}}; \bar{z}) = e^{-i\mu_0\bar{z}}A_1(\bar{\mathbf{x}})$, where $A_1 \approx A_0 e^{-w_1 R^2}$ is the fundamental mode ($R = |\bar{\mathbf{x}}|/\bar{\sigma}_x = |\mathbf{x}|/\sigma_x$), and A_0 is the appropriate normalization coefficient. Thus, we can write the third nonlinear harmonic $a_3(\bar{\mathbf{x}}; \bar{z}) = e^{-3i\mu_0\bar{z}}A_3(\bar{\mathbf{x}})$ with the transverse profile A_3 given by [3]

$$\left[-3i(\mu_0 - \bar{v}_0) + \frac{\bar{\nabla}_\perp^2}{6i}\right]A_3(\bar{\mathbf{x}}) - \left(\frac{K_3}{K_1}\right)^2 \int \mathrm{d}^2\bar{p}$$

$$\times \int \mathrm{d}\bar{\eta} \int_{-\infty}^0 \mathrm{d}\tau_1 e^{3i(\phi-\mu_0)\tau_1}A_1(\bar{\mathbf{x}}^{(0)}(s_1))\frac{\partial}{\partial\bar{\eta}}$$

$$\times \int_{-\infty}^0 \mathrm{d}\tau_2 e^{2i(\phi-\mu_0)\tau_2}A_1(\bar{\mathbf{x}}^{(0)}(s_2))\frac{\partial}{\partial\bar{\eta}}$$

$$\times \int_{-\infty}^0 \mathrm{d}\tau_3 e^{i(\phi-\mu_0)\tau_3}A_1(\bar{\mathbf{x}}^{(0)}(s_3))\frac{\partial\bar{f}_0}{\partial\bar{\eta}} \quad (17)$$

where $\tau_m = \bar{s}_m - \bar{s}_{m-1}$ for $m = 1, 2, 3$, and $\bar{s}_0 = \bar{z}$. We have extended the lower limit of the integral $\int \mathrm{d}\tau_m$ to $-\infty$ due to the exponential growth. Solving Eq. (17) with the Hankel transformation, we obtain [3]

$$A_3(R) = \left(\frac{K_3}{K_1}\right)^2 A_0^3 \int_0^\infty Q\,\mathrm{d}Q J_0(QR)H(Q), \quad (18)$$

where

$$H(Q) = \frac{-1/w_1^2}{12(\mu_0 - \bar{v}_0) - 2Q^2/(3\bar{\sigma}_x^2)}$$

$$\times \int_{-\infty}^0 \mathrm{d}\tau_1 \int_{-\infty}^0 \mathrm{d}\tau_2$$

$$\times \int_{-\infty}^0 \mathrm{d}\tau_3 \frac{3\tau_1(3\tau_1 + 2\tau_2)(3\tau_1 + 2\tau_2 + \tau_3)}{U}$$

$$\times \exp\left[-\frac{\bar{\sigma}_\eta^2}{2}(3\tau_1 + 2\tau_2 + \tau_3)^2\right.$$

$$\left. - i\mu_0(3\tau_1 + 2\tau_2 + \tau_3) - \frac{Q^2}{4w_1 S}\right] \quad (19)$$

and

$$S = \frac{U}{V + \sum_{m=1,2,3} \sin^2(\bar{k}_\beta \sum_{l=1}^m \tau_l)}$$

$$U = \left[\sum_{m=1}^3 \sin^2\left(\bar{k}_\beta \sum_{l=1}^m \tau_l\right)\right]$$

$$\times \left[\sum_{m=1}^3 \cos^2\left(\bar{k}_\beta \sum_{l=1}^m \tau_l\right)\right]$$

$$- \left[\sum_{m=1}^3 \sin\left(\bar{k}_\beta \sum_{l=1}^m \tau_l\right)\cos\left(\bar{k}_\beta \sum_{l=1}^m \tau_l\right)\right]^2$$

$$+ V^2 + 3V$$

$$V = \frac{1}{2w_1} + \frac{i}{2w_1}\bar{k}_\beta^2\bar{\sigma}_x^2(3\tau_1 + 2\tau_2 + \tau_3). \quad (20)$$

Eq. (18) can be computed using a discrete Hankel transformation. In general, the third-harmonic radiation is also transversely coherent with a Gaussian-like profile and a narrower spot size than the fundamental.

For a SASE FEL, the fundamental radiation starts with a white noise spectrum and has a finite gain bandwidth. In the time domain, the temporal structure of the fundamental is chaotic with many random spikes. Due to the nonlinear generation mechanism, the temporal structure of the third-harmonic radiation is similar to the fundamental, but with more fluctuation from spike to spike. It can be shown that [3]

$$A_3(\theta, R; \bar{z}) \approx \left(\frac{K_3}{K_1}\right)[G_1(\theta; \bar{z})]^3 A_0^3 H_0 e^{-w_3 R^2} \quad (21)$$

where $G_1(\theta; \bar{z})$ is a Gaussian random variable in θ and a slowly varying function in $\bar{z}$ for SASE ($G_1 = 1$ for CA), $H_0 = (K_3/K_1) \int Q \, dQ H_3(Q)$, and w_3 characterizes the transverse profile of the third-harmonic radiation. The average radiation power can be obtained by integrating over the transverse intensity profile and averaging over the temporal fluctuation. Thus, we have [3]

$$\left(\frac{P_3}{\rho P_{beam}}\right) \approx |H_0|^2 \frac{16 w_{1r}^3}{w_{3r}} \left(\frac{P_1}{\rho P_{beam}}\right)^3 \begin{cases} 1 & CA \\ 6 & SASE. \end{cases} \tag{22}$$

Here $P_{beam} = 2\pi \sigma_x^2 \gamma mc^3 n_0$ is the total electron beam power, and w_{1r} and w_{3r} are the real parts of w_1 and w_3. Thus, the third-harmonic radiation for a SASE FEL has a power level roughly 6 times larger than the corresponding steady-state case, but with more shot-to-shot fluctuations compared to the fundamental [3]. The third-harmonic bunching parameter is obtained by averaging $\langle e^{-3i\theta} f_3 \rangle$ over the 6-D phase-space volume and taking the absolute value [3]:

$$b_3 = 8 w_{1r}^{3/2} \left(\frac{P_1}{\rho P_{beam}}\right)^{3/2} \left| \int_{-\infty}^0 d\tau_1 \int_{-\infty}^0 d\tau_2 \right.$$

$$\times \int_{-\infty}^0 d\tau_3 \frac{3\tau_1(3\tau_1 + 2\tau_2)(3\tau_1 + 2\tau_2 + \tau_3)}{4 w_1^2 U}$$

$$\times \exp\left[-\frac{\bar{\sigma}_\eta^2}{2}(3\tau_1 + 2\tau_2 + \tau_3)^2 \right.$$

$$\left. \left. - i\mu_0(3\tau_1 + 2\tau_2 + \tau_3) \right] \right| \begin{cases} 1 & CA \\ 1.3 & SASE. \end{cases} \tag{23}$$

For example, using the design parameters (see Table 1) for the low-energy undulator test line (LEUTL) FEL at the Advanced Photon Source [8] and the proposed Linac Coherent Light Source (LCLS) at Stanford Linear Accelerator Center [9], we can compute the transverse profile of the third harmonic through Eqs. (18), (19) and (20) see Figs. 1 and 3 of Ref. [3]). The third-harmonic power and the bunching parameter are also calculated according to Eqs. (22) and (23). Table 1 lists the results when the fundamental power reaches one-half of the saturation power, when the exponential growth process is supposed to stop. We have compared the evolution of the third-harmonic

Table 1
Nonlinear harmonic generation for SASE FELs

	LEUTL	LCLS
E-beam and undulator		
Energy	220 MeV	14.4 GeV
Peak current	150 A	3400 A
Normalized emittance	5 μm	1.5 μm
Energy spread	0.1%	0.02%
Average beta function	1.5 m	18 m
Undulator period	3.3 cm	3 cm
Undulator strength	3.1	3.71
Fundamental radiation		
Resonant wavelength	518 nm	1.5 Å
Power gain length	0.67 m	6.1 m
Saturation power (P_{sat})	70 MW	8 GW
Harmonics at $P_1 = P_{sat}/2$		
3rd-harmonic power	3.6 MW	15 MW
3rd-harmonic bunching	0.39	0.018
2nd-harmonic power	550 kW	15 kW
2nd-harmonic bunching	0.47	0.056

power for the LEUTL FEL with the steady-state MEDUSA simulation [2], and the third-harmonic bunching for the LCLS case with the steady-state GINGER simulation [3]. Good agreement for both cases have been found.

4. Second-harmonic radiation

The second-harmonic radiation can be calculated from Eq. (10) with $h = 2$ and f replaced by f_2 of Eq. (15). One can follow the same procedure as in Section 3 to solve for the second-harmonic field for CA and SASE, except that the Hankel transformation should be replaced by the 2-D Fourier transformation in x and y because the radiation profile has the odd symmetry in x. Since the wiggling amplitude $K/(\gamma_0 k_u)$ is usually much smaller than the rms beam size σ_x, the power of the second-harmonic radiation is less than that of the third harmonic. We can estimate the power of the second-harmonic radiation by

$$\frac{P_2}{b_2^2} \approx \left(\frac{K}{\gamma_0 k_u \sigma_x}\right)^2 \left(\frac{K_2}{K_3}\right)^2 \frac{P_3}{b_3^2}. \tag{24}$$

Here the second harmonic bunching parameter b_2 is obtained by averaging $(e^{-2i\theta}f_2)$ over the 6-D phase-space volume and taking the absolute value

$$
b_2 = 4w_{1r}\left(\frac{P_1}{\rho P_{beam}}\right)\Bigg|\int_{-\infty}^{0}d\tau_1
$$
$$
\times \int_{-\infty}^{0}d\tau_2\frac{2\tau_1(2\tau_1+\tau_2)}{4w_1^2 U_2}
$$
$$
\times \exp\left[-\frac{\bar{\sigma}_\eta^2}{2}(2\tau_1+\tau_2)^2 - i\mu_0(2\tau_1+\tau_2)\right]\Bigg| \quad (25)
$$

where

$$
U_2 = \left[\sum_{m=1}^{2}\sin^2\left(\bar{k}_\beta\sum_{l=1}^{m}\tau_l\right)\right]
$$
$$
\times \left[\sum_{m=1}^{2}\cos^2\left(\bar{k}_\beta\sum_{l=1}^{m}\tau_l\right)\right]
$$
$$
- \left[\sum_{m=1}^{2}\sin\left(\bar{k}_\beta\sum_{l=1}^{m}\tau_l\right)\cos\left(\bar{k}_\beta\sum_{l=1}^{m}\tau_l\right)\right]^2
$$
$$
+ V_2^2 + 2V_2
$$

$$
V_2 = \frac{1}{2w_1} + \frac{i}{2w_1}\bar{k}_\beta^2\bar{\sigma}_x^2(2\tau_1+\tau_2). \quad (26)
$$

Using the LEUTL and LCLS examples, we calculate the second-harmonic bunching and estimate the second-harmonic power by Eq. (24). From Table 1, we see that a significant amount of second-harmonic radiation can be generated in the LEUTL FEL because the wiggling amplitude (proportional to $1/\gamma_0$) is about one-third of the beam size. However, for X-ray FELs employing a high-energy electron beam, such as the LCLS, the second-harmonic radiation is much reduced.

5. Conclusion

We have presented a perturbation scheme to analyze the 3-D evolution of the nonlinear harmonic radiation in coherent amplification and self-amplified spontaneous emission, with explicit calculation of second-harmonic and third-harmonic radiation based on current high-gain FEL projects. The transverse coherence and the substantial power level of the third harmonic could be useful in extending the short wavelength reach of a high-gain FEL.

Acknowledgements

We thank S. Biedron, Y. Chae, W. Fawley, H. Freund, S. Milton, and C. Pelligrini for useful discussions. This work was supported by the U.S. Department of Energy, Office of Basic Energy Sciences, under Contract No. W-31-109-ENG-38.

References

[1] R. Bonifacio, L. De Salvo, P. Pierini, Nucl. Instr. and Meth. A 293 (1990) 627.
[2] H.P. Freund, S.G. Biedron, S.V. Milton, IEEE J. Quantum Electron. QE-36 (2000) 275; H.P. Freund, S.G. Biedron, S.V. Milton, Nucl. Instr. and Meth. A 445 (2000) 53.
[3] Z. Huang, K.-J. Kim, Phys. Rev. E 62 (2000) 7259.
[4] W.B. Colson, IEEE J. Quantum Electron. QE-17 (1981) 1417.
[5] M.J. Schmitt, C.J. Elliott, Phys. Rev. A 34 (1986) 4843; M.J. Schmitt, C.J. Elliott, Phys. Rev. A 41 (1990) 3853; M.J. Schmitt, C.J. Elliott, Nucl. Instr. and Meth. A 296 (1990) 394.
[6] R. Bonifacio, C. Pellegrini, L.M. Narducci, Opt. Commun. 50 (1984) 373.
[7] Z. Huang, K.-J. Kim, Nucl. Instr. and Meth. A 475 (2001) 59, These Proceedings.
[8] S.V. Milton, et al., Nucl. Instr. and Meth. A 407 (1998) 210.
[9] Linac Coherent Light Source Design Study Report, SLAC-R-521, 1998.

ELSEVIER

Nuclear Instruments and Methods in Physics Research A 475 (2001) 118–126

**NUCLEAR
INSTRUMENTS
& METHODS
IN PHYSICS
RESEARCH**
Section A

www.elsevier.com/locate/nima

Nonlinear harmonics in the high-gain harmonic generation (HGHG) experiment

S.G. Biedron[a,b,*], H.P. Freund[c], S.V. Milton[a], L-.H. Yu[d], X.J. Wang[d]

[a] *Advanced Photon Source, Argonne National Laboratory, 9700 S.Cass Avenue, Argonne, IL 60439, USA*
[b] *MAX-Laboratory, University of Lund, Lund S-22100, Sweden*
[c] *Science Applications International Corp., McLean, VA 22102, USA*
[d] *Brookhaven National Laboratory, Upton, NY 11973, USA*

Abstract

We have previously performed rigorous analyses of the nonlinear harmonics in self-amplified spontaneous emission (SASE) free-electron lasers (FELs) using a 3D simulation code. To date, we have presented only preliminary results of these higher harmonics resulting in the high-gain harmonic generation (HGHG) process. A single-pass, high-gain FEL experiment based on the HGHG theory is underway at the Accelerator Test Facility (ATF) at Brookhaven National Laboratory (BNL) in collaboration with the Advanced Photon Source (APS) at Argonne National Laboratory (ANL). Using the above experiment's design parameters, the specific case of the harmonic output from the HGHG experiment will be examined using a 3D simulation code. The sensitivity of nonlinear harmonic output for this HGHG experiment as functions of emittance, energy spread, and peak current in both cases, and for the dispersive section strength and input seed power in the HGHG case, will be presented. © 2001 Elsevier Science B.V. All rights reserved.

PACS: 41.60Cr; 42.65.Ky; 42.25.Kb; 52.59.−f

Keywords: Free-electron lasers; Harmonic generation; Frequency conversion; Coherence; Intense particle beams; Radiation sources

1. Introduction

A single-pass, high-gain free-electron laser (FEL) experiment based on the high-gain harmonic generation (HGHG) theory is underway at the Accelerator Test Facility (ATF) at Brookhaven National Laboratory (BNL) in collaboration with the Advanced Photon Source (APS) at Argonne National Laboratory (ANL) [1]. This project achieved first lasing in August 1999, when gain of 10^7 above spontaneous was measured [2,3]. It has been shown through both the 1D [4–7] and

3D [8] analytical models and in 3D simulations [5–7] that the higher nonlinear harmonics are expected to grow at rates that scale in inverse proportion to the harmonic number in all single-pass, high-gain FELs. Also, the powers in these harmonics are expected to be substantial. An HGHG FEL configuration already implements frequency up-conversion (as explained briefly in the next section), and, in such a system, these higher nonlinear harmonics are unusually interesting, as even shorter wavelengths can be exploited simultaneously. The specific case of the simulated higher nonlinear harmonic output for the BNL/APS HGHG experiment will be examined here.

---

*Corresponding author.

E-mail address: biedron@aps.anl.gov (S.G. Biedron).

0168-9002/01/$ - see front matter © 2001 Elsevier Science B.V. All rights reserved.
PII: S 0 1 6 8 - 9 0 0 2 (0 1) 0 1 5 5 4 - 6

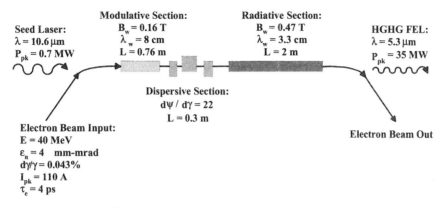

Seed Laser:
$\lambda = 10.6\,\mu m$
$P_{pk} = 0.7\,MW$

Modulative Section:
$B_w = 0.16\,T$
$\lambda_w = 8\,cm$
$L = 0.76\,m$

Radiative Section:
$B_w = 0.47\,T$
$\lambda_w = 3.3\,cm$
$L = 2\,m$

HGHG FEL:
$\lambda = 5.3\,\mu m$
$P_{pk} = 35\,MW$

Dispersive Section:
$d\psi / d\gamma = 22$
$L = 0.3\,m$

Electron Beam Input:
$E = 40\,MeV$
$\varepsilon_n = 4\quad mm\text{-}mrad$
$d\gamma/\gamma = 0.043\%$
$I_{pk} = 110\,A$
$\tau_e = 4\,ps$

Electron Beam Out

Fig. 1. The BNL/APS HGHG experimental layout.

2. High-gain harmonic generation (HGHG) theory

In a self-amplified spontaneous emission (SASE) FEL, the spontaneous emission "noise" serves to ignite the FEL bunching process, resulting in a noisy output. In a seeded (amplified) FEL, however, the output radiation quality is related to the coherence properties of the seed. In both cases, the process saturates and maximum power is reached. In practical design, this saturation point should be made to occur in the shortest possible distance. It is well known that in order to achieve this, both SASE and amplifier schemes require a high peak current, low emittance, and low energy-spread electron beam, as well as a properly shimmed undulator, to reduce magnetic error effects on the electron beam trajectory. From a machine user's standpoint, the amplifier scheme is more desirable due to the quality of the output radiation. To reach ultrashort wavelengths in this mode of operation, however, a seed laser already at the desired short wavelength is required. In the hard X-ray regime, for example, no seed is currently available.

Another possible mode of FEL operation capable of providing a very desirable output light beam is HGHG [9,10]. Here, a coherent radiation source, at a subharmonic of the desired output radiation wavelength overlaps an electron beam resonant with the seed radiation. Energy modulation of the electron beam with a period equal to the seed radiation is the result. Following this

modulation, a dispersive section is traversed. Spatial bunching is induced with a strong higher harmonic content. The beam then enters a second undulator, the radiative section, tuned to resonance at the desired harmonic. (In the BNL/APS HGHG experiment, the second harmonic is of interest.) Coherent radiation and ultimately saturation at this higher harmonic is then achieved within a reasonable number of undulator periods and with an excellent radiation beam quality, as compared to the SASE process. Recall that this better radiation beam quality is defined by the coherent seed radiation source. Although this method has been proven at $5.3\,\mu m$, it could quite possibly be extended to higher photon energies, where the radiator is tuned to a much higher harmonic, and could possibly achieve saturation in the UV, VUV, or X-ray regimes. A schematic of the HGHG process for the specific design case of the BNL/APS collaborative experiment is provided in Fig. 1.

3. Nonlinear harmonic generation in single-pass, free-electron lasers

The nonlinear harmonics grow as bunching at the fundamental occurs. In a planar undulator, which will be considered here, the natural motion of the electron beam causes the odd harmonics to be most significant in the forward direction. It is important to observe that all the harmonic growth rates are faster than the fundamental growth rate; the gain length scales inversely with the harmonic number [4–8].

In order to illustrate this scaling, consider Maxwell's equation in 1D:

$$\left(\frac{\partial^2}{\partial z^2} - \frac{1}{c^2}\frac{\partial^2}{\partial t^2}\right)E_x(z,t) = \frac{4\pi}{c^2}\frac{\partial}{\partial t}J_x(z,t), \tag{1}$$

where the electric field is given by

$$E_x(z,t) = \frac{1}{2}\sum_h \hat{E}_h(z)\exp[ihk_0(z-ct)] + \text{c.c.} \tag{2}$$

and the source current is given in terms of a nonlinear conductivity, $\sigma_h^{(NL)}$,

$$J_x(z,t) = \frac{1}{2}\sum_h [\sigma_h^{(L)}\hat{E}_h + \sigma_h^{(NL)}\hat{E}_1^h]$$

$$\times \exp[ihk_0(z-ct)] + \text{c.c.}, \tag{3}$$

where $\sigma_h^{(L)}$ describes the linear interaction. The detailed form for the nonlinear conductivity is not specified here, yet it is included implicitly in the nonlinear formulation. As stated above, these nonlinear terms are driven by bunching at the fundamental, and so we can write Maxwell's equations to lowest order as

$$\frac{d}{dz}\hat{E}_h \cong -\frac{2\pi}{c}(\sigma_h^{(L)}\hat{E}_h + \sigma_h^{(NL)}\hat{E}_1^h). \tag{4}$$

Therefore, if the fundamental grows as $\hat{E}_1 \approx \exp[\Gamma z]$, then the harmonics will grow as $\hat{E}_1^h \approx \exp[h\Gamma z]$, and the gain length scales as $L_g \approx (2h\Gamma)^{-1}$. The scaling of the gain length with the harmonic number is characteristic of the nonlinear mechanism. This is a well-known phenomenon in traveling wave tubes (TWTs) [11], has been seen in a 1D analysis for the first and third harmonics in an FEL [4–7], and was exemplified with a 3D simulation code [5–7]. More recently, a 3D analytical model has been developed

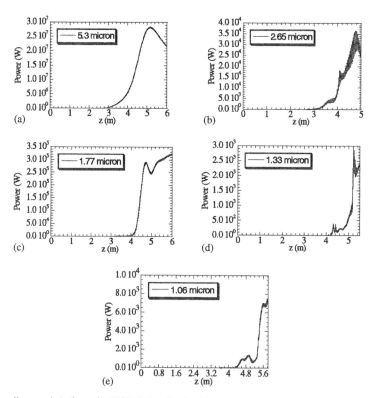

Fig. 2. Power (W) versus distance (m) along the HGHG line for the (a) 5.3 μm, (b) 2.65 μm, (c) 1.77 μm, (d) 1.33 μm, and (e) 1.06 μm output radiation.

and shows very good agreement with the previous work [8].

4. Harmonic output in the BNL/APS HGHG case

The output power of HGHG at the fundamental, with respect to the resonance condition of the radiative section of 5.3 μm, has been estimated [1,12] and measurements have been made [2,3]. The theoretical calculations were performed using the electron beam, seed laser, and magnetic parameters listed in Fig. 1.

The HGHG simulations including the higher, nonlinear harmonics relative to the radiative

section, were performed using the 3D, polychromatic, nonundulator averaged code, MEDUSA [13,14,5–7]. Again, the design parameters shown in Fig. 1 were used. For a dispersive strength of 1.7 kG ($R_{56} = 0.97$ mm), an input seed laser power of 0.7 MW ($E_0 = 1.62 \times 10^7$ V/m), an incoherent electron beam energy spread of 0.043%, a peak current of 110 A, and a normalized emittance of 4π mm-mrad, the powers as a function of distance along the HGHG line are shown in Fig. 2 for the fundamental of the radiative section (5.3 μm) and its four higher harmonics. Note the radiative section begins at $z = 2.65$ m. The higher harmonics are the second through fifth harmonics to the radiative section resonant wavelength (5.3 μm),

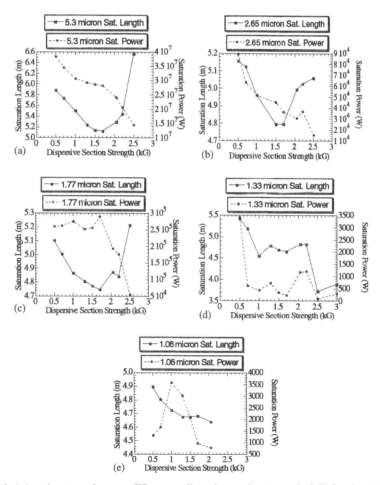

Fig. 3. Saturation length (m) and saturated power (W) versus dispersive section strength (kG) for the (a) 5.3 μm, (b) 2.65 μm, (c) 1.77 μm, (d) 1.33 μm, and (e) 1.06 μm output radiation.

but note they are the fourth, sixth, eighth, and tenth harmonics of the input seed laser (10.6 μm). This multiplicative increase in the frequency up-conversion based on the ratio of the resonant wavelengths of the radiative and modulative sections in HGHG makes these nonlinear harmonics even more fruitful than those found in SASE and amplifier schemes. In other words, in HGHG the shorter wavelengths are attainable more readily and with more control than in the other schemes.

Note that the case described above has slightly better performance than the nominal beam case for the fundamental as well as the harmonics and was found by lowering the dispersive section from

2.30 to 1.7 kG (reducing R_{56} from 1.67 to 0.97 mm). In addition to the simulations at this case, parameter scans were performed by varying dispersion section strength, seed laser power density, emittance, energy spread, and peak current, about the nominal design case. Figs. 3–7 show the output power and saturation length for each of these scans for the fundamental of the radiative section (5.3 μm) and the four higher harmonics (2.65, 1.77, 1.33, and 1.06 μm).

Upon examination of the parameter scans of dispersive section strength and the input seed power, some comments can be made. Both of these parameters contribute to the efficiency of the HGHG process, and the results of the scans

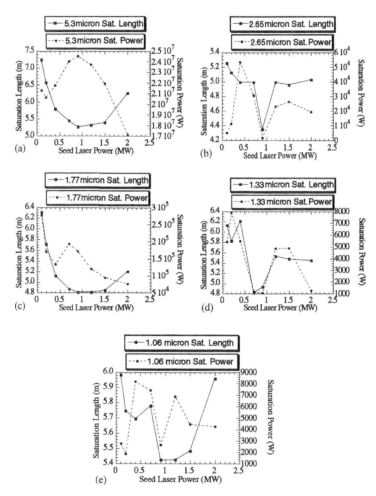

Fig. 4. Saturation length (m) and saturated power (W) versus input seed laser power (MW) for the (a) 5.3 μm, (b) 2.65 μm, (c) 1.77 μm, (d) 1.33 μm, and (e) 1.06 μm output radiation.

resemble each other greatly. For example, reduced HGHG efficiency will result from too little energy modulation; alternatively, too little phase-space rotation in the dispersive section leads to an underbunched electron beam at the entrance of the radiative section. Reduced HGHG efficiency will also result from too much energy modulation or too much phase-space rotation in the dispersive section as the beam is overbunched when entering the radiative section. Basically, there is a balance between saturated output power and saturation length based on the dispersive section strength and the degree of energy modulation. The optimum HGHG efficiency, if based solely on the degree of

energy modulation and dispersive strengths, is the shortest distance to saturation in the radiative section. The equation governing this optimum is expressed as

$$\left.\frac{d\Psi}{d\gamma}\right|_{\text{DispersiveSection}} = \frac{1}{\left.d\gamma/d\Psi\right|_{\text{MaximumLaserInducedModulation}}}$$

The saturated power decreases fairly readily with degraded electron beam energy spread, peak current, and emittance, and increases with beam quality. The saturation length also increases with electron beam degradation and is reduced as the beam quality sharpens.

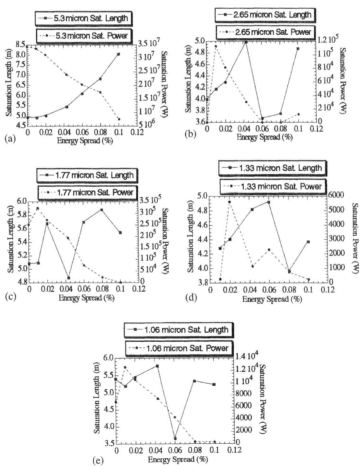

Fig. 5. Saturation length (m) and saturated power (W) versus energy spread (%) for the (a) 5.3 μm, (b) 2.65 μm, (c) 1.77 μm, (d) 1.33 μm, and (e) 1.06 μm output radiation.

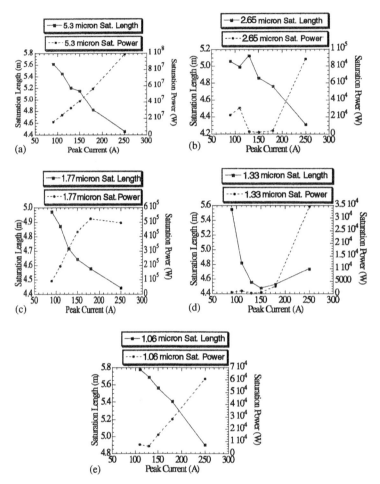

Fig. 6. Saturation length (m) and saturated power (W) versus peak current (A) for the (a) 5.3 μm, (b) 2.65 μm, (c) 1.77 μm, (d) 1.33 μm, and (e) 1.06 μm output radiation.

5. Conclusion

The nonlinear harmonic output in all single-pass, high-gain FELs is important, as shorter wavelengths are achievable. The harmonic output from HGHG is particularly interesting, however, since even shorter wavelengths can be achieved in such a configuration, and they are imprinted for coherence unlike the noisy SASE case. (The same is true when using the (seeded) Bonifacio et al. two-undulator scheme [15] as well.) In this presentation, the harmonic output from HGHG was shown in simulation by sensitivity studies of the fundamental and four higher

harmonics of the radiative section of the BNL/APS HGHG experiment. Harmonic measurements are ongoing at this experiment and have proven comparable to theory/simulation [16].

This harmonic experiment will also be performed for the fundamental and second harmonic with wavelengths of 530 and 265 nm, respectively, corresponding to an electron beam energy of ≈ 217 MeV in a purely SASE mode of operation at the Advanced Photon Source's SASE FEL at Argonne National Laboratory [7,17].

It is the desire of the light source community to achieve shorter wavelengths, on the order of 1 Å,

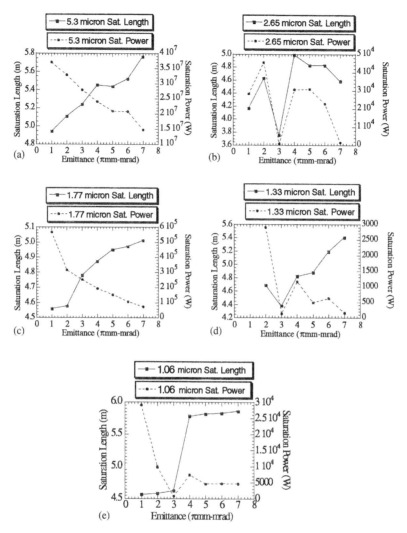

Fig. 7. Saturation length (m) and saturated power (W) versus emittance (π mm-mrad) for the (a) 5.3 μm, (b) 2.65 μm, (c) 1.77 μm, (d) 1.33 μm, and (e) 1.06 μm output radiation.

and various prototypical experiments other than HGHG are underway [18] using the fundamental radiation. To achieve even shorter wavelengths with lower electron beam energies, the nonlinear harmonics, which are generated in self-amplified spontaneous emission (SASE), amplifier, and HGHG FEL schemes, will be important. In addition, the use of soft X-ray seed lasers [19] along with combinations of the above schemes will also be important in obtaining these shorter wavelengths [20–22].

Acknowledgements

Many thanks are extended to our HGHG collaboration friends Marcus Babzien, Adnan Doyuran, and Timur Shaftan for helping with the experimental measurements.

This work is supported by the U.S. Department of Energy, Office of Basic Energy Sciences, under Control No. W-31-109-ENG-38.

The activity and computational work for H.P. Freund is supported by Science Applications

International Corporation's Advanced Technology Group under IR&D sub-project 01-0060-73-0890-000.

References

[1] L.-H. Yu, et al., The status of a high-gain harmonic generation experiment at the accelerator test facility, Proceedings of the IEEE 1999 Particle Accelerator Conference, 1999, p. 2470.

[2] L.-H. Yu, et al., Nucl. Instr. and Meth. A 445 (2000) 301.

[3] L.-H. Yu, et al., Science 289 (2000) 932.

[4] R. Bonifacio, L. DeSalvo, P. Pierini, Nucl. Instr. and Meth. A 293 (1990) 627.

[5] H.P. Freund, S.G. Biedron, S.V. Milton, IEEE J. Quant. Electron. 36 (2000) 275.

[6] S.G. Biedron, H.P. Freund, S.V. Milton, 3D FEL code for the simulation of a high-gain harmonic generation experiment, in: H. E. Bennett, D. H. Dowell (Ed.), Free-Electron Laser Challenges II; Proc. SPIE 3614 (1999) 96.

[7] H.P. Freund, S.G. Biedron, S.V. Milton, Nucl. Instr. and Meth. A 445 (2000) 53.

[8] Z. Huang, K.-J. Kim, Phys. Rev. E 62 (2000) 7295.

[9] L.-H. Yu, Phys. Rev. A 44 (1991) 5178.

[10] I. Ben-Zvi, et al., Nucl. Instr. and Meth. A 318 (1992) 208.

[11] J.W. Hansen, G.A. Lange, A.S. Rostad, R.L. Woods, System aspects of communications TWTAs on how to deal with the tube manufacturer to your best advantage, Hughes Aircraft Company Electron Dynamics Division Applications Note, August 1992.

[12] HGHG Design Handbook, BNL Report, (1996).

[13] H.P. Freund, Phys. Rev. E 52 (1995) 5401.

[14] S.G. Biedron, H.P. Freund, L.-H. Yu, Nucl. Instr. and Meth. A 445 (2000) 95.

[15] R. Bonifacio, L. de Salvo Souza, P. Pierini, E.T. Scharlemann, Nucl. Instr. and Meth. A 296 (1990) 787.

[16] A. Doyuran, et al., Nucl. Instr. and Meth. A 475 (2001) 260, These Proceedings.

[17] S.V. Milton, et al., Phys. Rev. Lett. 85 (2000) 988.

[18] S.V. Milton, et al., FEL development at the APS: the APS SASE FEL, in Free-Electron Laser Challenges II; H. E. Bennett, D. H. Dowell (Ed.), Proc. SPIE 3614 (1999) 86.
The VISA Collaboration (BNL, LANL, LLNL, SLAC, UCLA), VISA Proposal, Proposal for a SASE-Free-electron laser experiment, VISA, at the ATF Linac, April 1998.
A VUV Free Electron Laser at the TESLA Test Facility at DESY: Conceptual Design Report, DESY Print, TESLA-FEL 95-03, June 1995.
LCLS Design Study Group, Linac Coherent Light Source (LCLS) Design Study Report, SLAC-R-521, 1998).

[19] S.G. Biedron, H.P. Freund, S.V. Milton, Y. Li, Simulation of the fundamental and nonlinear harmonic output from an FEL amplifier with a soft X-ray seed laser, Proceedings of seventh European Accelerator Conference (EPAC), 26–30 June, 2000, Vienna, p. 729.

[20] G. Dattoli, P.L. Ottaviani, J. Appl. Phys. 86 (1999) 5331.

[21] S.G. Biedron, S.V. Milton, H.P. Freund, Nucl. Instr. and Meth. A 475 (2001) 401, These Proceedings.

[22] J. Wu, Nucl. Instr. and Meth. A 475 (2001) 104, These Proceedings.

ELSEVIER

Nuclear Instruments and Methods in Physics Research A 475 (2001) 127–131

NUCLEAR
INSTRUMENTS
& METHODS
IN PHYSICS
RESEARCH
Section A

www.elsevier.com/locate/nima

Two-color operation in high-gain free-electron lasers

H.P. Freund[a],*, P.G. O'Shea[b]

[a] *Science Applications International Corporation, 1710 Goodridge Drive, McLean, VA 22102, USA*
[b] *Department of Electrical and Computer Engineering, and Institute for Plasma Research, University of Maryland College Park, MD 20742, USA*

Abstract

Two-color operation in free-electron laser (FEL) amplifiers is studied using a 3D polychromatic simulation. We study the growth of two seed signals at closely spaced wavelengths within the gain band as well as a discrete spectrum of sidebands that are outside the gain band. The sidebands grow parasitically with growth rates higher than the seeded waves due to bunching in the seeded waves. An example of an X-ray FEL is discussed; however, the physics is applicable to all spectral ranges. © 2001 Elsevier Science B.V. All rights reserved.

PACS: 41.60.Cr; 52.75.Ms

Keywords: Free-electron lasers; Side bands; Intermodulation

Two-color operation of free-electron lasers (FELs) at closely spaced wavelengths is of considerable interest. Applications exist over a broad range of wavelengths such as pump–probe experiments [1], multiple wavelength anomalous scattering [2], or any process where there is a large change in cross-section over a narrow wavelength range [3]. Previous demonstrations of two-color operation were low gain oscillators in the infrared, and fell into two categories: (1) pulse switching mode, and (2) segmented undulator. In the pulse switching case [4], two different undulators are used and the switching rate is typically on the order of 10 Hz between wavelengths. The alternative scheme involves using a two part undulator [5] with a different gaps and undulator parameters allowing simultaneous operation at two wavelengths.

We consider a high-gain FEL amplifier seeded at two closely spaced two wavelengths using a single-segment undulator. These two seeds fall within the gain-bandwidth of the FEL, and each is assumed to have a bandwidth much less than the gain-bandwidth. Because the seeds can be controlled independantly, two-color operation can occur not only simultaneously, but also sequentially in any temporal combination determined by the switching mode of the seeds and the electron beam pulse structure. Both colors will travel along the same path at the FEL exit. For some applications this feature may be undesirable; however, the colors can be separated using dispersive elements.

The specific example considered is the Linac Coherent Light Source (LCLS) at the Stanford Linear Accelerator Center [6]; hence, the analysis has implications for future 4th generation light

---

*Corresponding author. Tel.: +1-202-767-0034; fax: +1-202-734-1280.

E-mail address: freund@mmace.nrl.navy.mil (H.P. Freund).

0168-9002/01/$ - see front matter © 2001 Elsevier Science B.V. All rights reserved.
PII: S 0 1 6 8 - 9 0 0 2 (0 1) 0 1 5 5 5 - 8

sources. Note that X-ray lasers are not available at the wavelengths of interest for the LCLS; however, an alternate scheme for generating the seeds is possible that relies on a proposal [7] in which the undulator is broken into two segments. X-rays across the entire FEL gain band are generated in the first segment by the amplification of noise (the so-called SASE or self-amplified spontaneous emission process). A monochromator selects a narrow band, which is reintroduced in synchronism with the electron beam in the second undulator segment. This proposal can be generalized to include a pair of monochromators that would select two wavelengths within the FEL gain band. The concept is relevant to all spectral ranges, however, and applications in the infrared (where seed lasers are available) may be of equal or greater interest.

The intermodulation of the electron beam by the two seeded waves gives rise to a nonlinear interaction that generates a discrete spectrum of sidebands extending beyond the FEL gain band, and which have growth rates higher than that of the seeded waves. This is a well-known phenomenon in traveling wave tubes used in communications [8] that operate by a similar axial bunching mechanism, where the intermodulation between multiple signals gives rise to unwanted nonlinearities. While these nonlinearities may or may not be important for two-color operation in FELs, the effects on the output spectrum are important and must be considered.

The 3D nonlinear, polychromatic simulation code MEDUSA [9,10] is used to simulate the nonlinear interaction of two narrow-band closely spaced driven signals. MEDUSA employs planar wiggler geometry and treats the electromagnetic field as a superposition of Gauss–Hermite modes using a source-dependent expansion [11]. Different wavelength components are constrained to be harmonics of a fundamental frequency ω_0. The field equations are integrated simultaneously with the 3D Lorentz force equations for an ensemble of electrons. No wiggler average is imposed on the orbit equations. A complete description of the formulation is given in Refs. [9,10].

We use ω_1 and ω_2 to denote the frequencies of the two seeds ($\omega_2 > \omega_1$), and choose ω_0 such that

$\omega_1 = N_1\omega_0$ and $\omega_2 = N_2\omega_0$ where N_1 and N_2 are integers. Since the two frequencies must be closely spaced in order for both waves to be within the gain band, N_1 and N_2 are typically much greater than unity. Since nonlinear intermodulations between the two waves are possible, we must also include a spectrum of beat waves separated from the seeded waves in frequency by harmonics of $\Delta\omega = \omega_2 - \omega_1$. In practice this spectrum must be truncated, and we include only the two nearest neighbor beat waves above and below the seeded waves. The beat waves at $2\omega_2 - \omega_1$ and $2\omega_1 - \omega_2$ are referred to as 3rd order intermodulation products, and those at $3\omega_2 - 2\omega_1$ and $3\omega_1 - 2\omega_2$ are 5th order intermodulation products. Since the beat waves are generated by nonlinear interactions between the driven waves, the intermodulation products grow rapidly. For driven signals with comparable grow rates, the gain lengths of the nth order intermodulation products scale as $1/n$ times the gain lengths of the seeded waves. This is similar to the scaling found in the nonlinear generation of harmonics in FELs [9]. While these intermodulation products have important implications on the linearity of the traveling wave tube amplifiers, the issue of linearity is not of prime importance in 4th generation light sources. However, the saturated powers in the seeded waves and their intermodulation products are of interest. For the example of interest, the electron beam has a 14.35-GeV energy, a 3400 A peak current, a normalized emittance of 1.5π mm mrad, and an energy spread of 0.02%. The peak, on-axis wiggler field strength is 13.2 kG with a period of 3.0 cm. We consider a matched electron beam that has been shown to result in the optimal growth rate [12]. The resonance wavelength for these parameters is in the vicinity of 1.5 Å. While the LCLS employs a multi-segment wiggler design, we choose a single segment, parabolic-pole-face wiggler model in the analysis for simplicity. This presents no handicap, however, since this is not a detailed design study. A seeded FEL requires drive powers in excess of the spontaneous synchrotron radiation. The spontaneous emission is ≈ 44 W per gain length [13]; hence, the seeded waves require powers of the order of several kW. A discussion of the monochromatic case is a necessary prelude to the multi-color analysis. The gain length versus wavelength is shown in Fig. 1 for energy spreads of 0%, 0.01% and 0.02%. The minimum gain length and the corresponding wavelength varies with energy spread from 6.21–8.04 m and 1.4920–1.4922 Å, respectively. The gain bandwidth is $\Delta\omega/\omega \approx 0.15\%$. The saturated power is plotted versus wavelength in Fig. 2. The saturated power depends sensitively on both the energy spread and the wavelength. Indeed, the maximum saturated power decreases from about 25 GW for a vanishing energy spread to 22 GW when the energy spread is 0.01% and finally to 8 GW at 0.02%. It is clear from Figs. 1 and 2 that the gain length begins to increase rapidly and the saturated power to decrease rapidly as the energy spread increases from 0.01% to 0.02%.

We consider two seeds with wavelengths of 1.4918 and 1.4925 Å, as well as the nearest neighbor beat waves at 1.4911, 1.4933, 1.4903, and 1.4940 Å. For convenience, we assume that seeded powers of 300 kW which overwhelms the spontaneous emission. We chose these two seed wavelengths without loss of generality. If we had chosen a different pair within the gain band, the results would be similar, except that the spacing of the sidebands would be altered. Note that, as shown in Fig. 1, with an energy spread of 0.02% the beat waves are outside the FEL gain band and would not normally grow without the nonlinear beating of the seeded waves. The evolution of the

seeded waves and the intermodulation products are shown in Fig. 3.

The gain lengths and saturated powers for each wavelength are shown in Table 1. The gain lengths for the two seeded waves of 9.373 m (1.4918 Å) and 9.013 m (1.4925 Å) correspond to the monochromatic case shown in Fig. 1. However, the saturated powers for the seeds are lower as would be expected in the case of the simultaneous growth of two waves. The interesting feature is the growth of the intermodulation products. If Γ_1 and Γ_2 denote the growth rates for the seeded waves, then the 3rd (5th) order intermodulation products will have growth rates of $2\Gamma_1 + \Gamma_2$ and $2\Gamma_2 + \Gamma_1$

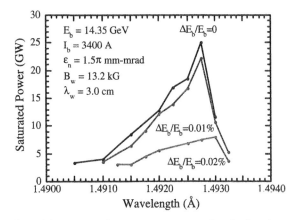

Fig. 2. The saturated power versus wavelength for three choices of the energy spread.

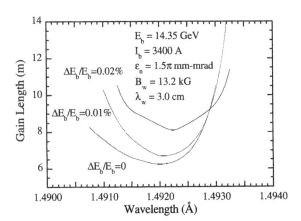

Fig. 1. The gain length versus wavelength for three choices of the energy spread.

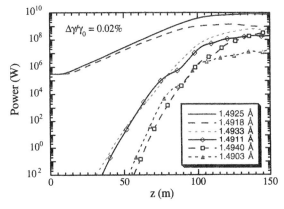

Fig. 3. Evolution of the power in the seeded waves and the beat waves.

Table 1
Gain lengths and saturated powers for the seeded and beat waves

Wavelength (Å)	Gain length (m)	Saturated power
1.4903	1.727	14.4 MW
1.4911	2.459	198 MW
1.4918	9.277	1.03 GW
1.4925	8.982	8.02 GW
1.4933	3.106	636 MW
1.4940	1.755	227 MW

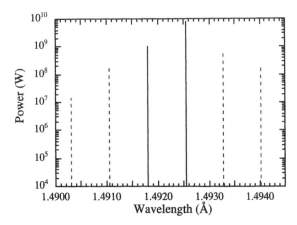

Fig. 4. Powers in the two driven wavelengths as well as the beat waves after 150 m. The solid lines correspond to the seeded waves and the dashed lines refer to the sidebands.

$(3\Gamma_1 + 2\Gamma_2$ and $3\Gamma_2 + 2\Gamma_1)$, and this is approximately reproduced in simulation. Observe that if $\Gamma_1 = \Gamma_2$ then the gain length of the nth order scales as $1/n$ times the gain length of the seeded waves.

The scaling law for the beat wave growth rates can be illustrated mathematically by noting that the seeded waves nonlinearly give rise to an oscillatory current. If the electric field of the seeded waves is $E_{\text{seed}} = E_1(z)\cos(k_1 z - \omega_1 t) + E_2(z)\cos(k_2 z - \omega_2 t)$, then the beating of these waves drives a nonlinear current

$$J_{NL} = \sum_{k,l=1}^{2} \sum_{m,n=1}^{\infty} \sigma_{k,l,m,n}^{(NL)} E_k^m E_l^n \\ \times \cos^m(k_k z \pm \omega_k t)\cos^n(k_l z \pm \omega_l t) \quad (10)$$

where $\sigma^{(NL)}$ is the nonlinear conductivity whose detailed form is not required to obtain the scaling law. The 3rd order intermodulation products arise from the terms in $E_1^2 E_2^1$ and $E_1^1 E_2^2$ that yield oscillatory components varying in frequency as $2\omega_1 - \omega_2$ and $2\omega_2 - \omega_1$ and, if $E_1 \approx \exp(\Gamma_1)$ and $E_2 \approx \exp(\Gamma_2)$, have growth rates of $2\Gamma_1 + \Gamma_2$ and $2\Gamma_2 + \Gamma_1$.

The power levels achieved by the intermodulation products are substantial. The beat waves exhibit a complex variation with z reflecting details of the bunching process. Hence, the beat waves saturate at different positions making a detailed comparison of the power levels difficult. However, at the saturation point of the fastest growing seed (i.e., $z = 138$ m), the 3rd order intermodulation products achieve peak power levels of 636 and 198 MW while the 5th order intermodulation products reach 227 and 14.4 MW. A comparison of these powers is shown in Fig. 4. It is evident that

while the seeds dominate, the beat waves must be taken into account in any two-color experiment.

In summary, we have studied two-color operation in high-gain FEL amplifiers using a 3D nonlinear polychromatic simulation code. The growth of two closely spaced seeds with wavelengths within the FEL gain band was analyzed along with a discrete spectrum of sidebands (not within the gain band) that grow due to the bunching caused by the seeded waves. For the example presented, the sidebands grow rapidly and reach substantial power levels; hence, the presence of these sidebands must be considered in any two-color experiment. We reiterate that while an X-ray example has been presented the basic physical mechanism is relevant to FEL operation in all spectral ranges, and applications to two-color experiments in the infrared may be of equal or greater interest.

This work was supported by a grant (N0017300P3053) from the US Navy. The computational work was supported by Science Applications International Corporations Advanced Technology Group under IR&D subproject 01-0060-73-0890-000.

References

[1] R. Prazeres, et al., Nucl. Instr. and Meth. A407 (1998) 464.

[2] J. Drenth, Principles of Protein X-ray Crystallography, 2nd Edition, Springer, New York, 1999.

[3] E. Johnson, personal communication.

[4] T.I. Smith, et al., Nucl. Instr. and Meth. A407 (1998) 151.

[5] D.A. Jaroszynski, et al., Phys Rev. Lett. 72 (1994) 2387.

[6] M. Cornacchia, et al., Free-electron laser challenges. in: P.G. O'Shea, H. Bennett (Eds.), Proceedings. SPIE, Vol. 2988, 1997, p. 5.

[7] J. Feldhaus, et al., Opt. Commun. 140 (1997) 341.

[8] A.S. Gilmour Jr., Principles of Traveling Wave Tubes, Artech House, Boston, 1994.

[9] H.P. Freund, T.M. Antonsen Jr., Principles of Free-electron Lasers, 2nd Edition, Chapman & Hall, London, 1996.

[10] H.P. Freund, S.G. Biedron, S.V. Milton, IEEE J. Quantum Electron 36 (2000) 275.

[11] P. Sprangle et al., Phys. Rev. Lett. 59 (1987) 202; Phys. Rev. A 36 (1987) 2773.

[12] H.P. Freund, P.G. O'Shea, Phys. Rev. Lett. 80 (1998) 520.

[13] M. Xie, Proc. IEEE 1995 Particle Accelerator Conference, Vol. 183, IEEE Cat. No. 95CH35843, 1995.

ELSEVIER

Nuclear Instruments and Methods in Physics Research A 475 (2001) 132–136

NUCLEAR
INSTRUMENTS
& METHODS
IN PHYSICS
RESEARCH
Section A

www.elsevier.com/locate/nima

Tunable non-relativistic FEL

A.A. Silivra*

6340 Quadrangle, Suite 210, BOPS Inc., Chapel Hill, NC 27514, USA

Abstract

A super broadband regime of operation of a long wavelength Free Electron Laser (FEL) has been studied. It has been shown that because of the physical nature of the underlying instability, a frequency region within which amplification and/or generation of the electromagnetic waves occurs, ranges from frequencies slightly below to many times above the resonant FEL frequency. The upper limit of the frequency band is imposed by the thermal spread in an electron beam. Therefore, in this regime, the FEL operating frequency is determined by the frequency characteristics of a FEL resonator, and can be tuned over a broad band without changing the electron beam energy or wiggler period. A theory of this regime has been developed and available experiments have been discussed. The most striking result of the study is that, although it cannot be understood (and, consequently, was not discovered) without using the relativistic equation of motion, the regime does not rely upon relativism of an electron beam. A non-relativistic implementation of this FEL regime in a submillimeter/THz device has been proposed. © 2001 Elsevier Science B.V. All rights reserved.

PACS: 52.75.Ms; 52.75.Vs; 41.60.Cr

Keywords: Coherent; Tunable; Broad band; Sub-millimeter; Generation; Non-relativistic electron beam

1. Introduction

A Free-Electron Laser (FEL) is the only vacuum electron device that can efficiently operate in the submillimeter–far-infrared wavelength band. Unfortunately, making an electron beam relativistic is a huge toll for obtaining electromagnetic radiation at the wavelengths mentioned above. However, if a certain relation between FEL parameters (such as the guiding and wiggler magnetic fields) holds, the operating frequency band of a FEL becomes extremely broad and, consequently, the operating frequency is no longer determined by the conventional FEL relation. A significant feature of this regime is that the frequency band expands towards higher frequencies. As a result, a much higher than conventional FEL frequency can be achieved without increasing the electron beam energy. In fact, the energy of the electron beam can be lowered to a non-relativistic level, which, thus, transforms the whole device into a device of traditional vacuum electronics.

The existence of unstable solutions in a broad frequency band above the FEL resonance frequency was first found numerically when the FEL dispersion relation was analyzed in Ref. [1]. At

---
*Tel.: +1-919-403-6757; fax: +1-919-402-9442.

E-mail address: silivra@ieee.org (A.A. Silivra).

0168-9002/01/$ - see front matter © 2001 Elsevier Science B.V. All rights reserved.
PII: S 0 1 6 8 - 9 0 0 2 (0 1) 0 1 6 9 6 - 5

about the same time broadband radiation from a FEL was registered experimentally [2]. In ;the experiment, a flat output radiation spectrum over a wavelength band from 6 to 0.9 mm (50–350 GHz) was observed. The results of this experiment were misinterpreted as a consequence of bad electron beam quality. Later, the broadband regime of FEL operation was found theoretically for the FEL with the so-called electromagnetic wiggler [3]. In the follow-up experimental generator [4], a practically flat spectrum of output radiation from 12 to 3.4 mm was observed with an electron efficiency of interaction of about 10%. For the chosen magnitude of the wiggler field, the broadband FEL regime existed within a certain region of the guide field magnitude close to the so-called cyclotron resonance. It should be noted that generally accepted FEL theory predicted a resonant regime of operation for the above mentioned device in the vicinity of 8 mm (38 GHz) and was totally unable to explain radiation with the wavelength shorter than 7 mm with any realistic assumptions about beam parameters. There are strong indications that in several other independent experiments different features related to this FEL regime have been observed although never understood.

2. Model and main equations

To study the underlying physical phenomena, let us consider a model of a transverse-uniform electron beam moving in a combined uniform axial magnetic field and periodic transverse magnetic (wiggler) field $\mathbf{B} = \mathbf{B}_0 + \mathbf{B}_\perp$, $\mathbf{B}_\perp = B_w[\mathbf{e}_x \cos(k_w z) + \mathbf{e}_y \sin(k_w z)]$. As is well known [1], in such a field electrons move along a steady-state trajectory $\mathbf{v} = v_\perp[\mathbf{e}_x \cos(k_w z) + \mathbf{e}_y \sin(k_w z)] + v_\| \mathbf{e}_z$. The constants of motion, or the steady-state transverse $v_\perp$ and longitudinal velocity $v_\|$, are determined as a solution to the following system of equations:

$$v_\perp = \frac{\Omega_w}{\Omega_0 - \gamma k_w v_\|} v_\|, \qquad v_\perp^2 + v_\|^2 = c^2(1 - \gamma^{-2})$$

$$(1)$$

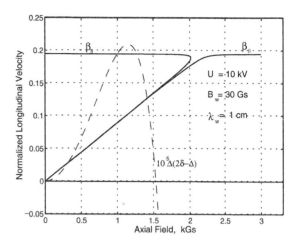

Fig. 1. Normalized longitudinal velocity and waves' coupling parameter vs. axial guide field.

where $\Omega_{0,w} = eB_{0,w}/(mc)$, e, m and γ are the electron charge, mass and relativistic factor, respectively, c is the speed of light.

The system of equations (1) has four solutions in a general case. Realizable solutions that satisfy a condition $v_\| > 0$ are shown in Fig. 1. Further, we will only be interested in the region where $\Delta = \Omega_0/\gamma - k_w v_\| \ll \Omega_0/\gamma$ which is represented by a steeper part of the most right velocity curve in Fig. 1.

Since electrons move along the steady-state trajectory, they form a flow that can be described by the hydrodynamic relativistic equation of motion (or in Euler's coordinates). The electromagnetic field is described by Maxwell's equations, which are coupled with the equation of motion through the electron current term $\mathbf{j} = -en\mathbf{v}$, n is the density of the electron beam.

3. Wave analysis

To find the waves that can exist in the model, the linearization procedure is used. The resulting system of equations has a simpler form if transverse components of the electromagnetic field and electron velocity are expressed via partial amplitudes of right-hand ($A_+ = A_x + iA_y$) and left-hand ($A_- = A_x - iA_y$) circular polarized waves.

The equations for the transverse components have the form

$$\frac{\partial \mathscr{E}_\pm}{\partial z} \mp \frac{i}{c}\frac{\partial \mathscr{B}_\pm}{\partial t} = 0$$

$$\frac{\partial \mathscr{B}_\pm}{\partial z} \pm \frac{i}{c}\frac{\partial \mathscr{E}_\pm}{\partial t} \mp \frac{4\pi e}{c} i n_0 v_\pm = \pm \frac{4\pi e}{c} i \tilde{n} v_\perp e^{\pm i k_w z}$$

$$\left(\frac{\partial}{\partial t} + v_\parallel \frac{\partial}{\partial z}\right) v_\pm \mp \frac{i\Omega_0}{\gamma} v_\pm \pm i k_w v_\parallel \frac{\gamma^2 v_\perp^2}{2c^2} v_\pm$$

$$+ \frac{e}{m\gamma}\left(\left(1 - \frac{v_\perp^2}{2c^2}\right)\mathscr{E}_\pm \pm \frac{iv_\parallel}{c}\mathscr{B}_\pm\right)$$

$$= \frac{ev_\perp^2}{2m\gamma c^2}e^{\pm 2ik_w z}\mathscr{E}_\mp \mp i\frac{\gamma^2 v_\perp^2}{2c^2}k_w v_\parallel e^{\pm 2ik_w z} v_\mp$$

$$\mp ie^{\pm ik_w z}\left(\frac{\Omega_w}{\gamma}v_z + \frac{\gamma^2 v_\parallel v_\perp}{c^2}k_w v_\parallel v_z \pm \frac{iev_\parallel v_\perp}{m\gamma c^2}E_z\right).$$

$$(2)$$

The equations for the longitudinal components have the form

$$\frac{\partial E_z}{\partial t} - 4\pi e\tilde{n} v_\parallel - 4\pi e n_0 v_z = 0$$

$$\frac{\partial E_z}{\partial z} + 4\pi e\tilde{n} = 0$$

$$\left(\frac{\partial}{\partial t} + v_\parallel\frac{\partial}{\partial z}\right)v_z + \frac{e}{m\gamma}(1 - v_\parallel^2/c^2)E_z$$

$$= \frac{ie}{2m\gamma c}\left[v_\perp\left(\mathscr{B}_+ - \frac{iv_\parallel}{c}\mathscr{E}_+\right) - B_w v_+\right]e^{-ik_w z}$$

$$- \frac{ie}{2m\gamma c}\left[v_\perp\left(\mathscr{B}_- + \frac{iv_\parallel}{c}\mathscr{E}_-\right) - B_w v_-\right]e^{ik_w z}.$$

$$(3)$$

It is clearly seen that the system of equations (2) and (3) is consistent if the solution for the transverse waves is proportional to $\exp i[\omega t - (k \mp k_w)z]$ and for the longitudinal waves to $\exp i(\omega t - kz)$. Assuming that coupling coefficients are small enough to be neglected, participating waves can be easily identified.

Thus, the transverse waves are forward and backward electromagnetic waves (of two different polarizations) passively coupled with the fast (wave index +) or slow (wave index −) cyclotron modes of the electron beam. The corresponding

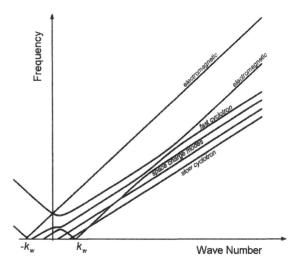

Fig. 2. Dispersion of waves participating in a FEL interaction.

dispersion relation has the form

$$D_\pm \equiv [\omega - (k \mp k_w)v_\parallel \mp \Omega_0/\gamma \pm k_w v_\parallel \gamma^2 \beta_\perp^2/2]$$
$$\times [\omega^2 - c^2(k \mp k_w)^2] - \omega_b^2/\gamma[\omega(1 - \beta_\perp^2/2)$$
$$- (k \mp k_w)v_\parallel] = 0 \qquad (4)$$

where $\omega_b = (4\pi n_0 e/m)^{1/2}$ is the plasma frequency of the electron beam. Note, that because of the transverse velocity modulation caused by the wiggler field, the wave number of the transverse waves has a parametric shift $\mp k_w$.

The longitudinal waves are space charge waves of the electron beam. The dispersion relation for them has the form $(\omega - kv_\parallel)^2 - (1 - v_\parallel^2/c^2)\omega_b^2/\gamma = 0$, which in a non-relativistic case obviously reduces to the well known $(\omega - k_w v_\parallel)^2 = \omega_b^2$. The dispersion relations for all eight waves are sketched in Fig. 2 to make the following consideration clearer.

4. Broadband instability

Besides the conventional FEL instability (developing under synchronism of the electromagnetic mode with the slow space charge mode) and FEL on the slow cyclotron wave [5], there is one more absolutely unique possibility to realize an instability in the system via bringing into synchronism the fast and slow cyclotron waves of the electron beam. It is possible in the vicinity of the so-called cyclotron resonance of the steady-state transverse velocity $v_\perp$, when the electron cyclotron frequency, Ω_0/γ, is close to the bounce frequency of electrons in the wiggler field, $k_w v_\parallel$. In this case, the denominator in the first formula (1) is small, $\Delta = \Omega_0/\gamma - k_w v_\parallel \ll k_w v_\parallel$, and the transverse velocity $v_\perp$ is relatively high (although a parameter $\beta_\perp^2 = v_\perp^2/c^2$ is always small $\beta_\perp^2 \ll 1$) that has a profound impact on the dispersion of the cyclotron waves. Indeed, under these same circumstances, a relativistic correction to the dispersion of cyclotron waves, $\delta = k_w v_\parallel \gamma^2 \beta_\perp^2 /2$, becomes large enough to significantly reduce or even fully compensate the offset of the cyclotron waves and bring them into synchronism which, in its turn, causes a broadband instability of the waves to develop.

The simplest way to verify this statement is to analyze the system (2) and (3) at asymptotically high frequencies where the principal waves' component is $v_\pm$. Neglecting in Eq. (2) other wave components as well as coupling between transverse and longitudinal waves, one can obtain the following system of equations:

$$(\omega - k v_\parallel - \Delta + \delta)v_+ = -\delta e^{2ik_w z} v_-$$
$$(\omega - k v_\parallel + \Delta - \delta)v_- = \delta e^{-2ik_w z} v_+. \qquad (5)$$

The corresponding dispersion equation $(\omega - k v_\parallel - \Delta + \delta)(\omega - k v_\parallel + \Delta - \delta) = -\delta^2$ has unstable roots if $\delta^2 > (\Delta - \delta)^2$. Or, equivalently, if $k_w v_\parallel \gamma^2 \beta_\perp^2 > \Omega_0/\gamma - k_w v_\parallel$, the fast and slow cyclotron waves of the electron beam are brought into synchronism which results in instability with the temporal rate $-\mathrm{Im}(\omega) = \sqrt{(2\delta - \Delta)\Delta}$ that does not depend on the wave frequency.

As can be seen from the sketch in Fig. 2, if the synchronism of waves results in an instability, the instability is extremely broadband. A numerical solution of the non-reduced dispersion equation obtained by solving the full system (2) and (3) fully justifies this conclusion.

6. Discussion and conclusion

Inspection of the system (5) and its solution shows that it does not exploit beam relativism. In the non-relativistic limit ($\gamma \to 1$) the criterion for this instability reduces to

$$k_w v_\parallel \beta_\perp^2 > \Omega_0 - k_w v_\parallel \qquad (6)$$

which still contains a small relativistic parameter $\beta_\perp^2 \ll 1$. This is why the relativistic equation of motion was needed to describe the process. In fact, unlike the traditional FEL theory, only the transverse motion of electrons must be treated as relativistic, while relativism of the longitudinal motion is not essential at all in this case. The magnitude of the parameter $(2\delta - \Delta)\Delta/(k_w c)^2$ that determines the instability rate, was computed for the non-relativistic beam and magnetic fields from an example given below. It is shown in Fig. 1 by the dashed line. Note, that it has a distinct maximum that is fully consistent with the experimental results [4].

The numerical solution of the non-reduced dispersion equation obtained from Eqs. (2) and (3) is needed to investigate the instability in more detail, in particular, the dependence of the instability rate on the beam plasma frequency shown in Fig. 3.

It should be noted that the instability extends toward indefinitely high frequencies only within the frame of the hydrodynamic consideration performed above. As is well known, this approach fails at frequencies $\omega_{\mathrm{lim}} \approx \omega_p \, v_\parallel/V_T$, where V_T is the thermal velocity spread in the electron beam.

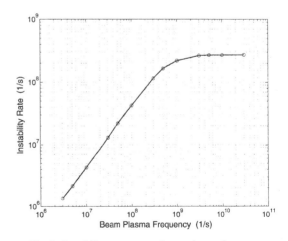

Fig. 3. Instability rate versus beam-plasma frequency.

For high quality non-relativistic electron beams, the limiting wavelength can be as short as 10 μm.

As an example of non-relativistic implementation, let us consider a system with the following parameters: beam voltage $U = 10$ kV, beam current $I = 100$ mA, axial magnetic field $B_0 = 1$ kGs, wiggler magnetic field $B_w = 30$ Gs, wiggler period 1 cm ($k_w = 6.28$ cm^{-1}). Under these parameters the normalized transverse velocity, $\beta_\perp = 0.17$ and longitudinal velocity, $\beta_\| = 0.09$, while the radius of the corresponding steady-state trajectory, $R = 3$ mm. Note, that the radius, R of the electron trajectory is rather large, so that in a real device the electron beam forms a single helix winding around the device axis. Nevertheless, the electron beam occupies the area in which magnetic fields are close to the idealistic model (or, in other words, dependence of the magnetic fields on the transverse coordinates is not significant). For $\omega_b =$ 10^9 s^{-1}, the instability rate is $-\mathrm{Im}\,\omega \simeq 10^{-3}$ $k_w c \simeq 2.2 \times 10^8$ s^{-1} and is suitable for implementation in generators and amplifiers.

Acknowledgements

The author would like to acknowledge support from BOPS, Inc for putting this paper together and filing a corresponding patent.

References

[1] I.B. Bernstein, L. Friedland, Phys. Rev. A-23 (1981) 816.
[2] K.L. Felch, et al., IEEE Quantum Electronics QE-17 (1981) 1354.
[3] N.Ya. Kotsarenko, A.A. Silivra, Izv. VUZov, Radiofizika 30 (1988) 1279 (in Russian).
[4] Ju.B. Victorov, et al., Opt. Commun. 79 (1) (1990) 81.
[5] A.A. Silivra, Nucl. Instr. and Meth. A 375 (1996) 248.

ELSEVIER

Nuclear Instruments and Methods in Physics Research A 475 (2001) 137–142

NUCLEAR
INSTRUMENTS
& METHODS
IN PHYSICS
RESEARCH
Section A

www.elsevier.com/locate/nima

On the anti-resonance effect in the reversed field free electron laser

J.T. Donohue[a,*], J.L. Rullier[b]

[a] Centre d'Etudes Nucléaires de Bordeaux-Gradignan, BP 120, 33175 Gradignan, France
[b] Commissariat à l'Energie Atomique Centre d'Etudes d'Ile de France, BP 12, 91680 Bruyères-le-Chatel, France

Abstract

In their reversed field experiment, Conde and Bekefi observed a dramatic decrease in RF power output when the anti-resonance condition was satisfied, i.e., cyclotron frequency = −FEL frequency. They further noticed that even with no injected RF signal, beam transmission also fell at the anti-resonance. We show here that the true cause of the observed orbital instability is not the anti-resonance, but rather the fact that the two independent oscillations about the ideal helical trajectory have frequencies whose ratio is −2. That this would be a source of trouble was pointed out by Cherry in 1925. In the Conde-Bekefi experiment, the condition for anti-resonance coincided with the 2: −1 ratio. By choosing wiggler parameters such that the two effects are dissociated, we demonstrate convincingly that the 2: −1 condition causes the instability, and that a generalized version of Cherry's model provides an adequate analytic description of our numerical results. © 2001 Elsevier Science B.V. All rights reserved.

Keywords: Free electron laser; Synchronization; Coupled oscillators

1. Introduction

In the period 1991–1992, Conde and Bekefi (henceforth CB) [1,2] performed an experiment using a microwave FEL equipped with a helical wiggler and an axial magnetic field. They investigated what happens when the axial field is anti-parallel to the beam, such that the cyclotron and wiggler motions have opposite senses. Their results are discussed by Freund and Antonsen [3], who devote Chapter 13 to the subject, and to which we

refer the reader for details. CB found that in the amplifier regime at 33 GHz, this configuration provided both high power and high conversion efficiency. However, for a small range of axial field, at constant wiggler field, the output power fell dramatically. Numerical calculations using standard FEL trajectory codes predicted that some, but not all, electron trajectories were erratic for this value of the field, and the electrons often struck the walls of the 5.1-mm radius waveguide. This occurred when,

$$eB_0/\gamma mc = -v_z k_w \quad (1)$$

where B_0 is the axial field and v_z the constant axial velocity of an electron tracing an ideal helical trajectory. By that we refer to motion in which the

*Corresponding author. Tel.: +33-5-57-12-07-73; fax: +33-5-57-12-07-77.
E-mail address: donohue@cenbg.in2p3.fr (J.T. Donohue).

0168-9002/01/$ - see front matter © 2001 Elsevier Science B.V. All rights reserved.
PII: S 0 1 6 8 - 9 0 0 2 (0 1) 0 1 5 5 6 - X

electron moves at fixed ρ and in such a way that $\phi - k_w z$ remains constant (i.e., $\dot{\phi} = v_z k_w$), where (ρ, ϕ, z) denote the cylindrical coordinates of the electron. The minus sign indicates that the axial field and the axial velocity are anti-parallel. When this condition is satisfied in the standard operating mode, without the minus sign, one speaks of the gyro-magnetic resonance, whence the name anti-resonance here. In principle, there is no stability problem in the reversed field configuration, in the sense that the oscillation frequencies Ω_+ and Ω_- are neither imaginary nor small (explicit expressions for Ω_+^2 and Ω_-^2 may be found in Ref. [3], Eq. (2.62)). In particular, at the anti-resonance, neither the exactly soluble one-dimensional analysis, nor the stability analysis given in Ref. [3] predict any anomalies in the trajectories. Nevertheless, both experiment and numerical calculations show clear evidence that at the anti-resonance, a substantial fraction of the electron trajectories are perturbed.

Over the past few years, several articles have appeared which argue that the anti-resonance condition is the cause of the anomalous trajectories [4–6]. As a justification for their arguments, the authors show numerically calculated trajectories, which clearly exhibit the observed anomalous behavior. Here we assert that this is fortuitous, and that the true explanation of the loss of electrons lies elsewhere, namely in the fact that Ω_+/Ω_- is approximately -2. In previous work (called DR hereafter) Ref. [7], we showed that Ω_+ is always positive, while in the reversed field $\Omega_- < 0$. Negative frequency means that an increase in the amplitude of oscillation corresponds to a decrease in the energy. When this occurs, a small perturbation can lead to simultaneous growth of both oscillations even though the total energy is conserved. Furthermore, one may show that whenever Eq. (1) is approximately true,

$$\Omega_+/\Omega_- = -1 + eB_0/\gamma m c v_z k_w + O(B_w/B_0)^2 \quad (2)$$

which means that if the anti-resonance condition is satisfied, the ratio Ω_+/Ω_- will be approximately -2, since typically, $B_w \ll B_0$. When this ratio is precisely -2, as was shown by Cherry [8] in 1925, an arbitrarily small cubic coupling can totally

de-stabilize two harmonic oscillators. Whittaker's book [9] presents a brief discussion of Cherry's results. Here we generalize Cherry's model to two arbitrary frequencies, while retaining integrability. We get a powerful predictive tool, and use it to prove our claim that it is the frequency ratio $= -2$, and not anti-resonance, which causes the anomalous trajectories. For brevity, we henceforth use the symbols "$2:-1$" to denote the situation where the frequency ratio is approximately -2.

We have seen from Eq. (2) that if the anti-resonance condition is satisfied the ratio of the frequencies is approximately -2 up to corrections of order $(B_w/B_0)^2$. In the experiment of CB, the value of B_w was 0.147 T. Given the beam energy (0.75 MeV) and wiggler step size (3.18 cm), the anti-resonance occurs for $|B_0| = 0.756$ T, while the $2:-1$ condition is satisfied at $|B_0| = 0.745$ T. These values of the axial field are so close that it would be difficult to attribute any putative effects to one cause or the other. To illustrate this, we show in Fig. 1 the curves in the (B_0, B_w) plane for which the anti-resonance (dotted) and $2:-1$ (solid) occur. For larger values of B_w, the two curves separate enough for a meaningful difference in their numerical predictions of electron trajectories to be observable. The horizontal lines indicate the value of B_w, 0.147 T, used by CB, and our choice, $B_w = 0.260$ T, which we used in our simulation.

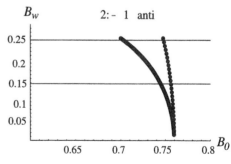

Fig. 1. Curves in the (B_0, B_w) plane along which the anti-resonance (dotted) and $2:-1$ (solid) conditions hold, for beam energy = 0.75 MeV. Horizontal lines indicate the CB value of 0.147 T, and our test value, 0.26 T.

2. Cherry's Hamiltonian

In Whittaker's treatise, one finds an account of work by Cherry. The Hamiltonian may be written as

$$H_C = \omega\big((p_1^2 + q_1^2)/2 - p_2^2 - q_2^2\big)$$

$$+ \alpha\big(q_2(q_1^2 - p_1^2) - 2p_2 q_1 p_1\big) \tag{3}$$

where the p_j and q_j denote canonical momenta and coordinates, ω is the natural frequency of one oscillator, -2ω that of the other and α is a cubic coupling. Cherry found a particular solution

$$q_1 = \sqrt{2}\sin(\omega t + \eta)/\alpha(t + \delta),$$

$$p_1 = \sqrt{2}\cos(\omega t + \eta)/\alpha(t + \delta)$$

$$q_2 = \sin(2\omega t + 2\eta)/\alpha(t + \delta),$$

$$p_2 = -\cos(2\omega t + 2\eta)/\alpha(t + \delta) \tag{4}$$

where δ and η are arbitrary constants. This solution diverges as $t \to -\delta$, whereas for $t \to \pm\infty$, it approaches the fixed-point of the Hamiltonian, $p_j = q_j = 0$. Note that the amplitudes of both oscillators can increase without bound while maintaining a fixed total energy ($H_C = 0$ in this example).

Although the example presented by Cherry might appear contrived, it is possible to relax the exact $2:-1$ condition on the frequencies and still find an exact solution. To do so we generalize Cherry's Hamiltonian H_C to the following expression:

$$H_G = H_C + \varepsilon(p_2^2 + q_2^2) \tag{5}$$

where the unperturbed frequency of oscillator 2 is now $2(\varepsilon - \omega)$. To solve the equations of motion we make the following canonical transformation:

$$P_1 = (p_1^2 + q_1^2)/2 - p_2^2 - q_2^2, \quad Q_1 = -\arctan(p_1/q_1)$$

$$P_2 = (p_2^2 + q_2^2)/2, \quad Q_2 = -2\arctan(p_1/q_1)$$

$$- \arctan(p_2/q_2) \tag{6}$$

where the Hamiltonian may be written in terms of the new variables as

$$H_G = \omega P_1 + 2\varepsilon P_2 + \alpha\sqrt{2P_2}(P_1 + 2P_2)\cos Q_2. \tag{7}$$

Since it depends neither on Q_1 nor t, there exist two conserved quantities, P_1 and H_G, and the system is integrable. Hamilton's equation for the quantity P_2 may be written as

$$\frac{dP_2}{dt} = \pm\alpha\sqrt{2}((P_1 + 2P_2)^2 P_2$$

$$- (H_G - \omega P_1 - 2\varepsilon P_2)^2/2\alpha^2)^{1/2} \tag{8}$$

where the sign is fixed by $\sin Q_2/|\sin Q_2|$. The solution to this differential equation is

$$P_2(t) = \wp\left(\pm\sqrt{2}\alpha t + C, g_2, g_3\right)$$

$$+ \varepsilon^2/6\alpha^2 - P_1/3 \tag{9}$$

where $\wp$ denotes the Weierstrassian elliptic function (see, for example Ref. [10]), and C is a constant. The parameters g_2 and g_3 are given by

$$g_2 = \big(P_1^2 - 6\varepsilon(H_G - \omega P_1)/\alpha^2 - 4\varepsilon^2 P_1/\alpha^2 + \varepsilon^4/\alpha^4\big)/3$$

$$g_3 = \big(2P_1^3 + 15\varepsilon^2 P_1^2/\alpha^2 - 12\varepsilon^4 P_1/\alpha^4 + 2\varepsilon^6/\alpha^6\big)/54$$

$$+ (H_G - \omega P_1)((4\varepsilon P_1 + 3(H_G - \omega P_1))/\alpha_2$$

$$- 2\varepsilon^3/\alpha^4)/6. \tag{10}$$

The behavior of the Weierstrassian elliptic function depends crucially on the sign of the discriminant $\Delta = g_2^3 - 27g_3^2$. If $\Delta < 0$, then $e_1 < \wp < \infty$ for real arguments, where e_1 denotes the unique real root of the cubic equation $4x^3 - g_2 x - g_3 = 0$. The quantity P_2 will ultimately increase without bound. However, if $\Delta > 0$, then there are three real roots $e_1 > e_2 > e_3$, and one may either have $\wp$ oscillating in the interval (e_3, e_2), or describing an unbounded trajectory in (e_1, ∞). Which of these alternatives occurs depends on $P_2(0)$. The time required for P_2 to increase from its minimum value to ∞ or to oscillate between e_3 and e_2 is fixed by the real half-period of $\wp$, and may be computed using the *Mathematica* [11] program Weierstrass half-periods. We thus have a complete analytic solution of the generalized Cherry problem.

3. Trajectories in the helical FEL and Cherry's problem

The connection between the preceding section and electron trajectories in a helical FEL with an axial guide field is furnished by DR, and we summarize briefly the main results. For an arbitrary trajectory, there is a conserved quantity, called the helical invariant, P_z. For each value of P_z, the electron's Hamiltonian, $H(\vec{p}, \vec{x}) = \sqrt{1 + (\vec{p} + e\vec{A}(\vec{x}))^2}$, (dimensionless units are used here) has a unique stable fixed-point (where its first derivatives vanish). After performing suitable canonical transformations DR obtained the expression

$$H = H_f + \Omega_+ |A_+|^2 + \Omega_- |A_-|^2 + \cdots \qquad (11)$$

where the ellipsis indicates higher order terms, and the fixed-point energy H_f, the frequencies $\Omega_\pm$ and the normal mode coefficients $A_\pm$ are known functions of P_z. In order to relate this approach to the generalized Cherry Hamiltonian, we make the following identification:

$$A_+ = \sqrt{P_2}\,e^{i(Q_2 - 2Q_1)}, \quad \Omega_+ = 2(\varepsilon - \omega)$$

$$A_- = \sqrt{P_1 + 2P_2}\,e^{iQ_1}, \quad \Omega_- = \omega. \qquad (12)$$

All that remains is to determine the coefficient of the appropriate cubic term, which is of the form $\sqrt{8}\alpha\Re(A_+ A_-^2)$. It then follows from Eqs. (5) and (12) that

$$\alpha = \frac{\partial^3 H}{\sqrt{8}\partial A_+ \partial A_-^2} \qquad (13)$$

where the derivative is to be evaluated at the fixed-point. In this way, we can take an arbitrary trajectory and calculate the initial values for the generalized Cherry Hamiltonian. Knowing the exact solution of the latter, we can predict whether or not the trajectory will be unstable.

In the CB experiment the following relation is a good approximation,

$$x + iy = \rho_f e^{iz} + \sqrt{2/\Omega_0}\left(\sqrt{P_2}\,e^{i(z + Q_2 - 2Q_1)}\right.$$

$$\left. - \sqrt{P_1 + 2P_2}\,e^{i(z - Q_1)}\right) \qquad (14)$$

where z, P_2, Q_1, and Q_2 are known functions of time. In addition, both $z - Q_1$ and Q_2 are slowly varying with time, which leads to a simple description of the motion in the transverse plane. The first term represents the ideal fixed-radius helical motion, the second a circular motion of similar angular speed but rotating in the opposite direction with a time varying radius, and the third is a very slow circular motion, again with a time-varying radius. The resulting motion is a rotating ellipse whose axes and center vary slowly with time. If the quantity P_2 becomes sufficiently large, the electron will strike the beam tube.

4. A numerical test

To justify our claim, we have performed a numerical test using the NDSolve program of *Mathematica* to solve the Lorentz force equations for the motion of the electrons. We chose the on-axis wiggler field $B_w = 0.260\,\mathrm{T}$, and we used two different axial fields, $B_0 = 0.7477\,\mathrm{T}$ for the anti-resonance, and $0.7029\,\mathrm{T}$ for $2:-1$. The hypothetical beam is made up of 49 electrons, with seven different initial transverse positions and seven different transverse velocities. These are shown in Fig. 2, where the center of each hexagon represents the ideal helical orbit (for the beam kinetic energy of $0.75\,\mathrm{MeV}$), and the offsets in transverse position and velocity are 0.15 and 0.05, respectively. The circle in Fig. 2a is the beam tube. For each electron the initial values were used to compute the parameters of the generalized Cherry Hamiltonian, and thus g_2 and g_3. For anti-resonance, 14 of

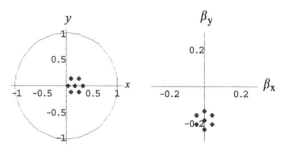

Fig. 2. Initial conditions in the transverse position and velocity planes for 49 test electrons.

the 48 non-ideal trajectories are predicted to be unbounded while for the $2:-1$ condition all 48 are predicted by the model to be unbounded. To test these predictions, both hypothetical beams were run through the wiggler for $0 < t < 350$, which corresponds to roughly 50 periods, the length of the wiggler used by CB. For anti-resonance, no electrons struck the wall. In contrast, only 7 of the 48 non-ideal electrons survived the trip when the $2:-1$ condition held. We conclude that the analytic model errs on the pessimistic side, in that some predicted failures do not occur, but not conversely.

We show in Fig. 3 a section of the hypothetical beams at four different longitudinal positions, for anti-resonance (top) and $2:-1$ (bottom). In the former the original hexagonal shape of the beam is reproduced periodically out to 45 periods. This demonstrates conclusively that the anti-resonance condition, once it is sufficiently removed from the $2:-1$ instability, does not cause electron loss. In contrast, for the $2:-1$ condition, the beam steadily loses electrons as it progresses down the wiggler. In Fig. 4 we show the projection on the $x-y$ plane of two trajectories with identical initial offsets, one at anti-resonance, and the other at

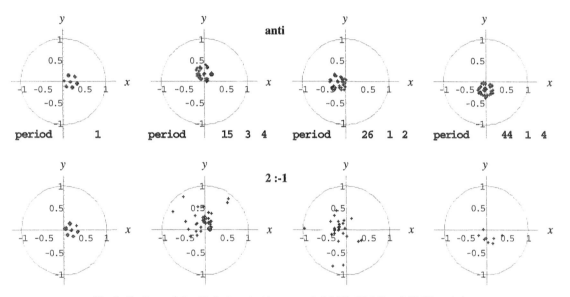

Fig. 3. Sections of the 49 electron test beams at 1, 15 3/4, 26 1/2 and 44 1/4 periods.

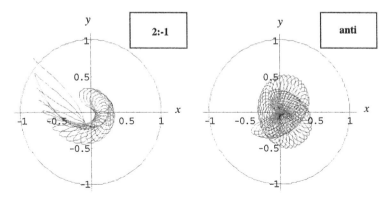

Fig. 4. Examples of a projected $x-y$ trajectory for anti-resonance and $2:-1$ frequency ratio.

$2:-1$. The behavior in both cases is similar to that predicted by Eq. (14), with ellipses of time varying eccentricity and elongation precessing about a slowly moving center. However, in one case the axis of the ellipse oscillates, whereas in the other it increases until the electron strikes the wall. Of 48 non-ideal $2:-1$ trajectories 41 are qualitatively similar to that shown here, as are all 48 anti-resonance trajectories. To conclude, we believe that our numerical simulation justifies our claim that the true cause of the phenomena seen by CB is not the anti-resonance.

References

[1] M.E. Conde, G. Bekefi, Phys. Rev. Lett. 67 (1991) 3082.
[2] M.E. Conde, Ph.D. Dissertation, Mass. Inst. of Tech., 1992, unpublished.
[3] H.P. Freund, T.M. Antonsen Jr., Principles of Free Electron Lasers, 2 Edition, Chapman & Hall, London, 1996.
[4] K.R. Chu, A.T. Lin, Phys. Rev. Lett. 67 (1991) 3235.
[5] X. Shu, Phys. Lett. A 162 (1992) 247.
[6] S.C. Zhang, J. Elgin, Phys. Rev. E 55 (1997) 4864.
[7] J.T. Donohue, J.L. Rullier, Phys. Rev. E 49 (1994) 766.
[8] T.M. Cherry, Trans. Camb. Phil. Soc. XXIII (1925) 199.
[9] E.T. Whittaker, Analytical Dynamics of Particles and Rigid Bodies, Cambridge Univ. Press, Cambridge, U.K. 1960, p. 412.
[10] E.T. Whittaker, G.N. Watson, Modern Analysis, Cambridge Univ. Press, Cambridge, U.K, 1962.
[11] Wolfram Research, Inc., Mathematica, Version 4, Champaign, IL, 1999.

ELSEVIER

Nuclear Instruments and Methods in Physics Research A 475 (2001) 143–146

NUCLEAR
INSTRUMENTS
& METHODS
IN PHYSICS
RESEARCH
Section A

www.elsevier.com/locate/nima

Modulated desynchronism in a free-electron laser oscillator

Oscar G. Calderón, Takuji Kimura*, Todd I. Smith

Stanford Picosecond FEL Center, Hansen Experimental Physics Laboratory, Stanford University, Stanford, CA 94305-4085, USA

Abstract

We study experimentally and theoretically, the effects of desynchronism modulation on short pulse free-electron laser (FEL) oscillators. We find that the output power and the micropulse length of the FEL beam oscillate periodically at the modulation frequency and the minimum micropulse length can be significantly shorter than that obtained without modulation. The FEL can operate during part of the modulation cycle in the normally inaccessible portion of the output power curve where the FEL gain is less than the cavity loss. © 2001 Elsevier Science B.V. All rights reserved.

PACS: 41.60.Cr; 42.60.Jf; 42.65.Re; 42.65.Sf

Keywords: Free-electron laser dynamics; Modulated desynchronism; Short pulse generation; Inaccessible region

1. Introduction

During the past few years there has been much interest in the topic of generating and controlling short FEL pulses [1–3]. To date, short pulse generation has been achieved while maintaining the two fundamental FEL parameters (the cavity detuning δL, and the losses α_0) constant. In this work we show a new method of controlling the pulse length of an FEL by modulating the cavity detuning [4,5]. We begin with the experimental observation that when δL is modulated by a few microns at a rate of 40 kHz, our FEL micropulse length varies from a minimum of 300 fs to a maximum of 800 fs. When the modulation is turned off, the micropulse length is 700 fs.

Throughout this paper the FEL wavelength is taken to be 5 μm.

2. Experimental results

The Stanford FEL parameters are shown in Table 1. Here the optical power signal is recorded as a function of time in Fig. 1(a) and (b) for different modulation levels. The flat signals were taken without modulation and are shown for comparison. At high modulation level ($\delta L_m = 3.7$ μm), the measured micropulse length ranges from 300 to 800 fs (FWHM).

The magnetic chicanes in the FEL beam line are non-isochronous, i.e., electron pulses with a higher energy will have a shorter transit time through the chicanes than lower energy pulses. The effect is calculated to be 0.03 ps/keV. Therefore modulation of the beam energy is translated into modulation of the electron bunch repetition

*Corresponding author. Tel.: +1-650-725-0450; fax: +650-725-8311.

E-mail address: takuji@stanford.edu (T. Kimura).

Table 1
Stanford FEL paramaters

Undulator periods, N_w	72
Undulator wavelength, λ_w	3.1 cm
RMS wiggler parameter, a_w	0.83
Radiation wavelength, λ	5 μm
Slippage, λN_w	360 μm
Peak electron current, I_b	13 A
Electron pulse length, L_b	390 μm
Repetition frequency, f_{FEL}	11.8 MHz
Cavity loss, α_0	3%
CW gain parameter, g_0	0.26
Gain parameter, $\gamma = (L_b/L_s)g_0$	0.283
Normalized loss, $\alpha = \alpha_0/\gamma$	10.6%

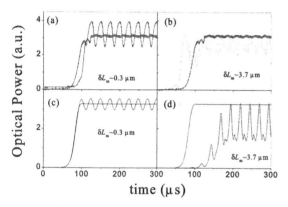

Fig. 1. (a,b) Experimental and (c,d) theoretical optical power signals as a function of time with different levels of energy modulation. The average desynchronism is $\delta L_0 = 1.5\,\mu m$ ($v_0 = 0.03$). The flat curves correspond to situations without energy modulation.

frequency. Since the change in the repetition frequency has the same effect as the cavity length detuning, the total cavity detuning can be written as $\delta L = \delta L_0 + (\delta L_m/2)\sin(\omega_m t)$, where ω_m is the modulation frequency and δL_m the modulation amplitude [4]. Typical energy modulation that is applied during the measurement corresponds to an effective cavity length modulation of about 3 μm peak to peak. This is comparable to the entire length of the cavity detuning curve for our FEL, and is likely to be responsible for the modulation of the micropulse length.

3. Theoretical model

3.1. Sinusoidal modulation of cavity detuning

$$\frac{\partial A}{\partial \tau} - v\frac{\partial A}{\partial \xi} + \frac{1}{2}\frac{\partial^2 A}{\partial \tau^2} + \frac{\gamma v^2}{2}\frac{\partial^2 A}{\partial \xi^2} + \frac{\alpha}{2}A = B \tag{1}$$

$$\frac{\partial B}{\partial \xi} = -iD, \quad \frac{\partial D}{\partial \xi} = -A - iSB - 2iQD + 2iQ^2B$$

$$\frac{\partial Q}{\partial \xi} = -[AB^* + \text{c.c.}], \quad \frac{\partial S}{\partial \xi} = -[AD^* + \text{c.c.}].$$

We use the short pulse FEL oscillator equations [3,5].

In these equations A is the optical field, $\xi = (ct - z)/L_s$ is the position within the optical pulse in units of the slippage length, $\tau = \gamma n$, where n is the pass number, and $v = 2\delta L/(L_s\gamma)$ is the normalized cavity detuning which can be written as $v = v_0 + (\Delta v_m/2)\sin(\omega_m\tau/(f_{FEL}\gamma))$. We have used in these equations a collective variable description [6]: $B = \langle \exp(-i\theta)\rangle$, $D = \langle p\exp(-i\theta)\rangle$, $Q = \langle p\rangle$, and $S = \langle p^2\rangle$, with $p = \partial_\xi\theta$, and θ the electron phase. The second derivative terms in Eq. (1) were introduced in a previous work [5], and they were found to be relevant to explaining the behavior of the system when the cavity detuning was modulated with large amplitude.

Eqs. (1) have been solved numerically. The initial beam is assumed to be monochromatic and unbunched, and the initial beam energy satisfies the FEL resonant condition. Hence the initial conditions are $B = D = Q = S = 0$ at $\xi = 0$. For $v < 0$ or > 0 we assume $A(\xi = 0, \tau) = A_0$ or $A(\xi = 1, \tau) = A_0$, where A_0 characterizes the level of spontaneous emission, $A(\xi, \tau = 0) = A_0$. We define the dimensionless output energy P as

$$P(\tau) = \int_{-\infty}^{+\infty} d\xi |A(\xi, \tau)|^2. \tag{2}$$

The theoretical optical power signal is shown in Figs. 1(c) and (d) as a function of time for two different modulation levels. These signals exhibit a periodic oscillation with the same frequency as the modulation. At $\delta L_m = 3.7\,\mu m$ calculations show that the micropulse length varies between 300 and 800 fs. This is in good agreement with the experimental result.

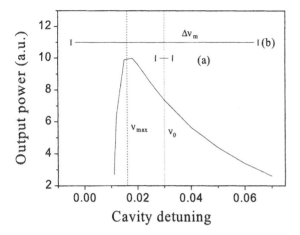

Fig. 2. Output power versus cavity detuning. The modulation amplitudes (a) $\Delta v_m = 0.006$, (b) $\Delta v_m = 0.07$ used in the simulation, the average cavity detuning $v_0 = 0.03$ and the maximum power cavity detuning v_{max} are shown.

In order to understand the role of the modulated desynchronism in FEL dynamics, we show in Fig. 2 the output power versus desynchronism curve obtained without modulation. The horizontal lines represent the range of detuning values due to various modulation amplitudes. When modulation is applied, the FEL must evolve as it attempts to follow the changing conditions. It is reasonable to assume that it moves toward the steady condition corresponding to the instantaneous detuning value. Therefore, for small modulation around detuning v_0 the power will decrease when $v > v_0$ and will increase when $v < v_0$. This explains the periodicity observed in the output power. When the modulation amplitude becomes large enough such that the detuning reaches very small values ($v < v_{max}$) outside the emission region, the equilibrium power is no longer monotonic in v, and the output power curve develops higher frequency structures (see Figs. 1(b) and (d)). The transient evolution in a short pulse FEL oscillator with perfect synchronism ($v = 0$) has been analyzed by Piovella [7].

3.2. Operation in the inaccessible region

The dynamics of the FEL operating in the normally inaccessible region ($v < v_{max}$) are very complex. To qualitatively understand these

dynamics, we investigate numerically the case where the cavity detuning changes as a step function. The detuning is initially set to a constant detuning $v_1 > v_{max}$. After the FEL reaches a steady state, the detuning is abruptly changed to a value $v_2 < v_{max}$. Fig. 3 shows the evolution of the optical micropulse shape during this process. As the change is applied, there is a strong tendency for all the optical energy to concentrate at the trailing edge of the electron bunch, causing the micropulse length to decrease. An interesting observation is that during a short period of time of approximately 6 μs, while the micropulse length is decreasing, the micropulse energy rises briefly above the steady state value before decreasing to zero. After choosing different combinations of the initial and final detuning values, we find that the highest micropulse energy occurs when the initial detuning is set to v_{max}, which corresponds to the maximum steady-state power, and the final detuning is set close to zero. Specifically for our case, the initial and final values are $v_1 = 0.017$ and $v_2 = 0.0001$, respectively. The evolution of the micropulse energy is shown in Fig. 4. The initial increase of the micropulse energy is clearly visible.

This phenomenon provides us the possibility to operate the FEL with average micropulse energy higher than that obtainable in the steady-state operation. A possible scheme is to employ a periodic step function as shown in Fig. 5(a).

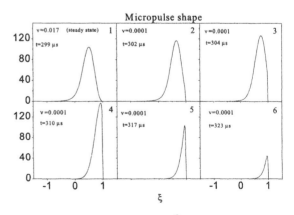

Fig. 3. Optical micropulse shape, $|A|^2$, versus the optical pulse position, ξ, recorded at different instances within a modulation cycle.

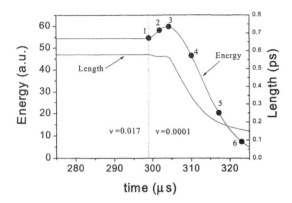

Fig. 4. Evolution of the micropulse energy and the micropulse length. The circles correspond to the cases plotted in Fig. 3.

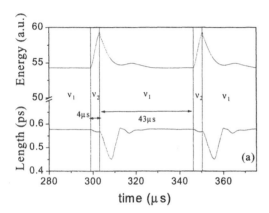

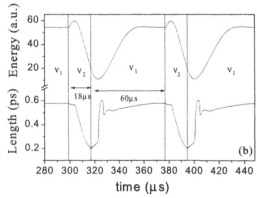

Fig. 5. Evolution of the micropulse energy and the micropulse length for a periodic step function with detuning values of $v_1 = 0.017$, $v_2 = 0.0001$.

Plotted are the simulated results of the micropulse energy and the macropulse length. Here the operation time in the normally inaccessible high-power region is about $4\,\mu s$, while the operation

time in the steady-state region is about $43\,\mu s$. Another interesting application will be the possibility to obtain a very short micropulse with a relatively high energy (see Fig. 5(b)). In that case a longer operation time in the normally inaccessible region is required.

4. Conclusions

We demonstrated a new method of controlling the micropulse length of a short pulse FEL oscillator by modulating the cavity detuning. Numerical simulations based on short pulse FEL equations show good agreement between experiment and theory. The dynamics of the FEL micropulse in the normally inaccessible region is investigated by studying a step-function change in cavity detuning. A correlation is found between optical pulse duration and optical energy when modulation is applied, i.e., shorter pulses are possible at the expense of reduced optical energy. Possible application of this technique is discussed.

Acknowledgements

This work was supported in part by ONR Grant N000140-94-1-1024. O.G.C. is supported by Becas Complutense "Flores Valles" and Becas Complutense "del Amo".

References

[1] D.A. Jaroszynski, A.F.G. van der Meer, P.W. van Amersfoort, W.B. Colson, Nucl. Instr. and Meth. A 296 (1990) 480.
[2] R.J. Bakker, G.M.H. Knippels, A.F.G. van der Meer, D. Oepts, D.A. Jaroszynski, P.W. van Amersfoort, Phys. Rev. E 48 (1993) R3256.
[3] N. Piovella, P. Chaix, G. Shvets, D.A. Jaroszynski, Phys. Rev. E 52 (1995) 5470.
[4] T. I. Smith, T. Kimura, In: G. R. Neil, S. V. Benson (Eds.), Proceedings of the 20th International Free-Electron Lasers Conference, Virginia, 1998, Elsevier Science BV, Amsterdam, 1999, p. II–9.
[5] O.G. Calderón, T. Kimura, T.I. Smith, Phys. Rev. ST Accel. Beams 3 (2000) 090701.
[6] R. Bonifacio, F. Casagrande, L. De Salvo Souza, Phys. Rev. A 33 (1986) 2836.
[7] N. Piovella, Phys. Rev. E 51 (1995) 5147.

ELSEVIER

Nuclear Instruments and Methods in Physics Research A 475 (2001) 147–152

NUCLEAR
INSTRUMENTS
& METHODS
IN PHYSICS
RESEARCH
Section A

www.elsevier.com/locate/nima

Model and simulation of wide-band interaction in free-electron lasers

Yosef Pinhasi*, Yuri Lurie, Asher Yahalom

Department of Electrical and Electronic Engineering, Faculty of Engineering, The College of Judea and Samaria, P.O. Box 3, Ariel 44837, Israel

Abstract

A three-dimensional, space-frequency model for simulation of interaction in free-electron lasers (FELs) is presented. The model utilizes an expansion of the total electromagnetic field (radiation and space-charge waves) in terms of transverse eigenmodes of the waveguide, in which the field is excited and propagates. The mutual interaction between the electron beam and the electromagnetic field is fully described by coupled equations, expressing the evolution of mode amplitudes and electron beam dynamics.

Based on the three-dimensional model, a numerical particle simulation code was developed. A set of coupled-mode excitation equations, expressed in the frequency domain, are solved self-consistently with the equations of particles motion. Variational numerical methods were used to simulate excitation of backward modes. At present, the code can simulate FELs operation in various modes: spontaneous (shot-noise) and self-amplified spontaneous emission, super-radiance and stimulated emission, all in the non-linear Compton or Raman regimes. © 2001 Elsevier Science B.V. All rights reserved.

1. Introduction

Several numerical models have been suggested for three-dimensional simulation of the free-electron laser (FEL) operation in the non-linear regime [1–10]. Unlike previously developed steady-state models, in which the interaction is assumed to be at a single frequency (or at discrete frequencies), the approach presented in this paper considers a continuum of frequencies, enabling solution of non-stationary, wide-band interactions in electron devices operating in the linear (small-signal) and non-linear (saturation) regimes. Solu-

tion of excitation equations in the space–frequency domain inherently takes into account dispersive effects arising from cavity and beam loading. The model is based on a coupled-mode approach expressed in the frequency domain [11] and used in the WB3D particle simulation code to calculate the total electromagnetic field excited by an electron beam drifting along a waveguide in the presence of a wiggler field of FEL.

The unique features of the present model enable one to solve non-stationary interactions taking place in electron devices such as spontaneous and super-radiant emissions in a pulsed beam FEL, shown in Fig. 1. We employed the code to demonstrate spontaneous and super-radiant emissions excited when a bunch of electrons passes through a wiggler of an FEL. Calculations of the

*Corresponding author. Tel.: +972-3-9066197; fax: +972-3-9066238.

E-mail address: yosip@eng.tau.ac.il (Y. Pinhasi).

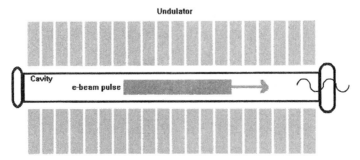

Fig. 1. The FEL scheme.

power and energy spectral distribution in the frequency domain were carried out. The temporal field was found by utilizing a procedure of inverse Fourier transformation. Super-radiance in the special limit of "grazing" (where dispersive waveguide effects play a role) was also investigated.

2. Dynamics of the particles

The state of the particle i is described by a six-component vector, which consists of the particle's position coordinates $\mathbf{r}_i = (x_i, y_i, z_i)$ and velocity vector $\mathbf{v}_i$. Here (x, y) are the transverse coordinates and z is the longitudinal axis of propagation. The velocity of each particle, in the presence of electric $\mathbf{E}(\mathbf{r}, t)$ and magnetic $\mathbf{B}(\mathbf{r}, t) = \mu \mathbf{H}(\mathbf{r}, t)$ fields, is found from the Lorenz force equation:

$$\frac{\mathrm{d}}{\mathrm{d}t}(\gamma_i \mathbf{v}_i) = -\frac{e}{m}[\mathbf{E}(\mathbf{r}_i, t) + \mathbf{v}_i \times \mathbf{B}(\mathbf{r}_i, t)] \qquad (1)$$

where e and m are the electron charge and mass, respectively. The fields represent the total (DC and AC) forces operating on the particle, and they include also the self-field due to space-charge. The Lorentz relativistic factor γ_i of each particle is found from the equation for kinetic energy:

$$\frac{\mathrm{d}\gamma_i}{\mathrm{d}t} = -\frac{e}{mc^2}\mathbf{v}_i \cdot \mathbf{E}(\mathbf{r}_i, t) \qquad (2)$$

where c is the velocity of light.

The equations are rewritten, such that the coordinate of the propagation axis z becomes the independent variable, by replacing the time derivative $\mathrm{d}/\mathrm{d}t = v_{zi}\,\mathrm{d}/\mathrm{d}z$. This defines a transformation of variables for each particle, which enables one to write the three-dimensional equations of motion in terms of z

$$\frac{\mathrm{d}\mathbf{v}_i}{\mathrm{d}z} = \frac{1}{\gamma_i}\left\{ -\frac{e}{m}\frac{1}{v_{zi}}[\mathbf{E}(\mathbf{r}_i, t) + \mathbf{v}_i \times \mathbf{B}(\mathbf{r}_i, t)] - \mathbf{v}_i\frac{\mathrm{d}\gamma_i}{\mathrm{d}z}\right\} \qquad (3)$$

$$\frac{\mathrm{d}\gamma_i}{\mathrm{d}z} = -\frac{e}{mc^2}\frac{1}{v_{zi}}\mathbf{v}_i \cdot \mathbf{E}. \qquad (4)$$

The time it takes a particle to arrive at a position z, is a function of the time t_{0i} when the particle entered at $z = 0$, and its instantaneous longitudinal velocity $v_{zi}(z)$ along the path of motion:

$$t_i(z) = t_{0_i} + \int_0^z \frac{1}{v_{zi}(z')}\,\mathrm{d}z'. \qquad (5)$$

3. The driving current

The distribution of the current in the beam is determined by the position and the velocity of the particles

$$\mathbf{J}(\mathbf{r}, t) = -\sum_i q_i\, \mathbf{v}_i\, \delta(x - x_i)\delta(y - y_i)\delta[z - z_i(t)]$$

$$= -\sum_i q_i\left(\frac{\mathbf{v}_i}{v_{zi}}\right)\delta(x - x_i)\delta(y - y_i)\delta[t - t_i(z)] \qquad (6)$$

here q_i is the charge of the ith macro particle in the simulation. The Fourier transform (in the positive frequency domain) of the current density is given by

$$\tilde{\mathbf{J}}(\mathbf{r}, f) = 2u(f) \int_{-\infty}^{+\infty} \mathbf{J}(\mathbf{r}, t)e^{-j2\pi ft}\, dt$$

$$= -2u(f) \sum_i q_i \left(\frac{\mathbf{v}_i}{v_{z_i}}\right)$$

$$\times \delta(x - x_i)\delta(y - y_i)e^{-j2\pi ft_i(z)} \tag{7}$$

here

$$u(f) = \begin{cases} 1, & f \geq 0 \\ 0, & f < 0 \end{cases}$$

is the step function.

This Fourier transform of the current (7) is substituted in the following excitation equations to find the evolution of the electromagnetic fields.

4. The electromagnetic field

The Fourier transform of the transverse component of the total electromagnetic field is given at the frequency domain as a superposition of waveguide transverse eigenmodes

$$\tilde{\mathbf{E}}_\perp(\mathbf{r}, f) = \sum_q \{C_{+q}(z, f)e^{-jk_{zq}z} + C_{-q}(z, f)e^{jk_{zq}z}\}$$

$$\times \mathscr{E}_{q\perp}(x, y)$$

$$\tilde{\mathbf{H}}_\perp(\mathbf{r}, f) = \sum_q \{C_{+q}(z, f)e^{-jk_{zq}z} - C_{-q}(z, f)e^{jk_{zq}z}\}$$

$$\times \mathscr{H}_{q\perp}(x, y) \tag{8}$$

and the expression for the longitudinal component of the electromagnetic field is found to be

$$\tilde{E}_z(\mathbf{r}, f) = \sum_q \{C_{+q}(z, f)e^{-jk_{zq}z} - C_{-q}(z, f)e^{+jk_{zq}z}\}$$

$$\times \mathscr{E}_{qz}(x, y) + \frac{j}{2\pi f\varepsilon}\tilde{J}_z(\mathbf{r}, f)$$

$$\tilde{H}_z(\mathbf{r}, f) = \sum_q \{C_{+q}(z, f)e^{-jk_{zq}z} + C_{-q}(z, f)e^{+jk_{zq}z}\}$$

$$\times \mathscr{H}_{qz}(x, y) \tag{9}$$

where

$$k_{z_q} = \sqrt{\left(\frac{2\pi f}{c}\right)^2 - k_{\perp_q}^2}$$

($k_{\perp_q}$ is the cut-off wave number of mode q and $C_{+q}(z, f)$ and $C_{-q}(z, f)$ are the qth mode's amplitude corresponding to the forward and backward waves, respectively. Eqs. (8) and (9) describe the total transverse and longitudinal electromagnetic field (radiation and space-charge waves) [11].

The evolution of the qth mode amplitudes $C_{\pm q}(z, f)$ is found after substitution of the current distribution (7) into the scalar differential excitation equation:

$$\frac{d}{dz}C_{\pm q}(z, f) = \mp \frac{1}{2\mathscr{N}_q(f)}e^{\pm jk_{zq}z}$$

$$\times \int\int \left[\left(\frac{Z_q}{Z_q^*}\right)\tilde{\mathbf{J}}_\perp(\mathbf{r}, f) + \hat{\mathbf{z}}\tilde{J}_z(r, f)\right]$$

$$\cdot \mathscr{E}_{\pm q}^*(x, y)\, dx\, dy$$

$$= \pm \frac{1}{\mathscr{N}_q(f)}e^{\pm jk_{zq}z} \sum_i q_i e^{-j2\pi ft_i(z)}$$

$$\times \left\{\frac{\varsigma_q}{v_{z_i}}\mathbf{v}_{\perp_i} \cdot \mathscr{E}_{\pm q\perp}^*(x_i, y_i) + \mathscr{E}_{\pm qz}^*(x_i, y_i)\right\} \tag{10}$$

here

$$\mathscr{N}_q(f) = \int\int \left[\mathscr{E}_{q\perp} \times \tilde{\mathscr{H}}_{q\perp}^*\right] \cdot \hat{z}\, dx\, dy$$

is the power normalization of mode q, and

$$\varsigma_q \equiv \frac{Z_q}{Z_q^*} = \begin{cases} +1, & \text{propagating modes} \\ -1, & \text{cut-off modes.} \end{cases}$$

The total electromagnetic field is found by inverse Fourier transformation of (8) and (9)

$$\mathbf{E}(\mathbf{r}, t) = \mathscr{R}\left\{\int_0^\infty \tilde{\mathbf{E}}(\mathbf{r}, f)e^{+j2\pi ft}df\right\}$$

$$\mathbf{H}(\mathbf{r}, t) = \mathscr{R}\left\{\int_0^\infty \tilde{\mathbf{H}}(\mathbf{r}, f)e^{+j2\pi ft}df\right\}. \tag{11}$$

The energy flux spectral distribution (defined in the positive frequency domain $f \geq 0$) is given by

$$
\frac{dW(z)}{df} = \frac{1}{2}\Re\left\{ \int\int \left[\tilde{\mathbf{E}}(\mathbf{r},f) \times \tilde{\mathbf{H}}^*(\mathbf{r},f) \right] \cdot \hat{\mathbf{z}}\, dx\, dy \right\}
$$

$$
= \frac{1}{2} \overset{\text{Propagating}}{\sum_q} \left[|C_{+q}(z,f)|^2 - |C_{-q}(z,f)|^2 \right]
$$

$$
\times \Re\{\mathcal{N}_q(f)\}
$$

$$
+ \overset{\text{Cut-off}}{\sum_q} \Im\left\{ C_{+q}(z,f)C^*_{-q}(z,f) \right\}
$$

$$
\times \Im\{\mathcal{N}_q(f)\}. \tag{12}
$$

5. The variational principle

The solution of Eqs. (3), (4) and (10) for forward waves is done by integrating the equations in the positive z-direction for a given boundary conditions at the point $z = 0$. For backward waves the natural physical boundary conditions are given at the end of the interaction region $z = L_{\mathrm{w}}$ and the direction of the integration is the negative z-direction.

In order to take into account excitation of both forward and backward waves, we introduce a variational functional

$$
F = \int_0^{L_{\mathrm{w}}} \left[C_{-q}(z,f)\frac{dC_{+q}(z,f)}{dz} - C_{+q}(z,f)\alpha_q(z,f) \right.
$$
$$
\left. + C_{-q}(z,f)\beta_q(z,f) \right] dz \tag{13}
$$

where

$$
\alpha_q(z,f) \equiv -\frac{1}{2\mathcal{N}_q(f)}e^{-jk_{zq}z}
$$

$$
\times \int\int \left[\left(\frac{Z_q}{Z^*_q}\right)\tilde{\mathbf{J}}_\perp(\mathbf{r},f) + \hat{\mathbf{z}}\,\tilde{J}_z(\mathbf{r},f) \right]
$$

$$
\cdot \mathscr{E}^*_{-q}(x,y)\, dx\, dy
$$

$$
\beta_q(z,f) \equiv +\frac{1}{2\mathcal{N}_q(f)}e^{+jk_{zq}z}
$$

$$
\times \int\int \left[\left(\frac{Z_q}{Z^*_q}\right)\tilde{\mathbf{J}}_\perp(\mathbf{r},f) + \hat{\mathbf{z}}\,\tilde{J}_z(\mathbf{r},f) \right]
$$

$$
\cdot \mathscr{E}^*_{+q}(x,y)\, dx\, dy. \tag{14}
$$

The variational derivative of the above functional is

The variational derivative of the above functional is

$$
\delta F = \int_0^{L_{\mathrm{w}}} \left[\delta C_{-q}(z,f)\left(\frac{dC_{+q}(z,f)}{dz} + \beta_q(z,f)\right) \right.
$$
$$
- \delta C_{+q}(z,f)\left(\frac{dC_{-q}(z,f)}{dz} + \alpha_q(z,f)\right)
$$

$$
\left. + \frac{d(C_{-q}(z,f)\delta C_{+q}(z,f))}{dz} \right] dz. \tag{15}
$$

For arbitrary variations $\delta C_{\pm q}(z,f)$ the functional minimizes (i.e., $\delta F = 0$) if and only if Eqs. (10) are satisfied, and the boundary term is

$$
\delta F_B \equiv \int_0^{L_{\mathrm{w}}} \frac{d(C_{-q}(z,f)\delta C_{+q}(z,f))}{dz} dz
$$
$$
= C_{-q}(z,f)\delta C_{+q}(z,f)\big|_0^{L_{\mathrm{w}}} = 0 \tag{16}
$$

resulting in

$$
C_{-q}(0,f)\delta C_{+q}(0,f) = C_{-q}(L_{\mathrm{w}},f)\delta C_{+q}(L_{\mathrm{w}},f). \tag{17}
$$

This enables solving an amplifier scheme in which the boundary conditions are $C_{-q}(L_{\mathrm{w}},f) = 0$ and $C_{+q}(0,f) = 0$, as well as an oscillator configuration where the boundary conditions are $C_{-q}(0,f) = C_{-q}(L_{\mathrm{w}},f)$ and $C_{+q}(0,f) = C_{+q}(L_{\mathrm{w}},f)$.

Table 1
The operational parameters of millimeter wave free-electron maser

Accelerator	
Electron beam energy	$E_k = 1 \div 3\,\mathrm{MeV}$
Electron beam current	$I_0 = 1\,\mathrm{A}$
Pulse duration	$T = 1\,\mathrm{pS}$
Wiggler	
Magnetic induction	$B_{\mathrm{w}} = 2000\,\mathrm{G}$
Period	$\lambda_{\mathrm{w}} = 4.444\,\mathrm{cm}$
Number of periods	$N_{\mathrm{w}} = 20$
Waveguide	
Rectangular waveguide	$1.01\,\mathrm{cm} \times 0.9005\,\mathrm{cm}$
Mode	TE_{01}

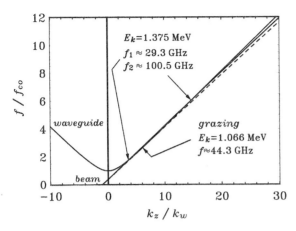

Fig. 2. FEL dispersion curves.

6. Numerical results

We shall use the code to investigate super-radiant emission radiated when an ultra short *e*-beam bunch (with duration of 1 ps, much shorter than the temporal period of the signal) passes through the wiggler of an FEL having operational parameters as given in Table 1. In this case, the power of super-radiant (coherent) emission is much higher than that of the incoherent spontaneous emission [12].

Fig. 2 shows two cases of dispersion relations: when the beam energy is set to 1.375 MeV, there are two separated intersection points between the beam and waveguide dispersion curves, corresponding to the "slow" ($v_{g1} < v_{z0}$) and "fast" ($v_{g2} > v_{z0}$)

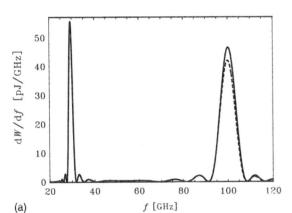

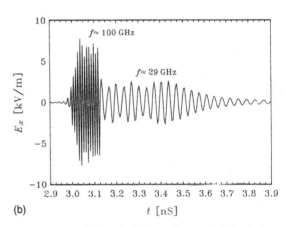

Fig. 3. Super-radiant emission from an ultra short bunch: (a) energy spectrum (analytic calculation and numerical simulation are shown by solid and dashed lines, respectively); (b) temporal wavepacket.

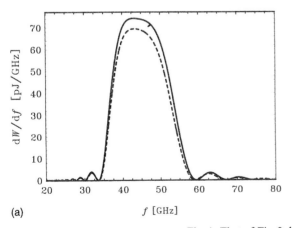

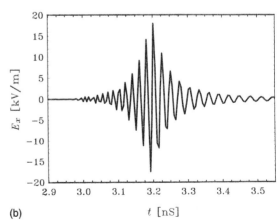

Fig. 4. That of Fig. 3, but in the grazing limit.

synchronism frequencies 29 and 100 GHz, respectively. Lowering the beam energy to 1.066 MeV results in a single intersection (at 44 GHz), where the beam dispersion line is tangential to the waveguide dispersion curve ($v_g = v_{z0}$—"grazing limit").

The calculated spectral density of energy flux in the case of two well-separated solutions is shown in Fig. 3a. The spectrum peaks at the two synchronism frequencies with main lobe bandwidth of

$$\Delta f_{1,2} \approx \frac{1}{\tau_{sp1,2}}, \quad \text{where } \tau_{sp1,2} \approx \left| \frac{L_{\mathrm{W}}}{v_{z_0}} - \frac{L_{\mathrm{W}}}{v_{g1,2}} \right|$$

is the slippage time. The corresponding temporal wave-packet (shown in Fig. 3b) consists of two "slow" and "fast" pulses with durations equal to the slippage times modulating carriers at their respective synchronism frequencies. The spectral bandwidth in the case of grazing, shown in Fig. 4a, is determined by dispersive effects of the waveguide taking into account by the simulation. The corresponding temporal wavepacket is shown in Fig. 4b.

Acknowledgements

The research of the second author (Yu. L.) was supported in part by the Center of Scientific Absorption of the Ministry of Absorption, State of Israel.

References

[1] W.M. Fawley, D. Prosnitz, E. T. Scharlemann, Phys. Rev. A 30 (1984) 2472.

[2] B.D. McVey, Nucl. Instr. Meth. Phys. Res. A 250 (1986) 449.

[3] A.K. Ganguly, H.P. Freund, Phys. Fluids 31 (1988) 387.

[4] S.Y. Cai, A. Bhattacharjee, T.C. Marshall, Nucl. Instr. Meth. Phys. Res. A 272 (1988) 481.

[5] T.M. Tran, J.S. Wurtele, Comput. Phys. Commun. 54 (1989) 263.

[6] T.-M. Tran, J.S. Wurtele, Phys. Reports 195 (1990).

[7] M. Caplan, Nucl. Instr. Meth. Phys. Res. A 318 (1992) 655.

[8] M. Caplan, et al., Nucl. Instr. Meth. Phys. Res. A 331 (1993) 243.

[9] Pallavi Jha, J.S. Wurtele, Nucl. Instr. Meth. Phys. Res. A 331 (1993) 243.

[10] Y. Pinhasi, A. Gover, V. Shterngartz, Phys. Rev. E 54 (1996) 6774.

[11] Y. Pinhasi, A. Gover, Phys. Rev. E 48 (1993) 3925.

[12] Y. Pinhasi, Yu. Lurie, "Generalized theory and simulation of spontaneous and super-radiant emissions in electron devices and free-electron lasers", Phys. Rev. E, in press.

ELSEVIER

Nuclear Instruments and Methods in Physics Research A 475 (2001) 153–157

NUCLEAR
INSTRUMENTS
& METHODS
IN PHYSICS
RESEARCH
Section A

www.elsevier.com/locate/nima

Electromagnetic wave pumped ion-channel free electron laser ☆

Liu Shenggang[a],*, R.J. Barker[b], Gao Hong[a], Yan Yang[a]

[a] University of Electronic Science and Technology of China, Sichuan, Chengdu 610054, China
[b] AFOSR, Washington DC, USA

Abstract

Theoretical study of electromagnetic wave pumped ion-channel free-electron laser (EPIC-FEL) is presented. The physical mechanism responsible for the generation of coherent radiation in the EPIC-FEL is described and the fundamental role of the ponderomotive wave in bunching and trapping the beam is emphasized. The dispersion relation of the EPIC-FEL has been obtained and growth rates are calculated for different parameters. It shows that EPIC-FEL has a very bright future. © 2001 Elsevier Science B.V. All rights reserved.

Keywords: Free electron laser; Ion-channel; Electromagnetic wave pump; Backward wave scattering

1. Introduction

Ion-channel guiding of an electron beam in a free-electron laser can eliminate the need for conventional focusing magnets, thereby reducing the capital and running cost [1]. Moreover, the presence of the ion-channel allows beam currents higher than the vacuum limit and also helps radiation guiding [2]. So there should be higher efficiencies and lower wavelengths for ion-channel guiding FEL. Comparing with wiggler FELs, there is a shorter period in an electromagnetic wave pumped FELs. So under the same injection energy of electrons, one can get a shorter laser wavelength output. Motivated by the benefits, the study of electromagnetic wave pumped ion-channel free-

electron laser (EPIC-FEL) will be more useful. Radiation generation in an ion-channel is one of the most attractive topics in science and technology since 1990 [3–9].

In this paper, we present the linear theory of EPIC-FEL. In the following discussions we also mainly consider the axial bunching for simplicity. Section 2 shows that there is an unstable electron orbit regime when $\Omega_0 = \omega_i$, while there is an ion-channel instability when $\Omega_1 = \omega_i$. Electron axial bunching is introduced in Section 3 and the dispersion relation is obtained in Section 4. Section 5 is the discussions and calculation results. The conclusion is given in Section 6.

2. Electron transverse motion

Consider a relativistic electron (charge $-|e|$, rest mass m, relativistic factor γ_0) moving along the z-axis. A backward scattering electromagnetic (EM)

☆ Work supported partly by AFOSR, and partly by Chinese National Science Foundation.

*Corresponding author. Tel.: +86-28-320-2868; fax: +86-28-325-4131.

E-mail address: liusg@uest.edu.cn (L. Shenggang).

wave as a pump wiggler field is described by

$$A_0 = A_0[e^{i(\omega_0 t + k_0 z)}e_+ + e^{-i(\omega_0 t + k_0 z)}e_-] \tag{1}$$

where A_0 is the amplitude of EM wave field, $k_0 = 2\pi/\lambda_0$, λ_0 is the EM wave wavelength, $e_\pm = (e_x \pm i e_y)/2$. The transverse static field caused by the ion-channel has the form

$$E_i = A_i(x, y, 0) \tag{2}$$

where $A_i = 4\pi|e|n_i$, n_i is the plasma density. In a form similar to the representation of the wiggler field (1) we represent the forward stimulated radiation by

$$A_R(z, t) = A_R[e^{i(kz - \omega t)}e_- + e^{-i(kz - \omega t)}e_+] \tag{3}$$

where A_R is the amplitude, k is the complex wave number, and ω is the real frequency. So, the total electric field and magnetic field are

$$E(z, t) = -\frac{\partial \Phi}{\partial z}e_z - \frac{1}{c}\frac{\partial(A_0 + A_R)}{\partial t} + A_i\chi \tag{4a}$$

$$B(z, t) = -\frac{\partial}{\partial z}[e_z \times (A_0 + A_R)] \tag{4b}$$

where χ represents the transverse displacement, Φ is the space-charge potential with the perturbed density δn, and is given by

$$\frac{\partial^2 \Phi}{\partial z^2} = 4\pi|e|\delta n. \tag{4c}$$

For the electron stable motion, assuming $\chi_{\perp 0} \propto e^{i\Omega_0 t} = e^{i(\omega_0 + k_0 v_{z0})t}$, we have

$$\chi_{\perp 0} = -\frac{i}{\Omega_0}V_{\perp 0}$$

$$V_{\perp 0} = \frac{|e|}{m\gamma_0 c} \cdot \frac{A_0}{\Omega_{i0}} \tag{5}$$

where $\Omega_{i0} = 1 - \omega_i^2/\Omega_0^2$, $\omega_i = (4\pi|e|^2 n_i/m\gamma_0)^{1/2}$ is the plasma frequency. We can see that for the electron steady-state motion, there is an unstable value when $\Omega_0 = \omega_i$. The electron orbit was similar with Ref. [3], i.e., it also has two stable regimes.

For the perturbed motion of electron, assuming $\chi_{\perp 1} \propto e^{-i\Omega_1 t} = e^{-i(\omega - k v_{z0})t}$, then

$$\chi_{\perp 1} = \frac{i}{\Omega_1}V_{\perp 1}$$

$$V_{\perp 1} = \frac{|e|}{m\gamma_0 c} \cdot \frac{A_R}{\Omega_{i1}} \tag{6}$$

where $\Omega_{i1} = 1 - \omega_i^2/\Omega_1^2$. In the above derivation we used the approximate of γ_0 instead of γ. This means we mainly consider the axial bunching. From Eq. (8), the ion-channel instability will occur when $\Omega_1 = \omega_i$.

3. Electron axial motion and bunching

From charge conservation, the perturbed beam density is given by

$$|e| \cdot \frac{\partial \delta n}{\partial t} = \frac{\partial \delta J_z}{\partial z} \tag{7}$$

where δJ_z is the perturbed axial beam current given by

$$\delta J_z(z, t) = -|e|(n_0 \delta v_z + \delta n v_{z0}). \tag{8}$$

In Eq. (8) δv_z and v_{z0} are the perturbed and unperturbed axial electron velocities. Combining Eqs. (7) and (8) yields the following expression for the perturbed density:

$$\frac{d\delta n}{dt} = -n_0 \frac{\partial \delta v_z}{\partial z}. \tag{9}$$

Taking the axial component of the motion equation, Eq. (5), and using the relation $d\gamma/dt = -|e|(v \cdot E)/mc^2$, we find that

$$\frac{dv_z}{dt} = -\frac{|e|}{m\gamma_0}\left[-\frac{\partial \Phi}{\partial z} + \frac{1}{c}(v \times B) \cdot e_z - \frac{v_z}{c^2}(v \cdot E)\right]. \tag{10}$$

Linearizing Eq. (10) by keeping terms to first order in the radiation field yields

$$\frac{d\delta v_z}{dt} = \frac{|e|}{m\gamma_0}\left\{-\gamma_{z0}^{-2}\frac{\partial \Phi}{\partial z} + i\left[\left(\frac{k_0}{\Omega_{i1}} + \frac{k}{\Omega_{i0}}\right) - \frac{\omega \beta_{z0}}{c\Omega_{i0}} + \frac{\omega_0 \beta_{z0}}{c\Omega_{i1}} - \frac{\omega_i^2}{\Omega_{i1}\Omega_{i0}\Omega_0}\left(\frac{\Omega_0}{\Omega_1} - 1\right)\right]\Phi_p(z, t)\right\} \tag{11}$$

where

$$\Phi_p(z, t) = \frac{-|e|}{m\gamma_0 c^2}A_0 A_R e^{i[(k + k_w)z - (\omega - \omega_0)t]} \tag{12}$$

is the ponderomotive potential and $\gamma_{z0} = (1 - \beta_{z0}^2)^{-1/2}$, $\beta_{z0} = v_{z0}/c$. Taking the convective time derivative of both sides of (9) and employing (11) and (4c) yields

$$\left(\frac{d^2}{dt^2} + \frac{\omega_b^2}{\gamma_0\gamma_{z0}^2}\right)\delta n = -\frac{in_0|e|}{m\gamma_0}\left[\left(\frac{k_0}{\Omega_{i0}} + \frac{k}{\Omega_{i0}} - \frac{\omega\beta_{z0}}{c\Omega_{i0}}\right.\right.$$

$$\left.\left. + \frac{\omega_0\beta_{z0}}{c\Omega_{i1}} - \frac{\omega_i^2}{\Omega_{i1}\Omega_{i0}\Omega_0}\left(\frac{\Omega_0}{\Omega_1} - 1\right)\right)\right]\frac{\partial\Phi_p(z,t)}{\partial z} \quad (13)$$

where $\omega_b = (4\pi|e|^2n_0/m)^{1/2}$ is the beam electron frequency, n_0 is the beam electron density. Eq. (15) shows that the perturbed charge density is driven by the ponderomotive potential wave.

4. Dispersion relation

The wave equation of radiation field is

$$\left(\frac{\partial^2}{\partial z^2} - \frac{1}{c^2}\frac{\partial^2}{\partial t^2}\right)A_R = -\frac{4\pi}{c}J_\perp \quad (14)$$

where $J_\perp = -|e|(n_0V_{\perp1} + \delta nV_{\perp0})$, substituting Eqs. (7) and (8) into Eq. (16) yields

$$\left(\frac{\partial^2}{\partial z^2} - \frac{1}{c^2}\frac{\partial^2}{\partial t^2} - \frac{\omega_b^2}{\gamma_0\Omega_{i1}}\right)A_R = \frac{4\pi|e|^2}{m\gamma_0}\frac{A_0}{\Omega_{i0}}\delta n. \quad (15)$$

Since the phase of the ponderomotive wave is $[(k+k_0) - (\omega-\omega_0)t]$, we see from Eq. (13) that the perturbed density should have a similar dependence in the time asymptotic limit, hence we write

$$\delta n(z,t) = \delta\tilde{n}e^{i[(k+k_w)z-(\omega-\omega_0)t]}. \quad (16)$$

Using Eq. (16) together with Eqs. (12) and (13), by eliminating $\delta\tilde{n}$ and A_R, Eq. (15) becomes

$$\left(\omega^2 - c^2k^2 - \frac{\omega_b^2}{\gamma_0\Omega_{i1}}\right)\cdot\left\{[(\omega-\omega_0) - (k+k_0)v_{z0}]^2\right.$$

$$\left. - \frac{\omega_b^2}{\gamma_0\gamma_{z0}^2}\right\} = \frac{\omega_b^2}{\gamma_0}\frac{V_0^2}{\Omega_{i0}}(k+k_0)$$

$$\times\left[\frac{k_0}{\Omega_{i1}} + \frac{k}{\Omega_{i0}} - \frac{\omega\beta_{z0}}{c\Omega_{i0}} + \frac{\omega_0\beta_{z0}}{c\Omega_{i1}}\right.$$

$$\left. - \frac{\omega_i^2}{\Omega_{i1}\Omega_{i0}\Omega_0}\left(\frac{\Omega_0}{\Omega_1} - 1\right)\right] \quad (17)$$

where $V_0 = |e|A_0/m\gamma_0c$ is the EM wave wiggle velocity. The above equation is the EPIC-FEL

dispersion relation. When $\omega_i = 0$, we may get the vacuum case. In this case, $\Omega_{i0} = \Omega_{i1} = 1$, the Eq. (17) reduces to the conventional EM wave pumped FEL dispersion relation [10,11].

5. Numerical calculations

Note that the phase velocity of the wave should be synchronized with electron beam, i.e.: $v_{ph} \approx v_{z0}$. The axial phase velocity is

$$v_{ph} = \frac{\omega - \omega_0}{k + k_0}. \quad (18)$$

So we can get

$$k = \frac{(1+\beta_{z0})^2}{(1-\beta_{z0}^2)}k_0 \quad \text{and} \quad \omega = \frac{(1+\beta_{z0})^2}{(1-\beta_{z0}^2)}\omega_0.$$

We can see it is difficult to satisfy this condition for a forward wave to amplify the wave.

Under the conditions of resonance for backward wave pumping, i.e. $k = [(1+\beta_{z0})^2/(1-\beta_{z0}^2)]k_0$, we can get $\Omega_0 = \Omega_1$ and $\Omega_{i0} = \Omega_{i1}$. Then the Eq. (17) is reduced to

$$\left(\omega^2 - c^2k^2 - \frac{\omega_b^2}{\gamma_0\Omega_{i0}}\right)\cdot\left\{[(\omega - \omega_0) - (k+k_0)v_{z0}]^2\right.$$

$$\left. - \frac{\omega_b^2}{\gamma_0\gamma_{z0}^2}\right\} = \frac{\omega_b^2}{\gamma_0}\frac{V_0^2}{\Omega_{i0}^2}4kk_0. \quad (19)$$

In obtaining Eq. (19) the use of the approximations, $\omega \approx ck$, $\omega_0 \approx ck_0$, $\beta_{z0} \approx 1$ and $k \gg k_0$ has been made. The above equation has been solved on a computer with $\gamma_0 = 6.0$, $k_0 = 2\pi/3 \text{ cm}^{-1}$, $\beta_0 = V_0/c = 0.2/\gamma_0$, $\beta_{z0} = 0.98$ and $\omega_0 = ck_0$. The growth rates (Im k) are plotted versus frequency for appropriate values of x (ω_i/Ω_0) for two stable regimes in Figs. 1 and 2, respectively [3]. The parameters used here are all listed in the figure captions. In order to satisfy the Budker condition for propagation of an electron beam in the ion-focused regime, the beam electron frequencies are different for regimes I and II in Figs.1 and 2. A plot of peak growth rate with x for regimes I and II (for the same parameters as in Figs. 2 and 3) is shown in Fig. 3. In regime I, with an increase in x

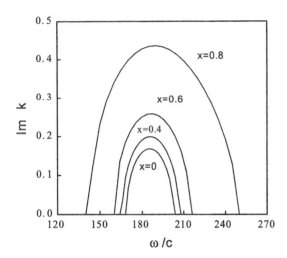

Fig. 1. Spatial growth rate $\mathrm{Im}\,k$ (cm^{-1}) vs. ω/c (cm^{-1}) for regime I, orbit, $\omega_b = 1.8 \times 10^{11}$ s^{-1}.

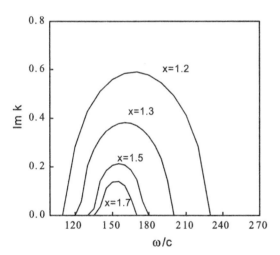

Fig. 2. Spatial growth rate $\mathrm{Im}\,k$ (cm^{-1}) vs. ω/c (cm^{-1}) for regime II, orbit, $\omega_b = 4.4 \times 10^{11}$ s^{-1}.

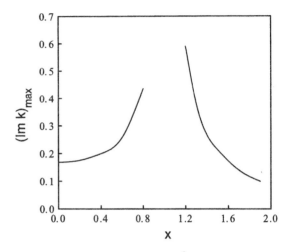

Fig. 3. Peak growth rate $(\mathrm{Im}\,k)_{\max}$ (cm^{-1}) as a function of x for regimes I and II.

the peak growth rate increases monotonically up to the singularity at the orbital stability boundary ($x \approx 1$). In regime II, the peak growth rate decreases slowly as ω_i becomes large. The results of the calculations are very interesting. Our results show that the values of the growth rates are lager than that of the magnetic wiggler case and the operating frequencies are much higher than those

in magnetic wiggler case. The growth rate in this case enhances as $\omega_i \approx \Omega_0$.

6. Conclusions

The presence of an ion-channel may lead to substantial enhancements in growth rate as $\omega_i \approx \Omega_0$. These agree with results obtained in Ref. [3]. But the values of the peak growth rates and resonant frequencies are larger than those of the former case. For all kinds of FELs, we can derive $\lambda \approx \lambda_0 / 2\gamma_z^2$. In order to operate FELs at low wavelengths, two methods have been used. First, reducing the wiggler periods. We know it is very difficult to obtain for conventional wiggler FEL case. However it is no problem for EM wave pumped FEL. Second, increasing the beam energy, but it may cause lower efficiencies. However, in ion-channel FELs, the ion-channel can improve the electron beam focusing and can compensate the efficiency loss by modifying the plasma frequency (Eq. (22)), so in the high beam energy case this kind of FEL can be operated at very low radiation wavelengths without losing much efficiency; meanwhile, the decreasing of the pump wavelength is very easy to accomplish for EPIC-FELs. With such benefits, we believe that EPIC-FELs will have a very bright future.

References

[1] K. Takayama, S. Hiramatsu, Phys. Rev. A 37 (1988) 173.

[2] Y. Seo, V.K. Tripathi, C.S. Liu, Phys. Fluids B 1 (1989) 221.

[3] P. Jha, P. Kumar, Phys. Rev. E 57 (1998) 2256.

[4] D.H. Whittum, A.M. Sessler, Phys. Rev. Lett. 64 (1990) 2511.

[5] K.R. Chen, T.C. Katsouleas, J.M. Dawson, IEEE Trans. Plasma Sci. 18 (1990) 837.

[6] D.H. Whittum, Phys. Fluids B 4 (1992) 730.

[7] K.R. Chen, J.M. Dawson, Phys. Rev. Lett. 68 (1992) 29.

[8] C.J. Tang, P.K. Liu, S.G. Liu, J. Phys. D 29 (1996) 90.

[9] S.G. Liu, R.J. Barker, 23rd International Conference on IR/MM waves, Essex Univ., September 7–11, 1998.

[10] K.R. Chen, J.M. Dawson, A.T. Lin, T. Katsouler, Phys. Fluids. B 3 (1991) 1270.

[11] V.L. Braman, G.G. Denisov, N.S. Ginzburg, Int. J. Electron. 59 (1985) 247.

ELSEVIER

Nuclear Instruments and Methods in Physics Research A 475 (2001) 158–163

NUCLEAR
INSTRUMENTS
& METHODS
IN PHYSICS
RESEARCH
Section A

www.elsevier.com/locate/nima

Self-amplified spontaneous emission in Smith–Purcell free-electron lasers

Kwang-Je Kim, Su-Bin Song*

Advanced Photon Source, Argonne National Laboratory, Argonne, IL 60439, USA

Abstract

We present an analysis of a Smith–Purcell system in which a thin current sheet of electrons moves above a grating surface in the direction perpendicular to the grating grooves. We develop a consistent theory for evolution of the electromagnetic field and electron distribution in the exponential growth regime starting from the initial electron noise and the incoming amplitude. The dispersion relation for the complex growth rate for this system is a quadratic equation. © 2001 Published by Elsevier Science B.V.

PACS: 41.60.Cr; 41.60.-m; 42.79.Dj; 52.75.Ms

Keywords: Smith–Purcell; FEL; SASE; Grating

1. Introduction

Beginning with the work by Smith and Purcell [1], the radiation generated by an electron beam passing over a grating surface has been studied for decades [2–4]. Recently, there has been a renewed interest in the Smith–Purcell system with the observation of possible exponential gain in an experiment using electron microscope beams [5,6].

Spontaneous emission in 2-D Smith–Purcell system in which electrons are moving line charges parallel to the grating surface was studied by Van den Berg based on a set of plane wave modes [2]. The set consists of the fundamental mode, which has the same phase velocity as electrons, and higher-order modes with additional spatial frequencies at harmonics of the grating frequency.

The grating surface is characterized by a scattering matrix connecting the incoming mth mode to the out going nth mode [7]. The electron beam generates the fundamental mode. Although the fundamental mode cannot itself propagate because it decays perpendicular to the grating surface, it may "scatter" at the grating surface to be converted into one or several propagating modes. The propagating modes then appear as the far-field Smith–Purcell radiation.

In the spontaneous emission problem discussed in the above, the amplitude of the fundamental mode was determined from Maxwell's equations where the electron current is regarded as fixed. If the mode amplitude became strong, it may act on the electron beam and induce density modulation. The evolution of the coupled radiation-electron beam system then is similar to the case of the usual free-electron laser in the exponential gain regime [8]. In this paper, we study the exponential growth

---

*Corresponding author.

E-mail address: sbsong@aps.anl.gov (S.-B. Song).

0168-9002/01/$ - see front matter © 2001 Published by Elsevier Science B.V.
PII: S 0 1 6 8 - 9 0 0 2 (0 1) 0 1 5 7 6 - 5

and self-amplified spontaneous emission in the 2-D Smith–Purcell system. The electron beam is assumed to be thin but has an arbitrary energy spread. We have derived a dispersion relation for the growth rate, which becomes a quadratic equation for a rectangular energy distribution. We are therefore able to derive an exact expression for the growth rate and also the criteria for an exponential growth in terms of the electron energy spread. As far as, we are aware these results have not been obtained before.

A calculation of the growth rate in a Smith–Purcell system was also carried out by Schächter and Ron [3] who discussed a cubic equation for the growth rate. However, they invoked an additional requirement that the scattering matrix be singular. The requirement is neither meaningful nor necessary.

2. Smith–Purcell FEL equation

2.1. Current

Fig. 1 shows the Smith–Purcell system studied in this paper. The surface of the metallic grating consists of a perfect conductor whose grooves are parallel and uniform in the y direction. Electrons move parallel to the z-axis in a thin sheet along the

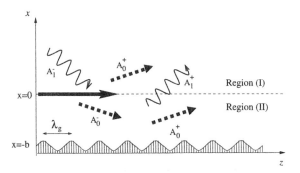

Fig. 1. Wave interaction in a Smith–Purcell system The system is translationally invariant in the y direction (perpendicular to the page). The wave lines, the dotted lines, and the solid line at $x = 0$ are, respectively, the $n = 1$ mode, the synchronous evanescent mode, and the electron beam. The electron beam is infinitely thin in the x direction. The grating surface is at $x = -b$.

grating surface. The current density is therefore

$$\mathbf{J}(x, z, t) = \frac{q}{\Delta y}\delta(x)\sum_i \delta(t - t_i(z))\hat{\mathbf{z}} \qquad (1)$$

where q is electron charge, Δy is the length in the y direction, $t_i(z) = t_i(0) + \int(1/v_i)\,\mathrm{d}z$ is the time when the ith electron passes through z, v_i is the velocity of the ith electron, and $\hat{\mathbf{z}}$ is the unit vector in the z direction. The Fourier transform of the current density is given by

$$\mathbf{J}(x, z; \omega) = \delta(x)K(z, \omega)\mathrm{e}^{\mathrm{i}\alpha_0 z} \equiv J_z\hat{\mathbf{z}} \qquad (2)$$

where

$$K(z, \omega) = \frac{q}{\Delta y}\sum \mathrm{e}^{\mathrm{i}\omega(t_i(z) - z/v_0)} \qquad (3)$$

and $\alpha_0 = k/\beta_0$, $\beta_0 = v_0/c$ is the normalized average velocity, $k = \omega/c$ is the wave number, and c is the speed of light. Since the overall phase factor $\mathrm{e}^{\mathrm{i}\alpha_0 z}$ is taken out, we expect that $K(z, \omega)$ is a slowly varying function of z.

2.2. Modes

From the Maxwell equation it follows that the magnetic field $\mathbf{H}$ is in the y direction for the present system [2] and that H_y satisfies the following Helmholtz equation:

$$\left(\frac{\partial^2}{\partial x^2} + \frac{\partial^2}{\partial z^2} + k^2\right)H_y(x, z; \omega) = \frac{\partial}{\partial x}J_z(x, z; \omega). \qquad (4)$$

The plane wave solution of the Helmholtz equation is of the form [2]

$$\mathrm{e}^{\mathrm{i}(\alpha_n z \pm p_n x)}, \qquad (5)$$

where $p_n = \sqrt{k^2 - \alpha_n^2}$. For the problem we are considering here, we may take $\alpha_n = \alpha_0 - nG$ ($G = 2\pi/\lambda_g$), where λ_g is the period of the grating grooves, and n is an arbitrary integer.

When p_n is real, Eq. (5) represents propagating waves. In this case, we will assume p_n to be positive. Therefore, the upper (lower) sign in Eq. (5) corresponds to an outgoing (incoming) wave to (from) $x = \infty$. However, p_n may be imaginary. In that case, we introduce the real quantity $p_n = \mathrm{i}\Gamma_n$, $\Gamma_n = |\sqrt{\alpha_n^2 - k^2}| > 0$. The corresponding waves are

$$\mathrm{e}^{\mathrm{i}\alpha_n z \mp \Gamma_n x}. \qquad (6)$$

Here, the upper (lower) sign corresponds to an evanescent wave "propagating" to (from) $x = \infty$. In particular, the fundamental mode $n = 0$ is evanescent: $\Gamma_0 = k/\beta_0\gamma_0$, where $\gamma_0 = 1/\sqrt{1 - v_0^2/c^2}$. Higher-order modes with $n < 0$ are also evanescent. The mode $n > 0$ may or may not be propagating. We assume that the $n = 1$ mode is propagating.

The evanescent $n = 0$ mode is strongly coupled to the current. Modes with $n \neq 0$ are either present as the incoming mode or generated at the grating surface. For simplicity, we consider the modes $n = 0$ and 1 only. Including other modes will be straightforward once the principle is understood.

In region (I) the magnetic field can be written as

$$H_y^{(I)} = A_{0,I}^+ e^{i\alpha_0 z - \Gamma_0 x} + A_1^- e^{i(\alpha_1 z - p_1 x)}$$
$$+ A_1^+ e^{i(\alpha_1 z + p_1 x)}. \tag{7}$$

The first term is an evanescent mode generated either directly from the electron beam or at the grating surface, the second term is an incoming mode from an external source, and the third term is a radiating mode generated at the grating surface.

In region (II) the magnetic field is

$$H_y^{(II)} = A_0^- e^{i\alpha_0 z + \Gamma_0 x} + A_{0,II}^+ e^{i\alpha_0 z - \Gamma_0 x}$$
$$+ A_1^- e^{i(\alpha_1 z - p_1 x)} + A_1^+ e^{i(\alpha_1 z + p_1 x)}. \tag{8}$$

The first term in Eq. (8) is an evanescent mode generated directly from the electron beam, the second term is an evanescent mode generated at the grating surface, the third term is an incoming radiating field from an external source, and the fourth term is a radiating mode generated at the grating surface.

2.3. Slowly varying amplitude approximation

Taking into account the interaction with the current, the amplitudes $A_n^\pm$, except A_1^-, cannot be regarded as constants. The incoming wave A_1^- is given by an external source and may be regarded as a constant. For other waves, we expect that the amplitudes are slowly varying functions of x and z. In order for the quantity $A_n^\pm(x, z)e^{i(\alpha z \pm px)}$ to satisfy the Helmholtz equation to the first order derivatives of $A_n^\pm$, we must require $(\alpha_n \partial/\partial z \pm p_n \partial/\partial x)A_n(x, z) = 0$. Therefore, the mode amplitude $A_n^\pm$ is a function of a particular combination of z and x:

$$A_n^\pm(x, z) = A_n^\pm(z \mp \alpha_n x/p_n). \tag{9}$$

2.4. Boundary conditions

By using the boundary conditions at the grating surface and at the current sheet, the amplitudes of the outgoing modes can be completely determined in terms of incoming wave amplitude and surface current K. First, the discontinuity in H_y between regions (I) and (II) at $x = 0$ is given by the surface current K:

$$A_{0,I}^+(z) - \{A_0^-(z) + A_{0,II}^+(z)\} = K(z). \tag{10}$$

Note that only the $n = 0$ modes couple to the current since the phase velocity of these modes are the same as the average electron velocity.

Second, E_z must be continuous across regions (I) and (II). Using $E_z = (\partial H_y/\partial x)/(-i\varepsilon_0\omega)$, we have

$$\frac{i\alpha_0}{\Gamma_0}\frac{\partial}{\partial z}A_{0,I}^+(z) - \Gamma_0 A_{0,I}^+(z)$$
$$= -\frac{i\alpha_0}{\Gamma_0}\frac{\partial}{\partial z}A_0^-(z) + \frac{i\alpha_0}{\Gamma_0}\frac{\partial}{\partial z}A_{0,II}^+(z)$$
$$+ \Gamma_0 A_0^-(z) - \Gamma_0 A_{0,II}^+(z). \tag{11}$$

From Eqs. (10) and (11) we can derive that

$$A_0^-(z) = -\tfrac{1}{2}K(z) \tag{12}$$

because $A_0^- = 0$ when $K = 0$.

The third and final boundary condition is at the grating surface $x = -b$, where the incident fields are linearly related to the reflected fields:

$$A_m^+ = \sum_n e_{mn}A_n^- \tag{13}$$

where $e_{mn}(\omega)$ is a reflection matrix of the grating. The elements of the matrix depend on the particular geometry of the grating. Accurate calculation of these coefficients usually requires a significant amount of numerical calculation [7], and is not treated here.

Eq. (13) becomes, in our problem,

$$A_{0,II}^+(z - i\alpha_0 b/\Gamma_0)e^{\Gamma_0 b}$$
$$= e_{00}A_0^-(z + i\alpha_0 b/\Gamma_0)e^{-\Gamma_0 b} + e_{01}A_1^-, \tag{14}$$

$$A_1^+(z + \alpha_1 b/p_1)$$
$$= e_{10}A_0^-(z + i\alpha_0 b/\Gamma_0)e^{(-\Gamma_0 + ip_1)b} + e_{11}A_1^-. \tag{15}$$

Inserting Eq. (14) into Eq. (10) yields, invoking the slowly varying approximation $(b|(\partial/\partial z)A_0^-| \ll (\Gamma_0/2\alpha_0)A_0^- = (1/2\gamma_0)A_0^-)$,

$$A_{0,I}^+(z) = \frac{1}{2}K(z) - \frac{e_{00}e^{-2\Gamma_0 b}}{2}K(z) + e_{01}A_1^- e^{-\Gamma_0 b}.$$
(16)

In Eq. (16), the first term corresponds to the direct field of electrons, the second term to the $n=0$ reflected mode of the grating, and the last term to the external field reflected by the grating.

The $n=1$ mode can be determined from Eqs. (12) and (15):

$$A_1^+(z) = -\frac{e_{10}}{2}K(z)e^{(-\Gamma_0 + ip_1)b} + e_{11}A_1^- e^{2ip_1 b}.$$
(17)

Eqs. (16) and (17) determine the outgoing fields $A_{0,I}^+$ and A_1^+ in terms of the input field A_1^- and the surface current K.

2.5. Electron motion

The electron motion is affected by the longitudinal electric field at the boundary between two regions $(x=0)$ that can be obtained:

$$E_z(\omega) = \frac{1}{i\varepsilon_0 \omega}\left\{\Gamma_0 A_{0I}^+(z) - \frac{i\alpha_0}{\Gamma_0}\frac{\partial}{\partial z}A_{0I}^+(z)\right\}e^{i\alpha_0 z}.$$
(18)

The second term in Eq. (18) can be ignored invoking the slowly varying approximation. From Eqs. (16) and (18), the longitudinal electric field of the $n=0$ mode at $x=0$ is obtained, in the time domain, as follows:

$$E_{0z}(0,z;t) = \frac{1}{2\pi i\varepsilon_0 c\gamma_0\beta_0}\int d\omega \left[\frac{1}{2}K(z)\right.$$
$$\left. - \frac{e_{00}}{2}e^{-2\Gamma_0 b}K(z) + e_{01}A_1^- e^{-\Gamma_0 b}\right]e^{-i\omega(t-z/v_0)}.$$
(19)

The term $\frac{1}{2}K(z)$ in the square brackets of Eq. (19) gives rise to a term in the longitudinal electric field that is proportional to $\sum_i \delta(t - t_i(z))$. This is the contribution from the self-field and should be removed in computing the force acting on the electrons.

As in conventional FEL equations of motion, we can define the electron phase and construct a set of electron equations of motion. The electron

phase in this system is defined as

$$t - z/v_0 = \xi$$
(20)

whose derivative is

$$\frac{d\xi}{dz} = (1/v - 1/v_0) = -\frac{\eta}{c\beta_0^3\gamma_0^2}$$
(21)

where $\eta = (\gamma - \gamma_0)/\gamma_0$, and $\beta_0 = v_0/c_0$.

After removing the self-field contribution from Eq. (19), the equation describing electron's energy change is

$$\gamma_0 mc^2\frac{d\eta}{dz} = \frac{qi}{4\pi\varepsilon_0 c\beta_0\gamma_0}\int d\omega$$
$$\times [e_{00}e^{-2\Gamma_0 b}K(z) - 2e_{01}A_1^- e^{-\Gamma_0 b}]e^{-i\omega\xi}.$$
(22)

Eqs. (3), (21), and (22) form a complete set of Smith–Purcell FEL equations. The energy is conserved between the field and the electron bunch in this system.

3. FEL evolution

3.1. Distribution function

We write the Klimontovich distribution function of electrons in phase space (ξ, η) as

$$f(\xi,\eta;z) = \frac{1}{v_0(dN_e/dz)}\sum_i \delta(\xi - \xi_i(z))\delta(\eta - \eta_i(z))$$
(23)

where dN_e/dz is the line density. This distribution satisfies the continuity equation in phase space:

$$\frac{\partial}{\partial z}f + \dot{\xi}\frac{\partial}{\partial\xi}f + \dot{\eta}\frac{\partial}{\partial\eta}f = 0$$
(24)

where dot represents the total derivative with respect to z. The distribution function can be separated into the smooth background $\bar{f}$ and the perturbation $\tilde{f}$:

$$f(\eta,\xi;z) = \bar{f}(\eta) + \tilde{f}(\eta,\xi;z).$$
(25)

We have assumed that the smooth background depends only on the electron energy η. Note that Eq. (23) implies a normalization such that $\int \bar{f}(\eta)\,d\eta = 1$.

Using Eq. (21) and Eq. (22), Eq. (24) becomes, after linearization,

$$\frac{df_\omega}{dz} + \frac{i\omega\eta}{c\beta_0^3\gamma_0^2}f_\omega + i[d_1 K(z,\omega) - d_2 A_1^-]\frac{\partial}{\partial\eta}\bar{f} = 0 \quad (26)$$

where $f_\omega = \int f e^{i\omega\xi}\,d\xi$, $d_1 = q e_{00} e^{-2\Gamma_0 b}/2\varepsilon_0\beta_0\gamma_0^2 mc^3$, and $d_2 = q e_{01} e^{-\Gamma_0 b}/\varepsilon_0\beta_0\gamma_0^2 mc^3$. Here the energy change of electrons is regarded as a first-order quantity. Eq. (26) can be solved using the Laplace transformation:

$$K(z,\omega) = \frac{q}{\Delta y}\oint \frac{d\mu\; e^{\mu z}}{2\pi i D(\mu)}$$
$$\times \left\{ \sum_i \frac{e^{i\omega\xi_i}}{\mu + i\omega\eta_i/c\beta_0^3\gamma_0^2} + iA_1^- d_2 v_0 \frac{dN_e}{dz} \right.$$
$$\left. \times \int d\eta \frac{\partial\bar{f}/\partial\eta}{\mu(\mu + i\omega\eta/c\beta_0^3\gamma_0^2)} \right\} \quad (27)$$

where we have defined the dispersion function $D(\mu)$ as

$$D(\mu) = 1 + \frac{iqd_1 v_0 dN_e}{\Delta y\quad dz}\int d\eta \frac{\partial\bar{f}/\partial\eta}{\mu + i\omega\eta/c\beta_0^3\gamma_0^2}. \quad (28)$$

The first term on the right-hand side of Eq. (27) contains a sum of stochastic phase factors and describes the process of self-amplified spontaneous emission, while the second term describes the amplification of the coherent input signal A_1^-.

3.2. Dispersion relation and exponential gain

The contour integral in Eq. (27) contains all singularities in the complex μ plane. In addition to poles of kinematic origin, the singularities as the solution of the dispersion relation

$$D(\mu) = 0 \quad (29)$$

determine the dynamics of the system [8]. For rectangular distribution,

$$\bar{f}(\eta) = \begin{cases} \dfrac{1}{\Delta\eta} & \text{if } |\eta| < \Delta\eta/2 \\ 0 & \text{otherwise} \end{cases} \quad (30)$$

the dispersion relation is reduced to a second-order equation

$$\mu^2 = \frac{q^2 k e_{00} e^{-2\Gamma_0 b}}{2\varepsilon_0 mc^2 \Delta y\beta_0^3\gamma_0^4}\frac{dN_e}{dz} - \left(\frac{k\Delta\eta}{2\beta_0^3\gamma_0^2}\right)^2. \quad (31)$$

For vanishing energy spread ($\Delta\eta = 0$), there is a mode growing exponentially with the growth rate

$$\mu = \frac{1}{\gamma_0^2\beta_0^2}\sqrt{\frac{2\pi k e_{00} e^{-2\Gamma_0 b}}{\Delta y}\frac{I}{I_A}} \quad (32)$$

where $I_A = 4\pi\varepsilon_0 mc^3/e = 17\text{ kA}$ is the Alfvén current. For a large energy spread

$$\Delta\eta \gtrsim \Delta\eta_{max} = \sqrt{\frac{2q^2 e_{00} e^{-2\Gamma_0 b}\beta_0^3}{\varepsilon_0 mc^2 \Delta yk}\frac{dN}{dz}} \quad (33)$$

the two solutions of the dispersion relation are both purely imaginary corresponding to an oscillatory amplitude.

For a growing mode, the behavior of the outgoing amplitudes A_0^+ and A_1^+ for $x, z \to \infty$ are

$$A_0^+(x,z) \propto e^{\mu(z+i\alpha_0 x/\Gamma_0)} \quad (34)$$

$$A_1^+(x,z) \propto e^{\mu(z-\alpha_1 x/p_1)}. \quad (35)$$

4. Discussion

The dispersion relation in Eq. (31) is notable since it is a quadratic equation in μ in contrast to the cubic equation in the usual FELs employing magnetic undulators [8]. This is probably due to the assumed translational invariance in the y direction. A different dispersion relation, which is cubic in μ, was derived in reference [3]. However they used an additional requirement that the reflection coefficient e_{00} be singular "*to support the synchronous wave*". We find that the requirement is neither meaningful nor necessary. The correct dispersion relation is that given by Eq. (31).

An experiment on the Smith–Purcell system was recently carried out at Dartmouth college [5]. The parameters of this experiment are: $I \lesssim 1\text{ mA}$, $\beta_0 \approx 0.35$, $\lambda \approx 300\,\mu\text{m}$. The electron beam was cylindrical with diameter $a \approx 20\,\mu\text{m}$. Taking $\Delta y \approx a$ and setting $e_{00} e^{-2\Gamma_0 b} \approx 1$, we obtain the gain length

$\mu^{-1} \gtrsim 8.6$ mm. Thus, the gain in the experiment is marginal since the grating length is 13 mm.

We have analyzed an idealized Smith–Purcell system in the exponential gain regime. The theory is based on the plane wave analysis of the system in two-dimensional geometry. Dispersion relation for the growth rate was derived from the coupled Maxwell–Klimontovich equation. The dispersion relation in this case is shown to be a quadratic equation, in contrast to the cubic dispersion relation of undulator FELs.

Acknowledgements

We thank Zhirong Huang for helpful discussions. This work is supported by the U.S. Department of Energy, Office of Basic Energy Sciences, under Contract No. W-31-109-ENG-38.

References

[1] S.J. Smith, E.M. Purcell, Phys. Rev. 92 (1953) 1069.
[2] P.M. van den Berg, J. Opt. Soc. Am. 63 (1973) 689.
[3] L. Schächter, A. Ron, Phys. Rev. A 40 (1989) 876.
[4] J.H. Brownell, J. Walsh, Phys. Rev. E 57 (1998) 1075.
[5] J. Urata, M. Goldstein, M.F. Kimmit, A. Naumov, C. Platt, J.E. Walsh, Phys. Rev. Lett. 80 (1998) 516.
[6] J.E. Walsh, J.H. Brownell, J. Swartz, S. Trotz, M. Kimmitt, G. Doucas, R. Fernow, H. Kirk, V. Yakimenko, Nucl. Instr. and Meth. A, these proceedings.
[7] R. Petit (Ed.), Electromagnetic Theory of Gratings, Springer, Berlin, 1980.
[8] K.-J. Kim, Nucl. Instr. and Meth. A 250 (1986) 396.

ELSEVIER

Nuclear Instruments and Methods in Physics Research A 475 (2001) 164–172

NUCLEAR
INSTRUMENTS
& METHODS
IN PHYSICS
RESEARCH
Section A

www.elsevier.com/locate/nima

Experimental and theoretical study of 2D Bragg structures for a coaxial free electron maser

A.W. Cross[a], W. He[a], I.V. Konoplev[a],*, A.D.R. Phelps[a], K. Ronald[a],
G.R.M. Robb[a], C.G. Whyte[a], N.S. Ginzburg[b], N.Yu. Peskov[b], A.S. Sergeev[b]

[a] *Department of Physics and Applied Physics, University of Strathclyde, Glasgow, G4 0NG, UK*
[b] *Institute of Applied Physics, Russian Academy of Sciences, Nizhny Novgorod 603600, Russia*

Abstract

Experimental results are presented of the first observation of two-dimensional distributed feedback in a coaxial 2D-Bragg structure. The use of two-dimensional distributed feedback has been proposed as a method of producing spatially coherent radiation from extremely powerful large sized relativistic electron beams. To obtain coherent high-power (GW) microwave and millimetre wave radiation, 2D Bragg resonators are needed to overcome the problems of mode selection and synchronisation of radiation from different parts of an oversized beam. The design and cold microwave measurements of a 2D Bragg coaxial structure for use in a Free Electron Maser (FEM) driven by an annular electron beam of circumference 25 times larger than the radiation wavelength is presented. The cavity for the 2D Bragg FEM consists of two 2D Bragg coaxial reflectors separated by a regular coaxial waveguide. The formation of pure two-dimensional distributed feedback without a mixture of one-dimensional parasitic feedback was demonstrated for the first time in this series of experiments. The first experimental comparison of 2D and 1D Bragg structures was also conducted and good agreement between experimental results and theoretical predictions was observed. The eigenmodes of the two-mirror cavity were calculated and it was shown that a single-mode steady-state operation regime could be obtained in a FEM based on such a novel cavity. © 2001 Elsevier Science B.V. All rights reserved.

PACS: 41.60.Cr; 41.60.Bq; 41.60.m; 42.50.Fx; 52.75.Ms

Keywords: Free-electron laser and maser; Distributed feedback; Bragg structure; High-power microwaves

1. Introduction

To date, most of the studies of the dynamics of multi-mode relativistic electron oscillators, such as free electron lasers have been concerned with systems involving Fabry–Perot like resonators composed from conventional mirrors which have a quasi-equidistant spectrum of eigenmodes with approximately the same Q-factors. In this system a single operating mode can be established due to non-linear mode competition [1,2]. However many FEL experiments involving microwave radiation use a different mirror system which selects the operating mode by distributed coupling of counterpropagating waves on a corrugated surface. Such mirrors are known as Bragg reflectors, which

*Corresponding author. Tel.: +44-141-548-4272; fax: +44-141-552-2891.

E-mail address: acp96115@strath.ac.uk (I.V. Konoplev).

0168-9002/01/$ - see front matter © 2001 Elsevier Science B.V. All rights reserved.
PII: S 0168-9002(01)01577-7

have been constructed from sections of waveguide with periodic one-dimensional (1D) corrugations along the direction of beam propagation [3]. This arrangement encourages single-mode operation due to the resonator being of high quality only for those modes, which lie inside the frequency band defined by the wave coupling coefficient, close to the Bragg frequency [4].

An attractive way to increase microwave radiation power for mildly relativistic electron beams is to use a high current electron beam of large transverse size of sheet or annular geometry which allows a total increase in beam current without increasing the current and radiation densities. This circumvents high-density beam instabilities and RF breakdown inside the interaction space. In a FEM driven by such an electron beam an oversized microwave system must be used. However, traditional 1D Bragg resonators, which were successfully utilised in many FEM experiments at Ka-band (26.5–40 GHz) frequencies [3–7], lose mode selectivity and spectral purity when the diameter of the microwave system (D) exceeds the wavelength of the radiation (λ) by more than $D/\lambda \sim 2$ to 4. To overcome these problems the use of two-dimensional (2D) distributed feedback has recently been proposed [8–12]. The novel feedback mechanism can be realised in 2D Bragg cavities of planar or coaxial geometry having a doubly periodic corrugation in two directions at an angle to each other. This corrugation provides transverse (with respect to the propagation of the electron beam) electromagnetic energy fluxes synchronising the radiation from all parts of the oversized electron beam.

The aim of the experiments at the University of Strathclyde is to demonstrate that 2D distributed feedback can be used to create a highly mode selective coaxial cavity which can synchronise radiation from an oversized annular electron beam of circumference 25 times larger than the radiation wavelength. A 37-GHz FEM exploring the novel feedback mechanism is under construction and was reported at the FEL-99 Conference [13].

In previous experimental studies of 2D Bragg structures the corrugations used were based on a rectangular grooves and it was shown that such a structure can be considered as a superposition of 1D and 2D Bragg corrugations [13–15]. This was supported by the measured transmission zones where 2D as well as 1D scattering was observed [15]. This paper presents results of theoretical and first experimental studies of a novel 2D Bragg coaxial structure based on a corrugation consisting of a "chessboard" pattern [14,15]. Measurement of the transmission characteristics of a 2D Bragg coaxial reflector based on a "chessboard" corrugated pattern resulted in the observation of pure 2D scattering. Good agreement between experimental measurements and theoretical predictions confirm this pattern to be the optimum substitute to an "ideal" corrugation [8–12] for operation of two-dimensional feedback. Computer simulations have been used to study operation of an FEM based on a 2D Bragg coaxial cavity and conditions for single mode oscillation are discussed.

2. Theoretical study of the 2D Bragg coaxial structure

A schematic outline of a coaxial 2D Bragg cavity consisting of two conductors of radii $r_{in,out}$, two 2D Bragg reflectors of length $l_{1,2}$ with a double-periodical corrugation in two directions at an angle to each other and a regular part of the waveguide of length l_0 is shown in Fig. 1(a). Let us note that in Fig. 1a an "ideal" corrugation is presented which was previously considered for theoretical studies [8–12]. However in this paper we studied a 2D "chessboard" structure as shown in Fig. 1b, which is easier to manufacture. The corrugated surface of the 2D "chessboard" structure can be presented as a multiplication of two periodic functions $a(\varphi, z) = a_1 g_\varphi(\varphi) g_z(z)$, where

$$g_\xi(\xi) = \begin{cases} -1 & 0 < \xi < d_\xi/2 \\ 1 & d_\xi/2 < \xi < d_\xi \end{cases},$$

$$g_\xi(\xi + d_\xi) = g_\xi(\xi),$$

d_ξ is the corrugation period over the ξ-coordinate. This corrugation can be described by the Fourier expansion and for a small depth a_1 it was

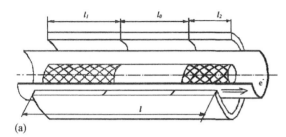

(a)

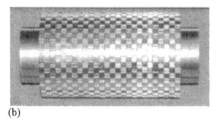

(b)

Fig. 1. (a) Schematic diagram of a two mirror coaxial 2D Bragg cavity. (b) Photograph of the corrugated inner conductor of the coaxial waveguide.

discovered that for the "chessboard" structure other harmonics in the spectrum were much smaller than the fundamental harmonic [14] and thus the corrugation can be approximated by the following expression:

$$a(\varphi, z) \cong \frac{16a_1}{\pi^2}\cos(\bar{m}\varphi)\cos(\bar{k}z) = \frac{8a_1}{\pi^2}[\cos(\bar{m}\varphi - \bar{k}z) + \cos(\bar{m}\varphi + \bar{k}z)] \quad (1)$$

where $\bar{k} = 2\pi/d_z$, d_z, is the period of the structure over the z-coordinate and $\bar{m}$ is the number of variations of the corrugation over the azimuthal coordinate φ. Therefore, despite the difference in appearance of the corrugations between Fig. 1a and b, it was found that the "chessboard" corrugation approximated well to an "ideal" 2D Bragg structure.

Corrugation (1) provides distributed coupling for four partial waves propagating over the longitudinal z and azimuthal coordinates:

$$\vec{E} = \text{Re}\{[\vec{E}_p^0(r)((B_+(\varphi, z)\,e^{-iM\varphi} + B_-(\varphi, z)\,e^{iM\varphi}) + \vec{E}_q^0(r)(A_+(\varphi, z)\,e^{-ik_z z} + A_-(\varphi, z)\,e^{ik_z z})]\,e^{i\omega t}\} \quad (2)$$

if the geometrical parameters of the corrugation $\bar{k}$ and $\bar{m}$ satisfy the Bragg conditions, with the

longitudinal wavenumber k_z and the azimuthal index M of the partial waves given by:

$$k_z = \bar{k}, \quad M \approx \bar{m}. \quad (3)$$

In (2) the amplitudes $A_\pm(\varphi, z)$, $B_\pm(\varphi, z)$ are the slow functions of the φ and z coordinates, the function $\vec{E}_{p,q}(r)$ describe the transverse structure of the modes over the radial r-coordinate and coincide with one of the eigenmodes of a coaxial waveguide. For a small corrugation depth $\bar{k}a_1 \ll 1$ and $r_{\text{in,out}} \gg r_{\text{out}} - r_{\text{in}}$, $r_{\text{in,out}} \gg \lambda$ where λ is the operating wavelength the dispersion equation for the eigenwaves of the coaxial waveguide can be reduced to that obtained for the eigenwaves of a planar waveguide [16]. This allows the small curvature of the cavity surface to be neglected enabling the planar co-ordinate system to be adopted. The amplitudes of the partial waves $A_\pm$, $B_\pm$ can therefore satisfy the coupled-wave equations described in Refs. [8–11].

A coaxial 2D Bragg cavity (Fig. 1a) can be realised in the form of a two-mirror cavity composed from two 2D Bragg coaxial structures acting as reflectors separated by a regular section of coaxial waveguide. Therefore, to study the eigenmodes' spectrum of such a cavity it is useful to obtain the reflection and transmission coefficients from the coaxial 2D Bragg structure. Taking into account the cyclic boundary condition i.e. presenting the partial waves field amplitudes as

$$A_\pm(\varphi, z) = \sum_{m=-\infty}^{\infty} A_\pm^m(z)e^{im\varphi}$$

$$B_\pm(\varphi, z) = \sum_{m=-\infty}^{\infty} B_\pm^m(z)e^{im\varphi}$$

and considering each Fourier term as a waveguide eigenmode with the azimuthal index m, assuming that one of the harmonics with an arbitrary index m of the incident wave $A_+(\varphi, z)$ has non zero amplitude while $A_-(\varphi, z)$ has zero amplitude on the right hand side of the structure i.e. $A_+^m(z = 0) = A_0^m$, $A_-^m(z = l_z) = 0$, from the 2D coupled-wave equations [8,10] one can obtain the following expressions for the reflection R_m and transmission T_m coefficients for the Bragg structure:

$$R_m = \frac{q_m}{i\Lambda_m \cot(\Lambda_m l_Z) - p_m}$$

$$T_m = \frac{i\Lambda_m}{\sin(\Lambda_m l_z)(i\Lambda_m \cot(\Lambda_m l_Z) - p_m)} \quad (4)$$

where

$$q_m = -\frac{2\alpha^2\delta}{\delta^2 - m^2 s^2}, \quad p_m = \delta\left(1 - \frac{2\alpha^2}{\delta^2 - m^2 s^2}\right)$$

$$\Lambda_m = \sqrt{\delta^2 - 4\alpha^2 - \frac{4\alpha^2 m^2 s^2}{\delta^2 - m^2 s^2}}$$

$\delta = \delta_0 - i\sigma$, $\delta_0 = (\bar{\omega} - \omega)/c$ is the frequency mismatch from Bragg resonance, $\bar{\omega}$ is the Bragg frequency which is determined by relations (3) and a waveguide modes dispersion, ω is the incident wave frequency, σ is the parameter of ohmic losses, c is the speed of light, α is the wave coupling coefficient on a 2D corrugation [8–12], $s = 2\pi/l_x$ and l_x is the mean circumference of the coaxial waveguide. Analysis of expressions (4) shows that for each mode of azimuthal index m the 2D Bragg mirror provides an effective reflection zone (Fig. 2) inside a frequency interval defined by condition $\Lambda_m^2 \leqslant 0$, i.e.

$$ms \leqslant \delta_0 \leqslant \sqrt{4\alpha^2 + m^2 s^2}. \quad (5)$$

As it can be seen from (4) and Fig. 2 the maximum reflection is achieved for azimuthally symmetric mode $m = 0$ at the frequency of precise Bragg resonance $\omega = \bar{\omega}$ (i.e. $\delta = 0$). This maximum reflection value does not depend on the mirror length parameter αl_z in contrast to the reflection coefficient for a 1D Bragg mirror, where the maximum reflection is defined as $R^{\max} \approx \tanh(\alpha l_z)$ [3]. As a result, for a 2D Bragg mirror it is possible to control the effective width of the reflection zone by adjusting parameter αl_z while maintaining the high reflection at the central (Bragg) frequency (compare Fig. 2a and b).

The analysis of (4) and (5) shows that for a two-mirror 2D Bragg resonator, when the length of the mirrors $l_{1,2}$ are much smaller than the length of the regular section l_0, the frequencies and Q-factors of the eigenmodes are described by the

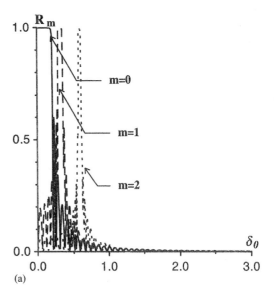

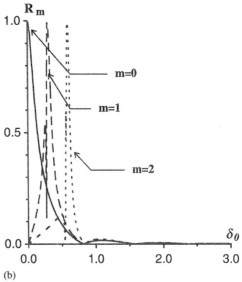

Fig. 2. Dependence of reflection coefficient R_m on the frequency detuning δ_0 for modes with azimuthal indexes $m = [0, 2]$ ($m = 0$ solid, $m = 1$ dashed and $m = 2$ dotted lines) when (a) $\alpha l_z = 4$ and (b) $\alpha l_z = 0.4$ ($r_0 = 3.5$ cm, $\sigma = 10^{-3}$ cm^{-1} and $\alpha/\sigma = 100$).

following expressions:

$$\mathrm{Re}\, \delta_{m,n} \approx \frac{\pi n}{l_0} + ms,$$

$$Q_{m,n} \cong \frac{2\pi l_0}{\lambda(1 - |R_1(\mathrm{Re}\,\delta_{m,n})R_2(\mathrm{Re}\,\delta_{m,n})|)} \quad (6)$$

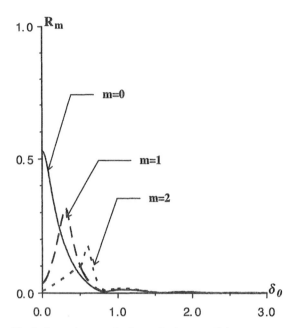

Fig. 3. Dependence of the reflection coefficient R_m on the frequency detuning δ_0 for modes with azimuthal indexes $m = [0, 2]$ ($m = 0$ solid, $m = 1$ dashed and $m = 2$ dotted lines) when $\alpha l_z = 0.4$, $r_0 = 3.5$ cm and $\alpha/\sigma = 2$.

where n is the longitudinal index of the modes along the regular part of the cavity. It is very important to note that high mode selectivity over both indexes m and n exists in such a cavity. The selection of the mode over azimuthal index m can be achieved by optimising parameter α/σ. It is easy to see from comparison of Fig. 2 with Fig. 3 that increasing parameter α/σ results in a dramatic decrease of the reflection of the azimuthal non-symmetric modes from such a structure. Thus taking into account (6) it is obvious that the Q-factor for these modes ($m \neq 0$) will also decrease significantly. From relations (5) and (6) one can conclude that by shortening the length of the regular section of the cavity or by decreasing the coupling coefficient such that:

$$l_0 < \frac{\pi}{2\alpha} \qquad (7)$$

it is possible to obtain a condition where only one eigenmode with $n = 0$ is located inside each reflection zone $\Lambda_m^2 \leqslant 0$. Thus analysing Figs. 2

and 3 it is easy to see that the azimuthally symmetric mode $m = 0$ and $n = 0$ at the frequency of the precise Bragg resonance $\omega = \bar{\omega}$ possesses the highest Q-factor. The electrodynamic suppression of the modes with index $m \neq 0$ and $n \neq 0$ will result in single-mode oscillation of the FEM.

To study build-up oscillations in a FEM with a two mirror coaxial 2D Bragg cavity computer simulations were carried out. A time-domain analysis described in Refs. [10,11], which takes into consideration the dispersion properties of 2D Bragg structures and ohmic losses, was used. In Fig. 4 the time-dependence of the FEM efficiency is presented for the set of parameters when condition (7) is fulfilled (Fig. 4a) and violated (Fig. 4b), respectively. The spectrum of the output radiation (Fig. 4a) proves single-mode operation for optimal FEM parameters.

3. Experimental study of the 2D Bragg coaxial structure

For the Strathclyde FEM experiment the two-mirror 2D Bragg cavity was machined with radii $r_{in} = 3.02$ cm, $r_{out} = 3.9$ cm and was designed to operate near 37.5 GHz. The 2D Bragg corrugation of depth $a_1 = 0.08$ cm, period $d_z = 0.8$ cm and number of azimuthal variations $\bar{m} = 24$ was made on the outer surface of the brass inner conductor only. It is important to note that for this number of variations $\bar{m}$ the corrugation period over azimuthal coordinate $d_\varphi = 2\pi r_{in}/\bar{m}$ is close to $d_\varphi \cong d_z$. In accordance with (3) this corrugation was chosen for operation at the designed frequency band with the TEM-mode ($A_\pm$ partial waves) and a $TE_{24,0}$-mode ($B_\pm$ partial waves). One notes that for the corrugation parameters chosen the radial variation index of the incident and scattered waves should be the same (see condition (3)). The 2D mirrors for the FEM were made of lengths 10.4 cm (up-stream) and 4 cm (down-stream), respectively.

The microwave parameters of the resonators were measured using a scalar network analyser in the frequency range 26–40 GHz with a resolution of up to 12.5 MHz. For excitation of the Bragg structures a wave-beam in the form of a TEM-

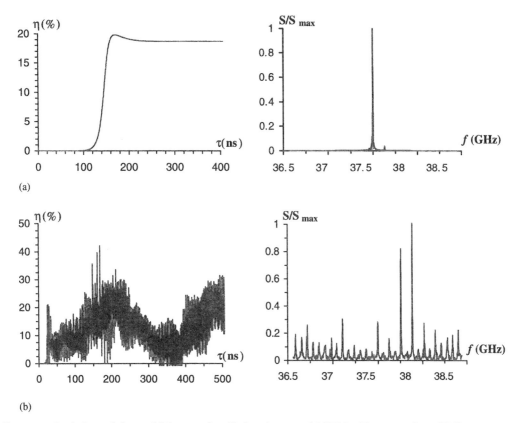

Fig. 4. Computer simulations of the establishment of oscillations in a coaxial FEM with a two-mirror 2D Bragg resonator. Time dependence of efficiency and spectrum of radiation in the stationary regime of oscillations: (a) $l_1 = 10$ cm, $l_0 = 22$ cm, $l_2 = 4$ cm and (b) $l_1 = 16$ cm, $l_0 = 50$ cm, $l_2 = 16$ cm ($\alpha = 0.1$ cm^{-1}, $C = 0.008$, $r_0 = 3.5$ cm, $\sigma = 6 \times 10^{-4}$ cm^{-1}).

mode of a coaxial waveguide was formed at the input of the structure. To produce such a wave-beam an additional transmission line was constructed and included two mode converters, which provide the required mode transformation in the operating frequency-band: from the launched TE$_{1,1}$ wave of a single-mode circular waveguide to a TM$_{0,1}$ wave of circular waveguide (the first mode converter) and to a TEM wave of a coaxial waveguide (the second mode converter). An additional coaxial slowly up-tapered waveguide horn of length ~ 60 cm and an opening angle of 3° were used to connect from the converters to the oversized coaxial cavity.

In accordance with calculations the effective zone of Bragg scattering was observed in the designed frequency region. Results of measure-

ments of the transmission coefficients from the 2D Bragg structures are presented in Fig. 5. As discussed previously the 2D Bragg structures demonstrate totally different behaviour of the transmission and reflection coefficients on the parameter αl_z as compared with 1D Bragg structures (see Fig. 7). For the 2D Bragg structures of different lengths the only difference was in the shape of the reflection zone while approximately the same minimum value of the transmission ($T \approx -20$ dB) at the Bragg frequency was maintained despite changes in the length of the reflector. To prove operation of the 2D feedback cycle via the mutual scattering of four partial waves the transverse electromagnetic energy fluxes (associated with partial waves $\mathbf{B}_{\pm}$) were also measured (Fig. 6) using a slit in the outer

(a)
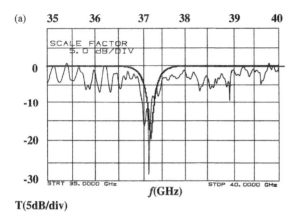

T(5dB/div)

(b)
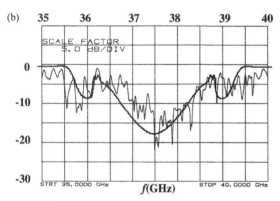

T(5dB/div)

Fig. 5. Frequency dependence of the transmission coefficient from a coaxial 2D Bragg structure of different lengths: thin line corresponds to the "cold" microwave measurements and thick line corresponds to calculations when $d_z = 0.8$ cm, $\bar{m} = 24$, $r_{in} = 3.02$ cm, $r_{out} = 3.95$ cm, $\alpha = 0.1$ cm^{-1}, $\sigma = 6 \times 10^{-4}$ cm^{-1} and (a) $l_z = 4$ cm; (b) $l_z = 10.4$ cm.

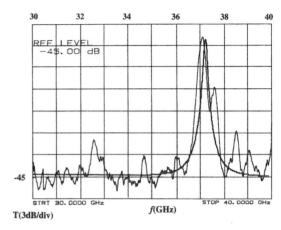

Fig. 6. Frequency dependence of the amplitude of the transverse electromagnetic energy fluxes for the coaxial 2D Bragg structure with length $l_z = 4$ cm (all other parameters are the same as for Fig. 5): thin line-"cold" microwave measurements, thick line-calculations.

conductor wall. This slit allowed a standard waveguide probe to be attached to the cavity without disturbing the field structure. It is important to note that it was only possible to measure distinct transverse electromagnetic fluxes as compared to a background RF signal when the waveguide probe was positioned tangentially to the waveguide surface. Thus a reflection zone associated with pure 2D Bragg scattering was observed for the first time. Comparing results with those obtained in previous experimental studies of 2D corrugated structures [13–15] a parasitic reflection zone corresponding to 1D Bragg scatter-

ing was measured and the centre of the reflection zone associated with 2D Bragg scattering was down shifted from the exact Bragg frequency. It is important to note that in the case of the previous 2B Bragg structure the existence of a parasitic zone of reflection may lead to unstable FEM operation [14].

During the experiments conducted with the "chessboard" pattern structure some small side bands were also observed near the main transmission zone (see Fig. 5). The presence of these "parasitic" zones may be caused by a small violation in the axial symmetry of the system and the presence of a small mixture of a non-symmetric TE$_{1,0}$ mode at the input to the Bragg structure. From the results of the "cold" experiments the mixture of the "parasitic" modes and the "parasitic" reflection may be estimated as less than 10%. It is important to note that the small downshift of the reflection zones from the central frequency can be attributed to an error in the period of the structure due to a tolerance accuracy of $\sim +0.05$ mm in the manufacture of the 2D Bragg reflector.

To increase understanding of 2D Bragg structures "cold" microwave measurements of a coaxial 1D Bragg structure was performed. The 1D Bragg mirror was made to operate at the same frequency band with length 30 cm, period 0.4 cm and corrugation depth 0.04 cm. In accordance with calculations two zones of effective scattering for the launched TEM-wave was observed in the operating frequency region: scattering into the backward wave of the same TEM-type wave near the frequency of 37.5 GHz and scattering into a backward $TM_{0,1}$ type wave of a coaxial waveguide near the frequency of 39 GHz. The widths of the zones, which is determined by the wave coupling parameter of the 1D Bragg structure [3], observed in the experiment correspond well with the results of calculations (Fig. 7). Note that the minimum value of the transmission coefficient at the frequency of precise Bragg resonance was close to $T^2 \approx 1 - \tanh^2(\alpha l_z)$. It is important to underline that no distinguishable (from the background signal) transverse electromagnetic fluxes along the azimuthal co-ordinate was measured for the 1D Bragg structure for both transmission zones. The signal was only obtained when the waveguide probe was located normal to the surface of the waveguide for the frequency band corresponding to the second zone (TEM→$TM_{0,1}$) Fig. 7. This proves that direct scattering in the longitudinal direction of the forward $\mathbf{A}_+$ and backward $\mathbf{A}_-$ partial waves occurred on the 1D Bragg structure. The observation of transverse waves for the 2D Bragg reflector in contrast with the

results obtained for 1D structures proves for the first time operation of two-dimensional distributed feedback via the transverse waves $\mathbf{B}_\pm$ which act to complete the 2D feedback cycle.

4. Conclusion

A theoretical and first experimental study of a novel two-dimensional coaxial Bragg structure with a "chessboard" pattern has been carried out. A reflection band near the designed frequency of 37.5 GHz for the fundamental TEM mode was observed. In the frequency band studied (26–40 GHz) only one reflection zone associated with 2D Bragg scattering was measured with no additional parasitic 1D Bragg reflection bands detected. Measurement of the transmission characteristics of the 2D Bragg coaxial reflector resulted in observation of pure 2D scattering as well as confirmation of the 2D feedback cycle. The results obtained from these experiments were compared with theoretical predictions under the assumption that the 2D Bragg structure could be considered an "ideal" corrugation. Good agreement was obtained between theoretical predictions and experimental results. This has confirmed the optimum corrugation geometry for two-dimensional feedback to be a "chessboard" pattern. First experimental comparison of the novel 2D Bragg structures with traditional 1D Bragg structures was carried out and the high selective properties of the 2D "chessboard" structure was demonstrated. The conditions for establishing single-mode oscillation were studied using a time-domain analysis and an efficiency of ~20% in the single-frequency regime was predicted.

Acknowledgements

The authors would like to thank Prof. G.G. Denisov for useful discussions and for his help with the experiments conducted. The authors would like to thank the Royal Society of UK, EPSRC, the Russian Foundation for Basic Research and the UK DERA for partial support of this work. I.V. Konoplev would like to thank the

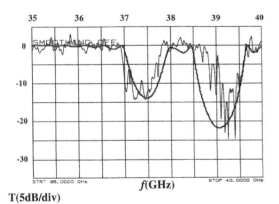

T(5dB/div)

Fig. 7. Frequency dependence of transmission coefficient from a coaxial 1D Bragg structure ($l_z = 30$ cm, $d_z = 0.4$ cm, $r_{in} = 3.02$ cm, $r_{out} = 3.95$ cm, $\alpha = 0.08$ cm^{-1}): thin line-"cold" microwave measurements, thick line-calculations.

Committee of Vice Chancellors and Principals of the Universities of the United Kingdom and Strathclyde University for support.

References

[1] T.M. Antonsen, B. Levush, Phys. Rev. Lett. 62 (1989) 1488.

[2] N.S. Ginzburg, M.I. Petelin, Int. J. Electron. 59 (1985) 291.

[3] V.L. Bratman, G.G. Denisov, N.S. Ginzburg, M.I. Petelin, IEEE J. Quantum Electron. QE-19 (1983) 282.

[4] T.S. Chu, F. Hartmann, B.G. Danly, R.J. Temkin, Phys. Rev. Lett. 72 (1994) 2391.

[5] M. Wang, Z. Wang, J. Chen, Z. Lu, L. Zhang, Nucl. Instr. and Meth. A 304 (1991) 116.

[6] P. Zambon, W.J. Witteman, P.J.M. Van der Slot, Nucl. Instr. and Meth. A 341 (1994) 88.

[7] N.S. Ginzburg, A.A. Kaminsky, A.A. Kaminsky, N.Yu. Peskov, S.N. Sedykh, A.P. Sergeev, A.S. Sergeev, Phys. Rev. Lett. 84 (2000) 3574.

[8] N.S. Ginzburg, N.Yu. Peskov, A.S. Sergeev, Opt. Commun. 112 (1994) 151.

[9] N.S. Ginzburg, N.Yu. Peskov, A.S. Sergeev, A.V. Arzhannikov, S.L. Sinitsky, Nucl. Instr. and Meth. A 358 (1995) 189.

[10] N.S. Ginzburg, I.V. Konoplev, A.S. Sergeev, Tech. Phys. (Zhurnal Tekhnicheskoi Fiziki) 66 (5) (1996) 108.

[11] N.S. Ginzburg, A.S. Sergeev, N.Yu. Peskov, G.R.M. Robb, A.D.R. Phelps, IEEE Trans. Plasma Sci. 24 (1996) 770.

[12] N.S. Ginzburg, N.Yu. Peskov, A.S. Sergeev, A.D.R. Phelps, I.V. Konoplev, G.R.M. Robb, A.W. Cross, A.V. Arzhannikov, S.L. Sinitsky, Phys. Rev. E 60 (1999) 935.

[13] I.V. Konoplev, A.W. Cross, W. He, A.D.R. Phelps, K. Ronald, G.R.M. Robb, C.G. Whyte, N.S. Ginzburg, N.Yu. Peskov, A.S. Sergeev, Nucl. Instr. and Meth. A 445 (2000) 236.

[14] A.V. Arzhannikov, N.V. Agarin, V.B. Bobylev, N.S. Ginzburg, V.G. Ivanenko, P.V. Kalinin, S.A. Kuznetsov, N.Yu. Peskov, A.S. Sergeev, S.L. Sinitsky, V.D. Stepanov, Nucl. Instr. and Meth. A 445 (2000) 222.

[15] N.Yu. Peskov, N.S. Ginzburg, G.G. Denisov, A.S. Sergeev, A.V. Arzhannikov, P.V. Kalinin, S.L. Sinitsky, V.D. Stepanov, Techn. Phys. Lett. (Pis'ma v ZhTF) 26 (4) (2000) 348.

[16] A.F. Harvey, Microwave Engineering, Academic Press, New York, 1963.

Nuclear Instruments and Methods in Physics Research A 475 (2001) 173–177

NUCLEAR
INSTRUMENTS
& METHODS
IN PHYSICS
RESEARCH
Section A

www.elsevier.com/locate/nima

Novel scheme of a multi-beam FEL synchronised by two-dimensional distributed feedback

N.S. Ginzburg[a], N.Yu. Peskov[a,*], A.S. Sergeev[a], A.V. Arzhannikov[b], S.L. Sinitsky[b]

[a] *Institute of Applied Physics, Russian Academy of Sciences (RAS), 46 Uljanov St., Nizhny Novgorod 603600, Russia*
[b] *Budker Institute of Nuclear Physics RAS, Novosibirsk 630090, Russia*

Abstract

A novel scheme of a multi-beam generator operated with two-dimensional distributed feedback has been proposed. The generator consists of different planar FEL-units, each one utilising a two-dimensional Bragg resonator and is driven by a sheet electron beam. Synchronisation of radiation from different units would be achieved via transverse electromagnetic energy fluxes taking place in 2-D Bragg resonators. © 2001 Elsevier Science B.V. All rights reserved.

PACS: 41.60.Bq; 42.50.Fx; 41.60.Cr; 52.75.Ms; 41.60.m

Keywords: FEL; Spatial coherence; Distributed feedback; Bragg resonator; High-power microwaves

1. Introduction

The possibility to use two-dimensional distributed feedback for generation of powerful microwave radiation from large-size sheet and annular relativistic electron beams is actively being studied both theoretically [1–3] and experimentally [4–6]. This feedback mechanism may be realised in two-dimensional Bragg resonators of planar and coaxial geometry. Computer simulations predicted the possibility to achieve a spatially coherent single-mode operating regime in both FEL schemes when the transverse size of the beam exceeds the wavelength by orders of magnitude. The first operation of this novel feedback mechan-

ism was demonstrated experimentally in a 75-GHz planar FEL based on accelerator ELMI (INP RAS, Novosibirsk). In these experiments the FEL was driven by a sheet electron beam with energy of 1 MeV, pulse duration of 5 μs, linear beam current density of 200 A/cm and transverse cross-section of $0.3 \times 20 \, \text{cm}^2$. As a result, in the experiments a 100 MW-level RF-pulse of microsecond duration was obtained [4,5].

A potential development of these experiments is to increase the total power in the microwave pulse by expanding the interaction space over both transverse coordinates. The use of 2-D distributed feedback makes it possible to design a multi-beam generator consisting of several planar FEL-units (modules) joined together (Fig. 1a) and working simultaneously in parallel. Each unit would be driven by a sheet electron beam moving in a planar undulator and based on a 2-D Bragg resonator.

*Corresponding author. Tel.: +7-8312-384-575; fax: +7-8312-362-061.
E-mail address: peskov@appl.sci-nnov.ru (N.Yu. Peskov).

(a)

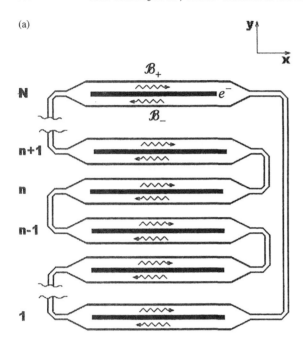

(b)

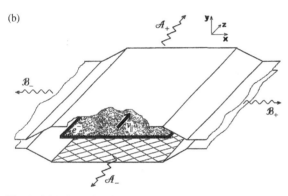

Fig. 1. (a) Schematic of multi-beam generator consisting of N planar FEL-units (in each unit the transverse cross-section of the electron beam and the transverse e.m. energy fluxes are shown). (b) Schematic of single planar FEL-unit based on a 2-D Bragg resonator driven by a sheet relativistic electron beam.

Such resonators would be coupled by a special waveguide system (Fig. 1a), and transverse (with respect to the electron beam) electromagnetic energy fluxes arising in these resonators would synchronise the whole generator and provide spatial coherence of the radiation.

It is possible to design both oscillator and amplifier schemes of the device. In the oscillator scheme the coupling waveguide system must form the circuit (Fig. 1a). In the amplifier scheme the coupling waveguide system would need to be opened in order to input the RF-signal. The present paper is devoted to theoretical analysis of the FEL-oscillator scheme. The project for experimental study of a dual-beam FEL is considered and will be based on ELMI accelerator.

2. Model and basic equations

A schematic of the multi-beam FEL-oscillator is presented in Fig. 1a. Each unit (Fig. 1b) includes the 2-D Bragg resonator consisting of two metal plates corrugated as

$$a(x, z) = a_1(\cos(\bar{h}x - \bar{h}z) + \cos(\bar{h}x + \bar{h}z)) \quad (1)$$

where $\bar{h} = \sqrt{2}\pi/d$, d is the corrugation period and $2a_1$ is the corrugation depth. Each unit is driven by a sheet relativistic electron beam moving in the $+z$ direction, which is guided by an axial magnetic field, and which oscillates in a planar wiggler covering each FEL-module. Different modules are coupled by a special waveguide system via the transverse electromagnetic energy fluxes arising in these resonators when excited by the electron beams. To complete the feedback circuit the first and last resonators must be connected to each other (Fig. 1a).

The electromagnetic field within the unit with number n may be presented in the form of the four partial waves: two partial waves $(\mathbf{A}_{\pm})$ propagating in forward and backward directions with respect to the electron beam and two partial waves $(\mathbf{B}_{\pm})$ propagating in the transverse directions

$$\vec{E} = \vec{E}_0 \operatorname{Re}\lfloor (\mathbf{A}_{+,n} e^{-i\bar{h}z} + \mathbf{A}_{-,n} e^{i\bar{h}z} + \mathbf{B}_{+,n} e^{-i\bar{h}x} + \mathbf{B}_{-,n} e^{i\bar{h}x}) e^{i\bar{\omega}t} \rfloor \quad (2)$$

where $\mathbf{A}_{\pm}(x, z, t)$, $\mathbf{B}_{\pm}(x, z, t)$ are slowly varying amplitudes along the coordinates and with respect to time, $\bar{\omega} = \bar{h}c$ is the Bragg frequency (frequency of precise Bragg resonance), which has been chosen as the reference frequency. The excitation of the nth unit by the electron beam and the build-up of oscillations can be described by the following system of equations (cf. with Refs. [2,3]):

$$\left(\frac{\partial}{\partial Z} + \beta_{gr}^{-1}\frac{\partial}{\partial \tau}\right)A_{+,n} + i\alpha(B_{+,n} + B_{-,n}) = J_n,$$

$$J_n = \frac{1}{\pi}\int_0^{2\pi} e^{-i\theta_n}\, d\theta_{0n} \qquad (3a)$$

$$\left(-\frac{\partial}{\partial Z} + \beta_{gr}^{-1}\frac{\partial}{\partial \tau}\right)A_{-,n} + i\alpha(B_{+,n} + B_{-,n}) = 0 \qquad (3b)$$

$$\left(\pm\frac{\partial}{\partial X} + \beta_{gr}^{-1}\frac{\partial}{\partial \tau}\right)B_{\pm,n} + i\alpha(A_{+,n} + A_{-,n}) = 0 \quad (3c, d)$$

$$\left(\frac{\partial}{\partial Z} + \beta_\parallel^{-1}\frac{\partial}{\partial \tau}\right)^2 \theta_n = Re(A_{+,n}e^{i\theta_n}) \qquad (3e)$$

where $Z = \bar{h}zC$, $X = \bar{h}xC$, $\tau = \bar{\omega}tC$, $(\hat{A}_\pm, \hat{B}_\pm) = e\kappa\mu(\mathbf{A}_\pm, \mathbf{B}_\pm)/mc\bar{\omega}\gamma_0 C^2$, $K \approx \beta_\perp/2\beta_\parallel$ is the parameter describing coupling between the wave and electrons, $\mu \approx \gamma_0^{-2}$ is the inertial bunching parameter, $v_{\parallel,\perp} = \beta_{\parallel,\perp}c$ is the longitudinal and transverse electron velocity, respectively, $v_{gr} = \beta_{gr}c$ is the wave group velocity, $\theta = \bar{\omega}t - hz - h_w z$ is the electron phase with respect to the synchronous wave, $h_w = 2\pi/d_w$, d_w is the wiggler period,

$$C = \left(\frac{e\hat{I}_0\,\lambda^2 K^2\mu}{mc^3 8\pi\gamma_0 a_0}\right)^{1/3}$$

is the gain parameter, and $\hat{I}_0$ is the unperturbed electron current per unit transverse size (linear current density), a_0 is the distance (gap) between plates and α is the wave coupling parameter on the 2-D Bragg structure proportional to the corrugation depth a_1 [1,2].

Boundary conditions for Eq. (3) for the partial waves propagating in longitudinal directions have the form

$$A_{+,n}|_{Z=0} = 0, \qquad A_{-,n}|_{Z=L_z} = 0 \qquad (4)$$

(L_z is the length of the FEL-units) and correspond to the e.m. energy fluxes from outside the resonator being absent and the partial waves not reflecting from the resonator edges. Boundary conditions for a monoenergetic unmodulated electron beam take the form

$$\theta_n|_{Z=0} = \theta_{0n} \in [0, 2\pi), \quad \left(\frac{\partial}{\partial Z} + \beta_\parallel\frac{\partial}{\partial \tau}\right)\theta_n\bigg|_{Z=0} = \Delta_n \qquad (5)$$

where $\Delta_n = (\bar{\omega} - hv_\parallel - h_w v_\parallel)/\bar{\omega}C$ is the initial mismatch from resonance in the nth unit. For the partial waves propagating in the transverse x directions and providing coupling between different FEL-units boundary conditions may be presented in the form

$$[B_{+,n}(\tau) = \Gamma B_{-,n-1}(\tau - T)]_{X=0},$$

$$[B_{-,n}(\tau) = \Gamma B_{+,n+1}(\tau - T)]_{X=L_X}$$

$$[B_{-,N}(\tau) = \Gamma^N B_{+,1}(\tau - NT)]_{X=L_X},$$

$$[B_{-,n}(\tau) = \Gamma^N B_{+,N}(\tau - NT)]_{X=L_X} \qquad (6)$$

where L_x is the width of the FEL-units, T is the time delay for the partial waves when propagating between two neighbouring units, $\Gamma < 1$ is the ohmic losses coefficient inside the coupling waveguide. We also assume that the time delay between the first and the last FEL-modules is equal to NT, where N is the total number of the FEL-modules. It is interesting to note that under conditions (6) all the modules represent a complete circuit and the process of synchronisation of oscillations in the multi-beam FEL may be considered analogous to the process in a coaxial FEL utilising 2-D feedback [3]. Similar to the coaxial 2-D Bragg FEL scheme, introducing the ohmic losses is necessary for obtaining steady-state oscillations.

3. Results of simulations

The simulations of the multi-beam FEL operation were carried out for the parameters close to conditions of the ELMI experiments (see Refs. [4,5] for details]). The oscillation transverse velocity into the beam $\beta_\perp \approx 0.2-0.25$ was pumped in a planar wiggler of 4 cm period, and the amplitude of transverse component at the axis up to 0.2 T with a guide magnetic field ~ 1 T. The width of each sheet beam was taken to be 15 cm with a linear current density ~ 200 A/cm. As a result for the designed frequency 75 GHz for the FEL operation we get a gain parameter $C \approx 4 \times 10^{-3}$. Figs. 2 and 3 present results of simulations for the generator made up from 6 FEL-units having the dimensionless width $L_x = 1.2$ (for the conditions

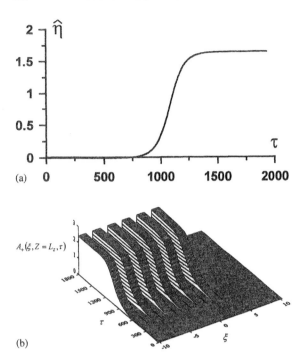

Fig. 2. Computer simulations of build-up oscillations in 6-modules ($N = 6$) FEL-generator. Time dependence of (a) normalised electron efficiency and (b) RF-field structure at the output of the generator ($Z = L_z$) for the parameters close to the conditions of the experiments at the ELMI accelerator ($L_z = 6.8$, $L_x = 1.2$, $T = 1.2$, $\alpha = 0.1$, $\Delta = -0.2$, $\Gamma = 0.95$).

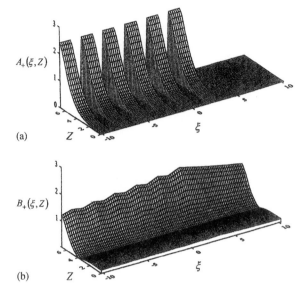

Fig. 3. Spatial profiles of the amplitudes of the partial waves (a) $\mathbf{A}_+$ and (b) $\mathbf{B}_+$ at the stationary regime of generation for the parameters given in Fig. 2.

of ELMI experiments this corresponds to $l_x = 20$ cm) and the dimensionless length $L_z = 6.8$ ($l_z = 110$ cm). The delay time between the units was taken as $T = 1.2$ (0.67 ns) and the ohmic loss inside the coupling waveguide was $\Gamma \approx 0.95$. Fig. 2 illustrates the process of build-up oscillations and establishment of stationary regime of generation: time dependence of electron efficiency summarised over all the units is shown in Fig. 2a, time dependence of the structure of output radiation (i.e. the partial wave $\mathbf{A}_+$) at $Z = L_z$ is shown in Fig. 2b. At this regime the oscillations established at the frequency coincided with the Bragg resonance frequency $\bar{\omega}$. Note that in the figures we introduce the transverse coordinate ξ, which describes the partial wave profile over the whole generator including all the FEL-units and the coupling waveguides. Fig. 3 presents the spatial profiles of the partial wave $\mathbf{A}_+$, which is synchronous with electrons, and one of the transversely propagating waves ($\mathbf{B}_+$), which provide synchronisation of oscillations from different units over the whole generator.

It is important to underline that the synchronisation of oscillations was observed even in the case when the different units were driven by electron beams with slightly different energies and, thus, possess different initial mismatches from the synchronism Δ_n. For the parameters of the system considered, the establishment of a stationary regime of spatially coherent generation took place up to the point where the difference in the mismatches Δ_n amounted to 50% from the zone of self-excitation. For a higher difference in mismatches, excitation of the generator was not observed. The regimes when the different modules generated different frequencies, i.e. when the synchronisation of oscillations was not realised, were not obtained in simulations that were performed.

Acknowledgements

The authors would like to thank the Russian Foundation for Basic Research, ISTC, and the UK Royal Society for partial support of this work.

References

[1] N.S. Ginzburg, N.Yu. Peskov, A.S. Sergeev, A.V. Arzhannikov, S.L. Sinitsky, Nucl. Instr. and Meth. A 358 (1995) 189.

[2] N.S. Ginzburg, N.Yu. Peskov, A.S. Sergeev, A.D.R. Phelps, I.V. Konoplev, G.R.M. Robb, A.W. Cross, A.V. Arzhannikov, S.L. Sinitsky, Phys. Rev. E 60 (1999) 935.

[3] N.S. Ginzburg, N.Yu. Peskov, A.S. Sergeev, Opt. Commun. 112 (1994) 151.

[4] A.V. Arzhannikov, N.V. Agarin, V.B. Bobylev, N.S. Ginzburg, V.G. Ivanenko, P.V. Kalinin, S.A. Kuznetsov, N.Yu. Peskov, A.S. Sergeev, S.L. Sinitsky, V.D. Stepanov, Nucl. Instr. and Meth. A 445 (2000) 222.

[5] A.V. Arzhannikov, N.V. Agarin, V.B. Bobylev, A.V. Burdakov, N.S. Ginzburg, E.V. Diankova, V.G. Ivanenko, P.V. Kalinin, S.A. Kuznetsov, N.Yu. Peskov, P.V. Petrov, A.S. Sergeev, S.L. Sinitsky, V.D. Stepanov, Abstracts of 22nd International FEL Conference, Durham, USA, 2000, p. 33.

[6] A.W. Cross, N.S. Ginzburg, W. He, I.V. Konoplev, N.Yu. Peskov, A.D.R. Phelps, G.R.M. Robb, K. Ronald, A.S. Sergeev, C.G. Whyte, Abstracts of 22nd International FEL Conference, Durham, USA, 2000, p. 176.

ELSEVIER

Nuclear Instruments and Methods in Physics Research A 475 (2001) 178–181

NUCLEAR
INSTRUMENTS
& METHODS
IN PHYSICS
RESEARCH
Section A

www.elsevier.com/locate/nima

Simulations of the TJNAF 10 kW free electron laser

R.D. McGinnis, J. Blau, W.B. Colson*, D. Massey, P.P. Crooker[1],
A. Christodoulou, D. Lampiris

Department of Physics, Naval Postgraduate School, Monterey, CA 93943, USA

Abstract

The TJNAF Free Electron Laser (FEL) will be upgraded to operate at 10 kW average power in the near future. Multimode simulations are used to analyze the operation describing the evolution of short optical pulses in the far infrared wavelength regime. In an FEL that recirculates the electron beam, performance can depend on the electron beam distribution exiting the undulator. The effects of varying the undulator field strength and Rayleigh length of the resonator are explored, as well as the possibility of using an optical klystron. The simulations indicate that the FEL output power can reach the design goal of 10 kW. © 2001 Elsevier Science B.V. All rights reserved.

PACS: 41.60Cr

Keywords: Frec-electron laser; Klystron; Rayleigh length

1. Introduction

The Thomas Jefferson National Accelerator Facility (TJNAF) has plans to modify their 2 kW [1] free electron laser (FEL) to operate at increased power of 10 kW [2]. The modification will involve adding an accelerator section to increase the electron beam energy up to 150 MeV. The accelerator will provide half picosecond long electron pulses at a repetition rate of 75 MHz with 270 A peak current, only 0.17% energy spread, and an electron beam radius of 350 μm. The average power in the electron beam will be 1.5 MW. To obtain a 10 kW beam from the FEL an extraction efficiency of approximately 0.7% is required. The undulator has $\lambda_0 = 20$ cm wavelength with rms undulator parameter $K = 2$ over $N = 24$ periods resulting in $\lambda = 5.8$ μm wavelength radiation. The undulator includes a dispersive section so that a klystron interaction can be explored. The optical resonator provides a Rayleigh length of 2 m and has 15% output coupling.

It is useful to introduce dimensionless parameters that describe the physics of the FEL design [3]. The electron pulse length l_e is nearly equal to the slippage distance $N\lambda$ and is described by $\sigma_z = l_e/N\lambda$. When σ_z is close to unity, the FEL interaction is affected by the continuously changing overlap of the short optical pulse slipping over the short pulse of bunching electrons. The dimensionless current density j is taken at the peak of the electron pulse, which is assumed to be

---

*Corresponding author. Tel.: +1-831-656-2765; fax: +1-831-656-2834.

E-mail address: colson@nps.navy.mil (W.B. Colson).

[1] On leave from Department of Physics and Astronomy, University of Hawaii, Honolulu, HI 96822, USA.

Table 1
TJNAF FEL parameters for various values of K

	$K = 1$	$K = 2$	$K = 3$	$K = 4$
λ (µm)	2.3	5.8	12	20
σ_z	2.7	1.1	0.54	0.32
jF	2.9	3.7	3.8	3.8
F	0.06	0.02	0.01	0.007

parabolic in shape. The resonator design determines the Rayleigh length, the transverse optical mode size, and the filling factor F averaged along the undulator. The desynchronism $d = -\Delta S/N\lambda$ is the shortening of the resonator cavity length by ΔS compared to the slippage distance $N\lambda$, and over many passes n, is crucial in determining the steady-state FEL gain, power, optical pulse shape, spectrum, and induced energy spread. The strength of the klystron dispersive section is measured by D, which is the equivalent drift length divided by the undulator length.

The undulator parameter is defined by $K = eB\lambda_0/2\pi mc^2$ where the rms field strength B of the electromagnetic undulator can be varied. Table 1 shows how various parameters are affected as K is increased from 1 to 4. The FEL wavelength increases from 2.3 to 20 µm due to the resonance condition, $\lambda = \lambda_0(1 + K^2)/2\gamma^2$. The slippage distance $N\lambda$ is increased, so the dimensionless electron pulse length drops from $\sigma_z = 2.7$ to 0.32. Longer wavelengths produce a large optical mode, so the filling factor is reduced from $F = 0.06$ to 0.007. The dimensionless current density j increases as K^2, but the product jF used in the simulations varies only slightly, from $jF = 2.9$ to 3.8.

2. Optical power

Fig. 1 shows the results of longitudinal multimode simulations with no klystron ($D = 0$). The average output power is plotted vs. desynchronism d. The maximum power of 14.5 kW is achieved with undulator parameter $K = 2$, and similar results are obtained for $K = 1$. The simulations also predict a short optical pulse of about 0.5 ps and a narrow optical spectrum. The final electron

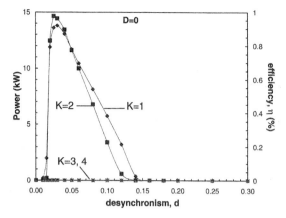

Fig. 1. Output power and efficiency vs. desynchronism without klystron ($D = 0$). The power is greater than the design goal of 10 kW for $K = 1, 2$. For $K = 3, 4$ there is no output due to small F and σ_z.

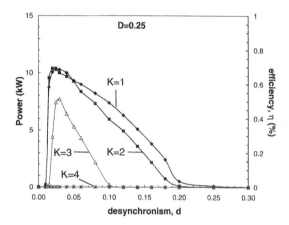

Fig. 2. Output power and efficiency vs. desynchronism with weak klystron ($D = 0.25$). The power is greater than the design goal of 10 kW for $K = 1, 2$. For $K = 3$, the maximum power is 7.5 kW, while for $K = 4$ there is no output due to small F and σ_z.

energy spread of $\sim 5.5\%$ is within the requirements for recirculation of the electron beam [2]. No power is achieved for $K = 3$ and 4 due to the smaller filling factor F and shorter electron pulse length σ_z.

Fig. 2 shows that the introduction of a klystron section of dispersive strength $D = 0.25$ should reduce the average power to slightly above 10 kW for $K = 1$ and 2. The weak klystron also provides enough gain to the FEL to enable operation with $K = 3$, but at an output power level of only 7.5 kW. In this case, the final electron energy spread is reduced to about 4%.

With stronger klystron dispersive values above $D = 0.5$, all four undulator parameter values produce some output power, but less than the desired value of 10 kW average power.

3. Transverse mode effects

Fig. 3 shows a three dimensional simulation in x, y, and τ of the proposed laser. The dimensionless parameters corresponding to undulator parameter $K = 2$ are shown in the upper right corner of the graph, along with the grey scale for the intensity plots. The transverse dimensions are normalized to $(L\lambda/\pi)^{1/2}$, and the longitudinal dimensions are normalized to the undulator length L. The electron beam radius is $\sigma_e = 0.12$ and the Rayleigh length is $z_0 = 0.4$. The dimensionless optical field amplitude is indicated by $|a|$. At each

pass, the electrons are injected with an initial Gaussian spread in phase velocity of width $\sigma_G = 0.5$, centered on $v_0 = 9$. Cavity losses are given by $Q = 6$ corresponding to approximately 15% mirror transmission and edge losses around the mirrors of 0.1% per pass. The mirror radius is $r_m = 3.2$, with radius of curvature $r_c = 1.2$, and the optical waist is located in the center of the undulator, $\tau_w = 0.5$. The mirror separation was artificially shortened to twice the undulator length for numerical convenience. The simulation was started in weak fields, $a_0 = 0.1$, and allowed to evolve until steady-state power was obtained in a stable resonator mode.

The upper-left plot $|a(x, n)|$ shows the evolution of a slice through the optical mode over $n = 500$ passes. The upper-middle plot $|a(x, y)|$ shows the final optical wavefront at the undulator exit. The center plot $|a(x, \tau)|$ shows a slice through the optical wavefront during the final pass. The lower-left plot $f(v, n)$ shows the evolution of the electron

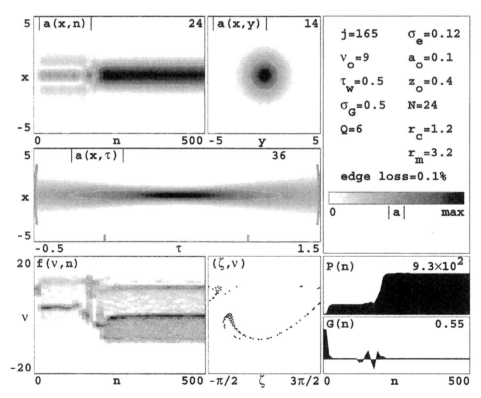

Fig. 3. Three dimensional simulation in x, y, and τ over many passes n, for normalized Rayleigh length $z_0 = 0.4$. The graphs and parameters are explained in the text.

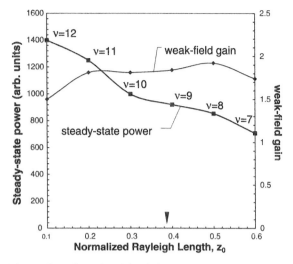

Fig. 4. Three dimensional simulation results for weak-field gain and steady-state power vs. z_0. The optimum resonance parameter v is indicated at each point.

phase velocity distribution, and next to it is the final electron phase space plot. In the lower-right is the evolution of the electron power $P(n)$ and gain $G(n)$.

In weak optical fields, $a \lesssim \pi$, the small electron beam distorts the resonator mode, reducing the filling factor and enhancing the gain in this example by about $\times 4$ compared to a simple theoretical estimate. As the field grows to saturation in strong fields, the figure shows that gain and mode distortion are reduced. Notice that at about $n = 200$ passes, the optical mode narrows and the power grows rapidly again, finally developing a more symmetric Gaussian mode with steady power.

The Rayleigh length affects the filling factor along the length of the cavity, and hence the FEL gain and efficiency. Fig. 4 shows the results of three dimensional simulations for weak-field gain and steady-state power exploring different Ray-leigh lengths z_0. The gain varies only slightly over the range $z_0 = 0.1{-}0.6$, with a peak around 0.5. The steady-state power increases approximately linearly as z_0 decreases. For $z_0 = 0.1$, the power is about 50% greater than at the original design value of $z_0 = 0.4$ (indicated by the triangular tick mark on the horizontal axis). Also shown at each point is the optimum resonance parameter v, which increases with shorter z_0 [3]. Another obvious advantage of a shorter z_0 is that the laser spot size at the mirrors increases, reducing the power density on the mirrors.

Three dimensional simulations were also used to look at mode distortion effects in the optical klystron. As expected, the weak-field gain is enhanced by the klystron, but the saturated power drops significantly. For $D = 0.25$, the power is reduced by $\times 4$ compared to the results with no klystron. For larger D the final power is also reduced significantly.

Acknowledgements

The authors are grateful for the support by the Naval Postgraduate School, and S. Benson for many useful discussions.

References

[1] G.R. Neil, et al., First Operation of an FEL in Same-cell Energy Recovery Mode, Nucl. Instr. and Meth. A 445 (2000) 192.

[2] D. Douglas, S.V. Benson, G.A. Krafft, R. Li, L. Merminga, B.C. Yunn, Driver accelerator design for the 10 kW upgrade of the Jefferson lab IR FEL, Linear Accelerator Conference, August 2000.

[3] W.B. Colson, in: W.B. Colson, C. Pellegrini, A. Renieri (Eds.), Laser Handbook, Vol. 6, North-Holland, Amsterdam, 1990 (Chapter 5).

ELSEVIER

Nuclear Instruments and Methods in Physics Research A 475 (2001) 182–186

**NUCLEAR
INSTRUMENTS
& METHODS
IN PHYSICS
RESEARCH**
Section A

www.elsevier.com/locate/nima

Simulations of the TJNAF FEL with tapered and inversely tapered undulators

A. Christodoulou[a], D. Lampiris[a], W.B. Colson[a],[*], P.P. Crooker[a],[1], J. Blau[a],
R.D. McGinnis[a], S.V. Benson[b], J.F. Gubeli[b], G.R. Neil[b]

[a] *Physics Department, Naval Postgraduate School, 833 Dyer Road, Monterey, CA 93943, USA*
[b] *FEL Department, Thomas Jefferson National Accelerator Facility, Newport News, VA 23606, USA*

Abstract

Experiments using the TJNAF FEL have explored the operation with both tapered and inversely tapered undulators. We present here numerical simulations using the TJNAF experimental parameters, including the effects of taper. Single-mode simulations show the effect of taper on gain. Multimode simulations describe the evolution of short optical pulses in the far infrared, and show how taper affects single-pass gain and steady-state power as a function of desynchronism. A short optical pulse presents an ever-changing field strength to each section of the electron pulse so that idealized operation is not possible. Yet, advantages for the recirculation of the electron beam can be explored. © 2001 Elsevier Science B.V. All rights reserved.

PACS: 41.60Cr

Keywords: Taper; Inverse taper; Free-electron laser; Simulation

1. Introduction

For any free-electron laser (FEL), the undulator design determines the physics of the interaction between the relativistic electrons and the co-propagating light pulse [1]. One way to alter the undulator design is to introduce taper. Taper may be achieved by varying the magnetic field along the undulator, and it has been shown that positively and negatively tapered undulators have characteristics that can be advantageous for particular

applications in strong optical fields [2–8]. Recently, Thomas Jefferson National Accelerator Facility (TJNAF) has experimentally explored many properties of the same FEL with no taper, positive taper, and negative taper [9]. The purpose of this paper is to numerically simulate those experiments with a view to theoretically understanding and extending the TJNAF experiments.

2. The TJNAF free-electron laser

The TJNAF FEL has $N = 41$ undulator periods of length $\lambda_0 = 2.7$ cm each. It was operated with short 0.5 ps electron pulses (length $l_e = 150$ μm) of total energy $E = 34.5$ MeV with a 0.25% energy

---

*Corresponding author. Tel.: +1-831-656-2765; fax: +1-831-656-2834.

E-mail address: colson@nps.navy.mil (W.B. Colson).

[1] On leave from Department of Physics and Astronomy, University of Hawaii, Honolulu, HI 96822, USA.

spread, $I = 50$ A peak current, $\lambda = 6$ μm optical wavelength, and a resonator $Q = 10$. The dimensionless undulator parameter $K = eB_{rms}\lambda_0/2\pi mc^2 = 0.98$, with linear taper rates of $\Delta K/K = \pm 5\%, \pm 7.5\%$, and $\pm 10\%$. From these values, we characterize the TJNAF FEL using dimensionless parameters: the current $j = 8N(\pi eKN\lambda_0)^2\rho F/\gamma^3 mc^2 = 10$, where ρ is the beam electron density, and the electron pulse length $\sigma_z = l_e/N\lambda = 1$. The taper is contained in the pendulum equation torque $\delta = -[4\pi NK^2/(1+K^2)](\Delta K/K) = 0, \pm 4\pi, \pm 6\pi, \pm 8\pi$; and for a cavity of length S, the desynchronism is given by $d = -\Delta S/N\lambda$ and is varied from 0 to 0.4.

3. Single mode results

Single mode behavior was explored by examining the gain for a range of initial phase velocities v_0 and initial fields a_0. It is interesting here to contrast the general properties of positive and negative taper. For *positive* taper, the optimum phase velocity in strong fields is at resonance ($v_0 = 0$) where bunching occurs for electrons initially trapped in the closed orbit region of phase space. For *negative* taper, the optimum phase velocity is above resonance ($v_0 > 0$), where bunching occurs as electrons travel around the closed orbit region of phase space. The motivation for using either positive or negative taper is to improve performance in strong fields ($a_0 \gg \pi$).

Fig. 1 shows the gain plotted against both the initial optical field a_0 and the initial phase velocity v_0 for the cases of $\delta = \pm 8\pi$ and $j = 10$. For reference, the untapered ($\delta = 0$, not shown) weak-field ($a_0 \lesssim \pi$) gain peaks at $G = 176\%$, and decreases to 12% in strong fields ($a_0 \gtrsim \pi$) [1]. As the taper increases, the curves are initially shifted along the v_0 axis by $-\delta/2$, the weak-field peak is reduced, and the strong field peak is increased. At large taper, as shown in Fig. 1, the gain spectrum becomes highly distorted. For $\delta = +8\pi$, the single $\delta = 0$ peak is replaced by two comparably sized peaks, which decrease from 29% in weak fields to 17% in strong fields. Two comparably sized peaks also appear for $\delta = -8\pi$, but in that case the peaks

do not merge at strong fields, decreasing from 27% in weak fields to 17% in strong fields. In strong fields the gain with either taper is better than that with no taper. Also note that the two curves in Fig. 1 are related to each other in the sense that $G(v_0, \delta, a_0) \simeq -G(-v_0, -\delta, a_0)$.

4. Multimode results

Multimode behavior is simulated by introducing a short parabolic electron pulse into the optical resonator and examining the evolution of the pulse and the optical mode as a function of the number of round trips n the optical pulse has made through the resonator. Fig. 2 shows the results after 2000 passes through an undulator with positive taper ($\delta = +8\pi$). The upper graphs give the optical field shape $|a(z,n)|$, the optical power spectrum $P(v,n)$, and the electron spectrum $f(v,n)$ at the final pass. The shading in the middle graphs shows how these quantities have evolved with n. On the bottom left, the longitudinal profile of the current density $j(z - \tau)$ is shown for reference at dimensionless times $\tau = 0$ and 1. The bottom center graph shows the weak-field gain spectrum for reference, and the right-hand bottom graph shows the evolution of the total power P as the pass number n increases. Several parameters are printed across the top; the peak current j, the resonator Q, the pulse width σ_z, δ, N, and the standard deviation of a small fluctuation $\delta\zeta$ in the initial electron phases. Also listed is the desynchronism $d = -\Delta S/N\lambda$, where ΔS is a reduction in resonator length from perfect synchronism. The displacement between the electron and optical pulses at $\tau = 0$ on each pass is determined by d. In addition to showing many interesting details of the behavior, the simulations also print out the total steady-state power P at saturation and the weak-field, steady-state gain G, both of which are strong functions of desynchronism. These quantities provide a good general assessment of the effect of taper on the FEL behavior as shown in Figs. 3 and 4.

Figs. 3 and 4 summarize the results of many simulations. The steady-state power P and weak-field gain G are plotted versus desynchronism for

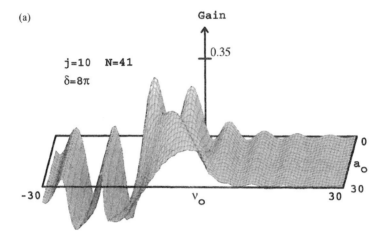

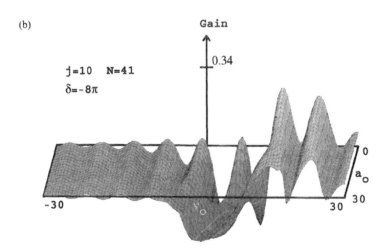

Fig. 1. FEL gain spectrum $G(v_0, a_0)$ for large tapers. (a) $\delta = +8\pi$; (b) $\delta = -8\pi$.

$\delta = 0, \pm 4\pi, \pm 6\pi, \pm 8\pi$ and $j = 10$. The most general feature is that the operating range decreases as δ increases in magnitude. [The operating range is also a function of the current: a taper rate of $\delta = \pm 8\pi$ and $j = 6$ (not shown) will not work at all.] For $d < 0.004$ the laser will not operate, but the power rises sharply before $d \simeq 0.02$. Simultaneously, the weak-field gain increases. In the start-up region, the number of passes n required to achieve the final power is large, but for higher d the number of passes is greatly reduced. At small values of desynchronism, the optical pulse is short (0.5 ps) with a broad spectrum. For $\delta = 0$ and negative tapers, the sharp peaks in power are accompanied by evidence of the trapped-particle

instability. This effect is reduced for positive tapers and vanishes for $\delta = 8\pi$. This instability is caused by electrons in the presence of strong optical fields becoming trapped in potential wells in phase space and oscillating at the trapped-particle synchrotron frequency $v_s = (|a|^2 - \delta^2)^{1/4}$ [1].

For larger d and positive δ, the power curves flatten, with the power diminishing significantly as the taper increases. For negative δ, flattening does not occur and the power declines uniformly to zero with increasing d. In the same region, the steady-state gain is approximately parabolic for all tapers, but with reduced range in d as δ increases in magnitude. Note that the untapered power exceeds the positively tapered power for most of the range of d. For $\delta = -4\pi$, however, tapering provides power greater than the untapered case up to $d \simeq 0.18$, while for $\delta = -6\pi$ the improvement ends at $d \simeq 0.09$. Finally, when d is very large, the optical and electron pulses fail to overlap, and the power is reduced to zero. In that region, the

optical pulses become very long, as much as $5\sigma_z$, with a long leading edge. Note also that the gain is approximately the same for both positive and negative tapers.

For $d < 0.1$, the final power, gain, and electron spectrum may oscillate regularly — up to 50% modulation of the average power in some cases and over hundreds of undulator passes. For these regions, shown by the large circles and squares on the power and gain curves in Figs. 3 and 4, only the peak values of the steady-state power and the average values of gain are shown. We attribute these non-steady effects to limit-cycle behavior, caused when trapped particles in strong fields combine with short optical pulses. The modulation, caused by the oscillation of the trapped current, continually modifies the shape of the short optical pulse. The different pulse shapes have different powers and spectra, causing oscillations as subpulse structures "march" through the pulse envelope. Fig. 2 shows an example for $\delta = +8\pi$ and $d = 0.026$, where one finds limit-cycle oscillations in $|a(z,n)|$, $P(v,n)$, $f(v,n)$, and $P(n)$.

From the electron spectrum, we have also found the electron energy spread $\Delta\gamma/\gamma$ as a function of d and δ. The curves (not shown) appear much like the power curves of Figs. 3 and 4, with initial

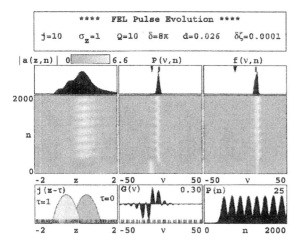

Fig. 2. Multimode simulation for $j = 10$, $d = 0.026$, and $\delta = +8\pi$. The various quantities are explained in the text. In this case, the oscillations in $|a(z,n)|$, $P(v,n)$, $f(v,n)$, and $P(n)$ are evidence for limit-cycle behavior.

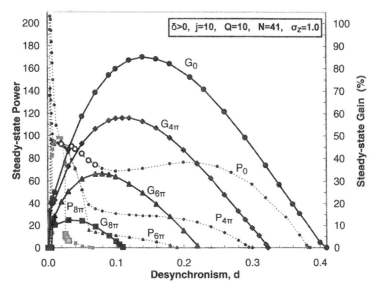

Fig. 3. Steady-state, saturated power P and weak-field, steady-state gain G versus desynchronism for positive taper rates $\delta = 0, +4\pi, +6\pi$, and $+8\pi$.

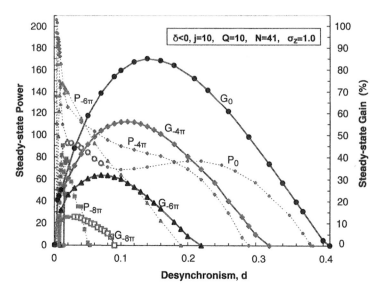

Fig. 4. Steady-state, saturated power P and weak-field, steady-state gain G versus desynchronism for negative taper rates $\delta = 0, -4\pi, -6\pi,$ and -8π.

sharp peaks (up to 7%) near $d = 0.01$. For $d > 0.05$ however, they are all less than 4%.

Finally, when exploring these parameters, we find that at slightly larger values of Q (above $Q = 12$) and negative taper ($\delta = -8\pi$), gain oscillations are observed in weak fields over many passes. The optical pulse becomes spatially modulated at the slippage distance, but there are no trapped electrons. This is a new effect caused by mode competition between two competing peaks in the gain spectrum.

Acknowledgements

The authors are grateful for support by the Naval Postgraduate School.

References

[1] W.B. Colson, in: W.B. Colson, C. Pellegrini, A. Renieri (Eds.), Laser Handbook, Vol. 6, North-Holland, Amsterdam, 1990 (Chapter 5).

[2] G.R. Neil, S. Benson, G. Biallas, C.L. Bohn, D. Douglas, H.F. Dylla, R. Evans, J. Fugitt, J. Gubeli, R. Hill, K. Jordan, G. Krafft, R. Li, L. Merminga, D. Oepts, P. Piot, J. Preble, M. Shinn, T. Siggins, R. Walker, B. Yunn, First operation of an FEL in same-cell energy recovery mode, Proceedings of the 21st International FEL Conference, Hamburg, Germany, August, 1999. Nucl. Instr. and Meth. A 445 (2000) 192.

[3] N.M. Kroll, P.L. Morton, M.N. Rosenbluth, Physics of Quantum Electronics, Vol. 7, 89, 1980.

[4] D.A. Jaroszynski, R. Prazeres, F. Glotin, J.M. Ortega, Nucl. Instr. and Meth. A 358 (1995) 224.

[5] D.A. Jaroszynski, R. Prazeres, F. Glotin, J.M. Ortega, D. Oepts, A.F.G. van der Meer, G. Knippels, P.W. van Amersfoort, Nucl. Instr. and Meth. A 358 (1995) 228.

[6] D.A. Jaroszynski, R. Prazeres, F. Glotin, O. Marcouille, J.M. Ortega, D. Oepts, A.F.G. van der Meer, G. Knippels, P.W. van Amersfoort, Nucl. Instr. and Meth. A 375 (1996) 647.

[7] E.L. Saldin, E.A. Schneidmiller, M.V. Yurkov, Nucl. Instr. and Meth. A 375 (1996) 336.

[8] W.B. Colson, R.D. McGinnis, Nucl. Instr. and Meth. A 445 (2000) 49.

[9] S. Benson, J. Gubeli, G.R. Neil, Nucl. Instr. and Meth. A 475 (2001) 276, these proceedings.

ELSEVIER

Nuclear Instruments and Methods in Physics Research A 475 (2001) 187–189

NUCLEAR
INSTRUMENTS
& METHODS
IN PHYSICS
RESEARCH
Section A

www.elsevier.com/locate/nima

Numerical analysis of radiation build-up in an FEL oscillator

Z. Dong*, K. Masuda, T. Kii, T. Yamazaki, K. Yoshikawa

Institute of Advanced Energy, Kyoto University, Gokasho, Uji, Kyoto 611-0011, Japan

Abstract

A 3D time dependent FEL oscillator simulation code has been expanded to include longitudinal mode competition. In this paper, FEL evolution from spontaneous emission to saturation is analysed. Simulation shows that the cavity transverse mode competition has little effect on the longitudinal mode competition and start-up process. © 2001 Elsevier Science B.V. All rights reserved.

PACS: 41.60.C

Keywords: 3D space-time interaction; Start up; Longitudinal mode

1. Introduction

Unlike an FEL amplifier in which radiation is locked and grows up from a seeded signal, in an FEL oscillator, the radiation starts up from spontaneous emission in the undulator. But in most FEL oscillator simulation codes, simulations usually start from a steady frequency signal, whose power is low enough to assure initial operation in the linear gain regime but high enough to avoid tedious computation in the competition of long-itudinal modes, i.e., it is assumed that the long-itudinal mode with its eigen-frequency nearest the maximum gain frequency has won in the mode competition and survived as the steady state. This means that the Start-up process is totally omitted. Previously published papers on the start-up process in an FEL oscillator have either been based on the mode dynamics equations for a long pulse FEL [1], or been limited to one spatial

dimension [2], which cannot answer whether the 3D effects or transverse mode competition play a role in the start-up process. In this paper, a 3D time dependent FEL code FELTMSM is expanded to include the Start-up process. FELTMSM code is developed by the transverse mode spectral method [4]; its degenerated 1D results show an agreement with the results of a previously developed 1D, multi-frequency code [2,3].

2. Computational model

In the expansion of FELTMSM [5] to Start-up region, the following equations are adopted.

2.1. Electron motion

The equations of motion for the jth electron are as follows:

$$\left(\frac{\partial}{\partial z} + \frac{1}{v}\frac{\partial}{\partial t}\right)\gamma_j$$
$$= \frac{1}{2\gamma_j}F(\mu_j)a_u(\vec{r}_j)k_L \bullet \langle a_{qmn}(\vec{r}_j, \tau)\cos\psi_{qj} \rangle_{qmn}, \quad (1)$$

*Corresponding author. Tel.: +81-774-38-3443; fax: +81-774-38-3449.

E-mail address: dong@iae.kyoto-u.ac.jp (Z. Dong).

$$\left(\frac{\partial}{\partial z} + \frac{1}{v}\frac{\partial}{\partial t}\right)\theta_j$$

$$= k_w - \frac{k_{qs}}{2\gamma_j^2}\left[1 + \frac{1}{2}a_w^2(\vec{r}_j)\right.$$

$$- F(\mu_j)a_w(\vec{r}_j) \cdot \langle a_{qmn}(\vec{r}_j, \tau)\sin\psi_{qj}\rangle_{qmn}$$

$$\left.+ \frac{1}{2}\langle a_{qmn}^2(\vec{r}_j, \tau)\rangle_{qmn} + \langle\gamma_j^2\beta_{\perp bj}^2\rangle\right], \qquad (2)$$

where $\langle \bullet\bullet\bullet \rangle$ means the sum over all TEM$_{qmn}$ modes. The detail definitions for the various parameters can be found in Ref. [5]. ψ_{qj}, the ponderomotive phase of the jth electron for the qth radiation mode can be determined from $\psi_{qj} = (1 - k_q/k_0)k_u z + (k_q/k_0)\theta_j$, where q labels the radiation mode number, k_0 is the wave number of the ω_0 mode, and the frequency of the qth mode of the radiation is $\omega_q = \omega_0(1 \pm q/N)$, $k_q = \omega_q/c$. N is a large number; usually the largest q and N are chosen by the CPU time and memory size limit.

2.2. Wave motion

The radiation field for TEM$_{qmn}$ can be expressed as

$$\left(\frac{\partial}{\partial z} + \frac{\partial}{c\partial t}\right)a_{qmn}$$

$$= iC\langle F(\mu_j)a_u(\bar{x}_j, \bar{y}_j, z)G_{qmn}^* e^{-i\psi_{qj}}/\gamma_j\rangle, \qquad (3)$$

where $C = 1/2k_s(\omega_p^2/c^2)\sigma_b$, ω_p is the electron plasma frequency, σ_b is the effective area of the electron beam. a_{qmn} is $N\lambda_{pon}$ periodic.

3. Simulation results

In this section, we show the numerical examples using the parameters that correspond to the FEL experiments at FELI [6]. In the simulation, a periodic boundary was assumed since the electron bunch is longer than the slippage distance, the particle number for each "electron slice" was chosen to be 256, and the cavity detuning was assumed to be $0.5\lambda_s$. We calculated 21 modes (q) with $N = 100$, the modes having in turn been assumed to be under the small signal gain curve.

The simulation was performed with the first 4 transverse modes (TEM$_{00}$, TEM$_{01}$, TEM$_{10}$, and TEM$_{11}$) to save the CPU time and memory size. The relative intensities of the q longitudinal modes at various passes are shown in Fig. 1, where a noise signal of spontaneous emission level [3] is input. From the simulation, we can conclude that: (1) In saturation, the mode (here $q = 9$) having its eigen-frequency nearest to the maximum gain wins the competition among the longitudinal modes. The contribution from the other longitudinal modes is $<10\%$ of the total intensity of the system at saturation, so we can say that the system is operated in the single longitudinal mode before post saturation (sideband occurs). The largest five

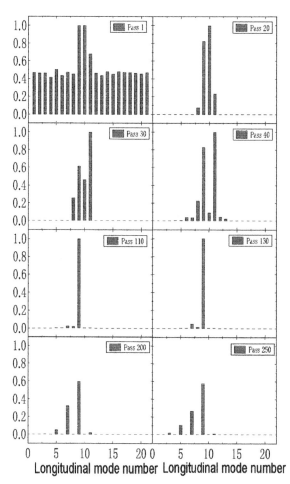

Fig. 1. The relative intensity as a function of the longitudinal mode numbers.

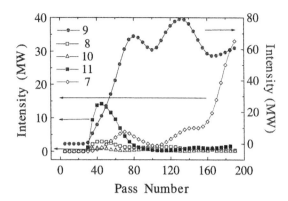

Fig. 2. The largest five longitudinal modes' intensity ($q = 7, 8, 9, 10, 11$). The mode $q = 9$ having its eigen-frequency nearest to the maximum gain wins the competition among the longitudinal modes.

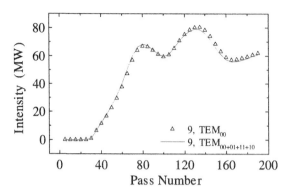

Fig. 3. The intensities of the eigen-frequency mode as a function of pass number. (a) Triangles for fundamental transverse mode only, (b) solid line for multiple transverse modes.

modes' intensities are shown in Fig. 2. (2) The competition among the longitudinal modes exists throughout the whole FEL process, not only in the start-up regime. (3) In the post saturation regime, sideband instability can develop, and the longitudinal mode powers ($q = 7, 5, \ldots$) in the longer wavelength increase. (4) The contribution from the higher transverse modes in the whole FEL process is $< 5\%$ for the 5 largest longitudinal modes. In order to investigate the effects of the transverse mode competition on the start-up process, we made a simulation with the fundamental mode TEM_{00} only; the power of the eigen-frequency mode ($q = 9$) is similar to the one with the first four transverse modes, as shown in Fig. 3. It is concluded that competition among the transverse modes has almost no effect on the start-up process.

4. Conclusions

Little effect of the transverse mode competition on the longitudinal coupling is concluded, i.e. the TEM_{00} mode is found dominant over other transverse modes either in the start-up regime, the linear growth regime or the deep saturation regime.

References

[1] I. Kimel, L.R. Elias, Nucl. Instr. and Meth. A 341 (1994) 191.
[2] Shin-ichiro Kuruma, et al., Nucl. Instr. and Meth. A 341 (1994) 289.
[3] Shin-ichiro Kuruma, et al., Nucl. Instr. and Meth. A 331 (1993) 421.
[4] C. Mei Tang, et al., IEEE J. Quantum Electron. QE-21 (7) (1985) 970.
[5] Z.W. Dong, et al., Nucl. Instr. and Meth. A 445 (2000) 101.
[6] A. Kobayashi, et al., Nucl. Instr. and Meth. A 393 (1997) 280.

ELSEVIER

Nuclear Instruments and Methods in Physics Research A 475 (2001) 190–194

NUCLEAR
INSTRUMENTS
& METHODS
IN PHYSICS
RESEARCH
Section A

www.elsevier.com/locate/nima

Requirements for a laser pumped FEL operating in the X-ray regime

D.F. Gordon[a], P. Sprangle[b], B. Hafizi[c], C.W. Roberson[d],*

[a] National Research Council Associate, Naval Research Laboratory, Washington, DC 20375, USA
[b] Plasma Physics Division, Naval Research Laboratory, Washington, DC 20375, USA
[c] Icarus Research, Inc., P.O. Box 30780, Bethesda, MD 20824-0780, USA
[d] Physical Sciences Division, Office of Naval Research, 800 N. Quimcy Street, Arlington, VA 22217, USA

Abstract

A laser pumped free electron laser (FEL) can in principle generate coherent, polarized, high power X-rays using a much lower energy electron beam than would be required by a conventional FEL. This is due to the dramatic reduction in the wiggler period, which in the case of a laser pumped FEL is the laser wavelength. However, a number of practical obstacles could prevent the realization of a laser pumped FEL in the laboratory. Foremost amongst these is the requirement that the axial velocity spread on the electron beam be less than the difference between the average velocity of the beam and the velocity of the ponderomotive wave. This requirement places a severe limitation on both the beam emittance and energy spread. Conditions are given on the emittance, energy spread, current density, and laser intensity needed to achieve lasing in the X-ray regime. © 2001 Elsevier Science B.V. All rights reserved.

PACS: 42.55.Vc; 41.60.Cr; 41.50.+h

Keywords: X-ray FEL; Laser pumped FEL

1. Introduction

Advances in laser technology have resulted in a number of new research areas. The newly-developed ultra-high power (>TW), ultra-short pulse (⩽1 ps) lasers [1] may have applications in a number of areas ranging from advanced high gradient particle accelerators to advanced sources of short wavelength radiation. Combining the advances in generating ultra high intensity laser beams with the steady improvements in generating high quality, high current density electron beams,

make a laser pumped X-ray FEL an interesting and possibly timely idea to reconsider.

In a conventional FEL, an electron beam is propagated through a static periodic magnetic (wiggler) field [2,3]. In a laser pumped FEL (LPFEL), an intense laser field replaces the wiggler field. The laser beam is directed anti-parallel to the direction of the relativistic electron beam and the stimulated radiation propagates parallel to the electron beam. In a LPFEL, the output radiation wavelength is given by $\lambda = \lambda_0/\mu_0$ where

$$\mu_0 = 4\gamma_0^2/(1 + a_0^2) \tag{1}$$

is the LPFEL frequency multiplication factor, λ_0 is the pump laser wavelength, a_0 is the laser strength

---

*Corresponding author. Tel.: +1-703-696-4222.
E-mail address: roberson@onr.navy.mil (C.W. Roberson).

parameter and γ_0 is the relativistic mass factor of the electrons. Here, it has been assumed the electron beam is highly relativistic. As an example, using a pump laser with $\lambda_0 = 1\ \mu m$ and $a_0 = 1$, the output radiation is in the XUV with $\lambda \approx 40\ \text{Å}$, yet the required electron beam energy is only 5 MeV (cf. [4]).

2. Growth rate and efficiency

In one dimension, for a cold electron beam in the high gain regime the spatial growth rate is [2,3]

$$\Gamma[\mu m^{-1}]$$

$$= 6.24 \times 10^{-3} f_x^{1/3} \left(\frac{J_b[kA/cm^2]}{\lambda_0[\mu m]} \right)^{1/3} \frac{\mu_0^{-1/2} a_0^{2/3}}{(1 + a_0^2)^{1/2}} \tag{2}$$

and the intrinsic efficiency is

$$\eta[\%] = 4.6 \frac{\lambda_0[\mu m] \Gamma[\mu m^{-1}]}{(1 - 1/\gamma_0)} \tag{3}$$

where $f_x = r_b^2 / r_x^2$ is the filling factor, $J_b = I_b / \pi r_b^2$ is the current density of a square profile electron beam in units of kA/cm^2, and a_0 is the normalized (dimensionless) vector potential of the pump laser (laser strength parameter). For a circularly polarized pump the laser strength parameter is given by

$$a_0 = 6 \times 10^{-10} \lambda_0[\mu m] (I_0[W/cm^2])^{1/2}$$

where $I_0 = 2P_0 / \pi r_0^2$ is the pump laser intensity in units of W/cm^2, λ_0 is the pump laser wavelength in units of μm, and r_0 is the laser spot size. For a fixed frequency multiplication factor the growth rate and efficiency are maximized for $a_0 = \sqrt{2}$.

3. Limitations due to beam quality

The expressions for the growth rate and efficiency are based on a one dimensional, cold beam model. A cold beam is defined by $S \ll 1$, where $S = \delta v_z / (v_{z0} - v_{pond})$, δv_z is the axial velocity spread, v_{z0} is the average beam velocity, and v_{pond} is the phase velocity of the ponderomotive wave [2]. If $S \gg 1$, lasing will not occur and X-rays will be generated incoherently [5–8]. The parameter S therefore provides a measure of beam quality.

The requirement $S \ll 1$ can be expressed in the form [3]

$$\frac{\delta v_z}{c} \ll \frac{\lambda}{L_e} \tag{4}$$

where λ is the operating wavelength and $L_e = 1/\Gamma$ is the e-folding length. The velocity spread δv_z consists of a contribution from the energy spread, $\delta \gamma$, and a contribution from the normalized transverse emittance, ε_n. Since these two quantities are generally uncorrelated, their combined effect may be found by summing each contribution in quadrature:

$$\frac{\delta v_z}{c} = \gamma_0^{-2} \left[(1 + a_0^2)^2 \left(\frac{\delta \gamma}{\gamma_0} \right)^2 + \frac{1}{4} \left(\frac{\varepsilon_n}{r_b} \right)^4 \right]^{1/2}. \tag{5}$$

Substituting Eq. (5) into the inequality (4) we find that the thermal effects can be neglected and the cold beam model is valid provided

$$\Gamma \gg \frac{4}{\lambda_0 (1 + a_0^2)} \left[(1 + a_0^2)^2 \left(\frac{\delta \gamma}{\gamma_0} \right)^2 + \frac{1}{4} \left(\frac{\varepsilon_n}{r_b} \right)^4 \right]^{1/2}. \tag{6}$$

The minimum emittance for a given beam current is determined from

$$(\varepsilon_n)_{min} \approx \left(\frac{I_b}{J_c \pi} \right)^{1/2} \left(\frac{k_B T_c}{mc^2} \right)^{1/2} \tag{7}$$

where I_b is the total beam current, J_c is the maximum current density at the cathode, T_c is the effective temperature of the cathode, and k_B is the Boltzmann constant. The minimum energy spread on the beam is given by $\delta \gamma = v$, where $v = (\omega_b r_b / 2c)^2 = I_b[kA] / 17\beta_0$ is Budker's parameter. This energy spread results from the potential drop across the beam due to self fields. If the expressions for the minimum emittance and minimum intrinsic energy spread are inserted into the inequality in Eq. (6), one finds that there is an *upper* bound on the electron beam current density:

$$J_b[kA/cm^2] \ll 1.74 \times 10^{-4} f_x^{1/2} (1 + a_0^2)^{3/4} a_0 \lambda_0[\mu m]$$

$$\times \left[\mu_0 \left(\frac{k_B T_c / mc^2}{J_c[kA/cm^2]} \right)^2 + 0.55 \times 10^{-16} \right.$$

$$\left. \times (1 + a_0^2) r_b^4[\mu m] \right]^{-3/4}. \tag{8}$$

The inequality in Eq. (8) is a manifestation of the fact that as the beam current is increased, the LPFEL growth rate does not increase rapidly enough to make up for the loss in beam quality.

4. Requirements for high output

Realization of a high output LPFEL requires that the number of e-foldings in a Rayleigh length of the pump laser, $N_e = \Gamma z_R$, and the X-ray power at saturation, $X = \eta I_b (\gamma - 1) mc^2/e$, both be large. From Eqs. (2) and (3), we find

$$N_e = 1.28 \mu_0^{-1/2} \left(\frac{I_b[\text{kA}]}{17} \right)^{1/3} \left(\frac{P_0[\text{GW}]}{a_0(1 + a_0^2)^{3/4}} \right)^{2/3}$$

(9)

$$X[\text{MW}] = 210 I_b[\text{kA}] \left(\frac{I_b[\text{kA}]}{17} \right)^{1/3} \left(\frac{a_0^4}{P_0[\text{GW}]} \right)^{1/3}.$$

(10)

The dependence on the laser parameters P_0 and a_0 is such that increasing N_e tends to lower X. However, as will be shown graphically below, it is possible to choose a direction in the (P_0, a_0) plane which increases both N_e and X simultaneously.

Increasing the beam current, I_b, increases both N_e and X. However, the beam current cannot be chosen arbitrarily due to the fact that the beam quality limitation imposed by Eq. (8) places an upper limit on the beam current.

5. Example: pumping with a CO$_2$ laser

Next, we consider the particular case of generating 100 Å radiation using a 10 μm laser as the pump. This implies $\mu_0 = 1000$. For the cathode, we take the typical photo-injector parameters $T_c \approx 0.2$ eV and $J_c = 600$ A/cm^2.

In Fig. 1 we plot the e-folding length, L_e, and the intrinsic efficiency, η, as functions of the current density J_b for three values of the laser strength parameter a_0. The e-folding length can be as little as 1 mm, although the intrinsic efficiency is low.

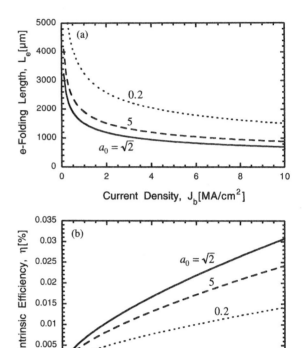

Fig. 1. LPFEL amplification with $\lambda_0 = 10$ μm and $\mu_0 = 1000$ for laser strength parameters of $a_0 = \sqrt{2}$, $a_0 = 0.2$, and $a_0 = 5$. (a) e-Folding length $L_e = 1/\Gamma$ vs. current density J_b. Minimum L_e occurs for $a_0 = \sqrt{2}$. (b) Intrinsic Efficiency η vs. current density J_b.

In Fig. 2 we plot the number of e-foldings in X-ray power, N_e, the saturated X-ray power, X, and the maximum current, I_{max}, as limited by the beam quality requirement of Eq. (8). These are plotted as iso-contours in the plane of laser power and laser strength parameter (P_0, a_0). It is desired that all three quantities (N_e, X, I_{max}) be large. By moving along a vector in the (P_0, a_0) plane with a slope less than that of the N_e contour but greater than that of the X contour all three quantities (N_e, X, I_{max}) can be simultaneously increased. In other words, it is advantageous to increase both P_0 and a_0, but only if it is done in the right proportion.

As an example of a possible operating condition, we take $P_0 = 10$ TW and $a_0 = 1$. Then $I_{max} \approx 540$ A. Taking $I_b = I_{max}/10 \approx 50$ A would

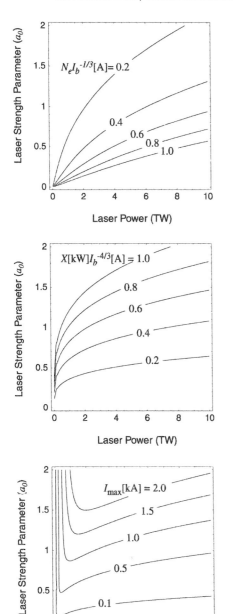

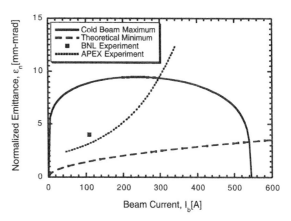

Fig. 3. Normalized emittance requirement for a pump laser with $\lambda_0 = 10$ μm, $P_0 = 10$ TW, $a_0 = 1$, and $r_0 = 150$ μm. The frequency multiplication factor is $\mu_0 = 1000$. The solid curve denotes the maximum normalized emittance allowed by the cold beam FEL model as determined from Eq. (11), assuming space charge limited energy spread. The dashed line is the minimum emittance of a photocathode as determined from Eq. (7) with $T_c = 0.2$ eV and $J_c = 600$ A/cm². The dotted line is an empirical fit to experimental data from the Los Alamos APEX photoinjector [9]. The solid square is the experimentally measured emittance at 110 A of the Brookhaven photoinjector [10].

Fig. 2. Total X-ray production with $\lambda_0 = 10$ μm and $\mu_0 = 1000$. (a) Contours of $N_e I_b^{-1/3}$[A], where N_e is the number of e-foldings in a Rayleigh length and I_b is the beam current, see Eq. (9). (b) Contours of X[kW]$I_b^{-4/3}$[A], where X is the saturated X-ray power, see Eq. (10). (c) Contours of I_{max}[kA], where I_{max} is the maximum current allowed by the cold beam FEL model assuming $T_c = 0.2$ eV and $J_c = 600$ A/cm², see Eq. (8).

give $2N_e \approx 4$ and $X \approx 70$ kW. The spot size would be 150 μm, the Rayleigh length would be $Z_R = 8$ mm, and the current density would be $J_b = 0.06$ MA/cm². The beam energy would be about 10 MeV.

To put the requirements on beam quality in a realistic context, we take the laser parameters from the above example and compare the maximum emittance allowed by the cold beam FEL model with the emittances of two actual photoinjectors. Using Eq. (6), one finds that the maximum emittance is

$$(\varepsilon_n)_{max} = 2^{1/2}(1 + a_0^2)^{1/2}r_b\left[\frac{\Gamma^2\lambda_0^2}{16} - \left(\frac{\delta\gamma}{\gamma_0}\right)^2\right]^{1/4}. \tag{11}$$

Fig. 3 shows $(\varepsilon_n)_{max}$ vs. I_b plotted alongside two experimentally determined emittances from two actual photoinjectors [9,10]. We see that the experimental emittances are less than $(\varepsilon_n)_{max}$ over a wide range of currents. Also plotted is the thermal limit on emittance, $(\varepsilon_n)_{min}$, as given by Eq. (7) using $T_c = 0.2$ eV and $J_c = 600$ A/cm^2. As expected, the experimental emittances are somewhat larger than $(\varepsilon_n)_{min}$. Note also that $(\varepsilon_n)_{min} = (\varepsilon_n)_{max}$ when $I_b = I_{max}$.

6. Conclusions

Using a multi-terawatt CO_2 laser, one might be able to produce tens of kilowatts of coherent 100 Å radiation using a 10 MeV electron beam as the lasing medium. The success of such a scheme would depend strongly on the quality of the electron beam. In particular, the beam emittance must be much less than the maximum emittance of Eq. (11). At the present time, the emittances of real electron beams with substantial currents are factors of a few less than the maximum emittance. To what extent this is sufficient can only be answered by a kinetic calculation of the gain coefficient. However, it appears that existing technology brings the idea of producing X-rays using a laser pumped free electron laser into the realm of reasonable discussion.

Acknowledgements

The authors are grateful to Drs. P. Serafim and A. Ting for many discussions related to this work. This work was supported by the Office of Naval Research.

References

[1] D. Strickland, G. Mourou, Opt. Comm. 56 (3) (1985) 219.
[2] C.W. Roberson, P. Sprangle, Phys. Fluids B 1 (1989) 3.
[3] P. Sprangle, C. Tang, W. Manheimer, Phys. Rev. A 21 (1980) 302.
[4] G. Dattoliand, P. Ottaviani, J. Appl. Phys. 86 (1999) 5331.
[5] P. Sprangle, B. Hafizi, F. Mako, Appl. Phys. Lett. 55 (1989) 2559.
[6] P. Sprangle, A. Ting, E. Esaray, A. Fisher, J. Appl. Phys. 72 (1992) 5032.
[7] C.M. Tang, B. Hafizi, S. Ride, Nucl. Instr. and Meth. A 331 (1993) 371.
[8] A. Ting, et al., J. Appl. Phys. 78 (1995) 575.
[9] P. O'Shea, et al., Phys. Rev. Lett. 71 (1993) 3661.
[10] I. Ben-Zvi, et al., Nucl. Instr. and Meth. A 318 (1992) 208.

ELSEVIER

Nuclear Instruments and Methods in Physics Research A 475 (2001) 195–204

**NUCLEAR
INSTRUMENTS
& METHODS
IN PHYSICS
RESEARCH**
Section A

www.elsevier.com/locate/nima

Operation of the OK-4/Duke storage ring FEL below 200 nm ☆

Vladimir N. Litvinenko[a,*], Seong Hee Park[b], Igor V. Pinayev[a], Ying Wu[c]

[a] *FEL Laboratory, Duke University, P.O. Box 90319, Durham, NC 27708, USA*
[b] *KAERI, 159 Deokjin-Dong Yusung-ku, Taejon, South Korea*
[c] *Lawrence Berkeley National Laboratory, 1 Cyclotron Road, Berkeley, CA 94720, USA*

Abstract

For a number of years the wavelength of 200 nm was a psychological barrier for FEL oscillators. The progress towards short wavelength was marginal since the OK-4/VEPP-3 storage ring FEL lased at 240 nm in 1988. After 10 years, in 1998, the OK-4/Duke FEL and the NIJI-IV FEL group moved the limit to 217 and 212 nm, respectively. Improvements of the OK-4/Duke storage ring FEL gain above 10% and the use of custom manufactured mirror coatings brought the success in August 1999. The OK-4 FEL lased in the range from 193.7 to 209.8 nm using electron energies from 500 to 800 MeV. In this paper, we present the description of the OK-4/Duke FEL up-grades and the lasing results below 200 nm obtained in August and October of 1999. © 2001 Elsevier Science B.V. All rights reserved.

PACS: 41.60.Cr; 29.20D; 78.66; 52.75

Keywords: Free electron laser; Storage ring; Optical klystron; Deep ultraviolet; Coherence

1. Introduction

The 200 nm was a serious barrier for FEL oscillators. Absence of robust highly reflective mirrors and rather low FEL gain were the premier showstoppers. The recent progress with increasing the FEL gain and improving mirror technology made it possible to break the barrier. The OK-4/Duke storage ring FEL (SR FEL) was the first FEL to lase below 200 nm in August 1999. This record was bitten by the rapidly developing TESLA Self-Amplified-Spontaneous-Emission (SASE) FEL, which lased in the range from 181 to 80 nm in April 2000 [1]. Both the FEL oscillators and the SASE FELs have their own flaws, advantages, as well as their unique places in family of coherent light sources. Presently, the OK-4/Duke SR FEL remains the only FEL oscillator which lased below 200 nm.

After a brief discussion of the prior progress towards the short wavelength FELs in Section 2, we present the description of the up-grades of the OK-4/Duke SR FEL systems which provided to lasing at and below 200 nm in August and October of 1999. In Section 4, we report experimental results obtained with the OK-4 FEL in this wavelength range. We conclude by discussing the potential of SR FELs and the future progress towards the VUV and soft-X-ray FEL oscillators.

☆ Work is supported by the ONR MFEL program under Contract #N00014-94-1-0818.

*Corresponding author. Tel.: +1-919-660-2568; fax: +1-919-660-2671.

E-mail address: vl@phy.duke.edu (V.N. Litvinenko).

2. Progress towards short wavelength FELs

The use of an Optical Klystron (OK) driven by a storage ring continues to be the mainstream approach for short wavelength FEL oscillators. The optical klystron—an advanced version of free-electron laser—was suggested and theoretically studied by Vinokurov and Skrinsky [2] in 1977. The traditional OK [2] has two wigglers separated by a buncher, which provides a substantial gain enhancement compared with a traditional FEL. The Budker Institute of Nuclear Physics (BINP, Novosibirsk, Russia) built and operated four OKs, simply named OK-1, OK-2, OK-3 and OK-4, driven by the VEPP-3 storage ring. In May 1995, the OK-4 FEL was transferred to the Duke FEL Laboratory (DFELL) where it was installed on the Duke storage ring and commissioned in 1996. Presently, DFELL and BINP are constructing an OK-5 FEL, a first distributed OK with four electromagnetic wigglers and three bunchers [3].

The first OK, the OK-1 (1979 demonstrated gain of 0.5% at 630 nm in 1980 [4]. The OK-2 (1981) and the OK-3 (1984) FELs had the gains of 1% and 2.5% per pass, respectively [5,6]. Inaccessibility to high reflectivity robust mirrors[1] forestalled the lasing with these FELs.

The ACO SR FEL (LURE, France) was the first visible FEL, which lased at 635 nm in June 1983 [7]. By 1987, it demonstrated lasing at 463 nm with the gain ~0.2% [8]. In 1988, the OK-4/VEPP-3 SR FEL lased in the range from 240 to 690 nm [9–12]. It had gains of 10% at 600 nm and of 4% at 240 nm. Discovery of the radiation resistive HfO/SiO_2 mirror coatings [11,12] made lasing at 240 nm possible for an FEL with gain of 2% per pass or higher.[2]

---

[1] The Soviet Union still existed and the "iron curtain" worked.

[2] According to the experience at Novosibirsk and Duke, initial losses of these mirrors can be as low as 0.5% per mirror. After months of exposure from the OK-4 FEL operating at energies from 500 to 800 MeV and K as high as 4.6, the losses stabilize at level of 1–1.1% per mirror at 237 nm. The reflectivity curve of the downstream mirror, strongly heated by spontaneous radiation, shifted for about 7 nm towards shorter wavelength. We flipped the downstream and upstream mirrors to match their annealing processes. Duke FEL Laboratory, Internal materials.

The further progress was very slow till 1998. The attack on 193 nm with the OK-4/Duke SR FEL in 1996 was inconclusive [13]. The UVSOR helical OK was the first to move the barrier to 239 nm in 1997 [14,15]. In 1998, the NIJI-IV and the OK-4/Duke SR FELs advanced the limit to 226 nm in April [16,17], to 217 nm in August [16], and to 212 nm in October [18,19]. The mirrors with Al_2O_3/SiO_2 coatings were used for 217 nm and 212 nm lasing. The 1998's attempt to lase with the OK-4 FEL at 193 nm with SiO_2/MgF coatings failed—the losses grew to 35% almost instantaneously. Finally, on August 10, 1999, the OK-4/Duke SR FEL lased in 193.7–209 nm range. This result was followed by lasing in a new ELETTRA SR FEL at 217.9 nm in 2000 [20] and abovementioned TESLA SASE FEL at 80 nm [1]. Three SR FEL oscillators, the OK-4/Duke, the NIJI-IV and the ELETTRA, promise to lase at 150 nm and below in near future [16,21].

3. The OK-4/Duke SR FEL up-grades

We made a number of important up-grades to the OK-4/Duke SR FEL systems to attack the 200 nm barrier. First, we installed and commissioned the Duke storage ring Beam-Position Measurement (BPM) system based on the Bergoz's electronics [22]. The use of this system allowed us to correct orbit and to shift the operational point closer to designed values of $Q_x = 9.11$, $Q_y = 4.18$ [23] from previously used $Q_x = 9.311$, $Q_y = 4.266$ [16]. This shift provided for better matching of electron beam and for its better stability. The main limiting factors for the single bunch beam current in the Duke storage ring are the low energy of injection (270 MeV) and the vertical instability. The threshold of this instability is very sensitive to the high order transverse modes of the Duke ring's RF cavity. The use of the new operational point allowed us to de-tune from most dangerous modes and to increase stored current to 17 mA per bunch. Overall performance of the injector was also improved after installation of new low-jitter timing system and new nitrogen laser for linac's photocathode. In 1999, the circulator for the RF system was repaired and the RF voltages up to 650 kV were achievable [24] compared with 200 kV limitation in 1998 [16].

Second, we enhanced the capabilities of the Duke storage ring and the OK-4 FEL diagnostic systems. A new RF master oscillator and a new betatron tune measurement system facilitated the creation of reliable ramping processes. We completed the commissioning of Hamamatsu dual-sweep streak-camera C5680 with deep-UV optics, the MgF window (extending the range to 115 nm), and time resolution better than 1.6 ps [25–27]. Its use provided for expeditious tuning of the optical cavity length, measurements of the e-bunch lengthening and estimation of the cavity losses. We installed the deep-UV Lumogen coated CCD-array system (TE/CCD-1024-EK, Princeton Scientific Inc.) on the second output window of the existing UV monochromator. This system provided for "in situ", live tuning of the OK-4 FEL using instantaneously measured spectra.

Third, we removed the NIST wiggler, which had a questionable quality of magnetic field, and its vacuum chamber from the northern straight section. This also improved the impedance budget of the storage ring.

Fourth, after lengthy discussions with manufacture of our mirrors (Lumonics Optics Group), we found a compromise on the composition of the mirror coatings. The main problem for 200 nm (and below) coatings using SiO_2/AlO_3 pair is a small difference in indexes of refraction. The other pair, SiO_2/MgF, has good indexes ratio, but is notorious for fast degradation. The Lumonics Optics Group found the way to combine the best features of these two pair and to make mirrors with composite coating and with the central wavelength of 200 nm. These mirrors turnout to be reasonably stable and had rather high reflectivity.[3]

All of the above improvements enhanced the OK-4 FEL gain well above 10% per pass and made it possible to lase at and below 200 nm. In 1999, we carried detailed studies of the e-beam quality in the Duke storage ring with special attention on the e-beam emittance growth via

Intra-Beam Scattering (IBS) and the energy spread growth caused by the microwave instability and the FEL lasing [25,26]. The measured bunch lengthening, shown in Fig. 1, indicated a substantial (almost 50%) reduction of the vacuum chamber impedance compared with that measured in August 1996. Bunch shapes did not deviate substantially from Gaussian in all operational ranges.

The broad impedance model approximates the bunch lengthening effects caused by Saturated Microwave Instability (MWI) as following:

$$2\pi \frac{\sigma_t^2 - \sigma_{t0}^2}{T_0^2} = \left(\frac{I_b Z_n/n}{2\pi V_{RF} h_{RF}} \right)^{2/3} \quad (1)$$

where $T_0 = l/f_0$ is the e-beam revolution time, Z_n/n is the effective impedance of the vacuum chamber [28], V_{RF} and $h_{RF} = 64$ are the voltage and the harmonic of the RF cavity, and σ_{t0}, is the natural bunch length. Based on this model, the calculated gain dependencies on the beam current and on the beam energy are shown in Figs. 2 and 3. For large currents, the MWI increases the relative energy spread $\sigma_{\varepsilon MW}$ well above its natural value $\sigma_{\varepsilon 0}$. When $\sigma_{\varepsilon 0} < \sigma_{\varepsilon MW} < 1/4\pi N_w$, where N_w is the number of wiggler periods, the gain of storage

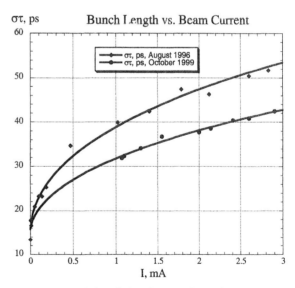

Fig. 1. Measured bunch lengths vs. e-beam bunch current. Both the 1996 and the 1999 measurements were at energy of 500 MeV with an RF voltage of 500 kV. Fits correspond to $Z/n \sim 3\,\Omega$ in 1996 and $Z/n \sim 1.6\,\Omega$ in 1999.

---

[3] The exact composition of coatings is the property of Lumonics Optics Group, 39 Auriga Drive, Nepean, Ont., Canada, K2E 7Y8.

ring OK is given by [12,29]

$$G_{peak} \cong \sqrt{\frac{\alpha_c}{\gamma}} \left(\frac{I_b}{I_A}\right)^{1/3} \left(\frac{Z_0}{Z_n/n}\right)^{2/3} \left(\frac{\pi V_{RF} h_{RF}}{4 I_A Z_0}\right)^{1/6}$$
$$\times N_w g(K_w) e^{-1/2} FF(\xi_{x,y}) f(\beta_{0,x,y}/L_w) \quad (2)$$

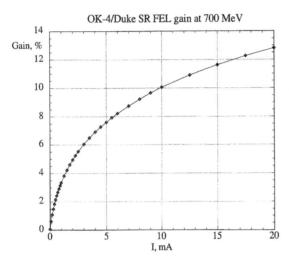

Fig. 2. Predicted OK-4/Duke SR FEL gain at $\lambda = 200$ nm, the beam energy $E_e = 700$ MeV and $V_{RF} = 600$ kV. Simulations include both MWI and IBS.

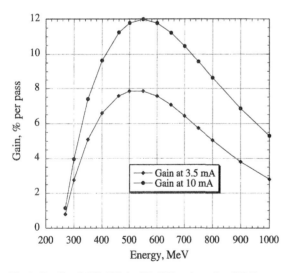

Fig. 3. Predicted OK-4/Duke SR FEL gain at $\lambda = 193.7$ nm as function of the beam energy ($V_{RF} = 600$ kV) for beam currents of 3.5 mA and 10 mA per bunch. Simulations include both the bunch lengthening and energy spread growth via MWI as well as emittance growth via IBS (i.e. the multiple Touschek effect).

where I_b is the current per bunch, $I_A = mc^3/e \sim 17$ kA is Alfen current, $Z_0 = 4\pi/c \sim 377 \, \Omega$ is the impedance of vacuum, α_c is the momentum compaction factor of the ring, $\gamma = E_0/mc^2$ is the relativistic factor of electron, K_w is the wiggler parameter, $g = 8\pi K_w^2/(1 + K_w^2)$ for helical wiggler and $g(K_w) = 4\pi\varsigma(J_0(\varsigma/2) - J_1(\varsigma/2))^2$; $\varsigma = K_w^2/(2 + K_w^2)$ for planar wiggler. The $FF_{x,y}(\xi_{x,y}) \cong (1 + \xi_{x,y} + \xi_{x,y}^2)^{1/2}$ takes into account 3D emittance effects [12,30], where $\xi_{x,y} = 4\pi\varepsilon_{x,y}/\lambda$ and $\varepsilon_{x,y}$ are the transverse emittances. The last parameter, $f(\beta_{0,x,y}/L_w)$ in Eq. (2) takes into account matching of the electron and optical beams with the length of the wiggler $L_w = N_w \lambda_w$, where $\beta_{0,x,y}$ are the Rayleigh range and, x and y β-functions [31]. For a properly optimized storage ring FEL with $\beta_{0,x,y} \sim L_w, f \sim 1$ [12,29,30]. Detailed gain calculations for the OK-4/Duke SR FEL were a bit more elaborate and involved the self-consistent code fel3D [32,33] as well as the emittance growth via IBS. The IBS plays very important role at low energies, high peak currents and short wavelengths.

The measured OK-4 FEL gains were, to our surprise, slightly higher than we predicted. For example, we lased at wavelength 194.5 and 206.27 nm using 2.6 mA in single bunch at 500 MeV. The predicted OK-4 FEL gain was only 6.92% per pass, while the measured optical cavity losses at this wavelength were 7.4±0.1%.

We were measuring the losses of the optical cavity using the traditional ring-down technique [34,8]. Typical measured ring down at $\lambda = 204$ nm is shown in Fig. 4. Due to the finite response time of the detector, this technique yields losses slightly lower than the real one. The only explanation of higher measured gain we can suggest at present time is following: we probably tuned some of high harmonics of the RF cavity to a resonance which increase the effective dV/dt and, therefore, the gain of the OK-4 FEL. The induced dV/dt should correspond to at least 300 kV of RF voltage at 64th harmonic to account for observed gain increase.

We were pleased with the results of up-grades and with excellent performance of the OK-4/Duke SR FEL during these runs. We chose 500 MeV as a starting energy to keep K_w, reasonable ($K_w = 2.4$ for 500 MeV and $K_w = 4.2$ at 800 MeV) to prevent

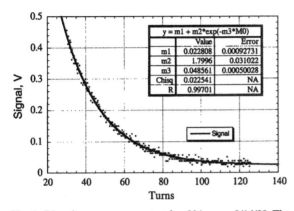

Fig. 4. Ring-down measurement at $\lambda = 204$ nm on 8/16/99. The measured losses were $4.86 \pm 0.05\%$.

fast initial mirror's degradation. The experiments started on August 9, 1999 and finished on October 21, 1999.

4. Lasing around 200 nm with the OK-4 FEL

It took 2 days to obtain first lasing. After installation and alignment of the mirrors, we measured the length of the optical cavity using low current and low energy e-beam (270 MeV) and the Hamamatsu streak camera. After correcting the measured mismatch of 1.9 mm, we obtained first lasing at central wavelength of 200.3 nm.

The OK-4/Duke SR FEL tunability around 200 nm

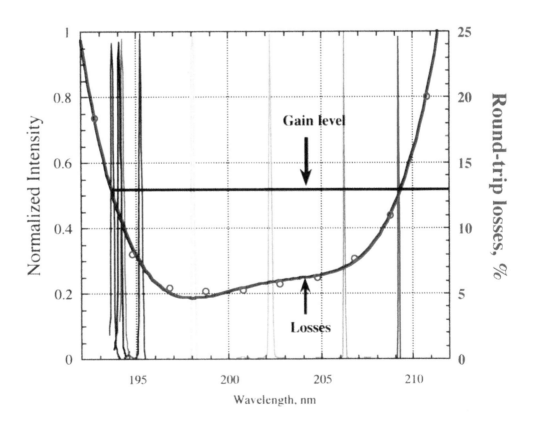

Fig. 5. The tunability measured at $E_e = 500$ MeV and beam current of 3.5–10 mA. Some of the measured lasing lines with peak intensities normalize to unity are shown. The brown curve is the fit of the measured round-trip losses of the optical cavity (circles). The stop-bands, i.e. the minimum and the maximum lasing wavelengths, provide information on the FEL gain. This measurement shows that gain was at least 13% per pass.

During next 8 hours, the OK-4 lased in the range from 193.68 to 209.8 nm. Typical tunability curve is shown in Fig. 5. We operated the OK-4 FEL using the same set of mirrors for about 3 weeks in August and October 1999. After initial lasing at 500 MeV, we studied the lasing spectra, temporal structure and the lasing power using a range of energies from 500 to 800 MeV. The best tunability range was obtained using beam with energies of 500–650 MeV. The maximum lasing power was demonstrated at 800 MeV with tunability range reduced to 194.5–208 nm. The large losses made the lasing at shortest wavelength (Fig. 6) rather weak with RMS line width of 0.067 nm ($\sigma_\lambda/\lambda = 3.5 \times 10^{-4}$).

The lasing at intermediate wavelengths was reasonably powerful and laser light had better quality with RMS line widths of 0.01–0.02% (see Fig. 7). The RMS laser micropulse duration was from 7 to 10 ps. Fig. 8 shows the streak camera image in synchroscan mode and the measured profiles of the laser and electron beam. The product of the measured line width and duration of the laser pulses was 14 times larger than the Fourier limit. The OK-4 FEL is demonstrating the Fourier limited FEL pulses since 1996 and we report these results in a separate paper [27].

Because of high absorption losses, the extraction efficiency was rather low (from 5% to 15% depending on the wavelength). At 500 MeV extracted power was at sub-mW level. For example, 3 mA beam generated 4 mW of lasing

power at $\lambda = 198$ nm, of which 0.33 mW were extracted. We measured 0.16 mW extracted through Downstream Mirror (DM) and 0.17 mW through Upstream Mirror (UM). Typical lasing power with 800 MeV e-beam was from 10–100 mW. For example, with 800 MeV, 4.7 mA beam current lased at 204 nm we measured out-coupled power of 2.95 mW (1.35 mW at UM and 1.6 mW at DM). Table 1 lists the parameters for "8 mA, 800 MeV, 198 nm" test-run, both the predictions and the measurements. A "simple-minded" prediction (Case a) includes the 3D emittance effects, the IBS, but assumes that the lasing completely suppresses the MWI. This assumption worked well for the results obtained with the OK-4/Duke SR FEL when the gain was much higher than the losses. The energy spread induced by the lasing was 2–5 times larger compared to the threshold of the MWI and suppressed the instability completely. In this case, the optimized lasing power can be calculated using formula [12,29]:

$$P_l = J_\varepsilon P_{SR} \frac{\sigma_{\varepsilon\mathrm{ind}}^2 - \sigma_{\varepsilon 0}^2}{\sigma_{\varepsilon\mathrm{ind}}} e^{-1/2} \mathrm{FF}\left(\frac{4\pi\varepsilon_{x,y}}{\lambda}\right); \; \sigma_\varepsilon = \frac{\sigma_E}{E_0}; \tag{3}$$

where P_{SR} is the power of synchrotron radiation, $\sigma_{\varepsilon\mathrm{ind}}$, $\sigma_{\varepsilon 0}$ are the induced and the natural relative

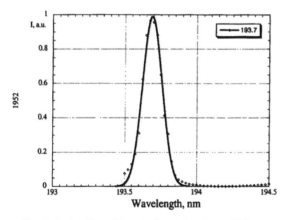

Fig. 6. Lasing line with central wavelength of 193.7 nm.

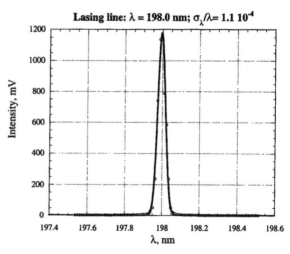

Fig. 7. Measured lasing line at 198 nm with RMS line width is 0.011%.

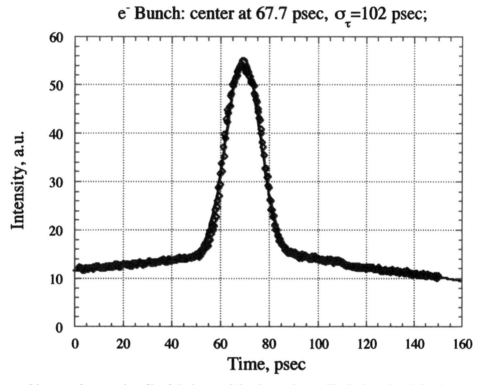

FEL Pulse: center at 69.6 psec, $\sigma_\tau = 7.07$ psec;

e⁻ Bunch: center at 67.7 psec, $\sigma_\tau = 102$ psec;

Fig. 8. Measured image and temporal profile of the laser and the electron beams. The horizontal scale for the image is 150 ps. The central (bright) spot is the laser beam with RMS duration of 7.07 ps, the long (blue) tails are from spontaneous radiation of the e-beam with RMS duration of 102 ps. $\lambda = 198$ nm, $\sigma_\lambda/\lambda = 1.1 \times 10^{-4}$.

energy spreads, and J_ε is energy partition number [31] ($J_\varepsilon = 2.07$ for the Duke ring). When the losses are comparable with the peak gain, the induced energy spread may be insufficient to suppress MWI (Case b). Lasing at 198 nm (8 mA, 800 MeV) induced the energy spread of $\sigma_{\varepsilon\text{ind}} \cong 1.22\sigma_{\varepsilon\text{MW}}$. In this case, $\sigma_{\varepsilon 0}$ should be replaced in Eq. (3) by

$\sigma_{\varepsilon\text{MW}} \cong 2.1\sigma_{\varepsilon 0}$ [29]. This estimation is closer to the measured results.

Still there was unknown loss of 1.7 mW of power: 0.5 mW (15%) at UM and 1.2 mW (35%) at DM. The explanation was found when we removed and examined the downstream CaF_2 window. It was strongly damaged by years of

Table 1
OK-4/Duke SR power at 198 nm

Beam energy (MeV)	800	
Beam current (mA)	8.05	
Losses per turn (%)	5.15	
Transparency per mirror (%)	0.3	
Predicted power (mW)	Case a	Case b
Lasing power	175	69
Extracted power (per mirror)	10.2	4.0
At power meter (per side)[a]	8.5	3.4
Measured power (mW)		
Upstream side (UM)	2.9	
Downstream side (DM)	2.2	

[a] Losses in the window and lens were 16%.

exposure, had a dark-violet color and an additional absorption. Similar, but weaker, degradation is possible for the upstream window. Annealing of the DM and the downshift of its transparency curve can also explain the observed asymmetry of power out-coupling between UM and DM. After few days of operation at 800 MeV, this ratio was $P_{UM}/P_{DM} = 1.32$ at 198 nm and $P_{UM}/P_{DM} = 0.85$ at 204 nm. This assumption is consistent with the measurements of the removed mirrors and 5 nm downshift of DM transparency curve.

Degradation of the mirrors was faster with 800 MeV e-beam. In October 1999, after 3 weeks of operations, the cavity losses grew up to $\sim 7\%$ per pass at 200 nm, and we decided to finish these experiments. In addition to lasing with measurable powers at and below 200 nm, the OK-4/Duke SR FEL generated, for the first time, semi-monoenergetic γ-ray beams with energies up to 58 MeV via Compton back-scattering and collimation [17].

5. Discussions

Further advancement of SR FEL oscillators into the VUV and soft-X-ray regions will require substantial increase of the FEL gain well above 100% per pass, the low e-beam emittances, as well as the optical and wiggler set-ups providing for longer mirror's lifetime. High-K helical wigglers

provide both advantages by increasing the FEL gain by factor ~ 2 and by eliminating the high intensity harmonic radiation on the axis of the optical cavity. Low emittances at nm rad level and low impedance of vacuum chamber are critical for attainment of the high FEL gain. The rapid success of the ELLETTRA SR FEL sets an example of the advantages provided by the third generation storage rings. It has low emittances, smooth (low impedance) vacuum chambers and peak e-beam currents at a half-kA level [20]. The ELLETTRA-type storage rings would be perfect drivers for next generation of the VUV storage ring FELs. They will make SR FELs competitive in short-wavelength range with other types of FELs and lasers.

High gain SR FELs need long straight sections similar to that 35 m long of the Duke storage ring [35,36]. The advanced, flexible high gain SR FEL systems can be based on distributed OK scheme [37]. The first example of such system is the OK-5 FEL [38], which is in development since 1998 and is close to the completion. The commissioning of the OK-5 FEL with helical EM wigglers [39], planned for 2002, should eliminate the mirror degradation problems at Duke. The OK-5/Duke SR FEL would have the gain $\sim 50\%$, which is sufficient to operate below 200 nm with broadband mirrors. The rather large emittance of the "2-and-half generation" Duke storage ring and high impedance of its vacuum chamber would prevent the OK-5 from having gain measured at hundreds of percents below 100 nm. We are considering further improvements of the Duke storage ring [40] with emittance of 1 nm rad at 1 GeV and a smooth ELETTRA-type vacuum chamber. After these up-grades, the OK-5/Duke SR FEL would have a gain of 830% per pass at 100 nm [38]. This level of gain would allow further progress towards soft-X-ray FEL oscillator at Duke. The high gain and the flexibility of the OK-5 FEL system was also designed to provide efficient soft-X-ray harmonic generation [3].

Prior to installation of the new OK-5 FEL, we plan some modest advances to shorter wavelength with OK-4 FFL. We also plan to start generation of coherent tunable harmonics in the OK-4 FEL operating in giant pulse mode in early 2001.

6. Conclusions

Recent progress towards VUV with storage ring FELs is very encouraging. The storage rings seem to overcome a number of critical barriers for making a storage ring soft-X-ray FEL feasible in foreseen future. The OK-4/Duke SR FEL broke at least one psychological barrier on this road.

Acknowledgements

The authors are thankful to the Duke FEL light source operation group, Dr. Ping Wang, James Gustavsson and Marurice Pentico, engineers and technicians at FEL laboratory, Mark Emamian, Joe Faircloth, Steve Hartman, Nelson Hower, Marty Johnson, Peter Morcombe, Owen Oakeley, Janet Patterson and Gary Swift.

References

[1] J. Rossbach, et al., Observation of Self-Amplified Spontaneous Emission in the Wavelength range from 80 nm to 180 nm at the TESLA Test Facility FEL at DESY, Nucl. Instr. and Meth. A 475 (2001) 13, these proceedings.

[2] N.A. Vinokurov, A.N. Skrinsky, Preprint 77–59, BINP, Novosibirsk, Russia, 1977;
N.A. Vinokurov, A.N. Skrinsky, Theory of Optical Klystron, MW Relativistic Electronics, Gorky, 1981, p. 204;
N.A. Vinokurov, Optical Klystron: Theory and Experiment, Thesis, Novosibirsk, 1986.

[3] V.N. Litvinenko, S.F. Mikhailov, N.A. Vinokurov, N.G. Gavrilov, D.A. Kairan, G.N. Kulipanov, O.A. Shevchenko, T.V. Shaftan, P.D. Vobly, Y. Wu, The OK-5/Duke Storage Ring VUV FEL with Variable Polarization, Nucl. Instr. and Meth. A 475 (2001) 407, these proceedings.

[4] A.S. Artamonov, N.A. Vinokurov, P.D. Voblyi, E.S. Gluskin, G.A. Kornyukhin, V.A. Kochubei, G.N. Kulipanov, V.N. Litvinenko, N.A. Mezentsev, A.N. Skrinsky, Nucl. Instr. and Meth. 177 (1980) 247.

[5] G.A. Kornyukhin, G.N. Kulipanov, V.N. Litvinenko, N.A. Vinokurov, P.D. Vobly, Nucl. Instr. and Meth. 208 (1983) 189.

[6] G.A. Kornyukhin, G.N. Kulipanov, V.N. Litvinenko, N.A. Mezentsev, A.N. Skrinsky, N.A. Vinokurov, P.D. Voblyi, Nucl. Instr. and Meth. A 237 (1985) 281.

[7] M. Billardon, P. Elleaume, J.M. Ortega, C. Bazin, M. Berger, M. Velghe, Y. Petroff, D.A.G. Deacon, K.E. Robinson, J.M.J. Madey, Phys. Rev. Lett. 51 (1983) 1652.

[8] M. Billardon, P. Elleaume, J.M. Ortega, C. Bazin, M. Berger, M.E. Couprie, Y. Lapierre, R. Prazeres, M. Velghe, Y. Petroff, Europhys. Lett. 3 (1987) 689.

[9] N.G. Gavrilov, L.G. Isaeva, G.N. Kulipanov, V.N. Litvinenko, S.F. Mikhailov, V.M. Popik, I.G. Silvestrov, A.S. Sokolov, E.M. Trakhtenberg, N.A. Vinokurov, P.D. Vobly, Nucl. Instr. and Meth. A 282 (1989) 422.

[10] V.V. Anashin, M.M. Brovin, A.A. Didenko, I.B. Drobyazko, A.S. Kalinin. G.N. Kulipanov, V.N. Litvinenko, L.A. Mironenko, S.F. Mikhaylov, S.I. Mishnev, V.M. Popik, V.G. Popov, Yu.A. Pupkov, I.G. Silvestrov, A.N. Skrinsky, A.S. Sokolov, E.M. Trakhtenberg, N.A. Vinokurov, P.D. Vobly, VEPP-3 dedicated straight section for OK operation, Preprint of Institute of Nuclear Physics 89–126, Novosibirsk, 1989;
S.A. Belomestnykh, M.M. Brovin, V.N. Litvinenko, N.A. Vinokurov, P.D. Vobly, The up-grades of VEPP-3 storage ring for OK-4 XUV FEL, Proceedings of 11th National Conference on Charged Particle Accelerators, 1988, Dubna, Vol. 1, 1989, p. 410.

[11] G.N. Kulipanov, V.N. Litvinenko, I.V. Pinayev, V.M. Popik, A.N. Skrinsky, A.S. Sokolov, N.A. Vinokurov, Nucl. Instr. and Meth. A 296 (1990) 1;
I.B. Drobyazko, G.N. Kulipanov, V.N. Litvinenko, I.V. Pinayev, V.M. Popik, I.G. Silvestrov, A.N. Skrinsky, A.S. Sokolov, N.A. Vinokurov, Nucl. Instr. and Meth. A 282 (1989) 424;
N.A. Vinokurov, I.B. Drobyazko, G.N. Kulipanov, V.N. Litvinenko, I.V. Pinayev, V.M. Popik, I.G. Silvestrov, A.N. Skrinsky, A.S. Sokolov, Rev. Sci. Instrum. 60 (7) (pt. II) 1989, 1435.

[12] V.N. Litvinenko, The optical klystron on VEPP-3 storage ring bypass—lasing in the visible and the ultraviolet, Thesis, Novosibirsk, 1989.

[13] V.N. Litvinenko, et al., SPIE Proceedings of the Conference on Free-Electron Laser Challenges, San Jose, CA, February 13–14, Vol. 2988, 1997, p. 188.

[14] H. Hama, et al., Proceedings of third Asian Symposium on FELs, IONICS Pub. Co. Ltd., Tokyo 1997, p. 17.

[15] H. Hama, Recent Progress of the UVSOR-FEL, Proceedings of ICFA 17th Advanced Beam Dynamics Workshop on Future Light Sources, APS, ANL, Argonne, IL, USA, April 6–9, 1999, http://www.aps.anl.gov/conferences/FLSworkshop/proceedings, wg3-15.

[16] V.N. Litvinenko, S.H. Park, I.V. Pinayev, Y. Wu, M. Emamian, N. Hower, O. Oakeley, G. Swift, P. Wang, Nucl. Instr. and Meth. A 429 (1999) 151.

[17] V.N. Litvinenko, S.H. Park, I.V. Pinayev, Y. Wu, Performance of the Ok-4/Duke storage ring FEL, Proceedings of SR'2000 Conference, Novosibirsk, July 2000, Nucl. Instr. and Meth. A 470 (2001) 66.

[18] K. Yamada, N. Sei, T. Yamazaki, H. Ohgaki, V.N. Litvinenko, T. Mikado, S. Sugiyama, M. Kawai, M. Yokoyama, Nucl. Instr. and Meth. A 429 (1999) 159.

[19] K. Yamada, N. Sei, H. Ohgaki, T. Mikado, S. Sugiyama, T. Yamazaki, Nucl. Instr. and Meth. A 445 (2000) 173.

[20] R. Walker et al., Initial Performance of the European UV/VUV Storage Ring FEL at ELETTRA, Nucl. Instr. and Meth. A 475 (2001) 20, these proceedings.

[21] K. Yamada, N. Sei, H. Oghaki, T. Mikado, T. Yamazaki, Characteristics of the NIJI-IV UV-VUV FEL system: towards lasing down to 150 nm using a compact storage ring, Nucl. Instr. and Meth. A 475 (2001) 205, these proceedings.

[22] P. Wang, N. Hower, V.N. Litvinenko, M. Moalem, O. Oakeley, G. Swift, Y. Wu, Proceedings of the 1999 Particle Accelerator Conference, New York, NY, March 29–April 2,1999, p. 1841.

[23] Y. Wu, V.N. Litvinenko, E. Forest, J.M.J. Madey, Nucl. Instr. and Meth. A 331 (1993) 287.

[24] P. Wang, P. Morcombe, Y. Wu, G.Ya. Kurkin, Proceedings of the 1995 Particle Accelerator Conference, Dallas, TX, May 1–5, 1995, p. 2099.

[25] S.H. Park, Characteristics of the Duke/OK-4 Storage Ring and γ-Ray Source, Thesis, Duke University, January 2000; S.H. Park, V.N. Litvinenko, I.V. Pinayev, Y. Wu. S. Canon, J. Kelly, E.C. Schreiber, W. Tornow, H.R. Weller, Dependence of energy resolution of Compton backscattered γ-rays on incoming beam parameters in the Duke/OK-4 Storage Ring FEL, Nucl. Instr. and Meth. A 475 (2001) 425, these proceedings.

[26] V.N. Litvinenko, S.H. Park, I.V. Pinayev, Y. Wu, Time structure of the OK-4/Dukestorage ring FEL, Nucl. Instr. and Meth. A 475 (2001) 240, these proceedings.

[27] V.N. Litvinenko, Microtemporal structure in the OK-4/Duke storage ring UV FEL, Oral Presentation at 19th International FEL Conference, Beijing, China, August 18–21, 1997; V.N. Litvinenko, S.H. Park, I.V. Pinayev, Y. Wu, A. Lumpkin, Fourier limited micro-pulses in the OK-4/Duke storage ring FEL operating with Super-modes, Nucl. Instr. and Meth. A 475 (2001) 234, these proceedings.

[28] D. Boussard, CERN LABII/rf/INT/75-2 (1975), review: J.L. LaClare, Proceedings of 11th International Conference on High Energy Accelerators, Geneva, 1980, p. 526.

[29] V.N. Litvinenko, S.H. Park, I.V. Pinayev, Y. Wu, Average and Peak Power Limitations and Beam Dynamics in the Duke/OK-4 storage ring FEL, Nucl. Instr. and Meth. A 475 (2001) 65, these proceedings.

[30] V.N. Litvinenko, Nucl. Instr. and Meth. A 359 (1995) 50.

[31] See for example, J. Murphy, Synchrotron Light Source Data Book, BNL 42333, Version 4, Brookhaven, NY, 1996.

[32] V.N. Litvinenko, B. Burnham, J.M.J. Madey, Y. Wu, Nucl. Instr. and Meth. A 358 (1995) 334.

[33] V.N. Litvinenko, B. Burnham, J.M.J. Madey, Y. Wu, Nucl. Instr. and Meth. A 358 (1995) 369.

[34] N.A. Vinokurov, V.N. Litvinenko, Method of measuring reflection coefficients close to unity, Preprint INP 79-24, Novosibirsk, 1979.

[35] Y. Wu, V.N. Litvinenko, J.M.J. Madey, Instr. and Meth. A 341 (1994) 363.

[36] Y. Wu, V.N. Litvinenko, S.F. Mikhailov, N.A. Vinokurov, N.G. Gavrilov, O.A. Shevchenko, T.V. Shaftan, D.A. Kairan, Lattice Modification and Nonlinear Dynamics for Elliptically Polarized VUV OK-5 FEL Source at Duke Storage Ring, Nucl. Instr. and Meth. A 475 (2001) 253, these proceedings.

[37] V.N. Litvinenko, Nucl. Instr. and Meth. A 304 (1991) 463.

[38] V.N. Litvinenko, N.A. Vinokurov, O.A. Shevchenko, Y. Wu, Predictions and Expected Performance for the VUV OK-5/Duke Storage Ring FEL with Variable Polarization, Nucl. Instr. and Meth. A 475 (2001) 97, these proceedings.

[39] V.N. Litvinenko, S.F. Mikhailov, N.A. Vinokurov, N.G. Gavrilov, G.N. Kulipanov, O.A. Shevchenko, P.D. Vobly, Helical wigglers for the OK-5 storage ring VUV FEL at Duke, Nucl. Instr. and Meth. A 475 (2001) 247, these proceedings.

[40] S.F. Mikhailov, V.N. Litvinenko, Low Emittance Lattice for the Duke Storage Ring soft-X-Ray FEL, Nucl. Instr. and Meth. A 475 (2001) 417, these proceedings.

ELSEVIER

Nuclear Instruments and Methods in Physics Research A 475 (2001) 205–210

NUCLEAR
INSTRUMENTS
& METHODS
IN PHYSICS
RESEARCH
Section A

www.elsevier.com/locate/nima

Characteristics of the NIJI-IV UV-VUV FEL system—toward lasing down to 150 nm using a compact storage ring

K. Yamada[a,*], N. Sei[a], H. Ohgaki[a], T. Mikado[a], T. Yamazaki[b]

[a] *Photonics Research Institute (AIST), 1-1-1 Umezono, Tsukuba, Ibaraki 305-8568 Japan*
[b] *Institute of Advanced Energy, Kyoto University, Gokasho, Uji, Kyoto 611-0011, Japan*

Abstract

Improvement of the NIJI-IV free electron laser (FEL) system is in progress. Recent gain enhancement made the lasing more stable in the wavelength range from 216 to 211 nm. The temporal lasing characteristics were studied with a dual sweep streak camera and a relatively stable cw lasing with a short micro-pulse width of ~ 14 ps (FWHM) was observed at 214 nm. Numerical calculation for the bunch lengthening and the laser gain revealed that full operation of our new RF cavity and installation of low impedance ring chambers will drastically increase the laser gain up to 20% at 200 nm and 12% at 150 nm. We can expect FEL lasing below 200 nm and even at wavelengths down to 150 nm using the compact storage ring NIJI-IV. © 2001 Elsevier Science B.V. All rights reserved.

PACS: 41.60.Cr; 52.75.Ms; 52.75.Va

Keywords: Free-electron laser; Storage ring; Optical klystron; Dielectric multilayer mirror

1. Introduction

Storage-ring free electron lasers (FELs) have substantial advantages, such as good coherence, good focusing, and easy synchronization with the synchrotron radiation from bending magnets or other insertion devices, compared with linac-based SASE FELs. These features make them excellent tools for both excitation and measurement in photon-related physics and chemistry. Efforts to shorten the wavelength in storage ring FELs from deep UV to VUV have been made and lasing was recently obtained at 193.7 and 218 nm at Duke and

Elettra [1,2], respectively. At the Electrotechnical laboratory (ETL), a FEL experiment has been carried out using the compact storage ring NIJI-IV to study the feasibility of a small-scale FEL system in the ultraviolet (UV)–vacuum ultraviolet (VUV) wavelength range. By improving the performances in both storage ring and the cavity mirrors, the NIJI-IV FEL already reached 212 nm on October 8, 1998 [3]. Taking into account the state-of-the-art dielectric mirror technology, we are aiming our tentative goal for FEL lasing at 150 nm. The laser gain presently obtained is insufficient to reach lasing below 200 nm, considering the rapid reduction of the mirror reflectance through degradation of dielectric multilayer mirrors. To enhance the laser gain, we introduced a new RF system to shorten the bunch length [4] and started to design

---

*Corresponding author. Tel.: +81-298-61-5679; fax: +81-298-61-5683.

E-mail address: kyamada@elt.go.jp (K. Yamada).

0168-9002/01/$ - see front matter © 2001 Elsevier Science B.V. All rights reserved.
PII: S 0 1 6 8 - 9 0 0 2 (0 1) 0 1 5 7 5 - 3

low-impedance-type ring vacuum chambers to minimize anomalous bunch lengthening through longitudinal microwave instability. Here, we report the spectral and temporal characteristics of the NIJI-IV FEL recently obtained at wavelengths from 216 to 211 nm and some important improvements of the NIJI-IV system in progress. The expected performance toward lasing down to 150 nm will be also discussed.

2. Performance of the NIJI-IV FEL system

The NIJI-IV is a compact and simple racetrack-type storage ring dedicated to FEL research, whose circumference and typical beam energy are ~ 30 m and 310 MeV, respectively. A 6.3-m optical klystron is equipped in one of the long straight sections in the NIJI-IV and the optical cavity, whose length is ~ 15 m, is composed of low-loss dielectric multilayer mirrors. We have been upgrading the NIJI-IV to shorten the FEL wavelength toward the VUV. The most effective improvement until last year was the installation of thin sextupole magnets into the ring to suppress the head–tail beam instability through chromaticity correction [5]. As for the laser cavity, Al_2O_3/SiO_2 dielectric multilayer mirrors were adopted for the wavelength below 220 nm, instead of HfO_2/SiO_2 mirrors, to avoid large light absorption in HfO_2. Typical laser gain and cavity loss presently obtainable are $\sim 2\%$ and $\sim 0.3\%$ around 214 nm. These enabled us to shorten the FEL wavelength down to 212 nm. For further wavelength shortening, both enhancement of the gain and development of low-loss VUV mirrors are necessary.

At present, light absorption by the high index dielectric used for the cavity mirrors is one of the most critical problems restricting the lasing wavelength in the NIJI-IV FEL. Since the optical absorption edge originating in the energy band gap of Al_2O_3 lies around 8.5 eV [6], we can expect to use Al_2O_3 as a high index dielectric down to ~ 150 nm; the light absorption, however, is also determined by impurities and vacancies, as well as stoichiometric structure, in the deposited thin films. As a result, the optical absorption edge usually shifts to the longer wavelength side. In our

case, the loss of the cavity composed of two Al_2O_3/SiO_2 mirrors optimized at 200 nm is measured to be 2.2%, which is much larger than that optimized at 214 nm. This suggests that light absorption originating in some mechanisms other than the fundamental absorption due to the energy band gap already occurs. Such a light absorption can be reduced by carefully tuning the dielectric coating condition. Now we are preparing Al_2O_3/SiO_2 mirrors whose loss is much smaller than 1% at wavelengths below 200 nm. For wavelengths much shorter than 90 nm, it will be necessary to adopt fluoride multilayer mirrors. Since the FEL gain we can obtain at present is not enough to overcome an increasing loss below 200 nm, we are also upgrading the NIJI-IV to enhance the laser gain sufficiently. Expected NIJI-IV performance will be shown later.

3. Recent lasing characteristics in the deep UV range

Now the NIJI-IV FEL can lase at wavelengths from the visible to the deep UV. Since the laser gain was recently enhanced by about 30% by increasing the cavity RF power from 1 to 3.3 kW (see Fig. 4), lasing became more stable than before, especially in the deep UV. Fig. 1 shows the FEL lines recently obtained for the shortest wavelength region. The tunable range has been slightly extended and we can observe lasing from 211 to

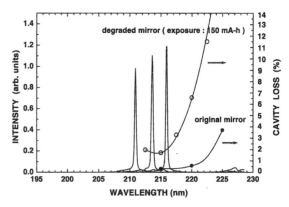

Fig. 1. Typical FEL line spectra obtained with Al_2O_3/SiO_2 mirrors optimized around 214 nm. Evolution of the cavity mirror degradation is also shown.

216 nm. The line intensity became much larger than before, especially at long and short wavelength fringes. The line width is measured to be 0.3 nm. Since this is almost the wavelength resolution of our measurement system, the actual width is expected to be less than 0.1 nm. The evolution of the cavity mirror degradation is also shown in the figure.

Fig. 2 shows the temporal FEL structure at 214 nm observed with a dual sweep streak camera. The upper and the lower traces correspond to a higher(~ 3 A) and a lower (~ 2 A) peak current condition. The cavity length is in almost optimal tuning condition for both cases. The upper trace indicates that we can obtain relatively stable cw

lasing with a very short-micro-pulse of ~ 14 ps in FWHM (~ 6 ps in rms), though its intensity is modulated with a few ms period for a longer time scale. On the other hand, the micro-pulse width becomes much larger for a lower current condition due to insufficient laser build-up. In this trace one can also find a slow temporal fluctuation with a frequency of 100 Hz. Although its origin has never been identified, it is probably caused by a mechanical vibration of the mirror vacuum chamber [7], because the base of our mirror holders has a slender structure due to limitation of the setting space. Although the amplitude of the slow fluctuation is ~ 23 ps, which is smaller than expected from such a slender structure like our

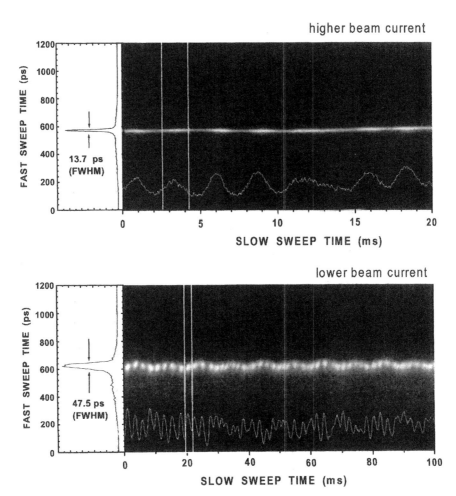

Fig. 2. Temporal FEL structure at 214 nm observed with a dual sweep streak camera. Upper and lower traces correspond to a higher(~ 3 A) and a lower (~ 2 A) peak current conditions.

mirror mounts, a mechanical stabilization as well as some electric feedback system will be needed for FEL application research, considering the short micro-pulse width (~ 14 ps). Faster intensity modulation with a few ms period is expected to be suppressed by enhancing the laser gain. The laser gain will be sufficiently enhanced through the upgrade of the NIJI-IV which is in progress, as shown in the following section.

4. Expected laser-gain enhancement toward the lasing down to 150 nm

To obtain a larger FEL gain in the VUV range, it is necessary to increase the peak current of the stored beam. For this purpose, the bunch length should be kept as short as possible. To shorten the natural bunch length and stabilize the beam, we upgraded the RF cavity. The natural bunch length is inversely proportional to the synchrotron oscillation frequency, which is roughly proportional to the 0.25th power of the net RF input to the cavity. The former cavity was an old one transferred from another machine and the allowed RF power input had been limited to only ~ 1 kW. We replaced the old cavity with a new one whose rated power input is 10 kW. Although the cavity is beginning to age and the present power input is 3.3 kW, a bunch shortening by a factor of 1.8 is expected in full power operation. Another critical problem to limit the FEL gain is bunch lengthening through potential-well distortion and longitudinal microwave instability. To suppress such bunch lengthening the ring impedance must be sufficiently small. In the NIJI-IV the broad band impedance is estimated to be $\sim 20\,\Omega$ from the threshold current for microwave instability, which is much larger than that in modern storage rings. Therefore, we started to design a new ring vacuum chamber whose impedance is smaller at least by an order of magnitude. With these improvements we can expect to suppress the bunch lengthening even in higher beam currents and to obtain sufficient FEL gain in the VUV range.

Fig. 3 shows the dependence of bunch length (in rms) on the average beam current in one bunch ring operation. Open and solid circles indicate the

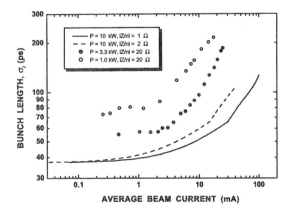

Fig. 3. Dependence of bunch length (in rms) on the average beam current in one bunch ring operation. Open and solid circles indicate the measured values for the cavity RF powers of 1 kW and 3.3 kW, respectively.

measured values for the cavity RF powers of 1 and 3.3 kW, respectively. One finds that the natural bunch length is effectively shortened from 80 to 55 ps with increasing RF power. However, an onset of anomalous bunch lengthening by microwave instability is clearly found at ~ 2 mA, thus checking the growth of peak beam current. The dashed and solid lines show the calculated bunch length for a ring impedance of 2 and $1\,\Omega$, respectively. The RF power was assumed to be 10 kW in both lines. The bunch length was estimated by $\sigma_1 = (\sigma_{1P}^2 + \sigma_{1M}^2)^{1/2}$ as a function of beam current I, where the bunch length under the influence of the potential well distortion σ_{1P} and that above threshold for microwave instability σ_{1M} were calculated with the following relations [8,9]:

$$I = \frac{\sqrt{2\pi}E}{e\alpha R^3}\left(\frac{f_s}{f_{rev}}\right)^2 \frac{1}{Im(Z/n)}(\sigma_{1P}^3 - \sigma_{10}^2\sigma_{1P}) \quad (1)$$

$$\sigma_{1M} = \left(\frac{\eta R^3 I|Z/n|_{bb}}{(2\pi)^{1/2}(E/e)(f_s/f_{rev})^2}\right)^{1/3}, \quad I \geq I_{th}. \quad (2)$$

Here, e and E are the electric charge and beam energy, f_s and f_{rev} are synchrotron oscillation frequency and ring revolution frequency, α and R are momentum compaction factor and ring average radius, σ_{10} and Z/n are natural bunch length and the ring impedance, η is the phase-slip factor defined as $\eta = \alpha - 1/\gamma^2$ and the threshold current for microwave instability I_{th} is expressed as follows;

$$I_{\mathrm{th}} = \frac{(2\pi)^{3/2} \eta f_{\mathrm{rev}} \sigma_{\mathrm{l}} cE}{e|Z/n|_{\mathrm{bb}}} \left(\frac{\sigma_\gamma}{\gamma}\right)^2 \qquad (3)$$

where c and σ_γ/γ are light velocity and relative beam energy spread.

From these curves one find that I_{th} increases up to ~ 15 and ~ 30 mA for an impedance of 2 and 1 Ω, respectively. This effectively suppresses the bunch lengthening and maintains a small value of σ_{l}, combined with the bunch shortening due to increased RF power. This will increase the peak current effectively and a drastic gain enhancement will be expected.

FEL gain at 200 nm was estimated by the well-known formula [5,10], taking into account the bunch length shown in Fig. 3. Fig. 4 shows the result. Here, the open and solid circles correspond to the presently obtainable gain for RF powers of 1 and 3.3 kW. Due to the large ring impedance, one finds that the gain tends to saturate at $\sim 2\%$, while it is enhanced by $\sim 30\%$ for RF power of 3.3 kW. On the other hand it should be noted that the gain is drastically enhanced up to more than 10% for $|Z/n| = 2\,\Omega$ and more than 20% for $|Z/n| = 1\,\Omega$ when the RF power is increased up to 10 kW. Since the reduction of the ring impedance down to 2–1 Ω is not difficult, we can expect 20% gain at 200 nm, which is sufficient to obtain lasing

below 200 nm. From this value, we can simply deduce the FEL gain at 150 nm to be 12%, considering the wavelength dependence. This seems to be a critical gain for lasing. However, we can reasonably expect to obtain lasing at 150 nm if the loss of the dielectric mirrors can be reduced down to less than a few%.

5. Conclusion

NIJI-IV FEL is being developed in the UV–VUV wavelength range to study the feasibility of a compact storage ring FEL system. The shortest lasing wavelength obtained was 211 nm, which was determined by the onset of a light absorption phenomenon probably due to some impurities or vacancies in the high index dielectric for the low-loss cavity mirrors. We obtained relatively stable cw lasing at 214 nm with a short micro-pulse width of ~ 14 ps in FWHM (~ 6 ps in rms). A slow temporal fluctuation of 100 Hz was also observed within the lasing. Although the amplitude of such a slow fluctuation is smaller than expected from the slender structure of our cavity mirror mounts, a mechanical stabilization as well as some electric feedback system will be needed for FEL application research, considering the short laser-pulse width obtained. To enhance the laser gain and obtain FEL lasing below 200 nm, we already introduced a new RF cavity and started to prepare low-impedance-type ring vacuum chambers. The numerical calculation for bunch lengthening and laser gain using analytic formulas revealed that such a ring improvement could increase the laser gain up to 20% at 200 nm. From this value, FEL gain at 150 nm is deduced to be 12%. These values are sufficient to obtain lasing below 200 nm and even at wavelengths down to 150 nm, considering the recent progress of low-loss mirror technology.

Acknowledgements

The authors would like to thank Dr. K. Eto of Japan Aviation Electronics Industry for his useful discussion about the low-loss dielectric mirrors. This work was supported by Peaceful Utilization

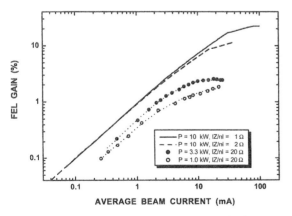

Fig. 4. Expected gain enhancement at 200 nm with improvement of the NIJI-IV.

Technology of Nuclear Energy from the Science and Technology Agency of Japan.

References

[1] V.N. Litvinenko, S.H. Park, I.V. Pinayev, Y. Wu, Nucl. Instr. and Meth. A 475 (2001) 65, these proceedings.

[2] R.P. Walker, J.A. Clarke, M.E. Couprie, G. Dattoli, M. Eriksson, D. Garzella, L. Giannessi, M. Marsi, D. Nolle, D. Nutarelli, M.W. Poole, H. Quick, E. Renault, R. Roux, M. Trov, S. Werin, K. Wille, Nucl. Instr. and Meth. A 475 (2001) 20.

[3] K. Yamada, N. Sei, H. Ohgaki, T. Mikado, S. Sugiyama, T. Yamazaki, Nucl. Instr. and Meth. A 445 (2000) 173.

[4] N. Sei, K. Yamada, H. Ohgaki, T. Mikado, S. Sugiyama, T. Yamazaki, Nucl. Instr. and Meth. A 445 (2000) 437.

[5] N. Sei, K. Yamada, H. Ohgaki, V.N. Litvinenko, T. Mikado, T. Yamazaki, Nucl. Instr. and Meth. A 429 (1999) 185.

[6] M.E. Innocenzi, R.T. Swimm, M. Bass, R.H. French, A.B. Villaverde, M.R. Kokta, J. Appl. Phys. 67 (1990) 7542.

[7] M. Hosaka, S. Koda, J. Yamazaki, H. Hama, Nucl. Instr. and Meth. A 445 (2000) 208.

[8] M. Furman, J. Byrd, S. Chattopadhyay, Beam instabilities, in: H. Winick (Ed.), Synchrotron Radiation Sources—a Primer, World Scientific, Singapore, 1994, pp. 306–343.

[9] H. Viedemann, Particle Accelerator Physics II, 2nd Edition, Springer, Berlin, 1999, p. 207.

[10] D.A.G. Deacon, J.M.J. Madey, K.E. Robinson, C. Bazin, M. Billardon, P. Elleaume, Y. Farge, J.M. Ortega, Y. Petroff, M. Velghe, IEEE Trans. Nucl. Sci. NS-28 (1981) 3142.

ELSEVIER

Nuclear Instruments and Methods in Physics Research A 475 (2001) 211–216

NUCLEAR
INSTRUMENTS
& METHODS
IN PHYSICS
RESEARCH
Section A

www.elsevier.com/locate/nima

Development of longitudinal feedback system for a storage ring free electron laser

S. Koda[a],*, M. Hosaka[a], J. Yamazaki[a], M. Katoh[a], H. Hama[b]

[a] *UVSOR Facility, Institute for Molecular Science, Myodaiji, Okazaki 444-8585, Japan*
[b] *Laboratory of Nuclear Science, Tohoku University, Mikamine, Taihaku-ku, Sendai 982-0826, Japan*

Abstract

A longitudinal feedback system for a storage ring free electron laser (SRFEL) has been developed at the UVSOR. Instantaneous temporal deviation of the FEL optical pulse with respect to the electron bunch is measured in the frequency domain by detecting a phase between higher harmonic components of respective revolution frequencies. The phase deviation is fed back to control the storage ring RF frequency so as to readjust the effective length of the optical cavity. Compensating for the temporal drift with the feedback system, synchronism between the FEL micropulse and the electron bunches was successfully maintained for a reasonably long time. © 2001 Elsevier Science B.V. All rights reserved.

PACS: 41.60.Cr

Keywords: Storage ring free electron laser; Longitudinal feedback; Phase detection

1. Introduction

Since the gain of SRFEL is generally small and evolution of the FEL optical pulse results from multiple interactions with the same electron bunches, maintaining synchronism between an optical pulse and an electron bunch with high accuracy is very important for stable laser oscillation. Even a very small detuning in the synchronism results in power fluctuation and time jitter of the micropulse. A longitudinal feedback system using a dissector at Super-ACO [1] and a

transverse feedback system at Duke university [2] has been developed to reduce various perturbations for the synchronism. In the UVSOR-FEL, the longitudinal detuning is often induced by environmental factors such as floor vibration due to vacuum pumps. Moreover, drift of the synchronism has been observed, probably caused by mechanical thermal changes of mirror holders or the mirror itself, so that readjustment of the RF frequency should be done frequently.

We have developed a longitudinal feedback system to compensate for modifications of the optical cavity length by means of changing the RF frequency according to temporal deviation of the micropulse from a reference position. The time deviation is measured as a phase between higher harmonics in longitudinal frequency spectra of the

---

*Corresponding author. Present address: Saga University, Synchrotron Light Application Center, 1 Honjo, Saga 840-8502, Japan. Tel.: +81-952-28-8854; fax: +81-952-28-8855.

E-mail address: koda@cc.saga-u.ac.jp (S. Koda).

micropulse and the electron bunch signals. The method enables us to measure the time deviation rapidly and accurately. Note that it is quite different from the method based on the dissector for the Super-ACO FEL [1]. In the following, we report the latest progress of the feedback system.

2. Detuning dependence of the UVSOR-FEL

Detuning dependence of the UVSOR-FEL is schematically shown in Fig. 1. Temporal structures of the optical pulse measured by a dual-sweep streak camera are shown in the upper figure, and the lower figure shows an amplitude proportional to the FEL power. Behavior against RF detuning is roughly divided into three regions: the cw, the pulse, and the quasi-cw [3]. In the cw region, where synchronism is perfectly tuned, the micropulse is located at the center of the electron bunch. The power fluctuation is smallest and the micropulse width is shortest. The synchronization range of the RF frequency to maintain the cw lasing is very small, within a relative range of $|\Delta f_{RF}/f_{RF}| < 10^{-8}$ in typical operation in the UVSOR-FEL. Passing through a transient region from the cw lasing, there are the regions of macropulse lasing at both sides of the cw region with $10^{-8} < |\Delta f_{RF}/f_{RF}| < 2.5 \times 10^{-7}$, where the FEL power fluctuates widely. In the larger detuning region $|\Delta f_{RF}/f_{RF}| > 2.5 \times 10^{-7}$, quasi-cw behavior is observed. The power fluctuation is small but the intensity is much lower than in the cw region. The micropulse is located far from the center of the electron bunch and the micropulse width is the largest among the operating regions.

The lasing at the cw region is probably of the highest quality because of high average power, small intensity fluctuation and short pulse width. But the lasing is not particularly stable, because mechanical perturbations from the environment can easily modify the optical cavity length outside the very narrow detuning range ($|\Delta f_{RF}/f_{RF}| = 10^{-8}$ corresponds to a cavity length change of 0.13 μm). To maintain the FEL oscillation in this region by readjusting the RF frequency is pretty difficult with manual operation.

3. Feedback system

Longitudinal signals of the FEL pulse and the electron bunch can be expanded by harmonics of a round-trip frequency in the optical cavity as follows:

$$f_{FEL}(t) = \sum_n F_{FEL}(2\pi n f_0) \exp(i2\pi n f_0) \qquad (1)$$

$$f_e(t) = \sum_n F_e(2\pi n f_0) \exp(i2\pi n f_0) \qquad (2)$$

where n is a harmonic number and, $F_{FEL}(\omega)$ and $F_e(\omega)$ are Fourier transforms for the pulse shape of the FEL micropulse and the electron bunch with angular frequency ω, respectively (normalization factors and responses of the detector and the circuits are included in the transforms). When the temporal deviation Δt of the micropulse signal occurs with respect to the electron bunch signal, the phase between the micropulse and the bunch is shifted by $\Delta \theta = 2\pi n f_0 \Delta t$ for nth harmonic component. The sensitivity for the time deviation is directly proportional to the harmonic number n in principle. Measuring the phase deviation between the micropulse and the electron bunch in certain higher harmonics n, we can derive the temporal deviation.

A schematic diagram of the feedback system is shown in Fig. 2. The detail of the phase detection is shown in a dotted box area in the figure. A photo-diode HAMAMATU S1722-02, which has a relatively fast rise time and a wide detection area, was employed to detect the FEL signal. The electron bunch signal is provided by a pick-up electrode in the storage ring. Both the FEL and the electron bunch signals are filtered at a higher harmonic frequency and the phase between those harmonic components is converted to an amplitude signal by a vector voltmeter HP8508A, which can be operated with a phase resolution of 0.1°. Taking a signal to noise ratio of the harmonics amplitude into account, we have chosen a frequency of 270 MHz, which is the 24th harmonic of the FEL round–trip frequency and corresponds to the third harmonic of the RF frequency. Phase angle of one degree in 270 MHz corresponds to a temporal deviation of 10 ps. The output phase signal from the vector voltmeter is sampled by a

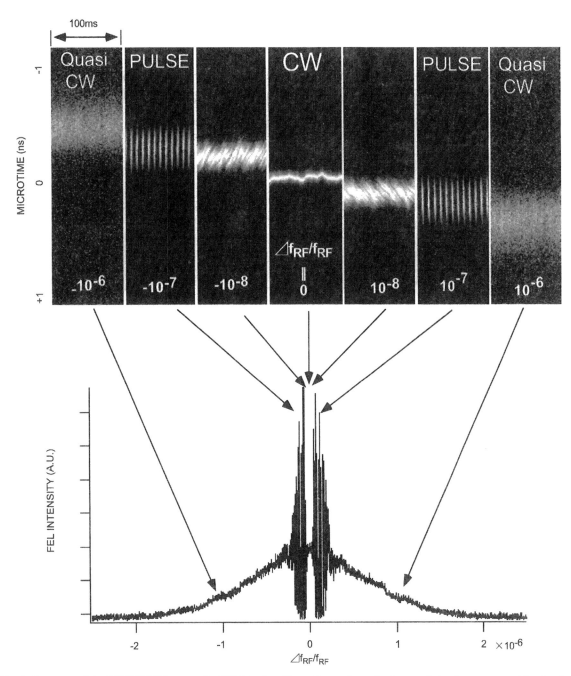

Fig. 1. Response of the UVSOR-FEL versus the RF detuning. Upper figure shows the temporal structures measured by the dual-sweep streak camera for respective RF detuning $\Delta f_{RF}/f_{RF}$. Lower figure shows the detuning dependence of the FEL intensity, which is measured as amplitude of 24th (270 MHz) harmonic component of the revolution frequency in the FEL signal. The nominal RF frequency was 90.10 MHz. The laser wavelength was 520 nm and the beam current was ~20 mA/2-bunch. The FEL net gain was expected to be 0.6%.

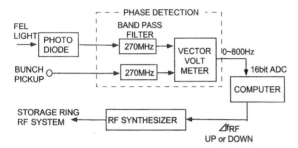

Fig. 2. Schematic flow of the feedback system.

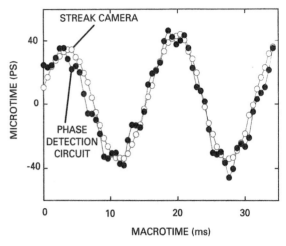

Fig. 3. Temporal deviations of the micropulse with respect to the electron bunch under a strong mechanical perturbation of 60 Hz. Data were simultaneously measured by the phase detection circuit (closed circle) and by the dual-sweep streak camera (open circle).

16 bit ADC with a sampling rate of 10^4 Hz, and averaged over 10 ms to reduce the noise. When the phase deviation from a reference phase exceeds a threshold angle $\Delta\theta_{TH}$, frequency correction with Δf_{RF} is applied to the master oscillator. The phase threshold $\Delta\theta_{TH}$ was chosen to be nearly equal to the micropulse width of 40 ps (FWHM) in the cw lasing region. Since, due to restriction of hardware, the minimum frequency change of the master oscillator is 0.1 Hz, the micropulse position can be stabilized within a region where an inclination of the phase response against the frequency change is $\Delta\theta/\Delta f_{RF} < 40$ ps/0.1 Hz.

4. Experimental result

In order to measure accuracy of the phase detection circuit, the detected phase was compared to temporal positions simultaneously measured by the dual-sweep streak camera. The measurement was performed under strong mechanical perturbation, so that the amplitude of the micropulse jitter of 60 Hz was much larger. The result is shown in Fig. 3. The difference between the temporal positions derived from the phase detection and the streak camera was about a few ps, which is in good agreement.

Typical phase response of the micropulse with respect to the RF detuning in the cw region is shown in the upper part of Fig. 4. It was found that the phase changed very rapidly, particularly at the center of the cw region. It seems to be difficult to stabilize the phase by controlling the RF frequency because the phase response $\Delta\theta/\Delta f_{RF}$ attained about 100 ps/0.1 Hz at this region. The

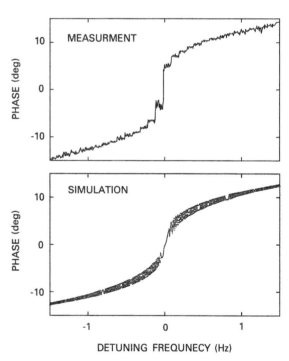

Fig. 4. Measured phase variation in a frequency of 270 MHz plotted as a function of the RF detuning (upper) with the frequency step of 0.1 Hz. Phase of 1° in the frequency is equivalent to 10 ps in the time domain. Lower figure shows a result of computer simulation using parameters nearly identical with the experiment.

response shows that the FEL can be very sensitive to a sudden jump of the micropulse position resulting from a very small change in the synchronism condition [4,5]. For practical stable control with the feedback system, the reference phase has to be chosen in a region of somewhat higher (or lower) phase than the perfect synchronism position. The lower part in Fig. 4 shows a result of computer simulation. In a model for the simulation, the gain profile is assumed to be proportional to the Gaussian shape of the electron bunch. In addition, assuming that the FEL induced energy spread for the electron bunch is proportional to total produced photon energy, the bunch lengthening due to potential-well distortion is taken into account. The simulation is qualitatively in agreement with the experimental result. However at the perfect synchronism position in the center of the cw region, variation of the measured phase is much steeper than that of the simulation. Theoretical analysis is highly desired to interpret the temporal response of the micropulse versus the detuning.

The feedback system was successfully demonstrated as shown in Fig. 5. Considering the steep phase response, the reference phase was shifted by 67 ps from the center of the cw region in the experiment. As one can see in the figure the power fluctuation and the micropulse jitter were reduced to less than 5% and 30 ps, respectively, during the feedback operation at 10 Hz. Once the feedback was switched off, the micropulse position began drifting and sometimes fluctuated widely, in addition to fluctuations in the power.

Another feedback action is shown in Fig. 6. When a person passed through near the mirror chamber, a very small detuning was induced. Probably the floor was slightly deformed by the weight of the person. The system could respond to such a momentary disturbance and compensate for the detuning in a shorter time.

5. Conclusion and summary

We have developed the longitudinal feedback system to compensate for the temporal deviation of the micropulse and then stabilize the FEL

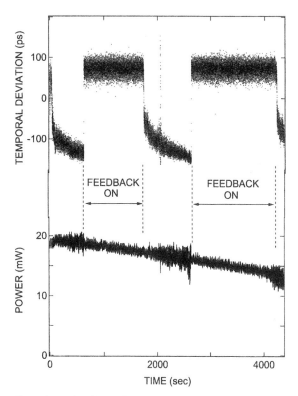

Fig. 5. Intensity fluctuation of the laser power (upper) and temporal variations of the micropulse (lower) recorded by applying feedback and no-feedback alternately. The laser power was linearly decreased due to decay of the beam current.

oscillation. The system readjusts the RF frequency to control the effective length of the optical cavity by detecting the phase between the micropulse and the electron bunch. The phase detection technique is very simple and the precise measurement of the time deviation can be continuously done. The feedback system can maintain the synchronism for a long time and reduce both the power fluctuation and the micropulse jitter, which may enable us to offer the FEL for user experiments. At present, the feedback system can respond to the longitudinal detuning varying slower than ~ 1 s because of the dull reaction of the master oscillator. Faster longitudinal instabilities will be suppressed by an improvement of the response time of the feedback system.

We have observed a phase response like a step function in the best synchronism region. This is one reason why the feedback system was operated

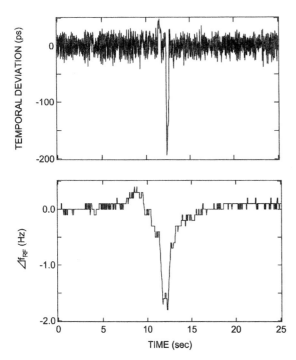

Fig. 6. An example of feedback action responded to an external mechanical perturbation, i.e., a person passed through near the forward mirror chamber during the feedback operation. Upper and lower figures show the micropulse phase and change of the RF frequency corrected by the feedback system, respectively.

with the reference phase set off-center in the cw region. There is a problem with operation for longer lasing durations, because the cw region narrows as the beam current decays, hence the FEL gain decreases, and the reference phase would be drifting into the pulse region. Consequently the reference phase should be readjusted sometimes in the practical operation. Another problem is that the detected phase of the electron bunch is not always the phase of the peak of the electron distribution if the bunch shape deviates from a symmetric one. It is well known that the resistive part of the ring impedance produces an asymmetric bunch shape (for example Ref. [6]). Fortunately, on the UVSOR ring, the electron bunch shape in the FEL oscillation is fairly

symmetric, and we have not yet experienced deviations caused by a difference between the bunch peak position and the detected phase.

In addition, from a view of the SRFEL dynamics, the steep response of the micropulse position for the very small change of the RF frequency is interesting. Although there is no qualitative explanation at the moment, the position stability of the micropulse against the detuning seems to be lost at the top of the electron distribution. At the best synchronism the temporal width of the micropulse is getting shorter and shorter, and moreover the longitudinal gain profile becomes much more flat due to the bunch lengthening. In this case the micropulse position might be just drifting according to the finite detuning because of no restoring force. Further study on the subject will be required.

Acknowledgements

Authors are grateful to staff of the UVSOR facility for their technical help. This work is supported by a Grant-in-Aid for Scientific Research from Ministry of Education, Science and Culture of Japan Contract No. 10450039.

References

[1] M.E. Couprie, D. Garzella, T. Hara, J.H. Codarbox, M. Billardon, Nucl. Instr. and Meth. A 358 (1995) 374.
[2] V.N. Litvinenko, et al., SPIE 2988 (1997) 188.
[3] M.E. Couprie, T. Hara, D. Gontier, P. Troussel, D. Garzella, A. Delboulbe, M. Billardon, Phys. Rev. E 53 (1996) 1871.
[4] M.E. Couprie, V. Litvinenko, D. Garzella, A. Delboulbe, M. Velghe, M. Billardon, Nucl. Instr. and Meth. A 331 (1993) 37.
[5] M.E. Couprie, D. Garzella, M. Hirsch, G. Moneron, L. Nahon, G. De Ninno, D. Nutarelli, E. Renault, Proceedings of the 7th European Particle Accelerator Conference, Austria Center Vienna, 2000, p. 735.
[6] H. Wiedemann, Particle Accelerator Physics II, Springer, Berlin, 1993.

ELSEVIER

Nuclear Instruments and Methods in Physics Research A 475 (2001) 217–221

NUCLEAR
INSTRUMENTS
& METHODS
IN PHYSICS
RESEARCH
Section A

www.elsevier.com/locate/nima

FEL induced electron bunch heating observed by a method based on synchronous phase detection

M. Hosaka[a],*, S. Koda[a], M. Katoh[a], J. Yamazaki[a], H. Hama[b]

[a] UVSOR Facility, Institute for Molecular Science, Myodaiji, Okazaki 444-8585, Japan
[b] Laboratory of Nuclear Science, Tohoku University, Mikamine, Taihaku-ku, Sendai 980-0826, Japan

Abstract

A new method for monitoring instantaneous bunch length has been developed on the UVSOR storage ring. Based on the bunch length dependence of the loss factor, the bunch length can be derived from the synchronous phase shift. Experiments using this method were performed to observe the bunch lengthening associated with the free electron laser (FEL) power variation. From the detuning dependence of the bunch heating, it was found that additional energy spread mostly has a constant ratio to the intracavity laser power normalized by the beam current. However, at the best synchronism region, the induced energy spread exhibits a strong non-linearity against the normalized FEL power and hence the power saturation is accomplished. © 2001 Elsevier Science B.V. All rights reserved.

PACS: 41.60.Cr; 41.75.Ht

Keywords: Free electron laser; Storage ring

1. Introduction

The saturation mechanism of laser power on a storage ring free electron laser (FEL) is qualitatively explained by the "bunch heating" process. In a storage ring FEL, the electron beam is recirculated many times in the interaction region, and the repetitive interactions cause enhancement of the energy spread and lengthening of the electron bunch; this process is called bunch heating. On the other hand, reduction of energy spreading occurs at the same time, due to the synchrotron radiation damping, and limits the enhancement of the energy spread. Since the FEL gain is directly related to the bunch length and the energy spread of the electron beam, these processes determine the FEL gain and consequently, lead to the power saturation. Although a number of experiments have been performed to examine the dynamics, some of the results have not been completely interpreted.

On the UVSOR-FEL, using a helical optical klystron [1], lasing stability has much improved after the remodeling of the optical cavity and the successful implementation of a longitudinal feedback system [2]. A precise measurement without external disturbances is now possible. We are carrying out experiments to investigate the FEL dynamics, especially the saturation dynamics. For this study, information on the bunch lengthening that reflects the energy spread is crucial.

Several methods to measure the bunch length have been developed; these are divided into two

---

*Corresponding author. Tel.: + 81-564-55-7402; Fax: + 81 + 564-54-7079.

E-mail address: hosaka@ims.ac.jp (M. Hosaka).

0168-9002/01/$ - see front matter © 2001 Elsevier Science B.V. All rights reserved.
PII: S 0 1 6 8 - 9 0 0 2 (0 1) 0 1 5 7 3 - X

classes: optical and electrical methods. The former corresponds to the measurement of the longitudinal profile of a bunch from synchrotron radiation and the most qualitative methods use a dual-sweep streak camera and a dissector [3]. They can provide a lot of information about electron distribution in a bunch over a wide dynamic range, but it takes at least several minutes to evaluate one numerical result and an instantaneous resolution is limited. In the electrical method, the longitudinal profile of the bunch is picked up using an electrode and the bunch length is obtained through the frequency spectrum [4]. Although information available is limited, real time continuous observation is possible by using the method.

On the UVSOR storage ring, we have developed a new method to measure an instantaneous bunch length continuously using the electrical method. The method is based on the bunch length dependence of the loss factor of the beam bunch. In this article, we introduce the loss factor and the principle of the measurement and show preliminary experimental results obtained by using the method.

2. Loss factor and bunch length

In a storage ring, the interactions of the electron bunch with the environment, such as the vacuum chamber components, lead to an energy loss, which in turn has to be compensated by the RF cavity field. This energy loss can be characterized through the loss factor, defined by

$$k = \frac{\Delta E}{e^2 N_b^2} \tag{1}$$

where ΔE, e and N_b are the energy loss per tune, the electron charge and the total number of electrons in the bunch. In the frequency domain, the loss factor is related to the resistive part of the longitudinal coupling impedance $Z_{res}(\omega)$ by

$$k = \frac{\pi}{e^2 N_b^2} \int_{-\infty}^{\infty} Z_{res}(\omega) |I(\omega)|^2 \, d\omega \tag{2}$$

where $I(\omega)$ is the current distribution at the frequency component ω [5].

The loss factor depends strongly on the bunch length. Assuming that the bunch distribution is

Gaussian with RMS bunch length σ_τ, the loss factor is

$$k = \frac{\pi I_0^2}{e^2 N_b^2} \int_{-\infty}^{\infty} Z_{res}(\omega) e^{-\omega^2 \sigma_\tau^2} \, d\omega \tag{3}$$

where I_0 is the total bunch current, and then one can get for example

$$k \propto \begin{cases} \sigma_\tau^{-1} & \text{for } Z_{res} \text{ is constant} \\ \sigma_\tau^{-2} & \text{for } Z_{res} \propto \omega \end{cases} \tag{4}$$

In the SPEAR storage ring, an overall loss factor dependence on bunch length has been measured and found to scale as $k \propto \sigma_\tau^{-1.21}$ [6]. The dependence has been also measured in BESSY [7], where the data was analyzed in term of the broad band resonator.

3. Measurement of loss factor and bunch length on the UVSOR storage ring

In order to obtain a scaling law between the loss factor and the bunch length, simultaneous measurements of the loss factor and the bunch length have been performed. On the UVSOR storage ring, the most efficient way to lengthen the electron bunch is by using FEL-induced bunch heating. As mentioned earlier, the FEL interaction causes enhancement of the energy spread of the electron bunch and one can vary the bunch length over a wide range by varying the FEL power.

Since the energy loss of the electron bunch is compensated by the RF cavity field, the loss factor can be deduced from the synchronous phase shift $\Delta\phi$ by

$$k = \frac{V_{RF} \left| \sin(\phi_{s0} + \Delta\phi) - \sin\phi_{s0} \right|}{e N_b} \tag{5}$$

where ϕ_{s0} corresponds to the synchronous phase for a zero beam current and V_{RF} is the peak accelerating voltage in the RF cavity. The setup for the synchronous phase detection is similar to the one developed for the longitudinal feedback (see Ref. [2]). A signal from the electron bunch, supplied by a button pickup electrode, is filtered by a band pass filter. In order to obtain the beam phase with high resolution, a frequency of 270 MHz corresponding to the third harmonics of the RF frequency has been chosen for the measurement. The cavity RF signal, which is used as a reference of the beam phase, is taken from a cavity pickup loop and is tripled in frequency. The phase between those signals is analyzed by a commercially available vector voltmeter at a sampling rate of 800 Hz and an output signal proportional to the phase is recorded through a digital voltmeter or an ADC. The overall resolution of the phase detection is estimated to be around 0.1° (full-width at half-maximum—FWHM) in 270 MHz frequency, which corresponds to about 1 ps in the time domain.

Simultaneous measurements of the synchronous phase shift and the bunch length have been carried out for a beam energy of 600 MeV, which is the nominal beam energy for the FEL experiment. A visible wavelength of 520 nm for the FEL oscillation was chosen because the FEL gain at that wavelength is rather high, and then the phase shift would be measured over a wide range of the bunch length. For the UVSOR-FEL, the storage ring is normally operated in two-bunch mode because the length of the optical cavity is a quarter of the ring circumference. For the measurement of the loss factor, the storage ring was operated with single bunch mode in order to minimize the multi-bunch effect. Multilayers of Ta_2O_5/SiO_2 films were used for the cavity mirrors and the measured round-trip cavity loss was 0.17%. At the same time, longitudinal bunch profiles have been taken by using a dual-sweep streak camera to calibrate the bunch length dependence of the loss factor. The bunch length has been stepped up with a higher beam current to 400 ps (RMS) from the natural bunch length of 115 ps. Within the beam current stored in the ring, however, the maximum bunch length without FEL induced bunch heating is less than 200 ps, an effect governed by potential-well distortion. We have also measured the synchronous phase at low beam currents without the FEL oscillation and deduced the phase for a zero beam current (ϕ_{s0} in Eq. (5)) by extrapolating the data.

The loss factors are deduced from the measured phase shifts using Eq. (5) and are plotted as a function of bunch length in Fig. 1. As seen in the figure, the relation between the loss factor and the bunch length does not completely obey the simple scaling law as in the SPEAR storage ring. It is probably due to the resistive-wall impedance,

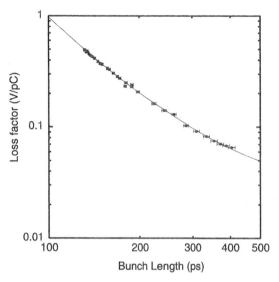

Fig. 1. Dependence of the loss factor on the bunch length. The solid line is the result of the fitting given in Eq. (6).

which plays an important role at the low frequency region. Accordingly, we introduce offsets for both the bunch length and the loss factor and perform a fitting with these parameters. The result of fitting is shown as a solid line in Fig. 1 and the fit is apparently good. The relation between the loss factor and the bunch length is written using the fitting parameters as

$$(k - 0.0234) = 3180(\sigma_\tau - 28.9)^{-1.90} \qquad (6)$$

where the loss factor k and the bunch length σ_τ are measured in units of [V/pC] and [ps], respectively. In the UVSOR storage ring, the resistive part of the longitudinal coupling impedance seems to depends linearly on the frequency (see Eq. (4)). There is not an exact basis in the empirical relation of the form given in Eq. (6) at the moment, because the contribution of the resistive-wall impedance to the broad band impedance is not clear. However, one can see that the relation is valid at least in the region of bunch length measured in the experiment (110–400 ps).

4. Observation of the detuning dependence of the bunch heating

One of the most interesting features of the storage ring FEL is strong detuning dependence in the lasing behavior [8]. The detuning of the synchronism between the optical pulse and the electron beam changes not only the temporal structure of the lasing but also the lasing power and hence the bunch heating. Our method to monitor the bunch length, in which continuous measurement is possible, is expected to be a powerful tool to observe the detuning behavior.

Experiments to observe the detuning dependence of the bunch heating have been carried out with almost the same condition as the one illustrated previously. In the experiment, the frequency of the RF cavity was swept at the rate of 1 Hz/s and the synchronous phase shift and the extracted laser power were recorded continuously. An example of the measurement performed with a beam current around 19 mA is shown in Fig. 2. The bunch length is deduced from measured synchronous phase shift using Eq. (6). Asymmetry of the synchronous phase shift versus the detuning seen in the figure is due to slight decay of the beam current during the measurement, on which the loss factor depends strongly (see Eq. (5)). At the same time the bunch lengths were also measured intermittently by using a streak camera and agree

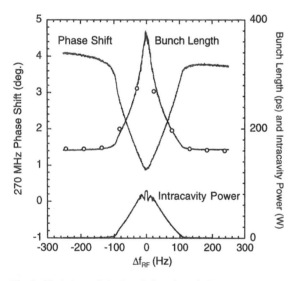

Fig. 2. Variation of the bunch length and the average power under RF detuning. Solid lines are measured synchronous phase and bunch length deduced using Eq. (6). Open circles are the bunch lengths measured by using a streak camera. Intracavity power was estimated with a transmission rate of 0.024% for a cavity mirror.

well with those measured by the phase detection methods as seen in the figure. The measured average laser power is also shown as a solid line. The maximum laser power was obtained in the "cw" zone around the perfect tuning. In the periodical pulse lasing zones, next to the cw zone, the fluctuation of the laser power was too large to measure properly using our system and the average power in the figure might underestimates the actual one. Therefore, in the following, we omit data in the periodical pulse lasing zone.

In order to obtain direct information about the FEL-induced bunch heating, we have deduced the energy spread from the measured bunch length. As far as a zero beam current, the energy spread is proportional to the bunch length. This is not the case, however, with a given non-zero beam current because of potential-well distortion due to the inductive ring impedance. We have numerically solved the Haissinski equation [9], and searched iteratively for a suitable energy spread (σ_γ) that reproduces a bunch length at the beam current. In the calculation, we have employed a pure inductive impedance of $L = 103$ nH, for which the beam current dependence of the bunch length without the laser interaction has been well reproduced. Subtracting the contribution from the natural energy spread ($\sigma_{\gamma 0}$) as $\Delta\sigma_\gamma^2 = \sigma_\gamma^2 - \sigma_{\gamma 0}^2$, the square of the additional energy spread due to the laser interaction is plotted as a function of the intracavity power for the detuning experiments at beam currents of 31, 19 and 10.5 mA in Fig. 3(a). As seen in the figure, the additional energy spread increases as the laser power increases and changes steeply near the cw zone.

It is clear that the additional energy spreads depend strongly on the beam current and so we plot them as a function of the intracavity power normalized by the beam current in Fig. 3(b). We have also performed a theoretical calculation based on the Renieri limit [10] and compared the experimental result. One can see that all the points for different beam currents lie on an identical curve and the tendency is almost reproduced by the calculation. Non-linear behavior, however, is getting stronger above 3 W/mA and the intracavity power seems to be saturated at 4–4.5 W/mA. This suggests that the increase rate of the energy spread

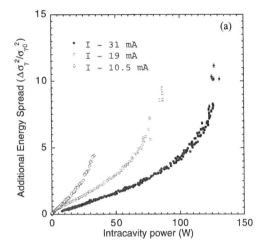

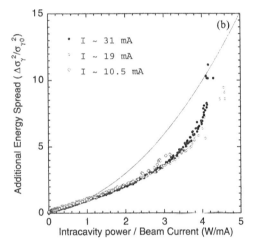

Fig. 3. Additional energy spread normalized by the natural energy spread ($\sigma_{\gamma 0}/\gamma = 3.4 \times 10^{-4}$) for the beam current around 31, 19 and 10.5 mA as a function of the intracavity power (a) and the one normalized by beam current (b). The solid line is a calculation based on the Renieri limit [10]. Data in the periodical pulse lasing zone are omitted.

is constant in a region of relatively low normalized intracavity power, while some variations in accomplishment of the equilibrium state occur at the higher intracavity power.

5. Conclusion

We have developed a new method to observe the bunch length in real time by the detection of the synchronous phase shift. In the method, the bunch length is deduced by using a scaling law for the loss factor determined experimentally. The exact spectrum of the resistive impedance of the storage ring has not been determined and if the bunch deviates much from the Gaussian distribution, the scaling law might not be applicable. It is clear, however, that the method is valid at least for a region of the bunch length in the present experiment.

Experiments using this method have been performed to observe the bunch lengthening associated with FEL power variation. From the detuning dependence of the bunch heating, it was found that additional energy spread mostly has a constant ratio to an intracavity laser power normalized by the beam current. A strong non-linearity of the increase rate was observed at higher intracavity power. More detailed experiment with various conditions for the FEL oscillation and theoretical investigations are in great demand to interpret the detuning behavior of the FEL power and the induced energy spread.

Acknowledgements

This work is supported by a Grant-in-Aid for Scientific Research from Ministry of Education, Science and Culture of Japan, Contract No. 10450039.

References

[1] H. Hama, Nucl. Instr. and Meth. A 375 (1996) 57.
[2] S. Koda, M. Hosaka, J. Yamazaki, M. Katoh and H. Hama, Nucl. Instr. and Meth. A 475 (2001) 211, these proceedings.
[3] E.I. Zinine, Nucl. Instr. and Meth. A 208 (1983) 439.
[4] T. Ieiri, Nucl. Instr. and Meth. A 375 (1996) 57.
[5] H. Wiedemann, Particle Accelerator Physics II, Springer, Berlin, 1993.
[6] P.B. Wilson, R.V. Servanckx, A.P. Sabersky, J. Gareyte, G.E. Fischer, A.W. Chao, IEEE Trans. NS-24 (1977) 1211.
[7] W. Anders, P. Kuske, Th. Westphal, HEACC Hamburg, 1992.
[8] M.E. Couprie, Nucl. Instr. and Meth. A 393 (1997) 13.
[9] H. Hama, M. Hosaka, Nucl. Instr. and Meth. A 429 (1999) 172.
[10] S.H. Park, Ph. D. Thesis, Department of Physics, Duke University, 2000, p. 56.

ELSEVIER

Nuclear Instruments and Methods in Physics Research A 475 (2001) 222–228

**NUCLEAR
INSTRUMENTS
& METHODS
IN PHYSICS
RESEARCH**
Section A

www.elsevier.com/locate/nima

Giant high-peak power pulses in the UV OK-4/Duke storage ring FEL using the gain modulator ☆

Igor V. Pinayev[a,*], Vladimir N. Litvinenko[a], Seong Hee Park[b], Ying Wu[c],
Mark Emamian[a], Nelson Hower[a], Janet Patterson[a], Gary Swift[a]

[a] FEL Laboratory, Duke University, P.O. Box 90319, Durham, NC 27708, USA
[b] KAERI, 159 Deokjin-Dong Yusung-Ku, Taejon, South Korea
[c] Lawrence Berkeley National Laboratory, 1 Cyclotron Road, Berkeley, CA 94720, USA

Abstract

We use the gain modulation technique to generate giant pulses in the OK-4/Duke storage ring FEL for applications requiring high peak power. This technique provides the increase of the peak power by several orders of magnitude. It is also very reliable, predictable and reproducible. The design, the parameters and the gain modulator performance are described. Comparison of expected and measured pulse forms is presented. Application of gain modulator for future harmonic generation experiments is also discussed. © 2001 Elsevier Science B.V. All rights reserved.

PACS: 41.60.Cr; 29.20D; 78.66; 52.75

Keywords: Free electron laser; Storage ring; Optical klystron; Peak power; Harmonics generation

1. Introduction

The maximum average lasing power for a short wavelength storage ring free-electron lasers (SR FEL) is limited [1,2] by effective "heating" of the electron beam (e-beam) participating in the lasing process. For high power applications the average storage ring FEL power can be redistributed into a series of giant pulses. In this case, peak power grows considerably while the average power is reduced only by about a factor 2. This mode of SR FEL operation can be achieved with the use of a conventional Q-modulation technique [3], the gain modulation technique [4,5], or modulation of RF frequency (used at Super-ACO SR FEL [6] and called the Q-switch). At Duke FEL Lab we use the gain modulation technique, i.e. the steering of electron beam orbit in transverse direction. This approach has several advantages:

- It eliminates insertion losses of the optical modulator and restrictions on the peak power.
- It provides for higher peak power, higher pulse energy, and better reproducibility.
- The synchrotron oscillations inevitably excited by modulation of the RF frequency are avoided.

In this paper, we present the evaluation, the design and the performance of the gain modulator

---

☆ Work is supported by the ONR MFEL program under Contract #N00014-94-1-0818.

*Corresponding author. Tel.: +1-919-660-2657; fax: +1-919-660-2671.

E-mail address: pinayev@fel.duke.edu (I.V. Pinayev).

for the OK-4/Duke storage ring FEL. We also discuss the use of the giant pulses for FEL diagnostic, user application and future harmonic generation experiments.

2. Choice of gain modulator

Detailed analysis reveals that the use of RF frequency modulation technique at Duke SRFEL will excite residual synchrotron oscillations with amplitude $\delta t > 40$ ps and $\delta E/E > 6 \times 10^{-4}$ [7]. The amplitude of these oscillations is comparable with the duration of the FEL micro-pulses and would impair advantages of the giant pulse mode. We found this technique to be unacceptable and used the gain modulation technique successfully tested with the OK-4 FEL at VEPP-3 storage ring in Novosibirsk [4]. We stop the lasing by moving the closed e-beam orbit away from the axis of the optical cavity, typically for $X_0 = 2.5$ mm in horizontal direction. To generate a giant pulse, we return the e-beam orbit onto the axis adiabatically. The cost-effective solution was to use one fast steering magnet driven by a critically damped circuit with time constant $\tau \approx 8.8$ μs. This time is significantly shorter than the duration of the giant pulse in the OK-4/Duke SR FEL, which is typically 50 μs to 100 μs long. The use of the critically damped circuit provides for the continuity of the first derivative and, therefore, for residual oscillation of less than $X_0/(\omega_x\tau)^2$. Here $\omega = 2\pi f_0 \nu_x$, $f_0 = 2.7898$ MHz is the e-beam revolution frequency (one turn takes 0.35845 μs), and ν_x is the non-integer part horizontal betatron tune. Detailed analysis shows that both coordinate and angle follow the shape of the driver with very small residual oscillations [7]. For typical Duke storage ring tunes ν_x from 0.11 to 0.17, the relative residual amplitude of betatron oscillations are slightly smaller than the estimate of $(2\pi\nu_x f_0 \tau)^{-2}$, and are 3.8×10^{-3} to 1.6×10^{-3}, respectively. A typical 2.5-mm displacement generates the residual oscillations with an amplitude of less than 10 μm at $\nu_x = 0.11$. These oscillations are much smaller than the RMS sizes of the electron and optical beams (~ 200 μm and ~ 250 μm) and do not affect the performance of the OK-4 FEL in giant pulse mode.

3. Design of the OK-4 FEL gain modulator

The OK-4 FEL, located in the dedicated straight section of the 1 GeV Duke storage ring, comprises of two electromagnetic wigglers and electromagnetic buncher. The lattice of the straight section and the optical cavity are optimized to maximize the OK-4 FEL gain. The configuration and controls are very flexible to optimize essential parameters of the OK-4 FEL. The detailed description of the OK-4 FEL and the Duke storage ring are published elsewhere [4,5,7–10].

The gain modulator is installed in the northern straight section of the Duke storage ring [8] near a focusing quadrupole. We chose the location to maximize the orbit displacement in the center of the OK-4 FEL. The main parameters of the fast steering magnet with ferrite yoke are listed in Table 1. The electronic circuit of the gain modulator is shown in Fig. 1. A high voltage MOSFET transistor connected in series with the coil serves as a fast switch. A clamping resistor and capacitor are connected in parallel with the coil. Together with the inductance of the steering magnet they form the critically damped LCR circuit. The predicted dependence of magnetic field is:

$$H/H_0 = 1\, t < 0,\ H/H_0 = (1 + t/\tau)\exp(-t/\tau)),\, t \geqslant 0.$$

The design of vacuum chamber for the gain modulator required major efforts. The solution was found in the design shown on Fig. 2. The cage of the aluminum rods has anodized ends providing for capacitive coupling between flanges. The cage has high impedance for the MHz range eddy

Table 1
Parameters of the gain modulator magnet

Length (ferrite) (cm)	20.3
Thickness (ferrite) (cm)	1.27
Length (magnetic) (cm), L_m	26.7
Magnet gap (cm)	14
Magnet width (cm)	15.2
Number of turns in the coil	60
Maximum current (A)	10
Maximum field integral (kGs cm)	1.44
Maximum field in ferrite (kGs)	0.88
Maximum kick @1 GeV (mrad)	0.43
Maximum X_0 @1 GeV (mm)	3.4

currents and allows the changes of the transverse magnetic field in the gain modulator at sub-μsec scale. At the same time, the image currents induced by a short e-beam in the walls of the vacuum chamber pass through the capacitance. This provides for low impedance of this vacuum vessel at high frequencies which is critical for microwave instability.

4. Performance of gain modulator

After manufacturing and assembling we tested the gain modulator on the bench. The voltage

induced in a pick-up coil by the decaying magnetic field was measured and integrated by digital oscilloscope LeCroy LC569A. The traces are shown in the Fig. 3. The measured shape and the time constant are in a good agreement with calculated from the values of the components.

After the installation, we measured the change in the equilibrium orbit of the real 500 MeV electron beam when the gain modulator was excited with 4 A current. The displacement of the electron beam was 1 mm at the entrance to the OK-4 and 2.5 mm at the end with $v_x = 0.17$. The Fig. 4 shows the action of the gain modulator on the e-beam orbit in the OK-4 FEL. We used position sensitive detector S2044 (PSD) (manufactured

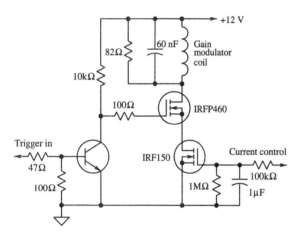

Fig. 1. The electronic circuit of the OK-4 FEL gain modulator. EPICS control system sets the DC current in the magnet via DAC voltage (current control input). The trigger pulse initiates the transition. The critically damped circuit comprises of the gain modulator coils inductance, the 82 Ω resistor and the 60 nF capacitor.

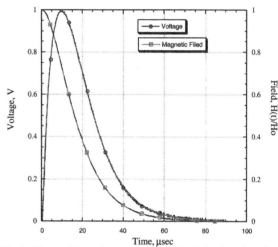

Fig. 3. Trace of the voltage induced in the B-dot loop installed inside the gain modulator magnet. The second curve gives the integral of the voltage and is proportional to the magnetic field in the magnet. Fits with functions of damped oscillator are almost perfect.

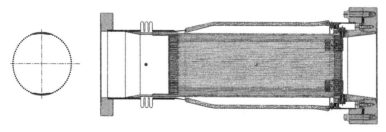

Fig. 2. The design of the vacuum chamber for the OK-4 FEL gain modulator (top view). The glass tube (6 "OD, 8 "long) provides the vacuum isolation. The inner side of the glass tube has thin chromium coating for the electrostatic discharge. On the left, the cross-section of the conducting cage with Al rods and rigidity "ribs".

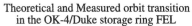

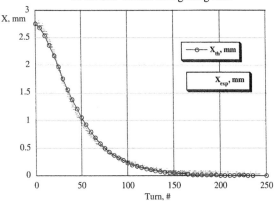

Fig. 4. The transition of the e-beam orbit in the OK-4 FEL. $E_e = 500$ MeV. $v_x = 0.11$, the gain modulator magnet current $I_{GM} = 4$ A. Theory—continuous line; dots—the measured PSD signal.

Table 2
The OK-4/Duke SRFEL in giant pulse mode

Wavelength (fundamental), (nm)	194–730
E-beam energy (GeV)	0.25–0.8
Current per bunch (mA)	up to 20
Giant pulse rep-rate (Hz)	1–60
Macropulse energy, (mJ)	0.05–3[a]
Macropulse duration, FWHM (μs)	40–200
Extracted peak power (MW)	0.1–3[b]
Peak intracavity power (GW)	0.1–1.2
Line-width, FWHM ($\delta\lambda/\lambda$)	$(1–2) \times 10^{-4}$
Micropulse duration $\sigma\tau$, (ps)	15–25
Micropulse separation (ns)	358.45–5.60
Spatial distribution	TEM_{00}
Peak spectral brightness (ph/sec/mm^2/mrad^2/10^3/BW)	$0.2–0.5 \times 10^{28}$ [b]

[a] Extracted, depends on wavelength and efficiency of extraction.

[b] Out-coupled per mirror, not optimized.

by Hamamatsu, with electronics similar to that described in Ref. [11], but with response time of ~0.5 μs) and digital oscilloscope LeCroy LC569A for these measurements. The agreement between the theoretical and experimental dependencies is very reasonable. We attribute small difference to non-linearity of PSD. We did not observe any measurable residual betatron oscillations. Overall, the performance of the gain modulator system was excellent.

5. Giant pulse mode of operation

In the giant pulse mode of OK-4/Duke SR FEL operation, the gain modulator periodically moves the closed electron beam orbit in the FEL region away from the axis of the optical cavity. This stops lasing and allows electron beam to reduce energy spread to the natural value. A pulse generator, installed in the control room, defines the sequence of the giant pulses with desirable repetition rate f_{rep}. Each pulse from generator initiates adiabatic transition of fresh electron beam onto the optical axis, and the FEL generates a giant pulse. The summary of the parameters obtained in the giant pulse mode is given in Table 2. In addition to boosting the peak power of the Duke deep-UV

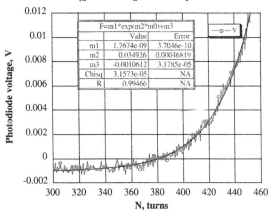

Fig. 5. Start-up of the OK-4 giant pulse shows the net gain of 3.5% (see the fit) which in combination with 2.2% cavity losses gives 5.7% FEL gain per pass.

FEL by up to four thousands times, the gain modulator turned out to be the ideal tool to tune the OK-4 FEL for maximum gain by observing the start-up of the lasing. We use it to study the OK-4 FEL as function of the e-beam current and energy, the RF voltage, and the strength of the buncher [12]. A typical start-up of the giant pulse, shown in Fig. 5, allows us to measure the net gain of the OK-4 FEL with high precision.

In combination with measured losses of the optical cavity, it gives the gain per pass. Our measurements are in good agreement with the SR FEL theory and our predictions [9,10,12]. For calculating the OK-4 FEL gain we take into account the bunch lengthening and the energy spread growth caused by microwave instability in the ring, as well as the finite emittances of electron beam. We did not observe any anomalies reducing the SR FEL gain, such as "dynamic gain or dynamic filling factor" suggested by Super-ACO group [13]. We studied the OK-4 FEL gain for medium beam currents from 3 to 15 mA and UV wavelength from 200 nm to 370 nm. The gain dependence on the e-beam energy E_e was proportional to $\sim E_e^{-1}$ at $E_e > 600$ MeV, and to $\sim E_e^{-0.25}$ at $E_e < 500$ MeV. The weaker dependence at low energies can be explained by the emittance growth via intra-beam scattering.

We used a Molectron integrating detector J3-09 to measure out-coupled macro-pulse energy and a silicon photodiode to measure the time structure of the optical power in macro-pulse. The photodiode signal passes through the second order low-pass filter with time constant of 0.3 μs. The filter smoothes 2.8 MHz structure of the FEL light but does not change the envelope of the giant pulse. Fig. 6 shows series of oscilloscope traces of optical

power with different electron current and buncher settings. After turning on the gain modulator minor changes of the electron orbit in the OK-4/Duke FEL region were needed to maximize optical power. We attribute this to slowly decaying fringe magnetic fields captured in the edges of vacuum chamber nearby the gain modulator.

When the repetition rate f_{rep} of giant pulses is much lower than the inverse damping time of the energy spread ($\tau_E = 8.5$ ms/E^3 (GeV) for the Duke case, i.e. $f_{rep} < f_c = 117$ Hz × E^3 (GeV)),the peak power and the shape of the giant pulses do not change and the average power is proportional to f_{rep}. At $f_{rep} \cong f_c$, the peak power in giant pulses is still higher that in CW mode by three orders of magnitude, while the average power is lower only by factor 2.3. Further increase of the rep-rate to $f_{rep} > f_c$ does not allow electron beam energy spread to "cool-down" completely and, therefore, reduces the peak power in the pulses. At the same time, the average lasing power in the giant pulse mode asymptotically grows to its CW level.

The observed repeatability of the giant pulses in the amplitude and the shape was reasonably good. It is our understanding that all variations in the amplitude and the shape of giant pulses were caused by ripples and drifts of the storage ring power supplies, some of which are of rather poor quality. The repeatability was the best (with variations at $\sim 1\%$ level) at high e-beam energies, where power supplies are more stable. We have in hand new high performance power supplies and plan to install them onto the Duke storage ring next spring. After this up-grade we expect to demonstrate repeatability of better than 1%.

In general, giant pulses can be optimized for the maximum start-up gain, maximum peak power or maximum energy per pulse by choosing proper setting of the current in the electromagnetic buncher. For the maximum gain G_p, the buncher should be set at $N_t = N_w + N_d = (4\pi\sigma_{Ee}/E_e)^{-1}$, where σ_{Ee} is initial ("cold beam") RMS energy spread, N_w is the number of wiggler periods, and $N_d = \Delta s/\lambda$ is the slippage in the buncher divided by the FEL wavelength [14]. For the OK-4 FEL, $N_w = 33.5$ and $N_d = 15.7$ $(I_B(\text{kA})/E_e(\text{GeV}))^2\lambda$ (μm), where I_B is the current in the buncher. With 3.25 mA beam current, $E_e = 500$ MeV, $V_{RF} =$

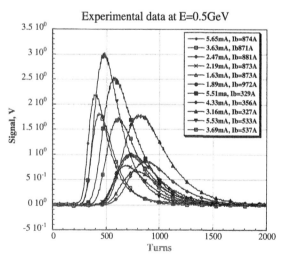

Fig. 6. Giant pulses measured with various e-beam and buncher currents. $\lambda = 370$ nm, $V_{RF} = 220$ kV.

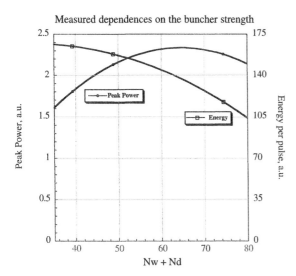

Fig. 7. Measured dependence of the peak power and energy in giant pulse at beam current of 3.25 mA as function of the buncher strength.

220 kV, $\lambda = 370$ nm (as on Fig. 7), the OK-4/Duke SR FEL has maximum gain of 5.8% per pass at $N_t \cong 87.25$.

Fig. 7 shows the measured dependencies of the out-coupled peak power and pulse energy as function of the buncher setting. For the maximum energy in the pulse, the buncher should be set at $N_t \cong (G_p/G_{th})^{-1/2} (4\pi\sigma_{Ee}/E_e)^{-1}$ [10], where G_{th} is the threshold gain defined by the losses in the optical cavity (measured to be 0.78% at $\lambda = 370$ nm for this set). It means that the pulse energy has maximum at $N_t \cong 32$, i.e. with buncher set close to zero. This conclusion is consistent with measured dependence shown on Fig. 7. To maximize the peak power, the setting should be between these two extremes ($N_t \cong 64$ for Fig. 7).

We used the self-consistent computer code [15,16] to simulate giant pulses in the OK-4/Duke SR FEL and compared them with experimental data. The essentially perfect fit of the theoretical and experimental curves indicated high accuracy of the SR FEL model used in the computer code [15,16], as well as, the flawless performance of the gain modulator. Detailed discussion of these results does not fit within this paper and is published elsewhere [7,12]. In addition to "proper" giant pulses (see Fig. 6) observed with "good"

synchronism, we also observed giant pulses with few maximums. This effect was observed only when the revolution frequency of the electron beam was substantially higher than synchronous. Even though these pulses are much weaker and less useful, the physics behind such behavior is interesting. The growing optical pulse increases local energy spread and suppresses lasing. The lasing re-occurs, when the synchrotron motion replaces the "spoiled" part of e-beam with a fresh one.

At the perfect or good synchronism, the duration of the giant pulse is insufficient to reach the Fourier limited FEL micro-pulses. Yet, the giant pulses from the OK-4/Duke storage ring FEL have a high degree of coherence (see Table 2) and are only 6 to 15 times short of the Fourier limit.

One of most important achievement with the OK-4 FEL gain modulator is the generation of intracavity peak powers at GW level in the deep-UV range. First, we can consider a fast optical kicker to extract this power for experiments with non-linear optical phenomena. Second, this level of power is at the super-pulse level [7,8], sufficient for efficient generation of harmonics within FEL itself with tunable wavelength [17]. The e-beam dynamics in the Super-pulse mode [8,18] provides for refreshing of the electron beam during the generation of the giant pulse. The fresh electron beam has the natural energy spread and will generate harmonics rather effectively. We observed a first indication of third harmonic while operating giant pulses at 370 nm. The set-up for this observation was not properly planned—the third harmonic radiation propagated through the mirror substrate, the view-port and the atmosphere. In addition, we used a standard (air-filled) monochromator. Currently, the UHV beam-line and vacuum soft-X-ray monochromator (manufactured by McPherson) are in the process of installation. The harmonics will be extracted through the 0.5-mm diameter hole in the downstream mirror of the optical cavity. We expect to generate the third and fifth harmonics of fundamental FEL wavelength, starting with 240 nm as fundamental. The basis for these experiments will be enhanced by planned installation of the distributed optical (the OK-5 FEL) with higher gain and with four helical wigglers and three bunches. One of the wigglers can be tuned on harmonic to enhance the efficiency [19].

6. Conclusions

The gain modulator installed on the Duke storage ring allows us to generate giant pulses in a stable and predictable manner. The performance of the OK-4/Duke FEL is well characterized and theoretical predictions are in a good agreement with experimental results. We plan to start the first experiments with harmonics generation in early 2001.

Acknowledgements

The authors are thankful to the Duke FEL light source operation group, Dr. Ping Wang, James Gustavsson and Maurice Pentico, engineers and technicians at FEL laboratory, Joe Faircloth, Steve Hartman, Marty Johnson, Peter Morcombe and Owen Oakeley for their help with this project.

References

[1] N.A. Vinokurov, A.N. Skrinsky, Power limitations for an optical klystron installed on a storage ring, Preprint INP 77–67, Novosibirsk, 1977.

[2] A. Renieri, Nuovo Cimento B 53 (1979) 160.

[3] O. Svelto, Principles of Lasers, Plenum Press, New York, 1998.

[4] G.N. Kulipanov, V.N. Litvinenko, I.V. Pinayev, V.M. Popik, A.N. Skrinsky, A.S. Sokolov, N.A. Vinokurov, Nucl. Instr. and Meth. A 296 (1990) 1.

[5] V.N. Litvinenko, S.H. Park, I.V. Pinayev, Y. Wu, M. Emamian, N. Hower, O. Oakeley, G. Swift, P. Wang, Nucl. Instr. and Meth A 429 (1999) 151.

[6] T. Hara, M.E. Couprie, A. Delboulbe, P. Troussel, D. Gontier, M. Billardon, Nucl. Instr. and Meth. A 341 (1994) 21.

[7] V.N. Litvinenko, I.V. Pinayev, S.H. Park, Y. Wu, Giant and Super-Pulses in the OK-4/Duke storage ring FEL-predictions and experimental results, Phys. Rev. Lett., in press.

[8] V.N. Litvinenko, B. Burnham, J.M.J. Madey, S.H. Park, Y. Wu, Nucl. Instr. and Meth. A 375 (1996) 46.

[9] V.N. Litvinenko, Nucl. Instr. and Meth. A 359 (1995) 50.

[10] V.N. Litvinenko, The optical klystron on VEPP-3 storage ring bypass—lasing in the visible and the ultraviolet, Thesis, Novosibirsk, 1989.

[11] I.V. Pinayev, M. Emamian, V.N. Litvinenko, S.H. Park, Y. Wu, System for control and stabilizing of OK4/Duke FEL optical cavity, Proc. Beam Instrumentation Workshop, AIP Conference Proceedings 451, Woodbury, NY, USA.

[12] V.N. Litvinenko, S.H. Park, I.V. Pinayev, Y. Wu, Nucl. Instr. and Meth. A 475 (2001) 65, these proceedings.

[13] D. Nutarelli, D. Garzella, M.E. Coupie, M. Billardon, Nucl. Instr. and Meth. A 393 (1997) 64.

[14] N.A. Vinokurov, A.N. Skrinsky, Optical klystron, Preprint INP 77–59, Novosibirsk, 1977.

[15] V.N. Litvinenko, B. Burnham, J.M.J. Madey, Y. Wu, Nucl. Instr. and Meth. A 358 (1995) 334.

[16] V.N. Litvinenko, B. Burnham, J.M.J. Madey, Y. Wu, Nucl. Instr. and Meth. A 358 (1995) 369.

[17] V.N. Litvinenko, X-ray storage ring FEL: new concepts and directions in: Proceedings of 10th ICFA Beam Dynamics Panel Workshop 4th Generation Light Sources, Grenoble, France, January 22–25, 1996, p. WG6-16.

[18] V.N. Litvinenko, B. Burnham, J.M.J. Madey, Y. Wu, SPIE V. 2521 (1995) 78.

[19] V.N. Litvinenko, S.F. Mikhailov, O.A. Shevchenko, N.A. Vinokurov, Nucl. Instr. and Meth. A 475 (2001) 97, these proceedings.

ELSEVIER

Nuclear Instruments and Methods in Physics Research A 475 (2001) 229–233

NUCLEAR
INSTRUMENTS
& METHODS
IN PHYSICS
RESEARCH
Section A

www.elsevier.com/locate/nima

Towards the Fourier limit on the super-ACO Storage Ring FEL

M.E. Couprie[a,b,*], G. De Ninno[a,b], G. Moneron[a,b], D. Nutarelli[b], M. Hirsch[a,b], D. Garzella[a,b], E. Renault[a,b], R. Roux[c], C. Thomas[d]

[a] Service Photons, Atomes et Molécules, CEA/DSM/DRECAM, Bât. 522, 91 191 Gif sur Yvette, France
[b] Laboratoire pour l'Utilisation du Rayonnement Electromagnétique (LURE), Centre Universitaire Paris-Sud,
Bât. 209D, BP 34, 91 898 Orsay, Cédex, France
[c] Sincrotrone Trieste, Exp. Division, Strada Statale 14, km 163,5 Basovizza, 34 0112, Italy
[d] Eindhoven University of Technology, The Netherlands

Abstract

Systematic studies on the Free Electron Laser (FEL) line and micropulse have been performed on the Super-ACO storage ring FEL with a monochromator and a double-sweep streak camera under various conditions of operation (detuning, "CW" and Q-switched mode). From these data, it appears that the FEL is usually operated very close to the Fourier limit. © 2001 Elsevier Science B.V. All rights reserved.

PACS: 41.60 Cr

Keywords: Storage ring free-electron laser; Fourier limit

1. Introduction

Following the first operation of a Storage Ring Free Electron Laser (FEL) on ACO in 1983 in Orsay (France) [1], the Super-ACO FEL [2] started to be used by the scientific community in various scientific domains since 1993. Since then, this type of coherent light source has reached a new stage of maturity. By taking advantage of their natural synchronisation with synchrotron radia-

tion, of their tunability, high repetition rate, high average power and coherence, storage ring FELs are unique sources for pump–probe experiments [3–5]. Recent developments on the DUKE FEL show that the tunability of Storage Ring FELs can be extented below 200 nm [6]. The rapid and successful operation of the FEL on the third generation storage ring ELETTRA in Italy confirms the reliability of these devices [7]. Consequently, it is particularly important to control the temporal coherence of the FEL sources, and to maintain them close to the Fourier limit for the FEL users. The work presented here reports about systematic measurements on the spectral and temporal profiles of the Super-ACO FEL [8]. The narrowing process is also analysed.

---

*Corresponding author. Laboratoire pour l'Utilisation du Rayonnement Electromagnétique (LURE), Centre Universitaire Paris-Sud, Bât. 209D, BP 34, 91 898 Orsay, Cédex, France. Tel.: + 33-1-64-46-80-44; fax: + 33-1-64-46-41-48.

E-mail address: couprie@lure.u-psud.fr (M.E. Couprie).

2. The Super-ACO FEL operation mode

The Super-ACO storage ring is now operated at 800 MeV with two cavities, the main at 100 MHz and a harmonic cavity at 500 MHz in the bunch narrowing mode, providing thus shorter electron bunches and a higher electronic density. The positron bunch distribution is slightly asymmetrical, with a sharp-rising edge because of the interaction of the electric field radiated by the head-on electrons in interaction with the vacuum chamber (see Fig. 1). The bunch length enhancement versus the ring current results from the microwave instability, and it could be explained theoretically with an impedance model of the vacuum chamber. For the studies presented here, the FEL was operated in the UV with various sets of mirrors, with an average output power in the 50–300 mW range. The Super-ACO FEL pulses present a microtemporal structure, reproducing the recurrence of the electron bunches at a high repetition rate. In addition, the FEL also presents a so-called "macro-temporal structure" at the ms scale (see Fig. 2), depending on the particular synchronisations between the electrons circulating in the storage ring and the optical pulses bouncing back and forth in the optical resonator (detuning condition). At perfect tuning, the FEL exhibits a

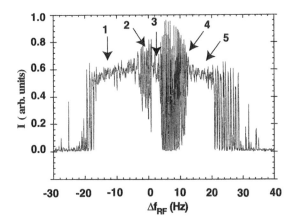

Fig. 2. Detuning curve of the Super-ACO FEL operated with two RF cavities. The fine-tuning of the synchronisation is changed by a modification of the RF frequency, whereas the rough tuning is performed with a cavity length change (1 kHz corresponds to 180 μm).

"cw" regime (zone 3 in Fig. 2) at the ms time scale. In the presence of a slight desynchronisation, the FEL power is noisy and shows a pulsed structure at the ms scale (zones 2 and 4). With a larger detuning, it becomes again CW. With the harmonic cavity, some additional pulsed zones can appear for larger detunings. The asymmetry of the detuning curve results from the shape of the electronic distribution. A longitudinal feedback system has been developed in order to maintain the FEL pulse in zone 3, and to reduce the jitter [9]. The displacement of the FEL pulse with respect to a reference position is deduced from the measurement of the FEL temporal profile with a stroboscopic detector [10], and the delay is converted with an electronic device into a voltage which is applied to the RF pilot for the adjustment of the main RF frequency. Various improvements of the electronic set-up allow the feedback system to correct rapid drifts (up to 150 μs), as illustrated in Fig. 3. It also stabilises the intensity fluctuations and the spectral drifts of the FEL.

The storage ring FEL evolution versus detuning has been reproduced using an iterative multipass model based on the pass-to-pass equations of evolution involving the laser intensity inducing the

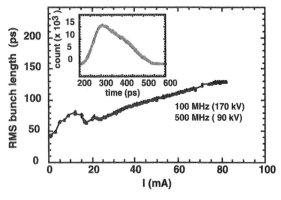

Fig. 1. Electron bunch profile (measured with a Hamamatsu double-sweep streak camera at 20 mA) and RMS positron bunch length versus current for Super-ACO operated with two cavities: 100 MHz (170 kV) and 500 MHz (90 kV). Data are analysed with an in-house software based on the moments' method.

Fig. 3. Compensation of drifts of the FEL with the longitudinal feedback. The FEL distribution is measured with a double-sweep streak camera with the fast sweep on the vertical axis and the slow sweep on the horizontal axis. The two displacements of the centre of mass of the FEL distribution are very rapidly corrected.

enhancement of the electron beam energy spread, the position of the laser micropulse [11] both in the time and frequency domains. Depending on the tuning condition and on the respective values of the laser rise time τ_r and the synchrotron damping time τ_s, the model can reproduce the macrotemporal structure for the different zones, the detuning curves and the FEL establishment regimes versus detuning. The zones where the FEL is pulsed appear for intermediate detunings. For a detuned FEL, besides the bunch heating process, the stored emission is shifted with respect to the synchronous electron and it interacts with the bunch in different positions in the electronic density. Thus the gain becomes smaller. The detuning of the FEL micro-pulse can then be partly responsible for the FEL saturation on Super-ACO. At the equilibrium, the displacement of the FEL micropulse with respect to the electronic density at each pass compensates the modification of the centre of mass of the FEL distribution because the part located towards the centre of the electronic density is much more amplified than the other edges. Because of the drift of the electrons with respect to the stored light, the spectral and temporal narrowing of the stored radiation cannot be so efficient, whereas the Fourier limit can be reached for perfect tuning.

3. Spectral characteristics of the Super-ACO FEL

The FEL line starts from the spectrum of the optical klystron, grows on one or several fringes of the optical klystron and significantly narrows. Storage ring FELs generally present a high spectral resolution, roughly given by: $\Delta\lambda/\lambda =$ $(1/\pi)\sqrt{\lambda/(N+N_d)}\sigma_\ell$ with σ_ℓ being the electron bunch length, N the number of periods of the undulator and N_d the interference order due to the dispersive section. The simulations show that for a perfectly tuned FEL [11], the laser line can continue to narrow even when the saturation is almost completely established, provided the beam is stable. The phenomenon has been called "adiabatic narrowing". The pulse narrowing is very quickly limited when the FEL is detuned. Any perturbation occurring during this process prevents the FEL from reaching the smallest spectral width, and the process restarts from the beginning. Fig. 4 shows the Super-ACO FEL line-width measured for various conditions of FEL operation. Experimentally, the spectral narrowing is more efficient for the FEL close to perfect tuning (zone 3) than for the detuned FEL (zone 5), which is in agreement with the theoretical predictions in Ref. [11]. The influence of perturbations on the laser narrowing process can explain why the points are scattered and why the whole system is not fully reproducible. The line width of the FEL operated in the Q-switched mode is generally slightly larger than in zone 3, probably because the complete narrowing cannot completely take place during the macropulse duration. Even if the longitudinal

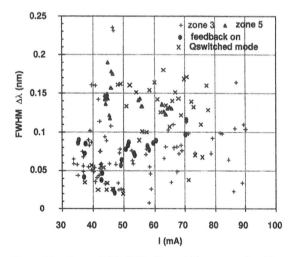

Fig. 4. The Super-ACO FEL line width measured with a Princeton monochromator (resolution of 0.1 Å) versus current, for the natural operation of the FEL in zone 3 (+) and in zone 5 (triangles), in the Q-switched mode (×).

feedback system was developed with the purpose of counteracting the jitter and the drift of the FEL pulse, it can also sometimes lead to a reduction of the FEL line width. As compared with previous measurements performed several years ago [12], the spectral width is in general smaller, probably because of the improvement of the feedback system and the slightly higher gain. There is no clear dependance of the FEL line width versus the stored current. In addition, some internal substructures in the FEL line can sometimes be observed, as already reported on the UVSOR FEL [13].

4. Temporal characteristics of the Super-ACO FEL

The storage ring FEL reproduces the micro-temporal structure of bunches at a high repetition rate (MHz). With the gain proportional to the electronic density, the FEL micropulse builds up to the maximum gain at perfect tuning. With respect to the initial positron distribution, the FEL pulse significantly narrows during the amplification process. Measurements of the FEL pulse duration on Super-ACO are shown in Fig. 5. In agreement with the simulations in Ref. [11], the FEL pulse narrows much more around perfect tuning (zone 3) than around well desynchronised

(zone 5). One also observes a dependence versus current, probably related to the one of the positron distributions itself. The FEL pulse width is also slightly higher in the Q-switched mode than in the natural regime in zone 3, again probably because the macropulse is not long enough for the quasi-"adiabatic" narrowing to proceed. Further data analysis in the Q-switched mode shows that the longer the macropulse, the shorter the FEL pulse. The influence of the longitudinal feedback system on the FEL pulse duration depends on the beam injections and on the natural value of the FEL. We could also observe a reduction of the FEL pulse in presence of the longitudinal feedback system. Besides, the pulse duration seems to be rather independent of the gap of the dispersive section. Some internal substructure can also be observed.

5. Analysis of the Fourier limit

Fig. 6 displays the FEL spectral width versus the FEL pulse duration, for data taken simultaneously. Clearly, FEL operation in zone 3 is much closer to the Fourier limit than for zone 5 (a factor of 2 above the best values), because the narrowing processes are limited for a detuned FEL. In zone 5, the FEL width is larger and the FEL pulse duration is shorter with the 500 MHz cavity than

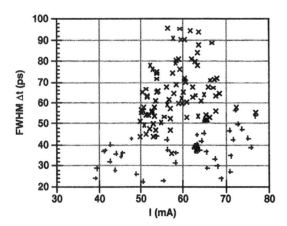

Fig. 5. The Super-ACO FEL temporal pulse measured with the Hamamatsu double-sweep streak camera versus current, for the natural operation of the FEL in zone 3 (+) and in zone 5 (triangles), in the Q-switched mode (×).

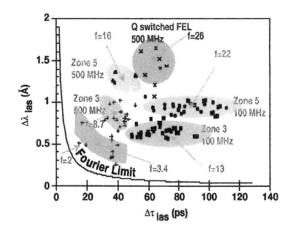

Fig. 6. The Fourier limit: FEL pulse width versus the FEL pulse duration. The new data are compared to these acquired previously. f gives the ratio of the product of the line width and the pulse duration with respect to the Fourier limit case.

with the 100 MHz cavity, since the FEL starts from a longer electron bunch in 100 MHz case. We also observe an improvement with respect to the Super-ACO FEL operation in 1994 with the 100 MHz cavity alone, possibly because of the present higher gain and the better stability control. The perfect control of the temporal coherence of the Super-ACO FEL has not yet been completely achieved, since some perturbations acting on the FEL (modulations and sudden jumps of beam parameters, or mechanical vibrations on the optical resonator) have not yet been permanently removed. Some of them are, nevertheless, compensated owing to our longitudinal feedback system.

6. Conclusion

The Super-ACO FEL is operating very close to the Fourier limit, even though some particular cases of internal substructure, multiline lasing or external modulations require further investigations. The temporal coherence achieved here is very good in terms of source quality for the user applications. This feature, coupled with a very high transverse coherence resulting from the laser mode build-up in a very long optical resonator, allows the Super-ACO FEL to be a unique tool for the scientific community.

Acknowledgements

A part of this study was supported by the European Community (ERBFMRX-C980245 and ERBFMGE-CT98-0102 contracts).

References

[1] M. Billardon, P. Elleaume, J.M. Ortéga, C. Bazin, M. Bergher, M. Velghe, Y. Petroff, D.A. Deacon, K.E. Robinson, J.M.J. Madey, Phys. Rev. Lett. 51 (1983) 1652.

[2] M.E. Couprie, P. Tauc, F. Merola, A. Delboulbé, D. Garzella, T. Hara, M. Billardon, Rev. Sci. Inst. 65 (5) (1994) 1485.

[3] M. Marsi, et al., Phys. Rev. B 61 (2000) R5070.

[4] L. Nahon, et al., Proc. SPIE 3775 (1999) 145; International Symposium on Optical Sciences: Accelerator-Based Sources of Infra-Red and Spectroscopic Applications II, Denver, USA, 18–23 juillet 1999.

[5] E. Renault, Nucl. Instr. and Meth., These Proceedings; E. Renault, L. Nahon, D. Nutarelli, D. Garzella, M.E. Couprie, F. Mérola, Tansient absorption spectroscopy using the super-ACO storage ring FEL, in: G.S. Edwards/ J.C. Sutherland (Eds.), Proceedings SPIE, Biomedical Applications of Free-Electrons Lasers, Vol. 3925, 2000, pp. 29–39.

[6] V. Litvinenko, et al., Nucl. Instr. and Meth. A 429 (1999) 151; V. Litvinenko, et al., Proceedings 21th FEL Conference, Hamburg, 1999.

[7] R. Walker, Nucl. Instr. and Meth. A 475 (2001) 20, These Proceedings; R.P. Walker, et al., Nucl. Instr. and Meth. A 429 (1999) 179.

[8] M.E. Couprie, R. Roux, D. Nutarelli, E. Renault, M. Billardon, Nucl. Instr. and Meth. A 429 (1999) 159.

[9] M.E. Couprie, D. Garzella, T. Hara, J.H. Codarbox, M. Billardon, Nucl. Instr. and Meth. A 358 (1995) 374.

[10] M.E. Couprie, D. Garzella, A. Delboulbé, M. Velghe, M. Billardon, Nucl. Instr. and Meth. A 331 (1993) 37.

[11] T. Hara, M.E. Couprie, M. Billardon, Nucl. Instr. and Meth. A 375 (1996) 67.

[12] M.E. Couprie, Laser à Électrons Libres Ultra-violet sur anneau de stockage, Habilitation à Diriger des Recherches, Univ. Paris-Sud, 17 September 1997.

[13] H. Hama, K. Kimura, J. Yamazaki, S. Takano, T. Kinoshita, M.E. Couprie, Nucl. Instr. and Meth. A 375 (1996) 32.

ELSEVIER

Nuclear Instruments and Methods in Physics Research A 475 (2001) 234–239

NUCLEAR
INSTRUMENTS
& METHODS
IN PHYSICS
RESEARCH
Section A

www.elsevier.com/locate/nima

Fourier limited micro-pulses in the OK-4/Duke storage ring FEL ☆

V.N. Litvinenko[a],*, S.H. Park[a], I.V. Pinayev[a], Y. Wu[a], A. Lumpkin[b], B. Yang[b]

[a] FEL Laboratory, Department of Physics, Duke University, Box 90319, Durham, NC 27708-0319, USA
[b] APS, Argonne National Laboratory, Argonne, IL 60439, USA

Abstract

The Super-modes are Fourier limited FEL micro-pulses predicted by Dattoli and Renieri in 1980. The OK-4 FEL at Duke, operating in the wavelength range from 193.7 nm to 730 nm, was the first to observe the Super-modes in a storage ring FEL in 1996. Since 1996, the up-graded diagnostics and improved control of the RF frequency allowed generation of Super-modes on a regular basis and systematic study of them. The Gaussian FEL micro-pulses with 1.3 ps RMS duration and Fourier limited RMS linewidth were generated in 1998–1999. In this paper we present the results of our studies and the comparison with theoretical predictions. We also present practical criteria for operating SR FELs with Super-modes. © 2001 Elsevier Science B.V. All rights reserved.

PACS: 41.60.C; 29.20.D; 78.66; 32.20; 33.30; 78.40.D

Keywords: Free-electron laser; Storage ring; Ultraviolet; Fourier limited spectra

1. Introduction

Fourier limited FEL micro-pulses in storage ring FELs (SR FELs) were predicted by Dattoli and Renieri, and named "Super-modes" [1]. Super-modes are steady-state solutions of the Schrödinger-type linear differential equation. They are not directly applicable for the linac driven FELs, where non-linear effects are responsible for the saturation of FEL power.

SR FELs easily generate laser beams in single TEM$_{00}$ mode with complete transverse coherence. In contrast, the attainment of full longitudinal

(time) coherence in SR FELs is rather challenging. To measure of the longitudinal coherence of an FEL pulse we will use the number of modes in longitudinal "phase-space" defined as

$$M = 2\sigma_s \sigma_k = \frac{4\pi c}{\lambda} \sigma_\tau \frac{\sigma_\lambda}{\lambda} \qquad (1)$$

where $\sigma_{s,\tau,k,\lambda}$ are RMS values of the FEL pulse length (s or τ) and its spectral line, λ is its wavelength and $k = 2\pi/\lambda$ is the wave vector. For a perfect Gaussian FEL pulse amplitude and constant phase the number of modes reaches minimum of $M = 1$, i.e. the pulse is Fourier limited. The use of spatial and spectral intensity in Eq. (1), instead of the amplitude of the field of the wave, causes the appearance of the factor 2.

Longitudinal profile of FEL *amplitude* gain $g(s) \equiv \sqrt{1 - G(s)} - 1$, where G is the FEL *power* gain, is *a real function* of the coordinate along the

☆ This work is supported by Office of Naval Research Contract #N00014-94-1-0818.

*Corresponding author. Tel.: +1-919-660-2658; fax: +1-919-660-2671.

E-mail address: vl@phy.duke.edu (V.N. Litvinenko).

beam (s). In CW mode of SR FEL operation, the $g(s)$ is a smooth bell-shape function with maximum in the center of the e-bunch ($s = 0$) [2,3]. It can be expanded around the origin: $g(s) \cong g_0 (1 - s^2/\sigma_g^2)$.

In a SR FEL oscillator, the de-tuning from exact synchronism, $\delta = cT_0 - 2L$, is defined by the beam revolution frequency $f_0(T_0 = 1/f_0)$ and the length of the optical cavity (L). Near the synchronism, the lengths of the optical pulses (σ_s) is much smaller than σ_g. The detailed analysis [2,4] shows that Super-modes exist only within a small tuning range with negative δ:

$$-1 < Y < 1; \quad Y = 1 + \frac{\delta}{\Delta g_0}. \tag{2}$$

Outside of this range, the stationary solutions do not exist, and SR FELs operate in so-called "general" lasing mode [2,4]. For the ideal tuning with $Y = 0$, the SMs are described by well-known functions [1–3],

$$a_n(s) = \alpha_n H_n\left(\frac{s}{\sigma_0}\right) \exp\left(-\frac{s^2}{2\sigma_0^2}\right); \quad \sigma_0 = \sqrt{\Delta}\sigma_g \tag{3}$$

where H_n is nth order Hermite polynomial, $n = 0, 1, 2, \ldots, \Delta$ is the slippage in the FEL. For an optical klystron, $\Delta = (N_w + N_d)\lambda$, where N_w is the number of wiggler periods and N_d is dimensionless dispersion of its buncher [2,3]. In modern SR FELs with $\Delta\sigma_g$ and $\sigma_0 \ll \sigma_g$ which justify the use of the gain expansion [1,2,5]. The OK-4/Duke SRFEL has $\Delta/\sigma_g \sim 0.003$ and $\sigma_0/\sigma_g \sim 0.05$. Increment (decrement) for the nth SMs (3) is

$$\gamma_n = g_0\left(1 - \frac{\Delta}{\sigma_g}(n + 1/2)\right) - 1 + r \tag{4}$$

where $R = r^2$ is reflectivity the optical cavity. In the steady-state mode, all γ_n must be equal or less than zero, i.e. $g_0 \cong 1 - r$. For small $\Delta\sigma_g$, difference in the increments (decrements) between supermodes is very small:

$$\delta\gamma = \gamma_{n+1} - \gamma_n = g_0\frac{\Delta}{\sigma_g} \cong (1 - r)\frac{\Delta}{\sigma_g}. \tag{5}$$

The main mode (a_0) *may* dominate the pulse only after $N \sim 1/\delta\gamma$ interactions. With low loss optical cavity ($1 - r \sim 1\%$), the OK-4 /Duke SR FEL needs at least 30,000 turns (~ 10 ms) to get into this mode. With higher loss cavity $\sim 10\%$ this time

decreases to ~ 1 ms. Further details and discussions can be found elsewhere [1–6].

For many years, SR FELs operated rather far from the Fourier limit [2,3] with M ranging from 10 to 100. The main reason behind the difficulty of obtaining Fourier limited pulses in a SR FEL is very short wavelengths (sub-μm) compared with the e-beam bunch-length (a few cm). Another reason is the macro-temporal structure [5], which often appeared in SR FELs because of imperfect tuning or/and small modulations of the FEL gain caused by ripple in the power supplies and e-beam instabilities [4]. Typical macro-temporal structure has 100% modulation of the FEL beam intensity and may have periodic or chaotic behavior [7]. The Super-ACO SR FEL was first to demonstrate stable CW lasing without significant macro-temporal structure [8]. Nevertheless, the Super-ACO SR FEL is still on its way to demonstrating Fourier limited FEL pulses [9].

2. Early OK-4/Duke SR FEL results

The OK-4/Duke storage ring FEL was commissioned in November 1996 with initial lasing in the near-UV range (345–413 nm) [10]. An UV Hamamatsu dual-sweep streak-camera was brought from APS, ANL (Argonne, IL) for the initial e-beam and FEL beam time-resolved diagnostics [11]. Typical RMS duration of FEL micro-pulses was ~ 8 ps (FWHM ~ 20 ps) with revolution frequency tuned close to exact synchronism.

High gain of the OK-4 FEL allowed us to operate with low beam currents which resulted in rather short e-bunch lengths. We were fortunate with the first observation of the short pulses from the OK-4/Duke SR FEL. At that time, the RF frequency was controlled by a surface-acoustic-wave (SAW) oscillator [12]. The accuracy of the control was insufficient for tuning into the Super-mode range. While studying the dependence of the micro- and macro-temporal structures of the OK-4 FEL beam, we suddenly observed a very short ~ 3 ps FEL micro-pulses. The duration of these pulses was consistent with the fundamental Super-mode [13]. The detailed analysis of the multiple pulse profiles revealed that the ratio between the amplitudes of

SM to the background is rather steady and varies from 2 to 1.7. We did not carry out spectral measurements in parallel with these studies.

Fig. 1 shows structure of FEL pulses recorded on November 15, 1996. The fit (on Fig. 1 (b)) shows a clear Gaussian peak with RMS duration of ~3 ps on the top of the background. The OK-4/DSR FEL operated at $\lambda = 388$ nm with 0.68 mA single bunch, 500 MeV e-beam. The RF voltage was 500 kV and the RMS e-beam bunch length was 38 ps. The OK-4 FEL buncher was set at $I_b = 1.27$ kA providing $\Delta = 0.155$ ps (i.e. 46.6 µm, $N_w = 33.5$ and $N_d = 86.25$). Theory predicts SR duration of $\sigma_\tau = 1.72$ ps which does not contradicts the measured duration. Simultaneous measurement of the spectrum was required to confirm that these pulses are Fourier limited.

3. Recent OK-4/Duke SR FEL results

Presently, the OK-4/Duke storage ring FEL is operating in the wavelength range from 193.7 to 730 nm [14]. The OK-5/Duke SR FEL has better stability when operating at higher energy, up to 800 MeV and short wavelengths. The RF frequency control is the most important and most precise tool for adjusting synchronization between the electron and laser micro-pulses. Since 1996, we substantially improved the accuracy of the RF frequency control as well as up-graded the diagnostics of both FEL and electron beams. We acquired our own Hamamatsu C5680 dual-sweep streak-camera with FWHM time resolution of ~1.5 ps (i.e. RMS time resolution of ~0.7 ps) and spectral range from the visible to 120 nm. The computer controlled MDR-23 monochromator allows us to measure time average spectra with RMS wavelength resolution of 0.0128 nm. These improvements allowed us to generate SMs on regular basis and to study them systematically with parallel time-resolved and spectral measurements.

We observed short FEL micro-pulses, consistent with SM, when we tuned the RF frequency with extremely high precision and stabilized the FEL operation. The suitable tuning range for SM observations was about ~0.05–0.002% of the full OK-4/DSR FEL tuning range [14]. With the optical cavity loss ~1% per pass around $\lambda = 380$ nm, very short FEL micro-pulses were observed within the range of $|\delta f/f| \leqslant 1.5 \times 10^{-9}$ ($|\delta f_0| \leqslant 0.005$ Hz). The study of SMs was simplified when we operated with low beam currents or with high loss mirrors. With 6% loss per turn at 218 nm [15], the suitable SM range was ~3 times wider. Within the suitable SM range, the fine adjustments with steps as small as $\delta f/f = 1 \times 10^{-10} (\delta f_0 \sim 0.0003$ Hz) were required to lase with Fourier limited micro-pulses comprising of a single SM.

Getting SMs at lower energies was less trivial because of less stable beam. The RF-noise and small coherent synchrotron oscillations of the e-beam also complicated the task. Nevertheless, we obtained the shortest FEL pulses when operated with very low e-beam current and lower e-beam energy. The averaged value of RMS FEL pulse duration was ~1.3 ps. The FEL time structure and the FEL spectrum are shown in Fig. 2. The parameters for this set were $E = 460$ MeV, $V_{rf} = 560$ kV; $I_{beam} = 0.3$ mA; $N_d = 72.75$. The RMS measured duration of e-bunch pulse was 33 ± 1 ps [6].

The deviation from the perfect Gaussian shapes in both the temporal and spectral profiles is rather small. Most probably it is caused by "contamination" by higher order SMs, which was as low as 5–10%.

We did not attempt to observe SMs when we operated below 200 nm and in the visible. Below 200 nm, we concentrated at studies of the OK-4 FEL gain. Nevertheless, with rather preliminary tuning, the OK-4 FEL generated micro-pulses ~4 ps RMS and $M \sim 5$ at 199 nm [14].

Fig. 1. The first observation of the SM. The dual-sweep image (a) from the APS steak-camera with the vertical (fast) time scale of 370 ps, and with the horizontal (slow) time scale of 0.5 ms. The pulse profiles, taken at $t \sim 0.2$ ms, and its fit with the fundamental SM and a broader background is shown in (b). The measured RMS width of 3 ± 0.35 ps was limited by the resolution of the streak-camera for this run ~2.9 ps. The quadrature formula gives 1.7 ps upper bound for SM RMS duration.

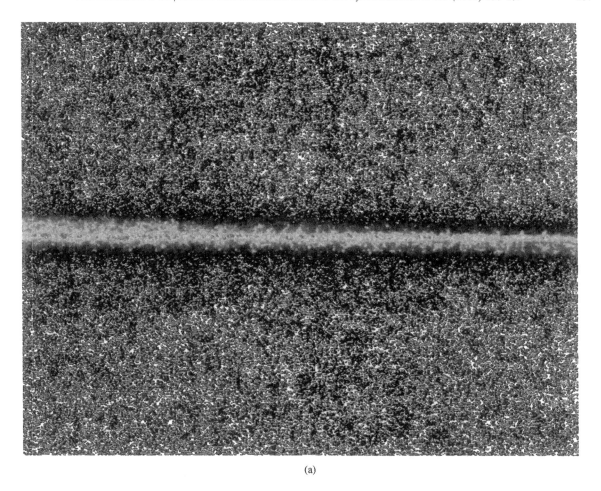

(a)

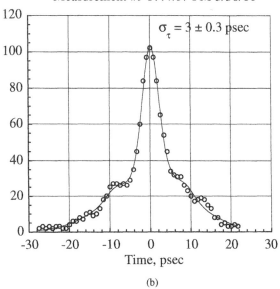

(b)

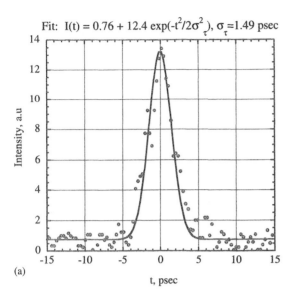

(a)

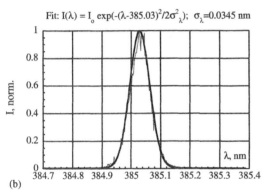

(b)

Fig. 2. Measured profiles of the FEL pulse (a) and is spectrum (b). The RMS duration of the FEL pulse is 1.31 ± 0.05 ps with the streak-camera RMS resolution of 0.7 ps taken into account. The RMS FEL line-width is 0.032 nm with RMS resolution of monochromator of 0.0128 nm taken into account.

4. Comparison of the experimental results and the theory

We used reasonably high loss mirrors $\sim 6\%$ per pass when the OK-4/DSR lased in the deep-UV. High losses and high energy of operation (700 MeV) made the attainment of SM an easier task. With SM, narrow spectra were observed in the deep-UV (Ref. [15], RMS linewidth of 4.12×10^{-5} at $\lambda = 218.65$ nm, $I_{beam} = 2.05$ mA,

$E = 700$ MeV, $V_{rf} = 200$ kV). The measured FEL micro-pulses had a shape close to the Gaussian with duration of 1.5 ± 0.1 ps. These measurements gave $M = 1.08 \pm 0.1$, which is in good agreement with the Fourier limit. For these parameters, the SM theory (3) predicts $\sigma_g \sim 63 \pm 7$ ps, which is in reasonable agreement of the measured RMS e-beam duration of 68 ± 5 ps.

The result at $\lambda = 385$ nm (see Fig. 2) is also very close to the Fourier limit. With the resolutions of the streak-camera and the monochromator taken into account, we got $M = 1.07 \pm 0.9$. Therefore, we conclude that we observed Fourier limited laser micro-pulses from the OK-4/Duke SR FEL.

According to the theory [2,4], SMs do not exist outside of the very narrow tuning range. The observed frequency ranges of $\delta f/f \sim 1-5 \times 10^{-9}$ are in reasonable agreement with the theory. The necessity of very fine-tuning with $\delta f_0 \sim 0.0005$ Hz (i.e. ~ 10 nm in the length of the optical cavity) needs further theoretical studies.

Overall, our experimental results are in good agreement with the theory within the accuracy of the measurements. We plan to use a monochromator with better resolution for improving the accuracy.

5. Conclusions and Acknowledgements

For the first time we demonstrated the operation of the SR FEL with Fourier limited micro-pulses. Fourier limited performance of the OK-4 FEL is needed for some user experiments. The forthcoming reliability up-grade of the Duke storage ring power supplies will improve the stability and will be the keystone toward the use of the OK-4 FEL for pump-probe experiments with time resolution ~ 2 ps.

The coherent synchrotron oscillations and the phase noise in the RF system are additional obstacles for the pump-probe experiments. We plan to improve further the stability of the RF system and to employ longitudinal feedback, if needed. As was demonstrated at UVSOR SR FEL, small vibration of the optical cavity mirrors can be a cause of the additional jitter and instability [16].

We plan to address these problems with a new design of the optical cavity.

The authors are grateful to the Office of Naval Research MFEL program for financial support. The authors would like to thank the staff and engineers of the Duke FEL laboratory, especially Dr. Ping Wang, for their help in these experiments.

References

[1] G. Datolli, A. Rinieri, Nuovo Cimento B 59 (1980) 1.

[2] V.N. Litvinenko, N.A. Vinokurov, Nucl. Instr. and Meth. A 304 (1991) 66.

[3] P. Elleaume, IEE J. Quantum Electron. QE-21 (1985) 1012.

[4] V.N. Litvinenko, Thesis, Novosibirsk, 1989.

[5] P. Elleaume, Nucl. Instr. and Meth. A 237 (1985) 28.

[6] V.N. Litvinenko, S.H. Park, Super-modes in storage ring FEL oscillators: theory and experiment, Phys. Rev. Lett., submitted.

[7] M. Billardon, Nucl. Instr. and Meth. A 304 (1991) 37.

[8] M.E. Couprie, FELs using Storage ring, Proc. of European Accelerator Conference, Sitges, Spain, 10–14 June, 1996.

[9] M.E. Couprie et al, Nucl. Instr. and Meth. A 475 (2001) 229, these proceedings.

[10] V.N. Litvinenko, et al., Nucl. Instr. and Meth. A 407 (1998) 8.

[11] A.H. Lumpkin, et al., Nucl. Instr. and Meth. A 407 (1998) 388.

[12] P. Wang, et al., Proceedings of the 1995 Particle Accelerator Conference, Dallas, TX, May 1996, p. 1841.

[13] V.N. Litvinenko, et al., Micro-temporal structure in the OK-4/Duke SR UV FEL: theory and experiment, presented at the FEL'97, Beijing, China, unpublished.

[14] V.N. Litvinenko, S.H. Park, I.V. Pinayev, Y. Wu, Operation of the OK-4/Duke storage ring FEL below 200 nm, Nucl. Instr. and Meth. A 475 (2001) 195, these proceedings.

[15] V.N. Litvinenko, S.H. Park, I.V. Pinayev, Y. Wu, M. Emamian, N. Hower, O. Oakeley, G. Swift, P. Wang, Nucl. Instr. and Meth. A 429 (1999) 151.

[16] M. Hosaka, S. Koda, J. Yamazaki, H. Hama, Nucl. Instr. and Meth. A 445 (2000) 208.

ELSEVIER

Nuclear Instruments and Methods in Physics Research A 475 (2001) 240–246

NUCLEAR
INSTRUMENTS
& METHODS
IN PHYSICS
RESEARCH
Section A

www.elsevier.com/locate/nima

Time structure of the OK-4/Duke storage ring FEL ☆

V.N. Litvinenko[a],*, S.H. Park[b], I.V. Pinayev[a], Y. Wu[c]

[a] *FEL Laboratory, Department of Physics, Duke University, Box 90319, Durham, NC 27708-0319, USA*
[b] *KAERI, 159 Deokjin-Dong Yusung-Ku, Taejon, South Korea*
[c] *Lawrence Berkeley National Laboratory, 1 Cyclotron Road, Berkeley, CA 94720, USA*

Abstract

In this paper, we present results of experimental and theoretical studies of macro- and micro-temporal of dynamics of the OK-4/Duke storage ring FEL (SR FEL) and electron beams. The experimental part of these studies utilized the Hamamatsu C5680 dual-sweep streak-camera with 1.2 ps resolution. We use both numerical and analytical tools for theoretical analysis of the FEL and e-beam distributions without any pre-imposed limitations. Our experimental results are in good agreement with the theoretical predictions. © 2001 Elsevier Science B.V. All rights reserved.

PACS: 41.60.Cr; 29.20D; 78.66; 52.75

Keywords: Free electron laser; Storage ring; Optical Klystron; Time structure; Peak power

1. Introduction

The macro- and micro-temporal structure of SR FEL beam is well studied experimentally [1–5]. In this paper, we review the experimental results obtained with the OK-4/Duke storage ring FEL and compare them with our theoretical models and predictions. We use the self-consistent, 3D model of the interaction between the electron and the optical beams in the FEL for the description of the evolution of the system. The set of two 3D codes #felTD and #vuvFEL, based on the analytical 3D SR FEL theory [6,7], was developed in 1994 [8]. This set turned out to be a universal

tool for an accurate description of SR FEL dynamics in the linear regime, which is typical for SR FELs. We used it for simulation of the gain [9], of the giant pulse mode [10], as well as of the average lasing power in the OK-4/Duke storage ring [11]. The codes are very efficient, allowing 30,000 macro-particles and runs for hundreds and thousands of turns [8,10].

In addition, this set of codes reveals complete dynamics of the FEL pulse and the phase-space distribution of the e-beam [9]. The laser light is represented by a wave-packet with slowly varying amplitude and phase (to be exact, the *real* and *imaginary* parts of the amplitude) evolving from turn to turn with all known effects taken into account. The incomplete list of effects on the FEL pulse includes the local complex gain with the slippage, the local interaction, the spontaneous radiation and the effect of the optical cavity. It is important to note that the codes are of

_____

☆This work is supported by ONR Contract N00014-94-1-0818.

*Corresponding author. Tel.: +1-919-660-2658; fax: +1-919-660-2671.

E-mail address: vl@phy.duke.edu (V.N. Litvinenko).

multi-frequency with the bandwidth as large as 1.5% [9]. This set of programs provides both time and spectral information on the FEL pulse. In the present state, this set of codes cannot handle nonlinear effects in FEL, such as harmonics generation.

In this paper, we focus on the temporal structure of the FEL pulses operated in so-called "general lasing mode" (GLM), while information on the other modes of operation is published separately [10,12]. First, we discuss the macro-temporal structure of the OK-4/Duke storage ring FEL and effects of the de-tuning. Second, we focus on details of the GLM micro-temporal structure and its specific features.

2. Tuning curve and temporal structure in the GLM

2.1. GLM

The de-tuning from exact synchronism, $\delta = cT_0 - 2L$, is defined by the beam revolution frequency $f_0 = 1/T_0$ and the length of the optical cavity L. We will use a dimensionless de-tuning parameter Y defined as [7,12]

$$Y = 1 + \frac{\delta}{\varDelta g_0}$$

where $\varDelta = (N_w + N_d)\lambda$, λ is the FEL wavelength, N_w is the number of wiggler periods and N_d is dimensionless dispersion of its buncher [12], and g_0 is the peak FEL gain. The de-tuning from the synchronism strongly affects practically all parameters of an SR FEL from its power to its temporal structure. When $|Y| < 1$, SF FELs operate with Fourier-limited laser pulses described by stationary solutions, known as Super modes [13]. Outside this range, i.e. with $|Y| > 1$, stationary solutions for FEL pulse do not exist, and SF FELs mode is the GLM [14]. Further, in this paper, we consider only the GLM. The GLM in storage ring FELs is similar to the SASE FEL mode. The main difference between them is that in SR FELs, the e-beam is losing phase information at λ-scale after each turn. Nevertheless, the statistics and the spectral features are similar to those in SASE FEL. For example, the GLM has

spikes (seen as stripes in Fig. 4) in the FEL pulse, which are similar to those in SASE FELs. In the GLM, spontaneous radiation, generated by the e-beam and captured in the optical cavity, interacts with the e-beam while drifting through it for $\Delta s \approx \delta$ at each turn. The sign of δ defines the direction of the drift (from head to tail or vice versa). This process is depicted in Fig. 1. When the radiation passes through the center of the beam, where gain is larger than the optical cavity loss, the radiation is amplified. The power reaches maximum at the point where the net gain is equal to zero. When the radiation drifts further to the wing of the e-beam, where the loss is larger than gain, it slowly decays.

For GLM and CW operation, the theory predicts the dependencies for the FEL-pulse width, the FEL spectral width, as well as the location of the FEL pulse with respect to the e-beam [7,14]. The predicted RMS FEL-pulse length σ_0 is [14]

$$\sigma_0 \cong \sigma_g \left| \frac{\delta}{(1-R)\sigma_g} \right|^{1/3} \left\{ \frac{3}{2} \ln \left(P_{FEL}/P_{SR} \right) \right\}^{-1/6} \quad (1)$$

where σ_g is the gain length (which is close to the e-bunch RMS length, see Ref. [12]), R is the reflectivity of the optical cavity, P_{FEL} and P_{SR} are FEL power and power of spontaneous radiation into FEL mode [7,14]. A typical value of $\ln(P_{FEL}/P_{SR})$ is ~ 12 for the OK-4/Duke storage ring FEL operating at 400–700 MeV.

2.2. Tuning curve

The tuning curve of an FEL, i.e. the dependence of the FEL power on δ, provides important information on the gain and the e-bunch length. We change the revolution frequency around the synchronous value ($\Delta f_0 = 0$) for measuring the tune curve instead of scanning the length of the optical cavity (δ). Both the methods are equivalent when the ratio $\delta/(cT_0) = -\Delta f_0/f_0$ is taken into account. The spectrum, the macro- and micro-temporal structures depend strongly on δ. It was found in earlier experiments that SR FELs have distinct macro-temporal structure with 100% modulation of the FEL power when tuned close to the synchronism [1,2]. It was also found that at both the wings of the tuning curve, SR FELs lase in CW mode. A number of later studies using a

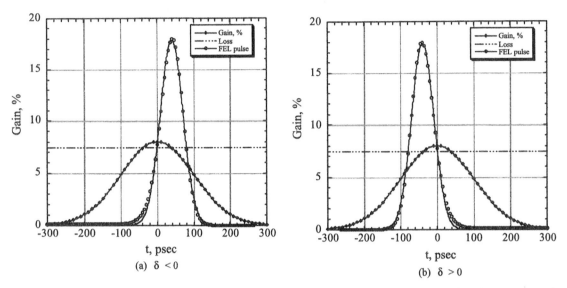

Fig. 1. The calculated "time average" profiles of the FEL beam (○) for gain curve (◇) with RMS duration of $\sigma_g = 100$ ps and 8% peak gain, and for 7.4% loss cavity mirrors (dashed line). De-tuning is $\delta \pm 25\,\mu$m (i.e. $\delta/c = 84$ fs). The FEL pulse profiles are slightly different from Gaussian fit and are located 39.4 ps left or right off center where the gain curve crosses the loss level. The FEL profiles (a) and (b) are neraly mirror images of each other and have RMS duration of 30.7 ps.

streak-camera provided more detail on the FEL power structure. The Super-ACO team found a stable CW region around $\delta = 0$, which they called "zone-3" [15]. The UVSOR team found very distinct stripes in the dual-sweep images, that are characteristic of the general lasing mode [16].

The tuning curve for the OK-4/Duke storage ring FEL is shown in Fig. 2. It has features similar to that of the Super-ACO SR FEL: five zones, three of which are CW zones. Tuning curves measured for the OK-4 FEL had typical FWHM of $\delta \sim 100$–$300\,\mu$m (related to the length of the optical cavity). This range translates into FWHM of 3–8 Hz in the beam revolution frequency f_0, or into FWHM ~ 190–500 Hz in the RF frequency, f_{rf}, which is the 64th harmonic of f_0. Both the "CW, zone 3" (~ 0.1 Hz) and the unstable zones 2 and 4 (~ 0.4 Hz) are rather narrow in our case. The Super-mode zone is even narrower (~ 0.01 Hz [12]).

We found a good fit for the measured average power to be $P(\Delta f) = P_0 \exp(1 - \sqrt{1 + \Delta f^2/\sigma_f^2})$ ($P_0 = 13.1$ mW and $\sigma_f = 1.162$ Hz in Fig. 2). At the wings ($|\Delta f_0| > 1.5$ Hz), the average power

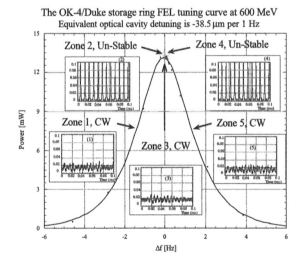

Fig. 2. Typical tuning curve of the OK-4/Duke storage ring FEL average power as function of revolution frequency ($f_0 \approx 2.7898$ MHz). 73% of average FEL power, out-coupled through up-stream mirror, was measured by Molectron power meter. The parameters are $E = 600$ MeV, $I_b = 3.6$ mA, $\lambda = 385$ nm, V_{rf} (178.55 MHz) = 600 kV. Clips show 0.1-s-long oscilloscope traces of macro-temporal FEL power structure typical for each of five zones.

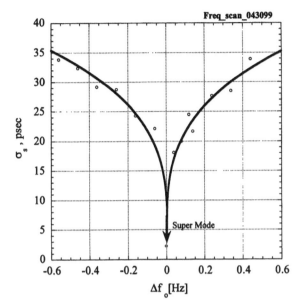

Fig. 3. Measured dependence of the RMS FEL pulse duration on the de-tuning. The fit is $\sigma_t = 5\,\text{ps} + |\Delta f_0/\sigma_f|^{1/3}$ with $\sigma_f = 0.00337\,\text{Hz}$.

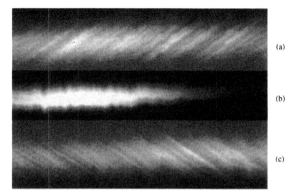

Fig. 4. Three dual-scan images of the OK-4 FEL beam operated in zones 1, 4, and 5, combined together: zones 1—bottom image (c), zone 4—the middle image (b) and zone 5—top image (a). The horizontal (slow) and vertical (fast) time scales are the same for all images: slow—5 ms, fast $\sim 600\,\text{ps}$ for combined image. $|\Delta f_0| \approx 0.6\,\text{Hz}$ for the top and bottom images and $\Delta f_0 \approx 0.05\,\text{Hz}$ for the middle. The streak camera resolution at this setting was $\sim 2.1\,\text{ps}$ RMS.

drops exponentially with $|\Delta f_0|$. Fig. 3 shows one of the measured dependencies of the FEL pulse length on the de-tuning. The fit of the theoretical curve (1) with the measured data is rather good. As we expected from the theory, at the wings of the tuning curve, the FEL pulse length σ_t is proportional to the cubic root of the de-tuning in the GLM [7,13]. In addition, the measured RMS e-bunch length of $120 \pm 5\,\text{ps}$ is in excellent agreement with the theoretical prediction from Formula (1): $\sigma_g = 121 \pm 2\,\text{ps}$. Relative locations of the FEL pulse and the e-beam were also in good agreement with theory.

3. Details of the temporal structure in general lasing mode

Fig. 4 shows three characteristic dual-scans of the temporal structures in GLM, similar to those observed at UVSOR [16,17]. The slow, horizontal scans allow to see the FEL bunch microstructure evolving from turn to turn. The characteristic "stripes", i.e. short wave-packets (SWP), are typical for dual-scan images for GLM SR FELs, where the SWP "slips" through the e-bunch, as described in the previous chapter. The lifetime of an SWP depends on the de-tuning, and can be estimated as $\tau = T_0 \upsilon$, where $\upsilon \sim \sigma_e/|\delta|$ is number of turns required for the SWP to pass the RMS e-beam bunch length of σ_e. The lifetime of SWP is longer and they are more powerful in zones (2) and (4). Their life is short in zones (1) and (5) and they co-exist within a single FEL micro-pulse. The correlation length of an SWP

$$s_c \propto \Delta \sqrt{\upsilon g_0} \approx \Delta \sqrt{\ln(P_{\text{FEL}}/P_{\text{SR}})}$$

gives an estimation of its typical duration (see below).

In zones (2) and (4), individual wave-packets evolve for rather longer time and become very intense. It might explain the unstable SR FELs behavior in these zones. The net gain along the pass is sufficient for such a pulse to accumulate the energy sufficient to blow up the energy spread of the e-beam. Such a pulse reduces the FEL gain below threshold and the e-beams need a time about the energy damping time to recover. As soon as this happens, the next pulse strikes.

Detailed analysis of images shows that for Fig. 4(a–c), the RMS duration of the e-beam are

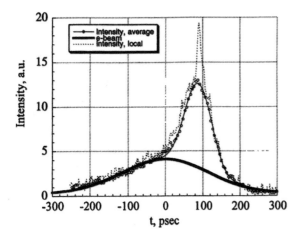

Fig. 5. The profiles of the FEL and e-beam for the image Fig. 4(c), $\Delta f_0 \approx 0.6$ Hz. The intensity of the FEL beam was reduced by a narrow-band filter to bring up the level of spontaneous radiation. Spontaneous radiation has the time profile of the e-bunch. Therefore, the image is a combination of shorter FEL pulse and longer e-pulse images. Open circles (O) show the measured profile average over turns and the thin continuous line shows a fit comprised by two Gaussian with adjustable amplitudes, width and centers, and a constant background.

118, 136 and 116 ps correspondingly. The RMS FEL pulse lengths (averaged over turns) are 33 ps for Fig. 4(a) and 18 ps for Fig. 4(b). The FEL pulses are shifted with respect to the e-beam center for −76 ps for Fig. 4 (a) and +47 ps for Fig. 4(b). Fig. 5 provides details of the image in Fig. 4(c).

Parameters for the fit in Fig. 5 are: amplitudes, 9.43 (FEL pulse) and 3.83 (e⁻ bunch), RMS widths are 35.3 ps (FEL pulse) and 116 ps (e⁻). The time is shifted for the e-beam to be in the center. The FEL pulse center is at +88.8 ps. The thick continuous curve shows the fitted e-beam profile. This graph has all the features of Fig. 1(a). In addition, the dashed curve shows a local profile, i.e. instant snap-shot, of the FEL micro-pulse with clear evidence of SWPs. In zones 1 and 5, SWPs have statistics distribution in amplitudes typical for a non-saturated SAFE FEL [18]. The SWP on the top in Fig. 5 has an RMS duration ~2.5 ps. The finite dynamic range and the resolution of the streak camera do not allow seeing finer structures in the pulse.

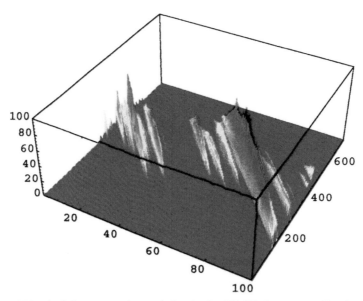

Fig. 6. 1000 turns ($t_1 - t_2 = 0.36$ ms) of the wave-packet evolution in the OK-4/Duke storage. The simulations were run for ~5 damping times to establish CW mode prior to the shown time interval. The 3-D output is composed of 100 wave-packets (slices) simulated and recorded by #vuvfel with interval of 10 turns (3.6 μs). The horizontal axis is the slice number (i.e. turn number/10). Vertical axis is the local wave power (intracavity, in kW). The third axis is fast time in wave packet bins (see [8])—the full scale is 131 ps. We show only 3% of the wave packets, the arrays with 20,000 bins. Other parameters are: $E = 600$ MeV, $I = 2.2$ mA, $V_{rf} = 500$ kV, $\delta = -45.8$ μm (−0.153 ps), $R = 99.3\%$. The resulting e-beam has RMS duration of 67 ps and RMS energy spread of 0.13%. The averaged FEL pulse had RMS duration of 39 ps, which coincides with the measurements.

The positions and amplitudes of SWPs change from turn to turn. We used the set of our codes (see Introduction) to simulate various modes of the OK-4 FEL operation including the GLM. The numerical model is based on macro-particles statistically representing electrons in an electron bunch. The interaction between the electron beam and the FEL wave-packet is both local and energy dependent. Most importantly, for GLM, the code includes correctly the spontaneous radiation, the slippage and the phase of the induced radiation of the e-beam in the FEL. The trivial part of the code takes care of the wave-packet propagation through the optical cavity and the de-tuning from synchronism. We are starting simulation with zero FEL field and its build up from the spontaneous radiation amplified by the e-beam itself. The simulation of the GLM showed all essential features observed in experiments: SWPs (or stripes) with correct propagation angle, correct shape and duration, and correct spectra. Fig. 6 shows results of one of these simulations with clearly visible SWPs. In contrast with the measurements, the simulations provide essentially unlimited time resolution. In the OK-4/Duke storage ring FEL, SWP can be ~ 0.1 ps RMS.

Short SWPs are present in the FEL pulse when the OK-4/Duke storage ring FEL operates in giant pulse (GPM) mode [10]. The GPM is a perfect example of the amplified spontaneous radiation and has structures similar to the GLM. The same set of programs was used to simulate the dynamics of the OK-4 FEL in GPM. Fig. 7 shows the wave-packet structure of the OK-4 FEL micro-pulse 180 μs after the start of the giant pulse. The streak camera images (see Fig. 8) have finite resolution and cannot show SWPs with RMS duration of < 0.5 ps.

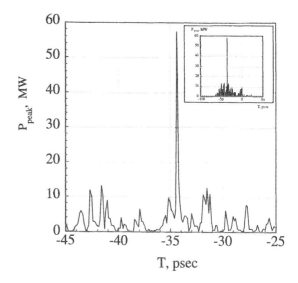

Fig. 7. The output of the #vuv at the turn 500 after the start of the giant pulse. The vertical axis is the peak power in MW. The clip at the up-right corner shows the complete micro-pulse, while the main plot shows only the center part. The short intense wave-packet has FWHM duration of 172 fs. The parameters for the run are $E = 500$ MeV, $I_b = 5.56$ mA, $\lambda = 360$ nm, $N_d = 42.75$, $V_{rf} = 220$ kV, cavity loss is 1.2% [10,11].

Fig. 8. Dual-sweep image of a typical giant pulse in the OK-4/Duke storage ring FEL with SWPs. Vertical scale is ~ 100 ps, the horizontal scale is 0.5 ms.

4. Conclusions

The experimentally observed temporal structures in the OK-4/Duke storage ring FEL are in good agreement with existing theory and with the simulations. The set of codes, we have developed, provides all essential information on the temporal behavior of the OK-4 FEL beam. The additional information on the FEL and beam structures available from these codes complements our experimental results. We were pleased that the predictions of our codes are in excellent agreement with the real OK-4 FEL performance. We plan to extend the capability of this set of codes to include more complex FEL structures, such as the OK-5

FEL, and a high gain, nonlinear regime in SR FELs.

The results of this paper are mostly focused on the GLM mode of the OK-4 FEL. The GLM in SR FEL provides FEL beams with comparable levels of average power but with longer macro-pulses and wider line-widths. It also provides for reliable CW mode of SR FEL operation. These features are attractive for experiments, where high peak power is not needed or is even harmful and where lasing lines $\sim 0.1\%$ is sufficient. For example, we use the general lasing mode for photo-emission electron microscopy (PEEM) user experiments, where the space charge effect limits the usable peak intensities [19].

Acknowledgements

The authors are grateful to the Office of Naval Research MFEL program for financial support. The authors would like to thank the staff, the engineers of the Duke FEL laboratory, and especially the accelerator operation group led by Dr. Ping Wang, for their help in these experiments.

References

[1] M. Billardon, et al., Nucl. Instr. and Meth. A 237 (1985) 224.

[2] I.B. Drobyazko, et al., Nucl. Instr. and Meth. A 282 (1989) 424.

[3] H. Hama, et al., Nucl. Instr. and Meth. A 375 (1996) 32; K. Kimura, Nucl. Instr. and Meth. A 375 (1996) 62.

[4] T. Hara, et al., Nucl. Instr. and Meth. A 375 (1996) 67.

[5] A.H. Lumpkin, et al., Nucl. Instr. and Meth. A 407 (1998) 388.

[6] V.N. Litvinenko, et al., Nucl. Instr. and Meth. A 359 (1995) 50.

[7] V.N. Litvinenko, Thesis, Novosibirsk, 1989.

[8] V.N. Litvinenko, et al., SPIE 2571 (1995) 78; V.N. Litvinenko, B. Burnham, J.M.J. Madey, Y. Wu, Nucl. Instr. and Meth. A 358 (1995) 334; V.N. Litvinenko, B. Burnham, J.M.J. Madey, Y. Wu, Nucl. Instr. and Meth. A 358 (1995) 369.

[9] V.N. Litvinenko, et al., Nucl. Instr. and Meth. A 407 (1998) 8; V.N. Litvinenko, et al., Nucl. Instr. and Meth. A 407 (1999) 151; V.N. Litvinenko, S.H. Park, I.V. Pinayev, Y. Wu, Operation of the OK-4/Duke storage ring FEL below 200 nm, Nucl. Instr. and Meth. A 475 (2001) 195, these proceedings.

[10] I.V. Pinayev, et al., Giant High-Peak Power Pulses in the UV OK-4/Duke Storage Ring FEL using the Gain Modulator, Nucl. Instr. and Meth. A 475 (2001) 222, these proceedings.

[11] V.N. Litvinenko, S.H. Park, I.V. Pinayev, Y.Wu, Power limitations and beam dynamics in the OK-4/Duke storage ring FEL, Nucl. Instr. and Meth. A 475 (2001) 65, these proceedings.

[12] V.N. Litvinenko, S.H. Park, I.V. Pinayev, Y. Wu, A. Lumpkin, Fourier limited micro-pulses in the OK-4/Duke storage ring FEL, Nucl. Instr. and Meth. A 475 (2001) 234, these proceedings.

[13] G. Datolli, A. Rinieri, Nuovo Cimento B 59 (1980) 1.

[14] V.N. Litvinenko, N.A. Vinokurov, Nucl. Instr. and Meth. A 304 (1991) 66.

[15] T. Hara, et al., Nucl. Instr. and Meth. A 341 (1994) 21; D. Garzela, et al., Nucl. Instr. and Meth. A 341 (1994) 24.

[16] H. Hama, et al., Nucl. Instr. and Meth. A 358 (1995) 365.

[17] S.H. Park, Thesis, Duke University, 2000.

[18] Claudio Pellegrini,, Nucl. Instr. and Meth. A 475 (2001) 328, these proceedings. J. Rossbach, et al., Observation of self-amplified spontaneous emission in the wavelength range from 80 nm to 180 nm at the TESLA Test Facility FEL at DESY, Nucl. Instr. and Meth. A 475 (2001) 13, these proceedings.

[19] H. Ade, Y. Wang, S.L. English, J. Hartman, R.F. Davis, R.J. Nemanic, V.N. Litvinenko, I.V. Pinayev, Y. Wu, J.M.J. Madey, Surf. Rev. Lett. 5 (6) (1998) 1257.

ELSEVIER

Nuclear Instruments and Methods in Physics Research A 475 (2001) 247–252

NUCLEAR
INSTRUMENTS
& METHODS
IN PHYSICS
RESEARCH
Section A

www.elsevier.com/locate/nima

Helical wigglers for the OK-5 storage ring VUV FEL at Duke ☆

V.N. Litvinenko[a],*, S.F. Mikhailov[a], N.A. Vinokurov[a], N.G. Gavrilov[b], G.N. Kulipanov[b], O.A. Shevchenko[b], P.D. Vobly[b]

[a] FEL Laboratory, Duke University, P.O Box 90319, Durham, NC 27708-0319, USA
[b] Budker Institute of Nuclear Physics, Novosibirsk, Russia

Abstract

In this paper, we present the design and parameters of electromagnetic wigglers with controllable polarization for the VUV OK-5/Duke storage ring FEL. The OK-5 FEL, the first distributed optical klystron, is comprised of four wigglers and three matching sections with individually controlled quadrupoles and bunchers. The geometry and the relative strength of horizontal and vertical fields determine the polarization of the radiation from the OK-5 wigglers. We compare the predicted and measured quality of the wiggler fields. © 2001 Elsevier Science B.V. All rights reserved.

PACS: 41.60.Cr; 29.20D; 78.66; 52.75

Keywords: Free electron laser; Storage ring; Optical klystron; Deep ultraviolet; Coherence

1. Introduction

The OK-5 FEL will replace the OK-4 FEL at the Duke storage ring, which successfully operated in a the wide spectrum region from the near IR to the deep UV [1]. The OK-5 FEL will be a first implementation of distributed optical klystron [2] with reasonably high gain in the UV and the VUV ranges of spectrum [3]. Detailed description of the OK-5/Duke storage ring FEL is presented elsewhere [4,5]. The main components of the OK-5

FEL are four electromagnetic wigglers (EM) with controllable polarization. The main advantages of the helical wigglers are the increase of the gain (almost by a factor of two at fixed period and wavelength) and the absence of high harmonics of radiation on its axis. The later feature provides for substantial reduction of the downstream mirror heating and degradation. The control of the polarization in the OK-4 FEL is critical for a number of experiments under consideration. First, we plan to use the OK-5/Duke SR FEL for effective generation of the beams of mono-energetic polarized γ-rays [4] in the scheme similar to that used with the OK-4 FEL [6]. The nuclear physics experiments with polarized targets require switchable circular and linear polarization of γ-rays [7,8]. Second, we plan to use left-and-right circular polarization in the 150–250 nm range for

☆ This work is supported by ONR Contract #N00014-94-1-0818.

*Corresponding author. Tel : + 1-919-660-2658; fax: + 1-919-660-2671.

E-mail address: vl@phy.duke.edu (V.N. Litvinenko).

studies of the dichroizm of biological objects for increasing the contrast of the their images. Third, we plan to conduct the experiment on the parity violation in atomic transitions 1S–2S in atomic hydrogen [9].

In Section 2, we describe the design and parameters of the helical EM wigglers. Section 3 is dedicated to the description of the 3D magnetic filed and its model. In Section 4, we compare the predicted and the measured magnetic fields as well as the current status of the devices. We conclude with the discussion of our plans.

2. Design and parameters of the OK-5 wigglers

The magnetic system of the OK-5 FEL will extend for 24.2 m and will occupy most of the 34-m South straight section (SS) of the Duke storage ring [10]. The OK-5 FEL is comprised of four EM wigglers separated by three 2.68 m long midsections [4]. The length of the segment (the wiggler + midsection) is equal to $\frac{1}{8}$ of the ring circumference. This configuration provides for effective generation of γ-rays using 8 e-bunches and three collision points [4].

Each helical wiggler is composed of two planar wigglers shifted with respect to each other on a quarter of the wiggler period. Fig. 1 shows the regular part of the wiggler with the top quarter taken off. The main wiggler parameters are presented at Table 1. Each component of the field is independently controllable by the coil current in vertical and horizontal arrays. Each coil consists of four bent water-cooled copper busses, which surround the poles in "snake-like" fashion, similar to the OK-4 wiggler design [11,12]. The coils are connected at the end of the wiggler. Two layers make one complete turn around each pole. The 4-m quarter-yokes of the wiggler are made from soft magnetic steel on high precision milling machine. The pole tips are made independently and pinned to the yokes. The final accuracy of the assembly is about 20 μm.

The requirement for switching the helicity of radiation leads to violation of the helical symmetry of the wiggler design. As the result, the additional steps should be taken to compensate asymmetry of the magnetic field. Special design of the wiggler terminations is also required to provide for the adiabatic entrance of particles into the regular part of the wiggler.

3. Description of the 3D magnetic field

When the steel is not saturated, the magnetic field of the wiggler can be presented as direct superposition of the fields produced by horizontal and vertical arrays. In the vacuum the field can be described by scalar potential: $\vec{B} = \nabla \psi$, where

$$\psi(x, y, z) = A_x \psi_0(x, y, z) + A_y \psi_0\left(y, x, z - \frac{\lambda_w}{4}\right). \tag{1}$$

Periodic part of the potential can be expanded into series

$$\psi_0(x, y, z) = \sum_{q=0}^{\infty} f_q(x, y) \cos(q k_w z) \tag{2}$$

with condition forced by Laplace equation $\Delta \psi_0 = 0$:

$$\frac{\partial^2 f_q}{\partial x^2} + \frac{\partial^2 f_q}{\partial y^2} - q^2 k_w^2 f_q = 0. \tag{3}$$

The f_q can be expanded into Tailor series for further studies:

$$f_q(x, y) = \sum_{m,n=0}^{\infty} b_{m,n,q} x^m y^n. \tag{4}$$

We used design with the plane symmetry, which provides for $\psi_0(x, y, z) = -\psi_0(x, -y, z)$, and therefore $b_{m,2l,q} = 0$. According to Eq. (3), coefficients $b_{m,n,q}$ obey the relations

$$(m+2)(m+1) b_{m+2,n,q} + (n+2)(n+1) b_{m,n+2,q}$$
$$= q^2 k_w^2 b_{m,n,q}.$$

The yokes and poles have both plane and left–right symmetry (see Fig. 2). The left–right symmetry of the field is slightly violated by asymmetric coil. The additional symmetry in the design provides for:

$$\psi_0(x, y, z) = -\psi_0\left(-x, y, \frac{\lambda_w}{2} - z\right).$$

Fig. 1. Regular part of the wiggler.

The asymmetry of the coils allows the presence of small average quadrupole and octupole fields:

$$f_0(x, y) = B_0^{(1)} xy + \frac{B_0^{(3)}}{6} \left(x^3 y - xy^3\right).$$

The magnetic field in the regular part of the wiggler was simulated with the 3D code "MER-MAID" [13]. Computations were made for the half period of the wiggler with current only in the coils exciting vertical field. The wiggler pole geometry is shown in Fig. 2.

Table 1
Parameters of OK-5 wiggler

Period λ_w, cm	12
Number of regular periods	30
Wiggler length, m	4.04
Wiggler gap, cm	4 × 4
Maximum designed field, kGs	3
Amplitude of the fundamental harmonic (at the bus current 2 kA), kGs	2.07
Relative value of the 3d field harmonic, %	0.6

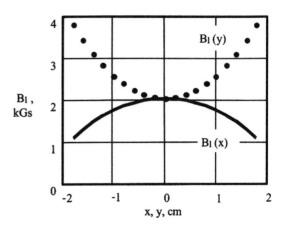

Fig. 3. Calculated dependence of fundamental harmonic of magnetic field on x and y. The current is 2 kA. The solid curve corresponds to the dependence on x at $y=0$ and the dotted curve—on y at $x=0$.

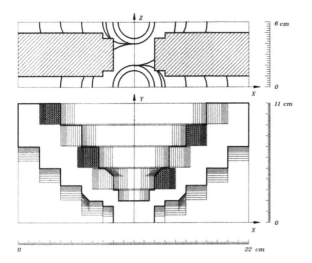

Fig. 2. 3D geometry of the wiggler half period used in MERMAID. The top fig shows a horizontal cross-section. The bottom fig shows 3D side view of the poles and coils. The dark squares show the coil cross-section. Coils are obviously asymmetric.

Because of the plane symmetry, we calculate the magnetic field only above the median plane. The boundary conditions are:

- $B_\tau = 0$ at median plane and outside boundaries.
- $B_n = 0$ at $z=0$ and 6 cm planes.

The calculated dependencies of the amplitude of fundamental harmonic of the magnetic field on transverse coordinates x and y are shown in Fig. 3. Numerical simulations confirmed the presence of the average quadrupole and octupole terms in the regular part of the wiggler. With 2 kA current in the coils, their amplitudes were $G = 5.64$ Gs/cm

and $O = -2.76$ Gs/cm³, respectively. These average terms can be compensated locally by small asymmetric cuts in the pole tips. Their integral compensation is also possible by special poles at the ends of the wigglers. We are testing both approaches.

4. Magnetic measurements

All four wigglers have been manufactured and magnetic measurements are under way. A Hall probe array with five horizontally spaced and five vertically spaced probes is employed for these measurements. These Hall probes provide accuracy of ±0.2 Gs.

Preliminary magnetic measurements show good agreement with the calculations. Fig. 4 shows the dependence of the vertical magnetic field along wiggler axis. The central particle trajectory based on the measured field is shown in Fig. 5. The nonzero coordinate and angle of the particle at the wiggler exit will be corrected by the adjustment of the field structure in the wiggler terminations. A relatively small deviation of the trajectory in the regular part of the wiggler from the ideal indicates the good quality of manufacturing.

The other indicator of the field quality is the spectrum of spontaneous radiation. The spectrum

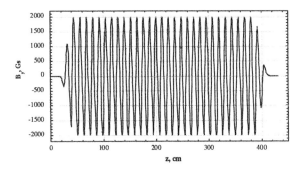

Fig. 4. Measured dependence of the vertical magnetic field along the wiggler axis.

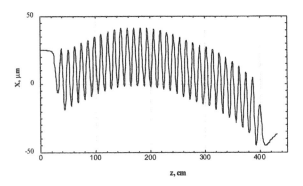

Fig. 5. Calculated trajectory of the central particle at $E = 1\,\mathrm{GeV}$.

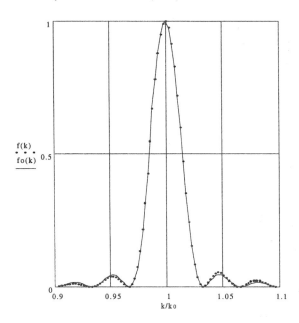

Fig. 6. Spectrum of spontaneous radiation from one wiggler. The solid curve is the $f_0 = \mathrm{sinc}(a(k-k_0))^2$ fit with $a = 1.35 \times 10^{-4}$ cm, $k_0 = 7.02 \times 10^5\,\mathrm{cm}^{-1}$.

of the radiation based on the measured magnetic field is shown in Fig. 6. The trajectory was calculated from the measured magnetic field (Fig. 4) for helical configuration. Using this trajectory, we calculated the retarded time and the value of vector potential on the wiggler axis as a function of observation time with even intervals. The time dependence of the radiated field was calculated by direct use of Lienard–Wiechert potential [14]. The radiated field was Fourier transformed into the radiation spectrum. The spectrum width corresponds to the effective number of wiggle periods $N = 30.2$, as expected.

The measured values of average gradient and octupole components are in reasonable agreement with the calculations. We are testing the pole tips, which should eliminate the average gradient and octupole completely.

The results of magnetic measurements will determine the choice of magnetic field structure in the wiggler terminations. We expect to complete modifications and magnetic measurements by Spring, 2001.

5. Conclusions

The wigglers for the OK-5 FEL are undergoing final adjustments to satisfy our requirements. We expect that all four wigglers will be ready by the end of the Spring 2001. The 3D simulations and preliminary magnetic measurements are in good agreement and demonstrate high quality of the regular part of magnetic field. We plan to install the wigglers and the rest of the OK-5 system into the Duke storage ring in 2001.

References

[1] V.N. Litvinenko, S.H. Park, I.V. Pinayev, Y. Wu, M. Emamian, N. Hower, O. Oakeley, G. Swift, P. Wang, Nucl. Instr. and Meth A 429 (1999) 151;

V.N. Litvinenko, S.H. Park, I.V. Pinayev, Y. Wu, Nucl. Instr. and Meth. A 470 (2001) 66.

[2] V.N. Litvinenko, Nucl. Instr. and Meth. A 304 (1991) 463.

[3] V.N. Litvinenko, N. A. Vinokurov, O.A. Shevchenko, Y. Wu, Predictions and Expected Performance for the VUV OK-5/Duke Storage Ring FEL with Variable Polarization, Nucl. Instr. and Meth. A 475 (2001) 97, these proceedings.

[4] V.N. Litvinenko, S.F. Mikhailov, N.A. Vinokurov, N.G. Gavrilov, D.A. Kairan, G.N. Kulipanov, O.A. Shevchenko, T.V. Shaftan, P.D. Vobly, Y. Wu, Nucl.Instr. and Meth., The OK-5/Duke Storage Ring VUV FEL with Variable Polarization, Nucl. Instr. and Meth. A 475 (2001) 407, these proceedings.

[5] Y. Wu, V.N. Litvinenko, S.F. Mikhailov, N.A. Vinokurov, N.G. Gavrilov, O.A. Shevchenko, T.V. Shaftan, D.A. Kairan, Lattice Modification and Nonlinear Dynamics for Elliptically Polarized VUV OK-5 FEL Source at Duke Storage Ring, Nucl. Instr. and Meth. A 475 (2001) 253, these proceedings.

[6] V.N. Litvinenko, et al., Phys. Rev. Lett. 78 (24) (1997) 4569.

[7] W. Tornow, V.N. Litvinenko, H. Weller, High Intensity γ-Ray Source for Nuclear Physics, Proposal to the US Department of Energy, August 2000.

[8] E.C. Schrieber, et al., Phys. Rev. C 61 (2000) 061604 (Rapid Comm.).

[9] T. Burt, V.N. Litvinenko, Feasibility studies of the parity violation experiment in 1S-2S transitions in atomic hydrogen using the OK-5/Duke storage ring FEL, unpublished.

[10] Y. Wu, V.N. Litvinenko, J.M.J. Madey, Instr. and Meth. A 341 (1994) 363.

[11] N.G. Gavrilov, et al., Nucl. Instr. and Meth. A 282 (1989) 422.

[12] I.B. Drobyazko, et al., Nucl. Instr. and Meth. A 282 (1989) 424.

[13] A.N. Dubrovin, MERMAID 2D/3D user's guide, 1998.

[14] L.D. Landau, E.M. Lifshitz, The Classical Theory of Fields, 4th Revised Edition, Pergamon Press, Oxford 1975, p.123.

ELSEVIER

Nuclear Instruments and Methods in Physics Research A 475 (2001) 253–259

NUCLEAR
INSTRUMENTS
& METHODS
IN PHYSICS
RESEARCH
Section A

www.elsevier.com/locate/nima

Lattice modification and nonlinear dynamics for elliptically polarized VUV OK-5 FEL source at Duke storage ring

Y. Wu[a],[*][1], V.N. Litvinenko[b], S.F. Mikhailov[b], O.A. Shevchenko[b],[2],
N.A. Vinokurov[c], N.G. Gavrilov[c], T.V. Shaftan[c], D.A. Kairan[c]

[a] *Lawrence Berkeley National Laboratory, 1 Cyclotron Road, Berkeley, CA 94720, USA*
[b] *FEL Laboratory, Department of Physics, Duke University, Durham, NC 27708, USA*
[c] *Budker Institute of Nuclear Physics, Novosibirsk, Russia*

Abstract

The Duke storage ring is a light source ring optimized for driving the Free Electron Lasers (FELs). To take advantage of the high brightness of the electron beam in the storage ring, a next generation FEL source — an elliptically polarized OK-5 FEL system, has been designed to match the electron beam quality of the Duke storage ring. In this paper, we present the storage ring lattice modifications which are necessary to accommodate the 24 m long OK-5 FEL. Because of its length and strong nonlinear focusing, the OK-5 FEL is expected to have a significant impact on the electron beam dynamics. We also present the preliminary results on the dynamic aperture calculation for the OK-5 FEL operated in different modes. © 2001 Elsevier Science B.V. All rights reserved.

PACS: 29.20.Dh; 29.27.Bd

Keywords: Storage ring; FEL; Dynamic aperture; Wiggler

1. Introduction

The Duke Free Electron Laser (FEL) storage ring is designed with a small emittance ($\varepsilon_x = 18$ nm at 1 GeV), a large dynamic aperture,

---

*Corresponding author. Present address: FEL Laboratory, Department of Physics, Duke University, Box 90319, LaSalle Street Extension, Durham, NC 27708-0319, USA. Tel.: +1-919-660-2654; Fax: +1-919-660-2671.

E-mail address: wu@fel.duke.edu (Y. Wu).
[1] Work supported by the Director, Office of Energy Research, Office of Basic Energy Sciences, Material Sciences Division, U.S. Department of Energy, under Contract No. DE-AC03-76SF00098.
[2] Work supported by Office of Naval Research Grant #N00014-94-1-0818.

and a long straight section for FEL wigglers. The first operational FEL on the Duke ring is the OK-4 FEL [1], originally designed for the VEPP-3 storage ring at Budker Institute of Nuclear Physics, Russia. The 8 m long OK-4 system is comprised of two planar wigglers with 10 cm periods and a three-pole buncher. Since its commissioning in 1995, the OK-4/Duke storage ring FEL has been successfully operated in a wide spectrum region from the near IR to deep UV. In 1999, it successfully lased at 194 nm, breaking the 200 nm wavelength barrier for FELs for the first time. However, the relatively small gain of this linearly polarized FEL has prevented us from taking advantage of the high brightness electron beam available in the Duke storage ring. In particular, the small FEL gain has limited the shortest wavelength that the OK-4 FEL can operate at, and the strong on-axis synchrotron radiation from the linearly polarized OK-4 wigglers has caused significant mirror damage, resulting in a shortened mirror lifetime.

To overcome these limitations, a next generation FEL source — an elliptically polarized OK-5 FEL system with 16 m long wigglers, has been specially designed to increase the FEL gain, to reduce the on-axis radiation, and to match the high quality of the electron beam in the storage ring [2]. The OK-5 FEL system [3] consists of four 4 m long wigglers and three bunchers. Each wiggler is comprised of a vertical and a horizontal electro-magnetic pole-tip array shifted for a quarter of the wiggler period with respect to each other. By controlling the horizontal and vertical magnetic fields independently via two power suppliers, the OK-5 wigglers can be configured to produce radiation with arbitrary polarizations, from various linear and elliptical polarizations to pure left and right circular polarizations. In Table 1, we compare the OK-5 and OK-4 wiggler parameters.

The variable polarization of the light source is essential for many experiments using the UV radiation and γ-rays. For example, the circularly polarized UV light is essential for studying the surface properties of the magnetic samples using a photoemission electron microscope (PEEM). The image contrast comes from the different absorption of the left- and right-handed circularly polarized photons by magnetic domains at the sample. Because the Compton scattered γ-rays inherit the polarization property of the FEL, the

γ-ray source driven by the OK-5 FEL will be a highly polarized radiation source with variable polarizations. Such a source is critical for a number of important nuclear physics experiments studying parity violations.

2. Lattice modification for the OK-5 FEL

The Duke FEL storage ring is a race-track structure comprised of two compact $180°$ arcs and two 34.21 m long straight sections. One of the two straight sections is dedicated to driving FEL light sources. The first FEL, the 8 m long OK-4 FEL, is located in the center of this bilaterally symmetric straight section [4]. The β-function matching between the OK-4 FEL and the arcs is provided by two pairs of quadrupole doublets on each end of the straight.

To achieve the flexibility in the lattice design as required for various operation modes of the OK-5 FEL, we elect to use a lattice structure with a combination of quadrupole doublets and triplets. We design the OK-5 straight section with two end matching cells (5.02 m long each) and a 24 m long FEL section in the middle (see Fig. 1). The matching cell employs two quadrupole doublets to match the β-functions between the arcs and the OK-5 wigglers. The OK-5 FEL portion of the lattice has three standard cells and a fourth wiggler. Each standard cell includes a quadrupole triplet (QF–QD–QF), a buncher, and an OK-5 wiggler. The three quadrupoles in the triplet are individually powered to provide compensations for linear focusing changes due to different OK-5 settings.

The exact length of the repetitive cell is chosen to optimize the production of γ-ray beams. In order to have three collision points at the same relative locations within each cell for eight evenly spaced electron bunches, we choose the cell length to be exactly one-sixteenth of the storage ring circumference (6.7 m). A total of five collision points are possible in this straight, three in the standard cells at the center of the defocusing quadrupole and two in the matching sections. Four sets of horizontal and vertical trim correctors per cell will be used for the fine adjustment of the

Table 1
Comparison of the OK-4 and OK-5 wiggler parameters

	OK-4 FEL	OK-5 FEL
Total wiggler length (m)	6.7	16.08
Number of wigglers	2	4
Number of periods per wiggler	33.5	32 (30+ transition)
Wiggler periods (cm)	10	12
Wiggler gap (mm)	22	40×40
Peak magnetic field (k Gs)	5.3	3.0
Max. wiggler $K = (eB_0)/k_w m_e c^2$	4.9	3.4

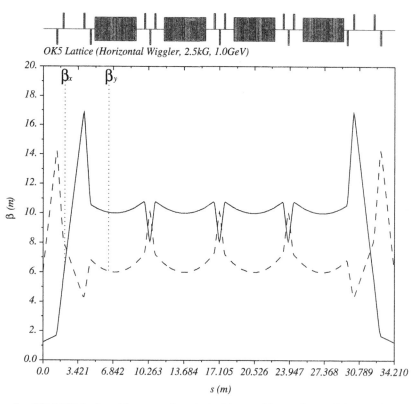

Fig. 1. The dispersion-free OK-5 FEL lattice with two quadrupole doublet matching sections and three triplet standard cells. The OK-5 wigglers are powered with horizontal polarization at 2.5 kG s. The solid line is β_x and the dashed line is β_y. At the center of the OK-5 wiggler, $\beta_x = 10$ m, $\beta_y = 6$ m.

electron beam positions and angles at the collision point. These trims will be used to move the electron beam off the FEL axis at the collision point; therefore, stopping the γ-ray production when operating for FEL user experiments.

Another design requirement for the OK-5 lattice is to allow a wide range of different β-function combinations at the center of wigglers for various operation conditions of the OK-5 FEL. The triplet cell lattice has been chosen for this purpose. As a result, β_x and β_y can be tuned from 4 to 10 m in arbitrary combinations. The two doublets matching sections are capable of providing good matching between the arc and triplet cells in all cases. We list some sample β-function combinations and phase advances of the OK-5 straight in Table 2. Adjusting the straight section quadrupoles, we can fully compensate the betatron

beating due to the operation of the OK-5 wigglers. After compensations, the residual tune changes in both planes are negligibly small, $\max(\mathrm{d}\nu_{x,y}) < 0.007$.

As seen in Table 2, the OK-5 lattices with different β-functions have significantly different phase advances in the straight ($\max(\Delta\nu_{x,y}) \approx 0.5$). To keep the overall storage ring tunes to their design values, we need a very flexible betatron tuning section. The other 34.21 m long straight is ideal for this purpose. Besides hosting the injection section and RF cavity, a large portion of this straight is free to be modified to compensate for the global betatron tunes. Using eight quadrupoles in this straight, we have developed a flexible global tuning scheme to keep the global tunes at the design values of $\nu_x = 9.11$ and $\nu_y = 4.18$, while allowing the horizontal and vertical β functions in

Table 2
β-functions at the centers of OK-5 wigglers vs. the phase advances of the south straight section lattice[a]

β_x (m)	β_y (m)	$dv_x(2\pi)$	$dv_y(2\pi)$
4	4	1.398	1.087
6	6	1.150	0.823
8	8	1.000	0.663
10	10	0.904	0.556
10	8	0.903	0.657
10	6	0.903	0.811
10	4	0.902	1.065

[a] In all the cases, the OK-5 wigglers are off.

the OK-5 straight to vary from 4 to 10 m. This tune compensation straight is shown in Fig. 2 for two different OK-5 settings.

3. Wiggler model

As we have found in the previous studies, long wigglers have a significant influence on the single particle dynamics [5]. The strong nonlinearities of long FEL wigglers will modify the nonlinear resonance strength and change the nonlinear tune shift with amplitude, resulting in particle losses at different nonlinear resonances. The Duke storage ring was designed with localized second-order geometric aberrations compensation in each of the two arcs [5]. Due to this compensation, the addition of the FEL wigglers did not change the lowest nonlinearities (i.e. the third-order resonances). In fact, previous studies had indicated that the Duke storage ring dynamic aperture was limited by the very strong high-order multipole components in the straight section quadrupoles. The inclusion of either the 6.8 m long OK-4 FEL wigglers or certain fictitious 26 m long horizontal FEL wigglers did not further reduce the dynamic aperture.

However, the OK-5 FEL wigglers differ from the previously studied wigglers in the following two ways: (1) the 16 m OK-5 wigglers are much longer than the 6.8 m OK-4 wigglers; (2) the OK-5 wigglers has notable 3D effects compared with 2D planar OK-4 wigglers for they are designed with two sets of wiggler arrays and as a result, different kinds of nonlinearities are excited. To study the effects of the OK-5 wiggler on the beam dynamics, we have developed a proper magnetic field model and a beam dynamics model for the arbitrarily polarized wiggler magnet. In particular, this model is capable of taking into account the finite pole-width effect of the wiggler.

The following 3D magnetic model has been used to represent the field of a linearly polarized wiggler. Assuming the mid-plane symmetry with respect to the y-plane and a mirror symmetry with respect to the x-plane, the magnetic field of a horizontal wiggler, $\vec{B}_H(x,y,z) = (B_x, B_y, B_z)$, can be expressed as follows:

$$B_x = -B_0 \sum_{m,n} C_{mn} \frac{k_{x,l}}{k_{y,m}} \sin(k_{x,l}x)$$
$$\times \sinh(k_{y,m}y)\cos(k_{z,n}z + \theta_{z,n})$$

$$B_y = B_0 \sum_{m,n} C_{mn} \cos(k_{x,l}x)\cosh(k_{y,m}y)$$
$$\times \cos(k_{z,n}z + \theta_{z,n})$$

$$B_z = -B_0 \sum_{m,n} C_{mn} \frac{k_{z,n}}{k_{y,m}} \cos(k_{x,l}x)$$
$$\times \sinh(k_{y,m}y)\sin(k_{z,n}z + \theta_{z,n}) \quad (1)$$

where $k_{y,m}^2 = k_{x,l}^2 + k_{z,n}^2$, $k_{z,n} = n2\pi/\lambda_u$, and λ_u is the wiggler period. We decide to use $\cos(k_x x)\cosh(k_y y)$ combinations with a real k_x in B_y to present the reduced vertical field strength due to the finite pole-width in the horizontal plane. The magnetic field of a vertical wiggler, $\vec{B}_V(x,y,z)$, can be expressed in the similar manner by interchanging the arguments x and y in the above

Fig. 2. Tune compensation straight for two different OK-5 operation conditions to preserve the global tunes, $v_x = 9.11$ and $v_y = 4.18$. The OK-5 wigglers are powered with vertical polarization at 2.5 kGs. The solid line is β_x and the dashed line is β_y. At the center of the OK-5 wigglers: (a) $\beta_x = \beta_y = 4$ m; (b) $\beta_x = 10$ m and $\beta_y = 6$ m.

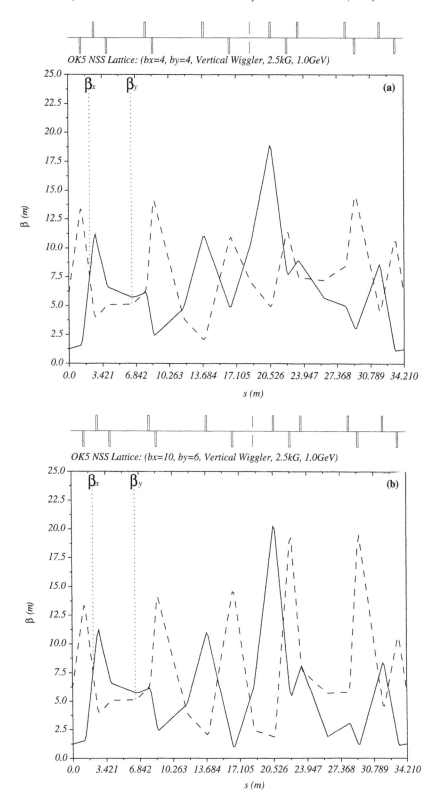

expression. When the iron saturation is not significant, the magnetic field of an arbitrarily polarized wiggler such as the OK-5 wiggler can be expressed by a linear superposition of the two linearly polarized wiggler fields:

$$\vec{B}(x,y,z) = a\vec{B}_H(x,y,z) + b\vec{B}_V\left(x,y,z+\frac{\lambda_u}{4}\right) \quad (2)$$

where $-1 \leqslant a \leqslant 1$ and $-1 \leqslant b \leqslant 1$.

In order to facilitate the development of the higher-order symplectic integrator, we choose a special field gauge for the vector potential such that $\vec{A} = (0, A_y, A_z)$. The following normalized Hamiltonian is then used to describe the electron beam dynamics in a horizontal wiggler:

$$
\begin{aligned}
&H(x, y, \delta, p_x, p_y, l; z) \\
&= -\sqrt{(1+\delta)^2 - p_x^2 - (p_y - a_y)^2} - a_z(x,y,z) \\
&\approx \left\{ -\delta + \frac{p_x^2 + p_y^2}{2(1+\delta)} \right\} + \left\{ \frac{a_y^2(x,y,z)}{2(1+\delta)} - a_z(x,y,z) \right\} \\
&\quad - \frac{p_y a_y}{1+\delta} \approx H_1(p_x, p_y, \delta) + H_2(x, y, \delta; z) \\
&\quad + H_3(x, y, \delta, p_y; z), \quad (3)
\end{aligned}
$$

where $a_{y,z} = (eA_{y,z}/P_0 c)$ are normalized vector potential components, $p_{x,y} = (P_{x,y})/P_0 c$ are normalized momenta, P_0 is the nominal momentum, $\delta = (P - P_0)/P_0$ is the relative momentum deviation, l is the relative path length. Since in $H_{1,2}$, the momenta and positions belonging to the same canonical pairs are clearly separated, $H_{1,2}$ can be integrated independently using an explicit high-order symplectic integrator. The small correction term in the Hamiltonian, $H_3(x, y, \delta, p_y; z)$, containing the coupling between p_y and a_y, can be integrated exactly using a first-order symplectic integrator and included in the tracking code if necessary.

A similar Hamiltonian is used to model a vertical wiggler. Applying the superposition principle, an arbitrarily polarized wiggler can be modeled by a linear combination of a horizontal and a vertical Hamiltonian.

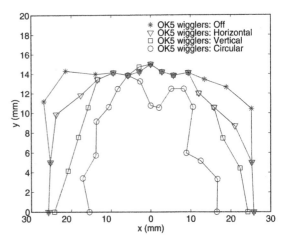

Fig. 3. Dynamic apertures for various settings of OK-5 wigglers in the straight section where $\beta_x = \beta_y = 6$ m. The electron beam energy is 1 GeV and the wiggler field is 2.5 kGs. Both the horizontal and vertical wigglers are energized at 2.5 kGs when the OK-5 is circularly polarized.

4. Preliminary results on dynamic aperture studies

We implemented the above Hamiltonian model in a storage ring simulation code, TRACY [6]. Using this model, we computed the Duke storage ring dynamic apertures for various modes of the OK-5 operation. In all calculations, only the fundamental harmonic of the magnetic field was taken into account. The storage ring energy was 1.0 GeV and the peak wiggler field was assumed to be 2.5 kGs.

Fig. 3 shows the calculated dynamic apertures with the OK-5 system. When the horizontal wigglers are energized, the dynamic aperture is slightly reduced as compared to the case when the OK-5 wigglers are off. Apparently, the nonlinearities of the strong sextupoles in the arcs still dominate in this case. The dynamic aperture reduction is more evident when the vertical OK-5 wigglers are turned on. This indicates that the vertical wigglers have a larger impact on the beam dynamics. By operating both sets of horizontal and vertical wiggler arrays to produce circularly polarized radiation, the dynamic aperture is significantly reduced. In fact, the effective magnetic field in this case is increased from 2.5 to

3.5 kGs due to the superposition of the two sets of wiggler fields. It is worth pointing out that even in this worst case, the available dynamic apertures (16 mm horizontally and 10 mm vertically) are still more than adequate for a high injection efficiency and good beam lifetime.

5. Conclusion

We have redesigned the Duke storage ring straight section lattice for the 24 m long OK-5 FEL system. The new lattice is very flexible to accommodate a large set of the OK-5 FEL operation modes. We have developed a magnetic and dynamics model to study the nonlinear dynamics of the arbitrarily polarized wigglers. Implementing this model in a tracking code, we have also studied the storage ring dynamics for various OK-5 FEL settings. The dynamics simulations indicate that we would have an adequate dynamic aperture for all cases. Further studies will be needed to understand the impact of wiggler harmonics, alignment errors, magnet power supply errors, and high-order multipole errors on the beam dynamics.

References

[1] I.B. Drobyazko, et al., Nucl. Instr. and Meth. A 282 (1989) 424.
[2] V.N. Litvinenko, et al., Predictions and Expected Performance for the VUV OK-5/Duke Storage Ring FEL with Variable Polarization, Nucl. Instr. and Meth. A 475 (2001) 97, these proceedings.
[3] V.N. Litvinenko, et al., Helical Wigglers for the OK-5 Storage Ring VUV FEL at Duke, Nucl. Instr. and Meth. A 475 (2001) 247, these proceedings.
[4] Y. Wu, et al., Proceedings of Particle Accelerator Conference, Washington, DC, 1993, pp. 218–220.
[5] Y. Wu, et al., Nucl. Instr. and Meth. A 341 (1994) 363.
[6] J. Bengtsson, H. Nishimura, TRACY, Lawrence Berkeley National Lab, private communication.

ELSEVIER

Nuclear Instruments and Methods in Physics Research A 475 (2001) 260–265

NUCLEAR
INSTRUMENTS
& METHODS
IN PHYSICS
RESEARCH
Section A

www.elsevier.com/locate/nima

New results of the high-gain harmonic generation free-electron laser experiment ☆

A. Doyuran[a],*, M. Babzien[a], T. Shaftan[a], S.G. Biedron[b], L.H. Yu[a], I. Ben-Zvi[a], L.F. DiMauro[a], W. Graves[a], E. Johnson[a], S. Krinsky[a], R. Malone[a], I. Pogorelsky[a], J. Skaritka[a], G. Rakowsky[a], X.J. Wang[a], M. Woodle[a], V. Yakimenko[a], J. Jagger[b], V. Sajaev[b], I. Vasserman[b]

[a] Brookhaven National Laboratory, Upton, NY 11973, USA
[b] Advanced Photon Source, Argonne National Laboratory, Argonne, IL 60439, USA

Abstract

We report on the experimental investigation of high-gain harmonic generation carried out at the Accelerator Test Facility at Brookhaven National Laboratory. A seed CO_2 laser at a wavelength of 10.6 μm was used to generate FEL output at a 5.3-μm wavelength. The duration of the output pulse was measured using a second-harmonic intensity autocorrelator, and the coherence length was measured using an interferometer. We also measured the energy distribution of the electron beam after it exited the second undulator, observing behavior consistent with that is expected at saturation. The intensity of the harmonic components of the output at 2.65 and 1.77 μm was determined relative to that of the 5.3-μm fundamental. Finally, using a corrector magnet upstream of the radiator, steering effects on the trajectories of the electron and light beams were studied. © 2001 Elsevier Science B.V. All rights reserved.

PACS: 41.60.Cr; 42.25.Kb; 42.55.Vc; 42.65.Ky

Keywords: Free-electron laser

1. Introduction

First lasing of a high-gain harmonic-generation (HGHG) free-electron laser (FEL) was reported in Refs. [1,2]. Here, the latest results characterizing the HGHG FEL output are presented. The experiment was carried out at the Accelerator Test Facillity (ATF) at Brookhaven National Laboratory in collaboration with the Advanced Photon Source at Argonne National Laboratory. A schematic of the HGHG layout is shown in Fig. 1. A coherent 10.6-μm seed provided by a CO_2 laser interacts with the electron beam in the first (energy-modulating) undulator, which is tuned to be resonant to 10.6-μm. The resulting energy modulation is then converted to spatial bunching while the electron beam traverses a dispersive section (a three-dipole chicane). In the second undulator, tuned to be resonant at 5.3 μm, the microbunched electron beam initially emits

☆ Work supported by US Department of Energy, Office of Basic Energy Sciences, under Contracts Nos. DE-AC02-98CH10886 and W-31-109-ENG-38 and by Office of Naval Research Grant no. N00014-97-1-0845.

*Corresponding author.

E-mail address: adoyuran@ic.sunysb.edu (A. Doyuran).

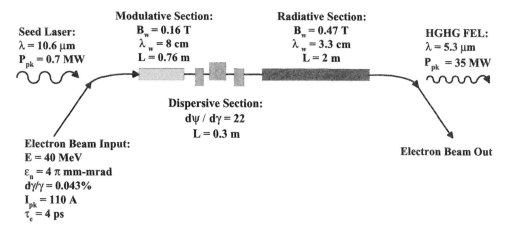

Fig. 1. HGHG experiment schematic and design parameters.

coherent radiation and then amplifies it exponentially until saturation is achieved. We have carried out measurements of the radiation at the radiator fundamental wavelength (5.3 μm), as well as at the second harmonic (2.65 μm) and the third harmonic (1.77 μm).

In previous work [2], the spectral distribution of the FEL output in the neighborhood of 5.3 μm was characterized using a single-shot imaging technique. The bandwidth of the HGHG output was found to be 15 nm FWHM, much narrower than the self-amplified spontaneous emission (SASE) bandwidth. In this note, we report results of measurements of the output pulse duration using a second-harmonic intensity autocorrelator, as well as the coherence length using an interferometer.

The electron beam parameters exhibit some small daily variation about the working point. Typical parameters are 0.8 nC charge, 6 ps FWHM pulse length, and 5 mm-mrad normalized emittance. The repetition rate of the ATF is 1.5 Hz, and the repetition rate of the CO_2 laser, limited only by its pulse forming network, is once every 15 s. The remaining FEL parameters are given in Fig. 1.

2. Autocorrelation

We constructed a background-free intensity autocorrelator using second-harmonic generation

in a $5 \times 5 \times 1$ mm^3 AgGaSe$_2$ nonlinear crystal. The group velocity mismatch [3] and geometric beam overlap in the crystal allow better than 0.5-ps resolution. The resolution of 1 ps is predominantly determined by the fluctuations of the electron beam and the CO_2 laser performance. The output radiation is measured with a single-element InSb photoconductive detector. The signal versus delay time (between the two arms of the autocorrelator) is shown in Fig. 2. Assuming a Gaussian pulse shape, the pulse duration is found to be $8.4/\sqrt{2} = 5.9$ ps. For an HGHG output pulse with energy 100 μJ and pulse length 5.9 ps, the output power is 17 MW, within a factor of two of the theoretical prediction of 35 MW.

3. Interferometer

An interferometer was used to investigate the temporal coherence of the HGHG output. The retroreflecting mirror in one arm was tilted off-axis and translated while the fringe contrast of the interference pattern was recorded on a thermal imaging camera, as shown in Fig. 3. Note that in order to collect more light we added a cylindrical mirror to produce a line-type image on the thermal camera. The variation of fringe contrast as a function of mirror displacement, plotted in Fig. 4, is a measure of the coherence length of the pulse. The optical coherence length is measured as

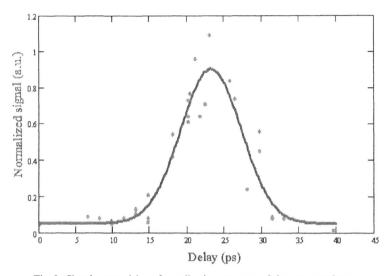

Fig. 2. Signal versus delay after adjusting one arm of the autocorrelator.

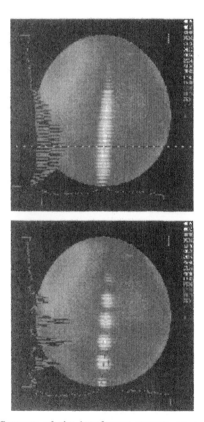

Fig. 3. Contrast of the interference pattern on a thermal imaging camera from the interferometer.

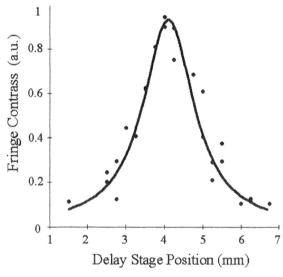

Fig. 4. Variation of fringe contrast as a function of mirror displacement revealing the coherence length of the HGHG radiation pulse.

1.6 mm, or 5.3 ps, based upon the delay change. The close agreement between pulse duration and coherence length indicates nearly full longitudinal coherence.

4. Steering

Using a small corrector magnet located 18.4 cm upstream of the radiator, we studied steering effects on electron and light beam trajectories. Five cerium-doped yttrium aluminum garnet (YAG) beam position measurement screens located at regular spacing through the radiator were used to observe the electron trajectory. A thermal imaging camera 1.7 m downstream of the end of the radiator was used to measure the deviation of the light beam. Steering correctors were varied separately in each plane to induce a maximum electron beam offset of 1.0 mm (horizontal) and 0.5 mm (vertical) on the fifth YAG screen, situated 18 cm downstream of the exit of the radiator. The electron beam steering produced a maximum HGHG output centroid displacement of 2.5 mm (horizontal) and 1.0 mm (vertical) at the thermal camera.

The horizontal steering of the electron beam did not cause observable intensity reduction. Fig. 5 shows the horizontal (top) and vertical (bottom) trajectories for the two cases. The measured shift of the optical HGHG output position at the thermal camera and the shift of the electron beam position on the fifth YAG screen suggest that HGHG output follows the electron beam.

When we steered the electron beam vertically by 0.5 mm at the fifth YAG screen, we observed about 1 mm displacement at the thermal camera. In this case, the intensity of the output was reduced because the radiator is a planer wiggler, i.e., the magnetic field increases as one moves out of the midplane in the vertical direction. Due to the change in magnetic field, the resonant wavelength shift is about 25 nm. Since the bandwidth of the HGHG is 20 nm, this field change causes the electrons to lose resonance with the radiation generated earlier in the wiggler.

5. Electron beam modulation and output radiation harmonics

The electron beam energy modulation was determined using the electron energy spectrometer after the radiator section. In the HGHG process,

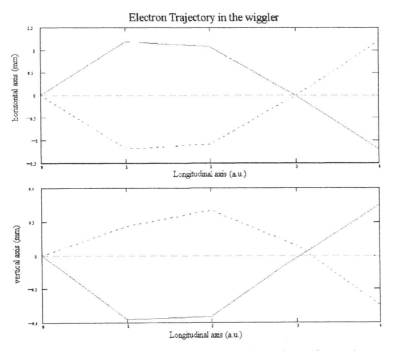

Fig. 5. Horizontal and vertical trajectory of the electron beam after steering.

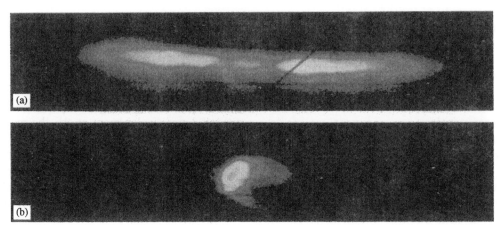

Fig. 6. The energy modulation of the electron beam image after the energy spectrometer (a) with and (b) without the CO_2 beam present.

energy modulation of the electron beam is generated in two ways: (1) through the initial interaction of the seed laser and electron beams in the modulator; and (2) through the HGHG FEL interaction itself in the radiator. The energy modulation produced in the radiator dominates. The electron beam images after the spectrometer (a) with CO_2 laser on, and (b) with the CO_2 laser off, are shown in Fig. 6. With the CO_2 laser on, the energy modulation is seen to be 2.5%.

The amount of modulation, as well as the strength of the higher harmonics, indicates that saturation or near-saturation has been reached. The higher harmonics have been studied theoretically in Refs. [4–7]. Here, we have measured the fundamental (5.3 µm), second (2.65 µm), and third (1.77 µm) harmonics relative to the radiator as a function of electron energy modulation, using the InSb detector in conjunction with the appropriate bandpass and neutral density filters, to produce similar signal levels on the detector. The responsivity of the InSb detector is a factor of 0.67 less at 2.65 µm and a factor of 0.5 less at 1.77 µm as compared to the 5.3-µm responsivity. For the fundamental (5.3 µm), a 10-nm bandpass filter with 80% transmission through this band was used with 1×10^6 attenuation. For the second harmonic (2.65 µm), a 35-nm bandpass filter with 75% transmission through this band was used with 1×10^3 attenuation. For the third harmonic

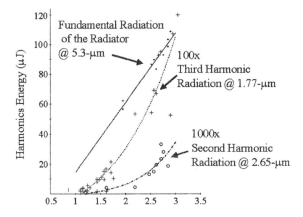

Fig. 7. Harmonic content (µJ) versus electron beam energy modulation (%).

(1.77 µm), a 35-nm bandpass filter with 60% transmission through this band was used with 1×10^4 attenuation.

In Fig. 7, we plot the harmonic contents (µJ) of the first three harmonics versus electron beam energy modulation (%). The rapid increase of the harmonic intensities at 2.65 and 1.77 µm as the electron energy modulation approaches 2.5% is strong evidence of saturation. In Table 1, the theoretical predictions and experimental measurements of the ratios of the harmonic-to-fundamen-

Table 1
The theoretical and experimentally measured harmonic-to-fundamental ratios

Wavelength (µm)	Simulation	Experiment
2.65	6×10^{-4}	2×10^{-4}
1.77	1×10^{-2}	0.8×10^{-2}

tal energies are presented for a 2.5% electron beam energy modulation. One sees that there is good agreement between experiment and theory.

6. Conclusion

New results in the HGHG experiment have provided additional confirmation of theory. For the HGHG output, we used an autocorrelator to measure the intensity pulse length and an interferometer to measure the coherence length of the HGHG output. The agreement of these two measurements indicates full longitudinal coherence. Using a spectrometer, we have previously determined [2] the single-shot spectrum in the neighborhood of 5.3 µm to have a 15-nm FWHM bandwidth. The measured pulse duration of 5.9 ps and bandwidth of 15 nm correspond to a time–bandwidth product that is a factor of ~1.7 larger than the minimum. However, the bandwidth and pulse duration measurement were not carried out on the same day, therefore the e-beam pulse shape might not be identical for the two measurements. To more precisely investigate the time–bandwidth product, we plan to carry out the spectral and pulse duration measurements in a single run. The intensity of the second (2.65 µm) and third (1.77 µm) nonlinear harmonics relative to the radiator section resonant wavelength (5.3 µm)

were measured to be in reasonable agreement with theory, and their nonlinear dependence on electron energy modulation provides strong evidence of saturation. Finally, the ability to steer the HGHG output will be useful in the future as a technique to align cascaded HGHG sections in an X-ray FEL [8–10].

References

[1] L.-H. Yu, M. Babzien, I. Ben-Zvi, L.F. DiMauro, A. Doyuran, W. Graves, E. Johnson, S. Krinsky, R. Malone, I. Pogorelsky, J. Skaritka, G. Rakowsky, L. Solomon, X.J. Wang, M. Woodle, V. Yakimenko, S.G. Biedron, J.N. Galayda, E. Gluskin, J. Jagger, V. Sajaev, I. Vasserman, Nucl. Instr. and Meth. A 445 (2000) 301.
[2] L.-H. Yu, M. Babzien, I. Ben-Zvi, L.F. DiMauro, A. Doyuran, W. Graves, E. Johnson, S. Krinsky, R. Malone, I. Pogorelsky, J. Skaritka, G. Rakowsky, L. Solomon, X.J. Wang, M. Woodle, V. Yakimenko, S.G. Biedron, J.N. Galayda, E. Gluskin, J. Jagger, V. Sajaev, I. Vasserman, Science 289 (2000) 932.
[3] V.G. Dmitriev, G.G. Gurzadyan, D.N. Nikogosyan, Handbook of Nonlinear Optical Crystals, 2nd Edition, Springer, Berlin, 1997.
[4] R. Bonifacio, L. De Salvo, P. Pierini, Nucl. Instr. and Meth. A 293 (1990) 627.
[5] H.P. Freund, S.G. Biedron, S.V. Milton, IEEE J. Quantum Electron. 36 (2000) 275.
[6] Z. Huang, K.J. Kim, Nucl. Instr. and Meth. A 475 (2001) 112, these proceedings.
[7] S.G. Biedron, H.P. Freund, X.J. Wang, L.-H. Yu, Nucl. Instr. and Meth. A 475 (2001) 118, these proceedings.
[8] L.H. Yu, Harmonic generation of hard X-rays, In: C.E. Eyberger, (Ed.), Proceedings of the IFCA Advanced Beam Dynamics Workshop on Future Light Sources, Argonne National Laboratory, Argonne, IL, 1999, URL: http://www. aps.anl.gov/conferences/ FLSworkshop/proceedings/papers/wg1-01.pdf.
[9] Juhao Wu, L.H. Yu, Nucl. Instr. and Meth. A 475 (2001) 104, these proceedings.
[10] S.G. Biedron, S.V. Milton, H.P. Freund, Nucl. Instr. and Meth. A 475 (2001) 401, these proceedings.

ELSEVIER

Nuclear Instruments and Methods in Physics Research A 475 (2001) 266–269

NUCLEAR
INSTRUMENTS
& METHODS
IN PHYSICS
RESEARCH
Section A

www.elsevier.com/locate/nima

High extraction efficiency observed at the JAERI free-electron laser

N. Nishimori*, R. Hajima, R. Nagai, E.J. Minehara

Free Electron Laser Laboratory, Advanced Photon Research Center, Kansai Research Establishment, Japan Atomic Energy Research Institute (JAERI), 2-4 Shirakata-Shirane, Tokai, Naka, Ibaraki 319-1195, Japan

Abstract

A high power Free-Electron Laser (FEL) has lased at a wavelength of 22 μm at the Japan Atomic Energy Research Institute (JAERI). The maximum power on a macro-pulse average is 1.7 kW, and it corresponds to an FEL energy of 160 μJ/micro-pulse. Extraction efficiency from the electron beam to the FEL radiation was measured to be 5.3% by an energy analyzer, when the maximum FEL power was coupled out. The rms wavelength spread was measured to be 4.6% at the same time. The extraction efficiency, in general, has a maximum value near the zero detuning length of an optical cavity, where (in contrast) the single-pass gain becomes smallest. A high peak current and a long macro-pulse duration are therefore indispensable for realizing high efficiency. The electron beam energy is 16.5 MeV, and the average current is 5.3 mA at a micro-pulse repetition rate of 10.4 MHz. The macro-pulse duration is 500 μs (5000 micro-pulses), long enough to reach saturation near the zero detuning length. The width and the peak current of the electron bunch are 5 ps FWHM and 100 A, respectively, at the undulator. © 2001 Elsevier Science B.V. All rights reserved.

PACS: 41.60.Cr; 52.75.Ms

Keywords: Free-electron laser; Superconducting linac; High power

1. Introduction

One of the attractive features of a Free-Electron Laser (FEL) is its capability for extracting high power from electron beam as radiation. An extraction efficiency over 3% has been observed at FELIX [1], ELSA [2], and Los Alamos [3]. Many interesting phenomena such as superradiance [1], spiking [4], limit-cycle oscillations [5], and spectral broadening due to universal brightness [1,2] have been observed at the FELs with high efficiency.

A high power FEL is useful for industrial applications. The Thomas Jefferson National Accelerator Facility FEL achieved 1.7 kW lasing in CW operation at a wavelength of 3 μm by the same cell energy recovery mode with an efficiency of 0.8% in 1999 [6]. The Japan Atomic Energy Research Institute (JAERI) FEL driven by a superconducting linear accelerator has also been developed to realize high power lasing in quasi-CW operation. As the wavelength is around 22 μm at JAERI-FEL, out-couplers of an optical cavity are restricted to a center hole on the cavity mirror

---

*Corresponding author. Tel.: +81-29-282-6752; fax: +81-29-282-6057.

E-mail address: nisi@milford.tokai.jaeri.go.jp (N. Nishimori).

0168-9002/01/$ - see front matter © 2001 Elsevier Science B.V. All rights reserved.
PII: S 0 1 6 8 - 9 0 0 2 (0 1) 0 1 6 3 8 - 2

or a scraper mirror. In either method, the out-coupled FEL power is approximately a half of the extracted power from the electron beam. This is because of the reflection loss on the cavity mirrors and the diffraction loss at the out-couplers and an undulator duct. High efficiency is therefore necessary for producing high power from the JAERI-FEL.

The JAERI-FEL has two unique features. One is a long macro-pulse inherent in superconducting linear accelerators. While the maximum macro-pulse length is limited up to 1 ms by a capacity of an electric power supplier, it is long enough to reach saturation even near the zero detuning length (δL) of an optical cavity, where the efficiency is maximum but the single-pass gain is minimum. The other unique feature is a high bunch charge of 0.51 nC, which compensates for the gain decrease near $\delta L = 0$ μm and exceeds total optical cavity loss. Hence, the JAERI-FEL has a great advantage for reaching strong saturation near $\delta L = 0$ μm.

The efficiency of JAERI-FEL in 1999 was consistent with the value by $1/4N_W$ (0.48% in our case), which is the theoretical efficiency of a non-bunched FEL [7]. The peak current and the pulse width at the undulator were 12 A and 50 ps, respectively, which were almost similar to originally designed values [8]. A recent improvement in our injection system, however, allowed us to transport 10 times longitudinally brighter beams inside the undulator. A strong FEL oscillation was observed near $\delta L = 0$ μm.

In the following section, the essence of improvements of the injection system and resultant electron beam performances are presented. In the last section, results of our FEL experiments are described.

2. Improved electron beam performance

Two modifications mainly contributed to the production of the high brightness electron beam with a long macro-pulse. One is an improvement of an electron gun, and the other is an optimum adjustment of a drift length between a 6th sub-harmonic buncher (SHB) and the first cell of the superconducting accelerator.

The electron gun at JAERI-FEL is a DC gun with a thermionic cathode driven by a grid pulser, and its DC voltage is 230 kV. The difficulty arises from the high voltage, although it is very useful not only for reducing space charge effects but also for keeping a high speed of $0.71c$ in the low energy region. In order to keep the high voltage against a discharge, the accelerator tube and the grid pulser are contained in a tank filled with SF_6 insulation gas. The absolute pressure of the gas is 2.5 atm. An input signal into the grid pulser on the high voltage terminal is fed through an optical fiber and has to go across an extra connector attached to the tank. As the input signal is attenuated too much at the connector, the grid pulser is affected by a slight error source such as a floating capacitance. This means that production of a stable beam is difficult. We assembled the grid pulser in such a way as to be unaffected by the slight error sources, and we installed it into the accelerator tube carefully. The detail of the improvement of the gun is described in Ref. [9]. Although the pulse width and the bunch charge are almost the same as those measured in 1999 [7], instabilities such as amplitude fluctuation and time jitter were reduced recently [9].

The time jitter at the gun causes arrival time fluctuation of the electron bunch at the undulator. The fluctuation corresponds to the shift of the effective optical cavity length. The fluctuation should be reduced to get high efficiency, because the efficiency is very sensitive to a shift of the detuning length near $\delta L = 0$ μm [10]. A simulation shows that the jitter inside the undulator should be smaller than 100 fs (rms) in order to reproduce the shape of the detuning curve shown in Fig. 1 [11]. In fact, halving the time jitter induced a 1.6-fold increase of the efficiency near $\delta L = 0$ μm at JAERI-FEL.

The drift length after the SHB must be determined to let the bunch be effectively compressed at JAERI-FEL. In 1999, the drift length was shortened from 7.4 to 4.5 m, and the bunch width decreased from 50 to 33 ps [7].

The FEL power was gradually increased by adjusting accelerator phases, optimizing out-coupling

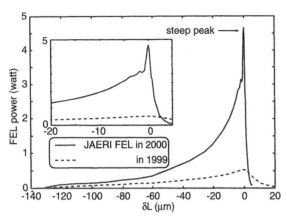

Fig. 1. Detuning curves obtained by a power meter with 0.4 ms macro-pulse at 10 Hz. Solid line is the present curve, and dotted line is obtained before modifications [7]. The enlargement around $\delta L = 0$ μm is shown above left.

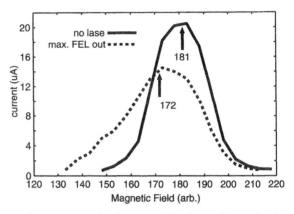

Fig. 2. Momentum distributions of the electron beam after the undulator when FEL was off (solid line), and the maximum FEL power was coupled out (dotted line) with 0.4 ms macro-pulse at 10 Hz. Their mean values are 181 and 172, respectively.

methods [12], reducing the time jitter of the gun, and removing frequency fluctuation of a master oscillator. Finally, the FEL output power far exceeded the value estimated from previous beam parameters measured in 1999. The bunch width was measured by a synchroscan streak camera (Hamamatsu M1955 with M1954-10) at the undulator center, and a value of 5 ps FWHM was obtained. The simulation with new parameters for the electron beam reproduces the observed FEL efficiency well [11].

Accelerator phases were obtained by measuring the beam load of accelerators and used as input parameters for PARMELA simulation. The simulation almost reproduces experimental results such as the emittance and the bunch width.

3. FEL performance

The FEL was coupled out by a gold-coated scraper mirror 20 mm in diameter through a KRS5 window, and its power was measured by a power meter near the window. Both sides of the optical cavity are gold-coated copper mirrors with reflectivity of 99.4%. Fig. 1 shows a detuning curve of the FEL power obtained by moving the downstream mirror along the cavity axis. A steep peak is seen in the detuning curve from $\delta L = -1$

to 0 μm. Here $\delta L = 0$ μm is defined as the position where the peak power was obtained. The steep peak is a feature not commonly observed at other FELs. The shape of the detuning curve including the steep peak is well reproduced by the simulation described in Ref. [11]. The FEL average power was obtained as 1.7 kW from the slope of the power to macro-pulse duration. It corresponds to 160 μJ/ micro-pulse. Because the transmittance of the KRS5 window is 70% at a normal incidence, the average power is estimated as 2.3 kW at the vacuum side of the window. The estimated FEL loss in the cavity is about 50% [12], so that the estimated efficiency is 5.2% from the observed FEL power.

The extraction efficiency was also measured by a bending magnet with a Faraday cup placed after the undulator. Energy spectra of the electron beam are shown in Fig. 2. The energy resolution is not good, because the analyzer was not designed and constructed as an energy analyzer but as a beam dump. The mean value of the momentum distribution is, however, reliable. The momentum distribution on lasing was measured with the maximum FEL power kept around $\delta L = 0$ μm, and it measured at $\delta L = +100$ μm off lasing. The position of the scraper mirror was determined for the out-coupled FEL power to be maximum, and the measured total optical cavity loss was 5% at

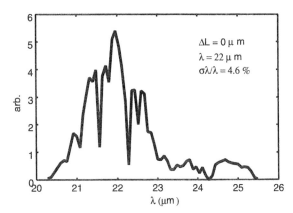

$\Delta L = 0\,\mu m$
$\lambda = 22\,\mu m$
$\sigma\lambda/\lambda = 4.6\,\%$

Fig. 3. FEL spectrum obtained at $\delta L \sim 0\,\mu m$. The mean value of λ is 22 μm and the rms wavelength spread is 4.6%.

that time [12]. The difference in the mean values between the two distributions corresponds to the converted power of the electron beam into the FEL, and is 4.6% in Fig. 2. The FEL is saturated after 50 μs from the beginning of a macro-pulse near $\delta L = 0\,\mu m$, and the macro-pulse duration was 400 μs at the measurement. The observed efficiency, therefore, should be multiplied by a corrective factor of $\frac{8}{7}$ to compensate for the non-lasing part, and an efficiency of 5.3% is obtained. It is consistent with the value of 5.2% derived from the average FEL power.

The FEL spectrum shown in Fig. 3 was obtained by a monochromator with a pyroelectric detector at a monitor room placed after optical transport of 14 m length in the air. The data was not corrected against vapor absorption. The rms wavelength (λ) spread $\sigma\lambda/\lambda$ is 4.6%. The optical pulse width is expected to be shorter than 0.9 ps FWHM from the assumption of the Fourier transform limit.

4. Summary

A high power FEL has been demonstrated at JAERI, lasing at a wavelength of 22 μm. The maximum power on a macro-pulse average is 1.7 kW, and it corresponds to the FEL energy of 160 μJ/micro-pulse. The extraction efficiency and the rms wavelength spread were 5.3% and 4.6%, respectively, when the maximum FEL power was coupled out. The high power FEL was obtained by means of a high brightness electron beam. The width and the peak current of the electron bunch are 5 ps FWHM and 100 A, respectively at the undulator. The high brightness beam was obtained by improving the injection system at JAERI-FEL.

References

[1] D.A. Jaroszynski, et al., Phys. Rev. Lett. 78 (1997) 1699; D.A. Jaroszynski, et al., Nucl. Instr. and Meth. A 393 (1997) 332.

[2] D. Iracane, et al., Phys. Rev. Lett. 72 (1994) 3985.

[3] R.W. Warren, et al., Nucl. Instr. and Meth. A 285 (1989) 1;
D.W. Feldman, et al., Nucl. Instr. and Meth. 285 (1989) 11.

[4] R.W. Warren, J.C. Goldstein, B.E. Newnam, Nucl. Instr. and Meth. A 250 (1986) 19.

[5] D.A. Jaroszynski, et al., Phys. Rev. Lett. 70 (1993) 3412;
D.A. Jaroszynski, et al., Nucl. Instr. and Meth. A 331 (1993) 52.

[6] G.R. Neil, et al., Phys. Rev. Lett. 84 (2000) 662;.
G.R. Neil, et al., Nucl. Instr. and Meth. A 445 (2000) 192.

[7] N. Nishimori, et al., Nucl. Instr. and Meth. A 445 (2000) 432.

[8] M. Sawamura, et al., Nucl. Instr. and Meth. A 318 (1992) 127.

[9] N. Nishimori, et al., Proceedings of 7th European Particle Accelerator Conference, Austria Center, Vienna, 2000, p. 1672.

[10] N. Piovella, P. Chaix, G. Shvets, D.A. Jaroszynski, Phys. Rev. E 52 (1995) 5470; N. Piovella, P. Chaix, G. Shvets, D.A. Jaroszynski, Nucl. Instr. and Meth. A 375 (1996) 156.

[11] R. Hajima, N. Nishimori, R. Nagai, E.J. Minehara, Nucl. Instr. and Meth. A 475 (2001) 270, these proceedings.

[12] R. Nagai, et al., Nucl. Instr. and Meth. A 475 (2001) 519, these proceedings.

ELSEVIER

Nuclear Instruments and Methods in Physics Research A 475 (2001) 270–275

NUCLEAR
INSTRUMENTS
& METHODS
IN PHYSICS
RESEARCH
Section A

www.elsevier.com/locate/nima

Analyses of superradiance and spiking-mode lasing observed at JAERI-FEL

R. Hajima*, N. Nishimori, R. Nagai, E.J. Minehara

Japan Atomic Energy Research Institute, 2-4 Shirakata-Shirane, Tokai-mura, Ibaraki 319-1195, Japan

Abstract

Japan Atomic Energy Research Institute (JAERI)-FEL has achieved quasi-CW lasing with an average power of 1.7 kW, the initial goal of the R&D program. The FEL extraction efficiency obtained completely exceeds the well-known limit for non-bunched beam, which is determined by the number of undulator periods. We have conducted numerical studies to characterize lasing dynamics observed at JAERI-FEL. Cavity-length detuning curves numerically obtained show good agreement with experimental results. Lasing behavior numerically obtained exhibits chaotic spiking-mode and superradiance as the cavity-length detuning approaches zero. Broadening of lasing spectrum observed in the experiments is explained by these lasing dynamics. The extraction efficiency becomes maximal at the perfect synchronization of the cavity length, where the lasing is quasi-stationary superradiance. We also compare these results with analytical theory previously reported. © 2001 Elsevier Science B.V. All rights reserved.

PACS: 41.60.Cr; 42.50.Fx; 29.17.+w

Keywords: Free-electron laser; Superconducting linac; Superradiance

1. Introduction

It is known that lasing dynamics in FEL oscillators with a bunched electron beam depend much on cavity-length detuning. In the single-supermode regime, optical pulses have smooth temporal profiles and show frequency spectra as narrow as the Fourier transform limit. As the FEL gain becomes much larger than the cavity loss, optical pulses with spiking structure appear at small cavity-length detuning, and the lasing spectrum has sidebands or broadbands. Extraction

efficiency increases as the broadening of the lasing spectrum and overcomes the well-known efficiency limit determined by the number of undulator periods: $\eta \sim 1/4 N_w$. This lasing behavior has been studied as spiking-mode, limit-cycle, chaotic turbulence and superradiance [1–4].

A high-power FEL developed at Japan Atomic Energy Research Institute (JAERI) achieved quasi-CW lasing with an average power of 1.7 kW recently [5]. We have not measured the temporal structure of the FEL micro-pulses, but we observed broadening of lasing spectrum near zero-detuning in the experiment which obeys the universal scaling law of superradiance between the extraction efficiency (η) and the rms bandwidth of lasing spectrum (σ_ω/ω): $\eta = 0.85(\sigma_\omega/\omega)$. The

---

*Corresponding author. Tel: +81-29-282-6315; fax: +81-29-282-6057.

E-mail address: hajima@popsvr.tokai.jaeri.go.jp (R. Hajima).

0168-9002/01/$ - see front matter © 2001 Elsevier Science B.V. All rights reserved.
PII: S 0 1 6 8 - 9 0 0 2 (0 1) 0 1 6 2 1 - 7

FEL lasing in our experiment, therefore, is considered as spiking-mode or superradiance. In the present study, characterization of the lasing dynamics in JAERI-FEL is conducted using numerical simulations.

2. Numerical simulations

Since our experiments show a large extraction efficiency corresponding to FEL lasing with deep saturation, nonlinear analyses are required for the characterization of the experimental results. We decided to make numerical simulations for the nonlinear analyses of lasing dynamics in JAERI-FEL. The numerical study also enables one to investigate inhomogeneous effects: energy spread, timing jitter, shot noise and radiation focusing.

For the analyses of lasing dynamics in JAERI-FEL, we must calculate the FEL pulse-propagation up to several thousands of round-trips, and a 3-D time-dependent simulation is not practical due to long computing time. We decided to use a 1-D time-dependent FEL code in the present study. Lasing dynamics in FEL oscillators are primarily dominated by longitudinal coupling between the optical pulse and the electron bunch, and in homogeneity of transverse dimension can be reduced to a filling factor.

Although characterization of nonlinear lasing dynamics is available using numerical simulations based on the collective variables [6,11], we make simulations with the macro-particle model. It is because the macro-particle model easily deals with shot noise of electron bunches. We will see that the shot noise has an important role in the lasing dynamics in the superradiant regime.

FEL equations expressed in dimensionless parameters are [12]

$$\frac{d\zeta_i}{d\tau} = -1, \quad \frac{d\mu_i}{d\tau} = Re[ia\exp(i\Psi_i)], \quad \frac{d\Psi_i}{d\tau} = \mu_i \quad (1)$$

where subscript i means i-th macro-particle, $\zeta = (z - ct)/N_w\lambda$ is the longitudinal position of a particle referred to the moving frame of the laser field and normalized by slippage distance, $\tau = ct/L_w$ is the dimensionless time variable, $\mu = 4\pi N_w(\gamma - \gamma_0)/\gamma_0$ is the dimensionless momentum

of a particle, $a = 4\pi a_w N_w^2[JJ]\lambda_w eE/\gamma_0 mc^2$ is the dimensionless complex laser field and $\Psi = (k_w - k)z - \omega t$ is particle phase. Evolution of the complex laser field can be calculated as

$$\frac{da(\zeta)}{d\tau} = ij_0(\zeta)\langle\exp(-i\Psi_i)\rangle_{\zeta_i=\zeta} \quad (2)$$

where $j_0 = 2(4\pi N_w\rho)^3$ is the dimensionless beam current and ρ is the FEL parameter.

The cavity-length detuning can be taken into account by shifting the optical field forward or backward every round-trip. Detuning smaller than the numerical grid interval can be made by interpolating field on the two neighbor grids. We also introduce shot noise according to the Penman's method [7], which gives the equivalent shot noise of the electron bunch.

All the parameters used in the following simulations are derived from our experiment: dimensionless peak current $j_0 = 33$, total loss of the optical cavity $\alpha_0 = 5\%$, FEL wavelength $\lambda = 22\,\mu m$, the number of undulator period $N_w = 52$. The electron bunch is assumed to be a triangle shape with 5 ps (FWHM) length. The slippage distance is calculated as $L_s = 1.1\,mm$, which is almost comparable to the bunch length.

We calculate the evolution of optical pulses up to 4000 round-trips, using Eqs. (1) and (2), and we plot the extraction efficiency obtained as a function of cavity-length detuning to simulate an experimental detuning-curve with macro-pulse duration of 400 μs. Experimental and numerical results are shown in Fig. 1. A sharp peak near zero-detuning appearing in the experiment is completely simulated in the numerical results.

Figs. 2 and 3 show the evolution of optical pulses calculated for two different cavity-length detuning $\delta L = -0.1$ and $-10\,\mu m$, respectively. In the case of $\delta L = -10\,\mu m$, optical spikes are generated at the trailing edge of the electron bunch and grow to large amplitude while moving forward every round trip by the cavity-length shortening. The optical spikes decay according to the total loss of the optical cavity after they are pushed away from the electron bunch. The generation of the optical pulses is irregular and chaotic. This is known as chaotic spiking-mode lasing, which occurs in large gain FEL oscillators

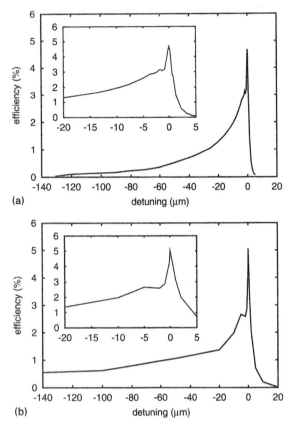

Fig. 1. Extraction efficiency averaged over a mucro-pulse as a function of cavity-length detuning. (a) Experimental result with macro-pulse of 400 µs. (b) Numerical result for 4000 passes.

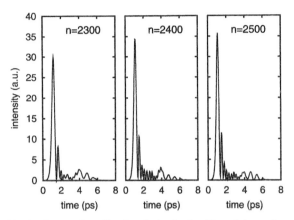

Fig. 2. Temporal profile of optical pulse for $\delta L = -0.1$ µm after round-trips of 2300, 2400 and 2500. The center of electron bunch is at 0 ps before the undulator and slips backword to 3.9 ps at the end of the undulator.

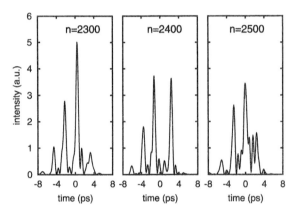

Fig. 3. Temporal profile of optical pulse for $\delta L = -10$ µm after round-trips of 2300, 2400 and 2500.

with small cavity-length detuning. Each of the spikes is superradiance. Lasing at $\delta L = -0.1$ µm shows different aspect. A single optical spike has large amplitude and keeps almost the same position in the moving frame, though the amplitude and the position of spike show small fluctuation every round trip. Since the optical spike also has the characteristics of superradiance, we can call it quasi-stationary superradiance. This type of lasing is obtained in our simulations with detuning range: -0.5 µm $< \delta L < 5$ µm, which corresponds to the sharp peak of the detuning curve observed in Fig. 1. The discontinuous jump of extraction efficiency in the experimental detuning curve is, therefore, considered as the transition of lasing dynamics from the chaotic spiking-mode to

quasi-stationary superradiance. The broadening of the lasing spectrum for small cavity-length detuning measured at the experiments [5] is consistent with the numerically obtained optical pulses much shorter than the electron bunch.

In our calculations, the shot noise of electron bunches is simulated using Penman's formula. Variation of noise factor alters effective power of seed light in the simulation. The lasing behavior in the early stage of starting-up, therefore, depends largely on the amount of noise factor. To the contrary, lasing dynamics after the onset of saturation are less affected by shot noise, because the amplitude of coherent component in the shot noise is much smaller than the saturated radiation stored in the optical cavity. For example, the coherent component of radiation excited by shot noise with Penman's formula is as small as 10^{-11} of saturated power in our simulation.

In spite of the above discussion, we have found the surprising result that the lasing dynamics in the quasi-stationary superradiant regime are largely affected by the shot noise even after the onset of saturation. Fig. 4 shows detuning curves near the perfect synchronism calculated for two different noise factor: (a) noise factor determined by Penman's formula, (b) 10^{-12} of that. We can see that the simulation with Penman's noise gives FEL lasing with large extraction efficiency near zero-detuning similar to the experimental result, but FEL lasing does not occur with detuning range of $\delta L > -0.2\,\mu m$ for the smaller shot noise. These results suggest that the lasing dynamics in the quasi-stationary superradiance are closely connected with a successive small fluctuation introduced by shot noise.

It is a well-known feature of FEL oscillators that the energy spread of electron bunches degrades FEL interaction. A handy gain-reduction factor due to the energy spread is derived as $f_E = 1/(1 + 1.7(4\sigma_E N_w)^2)$, where σ_E is the rms energy spread in an electron bunch and N_w is the number of undulator periods [8]. This formula is valid for

FEL lasing in which an electron bunch interacts with an optical pulse continuously through the whole length of the undulator. This situation appears in an FEL oscillator with single-super-mode regime. If an FEL oscillator is operated in the spiking-mode or the superradiant regime, electron bunches interact with optical spikes shorter than the slippage distance and lose their energy during a small number of undulator periods. This means that the effective number of undulator periods becomes smaller than the real undulator and the effect of the energy-spread is relaxed consequently.

Detuning curves calculated for $\sigma_E = 0.8\%$ and 0.5% are shown in Fig. 5 and compared with the reference case without energy-spread, shown in Fig. 1(b). It can be clearly seen that the detuning range becomes narrower for the larger energy-spread. This is explained by the FEL gain reduction due to the energy-spread in the single-supermode and the weak sideband regime. In the superradiant regime near the perfect synchronism, however, the detuning curves have almost the same profile for all cases. Detailed analyses show that the small reduction of peak efficiency observed in the detuning curve with large energy-spread is due to the longer starting-up period in a macro-pulse. They also show that the extraction efficiency after complete saturation is independent of the energy-spread.

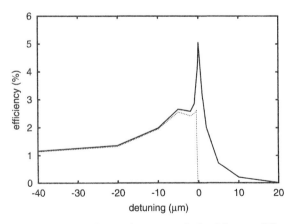

Fig. 4. Cavity-length detuning curve calculated for two different noise-factor: shot noise determined by Penman's formula (solid line) and 10^{-12} of that (dashed line).

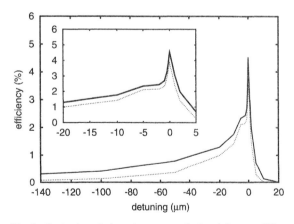

Fig. 5. Cavity-length detuning curve calculated for two different energy-spread: $\sigma_E = 0.5\%$ (solid line) and $\sigma_E = 0.8\%$ (dashed line).

In an electron linac, the energy spread is generally introduced by several sources, which include the residual component of energy modulation applied at an injector, longitudinal phase spread during acceleration and coherent synchrotron radiation at circular paths. In JAERI-FEL, the dominant source of energy-spread is coherent synchrotron radiation at a 180-degree arc before the undulator. Although we have not measured the energy spread at the undulator, numerical simulations suggest that the coherent synchrotron radiation causes rms energy-spread of 0.4% in a bunch. The detuning curves obtained in the experiment and the calculations suggest an rms energy-spread of 0.8%, which may include energy-spread introduced before the arc.

In the numerical studies, the timing jitter of electron bunches can be simulated by jitter in the cavity-length detuning. Fig. 6 shows detuning curves calculated for rms timing-jitter of 100 and 500 fs, where we assume the timing jitter has a Gaussian distribution. It is shown that the detuning curve becomes dull and that the maximum extraction efficiency decreases for the larger timing jitter as we expected. The cavity-length detuning curves in the experiments has a strict peak near zero detuning, the width of which is around 1 μm. If we assume a shift of cavity-length detuning equal to $-1\,\mu m$ and total loss of the cavity $\alpha_0 = 5\%$, a specific position at the optical spikes

moves 40 μm before it disappears by cavity loss. Acceptable timing jitter can be estimated from the above consideration and found to be 130 fs, which is almost consistent with the numerical results.

We measured timing jitter of electron bunches just downstream of the gun and obtained an rms value of 23 ps [9]. Since the electron bunches of 870 ps at the gun are compressed into 5 ps at the undulator, the jitter can be estimated as 140 fs at the undulator with the assumption of linear compression of the longitudinal phase space.

3. Discussions

For an FEL oscillator with short electron bunches operated in the superradiant regime, the maximum extraction efficiency is analytically derived as $\eta \simeq 1.43\,\rho\sqrt{L_b/\alpha_0 L_c}$, where L_b is the bunch length and $L_c = \lambda/4\pi\rho$ is the cooperation length [11]. In JAERI-FEL, electron bunches are slightly longer than the slippage distance and L_b in the above formula must be replaced by slippage distance L_s. The extraction efficiency for JAERI-FEL is found to be $\eta \simeq 1.43\sqrt{4\pi N_w\rho^3/\alpha_0} = 4.0\%$. This is almost consistent with our experimental and numerical results.

A linear analysis by Piovella also shows that a short bunch FEL oscillator operated at the perfect cavity-length synchronism has a transient solution of superradiant start-up [10]. This superradiance, however, ceases to lase before reaching saturation and does not keep its amplitude over a number of round-trips. Our numerical simulations give a similar result in the limit of small shot noise. In the simulations with appropriate shot noise, however, a superradiant spike generated at the linear transient regime grows, moreover, into the nonlinear saturating regime and keeps quasi-stationary lasing with chaotic irregular fluctuation. It is considered that lasing at zero-detuning and small reverse-detuning observed in our calculations has strong connection with shot noise.

The effect of shot noise in an FEL oscillator with reverse-detuning was previously reported by Benson and Madey [13], and Brau [12], who showed that the lasing behavior in the reverse-detuning is affected by shot noise. The enhance-

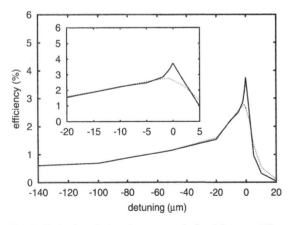

Fig. 6. Cavity-length detuning curve calculated for two different timing-jitter of electron bunches: 100 fs (solid line) and 500 fs (dashed line).

ment of the FEL efficiency by shot noise at the perfect synchronism has never been reported, however.

4. Conclusions

We have conducted numerical analyses to characterize the nonlinear lasing dynamics in JAERI-FEL, in which extraction efficiency has a large value near the zero-detuning. The discontinuous jump of extraction efficiency observed in experimental detuning-curves can be explained by transition of lasing dynamics from chaotic spiking-mode to quasi-stationary superradiance. Our numerical consideration also shows that the lasing dynamics in quasi-stationary superradiance is strongly related to shot noise of electron bunches which introduces successive small fluctuation into the system.

References

[1] R.W. Warren, et al., Nucl. Inst. and Meth. A 285 (1989) 1.
[2] D.A. Jaroszynski, et al., Phys. Rev. Lett. 70 (1993) 3412.
[3] D. Iracane, et al., Phys. Rev. Lett. 72 (1994) 3985.
[4] D.A. Jaroszynski, et al., Phys. Rev. Lett. 78 (1997) 1699.
[5] N. Nishimori, et al., Nucl. Instr. and Meth A 475 (2001) 266, these proceedings.
[6] N. Piovella, et al., Phys. Rev. E 52 (1995) 5470.
[7] C. Penman, B.W.J. McNeil, Opt. Commun. 90 (1992) 82.
[8] G. Dattoli, et al., IEEE-QE 20 (1984) 637.
[9] N. Nishimori, et al., in Proceedings of the EPAC-2000, Vienna, Austria, 2000, pp.1672–1674.
[10] N. Piovella, Phys. Rev. E 51 (1995) 5147.
[11] P. Chaix, et al., Phys. Rev. E 59 (1999) 1136.
[12] C.A. Brau, Free-Electron Lasers, Academic Press, New York, 1990.
[13] S. Benson, M.J. Madey, Nucl. Inst. and Meth. A 237 (1985) 55.

ELSEVIER

Nuclear Instruments and Methods in Physics Research A 475 (2001) 276–280

NUCLEAR
INSTRUMENTS
& METHODS
IN PHYSICS
RESEARCH
Section A

www.elsevier.com/locate/nima

An experimental study of an FEL oscillator with a linear taper

S. Benson*, J. Gubeli, G.R. Neil

Jefferson Lab, MS 6A, 12000 Jefferson Ave., TJNAF, Newport News, VA 23606, USA

Abstract

Motivated by the work of Saldin, Schneidmiller and Yurkov, we have measured the detuning curve widths, spectral characteristics, efficiency, and energy spread as a function of the taper for low and high Q resonators in the IR Demo FEL at Jefferson Lab. Both positive and negative tapers were used. Gain and frequency agreed surprisingly well with the predictions of a single mode theory. The efficiency agreed reasonably well for a negative taper with a high Q resonator but disagreed for lower Q values both due to the large slippage parameter and the non-ideal resonator Q. We saw better efficiency for a negative taper than for the same positive taper. The energy spread induced in the beam, normalized to the efficiency is larger for the positive taper than for the corresponding negative taper. This indicates that a negative taper is preferred over a positive taper in an energy recovery FEL. © 2001 Elsevier Science B.V. All rights reserved.

PACS: 29.27.−a; 29.17.+W; 41.60.Cr

Keywords: High power; Tapered FEL; Efficiency enhancement

1. Introduction

In the early days of free-electron lasers (FELs), Kroll, Morton, and Rosenbluth, using analogies to accelerator physics [1], suggesting the use of variable parameter wigglers to enhance the efficiency of FELs. The theory of the tapered wiggler FEL [2] assumed a single frequency, plane wave input and was thus applicable to amplifiers. The phase displacement scheme [3] also assumed a plane wave, though pulsed effects were considered. The two schemes share one common feature. They assume a well-defined bucket as an initial condi-

tion. However, the small-signal gain at the optimal saturated frequency was smaller than the saturated gain, so oscillators had a startup problem [4]. The first tapered oscillator, therefore, used a multi-component design that enhanced the small-signal gain at the appropriate frequency and produced a good trapping fraction [5]. Later, high efficiency (up to 45%) was achieved in an amplifier experiment at Livermore [6]. In 1995, experimental studies at FELIX and Orsay [7] and theoretical studies by Saldin et al. [8] showed that a mild negative taper should produce better extraction efficiency than a positive taper. The authors in Ref. [8] showed that there was a mismatch between the optimal frequency for small-signal gain and the optimal frequency for saturated lasing for the case of a positive tapered wiggler, leading to poor

---

*Corresponding author. Tel.: +1-757-269-5026; fax: +1-757-269-5519.

E-mail address: felman@jlab.org (S. Benson).

performance for mildly tapered FEL oscillators. There is no such mismatch for a moderately inverse tapered oscillator allowing it to operate more efficiently.

In many experiments, inserting a wedge into the wiggler gap, leaving the untapered gap at either the front or back end, provides the taper. In this case, the ratio of the small-signal gain of the tapered wiggler to that of the untapered wiggler is independent of which end is opened. Similarly, the frequency for maximum small-signal gain is the same for the two cases. This frequency decreases for both positive and negative tapers since the maximum field is being held constant.

The analysis in Ref. [8] motivated us to look at the behavior in the IR Demo FEL [9] at Jefferson Lab with a linear taper. We created the taper by introducing a linear gap change. The taper is only slightly non-linear and the magnetic field quality is still excellent. One major difference between our FEL and the analysis in Ref. [8] is the assumption of a single frequency. The IR Demo FEL has quite short electron pulses and operates in the high slippage regime with broad spectra. In this regime, the FEL tends to develop very short micropulses that pass over the electrons in much less than one pass through the laser. An electron, therefore, effectively sees a shorter wiggler. This can enhance the efficiency of the untapered FEL and can negate the effects of a taper since the electrons do not see a significant taper during the time that the optical field is present. Another difference is that the maximum wiggler field is a constant vs. taper while, in Ref. [8], the entrance wiggler field is constant. Finally, the resonator Q for most of our data was lower than that in Ref. [8]. This should result in lower efficiencies for the IR Demo for most tapers.

Since the IR Demo FEL utilizes energy recovery, the overall performance is sensitive to the *total* energy spread at the output of the FEL. Experiments with step-tapered oscillators indicated that the r.m.s. energy spread was smaller for an inverse step taper than for an untapered FEL for the same power [7]. We were, therefore, quite interested to see if an inverse taper could provide enhanced efficiency while maintaining or even decreasing the exhaust energy spread.

2. Description of the experiment

The IR Demo FEL and accelerator are described in Ref. [9]. Two experimental runs with the laser were carried out. In the first, the laser was operated at wavelengths near 3 μm with a resonator Q of 10. Three tapers of both signs were studied. Exhaust energy profiles were obtained by looking at a viewer downstream of the FEL when lasing with pulsed beam. In the second set of experiments, the laser was operated near 6 μm. One positive taper and two negative tapers were studied with a resonator Q of 10 and one of each was studied with a resonator Q of 50. In each case, slit scans of the exhaust beam energy spread enabled us to measure the exhaust energy spectrum of the laser.

For each taper, the laser was optimized with pulsed lasing conditions (typically 1 ms pulses at 60 Hz). At this power level, slightly higher electron losses could be tolerated and mirror-heating effects were negligible compared to CW operation. With almost 19,000 cavity round trips during the macropulse the laser had plenty of time to reach an equilibrium state. The average power as a function of cavity length detuning was measured. For the case of no taper or a weak taper, the detuning curve was also measured for a 9.4 MHz micropulse frequency where two resonator passes are required for each gain pass. For each taper, we measured spectra at the peak of the detuning curve and one-third and two-thirds of the way out on the curve. The exhaust energy spreads were measured at several points on the detuning curve. Finally, the performance with CW lasing was optimized and measured.

Tapers were obtained by the insertion of precision shims at either end of the wiggler. Dial gauges on either side of the wiggler measured the position and gap of the wiggler for each taper. The variation of the wiggler field with respect to gap is known so the field taper can be calculated from the gap taper. The actual taper will be exponential instead of linear, but with a taper of only 10% the linear taper is an excellent approximation to the exponential taper. The resonant energy prediction will differ by less than 0.1% from the actual resonant energy for a resonant energy taper of 5% (10% field taper).

3. Experimental results

Dattoli has shown that the total length of the detuning curve should be linearly proportional to the gain [10]. We have found that this is not true when the gain-to-loss ratio is very large, but we can increase the losses by doubling the number of passes for each gain pass. If this is done, the results are fairly consistent with Dattoli's theory. The detuning curve length vs. the normalized taper strength is shown in Fig. 1 along with the relative gain vs. taper calculated from theory. The normalized taper strength b as defined in Ref. [8] is the change in the normalized resonant energy, i.e. $b = v_f - v_i$, where v_i and v_f are the normalized resonant energies at the beginning and end of the wiggler, respectively. The agreement is quite good considering the inaccuracies of this method. Also note that the gain was independent of the sign of the taper as expected from theory.

The normalized efficiency $2N\eta$ is shown in Fig. 2 for a negative taper, along with the dependence from Ref. [8] using the same dimensionless parameters. The efficiency for no taper is actually much higher than for the single frequency theory, due to short pulse effects similar to those described in Ref. [11]. This effect persists as the taper is increased but the efficiency goes down. This is presumably due to the fact that the gain-to-loss ratio is decreasing. This lowers the fraction of

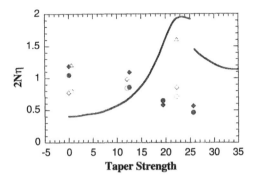

Fig. 2. The experimental normalized efficiency $2N\eta$ vs. taper strength is shown. The curve is scaled from the one in Ref. [8]. The circles are CW efficiencies, the diamonds are pulsed efficiency and the triangles are the pulsed efficiency for the case of $Q = 50$. The filled symbols are taken at 3 μm and the empty ones at 6 μm. Two of the symbols are slightly offset at $b = 0$ for clarity.

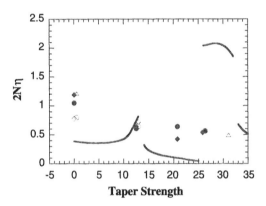

Fig. 3. The experimental normalized efficiency vs. taper strength for a positive taper is shown. The curve is scaled from Ref. [8]. The symbol definition is the same as for Fig. 2.

electrons contributing to the interaction and lowers the effectiveness of the short pulse effects and the tapered operation. When the cavity Q is raised so that the gain-to-loss is the same as that of the untapered case, we find that the efficiency agrees quite well with the single frequency model. No CW data are shown for the $Q = 50$ resonator since the mirror heating greatly limited the efficiency.

In Fig. 3, we show the efficiency vs. taper for the positive taper runs. The general behavior is similar to the negative taper runs. The efficiency is highest for an untapered configuration and falls off as the taper increases. The efficiency is uniformly less than the negative taper for all values of the taper.

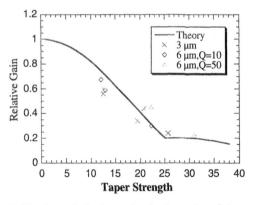

Fig. 1. The theoretical gain vs. the absolute value of the taper and the experimental detuning curve length relative to that for no taper are shown. The gain is the same for both positive and negative tapers. The kink in the curve occurs when the gain peak shifts from peak to peak as described in Ref. [8].

The efficiency for the case of the high Q resonator is surprisingly low. Higher efficiency was limited by the inability to recirculate beam due to the occasional strong lasing accompanied by an extremely large exhaust energy spread. The optical spectrum was at least 3% in width but was extremely noisy at any one wavelength as if the wavelength was jumping around from pulse to pulse. There was also some evidence for synchrotron detrapping at some cavity lengths.

As noted above, the inverse taper is expected to have a smaller energy spread at the FEL exit. In addition, the distributions for a positive taper, a negative taper, and no taper should be qualitatively different. We took slit scans and viewer images at a dispersed location to study the distributions. We found that the untapered case has a nearly top hat energy distribution. In general, the energy distributions varied with the cavity length and the degree of taper. The general trend was for the positive taper to have a low-energy tail and the negative taper to have a high-energy tail and a very sharp low-energy edge. The positive taper typically has a double humped distribution with a decelerated bucket and a second peak corresponding to the untrapped electrons. The positive tapers usually showed evidence of poor trapping efficiency.

If the full width of the energy distribution for each taper at the peak of the detuning curve is divided by the laser efficiency, one has a measure of how appropriate the wiggler is for energy recovery (lower ratios are preferred). The ratios for all our configurations are shown in Fig. 4. The lowest numbers are for an untapered wiggler, a mild negative taper, or the strong negative taper with a high cavity Q. The general trend is towards a higher ratio for the positive taper and a lower ratio for the corresponding negative taper. One might expect the positive taper ratio to be much lower than it is. The problem is that the trapping efficiency is rather poor as noted above.

As pointed out in Ref. [8], the wavelength for maximum small signal gain is not the correct wavelength for optimum large signal operation with a positive taper, leading to poor efficiency. One might think, however, that the large spectral bandwidth present in an FEL with a large slippage

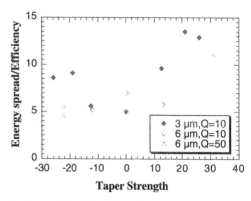

Fig. 4. Ratio of the full-width of the exhaust energy spectrum to the efficiency at the peak of the cavity length detuning curve. Only pulsed efficiencies were used since the energy spreads could only be measured using pulsed beam.

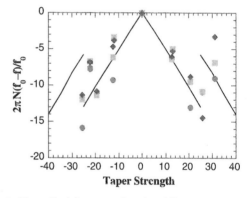

Fig. 5. Normalized frequency detuning shift vs. taper strength. The curves correspond to the frequency expected if the frequency varies with the peak of the small signal gain. The circles indicate the shift at the peak of the detuning curve. The squares and diamonds indicate spectra one-third and two-thirds of the way out on the detuning curve.

parameter might allow the saturated signal to move over to a frequency where the saturated gain is higher. That this was not the case is evident in Fig. 5. In this figure, we show the shift in frequency from that of an untapered wiggler for a similar point in the cavity length detuning curve. The curves are the shift in the frequency for the maximum small signal gain as a function of taper. The points taken with a very large taper shows that the jump at $b = 26$ occurs in the experiment. The spectrum does not approach the wavelength necessary for the tapered FEL to operate most efficiently. The points for $b = 31$ have a large uncertainty since the wavelength jumped around rapidly with time.

4. Conclusions

Some of the predictions of the single frequency model for linearly tapered oscillators are borne out by the experimental results. The gain is independent of the sign of the taper and the reduction of the gain, as measured by the length of the cavity length detuning curve, matches well with the predictions. The negative taper is clearly superior to the positive with respect to efficiency and exhaust energy spread. Finally, the saturated wavelength is close to the wavelength for largest small signal gain even for a system with very broad bandwidth. This is really surprising given the broadband nature of the IR Demo FEL.

We did not expect the experimental efficiency to be the same as that predicted in Ref. [8] due to the relatively low Q and the large slippage parameter. The behavior seen as the gain decreases is what one might expect for an untapered system with a large slippage. When the Q was raised to match the gain-to-loss ratio of the untapered configuration, however, the efficiency of the negative taper matched that of Ref. [8] well. The surprise is that the high Q positive taper did not work well. This was due to a very poor trapping fraction, which appears to have been due to synchrotron–betatron detrapping. The stochastic nature of this led to an extremely unstable operation in both power and wavelength.

How appropriate is a tapered wiggler for an FEL with energy recovery? The positive taper is clearly poorly suited due to its unstable operation, its low efficiency, and its large exhaust energy spread. The negative taper worked as well as the untapered wiggler but, in the case of the optimal taper, the Q had to be raised. This limited the average power of the device due to mirror heating. It should be noted, however, that the fact that the negative taper has a high-energy tail and a sharp low-energy edge means that the system can be pushed very hard without leading to a beam-loss instability in the RF system. A recirculating linac is stable to losses on the high-energy edge. A system with longer micropulses would also benefit from a negative taper since the efficiency of the laser in the absence of a taper would not be as high.

Acknowledgements

Many thanks to all the technical staff at Jefferson Lab who helped make this work possible, especially Rich Hill, Kevin Jordan, and Jim Coleman who got the slit scanner to work. This work was supported by the U.S. Department of Energy under contract DE-AC05-84-ER40150, the Office of Naval Research, the Commonwealth of Virginia, and the Laser Processing Consortium.

References

[1] P.L. Morton, Phys. Quantum Electron. 8 (1981) 1.
[2] N.M. Kroll, P.L. Morton, M.N. Rosenbluth, IEEE J. Quantum Electron. QE-17 (1981) 1436.
[3] N.M. Kroll, P.L. Morton, M.N. Rosenbluth, Proceedings of the Telluride Conference, Phys. Quantum Electron. 7 (1979) 104.
[4] C.-C. Shih, M.Z. Caponi, Phys. Quantum Electron. 8 (1981) 381.
[5] J.A. Edighoffer, G.R. Neil, C.E. Hess, T.I. Smith, S.W. Fornaca, H.A. Schwettman, Phys. Rev. Lett. 52 (1984) 344.
[6] T.J. Orzechowski, et al., Phys. Rev. Lett. 57 (1986) 2172.
[7] D.A. Jaroszynski, et al., Nucl. Instr. and Meth. A 358 (1995) 228.
[8] E.L. Saldin, E.A. Schneidmiller, M.V. Yurkov, Opt. Commun. 103 (1993) 297.
    ibid., Nucl. Instr. and Meth. A 375 (1996) 336.
[9] G.R. Neil, et al., Phys. Rev. Lett. 84 (2000) 662.
[10] G. Dattoli, T. Hermsen, L. Mezi, A. Torre, J.C. Gallardo, Nucl. Instr. and Meth. A 272 (1988) 351.
[11] N. Piovella, P. Chaix, G. Shvets, D.A. Jaroszynski, Phys. Rev. E 52 (1995) 5470.

ELSEVIER

Nuclear Instruments and Methods in Physics Research A 475 (2001) 281–286

NUCLEAR
INSTRUMENTS
& METHODS
IN PHYSICS
RESEARCH
Section A

www.elsevier.com/locate/nima

Experimental study of compact FEL with micro wiggler and electrostatic accelerator

S. Fujii[a],*, T. Fujita[b], T. Mizuno[c], T. Ohshima[c], M. Kawai[d],
H. Saito[c], S. Kuroki[b], K. Koshiji[a]

[a] Science University of Tokyo, 2641 Yamazaki, Noda, Chiba 278-8510, Japan
[b] Soka University, 1-236 Tangi, Hachioji, Tokyo 192-8577, Japan
[c] Institute of Space and Astronautical Science, 3-1-1 Yoshinodai, Sagamihara, Kanagawa 229-8510, Japan
[d] Kawasaki Heavy Industries Ltd., 118 Futatsuzuka, Noda, Chiba 278-8585, Japan

Abstract

A compact FEL for submillimeter and far infrared regions is studied at the Institute of Space and Astronautical Science. The FEL can be compact by using an electrostatic accelerator and a micro wiggler. The electrostatic accelerator (DISKTRON) with a diameter of 1 m can generate up to 1 MV continuously. The micro wiggler is fabricated using permanent magnets made from Nd–Fe–B. (period: 8 mm, total length: 248 mm, gap: 2–10 mm, K parameter: 0.07–0.7) An electron beam of high quality is generated by means of a photo cathode. (731 kV, 1.5 A, 25 ns, 2 mmϕ, $\Delta E/E$:0.18%) In the preliminary phase, detection of the FEL at the millimeter wave region of 96 GHz is conducted. The electron beam is injected into a resonator with Distributed Bragg Reflector. A small millimeter wave signal has been detected. © 2001 Elsevier Science B.V. All rights reserved.

PACS: 41.60.cr; 07.57.Hm

Keywords: Free electron laser; Micro wiggler; Electrostatic accelerator

1. Introduction

Recently, research on FELs is popular in two frequency regions, which are the VUV region and the submillimeter (far infrared) region. It is difficult for other wave sources to generate high power and coherent radiation in these regions. The technology of submillimeter and far infrared is at the border between a microwave device and a photonic device [1]. FELs of frequency from 0.1 to 10 THz are very useful for various studies. There are 8–10 FEL user facilities in the world [2]. The facilities of UCSB (MM-FEL, FIR-FEL) and FOM (FELIX FEL-1) provide the radiation of this frequency region. However, their accelerators (6 MeV Van de Graaff and 25 MeV linac) are still huge and expensive, and total lengths of wigglers are several meters; therefore whole sizes of apparatuses are very large. Users of FELs desire a compact FEL, which is tabletop size. The compact FEL for submillimeter and far infrared regions using the electrostatic accelerator and the micro wiggler is studied at the Institute of Space

*Corresponding author. Tel.: +81-42-759-8365.
E-mail address: reishi@koshijia.ee.noda.sut.ac.jp (S. Fujii).

0168-9002/01/$ - see front matter © 2001 Elsevier Science B.V. All rights reserved.
PII: S 0 1 6 8 - 9 0 0 2 (0 1) 0 1 6 1 9 - 9

and Astronautical Science, Japan. The FEL has tabletop size with a total length of 2.5 m. In the preliminary phase, we try to oscillate the FEL in the millimeter wave of 96 GHz since measuring instruments are easy to obtain in this region.

2. Design and apparatus of FEL

The schematic diagram of experimental apparatus is shown in Fig. 1. The FEL system consists of an electron gun with a photo cathode, the electrostatic accelerator (DISKTRON), a high gradient accelerating tube, a beamline with electron beam control system, the micro wiggler and a waveguide resonator with Distributed Bragg Reflector (DBRs) [3]. The micro wiggler is in the vacuum chamber, and the resonator is installed in the narrow gap of the micro wiggler. Since the width of the wiggler gap is only 3 mm and the length is 248 mm, a waveguide resonator is applied as a resonator. DBRs, which the electron beam can pass through, are used as reflecting mirrors of resonator.

A millimeter wave of 96 GHz is selected as a frequency of the preliminary experiment. Designed parameters of the FEL at 96 GHz are listed in Table 1. The FEL is designed to obtain interaction between the electron beam with the kinetic energy of 731 keV and the electromagnetic wave in the waveguide with the width of 2.25 mm. The electromagnetic wave experiences 13 round trips in the resonator during pulse width of the electron beam.

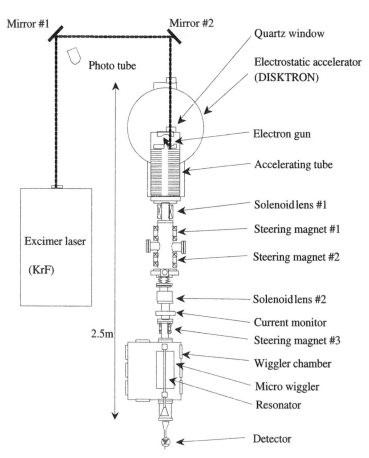

Fig. 1. Schematic diagram of the experimental apparatus.

Table 1
Parameters of FEL system for 96 GHz

Micro wiggler	
Type	Planer Halbach
Period	8 mm
Number of periods	29
Total length	248 mm
Gap	3 mm
K-parameter	0.48
Magnets	Nd–Fe–B
Resonator	
Type	Rectangular wave guide
Width	2.25 mm
Height	3.5 mm
Length	290 mm
Q	2230
Reflectors	
Type	Distributed Bragg reflector
Reflection index	0.91
Center frequency	96.2 GHz
FWHM	8.5 GHz
FEL	
Growth rate	6.0 m^{-1} (0.5 A)

There are two DBRs on both ends of the waveguide. The reflection index of the DBRs is designed to be 0.91. Then, the Q parameter of the resonator is 2330. Since the following condition [4] is satisfied, the FEL is expected to operate in the strong pump Compton regime.

$$\beta_{0\perp} \gg \beta_{crit} = 4(\xi/\gamma^3)^{1/2} \qquad (1)$$

where, $\beta_{0\perp}$, $\xi(= \omega_b/\gamma^{1/2} ck_0)$, ω_b, k_0, and γ are the transverse beam velocity, the beam strength parameter, the plasma frequency, the wave number, the Lorenz factor, respectively. The gap (3 mm) of the micro wiggler is not so much wide compared to the diameter of the electron beam, and the betatron motion may play an important role. The betatron motion of the electron at the edge of the electron beam may decrease the growth rate of the FEL. If, however, we ignore the influence of the betatron motion, the growth rate, Γ of the FEL is

$$\Gamma = \frac{\sqrt{3}}{2^{4/3}} (\xi\beta_{0\perp})^{2/3} k_0. \qquad (2)$$

The total gain of the FEL is expressed as [5]

$$G = \left[\frac{1-\alpha}{m^2} \left(e^{2\Gamma L} - e^{-2\Gamma L} \right) \right]^n, \qquad (3)$$

where α, L, and n are the loss of the resonator per a pass, the length of the interaction region, and the number of round trips in the resonator, respectively. The quantity m is equal to 3 in the Compton regime. In the case of the electron beam of 0.3 A with the diameter of 2 mm, the growth rate of the FEL is 5.1 m^{-1} and the total gain of the FEL is 0 dB. In the case of the electron beam with current of 0.5 A and diameter of 2 mm, the growth rate of the FEL is 6.0 m^{-1} and the gain of the FEL is estimated about 20 dB.

The equivalent circuit of DBR provides the amplitude index of reflection R as [6]

$$R = \frac{A + B/Z_0 - CZ_0 - D}{A + B/Z_0 + CZ_0 + D} \qquad (4)$$

where

$$
\begin{pmatrix} A & B \\ C & D \end{pmatrix} = \left[\begin{pmatrix} 1 & 0 \\ jB_a & 1 \end{pmatrix} E \begin{pmatrix} 1 & j/B_b \\ 0 & 1 \end{pmatrix} \right.
$$
$$
\times E \begin{pmatrix} 1 & 0 \\ jB_a & 1 \end{pmatrix} E \begin{pmatrix} \cos f\phi & jZ_{0\sin} f\phi \\ j\frac{1}{Z_0}\sin f\phi & \cos f\phi \end{pmatrix}^{N-1}
$$
$$
\left. \times E \begin{pmatrix} 1 & 0 \\ jB_a & 1 \end{pmatrix} E \begin{pmatrix} 1 & j/B_b \\ 0 & 1 \end{pmatrix} E \begin{pmatrix} 1 & 0 \\ jB_a & 1 \end{pmatrix} \right) \qquad (5)
$$

where B_a, B_b, Z_0, ϕ, and N are the admittance of the equivalent circuit, the impedance of the equivalent circuit, characteristic impedance of the equivalent circuit, the phase constant of the equivalent circuit, and the number of the periodic structure, respectively.

The power reflection index $|R|^2$ of DBR is calculated by using Eq. (4). The result of the calculation is shown in Fig. 2. The electromagnetic field in the same structure was calculated by using KCC micro stripes that provide 3D electromagnetic analysis. The dashed line of Fig. 2 shows the result of the calculation. From this figure, we can find that the estimate by Eq. (4) agrees with the result of the KCC micro stripes.

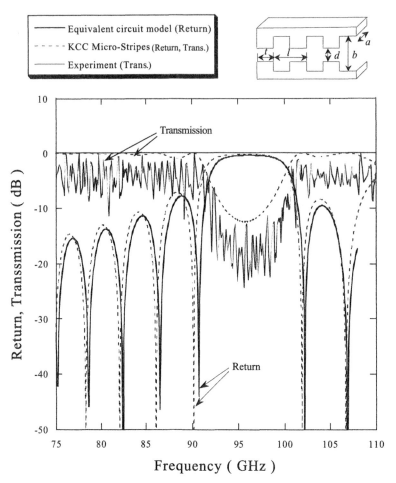

Fig. 2. Characteristics of the DBR. Return of the DBR calculated by using the equivalent circuit model (bold solid). Return and transmission of DBR calculated by the KCC micro-stripes (dashed). Transmission of the DBR measured by the scalar network analyzer (solid). $a = 2.25$ mm, $b = 3.5$ mm, $d = 2.9$ mm, $l = 2.1$ mm, $t = 1.05$ mm, $N = 10$.

DBRs were fabricated based on the design by means of electric casting. The transmission of a DBR is measured by using a scalar network analyzer. The result is shown in Fig. 2. The DBR can reflect the electromagnetic wave at the center frequency of 96.2 GHz and the width of 8.5 GHz (FWHM) as designed.

3. Characteristics of electron beam

The trajectories of the electron in the electron gun and accelerating tube are calculated by the simulation code E-GUN to optimize the shape of

electrode in the electron gun. The control system of the electron beam at the beamline is designed by TRACE 3-D which calculates the envelop of the electron beam [3]. The electron gun and the beamline are fabricated based on these designs. The peak current, the diameter and the energy spread of the electron beam are measured by using a current monitor, a CCD camera, and an energy analyzer, respectively. Characteristics of the electron beam obtained by the measurements are listed in Table 2.

The upper part of Fig. 3 shows the schematic diagram of the energy analyzer. The tungsten slit whose width is 0.1 mm is positioned at the

Table 2
Characteristics of electron beam

Energy	731 keV
Extract voltage	49 kV
Peak current	1.5 A (entrance of resonator)
	0.12 A (exit)
Pulse width	25 ns
Repetition rate	1 Hz
Beam diameter	3.1 mm (entrance of resonator)
	2.5 mm (center)
	2.9 mm (exit)
Current density	31 A/cm^2
Energy spread	0.18%
Normalized emittance	4.38 π mm mrad (designed)

entrance of the energy analyzer. Electrons that can pass through the slit are bent by 180° by the magnetic force in the analyzer. The digital video camera takes a picture of Cherenkov radiation from the PYREX glass plate at the end of the energy analyzer. The energy spread of the electron is expressed by Eq. (6).

$$\frac{\Delta E}{E} = \left(1 + \frac{1}{\gamma}\right)\left(\frac{S_1' - S_1}{2R}\right) \tag{6}$$

where S_1, S_1', and R are the width of the entrance slit, the width of the electron distribution at the end of the energy analyzer, and the Larmor radius of the electron, respectively. The lower part of Fig. 3 shows a typical distribution of the electron obtained by plotting the brightness of pixels on the picture. The width of the electron distribution is defined as the full width at 20% of peak brightness. The width of the electron distribution is 0.36 mm, taking the average of 10 shots; therefore the energy spread of the electron is calculated to be 0.18% by Eq. (6). The value of 0.18% corresponds to the energy spread $\Delta E = 1.3$ keV. The electric capacity of the accelerator is about 64 pF. The drop in the accelerating voltage during the pulse width of the electron beam (25 ns) is calculated to be 0.59 kV assuming the electron beam of 1.5 A. The potential difference between the electrons at the center of the beam and the electron at the edge of the beam is calculated as 0.043 kV assuming an electron beam of 1.5 A. The sum of two voltages is 0.63 kV, which is still less than the voltage obtained by measuring of the energy spread. The

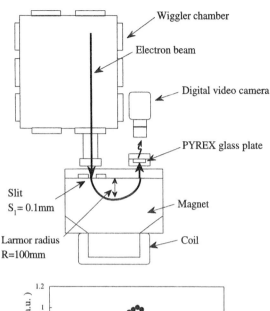

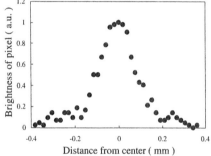

Fig. 3. Schematic diagram of the energy analyzer and the distribution of electrons.

other cause of the discrepancy may be the electron beam divergence. We plan to measure this effect in the next experiments.

4. Experiment and conclusion

The electron beam was injected into the resonator with the micro wiggler to oscillate the FEL at 96 GHz. A horn antenna in the vacuum chamber guides the output from the waveguide resonator. Then, the radiation passes through the acrylic vacuum window. Another horn antenna outside the chamber receives the output. A broadband crystal detector is used to detect the output. Fig. 4 shows one of

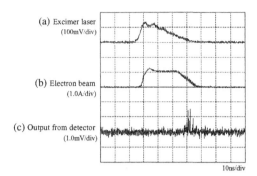

Fig. 4. Typical waveforms of oscilloscope. (a) Waveform of the excimer laser. (b) Current of the electron beam. (c) Signal of output detection.

the typical waveforms. The detected signal is obtained at a kinetic energy of 630 keV, which is different from the expected energy of 731 keV for FEL oscillation of 96 GHz. Fig. 4(a),(b) and (c) are the pulse of the excimer laser by photo tube, the beam current, and the detected RF signal, respectively. The spike-like, weak signals are detected at the end of the electron beam pulse. It is confirmed that this signal is not the noise signal but the signal of the millimeter wave. The power received by the crystal detector is 15 μW. The fabrication error of 0.1 mm for the waveguide width could be a reason for the difference of the electron energy between the expected value and the experimental value where the output is obtained.

Presently, the wavelength of the output is being measured to confirm that the output is the oscillation of the FEL. We plan to improve the intensity and stability of the output as described below. The first step is extending of the pulse width. As shown in Fig. 4, rising timing of the output corresponds to the very end of the beam pulse. A rapid increase in the FEL gain is expected by extending of the pulse width. Secondly, increase in the beam current that passes through the waveguide resonator is effective. Only about 10% of the electron beam can pass through the waveguide resonator presently. Thirdly, increase of the resonator Q by means of increasing the number of periods in the DBR can increase the intensity of the output.

References

[1] J. Allen, B. Gwinn, G. Ramain, M. Sherwin, Materials Research with the UCSB FELs at the CTST, 1997.
[2] Committee on Free Electron Lasers and Other Advanced Coherent Light Sources, Free Electron Lasers and Other Advanced Sources of Light, National Academy Press, Washington, DC, 1994.
[3] S. Fujii, T. Mizuno, H. Saito, et al., Nucl. Instr. and Meth. A 445 (2000) II-9.
[4] K.J. Button, Infrared and Millimeter Waves, Vol. 1, Academic Press, New York, 1979.
[5] H. Saito, et al., Free Electron Laser and its Application, Corona Publishing, 1990.
[6] MIT Radiation Laboratory Series, Waveguide Handbook, McGraw-Hill, New York, 1951.

ELSEVIER

Nuclear Instruments and Methods in Physics Research A 475 (2001) 287–295

**NUCLEAR
INSTRUMENTS
& METHODS
IN PHYSICS
RESEARCH**
Section A

www.elsevier.com/locate/nima

Progress in development of high-power FELs with two-dimensional Bragg resonators

N.S. Ginzburg[a], N.Yu. Peskov[a,*], A.S. Sergeev[a], A.V. Arzhannikov[b], S.L. Sinitsky[b], P.V. Kalinin[b], A.D.R. Phelps[c], A.W. Cross[c]

[a] *Institute of Applied Physics, Russian Academy of Sciences, 46 Uljanov St., Nizhny-Novgorod 603600, Russia*
[b] *Budker Institute of Nuclear Physics RAS, Novosibirsk 630090, Russia*
[c] *University of Strathclyde, Glasgow G40NG, UK*

Abstract

Theoretical and experimental studies of novel free-electron laser (FEL) schemes with 2-D Bragg resonators of planar and coaxial geometry are discussed. The highly selective properties of 2-D Bragg resonators are described including comparison with results of "cold" microwave measurements. First results of experimental studies of a planar W-band FEL based on the accelerator ELMI (INP RAS) are presented. Progress in a coaxial Ka-band FEL experiment under construction at the University of Strathclyde is described. A new project at the Budker Institute investigating a multi-beam FEL where 2-D distributed feedback is used to synchronise several FEL units of planar geometry is also discussed. © 2001 Elsevier Science B.V. All rights reserved.

PACS: 41.60.Bq; 42.50.Fx; 41.60.Cr; 52.75.Ms; 41.60.m

Keywords: FEL; Spatial-coherence; Distributed feedback; Bragg resonator; High-power microwaves

1. Introduction

In recent years, two-dimensional (2-D) distributed feedback, which can be realised in a 2-D Bragg resonator of planar and coaxial geometry, has been proposed [1–3] as a method to provide spatial-coherent radiation from a free-electron laser (FEL) when driven by a large size sheet or annular electron beam. Computer simulations demonstrated the possibility to achieve single-mode single-frequency operation in such an FEL when the transverse size of the interaction region exceeds the radiation wavelength by up to two or three orders of magnitude. The first operation of a planar FEL with the novel feedback mechanism has been observed [4] using the ELMI accelerator (INP RAS, Novosibirsk), which forms a small-scale sheet electron beam.

This paper is devoted to the progress in theoretical and experimental studies of novel FEL schemes with 2-D distributed feedback. The selective properties of 2-D Bragg resonators are described including comparison with the results of "cold" microwave measurements. Experimental results of a planar W-band FEL with 2-D distributed feedback carried out using the

*Corresponding author. Tel: +7-8312-384-575; fax: +7-8312-362-061.
E-mail address: peskov@appl.sci-nnov.ru (N.Y. Peskov).

0168-9002/01/$ - see front matter © 2001 Elsevier Science B.V. All rights reserved.
PII: S 0 1 6 8 - 9 0 0 2 (0 1) 0 1 6 1 8 - 7

ELMI-accelerator are presented. Further developments are discussed, including a coaxial FEL experiment, which is currently under construction at the University of Strathclyde (Glasgow, UK), as well as a multi-beam FEL project at the Budker INP RAS, where 2-D feedback would be used to synchronise several FEL units of planar geometry.

2. Eigenmodes of a 2-D planar Bragg resonator

A two-dimensional planar Bragg resonator (Fig. 1) consists of two metal plates, which are doubly corrugated so that the translational vectors of the corrugations have an angle φ to each other. Operation of a 2-D Bragg resonator is based on scattering of four perpendicular-propagating partial waves on this corrugation (see Fig. 1):

$$\vec{E} = \mathrm{Re}\left[\left(A_+\vec{E}_{a+}e^{-ih_{a+}z} + A_-\vec{E}_{a-}e^{ih_{a-}z}\right\}\right.$$
$$\left. + B_+\vec{E}_{b+}e^{-ih_{b+}x} + B_-\vec{E}_{b-}e^{ih_{b-}x}\right)e^{i\omega t}\right]. \quad (1)$$

In the general case the partial waves may possess different wavenumbers (h_j) and different transverse (over y-coordinate) structures $\left(\vec{E}_j\right)$, which coincide with the structures of the modes of a planar waveguide. One of the partial waves (let us assume A_+) propagates along the electron beam and interacts with the electrons. Thus, two other partial waves $B_\pm$ propagate in transverse directions and provide synchronisation of radiation from different parts of a wide sheet electron

beam. The backward wave A_- forms the feedback cycle.

A 2-D Bragg structure may be produced in several different ways and Fig. 2 shows three possible versions of the structures. In the "ideal" case the corrugation surface should be represented by a sine-function [1–3] (Fig. 2a)

$$a(x,z) = a_1\left(\cos\left(\bar{h}x - \bar{h}z\right) + \cos\left(\bar{h}x + \bar{h}z\right)\right). \quad (2a)$$

However, the simplest Bragg structure from the technological point of view is a set of rectangular grooves cut in two perpendicular ($\varphi = 90°$) directions (Fig. 2b). Such a corrugated surface may be presented as a multiplication of two periodic functions [5]:

$$a(x,z) = a_1 f(x + z)f(x - z) \quad (2b)$$

where

$$f(\xi) = \begin{cases} 0 & 0 < \xi < d_\xi/2 \\ 1 & d_\xi/2 < \xi < d_\xi \end{cases}$$

$f(\xi + d_\xi) = f(\xi)$, d_ξ is the corrugation period over the ξ-coordinate. For a small depth a_1 this corrugation may be approximated by the Fourier expansion:

$$a(x,z) = \frac{a_1}{\pi}\left(\cos\left(\bar{h}x - \bar{h}z\right) + \cos\left(\bar{h}x + \bar{h}z\right)\right.$$
$$\left. + \frac{2}{\pi}\cos(2\bar{h}x) + \frac{2}{\pi}\cos(2\bar{h}z)\right) \quad (3)$$

where $\bar{h} = \sqrt{2}\pi/d$, d is the corrugation period over the diagonals. As a result, this corrugation provides two feedback circuits. The first two terms in Eq. (3) are responsible for a 2-D coupling of

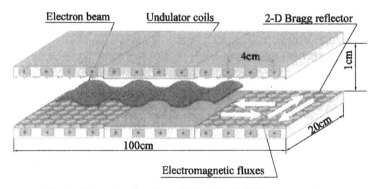

Fig. 1. Schematic of a planar FEL with a 2-D Bragg resonator.

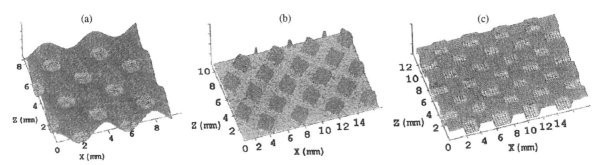

Fig. 2. Schematic of a 2-D Bragg grating with (a) sinusoidal, (b) "rectangular grooves" and (c) "chessboard pattern" corrugation profiles.

four partial waves (i.e. $A_+ \leftrightarrow B_\pm \leftrightarrow A_-$) which for the case $\varphi = 90°$ should have the same wavenumbers [1–3]:

$$h_{a\pm} = h_{b\pm} \approx \bar{h}. \qquad (4)$$

Simultaneously, the two last terms in Eq. (3) provide a traditional 1-D Bragg coupling (direct scattering $A_+ \leftrightarrow A_-$ and $B_+ \leftrightarrow B_-$) [6,7] when two counter propagating partial waves meet the resonance condition:

$$h_{a+} + h_{a-} \approx 2\bar{h} \quad h_{b+} + h_{b-} \approx 2\bar{h}. \qquad (5)$$

These waves may have different transverse structures and wavenumbers. Obviously, if the partial waves satisfy the resonance condition (4) then they will simultaneously satisfy the resonance condition (5). At the same time, from the fulfilment of resonance condition (5) it does not necessarily follow that condition (4) will also be fulfilled.

Scattering of the partial waves Eq. (1) in the condition of simultaneous Bragg resonances (4) and (5) was studied in Ref. [5]. It was shown that both resonators (an "ideal" resonator and a resonator with a corrugation profile given by Eq. (2b)) possess selectivity over both longitudinal and transverse mode indexes. At the same time, direct scattering of the forward and backward partial waves (i.e. $A_+ \leftrightarrow A_-$ and $B_+ \leftrightarrow B_-$) destroys symmetry of the both eigenmodes spectrum and integral reflection coefficient, as compared to the case of an "ideal" resonator (compare solid and dashed lines in Fig. 3a). Simultaneously, the spectrum is shifted to a lower frequency region (see Ref. [5] for details).

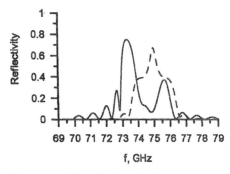

Fig. 3. Frequency dependencies of the integral reflection coefficient R for the 2-D Bragg structures with an "ideal" sinusoidal (dashed line) and a "rectangular grooves" (solid line) corrugation profile ($l_x = 20$ cm, $l_z = 18$ cm, $d = 0.282$ cm, $a_1 = 0.2$ mm).

The frequencies and Q-factors of the eigenmodes may be found when modeling its excitation by a plane wave entering the resonator through one of its edges. Then, the eigenmodes positions correspond to the minimums in the reflection coefficient. This modeling is useful for interpretation of "cold" microwave measurements of the resonator. Figs. 3 and 4 give the frequency dependence of the integral coefficients of reflection R for 2-D Bragg resonators with corrugations of different profiles. Parameters used in the simulations are close to the parameters of the resonators used in "hot" experiments (see Section. 4).

The "cold" microwave measurements of the reflection coefficients were done using a scalar network analyser. An additional quasi-optical transmission line was used for excitation of the

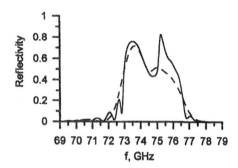

Fig. 4. Frequency dependencies of the integral reflection coefficient R for the resonator with the "chessboard pattern" corrugation profile: $l_x = 20$ cm, $d = 0.282$ cm, $a_1 = 0.3$ mm, $l_z = 18$ cm (solid line) and $l_z = 6$ cm (dashed line).

resonator. It was made from two parallel metal plates with a parabolic mirror placed between them. A standard W-band rectangular waveguide irradiated this mirror at its focus and, as a result, a wave-beam with a plane phase profile was formed at the transmission line output.

Results of "cold" microwave measurements of the frequency dependence of reflection from the 2-D Bragg mirrors used for operation of the first planar FEL based on 2-D feedback (presented in Ref. [4]) are given in Fig. 5a. Note that in the first experiments mirrors with the simplest corrugation profile shown in Fig. 2b were manufactured. In accordance with calculations three zones of effective reflection were observed in the frequency regions near 71–73, 75–76 and 77–78 GHz (see Fig. 5a). The zones near 71–73 and 75–76 GHz correspond to the mutual scattering of four partial TEM-type waves (compare with Fig. 3, solid line). In addition the effective Bragg reflection zone near the frequency of 78 GHz was also observed. This zone corresponded to the mutual scattering of the waves TEM $\leftrightarrow$ TE$_2$ under the resonance condition (5). In this frequency zone only the direct scattering of the forward (A_+) and backward (A_-) partial waves took place, which was corroborated in the experiment with the measurement of zero scattering of radiation in the transverse directions.

The presence of several reflection zones can be considered a drawback. In fact, in the first "hot" experiments [4] a FEL based on such a resonator

generated simultaneously several frequencies corresponding to different zones of reflection (see Section 4). To suppress such spurious modes of generation new mirrors with a so-called "chessboard pattern" profile (Fig. 2c) have been manufactured. This profile approximates much better the ideal sine-function (i.e. the harmonic amplitude in the Fourier expansion corresponding to 1-D coupling is much smaller than those given by Eq. (3)). As shown in Fig. 5b the zone of spurious reflection around 77–78 GHz for the "chessboard pattern" corrugation is practically zero. Fig. 4 presents the calculated value of the reflection coefficient for mirrors of different lengths and the admixture of "parasitic" 1-D scattering on the level of 30% (this level of 1-D coupling on the grating was estimated from the value of reflection near 77–78 GHz). Comparison of Figs. 4 and 5b demonstrates the good agreement between the calculated and measured data. Note, however, that the new gratings were made with a corrugation of depth of 0.3 mm, which was larger than in the previous experiments (where a corrugation of depth of 0.2 mm was used). This caused an increase in the wave coupling coefficient and resulted in a widening of the Bragg reflection zone (compare Fig. 5a and b).

Similar tests of a coaxial 2-D Bragg mirror having a "chessboard pattern" corrugation profile have been performed at the University of Strathclyde and also showed good agreement with the results of simulations [8].

3. Build-up of oscillations in a FEL with a 2-D Bragg resonator

Preliminary dynamics of a FEL with 2-D distributed feedback were studied theoretically. A schematic diagram of the FEL with a two-mirror planar 2-D Bragg resonator driven by a sheet electron beam is shown in Fig. 1. In the experiments performed such a resonator scheme was utilised and consisted of two 2-D Bragg reflectors of lengths $l_{Br1,2}$ separated by a regular waveguide of length l_0.

Time dependence of the efficiency for the range of parameters corresponding to the experiments

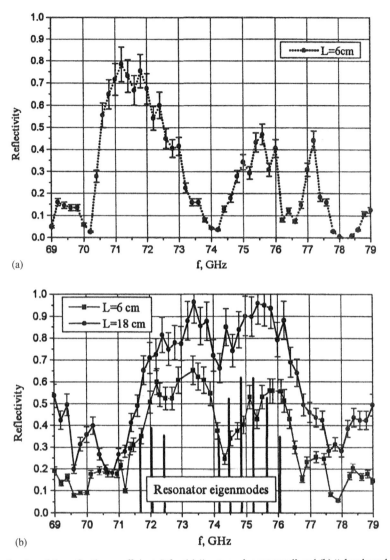

Fig. 5. Frequency dependencies of the reflection coefficient R for (a) "rectangular grooves" and (b) "chessboard pattern" corrugation profiles measured in "cold" microwave experiments.

are presented in Fig. 6. In the simulations a time domain analysis taking into consideration the dispersion properties of the Bragg reflectors was used [2,3]. As it was shown in the simulations a stationary regime of oscillation was established at the frequency of the precise Bragg resonance (determined by Eq. (4)), which corresponded to the frequency of the fundamental mode of the resonator (for parameters of the experiments using the ELMI-accelerator the Bragg frequency was

equal to 75 GHz). Profiles of the partial waves in the stationary regime are given in Fig. 7. It should be noted that the amplitudes of the partial waves $B_{\pm}$ are rather high inside the output (collector-side) mirror. This can lead to the development of RF breakdown and can therefore be considered a drawback of the two-mirror 2-D Bragg resonator scheme. To solve this problem at the next stage of the experiment with a planar FEL we intend to replace the 2-D collector-side mirror with a 1-D

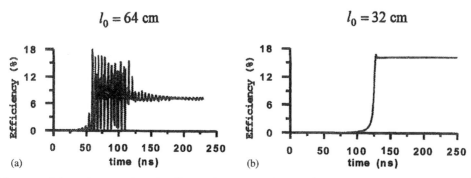

$l_0 = 64$ cm $l_0 = 32$ cm

(a)   time (ns)   (b)   time (ns)

Fig. 6. Establishment of the stationary regime of oscillations in a FEL-oscillator with two-mirror 2-D Bragg resonator for parameters close to the conditions of the experiments based on ELMI accelerator ($l_x = 20$ cm, $l_{Br1} = 18$ cm, $l_{Br2} = 6$ cm and (a) $l_0 = 64$ cm, (b) $l_0 = 32$ cm).

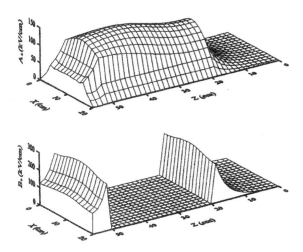

Fig. 7. Spatial structures of the partial wave amplitudes in the stationary regime of oscillations for parameters given in Fig. 6a.

mirror. Simulations of the FEL with this hybrid resonator consisting of a 2-D cathode-side mirror and a 1-D collector-side mirror showed stable single-mode operation [9].

4. Experiments with planar FEL

The experimental studies of the planar FEL with 2-D feedback were carried out using the ELMI-accelerator (see Ref. [4] for further details). The ribbon diode produced a beam with an electron energy of ~1 MeV, current of ~5 kA, pulse duration of ~5 μs and cross section of 1 cm × 20 cm. The undulator was located inside a slit vacuum channel and coils which produced a uniform longitudinal magnetic field (B_0) of up to 1.4 T. The undulator magnetic field (B_{und}) used for pumping the transverse electron velocity had a period of 4 cm and a maximum on axis value of 0.2 T. The total length of the undulator was ~1 m with an adiabatic entrance over the first 3 periods. After passing the resonator the beam was dumped on a large area fibrous graphite collector by using a pulse deflector.

Inside the resonator the sheet beam had a cross-section of 0.4 cm × 12 cm. The planar resonator with cross-section 0.85 cm × 20 cm consisted of two Bragg reflectors separated by a section of regular waveguide. The length of the cathode-side Bragg reflector was 18 cm, the collector-side 2-D Bragg reflectors had lengths of 10 and 6 cm for different series of experiments. The length of the regular section was varied from 64 to 32 cm to optimise the interaction.

To measure the absolute value of the microwave radiation power calibrated semiconductor "hot"-carrier detectors were used. For spectral analysis of the W-band radiation a set of detectors (traditional semiconductor diodes) equipped with band-pass filters were used. The radiation was received by these detectors inside a vacuum chamber at a distance of 1 m from the oscillator output. To obtain the average spectral density at a given frequency band the signals of the diodes were divided by the filter bandwidth.

In the first series of experiments the Bragg gratings had a "rectangular groove" profile (Fig. 2b) with a period of 2.82 mm and depth of 0.2 mm and were separated by a 64 cm long regular section [4]. The highest value of spectral density was registered in the 74–76 GHz frequency band that corresponded to the 2-D feedback operating mode. About 20% of output power was measured in the upper 76.5–78.5 GHz frequency band and was related to the influence of spurious 1-D feedback cycle as a consequence of the corrugation profile that was used (see Ref. [4] for details).

In the last series of the experiments 2-D gratings with the improved "chessboard pattern" profile were used. These gratings had a depth of 0.3 mm and a 4 mm period along the x- and z-coordinates (i.e. a 2.83 mm period over the diagonals). As discussed before, the 1-D Bragg scattering was essentially suppressed for this "chessboard pattern" profile. Also, the long regular section was replaced by a short section of length 32 cm and the length of output reflector was reduced to a minimum value of 6 cm [10]. The beam current for this series of experiments was ~ 1.5 kA and the diode voltage was varied from 0.86 to 0.96 MV. Fig. 8a shows the dependence of power on the undulator field. The different symbols used to show the experimental data refer to the shots in which generation of microwave pulses occurred at values of the electron energy within the values indicated above. The three solid lines represent the theoretical dependence of electron efficiency on the undulator field for three values of the electron energy: 0.8, 0.9 and 1.0 MeV, respectively, and are in good agreement with the experimental results. Both theoretical curves and experimental points exhibit similar behaviour at different voltages.

It is important to note that for the FEL with the new grating all the radiation spectrum was concentrated inside a frequency band near 75 GHz. Without this band the signals were practically zero (Fig. 8b) indicating that the generation of spurious 1-D feedback was suppressed. However the RF-pulse duration of ~ 100–200 ns

became much shorter than in previous experiments (cf. Ref. [4]). This may be explained by the fact that the depth of the corrugation was too large resulting in the initiation of RF breakdown.

The absolute power of the 4 mm-radiation was measured at the 100 MW-level for both series of experiments. As observed in the experiments this level of output power was limited by pulse shortening. The electron efficiency in the last experiments was estimated to be ~ 10%.

5. New developments

A new FEL experiment, which will employ two-dimensional distributed feedback, driven by an annular electron beam powered by a high-current accelerator at the University of Strathclyde, is currently in progress [8]. The experiment is based on a high-current explosive emission electron diode, which will generate a 250 ns duration electron beam pulse with energy of 0.5 MeV and current of ~ 1.2 kA. The average diameter of the annular electron beam will be 7 cm. Transportation of the electron beam will be achieved in a pulsed solenoid and an azimuthally symmetrical wiggler of 4 cm period. The cavity will be made from coaxial waveguide sections consisting of an 8 cm outer conductor and a 6 cm diameter inner conductor having a corrugation of period $d_z = 0.8$ cm with the number of variations over the azimuthal direction $\bar{m} = 35$ and depth of 0.04 cm to provide selective operation of the FEM at a frequency of ~ 38 GHz. Computer simulations show that in the two-mirror cavity of total length ~ 36 cm spatially-coherent single-mode oscillation with an output power of ~ 100–150 MW and an efficiency of about 15–20% may be realized. The aim of these experiments is to demonstrate operation of the novel 2-D feedback mechanism for an annular electron beam of oversized dimension $D_{beam}/\lambda \approx 10$ (see Ref. [8] for more details).

As demonstrated, 2-D distributed feedback provides the possibility for practically unlimited expansion of one of the transverse dimensions of the interaction space while maintaining spatially coherent radiation. Indeed at the U-2 accelerator

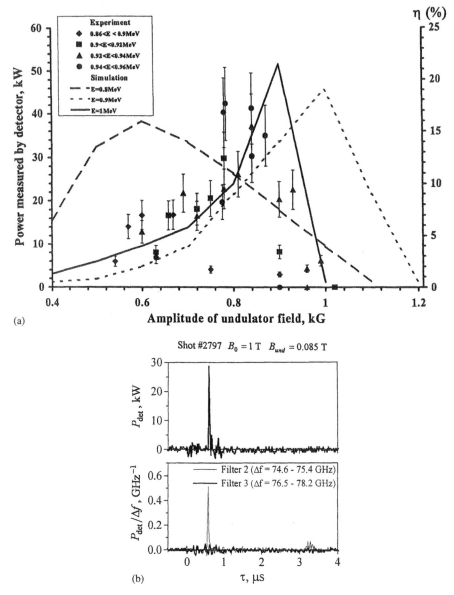

Fig. 8. (a) Dependence of the radiation power on the undulator field and (b) RF-power received through band-pass filters for a FEL with resonator having a "chessboard pattern" corrugation profile.

(Budker INP RAS, Novosibirsk) a microsecond relativistic sheet electron beam with an electron energy of 1 MeV, current per unit transverse size (linear current density) of 1 kA/cm and a transverse size of up to 140 cm was generated [11,12]. The power of this beam is tens of gigawatts and its total energy is up to 0.5 MJ. Based on the principle of 2-D distributed feedback it is possible for an FEL to generate an output power of up to 5 GW when driven by the full-scale sheet electron beam formed by the U-2 accelerator [3].

Another possibility is to expand the interaction space over both transverse coordinates. According to theoretical analysis [13] the use of 2-D distributed feedback makes it possible to design a multi-beam generator consisting of several planar FEL-units (modules) joined together and working simultaneously in parallel as a whole device. Each unit would be driven by a sheet electron beam moving in a planar undulator and would be based on a 2-D Bragg resonator. Such resonators would be coupled by a special waveguide system (Fig. 1. in Ref. [13]) and transverse (with respect to the electron beams) e.m. energy fluxes arising in these resonators would synchronize the whole generator and provide spatial coherence of the radiation. In fact, such a multi-beam oscillator may be considered analogous to a coaxial FEL utilising 2-D feedback. The design of a dual-beam FEL based on the ELMI accelerator is currently under progress.

Acknowledgements

The authors would like to thank Dr. P.V. Petrov (VNIITF, Snezhinsk, Russia), Dr. K. Ronald, Dr. W. He and Mr. I.V. Konoplev (University of Strathclyde, Glasgow, UK), Mr. N.V. Agarin, Mr. V.B. Bobylev, Mr. V.G. Ivanenko, Mr. S.A. Kuznetsov and Mr. V.D. Stepanov (INP RAS, Novosibirsk, Russia) for their active participation in this work. The authors would like to thank the Russian Foundation for Basic Research, ISTC, DERA and the UK Royal Society for partial support of this work.

References

[1] N.S. Ginzburg, N.Yu. Peskov, A.S. Sergeev, Opt. Commun. 96 (1993) 254.

[2] N.S. Ginzburg, N.Yu. Peskov, A.S. Sergeev, A.V. Arzhannikov, S.L. Sinitsky, Nucl. Instr. and Meth. A 358 (1995) 189.

[3] N.S. Ginzburg, N.Yu. Peskov, A.S. Sergeev, A.D.R. Phelps, I.V. Konoplev, G.R.M. Robb, A.W. Cross, A.V. Arzhannikov, S.L. Sinitsky, Phys. Rev. E 60 (1999) 935.

[4] A.V. Arzhannikov, N.V. Agarin, V.B. Bobylev, N.S. Ginzburg, V.G. Ivanenko, P.V. Kalinin, S.A. Kuznetsov, N.Yu. Peskov, A.S. Sergeev, S.L. Sinitsky, V.D. Stepanov, Nucl. Instr. and Meth. A 445 (2000) 222.

[5] N.Yu. Peskov, N.S. Ginzburg, G.G. Denisov, A.S. Sergeev, A.V. Arzhannikov, P.V. Kalinin, S.L. Sinitsky, V.D. Stepanov, Pis'ma v ZhTF 26 (8) (2000) 72 (in Russian);
N.Yu. Peskov, N.S. Ginzburg, G.G. Denisov, A.S. Sergeev, A.V. Arzhannikov, P.V. Kalinin, S.L. Sinitsky, V.D. Stepanov, Sov. Techn. Phys. Lett. 26 (4) (2000) 348;
N.Yu. Peskov, N.S. Ginzburg, G.G. Denisov, A.S. Sergeev, A.V. Arzhannikov, P.V. Kalinin, S.L. Sinitsky, V.D. Stepanov, Opt. Commun. 187 (2001) 311.

[6] N.F. Kovalev, I.M. Orlova, M.I. Petelin, Izv. VUZov: Radiofizika 11 (1968) 783 (in Russian);
N.F. Kovalev, I.M. Orlova, M.I. Petelin, Radiophys. Quantum Electron. 11 (1968) 449.

[7] V.L. Bratman, G.G. Denisov, N.S. Ginzburg, M.I. Petelin, IEEE J. Quantum. Electron. QE-19 (1983) 282.

[8] A.W. Cross, N.S. Ginzburg, W. He, I.V. Konoplev, N.Yu. Peskov, A.D.R. Phelps, G.R.M. Robb, K. Ronald, A.S. Sergeev, C.G. Whyte, Abstracts of the 22nd International FEL Conference, Durham, USA, 2000, p. 176.

[9] N.S. Ginzburg, N.Yu. Peskov, A.S. Sergeev, A.V. Arzhannikov, S.L. Sinitsky, Pis'ma v ZhTF 26 (16) (2000) 8 (in Russian).

[10] A.V. Arzhannikov, N.V. Agarin, V.B. Bobylev, A.V. Burdakov, V.S. Burmasov, E.V. Diankova, N.S. Ginzburg, V.G. Ivanenko, P.V. Kalinin, S.A. Kuznetsov, N.Yu. Peskov, P.V. Petrov, A.S. Sergeev, S.L. Sinitsky, V.D. Stepanov, Abstracts of International Conference "Euro Electromagnetics" (EUROEM-2000), Edinburgh, UK, 2000, p. 99.

[11] A.V. Arzhannikov, V.S. Nikolaev, S.L. Sinitsky, M.V. Yushkov, J. Appl. Phys. 72 (1992) 1657.

[12] A.V. Arzhannikov, V.B. Bobylev, V.S. Nikolaev, S.L. Sinitsky, A.V. Tarasov, Proceedings of the 10th International Conference On High-Power Particle Beams, San Diego, USA, 1994, Vol. 1, p. 136.

[13] N.S. Ginzburg, N.Yu. Peskov, A.S. Sergeev, A.V. Arzhannikov, S.L. Sinitsky, Techn. Tech. Phys. Lett., 27 (2001) 240;
Abstracts of the 22nd International FEL Conference, Durham, USA, 2000, p. 190.

ELSEVIER

Nuclear Instruments and Methods in Physics Research A 475 (2001) 296–302

NUCLEAR
INSTRUMENTS
& METHODS
IN PHYSICS
RESEARCH
Section A

www.elsevier.com/locate/nima

Long wavelength compact-FEL with controlled energy–phase correlation

A. Doria, V.B. Asgekar[1], D. Esposito[2], G.P. Gallerano*, E. Giovenale,
G. Messina, C. Ronsivalle

*ENEA, INN-FIS , C.R. Frascati, Via Enrico Fermi 45,
P.O. Box 65, 00044 Frascati, Italy*

Abstract

A radio-frequency modulated electron beam passing through a magnetic undulator generates coherent spontaneous emission (CSE) when the electron bunch length is comparable to the FEL resonant wavelength. At long wavelengths, CSE can be significantly enhanced by a proper ramping of the electron energy within the bunch duration, since this allows the contributions of the individual electrons to be added in phase. In this paper, the physical principles of the coherent spontaneous emission are reviewed together with its main experimental features. An accelerating structure, constructed at ENEA-Frascati for the systematic investigation of energy–phase correlation effects, is described together with an analysis of the expected performance. © 2001 Elsevier Science B.V. All rights reserved.

PACS: 41.60.Cr

Keywords: Free electron laser; Coherent emission; Longitudinal phase space

1. Introduction

A beam of relativistic electrons passing through a magnetic undulator emits electromagnetic radiation that is partially coherent and is called 'spontaneous emission'. When a radio-frequency (RF) accelerator is used as an electron source, the electron beam is delivered in the form of bunches. The electromagnetic radiation emitted by electrons

in the bunch adds up coherently at wavelengths of the order of the bunch size. Such a radiation is termed coherent spontaneous emission (CSE). The various methods employed to calculate the properties of the radiation [1–3] make use of the trajectories of the electrons in the undulator magnetic field. The phase space distribution of the electrons strongly affects the properties of the electromagnetic radiation. The conditions on the phase space distributions of the electrons in the transverse planes are well known and are expressed in terms of the beam quality requirements. It has been proved theoretically that the distribution of the electrons in the longitudinal phase space also plays a crucial role in determining the

---

*Corresponding author. Tel.: +39-06-9400-5223; fax: +39-06-9400-5334.

E-mail address: gallerano@frascati.enea.it (G.P. Gallerano).

[1] Department of Physics, University of Pune, 411 007 Pune, India.

[2] ENEA student.

coherence properties of the radiation [4]. An appropriate redistribution of electrons in the longitudinal phase space not only enhances the intensity of the fundamental component but also of its harmonics [5]. Experimentally, the redistribution of electrons in the longitudinal phase space can be achieved by passing the electrons delivered by an LINAC through a suitable RF structure [6].

In this paper, after a brief review of the main theoretical and experimental features of CSE, we describe the experiment being carried out at ENEA-Frascati to study the enhancement of the coherent radiation at millimeter and sub-millimeter wavelengths due to the longitudinal phase space manipulation of the electron beam.

2. Theoretical and experimental features of CSE

The electron beam produced by an RF accelerator has a periodic temporal structure. As a consequence, the transverse current due to the wiggling electron motion in the undulator carries a modulation at harmonics of the fundamental $\omega_l = 2\pi l/T_{RF}$, where l is a positive integer and T_{RF} is the RF period, with a relative amplitude that depends on the electron bunch shape [1]. If a rectangular waveguide of cross-section $\Sigma = ab$, placed within the undulator of length L, is considered, the transverse current acts as a localized source inside the waveguide and excites a number of waveguide modes. Utilising the physical model described in Ref. [1] it is possible to calculate the CSE field amplitude radiated in a given $TE_{0,n}$ waveguide mode by integrating the $\vec{J} * \vec{E}$ product, that represents the energy exchange rate between the electron beam current and the radiation field, over the interaction volume. Following the analysis reported in Ref. [4] it is possible to treat each electron bunch at the undulator entrance as an ensemble of N_e particles distributed in the longitudinal phase space (ψ, γ), each having energy γ_j and phase $\psi_j = \omega_{RF} t_j$ with respect to a reference charge injected at $t = 0$ with velocity β_{z0}. The radiated power at the harmonic ω_l is then calculated as a sum of the single electron contributions:

$$P_{l,0,n} = \frac{\beta_{gl}}{2Z_0}|A_{l,0,n}|^2$$

where

$$A_{l,0,n} = -\frac{Z_0}{\beta_{gl}}\frac{e}{T_{RF}}\frac{KL}{\sqrt{ab}}F\sum_{j=1}^{N_e}\frac{1}{\beta_j\gamma_j}\frac{\sin(\theta_j/2)}{\theta_j/2}\mathrm{i}e^{\mathrm{i}(\theta_j/2+l\psi_j)}$$

(1)

where $\theta_j(l) = (\omega_l/c\beta_{z,j} - k_u - k_{0,n})L$ is the usual definition of the phase shift parameter describing the FEL resonance condition in a waveguide, $Z_0 = 377\,\Omega$ is the free space impedance, β_g is the normalized group velocity of the waveguide mode at the frequency ω_l with wave vector $k_{0,n}$, K is the undulator parameter $K = eB_0\lambda L_u/(2\sqrt{2}\pi m_0 c^2)$, and F is a form factor describing the overlapping between the e-beam transverse distribution and the waveguide mode.

Expression (1) represents a multiple line spectrum with a radiated power at the harmonic frequency ω_l that is proportional to the square of the electron current and to the square of the undulator length. The frequency dependence of the radiated power has the usual behaviour of the undulator spontaneous emission. The width of the individual lines is related to the number of correlated bunches. If correlation between bunches is assumed to occur over the whole macropulse duration T_M, typically of the order of 1 μs, a relative linewidth $\Delta\upsilon_l/\upsilon_l \approx T_{RF}/T_M$ is expected. The quadratic dependence of the emitted power on the electron beam current and the "line-structure" of the CSE spectrum, shown in Eq. (1), were clearly observed in a number of experiments, and the limits on the relative width of the individual harmonics were also investigated in detail by means of a high resolution Fabry–Perot interferometer [7]. A recent analysis of the CSE spectrum showed that the linewidth of the individual harmonics can be as small as $\Delta\upsilon_l/\upsilon_l \approx 5\times10^{-4}$ (see Fig. 1).

Another interesting feature of CSE is the dependence of the phase of the radiated field on the electron drift velocity, which is implicit in $\theta_j(l)$, and on the phase of the electron at the undulator entrance, as it is shown by the factor $e^{\mathrm{i}(\theta_j/2+l\psi_j)}$ in Eq. (2). One observes that the emitted power $P_{l,0,n}$ is maximised when the single electron contribu-

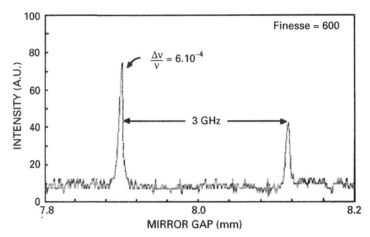

Fig. 1. Detail of a Fabry–Perot interferogram of CSE taken at a central wavelength of 2.6 mm. Two harmonics of the fundamental RF are shown in the figure. The FP finesse is $F = 600$, which corresponds to an instrumental resolution of 2.5×10^{-4}. The measured relative bandwidth is 6×10^{-4}, which, after deconvolution, implies a linewidth of about 5×10^{-4}.

tions in the bunching factor shown by the sum in the expansion coefficient $A_{l,0,n}$ interfere constructively with each other. This happens when the electrons are distributed in the longitudinal phase space (LPS) as close as possible to the "phase-matching" curve:

$$\psi = -\pi \frac{L}{cT_{RF}} \left(\frac{1}{\beta_z(\gamma)} - \frac{1}{\beta_{z_0}} \right). \tag{2}$$

A detailed analysis of this process has shown that the emission from a "correlated" distribution of electrons in the (ψ, γ) plane, satisfying the phase-matching condition (2), can be more than one order of magnitude higher than that from an "uncorrelated" distribution [6]. There is therefore an interest in investigating experimentally the emission behaviour as a function of the LPS distribution of the electron bunches. An experimental device has been designed and constructed to this purpose and will be described in the next sections.

3. Layout of the experiment and expected performance

A block diagram of the experiment is shown in Fig. 2. A pulsed triode gun, equipped with a 7.7 mm diameter osmium treated dispenser cathode, produces 10 µs long 1 A electron pulses at 13 kV. Following the gun, the beam is injected into an accelerating structure composed by two modules:

1. A β-graded on axis S band (2998 MHz) LINAC, with three full cells and two half end-cells, operating in the $\pi/2$ mode. This section accelerates electrons up to a kinetic energy of 1.8 MeV, with a macropulse current of 0.40 A.

2. A phase matching section (PMS) placed 4 cm downstream of the linac output. The drift space and the phase shift of the PMS are set so as to have the reference electron from the LINAC crossing the center of the PMS at a phase close to zero and with the proper slope of the electric field. The PMS is composed by three on axis coupled cavities (one full and two half cells) operating in the $\pi/2$ mode and tuned at the same frequency of the linac. This is accomplished by means of two motorized plungers, inserted at the end cavities (see Fig. 3) and driven in a closed loop. The measured tuning range of the PMS is shown in Fig. 4. A cooling system keeps both LINAC and PMS at the fixed temperature of 30°C ± 0.05°C.

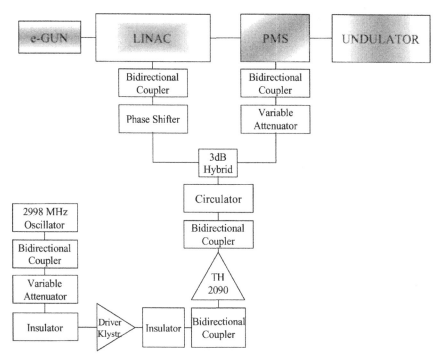

Fig. 2. Block diagram of the experiment.

Fig. 3. (a) Photograph of the LINAC (1) and PMS (2). (b) View of the frequency tuning plungers (3).

The emittance of the beam generated by the gun is matched to the acceptance of the LINAC by a solenoidal magnetic lens. Numerical simulations, performed using the E-GUN and PARMELA codes, show that the electron capture and the beam emittance are optimized at a field of about

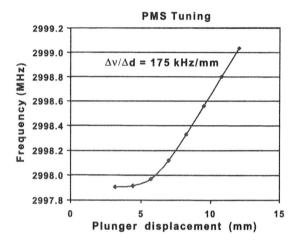

Fig. 4. PMS tuning curve.

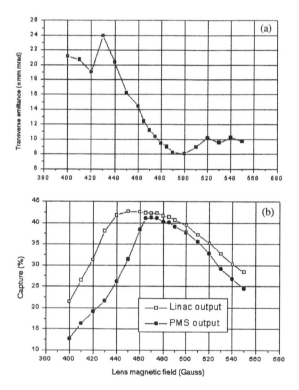

Fig. 5. (a) Beam emittance and (b) electron capture as a function of the lens magnetic field.

480 G in the magnetic lens (see Fig. 5). The distribution of the bunched electrons in the longitudinal phase space at the PMS output can be changed by varying the phase and the amplitude of

the RF field driving the PMS with respect to the linac by a variable phase shifter and a high power attenuator, which are part of the RF system [6], as shown in Fig. 2.

In the PMS electrons with energy higher than the reference particle are decelerated, whereas those with lower energy are accelerated to provide the required energy ramping. Due to the flexibility of operation of the RF system, in addition to changing the phase space distribution of the electrons, the PMS can also be used, if required, to further accelerate the beam to about 4 MeV.

Steering coils and a quadrupole triplet are placed between the PMS and the undulator to control the beam position and focus at the undulator entrance. The permanent magnet undulator is a 16 period long Halbach structure with 4 periods per block and 2.5 cm periodicity. It provides a magnetic field amplitude of 6 kG corresponding to $K = 1$. A rectangular waveguide WR42 with internal dimensions $10.668 \times 4.318\,\text{mm}^2$ is placed inside the undulator gap. A microwave horn at the end of the undulator collects the radiation which is then transported, by means of a copper light-pipe, to a Fabry–Perot interferometer for the spectral analysis.

A detailed parametric study has shown that the emitted power in the FEL resonance band, or at a specific harmonic frequency, can be optimized by a proper choice of the phase shift and field amplitude in PMS [6]. Numerical simulations of the beam dynamics were performed using the PARMELA code. Without entering the detail of these simulations, we recall here that a net increase in the CSE is expected when the electric field in the PMS is set to 50% of the electric field in the linac ($E_{\text{linac}} = 25\,\text{MV/m}$) and the relative phase is adjusted in order to have the proper correlation at the undulator input.

4. Electron beam phase space measurement

A suitable diagnostics is needed to verify the fulfilment of the phase matching condition. To this purpose, a system has been designed to measure the actual electron beam longitudinal phase space distribution. The chosen approach [8] makes use of a static magnetic field B_0 to induce a spatial energy dispersion in the plane perpendicular to its direction, and an RF electric field $E_0 \sin(2\pi\omega_{RF}t + \phi)$ perpendicular to B_0, to disperse electrons having a different longitudinal position in the bunch. In spite of the conceptual simplicity of this approach, there are a number of practical issues to consider. First of all, to reach an adequate spatial resolution of the measuring device proper values of the amplitude of the electric and magnetic fields and a proper value of the distance of the target screen from the region where the fields are applied have to be chosen (see Fig. 6). Secondly, the finite transverse size of the electron beam has to be kept much smaller than the spatial dispersion induced by the field. Finally the natural "diffraction" of the e-beam, governed by the transverse emittance, has also to be taken into account when designing the measuring device.

The measuring technique consists in taking an image of the e-beam on an $Al_2O_3 : Cr$ fluorescent target as shown in the block diagram of Fig. 6 with and without the applied fields. The measured

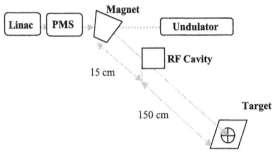

Fig. 6. Block diagram of the LPS diagnostic system.

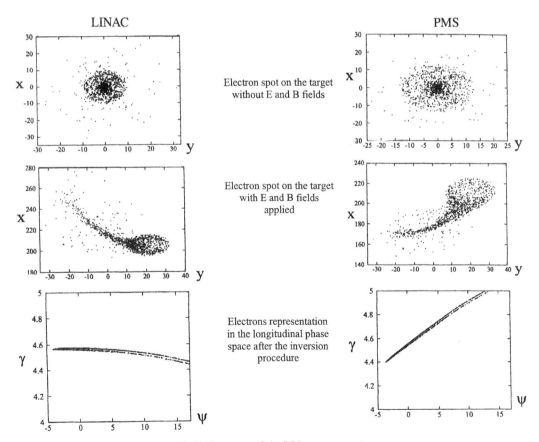

Fig. 7. Summary of the LPS measurements.

image intensity is proportional to the density of electrons impinging on the target. Because of the contribution of the finite size of the beam and of the emittance to the spot on the target, an inversion procedure, which will be described in more detail elsewhere, is then applied to extract the LPS information. Briefly, the inversion procedure makes use of the following equations:

$$v_{zn} = \frac{1}{\sqrt{1/c^2 + (m_0\Delta x_n)^2/(d_1 eB_0\Delta b)^2}} \qquad (3)$$

$$\frac{\Delta y_{yn}}{d}v_{zn} - e \cdot \frac{\int_{t_n}^{t_n+t_e} E_0\sin(2\pi\omega_{RF}t + \phi)\, dt}{m_0\gamma_n} = 0 \qquad (4)$$

where m_0 is the electron rest mass and t_e is the electron transit time through the RF field region.

Measuring the position Δx of the nth electron on the target and knowing the distance d_1 between the target and the constant magnetic field region of dimension Δb, the electron velocity v_{zn} is obtained from Eq. (3) and thus its energy γ_n. Solving Eq. (4) the time t_n of the nth electron in the microbunch is then obtained by measuring its position Δy on the target, placed at a distance d from the dispersing region, and knowing its velocity from Eq. (3). Applying an RF electric field of $3\,MV/m$ amplitude at $3\,GHz$, it is possible to achieve a temporal dispersion of $1\,ps/mm$ at the target with $d = 1.5\,m$. Analogously, an energy dispersion of $25\,keV/mm$ can be achieved with a static magnetic field of $4.5\,kG$ and a distance $d_1 = 1.65\,m$. The results are summarized in Fig. 7, where a simulation is shown using the output of a PARMELA code applied to the LINAC and PMS output.

5. Conclusions

The main features of coherent spontaneous emission by an RF modulated electron beam have been described in this paper. A device for the systematic investigation of the manipulation of the longitudinal phase space distribution has been constructed at ENEA-Frascati and is currently under test. It will provide a tool for the realization of high efficiency generators of coherent radiation in the mm-wave region.

Acknowledgements

One of the authors, V.B. Asgekar, undertook this work with the support of the "ICTP Programme for Training and Research in Italian Laboratories, Trieste, Italy".

References

[1] A. Doria, R. Bartolini, J. Feinstein, G.P. Gallerano, R.H. Pantell, IEEE J. Quantum Electron. 29 (1993) 1428 and references therein.

[2] D.A. Jaroszynski, R.J. Bakker, C.A.J. van der Geer, D. Oepts, P.W. van Amersfoort, Phys. Rev. Lett. 71 (1993) 3798.

[3] P. Kung, H. Lihn, H. Wiedemann, D. Bocek, Phys. Rev. Lett. 73 (1994) 967.

[4] A. Doria, G.P. Gallerano, E. Giovenale, S. Letardi, G. Messina, C. Ronsivalle, Phys. Rev. Lett. 80 (1998) 2841.

[5] G. Dattoli, L. Giannessi, P.L. Ottaviani, Nucl. Instr. and Meth. A 432 (1999) 501.

[6] E. Giovenale, A. Doria, G.P. Gallerano, S. Letardi, G. Messina, C. Ronsivalle, A. Vignati, Nucl. Instr. and Meth. A 437 (1999) 128.

[7] G.P. Gallerano, A. Doria, E. Giovenale, G. Messina, Nucl. Instr. and Meth. A 358 (1995) 78.

[8] D. Esposito, Thesis, La Sapienza—Electrical and Electronic Engineering Faculty, University of Rome, (2000).

ELSEVIER

Nuclear Instruments and Methods in Physics Research A 475 (2001) 303–307

NUCLEAR
INSTRUMENTS
& METHODS
IN PHYSICS
RESEARCH
Section A

www.elsevier.com/locate/nima

Enhancement of FEM radiation by prebunching of the e-beam (stimulated super-radiance)

M. Arbel[a],*, A.L. Eichenbaum[a,b], H. Kleinman[a], I.M. Yakover[a], A. Abramovich[b], Y. Pinhasi[b], Y. Luria[b], M. Tecimer[a], A. Gover[a]

[a] Tel-Aviv Department of Physical Electronics, Tel Aviv University, P.O.B. 39040, 69978 Tel Aviv, Israel
[b] Ariel - Department of Electrical and Electronic Engineering, The College of Judea and Samaria, 69978 Tel Aviv, Israel

Abstract

An electron beam (e-beam) prebunched at the synchronous FEM frequency and traversing through a waveguide, located coaxially with a magnetic undulator, emits coherent radiation at the bunching frequency. Introduction of both a premodulated e-beam and a radio-frequency (r.f.) signal at the same frequency at the input of the waveguide can lead to more efficient interaction, and thus more power can be extracted from the electron beam. In order to achieve this, the density modulation of the electron beam should be at an appropriate phase with respect to the r.f. signal.

We report a first experimental demonstration of the influence of the phase difference between the r.f. input signal and the fundamental component of the density modulation of the e-beam on the radiated power in a Free-Electron Maser (FEM). Our experimental system allows control of the current density modulation, of the r.f. input power level, in the undulator region and of the phase between that r.f. input and the modulation of the e-beam.

A comparison between measured radiation power with that predicted by theory for various phase differences, current density modulation, and r.f. signal levels, was made. Good correlation was obtained. © 2001 Elsevier Science B.V. All rights reserved.

1. Introduction

In the usual FEL operation, an electron beam interacts with an electromagnetic (EM) wave in an undulator. A bunching force resulting from FEL interaction density modulates the electron beam as it transits the undulator. In order to extract considerable energy from the e-beam, electron bunches must be located in the decelerating regions of the ponderomotive EM wave. The EM wave is thus amplified and coherent FEL radiation

is obtained [1]. It was shown that coherent radiation from an e-beam can also be obtained if the e-beam is modulated prior to its entrance to the interaction region, even when no EM wave is introduced at the input to the wave interaction region [2–7]. This radiated power is called "Super-radiance" (or "Prebunched Beam power"—PB). The radiation process can be substantially enhanced under certain conditions if a modulated e-beam and an EM wave (r.f. signal) are introduced simultaneously into the interaction region. This process is called "Stimulated Super-radiance" (or "Stimulated Prebunched Beam power"—SPB). In order to obtain maximal radiation power, the r.f. signal and the modulation of the e-beam must be

---

*Corresponding author. Tel.: +972-3-640-8148; fax: +972-3-642-3508.
E-mail address: arbel@eng.tau.ac.it (M. Arbel).

phase-matched [2–4]. The radiated power in a Free-Electron Maser (FEM) as a function of such a phase matching is reported in the following. As far as we know such an experimental demonstration has not been reported previously.

2. Analytical model

In the general case, the total radiated power is the sum of the three radiation processes described above [2,3]:

$$
\begin{aligned}
P(L) &= P_s |C(L)|^2 \\
&= P(0) F_{\mathrm{FEL}}(\bar{\theta}, \bar{\theta}_{\mathrm{pr}}) + P_B F_{\mathrm{PB}}(\bar{\theta}, \bar{\theta}_{\mathrm{pr}}) \\
&\quad + \sqrt{P(0) P_B} F_{\mathrm{SPB}}(\bar{\theta}, \bar{\theta}_{\mathrm{pr}}, \phi)
\end{aligned} \tag{1}
$$

where

P_s	the normalization power of the waveguide mode
$C(L)$	the amplitudes of the EM field at $z = L$
$P(0)$	the r.f. signal input power
$P_B = \dfrac{1}{32} \left(\dfrac{a_w}{\gamma\beta}\right)^2 \dfrac{Z_{\mathrm{mode}}}{A_{\mathrm{em}}} I_o^2 L^2$	e-beam power parameter
a_{w}	the wiggler parameter
$Z_{\mathrm{mode}}, A_{\mathrm{em}}$	Impedance and effective area of the EM mode, respectively
I_o, L	e-beam DC current and interaction length, respectively

The three detuning functions in Eq. (1) (assuming zero velocity modulation at the input wiggler) are defined as [2]

$$
F_{\mathrm{FEL}} = 1 + \frac{\bar{Q}}{2\bar{\theta}_{\mathrm{pr}}} \left\{ \mathrm{sinc}^2 \left(\frac{\bar{\theta} + \bar{\theta}_{\mathrm{pr}}}{2}\right) - \mathrm{sinc}^2 \left(\frac{\bar{\theta} - \bar{\theta}_{\mathrm{pr}}}{2}\right) \right\} \tag{2}
$$

$$
\begin{aligned}
F_{\mathrm{PB}}(\bar{\theta}, \bar{\theta}_{\mathrm{pr}}) &= M_j^2 \frac{\bar{\theta}^2}{(\bar{\theta}^2 - \bar{\theta}_{\mathrm{pr}}^2)^2} \\
&\times \left(1 + \cos^2 \bar{\theta}_{\mathrm{pr}} + \left(\frac{\bar{\theta}_{\mathrm{pr}}}{\bar{\theta}}\right)^2 \sin^2 \bar{\theta}_{\mathrm{pr}} \right. \\
&\quad \left. -2 \cos \bar{\theta} \cos \bar{\theta}_{\mathrm{pr}} - 2\bar{\theta}_{\mathrm{pr}} \sin \bar{\theta}_{\mathrm{pr}} \frac{\sin\bar{\theta}}{\bar{\theta}} \right)
\end{aligned} \tag{3}
$$

$$
\begin{aligned}
F_{\mathrm{SPB}}(\bar{\theta}, \bar{\theta}_{\mathrm{pr}}, \phi) &= M_j \left\{ \mathrm{sinc}\left(\frac{\bar{\theta} + \bar{\theta}_{\mathrm{pr}}}{2}\right) \right. \\
&\times \cos\left(\frac{\bar{\theta} + \bar{\theta}_{\mathrm{pr}}}{2} + \phi\right) + \mathrm{sinc}\left(\frac{\bar{\theta} - \bar{\theta}_{\mathrm{pr}}}{2}\right) \\
&\left. \times \cos\left(\frac{\bar{\theta} - \bar{\theta}_{\mathrm{pr}}}{2} + \phi\right) \right\}
\end{aligned} \tag{4}
$$

where

$\bar{\theta} = \left(\dfrac{\omega}{v_z} - k_z - k_{\mathrm{w}}\right) L$	normalized detuning parameter
$\bar{\theta}_{\mathrm{pr}} = \dfrac{\omega_p}{v_z} L$	normalized plasma frequency parameter
$\bar{Q}$	normalized gain parameter
$M_J e^{\mathrm{i}\phi} = \tilde{J}(0)/J_0$	current modulation index
ϕ	phase of the fundamental component of current density modulation with respect to the ponderomotive wave phase.

For a tenuous e-beam ($\bar{\theta}_{\mathrm{p}} \ll \pi$), the detuning functions are reduced to the following form [2,3]:

$$
F_{\mathrm{FEL}} = 1 + \bar{Q} \frac{\mathrm{d}}{\mathrm{d}\bar{\theta}} \mathrm{sinc}^2(\bar{\theta}/2) \tag{5}
$$

$$
F_{\mathrm{PB}} = M_j^2 \mathrm{sinc}^2(\bar{\theta}/2) \tag{6}
$$

$$
F_{\mathrm{SPB}} = 2M_J \mathrm{sinc}(\bar{\theta}/2)\cos(\bar{\theta}/2 + \phi). \tag{7}
$$

Eq. (5) expresses the FEL gain in the absence of prebunching. The prebunched FEM radiation (expressed by Eq. (6)) was investigated by us previously [5]. The SPB radiation given by Eq. (7) has a sinusoidal dependence on the phase

Table 1
Operational parameters of the compact FEM

Electron accelerator:	Pierce gun + Electrostatic accelerator
Electron beam energy and current	70 keV, 0.7 A
Electron beam prebuncher:	Traveling wave type
Prebuncher frequency band	$3\,\text{GHz} \leqslant f_m \leqslant 12\,\text{GHz}$
Prebuncher input power	$O \leqslant P_{\text{buncher}} \leqslant 2\,\text{W}$
Prebuncher current modulation at undulator input	$O \leqslant M_j \leqslant 0.26$
Wiggler type:	Rectangular
Magnetic induction	300 G
Length of period	4.44 cm
Number of periods	$N_w = 17$
Interaction length	$L_w = 85\,\text{cm}$
Waveguide type:	WR-187
Cross-section	2.215 cm × 4.755 cm
Mode	TE_{10}

difference ϕ between the current modulation and the injected r.f. signal.

For the set of parameters of our prebunched FEM (see Table 1) Eqs. (5)–(7) described the experimental scenario fairly well [5].

The total radiated power (Eq. (1)) is the sum of the three terms given by Eqs. (5)–(7). F_{FEL} and F_{PB} (Eqs. (5) and (6)) do not depend on phase ϕ. Since the SPB term (Eq. (7)) is sinusoidal vs. phase ϕ, the total radiated power (Eq. (1)) also varies sinusoidally with ϕ and has an average power determined by the amplified input signal (FEL power) and by the PB power (first two terms of Eq. (1)).

We investigated experimentally the influence of the relative phase between the fundamental frequency of current density modulation of the e-beam and the r.f. signal wave introduced to the waveguide input; i.e. we investigated the third term of Eq. (1) (the SPB power).

3. Experimental setup and measurements

The experimental demonstration was made with the aid of a table-top prebunched beam FEM developed at Tel-Aviv University. The use of prebunching in operation as an oscillator permits mode selection and single frequency operation [9].

Efficiency enhancement of the FEM oscillator was made possible by selection of an appropriate eigenfrequency as described in [10]. Possible uses as a frequency agile oscillator (on a pulse to pulse basis) were described in [11]. An experimental study of super-radiance both at the upper and at the lower synchronous frequency was reported in [5].

A schematic illustration of the experimental setup is shown in Fig. 1. The premodulated e-beam is derived from a traveling-wave-type prebuncher [8]. The fundamental component of the e-beam current modulation frequency is simply the input frequency to the traveling wave (TW) prebuncher. The prebunching modulation index "M_J" at the fundamental bunching frequency can be varied by the adjustment of the prebuncher r.f. input power (P_{bunch}).

The premodulated e-beam, derived from the prebuncher, traverses a rectangular waveguide located in the "wiggler" section. At 70 keV beam energy, the predicted upper synchronous frequency in the TE_{10} mode is about 4.9 GHz. The r.f. output port of the FEM is terminated in a matched load. A power divider placed at the output of the r.f. signal generator splits the source power into two separate channels. One channel is used to provide the r.f. signal wave into the waveguide. The other channel is used to provide

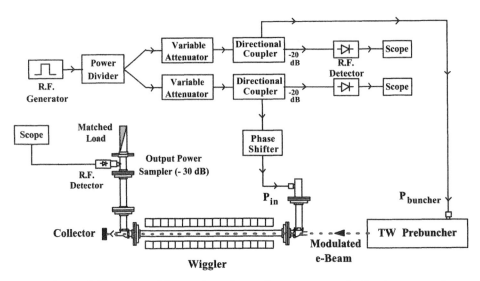

Fig. 1. Schematic illustration of the experimental setup for measurment of radiated power vs. phase.

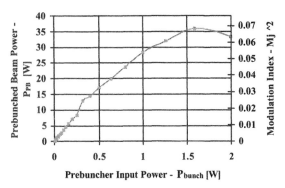

Fig. 2. Super-radiant (PB) power and current density modulation index M_J^2 vs. r.f. input power to TW prebuncher.

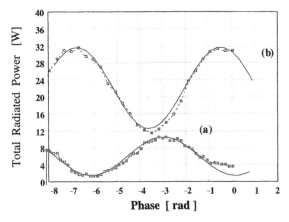

Fig. 3. Comparison of measured to calculated radiated power vs. phase (r.f. input power to waveguide – 1 W; prebuncher r.f. input power 0.25 W (curve a) and 0.8 W (curve b)).

r.f. input power to the prebuncher. A variable attenuator in each channel allows the adjustment and r.f. power setting at the inputs to the waveguide and to the TW prebuncher. A directional coupler in each channel allows the measurement of these power levels. The calibrated phase-shifter allows the change of relative phase between the r.f. signal wave and the modulated e-beam (at least over a range of 2π radians with a resolution of $0.36°$). The operational parameters of the compact FEM are shown in Table 1.

Fig. 2 shows that the PB radiated power is nearly proportional to prebuncher input power in

the linear regime ($P_{bunch} < 1$ W) as per Eq. (6). Therefore, M_J^2 evaluated from this curve is proportional to the prebuncher input power. Thus, by controlling P_{bunch} and the buncher input frequency we control "M_J" and the bunching frequency, respectively.

Fig. 3 shows measurements of the total radiated power vs. phase for an input r.f. power of 1 W to the waveguide and for two prebuncher input power levels: 0.25 W (curve a) and 0.8 W (curve b). For a constant input power $P(0) = 1$ W, the average radiated power is greater for $P_{bunch} = 0.8\,W$ ($M_J = 0.19$) as compared to $P_{bunch} = 0.25\,W$ ($M_J = 0.1$). For the higher M_J we note that the PB radiation power (proportional to M_J^2) is higher than the SPB radiation power (proportional to M_J). These experimental results agree with the predictions of the analytic model [2]. For other experimental r.f. input power levels, agreement with theory was also good.

4. Conclusion

We demonstrated experimentally the existence of a periodic variation of radiation power vs. phase difference between the EM wave introduced into the wiggler region and the phase of the fundamental current of a prebunched e-beam. The radiated power is maximized (for phase matching) and minimized (for antiphase conditions). The amplitude of the variable component of radiation power corresponds closely to that predicted theoretically. The maximum total radiated power also corresponds to theoretically predicted values and under optimal phase conditions, the FEM radiation power is enhanced considerably. The measured variation of radiated power as a function of phase difference, modulation index, and r.f.

input power into the wiggler region corresponds well to the theoretical predictions confirming that both theory and experimental tests are valid.

References

[1] J.M.J. Madey, Stimulated emission of Bremsstrahlung in periodic magnetic fields, J. App. Phys. 42 (1970) 1906.

[2] I. Schnitzer, A. Gover, The prebunched free-electron laser in various operating gain regimes, Nucl. Instr. and Meth. A 237 (1985) 124.

[3] M. Cohen, Ph.D. Thesis, Tel-Aviv University 1995.

[4] A. Doria, et al., Coherent emission and gain from a bunched electron beam, IEEE J. Quantum Electron. QE-29 (1993) 1428.

[5] M. Arbel, et al., Super-radiance in a prebunched beam free electron maser, Nucl. Instr. and Meth A. 445 (2000) 247.

[6] S. Mayhew, et al., A tunable pre-bunched CW-FEM, Nucl. Instr. and Meth. A 393 (1997) 356.

[7] Y. Pinhasi, Ph.D. Thesis, Tel-Aviv University 1995.

[8] A. Eichenbaum, Traveling wave prebunching of electron beams for Free Electron Masers, IEEE Trans. Plasma Sci. 27 (2) (1999) 568.

[9] M. Cohen, et al., Masing and single-mode locking in a free electron maser employing prebunched electron beam, Phys. Rev. Lett. 74 (1995) 3812.

[10] A. Abramovich, et al., Appl. Phys. Lett. 76 (2000).

[11] A. Eichenbaum, et al., A novel free-electron maser as a high power microwave source of sophisticated signals, 18th Convention of Electrical and Electronics Engineers in Israel, IEEE 1995, pp. 444/1–5, New York, NY.

ELSEVIER

Nuclear Instruments and Methods in Physics Research A 475 (2001) 308–312

NUCLEAR
INSTRUMENTS
& METHODS
IN PHYSICS
RESEARCH
Section A

www.elsevier.com/locate/nima

An experimental investigation into the evolution of the output of a free electron maser oscillator

C.C. Wright*, R.A. Stuart, J. Lucas, A.I. Al-Shamma'a

University of Liverpool, Department of Electrical Engineering and Electronics, Brownlow Hill, Liverpool L69 3GJ, UK

Abstract

We have constructed a Free Electron Maser (FEM) oscillator that can be operated in single mode in the X-band range (9–11 GHz). The electron gun has two anodes, the first being at ground while the second is held at $+60\,kV$. Electron beam current flows from the thermionic cathode when the cathode is pulsed with up to $-3\,kV$. With this arrangement we have found that it is possible to generate an electron beam current pulse of up to $50\,mA$ with extremely fast rise and fall times (45 ns) with less than $100\,V$ droop of the accelerating voltage over a $100\,\mu s$ pulse. As such this device can be used to make accurate measurements of the growth of the microwave output as a function of time. We have measured the exponential gain of a longitudinal mode up to saturation at a fixed accelerating voltage and found that the spread of times taken to reach a given power level can only be explained if it is assumed that there is an initial phase during which the cavity is filling with spontaneous radiation. For the low small signal gains achievable with this system, this phase lasts for about a microsecond. © 2001 Elsevier Science B.V. All rights reserved.

PACS: 41.60.Cr

Keywords: FEM; Noise; Gain; Growth

1. Introduction

There has been renewed interest recently in long pulse FELs with the construction and first operation of the devices at FOM, Israel and Korea for example [1–3]. These FELS suffer from the same problems that were highlighted by Elias and his co-workers [4], namely:

1. A statistical variation of output from pulse to pulse said to be due to the instability of the high voltage regulation system.

2. Multimode competition because of the high gain bandwidths employed. The time taken for one cavity mode to eventually dominate and suppress the others is claimed to be short by some theoreticians [5] and long compared to typical pulse duration of tens of microseconds by others [6].

3. The droop of the accelerating voltage due to the discharge of the high voltage terminal capacitance by the electron beam current allows the output to hop from one mode to another [7].

We have constructed a flexible, low voltage, low output, long pulse FEM with which it was hoped to model these problems of higher energy devices. The device has been described in a previous paper

---
*Corresponding author. Tel.: +44-151-794-4600; fax: +44-151-794-4540.

E-mail address: ccwright@liv.ac.uk (C.C. Wright).

0168-9002/01/$ - see front matter © 2001 Elsevier Science B.V. All rights reserved.
PII: S 0 1 6 8 - 9 0 0 2 (0 1) 0 1 6 1 5 - 1

Table 1
FEM operating parameters

Electron gun	
Cathode	−3 kV pulsed
First anode	Grounded
Second anode	50–70 kV
Beam current	50 mA max.
Undulator magnet	
Material	NdFeB
Period	1.9 cm
Number of periods	31
Length	58.9 cm
Gap	22–24 mm
Peak field strength	320 G
Waveguide cavity	
Guide type	WR75
Dimensions	$a = 1.905$ cm, $b = 0.9525$ cm
Length of cavity	71 cm
Longitudinal mode spacing	90–140 MHz over output frequency range 9–11 GHz
Small signal gain less round trip losses in guide	1–2%

[8]. A summary of our FEM operating parameters is given in Table 1. In this paper we report on measurements we have carried out to investigate point 1 above, the statistical variation in the output from shot to shot.

2. The growth of microwave output with time

When the accelerating voltage is adjusted so that only one of the longitudinal modes is amplified, then the power P in the cavity at time t after the start of the electron beam pulse would be expected to be given by

$$P = P_n \exp(gt/t_r) \qquad (1)$$

where t_r is the time taken for the power at the oscillating frequency to make one round trip of the cavity, g is the net small signal gain (electronic gain less round trip cavity loss) and P_n is the starting spontaneous noise power in the cavity.

Fig. 1 shows the growth of microwave power in the cavity to the saturated level for several shots of the FEM at ostensibly identical operating conditions. Even though the accelerating voltage and the beam current were highly stable, for example

the pulse to pulse variation of beam current was < 0.01%, there was a large spread of about ± 7% in the time taken for the microwave power to reach half the saturation level.

The slopes in the exponential gain region were measured and found to be the same to within ±1%. This confirmed that the net gain (g) was constant as expected from the constancy of the supplies. It was concluded that the spread in times to reach a given power level must be due to variations in the initial noise power in the cavity, P_n.

To investigate this effect further, measurements were made of the time taken to reach a level of 165 W (the point in the exponential growth region at which gave a convenient 10 mV output from our detector) for 1000 single shots of the FEM. A spacing of 15 s, approximately five times the RC recharging time constant of the second anode voltage, was allowed between shots to ensure constancy of operating conditions. From the times recorded, using a measured value of $g = 1.58\%$, a set of values for P_n were obtained using Eq. (1). Fig. 2 shows the distribution of the apparent noise power levels.

For the 50 mA electron beam current used, the shot noise generated spontaneous noise power is expected to be approximately 3 nW per round trip. Fig. 2 shows that the observed noise was much greater than expected. It has a roughly exponential distribution with a decay constant of 25 nW. It is well known that initially the cavity traps, and therefore fills up with, spontaneous radiation for a short time after the electron beam is switched on. It was considered that this process might explain our observations.

Suppose we consider the first section of electron beam to enter the cavity. Let its length (L) be such that the time it takes to pass through the cavity is (t_r), the radiation round trip time. This length of electron beam adds an amount of spontaneous power, P_{spon}, to the cavity. The next length L of electron beam might be considered to add another P_{spon} of power to the cavity and so on. The cavity power is initially increasing at a much greater rate than would be caused by our low level of electronic gain. This process of filling the cavity with spontaneous power will continue until the cavity power reaches a level where the increase due to the

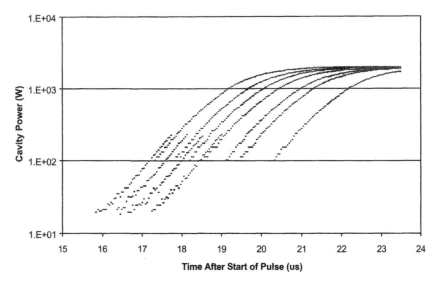

Fig. 1. A plot of the microwave power output for several electron beam pulses taken under identical operating conditions with microwave power plotted logarithmically (vertical axis). The short traces were taken at greater oscilloscope sensitivity to extend the exponential growth region for measurement of gradient.

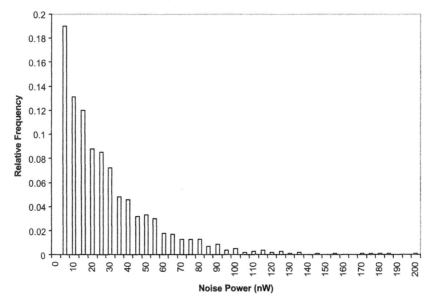

Fig. 2. A histogram showing the distribution of initial noise power calculated from 1000 measurements of the time taken to reach 165 W.

addition of another P_{spon} is masked by that due to electronic gain.

This transition will occur after n round trips, where n is given by

$$gnP_{spon} \approx P_{spon}$$

i.e. when

$$n \approx 1/g.$$

At this point the power will be essentially growing exponentially and the spontaneous noise contribu-

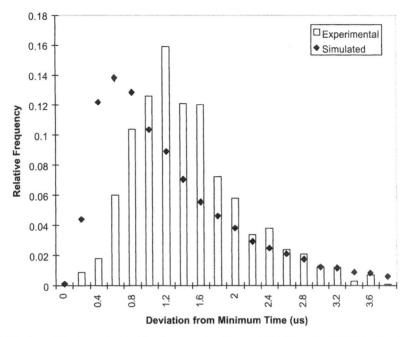

Fig. 3. Frequency distribution of the time calculated for the power level to reach $(1/g)P_{spon}$ for 20,000 simulated pulses, superimposed on the experimental data for 1000 measurements of the time taken to reach 165 W.

tion can be neglected. For the measured gain of 1.58%, a total of $n = 64$ round trips would be required to fill the cavity. The initial power from which microwave oscillations build up could then be considered to be 64 times larger than would be expected on the basis of a spontaneous power calculation, which would simply be P_{spon}.

The argument above would only apply in the unrealistic case that each length L of electron beam added the same power P_{spon} to the cavity. In reality these additions would be of random phase and have a spread of magnitude because of the linewidth of the spontaneous radiation.

Assuming, for simplicity, that each noise contribution has the same magnitude but has a random phase between 0 and 2π, we have calculated the distribution of times[1] at which the radiation should reach the level $(1/g)P_{spon}$.

Fig. 3 shows the spread of times obtained by these calculations superimposed on the histogram of experimentally measured times to reach 165 W.

The two distributions should be of the same width, being offset by the constant time taken for the power to grow from $(1/g)P_{spon}$ to 165 W. The theoretical and experimental results show good agreement in view of the simplifying assumptions made. The apparent skewing of the peak of the experimental data to the right of the peak of the experimental points may be due to the assumption of constant noise magnitude or due to the fact that even when the noise level has reached $(1/g)P_{spon}$, it is still possible for the cavity power to reduce if several successive noise contributions are in antiphase to the cavity radiation.

However, there seems to be little doubt that the results obtained indicate that the spread in the observed onset times is indeed due to the filling of the cavity with spontaneous radiation.

The average spontaneous noise level P_{spon} predicted by this model is 0.64 nW. Colson's formula [9] for P_{spon} with our FEM parameters gives a value of 3.0 nW, but this applies to a

---

[1] This simulation involves the repeated solution of the equation $1/g = \left| \sum_{k=1}^{n} \exp(j\theta_k) \right|^2$ for the number of round trips, n. θ_k is a random, uniformly distributed phase between 0 and 2π.

standard FEL rather than a waveguide system. Other calculations for the spontaneous noise in a waveguide [10] do not lend themselves to a simple calculation of P_{spon} but would be expected to give a value of the same order of magnitude.

3. Conclusions

The time taken for the microwave output of an FEM oscillator to build up has been measured over many electron beam pulses and it has been found that the spread in times is due to an initial phase during which the cavity gradually fills with spontaneous noise generated by the shot noise on the electron beam. The transition from this filling stage to the exponential gain regime occurs when the spontaneous noise trapped in the cavity reaches such a level that, on the next round trip, the power increase due to electronic gain exceeds the increase due to spontaneous noise on average.

References

[1] W.H. Urbanus, et al., Nucl. Instr. and Meth. A 331 (1993) 235.
[2] Abramovich, et al., Nucl. Instr. and Meth. A 407 (1998) 81.
[3] Byung Cheol Lee, et al., Nucl. Instr. and Meth. A 375 (1996) 28.
[4] A. Amir, et al., Appl. Phys. Lett. 47 (1985) 1251.
[5] I. Kimel, L.R. Elias, Nucl. Instr. and Meth. A 341 (1994) 191.
[6] T.M. Antonsen, B. Levush, Phys Fluids B 1 (1989) 1097.
[7] G. Ramian, Nucl. Instr. and Meth. A 318 (1992) 225.
[8] C.C. Wright, et al., Nucl. Instr. and Meth. A 445 (2000) 197.
[9] W. Colson, Ph.D.Thesis, Stanford University, 1977, p. 26.
[10] A. Amir, et al., Phys. Rev. A 32 (1985) 2864.

ELSEVIER

Nuclear Instruments and Methods in Physics Research A 475 (2001) 313–317

NUCLEAR
INSTRUMENTS
& METHODS
IN PHYSICS
RESEARCH
Section A

www.elsevier.com/locate/nima

The dependence of the oscillation frequency of an FEM on beam voltage

R.A. Stuart*, C.C. Wright, J. Lucas, A.I. Al-Shamma'a

Department of Electrical Engineering and Electronics, The University of Liverpool, Brownlow Hill, Liverpool L69 3GJ, UK

Abstract

A waveguide free electron maser oscillator operating in the X-band frequency range has been constructed. It has been observed that the frequencies of oscillation, although close to the longitudinal modes of the cavity, vary slightly with electron beam voltage. It is proposed that this effect is due to the voltage-dependent variation in the phase of the radiation caused by the interaction with the electron beam. An initial comparison has been made between the observed variation of frequency with voltage and simple calculations carried out using the small signal approximation. © 2001 Elsevier Science B.V. All rights reserved.

Keywords: Free-electron maser

1. Introduction

We have constructed a waveguide free electron maser (FEM) oscillator which operates at X-band frequencies [1]. Whilst 'priming' this device using an external source, to perform experiments similar to those carried out by McCurdy [2] on gyrotrons, it was observed that the operating frequency of the FEM could be changed slightly by adjusting the electron beam accelerating voltage. This was unexpected, as it had been supposed initially that the operating frequency would be the same as one of the 'cold' cavity longitudinal modes, which are only dependent on the dimensions of the resonant cavity, and not apparently on the parameters of the electron beam.

One possible explanation that was considered was the heating effect of the electron beam on the cavity. If the cavity temperature increased with increasing electron beam power level, the cavity would expand and the resonant frequency would decrease. In fact, we observed that increasing the electron beam voltage caused the observed oscillation frequency to increase.

Our results suggest that this dependence of operating frequency on accelerating voltage can be explained by assuming there is a change in the phase of the electromagnetic radiation in the cavity as it is amplified. Such a phase variation has been predicted by various authors [3,4], and measured for an amplifier configuration [5]. In our oscillator configuration, there is also a need to achieve constructive interference over one trip of the radiation round the cavity. This, combined with the phase variation, implies a shift in the oscillation frequency from the cold cavity value by

_____
*Corresponding author.
E-mail address: ee98@liv.ac.uk (R.A. Stuart).

an amount dependent on the gain, and consequently the operating voltage. We have found experimentally that the frequency shift is approximately linearly related to the change in accelerating voltage.

2. Theory

An electromagnetic mode propagating in an empty, lossless, waveguide must satisfy the dispersion relationship between its angular frequency, ω, and its wavenumber, k, given by

$$\omega = (\omega_{co} + c^2 k^2)^{1/2} \tag{1}$$

where ω_{co} is the cut off frequency of the mode and c is the velocity of light in vacuo.

If the guide is used to construct a resonant cavity, length L_c, then propagation is further restricted to a set of discrete frequencies of the mode, ω_m, satisfying

$$k_m = \frac{m\pi}{L_c} \quad (m \text{ positvie inetger}). \tag{2}$$

In a guide with losses, the wavenumber is complex, but the real part still satisfies Eqs. (2) and (1) to a good approximation for frequencies well above cut off.

We now assume that such a waveguide cavity is incorporated into a long pulse FEM operating as an oscillator. When the electron beam is turned on, spontaneous radiation generated as electrons pass through the undulator is trapped in the cavity. The electron beam amplifies the radiation, which is simultaneously filtered by the cavity, until eventually saturation is reached and microwave power at just one particular longitudinal cavity resonance remains at a frequency ω_p.

The above argument that the oscillator should operate at one of the cold cavity resonance frequencies ignores the fact that as well as amplification of the electromagnetic field, there is also a phase shift of the radiation relative to vacuum propagation. The interaction between the electrons and the mode changes the wavenumber of the radiation that co-propagates with the electron beam. The imaginary part, which controls the gain, is most affected but the real part, k_f,

which controls the wavelength, is also altered. The electron beam does not influence the wavenumber of the radiation that is travelling against the electron stream, and its real part, k_b, therefore would not be expected to change compared with the cold cavity value. The condition which replaces Eq. (2) is

$$k_f + k_b = \frac{2\pi m}{L_c} \tag{3}$$

since a total phase change of $2\pi m$ per round trip is required for constructive interference to take place. The right-hand side of this equation is a constant, so any change in the real part of the forward wavenumber, k_f, must be compensated by an equal and opposite change in reverse wavenumber, k_b. The only way in which k_b can change, however, is if there is a change in the resonant frequency through Eq. (1). If the real parts of the wavenumbers are only slightly different from their cold cavity resonance values, then

$$k_f = \frac{m\pi}{L_c} + \delta k \quad \text{and} \quad k_b = \frac{m\pi}{L_c} - \delta k. \tag{4}$$

The 'hot' resonance frequency can be approximated by

$$\omega = \omega_p - v_{gr} \, \delta k \tag{5}$$

where v_{gr} is the group velocity of the cold mode at $k = m\pi/L_c$.

It should be noted that the value of k_f used above is an average taken along the length of the cavity. In fact, as the forward wave propagates through the wiggler, the wavelength, and hence k_f changes continuously as the electrons become more bunched. The additional total phase change from the entrance of the wiggler to the exit has been calculated in the small signal, Compton regime [4] and simulated numerically for certain FELs operating in saturation [3,5]. If we assume this additional phase is Φ, then δk is given by

$$\delta k = \frac{\Phi}{L_c}. \tag{6}$$

Fig. 1 illustrates the variation of the additional phase as predicted by Brau [4] compared with the small signal gain for the Compton regime for the parameters of the Liverpool FEM.

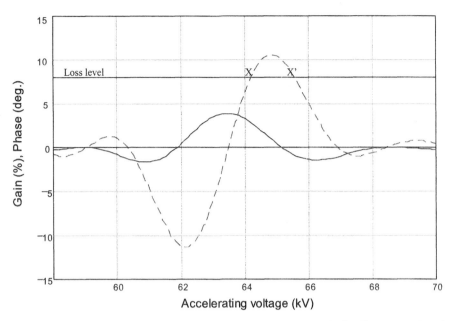

Fig. 1. FEM phase shift and small signal gain vs. accelerating voltage. The points X and X' indicate the region where gain exceeds losses of 8% and oscillation is possible.

The FEM will only oscillate in the range of electron energy where the small signal gain is greater than the total losses. In Fig. 1, such a region lies between X and X'. Within this range, where the gain has a maximum, the variation of the additional phase with accelerating voltage is approximately linear. This would be expected on the basis of the Kramers–Kronig relations. Consequently, we expect that the oscillation frequency, f, will also vary linearly with accelerating voltage, V, since differentiating Eq. (5) gives

$$\frac{df}{dV} = \frac{1}{2\pi}\frac{d\omega}{dV} = -\frac{1}{2\pi}\frac{v_{gr}}{L_c}\frac{d\Phi}{dV}. \qquad (7)$$

Note that df/dV is positive since $d\Phi/dV < 0$. For the parameters of the Liverpool FEM, Eq. (7) gives a rate of change of 1500 kHz/kV. However, this theory is only applicable in the small signal regime, whereas our measurements of beat frequency were made after saturation. When saturation occurs, simulations performed by Colson (see Fig. 6 in Ref. [3]) indicate that the variation should still be approximately linear but with a significantly reduced gradient.

3. Experiment and results

The FEM consists basically of an electron gun, an undulator/waveguide-cavity assembly at elevated potential and a collector system [1]. Low power microwave radiation can be extracted from the cavity via a horn, connected to the main cavity by a -36 dB cross coupler. The microwave power is collected in another horn that feeds a mixer diode detector. The purpose of the horns is to isolate the detector system from the high voltage cavity. The diode was also supplied microwave power by a Marconi 6200 test set. The diode mixes the test set output with the FEM output producing a difference beat signal, which can be monitored on a high speed digital storage oscilloscope. An example of the diode output observed for a single pulse is shown in Fig. 2. This clearly shows the beat signal superimposed on top of the saturated FEM output. Fig. 3 shows the variation of the beat frequency as the test set was adjusted around the FEM output frequency. As expected, the variation of the beat depends linearly on the test set frequency, with a slope magnitude measured

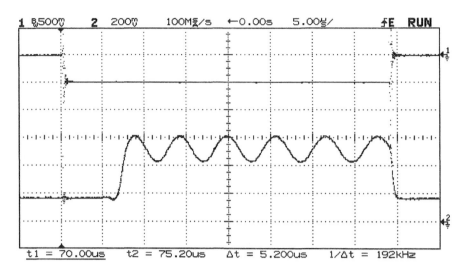

Fig. 2. Output of the mixer when electron gun is pulsed. Notice the beat wave superimposed on the microwave output. In this example the beat was at 192 kHz.

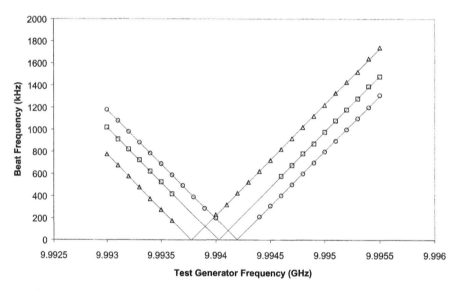

Fig. 3. The variation of the beat frequency with the frequency of the Marconi test generator for three values of electron beam voltage. △: 65.10 kV; □: 65.52 kV; ○: 65.74 kV.

experimentally at 1.000 ± 0.005. Since the beat has the same frequency whether the test set frequency is a constant amount above or below the operating frequency of the FEM, sweeping the test set was necessary to allow an unambiguous determination of the FEM frequency.

Sets of results are shown in Fig. 3 for three values of electron beam voltage. As can be seen

from the points where the beat frequencies are zero, the FEM operating frequency increases slightly with electron energy. To investigate this further, measurements were made over a wide range of applied voltage (see Fig. 4). It can be seen that the oscillator frequency changes linearly with accelerating voltage with a positive gradient of 660 kHz/kV. This result is in not unreasonable

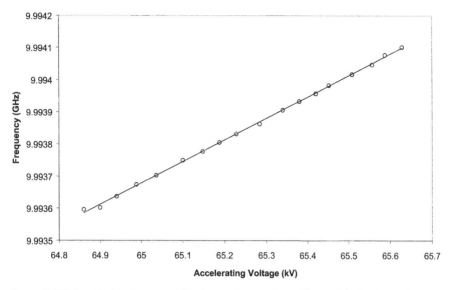

Fig. 4. The change in FEM oscillation frequency with electron beam voltage. The straight line has a slope of 660 kHz/kV.

agreement with the simple theory outlined earlier which predicted a positive slope of less than 1500 kHz/kV by an amount dependent on the degree of saturation of the FEM. The fact that experimental and theoretical values of the slope are of the same order is encouraging and it should be possible to achieve closer agreement in future by comparing these results with a non-linear theoretical model of our device.

4. Conclusions

We observe that the operating frequency of an FEM oscillator differs slightly from the expected cold cavity longitudinal resonance value by an amount which varies with electron energy. This effect has been ascribed to the change in total phase in the electromagnetic wave due to its interaction with the electron beam. In the small signal Compton regime, it is possible to predict the effect of this phase change, which is to make the FEM oscillator frequency increase linearly with electron beam voltage with a slope of 1500 kHz/kV

for the parameters of our device. As the oscillator saturates, simulations [6] suggest that the oscillator frequency should be less dependent on voltage. We have observed the voltage dependence in saturation for our FEM to be 660 kHz/kV which encourages us to believe that our explanation of the effect is correct. It is hoped to compare our results with a non-linear theory of oscillator evolution shortly in order to obtain better agreement.

References

[1] C.C. Wright, A.I. Al-Shamma'a, J. Lucas, R.A. Stuart, Nucl. Instr. and Meth. A 445 (2000) 197.
[2] A.H. McCurdy, C.M. Armstrong, IEEE Trans. Microwave Theory Technol. MTT-36 (1988) 891.
[3] W.B. Colson, IEEE J. Quantum Electron. QE-17 (1981) 1417.
[4] C.A. Brau, Free-Electron Lasers, Academic Press, New York, 1990.
[5] T.J. Orzechowski, E.T. Scharlemann, D.B. Hopkins, Phys. Rev. A 35 (1987) 2184.

ELSEVIER

Nuclear Instruments and Methods in Physics Research A 475 (2001) 318–322

NUCLEAR INSTRUMENTS & METHODS IN PHYSICS RESEARCH
Section A

www.elsevier.com/locate/nima

A metal-grating FEL experiment at the ENEA compact-FEL facility

A. Doria[a], G.P. Gallerano[a],*, E. Giovenale[a], G. Messina[a], V.B. Asgekar[b], G. Doucas[c], M.F. Kimmitt[d], J.H. Brownell[e], J.E. Walsh[e]

[a] *ENEA, Divisione Fisica Applicata, Via Enrico Fermi 45, 00044 Frascati, Italy*
[b] *Department of Physics, University of Pune, 411 007 Pune, India*
[c] *Particle Physics Laboratory, University of Oxford, Oxford OX1 3RH, UK*
[d] *Department of Physics, University of Essex, Colchester CO4 3SQ, UK*
[e] *Department of Physics & Astronomy, Dartmouth College, Hanover, NH 03755-3528, USA*

Abstract

We report the first results from an experiment aimed at a detailed study of the interaction of an electron beam with a metallic grating (the Smith–Purcell effect). The electron beam energy was 2.3 MeV (total) and the emitted far infrared radiation was observed at emission angles in the 35–45° range. The observed signal was about 3 orders of magnitude stronger than anticipated by incoherent emission theory. © 2001 Elsevier Science B.V. All rights reserved.

PACS: 41.60.Cr; 07.57.Hm

Keywords: Free electron laser; Electron beam; Metal grating; Smith–Purcell radiation

1. Physical principles

When an electron beam passes close to the surface of a grating, radiation is emitted because of the interaction of the beam with the periodic structure. The origin of this radiation can be understood in terms of image charges and, hence, currents induced on the surface of the grating by the passing electron and 'accelerated' by the periodic profile of the grating [1,2]. Although alternative approaches have been suggested [3,4], the surface current picture is easier to follow and

*Corresponding author. Tel.: + 39-06-9400-5223; fax: + 39-06-9400-5334.

E-mail address: gallerano@frascati.enea.it (G.P. Gallerano).

has received some experimental support, typically at energies of a few MeV [5].

The wavelength λ of the emitted radiation depends on the period l of the grating, the speed of the electron, expressed in terms of β, and also on the angle of observation θ relative to the beam direction, i.e.

$$\lambda = \frac{l}{n}\left(\frac{1}{\beta} - \cos\theta\right) \tag{1}$$

where n is the order of the radiation. This equation is well established experimentally. It is thus possible to select a wavelength by varying, for example, the angle of observation θ. It is also possible to select harmonics higher than 1.

Significant power can be radiated, especially in the case of highly relativistic beams. In order to calculate the radiated power, one starts with the formula for the energy emitted per unit solid angle $d\Omega$, per unit frequency $d\omega$ by a continuous distribution of charge [6]:

$$\frac{d^2 I}{d\omega\, d\Omega} = \frac{\omega^2}{4\pi^2 c^3}\left| \int_{-\infty}^{\infty} dt \int_{-\infty}^{\infty} \bar{n}\times(\bar{n}\times\bar{J})\exp\left[i\omega\left(t - \frac{\bar{n}\cdot\bar{r}}{c}\right)\right] dy\, dz \right|^2$$

(2)

where $\bar{n} = (\sin\theta\cos\phi, \sin\theta\sin\phi, \cos\theta)$ is the direction of observation, $\bar{r} = (x, y, z)$ is the position vector of the induced charge and $\bar{J} = (J_x, J_y, J_z)$ is the surface current density induced on the grating. It can then be shown that the spontaneous power dP emitted in a solid angle $d\Omega$, in a direction θ and an azimuthal angle ϕ, by a beam current I passing at a distance x_0 above a grating of length z is given by

$$\frac{dP}{d\Omega} = 2\pi e I \frac{Z}{l^2}\frac{\beta^3 n^2}{(1 - \beta\cos\theta)^3}\exp\left[-\frac{2x_0}{\lambda_e}\right] R^2.$$

(3a)

The quantity λ_e is called the 'evanescent wavelength' and is defined by

$$\lambda_e = \lambda\frac{\beta\gamma}{2\pi\sqrt{1 + \beta^2\gamma^2\sin^2\theta\sin^2\phi}}.$$

(4)

The term R^2 is a complicated expression that depends on the profile of the grating and the wavelength of the radiation. Although an analytic form can be written down for the case of a profile consisting of two straight facets only, it is best to evaluate R^2 numerically. If the beam is not a delta function at a height x_0 but has, say, a Gaussian distribution in the x direction, perpendicular to the grating, then the above expression becomes

$$\frac{dP}{d\Omega} = 2\pi e I \frac{Z}{l^2(1 - \beta\cos\theta)^3}\beta^3 n^2$$

$$\times R^2\frac{1}{\sqrt{2\pi}\sigma_x}\int_0^{\infty}\exp\left[-\frac{(x - x_0)^2}{2\sigma_x^2} - \frac{2x}{\lambda_e}\right] dx.$$

(3b)

The detailed analysis of the power distribution formula requires numerical methods. As a qualitative comment for relativistic beams, one would expect the radiation to be peaked in the forward

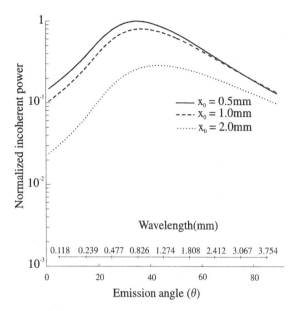

Fig. 1. Calculated spontaneous emission power as a function of the emission angle, for three different beam positions. The normalization factor is 1.3×10^{-6} W/sr (per cm of grating length, per 1 A of beam current). The grating period is 4 mm, its blaze angle 16°, the beam radius is 1.0 mm and its kinetic energy 1.8 MeV ($\gamma = 4.5$).

direction, but at an angle that can be adjusted to a certain degree by the selection of the grating parameters. The spontaneous power expected from our experimental parameters is shown in Fig. 1.

The above analysis is valid for a DC beam, in which case the individual electrons act as 'incoherent' oscillators and their contributions are simply added together. In the case of bunched beams it is possible to have 'coherent enhancement' of the radiation, especially in the case where the bunch length becomes comparable to the wavelength of the radiation. The coherent enhancement can increase the emitted power dramatically, up to a factor N_e, where N_e is the number of electrons in the bunch. The calculation of this enhancement requires knowledge of the distribution of electrons inside the bunch in the x, y and z directions. Its effects are critically dependent on the assumed longitudinal profile (z-direction) of the bunch.

As an example we report in Fig. 2 the calculations performed assuming that the bunch has a

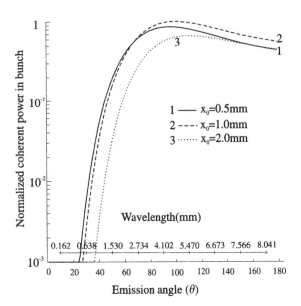

Fig. 2. Calculated coherent emission power as a function of emission angle, for three beam positions. The normalization factor is 28.6 W/sr (per cm of grating length, per bunch). See text for bunch details. Other parameters as in Fig. 1.

Table 1
Grating parameters for two different e-beam energies.

	l (mm)	θ_m (°)	θ_l (°)
10	2.5	28.5	14
4.6	4.0	32	16

at an electron beam energy of either 2.3 MeV (total) or 5 MeV and is capable of providing up to 4 A of peak current in the bunch. The pulse structure of the accelerator consists of 15 ps long bunches, spaced about 300 ps apart. The average current in the pulse train was, approximately, 200 mA. The normalized emittance of the beam is about 10π mm mrad.

Two gratings were constructed, each one optimized for one of the two possible machine energies. Their design parameters are listed in Table 1. The overall length and width of each grating are 100 and 20 mm, respectively.

In the experiments reported here only the 4 mm period grating was used. It was mounted in the multi-purpose interaction chamber of the facility (Fig. 3a), designed to host undulators, Cherenkov and grating FELs. The optical system (Fig. 3b) consists of three copper mirrors, held together in a rigid frame and rotating as a unit. It allows radiation to be collected at different emission angles θ, ranging from about 10° to 170°. Radiation is extracted from the side of the chamber and analysed by means of a Fabry–Perot interferometer, using an InSb detector cooled to liquid helium temperature.

profile that can be approximated by a 'double exponential', i.e. an exponential symmetrical around $z = 0$, of the form $\exp[-|z|/\sigma_z]$, this results in an enhancement in the region of $\lambda = 1$ mm (see Fig. 2). Comparison of Fig. 2 with Fig. 1 demonstrates clearly the effect of coherent enhancement: the emitted energy (or power) has increased dramatically and this is particularly obvious for the higher emission angles where the wavelength exceeds the pulse length. In plotting Fig. 2 we assume also that $\sigma_z = 0.889$ mm, so that for such an exponential distribution about 90% of the beam is contained within a pulse length of 14 ps. The ordinates of the graph express the output in terms of power in the bunch per steradian, per cm of grating length. We have also assumed 3.3×10^8 electrons/bunch.

2. Experimental arrangement

The experiment was carried out at the ENEA FEL Facility at Frascati [7]. The electron beam was provided by a Microtron, which can operate

3. Results and discussion

The emitted infrared radiation was put through a Fabry–Perot interferometer in order to determine its wavelength. The results of two scans are shown in Fig. 4a and b for emission angles $\theta = 45°$ and 36°, respectively. The emitted radiation peaks in the wavelength region between 1.2 and 1.6 mm for $\theta = 45°$ and at slightly shorter wavelength for $\theta = 36°$. In view of the uncertainty in the determination of the exact angle of the mirror assembly, the agreement with calculations (Fig. 1)

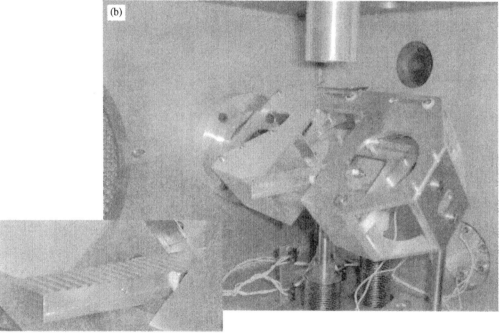

Fig. 3. (a) FEL layout: Microtron, transport channel, interaction chamber. (b) Grating and light-collection system, inside interaction chamber.

is reasonable. However, it was obvious from the start that the power levels observed were significantly higher (about 3 orders of magnitude) than those expected from consideration of the incoherent emission alone. This raised the possibility that there might be some coherent enhancement of the radiation.

In Fig. 5 we report the measured output power as a function of the height of the beam above the grating. In the same graph we also report the

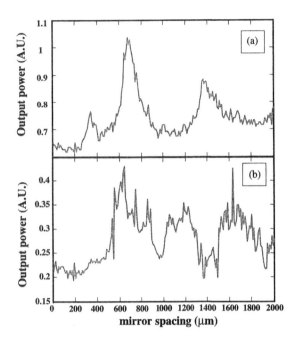

Fig. 4. Fabry–Perot interferogram of the grating output radiation taken at elevation angles 45° and 36°.

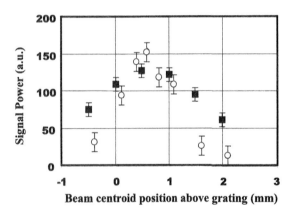

Fig. 5. Output power as a function of beam centroid position above grating. Circles: experimental results; Black squares: normalized theoretical predictions.

normalized theoretical behaviour. For calculations we have assumed that the beam has Gaussian distributions in the x and y directions with $\sigma_x = 1.41$ mm and $\sigma_y = 2.82$ mm. For the z direction, different bunch shapes result in a change of the relative amplitude between the experimental and theoretical curves, but do not affect the dependence of the normalized emitted power with

respect to the e-beam distance from the grating. We are currently investigating the emission patterns that arise out of different longitudinal beam profiles.

4. Summary and conclusions

Preliminary results obtained at the ENEA FEL Facility at Frascati indicate that the interaction of a 2.3 MeV (total) electron beam with a grating produces radiation in the far infrared that shows strong indications of coherent enhancement. The initial experimental results have been successfully compared to the predictions of a theoretical treatment based on the acceleration of charges induced on the grating surface. New tests are scheduled with different grating periods and at an electron beam of energy 5 MeV.

Acknowledgements

We are grateful to W. Linford at Oxford for the design and manufacture of the rotating mirror assembly. We are also grateful to A. Fastelli and R. Grossi for their assistance in setting up this experiment. Two of the authors (GD and MFK) would like to acknowledge the financial support of the Royal Society. One of the authors, V.B. Asgekar, undertook this work with the support of the "ICTP Programme for Training and Research in Italian Laboratories, Trieste, Italy".

References

[1] J.H. Brownell, J.E. Walsh, G. Doucas, Phys. Rev. E 57 (1998) 1075.
[2] S.R. Trotz, J.H. Brownell, J.E. Walsh, G. Doucas, Phys. Rev. E 61 (2000) 7057.
[3] O. Haeberl, P. Rullhusen, J.-M. Salom, N. Maene, Phys. Rev. E 49 (1994) 3340.
[4] A.P. Potylitsyn, Phys. Rev. Lett. A 238 (1998) 112.
[5] J.H. Brownell, G. Doucas, M.F. Kimmitt, J.H. Mulvey, M. Omori, J.E. Walsh, J. Phys. D: Appl. Phys. 30 (1997) 2478.
[6] J.D. Jackson, Classical Electrodynamics, Wiley, New York, 1975.(Chapter 4)
[7] G.P. Gallerano, A. Doria, E. Giovenale, A. Renieri, J. Infrared Phys. Technol. 40 (3) (1999) 161.

ELSEVIER

Nuclear Instruments and Methods in Physics Research A 475 (2001) 323–327

NUCLEAR
INSTRUMENTS
& METHODS
IN PHYSICS
RESEARCH
Section A

www.elsevier.com/locate/nima

Optimization of the design for the LCLS undulator line

E. Gluskin[a], N.A. Vinokurov[b], G. Decker[a], R.J. Dejus[a], P. Emma[c], P. Ilinski[a],
E.R. Moog[a,*], H.-D. Nuhn[c], I.B. Vasserman[a]

[a] *Advanced Photon Source, Argonne National Laboratory, XFD-401, 9700 S. Cass. Avenue, Argonne, IL 60439, USA*
[b] *Budker Institute of Nuclear Physics, 630090 Novosibirsk, Russian Federation*
[c] *Stanford Linear Accelerator Center, Stanford, CA 94309, USA*

Abstract

The Linac Coherent Light Source (LCLS) undulator line will consist of undulator segments separated by breaks of various lengths. Focusing quadrupoles, in a FODO lattice, and electron-beam diagnostics will be located in the breaks, and every third break will be longer to also accommodate photon diagnostics. The electron-beam beta function and the undulator period were selected to minimize the saturation length. The FEL simulation code RON has been used to optimize parameters such as the length of the undulators and the break lengths between undulators. Different break lengths after the first three undulators have been found to help reduce the overall undulator line saturation length. Tolerances for individual undulators have also been determined. © 2001 Elsevier Science B.V. All rights reserved.

PACS: 41.60. Cr

Keywords: Free-electron laser; SASE

1. Introduction

One possible way to create an X-ray free-electron laser (FEL) is with the self-amplified spontaneous emission (SASE) scheme. This scheme involves only the undulator and the electron beam propagating through it. The electron beam is unstable in that it bunches at the wavelength of the fundamental harmonic of the spontaneous undulator radiation. For small bunching, the dependence is linear, so the Fourier harmonics of the beam current at this frequency grow exponentially with distance traveled through the undulator. The gain length is the characteristic length where the squared magnitude of the fundamental Fourier harmonic increases by a factor of e. At the saturation length, the electron beam has become significantly bunched and there is no further growth. The coherent undulator radiation produced by the bunched beam is the FEL output. An advantage of this FEL scheme is the absence of mirrors, which are a serious problem for X-ray wavelengths. A disadvantage is that the radiation spectrum is relatively wide and the efficiency is low. The main problems in building such a device are obtaining a high-current low-emittance low-energy-spread electron beam to keep the saturation length reasonable (i.e., not much over 100 m) and meeting tight tolerances for field errors, misalignments and steering errors of the undulator.

---

*Corresponding author. Tel.: +1-630-252-5926; fax: +1-630-252-9303.

E-mail address: moog@aps.anl.gov (E.R. Moog).

The saturation length is typically about 20 gain lengths. For an FEL that is barely (or not) long enough to saturate, nearly all the output light comes from the end of the undulator line. Most of the line is devoted to bunching the electron beam by linearly amplifying the initial particle density fluctuations. Therefore, the goal in optimizing this part of the undulator line is to minimize the gain length.

2. Optimal period and focusing

The main parameters of the LCLS project [1] are listed in Table 1.

A challenging feature of the LCLS undulator is the high ratio of 4.5 between the beam emittance and the "minimum radiation emittance" for the light (i.e., wavelength divided by 4π). This means that only a small fraction of the particles will overlap with the light so as to be involved in the radiation amplification. Increasing the beam energy (for the same normalized emittance) would decrease the beam emittance. There are two limitations to increasing the energy, however: it would increase the energy spread due to quantum fluctuations of the radiation, and the linac energy for the LCLS is limited. This affects the possible choices for undulator period and beta function.

Table 1
Some parameters of the LCLS project

Radiation wavelength	0.15 nm
Beam energy	14.35 GeV
Normalized emittance	1.5 mm mrad
Beam peak current	3.4 kA
Energy spread (standard deviation)	3 MeV
Focusing	FODO
Undulator period	30 mm
Undulator parameter K	3.71
Undulator effective field	13.250 kG
Nominal magnetic gap	6 mm
Undulator length	3.36 m
Break length (short)	0.231 m
Break length (long)	0.463 m
Supercell length (6 undulators)	22.010 m
Number of undulators	33

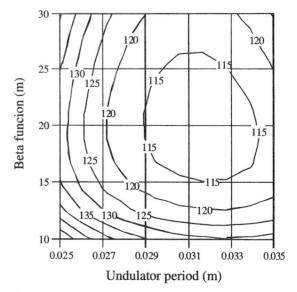

Fig. 1. Contours of constant saturation length. While the wavelength of the light is always 1.5 Å, the undulator magnetic field strength is assumed to change with the undulator period length following Halbach's formula [2]; the energy of the electron beam changes so as to preserve the wavelength of the light. The effect of the increase in the electron-beam energy-spread along the undulator line due to quantum fluctuations is included in the calculation. All numbers are in meters.

In Fig. 1, the dependence of the saturation length on the undulator period and matched beta function (lines of equal saturation length) for the planar permanent magnet undulator is shown. This dependence was obtained using the formulas of Halbach [2] and Ming Xie [3] and takes into account both the energy spread due to quantum excitation and the undulator "filling factor" (the fraction of the undulator line length occupied by undulators rather than by breaks between undulator sections). The wavelength of the output radiation is kept at 1.5 Å, but the magnetic field strength changes in accordance with Halbach's relation [2], and the electron beam energy must change as well to keep the wavelength constant. Fig. 1 shows that the design values of 0.03 m for the undulator period (corresponding to the 14.35 GeV energy) and 20 m for the beta function are close to optimal. For lower energy spread and emittance, the optimal undulator period decreases.

The use of a superconducting helical undulator was also considered. For a period of 0.24 m, a field of 1.3 T, and with other parameters the same as for the planar permanent magnet option, the saturation length is about 70 m. Although this saturation length is shorter than for a planar undulator, some as yet untested aspects to the mechanical design of a superconducting helical device remain. Since planar permanent magnet undulators are an established technology, they will be used for the LCLS project.

3. Irregularities and imperfections

The linear theory of high gain is well developed now (see, for example Ref. [4]). Nevertheless, the design of a real magnetic system for a short-wavelength high-gain FEL requires consideration of an inhomogeneous nonsymmetric magnetic system with separated focusing quadrupoles inserted into the breaks between undulator sections. Field, steering, and alignment errors must be considered. The linear time-independent code RON [5] was written for the optimization of such magnetic systems. It was used successfully in the design optimization of the Argonne FEL [6], which first used the separated-focusing approach and has tested and proved many features of the current LCLS design. This code has now also been used for the optimization of the LCLS undulator line.

The simplest way to provide proper focusing is to use a FODO lattice, and this choice has been made for the LCLS project. The magnetic system of the undulator line will consist of undulator sections with breaks between undulators where quadrupoles and beam position monitors will be installed. After every third undulator, the break will be longer so that photon diagnostics can be installed as well. This structure is geometrically similar to the existing APS FEL except that the photon diagnostics are only after every third undulator. Another lattice based on quadrupole triplets between undulators was considered and rejected because of very tight tolerances for the relative alignment of the three quadrupole centers.

The following parameter choices were made, based on the results of RON calculations:

1. The optimal undulator length was found to be near 3.4 m. For shorter lengths, the "filling factor" is less, increasing the effective gain length. (This assumes that the break length is kept at about 0.2 m, which is required by phasing conditions.) For longer undulator lengths, the gain length at a beam energy of 4.5 GeV[1] increases due to the variation of the beta function within the undulator. Longer lengths are also more difficult mechanically.
2. The optimal average value for the beta function was found to be 20 m. The focal lengths of the quadrupoles will be chosen accordingly.
3. The break lengths between undulators were optimized by calculating the corrections to the "resonance" break length due to the effect of finite emittance and diffraction.
4. An option that included magnetic bunchers between the undulator sections was considered and optimized. No significant improvement was found, so no magnetic bunchers are included in the undulator line design.
5. The effect of the residual quadrupole misalignment after simulated beam-based alignment [7] was calculated for the optimized undulator line. The increase of the saturation length was found to be about 10 m.
6. The effect of the spread of deflection parameters K in different undulator sections was simulated, to determine the corresponding tolerances.

4. Tolerances for the undulator section

The aim of optimization is to minimize the gain length and consequently the saturation length. There are tens of significant parameters in the system, and a deviation in any of these parameters will increase the gain length. A tolerance budget was worked out for the various parameters so that the overall gain length increase does not exceed 3%, which corresponds to a 4 m increase in

---

[1] The 4.5 GeV mode is planned for initial FEL commissioning, to produce a wavelength of 15 nm.

saturation length. Tolerances were set assuming simultaneous worst cases for all parameters. The overall tolerances for the undulator line were used to determine tolerances for a single undulator section.

The following requirements for the undulator section field errors were developed.

1. The trajectory walk-out from a straight line must not exceed 2 μm. The beam-based alignment technique will minimize deviations in the transverse beam coordinates near the beam position monitors (BPMs) between the undulator sections, so the trajectory walk-outs $x(z)$ and $y(z)$ with zero initial (at the upstream BPM) and final (at the downstream BPM) coordinates have to be specified:

$$x(z) = \frac{1}{\gamma} \int_0^z I_{1x}(z') \, dz', \quad y(z) = \frac{1}{\gamma} \int_0^z I_{1y}(z') \, dz'$$

where γ is the relativistic factor,

$$I_{1x}(z) = \frac{e}{mc^2} \left[\int_0^z B_x(z') \, dz' - \frac{1}{L} \int_0^L \int_0^{z'} B_x(z'') \, dz'' \, dz' \right],$$

$$I_{1y}(z) = \frac{e}{mc^2} \left[\int_0^z B_y(z') \, dz' - \frac{1}{L} \int_0^L \int_0^{z'} B_y(z'') \, dz'' \, dz' \right],$$

e and m are electron charge and mass, c is the velocity of light, B_x and B_y are the measured transverse components of magnetic field, and L is the cell length (the distance between BPMs). The 2 μm deviations in both the x and y directions gives an increase in the gain length of less than 0.2%, and can be achieved with present magnetic measurement and tuning techniques.

2. The reduction in spectral intensity of the zero-angle radiation must not exceed 4%. The spectral intensity of the zero-angle radiation is $e^2 k^2 |A|^2 / 2\pi c \gamma^2$, where k is the fundamental harmonic wavevector of the undulator radia-

tion, and

$$A = \int_0^L I_{1y}(z) e^{-ik/2\gamma^2 \left[z + \int_0^z I_{1x}^2(z') \, dz' + \int_0^z I_{1y}^2(z') \, dz' \right]} \, dz.$$

The "reduction" is as compared with an ideal undulator, but in practice the comparison can be with the best undulator, i.e., the one which gives the highest value of $|A|$. A 4% intensity reduction corresponds to an increase in the gain length by 1.1%.

3. The calculated particle phase deviation from the design value must be less than 10°. This phase is simply the particle-wave slippage

$$\varphi = \frac{k}{2\gamma^2} \left[L + \int_0^L I_{1x}^2(z) \, dz + \int_0^L I_{1y}^2(z) \, dz \right]$$

and the "design value" is an integer multiple of 2π. A 10° dephasing causes an increase in gain length of 1.7%.

4. The undulator median plane must be defined (and aligned) with an accuracy better than 50 μm vertically. If the beam is off-axis vertically by 50 μm, it will see a stronger undulator field, resulting in about 10° of additional phase slippage.

5. Magnetic and mechanical designs

The magnetic design of an undulator meeting the parameters of Table 1 has been completed. The undulators will rely on proven hybrid technology, using vanadium permendur poles and Nd–Fe–B permanent magnets. The grade of magnet material chosen has very high coercivity to increase the magnets' resistance to radiation-induced demagnetization. The poles and magnets will be rectangular rather than wedged, to help keep the mechanical design and fabrication of the magnetic structure simpler. These 30-mm-period undulators will be similar enough to the APS-standard 33-mm-period undulators that the tuning techniques developed for the latter should transfer directly. In fact, the tolerances presented in the previous section are already met by the undulators that were tuned magnetically for installation in the APS FEL.

Straightness of the trajectory is a significant requirement for the undulators, and a proper design for the ends of the undulators will help keep the trajectory straight. The sequence of pole strengths at the undulator ends will be 0.25, 0.75, 1. This gives an entrance into (or exit from) the undulator with no angle kick and no trajectory offset.

Proper phasing between undulators also demands proper tuning of the undulator ends. The magnetic phasing must match the physical distance between undulators. End phase tuning techniques were developed for the APS FEL that could tune the phasing by $\pm 38°$; these techniques will be applied to the LCLS undulators.

The mechanical design for the undulators is in progress.

Acknowledgements

This work is supported by the US Department of Energy, Office of Basic Energy Sciences, under Contract No. 31-109-Eng-38.

References

[1] LCLS Design Study Report, SLAC-R521 and UC-414, Rev. December 1998.
[2] K. Halbach, J. Phys. Collo. C1 44 (suppl.) (2) (1983) C1.
[3] M. Xie, IEEE Proceedings of the 1995 Particle Accelerator Conference, Dallas, TX, May 1–5, 1995, IEEE, 1996, p. 183.
[4] K.-J. Kim, M. Xie, Nucl. Instr. and Meth. A 331 (1993) 359.
[5] R. Dejus, O.A. Shevchenko, N.A. Vinokurov, Nucl. Instr. and Meth. A 429 (1999) 225.
[6] E.S. Gluskin, et al., Nucl. Instr. and Meth. A 429 (1999) 358.
[7] P. Emma, R. Carr, H.-D. Nuhn, Nucl. Instr. and Meth. A 429 (1999) 407.

ELSEVIER

Nuclear Instruments and Methods in Physics Research A 475 (2001) 328–333

NUCLEAR INSTRUMENTS & METHODS IN PHYSICS RESEARCH
Section A

www.elsevier.com/locate/nima

Optimization of an X-ray SASE-FEL

C. Pellegrini[a], S. Reiche[a,*], J. Rosenzweig[a], C. Schroeder[a], A. Varfolomeev[b],
S. Tolmachev[b], H.-D. Nuhn[c]

[a] *UCLA - Department of Physics & Astronomy, 405 Hilgard Ave., Los Angeles, CA 90095-1444, USA*
[b] *Russian Research Center, Kurchatov Institute, Moscow 123128, Russia*
[c] *Stanford Linear Accelerator Laboratory, Stanford, CA 94309, USA*

Abstract

The most important characteristics of an X-ray SASE-FEL are determined by the electron beam energy, transverse and longitudinal emittance, and by choice of the undulator period, field, and gap. Among them are the gain and saturation length, the amount and spectral characteristics of the spontaneous radiation, the wake fields due to the vacuum pipe. The spontaneous radiation intensity is very large in all X-ray SASE-FELs now being designed, and it contributes to the final electron beam energy spread, thus affecting the gain. It also produces a large background for the beam and radiation diagnostics instrumentation. The wake fields due to the resistivity and roughness of the beam pipe through the undulator, also affects the beam 6-dimensional phase space volume, and thus the gain and the line width. In this paper, we discuss ways to optimize the FEL when considering all these effects. In particular we consider and discuss the use of a hybrid iron-permanent magnet helical undulator to minimize some of these effects, and thus optimize the FEL design. © 2001 Elsevier Science B.V. All rights reserved.

PACS: 41.60.Cr

Keywords: FEL; Wake fields; X-rays; Helical undulator

1. Introduction

When moving towards shorter FEL wavelengths the demands on the electron beam parameters and undulator design becomes more severe. While the typical undulator length is longer, the tolerable emittance and energy spread are smaller. Therefore, the FEL performance is more sensitive to effects such as wake fields and spontaneous emission.

The emission of incoherent undulator radiation is the same for all electrons, and yields an energy loss, which can be compensated by tapering the undulator, and an increased energy spreads, which can reduce the gain. Wake fields effects change depending on the electron position within the bunch, and produce an energy loss from the entrance to the exit of the undulator which reduces the gain.

The wake fields effects can, however, be reduced by proper design of the vacuum chamber, increasing its diameter and reducing the resistivity and the roughness, and of the undulator, to reduce its length by maximizing the gain. The general

---

*Corresponding author. Tel.: +1-310-206-5584; fax: +1-310-825–8432.

E-mail address: reiche@stout.physics.ucla.edu (S. Reiche).

0168-9002/01/$ - see front matter © 2001 Elsevier Science B.V. All rights reserved.
PII: S 0 1 6 8 - 9 0 0 2 (0 1) 0 1 6 1 1 - 4

strategy to reduce wake fields effects is to increase the vacuum chamber diameter and to shorten the undulator length. Therefore, in this discussion we compare the planar LCLS undulator, considered as a reference, with helical undulators, in particular a novel design by the Kurchatov Institute for a helical undulator with very high magnetic field. With this design a shorter length can be achieved compared to planar undulator.

2. Undulator and beam parameters models

Both planar and helical undulators, have been successfully used for FELs in the past. In the models, presented in this paper, we consider helical undulators with large gaps, except for the LCLS reference case, which uses a planar hybrid undulator. The choice of high field helical undulator is motivated by the fact that with a new design helical undulators can give a short gain length, while the gap can be as large as in the planar case. With a larger gap the vacuum chamber size can be increased, thus reducing the impact of wake fields (see Section 4).

For the discussion we consider 4 cases

- a high field helical undulator (case A);
- a low field helical undulator (case B);
- a low charge, high field helical undulator (case C);
- the reference LCLS case, as presented in the CDR [1].

The exceptionally high field helical undulator is a permanent magnet system designed by the Kurchatov group. This undulator can provide a large field or a large gap (see Fig. 1). The choice has to be determined by considerations of wake field effects in the undulator vacuum pipe, and total undulator length. For this discussion a gap width of 8.5 mm is used providing the resonant wavelength at almost the same energy as the LCLS undulator.

In case B a different, conventional design for a helical undulator is used, providing a lower magnetic field and larger period length. This requires a lower electron beam energy to obtain the same resonance wavelength. The benefit of the

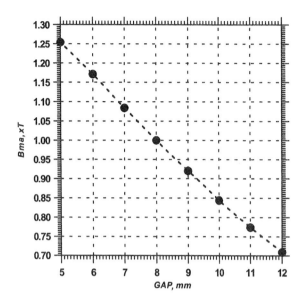

Fig. 1. On-axis field strength versus gap for the Kurchatov design of the helical undulator.

model is the rather simple design of the undulator. Beside combinations of permanent magnets and iron yokes the same field profile can be obtained by a double helix of current carrying copper embedded in an iron yoke. The required current does not exceed 2000 A, and the Ohmic losses can be cooled by liquid nitrogen.

Model C is a modified version of Model A, where the bunch charge and, thus, the bunch length and beam emittance are smaller [2]. The reduced emittance effects and the use of stronger focusing result in a shorter undulator.

The undulator and electron beam parameters for the four cases are presented in Table 1.

For completeness a fifth model could have been considered, using a planar undulator with larger gap while providing the same on-axis peak field as the LCLS case. The construction of this undulator was suggested by Kurchatov group also. The physics would have been same as for the LCLS case except that the wake fields are reduced to the level of the helical models. This impact is covered by Section 4.

Table 1
Parameters for LCLS (planar undulator) and alternative models based on helical undulators (cases A–C)

	LCLS	A	B	C
Undulator period (cm)	3	3	4	3
Undulator field (T)	1.3	0.96	0.48	0.96
Undulator (K)	3.7	2.7	1.8	2.7
Undulator gap (mm)	6.0	8.5	8.0	8.5
Focusing beta function (m)	18.0	17.7	20.5	5.0
Beam energy (GeV)	14.4	14.7	10.65	14.7
Total synchrotron radiation (GW)	90	50	11.6	10
Normalized emittance, mm mrad	1.1	1.1	1.1	0.3
Charge (nC)	0.95	0.95	0.95	0.2
Peak current (kA)	3.4	3.4	3.4	1.17
Relative energy spread at undulator entrance (10^{-5})	6	6	8	6
Resonant wavelength (Å)	1.5	1.5	1.5	1.5
FEL parameter ($\rho \times 10^4$)	5	6	6	10
Gain length (m)	4.2	2.8	4.2	1.84

3. Power and saturation length

For the simulation we used the 3D time-dependent FEL code GENESIS 1.3 [3], which has been benchmarked to various other FEL codes in the steady-state regime of an FEL [4]. To reduce the CPU time and to exclude many independent runs to exclude the fluctuation due to the SASE process all models have been simulated as an FEL amplifier. The initial power level has been estimated by the 1D theory [5], which is applicable for X-ray FELs because they are not dominated by diffraction effects.

The general performance of the different models is given by the solid lines in Fig. 2 (Section 4 discusses the wake fields effect, also shown in this plot).

For all models the saturation level is almost identical at 10 GW while the saturation length varies. Model C shows the best performance with a saturation length of 37 m due to the reduced emittance and stronger focusing. But even with the same electron beam parameters the performance of the high field undulator (case A) exceeds that of the LCLS undulator. The low field helical undulator has the longest saturation length. This is caused by the emittance effects which are stronger for a lower beam energy while keeping the undulator parameter the same [6].

In the conceptional design study of LCLS the ability to tune the resonant wavelength between 1.5 and 15 Å is an important feature for a wider range of experiments with a high brilliant X-ray beams. All models are able to fulfill this requirement by changing only the electron beam energy. At 15 Å the saturation length is below 30 m for all cases. The benefits of a reduced bunch charge and emittance are not as significant as for the 1.5 Å case, because the FEL performance is less affected by emittance effects at lower beam energy.

4. Spontaneous emission

The difference between a planar and a helical undulator becomes obvious when the spontaneous radiation is taken into account. Typically a planar undulator radiates at a higher power lever with a richer content of higher harmonics than a helical undulator. Since the FEL is driven by a lower beam energy, case B has the lowest radiation level considered in this discussion. In addition, spontaneous radiation of a helical undulator is always emitted at an angle with respect to the undulator axis, which makes it easier to separate the FEL radiation from the spontaneous emission.

Another effect for X-ray FELs is the fluctuation in the number of emitted photons in the high frequency part of spectrum of the radiation spectrum. It yields an increased energy spread of the electron beam [7], which cannot be compensated by field tapering as it is the case for the average energy loss due to the spontaneous emission.

The growth rate of the energy spread scales with the electron beam energy. Therefore case B benefits from the lower beam energy and the FEL is hardly affected. For the other cases the initial energy spread grows about 100% before it is dominated by the saturation of the FEL (see Fig. 3). Because the FEL dynamic is rather affected by the electron beam emittance than the energy spread the overall effect of the quantum fluctuation in the spontaneous emission is small. Even for the worst case the power level at saturation is degraded by less than 10%.

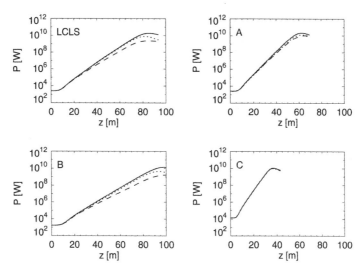

Fig. 2. Radiation power for the different cases including no wake fields, resistive wall wake fields and resistive wall + surface roughness wake fields (solid, dotted and dashed line, respectively).

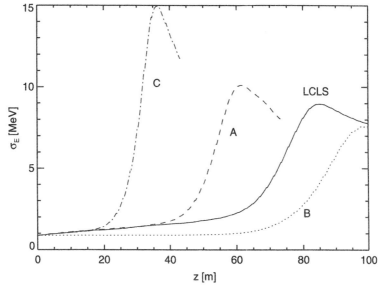

Fig. 3. Energy spread along the undulator including the effect due to the fluctuation of the emitted photons in the spectrum of the spontaneous emission.

5. Wake fields effects

Wake field effects within the undulator cannot be neglected due to the high peak current of the electron beam and the small diameter of the vacuum chamber, which enhance the wake field amplitude significantly. These effects have been the subject of much recent work [8–10]. The two main effects considered are those due to the resistivity and the imperfections of the vacuum pipe wall. A third source of wake field, produced by discontinuities in the vacuum chamber, is excluded in this discussion because an estimate of the wake field amplitude relies on an explicit design of the vacuum chamber including pumping ports, diagnostic sections and bellows.

Wake fields have longitudinal and transverse components. The latter are not taken into account because they are of higher order in the electron beam misplacement. Longitudinal wake fields cause a modulation of the electron beam energy along the bunch, and are commonly described by a wake potential.

For the resistive wall wake fields the wake potential [11] is

$$W_z(t) = -\frac{4ce^2Z_0}{\pi R^2} \left[\frac{1}{3}e^{t/\tau}\cos(\sqrt{3}t/\tau) - \frac{\sqrt{2}}{\pi}\int_0^\infty \mathrm{d}x \frac{x^2 e^{tx^2/\tau}}{x^6+8} \right] \quad (1)$$

where t is the longitudinal position of the test particle with respect to the particle generating the field, $\tau = [2R^2/Z_0\sigma c^3]^{1/3}$ is the characteristic scale of the wake potential, σ is the conductivity of the vacuum chamber, R is the chamber radius and $Z_0 \approx 377\,\Omega$ is the vacuum impedance.

For the effect of imperfection there are several models under consideration. A more conservative model [8] describes the surface roughness by a thin dielectric layer where the thickness δ is equivalent to the rms surface modulation.

The resulting wake potential is

$$W_z(t) = -\frac{ce^2Z_0}{\pi R^2}\cos(k_0 t) \quad (2)$$

with $k_0 \approx \sqrt{4/R\delta}$.

If the longitudinal characteristic length of the surface imperfection is much larger than the depth, the wake field amplitude is strongly reduced [9] and negligible compared to the resistive wake field.

Fig. 4 shows the wake potential for the LCLS parameter using a beam pipe radius of 2.5 mm. For larger radii, as it is the case for all helical undulators (A–C), the amplitude is noticeable reduced and the FEL performance is less degraded (Fig. 2). The wake fields are further reduced by using a low charge beam (case C), where the resulting effects are almost negligible. The benefit of a larger beam pipe is partially removed by a longer saturation length for case B because the energy modulation is accumulated over a longer distance.

The FEL amplification is less influenced in regions of the electron bunch where the total wake potential is zero. The average temporal radiation profiles for the LCLS case and case A are shown in

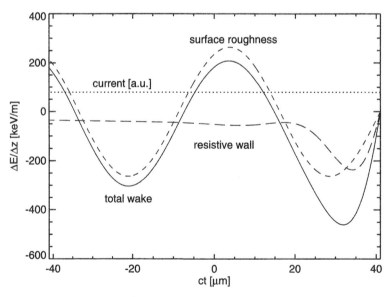

Fig. 4. Wake potentials for the LCLS undulator.

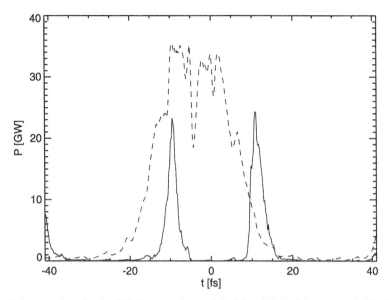

Fig. 5. Average radiation envelope for the LCLS case and case A (solid and dashed line, respectively) close to saturation.

Fig. 5. Again, the helical case exhibits superior performance regarding the effective length of the radiation pulse, which is approximately 40% for case A but 10% for LCLS, compared to the electron bunch length.

6. Conclusion

The impact of spontaneous emission and wake fields on the X-ray FELs gain has to be taken into account during the design and manufacturing phase of the undulator. To reduce the amplitude of wake fields a larger chamber size and a shorter undulator are desired. This can be fulfilled by a novel designs of helical undulators with a very high magnetic field, leading to a consideration of these undulator for an X-ray FEL, instead of planar undulators based on a more conventional design. An additional benefit of helical undulators is the easier separation between spontaneous emission and FEL radiation, thus facilitating the diagnostic for the system. This feature should not be underestimated during the commissioning phase of the FEL.

The energy loss by spontaneous emission can be compensated by field tapering and does not depend on the undulator type. Similarly, the increase of the energy spread is mainly determined by the beam energy, and puts a limit to the shortest wavelength obtainable for SASE-FELs.

In conclusion, our results show that the novel helical undulator design of the Kurchatov group, providing an unusual high magnetic field, has important advantages compared to planar undulators for an X-ray FEL.

References

[1] LCLS Design Study Report, Report SLAC-R-521, 1998.
[2] J.B. Rosenzweig, E. Colby, Proc. Conf. Adv. Accel. Conc. AIP 335 (1995) 724.
[3] S. Reiche, Nucl. Inst. and Meth. A 429 (1999) 243.
[4] S.G. Biedron, et al., Nucl. Inst. and Meth. A 445 (2000) 110.
[5] R. Bonifacio, et al., Phys. Rev. Lett. 73 (1994) 70.
[6] E.L. Saldin, et al., DESY Print, TESLA-FEL, Hamburg, DESY, 1995.
[7] E.L. Saldin, et al., Nucl. Inst. and Meth. A 381 (1996) 545.
[8] A. Novokhatski, A. Mosier, Proceedings of the PAC97 Conference, Vancouver, 1997.
[9] G. Stupakov, et al., Phys. Rev. Special Top. — Accel. Beams 2 (1999) 060 701.
[10] L. Palumbo, et al., Proceedings of the EPAC00 Conference, Vienna, 2000.
[11] K. Bane, SLAC Report AP-87, 1991.

ELSEVIER

Nuclear Instruments and Methods in Physics Research A 475 (2001) 334–338

NUCLEAR
INSTRUMENTS
& METHODS
IN PHYSICS
RESEARCH
Section A

www.elsevier.com/locate/nima

Wavelength spectrum of self-amplified spontaneous emission in the far-infrared region

R. Kato, T. Okita, R.A.V. Kumar, T. Igo, T. Konishi, M. Kuwahara,
M. Fujimoto, S. Mitani, S. Okuda, S. Suemine, G. Isoyama*

Institute of Scientific and Industrial Research, Osaka University, 8-1 Mihogaoka, Ibaraki, Osaka 567-0047, Japan

Abstract

We are conducting experiments to generate self-amplified spontaneous emission (SASE) in the far-infrared region and to measure its characteristics, using a single bunch electron beam accelerated with the L-band linac at ISIR, Osaka University. We have measured the wavelength spectra of SASE in the wavelength region between 80 and 180 μm using a grating monochromator and a Ge : Ga detector. The measured spectral widths of the fundamental peak are slightly larger than those predicted by the one-dimensional model. We have also observed the second harmonic peak of SASE. © 2001 Elsevier Science B.V. All rights reserved.

PACS: 41.60.Cr; 42.25.Kb; 29.17. + w; 07.57.Hm

Keywords: Free-electron laser(s); SASE; High gain; Far-infrared; RF linac

1. Introduction

The laser is one of the greatest inventions of this century. Its wavelength is, however, restricted within the region from near infrared to vacuum ultraviolet, since it depends on the gain medium and the mirror efficiency. On the other hand, the X-ray is a powerful tool for modern science and the high brightness SR rings known as "third generation light sources" are running as strong X-ray sources. A laser producing coherent X-rays is awaited as the fourth generation light source. Self-amplified spontaneous emission (SASE) using the

high-gain single-pass free-electron laser (FEL) configuration is one of promising approaches to realize the X-ray laser.

A few SASE-FEL projects are now in progress in the short wavelength region [1,2]. SASE-FELs are extensively studied in theory and simulation, while experiments in the short wavelength region have just started in the visible and ultraviolet region. Meanwhile several experiments have been conducted in the infrared region. We observed SASE in 1991 at wavelengths of 20 and 40 μm using a single bunch beam accelerated with the 38 MeV, L-band linac at the Institute of Scientific and Industrial Research (ISIR), Osaka University [3]. Recently, we began to study SASE-FEL again in the wavelength region from far infrared to submillimeter, and observed strong far infrared light emitted when the single-bunch electron

---

*Corresponding author. Tel.: + 81-6-6879-8485; fax: + 81-6-6879-8489.

E-mail address: isoyama@sanken.osaka-u.ac.jp (G. Isoyama).

beam passed through a planar wiggler. We concluded it to have originated from SASE, judging from the intensity variation with the K-value of the wiggler [4]. Then we increased the injection current up to 18 A from a gun cathode into the linac and have measured the wavelength spectra of SASE up to 180 μm. In this paper, we report the recent progress of the SASE-FEL study at ISIR, Osaka University.

2. Experimental

The L-band linac is equipped with a three-stage sub-harmonic buncher (SHB) system composed of two $\frac{1}{12}$ and one $\frac{1}{6}$ SHBs in order to produce an intense single-bunch beam with charge up to 91 nC/bunch. For the single-bunch operation mode, the electron beam with a peak current up to 28 A (typically 18 A in our experiments) and a duration of 5 ns is injected from a thermionic gun with a cathode area of 3 cm^2 (EIMAC, YU-156) into the SHB system. After being compressed to a single bunch, the electron beam is accelerated to 11–32 MeV with the 1.3 GHz accelerating tube. The electron beam is transported via the achromatic beam transport line to the FEL system. The wiggler is a 32 period planar wiggler with a period length of 60 mm. The K-value can be varied from 0.013 to 1.472. The main characteristics of the electron beam and the wiggler are listed in Table 1.

In SASE experiments, the upstream mirror for the optical resonator of the FEL was removed and the downstream mirror was replaced with a mirror of an appropriate focal length. Light emitted by the single bunch beam passing through the wiggler was reflected by the downstream mirror and led to a far-infrared monochromator in the measurement room via the evacuated optical transport line. The monochromator is a cross Czerny-Turner type with a plane reflective grating. The effective aperture ratio is $f/4.0$. Collimating and focusing mirrors are spherical mirrors with a focal length of 500 mm and have the square shape of 120 × 120 mm^2. The grating is grooved with 7.9 lines/mm (Milton Roy) and has a square shape of 64 × 64 mm^2. Its blaze wavelength is 112.5 μm. The monochromator can be used in the wavelength

Table 1
Main parameters of the electron beam and the wiggler

Electron beam	
Accelerating frequency	1.3 GHz
Energy	11.96 MeV
Energy spread (FWHM)	2.08%
Charge/bunch	15.5 nC
Bunch length	20–30 ps
Peak current	0.5–0.8 kA
Normalized emittance	150–200π mm mrad
Repetition	60 Hz
Mode	Single bunch
Wiggler	
Total length	1.92 m
Magnetic period	60 mm
No. of periods	32
Magnet gap	120–30 mm
Peak field	0.37 T
K-value	0.013–1.472

region from 60 to 190 μm. Spectral resolution is dependent on the measured wavelength and the slit width, and it is about 1.5 μm in this wavelength region for the slit width of 6 mm. The monochromator is evacuated with a rotary pump through the optical transport line. The monochromatized light is taken out through a 2 mm thick single-crystal-quartz window, and detected with a Ge:Ga photo conductive detector cooled with liquid helium, which is connected to the window with another vacuum pipe. The Ge:Ga detector has the highest sensitivity around 105 μm and the sensitivity drops steeply in the longer wavelength side and slowly in the shorter. Though the sensitivity at 140 μm drops down to approximately 10% of the peak value, a non-zero value still remains over 170 μm. Therefore we can measure the optical signal around 180 μm if the intensity of light is very high. The sensitivity on the short wavelength side is limited to 50 μm by the short wavelength cut-off filter in the detector.

3. Experimental results

We conducted SASE experiments in the wavelength region from far infrared to submillimeter, in which very high power gain of SASE is expected.

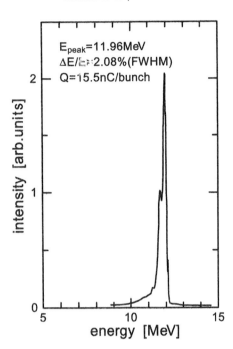

Fig. 1. The energy spectrum of the single bunch electron beam.

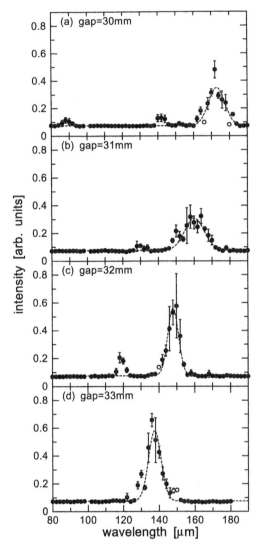

Fig. 2. Wavelength spectra of SASE emitted by the single bunch beam, measured for some wiggler gaps. The solid and the open circles denote the measured data points. The dashed lines were sums of a function for the main peak and a straight line for the background fitted to the data points denoted by the solid circles. See text for details.

The peak wavelength of SASE is given by $\lambda = \lambda_{\rm w}(1 + K^2)/2\gamma^2$, where $\lambda_{\rm w}$ is the period length of the wiggler magnet, γ the Lorentz factor and K the deflection parameter. The energy spectrum of the single bunch electron beam measured with a momentum analyzer magnet is shown in Fig. 1. In this experiment, the peak energy is 11.96 MeV and the energy width is 2.08% (FWHM). The average charge per bunch passing through the wiggler is 15.5 nC. By changing the wiggler gap from 30 to 33 mm, the K-value is decreased from 1.47 to 1.26. For the above energy, it is possible to change the peak wavelength from 173 to 141 μm with the fundamental peak. Fig. 2 shows wavelength spectra of SASE measured in these experimental conditions. The data points and the error bars show average values and standard deviations, respectively, of the five highest intensities in thirty successive optical pulses. These intensities are not corrected with detector sensitivity and grating efficiency, and therefore they are preliminary. Since the sensitivity and the efficiency have peaks at 105 and 112.5 μm, respectively, we thought that

a peak in the longer wavelength side had higher power.

4. Discussions

Two or three peaks can be seen in each spectrum in Fig. 2. The wavelength of the highest peak in each spectrum agrees quite well with the calculated wavelength using the electron energy and the K-value corresponding to the gap quoted in each panel. These peaks are, therefore, identified as fundamental peaks of SASE. The wavelength of the rightmost peak in Fig. 2(a) is 88 μm and it is precisely half the wavelength of the fundamental peak. The peak has been observed only for the larger K-value. It is theoretically predicted that the n-th harmonic of SASE grows more rapidly than the fundamental peak with a gain length equal to that of the fundamental SASE divided by the order of harmonic, n [5]. This peak disappeared when a short-wavelength cut filter of 100 μm was inserted between the monochromator and the detector, while the fundamental peak survived. These results exclude a possibility that it may be due to stray light of the fundamental peak. Thus the peak has been identified as the second harmonic of the fundamental SASE peak, which presumably originates from non-linear harmonic generation of SASE. The intensity of the second harmonic peak relative to the fundamental peak is roughly estimated to be 10^{-2}–10^{-3} using the intensity ratio in the spectrum together with the detector sensitivity and the monochromator efficiency.

The origin of the peak is not clear, appearing on the shorter wavelength side of the fundamental peak in Fig. 2. It can be seen in Fig. 2(a)–(c), but not in Fig. 2(d). Appearance of the peak and its intensity seem to depend on fine-tuning of the linac. Its wavelength is approximately 30 μm shorter than that of the fundamental peak in Fig. 2 and the intensity at the highest would be presumably two orders of magnitude weaker than that of the fundamental peak if efficiency of the detection system is taken into account. An impurity bunch in the electron beam, which has 10% higher energy than the main bunch, could make such a peak, if the intensity of the impurity bunch is strong enough to produce SASE. The intensity of SASE strongly depends on charge in a bunch; the electron gun provided a beam current of 18 A in the present measurement, while it is impossible to measure the wavelength spectrum with a current less than 16 A. The intensity of an impurity bunch should be comparable to that of the main bunch were it producing the satellite peak, but it is usually less than 1% of the main bunch, if it exists at all. We do not see any higher energy component in the energy spectrum of the electron beam shown in Fig. 1. Furthermore, there is a slit with a momentum acceptance of $\pm 1.5\%$ in the FEL beam line. All these facts exclude the possibility that the satellite peak may originate from an impurity bunch. It is not possible to exclude the possibility that the peak may be due to stray light in the monochromator, but it is not consistent with the fact that appearance and intensity are dependent on conditions other than the monochromator. Since this peak may be related to some physical process, we will continue to study it.

The spectral width of the fundamental peak was derived by the least square fit of a Gaussian function for the peak and straight line for the background to the measured spectrum as shown by the dashed lines in Fig. 2. Standard deviations are (a) 2.6%, (b) 3.4%, (c) 2.0% and (d) 2.5%, respectively, for the fundamental peaks shown in Fig. 2. The spectrum width predicted by the one-dimensional model is given by $\sigma_\omega/\omega = 0.91\sqrt{\rho/N}$, where ρ is the FEL parameter and N the number of periods [6]. The calculated spectrum width varies slightly with the wavelength from 2.2% to 2.3%. These values are slightly smaller than the average value of the measured spectral width 2.5%. If we take into account a contribution of the energy spread of the electron beam, 0.9% (r.m.s.), the theoretical values would come closer to the measured values. The measured values should also be compared with the spectral width of spontaneous emission calculated by $\sigma_\omega/\omega = 0.42 \times 1/N$ as 1.3% or the width calculated with the energy spread as 1.6%. The spectral width of SASE is larger than that of spontaneous emission, probably due to the short gain length in our experimental conditions.

5. Summary

We have measured the wavelength spectra of SASE in the wavelength region from far infrared to sub-millimeter. We observed the fundamental peak of SASE, and the second harmonic. The measured spectral widths of SASE were consistent with predictions of the one-dimensional model.

References

[1] The LCLS Design Study Group, SLAC-R-521, April 1998.

[2] J. Rossbach, Nucl. Instr. and Meth. A 375 (1996) 269.

[3] S. Okuda, J. Ohkuma, N. Kimura, Y. Honda, T. Okada, S. Takamuku, T. Yamamoto, K. Tsumori, Nucl. Instr. and Meth. A 331 (1993) 76.

[4] R. Kato, R.A.V. Kumar, T. Okita, S. Kondo, T. Igo, T. Konishi, S. Okuda, S. Suemine, G. Isoyama, Nucl. Instr. and Meth. A 445 (2000) 164.

[5] H.P. Freund, S.G. Biedron, S.V. Milton, IEEE Quantum Electron. QE-36 (2000) 275.

[6] K.J. Kim, Nucl. Instr. and Meth. A 250 (1986) 396.

ELSEVIER

Nuclear Instruments and Methods in Physics Research A 475 (2001) 339–342

NUCLEAR
INSTRUMENTS
& METHODS
IN PHYSICS
RESEARCH
Section A

www.elsevier.com/locate/nima

Initial gain measurements of an 800 nm SASE FEL, VISA

P. Frigola[a], A. Murokh[a], P. Musumeci[a], C. Pellegrini[a], S. Reiche[a], J. Rosenzweig[a],
A. Tremaine[a,*], M. Babzien[b], I. Ben-Zvi[b], E. Johnson[b], R. Malone[b],
G. Rakowsky[b], J. Skaritka[b], X.J. Wang[b], K.A. Van Bibber[c], L. Bertolini[c],
J.M. Hill[c], G.P. Le Sage[c], M. Libkind[c], A. Toor[c], R. Carr[d], M. Cornacchia[d],
L. Klaisner[d], H.-D. Nuhn[d], R. Ruland[d], D.C. Nguyen[e]

[a] UCLA, 405 Higard Avenue, Los Angeles, CA 90095, USA
[b] BNL, Upton, NY 11973, USA
[c] LLNL, Livermore, CA 94551, USA
[d] SLAC, Stanford, CA 94720, USA
[e] LANL, Los Alamos, NM 87545, USA

Abstract

The Visible to Infrared SASE Amplifier (VISA) FEL is designed to obtain high gain at a radiation wavelength of 800 nm. The FEL uses the high brightness electron beam of the Accelerator Test Facility (ATF), with energy of 72 MeV. VISA uses a novel, 4 m long, strong focusing undulator with a gap of 6 mm and a period of 1.8 cm. To obtain large gain the beam and undulator axis have to be aligned to better than 5 μm. Results from initial measurements on the alignment, gain, and spectrum will be presented and compared to theoretical calculations and simulations. © 2001 Elsevier Science B.V. All rights reserved.

PACS: 41.60.Cr; 41.60.Ap

Keywords: SASE; FEL; Undulator; High gain

1. Introduction

Current SASE FEL research is directed at developing shorter wavelength devices [1,2] and a current proposal [3] to generate high power, coherent X-rays at 1.5 Å has already received initial funding—Linac Coherent Light Source (LCLS). During the last few years, SASE FELs have demonstrated very high gain [2] and have agreed well with simulations in the far infrared. VISA, while R&D for the LCLS, will continue this study into the visible and harmonic UV wavelengths. In addition, VISA employs the use of a novel small gap 4 m long strong focusing undulator which not only makes the intra-undulator diagnostics challenging, but the alignment of the undulator to be very critical on FEL performance. These challenges are unique to VISA and an understanding of them is needed in order to build the next generation SASE FEL devices.

*Corresponding author. Tel.: +1-310-206-5584; fax: +1-310-206-1091.

E-mail address: tremaine@physics.ucla.edu (A. Tremaine).

0168-9002/01/$ - see front matter © 2001 Elsevier Science B.V. All rights reserved.
PII: S 0 1 6 8 - 9 0 0 2 (0 1) 0 1 6 0 9 - 6

Table 1
VISA experimental design parameters including output radiation, electron beam requirements, and undulator parameters

Peak power (saturation)	P_{Sat}	60 MW
Wavelength	λ_r	800 nm
Electron beam energy	E	71 MeV
Peak current	I_P	200 A
Normalized emittance	ε	2 mm mrad
Bunch length (FWHM)	τ	< 10 ps
Undulator period	λ_u	1.8 cm
On axis field	B	0.75 T
1-D FEL parameter	ρ	0.0085
Field gain length	L_g	0.35 m

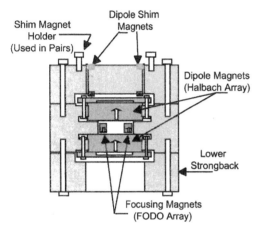

Fig. 1. Endview of the VISA undulator. Shown are the dipole magnets along with the quadrupole magnet array (FODO) which provides strong focusing. The electron beam comes out of the page.

2. Experimental setup

VISA has been commissioned and initial results have been obtained at the ATF [4] in BNL. The electron source is a 1.6 cell S-band photoinjector with an emittance compensating solenoid. A separate klystron powers two 3 m traveling wave linac sections after which the electron beam has an energy greater than 70 MeV. Added to the VISA beamline is a matching section that includes a focusing quadrupole triplet, beam position and profile monitors. This setup provides the high quality electron beam matching conditions (Table 1) necessary to achieve high gain.

The VISA undulator [5] is a strong focusing 4 m planar permanent magnet Halbach array design;

the parameters are given in Table 1 and an endview of the undulator is shown in Fig. 1. Dipole magnets with vertically oriented fields are put on either side of the gap providing strong quadrupole focusing [6]. These quadrupoles are arranged along the undulator in an FODO lattice design with a period of 24.75 cm giving four FODO cells per 1 m undulator section and an average β-function of 30 cm. According to simulations [7], implementation of the strong focusing decreases the gain length by up to 40% compared to a natural focusing scheme.

Simulations [7] have also shown that the electron beam and co-propagating SASE FEL radiation must remain within an rms beam diameter, 60 mm, of each other throughout the undulator in order for VISA to achieve high gain. This puts very rigid requirements on the measurements and alignment of the undulator. We have set the requirement on aligning the four 1 m sections to be within 30 μm. To do this, a laser interferometric alignment system [8] has been developed in which these strict alignment tolerances can be attained.

Eight steering magnets and eight diagnostic pop-in ports are evenly spaced (by 50 cm) down the length of the undulator. The steering might be necessary to help propagate the electron beam through the 4 m and prevent electron beam walk-off. Each actuated diagnostic port serves to make electron beam measurements using OTR and also to extract FEL light [9]. This setup gives VISA the ability to make accurate electron beam profile and SASE radiation measurements vs. distance along the undulator.

3. Experimental measurements

In order to study gain, the gun phase was set to compress the electron beam to 2 ps FWHM, which gave a peak current of 130 A and a beam emittance of about 3.5 mm mrad. It will be discussed further below that the undulator sections were not aligned to the specified tolerances given above and for a strong focusing system like VISA, a > 100 μm misalignment is very detrimental to SASE FEL performance [7].

A factory calibrated Molectron Joulemeter collects the undulator radiation and is placed

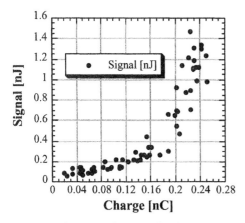

Fig. 2. Detector signal vs. charge at the down stream Joule-meter about 2 m past the exit of the undulator. The graph is highly non-linear indicating gain, which we calculate to be ~110.

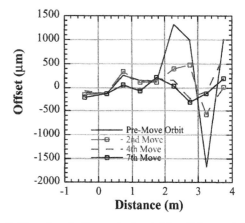

Fig. 3. Beam-based alignment of undulator. Before undulator is moved, the electron orbit has 3 mm amplitude. Trajectory model fitting and multiple movements of undulator sections reduced trajectory amplitude to 500 μm. $z = 0$ is the undulator entrance.

about 2 m from the exit of the undulator with an acceptance angle of 10 mrad. Charge measurements are made by a faraday cup located at the end of the beamline. Once SASE signal was seen, the intra-undulator corrector magnets, gun focusing solenoid, and matching sections quadruples were all used to peak the signal. Fig. 2 plots detector energy vs. charge. The plot is very non-linear indicating gain. Spontaneous radiation is linear with charge and for our experimental setup we calculate 0.68 nJ of expected spontaneous energy at 240 pC. Extrapolating the low charge linear portion of Fig. 3 (0–0.16 nC) which is purely spontaneous radiation, one can see we measured about 0.5 nJ of spontaneous energy at 240 pC. Thus, the signal detected above the 0.5 nJ at 240 pC is SASE and at 240 pC a peak SASE signal of 1 nJ was measured. We define gain as

$$g = \frac{I}{I_0} = \frac{1}{9} \exp\left(\frac{L_{und}}{L_{gain}}\right).$$

where I and I_0 are the SASE and spontaneous energy inside the coherency cone respectively, L_{und} is the undulator length and L_{gain} is the gain length. Calculations show for our setup $I_0 = 9$ pJ thus, giving a peak gain, $g = 110$, and a gain length, $L_{gain} = 58$ cm. As the gain signal increases, the SASE shot-to-shot signal is expected to fluctuate more due to the statistical nature of the SASE

FEL start-up process [10]. This is clearly the case in Fig. 2 where signal fluctuations near 240 pC are quite large and decrease as the charge/SASE signal lessens. At the moment there is no comparison with theory because of the uncertainty in the undulator section positions; however, in the future when the alignment becomes more certain, a comparison will be more relevant.

Electron beam trajectory studies through the undulator are necessary in order to understand and optimize the SASE FEL performance. If the four 1 m sections are misaligned (which is the case shown below), orbit kicks will develop in electron beam trajectories.

Without the use of the corrector magnets along the length of the undulator, the beam horizontal (wiggle plane) orbits were studied by changing the upstream (matching section) launch conditions into the undulator. Shown in Fig. 3 is the electron beam centroid positions at the BPM just before the undulator and inside the undulator at the eight BPM pop-in positions relative to our alignment laser where $z = 0$ is the undulator entrance. The alignment laser gives a straight-line path reference through the undulator. Fig. 3 shows a peak-to-peak trajectory amplitude of nearly 3 mm (pre-move orbit) and a significant kick between undulator Sections 2 and 3 before steps were taken for improvement.

After we were satisfied this was the best trajectory we could attain, we decided to do a beam-based alignment and move the undulator while under vacuum. The VISA system has the ability to horizontally (in the wiggle plane) move the undulator while under vacuum. An undulator monitoring system was implemented during installation in which the body of the undulator could be monitored to an accuracy of $<15\,\mu m$ giving us fairly precise control of the undulator's horizontal position. Using a trajectory simulator, we were able to predict which direction and how far to move the undulator sections, do another orbit study, and move the undulator again. To the accuracy of the simulator, we needed to move only the last two sections, Sections 3 and 4. After seven iterative undulator section moves, we had reduced the peak-to-peak oscillation from 3 mm to 500 μm. Fig. 3 shows only three trajectories after moving the undulator for clarity of picture. At first, Section 3 was moved until the oscillation amplitude through it was similar to the first two sections and then movement of Section 4 followed. The distance the sections needed to be moved was between 150 and 600 μm depending where the realignment was necessary. The limiting factor of this method is the fact that the trajectory model assumes only one undulator section to be misaligned which was clearly not the case. Fig. 3 shows great improvement in the electron beam trajectory using this method to align our undulator.

4. Conclusion

For strong focusing undulator systems, alignment is very critical. We demonstrated beam-based alignment on the VISA system by a series of trajectory studies and undulator movements. This method re-aligned the undulator from a peak-to-peak trajectory amplitude of 3 mm to 500 μm. Also, a significant gain, $g = 110$, was measured with gross misalignments of the undulator sections. Currently, work is in progress to bring the alignment to the several microns level using the interferometric system. Once aligned the trajectory orbit amplitude should be reduced further along with attaining much higher gain.

References

[1] J.B. Murphy, C. Pellegrini, Nucl. Instr. and Meth. A 237 (1985) 108.
[2] M. Hogan, et al., Phys. Rev. Lett. 80 (1998) 298.
[3] M. Cornacchia, Performance and design concepts of a free electron laser operating in the X-ray region, SLAC-PUB-7433, March 1997.
[4] X.J. Wang, et al., Brookhaven accelerator test facility energy upgrade, Proceedings of the 1999 Particle Accelerator Conference, New York, NY, 1999.
[5] M. Libkind, et al., Mechanical design of the VISA undulator, Proceedings of the 1999 Particle Accelerator Conference, New York, NY, 1999.
[6] A.A. Varvolomeev, A.H. Hairetdinov, Nucl. Instr. and Meth. A 341 (1994) 462.
[7] P. Emma, et al., FEL trajectory analysis for the VISA experiment, Proceedings of the 20th International FEL Conference (FEL98), Williamsburg, VA, USA, August 1998, SLAC-PUB-7913.
[8] R. Ruland, et al., Alignment of the VISA undulator, Proceedings of the 1999 Particle Accelerator Conference, New York, NY, 1999.
[9] A. Murokh, et al., in: V.N. Litvinenko (Ed.), Proc. 22nd FEL Conference, Elsevier Science B.V., Amsterdam, 2001, p. II–35.
[10] R. Bonifacio, et al., Phys. Rev Lett. 73 (1994) 70.

ELSEVIER

Nuclear Instruments and Methods in Physics Research A 475 (2001) 343–347

NUCLEAR
INSTRUMENTS
& METHODS
IN PHYSICS
RESEARCH
Section A

www.elsevier.com/locate/nima

Simulation studies of the proposed far-infrared amplifier at ISIR, Osaka University

Ravi A.V. Kumar*,[1], Ryukou Kato, Goro Isoyama

Institute of Scientific and Industrial Research, Osaka University, 8-1 Mihogaoka, Ibaraki, Osaka 5670047, Japan

Abstract

Computer simulation studies are currently underway for the proposed FEL amplifier experiment to be carried out using the existing 38 MeV, L-band linac based FIR FEL system at the Institute of Scientific and Industrial Research (ISIR), Osaka University. This paper describes the time-dependent and time-independent simulation results using Genesis code. Dependence of FEL amplifier characteristics are investigated such as the power spectrum and profile, frequency spectrum and FEL bunching parameter on parameters like the beam current, input power and interaction length. The operating current regime and input seed laser powers required for the optimum amplification are given. © 2001 Elsevier Science B.V. All rights reserved.

PACS: 41.60.Cr; 41.85.Lc; 42.25.Fx; 29.17.+w

Keywords: Free-electron laser(s); Far infrared; FEL amplifier

1. Introduction

The Free-Electron Laser (FEL) has been deemed the 'next generation light source'. With the extensive research and development in the field of single pass, Self-Amplified Spontaneous Emission (SASE) FELs, the accessible wavelength of the FEL has now been extended far beyond the UV region. The single-pass SASE is the best suited configuration because of problems pertaining to use of optical elements at shorter wavelengths.

*Corresponding author. Tel.: +81-6-6879-8487; fax: +81-6-6879-8489.

E-mail address: ravi25@sanken.osaka-u.ac.jp (R.A.V. Kumar).

[1] Currently on leave from Institute for Plasma Research, Gandhinagar, India.

SASE was first observed here at the Radiation Laboratory of the Institute of Scientific and Industrial Research (ISIR), Osaka University in 1991 during the course of developing the infrared FEL using the 38 MeV, L-band linac [1]. Detailed experimental study on SASE is currently being conducted using the wiggler for the infrared FEL. Although strong SASE has been observed, it is not possible to go into the saturation regime under the present experimental conditions due to the limitation of the comparatively short number of 32 wiggler periods. It was proposed that the above-mentioned FEL system could be used to conduct single-pass FEL amplifier experiments using an injected seed laser source, which would help understand the phenomenon of SASE in the strong signal regime. It is also necessary to know the behaviour of the system and its dependence on various experimental parameters before actually conducting the experiment. The one-dimensional (1-D) theory does offer an insight into the behaviour, but due to the fact that many factors are often neglected in 1-D estimations, the results may not correlate exactly with the experimental observations. Hence it is essential that a more detailed, realistic simulation is carried out, taking into account as many experimental parameters as possible. This paper attempts to study the effect of the beam current, input seed laser power and interaction length on the output characters and attempts to obtain optimum operating conditions for the experiment.

2. The FEL simulation code

Ever since the design and building of FELs began, several simulation codes have been developed [2], each one tailored to describe the phenomenon of FEL occurring in different regimes. Genesis has evolved from an earlier 3-D FEL simulation code TDA, and then TDA3D [3,4]. Genesis [5], like its predecessors, solves the paraxial FEL equations with the approximation of slow varying amplitude of the radiation field. Using Genesis, time-dependent simulation is also possible.

Table 1
Parameters used in the simulation studies

Data from on-going experiments	
Beam energy	12.42 MeV
Energy spread (FWHM)	3.0%
Wiggler K parameter	1.472
Wiggler period	6.0 cm
No. of periods	32
Simulation wavelength	166 μm
Normalized emittance (x)	150 π mm mrad
Normalized emittance (y)	93 π mm mrad
e-beam radius (x)	3.3 mm
e-beam radius (y)	1.1 mm
Raleigh length	0.6 m
Bunch length	6.0 mm
Peak current	500 A
Simulation parameters	
No. of slices	150
No. of particles	4096

In the proposed experiment, high-intensity, single bunch e-beam of charge up to 73 nC and a pulse length in the range 9–50 ps from the 38 MeV ISIR linac is traversed through a 32 period Halbach type, permanent magnet (Nd–Fe–B) planar wiggler of 60 mm period. The wiggler gap can be adjusted so that the wiggler constant K can be varied from 1.47 to 0.013. A seed laser source at appropriate wavelength and power would be injected along the axis of the beam and this would be amplified by the single-pass FEL process. The parameters used in these simulation studies are given in Table 1. These values were obtained from real-time, measured data from on-going SASE experiments [6].

3. Results and discussion

In this paper, the influence of three basic parameters viz. the beam current, interaction length and the input seed power on the output power and frequency spectrum are studied.

Fig. 1 shows the dependence of the output power on the peak beam current. It can be clearly seen that there are two distinct regimes of current indicated by the two different slopes in the graph. The output in the lower current regime (slope < 1 W/A) may not be wholly due to the SASE, but a major contribution is due to other processes such as spontaneous emission. But at higher currents (slope ∼ 5), the SASE process is dominant, which can also be inferred from the plots of the FEL bunching factor, as shown in Fig. 2. The bunching factor is a measure of the spatial modulation of the electrons in the wiggler. The bunching factor is not very well defined until the current approaches the second regime. The temporal profile of a typical electron bunch is also indicated by the dashed line in the graph (in this case, 2.5 kA peak current).

The dependence of the output power on the input seed power is shown in Fig. 3. The shot-noise power of the system is given by [7]

$$P_{shot} \approx \frac{3\sqrt{4\pi}\rho^2 P_b}{N_\lambda \sqrt{\ln\left(\frac{N_\lambda}{\rho}\right)}}. \qquad (1)$$

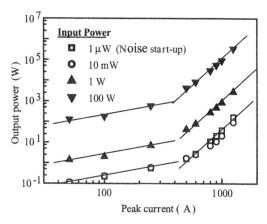

Fig. 1. The dependence of the output power on beam current.

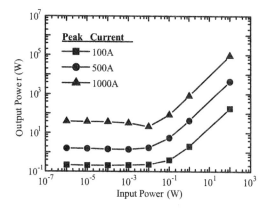

Fig. 3. Dependence of the output power on input seed laser power at different peak currents.

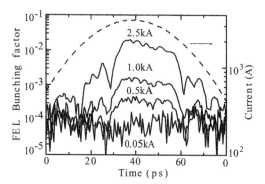

Fig. 2. The evolution of the FEL bunching factor in a single electron bunch at different peak currents.

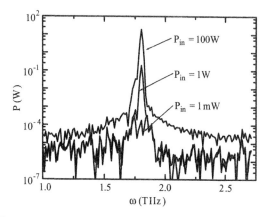

Fig. 4. The dependence of the power spectrum on input power.

where ρ is the FEL parameter, P_b is beam power and N_λ is the number of electrons in one wavelength. This power is typically in the 10^{-3}–10^{-2} W range for the parameters used in this simulation. It can be seen that for input powers less than the shot-noise power, the output is more or less constant. Thus, it is important to keep the input seed laser power above the shot-noise power level as expected, in order that the system effectively acts as an amplifier for the input seed laser. Fig. 4 shows the power spectra at different input seed laser powers. It can be seen that the power spectrum for higher input power is narrower as compared to the noise start-up.

The time-dependent simulation is used to study the spatial profile of the output radiation and its dependence on the peak current as well as input

power. Fig. 5 shows the near-field radiation pattern of the amplified output at the wiggler exit. The simulation uses a modest peak current of 500 A, noise start-up (Fig. 5a) which is the SASE process, and the amplifier mode using a 1.0 W seed laser input (Fig. 5b). It can clearly be seen that the spatial profile of the amplified radiation with a seed laser source is more pure than that from noise start-up.

The saturation of the system is studied as a function of the interaction length as well as beam current and input laser power and is shown in Fig. 6. Saturation occurs in the region of 60 periods for a I_{peak} of 1 kA (SASE) and P_{input} of ~ 1 W (amplifier) while it occurs at about 130 when current is 500 A and input power is 1 W. For lower currents and input power, the saturation

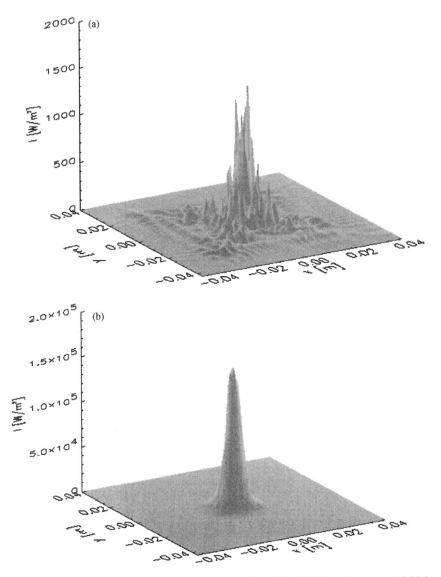

Fig. 5. The near-field spatial profile of the output radiation at the wiggler exit for 100 A peak current. (a) Noise start-up (SASE; $P_{in} = 1\,\mu W$), (b) amplifier mode ($P_{in} = 1\,W$).

progressively shifts to higher period numbers. By using the amplifier mode, it is possible to achieve saturation regimes even at lower currents, provided the interaction length and seed power are high enough. Hence it is experimentally possible to approach saturation by controlling any or all of these three parameters, viz., interaction length, beam current and the input power. It is thus planned to increase the number of periods of the

existing wiggler from the existing 32 to about 70 in the near future and this would enable operation of the amplifier at a regime closer to the saturation of the system.

4. Conclusions

The time-dependent 3-D simulations for the proposed FEL FIR amplifier experiments are

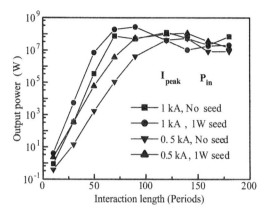

Fig. 6. The dependence of the saturation of the system on interaction length, peak current and input power.

being carried out using Genesis code. Dependence of the output power on beam current, input power and interaction length are presented. In order that amplification actually occur, it is essential that the input power be kept well above the shot-noise power and beam currents above the threshold value above which the SASE contribution to the output is dominant. It can be thus concluded that by using the single-pass amplifier mode of operation, it is possible to obtain a relatively stable and

spatially pure output radiation signal as compared to the SASE noise start-up process. This would allow a better understanding of the single pass SASE phenomenon. Also, with a peak current of 1 kA, a modest input power more than a few hundred mW (higher than the shot-noise power) or more in addition to an enhanced interaction length, it would be feasible to approach the saturation regime using the existing FEL set-up.

References

[1] S. Okuda, J. Ohkuma, N. Kimura, Y. Honda, T. Okada, S. Takamuku, T. Yamamoto, K. Tsumori, Nucl. Instr. and Meth. A 331 (1993) 76.

[2] T.M. Tran, J.S. Wurtele, Phys. Rep. 195 (1990) 1.

[3] T.M. Tran, J.S. Wurtele, Comput. Phys. Commun. 54 (1989) 263.

[4] S. Reiche, Nucl. Instr. and Meth. A 429 (1999) 243.

[5] GENESIS 1.3, A time dependent FEL simulation code, Version 0.1, 1998. http://www.desy.de/~reichesv/genesis/genesis.html.

[6] R. Kato, R.A.V. Kumar, T. Okita, S. Kondo, T. Igo, T. Konishi, S. Okuda, S. Suemine, G. Isoyama, Nucl. Instr. and Meth. A 445 (2000) 164.

[7] E.L. Saldin, E.A. Schneidmiller, M.V. Yurkov, Phys. Rep. 260 (1995) 187.

ELSEVIER

Nuclear Instruments and Methods in Physics Research A 475 (2001) 348–352

NUCLEAR
INSTRUMENTS
& METHODS
IN PHYSICS
RESEARCH
Section A

www.elsevier.com/locate/nima

Observation of longitudinal phase space fragmentation at the TESLA test facility free-electron laser

M. Hüning, Ph. Piot*, H. Schlarb

Deutsches Elektronen-synchrotron, DESY, D-22603 Hamburg, Germany

Abstract

It has been reproducibly observed that the energy distribution of the beam, when fully longitudinally compressed for SASE operation, breaks up into several peaks. In this paper a description of the experimental setup, beam operating conditions, and observations is presented to enable further theoretical studies of this effect. © 2001 Elsevier Science B.V. All rights reserved.

PACS: 29.27.Bd; 41.20.Bt; 41.60.Ap; 41.85.Lc

Keywords: Bunch compression; FELs; Microbunch; Collective effect

1. Introduction

High-gain single pass free-electrons lasers (FEL) require high charge (nC-level), ultrashort (ps-level) and low emittance (mm-mrd) electron bunches. Since state-of-art photoinjectors cannot directly produce the required short bunch, magnetic bunch compressors are generally used to achieve the picosecond-level bunch length needed. Compressing and then transporting these short bunches to the undulator section becomes challenging since significant phase space dilution due to collective effects (e.g. wakefields, coherent synchrotron radiation, etc) might arise. In this paper, we present a series of measurements taken at the TESLA Test Facility (TTF) FEL [1]. Our observations indicate that a bunch undergoing compression tends to have its energy profile

fragmented. It is worthwhile mentioning that similar features have been recently observed at other facilities [2,3].

2. Experimental background

A schematics of the TTF FEL driver accelerator is depicted in Fig. 1. We will only concentrate on the longitudinal phase space manipulation since the measurements reported hereafter pertain to this plane. The beam generation line consists of a L-band radio-frequency (RF) photoinjector coupled to a TESLA-type superconducting accelerating cavity that boosts the beam up to 17 MeV. The photocathode drive-laser is Gaussian-shaped with a rms time duration of 8 ps. The electron bunch then enters an accelerating section (Acc. #1) that consists of eight TESLA-type superconducting cavities. The injection phase of the beam in this latter accelerating section is chosen, under nominal

---
*Corresponding author. Tel.: + 49-0-40-8998-2755.
E-mail address: philippe.piot@desy.de (P. Piot).

0168-9002/01/$ - see front matter © 2001 Elsevier Science B.V. All rights reserved.
PII: S 0 1 6 8 - 9 0 0 2 (0 1) 0 1 5 3 7 - 6

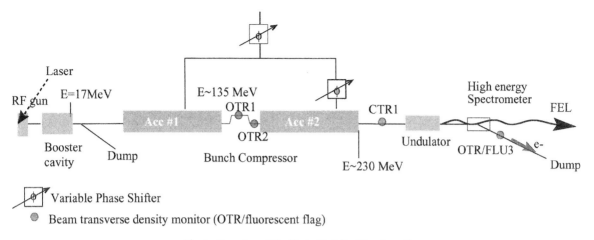

Fig. 1. Overview of the Tesla Test Facility phase I.

operating conditions, to impart the proper time–energy correlation to compress the bunch using the downstream magnetic chicane-based compressor. The beam transport downstream of the bunch compressor consists of a second accelerating section (Acc. #2, identical to Acc. # 1), nominally operated for maximum energy gain, followed by an undulator. Behind the undulator, the electron beam is separated from the FEL beam by a spectrometer dipole (that bends in the same plane as the compressor). The transfer line to the dump is instrumented with several diagnostics including a beam profile measurement station.

A detailed description of the magnetic compressor is presented in Ref. [4]. During our measurement, the bending angle in the compressor was set to 19°, which corresponds to a momentum compaction $R_{56} \simeq -0.18$ m. This results in an optimum operating phase for the first accelerating section of 10° off-crest approximately. Because of the rather long bunch length at the injector front-end, typically 2.7 mm (rms), the longitudinal phase space is strongly distorted via RF-induced curvature during its acceleration in Acc. #1. This distortion affects the compression process and limits the minimum attainable bunch length to 0.5 mm (rms) approximately (as inferred from multi-particle simulations and experimentally verified).

3. Experimental techniques

During our measurements, we used the different beam profile monitors showed in Fig. 1: the energy spread was measured using the optical transition radiation (OTR) viewers OTR1 and OTR/FLU3 (this latter viewer incorporates both an OTR radiator and a fluorescent screen). At the location of these two beam profile stations, the linear dispersion is estimated to $R_{16} = 0.31$ m (bunch compressor) and $R_{16} = -1.15$ m (spectrometer), respectively. The nonlinear dispersion, T_{166}, is found, for both locations, to have insignificant impact on the beam horizontal profile so that the horizontal coordinate of an electron, x, scales linearly (to first order) with its relative momentum offset, δ, following $x \simeq R_{16}\delta$. This scaling implicitly assumes that the contribution from the pure betatron term is insignificant. A true assumption since, for all measurements presented in this paper, a set of quadrupoles, located upstream of the bunch compressor or the spectrometer dipole, respectively, were tuned to minimize the horizontal beam spot at the observation point. This guarantee that the beam spot was dominated by the dispersive contribution. Based on the measurements of energy profiles at OTR/FLU3 for different phases of Acc. #2, a longitudinal tomography technique was implemented. The method to reconstruct the longitudinal phase space is based on the MENT [5] (Maximum ENTropy) algorithm which computes the best estimate of the phase space density by maximizing its entropy. At the compressor exit the transverse beam density can also be measured using OTR2. At that point the dispersion was found not to be zero, we believe because of spurious dispersion generated by nonzero value of upstream correctors. The diagnostics package also include a bunch length monitor located downstream of the second accelerating section. This device provides bunch length measurement by the mean of a sub-millimeter wave polarizing interferometry of coherent transition radiation [6] (CTR). We did not systematically measure the bunch length but rather inferred whether the first accelerating section (Acc. #1) was operated for maximum compression by simply "peaking" the CTR power detected by a pyroelectric detector. All of the measurements reported in the following were performed with a charge per bunch of 1 nC (within 10%). We have verified, using a streak camera setup, that our observations result from a single RF-bucket.

4. Observations

4.1. Energy profiles versus incoming time-energy correlation

In this series of measurement the compressor is operated, and the bunch compression is varied by operating Acc. #1 at different phase to act on the incoming longitudinal phase space slope, $d\delta/ds$: at maximum compression it is related to the bunch compressor momentum compaction by relation $d\delta/ds = -1/R_{56}$. We found, using the CTR signal, that the maximum compression was occuring $-10°$ w.r.t. the maximum energy gain phase. For this operating condition, the beam energy was about 135 MeV at the bunch compressor location and 230 MeV downstream Acc. #2. After each change of Acc. #1 phase, the phase of the Acc. #2 accelerating section was reset to maximize the energy at the machine front end. Fig. 2 depicts the evolution of the beam density recorded at the observation point (i.e. OTR/FLU3) for some phase of the Acc. #1. The maximum compression

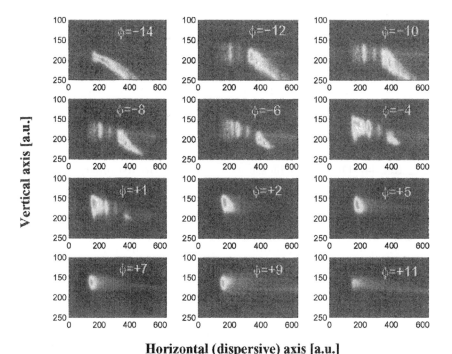

Fig. 2. Single bunch beam transverse density observed at the profile measurement station OTR/FLU3. These measurements were performed using the fluorescent screen.

occurs at $\phi \simeq -8°$[1] whereas the maximum energy gain through the whole linac is obtained for $\phi \simeq +2°$, From Fig. 2, we conclude that when the time–energy-induced correlation by Acc. #1 does not provide compression, i.e. $\phi > +2°$, the energy profile is as expected: it consists of a bright core with a long energy tail due to the RF-induced curvature because of the long incoming bunch. As the time–energy-induced correlation allows compression, $\phi < +2°$, the energy profile multiple fine structure which seems to separate into two main "islands" at maximum compression. Finally, in the over-compression regime, $\phi < -8°$, these multiple structures start disappearing and are hardly observable at phases below $-14°$.

4.2. Impact of the bunch compressor

For three cases (i) bunch compressor operated and accelerated section Acc. #1 operated for maximum compression, (ii) bunch compressor operated and Acc. #1 operated on-crest, and (iii) bunch compressor off with Acc. #1 operated approximately $-10°$ off-crest, we have investigated the evolution of the energy profile in the high energy spectrometer line for various phase of Acc. #2 ultimately to take tomographic data in order to recover the full longitudinal phase space density. The energy profiles for these three cases are presented in Fig. 3 (Acc. #2 was set for maximum energy gain). A structured energy profile, as the one presented previously, is observed only for the case where the linac operated to provide maximum compression. This feature, which has also been observed for other bunch charges, suggests the induced fragmentation of the energy profile related to the compression process. At the bunch compressor exit, some structure can already be noticed by observing the beam transverse density at OTR2, but none was observable at OTR1. For case (ii), the reconstructed longitudinal phase space at the compressor exit is shown in Fig. 4.

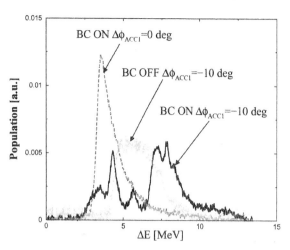

Fig. 3. Energy profile obtained for three different senarii of operation: bunch compressor ON with Acc. #1 on-crest ($\Delta\phi = 0°$), and the setup for minimum bunch length ($\Delta\phi = 0°$), and bunch compressor OFF with $\Delta\phi = -10°$.

4.3. Charge-dependence

To check whether the observation could be attributed to collective effects, we also measured the energy distribution for bunch charge ranging from 0.5 to 4 nC. The energy profile showed a weak dependence on the charge. Nevertheless the collective effects[2] hypothesis cannot be ruled out: the charge was varied by changing the photocathode drive laser intensity, which also impacts the bunch length. From multi-particle simulations we expect that increasing the charge from 0.5 to 4 nC would approximately double the bunch length (from 2 to 4 mm (rms)) principally because of longitudinal space charge force charge-dependence in the RF-gun and following drift section.

5. Discussion

An anomalous "beam break-up" of the longitudinal phase space has been observed at TTF-

---

[1] The phase value ϕ mentioned hereafter are arbitrary value read from the control system

[2] for the collective effects considered, the relative energy loss of one electron is of the form $\delta p/p \propto Q/\sigma_s^\alpha$, for standard geometric wakefield $\alpha \simeq \frac{1}{3}$, for coherent synchrotron radiation $\alpha = \frac{4}{3}$, with Q being the bunch charge. The charge density is assumed to be a Gaussian distribution with rms length σ_s.

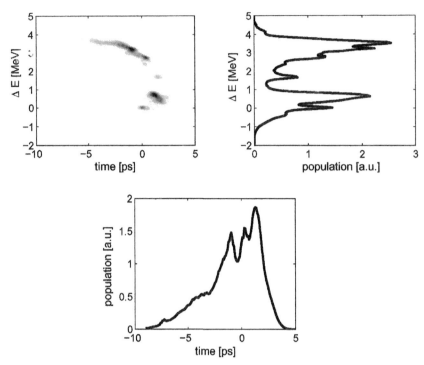

Fig. 4. Reconstructed longitudinal phase space downstream the bunch compressor (top left) and energy (top right) and time (bottom) charge density. (Time > 0 correspond to the bunch head).

FEL when the linac is operated so that the bunch compression is maximized. We discussed elsewhere [7] a possible mechanism, based on CSR self-interaction in the bunch compressor, that could account for our observation. In the TTF accelerator among the different source of wakefield that have been considered, CSR wakefield has the strength to yield an energy redistribution such as the one we have observed. Moreover, the CSR wake can be strongly enhanced due to local charge concentration, as pointed out in Ref. [8], e.g. due to longitudinal phase space curvature.

References

[1] Andruszkow, et al., Phys. Rev. Lett. 85 (2000) 3825.

[2] Ph. Piot, et al., Proceedings of EPAC 2000, Vienna, Austria, 2000, pp. 1546–1548.
[3] M. Borland, LEUTL-ANL, private communication, 2000.
[4] A. Kabel, et al., Proceedings of 1999 part. accel. conf., IEEE catalog #0-7803-5573-3/99, 1999, pp. 2507–2509.
[5] G. Minerbo, Comput. Graphics Image Process. 10 1970 48.
[6] M. Geitz, et al., Proceedings of the 1999 Part. Accel. Conference, IEEE Catalog #0-7803-5573-3/99, 1999, pp. 2174–2176.
[7] T. Limberg, Ph. Piot, E. Schneidmiller, An analysis of longitudinal phase space fragmentation at the TESLA test facility, Nucl. Instr. and Meth A 475 (2001) 353, these proceedings.
[8] R. Li, Proceedings of the EPAC 2000, Vienna, Austria, 2000, pp.1312–1314.

ELSEVIER

Nuclear Instruments and Methods in Physics Research A 475 (2001) 353–356

NUCLEAR
INSTRUMENTS
& METHODS
IN PHYSICS
RESEARCH
Section A

www.elsevier.com/locate/nima

An analysis of longitudinal phase space fragmentation at the TESLA test facility

T. Limberg, Ph. Piot*, E.A. Schneidmiller

DESY, Deutsches Elektronen-Synchrotron, D-22603 Hamburg, Germany

Abstract

It has been repeatedly observed, at TTF-FEL, that the energy profile of the fully longitudinally compressed beam breaks up into peaks. In this paper we analyze a potential cause of this effect. We study the bunch self-interaction via coherent synchrotron radiation, leading to a fragmentation of the longitudinal phase space. For our analysis, we use a simple model as well as the simulation code TraFiC^4 (Nucl. Instr. and Meth. A 445 (2000) 338) to evaluate the CSR-induced effects. © 2001 Elsevier Science B.V. All rights reserved.

PACS: 29.27.Bd; 41.20.Bt; 41.60.Ap; 41.85.Lc

Keywords: Bunch compression; FELs; Microbunch; Collective effect

1. Introduction

In the TESLA Test Facility (TTF) [1] free-electron laser (FEL), electron bunches of 2.7 mm length (rms) are emitted from an RF-gun and accelerated by a booster cavity before being injected into a standard TESLA superconducting accelerating section (Acc. #1). Downstream of the Acc. #1, at an energy of approximately 135 MeV, a magnetic bunch compressor chicane (BCC) (see Fig. 1) compresses the bunches to a length of about 0.5 mm (rms). Then a second accelerating section (Acc. #2) increases the energy to 230 MeV; afterwards the beam enters an undulator, followed by a spectrometer magnet and a dump. The spectrometer bends the beam in the same plane as the bunch compressor. The energy profile is

monitored using a fluorescent screen located between the spectrometer and beam dump. The energy distribution observed [2] can be approximately described as a two-peak structure with a separation of 3 MeV which appears when the phase of Acc. #1 is close to providing maximum compression. Thus a significant amount of particles must suffer an energy loss of the order of that separation. Similar effects have been observed at other laboratories (see for instance Ref. [3]). The longitudinal wakefields induced by the TESLA RF structures are, even for the case when very short ($< 50 \, \mu m$) sub-structures carry most of the charge, more than one order of magnitude too weak to account for the phenomenon. Coherent synchrotron radiation (CSR) induced effects, however, do have the strength to cause this kind of energy loss. At the bunch center, the longitudinal CSR-field, E_0 (which is nearly the maximum field), produced by a 1-D Gaussian bunch with rms length σ_s and

---

*Corresponding author. Tel.: +49-40-8998-2756.
E-mail address: philippe.piot@desy.de (Ph. Piot).

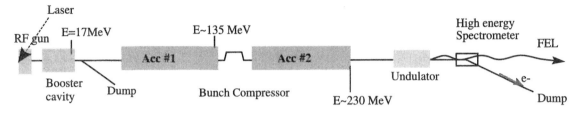

Fig. 1. Overview of the TESLA Test Facility phase I.

charge q_0 moving on an infinitely long trajectory with bending radius R_0 is given by

$$E_0 = \frac{-q_0}{4\pi\varepsilon_0}\left[\frac{2^{1/3}\Gamma\left(\frac{5}{6}\right)}{3^{1/3}\sqrt{\pi}R_0^{2/3}\sigma_s^{4/3}}\right] \tag{1}$$

where $\Gamma(\ldots)$ is the complete gamma-function, and ε_0 is the electric permittivity for vacuum. For instance if half of the total bunch charge would be (locally) concentrated in a (Gaussian) spike of $25\,\mu m$ (rms), the latter equation would yield a field $E_0 \simeq 3\,MV/m$. In the next sections we look into the possible origin of such narrow peaks and endeavor to explain the observed energy profile break-up as a consequence of CSR bunch self-interaction.

1.1. Nonlinear compression of long bunches

Firstly, we need to understand how a considerable amount of charge can be gathered in local narrow peak(s). In principle [4], an initial bunch modulation with a Fourier component of the charge density in the Fourier space $\tilde{\Lambda}^-(\omega) = \alpha\cos(\omega t)$ in the presence of an energy spread dilution mechanism characterized by the longitudinal impedance $Z_\parallel(\omega)$ can be enhanced when passing a beamline section with non-zero momentum compaction.[1] For the outgoing beam, the amplitude of the charge density component (after linearization) can be written as [4] $\tilde{\Lambda}^+(\omega) \simeq [\alpha R_{56}\omega Z_\parallel(\omega)]/[c\gamma_0]\cos(\omega t + \varphi)$, where φ represents a phase shift and γ_0 the average beam energy. The latter expression shows that depending on ω and $Z_\parallel(\omega)$ there can be significant amplification of the modulation.

We will hereafter consider the simpler case of a Gaussian charge density downstream of the injector, so that the peak has to be generated during the compression process itself. Indeed, a rather long bunch, in our case 2.7 mm (rms), accumulates enough curvature in longitudinal phase space due to the non-linear cosine shape of the RF-wave to require full compression even for final bunch lengths of around 0.5 mm (see Fig. 2). Consequently, the longitudinal profile after compression (see Fig. 3) consists of a high density peak (see also [5]) followed by a tail. The width of the leading peak is given by the initial uncorrelated energy spread; in the case shown, an initial spread of 15 keV (rms) results in an rms peak width of approximately $50\,\mu m$ which contains 50% of the charge. Since this is only the case towards the end of the third and through the fourth dipole of the BCC, we are just approaching the measured energy losses.

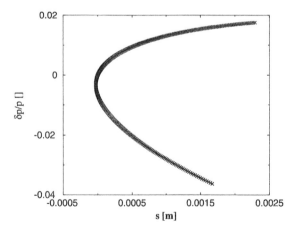

Fig. 2. A typical longitudinal phase space distribution after compression.

---

[1] We define the momentum compaction, R_{56}, as $R_{56} = \partial S/\partial(\delta p/p)$, i.e. as the linear dependence of path length S on relative energy spread $\delta p/p$.

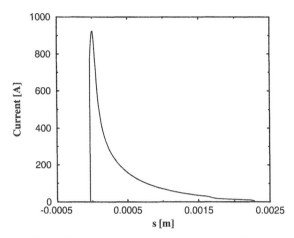

Fig. 3. Bunch charge density computed from Fig. 2.

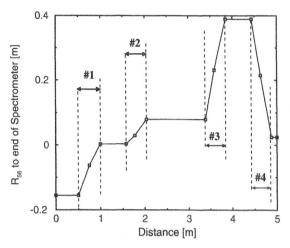

Fig. 4. Momentum compaction, R_{56}, to the end of the spectrometer magnet starting from different locations within BCC (zero corresponds to 0.5 m upstream of the first dipole of BCC). The labels #1–#4 indicate the location of BCC dipoles.

1.2. Bunch compression in the spectrometer dipole magnet

Let us now analyze the role of the spectrometer magnet, a dipole magnet of 1 m length, with a deflection angle comparable to the bunch compressor magnets. Any induced energy spread at a given point acts on beam parameters in the bending magnets downstream. Fig. 4 shows the momentum compaction, R_{56} up to the end of the spectrometer bending magnet for different positions in the bunch compressor beam line. The energy gradient provided by the CSR-field[2] in the end region of the third dipole of BCC, for instance, will cause a further local compression of the bunch in the spectrometer magnet, since the R_{56} to the end of that magnet is of opposite sign and twice as large as that for the whole bunch compressor.

1.3. TraFiC⁴ simulation calculations

The code TraFiC⁴ [7] was used to simulate the TTF-FEL beamline from the entrance of the compressor to 1 m downstream of the spectrometer dipole. The initial phase space was generated by tracking a phase space distribution through the injector using the multi-particle code ASTRA [6]. The beamline between the chicane and the spectro-

meter was modeled by a simple four-quadrupole telescope that provides the same transverse transfer functions as the nominal settings computed for the TTF-FEL and used during the experiment reported in Ref. [2]. Since the distortions of the longitudinal phase space are quite strong, the calculations were performed in a self-consistent manner: the generated CSR-fields are fed back to the source ensemble in a leap-frog algorithm. The incoming bunch was modeled with 320 3-D Gaussian sub-bunches of 25 μm (rms) length. A greater number of shorter sub-bunches would be required to model the long incoming bunch as a smooth charge distribution. The computing time scales with the square of the number of sub-bunches. At the present stage of our studies we are limited to bunch sub-structures of 25 μm and longer. Fig. 5 depicts the simulation results: downstream of the compressor the energy profile starts to fragment and downstream of the spectrometer the peaks get stronger. The peak separation, however, falls short by a factor of 2–3 when compared with the measurement. The underestimate of this effect is most likely due to the limited resolution used in the calculation which prevents us from accurately evaluating radiative fields of sub-structures undergoing over-compression in the spectrometer dipole.

---

[2] In our convention CSR-induced local time–energy correlation is compressed by positive R_{56} section.

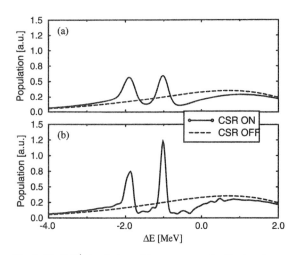

Fig. 5. TraFiC^4 simulation of beam energy distribution downstream of the compressor (A) and downstream of the analyzing spectrometer magnet (B).

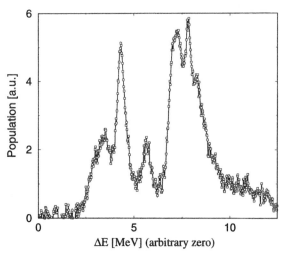

Fig. 6. An example of measured [2] energy distribution after the spectrometer (to be compared with Fig. 5(B)).

2. Conclusion and future plans

If non-linearities such as the curvature of the accelerating RF sections are significant, longitudinal bunch compression will produce non-Gaussian and locally peaked distributions. For such types of non-Gaussian distributions, CSR effects can be much stronger compared to Gaussian distributions with the same rms bunch length. Thus the modeling of the FEL driver-accelerator have to be performed with more realistic beam distributions. In the case of the TTF-FEL, we have shown that stronger CSR-induced energy redistribution can split up the beam energy profile similar to the measurements reported in Ref. [2]. The FEL operation will profit from a much higher peak current than that calculated with a linear model of bunch compression. On the other hand, CSR-fields will be stronger and lead to emittance dilution both in the transverse and longitudinal planes. Thus, the FEL process simulations have to be performed with realistic phase space distributions, i.e., obtained from tracking calculations which include CSR effects. Future projects, like the TESLA X-FEL, have to be simulated and optimized using the same approach.

As for experimental studies at the TTF-FEL, we are presently engaged in further experiments to prove that we are indeed observing a CSR-driven effect. We especially plan to vary the bunch charge and measure the energy and charge profiles simultaneously. We will evaluate the corresponding transverse emittance dilution that should occur in the bending plane. Finally, we plan to study the evolution of the structure of the energy profile for different in-coming dispersion into the spectrometer magnet by changing the upstream quadrupole magnets (Fig. 6).

References

[1] Andruszkov, et al., Phys. Rev. Lett. 85 (2000) 3825.
[2] M. Hüning, Ph. Piot, H. Schlarb, Observation of longitudinal phase space fragmentation at the TESLA test facility free-electron laser, Nucl. Instr. and Meth. A 475 (2001) 348, these proceedings.
[3] L. Grönig, et al., Proceedings of EPAC 2000, Vienna, Austria, 2000, pp. 472–473;
Ph. Piot, et al., pp. 1546–1548; R. Hajima, et al., Nucl. Instr. and Meth. A 429, (1999) 264;
S. Okuda, et al., Nucl. Instr. and Meth. A 445 (2000) 351.
[4] E. Saldin, E. Schneidmiller, M. Yurkov, Longitudinal phase space distortion in magnetic bunch compressor, Report TESLA-FEL 2001-09.
[5] R. Li, Proceedings of EPAC 2000, Vienna, Austria, (2000) pp. 1312–1314.
[6] K. Flöttmann, ASTRA User Manual, www.desy.de/~mpy-flo/ASTRA_dokumentation.
[7] M. Dohlus, A. Kabel, T. Limberg, Nucl. Instr. and Meth. A 445 (2000) 338.

ELSEVIER

Nuclear Instruments and Methods in Physics Research A 475 (2001) 357–362

NUCLEAR
INSTRUMENTS
& METHODS
IN PHYSICS
RESEARCH
Section A

www.elsevier.com/locate/nima

X-ray FEL with a meV bandwidth

E.L. Saldin[a], E.A. Schneidmiller[a],*, Yu.V. Shvyd'ko[b], M.V. Yurkov[c]

[a] *Deutsches Elektronen-Synchrotron (DESY), Notkestrasse 85, D-22607 Hamburg, Germany*
[b] *II. Institut für Experimentalphysik, Universität Hamburg, D-22761 Hamburg, Germany*
[c] *Joint Institute for Nuclear Research, Dubna, 141980 Moscow Region, Russia*

Abstract

A new design for a single pass X-ray Self-Amplified Spontaneous Emission (SASE) FEL was proposed by Feldhaus et al. (Opt. Commun. 140 (1997) 341) and named "two-stage SASE FEL". The scheme consists of two undulators and an X-ray monochromator located between them. For the Angström wavelength range the monochromator can be realized using Bragg reflections from crystals. We propose a scheme of monochromator with a bandwidth of 20 meV for the 14.4 keV X-ray SASE FEL being developed in the framework of the TESLA linear collider project. The spectral bandwidth of the radiation from the two-stage SASE FEL (20 meV) is determined by the finite duration of the electron pulse. The shot-to-shot fluctuations of energy spectral density are dramatically reduced in comparison with the 100% fluctuations in a SASE FEL. The peak and average brilliance are three orders of magnitude higher than the values which could be reached by a conventional X-ray SASE FEL. © 2001 Elsevier Science B.V. All rights reserved.

PACS: 41.60.Cr; 42.55.V; 07.85—m; 76.80+y

Keywords: Free electron laser; X-ray laser; SASE; X-ray monochromator; Mössbauer radiation

1. Introduction

A single pass X-ray SASE FEL [2,3] can be modified as proposed in Ref. [1] to significantly reduce the bandwidth and the fluctuations of the output radiation. The modified scheme consists of two undulators and an X-ray monochromator located between them. The first undulator operates in the linear regime of amplification starting from noise and the output radiation has the usual SASE properties. After the first undulator, the electron beam is guided through a bypass and the X-ray beam enters the monochromator, which selects a narrow band of radiation. At the entrance of the second undulator, the monochromatic X-ray beam is combined with the electron beam and is amplified up to the saturation level.

The electron micro-bunching induced in the first undulator should be destroyed prior to its arrival at the second one. This is achieved automatically due to the energy spread of the electron beam guided through the bypass. At the entrance of second undulator, the radiation power from the monochromator dominates significantly over the shot noise and the residual electron bunching. Also, the input signal bandwidth is small compared to the FEL amplifier bandwidth.

The monochromatization of the radiation is performed at a relatively low level of radiation

*Corresponding author. Tel.: +49-40-8998-2676; fax: +49-40-8998-4475.

E-mail address: schneidm@mail.desy.de (E.A. Schneidmiller).

power, which allows one to use conventional X-ray optical elements for the monochromator design. X-ray grating techniques can be used successfully down to wavelengths of several angstroms and at shorter wavelengths crystal monochromators could be used.

The proposed two-stage scheme possesses two significant advantages. First, it gives the possibility for achieving monochromaticity of the output radiation close to the limit given by the finite duration of the electron pulse and for increasing the brilliance of the SASE FEL. Second, shot-to-shot fluctuations of the energy spectral density could be reduced from 100% to <10% when the second undulator section operates at saturation. Since it is a single bunch scheme, it does not require any special time diagram for the accelerator operation.

The conditions that are necessary and sufficient for the effective operation of a two-stage SASE FEL were discussed in Ref. [1] and can be summarized as follows:

$$P_{in}^{(2)}/P_{shot} = G^{(1)}R_m(\delta\lambda/\lambda)_m/(\delta\lambda/\lambda)_{SASE} \gg 1 \qquad (1)$$

$$\lambda/\pi\sigma_z < (\delta\lambda/\lambda)_m \ll (\delta\lambda/\lambda)_{SASE} \qquad (2)$$

$$G^{(1)} \ll G_{sat}(SASE). \qquad (3)$$

Here $P_{in}^{(2)}$ is the input radiation power at the entrance to the second undulator, P_{shot} is the effective power of shot noise, $G^{(1)}$ is the power gain in the first undulator, R_m is the net reflection coefficient of the mirrors and the dispersive elements of the monochromator, $(\delta\lambda/\lambda)_m$ is the resolution of the monochromator, $(\delta\lambda/\lambda)_{SASE}$ is the radiation bandwidth of the SASE FEL at the exit of the first undulator, σ_z is the rms length of the electron bunch, and $G_{sat}(SASE)$ is the power gain of SASE FEL at saturation.

An application of such a two-stage scheme to 6 nm SASE FEL at the TESLA Test Facility at DESY [4] was discussed in Refs. [1,5]. Now, it is funded and is expected to be the main option for operation of the user facility. In this paper we consider a possible design of the two-stage FEL operating in the Angström range for the X-ray laboratory integrated into the TESLA linear collider project [6]. Specifically, we consider the

FEL optimized for the 14.4 keV X-rays (0.86 Å). There is special interest in the 14.4 keV X-rays because of additional possibilities which the powerful and diverse nuclear resonance scattering techniques which use the highly monochromatic 14.4 Mössbauer radiation [7] open for studies of structure and dynamics of solids, biological molecules, etc.

2. Parameters of the two-stage X-ray FEL

The main parameters of the electron beam and the undulators are presented in Table 1 and coincide with those of the usual SASE FEL at 14.4 keV being designed for the TESLA X-ray laboratory. The SASE FEL bandwidth at the exit of the first stage is about 7×10^{-4} and weakly depends on the gain. We require the monochromator FWHM bandwidth to be about 20 meV, or 1.4×10^{-6} (see Eq. (2)). The net reflection coefficient of all the crystals of the monochromator is expected to be in a range of 0.3–0.5. Requiring the excess of the input radiation power at the entrance to the second undulator $P_{in}^{(2)}$ over the effective power of the shot noise P_{shot} to be two orders of magnitude (see Eq. (1) and Ref. [5]) we end up with a required gain of 1.5×10^5 in the first undulator. The SASE FEL gain at the saturation would be about 4×10^6, i.e., condition (3) is satisfied.

Table 1
Parameters of the electron beam and the undulators

Electron beam	
Energy, $\mathscr{E}_0$	25 GeV
Peak current, I_0	5 kA
rms bunch length, σ_z	23 μm
Normalized rms emittance, ε_n	1.6π mm mrad
rms energy spread (entrance)	2.5 MeV
External β-function	45 m
Bunch separation	93 ns
Number of bunches per train	11315
Repetition rate	5 Hz
Undulator	
Type	Planar
Period, λ_w	4.5 cm
Peak magnetic field, H_w	9.5 kGs

Table 2
Parameters of the first and the second stages

First stage	
Wavelength, λ	0.860 Å
Effective power of shot noise, P_{shot}	5 kW
Length of undulator, $L_w^{(1)}$	140 m
FWHM bandwidth, $(\delta\lambda/\lambda)_{SASE}$	7×10^{-4}
Radiation spot size (FWHM)	50 μm
Angular divergence (FWHM)	1 μrad
Peak power	0.75 GW
Average power	4 W
Second stage	
Input power, $P_{in}^{(2)}$	0.5 MW
Length of undulator, $L_w^{(2)}$	170 m
FWHM bandwidth, $\delta\lambda/\lambda$	1.4×10^{-6}
Angular divergence (FWHM)	0.7 μrad
Radiation spot size (FWHM)	110 μm
Peak power	20 GW
Average power	110 W
Peak brilliance	3×10^{36} Phot./ (s × mrad^2 × mm^2 × 0.1% bandw.)
Average brilliance	2×10^{28} Phot./ (s × mrad^2 × mm^2 × 0.1% bandw.)

Table 3
Parameters of the monochromator and the electron beam bypass

Monochromator	
Nominal energy	14.4 keV
Bandwidth	20 meV
Tunability range	2–4 keV
Total reflection coefficient	0.3–0.5
Absorbed average power	<200 mW
Absorbed average power density	<50 W/mm^2
Electron beam bypass (chicane)	
Total length	40 m
Bending angle of magnets	1.3°
Path lengthening	1 cm

The parameters of the first and the second stages are presented in Table 2. They have been calculated with the FEL simulation code FAST [8]. When calculating these parameters we have taken into account the growth of energy spread in the electron beam due to the quantum fluctuations of undulator radiation [9,10]. The peak and average brilliance of the X-ray beam at the exit of the second stage are 500 times larger than in the case of the usual SASE FEL. The shot-to-shot fluctuations of the energy spectral density are reduced to the 10% level due to a nonlinear stabilization mechanism [5].

The distance between the two undulators is mainly determined by parameters of the electron beam bypass (chicane) that must compensate for a path delay of X-rays in the monochromator. The latter is on the order of 1 cm (cf. Section 3). For a bending angle of the chicane magnets of 1.3° the total length of the chicane is about 40 m. The electron beam microbunching is completely destroyed at the end of the bypass due to the uncorrelated energy spread in the beam and reasonable longitudinal dispersion of the chicane [1]. Because of the small angular divergence of the radiation coming out of the first undulator, the focusing of this radiation is not necessary. Indeed, the calculations show that the input coupling factor [11] to the eigenmode in the second undulator decreases by about 30% with respect to the case of optimal focusing. The main parameters of the monochromator and the electron beam bypass are presented in Table 3.

3. High energy-resolution, high heat-load, tunable X-ray monochromator

The main requirements for the X-ray monochromator of the two stage XFEL are

(i) degree of monochromatization: $\lambda/\delta\lambda = E/\delta E = 0.7 \times 10^6$;
(ii) tunability range: a few keV;
(iii) resistance to high heatload.

To reach the required value of monochromatization alone is not a problem. Nowadays, a monochromatization of 10^7 and more is possible. Bragg diffraction is the main tool used for such purposes. For a recent review of the techniques used and achievements in this field see, e.g., Ref. [12]. However, the combination of the three requirements renders the realization of such a monochromator less straightforward.

3.1. Spectral width of Bragg reflections and tunability range

Tunability of an X-ray monochromator for a given monochromaticity will be addressed first.

The relative energy width of a Bragg reflection in a thick nonabsorbing crystal (like silicon, diamond, etc.) is given in the dynamical theory of diffraction in perfect crystals (see, e.g., Ref. [13]) by

$$\frac{\delta E}{E} = \frac{\delta \lambda}{\lambda} = \frac{|\chi_{\mathbf{g}}|}{\sin^2 \theta}.$$

Here θ is the glancing angle of the radiation plane wave to the reflecting atomic planes $(h\,k\,l)$ with the interplanar distance $d_{h\,k\,l}$ and the related reciprocal vector $\mathbf{g}$, $|\mathbf{g}| = 2\pi/d_{h\,k\,l}$. The relation between the wavelength λ of the reflected X-rays and θ is given by the Bragg law $2d_{h\,k\,l}\sin\theta = \lambda$.

$$\chi_{\mathbf{g}} = -\frac{r_e \lambda^2}{\pi V} Zf\left(\frac{\sin\theta}{\lambda}\right) \exp\left(-\frac{\langle u^2 \rangle}{\lambda^2} 8\pi^2 \sin^2\theta\right)$$

is the Fourier component of the electric susceptibility corresponding to the reciprocal vector $\mathbf{g}$. The expression is valid for a single-atom crystal. The following notations are used: V is the volume of the crystal unit cell, Z is atomic number, r_e is the classical electron radius, $f(\ldots)$ is the atomic scattering formfactor, $\exp(\ldots)$ is the Debye–Waller factor, $\langle u^2 \rangle$ is mean square displacement of atoms due to thermal vibrations in the direction of the scattering vector $\mathbf{g}$. The combination of the both equations gives

$$\frac{\delta E}{E} = \frac{r_e \lambda_{h\,k\,l}^2}{\pi V} Zf(\lambda_{h\,k\,l}^{-1}) \exp\left(-8\pi^2 \frac{\langle u^2 \rangle}{\lambda_{h\,k\,l}^2}\right). \tag{4}$$

Here the Bragg wavelength $\lambda_{h\,k\,l} = 2d_{h\,k\,l}$ is introduced—the largest wavelength of X-rays allowed by the Bragg law to be reflected from the $(h\,k\,l)$ atomic planes.

An important and very favorable implication of Eq. (4) for our applications is that the relative spectral width for the given Bragg reflection $(h\,k\,l)$ is independent of the energy or glancing angle of X-rays and defined merely by properties of the crystal and reflecting atomic planes. In particular, it implies that the choice of a crystal, reflecting atomic planes and crystal temperature determines the

spectral resolution. Fig. 1 shows results of evaluations of the monochromaticity $E/\delta E$ of X-rays reflected from different atomic planes $(h\,k\,l)$ in diamond (C) and silicon (Si) single crystals at room temperature. The range of tunability is limited only by the lowest X-ray energy allowed by the Bragg law—the Bragg energy $E_{h\,k\,l} = hc/\lambda_{h\,k\,l}$.

The 1 μrad divergence (FWHM) of X-rays from the first undulator was also taken into account, which shows up in the decreasing monochromaticity with increasing X-ray energy. This occurs when the angular acceptance of Bragg reflections approaches the angular divergence of the incoming beam.

As it is seen from Fig. 1 the number of possible Bragg reflections which provide required monochromaticity and tunability range is rather limited. In case of diamond (C), these are $(1\,3\,7)$ or $(1\,1\,7)$ and equivalent ones. In case of silicon single crystals these are $(1\,3\,9)$ or $(3\,3\,9)$ and equivalent ones.

We are discussing here only silicon and diamond single crystals. There are two reasons for this. Si single crystals are the most perfect crystals available nowadays. This is an important feature which ensures the preservation of the coherent properties of the radiation from the first undulator. Diamond, although not so perfect as silicon, nevertheless, sufficiently large ($\approx 10\ \text{mm}^2$) perfect crystals are available [14,15]. The greatest advantage of diamond is its ability to withstand high heat load due to the extremely high thermal conductivity, low thermal expansion, small X-ray absorption, and high reflectivity [16].

3.2. Actual scheme

We have chosen a four-bounce scheme for the X-ray monochromator as shown in Fig. 2. This solution is advantageous because it allows one to keep the direction and the position of the X-ray beam at the exit the same as at the entrance of the monochromator.

This solution is also advantageous because it allows one to use the first two Bragg reflectors as a high-heat load premonochromator, which withdraws the major heat load from the actual high-energy-resolution monochromator—the third and the fourth crystals. In front of the first crystal one should install a collimator with a hole of diameter

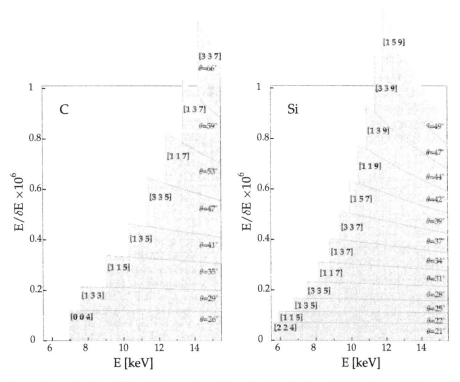

Fig. 1. Degree of monochromatization $E/\delta E$ of X-rays reflected from the atomic planes $(h\,k\,l)$ in diamond (C) and silicon (Si) single crystals at room temperature. The divergence of the incident X-rays is assumed to be 1 μrad. At the right end of each graph the glancing angle θ is given corresponding to the highest X-ray energy $E = 15.5$ keV considered.

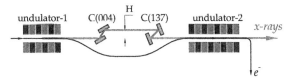

Fig. 2. The two-stage XFEL with the four-bounce X-ray monochromator: $C(004) \times C(004) - C(137) \times C(137)$. Additional path length acquired by the X-rays in the monochromator is $\delta L = 3.0H$. Alternative realization is $C(004) \times C(004) - Si(139) \times Si(139)$ with silicon crystals as a high-energy-resolution monochromator. The additional path length is $\delta L = 1.7H$.

about 100 μm. The spontaneous emission power load on the crystal will be then suppressed to a level much lower than the power of the SASE radiation. In the pre-monochromator part one can use, e.g., diamond crystal plates 100 μm thick and Bragg reflection $(0\,0\,4)$. Given that the crystal is perfect, it reflects 99% of the incident X-rays within a band of

132 meV. Only 5% of the off-band radiation is absorbed, and the rest passes through. The absorbed power is thus about 200 mW, nearly 20 times less than the incident power. The absorbed power density is about 50 W/mm^2. It is comparable with the power at the monochromators of the third generation synchrotron sources [17–20]. The radiation power which reaches the high-resolution monochromator crystals is $\simeq 1.3\%$ of the initial value ($\simeq 50$ mW). The latter can be reduced by a factor of two if one uses the reflection $C(1\,3\,3)$ with a bandwidth of 75 meV.

The final monochromatization to the required level takes place by a high-index reflection from the third and the fourth crystals. The required monochromatization $E/\delta E = 0.7 \times 10^6$ of the 14.4 keV X-rays can be achieved, according to Fig. 1, by only a very limited number of Bragg reflections. The final choice of the reflection should be dictated by the requirements of tunability and heat load. The Bragg reflections in diamond have a smaller tunability range. On the other hand, they have higher reflectivity and angular acceptance. Fine adjustment of the angular acceptance and the energy bandwidth can be performed by using asymmetric Bragg reflections.

An important technical issue is a path delay (with respect to the straight path) which the X-ray pulse acquires in the monochromator. The path delay equals

$$\delta L = H(\tan \theta_{H\,K\,L} + \tan \theta_{h\,k\,l}),\qquad(5)$$

where H is the beam shift, $\theta_{H\,K\,L}$ in the Bragg angle of the reflections in the high heat-load part (the first two crystals), and $\theta_{h\,k\,l}$ in the Bragg angle of the reflections in the high-energy-resolution part of the monochromator (the third and the fourth crystals). By varying H one can keep the delay δL constant in the whole tunability range of the monochromator. For the proposed monochromator schemes the actual values of δL are given in Fig. 2 caption and can be about 1 cm.

4. Conclusion

Our analysis shows that the construction of the high-brilliant two-stage SASE FEL in the Angström spectral range is feasible. We have considered an extreme case when the spectral bandwidth is defined by the length of the electron bunch (lower limit in Eq. (2)). By increasing the bandwidth one can increase the tunability range and reduce the power density incident on crystals of the monochromator. The final choice of parameters will be dictated by needs of potential users of intense monochromatic X-rays.

Acknowledgements

We thank G. Materlik, J. Pflüger and J. Rossbach for their interest in this work. We are grateful to J. Feldhaus and J.R. Schneider for many useful discussions on the two-stage SASE FEL.

References

[1] J. Feldhaus, E.L. Saldin, J.R. Schneider, E.A. Schneidmiller, M.V. Yurkov, Opt. Commun. 140 (1997) 341.
[2] Ya.S. Derbenev, A.M. Kondratenko, E.L. Saldin, Nucl. Instr. and Meth. 193 (1982) 415.
[3] J.B. Murphy, C. Pellegrini, Nucl. Instr. and Meth. A 237 (1985) 159.
[4] J. Rossbach, Nucl. Instr. and Meth. A 375 (1996) 269.
[5] E.L. Saldin, E.A. Schneidmiller, M.V. Yurkov, Nucl. Instr. and Meth. A 445 (2000) 178.
[6] R. Brinkmann, G. Materlik, J. Rossbach, A. Wagner (Eds.), Conceptual Design of a 500 GeV e+e− Linear Collider with Integrated X-ray Laser Facility, DESY 97-048, Hamburg, 1997.
[7] E. Gerdau, H. de Waard (Eds.), Nuclear Resonant Scattering of Synchrotron Radiation, Hyperfine Interactions 123/124 (1999) and 125 (2000).
[8] E.L. Saldin, E.A. Schneidmiller, M.V. Yurkov, Nucl. Instr. and Meth. A 429 (1999) 233.
[9] E.L. Saldin, E.A. Schneidmiller, M.V. Yurkov, Nucl. Instr. and Meth. A 374 (1996) 401.
[10] E.L. Saldin, E.A. Schneidmiller, M.V. Yurkov, Nucl. Instr. and Meth. A 381 (1996) 545.
[11] E.L. Saldin, E.A. Schneidmiller, M.V. Yurkov, The Physics of Free Electron Lasers, Springer, Berlin, 1999.
[12] T. Toellner, Hyperfine Interactions 125 (2000) 3.
[13] Z.G. Pinsker, Dynamical Scattering of X rays in Crystals, Springer, Berlin, 1978.
[14] T. Ishikawa, K. Tamasakua, M. Yabashib, S. Gotob, Y. Tanakaa, H. Yamazakib, K. Takeshitab, H. Kimurab, H. Ohashib, T. Matsushitab, T. Ohatab, Proc. SPIE 4145 (2000).
[15] J.P.F. Sellschop, S.H. Connell, R.W.N. Nilen, C. Detlefs, A.K. Freund, J. Hoszowska, R. Hustache, R.C. Burns, M. Rebak, J.O. Hansen, D.L. Welch, C.E. Hall, New Diamond Frontier Carbon Technol. 10 (2000) 253.
[16] A.K. Freund, Opt. Eng. 34 (1995) 432.
[17] A.K. Freund, Rev. Sci. Instr. 67 (1996) 5.
[18] D.M. Mills, J. Synchrotron Rad. 4 (1997) 117.
[19] M. Yabashi, H. Yamazaki, K. Tamasaku, S. Goto, K. Takeshita, T. Mochizuki, Y. Yoneda, Y. Furukawa, T. Ishikawa, Proc. SPIE 3773 (1999) 2.
[20] D.H. Bilderback, A.K. Freund, G.S. Knapp D.M. Mills, J. Synchrotron Rad. 7 (2000) 53.

ELSEVIER

Nuclear Instruments and Methods in Physics Research A 475 (2001) 363–367

NUCLEAR
INSTRUMENTS
& METHODS
IN PHYSICS
RESEARCH
Section A

www.elsevier.com/locate/nima

Development of a pump-probe facility combining a far-infrared source with laser-like characteristics and a VUV free electron laser

B. Faatz[a,*], A.A. Fateev[b], J. Feldhaus[a], J. Krzywinski[c], J. Pflueger[a],
J. Rossbach[a], E.L. Saldin[a], E.A. Schneidmiller[a], M.V. Yurkov[b]

[a] *DESY, Deutsches Elektronen-Synchrotron, Notkestrasse 85, D-22603 Hamburg, Germany*
[b] *Joint Institute for Nuclear Research, Dubna, 141980 Moscow Region, Russia*
[c] *Institute of Physics of the Polish Academy of Sciences, 02688 Warszawa, Poland*

Abstract

The TESLA Test Facility (TTF) at DESY is a facility producing sub-picosecond electron pulses for the generation of VUV or soft X-ray radiation in a free electron laser (FEL). The same electron pulses would also allow the direct production of high-power coherent radiation by passing the electron beam through an undulator. Intense, coherent far-infrared (FIR) undulator radiation can be produced from electron bunches at wavelengths longer than or equal to the bunch length. The source described in this paper provides, in the wavelength range 50–300 μm, a train of about 1–10 ps long radiation pulses, with about 1 mJ of optical energy per pulse radiated into the central cone. The average output power can exceed 50 W. In this conceptual design, we intend to use a conventional electromagnetic undulator with a 60 cm period length and a maximum field of 1.5 T. The FIR source will use the spent electron beam coming from the VUV FEL which allows one to significantly extend the scientific potential of the TTF without interfering with the main option of the TTF FEL operation. The pulses of the coherent FIR radiation are naturally synchronized with the VUV pulses from the main TTF FEL, enabling pump-probe techniques using either the FEL pulse as a pump or the FIR pulse as a probe, or vice versa. © 2001 Elsevier Science B.V. All rights reserved.

PACS: 41.60.Cr; 52.75.M; 07.57.Hm

Keywords: Free electron laser; High-gain FEL; Far-infrared source

1. Introduction

The FIR range of the electromagnetic spectrum is not well covered by intense sources except for a

*Corresponding author. Tel.: + 49-40-8998-4513; fax: + 49-40-8998-2787.

E-mail addresses: faatz@desy.de (B. Faatz), saldin@mail.-desy.de (E.L. Saldin), saldin@mail.desy.de (E.L. Saldin).

few operating FELs. The analysis of the parameters of existing FIR FELs shows that practical sources of broadly tunable, powerful, coherent FIR radiation remains essentially unavailable at wavelengths beyond 200 μm [1–3]. This situation, however, might change soon. The development of magnetic bunch compression systems, together with advances in superconducting accelerator technology and design, now offers the new possibility of laser-like sources in the far-infrared wavelength range. The generation of relativistic, sub-picosecond electron pulses allows the direct production of high-power, coherent, narrow-band, FIR radiation by passing the electron beam through an undulator. This provides a reliable and easily tunable powerful source of FIR radiation for scientific applications [4,5].

The TTF is a facility producing sub-picosecond electron pulses (50 μm rms) for the generation of VUV or soft X-ray radiation. Utilizing these sub-picosecond electron bunches can also provide broadband FIR source. Intense, coherent FIR radiation can be produced from sub-picosecond electron bunches at wavelengths longer than or equal to the bunch length. The total radiation from an electron bunch is the summation of the electric fields emitted by each individual electron and the total radiated energy is then equal to the square of the total electric field. The coherent radiation energy is proportional to the square rather than linearly proportional to the number of radiating electrons. Since there are 6×10^9 electrons in each bunch, the radiation intensity is enhanced by this large factor over the incoherent radiation. This paper describes such a coherent source, proposed as part of the TTF FEL user facility. The FIR source addresses the needs of the science community for a high-brightness, tunable source covering a broad region of the far-infrared spectrum — from 50 to 300 μm. The FIR radiator described in this paper provides a train of about 1–10 ps long pulses with up to 1 mJ of optical energy per pulse into the central cone. The average output power can exceed 50 W.

The pump-probe technique is one of the most promising methods for the application of a high power FIR source [6]. It is the aim of the present project to develop a user facility for pump-probe experiments in the picosecond regime, combining FIR and short-wavelength FEL radiation. The TTF will allow, for the first time, the integration of a far-infrared coherent radiation source and a VUV beamline. One type of experiment will use the VUV FEL beam as a pump and the far-infrared photon beam as a probe; in this mode, researchers will be able to study the vibrational structure of highly excited and superexcited molecules. The other mode (far-infrared beam pump and VUV beam probe) can be used to study cluster energetics and dynamics. The FIR radiation can be used to excite the clusters, which can subsequently be dissociated or ionized by the VUV radiation. Spectroscopic and structural information can thus be extracted. Spectroscopy of gas-phase free radicals will also benefit from the FIR beam pump and VUV beam probe experiments. In these experiments, a cold molecular beam containing a small concentration of radicals would be excited by the intense FIR beam, tunable across the absorption spectrum. Since the density of radicals in the beam is not high enough to allow the direct measurement of absorption, a VUV beam from the TTF FEL would be used to detect the infrared-exited states of molecules by selectively ionizing the vibrationally excited radicals.

2. Temporal coherent undulator radiation

The electron beam current is made up of moving electrons randomly arriving at the entrance to the undulator:

$$I(t) = (-e) \sum_{k=1}^{N} \delta(t - t_k)$$

where $\delta(\cdots)$ is the delta function, $(-e)$ is the charge of the electron, N is the number of electrons in a bunch, and t_k is the random arrival time of the electron at the undulator entrance. The electron bunch profile is described by the profile function $F(t)$. The beam current averaged over an ensemble of bunches can be written in the form:

$$\langle I(t) \rangle = (-e)NF(t).$$

The probability of arrival of an electron during the time interval $t, t + dt$ is equal to $F(t) \, dt$.

The radiation power at the frequency ω, averaged over an ensemble, is given by the expression:

$$\langle P(\omega) \rangle = p(\omega)[N + N(N - 1)|\bar{F}(\omega)|^2]$$

where $p(\omega)$ is the radiation power from one electron and $\bar{F}(\omega)$ is the Fourier transform of the bunch profile function. For wavelengths shorter than the bunch length the form factor reduces to zero and approaches unity for longer wavelengths.

To optimally meet the needs of basic research with FIR coherent radiation, it is desirable to provide specific radiation characteristics. To generate these characteristics, radiation is produced from an undulator installed along the electron beam path. The undulator equation

$$\omega = 2ck_w\gamma^2\left[1 + \frac{K^2}{2} + \gamma^2\theta^2\right]^{-1}$$

tells us the frequency of radiation as a function of the undulator period $\lambda_w = 2\pi/k_w$, undulator parameter K, electron energy γ, and polar angle of observation θ. Note that for radiation within the cone of half angle

$$\theta_{con} = \frac{\sqrt{1 + K^2/2}}{\gamma\sqrt{N_w}}$$

the relative spectral bandwidth is $\Delta\omega/\omega \simeq 1/N_w$, where N_w is the number of undulator periods. The energy radiated into the central cone, for a single electron, is given by

$$\Delta\mathscr{E}_{con} \simeq \pi e^2 A_{JJ}^2 \omega_0 K^2 / [c(1 + K^2/2)].$$

Here $\omega_0 = 2\gamma^2 k_w/(1 + K^2/2)$ is the resonant frequency, $A_{JJ} = [J_0(Q) - J_1(Q)]$, J_n is the Bessel function of nth order, and $Q = K^2/(4 + 2K^2)$. The coherent radiation enhances the energy radiated into the central cone by a factor of $N|\bar{F}(\omega_0)|^2$.

3. Facility description

In the far-infrared, beyond 200 μm, a source based on coherent undulator radiation has unique capabilities. In this paper, we propose to integrate such a source into the TESLA Test Facility at DESY [7,8]. This source will be able to deliver up to 800 μs long trains of FIR pulses at a separation of 111 ns with about 1–10 ps duration,[1] up to 1 mJ energy radiated into the central cone, and 50–300 μm wavelength. The superconducting linac

will operate at about 1% duty factor, and the average output power of coherent FIR radiation can exceed 50 W.

We propose to install an additional undulator after the VUV SASE FEL. Because the FIR source uses the electron beam coming from the VUV FEL, the proposed source operates in a "parasitic" mode not interfering with the main mode of the VUV FEL operation (see Fig. 1). Starting points of the design are the project parameters of the electron beam after the VUV FEL at the TTF (see Table 1). The planar undulator is an inexpensive electromagnetic device with 10 periods, each 60 cm long. At the operation wavelength of the FIR source around 300 μm, the peak value of the magnetic field is about 1.5 T.

Coherent radiation produced by the electron bunch strongly depends on the bunch profile and charge. Fig. 2 shows the expected axial profile of the electron bunch from the TTF accelerator.[2] Using such a bunch, we can generate powerful FIR radiation in a wide wavelength band, from 50 to 300 μm (see Fig. 3). A unique feature of the TTF accelerator consists of the possibility to control, in a wide limit, the axial bunch profile by means of bunch compressors. Running of the TTF accelerator with shorter bunches would allow to shift the short wavelength boundary of the FIR spectrum below 50 μm. Another unique feature of the TTF accelerator is that it is capable of accelerating electron bunches with a bunch charge of about 10 nC, thus providing the possibility to generate FIR pulses with peak power exceeding a gigawatt level.

Many practical applications require the control of the shape and time characteristics of the FIR pulse. For instance, closely spaced picosecond FIR pulses with controllable phase relations are needed for coherent multi-photon excitation and selective excitation of, for example, certain molecular vibrations. The proposed FIR source provides wide possibilities to control and modify, in a well-defined manner, the shape of the radiation pulse on a picosecond time scale (see Fig. 4). The electron beam passing a uniform undulator produces an FIR pulse with a rectangular profile.

---

[1] When the electron bunch moves along the undulator, the electromagnetic wave advances the electron beam by one wavelength at one undulator period.

---

[2] Calculations have been performed by T. Limberg.

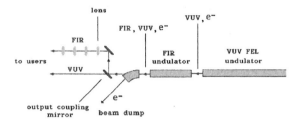

Fig. 1. Schematic layout of the FIR–VUV pump-probe facility.

Table 1
Parameters of the FIR coherent radiation source

Electron beam	
Energy	1000 MeV
Bunch charge	1 nC
Rms bunch length	50 μm
Normalized emittance	2π mm-mrad
Rms energy spread	2.5 MeV
Bunch repetition rate	9 MHz
Duty factor	1%
Undulator	
Type	Planar
Period	60 cm
Peak magnetic field	1.2–1.5 T
Number of periods	10
Output radiation	
Wavelength	50–300 μm
Bandwidth	transform-limited
Peak power	up to 100 MW
Average power	up to 50 W
Micropulse duration	1–10 ps
Micropulse energy	up to 1 mJ

FIR optical pulses can be shaped in a complicated manner by means of individual tuning of the magnetic field in each period.

The proposed FIR source is compatible with the layout of the TTF and the VUV FEL at DESY, and can be realized with minimal additional efforts. The undulator and outcoupling optical system can be installed in the unoccupied straight vacuum line used for the electron and VUV beamlines behind the dipole magnet separating the electron beam from the VUV beam (see Fig. 1). In order to make use of the FIR radiation an additional mirror is needed to couple out the

major fraction of the optical power in the central cone and to direct it to the experimental area. The distance between the mirror and the exit of the second undulator is 10 m, the distance between the mirror and the exit of the FEL undulator is about 16 m. The minimum size of the hole in the outcoupling mirror is defined by the condition that VUV radiation losses due to hole aperture limitation should be avoided. The fraction of the output FEL power passing the mirror hole is calculated from the angular distribution of the VUV radiation (20 μrad rms). For a hole diameter of 2 mm the fraction of VUV power directed into the experimental area is close to 100%. Due to the angle-frequency correlation of the coherent

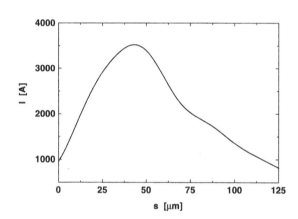

Fig. 2. Axial bunch profile of the electron bunch from TTF accelerator.

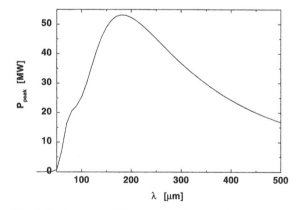

Fig. 3. Peak power of FIR source at the TESLA Test Facility produced by the bunch with axial profile plotted in Fig. 2.

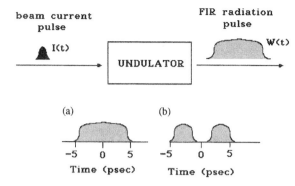

Fig. 4. FIR pulse profiles generated by the FIR undulator in the central cone (300 μm wavelength). Plot (a) illustrates the pulse produced by a uniform undulator (rectangular pulse of 10 ps duration). Plot (b) illustrates the case when four central poles are switched off (two pulses of 3 ps duration each delayed by 3 ps).

undulator radiation the required frequency bandwidth of radiation could be provided by angular selection. To provide a natural selection of coherent radiation in the central cone ($\theta_{con} \simeq 5$ mrad), the radius of the mirror should be equal to 6 cm at a distance of about 10 m between the exit of the second undulator and the mirror.

The operation of the proposed FIR source is insensitive to the emittance of the electron beam, since the condition of the optimal electron beam's transverse size is: $\sigma^2 \ll Lc/\omega \simeq 1$ cm^2, where L is the undulator length. The analysis of the parameters of the FIR source has shown that it will operate reliably even for an emittance exceeding the project value of 2π mm mrad by two orders of magnitude.

An undulator is a sequence of bending magnets where particles with different energies have different path length. As a result, the energy spread in the beam leads to the bunch lengthening in the undulator. When an electron bunch passes the FIR undulator, radiation interaction induces additional energy spread in the electron beam which can also lead to bunch lengthening. Recently, the problem connected with radiative interaction of the particles in a line-charge microbunch moving in an undulator has been investigated analytically [9]. In the case of a Gaussian

bunch profile the induced energy spread $\Delta \mathscr{E}_f$ is given by

$$\frac{\Delta \mathscr{E}_f}{\mathscr{E}} = \frac{r_e N K^2 L}{\sqrt{2\pi}\sigma_z^2 \gamma^3} G(p, K) \qquad (1)$$

where, r_e is the classical electron radius, $p = k_w \sigma_z \gamma^2 / [(1 + K^2/2)]$ is the bunch length parameter, and σ_z is the rms bunch length. Parameters of the FIR source project are: $N = 6 \times 10^9$, $\sigma_z = 50$ μm, $K = 70$ (for 200 μm wavelength), $L = 6$ m, and $\gamma = 2 \times 10^3$. The value of G is $G \simeq 0.5$. Substituting these values into (1), we obtain an induced correlated energy spread at the exit of undulator $\Delta \mathscr{E}_f / \mathscr{E} \simeq 2\%$. This leads to an increase of the bunch length:

$$\Delta l_b \simeq \frac{L(1 + K^2/2)\Delta \mathscr{E}_f}{2\gamma^2} \frac{\Delta \mathscr{E}_f}{\mathscr{E}} \simeq 20 \ \mu m.$$

Since this value is much less than the radiation wavelength $\lambda \simeq 100$ μm, we can conclude that bunch decompression in the undulator due to induced energy spread should not be a serious limitation in our case.

Acknowledgements

We thank D. Trines and J.R. Schneider for interest in this work and support, and A.F.G. van der Meer for useful discussions on scientific applications of coherent far-infrared sources.

References

[1] G. Ramian, Nucl. Instr. and Meth. A 318 (1992) 225.
[2] D. Oepts, A.F.G. van der Meer, P.W. van Amersfort, Infrared Phys. Tech. 36 (1995) 297.
[3] H.A. Schwettman, T.I. Smith, R.L. Swent, Nucl. Instr. and Meth. A 375 (1996) 662.
[4] D. Bocek, et al., SLAC-PUB-7016, Nucl. Instr. and Meth. A 375 (1996) 13.
[5] C. Settakorn, et al., SLAC-PUB-7812, May 1998.
[6] K.-J. Kim, et al., LBL preprint Pub-5335, 1992.
[7] A VUV Free Electron Laser at the TESLA Test Facility: Conceptual Design Report, DESY Print TESLA-FEL 95-03, Hamburg, DESY, 1995.
[8] J. Rossbach, Nucl. Instr. and Meth. A 375 (1996) 269.
[9] E.L. Saldin, E.A. Schneidmiller, M.V. Yurkov, Nucl. Instr. and Meth. A 417 (1998) 158.

ELSEVIER

Nuclear Instruments and Methods in Physics Research A 475 (2001) 368–372

**NUCLEAR
INSTRUMENTS
& METHODS
IN PHYSICS
RESEARCH**
Section A

www.elsevier.com/locate/nima

Development of a pump-probe facility with sub-picosecond time resolution combining a high-power ultraviolet regenerative FEL amplifier and a soft X-ray SASE FEL

B. Faatz[a], A.A. Fateev[b], J. Feldhaus[a], J. Krzywinski[c], J. Pflueger[a], J. Rossbach[a], E.L. Saldin[a], E.A. Schneidmiller[a,*], M.V. Yurkov[b]

[a] *Deutsches Elektronen-Synchrotron (DESY), Notkestrasse 85, D-22607 Hamburg, Germany*
[b] *Joint Institute for Nuclear Research, Dubna, 141980 Moscow Region, Russia*
[c] *Institute of Physics of the Polish Academy of Sciences, 02688 Warszawa, Poland*

Abstract

This paper presents the conceptual design of a high power radiation source with laser-like characteristics in the ultraviolet spectral range at the TESLA Test Facility (TTF). The concept is based on the generation of radiation in a regenerative FEL amplifier (RAFEL). The RAFEL described in this paper covers a wavelength range of 200–400 nm and provides 200 fs pulses with 2 mJ of optical energy per pulse. The linac operates at 1% duty factor and the average output radiation power exceeds 100 W. The RAFEL will be driven by the spent electron beam leaving the soft X-ray FEL, thus providing minimal interference between these two devices. The RAFEL output radiation has the same time structure as the X-ray FEL and the UV pulses are naturally synchronized with the soft X-ray pulses from the TTF FEL. Therefore, it should be possible to achieve synchronization close to the duration of the radiation pulses (200 fs) for pump-probe techniques using either an UV pulse as a pump and soft X-ray pulse as a probe, or vice versa. © 2001 Elsevier Science B.V. All rights reserved.

PACS: 41.60.Cr; 52.75.M; 42.72.Bj

Keywords: Free electron laser; High-gain FEL; Ultraviolet source

1. Introduction

Pump-probe techniques using either the soft X-ray pulse from the TTF FEL [1] as a pump and the visible–UV laser as a probe pulse, or vice versa, promise unprecedented insight into the dynamics of electronic excitations, chemical reactions and phase transitions of matter, from atoms, through organic and inorganic molecules and clusters, to surfaces, solids and plasmas. For applications in the visible and near-visible wavelength range a pump-probe facility based on a conventional quantum laser system will be available at the TTF.[1] The laser will provide in the visible spectral region between 750 and 900 nm a train of 150 fs

---

*Corresponding author. Tel.: + 49-40-8998-2676; fax: + 49-40-8998-4475.

E-mail address: schneidm@mail.desy.de
(E.A. Schneidmiller).

---

[1] Development of a pump-probe facility combining a high-power optical laser and a soft X-ray FEL: Proposal. Available at DESY by request only.

pulses with 100 μJ of optical energy per pulse, at the same repetition rate as the X-ray FEL. The synchronization of the optical laser with the soft X-ray FEL pulses to within 200 fs is the most challenging task of this project. The main problem is the time jitter (± 1 ps) of electron bunches which are synchronous with the soft X-ray FEL pulses.

In this paper, we describe the extension of the pump-probe facility into the ultraviolet wavelength range. Our approach is based on the idea to use a narrow band feedback between exit and entrance of a high gain FEL amplifier operating in multibunch mode (so-called regenerative FEL amplifier—RAFEL [2]). Such a feedback can be realized in the UV wavelength range using mirrors, lenses, and a grating as a dispersive element. The RAFEL output radiation has the same pulse format as the X-ray FEL and produces 200 fs micropulses with 2 mJ of radiation energy per micropulse and transform-limited spectral width.

2. Facility description

We propose to install an additional undulator after the soft X-ray FEL. A layout of the proposed RAFEL is shown in Fig. 1 and the main parameters are presented in Table 1. Our design makes use of the spent electron beam leaving the X-ray undulator. The SASE process in the X-ray FEL induces an additional energy spread in the electron beam. Nevertheless, the electron beam at the exit of the X-ray FEL is still a good "active medium"

Table 1
Parameters of the UV pump-probe facility (RAFEL option)

Electron beam	
Energy	1000 MeV
Charge per bunch	1 nC
rms bunch length	50 μm
rms emittance	2π mm mrad/γ
rms energy spread	2.5 MeV
Number of bunches	7200/train
Bunch spacing	111 ns
Repetition rate	10 Hz
Undulator	
Type	Planar
Period	7 cm
Peak magnetic field	1–1.4 T
Number of periods	85
Output radiation	
Wavelength	200–400 nm
Bandwidth	Transform-limited
Micropulse duration	200 fs
Micropulse energy	2 mJ

for an UV FEL amplifier. Since the RAFEL uses the spent electron beam, the proposed laser system operates in a "parasitic" mode not interfering with the main mode of the X-ray FEL operation. Since the X-ray and UV radiation pulses are generated by the same electron bunch, there is no problem of synchronization for pump-probe experiments with an accuracy close to the duration of the radiation pulses (200 fs). The RAFEL proposed here will provide intense, tunable and coherent radiation in the UV region of the spectrum between 200 and 400 nm as direct laser output.

The RAFEL undulator and outcoupling optical system proposed can be installed in the unoccupied straight vacuum line used to transfer the X-ray beam to the experimental area. The installation of the feedback is greatly facilitated by the fact that there is free space available for the input optical system. In order to get fully coherent X-ray radiation, a seeding option will be implemented into the X-ray FEL under construction at DESY.[2] The X-ray FEL seeding option consists of two undulators, an electron bypass and X-ray grazing

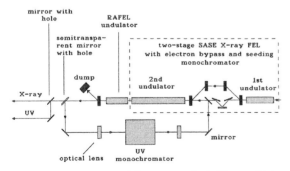

Fig. 1. Schematic layout of the pump-probe facility combining UV regenerative FEL amplifier and soft X-ray FEL at TTF.

[2] Seeding option for the VUV Free Electron Laser at DESY: Proposal. Available at DESY by request only.

incidence monochromator (see Fig. 1). The electron bypass is necessary to delay the electron beam by the same amount as the X-ray photon beam is delayed by the X-ray monochromator. The magnetic chicane has to deflect the electron beam out of the straight flight pass to make room for the X-ray monochromator and input optical elements of RAFEL.

3. Optical feedback system

The RAFEL operates as follows. The first bunch in a train of up to 7200 bunches amplifies shot noise and produces intense, but wide-band radiation. A fraction of the radiation is back-reflected by a semi-transparent output coupling mirror. The spherical grating, which is installed in the straight section of the feedback loop, disperses the light and focuses a narrow band of radiation back on the entrance of the undulator. The bandwidth of the feedback is chosen to produce a photon pulse length about ten times as long as the electron bunch length in order to avoid effects from a ± 1 ps time jitter. This requires a resolving power $\lambda/\Delta\lambda \simeq 3000$ at $\lambda = 200$ nm (photon pulse duration at the monochromator exit $l_{ph} \simeq \lambda^2/(c\Delta\lambda) \simeq 2$ ps).

After the undulator the electron and the radiation beams are separated. The electron beam is guided into the beam dump and the radiation enters the output coupling system. The distance between the feedback outcoupling mirror and the exit of the RAFEL undulator is 20 m, the distance between the mirror and the exit of the X-ray FEL undulator is about 27 m. At a diameter of the hole in the mirror of 3 mm the fraction of the X-ray power directed through the mirror is close to 100%. This mirror is semi-transparent for the UV radiation, and approximately 50% of the UV radiation power is transmitted through it and delivered to the experimental area. Calculations show that an alignment accuracy of about 10 μrad is sufficient for reliable operation of the optical feedback.

The monochromator for the RAFEL should be able to select any wavelength between 200 and 400 nm. We adopted the Namioka scheme where

tuning of the wavelength is performed by means of rotation of the grating, while entrance and exit slits are fixed. Commercially available holographic gratings allow one to focus the image of the entrance slit exactly on the exit slit at small values of coma and astigmatism in the 200–400 nm wavelength band. The feedback transmission factor can be written as $T_{fb} = K_{coupl} \times R_{loss}$, where K_{coupl} is the fraction of output radiation coupled out through the semi-transparent mirror, R_{loss} refers to the losses in the optical elements (mirrors, lenses, grating) of the feedback system. In addition, the grating reduces the peak power of the coherent signal further since it stretches the pulse longitudinally by a factor $\lambda^2/(\sigma_z\Delta\lambda)$ [3].

In the present design the lateral size of the photon beam focus, w, is completely determined by the fixed geometry of the feedback optical system (i.e. the focal distances of the mirrors and the aperture of the X-ray undulator vacuum chamber). In our numerical example for 400 nm wavelength the size of the photon beam focus is about $w \simeq 800$ μm, which is about 15 times larger than the rms electron beam size, σ. Such a mismatch, however, is not dramatic and will result in a reduction of the gain by a factor of about 3 only (see Section 5 for more details). Taking into account all the effects mentioned above, the overall loss factor for the feedback system is about 10^{-2}. When the power gain in the undulator, G, exceeds the relative losses of power in the optical feedback system, the output radiation power begins to grow, i.e. lasing takes place.

4. Undulator

The undulator is one of the central components of the RAFEL. The values of the peak field and the period length are given in Table 1. The required field strength can be achieved using a hybrid configuration. At a gap of 12 mm, a peak field up to 1.5 T is feasible for the undulator period of $\lambda_u = 7$ cm. To minimize the length of the matching section between the VUV/X-ray undulator and RAFEL undulator, we decided to use the same value of the beta function of 3 m. It is shown in Section 5 that the maximum value of the gain well exceeds the relative losses of peak power in the optical feedback system when the undulator is at least 6 m long. The wavelength can be tuned continuously by changing the undulator magnetic field and adjusting the monochromator wavelength by simple rotation of the grating.

5. Operation of the FEL amplifier

The main physical effects defining the operation of the FEL amplifier are the diffraction effects and the space charge effects. The longitudinal velocity spread was calculated using actual energy distribution in the electron beam after leaving the VUV/X-ray undulator. The distribution function of the electrons can be fitted well by a Gaussian distribution with the rms deviation $\sigma_E \simeq 2.5$ MeV. The energy spread does not influence too much the gain at chosen parameters (the power gain length is increased by 20% only with respect to the case of a "cold" electron beam). Calculations of the total power gain must take into account the details of focusing of the external radiation on the electron beam at the undulator entrance. A quantitative description may be performed in the following way. We assume that the seed radiation has a Gaussian radial intensity distribution which is characterized by the position of the focus, z_0, and the size of the waist in the focus, w. In the high-gain linear (steady-state) regime the radiation power grows exponentially with the undulator length: $G = P_{out}/P_{in} = A \exp[z/L_g]$, where P_{out} and P_{in} is the output and input power, respectively. The input coupling factor A depends on the focusing of the seed radiation and is a function of z_0 and w. It should be maximized by an appropriate choice of w and z_0. This problem has been studied in detail in Ref. [4] using the solution of the initial-value problem. It has been found that the value of A at $z_0 = 0$ does not differ significantly from its maximal value and the position of the Gaussian beam waist can be placed at the coordinate of the undulator entrance. However, due to the constraints of the present design we have no possibility of achieving the optimal value. The reason for this is that the focusing mirror can only be placed at a fixed position of 20 m apart

from the RAFEL undulator entrance. Also, there is an aperture limitation of 9 mm due to the diameter of the vacuum chamber in the VUV/X-ray undulator. Fortunately, this does not lead to significant degradation of the gain as it is illustrated in Fig. 2.

Fig. 3 presents the plots for the power gain versus undulator length for the FEL amplifier operating in the linear regime at 200 and 400 nm wavelength. It is seen that the gain only slightly

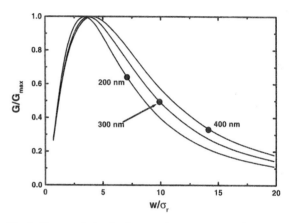

Fig. 2. Power gain in the linear regime versus spot size of the seeding radiation beam (the rms transverse size of the electron beam is $\sigma_r = 55 \,\mu$m).

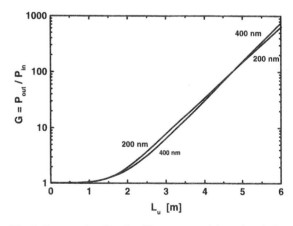

Fig. 3. Power gain, $G = P_{out}/P_{in}$ versus undulator length for the FEL amplifier operating at 200 and 400 nm wavelength (radiation beam size at the undulator entrance w is equal to 400 and 800 μm, respectively).

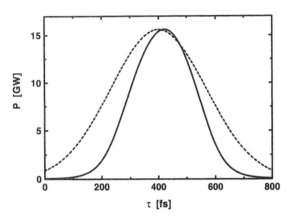

Fig. 4. Temporal structure of the radiation pulse at the exit of the RAFEL undulator. The dotted line represents the longitudinal profile of the electron beam current (the maximum corresponds to 2.5 kA).

FEL amplifier are presented in Table 1. Fig. 4 shows the time structure of the radiation pulse at the exit of the undulator. We obtain that the FEL amplifier operating at saturation produces 4 mJ pulses of about 200 fs pulse duration.

The analysis of the RAFEL parameters has shown that it will operate reliably even at a value of the energy spread exceeding the project value for the TTF by a factor of two. There is also a safety margin (by a factor of two) with respect to the value of the emittance.

Acknowledgements

We thank D. Trines and J.R. Schneider for their interest in this work and their support.

depends on the wavelength for an undulator length of 6 m. On the one hand, the focusing of radiation becomes less optimal at increasing wavelength (see Fig. 2). On the other hand, the power gain length decreases with the increase of the wavelength. At a relatively short undulator length these two effects compensate each other.

The final optimization of the FEL amplifier parameters has been performed with the nonlinear, three dimensional, time-dependent simulation code FAST [5]. The optimized parameters of the

References

[1] J. Rossbach, Nucl. Instr. and Meth. A 375 (1996) 269.
[2] J. Goldstein, D. Nguyen, R. Sheffield, Nucl. Instr. and Meth. A 393 (1997) 137.
[3] J. Sollid, et al., Nucl. Instr. and Meth. A 285 (1989) 147.
[4] E.L. Saldin, E.A. Schneidmiller, M.V. Yurkov, The Physics of Free Electron Lasers, Springer, Berlin, 1999.
[5] E.L. Saldin, E.A. Schneidmiller, M.V. Yurkov, Nucl. Instr. and Meth. A 429 (1999) 233.

ELSEVIER

Nuclear Instruments and Methods in Physics Research A 475 (2001) 373–376

NUCLEAR
INSTRUMENTS
& METHODS
IN PHYSICS
RESEARCH
Section A

www.elsevier.com/locate/nima

Nonlinear harmonic generation in distributed optical klystrons

H.P. Freund[a],*, G.R. Neil[b]

[a] *Science Applications International Corporation, 1710 Goodridge Drive, McLean, VA 22102, USA*
[b] *Thomas Jefferson National Accelerator Facility, Newport News, VA 23606, USA*

Abstract

A distributed optical klystron has the potential for dramatically shortening the total interaction length in high-gain free-electron lasers (INP 77-59, Novosibirsk, 1977; Nucl. Instr. and Meth A 304 (1991) 463) in comparison to a single-wiggler-segment configuration. This shortening can be even more dramatic if a nonlinear harmonic generation mechanism is used to reach the desired wavelength. An example operating at a 4.5 Å fundamental and a 1.5 Å harmonic is discussed. © 2001 Elsevier Science B.V. All rights reserved.

PACS: 41.60.Cr; 52.75.Ms

Keywords: Free-electron lasers; Harmonics; Optical klystron

The free-electron laser (FEL) is a candidate for 4th generation X-ray light sources [1–3], and are intended to operate in the Self-Amplified Spontaneous Emission (SASE) mode where noise is amplified to high levels in a single pass. However, the gain lengths are long and the wiggler length required can exceed 100 m. One alternative to this design is to use a distributed optical klystron employing a multi-segment wiggler in which dispersive magnetic elements are located in the gaps between the wigglers. High gain is achieved because the dispersive elements enhance the electron bunching [4–8]. The multi-segment design has been discussed in regard to interaction at the fundamental [5–8], and it is known that the bulk amplification can be much faster than exponential if a sufficiently high beam brightness can be

achieved. Here, we discuss the possibility of using the nonlinear harmonics, and employ dipole triplets (chicanes) as dispersive elements where the outermost dipoles have the same magnitudes (B_0) and lengths (L_d) and the center dipole has the same magnitude as the outermost dipoles but twice the length.

A unit cell of the structure is illustrated schematically in Fig. 1, and the overall structure is composed of many such unit cells. The 3D nonlinear, polychromatic simulation code MEDUSA [9,10] is used. MEDUSA uses a Gaussian electron beam distribution in energy and phase space, planar wiggler geometry, and treats the electromagnetic field as a superposition of Gauss–Hermite modes. The field equations are integrated simultaneously with the 3D Lorentz force equations for an ensemble of electrons. No wiggler average is imposed on the orbit equations; hence, MEDUSA is capable of propagating the electron beam through arbitrary magnetic structures (such

---

*Corresponding author. Tel.: +1-202-767-0034; fax:+1-202-734-1280.

E-mail address: freund@mmace.nrl.navy.mil (H.P. Freund).

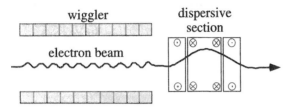

Fig. 1. Schematic illustration of a unit cell.

as wigglers, quadrupoles, dipoles, as well as FODO lattices and dipole chicanes) and includes the self-consistent interaction with the electromagnetic wave.

We consider a 1.5 Å source corresponding to the Linac Coherent Light Source (LCLS) [2] which employs a 14.35 GeV/3400 Å electron beam with an rms beam radius of 88 μm, and operates on the fundamental. Gun/linac simulations indicate that a slice emittance $\varepsilon_n = 1.0\pi - 1.5\pi$ mm mrad and an energy spread of 0.006% can be achieved. For simplicity, we assume parabolic-pole face (PPF) wiggler segments. In studying the harmonic interaction, we lower the beam energy to 8.265 GeV to obtain a fundamental resonance at 4.5 Å, and reduce the length of each wiggler segment to 2.25 m with respect to the parameters used in Ref. [8]. All other parameters are held fixed; in particular, the wiggler segments are separated by 2.0 m gaps, and the dipole chicanes are centered in the gaps. The individual dipoles have lengths of 0.4 m (for an overall chicane of 1.6 m). The number of unit cells and the dipole magnitudes are varied in simulation to obtain the optimal performance.

The length of the wiggler segments is less than the gain length for the 8.265 GeV/4.5 Å (fundamental) interaction. MEDUSA is in substantial agreement with the results of linear theory [11] for a single-segment wiggler from which we find gain lengths of 2.96 m in MEDUSA and 2.99 m gain length from the linear theory. It was pointed out in treating the fundamental interaction at 1.5 Å [8] that the growth in the distributed optical klystron can be substantially faster than exponential. This is because the enhanced bunching in a chicane gives rise to amplification in the next wiggler segment that is initially faster than exponential but

falls off to the expected exponential rate after about one gain length. Therefore, if the wiggler segments are shorter than a gain length, then the interaction is faster than exponential and the saturation length can be considerably shortened.

The interaction in an optical klystron is very sensitive to both beam energy and the magnetic field in the chicanes because the increase in the path length due to the chicane is $L = 2L_d^3 B_0^2/3\gamma^2 c^2$, so the change in path length with energy relative to the wavelength (λ) is

$$\frac{\Delta L}{\lambda} = \frac{4L_d^3 \Omega_0^2}{3\gamma^2 c^2 \lambda}\left(\frac{\Delta B_0}{B_0} - \frac{\Delta E_b}{E_b}\right) \tag{1}$$

where Ω_0 is the chicane cyclotron frequency. Optimal performance requires that $|\Delta L/\lambda| < 1$ or else overbunching occurs. This implies that

$$\frac{\Delta E_b}{E_b} < \frac{3}{8}\frac{c^2 \lambda_w}{L_d^3 \Omega_0^2}(1 + K^2/2) \tag{2}$$

for $\lambda \approx \lambda_w(1 + K^2/2)/2\gamma^2$ where K is the wiggler parameter. For the present example, optimal performance requires that $\Delta E_b/E_b < 0.01\%$ for a 2 kG field in the chicanes.

It is important to note that while the sensitivity to energy spread improves in a single-segment wiggler as the fundamental wavelength increases from 1.5 to 4.5 Å with a decrease in beam energy, this is not necessarily the case in the optical klystron. The reason for this is contained in Eq. (1) where the change in path length through the chicane depends upon the product of $\gamma^2\lambda$ that is independent of beam energy. Thus, the principal advantage of the nonlinear harmonics is in a shorter saturation length. However, we could have reduced the sensitivity to energy spread if, instead of using lower beam energy to reach 4.5 Å, we had increased the wiggler period.

Our principal purpose is to examine harmonic performance for the generation of 1.5 Å X-rays, and we begin by examining the fundamental performance. The power at the fundamental (4.5 Å) as a function of axial distance is shown in Fig. 2 for $\varepsilon_n = 1.5\pi$ mm mrad and $\Delta E_b/E_b = 0$, 0.001%, and 0.002%. The single-segment performance is also shown for comparison. In all cases, the initial noise power was chosen to be 226 W. Note that saturation occurs over a distance of

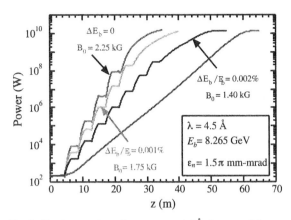

Fig. 2. Fundamental performance at 4.5 Å for $\varepsilon_n = 1.5\pi$ mm mrad.

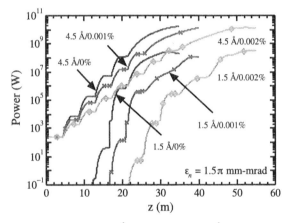

Fig. 3. Fundamental (4.5 Å) and harmonic (1.5 Å) performance for $\varepsilon_n = 1.5\pi$ mm mrad.

approximately 65 m for the single-segment wiggler. In the ideal case of a vanishing energy spread, this distance can be reduced to a total length of about 35 m including six wiggler segments with a total length of only 26.25 m. As the energy spread increases, the performance degrades; however, substantial advantages over the single-segment wiggler can still be obtained for energy spreads up to about 0.002–0.003%.

We now turn to the harmonic performance. The harmonics are undriven in the simulations and grow due to the bunching at the fundamental [10]. The evolution of the fundamental and harmonics for the preceding example for $\varepsilon_n = 1.5\pi$ mm mrad

is shown in Fig. 3. For $\Delta E_b/E_b = 0$ the peak power at the fundamental is 16.95 GW while the harmonic peaks at 265.5 MW for a power ratio of the harmonic to the fundamental of 1.6%. When the energy spread increases to $\Delta E_b/E_b = 0.001\%$, the fundamental power drops to 12.76 GW and the harmonic power to 124.2 MW for a power ratio of 1.0%. As the energy spread increases further to $\Delta E_b/E_b = 0.002\%$, the fundamental and harmonic powers reach 13.87 GW and 310.6 MW, respectively, for a power ratio of 2.2%. Note that the increased power at $\Delta E_b/E_b = 0.002\%$ relative to 0.001% is due to the fact that it is difficult to precisely find the optimum magnetic field.

The sensitivity to the chicane magnetic field is also derived from Eq. (1), and has been noted in the previous studies [8,9]. It follows from Eq. (1) that, for the parameters of interest, $\Delta L/\lambda \approx 4.9\Delta B_0$ for ΔB_0 in Gauss and $B_0 \approx 2$ kG. The performance, therefore, should be roughly periodic for integer changes in $\Delta L/\lambda$, and this implies performance variations as the magnetic field changes by about 0.2 G. This is clearly shown in Fig. 4, where we plot the fundamental and harmonic performance for $\Delta E_b/E_b = 0$ and an emittance of 1.5π mm mrad near the optimal B_0. The performance for 1.0π mm mrad is similar, and the evolution of the power at the fundamental and harmonic for $\Delta E_b/E_b = 0$, 0.002%, and 0.004% is shown in Fig. 5.

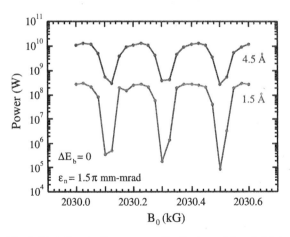

Fig. 4. Variation in the power in the fundamental and third harmonic with dipole field strength.

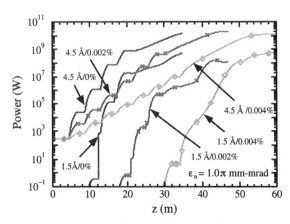

Fig. 5. Fundamental (4.5 Å) and harmonic (1.5 Å) performance for $\varepsilon_n = 1.0\pi$ mm mrad.

In summary, we have presented a study of the performance of a distributed optical klystron operating on the 3rd harmonic to reach a wavelength of 1.5 Å. This has the advantage that faster than exponential growth greatly reduces the saturation length, while the principal disadvantage is the stricter constraint on beam quality. However, it should be remarked that while the energy spread required for an optical klystron to be effective at a wavelength of 1.5 Å is below the currently projected slice energy spread (0.006%) of the LCLS, the discrepancy is within a factor of only 2–3. Hence, only a modest improvement in technology is required for this concept to be as effective as an alternative to the LCLS design.

Acknowledgements

Computational work was supported by the Advanced Technology Group at SAIC under IR&D subproject 01-0060-73-0890-000. One of us (GRN) was supported by DOE Contract DE-AC05-84ER40150 and the Commonwealth of Virginia.

References

[1] B. Faatz, et al., Nucl. Instr. and Meth. A 375 (1996) 441.
[2] NTIS Doc. No. DE98059292 (LCLS Design Group, "LCLS Design Report," April 1998), National Technical Information Service, Springfield, VA 22162.
[3] S.V. Milton, et al., Phys. Rev. Lett. 85 (2000) 988.
[4] N.A. Vinokurov, A.N. Skrinsky, preprint INP 77-59, Novosibirsk, 1977.
[5] V.N. Litvinenko, et al., Nucl. Instr. and Meth. A 304 (1991) 463.
[6] V.A. Bazylev, M.M. Pitatelev, Nucl. Instr. and Meth. A 358 (1995) 64.
[7] N.A. Vinokurov, Nucl. Instr. and Meth. A 375 (1996) 264.
[8] G.R. Neil, H.P. Freund, Nucl. Instr. and Meth. A 475 (2001) 381, this issue.
[9] H.P. Freund, T.M. Antonsen Jr., Principles of Free-electron Lasers, 2nd Edition, Chapman & Hall, London, 1996.
[10] H.P. Freund, et al., IEEE J. Quantum Electron. 36 (2000) 275.
[11] M. Xie, Proceedings of IEEE 1995 Particle Accelerator Conference (IEEE Cat. No. 95CH35843), 1995, p. 183.

ELSEVIER

Nuclear Instruments and Methods in Physics Research A 475 (2001) 377–380

NUCLEAR
INSTRUMENTS
& METHODS
IN PHYSICS
RESEARCH
Section A

www.elsevier.com/locate/nima

A comparison between the fundamental and third harmonic interactions in X-ray free-electron lasers

H.P. Freund*

Science Applications International Corp. 1710 Goodridge Drive, McLean, VA 22102, USA

Abstract

Free-electron lasers are under study as 4th generation light sources for 1.5 Å X-rays. However, the interaction is extremely sensitive to the beam quality, and it is useful to consider alternate schemes. The scheme discussed here is to generate 4.5 Å X-rays using the fundamental interaction and rely on nonlinear 3rd harmonic generation to reach 1.5 Å. The specific example discussed corresponds to the Linac Coherent Light Source proposed at the Stanford Linear Accelerator Center. © 2001 Elsevier Science B.V. All rights reserved.

PACS: 41.60.Cr; 52.75.Ms

Keywords: Free-electron lasers; Harmonics; Nonlinear harmonics

Studies are in progress for a free-electron laser (FEL) to serve as a 4th generation light source. The design goal of the Linac Coherent Light Source (LCLS) is to generate 1.5 Å X-rays using the fundamental interaction [1]. However, the interaction is extremely sensitive to the beam quality; hence, it is useful to consider alternate schemes. One alternate scheme considered here is to generate 4.5 Å X-rays using the fundamental and to rely on nonlinear 3rd harmonic generation [2,3] to reach 1.5 Å. At this longer wavelength, the fundamental and 3rd harmonic are less sensitive to beam quality.

These two schemes are compared for parameters similar to the nominal design parameters of the LCLS. The analysis is restricted to the case where

the 4.5 Å fundamental interaction is achieved by increasing the wiggler period while holding other parameters fixed. We then study the sensitivity of both configurations to increases in the energy spread of the beam. The advantage of using a longer wiggler period is that both the wiggler K-value and the gap can be increased simultaneously. The increase in K is responsible for the reduced sensitivity to beam quality while the increase in the gap permits the use of a larger drift tube which reduces the influence of wakefields on the beam. However, alternate schemes such as reducing the beam energy are also possible.

The MEDUSA simulation code [2,4] is used to study both schemes. MEDUSA is a three-dimensional, multi-frequency simulation code where the electromagnetic field is represented as a superposition of Gauss-Hermite modes. The electron beam is modeled as a Gaussian with a matched

*Tel.: +1-202-767-0034; fax: +1-202-734-1280.

E-mail address: freund@mmacc.nrl.navy.mi (H.P. Freund).

0168-9002/01/$ - see front matter © 2001 Elsevier Science B.V. All rights reserved.
PII: S 0 1 6 8 - 9 0 0 2 (0 1) 0 1 5 4 3 - 1

beam radius. The field equations are integrated simultaneously with the 3D Lorentz force equations. A comparison between MEDUSA and four other FEL simulation codes, as well a linear analytic theory, has shown good agreement [5].

The nonlinear harmonic generation mechanism is well-known in traveling-wave tubes and has been discussed in FELs [2,3,6–8]. As the fundamental grows, the beam develops microbunches that give rise to enhanced nonlinear harmonic growth and can become the dominant harmonic growth mechanism where the gain length scales inversely with the harmonic number. In addition, the harmonic powers generated by this process can be quite high and can reach 10% of the power in the fundamental. The essential point is that the harmonic growth and saturation are controlled by the fundamental. This contrasts with the linear harmonic instability that is more sensitive to beam and wiggler quality than the fundamental. Hence, any decreased sensitivity of the fundamental to beam and wiggler quality also results in a decreased sensitivity of the nonlinear harmonic mechanism.

A paper investigating the sensitivity of the nonlinear harmonic generation to wiggler imperfections is in preparation. The issue considered here is the sensitivity to beam energy spread. The nominal LCLS design employs a 14.35 GeV/ 3400 A electron beam with a normalized emittance of 1.5π mm-mrad and slice and global energy spreads of 0.006% and 0.02%, respectively. The wiggler has an amplitude of 13.2 kG and a period of 3.0 cm. We achieve operation on the fundamental at 4.5 Å by the simple expedient of increasing the wiggler period from 3.0 to 4.44 cm. Since the amplitude is fixed, this yields an increased $K = 5.47$. The electron beam is unchanged, although variations in the energy spread are considered for both cases. No further optimization of the 4.5 Å design is attempted. In particular, the wiggler amplitude can be increased by using longer wiggler periods, and this results in further increases in K; however, at the cost of increases in the beam energy.

One difference in this analysis from the LCLS design is that a single long parabolic-pole face (PPF) wiggler is considered for simplicity, since no additional focusing is required. However, the betatron period for the PPF wiggler is of the order of 72 m. The actual wiggler design makes use of multiple wigglers with strong focusing resulting in (1) a shorter betatron period, (2) shorter gain length and (3) a reduced sensitivity to beam quality relative to the single PPF wiggler model. The PPF model, however, provides a qualitative picture of the nonlinear harmonic mechanism.

First, consider a comparison of the performance of the 1.5 and 4.5 Å designs for the ideal limit of zero beam energy spread. While the LCLS is to be used in Self-Amplified Spontaneous Emission (SASE) mode, it is convenient to consider an amplifier model where a drive power of 30 kW is assumed at either 1.5 or 4.5 Å for the two cases. This permits the determination of the gain lengths and saturated powers to be expected in the SASE design. The results of the simulation are shown in Fig. 1. Note that although harmonics are also associated with the 1.5 Å fundamental case, these are not included in this case since our primary interest is the performance at 1.5 Å.

The saturated power for the nominal 1.5 Å design is about 12.8 GW with a gain length of 6.21 m. This is in reasonable agreement with a linear theory [9], which predicts a gain length of 5.40 m and a saturated power of 17.4 GW. The peak power at 4.5 Å is 26.8 GW, which is higher than the 1.5 Å design case due to the increased K

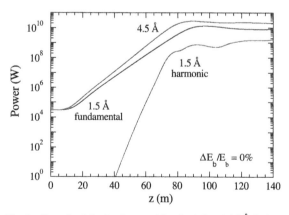

Fig. 1. Growth of the fundamental for the 1.5 and 4.5 Å designs and of the 3rd harmonic of the 4.5 Å interaction for zero energy spread.

value, and the gain length of 5.09 m is shorter. The gain length for the 3rd harmonic is about 1.67 m, which is approximately one-third that at the fundamental, as expected [2]. The harmonic power is a somewhat more ambiguous quantity. Typically, the nonlinear bunching at the harmonic causes the harmonic power to peak at a point just prior to the saturation of the fundamental. Subsequently, the harmonic power decreases due to overbunching but typically recovers and reaches its peak value somewhere downstream from the peak power point of the fundamental. For this example, the harmonic power peaks after a distance of 92 m at a power of 737 MW (about 2.8% of the fundamental power). However, the asymptotic value for the harmonic power can exceed this figure.

Because of this ambiguity in the optimal harmonic power point, it is more useful to illustrate the effects of increasing energy spread. To this end, the corresponding evolution of the powers in the 1.5 Å fundamental and the 4.5 Å fundamental and 1.5 Å harmonic for energy spreads of 0.01%, 0.02%, and 0.03% are shown in Figs. 2–4, respectively. First note that the gain lengths (saturation power levels) for the 1.5 Å fundamental are 6.64 m (12.2 GW), 8.04 m (8.35 GW), and 10.07 m (5.70 GW) for $\Delta E_b/E_b = 0.01\%$, 0.02%, and 0.03%, respectively. These are longer (lower) than the gain lengths

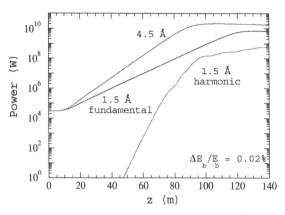

Fig. 3. Growth of the fundamental for the 1.5 and 4.5 Å designs and of the 3rd harmonic of the 4.5 Å interaction for a 0.02% energy spread.

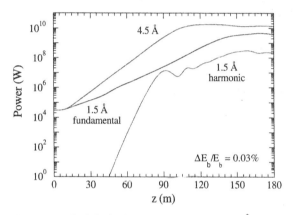

Fig. 4. Growth of the fundamental for the 1.5 and 4.5 Å designs and of the 3rd harmonic of the 4.5 Å interaction for a 0.03% energy spread.

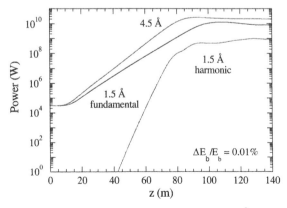

Fig. 2. Growth of the fundamental for the 1.5 and 4.5 Å designs and of the 3rd harmonic of the 4.5 Å interaction for a 0.01% energy spread.

(power levels) for the 4.5 Å fundamental which are 5.13 m (22.8 GW), 5.87 m (21.2 GW), and 7.15 m (17.5 GW) for $\Delta E_b/E_b = 0.01\%$, 0.02%, and 0.03%. The gain for the 1.5 Å harmonic is approximately one-third that of the 4.5 Å fundamental and are 1.71, 1.94 and 2.17 m for the three choices of energy spread. In terms of the output power, it is clear that the nominal LCLS design exceeds that of the 1.5 Å harmonic for energy spreads up to 0.03%. However, the saturation lengths for the nominal design for the 1.5 Å fundamental exceed those for the harmonic design, and the power levels for the 1.5 Å harmonic are appreciable and may become competitive if the achievable energy spread exceeds 0.03%.

In summary, this study indicates that the expected powers at 1.5 Å for the nominal LCLS design exceed those to be expected from harmonics for energy spreads below about 0.03%. This is for the case in which the beam energy and wiggler amplitude are held fixed while the wiggler period is increased; however, this conclusion may change if the harmonic is generated using lower beam energies or higher wiggler amplitudes. While the present analysis is based upon the natural focusing provided by a PPF wiggler, the basic physics applies to a strong focusing system as well.

Computational work was supported by the Advanced Technology Group at Science Applications International Corporations under IR&D subproject 01-0060-73-0890-000.

References

[1] NTIS Doc. No. DE98059292 (LCLS Design Study Group, "LCLS Design Report," April 1998), National Technical Information Services, US Department of Commerce, 5285 Port Royal Road, Springfield, VA 22161.

[2] H.P. Freund, et al., IEEE J. Quantum Electron. 36 (2000) 275.

[3] Z.H. Huang, K.J. Kim, Nucl. Instr. and Meth. A 475 (2001) 112, this issue.

[4] H.P. Freund, T.M. Antonsen Jr., Principles of Free-electron Lasers, 2nd Edition, Chapman & Hall, London, 1996.

[5] S.G. Biedron, et al., Nucl. Instr. and Meth. A 445 (2000) 110.

[6] R. Bonifacio, et al., Nucl. Instr. and Meth. A 293 (1990) 627.

[7] R. Bonifacio, et al., Nucl. Instr. and Meth. A 296 (1990) 787.

[8] R. Bonifacio, R. Corsini, P. Pierini, Phys. Rev. A 45 (1992) 4091.

[9] M. Xie, Proceedings of the IEEE 1995 Particle Accelerator Conference, IEEE Cat. No. 95CH35843, 1995, p. 183.

ELSEVIER

Nuclear Instruments and Methods in Physics Research A 475 (2001) 381–384

NUCLEAR
INSTRUMENTS
& METHODS
IN PHYSICS
RESEARCH
Section A

www.elsevier.com/locate/nima

Dispersively enhanced bunching in high-gain free-electron lasers

G.R. Neil[a], H.P. Freund[b],*

[a] Thomas Jefferson National Accelerator Facility, Newport News, VA 23606, USA
[b] Science Applications International Corp., 1710 Goodridge Drive, McLean, VA 22102, USA

Abstract

A free-electron laser based on a distributed optical klystron is studied using the 3D simulation code MEDUSA. The configuration consists of multiple wiggler segments separated by dipole chicanes. The chicanes ballistically enhance electron beam bunching and, in a properly designed system, can result in faster than exponential growth over the bulk of the interaction length. The principal constraint on the concept is the stringent requirements on beam quality. © 2001 Elsevier Science B.V. All rights reserved.

The free-electron laser (FEL) is the prime concept for future 4th generation X-ray light sources [1–3]. We propose a distributed optical klystron to reduce the overall length of both the system in comparison to the most common designs under study today. The design employs a multi-segment wiggler with dispersive magnetic elements located in the gaps between the wigglers. Here, these dispersive elements are dipole chicanes where the dipoles have the same magnitudes (B_0) and lengths and the center dipole has twice the length of the outer ones (L_d). The chicane enhances the bunching of the beam over that from the preceding wiggler segment. A unit cell of the structure is illustrated schematically in Fig. 1, and the overall structure is composed of many unit cells.

The optical klystron was first proposed for low-gain oscillators using two wigglers separated by a drift space by Vinokurov and Skrinsky [4], and has been implemented experimentally [5–8]. Optical klystrons in high-gain FELs operating on the fundamental [9–13] and a harmonic [14] have also been discussed. While it is not practicable for X-ray wavelengths, a low-gain oscillator using an optical klystron has been studied as a prebuncher for a beam that is then extracted from the optical cavity and injected into another wiggler for amplification [15]. The configuration described here differs from past work in that we are interested in X-ray wavelengths, and no harmonic interactions are involved. It is shown that when properly designed the growth rate in the distributed optical klystron can substantially exceed the exponential growth found in single-segment wiggler designs and the overall system length, as well as the cumulative length of the wiggler, can be greatly reduced.

The 3D nonlinear, polychromatic simulation code MEDUSA [16–18] is used to study the concept. MEDUSA uses planar wiggler geometry and treats the electromagnetic field as a superposition of Gauss–Hermite modes. The field

*Corresponding author. Tel.: +1-202-767-0034; fax: +1-202-734-1280.

E-mail address: freund@mmace.nrl.navy.mil (H.P. Freund).

0168-9002/01/$ - see front matter © 2001 Elsevier Science B.V. All rights reserved.
PII: S 0 1 6 8 - 9 0 0 2 (0 1) 0 1 5 4 4 - 3

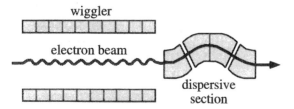

Fig. 1. Schematic illustration of a unit cell.

equations are integrated simultaneously with the 3D Lorentz force equations for an ensemble of electrons. No wiggler average is imposed on the orbit equations, and MEDUSA is capable of propagating the electron beam through both quadrupole and dispersive dipole magnets. A Gaussian electron beam distribution is used in energy and phase space.

Here, we consider the Linac Coherent Light Source (LCLS) [2] which employs a 14.35 GeV/ 3400 A electron beam with an 88 μm rms beam radius. The emittance and energy spread derived from gun/linac simulations for a single-wavelength "slice" of the beam indicate that a normalized emittance $\varepsilon_n = 1.0\pi$–1.5π mm mrad and an energy spread of 0.006% can be achieved. The LCLS wiggler uses multiple segments of a flat-pole face design with an amplitude of 13.2 kG and a period of 3.0 cm with strong focusing in the gaps. The resonant wavelength is near 1.5 Å. For simplicity, we assume parabolic-pole face (PPF) wiggler segments (eliminating the necessity of any external focusing) 4.8 m in length separated by 2.0 m gaps. The dipoles have lengths of 0.4 m (for an overall chicane of 1.6 m) and are centered in the gaps. We vary the number of unit cells and the dipole magnitudes in simulation to obtain the optimal performance. Here, we consider uniform dipole and wiggler parameters; however, further enhancements may be possible by varying the wigglers and dipoles in each unit cell.

The single-segment performance has been discussed previously [17]. For a PPF wiggler, the minimum exponential gain length is 6.21 m when $\Delta E_b/E_b = 0\%$ and 6.64 m for $\Delta E_b/E_b = 0.010\%$ but increases rapidly thereafter and rises to 8.04 m when $\Delta E_b/E_b = 0.020\%$. The saturated power shows a similar sensitivity to energy spread. Note

that this is for natural focusing in the PPF wiggler, and the strong focusing in the LCLS design yields a shorter gain length. However, the simplification associated with the PPF wiggler is suitable to illustrate the advantages/disadvantages of the concept.

Now consider the distributed optical klystron for $\varepsilon_n = 1.5\pi$ mm mrad and an energy spread $\Delta E_b/E_b = 0$. The power is plotted versus distance in Fig. 2 for the optical klystron and a single-segment wiggler for a drive power of 30 kW at a wavelength of 1.4920 Å. The optimal performance in the optical klystron (corresponding to $B_0 = 1.88$ kG) shows that saturation is at a higher power and shorter distance (15.5 GW over 50 m) than the single-segment wiggler (10.7 GW over 85 m). The essential point, however, is that there is no region of exponential gain in the optical klystron. Rather, the growth rates are faster than exponential.

The enhanced bunching in the chicane causes growth to be faster than exponential. The growth rate in a wiggler segment after a chicane is initially faster than exponential but decreases to the expected exponential growth rate after about one gain length. The gain length here exceeds 6 m which is longer than the wiggler segments (4.8 m), however, and the wiggler terminates before the growth rates roll over to the exponential rate. This process is repeated in each unit cell until the power approaches saturation. However, near saturation

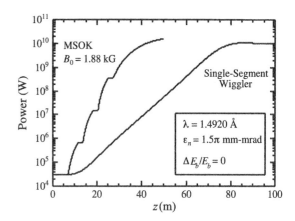

Fig. 2. Power growth for the distributed optical klystron and a single-segment wiggler for zero energy spread.

the enhanced bunching becomes less effective and the length of the final segment is chosen to reach saturation.

The variation in performance for 5 wiggler segments and 50 m total length is shown in Fig. 3 versus B_0. The optimal B_0 is 1.88 kG; however, there are large fluctuations because the ratio of the path length in the chicane to the wavelength is

$$\frac{L}{\lambda} = \frac{2}{3} \frac{L_d^3 \Omega_0^2}{\gamma^2 c^2 \lambda} \qquad (1)$$

for an electron propagating paraxially and where Ω_0 is the cyclotron frequency. Thus, $\Delta L/\lambda \approx 4.9 \Delta B_0$ for the parameters of interest and the path length changes by one wavelength when B_0 changes by about 0.20 G. The performance, therefore, oscillates as B_0 changes by this amount and requires careful design and control of the fields in the chicanes.

The principal disadvantage is that the distributed optical klystron is more sensitive to emittance and energy spread than a single-segment design. This is explained by noting that the change in path length in the chicane, ΔL, due to a change in energy is

$$\frac{\Delta L}{\lambda} = -\frac{4}{3} \frac{L_d^3 \Omega_0^2}{\gamma^2 \lambda} \frac{\Delta \gamma}{\gamma}. \qquad (2)$$

As a result, $|\Delta L/\lambda| \approx 8.82 \times 10^3 \Delta \gamma / \gamma$ for the parameters under study. Since $|\Delta L/\lambda| \ll 1$ is required for coherent bunching, this implies that the energy spread must be significantly less than 0.01%. Of

course, the actual criterion is more complicated because the path length also depends on the entry angle into the chicane.

This is illustrated in Fig. 4 where we plot the power versus distance for $\varepsilon_n = 1.5\pi$ mm mrad and for a single-segment wiggler and the optical klystron with $\Delta E_b/E_b = 0, 0.001\%$, and 0.002%. The power for a single-segment wiggler is virtually independent of energy spread over this range. There is little change in the performance of the optical klystron as $\Delta E_b/E_b$ increases to 0.001%. However, the performance degrades rapidly as $\Delta E_b/E_b$ increases beyond this value, and 7 wiggler segments and an overall length of 70 m is required to reach saturation for $\Delta E_b/E_b = 0.002\%$. This saturation length is only 17.6% shorter than that for the single-segment wiggler. The net benefit from the optical klystron vanishes almost entirely when the energy spread increases to 0.0025%. Note also that the optimal B_0 decreases from 1.88 to 1.82 to 1.47 kG as the energy spread increases from 0% to 0.001% to 0.002%.

The constraint on $\Delta E_b/E_b$ is less severe at lower emittance. This is illustrated in Fig. 5, where we plot the power versus distance for $\varepsilon_n = 1.0\pi$ mm mrad and for a single-segment wiggler and the distributed optical klystron for $\Delta E_b/E_b = 0, 0.002\%$, and 0.004%. It is clear that higher energy spreads can be tolerated for this lower emittance value, and the optical klystron performance will substantially exceed that of the single segment wiggler for $\Delta E_b/E_b$ below about

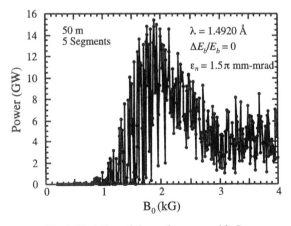

Fig. 3. Variation of the performance with B_0.

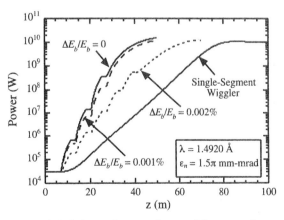

Fig. 4. Relative performance for $\varepsilon_n = 1.5\pi$ mm mrad.

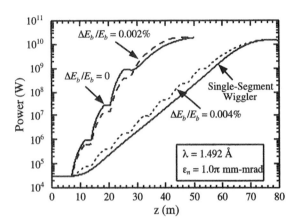

Fig. 5. Relative performance for $\varepsilon_n = 1.0\pi$ mm mrad.

0.003%. A decrease in the optimal B_0 is also seen for this emittance, and varies from 1.82 to 1.63 to 0.70 kG as $\Delta E_b / E_b$ increases from 0% to 0.002% to 0.004%.

In summary, a detailed analysis has been presented of distributed optical klystron for an X-ray application using the 3-D FEL simulation code MEDUSA. The detailed parameters studied corresponds to the LCLS. Our results indicate that stringent constraints are required. The dipole fields in the chicanes must be controlled to within about $\pm 0.01\%$ that we feel is possible with careful design. The beam quality requirements, however, are somewhat beyond what has been demonstrated. Simulations indicate that slice emittances and energy spreads in the range of 1.0π–1.5π mm mrad and 0.006%, respectively, are possible. Simulations indicate that for these emittances, the slice energy spread must be kept below about 0.003%. While this appears to be presently beyond the state-of-the-art, future improvements in beam quality may make this an attractive option for a 4th generation X-ray light source. It should also be remarked that the beam constraints will become less severe for longer wavelengths, and that the concept may already be an attractive option at wavelengths above 10 Å.

This paper represents only a preliminary study since we have used fixed length wigglers, gaps, and dipoles as well as uniform, repeated unit cells. We expect that the performance can be improved by step tapers in some or all of these parameters. In addition, our analysis considered monochromatic wave growth. MEDUSA, however, can simulate the full polychromatic interaction, and future work will extend the present analysis to this regime.

The computational work was supported by the Advanced Technology Group at Science Applications International Corporations under IR&D subproject 01-0060-73-0890-000. One of the authors (GRN) was supported by DOE Contract # DE-AC05-84ER40150, and by the Commonwealth of Virginia.

References

[1] B. Faatz, et al., Nucl. Instr. Meth. A 375 (1996) 441.
[2] NTIS Doc. No. DE98059292, LCLS Design Group, LCLS Design Report, April 1998, (Copies may be ordered from the National Technical Information Service) Springfield, VA 22162.
[3] S.V. Milton, et al., Phys. Rev. Lett., submitted for publication.
[4] N.A. Vinokurov, A.N. Skrinsky, Preprint INP 77-59, Novosibirsk, 1977.
[5] A.S. Artamonov, et al., Nucl. Instr. Meth. 177 (1980) 247.
[6] M. Billardon, et al., J. Physi.-Colloq. 44 (1983) C1–29.
[7] V.N. Litvinenko, et al., Nucl. Instr. Meth. A 429 (1999) 151.
[8] G.R. Neil, et al., Nucl. Instr. Meth. A 237 (1985) 199.
[9] J.C. Gallardo, C. Pellegrini, Nucl. Instr. Meth. A 296 (1990) 448.
[10] V.N. Litvinenko, Nucl. Instr. Meth. A 304 (1991) 463.
[11] V.A. Bazylev, M.M. Pitatelev, Nucl. Instr. Meth. A 358 (1995) 64.
[12] N.A. Vinokurov, Nucl. Instr. Meth. A 375 (1996) 264.
[13] J. Chen, et al., Nucl. Instr. Meth. A 375 (1996) 299.
[14] L.H. Yu, et al., Nucl. Instr. Meth. A 445 (2000) 301.
[15] S.J. Hahn, et al., Nucl. Instr. Meth. A 358 (1995) 167.
[16] H.P. Freund, T.M. Antonsen Jr., Principles of Free-electron Lasers, 2nd Edition, Chapman & Hall, London, 1996.
[17] H.P. Freund, et al., IEEE J. Quantum Electron. 36 (2000) 275.
[18] H.P. Freund, P.G. O'Shea, Phys. Rev. Lett. 84 (2000) 2861.

ELSEVIER

Nuclear Instruments and Methods in Physics Research A 475 (2001) 385–390

NUCLEAR
INSTRUMENTS
& METHODS
IN PHYSICS
RESEARCH
Section A

www.elsevier.com/locate/nima

Generation of powerful ultrashort electromagnetic pulses based on superradiance

N.S. Ginzburg[a],*, I.V. Zotova[a], Yu.V. Novozhilova[a], A.S. Sergeev[a],
A.D.R. Phelps[b], A.W. Cross[b], S.M. Wiggins[b], K. Ronald[b], V.G. Shpak[c],
M.I. Yalandin[c], S.A. Shunailov[c], M.R. Ulmaskulov[c]

[a] Institute of Applied Physics, RAS, Nizhny Novgorod, Russia
[b] University of Strathclyde, Glasgow G4 0NG, UK
[c] Institute of Electrophysics, RAS, Ekaterinburg, Russia

Abstract

Experimental results of the observation of superradiation from intense, subnanosecond electron bunches moving through a periodic waveguide and interacting with a backward propagating TM_{01} wave are presented. The ultra-short microwave pulses in Ka, W, and G band were generated with repetition frequencies of up to 25 Hz. Observation of RF breakdown of ambient air, as well as direct measurements by hot-carrier germanium detectors, leads to an estimate of the peak power as high as 60–120 MW for the 300–400 ps pulses at 38 GHz. The initial observation of 75 GHz 10–15 MW radiation pulses with duration less than 150 ps, and of 150 GHz microwave spikes with a risetime of 75 ps are also reported. Comparison with simulations is discussed as well. © 2001 Elsevier Science B.V. All rights reserved.

PACS: 41.60.Bq; 42.50.Fx; 41.60.Cr; 52.75.Ms

Keywords: Cherenkov emission; Coherence; Electron bunch

1. Introduction

Coherent emission from short electron bunches with lengths much smaller than the wavelength is well known and was first observed in the Smith–Purcell experiment. Another type of conventional coherent emission, which is used extensively in microwave electronics, arises from continuous or quasi-continuous electron beams with length much greater than the wavelength. In this case coherent emission is related to stimulated processes resulting in selfbunching of the electrons. For the intermediate case, i.e. for electron bunches with a size of several wavelengths, it was traditionally assumed that radiation can be related only to density fluctuations inside the electron bunch (incoherent component) or to coherent emission from the sharp edges of the bunch. Only recently it was recognised that much more intense coherent emission can occur from the entire volume of such bunches due to selfbunching and slippage of the wave over the electron pulse. Based on similar mechanisms investigated early in quantum electronics for ensembles of inverted atoms [1–3] coherent

*Corresponding author. Tel.: + 831-2-33-51-38; fax: + 7-831-2-33-51-38.

E-mail address: ginzburg@appl.sci-nnov.ru (N.S. Ginzburg).

emission of such bunches can be regarded as superradiance. According to theoretical considerations of superradiance from non-equilibrium ensembles of classical electrons undertaken in recent years [4–10] this phenomenon can be utilised to generate intense ultrashort electromagnetic pulses. Superradiance can be related to different mechanisms of stimulated emission: bremsstrahlung, cyclotron, Cherenkov, etc. All these mechanisms of SR have been observed recently experimentally [11–14]. For microwave band maximal peak power was obtained for the Cherenkov mechanism of SR [14] when electron bunches moving though the periodical corrugated waveguide interacted with a backward propagating wave (for the long pulse regime this type of interaction is traditionally used in BWO).

In this paper we present results of theoretical and experimental studies of Cherenkov type emission in corrugated waveguides. Reported experiments were directed mainly at increasing the peak power of the SR pulses as well as on increasing the carrier frequency of these pulses. As a result powerful SR pulses have been observed throughout the band of 38–150 GHz.

2. Basic description of Cherenkov SR in corrugated waveguide

Cherenkov SR in the slow-wave structure in the form of periodically corrugated metallic waveguides was observed under synchronism with a slow spatial harmonic of the backward wave

$$\omega = (-h + h_c)V_0 \qquad (1)$$

where ω and h are the wave frequency and longitudinal wave number of fundamental harmonic respectively, $h_c = 2\pi/\lambda_c$, and λ_c is the corrugated period. In this case the longitudinal component of the electric field of the synchronous spatial harmonic can be presented in the form

$$E_z = Re\left[E_z^s(r_\perp)A(z,t)\exp\left(i\omega\left(t - z/V_0\right)\right)\right]$$

where $E_z^s(r_\perp)$ describes the transverse distribution and $A(z,t)$ describes a temporal evolution of the longitudinal distribution of synchronous wave. The interaction of the electrons with the radiation

can be described by the following equations

$$\left(\frac{\partial}{\partial\zeta} - \frac{\partial}{\partial\tau}\right)a = Gf(\zeta)J, J = \frac{1}{\pi}\int_0^{2\pi}\exp(-i\theta)\,d\theta_0 \quad (2)$$

$$\frac{\partial^2\theta}{\partial\zeta^2} = Re[a\exp(i\theta)] \qquad (3)$$

Here we are using dimensionless variables:

$$\zeta = z\frac{\omega}{2c}\gamma_0^{-2}\beta_0^{-1}, \quad \tau = \frac{\omega}{2\gamma_0^2}\frac{(t - z/V_0)}{(1 + V_0/V_{gr})},$$

$$a = \frac{4e\gamma_0 AE_z^s(R_0)}{mV_0\omega}, \quad G = \frac{4\gamma_0^3 eI_0|Z|}{mc^2\beta_0^2}$$

I_0 is the electron current, R_0 is the radius of electron injection, Z is the coupling impedance of the TM_{01} mode, $\theta = \omega t - h_s z$ is the electron phase with respect to the synchronous wave, $h_s = h_c - h$, V_0 is the electron velocity, V_{gr} is the electromagnetic wave group velocity. The function $f(\tau)$ describes the unperturbed electron density $f(\tau) = 1, 0 < \tau < T$, where T is the dimensionless duration of the electron bunch. In the case when emission start up is related to electron density fluctuations, described by parameter $q \ll 1$, the boundary conditions can be presented in the form

$$\theta|_{\zeta=0} = \theta_0 + q\cos\theta_0, \quad \frac{\partial\theta}{\partial\zeta}\bigg|_{\zeta=0} = 0,$$

$$\theta_0 \in [0, 2\pi], a|_{\zeta=L} = 0$$

where L is the length of the interaction region.

In Fig. 1a the SR pulse profile is shown for different values of electron pulse duration for the parameter $G = 0.15$. As seen from Fig. 1b, the peak amplitude growth proportional to the electron pulse duration until this duration is rather short ($T < 10$). It corresponds to a squared dependence of the peak radiation power with respect to the total number of electrons. It means that the all electrons of bunch radiate coherently. Confirmation of this fact can be also found from the dependence of peak power on parameter G, i.e. total pulse current (Fig. 2). As it follows from Fig. 1b, saturation of the growth of peak amplitude occurs when electron pulse duration exceeds a certain value ($T > 10$) because the electron pulse becomes too long to provide coherent radiation from all over pulse length. It should be also noted

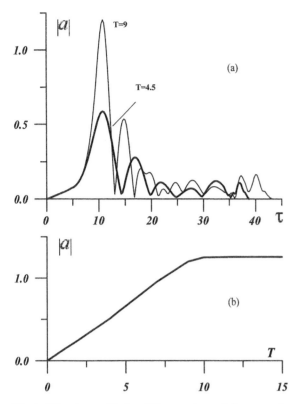

Fig. 1. SR pulse profile for different values of electron pulse duration (a) Peak amplitude vs pulse duration (b). $L = 34$, $G = 0.15$.

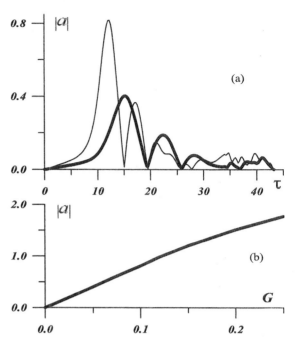

Fig. 2. SR pulse profile for different values of electron current (a) Peak amplitude vs total electron current (b). $L = 34$, $T = 9$.

that the effective SR pulse duration decreases with increasing its amplitude (see, Figs. 1a and 2a).

In physical variables for 500 ps, 1.5 kA, 200 keV electron bunch, moving in the corrugated wavegude with the parameters given below in Section. 3 we obtain SR duration about 250 ps and peak power 200 MW (for accurate power estimation instead pendulum Eq. (3) we have used full relativistic electron motion equations).

3. Experimental set-up

A compact pulsed accelerator [15,16] based on the RADAN 303 modulator equipped with a subnanosecond pulse sharpener was used to inject typically 0.5–4 ns, 1–2 kA, 200–250 keV electron bunches into the interaction region. These electron bunches were generated from a magnetically insulated coaxial vacuum diode, which utilised a cold, explosive emission cathode. High current electron pulses were transported through the interaction space which took the form of a corrugated waveguide of total length 1.5–10 cm, in a longitudinal guiding magnetic field created either by a pulsed solenoid ($B_Z \sim 2.5$–5 T) or superconducting magnet ($B_{Z\max} \sim 8.5$ T). The period of corrugation of the Ka-band slow-wave structure was approximately 3.5 mm, the corrugation depth was 0.75 mm with a mean radius of 3.75 mm. The W- and G-band structures dimensions were half and quarter those of the Ka-band structure, respectively. The cathode diameter was varied from 5.5 mm (Ka band) down to 1.2 mm (G-band). For measurement of the output radiation a hot-carrier germanium detector which had a transient response of ~ 100 ps was used.

4. PIC code simulation of superradiative emission

Additional simulation of the radiation from subnanosecond electron bunches passing through periodic waveguide structures under experimental

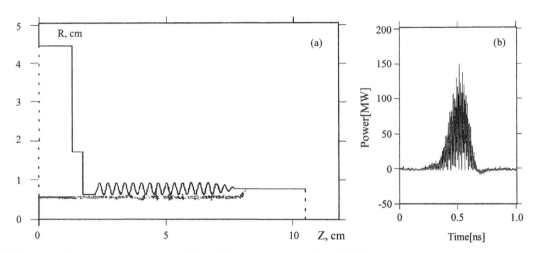

Fig. 3. Geometry of interaction region and positions of electrons at time $t = 1$ ns (a) Pointing vector integrated over transverse cross section as a function of time at $z = 9$ cm (b).

conditions was carried out using the particle-in-cell (PIC) code KARAT (see in more detail [14]). In Fig. 3a is shown the system geometry (all sizes are in centimetres) and electron bunch 1 ns after a 240 kV driving voltage pulse was imposed on the coaxial line. In the presented simulations the strength of the guide magnetic field was 5 T. The pulse duration of about 0.8 ns and peak current ~ 1.3 kA were close to the experimental measurements of the current pulse. In Fig. 3a we see modulation of the bunch density. The dependence of RF output power on time is presented in Fig. 3b. The microwave pulse duration is about 300 ps. The peak power reached 75 MW (after averaging over the high frequency it should be half of the peak magnitude presented in Fig. 3b) and the electron efficiency was about 7%. The spectrum of radiation with a central frequency about 38 GHz agrees with the experimental data and the frequency, which can be found from the resonance conditions. The distribution over radius of the radial component of the electric field corresponds to the excitation of the TM_{01} mode.

The dependence of peak power on the voltage pulse duration is shown in Fig. 4. Because the electron peak current was practically constant and the electron pulse duration corresponded to the voltage pulse duration, this diagram actually represents the dependence of peak power on the total charge in the bunch. Obviously this depen-

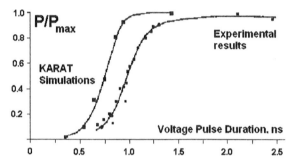

Fig. 4. Dependence of peak power on accelerating pulse duration.

dence is rather close to a square law when the pulse duration is close to the correlation time $T_c = L(1/V_0 + 1/V_{gr}) \sim 0.8$ ns.

Note that for small pulse duration the given interaction length of 6 cm is not sufficient for formation of SR spikes. For long electron pulses saturation of the growth of the peak power is obviously related to the fact that the pulse becomes too long to provide coherent radiation emission from the entire pulse length. Actually, subsequent microwave spikes are emitted when the pulse duration is increased. Alongside the investigation of the generation of sub-nanosecond microwave pulses at Ka-band, similar experiments were performed for a W-band and G-band backward-wave microwave structures (all geometrical sizes of slow-wave structures were reduced proportionally to the radiation wavelength). The PIC-code simulations have shown a peak power of up to 30 MW can be attained for the pulsewidth of 100–150 ps at the frequency of 75 GHz.

5. Experimental results

5.1. Ka-band experiments

All the experimental data for the Ka band pulses were obtained for a magnetic field of 5–7 T exceeded the cyclotron resonance value (~ 3 T).

A typical oscilloscope trace of the Ka band microwave signal is presented in Fig. 5a. The observed microwave spikes have duration of about 300 ps and a rise time of 200 ps. It was shown that a 6 cm interaction length is sufficient for the formation of a superradiance spike as neither the pulse waveform or peak power changed when the interaction length was increased up to 10 cm.

Frequency measurements at Ka the band have been made using a set of cut-off waveguide filters and showed that the main peak had a central frequency of approximately 38 GHz. The relative radiation spectrum bandwidth was about 5%. The measured radiation pattern corresponded with good accuracy to the excitation of the TM_{01} mode. This measurement allowed the absolute peak power to be calculated by integrating the signal from the detector over its radial position. The peak power estimated by this method was about 60 MW. A rather high level of radiation power was indicated also by the illumination of a neon bulb panel when the radiation signal irradiated the panel at a distance of 30 cm from the output horn.

Application of the technology of the energy compression for high voltage subnanosecond pulses generated by the modulator provided us with a 20–25% increase of the pulse amplitude. By attaining electron energies of 300 keV we found an essential rise of the power of Ka band pulses. Estimates show that the power level exceeds 150 MW for the 300–400 ps microwave spikes.

Control of the parameters of the accelerating voltage pulse (0.5–2.5 ns) provided the possibility for study of the dependence of the peak power on the duration of the beam current pulse and, hence,

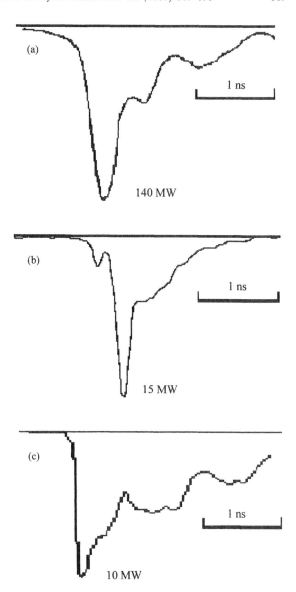

Fig. 5. Oscilloscope traces of SR pulses at (a) Ka, (b) G and (c) W-band.

on the total number of electrons participating in the interaction. The dependence of peak power on the electron bunch length is very close to square law (Fig. 4). This means that mostly from the entire pulse length electrons radiate coherently.

5.2. W-band experiments

Alongside the investigation of the generation of sub-nanosecond microwave pulses at Ka-band, similar experiments were performed for a W-band backward-wave microwave structure. The PIC-code simulations have shown a peak power of up to 30 MW can be attained for the half-amplitude pulsewidth of 100–150 ps at a frequency of 75 GHz. The risetime of the front of the pulse did not exceed 100 ps. In the experiments such pulses were clearly observed (Fig. 5b). The integral mode pattern scan has shown that the total peak power of the W band pulses was between 10 and 15 MW when the accelerating voltage at the cathode was 250 kV, e-beam current was of 800 A, and the axial B-field was 3.5 T. The W-band spikes possessed a sharp leading edge (<120 ps), e.g. rising two times faster than the Ka band pulses. Such a fast front corresponded to the limits of the transient response of the oscilloscope's cable delay line provided for the measurements (120 ps). Due to these bandwidth limitations of the measuring system we believe that the peak power was somewhat underestimated.

5.3. G-band experiments

For generation of G-band subnanosecond microwave spikes the energy of the electron beam was reduced down to 180–200 keV. This was required to prevent strong collimation of the current pulse at the inlet of the slow-wave structure which had a drift channel of 1.5 mm diameter. Reducing the accelerating voltage leads to a decrease of the transverse velocities of the electrons emitted from the cathode. It is important to note that in the G band experiment a pulsed axial magnetic field of 2.5 T was applied. Fig. 5c shows the 150 GHz microwave pulse obtained for the conditions when only electrons emitted at the peak of the accelerating pulse were in synchronism with the TM_{01} wave. The measured pulse rise time did not exceed 75 ps which was the limit of the transient characteristic of the 5 GHz oscilloscope, type C7–19. This corresponds well with the PIC-simulations which predict that the half-amplitude pulsewidth of the G band spike to be as short as 70 ps.

6. Conclusion

In conclusion we believe that these experiments carried out together with the simulations of the electron bunch/electromagnetic wave interaction give us reason to believe that the observed emission can be interpreted as superradiative. The microwave pulses generated by such a mechanism have unique short duration of 300 ps with a 150 MW peak power level measured in Ka-band. The initial observation of 75 GHz 10–15 MW radiation pulses with duration less than 150 ps and 150 GHz microwave spikes with a risetime of 75 ps are also reported. The achieved repetition frequency of 25 Hz also supports the conclusion that a novel source of powerful subnanosecond pulses has been developed. In this context an important factor is that the whole device is in the form of a table-top system.

It is promising to consider the application of subnanosecond multi-megawatt millimetre band pulses in areas such as novel diagnostics and the study of non-linear phenomena in plasmas and solids. Another area is related to using such pulses in radiotechnical applications. It would be interesting also to test the influence of such pulses on biological matter.

Acknowledgements

This work was supported by the Russian Foundation for Basic Research (grant 98-02-17308)

References

[1] R.H. Dicke, Phys. Rev. 99 (1954) 131.
[2] J.C. MacGillivray, M.S. Feld, Phys. Rev. A 14 (1976) 1169.
[3] N. Scribanowitz, et al., Phys. Rev. Lett. 30 (1973) 309.
[4] R. Bonifacio, et al., Opt. Commun. 68 (1988) 369.
[5] R. Bonifacio, et al., Phys. Rev. A 44 (1991) 3441.
[6] N.S. Ginzburg, Sov. Tech. Phys. Lett. 14 (1988) 197.
[7] G.R.M. Robb, et al., Phys. Rev. Lett. 77 (1996) 1492.
[8] N.S. Ginzburg, et al., Pis'ma v ZhETF 60 (7) (1994) 501.
[9] N.S. Ginzburg, et al., Sov. Tech. Phys. Lett. 22 (1996) 359.
[10] V. Piovella, et al., Phys. Rev. E 52 (1995) 5470.
[11] N.S. Ginzburg, et al., Phys. Rev. Lett. 78 (1997) 2365.
[12] D.A. Jaroszynski, et al., Phys. Rev. Lett. 78 (1997) 1699.
[13] N.S. Ginzburg, et al., Nucl. Instr. and Meth. A 393 (1997) 352.
[14] N.S. Ginzburg, et al., Phys. Rev. E 60 (3) (1999) 3297.
[15] M.I. Yalandin, et al., in: Proceedings of the 9th IEEE International Pulsed Power Conference, Albuquerque, NM, 1993, p. 388.
[16] M.I. Yalandin, et al., Pis'ma v ZhTF. 25 (10) (1999) 19.

ELSEVIER

Nuclear Instruments and Methods in Physics Research A 475 (2001) 391–396

NUCLEAR
INSTRUMENTS
& METHODS
IN PHYSICS
RESEARCH
Section A

www.elsevier.com/locate/nima

Design considerations of 10 kW-scale, extreme ultraviolet SASE FEL for lithography

C. Pagani[a], E.L. Saldin[b],*, E.A. Schneidmiller[b], M.V. Yurkov[c]

[a] INFN Milano-LASA, Via Cervi, 201, 20090 Segrate (MI), Italy
[b] Deutsches Elektronen-Synchrotron (DESY), Building 25F, Notkestrasse 85, Hamburg, Germany
[c] Joint Institute for Nuclear Research, Dubna, 141980 Moscow Region, Russia

Abstract

The semiconductor industry growth is driven to a large extent by steady advancements in microlithography. According to the newly updated industry road map, the 70 nm generation is anticipated to be available in the year 2008. However, the path to get there is not clear. The problem of construction of extreme ultraviolet (EUV) quantum lasers for lithography is still unsolved: progress in this field is rather moderate and we cannot expect a significant breakthrough in the near future. Nevertheless, there is clear path for optical lithography to take us to sub-100 nm dimensions. Theoretical and experimental work in Self-Amplified Spontaneous Emission (SASE) Free Electron Lasers (FEL) physics and the physics of superconducting linear accelerators over the last 10 years has pointed to the possibility of the generation of high-power optical beams with laser-like characteristics in the EUV spectral range. Recently, there have been important advances in demonstrating a high-gain SASE FEL at 100 nm wavelength (J. Andruszkov, et al., Phys. Rev. Lett. 85 (2000) 3821). The SASE FEL concept eliminates the need for an optical cavity. As a result, there are no apparent limitations which would prevent operating at very short wavelength range and increasing the average output power of this device up to 10-kW level. The use of super conducting energy-recovery linac could produce a major, cost-efficient facility with wall plug power to output optical power efficiency of about 1%. A 10-kW scale transversely coherent radiation source with narrow bandwidth (0.5%) and variable wavelength could be excellent tool for manufacturing computer chips with the minimum feature size below 100 nm. All components of the proposed SASE FEL equipment (injector, driver accelerator structure, energy recovery system, undulator, etc.) have been demonstrated in practice. This is guaranteed success in the time-schedule requirement. © 2001 Elsevier Science B.V. All rights reserved.

PACS: 41.60.Cr; 52.75.M; 42.62.Cf

Keywords: Free electron laser; Superconducting accelerator; Industrial applications

1. Introduction

As the free electron laser (FEL) [1] has several excellent features, such as high efficiency, high power and wavelength tunability, a very wide range of industrial applications contemplated [2]. The most useful and pertinent frequency ranges for industrial-use FELs are in the UV and VUV. Recently, an industrial UV FEL project has been launched by a consortium of industrial firms including DuPont, Xerox, and IBM [3]. In the near future, one can predict that FELs will be

*Corresponding author. Tel.: +49-40-8998-2676; fax: +49-40-8998-4475.
E-mail address: saldin@vxdesy.desy.de (E.L. Saldin).

0168-9002/01/$ - see front matter © 2001 Elsevier Science B.V. All rights reserved.
PII: S 0 1 6 8 - 9 0 0 2 (0 1) 0 1 5 4 6 - 7

widely introduced into high-technology industries. In particular, in the next decade lithography will be highly supported by short-wavelength FELs. Moore's law, postulated in 1965, predicted the exponential increase in the number of devices per chip which has driven the decrease of lithography dimensions. The exponential decrease in the minimum feature size sustained by optical lithography over the past several decades has enabled exponential increase in memory chips (from 1-kb chips for $10\,\mu m$ linewidth dimension to 1-Gb for $0.18\,\mu m$) [4]. Critical dimensions for use in high volume manufacturing are anticipated to decrease from 180 nm in the year 1999 to 100 nm in the year 2006 and to 50 nm in the year 2012 [4,5]. Nevertheless, now there is no clear path for optical lithography to take us to sub-100 nm dimensions. Uncertainty regarding the extendibility of optical lithography casts doubt on the ability of the semiconductor industry to continue exponential rate of progress.

In principle, optical lithography can cover the dimension range from 100 to 10 nm. Potential candidate technology for high volume manufacturing beyond the use of 193 nm wavelength ArF lasers is extreme ultraviolet lithography (EUVL) (see Ref. [5]). Based on multilayer coated reflective optics, it makes a jump in wavelength to the sub-100 nm region while maintaining the evolution of optical techniques and the industry investment therein. Fig. 1 shows schematically the basic elements of EUV lithography (see Ref. [5]). Transversely coherent radiation illuminates a multilayer-coated reflective mask that is over-coated with an absorber pattern. Multilayer-coated reduction optics are then used to replicate the pattern at nominal 4 : 1 reduction on a photoresist-coated wafer. In order to correct for aberrations across the relatively large field, and being limited to a few optical surfaces, one must turn to aspheric optics. The reduction optics must be highly corrected so as to print near diffraction-limited patterns at the wafer. This is a new challenge for mirrors and multilayers. It is also necessary to develop the new materials needed for photoresists and photomask.

The new short wavelength light sources must be developed for the EUV lithographic process. Significant efforts of scientists and engineers working in the field of conventional quantum lasers are directed towards the construction of powerful EUV laser for lithography. Nevertheless, this problem is still unsolved: progress in this field is rather moderate and we cannot expect a significant breakthrough in the near future. Uncertainty regarding the extendibility of convenient quantum lasers to EUV wavelength region casts doubt on the ability of the optical lithography to continue to dominate in the next decade.

In this paper we describe the approach being taken to extend the capability of light sources for lithography up to EUV region. Our approach is based on the idea to use SASE FEL for delivering extremely brilliant, coherent light with wavelength in the EUV range. Compared to the state-of-the-art EUV plasma lasers, one expects full transverse coherence, and up to 9–10 order of magnitude larger average brilliance. Since the wavelength of an SASE FEL is adjustable, selection of new materials needed for photoresists and photomask may be much easier than for the case of fixed-wavelength lasers. Recently there have been important advances in demonstrating a high-gain SASE FEL at 100 nm wavelength [1]. The experimental results presented in Ref. [1] have been achieved at the TESLA Test Facility (TTF) FEL at DESY. The goal of the TTF FEL is to demonstrate SASE FEL emission in the VUV and, in the second phase, to build a VUV- soft X-ray user facility [6].

We show that it is feasible to construct a 10-kW scale SASE FEL. The technical approach adopted

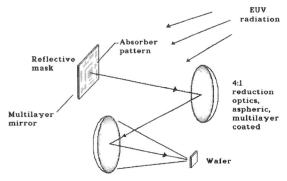

Fig. 1. The basic concept of EUV lithography.

in our design makes use of superconducting RF linear accelerator (SRF accelerator). With SRF linac, a SASE FEL would acquire high average power, thanks to the input beam continuous-wave (CW) nature. The energy recovery of most of the driver electron beam energy would further increase the power efficiency. The stringent electron beam qualities required for EUV SASE FEL operation can be met with a conservative injector design (using a conventional thermionic DC gun and subharmonic bunchers) and the beam compression and linear accelerator technology, recently developed in connection with high-energy linear collider and X-ray FEL programs [7,10].

2. Facility description

Fig. 2 shows the general scheme of the 10 kW-scale EUV SASE FEL driven by a 1300-MHz superconducting linear accelerator. In the acceleration sections, the superconducting cavities are designed to operate with nominal acceleration gradients of 10 MV/m. The electron beam originates in a 300-kV DC gun with a gridded thermionic cathode. The injector, which is practically identical to that designed at LBL for the CW-mode operation infrared FEL [8], includes two

subharmonic, room-temperature buncher cavities and 500-MHz accelerator buncher cavity. The injector produces 6.5 MeV electron pulses with a duration 33 ps (FWHM), at an average current of 12.2 mA (2 nC of charge, 6.1 MHz repetition rate). A 500-MHz single-cavity cryounit follows the injector which increases the beam energy to 12 MeV. The optimized beam parameters at the exit of the cryounit are: bunch charge 2 nC, bunch length 4.2 mm rms, normalized transverse emittance 8π mm mrad, and longitudinal rms emittance $300\,\pi$ keV mm. Accelerator buncher cavity and first accelerating cavity operating without energy recovery, will require about 170 kW RF power. The klystrons are two 75-kW TH2133 tubes, combined through a magic tee to provide the 120-kW of RF at the input coupler to each cryounit. Results of stability analysis of the injector are presented in Ref. [8]. The charge stability $<2\%$, bunch length stability $<2\%$ and bunch timing stability <3 ps are well within RF control system capability. The 12 MeV energy electron beam then enters the SRF linac for further accelerating up to an energy of 1000 MeV. The electron beam enters the undulator, yields EUV coherent radiation, and finally decelerates through an energy recovery pass in the SRF driver linac before its remaining energy is

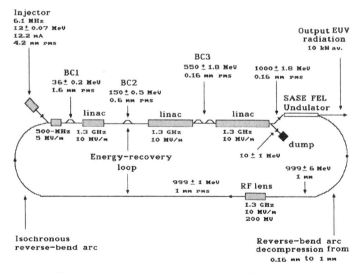

Fig. 2. Basic scheme of the high-power EUV SASE FEL.

absorbed in the beam dump at a final energy of about 10 MeV. In the present design the beam dump energy is below the photon–neutron production threshold, so the problem of radio-nuclide production in the dump does not exist.[1]

The SASE FEL provides a continuous train of 0.5 ps micropulses, with 2 mJ of optical energy per micropulse at a repetition rate 6.1 MHz. The average radiation output power can exceed 10 kW. The radiation from SASE FEL is spatially (or transversely) coherent. The temporal (or longitudinal) coherence, however, is poor due to the start-up from noise. The bandwidth of the output radiation would be about 0.5% (FWHM). A characteristic feature of multilayer mirrors is their rather small bandwidth of the reflected radiation. It is interesting to note that the radiation bandwidth of SASE FEL is close to the typical bandwidth of the multilayer mirror reflectivity.

A driver linac design requires considerable manipulation of the longitudinal and transverse beam dynamics in order to provide, on the one hand, the bunch parameters for effective generation of the SASE radiation, and on the other hand, to make effective energy recovery feasible. For the driver accelerator design we assume to use a three-stage compressor design. The compression performs in three steps: at 36 MeV (from 4.2 to 1.6 mm rms), 150 MeV (from 1.6 to 0.6 mm rms) and 550 MeV (from 0.6 to 0.16 mm rms). Between the first and second bunch compressors the curvature of the accelerating field would impose an intolerable nonlinear correlated energy distribution along the bunch. Thus, the use of third harmonic deceleration structure is foreseen in order to reduce the nonlinear energy spread. First (BC1) and second (BC2) bunch compressors are simple chicanes formed of four rectangular dipole magnets. The third (BC3) compressor is a sequence of two magnetic chicanes. The shorter bunch and higher energy allow for a much longer and more complicated design than BC2 and the complexity of a double chicane is required [7]. The

last part of the driver linac accelerates the bunch with an on-crest phase up 1000 MeV.

In our conceptual design we assume the use of an energy recovery system. Only about 0.1% of the electron energy is converted to light. The remainder undergoes energy recovery, being returned to the SRF cavities, where most of it is converted back into RF power at the cavities' resonant frequency. An off-crest deceleration phase should be tuned in order to minimize the RF power consumption by the accelerator. The decelerated beam is then dumped. The SRF linac must decelerate the bunches from an energy of about 999 MeV to about 10 MeV in the beam dump. The energy spread of the electron beam after leaving the undulator is pretty large, about $\Delta E \simeq \pm 6$ MeV. An important feature of our design is that a very short electron bunch (of about 0.16 mm rms) is used for the generation of the EUV radiation. Thus, the use of energy bunching is foreseen in order to reduce the energy spread.

Energy bunching is appropriate for the situation in which particles are bunched tightly in phase but have a large energy spread. The transformations are in the reverse order from those used for phase bunching. A relation is first established between phase and energy, creating a skew "ellipse" in longitudinal phase space. This is followed by a RF lens that reduces the energy spread by applying a reverse voltage that returns the "ellipse" to axis. Phase separation (i.e. linear correlation energy and longitudinal position) can be obtained in our case by the first, 180° bend of the recovery loop. A correlated energy spread in the bunch is canceled by passing a RF accelerator structure at 90° crossing phase(0° corresponding to running on-crest). We select 1300-MHz structure for the RF lens, based on SRF cavities operating with gradient 10 MV/m. For the chosen parameters of the EUV SASE FEL we get induced energy spread 6 MeV. Voltage which is sufficient to cancel the 6 MeV energy spread is equal to 200 MV. The transformed energy spread is about 1 MeV. It should be noted that energy bunching, in our case can be treated by single particle dynamic theory. This situation is in marked contrast to phase bunching, in which the space charge and wake field

---

[1] Recently Jefferson Laboratory energy-recovery SRF linac achieved 48 MeV of beam energy with 4 mA of average beam current [9].

effects determine the effective phase-space area occupied by the particles.

The wall plug power to output optical power efficiency of SASE FEL for industrial applications is an important criterion. For the present design we fixed on a rather conservative value of the ratio of the energy in the radiation pulse to the energy in the electron pulse of about 0.1%. Energy recovery of most of the driver electron beam energy would increase the power efficiency and we can reach RF power to radiation beam power efficiency of about 7%. Assuming the efficiency of the klystron modulator 80%, and electronic efficiency of the klystron 60%, we obtain that the AC wall plug power to output radiation power efficiency is about 3%. The present design requires cooling of about 25 cryomodules. To do this, we need a He refrigerator with net power consumption about 1 MW. As a result we obtain total efficiency of the proposed SASE FEL to about 1%.

A complete description of the SASE FEL can be performed only with three-dimensional (3-D) time-dependent numerical simulation code. With the design and construction of VUV and X-ray SASE FELs, many 3-D time-dependent codes have been developed over the years in order to describe FEL amplifier start-up from shot noise. Optimization of the parameters of the EUV SASE FEL in our case has been performed with the code FAST [11].

The optimized parameters of the EUV SASE FEL are presented in Table 1. Averaging over 100 simulation runs with statistically independent shot noise in the electron beam gives the radiation pulse shape, which is plotted in Fig. 3. The obtained duration of the radiation pulse and pulse energy are about 0.5 ps and 2 mJ, respectively. At the next step of calculation we find the spectral distribution of the radiation power for each angle in the far zone, and after integrating over all angles we obtain 0.5% (FWHM) as the integral spectrum width. An important characteristic of the radiation source is the degree of transverse coherence. Correspondings definitions for the degree of coherence in the high-gain linear regime of the SASE FEL can be found in Ref. [12]. Our simulations show that the degree of coherence in our case is close to unity ($\zeta \simeq 0.9$).

Table 1
Performance characteristics of the EUV SASE FEL

Electron beam	
Energy (MeV)	1000
rms energy spread (%)	0.18
Normalized emittance (π mm mrad)	8
Bunch charge (nC)	2
rms bunch length (mm)	0.16
Repetition rate (MHz)	6.1
Undulator	
Type	Planar
Period (cm)	4.5
Gap (mm)	11
Maximum peak field (kG)	11
External beta-function (cm)	100
Number of undulator periods	700
Output radiation	
Wavelength (nm)	70
Micropulse duration (ps) (FWHM)	0.5
Spectrum width (%) (FWHM)	0.5
Micropulse energy (mJ)	2
Peak power (GW)	3
Average power (kW)	10

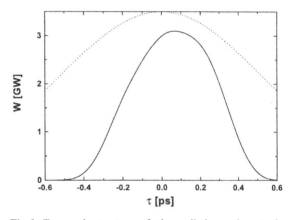

Fig. 3. Temporal structure of the radiation pulse at the undulator length of 34 m. The smooth curve is the radiation pulse profile averaged over large number of statistically independent runs. The dashed line presents the longitudinal profile of the electron beam current.

Analysis of parameters of a high-power EUV SASE FEL shows that its radiation wave-length range is clearly limited by the quality of the electron beam achievable with injector. For

10 kW-scale EUV SASE FEL operating in 10–20 nm wavelength range, a new approach for the injector has to be considered. In this context the R&D work on SRF photoinjector [13] looks very promising.

References

[1] J. Andruszkov, et al., Phys. Rev. Lett. 85 (2000) 3821.

[2] C. Yamanaka, Nucl. Instr. and Meth. A 318 (1992) 1.

[3] G.R. Neil, et al., Nucl. Instr. and Meth. A 358 (1995) 159.

[4] C.P. Ausschnitt, Microelectron. Eng. 41/42 (1998) 41.

[5] D. Attwood, Soft X-rays and Extreme Ultraviolet Radiation, Cambridge University Press, Cambridge, 1999.

[6] J. Rossbach, Nucl. Instr. and Meth. A 375 (1996) 269.

[7] The Linac Coherent Light Source (LCLS) Design Study Report, SLAC-R-521, 1998.

[8] K.-J. Kim, et al., LBL preprint Pub-5335, 1992.

[9] G.R. Neil, et al., Nucl. Instr. and Meth. A 445 (2000) 192.

[10] R. Brinkman, et al., (Eds), Conceptual Design of 500 GeV e^+e^- Linear Collider with Integrated X-ray Facility, DESY 1997-048, ECFA 1997-182, Hamburg, May 1997.

[11] E.L. Saldin, E.A. Schneidmiller, M.V. Yurkov, Nucl. Instr. and Meth. A 429 (1999) 233.

[12] E.L. Saldin, E.A. Schneidmiller, M.V. Yurkov, The Physics of Free Electron Lasers, Springer, Berlin, Heidelberg, New York, 1999.

[13] F. Gabriel, et al., Nucl. Instr. and Meth. A 429 (1999) II 91.

ELSEVIER

Nuclear Instruments and Methods in Physics Research A 475 (2001) 397–400

**NUCLEAR
INSTRUMENTS
& METHODS
IN PHYSICS
RESEARCH**
Section A

www.elsevier.com/locate/nima

Short wavelength free electron lasers in 2000

W.B. Colson*

Physics Department, Naval Postgraduate School, 833 Dyer Road, Monterey, CA 93943, USA

Abstract

Twenty-three years after the first operation of the short wavelength free electron laser (FEL) at Stanford University, there continue to be many important experiments, proposed experiments, and user facilities around the world. Properties of FELs operating in the infrared, visible, UV, and X-ray wavelength regimes are listed and discussed. © 2001 Elsevier Science B.V. All rights reserved.

PACS: 41.60.Cr

Keywords: Free-electron laser; Short wavelength

Table 1 lists existing and proposed relativistic free electron lasers (FELs) in 2000. The top part of the table lists existing FELs, and the bottom part of the table lists proposed FELs. Each FEL is identified by a location, or institution, followed by the FEL's name in parentheses. The table can be found at http://www.physics.nps.navy.mil/fel.html. Additions and corrections can be transmitted to us for inclusion in the table in future.

The first column of the table lists the operating wavelength λ, or wavelength range. The large range of operating wavelengths, six orders of magnitude, indicates the flexible design character-istics of the FEL mechanism. In the second column, σ_z is the electron pulse length divided by the speed of light c, and ranges from CW to short sub-picosecond pulse time scales. The expected optical pulse length can be 3–5 times shorter or longer than the electron pulse depending on the optical cavity Q, the FEL desynchronism, and the

FEL gain. If the FEL is in an electron storage ring, the optical pulse is typically much shorter than the electron pulse. Most FEL oscillators produce an optical spectrum that is Fourier transform limited by the optical pulse length.

The electron beam energy E and peak current I provided by the accelerator are listed in the third and fourth columns. The accelerator type is listed as the first entry in the last column with a code such as RF for the radio-frequency linac. While there are a variety of accelerators used, most are RF with some electron storage rings, microtrons, and electrostatic accelerators. Storage rings tend to be used for the short wavelength applications, while the electrostatic accelerators provide longer wavelengths.

The next three columns list the number of undulator periods N, the undulator wavelength λ_0, and the undulator parameter $K = eB\lambda_0/2\pi mc^2$, where e is the electron charge magnitude, B is the rms undulator field strength, and m is the electron mass. For an FEL klystron undulator, there are multiple undulator sections as listed in the

*Tel.: +1-831-656-2765; fax: +1-831-656-2834.

E-mail address: colson@nps.navy.mil (W.B. Colson).

0168-9002/01/$ - see front matter © 2001 Elsevier Science B.V. All rights reserved.
PII: S 0 1 6 8 - 9 0 0 2 (0 1) 0 1 6 8 7 - 4

Table 1
Relativistic short wavelength free electron lasers (2000)[a]

FELs	λ (μm)	σ_z	E (MeV)	I (A)	N	λ_0 (cm)	K (rms)	Acc. type [Ref.]
Existing								
UCSB (mm FEL)	340	25 μs	6	2	42	7.1	0.7	EA,O [1]
Dartmouth (FEL)	200	CW	0.04	0.001	50	300	—	SP,O [2]
Korea (KAERI-FEL)	97–150	25 ps	6.5	0.5	80	2.5	1.6	RF,O [47]
Himeji (LEENA)	65–75	10 ps	5.4	10	50	1.6	0.5	RF,O [3]
UCSB (FIR FEL)	60	25 μs	6	2	150	2	0.1	EA,O [1]
Osaka (ILE/ILT)	47	3 ps	8	50	50	2	0.5	RF,O [4]
Osaka (ISIR)	40	30 ps	17	50	32	6	1	RF,O [5]
Tokai (JAERI-FEL)	22	5 ps	16.5	100	52	3.3	0.7	RF,O [6]
Bruyeres (ELSA)	20	30 ps	18	100	30	3	0.8	RF,O [7]
Osaka (FELI4)	18–40	10 ps	33	40	30	8	1.3–1.7	RF,O [8]
UCLA-Kurchatov	16	3 ps	13.5	80	40	1.5	1	RF,A [9]
LANL (RAFEL)	15.5	15 ps	17	300	200	2	0.9	RF,O [10]
Stanford (FIREFLY)	15–65	1–5 ps	15–32	14	25	6	1	RF,O [11]
UCLA-Kurchatov- LANL	12	5 ps	18	170	100	2	0.7	RF,A [12]
Maryland (MIRFEL)	12–21	5 ps	9–14	100	73	1.4	0.2	RF,O [13]
Beijing (IHEP)	10	4 ps	30	14	50	3	1	RF,O [14]
Darmstadt (IR-FEL)	6–8	2 ps	25–50	2.7	80	3.2	1	RF,O [15]
BNL (HGHG)	5.3	6 ps	40	120	60	3.3	1.44	RF,A [16]
Osaka (FELI1)	5.5	10 ps	33.2	42	58	3.4	1	RF,O [17]
Tokyo (FEL-SUT)	5–16	2 ps	32	0.2	40	3.2	0.7–1.8	RF,O [48]
Nieuwegein (FELIX)	4–200	1 ps	50	50	38	6.5	1.8	RF,O [18]
Duke (MarkIII)	3	3 ps	44	20	47	2.3	1	RF,O [19]
Stanford (SCAFEL)	3–13	0.5–12 ps	22–45	10	72	3.1	0.8	RF,O [21]
Orsay (CLIO)	3–53	0.1–3 ps	21–50	80	38	5	1.4	RF,O [22]
Vanderbilt (FEL1)	2.0–9.8	0.7 ps	43	50	52	2.3	1	RF,O [23]
Osaka (FELI2)	1.88	10 ps	68	42	78	3.8	1	RF,O [17]
TJNAF (FEL)	1–6	0.4 ps	48	60	41	2.7	0.9	RF,O [20]
BNL (ATF)	0.5	6 ps	50	100	70	0.88	0.4	RF,O [24]
Dortmund (FELICITAI)	0.47	50 ps	450	90	17	25	2	SR,O [25]
ANL (APSFEL)	0.385	0.65 ps	354	184	648	3.3	2.2	RF,S [40]
Orsay (Super-ACO)	0.3–0.6	15 ps	800	0.1	2×10	13	4.5	SR,O [26]
Osaka (FELI3)	0.3–0.7	5 ps	155	60	67	4	1.4	RF,O [27]
Okazaki (UVSOR)	0.24	6 ps	607	10	2×9	11	2	SR,O [28]
Tsukuba (NIJI-IV)	0.2–0.6	160 ps	310	5	2×42	7.2	2	SR,O [29]
Italy (ELETTRA)	0.2–0.4	28	1000	150	2×19	10	4.2	SR,O [38]
Duke (OK-4)	0.1947	1.6 ps	800	35	2×33	10	0–4	SR,O [30]
DESY (TTF1)	0.109	0.5 ps	233	300	492	2.73	1.1	RF,S [42]
Proposed								
Florida (CREOL)	355	8 μs	1.3	0.13	185	0.8	0.1	EA,O [31]
Netherlands (TEUFEL)	180	20 ps	6	350	50	2.5	1	RF,O [32]
Rutgers (IRFEL)	140	25 ps	38	1.4	50	20	1	MA,O [33]
Moscow (Lebedev)	100	20 ps	30	0.25	35	3.2	0.8	MA,O [34]
Novosibirsk (RTM)	2–11	20 ps	98	100	4×36	9	1.6	RF,O [35]
TJNAF (UVFEL)	0.16–1	0.2 ps	160	270	72	3.3	1.3	RF,O [20]
Rocketdyne/Hawaii (FEL)	0.3–3	2 ps	100	500	84	2.4	1.2	RF,O [36]
Harima (SUBARU)	0.2–10	26 ps	1500	50	33,65	16,32	8	SR,O [37]
ANL (APSFEL)	0.12	1 ps	440	150	864	3.3	3.1	RF,S [40]
BNL (DUVFEL)	0.1	6 ps	230	1000	256	2.89	1.2	RF,A [39]
Frascati (COSA)	0.08	10 ps	215	200	400	1.4	1	RF,O [41]
Duke (VUV)	0.01–1	1 ps	1000	50	4×32	12	3	SR,O [43]
DESY (TTF2)	0.006	0.17 ps	1000	2500	981	2.73	0.9	RF,S [44]

Table 1 (*continued*)

FELs	λ (μm)	σ_z	E (MeV)	I (A)	N	λ_0 (cm)	K (rms)	Acc. type [Ref.]
SLAC (LCLS)	0.00015	0.07 ps	14 350	3400	3328	3	3.7	RF,S [45]
DESY (TESLA)	0.0001	0.08 ps	35 000	5000	1200	5	4.2	RF,S [46]

[a]Note: RF—RF linac accelerator; MA—microtron accelerator; SR—electron storage ring; EA—electrostatic accelerator; A—FEL amplifier; O—FEL oscillator; SP—smith-purcell oscillator; S—SASE FEL.

N-column. Note that the range of values for N, λ_0, and K are much smaller than for the other parameters indicating that most undulators are similar. Only a few of the FELs use the klystron undulator at present, and the rest use the conventional periodic undulator. The FEL resonance condition,

$$\lambda = \frac{\lambda_0(1 + K^2)}{2\gamma^2}$$

where γ^2 is the relativistic Lorentz factor $\gamma = E/mc^2$, provides a relationship that can used to derive K from λ, E, and λ_0. The middle entry of the last column lists the FEL type: "O" for oscillator, "A" for amplifier, etc. Most of the FELs are oscillators, but recent progress has resulted in short wavelength FELs using Stimulated Amplification of Spontaneous Emission (SASE) to produce 109 nm radiation. A reference describing the FEL is provided at the end of each line entry.

For the conventional undulator, the peak optical power can be estimated by the fraction of the electron beam peak power that spans the undulator spectral bandwidth, $1/4N$, or $P \approx EI/4eN$. For the FEL using a storage ring, the optical power causing saturation is substantially less than this estimate and depends on ring properties. For the high-gain FEL amplifier, the optical power at saturation can be substantially more. The average FEL power is determined by the duty cycle, or spacing between electron pulses, and is typically many orders of magnitude lower than the peak power. The TJNAF IR FEL has now reached an average power of 1.7 kW with the recovery of the electron beam energy in superconducting accelerator cavities.

In the FEL oscillator, the optical mode that best couples to the electron beam in an undulator of length $L = N\lambda_0$ has Rayleigh length $z_0 \approx L/\sqrt{12}$ and has a mode waist radius of $w_0 \approx \sqrt{N\gamma\lambda/\pi}$. The

FEL optical mode typically has more than 90% of the power in the fundamental mode described by these parameters.

The author is grateful for support by the Naval Postgraduate School.

References

[1] G. Ramian, Nucl. Instr. and Meth. A 318 (1992) 225.
[2] J. Urata, et al., Phys. Rev. Lett. 80 (1998) 516.
[3] T. Mochizuki, et al., Nucl. Instr. and Meth. A 393 (1997) II-47;
S. Miyamoto, T. Mochizuki, J. Jpn Soc. Infrared Sci. Technol. 7 (1997) 73.
[4] N. Ohigashi, et al., Nucl. Instr. and Meth. A 375 (1996) 469.
[5] S. Okuda, et al., Nucl. Instr. and Meth. A 341 (1994) 59.
[6] N. Nishimori, et al., Nucl. Instr. and Meth. A 445 (2000) 432.
[7] P. Guimbal, et al., Nucl. Instr. and Meth. A 341 (1994) 43.
[8] T. Takii, et al., Nucl. Instr. and Meth. A 407 (1998) 21.
[9] M. Hogan, et al., Phys. Rev. Lett. 80 (1998) 289.
[10] D.C. Nguyen, et al., Nucl. Instr. Meth. A 429 (1999) 125.
[11] K.W. Berryman, T.I. Smith, Nucl. Instr. and Meth. A 375 (1996) 6.
[12] M. Hogan, et al., Phys. Rev. Lett. 81 (1998) 4867.
[13] I.S. Lehrman, et al., Nucl. Instr. and Meth. A 393 (1997) 178.
[14] J. Xie, et al., Nucl. Instr. and Meth. A 341 (1994) 34.
[15] J. Auerhammer, et al., Nucl. Instr. and Meth. A 341 (1994) 63.
[16] I. Ben-Zvi, et al., Nucl. Instr. and Meth. A 318 (1992) 208.
[17] A. Kobayashi, et al., Nucl. Instr. and Meth. A 375 (1996) 317.
[18] D. Oepts, et al., Infrared Phys. Technol. 36 (1995) 297.
[19] S.V. Benson, et al., Nucl. Instr. and Meth. A 250 (1986) 39.
[20] S. Benson, et al., Nucl. Instr. and Meth. A 429 (1999) 27.
[21] H.A. Schwettman, et al., Nucl. Instr. and Meth. A 375 (1996) 662.
[22] J.M. Ortega, et al., Nucl. Instr. and Meth. A 375 (1996) 618.
[23] C. Brau, Nucl. Instr. and Meth. A 318 (1992) 38.
[24] K. Batchelor, et al., Nucl. Instr. and Meth. A 318 (1992) 159.

[25] D. Nolle, et al., Nucl. Instr. and Meth. A 341 (1994) ABS7;
Schmidt, et al., Nucl. Instr. and Meth A 341 (1994) ABS9.

[26] M.E. Couprie, et al., Nucl. Instr. and Meth. A 407 (1998) 215.

[27] T. Tomimasu, et al., Nucl. Instr. and Meth. A 393 (1997) 188.

[28] H. Hama, et al., Nucl. Instr. and Meth. A 341 (1994) 12.

[29] T. Yamazaki, et al., Nucl. Instr. and Meth. A 341 (1994) ABS3.

[30] V.N. Litvinenko, et al., Nucl. Instr. and Meth. A 407 (1998) 8.

[31] M. Tecimer, et al., Nucl. Instr. and Meth. A 341 (1994) A126.

[32] J.I.M. Botman, et al., Nucl. Instr. and Meth. A 341 (1994) 402.

[33] E.D. Shaw, et al., Nucl. Instr. and Meth. A 318 (1992) 47.

[34] K.A. Belovintsev, et al., Nucl. Instr. and Meth. A 341 (1994) ABS45.

[35] N.G. Gavrilov, et al., Status of Novosibirsk high power FEL Project, SPIE Proc. 2988 (1997) 23;.N.A. Vinokurov, et al., Nucl. Instr. and Meth. A 331 (1993) 3.

[36] R.J. Burke, et al., Proceedings of the SPIE: Laser Power Beaming, Vol. 2121, Los Angeles, January 27–28, 1994.

[37] S. Miyamoto, et al., Report of the Spring-8 International Workshop on 30 m Long Straight Sections, Kobe, Japan, August 9, 1997.

[38] R.P. Walker, et al., Nucl. Instr. and Meth. A 429 (1999) 179;.Twenty-Second International Free Electron Laser Conference, Duke University, Durham, NC.

[39] E.D. Johnson, Nucl. Instr. and Meth. A 393 (1997) II-12.

[40] S.V. Milton, et al., Nucl. Instr. and Meth. A 407 (1998) 210;.
Twenty-Second International Free Electron Laser Conference, Duke University, Durham, NC.

[41] F. Ciocci, et al., A. Torre, IEEE J. Quantum Electron. 31 (1995) 1242.

[42] W. Brefeld, et al., Nucl. Instr. and Meth. A 393 (1997) 119.

[43] V.N. Litvinenko, et al., Nucl. Instr. and Meth. A 358 (1995) 369.

[44] W. Brefeld, et al., Nucl. Instr. and Meth. A 375 (1996) 295.

[45] M. Cornacchia, Proc. SPIE 2998 (1997) 2.

[46] R. Brinkmann, et al., Nucl. Instr. and Meth. A 393 (1997) 86.

[47] Young Uk Jeong, et al., Nucl. Instr. and Meth. A 475 (2001) 47, these proceedings.

[48] M. Yokoyama, et al., Nucl. Instr. and Meth. A 475 (2001) 38, these proceedings.

ELSEVIER

Nuclear Instruments and Methods in Physics Research A 475 (2001) 401–406

NUCLEAR
INSTRUMENTS
& METHODS
IN PHYSICS
RESEARCH
Section A

www.elsevier.com/locate/nima

Modular approach to achieving the next-generation X-ray light source

S.G. Biedron[a,b,*], S.V. Milton[a], H.P. Freund[c]

[a] *Advanced Photon Source, Argonne National Laboratory, 9700 S. Cass Avenue, Argonne, IL 60439, USA*
[b] *MAX-Laboratory, University of Lund, S-221 00 Lund, Sweden*
[c] *Science Applications International Corporation, McLean, VA 22102, USA*

Abstract

A modular approach to the next-generation light source is described. The "modules" include photocathode, radio-frequency, electron guns and their associated drive-laser systems, linear accelerators, bunch-compression systems, seed laser systems, planar undulators, two-undulator harmonic generation schemes, high-gain harmonic generation systems, nonlinear higher harmonics, and wavelength shifting. These modules will be helpful in distributing the next-generation light source to many more laboratories than the current single-pass, high-gain free-electron laser designs permit, due to both monetary and/or physical space constraints. © 2001 Elsevier Science B.V. All rights reserved.

PACS: 41.60.Cr; 42.65.Ky; 42.25.Kb; 52.59.–f

Keywords: Free-electron lasers; Harmonic generation; Frequency conversion; Coherence; Intense particle beams and radiation sources

1. Introduction

It is the desire of the light source community to design and build a next-generation light source capable of producing pulses that have ultrashort pulse lengths exhibiting temporal coherence unachievable by the existing third-generation machines [1–3].

We present five examples using "modules" to build toward the fourth-generation light source, one of which will be examined more fully. In these examples, the electron beams never exceed 6 GeV, attempts are made to use the lowest number of new electron beams and the shortest possible radiation production lines (undulators, etc.), while applying the methods of single-pass, high-gain free-electron lasers (FELs).

2. The modular approach

Just as the user stations in today's third-generation X-ray machines are rapidly being occupied and additional machines are being built to try to accommodate the high demands, a similar trend is expected after the completion of the Linac Coherent Light Source (LCLS) [4] and TESLA [5] X-ray FEL facilities. Here, combinations of "modules" to produce next-generation X-ray light sources are used that may be more flexible and have better properties than currently expected.

*Corresponding author.
E-mail address: biedron@aps.anl.gov (S.G. Biedron).

The types of modules include the following: photocathode; radio-frequency (rf) guns and associated drive-lasers; linear accelerators; bunch compressors; seed lasers; planar undulators; two-undulator harmonic generation schemes (TUHGS) [6]; high-gain harmonic generation (HGHG), consisting of a modulative section, a dispersive section, and a radiative section [7–12]; nonlinear higher harmonics [13–20]; and wavelength shifting.

The modular approach alone, less the higher nonlinear harmonics, represents a powerful tool; but with these higher nonlinear harmonics, shorter wavelengths can be reached beyond those available from the fundamental in single-pass, high-gain FELs. In a self-amplified spontaneous emission (SASE) [21–26] or amplifier system, the nonlinear harmonics appear to be substantial and quite useful [15–17,20]. In TUHGS and HGHG schemes, however, the downstream undulator is tuned to a higher harmonic than the input seed laser, generating coherent output at the fundamental *and* at the higher nonlinear harmonics of this second undulator. In other words, in TUHGS and HGHG schemes, the shorter wavelengths are attainable more readily than in the SASE and amplifier schemes [19]. These nonlinear harmonics arise in all single-pass FELs based on the planar undulator designs and have been recently measured on the joint BNL/APS HGHG experiment [12] and plan to be measured in the APS SASE FEL at Argonne [15].

Many combinations of the modular path toward the next-generation light source exist, hence, we will present the five major representative examples. Along with the most necessary modules of linear accelerators, seed lasers, bunch compressors, and nonlinear harmonic generation, the following best exemplify the general techniques employed using radiation module combinations:

- multiple amplifier modules,
- multiple TUHGS modules or multiple HGHG modules,
- combinations of amplifier, TUHGS, and/or HGHG modules,
- tabletop soft X-ray seed laser with amplifier, TUHGS, and/or HGHG modules,

- wavelength shifting coupled with amplifier, TUHGS, and/or HGHG modules.

3. The five conceptual examples

3.1. Example I: four amplifier modules

This scheme is composed of four amplifier modules (AMP I–IV) tuned to the fundamental resonance with four fresh electron bunches. Here, a $\lambda_{seed} = 266$ nm, $P_{pk} = 100$ MW seed laser is used. The fifth nonlinear harmonic of the output radiation from AMP I-III each seed the next respective module. The final wavelength, the fifth nonlinear harmonic of AMP IV, is 4.256 Å with an electron beam energy of ~ 4.3 GeV.

3.2. Example II: two HGHG modules

Although either the TUHGS or the HGHG schemes could represent this type of combination, the case of multiple HGHG modules is shown here. This example is related to Yu's Cascading Stages of HGHG [27] and further extends its usefulness by utilizing the higher nonlinear harmonics in the system. This is also related to Dattoli's modular scheme [28]. Of our five examples, this scheme provides the shortest wavelength via a relatively low electron beam energy with the least number of new electron bunches as compared to the current designs.

As seen in Fig. 1, a laser of $\lambda_{seed} = 266$ nm with $P_{pk} = 1$ MW is used as the seed for the first HGHG module (HGHG I). This seed is the "original" fundamental wavelength that drives the entire resultant system. Recall that each HGHG module consists of modulative, dispersive, and radiative sections where the radiative section is tuned to the desired harmonic. Along with the seed laser, a ~ 700-MeV electron beam enters HGHG I, where the energy modulation and spatial bunching are induced in the modulative and radiative sections, respectively. Here, radiative section I (RAD I) is tuned to the seventh harmonic (38 nm) of the "original" fundamental (266 nm). The fifth nonlinear harmonic (7.6 nm) of the output from RAD I (which is the 35th harmonic

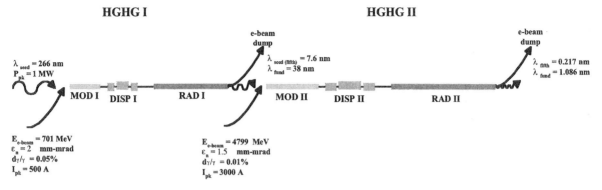

Fig. 1. Example II—two HGHG modules.

to the "original" fundamental) is used to seed the second HGHG section (HGHG II). The first electron beam is bent into a dump and a second, ∼4.8-GeV electron beam enters HGHG II along with this 7.6-nm seed. This grows in an amplifier mode of this second modulative section (MOD II) until enough energy modulation is imparted on the electron beam. The beam then passes through the dispersive section and through radiative section II (RAD II), tuned to the seventh harmonic of the 7.6-nm seed, which is 1.086 nm. Here, the longitudinally coherent output radiation in the fifth nonlinear harmonic has a wavelength of 2.1 Å.

3.3. Example III: two amplifier modules and one HGHG module

This third scheme employs the same seed laser described in Example I and uses three fresh electron bunches in two amplifier modules (AMP I and AMP II) and one HGHG module (HGHG), respectively. An alternative to an HGHG module is to use a TUHGS module.

As seen in Example I, the fifth nonlinear harmonic output from AMP I and AMP II serve as seeds for the following modules. The modulative section in the HGHG module is long enough to induce the desired energy modulation on the electron beam. The radiative section in HGHG is tuned to the seventh harmonic of the input seed from AMP II. The final wavelength in the fifth nonlinear harmonic emitted is 3.04 Å with an electron beam energy of ∼4 GeV.

3.4. Example IV: soft X-ray seed laser amplifier and one HGHG module

In this example, a tabletop, Ni-like molybdenum soft X-ray laser with $\lambda_{\text{seed}} = 18.9$ nm, $P_{\text{pk}} = 5$ GW is used as the seed amplifier module (AMP), of which the coherent power of the correct polarization is ∼0.1 MW. This tabletop, soft X-ray laser would be identical to the "COMET" laser that is currently operational at Lawrence Livermore National Laboratory (LLNL) [18]. The fifth nonlinear harmonic output from AMP would serve to seed an HGHG module (HGHG), where the radiator is tuned to the seventh harmonic of the input seed, requiring an electron beam energy of ∼6 GeV. Utilizing the fifth nonlinear harmonic of the output radiation yields 1.0 Å radiation.

3.5. Example V: wavelength shifting

In the wavelength shifting case, the acceleration and radiation-producing modules are more integrated than in the previous examples. Here, the simplest case is to use modules in the following order, as seen in Fig. 2. First, an electron beam is produced using a gun and a linear accelerator. A seed laser is introduced to the electron beam in an undulator, whose fundamental is tuned to the seed laser for the electron beam energy. This is performed to induce a specified amount of energy modulation. Then, the electron beam is overrotated in phase space using a bunch compressor and further introduced to an accelerating section whose phase is slightly off-crest to induce an

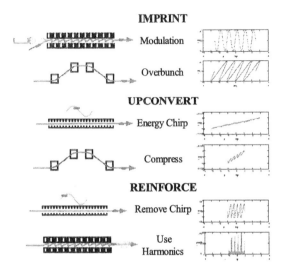

Fig. 2. Example V—wavelength shifting concept.

energy chirp. Next, the electron beam is compressed through a second bunch compressor and the chirp is removed in an additional accelerating section. After further acceleration, the electron beam is injected into an undulator, TUHGS, or HGHG module, where, since it is prebunched to a wavelength tuned by the compression process, the electron beam radiates in a fully coherent fashion. The resultant coherent output can be further introduced to more of the same wavelength-shifting modules or other modules.

As an aside, wavelength shifting allows one to generate arbitrary wavelengths independent of the seed after already imprinting its quality on an electron bunch. It can also be used to shift wavelengths up or down and so can be used for final wavelength tuning by passing wavelength-shifted, "saturated" beams through undulators tuned to the microbunch spacing.

4. Simulation code

MEDUSA [29] is a 3D, nonlinear polychromatic code based on the source-dependent expansion [30] of the Gauss–Hermite waveguide modes. It has been benchmarked at the fundamental against four other simulation codes and demonstrates good agreement [31,32]. It is capable of simulating

TUHGS, HGHG, and nonlinear harmonic generation [6]. The power in the third nonlinear harmonic has also been compared with the 3D analytical model and is in good agreement [20]. For the following modular cases, the output radiation at the desired nonlinear harmonic was fed into the next section with the "fresh" electron bunch in consecutive computer runs.

5. Numerical example: example II

Using MEDUSA, example II was treated in numerical simulation. The output power of the fifth nonlinear harmonic from the HGHG I radiative section is 8.6×10^4 W and serves to seed HGHG II along with a fresh electron bunch. The fifth nonlinear harmonic to the radiative section of HGHG II yields 2.2×10^6 W. The output saturation power of the fundamental and nonlinear harmonic output of radiative sections HGHG I and HGHG II are further summarized in Table 1.

We now further analyze the specific case that each linear accelerator module is composed of 3-m, SLAC-type accelerating structures, each capable of 50-MeV acceleration (assuming two SLED cavities and four 3-m structures per each modulator and klystron assembly). This translates into 1 GeV/60 m of linear accelerator. This conservative estimate is used to compare the "amount" of linear accelerator required for each of the five examples specified above. Table 2 lists the required

Table 1
Power after HGHG module in numerical example II for the odd harmonics up to $h = 7$

	Harmonic number	Radiation wavelength (nm)	Saturation point z(m)	Power (W)
HGHG I	1	38	19.7	5.3×10^8
	3	12.67	20.1	3.8×10^6
	5	7.6	18.5	8.6×10^4
	7	5.4	19.5	1.3×10^4
HGHG II	1	1.09	92.9	1.3×10^9
	3	0.36	93.2	7.4×10^6
	5	0.218	86.6	2.2×10^6
	7	0.156	90.6	5.8×10^5

Table 2
Length review of numerical example II

	Electron beam energy (GeV)	Total length of linear accelerator required (m)	Radiation wavelength (nm) of fifth harmonic to HGHG section	Total length of radiation production section required (m)
HGHG I	0.7	42	7.6	21
HGHG II	4.8	288	0.217	94
Total lengths	na	330	na	115

electron beam energy, radiation wavelength, and the length of the accelerator and radiation production sections (all-inclusive; undulators, drift, and dispersive sections) for example II.

6. Conclusions

The modular approach to achieving the next-generation X-ray light source (or any light source) allows: (1) imparting full longitudinal coherence on the output radiation by seeding at a much lower wavelength than that of the desired output wavelength with a coherent source, (2) the use of multiple seeding configurations to work toward an ultrashort wavelength based on substantial frequency up-conversion, (3) utilizing a much lower electron beam energy to produce the desired wavelengths, and (4) the option of building toward shorter radiation wavelengths by first starting with a modest system and then adding additional modules, as time, money, and space permit.

It was the intent to demonstrate the importance of the modular methods as well as nonlinear harmonic generation for building toward the next-generation light source. Although many other modular combinations do indeed exist, producing both longer and shorter wavelengths of varying powers, these five examples were chosen for discussion to promote a type of source that was attainable, within monetary and/or physical space constraints, by many more institutions than currently expected. In particular, new electron beams are not fully necessary, as multiple bunches can be generated in the linear accelerator and fast kicker magnets could simply gate bunches into specific modules. A thorough design review of each of the five examples described is underway involving the electron beam and undulator tolerances.

Acknowledgements

Thanks are extended to our friends for each of their unique contributions to this work: M. Babzien, Y.-C. Chae, R. Dejus, A. Doyuran, M. Eriksson, B. Faatz, W. Fawley, J. Galayda, Z. Huang, K.-J. Kim, Y. Li, I. Lindau, A. Lumpkin, H.-D. Nuhn, A. Nassiri, S. Reiche, T. Shaftan, L. C. Teng, X. Wang, S. Werin, B.X. Yang, and L.-H. Yu.

This work is supported by the U.S. Department of Energy, Office of Basic Energy Sciences, under Contract No. W-31-109-ENG-38. The activity and computational work for H.P. Freund was supported by Science Applications International Corporation's Advanced Technology Group under IR&D subproject 01-0060-73-0890-000.

References

[1] M. Cornacchia, Proceedings of the Workshop on Fourth-Generation Sources, February 24–27, 1992.
[2] Proceedings of the 10th ICFA Beam Dynamics Panel Workshop on Fourth-Generation Light Sources, Grenoble, January 22–25, 1996.
[3] Workshop on Future Light Sources, at the APS at Argonne National Laboratory, October 27–29, 1997.
[4] LCLS Design Study Group, Linac Coherent Light Source (LCLS) Design Study Report, SLAC-R-521, 1998.
[5] A VUV Free Electron Laser at the TESLA Test Facility at DESY: Conceptual Design Report, DESY Print, TESLA-FEL 95-03, June 1995.
[6] R. Bonifacio, L. De Salvo Souza, P. Pierini, E.T. Scharlemann, Nucl. Instr. and Meth. A 296 (1990) 787.
[7] L.-H. Yu, Phys. Rev. A44 (1991) 5178.
[8] I. Ben-Zvi, et al., Nucl. Instr. and Meth. A 318 (1992) 208.
[9] L.-H. Yu, et al., Proceedings of the IEEE 1999 Particle Accelerator Conference, 1999, p. 2470.
[10] L.-H. Yu, et al., Nucl. Instr. and Meth. A 445 (2000) 301.
[11] L.-H. Yu, et al., Science 289 (2000) 932.
[12] A. Doyuran, et al., Nucl. Instr. and Meth. A 475 (2001) 260, These Proceedings.
[13] R. Bonifacio, L. DeSalvo, P. Pierini, Nucl. Instr. and Meth. A 293 (1990) 627.

[14] S.G. Biedron, H.P. Freund, S.V. Milton, in: H.E. Bennett, D.H. Dowell (Eds.), Free-Electron Laser Challenges II, Proceedings of SPIE, Vol. 3614, SPIE, Bellingham, MA, 1999, p. 96.

[15] H.P. Freund, S.G. Biedron, S.V. Milton, IEEE J. Quantum Electron. 36 (2000) 275, and references therein.

[16] H.P. Freund, S.G. Biedron, S.V. Milton, Nucl. Instr. and Meth. A 445 (2000) 53.

[17] Z. Huang, K.-J. Kim, Phys. Rev. E 62 (2000) 7295.

[18] S.G. Biedron, H.P. Freund, S.V. Milton, Y. Li, Proceedings of the seventh European Accelerator Conference (EPAC), Vienna, 26–30 June 2000, p. 729, and references therein, in preparation.

[19] S.G. Biedron, H.P. Freund, S.V. Milton, X.J. Wang, L-.H. Yu, Nucl. Instr. and Meth. A 475 (2001) 118, These Proceedings.

[20] S.G. Biedron, et al., The use of harmonics to achieve coherent short wavelengths, Proc. of the IEEE 2001 Particle Accelerator Conference, to be published.

[21] R. Bonifacio, C. Pelligrini, L.M. Narducci, Opt. Commun. 50 (1984) 373.

[22] J.B. Murphy, C. Pelligrini, J. Opt. Soc. Am. B 2 (1985) 259.

[23] YaS. Derbenev, A.M. Kondratenko, E.L. Saldin, Nucl. Instr. and Meth. A 193 (1982) 415.

[24] K.-J. Kim, Phys. Rev. Lett. 57 (1986) 1871.

[25] L.H. Yu, et al., Phys. Rev. Lett. 64 (1990) 3011.

[26] M. Xie, Proceedings of the IEEE 1995 Particle Accelerator Conference, 1995, p. 183.

[27] L.H. Yu, J. Wu, Nucl. Instr. and Meth. A 475 (2001) 79, These Proceedings.

[28] G. Dattoli, P.L. Ottaviani, J. Appl. Phys. 86 (1999) 5331.

[29] H.P. Freund, Phys. Rev. E 52 (1995) 5401.

[30] P. Sprangle, A. Ting, C.M. Tang, Phys. Rev. A 36 (1987) 2773.

[31] S.G. Biedron, Y.C. Chae, R.J. Dejus, B. Faatz, H.P. Freund, S.V. Milton, H.-D. Nuhn, S. Reiche, Nucl. Instr. and Meth. A 445 (2000) 110.

[32] S.G. Biedron, Y.-C. Chae, R.J. Dejus, B. Faatz, H.P. Freund, S. Milton, H.-D. Nuhn, S. Reiche, Proceedings of the IEEE 1999 Particle Accelerator Conference, 1999, p. 2486.

ELSEVIER

Nuclear Instruments and Methods in Physics Research A 475 (2001) 407–416

NUCLEAR
INSTRUMENTS
& METHODS
IN PHYSICS
RESEARCH
Section A

www.elsevier.com/locate/nima

The OK-5/Duke storage ring VUV FEL with variable polarization ☆

V.N. Litvinenko[a,*], S.F. Mikhailov[a], O.A. Shevchenko[a], N.A. Vinokurov[b],
N.G. Gavrilov[b], G.N. Kulipanov[b], T.V. Shaftan[b], P.D. Vobly[b], Y. Wu[c]

[a] FEL Laboratory, Department of Physics, Duke University, P.O. Box 90319, Durham, NC 27708-0319, USA
[b] Budker Institute of Nuclear Physics, Novosibirsk, Russia
[c] ALS, Lawrence Berkeley National Laboratory, Berkeley, CA, USA

Abstract

The OK-5/Duke storage ring free electron laser (FEL) project was started in 1998. Presently, the components of the OK-5 FEL and the new South straight section are in the final stage of manufacturing. This paper describes the design and the main features of the OK-5/Duke storage ring FEL. The basic concepts and main compromises made in the design process are presented. Plans for the OK-5 FEL commissioning are discussed. © 2001 Elsevier Science B.V. All rights reserved.

PACS: 41.60.Cr; 29.20D; 78.66; 52.75

Keywords: Free electron laser; Storage ring; Optical klystron; Deep ultraviolet; Coherence

1. Introduction

The initial concept of the distributed optical klystron (DOK) [1] for the Duke storage ring was developed in 1993 [2]. Its present realization, named the OK-5 "Blue Devil" FEL, is close to completion and will replace the OK-4 FEL at the Duke storage ring in the near future. The OK-5 FEL has the same origin as the series of optical klystrons named the OK-1, OK-2, OK-3 and OK-4, which were developed in the Budker Institute of Nuclear Physics (BINP), Novosibirsk [3]. Hence, the OK-5 has its natural "serial number" with "Blue Devil" sub-title indicating its future location[1] and the UV/VUV operation range. The OK-5 FEL is the first implementation of the distributed optical klystron with reasonably high gain in the UV and the VUV ranges of spectrum [4].

The schematic layout of the OK-5/Duke storage ring FEL system is shown in Fig. 1. The 24.2-meter long magnetic system of the OK-5 is comprised of four 4-m long electromagnetic

☆ Work supported Office of Naval Research Grant #N00014-94-1-0818, by the Duke University School of Arts and Sciences, by the Director, Office of Energy Research, Office of Basic Energy Sciences, Material Sciences Division, U.S. Department of Energy, under Contract No. DE-AC03-76SF00098.

*Corresponding author. Tel.: +1-919-660-2568; fax: +1-919-660-2671.

E-mail address: vl@phy.duke.edu (V.N. Litvinenko).

[1] The "Blue Devil" basketball team and the deep blue color are popular symbols of the Duke University.

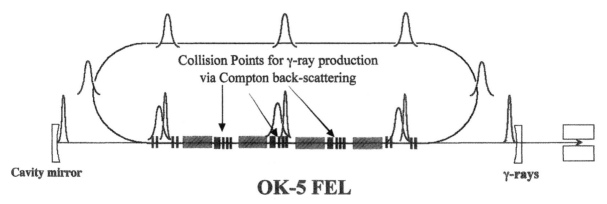

Fig. 1. Schematic layout of the OK-5/Duke storage ring FEL and γ-ray source. Four OK-5 wigglers (green) are separated by three midsections. Each midsection comprised of a triplet of quadrupoles, a buncher and eight dipole correctors. The length of the "OK-5 cell", i.e. the wiggler and midsection is equal to eight RF wavelength. With eight circulating electron bunches, this configuration provides three points where electrons and FEL photons collide head-on and generate a beam of γ-rays.

wigglers and three matching midsections between them. The design and the detailed description of the OK-5 electromagnetic wigglers are presented elsewhere [5]. These wigglers have the horizontal and vertical arrays of the poles shifted for a quarter of the period with respect to each other. The coils for the horizontal and vertical arrays are controlled independently. This feature provides the control of the polarization of the FEL light, which includes the left and the right circular polarization and the horizontal and the vertical linear polarization. The main mode of the OK-5 FEL will be based on the circular polarization. We plan to use the elliptical polarization of the wigglers for all OK-5 FEL operations. We will use the opposite helicity of the adjacent wigglers to lase with linear polarization in the OK-5 FEL. The use of the elliptical polarization will provide for strong reduction of the thermal load and the degradation of the downstream mirror compared with the straightforward use of plane wiggler field.

The control of the polarization in the OK-5 FEL is critical for a number of experiments under consideration. First, we plan to use the OK-5/Duke SR FEL for effective generation of mono-energetic polarized γ-ray beams in the scheme similar to that used with the OK-4 FEL [6]. The nuclear physics experiments with polarized targets require switchable, circular and linear polarization of γ-rays [7]. Second, we plan to use left-and-right

circular polarization in the 150–250 nm range for studies of the dichroizm of biological objects. The use of the polarization increases the contrast of the images. Third, we plan to conduct the experiment on the parity violation in atomic transitions 1S–2S in hydrogen [8]. We also plan to use the OK-5 FEL as the research tool to develop an effective scheme for coherent harmonics generation in the soft-X-ray range. The variety of anticipated applications and the requirements to support "top-off" injection, are reflected in the complexity of the OK-5/Duke storage ring FEL design. We designed the lattice for the OK-5 FEL to be flexible for a wide spectrum of applications [9] as well as with a potential for low emittance operations [10]. In both cases, the lattices provide sufficient dynamic aperture for the effective "top-off" injection and a reasonably long beam lifetime [9,10]. This paper is focused on the global aspects of the OK-5/Duke storage ring system while the details can be found in Refs. [4,5,9,10].

In Section 2 we describe the design of the OK-5/Duke storage ring FEL. Section 3 presents issues related to the optical cavity with emphasis on the downstream mirror heat-load and degradation. In Section 4 we discuss the spectrum and polarization of spontaneous radiation, and the features of the OK-5 FEL gain. We conclude with the summary of the main parameters and discussions of our plans.

2. Design of the OK-5/Duke storage ring FEL system

The 24.2-m long magnetic system of the OK-5 comprises four electromagnetic wigglers with controllable polarization and three midsections between them. Each mid-section has a buncher and a triplet of qudrupoles. The OK-5 magnetic system fits into the center of the 34.2-m long, dispersion free South straight section of the Duke storage ring. Two 5-m long bilaterally symmetric sections with four quadrupoles provide the matching between the arcs and the OK-5 system [9].

The design of the complete OK-5/Duke storage ring FEL system went through a number of iterations bringing together a set of contradicting requirements. The main compromise in the OK-5 FEL performance was imposed by the need for efficient generation of γ-ray beams. A Compton γ-ray source with good energy resolution requires a field-free collision point with low angular e-beam spread [11]. These points could not be located in the wigglers or between the quadrupoles, where $\alpha_{x,y}$ are large. Collision points are located in the triplet centers, where $\alpha_{x,y} = 0$ [9]. It was the natural choice to space them by the OK-5 FEL super-period (SP), i.e. the wiggler + the mid-section. The SP length must satisfy the collider-condition, i.e. must be equal to an integer number of the RF half-wavelengths, λ_{RF}. The Duke ring RF system operates on 64th harmonic with $\lambda_{RF} = 1.68$ meters. Given lengths of the OK-5 wigglers, the buncher and the quadrupoles reduce our choices of SP

length the either to $7/2 \, \lambda_{RF}$ or $4 \, \lambda_{RF}$. The "$7/2 \, \lambda_{RF}$" option provided for 15% higher OK-5 FEL gain but would reduce the effectiveness of the γ-ray production by a factor of ~ 3. The oddness of the numbers, 7 and 64, contradicts to the effective use of three collision points. The "$4 \, \lambda_{RF}$" with 2.68-m long SP option is a natural match with evenly spaced 8 bunches and provides 3×8 of collisions per turn. We decided that improved performance of the γ-ray source is worthy of the 15% reduction in the OK-5 FEL gain.

The final design of the OK-5 FEL midsection is shown in Fig. 2. The lattice of the midsection has bilateral symmetry with the triplet in the middle. The symmetry of the midsection is slightly violated by the buncher, which provides a weak vertical focusing. A small variation in the triplet setting compensates this asymmetry. Four x- and four y-dipole correctors provide the independent, local control of the "$x-y$" positions and angles in triplet center, i.e. in the collision point. A beam-position monitor (BPM) is located next to the collision point. These features are essential for attainment of the high γ-ray flux and low energy spread of the collimated γ-ray beam. When desired, the correctors will completely "turn-off" the γ-ray production by shifting the e-beam for ~ 5 mm away from the optical axis in the collision points, where RMS radius of FEL beam is less than 0.5 mm. In addition, we plan to use these dipole correctors for testing of the "e-beam" outcoupling technique [12] for harmonic generation.

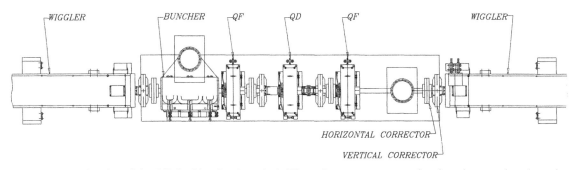

Fig. 2. The layout drawing of the OK-5 midsection (top-view). The main component are a buncher, three quadrupoles, a beam-position monitor of APS type (adjacent to the center quadrupole), two ion pumps (equipped with sublimation pumps) and eight dipole correctors.

We will use a simple design of the buncher. The 0.7-m long buncher is a compensated 3-pole EM wiggler with 40 mm gap and $K_{max} \sim 30$, similar to that of the OK-4 FEL [3] but twice longer. Longer buncher length lowers the peak value of magnetic field to 5.5 kGs at its maximal setting. Lower fields in the buncher reduces the power of synchrotron radiation and, therefore, the potential mirror damage. The downstream mirror is naturally exposed to the synchrotron radiation from the end-of-the arcs bending magnets, whose magnetic field is ~ 19 kGs at 1.2 GeV. The power density at the downstream mirror from the bending magnets exceeds that of three bunchers at least four folds. We are convinced that the use of helical bunchers for reducing mirror damage is not justified in our case.

The vacuum chamber in the OK-5 FEL has a uniform round pipe with an internal diameter of 37 mm. There are two water-cooled copper-masks at the entrance and at the middle of each midsection. The masks have an internal diameter of 35 mm and intercept only a small part of the radiation from the wigglers. The main goal of these masks is to protect fragile BPM buttons and bellows while having sufficient aperture for an effective injection. The masks also shield a part of the vacuum chamber in the wigglers. In most of the practical settings, a substantial part of the up-stream wiggler radiation will reach the walls of the vacuum chamber in the downstream wiggler. The vacuum chambers of the end-matching sections have 60-mm external and 58-mm internal diameters. These sections are connected with other vacuum chambers via smooth transitions. The main absorber combined with a smooth transition is located at the downstream end of the straight section. Other local masks protect fragile parts of the system.

The 4.04-m length of the OK-5 wigglers was determined by the manufacturing capability of the high precision and computer controlled milling machine at BINP. The wiggler period of 12 cm was chosen [13] to provide the desirable tunability range with rather large, 4 cm × 4 cm, gap required for effective injection [4,5,9]. A round stainless-steel vacuum chamber for the OK-5 wigglers has external diameter of 40 mm. The inside wall of the

vacuum pipes is electrochemically polished to reduce the out-gassing. Four cooling channels are attached to vacuum pipe with special thermo-conducting glue. This design provides sufficient cooling capacity to evacuate the heat generated by radiation in the vacuum chamber. We anticipate an initial period of operation with mediocre vacuum before the vacuum chamber will be cleaned by the radiation. The details of the OK-5 wiggler design and its magnetic field are published in a separate paper [5].

3. The optical cavity for the OK-5/Duke storage ring FEL system

We will use an improved version of the 53.73-m long OK-4 optical cavity system [14] with the feedback and the computer control incorporated into the Duke storage ring EPICS control system. For the future operations, we plan to design and to build a new optical cavity system with a vacuum loading dock for new mirrors and the capability for the in-vacuum change of mirrors.

The OK-5 FEL has maximum gain with Rayleigh range $\beta_0 \sim 4.5$ m, which requires mirror radii to be 27.62 m. Variation of the β_0 within the range of 3–6 m changes the relative value of the OK-5 gain by less than 10%. It means that initially we can use existing OK-4 mirrors with 22.27-m radii of curvature and $\beta_0 = 3.3$ m. It also means, that we can use $\beta_0 \sim 5.5$ m with mirror radii of 28 m with improved stability of the optical cavity with marginal loss in the OK-5 gain. The use of 28-m radii will reduce sensitivity to the angular misalignments of the mirrors by a factor of 2.7 compared with the OK-4 FEL system. It also will allow us to reduce the diameter of the mirror substrates from the present 5 to 3 cm for long wavelength and to 2-cm for short wavelengths. Even at $\lambda = 1$ µm the RMS FEL beam radius at the mirror will be ~ 3.4 mm ($\beta_0 = 5.5$ m) and the use of 1.5-cm radius (3-cm diameter) mirror substrates provide for very low diffraction losses. In the UV and the VUV, a 2-cm diameter mirrors will be sufficient. Presently, the vertical aperture for the OK-4 FEL beam is limited to the 2-cm in the end-of-the-arc bending magnets. This aperture translates into 3.1 cm at the mirrors. This limitation did not create any adverse effects for the OK-4 FEL operating at wavelengths from 193.7 to 730 nm. This experience made us confident and committed to the reduction of the mirror diameter in the near future.

Small mirrors have serious advantages when used in FEL with high-K helical or elliptical wigglers. These wigglers have very low intensity of the radiation on the axis while the most intense radiation is directed into the K/γ cone around the axis [15]. Fig. 3 shows the power of spontaneous radiation absorbed in a downstream mirror from 100 mA, 1 GeV e-beam and the OK-5 FEL wigglers. The 3 cm or 2 cm mirrors will absorb 5.1 W or 1.0 W, respectively, compared with 27.7 W for a 5 cm mirror.

In addition to the lower power load, the smaller mirrors are more rigid and allow effective cooling using indium brazing. We are testing this concept with the existing 5 cm mirrors. Mirrors with 28-m radii also provide wider range of the stability that currently used, and allow larger stresses. We are inclined to chose $\beta_0 \sim 5.5$ m for the OK-5 FEL optical cavity.

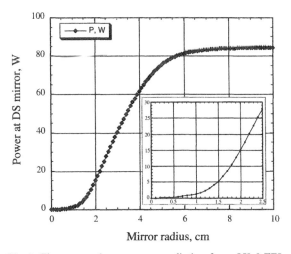

Fig. 3. The power of spontaneous radiation from OK-5 FEL wigglers deposited onto the downstream mirror as the function of its radius. The distance from the downstream mirror to end of the fourth wiggler is 14.81 m. The furthest distance to the beginning of the first wiggler is 38.92 m. $E = 1$ GeV, $I_b = 100$ mA, $\lambda = 100$ nm. Details for the center cone of radiation are shown in the right-down corner.

4. Spectrum and polarization of the OK-5 FEL spontaneous radiation

We intend to use spectral and polarization features of spontaneous radiation from the OK-5 FEL system for selected user applications [16], as well as for the FEL diagnostic. The flexibility of the OK-5 FEL system provides features which can be very unique. For example, in configuration in which adjacent wigglers have opposite helicity, we can generate the beam combining horizontally polarized photons at wavelength λ_1 and vertically polarized photons at wavelength λ_2. By adjusting the OK-5 bunchers, we can control the difference in the wavelength $\lambda_1 - \lambda_2$.

In the OK-5 wigglers, the horizontal and the vertical fields are controlled independently by their coil currents. This configuration allows controlling both the wavelength and polarization of radiation. The wavelength is determined by the values of the wiggler field $B_{x,y}$, the wiggler period λ_w and the e-beam relativistic factor $\gamma = E/mc^2$:

$$\lambda = \frac{\lambda_w}{2\gamma^2}\left(1 + a_x^2 + a_y^2\right), \qquad a_{x,y} = \frac{e\lambda_w\sqrt{\langle B_{y,x}^2 \rangle}}{2\pi mc^2} \quad (1)$$

where e and m are the charge and the mass of electron, c is the speed of the light. It means that RMS values of the fields determine the wavelength:

$$\lambda = \frac{\lambda_w}{2\gamma^2}\left(1 + K_w^2\right), \qquad K_w \equiv \sqrt{a_x^2 + a_y^2}. \quad (2)$$

The spectrum of spontaneous radiation from the OK-5 FEL differs from that of a conventional FEL and a conventional OK. For identical setting of the bunchers and a single electron with relativistic factor of γ_0, the spectrum at zero angle is a result of interference of radiation from four wigglers:

$$I_{DOK}(k) = I_0 \left(\frac{\sin\pi N_w(1-X)}{\pi N_w(1-X)}\right)^2 \times \left(\frac{\sin(4\pi(N_w+N_d)X)}{\sin(\pi(N_w+N_d)X)}\right)^2, \quad X = \frac{k}{k_0} \quad (3)$$

where $k_0 = 2\pi/\lambda_0$, N_w is the number of periods per wiggler, $N_d = \Delta s(\gamma_0)/\lambda_0$ is the dimensionless slippage in the bunching straight section. The first term in Eq. (3) is the well-known spectrum of the radiation from one wiggler. The second term is the interference term. Fig. 4(a) shows this spectrum for ideal electron beam for $N_d = 25$ and its envelope. The well-developed fringe pattern and clearly separated peaks are distinguishable features of DOK spectra. Finite energy spread and emittance of electron beam smooth those features making them less pronounceable. Formulae (3) can be easily integrated with the e-beam distribution with finite energy spread and emittance using, for example, Mathematica [17]. Figs. 4(b) and (c) show the effect on the spectra by finite energy spread and emittance of the electron beam. As expected, the DOK is very susceptible to the energy spread, which reduces the depth of the

fringes and smoothes other fine features. Similar to the standard OK, the measured depth of modulation can be used for measuring the e-beam energy spread and for tuning bunchers to maximize the FEL gain [18]. The finite angular spread shifts spectra towards longer wavelength and makes it asymmetric. At low level, the DOK gain is proportional to the negative derivative on the e-beam energy of the spontaneous radiation intensity into TEM_{oo} mode [19]. Even though the on-axis, i.e. zero-angle, spectra do not reflect all 3-D effect, they are practically useful for optimization of the FEL performance. Specifically, optimal buncher strengths (see [4] for specific examples) do not depend strongly on the beam emittance. Therefore, the maximization of the sharpness of

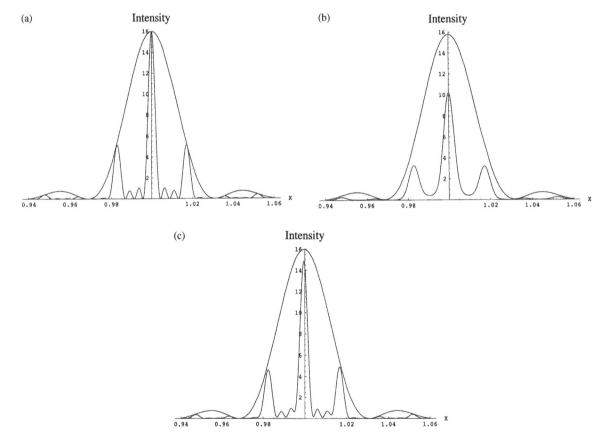

Fig. 4. On-axis spontaneous radiation spectra from the OK-5 FEL system with $N_d = 25$. The envelope is the spontaneous radiation spectrum from one wiggler scaled up by 16-fold. Vertical scale is in units of the radiation from one wiggler, horizontal scale, $X = k/k_0$, is the relative detuning from the FEL resonance. (a) for ideal e-beam; (b) for e-beam with 0.1% RMS energy spread; (c) for e-beam with 10 nm·rad emittance at $E = 700$ MeV, $\lambda = 200$ nm. .

the fringes provides a tool for optimizing DOK gain. Fig. 5 shows typical dependence of the DOK spectra on N_d which clearly indicates the best range for the gain around $N_d \sim 55$. The OK-5 FEL gain has fast oscillating dependencies on a number of parameters, including the relative strengths of three bunchers. The detailed studies are published separately [4]. The gain dependence for OK-5 FEL

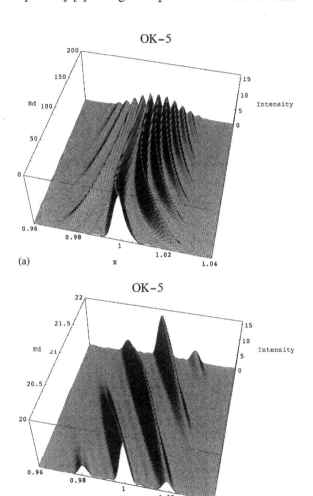

(a)

(b)

Fig. 5. The spectrum of spontaneous radiation from OK-5 FEL as the function of the detuning $X = k/k_0$ and the strength of the bunchers N_d for e-beam with the RMS energy spread of 0.05%. Vertical axis is the spectral intensity in the units of the radiation from one wiggler at $X = 0$. The horizontal axis is X, the third axis is N_d: (a) N_d in the range from 0 to 200 with an integer step of 2 for avoiding the fast oscillating features; (b) N_d in the range from 20 to 22 with a fractional step of 0.02 to show the fast oscillating features.

on the buncher strength differs from that of an FEL or conventional OK. Fig. 6 illustrates this difference. The conventional OK has the cos-like gain dependence with the maximum very close to $N_d = $ integer -0.25, i.e. exactly as it was predicted by OK inventors [20]. The OK-5 FEL gain has two closely space positive and negative peaks, separated by a "no-gain" flat zone. The maximum of the OK-5 gain is located at $N_d \approx$ integer -0.13. This tendency for DOK peak gain to shift towards $N_d \approx$ integer was also expected [1]. This feature is the direct consequence of the DOK spontaneous radiation spectrum and the Einstein relations between induced and spontaneous radiation.

Another interesting feature of the OK-5 is its flexible polarization determined by the geometry of the wiggler and the relative strength of horizontal and vertical fields. On the axis of the wiggler the polarization is

$$\rho_{xy} = \frac{1}{2} \begin{bmatrix} 1+\xi_3 & \xi_1 - i\xi_2 \\ \xi_1 + i\xi_2 & 1 - \xi_3 \end{bmatrix} = \frac{1}{a_x^2 JJ_+^2 + a_y^2 JJ_-^2}$$

$$\times \begin{bmatrix} a_x^2 JJ_+^2 & -ia_x a_y JJ_+ JJ_- \\ ia_x a_y JJ_+ JJ_- & a_y^2 JJ_-^2 \end{bmatrix}$$

$$JJ_\pm = J_0(\xi) \pm J_1(\xi), \qquad \xi = \frac{a_y^2 - a_x^2}{2(1 + a_y^2 + a_x^2)}$$

$$\xi_2 = \frac{2a_x a_y JJ_+ JJ_-}{a_x^2 JJ_+^2 + a_y^2 JJ_-^2}; \qquad \xi_3 = \frac{a_x^2 JJ_+^2 - a_y^2 JJ_-^2}{a_x^2 JJ_+^2 + a_y^2 JJ_-^2} \qquad (4)$$

where J_0 and J_1 are Bessel functions of the first kind and ξ_2 and ξ_3 are degrees of circular and linear polarization, respectively [21]. The OK-5 wiggler geometry defines the absence of linear polarization at $45°$ ($\xi_1 = 0$) in contrast with permanent magnet devices, where polarization is changed by lateral shift of the horizontal and vertical array with respect to each other [15,22]. The OK-5 light is generally 100% elliptically polarized with $\xi_2^2 + \xi_3^2 = 1$. When field amplitudes are equal ($|a_x| = |a_y|$), the polarization is circular with helicity defined by the sign of the product $a_x a_y$: $\xi_2 = 1$ for right, and $\xi_3 = -1$ for left circular polarization. Therefore, it is very straightforward to set the OK-5 FEL for circular polarization.

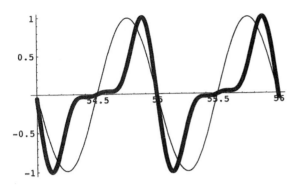

Fig. 6. Detailed dependence of the OK-5 (DOK) FEL and a conventional OK (COK) gains on the buncher(s) strength N_d (horizontal axis) in the range from 54 to 56. The gains (vertical axis) are calculated at the resonant wavelength ($X = 1$) and normalized to their maximal value in the range. The COK is a reduced OK-5 with two wigglers and one buncher. The e-beam RMS energy spread is 0.05%. The thick line is the OK-5 gain, the thin line is the COK gain. The OK-5 gain reaches maximum at $N_d = 54.8694$ and the COK gain has the peak gain at $N_d = 54.7758$.

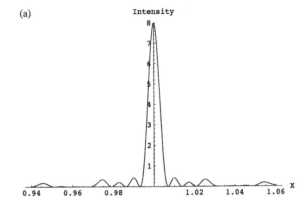

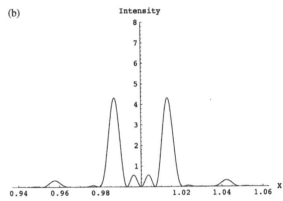

Fig. 7. Typical example of spectra for linearly polarized light with alternating helicity in the OK-5 wigglers by changing sign of the horizontal filed in adjacent wigglers. $N_d = 5$, (a) spectrum of horizontally polarized light; (b) spectrum of vertically polarized light.

When one of the field components is zero, the polarization is linear: $\xi_3 = 1$ for horizontal, and $\xi_3 = -1$ for vertical linear polarization. Nevertheless, this simple-minded approach of turning off one of the arrays has three major drawbacks. First, the radiation for the planar wiggler has strong harmonics on the axis, which can be harmful for the OK-5 mirrors. Second, in planar configuration the OK-5 wigglers will have reduced tunability range by a factor ~ 2. Third, some of our spontaneous radiation users need a clean narrow-band fundamental line in the VUV. The harmonics of the fundamental wavelength could be harmful for experiments. The use of a monochromator or a filter in the VUV means a big loss in the beam intensity and, often, makes experiment impossible. We plan to use helical (or almost circular) configuration of the wiggles with opposite the helicity in the adjacent wigglers. The spectra for this configuration for the horizontal and the vertical polarization are shown in Fig. 7. One of the low N_d settings (Fig. 7) provides the narrow line horizontally polarized light for user application, while vertically polarized light is shifted to the wings.

To lase in the OK-5 FEL with horizontal polarization, we will use higher N_d and will shift the spectra of the desirable linear polarization to the position of the maximum gain. Fig. 8 shows the relative values of gain for horizontally and vertically polarized light. The net difference in the gain of $\sim 8\%$ is sufficient for complete domination of the horizontal linear polarization in the lasing mode. The other choice of the N_d or/and other alternation scheme would give preference for the vertical polarization. We can also use a slight elliptical polarization with a few percents difference in the x or y coils to increase the gain difference. The OK-5 FEL operating with the linear polarization will have $\frac{1}{2}$ gain compared to that for the circular polarization. The OK-5 FEL

(a)
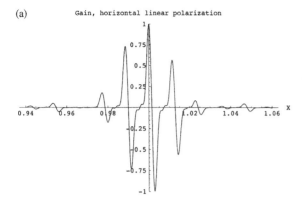

(b)
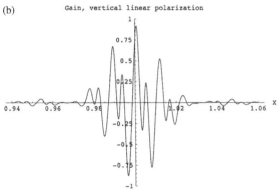

Fig. 8. Gain vs. detuning parameter X. The OK-5 has alternating helicity of the wigglers and the bunchers set at $N_d = 54.8694$. The gain is normalized to its maximum value for horizontally polarization. (a) gain for horizontally polarized light; (b) gain for vertically polarized light.

gain with planar wiggler and with alternating helical configurations will be at the same level.

In conclusion, the OK-5 can be tuned in the way that the horizontal and vertical polarization have the same level of gain, but at different wavelength. This unusual lasing mode will provide two-color FEL beam easily separable by polarizers.

5. Conclusions and acknowledgements

The OK-5 FEL/Duke storage ring FEL promises further advanced of FEL-oscillators into the UV and VUV range. It has substantial gain to operate below 200 nm. It will provide remarkable features for the users of spontaneous and coherent radiation in the UV and the VUV ranges of spectra. A pleasant side-product of the OK-5 FEL, the Compton back-scattered monochromatic, polarized γ-ray beams, will be used for nuclear physics experiments [4].

The real potential of the OK-5 FEL will be realized after modification of the Duke storage ring with lower emittance and low impedance vacuum chamber [10]. Gain $\sim 1000\%$ per pass [4] will allow the OK-5 to lase below 100 nm and to generate coherent soft-X-ray harmonics.

The final design of all components of the OK-5 FEL and 34 m straight sections was completed by mid-2000. Four OK-5 wigglers have been manufactured and are undergoing magnetic measurements, fine tuning and minor modification of the pole-tips. The quality of the wigglers seems to be excellent. The rest of the system is in the production stage and should be finished in the summer of 2001. The OK-5 FEL system will go through additional tests at Duke University prior to installation onto the Duke storage ring. We hope to report the first results attained with the OK-5/Duke storage ring in 2002.

Authors would like to acknowledge the financial support for this project by the Office of Naval Research and Duke University. Authors are grateful to Prof. Berndt Müller, the Dean of the Natural Sciences, Duke University for continuous financial and moral support of the project. Authors would like to thank colleagues from the Duke Free Electron Laser Laboratory, especially Mark Emamian and Gary Swift, for their thoughtful contributions into this project, and Prof. Glenn Edwards, the Director of FEL laboratory for his support of the project.

References

[1] V.N. Litvinenko, Nucl. Instr. and Meth. A 304 (1991) 463;.
N.A. Vinokurov, Nucl. Instr. and Meth. A 375 (1996) 264.
[2] DFELL booklet, Duke University, 1993.
[3] A.S. Artamonov, et al., Nucl. Instr. and Meth. 177 (1980) 247;
G.A. Kornyukhin, G.N. Kulipanov, V.N. Litvinenko, N.A. Mezentsev, A.N. Skrinsky, N.A. Vinokurov, P.D. Vobly, Nucl. Instr. and Meth. A 237 (1985) 281;
I.B. Drobyazko, et al., Nucl. Instr. and Meth. A 282 (1989) 424.

[4] V.N. Litvinenko, N.A. Vinokurov, O.A. Shevchenko, Predictions and Expected Performance for the VUV OK-5/Duke Storage Ring FEL with Variable Polarization, Nucl. Instr. and Meth. A 475 (2001) 97, these proceedings.

[5] V.N. Litvinenko, S.F. Mikhailov, N.A. Vinokurov, N.G. Gavrilov, G.N. Kulipanov, O.A. Shevchenko, P.D. Vobly, Helical wigglers for the OK-5 storage ring VUV FEL at Duke, Nucl. Instr. and Meth. A 475 (2001) 247, these proceedings.

[6] V.N. Litvinenko, et al., Phys. Rev. Lett. 78 (24) (1997) 4569.

[7] W. Tornow, V.N. Litvinenko, H. Weller, High Intensity γ-Ray Source for nuclear physics, Proposal to the US Department of Energy, August 2000;
E.C. Schrieber, et al., Phys. Rev. C 61 (Rapid Comm.) (2000) 061604.

[8] T. Burt, V.N. Litvinenko, Feasibility studies of the parity violation experiment in ^1S–^2S transitions in atomic hydrogen using the OK-5/Duke storage ring FEL, unpublished.

[9] Y. Wu, V.N. Litvinenko, S.F. Mikhailov, N.A. Vinokurov, N.G. Gavrilov, O.A. Shevchenko, T.V. Shaftan, D.A. Kairan, Lattice Modification and Nonlinear Dynamics for Elliptically Polarized VUV OK-5 FEL Source at Duke Storage Ring, Nucl. Instr. and Meth. A 475 (2001) 253, these proceedings.

[10] S.F. Mikhailov, V.N. Litvinenko, Y. Wu, Low Emittance Lattice for the Duke Storage Ring soft-X-Ray FEL, Nucl. Instr. and Meth. A 475 (2001) 417, these proceedings.

[11] V.N. Litvinenko, J.M.J. Madey, SPIE 2521 (1995) 55;
V.N. Litvinenko, J.M.J. Madey, Nucl. Instr. and Meth. A 375 (1996) 580.

[12] N.A. Vinokurov, Concept of the e-beam outcoupling, BINP, 1985.

[13] N.A. Vinokurov, Magnetic design of the helical undulators for the Duke FEL, Design Report, Novosibirsk, 1998.

[14] I.V. Pinayev, M. Emamian, V.N. Litvinenko, S.H. Park, Y. Wu, CP451, Beam Instrumentation Workshop, in: R.O. Hettel, S.R. Smith, J.D. Masek (Eds.), The American Institute of Physics, 1998, p. 545.

[15] H. Hama, K. Kimura, M. Hosaka, J. Yamazaki, T. Kinoshita, Nucl. Instr. and Meth. A 393 (1997) 23.

[16] H. Ade, Y. Wang, S.L. English, J. Hartman, R.F. Davis, R.J. Nemanic, V.N. Litvinenko, I.V. Pinayev, Y. Wu, J.M.J. Madey, Surf. Rev. and Lett. 5 (6) (1998) 1257.

[17] Mathematica 4, © Copyright 1988–1999 Wolfram Research, Inc.

[18] G.A. Kornyukhin, G.N. Kulipanov, V.N. Litvinenko, N.A. Mezentsev, A.N. Skrinsky, N.A. Vinokurov, P.D. Vobly, Nucl. Instr. and Meth. A 237 (1985) 281;.
N.A. Vinokurov, A.N. Skrinsky, Theory of the Optical Klystron, MW Relativistic Electronics, Gorky, 1981, p. 204.

[19] N.A. Vinokurov, Classical analog of Einstein relations, Preprint 81-02, BINP, Novosibirsk, 1981;
V.N. Litvinenko, N.A. Vinokurov, Nucl. Instr. and Meth. A 331 (1993) 440.

[20] N.A. Vinokurov, A.N. Skrinsky, Optical klystron, Preprint 77–59, BINP, Novosibirsk, 1977.

[21] L.D. Landau, E.M. Lifshitz, The Classical Theory of Fields, 4th Revised Edition, Pergamon Press, Oxford, 1975, p. 123.

[22] R. Walker, et al., Initial Performance of the European UV/VUV Storage Ring FEL at ELETTRA, Nucl. Instr. and Meth. A 475 (2001) 20, these proceedings.

ELSEVIER

Nuclear Instruments and Methods in Physics Research A 475 (2001) 417–424

NUCLEAR INSTRUMENTS & METHODS IN PHYSICS RESEARCH
Section A

www.elsevier.com/locate/nima

Low emittance lattice for the Duke storage ring soft X-ray FEL ☆

S.F. Mikhailov[a,*], V.N. Litvinenko[a], Y. Wu[b]

[a] *FEL Laboratory, Duke University, Box 90139, LaSalle St. ext., Durham, NC 27708-0319, USA*
[b] *ALS, LBNL, Berkeley, CA, USA*

Abstract

In this paper, we present a possible lattice for the Duke storage ring with horizontal emittance of 1.4 nm rad at an energy of 1 GeV. The new lattice was constrained to fit the layout of the existing storage ring and to re-use existing magnets. Within these constrains, the improvement of the emittance (from current 18 nm rad) is possible due to the use in the arcs of combined function bending magnets with strong dipole, quadrupole and sextuple fields. We present the results of 2D and 3D simulations of magnetic fields for these magnets. We discuss the choice of the arc's lattice cell and the per-cell tune advance. The lattice is based on the concept of local compensation of the non-linear geometrical aberrations. Preliminary studies of the dynamic aperture for this lattice are very encouraging. We briefly discuss the possibility further emittance reduction using new magnets for the arcs. © 2001 Published by Elsevier Science B.V.

PACS: 41.60.c; 29.20.D; 41.75.L; 07.85.N; 07.85

Keywords: Emittance; Storage ring; Free-electron-laser

1. Introduction

The 1 GeV Duke storage ring is designed to accommodate a variety of FELs [1–3]. Its 34-m long, dispersion-free straight sections and their flexible lattices provide the perfect environment for installing various FEL systems and optimising lattices for their best performance [4,5]. The fundamental principle providing simultaneously for this flexibility and for large dynamic aperture is the local compensation of the non-linear geome-

trical aberrations (LCNGA), developed and tested at Duke FEL laboratory [6]. The main source of non-linear aberrations are sextupoles in the arcs, which are required for the chromaticity compensation and control. LCNGA concept limits all significant geometrical aberrations to exist only within the arcs. Hence, the choice of the linear lattice in the straight section does not significantly affect the dynamic aperture of the ring. These feature proved to be very important for accommodating both the 7.5-m long OK-4 with plane wigglers and 28-m long OK-5 FEL with helical wigglers [4,5] without any changes in the arc lattice.

Even though the performance of the Duke storage ring and its lattice is rather remarkable [7], its natural horizontal (18 nm rad at 1 GeV)

☆ This work is supported by the Dean of Natural Sciences, Duke University.

*Corresponding author. Tel.: +1-919-660-2647; fax: +1-919-660-2671.

E-mail address: smikhail@fel.dike.edu (S.F. Mikhailov).

emittance is too large for very short wavelength FELs with wavelength at and below 100 nm. The existing arcs lattice was constrained by hardware, to be exact the vacuum chambers, which were designed and build in Stanford in mid-eighties. The hardware was brought to Duke University and imposed the FODO lattice for the ring's arcs [8]. The use of old vacuum chambers also defined rather large longitudinal impedance of the system. The alternative lattice with 3 nm rad emittance [9] was dismissed as too expensive. Therefore, the Duke storage ring did not reach emittance of the third generation light sources and should be classified as the "second-and-half" generation. In near future the Duke ring will undergo a reliability up-grade of the arcs. The new quadrupole coils and new power supplies will provide operation at energies from 0.25 to 1.2 GeV.

At present, the OK-4/Duke storage ring FEL reached a wavelength range where emittance could be the main limiting factor [10]. The installation of the advanced OK-5 FEL on the Duke storage ring makes it natural to advance

Table 1
Parameters of the existing and the new lattices

	Existing	New
Maximum energy (GeV)	1.2	
Nominal energy (GeV)	1.0	
Injection energy (GeV)	0.27	0.25–1.2
Circumference (m)	107.46	
RF frequency (MHz)	178.547	
Number of bunches	1–64	
Horizontal emittance $\varepsilon_x @ E = 1$ GeV (nm rad)	18	1.4
Losses per turn (keV)	42.0	34.6
Damping times ($E = 1$ GeV):		
Horizontal τ_x (ms)	18.3	12.2
Vertical τ_y (ms)	17.0	20.7
Longitudinal τ_E (ms)	8.2	16.0
Momentum compaction	0.0086	0.00225
Natural chromaticities:		
Horizontal $dQ_x/d\delta$	−10.0	−26.1
Vertical $dQ_y/d\delta$	−9.8	−11.5
Betatron tunes:		
Horizontal Q_x	9.11	12.11
Vertical Q_y	4.18	6.13
Longitudinal Q_s ($U_{RF} = 850$ kV)	0.0086	.0044
Longitudinal emittance ε_s (μm)	5.75	3.84

Table 2
Parameters of magnetic system and the lattices @ 1 GeV

	Existing			New	
Number of cells	20			24	
Number of dipoles	40			48	
Number of arc quads	42			48	
Parameters of the dipoles					
Dipole gap ($x = 0$) (cm)	2.4			2.5	
Magnetic length (m)		0.335			
Dipole field B_0 (kGs)	15.7			13.05	
Gradient G (kGs/cm)	0			1.175	
Sextupole $\int B_y'' ds$ (kGs/cm)	2.25			15.8	
Parameters of the quadrupoles	QF		QD	QF1	QF2
Bore diameter (cm)		4.0			
Magnetic length (m)	0.20		0.14	0.20	0.14
Gradient G (kGs/cm)	3.18		2.73	3.28	
Sextupole B_y'' (Gs/cm²)	266		228	372	
Beta functions and dispersion in arcs:	min		max	min	max
β_x (m)	0.43		2.48	0.15	4.28
β_y (m)	1.56		5.06	0.77	3.34
D_x (cm)	12.7		24.5	3.0	11.3
Tune advances per cell:					
ΔQ_x		3/10		5/12	
ΔQ_y		1/10		1/6	

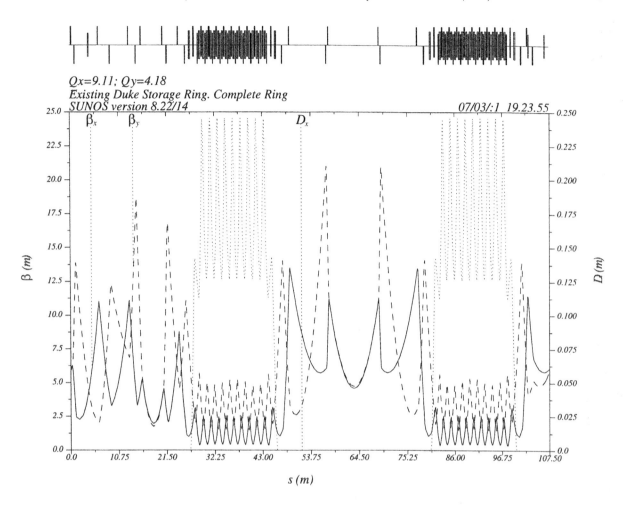

Fig. 1. The existing lattice of the Duke storage ring. Solid line—β_x; dash-line—β_y; dotted line—η.

other ring's parameters such as the emittance and the longitudinal impedance. These improvements would provide for the OK-5 FEL gain measured in hundreds of percents and for its effective operation in the VUV [11]. In this paper we focus on the new lattice providing low emittance while reusing the most of the existing magnets after modest modifications. Section 2 provides the details of the lattice design. Section 3 is focused on issues of the non-linear effects and the dynamic aperture. We conclude with the discussion of the results and possibility of emittance reduction below 1 nm rad.

2. Linear aspects of the new lattice

We constrained the new lattice to fit the layout of the existing Duke storage ring and to re-use as many of existing magnets as possible. The upgrade of the ring will require:

- Modification of the yokes in the existing dipole magnets;
- Reshuffling of the existing magnetic elements in the arcs;
- Modification of the supports to install two additional cells.

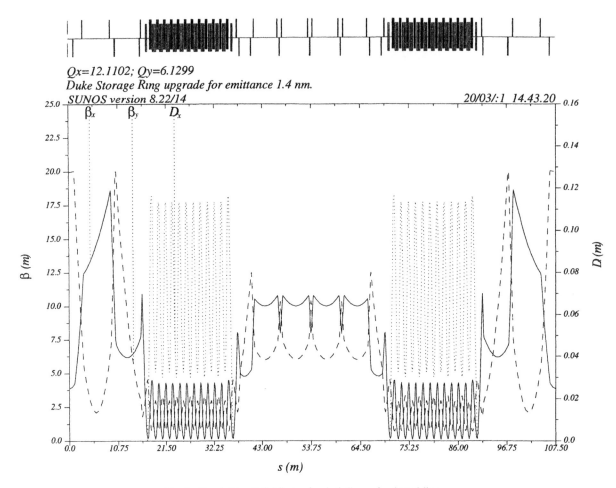

$Q_x=12.1102$; $Q_y=6.1299$
Duke Storage Ring upgrade for emittance 1.4 nm.
SUNOS version 8.22/14   20/03/:1 14.43.20

Fig. 2. New lattice. Solid line—β_x; dash-line—β_y; dotted line—η.

Tables 1 and 2 list the main parameters and compare the existing and the new lattices. Figs. 1 and 2 show detailed structure of β- and η-functions for the existing and the new lattices. The new lattice incorporates the new South straight section with the OK-5 FEL [5]. β-functions in the OK-5 straight section may be independently varied within the range of $\beta_{x,y} =4$–10 m. The change of the tune/phase advance in the OK5 FEL straight section may be compensated in the RF/injection straight section.

Detailed β- and η-functions for the existing and the new regular arc cells are shown in Figs. 3 and 4. The new lattice will reduce the horizontal beam emittance from 18 to 1.4 nm rad. This

is achieved with the use of combined function dipole magnets with strong quadrupole and sextupole components. It makes the regular cell more compact and provides for more cells per arc. The combination of the bending and vertical focusing functions on one element enables us to reduce the horizontal η-function, and therefore, the emittance. The concept is similar to that utilized in the Advanced Light Source at LBL [12]. Strong sextupole components in the dipoles and quadrupoles are necessary to compensate the vertical and horizontal chromaticity of the ring. The horizontal sextupole moment in the focusing quadrupoles is generated by asymmetric excitation of the coils. This method has been

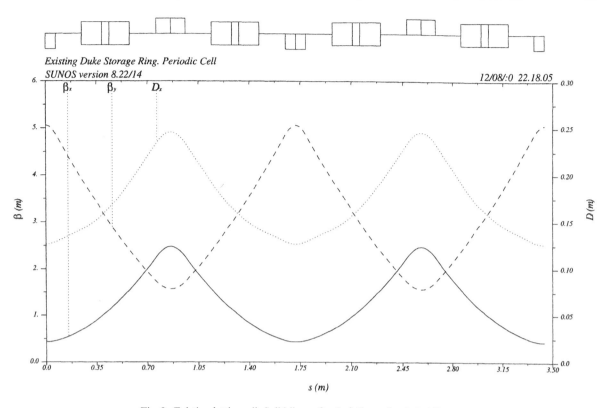

Fig. 3. Existing lattice cell. Solid line—β_x; dash-line—β_y; dotted line—η.

proved to be very efficient without serious side-effects [13].

The regular cell of the new lattice comprises two complete bends and two complete quadrupoles. Each complete 15° bend consists of two existing 7.5° dipoles, and each complete quadrupole consists of two existing quadrupoles. This arrangement provides for the re-use of the existing magnets (see Table 2). The choice of the tune advance per cell provides for the low emittance as well as for a natural compensation of the most of the non-linear geometrical aberrations. Stronger focusing in the arcs and in the OK-5 FEL increases the absolute value of horizontal and vertical chromaticities by factor 2.6 and 1.2, respectively. In combination with substantially reduced values of η-functions, the later requires strong sextupole fields in arc's dipoles and quadrupoles.

We used the MERMAID 2D and 3D codes [14] for magnetic design of the modified dipole shown in Fig. 5. The increase number of bending magnets in the new lattice to forty-eight provided for decrease of the magnetic field to 13 kGs. The lower value of the dipole filed enables us to obtain good quality of magnetic field in the aperture of $\Delta x = \pm 2$ cm. The challenging part of the design was the attainment of the high quality sextupole field. The 3D version of the MERMAID code proved to be very reliable and allows avoiding prototyping of the magnets.

The new lattice has new configuration of the North straight section, which is optimized for future installation of the modified OK-4 FEL and for the full energy injection from the future 1.2 GeV booster [15]. Similar to the existing lattice, the new arc lattice has η-matching end-cells providing for dispersion-free straight sections. The η-matching uses a slightly modified half-cell with adjusted quadrupole strengths.

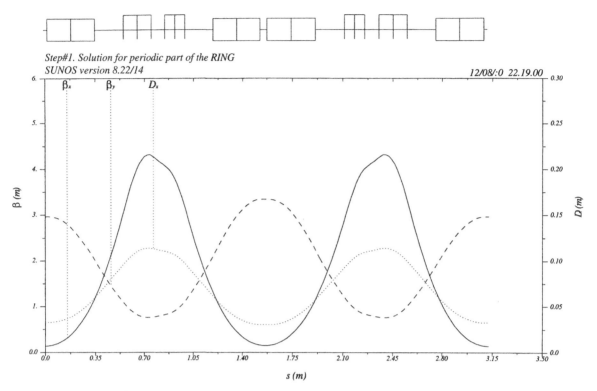

Fig. 4. New lattice cell. Solid line—β_x; dash-line—β_y; dotted line—η.

3. Non-linear aspects of the new lattice

The new lattice is based on the concept of local compensation of the second-order geometrical aberrations (SOGA). The SOGA concept is a modification of well-known second-order achromat [16]. This concept is used in the existing lattice and is very effective. Strong, "so-called" chromatic sextupoles in the arcs are the main non-linear elements of the Duke storage ring. In first-order perturbation theory [17] the non-linear geometrical aberrations induced by sextupole fields are completely described by five complex integrals:

$$A_n(s) = \int_s^{s+C} Se^{in\psi_x}\beta_x^{3/2}\,ds'; \quad n = 1,3; \quad B_m(s)$$
$$= \int_s^{s+C} Se^{i(m\psi_y+\psi_x)}\beta_x^{1/2}\beta_y\,ds'; \quad m = 0, \pm 2 \quad (1)$$

where S is the sextupole strength and $\psi_{x,y}$ are betatron phases. The same results can obtained

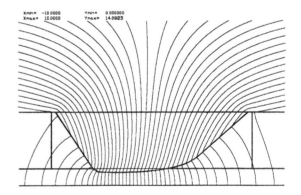

Fig. 5. The pole profile of the modified dipole magnet.

using non-linear Lie maps [8]. Making the integrals (1) over each arc to be zero,

$$A_n = \int_{\text{arc}} Se^{in\psi_x}\beta_x^{3/2}\,ds' = 0; \quad B_m(s)$$
$$= \int_{\text{arc}} Se^{i(m\psi_y+\psi_x)}\beta_x^{1/2}\beta_y\,ds' = 0$$

we cancel the SOGA in the straight sections and make the dynamic aperture less susceptible to the specific lattice of the straight sections. In the next order of perturbation theory, the sextupoles cause non-linear tune shifts. At large amplitudes of betatron oscillations, the non-linear tune shift causes a phase shift between the cells which violates exact compensation of the SOGA. This effect is most profound for the horizontal oscillations. We used one family of octupoles to compensate horizontal non-linear tune shift to increase the dynamic aperture of the ring.

For simulation of dynamic aperture of the bare lattice we used the MAD8.22 code. After correction of the non-linear tune shift, the horizontal dynamic aperture reached 55 mm mrad. The vertical dynamic aperture also increased to 20 mm mrad. The energy deviation of $\pm 2\%$ did not affect the dynamic aperture considerably. These preliminary results are very encouraging. We plan to continue simulation of the dynamic aperture using more sophisticated features of the TRACY and DESPOT codes [18].

4. Conclusions and acknowledgements

In this paper, we described the "cost-effective" solution for the low emittance lattice of the Duke storage ring. The results of these studies showed that such a lattice is feasible and is compatible with both the OK-4 and the OK-5 FELs. Nevertheless, the imposed conditions for the re-use of the existing components did not allow us to reach the best lattice and the best emittance. There are clear indications that the use of fourteen cells per arc provides additional cancellations of the higher-order resonances as well as for better emittance [19]. This more radical lattice is based on similar principles as described in this paper and has horizontal emittance < 1 nm rad at 1 GeV. It will require complete set of new magnetic elements.

With the proposed modifications, the Duke FEL storage ring will join the family of the medium energy, third generation light sources, such as ELETTRA, ALS, SLS and BESSY-II.

The authors would like to thank Prof. Berndt Müller, the Dean of Natural Sciences, Duke University, for the support of this project. The authors are grateful to Dr. Andrey Dubrovin, Budker Institute of Nuclear Physics, Novosibirsk, Russia for providing the up-dated version of the Mermaid code.

References

[1] V.N. Litvinenko, S.H. Park, I.V. Pinayev, Y. Wu, M. Emamian, N. Hower, O. Oakeley, G. Swift, P. Wang, Nucl. Instr. and Meth. A 429 (1999) 151.

[2] V.N. Litvinenko, S.F. Mikhailov, N.A. Vinokurov, N.G. Gavrilov, D.A. Kairan, G.N. Kulipanov, O.A. Shevchenko, T.V. Shaftan, P.D. Vobly, Y. Wu, Nucl. Instr. and Meth. A 475 (2001) 407, these proceedings.

[3] V.N. Litvinenko, B. Burnham, S.H. Park, Y. Wu, M. Emamian, R. Cataldo, J. Faircloth, S. Goetz, N. Hower, J.M.J. Madey, J. Meyer, P. Morcombe, O. Oakeley, J. Patterson, G. Swift, P. Wang, I.V. Pinayev, M.G. Fedotov, N.G. Gavrilov, V.M. Popik, V.N. Repkov, L.GI saeva, G.N. Kulipanov, G.Ya. Kurkin, S.F. Mikhailov, A.N. Skrinsky, N.A. Vinokurov, P.D. Vobly, E.I. Zinin, A. Lumpkin, B. Yang, Nucl. Instr. and Meth. A 407 (1998) 8.

[4] Y. Wu, V.N. Litvinenko, J.M.J. Madey, Nucl. Instr. and Meth. A 341 (1994) 363.
Y. Wu, V.N. Litvinenko, E. Forest, J.M.J. Madey, Nucl. Instr. and Meth. A 331 (1993) 287.

[5] Y. Wu, V.N. Litvinenko, S.F. Mikhailov, O.A. Shevchenko, N.A. Vinokurov, N.G. Gavrilov, T.V. Shaftan. D.A. Kairan, Nucl. Instr. and Meth. A 475 (2001) 253, these proceedings.

[6] V.N. Litvinenko, Y. Wu, B. Burnham, J.M.J. Madey, S.H. Park, Proceedings of the 1995 Particle Accelerator Conference, Dallas, TX, May 1–5, 1995, p.796;
Y. Wu, V.N. Litvinenko, J.M.J. Madey, Proceedings of the 1993 IEEE Particle Accelerator Conference Washington, DC, May 17–20, 1993, p. 218;
B. Burnham, N. Hower, V.N. Litvinenko, J.M.J. Madey, Y. Wu, Proceedings of the 1993 IEEE PAC, Washington, DC, May 17–20, 1993, p. 889;
Y. Wu, V.N. Litvinenko, B. Burnham, Proceedings of the 1995 Particle Accelerator Conference, Dallas, TX, May 1–5, 1995, p. 2877.

[7] Y. Wu, V.N. Litvinenko, B. Burnham, S.H. Park, J.M.J. Madey, IEEE Trans. Nucl. Sci. 44 (5) (1997) 1753;
Y. Wu, V.N. Litvinenko, B. Burnham, J.M.J. Madey, S.H. Park, Nucl. Instr. and Meth. A 375 (1996) 74.

[8] Y. Wu, Theoretical and experimental studies of the beam physics in the Duke FEL storage ring, Thesis, Duke University, August 1995.

[9] V.N. Litvinenko, Triplet lattice for the Duke ring, 1991, DFELL materials.

[10] V.N. Litvinenko, S.H. Park, I.V. Pinayev, Y. Wu, Operation of the OK-4/Duke storage ring FEL below 200 nm, Nucl. Instr. and Meth. A 475 (2001) 195, these proceedings.

[11] V.N. Litvinenko, O.A. Shevchenko, N.A. Vinokurov, Y. Wu, Predictions and Expected Performance for the VUV OK-5/Duke Storage Ring FEL with Variable Polarization, Nucl. Instr. and Meth. A 475 (2000) 97, these proceedings.

[12] 1–2 GeV synchrotron radiation source, Conceptual Design Report, July 1986, LBL, Berkeley, CA, Pub-5172 Rev.

[13] B. Burnham, V.N. Litvinenko, Y. Wu, Proceedings of the 1995 Particle Accelerator Conference, Dallas, TX, May 1–5, 1995, p. 524.

[14] Mermaid, the 2D/3D code for magnetic design, © A.N. Dubrovin, Novosibirsk, Russia.

[15] HIγS proposal to the US Department of Energy, Technical Description, September 2000.

[16] K.L. Brown, A second-order Magnetic Optical Achromat, SLAC Report 2257. SLAC, February 1979.

[17] N.N. Bogolyubov, Yu.A. Mitropolsky, Asymptotic methods in the theory of non-linear oscillations, Fizmatgiz, Moscow, 1977;
V.N. Litvinenko, Concept of local compensation of the second order non-linear geometrical aberrations for the Duke storage ring, DFELL materials, Duke University, 1991.

[18] DESPOT, © E. Forest; TRACY, © J. Bengtsson, H. Hishimira.

[19] S.F. Mikhailov, V.N. Litvinenko, Y. Wu, Lattice for the Duke storage ring with 1-nm horizontal emittance, in preparation.

ELSEVIER

Nuclear Instruments and Methods in Physics Research A 475 (2001) 425–431

NUCLEAR
INSTRUMENTS
& METHODS
IN PHYSICS
RESEARCH
Section A

www.elsevier.com/locate/nima

Spatial distribution and polarization of γ-rays generated via Compton backscattering in the Duke/OK-4 storage ring FEL ☆

S.H. Park[a,*], V.N. Litvinenko[b], W. Tornow[c], C. Montgomery[c]

[a] *Laboratory for Quantum Optics, Korea Atomic Energy Research Institute, Taejon 305-353, South Korea*
[b] *Free Electron Laser Laboratory, Department of Physics, Duke University, Durham, NC 27708, USA*
[c] *Department of Physics, Duke University and Triangle Universities Nuclear Laboratory, Durham, NC 27708, USA*

Abstract

Beams of nearly monochromatic γ-rays are produced via intracavity Compton backscattering in the OK-4/Duke storage ring FEL, the high-intensity γ-ray source (HIγS). Presently, HIγS generates γ-ray beams with an energy tunable from 2 to 58 MeV and a maximum flux of 5×10^7 γ-rays per second. The γ-rays are linearly polarized with a degree of polarization close to 100% (V.N. Litvinenko, et al., Predictions and expected performance for the VUV OK-5/Duke Storage Ring FEL with variable polarization, Nucl. Instr. and Meth. A, to be published in this proceeding) and they are collimated to pencil-like semi-monoenergetic beams with RMS energy spreads as low as 0.2%. The detailed theoretical and experimental studies of the γ-ray beam quality were conducted during the last two years (S.H. Park, Thesis, Duke University, Durham, NC, USA, 2000). In this paper, we present the theoretical analysis and the experimental results on the spatial distribution and polarization of γ-rays from the HIγS facility. © 2001 Elsevier Science B.V. All rights reserved.

PACS: 13.60.Hb; 29.17.+w; 42.60.Cr; 42.75.Ht

Keywords: Compton backscattering; Free-electron lasers; Gamma rays

1. Introduction

Compton backscattering of laser light from high-energy electrons for γ-ray production in storage rings is widely used. The need for nuclear physics applications drives the quest for intense, high quality, polarized γ-ray beams. The variable polarization of γ-rays is very important in studies of nuclear phenomena. Using external lasers, a polarizer can control the polarization of the photons. However, the use of external laser, limits the γ-ray flux well below that attainable via the FEL intracavity Compton backscattering. In the FELs, wigglers can control the polarization of photons. For example, helical wigglers provide for circular polarization while planar wigglers provide for linear polarization. The OK-5 FEL [1], the next Duke storage ring FEL, will provide high-quality γ-ray beams with variable polarization.

☆ This work is supported by ONR Contract #N00014-94-1-0818.

*Corresponding author. Tel.: +82-42-868-8337; fax: +82-42-861-8292.

E-mail address: shpark@nanum.kaeri.re.kr (S.H. Park).

The angular distribution of γ-rays generated via Compton backscattering depends on the polarization of the incoming beams. A linearly polarized γ-ray beam has an asymmetric distribution compared with an azimuthally symmetric circular polarized γ-ray beam. Hence, the spatial distribution of γ-rays provides information on their polarization. The Compton γ-rays from the Duke/OK-4 storage ring FEL are almost 100% linearly polarized [2]. We measured the spatial distribution using special X-ray films with γ-ray converters. The measurements are consistent with 100% linear polarization of the γ-ray beam.

Section 2 describes the relation between the scattering cross-section and the polarization. The experimental results are described in Section 3.

2. Scattering cross-section and polarization effects

Compton scattering is a second-order process. In the covariant QED theory it is described by the S-matrix [3]

$$S_{\text{fi}} = e^2 \int d^4x_1 \int_{t_2 < t_1} d^4x_2 \langle k'p' | \bar{\psi}(x_1)\gamma_\mu \psi(x_1)$$
$$\times A_\mu(x_1)\bar{\psi}(x_2)\gamma_\nu \psi(x_2) A_\nu(x_2) | kp \rangle \quad (1)$$

where $|kp\rangle$ and $\langle k'p'|$ represent the initial state of the photon with wave vector k and the electron with momentum p, and the final state of the photon with wave vector k' and the electron with momentum p'. The Feynman diagrams of the process are shown in Fig. 1. The detailed calculation of the scattering cross-section can be found in Refs. [3,4].

In the Duke/OK-4 storage ring FEL, the collision occurs between unpolarized electrons and 100% linearly (horizontally) polarized FEL

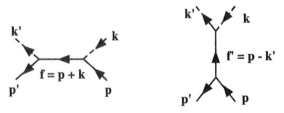

Fig. 1. Feynman diagrams for Compton scattering.

photons. The Compton scattering cross-section for polarized photons and unpolarized electrons is [4]

$$d\sigma = \frac{d\bar{\sigma}}{2} + \frac{2r_e^2 \, dy \, d\varphi}{x^2}$$
$$\times \left\{ -(\xi_3 + \xi_3')\left[\left(\frac{1}{x} - \frac{1}{y}\right)^2 + \left(\frac{1}{x} - \frac{1}{y}\right)\right] \right.$$
$$+ \xi_1 \xi_1' \left(\frac{1}{x} - \frac{1}{y} + \frac{1}{2}\right)$$
$$+ \frac{1}{2}\xi_2 \xi_2' \left(\frac{x}{y} + \frac{y}{x}\right)\left(\frac{1}{x} - \frac{1}{y} + \frac{1}{2}\right)$$
$$\left. + \xi_3 \xi_3' \left[\left(\frac{1}{x} - \frac{1}{y}\right)^2 + \left(\frac{1}{x} - \frac{1}{y}\right) + \frac{1}{2}\right]\right\} \quad (2)$$

where φ is the azimuthal angle, $\vec{\xi} = (\xi_1, \xi_2, \xi_3)$ and $\vec{\xi}' = (-\xi_1', -\xi_2', \xi_3')$ are the Stokes parameters of the incoming (initial) and outgoing (final) photons, and $d\bar{\sigma}$ is the scattering cross-section for unpolarized electrons and photons. The expressions for relativistic invariants x and y in the laboratory frame are:

$$x = \frac{2\gamma\hbar\omega(1 - \beta\cos\theta_i)}{mc^2}, \qquad y = \frac{2\gamma\hbar\omega'(1 - \beta\cos\theta_f)}{mc^2}$$
$$(3)$$

where ω and ω' are the frequencies of the incoming and scattered photons, θ_i and θ_f are their angles with respect to the incoming electron, and γ is the relativistic factor. The factor $\frac{1}{2}$ in Eq. (2) appears because there is no summation over the polarization of the final photons.

The directions of the photon polarization are determined by the components of the vectors $(e^{(1)}, e^{(2)}, \hat{k})$ and $(e'^{(1)}, e'^{(2)}, \hat{k}')$. The $e^{(1)}$ $(e^{(2)})$ and $e'^{(1)}(e'^{(2)})$ are the directions of the polarization perpendicular to the scattering plane (in the plane of the scattering). They are:

$$e^{(1)}, \; e'^{(1')} \| \hat{k} \times \hat{k}', \; e^{(2)} \| \hat{k} \times [\hat{k} \times \hat{k}'], \; e'^{(2)} \| \hat{k}' \times [\hat{k} \times \hat{k}'].$$

A sign change of $e'^{(2)}$ the photon density matrix is equivalent to a sign change of ξ_1' and ξ_2'. Hence, $\vec{\xi}' = (-\xi_1', -\xi_2', \xi_3')$. The density matrices of the initial and final photons (described by the unit four-vectors $e^{(1)}$ and $e^{(2)}$) are

$$\rho^{(k)} = \frac{1}{2}(1 + \vec{\xi} \cdot \vec{\sigma})$$

$$= \frac{1}{2}\begin{bmatrix} 1 + \xi_3 & \xi_1 - i\xi_2 \\ \xi_1 + i\xi_2 & 1 - \xi_3 \end{bmatrix}, \quad \xi_1^2 + \xi_2^2 + \xi_3^2 = 1$$

$$\rho^{(k')} = \frac{1}{2}(1 + \vec{\xi}' \cdot \vec{\sigma})$$

$$= \frac{1}{2}\begin{bmatrix} 1 + \xi_3' & -\xi_1' + i\xi_2' \\ -\xi_1' + i\xi_2' & 1 - \xi_3' \end{bmatrix}, \quad \xi_1'^2 + \xi_2'^2 + \xi_3'^2 = 1$$

where $\vec{\sigma}$ is the Pauli matrix. $\xi_3 = 1$ corresponds to the photons polarized in the direction perpendicular to the scattering plane. $\xi_3 = -1$ corresponds to photons polarized in the direction parallel to the scattering plane. $\xi_1 = 1$ describes 45° linear polarization. Finally, ξ_2 characterizes circular polarizations. The quantities ξ_2 and $\sqrt{\xi_1^2 + \xi_3^2}$ are invariant under Lorentz transformation.

For horizontally polarized incoming photons, the ratio between the scattering cross-sections for vertically and horizontally polarized outgoing photons can be expressed through the recoil parameter R. For the most energetic photons scattered at $\theta_f = 0$ in head-on collisions with $\theta = \theta_f = 0$, the ratio is given by

$$\frac{d\sigma_\perp}{d\sigma_\parallel}(\theta = 0) = \frac{R^2}{(2 + R)^2}, \quad R = \frac{2\gamma\hbar\omega(1 + \beta)}{mc^2}. \quad (4)$$

Therefore, for small recoil parameter R, the probability for scattering with vertical polarization is negligibly small. For example, the scattering of 3.3 eV photons off 600 MeV electrons gives $R = 0.03$ and $d\sigma_\perp/d\sigma_\parallel \cong 2.2 \times 10^{-4}$.

In the Duke/OK-4 FEL the optical beam is horizontally polarized. In the Cartesian coordinate system (Fig. 2) for head-on collisions it means that $\xi_1 = \sin 2(\pi/2 + \varphi)$ and $\xi_3 = \cos 2(\pi/2 + \varphi)$. We calculated the distributions of γ-rays 30 m downstream from the collision point in the Duke/OK-4 storage ring FEL. Fig. 3 shows the distributions for horizontally and vertically polarized γ-rays. The four lobes in the distribution for vertically polarized γ-rays indicate the nature of the process. The non-zero value of vertically polarized γ-rays on the axis is directly connected with the spin-flip of the electron [4].

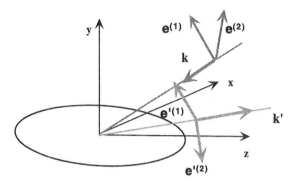

Fig. 2. Polarization vectors in Cartesian coordinates.

The X-ray film and converter used in our experiments do not distinguish between different polarization directions of the γ-rays. Hence, the film will provide us with a distribution of the total γ-ray flux. The sum over polarization gives a simple expression for the cross-section

$$d\sigma = d\bar{\sigma} - \frac{4r_e^2 \, dy \, d\varphi}{x^2} \xi_3 \left[\left(\frac{1}{x} - \frac{1}{y}\right)^2 + \left(\frac{1}{x} - \frac{1}{y}\right) \right]. \quad (5)$$

It depends on Stoke's parameter ξ_3 of the incoming photon. For a circular polarized photon as well as for an unpolarized photon, we have $\xi_3 = 0$. The distribution of γ-rays is axis-symmetric is these cases.

In the case of head-on collisions, $\theta_i = \pi$, a number of variables can be expressed through the angle of the final photon in the laboratory frame $\theta = \theta_f$. Writing the dy in terms of $\sin\theta \, d\theta$ we get:

$$dy = 2\left(\frac{\hbar\omega'}{mc^2}\right)^2 \sin\theta \, d\theta.$$

The differential cross-section in the lab frame becomes [5,6]

$$\frac{d\sigma}{d\Omega} = \frac{8r_e^2}{x^2}\left\{ (1 - \xi_3)\left[\left(\frac{1}{x} - \frac{1}{y}\right)^2 + \left(\frac{1}{x} - \frac{1}{y}\right) \right] \right.$$

$$\left. + \frac{1}{4}\left(\frac{x}{y} + \frac{y}{x}\right)\right\} \left(\frac{\hbar\omega'(\theta)}{mc^2}\right)^2 \quad (6)$$

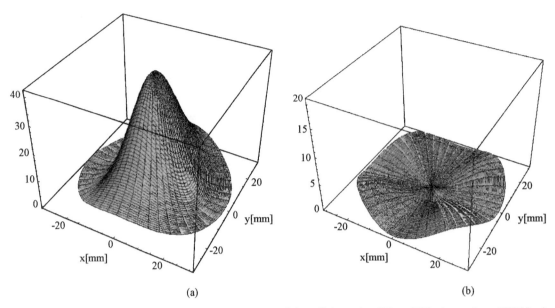

(a)   (b)

Fig. 3. Calculated distributions for γ-ray beams 30 m downstream of the collision point. 370 nm FEL photons have 100% horizontal linear polarization, and 600 MeV electrons are unpolarized. (a) Horizontally polarized γ-rays and (b) vertically polarized γ-rays.

where

$$x = \frac{2\gamma\hbar\omega(1+\beta)}{mc^2}$$

$$y = \frac{2\gamma\hbar\omega(1-\beta\cos\theta)}{mc^2}$$

$$\times\frac{(1+\beta)}{(1-\beta\cos\theta+(1+\cos\theta)\hbar\omega/\gamma mc^2)}.$$

The polarization-averaged distribution of γ-rays has azimuthal modulations for linearly polarized incoming photons. Fig. 4 shows the calculated density distributions of γ-rays for circularly and linearly polarized incoming photons. For linearly polarized FEL photons, one observes a nodal point at $\theta \sim 1/\gamma$. The origin of nodes is rather simple when we consider the process in the rest frame of the electrons. A horizontal electric field forces the electron to oscillate in the horizontal direction (with frequency $\varpi = \gamma\omega(1+\beta)$). The dipole radiation of the electron has two nodes in the direction of motion (i.e. $\vartheta = \pm\pi/2$ in the horizontal plane). The Lorentz boost into the laboratory frame transfers these nodes to be at $\theta_x \sim 1/\gamma$. For a 1 GeV electron and a distance of

$l = 30$ m from the collision point, the nodes are located at $\Delta x \sim 15.3$ mm to the left and to the right of the beam axis.

3. Experimental results

We used an electron beam with an energy of 600 MeV and the OK-4 FEL tuned at 375 nm (3.3 eV) to generate γ-rays with a peak energy of 17.7 MeV. X-ray films with a converter were used to measure density distributions of the γ-rays. The measurements were completed before the installation of the γ-ray beamline, which currently extends about 60 m downstream of the collision point into a dedicated shielded γ-ray vault [7]. Therefore, we were able to install the films adjacent to the downstream mirror and the outcoupling window at a distance of only 27.8 m from the collision point. The γ-rays passed through the 11 mm of quartz substrate of the downstream mirror and through the 5-mm CaF window before reaching the film. Fig. 5 shows the measured and the calculated spatial distributions of γ-rays at this

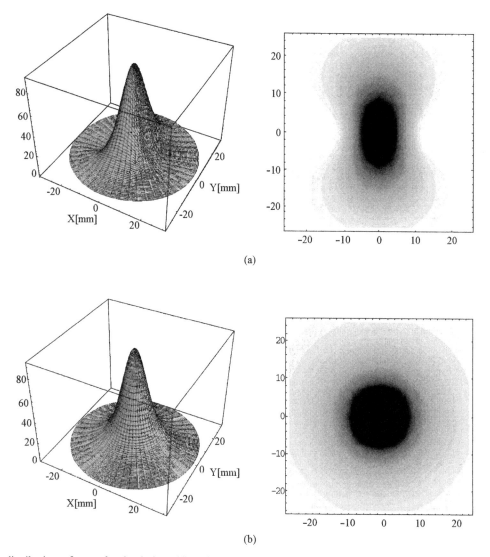

Fig. 4. The distributions of γ-rays for circularly and linearly polarized incoming photons at a location 30 m from the collision point. (a) Linearly polarized incoming photons of wavelength 120 nm and unpolarized electrons of energy 1 GeV and (b) circularly polarized incoming photons of wavelength 120 nm and unpolarized electrons of energy 1 GeV).

location. The X-ray films provided low contrast and the gray background resulted in limited experimental resolution. These are the reasons why the measured density plots are not as clear as the theoretical ones. Nevertheless, the effect of the linear polarization is clearly indicated on the exposed films. At electron beam energy of 600 MeV, the nodes were located at ±23.68 mm from the beam axis. The horizontal aperture for

the γ-ray beam was limited to 44 mm by the mirror holder. The vertical aperture was limited to ∼20 mm by the vacuum chamber in the bending magnet located 17.5 m downstream from the collision point. This translates into the vertical aperture of ∼30 mm seen on the film. Both photographs in Fig. 5 show two narrow dark horizontal lines above and below the main spot. These lines correspond to the gaps (which

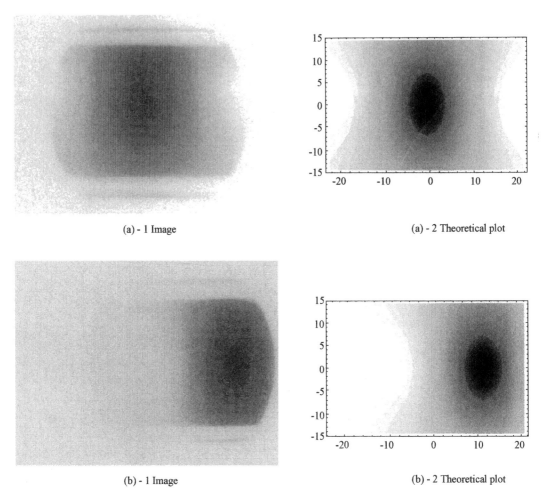

(a) - 1 Image   (a) - 2 Theoretical plot

(b) - 1 Image   (b) - 2 Theoretical plot

Fig. 5. Predicted and measured spatial distribution of γ-rays generated via Compton backscattering of FEL photons with 3.3 eV (376 nm) from 600 MeV electrons. (a) The γ-ray beam is aimed at the center of the horizontal aperture and (b) the γ-ray beam is aimed 12.9 mm to the right of the center of the horizontal aperture.

measures ∼0.25 mm on height) between the walls of the flat vacuum chamber and the poles of the dipole magnet. The flanges of the dipole vacuum chamber (∼5 cm think stainless steel) are partially transparent for γ-rays while the long walls and thick yoke of the dipole magnets efficiently absorb the γ-rays. Therefore, in addition to measuring the distribution of the γ-ray beam, we obtained the γ-ray image of our vacuum chamber.

Fig. 5(a) shows the image of the γ-ray beam with nodes located at ±23.7 mm just outside of the horizontal aperture of ±22 mm. It shows a clear

indication of the nodes as well as a vertical displacement caused by a 60 μrad vertical angle of the electron beam at the collision point. The observed similarity between the measured and the predicted images is reasonable, considering possible differences in the gray-scale.

In order to see one complete node within the available aperture size, we changed the horizontal angle of the electron and FEL beams at the collision point by 0.46 mrad, aiming the γ-ray beam axis 12.9 mm to the right of the mirror center. This procedure moved the left node to

−10.77 mm from the aperture center ("−"sign stands for the left side), as shown on Fig. 5(b). We also corrected the 60 μrad vertical angle to put the γ-ray beam axis onto the horizontal plane. The similarity between the measured and the predicted images is rather remarkable. The location of the measured node coincides with the predicted one within the accuracy of our measurements.

4. Conclusion

We measured for the first time, the distribution of γ-rays generated by linearly polarized FEL photons via Compton backscattering. The results are in good agreement with the theory considering the limited accuracy provided by the finite resolution of the films used in the present experiment. The shape of the measured spatial distribution clearly indicates the high degree of polarization of the γ-rays. The angle between the γ-ray beam center and the nodes is in good agreement with the predicted angle.

More accurate measurements will be possible when a γ-ray detector system with better spatial resolution becomes available. A strip-γ-ray detector is one of the potential candidates for future experiments. With the OK-5 system, we also plan to measure spatial distributions of γ-ray beams generated by FEL photons adjustable polarization.

References

[1] V.N. Litvinenko, et al., Predictions and expected performance for the VUV OK-5/Duke Storage Ring FEL with variable polarization, Nucl. Instr. and Meth. A 475 (2001) 97, these proceedings.

[2] V.N. Litvinenko, et al., Gamma-ray production in a storage ring free electron laser, Phys. Rev. Lett. 78 (1997) 4569.

[3] J.J. Sakurai, Advanced Quantum Mechanics, Benjamin/Cummings Publishing Company, Menlo Park, CA, 1967.

[4] L.D. Landau, E.M. Lifshitz, in: Quantum Electrodynamics Course of Theoretical Physics, Vol. 4, Pergamon Press, Oxford, 1982.

[5] V.N. Litvinenko, J.M.J. Madey, SPIE 2521 (1995) 55.

[6] S.H. Park, Thesis, 2000, Duke University, Durham, NC, USA.

[7] V.N. Litvinenko, M. Emamian, γ-Ray Beamline for HIγS Facility, DFELL Materials, 2000.

ELSEVIER

Nuclear Instruments and Methods in Physics Research A 475 (2001) 432–435

**NUCLEAR
INSTRUMENTS
& METHODS
IN PHYSICS
RESEARCH**
Section A

www.elsevier.com/locate/nima

Proposal for a IR waveguide SASE FEL at the PEGASUS injector

S. Reiche*, J. Rosenzweig, S. Telfer

UCLA - Department of Physics & Astronomy, 405 Hilgard Ave., Los Angeles, CA 90095-1444, USA

Abstract

Free Electron Lasers up to the visible regime are dominated by diffraction effects, resulting in a radiation size much larger than the electron beam. Thus the effective field amplitude at the location of the electron beam, driving the FEL process, is reduced. By using a waveguide, the radiation field is confined within a smaller aperture and an enhancement of the FEL performance can be expected. The PEGASUS injector at UCLA will be capable to provide the brilliance needed for an IR SASE FEL. The experiment Power Enhanced Radiation Source Experiment Using Structures (PERSEUS) is proposed to study the physics of a waveguide SASE FEL in a quasi 1D environment, where diffraction effects are strongly reduced as it is the case only for future FELs operating in the VUV and X-ray regime. The expected FEL performance is given by this presentation. © 2001 Elsevier Science B.V. All rights reserved.

PACS: 41.60.Cr

Keywords: FEL; GENESIS 1.3; Wave guides; PERSEUS

1. Introduction

Todays electron beam sources for SASE FELs, operating in the 500 nm or longer wavelength regime, provide a sufficient beam quality so that a beam size below the millimeter level is achieved within the undulator. The resulting diffraction of the emitted radiation field degrades the FEL. Even for a high gain (gain > 10^5) SASE FEL experiment at 12 μm [1] the ratio between the electron beam and radiation field size is 1 : 3 at its equilibrium state in the linear regime of the FEL amplification.

Diffraction is eliminated if the radiation field is enclosed by a waveguide. Waveguides reduces the diffraction effects and have successfully been used in FEL experiments in the millimeter wavelength region [2]. Based on the current research on IR waveguides the enhancement of the FEL efficiency can be extended to shorter wavelengths such as the FEL of the PEGASUS injector [3].

2. The PERSEUS experiment

Power Enhanced Radiation Source Experiment Using Structures (PERSEUS) is an extension of the FEL at the PEGASUS injector. The undulator itself has already been used for an IR SASE FEL experiment with a gain larger than 10^5 [1]. Table 1 lists all important parameters of the PERSEUS experiment.

---

*Corresponding author. Tel.: + 1-310-206-5584; fax: + 1-310-825-8432.

E-mail address: reiche@stout.physics.ucla.edu (S. Reiche).

0168-9002/01/$ - see front matter © 2001 Elsevier Science B.V. All rights reserved.
PII: S 0 1 6 8 - 9 0 0 2 (0 1) 0 1 6 3 4 - 5

Table 1
Parameters of the PERSEUS waveguide FEL

Electron beam	
Energy	17.9 MeV
Energy spread (rms)	0.15%
Normalized emittance	2π mm mrad
Charge	1 nC
Bunch length (rms)	0.9 mm
Undulator (planar)	
Period length	2.05 cm
Undulator parameter K	1.04
Undulator length	2 m
β-function	22 cm
Waveguide size	1 mm × 1 mm
Radiation	
Resonant wavelength	12.85 μm
Fundamental waveguide mode	TE_{01}
ρ-parameter (1D)	2.04×10^{-2}
Diffraction parameter B	0.061

The IR-waveguide is a Hollow Glass Waveguide (HGW) with a thin metal and dielectric layer deposit on the inside of the waveguide. Losses less than 0.2 dB/m at 10.3 μm have been measured for a waveguide radius of 500 μm and a layer thickness of 600 nm [4]. The same type of waveguide has been used to transport the radiation at the Vanderbilt FEL without any need for cooling [5]. The average power level of the PERSEUS experiment is 300 mW and thus lower than that for the Vanderbilt FEL. Therefore no FEL performance limitations are expected caused by the waveguide.

3. Waveguide modes

The resonance condition for a waveguide mode is

$$\beta_z = \frac{\omega_{mn}/c}{k + k_U} \quad (1)$$

where β_z is the longitudinal velocity of the electrons, normalized to c, k_U is the undulator wave number, k the wave number of the waveguide mode and ω_{mn} the frequency of the TE_{mn} or TM_{mn} mode.

Further we assume that the waveguide size is much larger than the radiation wavelength and, thus, the waveguide is overmoded. The frequency can be approximated as

$$\omega_{mn} = ck + \frac{c}{2k}\left[\left(\frac{m\pi}{l_x}\right)^2 + \left(\frac{n\pi}{l_y}\right)^2\right] \quad (2)$$

where l_x and l_y define the size of the waveguide in the x- and y-direction, respectively. A detailed description of the physics of waveguide FELs can be found in Refs. [6] or [7].

The 3D time-dependent code GENESIS 1.3 [8] has been modified to study the expected performance of the PERSEUS FEL. The radiation field is expanded into a series of empty waveguide modes for a rectangular waveguide.

The performance of the PERSEUS FEL for different sizes of the waveguide is shown in Fig. 1. Beyond 5 mm the FEL amplification is almost independent of the waveguide size and converges towards the results of the free-space FEL excluding waveguides.

Unfortunately, the radiation power does not saturates for the design waveguide size of 1 mm × 1 mm. As discussed in the next section the FEL performance is rather insensitive on variation in the beam size or emittance. This would allow to runs the PEGASUS injector with a higher bunch

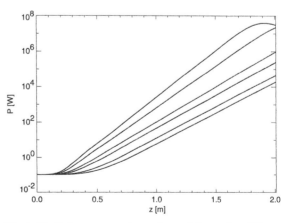

Fig. 1. Radiation power along the undulator for different waveguide sizes (Graphs from top to bottom correspond 0.8, 1, 1.5, 2, 2.5 and 5 mm waveguide size).

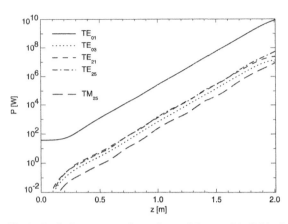

Fig. 2. Radiation power along the undulator of individual empty waveguide modes.

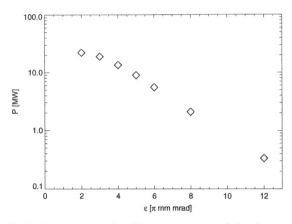

Fig. 3. Output power for different emittances of the electron beam.

charge, where the benefit of a higher peak current exceeds the degradation by a larger emittance.

As mentioned above the fundamental mode is the TE_{01} mode (Fig. 2). The next higher modes are the TM_{21} and the TE_{21} mode. The TE mode couples only half as strong as the corresponding TM mode [9].

All modes have the same growth rate and belong to the decomposition of the fundamental eigenmode of the FEL amplification into the eigenmode of the empty waveguide. The second largest mode (TE_{21}) is of the percent level compared to the dominant fundamental TE_{01} mode. Therefore the TE_{01} mode can be regarded as a good approximation of the fundamental FEL eigenmode.

4. Beam tolerances and diagnostics

For the PERSEUS experiment the tolerance of only two parameters are of particular interest: the emittance and the beam alignment.

The dependence on the emittance is shown in Fig. 3. Below 5π mm mrad the dependence is rather weak, which yields a relaxed tolerance on the emittance. The degradation above 5π mm mrad arises mainly from particle losses because the electron beam cannot be matched to the focusing of the undulator with the given aperture of the waveguide.

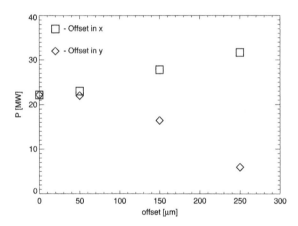

Fig. 4. Output power for different initial offsets of the electron beam.

Fig. 4 shows the output power for different initial beam offsets. The power drops for offsets in y because any offset within the undulator reduces the coupling between the electron beam and the fundamental TE_{01} mode. The reduction is 70% for an offset of 250 μm.

The coupling of the beam to the fundamental TE_{01} mode is independent on the x-position. The enhancement for an offset in x arises from the excitation of the next higher modes, namely the TE_{11} and TM_{11} modes. The coupling strength is identical for both modes and increases with increasing offset. If the electron beam is aligned to the waveguide center the coupling to these modes is zero. The excitation of these modes adds up coherently with the fundamental mode, increasing the bunching process and, thus, the FEL amplification. The phase advance of the TE_{11} and TM_{11} with respect to the electron beam is comparable to the betatron motion. Therefore the direction of the electric field of these modes remains almost constant at the position of the electron beam over the entire undulator.

The FEL performance might also be affected by the need for beam diagnostic within the undulator. The diagnostics consist mainly of insertable screens for OTR measurements. The waveguide is cut into smaller pieces, which are separated by cavities to provide the required space for the screens. This causes two effects: increased wake fields and coupling losses of the fundamental TE_{01} mode into the next waveguide segment.

The wake fields due to the resistance of the vacuum chamber and the dielectric layer are negligible compared to those caused by the change in the waveguide aperture. The resulting energy modulation has its maximum close to the beam center with an amplitude of 90 keV/m [10]. The reduction of the FEL output power is less than 1% and can be considered as negligible.

More stringent is the requirement for good coupling of the radiation field into the next segment of the waveguide. For a cavity length of 12 mm the loss per single cavity is 11% adding up to 30% power reduction for three cavities in total.

5. Conclusion

The usage of a waveguide for the FEL at the PEGASUS injector improves the FEL performance by reducing the saturation length of about 50%. Another improvement is the tolerance on the beam alignment, which is more relaxed compared to the free-space case. Therefore commercially available waveguides of sub-millimeter sizes can either reduces the saturation length of IR-SASE FELs or allows to operate the FEL with a reduced brilliance of the electron beam source.

References

[1] M. Hogan, et al., Phys. Rev. Lett. 81 (1998) 4867.
[2] T.J. Orzechowski, et al., Phys. Rev. Lett. 54 (1985) 889.
[3] C. Pellegrini, et al., Nucl. Instr. and Meth. A 475 (2001) 328, these proceedings.
[4] Y. Matsuura, et al., Appl. Opt. 34 (1995) 6842.
[5] H.S. Pratisto, et al., Clinical beam delivery of the Vanderbilt FEL: design and performance of a hollow waveguide-based handheld probe for neurosurgery, in: J. Harrington, A. Katzir (Eds.), Specialty Fiber for Medical Application, SPIE, Bellingham, 1999.
[6] E. Jerby, A. Gover, Nucl. Instr. and Meth. A 272 (1988) 380.
[7] H.P. Freund, T.M. Antonsen, Principles of Free-electron Lasers, Chapman & Hall, New York, 1996.
[8] S. Reiche, Nucl. Instr. and Meth. A 429 (1999) 243.
[9] R.E. Collin, Field theory of Guided Waves, IEEE Press, New York, 1991.
[10] A.W. Chao, Physics of Collective Beam Instabilities in High Energy Accelerators, Wiley, New York, 1993.

ELSEVIER

Nuclear Instruments and Methods in Physics Research A 475 (2001) 436–440

**NUCLEAR
INSTRUMENTS
& METHODS
IN PHYSICS
RESEARCH**
Section A

www.elsevier.com/locate/nima

Generation of ultra-short quasi-unipolar electromagnetic pulses from quasi-planar electron bunches

V.L. Bratman[a], D.A. Jaroszynski[b], S.V. Samsonov[a], A.V. Savilov[a],*

[a] *Institute of Applied Physics, Russian Academy of Sciences, 46 Ulyanov Str., Nizhny Novgorod, 603950 Russia*
[b] *University of Strathclyde, Department of Physics and Applied Physics, John Anderson Building, 107 Rottenrow, G4 ONG, Glasgow, UK*

Abstract

A method for the generation of quasi-unipolar pulses based on coherent synchrotron radiation from a quasi-planar electron bunch moving along a curved trajectory is proposed and theoretically studied. It is demonstrated that the experimental realization of this method at an existing installation (Terahertz to Optical Pulse Source) can result in generation of picosecond pulses with a peak power of up to 200 MW. © 2001 Elsevier Science B.V. All rights reserved.

PACS: 41.60.−m; 41.60.Ap

Keywords: Synchrotron radiation; Ultra-short broadband electromagnetic pulses

1. Introduction

Coherent spontaneous emission in free-electron lasers provides a method of generating short pulses of far-infrared radiation [1,2] However, to obtain ultra-broadband quasi-unipolar pulses a short undulator should be used to optimize the emission bandwidth. In the limit of half an undulator period a bandwidth of $\delta\lambda/\lambda \sim 1$ should be possible. In this paper, we examine methods of generating intense quasi-unipolar pulses using a "pancake" shaped relativistic electron beam acting as a phased antenna array in a half-period undulator.

Photoinjectors are commonly used as a front end in modern accelerators to produce low emittance high charge sub-picosecond bunches of

relativistic electrons at very high powers, up to 10^9–10^{13} W. The availability of ultra-short electron bunches opens up the attractive possibility of producing quasi-unipolar electromagnetic pulses with durations comparable or even shorter than the duration of the electron pulse. It is clear from general principles that the pulse shape follows that of the electron current pulse. Here, we consider unipolar synchrotron radiation from a relativistic electron bunch moving along a finite curved trajectory, whose form determines the field of the radiated pulse. To ensure high directionality of the radiation, a quasi-planar bunch with transverse dimensions much larger than its longitudinal dimension is necessary to provide a "phased antenna array" pattern of the emitted radiation. A short length of the bunch provides a coherent summation of fields (rather than powers) emitted by particles (coherent radiation). According to the theory, a powerful unipolar electromagnetic pulse

---

*Corresponding author. Tel.: +7-8312-384-318; fax: +7-8312-362-061.

E-mail address: savilov@appl.sci-nnov.ru (A.V. Savilov).

can be generated from a bunch emitting as it briefly executes an imposed transverse motion due to a magnetic field.

2. Basic principles

We begin with the well-known [3] example of radiation emitted by a single charged plane, which moves only along the x-direction for a finite interval of time with an imposed velocity $v_x(t)$ (Fig. 1a). The plane radiates a pulse composed of E_x and B_y components of electric and magnetic fields, which are described by the Maxwell equations,

$$\frac{\partial E_x}{\partial z} = -\frac{1}{c}\frac{\partial B_y}{\partial t}, \quad \frac{\partial B_y}{\partial z} = -\frac{1}{c}\frac{\partial E_x}{\partial t} - \frac{4\pi}{c}j_x(t,z) \quad (1)$$

where $j_x(t,z) = \sigma v_x(t)\delta(z)$, and σ is the surface charge density. The plane radiates two identical pulses propagating in the $\pm z$-directions, with spatio-temporal profiles reproducing the velocity "profile", $v_x(t)$:

$$E_x(t,z) = -\frac{2\pi\sigma}{c}\begin{cases} v_x(t - z/c), & z > 0, \\ v_x(t + z/c), & z < 0, \end{cases}$$

$$B_y(t,z) = \begin{cases} E_x(t,z) & z > 0 \\ -E_x(t,z) & z < 0. \end{cases}$$

Thus, monotonic motion of the charge plane in the x-direction (i.e. when the sign of $v_x(t)$ does not change) results in radiation of a unipolar pulse.

An analogous result is produced when the plane moves not only in the x-direction, but also along the z-coordinate with a constant longitudinal velocity, v_z (Fig. 1b) [4]:

$$E_x(t,z) = -\frac{2\pi\sigma}{c}\begin{cases} \dfrac{v_x(\zeta_+)}{1 - \beta_z}, & z > v_z t, \\ \dfrac{v_x(\zeta_-)}{1 + \beta_z}, & z < v_z t, \end{cases}$$

$$B_y(t,z) = \begin{cases} E_x(t,z), & z > v_z t, \\ -E_x(t,z), & z < v_z t. \end{cases} \quad (3)$$

Here $\beta_z = v_z/c$, and $\zeta_\pm = (t \mp z/c)/(1 \mp \beta_z)$. At $v_z \sim c$ the peak power of the forward pulse is significantly higher than that of the backward pulse. Due to the Doppler effect, the duration of the forward pulse is contracted, whereas the backward pulse is stretched.

Obviously, similar electromagnetic pulses can be generated from a quasi-planar moving electron bunch (Fig. 2), whose longitudinal ($z-$) dimension is much shorter than its transverse ($x-$ and $y-$) dimensions. Such a bunch radiates mainly in the z-direction, as from a phased antenna array. A

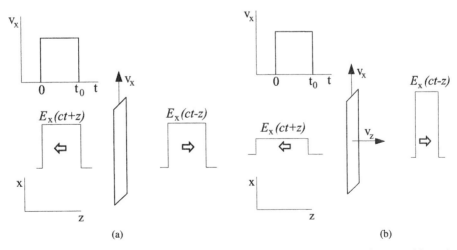

Fig. 1. Radiation of a moving charged plane: the plane does not move along z-coordinate (a) and moves with a velocity close to the speed of light (b).

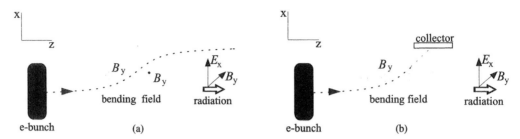

Fig. 2. Trajectories of a quasi-planar electron bunch moving through bending systems: two bending magnets (a), and a uniform magnetic field with a collector (b).

magnetostatic field bends the bunch trajectory so that during a finite interval of the time the bunch has a positive x-component of velocity. The magnetic field could be due to two bending magnets providing a translation of the bunch in the x-direction (Fig. 2a). Another possibility could be a single bending magnet followed by a collector (Fig. 2b); the collector is necessary to stop the motion of the electron bunch along the x-direction and, therefore, to control the duration and polarization of the radiated pulse. In both cases, the "pulse" of transverse motion of the bunch results in synchrotron radiation with linear (E_x, B_y) polarization. Since on the entire bunch trajectory the transverse velocity does not change its sign, the radiation has a unipolar character. The forward-pulse duration is estimated using the Doppler-converted time of the radiation,

$$t_p \approx L_b (1 - \bar{v}_z/c)/\bar{v}_z \qquad (4)$$

where L_b is the length of the bending system, and $\bar{v}_z$ is the characteristic transnational velocity of the bunch in the bending region. Thus, the use of an ultra-relativistic electron bunch enables the production of ultra-short electromagnetic pulses.

3. Simulations for the TOPS project

As a concrete example, we consider the ultra-short pulse electron source at the Terahertz to Optical Pulse Source (TOPS) developing at the Strathclyde University [5]. This facility will produce quasi-planar electron bunches with a total charge of the order of 1 nC and an energy

of 3–4 MeV. At an electron energy of 3.2 MeV, the Doppler conversion factor, $\Gamma = 1/(1 - v_z/c) \approx 2\gamma^2$, is of about 100. Here γ is the relativistic Lorentz-factor of the electrons. According to Eq. (4), the bending system length of few centimeters corresponds to radiation of a quasi-unipolar pulse with a duration $t_p \sim 1$ ps. Such a pulse has a central frequency of $f_c \approx 1/t_p \sim 1$ THz and a bandwidth of the order of f_c.

For the TOPS project, we simulate the radiation of a quasi-planar cylindrical electron bunch with a z-dimension of 0.3 mm and a transverse diameter of 1.4 mm, which initially moves rectilinearly along the z-coordinate. A finite-extent magnetostatic system provides a bending field, $\mathbf{B}_b = \mathbf{y}B_b$, driving the electrons along the x-coordinate. The electron motion is described by the following equations for the relativistic electron momentum, $\mathbf{p} = m\gamma\mathbf{v}$,

$$\frac{dp_x}{dt} = -eE_x + \frac{e}{c}v_z(B_y + B_b),$$

$$\frac{dp_z}{dt} = -eE_c - \frac{e}{c}v_x(B_y + B_b). \qquad (5)$$

Here E_x and B_y are electric and magnetic components of the radiated field, and E_c is the z-component of the Coulomb field. The Coulomb repulsion of electrons in transverse directions is neglected because it is partially compensated by the magnetic attraction of parallel electron currents. We assume that the bending field is weak enough to result in a negligibly small change in longitudinal electron velocity. Then the temporal derivative in Eq. (5) is transformed into the form $d/dt \approx \partial/\partial t + v_0 \partial/\partial z$, where v_0 is the initial electron velocity. The variables t and z are then substituted by new normalized variables, $\zeta = (z - v_0 t)/L_b$ and $\tau = ct/L_b$. This leads to the following equations for the normalized x- and z-components of the electron momentum,

$$\frac{\partial(\gamma\beta_x)}{\partial\tau} = -\hat{E}_x + \beta_z(\hat{B}_y + \hat{B}_b),$$

$$\frac{\partial(\gamma\beta_z)}{\partial\tau} = -\hat{E}_c - \beta_x(\hat{B}_y + \hat{B}_b) \quad (6)$$

and the following equations for the electron coordinates,

$$\frac{\partial\hat{x}}{\partial\tau} = \beta_x, \quad \frac{\partial\zeta}{\partial\tau} = \beta_z - \beta_0 \quad (7)$$

where $\hat{x} = x/L_b$, $\beta_{x,z} = v_{x,z}/c$ are the electron velocities normalized by the speed of light, $\beta_0 = v_0/c$, and $(\hat{E}, \hat{B}) = (eL_b/mc^2)(E, B)$ are the normalized fields.

The Maxwell equations (1) are transformed into the following form:

$$\frac{\partial\hat{E}_x}{\partial\zeta} = -\frac{\partial\hat{B}_y}{\partial\tau} + \beta_0\frac{\partial\hat{B}_y}{\partial\zeta}, \quad \frac{\partial\hat{B}_y}{\partial\zeta} = -\frac{\partial\hat{E}_x}{\partial\tau} + \beta_0\frac{\partial\hat{E}_x}{\partial\zeta}$$

$$-\frac{4\pi L_b^2}{I_A}j_x(\tau, \zeta) \quad (8)$$

where $I_A = mc^3/e$. The boundary conditions for Eqs. (8) can be formulated as radiation conditions in vacuum,

$$\frac{E_x}{B_y} = \begin{cases} 1, & \zeta > \zeta_{max} \\ -1, & \zeta < \zeta_{min} \end{cases} \quad (9)$$

where ζ_{max} and ζ_{min} are the extrema of the electrons in the bunch

We first consider a magnetostatic system in the form of two bending magnets (Fig. 2a), which can be modeled by the following y-component of the magnetic field,

$$B_b = B_0 \cos(\pi z/L_b), \quad 0 \leqslant z \leqslant L_b. \quad (10)$$

The results of simulations of Eqs. (5)–(9) for the bending field Eq. (10) and the length of the bending system $L_b = 1.2$ cm are shown in Fig. 3. For relatively small values of the bending field the forward radiated pulse is quasi-unipolar with duration of 1–2 ps. The power of the emitted radiation increases with the increase of the bending field, with a maximum of 200 MW achievable for a bending field magnitude of $B_0 = 8$ kGs. This is

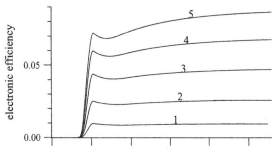

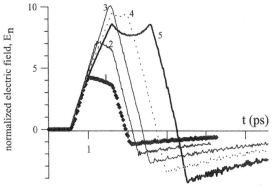

Fig. 3. Synchrotron radiation of an electron bunch during its motion through a two-magnet bending system. Electronic efficiency and radiated electric field versus time in the cases of magnitudes of the bending field, B_0, 3.3 kGs (1), 5.5 kGs (2), 8.0 kGs (3), 10.0 kGs (4), and 12.0 kGs (5). The normalized field, E_n, is connected with the radiated power by the formula $P = 2$ MW E_n^2.

accompanied by an increase of the pulse duration, which can be explained by two effects. The first one is an increase of the transverse electron velocity in the bending region, and, therefore, a decrease of the Doppler conversion factor, $\Gamma = (1 - \bar{v}_z/c)^{-1}$. The second effect is a perturbation of the electron motion by the radiated and Coulomb fields (radiation reaction and Coulomb repulsion). For bending field over 8 kGs, these effects lead to a saturation of the radiation process: any further increase of B_0 results in a decrease of the peak power accompanied by an increase in the duration of the radiated pulse.

At high bending fields the radiated pulse becomes bipolar. The electric field at the rear of the pulse reverses direction and results in a long "tail" of weak field having a "wrong" polarization. The origin of the negative "tail" is the transverse motion of electrons in the region after the bending system.

This phenomenon is caused by the collective effect of the interaction of different charge planes of the electron bunch, namely, the influence of the radiated field on the transverse motion of electrons. Actually, according to Eqs. (2) or (3), the radiated electric field counteracts the electron x-motion imposed by the bending field. Thus, the counter-force of the radiation acts to decrease the transverse velocity. Moreover, in the region after the bending system, the radiated electric field drives the electrons in the opposite direction, due to the retardation of the radiation. This results in a more complicated form of the radiated pulse: its duration increases and its rear experiences a reverse of the polarization. For weak bending fields, B_0, the collective effect is unimportant. In contrast, large values of B_0 modify the pulse shape into a bipolar form.

Evidently, the bipolar nature of the pulse can be avoided by collecting the bunch inside the bending region (Fig. 2b). As an example, we consider the bunch motion in a uniform magnetic field, $\mathbf{y}B_b$. The electron motion is abruptly stopped by a collector placed at a distance of $x = 0.5$ cm from the point of the bunch injection (in the simplest model we neglect the electron interaction with the collector, which could lead to undesired radiation). The output radiation represents an exactly uni-polar pulse with power and duration depending on the value of the bending field, B_b (Fig. 4). Magnetic fields up to 1 kGs shorten the output electromagnetic pulse due to a decrease of the characteristic bending length. However, any further increase of the bending field results in an increase of the duration of the output pulse due to a decrease in the Doppler conversion factor.

4. Conclusion

According to the theory, the proposed method enables the production of picosecond quasi-uni-polar pulses with a high peak power. It should be noticed that in this work we have not considered propagation of the quasi-unipolar pulses. This is justified because the radiation is produced in a short interaction region (bending system) by a "pancake" shaped electron bunch. Thus, we have used a model of radiating charge planes in vacuum that neglects both diffraction of the pulse and dispersion of the medium. The influence of these effects can be minimized using, for example, a weakly dispersive strip-line waveguide.

This work was supported by INTAS, Grant No. 97-32041.

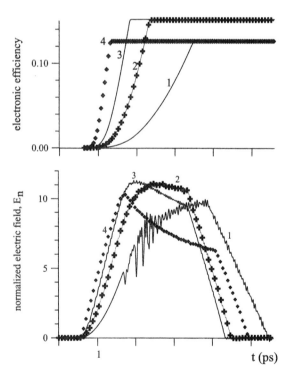

Fig. 4. Synchrotron radiation of TOPS electron bunch during its motion through a uniform bending field. Electronic efficiency and radiated electric field versus time in the cases of values of the bending field, B_b, 0.2 kGs (1), 0.5 kGs (2), 1.0 kGs (3), and 3.0 kGs (4). The normalized field, E_n, is connected with the radiated power by the formula $P = 2$ MW E_n^2.

References

[1] A. Doria, R. Bartolini, J. Feinstein, G.P. Gallernao, R.H. Pantell, IEEE J. Quantum Electron. 29 (1993) 1428.
[2] D.A. Jaroszynski, R.J. Bakker, A.F.G. van der Meer, D. Oepts, P.W. van Amersfoort, Phys. Rev. Lett. 71 (1993) 3798.
[3] R.P. Feynman, R.B. Leighton, M. Sands, in: The Feynman Lectures on Physics, The Electromagnetic Field, Vol. 2, Addison-Wesley, Reading, MA, 1965.
[4] V.L. Bratman, S.V. Samsonov, Phys. Lett. A 206 (1995) 337.
[5] D.A. Jaroszynski, B. Ersfeld, G. Giraud, et al., Nucl. Instr. and Meth. A 445 (2000) 317.

ELSEVIER

Nuclear Instruments and Methods in Physics Research A 475 (2001) 441–444

NUCLEAR
INSTRUMENTS
& METHODS
IN PHYSICS
RESEARCH
Section A

www.elsevier.com/locate/nima

Spontaneous and stimulated undulator radiation by an ultra-relativistic positron channeling in a periodically bent crystal

W. Krause*, A.V. Korol, A.V. Solov'yov, W. Greiner

Institut für Theoretische Physik der Johann Wolfgang Goethe-Universität, Postfach 11 19 32, 60054 Frankfurt am Main, Germany

Abstract

We discuss the radiation generated by positrons channeling in a crystalline undulator. The undulator is produced by periodically bending a single crystal with an amplitude much larger than the interplanar spacing. Different approaches for bending the crystal are described and the restrictions on the parameters of the bending are discussed. We also present numeric calculations of the spontaneous emitted radiation and estimate the conditions for stimulated emission. Our investigations show that the proposed mechanism could be an interesting source for high energy photons and is worth to be studied experimentally. © 2001 Elsevier Science B.V. All rights reserved.

PACS: 41.60

Keywords: Undulator radiation; Channeling positrons; Periodically bent crystal

1. Introduction

We discuss a new mechanism, initially proposed in Refs. [1,2], for the generation of high energy photons by means of the planar channeling of ultra-relativistic positrons through a periodically bent crystal. In this system there appears, in addition to the well-known channeling radiation, an undulator type radiation due to the periodic motion of the channeling positrons which follow the bending of the crystallographic planes. The intensity and the characteristic frequencies of this undulator radiation can be easily varied by changing the positrons' energy and the parameters of the crystal bending.

A periodically bent crystal is used as an undulator. The ultra-relativistic positrons channel between the bent planes. It is important to stress that we consider the case when the amplitude a of the bending is much larger than the interplanar spacing d ($\sim 10^{-8}$ cm) of the crystal ($a \sim 10\,d$) and is simultaneously much less than the period λ of the bending ($a/\lambda \sim 10^{-5}\ldots 10^{-4}$).

In addition to the spontaneous photon emission by the crystalline undulator, the scheme we propose leads to the possibility of generating stimulated emission. This is due to the fact that photons emitted at the points of the maximum curvature of the trajectory travel almost parallel to the beam and thus, stimulate the photon

---

*Corresponding author.

E-mail address: krause@th.physik.uni-frankfurt.de (W. Krause).

generation in the vicinity of all successive maxima and minima of the trajectory.

2. The bent crystal

The bending of the crystal can be achieved either dynamically or statically. In Refs. [1,2] it was proposed to use a transverse acoustic wave to dynamically bend the crystal. One possibility to couple acoustic waves to the crystal is to place a piezo sample atop the crystal and to use radio frequencies to excite oscillations. The usage of a statically and periodically bent crystal was discussed in Ref. [3]. The idea is to construct a crystalline undulator based on graded composition strained layers.

We now consider the conditions for stable channeling. The channeling process in a periodically bent crystal takes place if the maximum centrifugal force in the channel, $F_{cf} \approx m\gamma c^2 / R_{min}$ (R_{min} being the minimum curvature radius of the bent channel), is less than the maximal force due to the interplanar field, F_{int} which is equal to the maximum gradient of the interplanar field (see Ref. [2]). More specifically, the ratio $C = F_{cf}/F_{int}$ has to be smaller than 0.15, otherwise the phase volume of channeling trajectories is too small (see also Ref. [4]). Thus, the inequality $C < 0.15$ connects the energy of the particle, $\varepsilon = m\gamma c^2$, the parameters of the bending (these enter through the quantity R_{min}), and the characteristics of the crystallographic plane.

A particle channeling in a crystal (straight or bent) undergoes scattering by electrons and nuclei of the crystal. These random collisions lead to a gradual increase of the particle energy associated with the transverse oscillations in the channel. As a result, the transverse energy at some distance L_d from the entrance point exceeds the depth of the interplanar potential well, and the particle leaves the channel. One may follow the method described in Refs. [5,6] to estimate the quantity L_d, called the dechanneling length [7]. Thus, to consider the undulator radiation formed in a crystalline undulator, it is meaningful to assume that the crystal length does not exceed L_d.

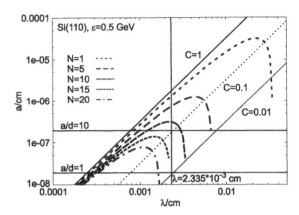

Fig. 1. The range of parameters a and λ for a bent Si(1 1 0) crystal at $\varepsilon = 500$ MeV.

Let us demonstrate how one can estimate for given crystal and energy ε the ranges of the parameters a and λ subject to the conditions formulated above. For doing this, we assume that the shape of the centerline of the periodically bent crystal is $a \sin(2\pi z/\lambda)$. Fig. 1 illustrates the above-mentioned restrictions in the case of $\varepsilon = 500$ MeV positrons channeling in Si between the (1 1 0) crystallographic planes. The diagonal straight lines correspond to various values (as indicated) of the parameter C. The curved lines correspond to various values (as indicated) of the number of undulator periods N related to the dechanneling length L_d through $N = L_d/\lambda$. The horizontal lines mark the values of the amplitude equal to d (with $d = 1.92 \times 10^{-8}$ cm being the (1 1 0) interplanar distance in Si) and to 10 d. The vertical line marks the value $\lambda = 2.335 \times 10^{-3}$ cm, for which the spectra (see Section 3) were calculated.

3. Spectra of the spontaneous emitted radiation

We calculated the spectra of the spontaneous emitted radiation using the quasiclassical method [8]. The trajectories of the particles were calculated numerically and then the spectra were evaluated [6]. The latter include both radiation mechanisms, the undulator and the channeling radiation.

The spectral distributions of the total radiation emitted in the forward direction for $\varepsilon = 500$ MeV

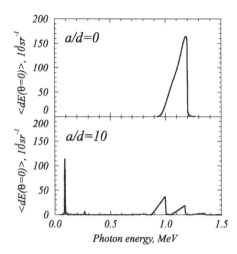

Fig. 2. Spectral distributions of the total radiation emitted in forward direction for $\varepsilon = 500$ MeV positrons channeling in Si between the (1 1 0) crystallographic planes for different a/d ratios.

positrons channeling in Si between the (1 1 0) crystallographic planes are plotted in Fig. 2. The wavelength is fixed at $\lambda = 2.335 \times 10^{-3}$ cm, while the ratio a/d is changed from 0 to 10. The length of the crystal is $L_{\rm d} = 3.5 \times 10^{-2}$ cm and corresponds to $N = 15$ undulator periods.

The first graph in Fig. 2 corresponds to the case of the straight channel ($a/d = 0$) and thus presents the spectral dependence of the ordinary channeling radiation only. In the case of 500 MeV positrons channeling between the (1 1 0) planes in Si the undulator parameter of the channeling radiation is $p_{\rm c} \approx 0.2 \ll 1$ and thus all the channeling radiation is concentrated within some interval in the vicinity of the energy of the first harmonic. The calculated spectrum is consistent with measured channeling radiation spectra.

Increasing the a/d ratio leads to modifications in the spectrum of radiation. The changes which occur manifest themselves via three main features, (i) the lowering of the ordinary channeling radiation peak, (ii) the gradual increase of the intensity of undulator radiation due to the crystal bending, (iii) the appearing of additional structure (the sub-peaks) in the vicinity of the first harmonic of the ordinary channeling radiation. A more detailed analysis of these spectra can be found in Ref. [6].

4. Discussion of stimulated photon emission

The mechanism described in this paper allows us to discuss the possibility to generate stimulated emission of high energy photons by means of a bunch of ultra-relativistic positrons moving in a periodically bent channel. Indeed, the photons emitted in the nearly forward direction at some maximum or minimum point of the trajectory by a group of particles of the bunch stimulate the emission of photons with the same energy by another (succeeding) group of particles of the same bunch when it reaches the next maximum/minimum.

In Ref. [2], estimates for the gain factor for the spontaneous emission in crystalline undulators were obtained. It was demonstrated that to achieve a total gain equal to 1 on the scale of the crystal length (equal to the dechanneling length), one has to consider volume densities n of the channeling positrons on the level of $10^{20}, \ldots, 10^{21}$ cm^{-3} for positron energies within the range 0.5...5 GeV. These magnitudes are high enough to question whether or not they can be really attained.

To estimate the volume density n of a positron bunch which can be achieved in modern colliders we use the data presented in Ref. [9, see p. 142] for a beam of 50 GeV positrons available at SLC (SLAC, 1989). The bunch length is 0.1 cm and the beam radius is 1.5 µm (II) and 0.5 µm (V), resulting in the volume of one bunch $V = 2.4 \times 10^{-9}$ cm^3. The number of particles per bunch is given as 4.0×10^{10}. Therefore, one obtains $n = 1.7 \times 10^{19}$ cm^{-3}.

This value, although lower by an order of magnitude than the estimates obtained in Ref. [2], shows that the necessary densities should be reachable in the future with accelerators optimized for high particle densities.

Finally, let us discuss the required transverse emittance of the beam. For doing this, we need to consider the angle between the particle's trajectory and the crystal plane. If this angle is larger than the Lindhard angle $\Psi_{\rm P}$, the particle will not be captured into the channeling mode and will leave the channel immediately [10]. For 5 GeV positrons channeling between the (1 1 0) planes of silicon, we have $\Psi_{\rm P} = 72$ µrad and for 50 GeV positrons it equals to $\Psi_{\rm P} = 23$ µrad.

One can compare these values with the divergence of the SLC beam, for which transverse emittance in vertical direction is given as $0.05\,\pi$ rad nm [9]. With a vertical beam radius of $0.5\,\mu$m we get for the vertical beam divergence $\Psi = 100\,\mu$rad. Thus, the divergence of the beam is about 4 times higher than the acceptance of the channel and so only a quarter of all particles will participate in the channeling process. Evidently, it is necessary to reduce the divergence of the beam, for example by increasing the beam radius. But then it is also necessary to reduce the bunch length to keep the particle density high enough. Fortunately, like for the particle density, the values achievable today differ only about one order of magnitude from the values estimated above for the stimulated emission.

5. Conclusion

To conclude, we point out that the use of crystalline undulators is a new and interesting way to produce high energy radiation. As we have shown above, the present accelerator technology is nearly sufficient to achieve the necessary conditions. Also, it appears that by means of modern technology it is possible to construct the required crystalline undulator, which, as compared to the undulators based on magnetic fields, will allow photon emission of much higher energy.

In our opinion, the effect described above is worth experimental study which, as a first step, may concentrate on measurements of the spontaneous undulator radiation spectra.

The related problems not discussed in this paper (but which are under consideration) include the investigation of the crystal damage due to the acoustic wave, photon flux and beam propagations.

Acknowledgements

The research was supported by DFG, GSI, and BMBF. AVK and AVS acknowledge the support from the Alexander von Humboldt Foundation.

References

[1] A.V. Korol, A.V. Solov'yov, W. Greiner, J. Phys. G. 24 (1998) L45.
[2] A.V. Korol, A.V. Solov'yov, W. Greiner, Int. J. Mod. Phys. E 8 (1) (1999) 49.
[3] U. Mikkelsen, E. Uggerhøj, Nucl. Instr. and Meth. B 160 (2000) 435.
[4] A.V. Korol, A.V. Solov'yov, W. Greiner, Int. J. Mod. Phys. E 9 (1) (2000) 77.
[5] V.M. Biruykov, Y.A. Chesnokov, V.I. Kotov, Crystal Channeling and its Application at High-Energy Accelerators, Springer, Berlin, 1996.
[6] W. Krause, A.V. Korol, A.V. Solov'yov, W. Greiner, J. Phys. G: Nucl. Part. Phys. 26 (2000) L87.
[7] D.S. Gemmel, Rev. Mod. Phys. 46 (1) (1974) 129.
[8] V.N. Baier, V.M. Katkov, V.M. Strakhovenko, High Energy Electromagnetic Processes in Oriented Single Crystals, World Scientific, Singapore, 1998.
[9] C. Caso, et al., Eur. Phys. J. C 3 (1998) 142.
[10] J. Lindhard, K. Dan, Viddensk. Selsk. Mat. Phys. Medd. 34 (1) (1965) 14.

ELSEVIER

Nuclear Instruments and Methods in Physics Research A 475 (2001) 445–448

NUCLEAR
INSTRUMENTS
& METHODS
IN PHYSICS
RESEARCH
Section A

www.elsevier.com/locate/nima

The research on two-color photon sources in infrared and X-ray ranges by compton scattering

Yu Zhao*, Jiejia Zhuang

Institute of High Energy Physics, P.O. Box 2732-16, Beijing 100080, China

Abstract

The generation of a two-color source of FEL light in both the infrared and soft X-ray ranges by intracavity Compton backscattering is demonstrated by the Beijing FEL facility. 1.20–1.35 keV soft X-rays are successfully extracted from the optical cavity of the FEL through a porous metallic mirror, while a 9–10 μm FEL laser is output in the other dielectric mirror simultaneously. The average output flux of X-ray is 10^2–10^3 photons/s when the average output FEL laser power is 6–20 mW. The experimental result allows us to envision a convenient way to expand the application areas of IR FEL facilities into X-ray or γ-ray ranges. © 2001 Elsevier Science B.V. All rights reserved.

PACS: 41.50.+h; 41.60.Cr

Keywords: Compton scattering; Free electron laser; X-ray

1. Introduction

A Compton scattering source, or Thomson scattering source, has been discussed frequently as a new kind of compact X- or γ-ray source in recent years [1–9]. Its theory is based on the Compton scattering of intense laser radiation from a counterstreaming electron beam. The wavelength and flux of scattering photons are, respectively, given by

$$\lambda_X = \frac{\lambda_L}{4\gamma^2}(1 + a_0^2/2 + \gamma^2\theta^2) \qquad (1)$$

$$F_X = \frac{\sigma}{4\pi w^2}N_e F_{ph} \qquad (2)$$

where $\sigma = 6.66 \times 10^{-29}\,\mathrm{m}^2$ is the cross-section of Thomson scattering, λ_L is the laser wavelength, θ is the observation angle, γ is the electron Lorentz factor, w is the size of laser spot, F_{ph} is the flux of incident laser pulse, N_e is the number of electrons in an electron bunch colliding with photons, and a_0 is the dimensionless laser strength parameter, which can be calculated as follows.

$$a_0 = 0.85 \times 10^{-9}\,\lambda_L(\mu m)I_0^{1/2}(\mathrm{W/cm}^2) \qquad (3)$$

where I_0 is the laser power density.

There are several schemes to build Compton scattering sources by means of different combinations of accelerators and lasers. Of these configurations, one simple and original method is to obtain the X-ray by Compton backscattering in an optical cavity of FEL facilities [10–12]. In fact, intracavity operation of IR FEL is a natural

*Corresponding author. Tel.: +86-10-62554583; fax: +86-10-62573660.

E-mail address: yuzhao@btamail.net.cn (Y. Zhao).

Compton scattering source and can be described briefly as follows:

First, the relativistic electrons collide with the virtual photons that are equivalent to the periodic magnetic field of undulator and the first phase of Compton scattering takes place. In this stage, the virtual photons, whose wavelength is in the centimeter range, are converted into real infrared photons that are stored in the optical cavity of FEL.

Second, the IR photons oscillate between two optical cavities. Electrons continuously interact with the photons and transform part of their kinetic energy into light energy by stimulated Compton scattering. Then laser is produced and amplified. The wavelength of the FEL laser is given by

$$\lambda_{FEL} = \frac{\lambda_u}{2\gamma^2}(1 + K^2/2) \qquad (4)$$

where λ_u is the wiggler period and K is the wiggler strength parameter.

Third, the backward infrared laser from the mirror of the FEL collides with the forward electrons driving the FEL and the second phase of Compton scattering happens. The high-frequency photons in X-ray band are generated in this stage. According to Eqs. (1) and (2), the wavelength of X-rays is given by

$$\lambda_X = \frac{\lambda_{FEL}}{4\gamma^2}(1 + a_0^2/2 + \gamma^2\theta^2)$$
$$= \frac{\lambda_u}{8\gamma^4}(1 + a_0^2/2 + \gamma^2\theta^2)(1 + K^2/2). \qquad (5)$$

According to the above analysis, there are two frequency-band photons produced in FEL: one is the infrared laser that is mainly considered in FEL research; the other comprises X-rays that are commonly neglected and are wasted in the mirror. So, if X-rays can be extracted from the optical cavity and the output of infrared laser is not affected at the same time, the FEL can be changed into a facility running in two wavelength ranges. The possibility of two-color photon sources had been envisioned in an experiment concerning intracavity Compton scattering on CLIO FEL [10].

2. Experimental setup

On the basis of the above idea, we designed an experiment to produce two-color photon sources in Beijing FEL (shown in Fig. 1). BFEL is a Compton region FEL oscillator operating in the mid-IR range which consists of a thermionic cathode RF gun, a 30 MeV linear accelerator, a 1.5 m NdFeB wiggler, and 2.5 m optical cavity with dielectric mirrors. Its electron peak current is about 15 A in 4 ps long bunches whose repetition is 2856 MHz during a 4 μs macropulse. The beam average normalized rms emittance is approximately 50π mm mrad, and its energy spread is below 1%. The wiggler has 50 periods and every period is 3 cm. The best radiation wavelength range is 9–11 μm.

The estimated wavelength of X-rays by Eq. (5) in BFEL is in the soft X-ray range. It is difficult to extract soft X-ray photons and reflect IR laser simultaneously with the cavity mirror because soft X-ray photons are very easily absorbed by mirror materials. To solve this problem, we designed a porous metallic mirror M1 to extract X-rays in the experiment. The central thickness of mirror M1 is 50 μm and many round apertures are punched in the central area of the mirror. According to high frequency theory, every aperture can be regarded as a round wave-guide and its maximum critical wavelength is given by

$$\lambda_C \cong 1.71d \qquad (6)$$

where d is the aperture diameter. The critical wavelength λ_C means that an electromagnetic

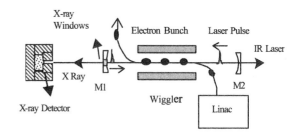

Fig. 1. Experimental setup of a two-color photon source in Beijing FEL.

wave cannot propagate in the round wave-guide and will be reflected completely by the mirror if its wavelength is longer than λ_C. By laser processing, the diameters of apertures are punched at 3–5 μm in this experiment and their critical wavelengths, calculated by Eq. (6), are shorter than 9 μm. Therefore, the FEL laser (9–11 μm) cannot pass through the porous metallic mirror and is reflected, but X-rays are easily output because their wavelength λ_X (<20 Å) is much shorter than λ_C.

To detect the X-ray signal, we developed a kind of gaseous proportional counter, with a 5 μm thick Be window, that is very sensitive to soft X-rays and is able to detect and distinguish photon energy from 600 eV to 20 keV. In experiment, the counter was located at 30 cm from the X-ray mirror M1 and the space between them was evacuated.

In our initial experiment, about 1% of the incident X-rays were extracted through apertures and 98.5% of the FEL laser was reflected in mirror M1. The infrared output mirror M2 is still a dielectric mirror (ZnSe) and its IR transmission is 1%. The calculated number of X-ray photons was 2.4/micropulse, 7.4×10^5/macropulse and about 10^6 photons/s in cavity when the intracavity peak power of FEL was 24 GW during the laser micropulse. The output flux is estimated at about 10^2–10^3 photons/s because there was 10^{-4} output loss due to X-ray emittance and 1% transmission of mirror M1.

3. Experimental result

In this experiment, although the whole counter was carefully shielded by lead bricks against ionizing radiation produced by the linac, the detected X-ray signal still contains the background radiation emanating from the FEL's linac. To extract the signal produced by the inverse Compton scattering, we first detect the total signal when the laser is on and then only detect background signal when the laser is off. The difference of the two kinds of signal, seen in Fig. 2, is the soft X-ray signal produced by Compton scattering. We can find that the X-ray signal in Fig. 2 has some

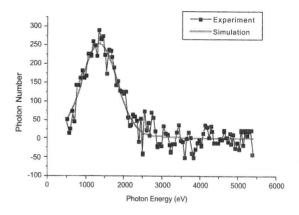

Fig. 2. Soft X-ray signal detected in experiment.

Table 1
The main experimental parameters of the two-color photon source

Electron energy (MeV)	26–27
Mirror reflectivity in 9–11 μm	M1: 98.5% M2: 99%
Laser wavelength (μm)	9–11
Laser macropulse duration (μs)	2.5
Rayleigh range (m)	0.7
Output laser-pulse energy (mJ/macropulse)	2–7
X-ray photon energy (keV)	1.20–1.35
X-ray average output flux (photons/s)	10^2–10^3
X-ray average angle spread (mrad)	5

fluctuation that is produced by the instability of the background radiation.

The main parameters of the two-color photon source in our experiment are summarized in Table 1. The calculated result was in accord with the experimental result in magnitude. The X-ray peak brightness is about 10^{10} photons/(s mm^2 mrad^2 0.1% bandwidth). There are some limitations in BFEL to generate the higher flux and brightness X-rays by Compton scattering at present, especially its very long Rayleigh length and low FEL power. Efforts at improvement, however, are underway in the BFEL laboratory. In current planning, the laser power, macropulse repetitions of electron beam, and X-ray transmission of mirror M1 will all be increased more than one order of magnitude, respectively, in the near future. Under these conditions, the flux of X-ray will reach more than 10^7 photons/s.

4. Conclusion

Based on the introduction and experiment detailed above, we can summarize the following three main advantages of two-color photon sources generated by IR FEL:

First, because IR FEL itself can provide high-power lasers and high-quality electron beams at the same time and there are no requirements for external lasers and accelerators, the cost of rebuilding FEL facilities to generate X-rays by Compton scattering is very low.

Second, intracavity Compton scattering can keep a good overlap between electron beams and photons naturally both in space and in time, which is difficult for the common method of external collision between electrons and lasers to realize the Compton scattering.

Third, the X-rays realized by Compton scattering extend the tunable range of IR FEL greatly and thus broaden the areas of application of IR FEL in X-ray areas.

At present, there are more than ten developed and developing IR FEL facilities based on an optical cavity in the world. Our method of two-color operation of FEL by means of intracavity Compton scattering is a simple and convenient way to expand the application areas of these FEL facilities into X-ray or γ-ray range.

Acknowledgements

We wish to thank Mingqi Cui for helpful advice concerning X-ray detection and also thank Yonggui Li, Gang Wu, Mingka Wang for their support in our experiment.

References

[1] P. Sprangle, A. Ting, E. Esarey, A. Fisher, J. Appl. Phys. 72 (11) (1992) 5032.
[2] C.A. Brau, Nucl. Instr. and Meth. A 318 (1992) 38.
[3] K.-J. Kim, S. Chattopadhyay, C.V. Shank, Nucl. Instr. and Meth. A 341 (1994) 351.
[4] I. Kimel, L.R. Elias, Nucl. Instr. and Meth. A 358 (1995) 20.
[5] R.W. Schoenlein, W.P. Leemans, A.H. Chin, P. Volfbeyn, T.E. Glover, P. Balling, M. Zolotorev, K.-J. Kim, S. Chattopadhyay, C.V. Shank, Science 274 (1996) 236.
[6] Jiejia Zhuang, Yu Zhao, High Energy Phys. Nucl. Phys. 24(9) (2000) 82.
[7] Eric Esarey, Sally K. Ride, Phillip Sprangle, Phys. Rev. E 48(4) (1993) 3003.
[8] Sally K.Ride, Eric Esarey, Michael Baine, Phys. Rev. E 52(5) (1995) 5425.
[9] Eric Esarey, et al., Phys. Rev. A 45(8) (1992) 5872.
[10] F. Glotin, J.-M. Ortega, R. Prazeres, G. Devanz, O. Marcouillé, Phys. Rev. Lett. 77 (15) (1996) 3130.
[11] F. Ciocci, G. Dattoli, L. Giannessi, P.L. Ottaviani, M. Quattromini, M. Carpanese, Nucl. Instr. and Meth. A 393 (1997) 536.
[12] M. Hosaka, H. Hama, K. Kimura, J. Yamazaki, T. Kinoshita, Nucl. Instr. and Meth. A 393 (1997) 525.

ELSEVIER

Nuclear Instruments and Methods in Physics Research A 475 (2001) 449–453

NUCLEAR
INSTRUMENTS
& METHODS
IN PHYSICS
RESEARCH
Section A

www.elsevier.com/locate/nima

Coherent amplification of volume plasmons

C.A. Brau*

Department of Physics, Vanderbilt University, Box 1807, station B, Nashville, TN 37235, USA

Abstract

It is shown that volume plasmons can be coherently amplified by a high-intensity electron beam. Because of natural damping of the plasmons and scattering of the electrons by ions in the metal the required electron-beam intensity is of the order of $10^{13}\,A/m^2$. Although the plasmons are confined inside the metal, they can be detected by their effect on the energy of the electrons emerging from the metal and by the enhancement of the transition radiation at the plasma frequency © 2001 Published by Elsevier Science B.V.

PACS: 71.45.Gm; 73.20.Mf; 79.2.Uv

Keywords: Plasmons; Electron beam; Plasmon excitation

1. Background

As the brightness of available electron beams has improved, the variety of free-electron lasers has proliferated and the laser wavelength has been extended into the ultraviolet. Nevertheless, all free-electron laser interactions excite transverse electromagnetic waves, although some are modified by the waveguides in which they propagate. It is proposed here that it should be possible to excite an entirely different type of coherent wave, specifically a volume plasmon in a metal, using a high-brightness, short-pulse electron beam.

A volume plasmon is a quantized electrostatic oscillation in the high-density plasma of a metal. The plasma frequency is

$$\omega_p = \sqrt{\frac{n_e e^2}{\varepsilon_0 m}} \qquad (1)$$

where n_e is the electron density, e the electron charge, ε_0 the permittivity of free space (SI units are used) and m the electron mass. Typically, the plasma frequency in metals is of the order of $\hbar\omega_p = O(10\,eV)$. Previously, it has been observed that high-energy electrons passing through a thin foil lose energy in quanta $\hbar\omega_p$, where $\hbar$ is Planck's constant, as they excite bulk plasmons in the foil [1]. The number of plasmons excited per unit length by an electron is

$$\frac{dn_p}{dz} = \frac{e^2 \omega_p}{4\pi\varepsilon_0 \hbar c^2 \beta_0^2} \ln\left(\frac{\beta_0 c}{a_0 \omega_p}\right) \qquad (2)$$

where $\beta_0 c$ is the electron velocity and a_0 the Bohr radius [2]. For 100-keV electrons passing through lithium the excitation rate is about 5 plasmons per μm of travel. The experiments are typically carried out with foils a few μm thick.

At low electron-beam intensity the plasmons are damped before they interact with another electron, but at sufficiently high intensity the process becomes nonlinear. In this case the electron

Tel.: +1-615-322-2559; fax: +1-615-343-1103.
E-mail address: c.a.brau@vanderbilt.edu (C.A. Brau).

0168-9002/01/$ - see front matter © 2001 Published by Elsevier Science B.V.
PII: S 0 1 6 8 - 9 0 0 2 (0 1) 0 1 6 3 0 - 8

velocity modulation caused by the electric fields of the plasmons can lead to bunching of the electrons and enhanced excitation of the plasmons. In a collisionless plasma the electrostatic waves do not propagate (the group velocity vanishes), and this is approximately true of plasmons in a metal. Nevertheless, the electrons in the beam propagate the coherence, and gain on the interaction is predicted as described below.

Although the plasmons are confined inside the metal, they can be detected by their effect on the energy of the electrons emerging from the foil. In addition, when the process saturates the electron bunching will cause the transition radiation emitted by the emerging electrons to become coherent, with a strong enhancement at the plasma frequency.

2. Gain

To compute the gain we must calculate the excitation of the plasmons by the bunched electron beam and the bunching of the electrons by the plasmons. Since the plasmon energy is typically small compared with the electron energy $O(100\,\mathrm{keV})$, we can use classical mechanics to describe the process. Since electrostatic oscillations have no magnetic field, the excitation of the plasma waves is found from Gauss's law

$$\nabla \cdot \mathbf{D} = \rho \tag{3}$$

where $\mathbf{D}$ is the electric displacement and ρ the charge density of the electron beam. The net charge density of the free electrons in the metal, as modulated by the plasma oscillations, is regarded as bound charge in this description and is included in the displacement vector $\mathbf{D}$. The electron-beam charge density is found from the continuity equation

$$\frac{\partial \rho}{\partial t} + \rho \nabla \cdot \mathbf{v} + \mathbf{v} \cdot \nabla \rho = 0 \tag{4}$$

and the velocity of the electrons is found from the equation of motion

$$\frac{\partial \gamma m \mathbf{v}}{\partial t} + (\mathbf{v} \cdot \nabla)(\gamma m \mathbf{v}) = e\mathbf{E} \tag{5}$$

where γ is the usual relativistic parameter of the electrons and $\mathbf{E}$ the electric field.

To solve these equations we linearize the continuity equation and the equation of motion about the quiescent values ρ_0 and $\mathbf{v}_0$ to get

$$\frac{\partial \rho'}{\partial t} + \mathbf{v}_0 \cdot \nabla \rho' = 0 \tag{6}$$

$$\gamma_0^3 \frac{\partial \mathbf{v}'}{\partial t} + \gamma_0^3 (\mathbf{v}_0 \cdot \nabla)\mathbf{v}' = \frac{e\mathbf{E}}{m} \tag{7}$$

where primes refer to perturbations from the quiescent values. We ignore for now the effects of the overall space charge of the beam, so that the charge density ρ in Gauss's law represents just the modulation of the electron-beam density by the plasma waves, and we write

$$\nabla \cdot \mathbf{D} = \rho'. \tag{8}$$

If we Fourier transform Eqs. (6)–(8) with respect to time (indicated by a tilde) and space (indicated by the subscript $\mathbf{k}$) we get

$$i\varepsilon(\omega)\mathbf{k} \cdot \tilde{\mathbf{E}}_{\mathbf{k}} = \tilde{\rho}'_{\mathbf{k}} \tag{9}$$

$$(\mathbf{v}_0 \cdot \mathbf{k} - \omega)\tilde{\rho}'_{\mathbf{k}} = -\rho_0 \mathbf{k} \cdot \tilde{\mathbf{v}}'_{\mathbf{k}} \tag{10}$$

$$i(\mathbf{v}_0 \cdot \mathbf{k} - \omega)\mathbf{k} \cdot \tilde{\mathbf{v}}'_{\mathbf{k}} = \frac{e}{\gamma_0^3 m}\mathbf{k} \cdot \tilde{\mathbf{E}}_{\mathbf{k}} \tag{11}$$

where $\varepsilon(\omega)$ is the permittivity of the conduction-electron plasma at the frequency ω, so that the displacement is

$$\tilde{\mathbf{D}}(\omega) = \varepsilon(\omega)\tilde{\mathbf{E}}(\omega). \tag{12}$$

Combining Eqs. (9)–(12) we obtain the dispersion relation

$$(\mathbf{v}_0 \cdot \mathbf{k} - \omega)^2 = \omega_b^2 \frac{\varepsilon_0}{\varepsilon(\omega)} \tag{13}$$

where

$$\omega_b = \sqrt{\frac{\rho_0 e}{\gamma_0^3 \varepsilon_0 m}} \tag{14}$$

is the effective plasma frequency of the electron beam.

To evaluate the permittivity $\varepsilon(\omega)$ we can use the Drude model, for which

$$\frac{\varepsilon_0}{\varepsilon(\omega)} = \frac{\omega^2 + i\gamma_p \omega}{\omega^2 + i\gamma_p \omega - \omega_p^2} \tag{15}$$

where γ_p is the damping rate for the plasmons [3]. If we introduce the reduced variables

$$\mathbf{v}_0 \cdot \mathbf{k} - \omega = \omega_b(\kappa' + i\kappa'') \tag{16}$$

$$\Omega = \frac{\omega}{\omega_p} \tag{17}$$

then the dispersion relation becomes

$$(\kappa' + i\kappa'')^2 = \frac{\Omega - i\omega_p\tau_p\Omega^2}{\Omega - i\omega_p\tau_p(\Omega^2 - 1)} \tag{18}$$

where $\tau_p = 1/\gamma_p$ is the plasmon lifetime. The largest growth rate for the instability, that is, the maximum value of κ'', occurs near the plasma frequency, $\Omega \approx 1$. For weakly damped plasmons $(\omega_p\tau_p \gg 1)$ we get

$$\kappa' + i\kappa'' \approx \sqrt{1 - i\omega_p\tau_p} \approx \sqrt{\frac{\omega_p\tau_p}{2}}(1 - i). \tag{19}$$

The spatial growth rate of the amplitude of the instability is then

$$k'' \approx \frac{\omega_b}{\beta_0 c}\sqrt{\frac{\omega_p\tau_p}{2}} \tag{20}$$

where $\beta_0 = v_0/c$.

3. Limitations

Several limitations make it difficult to achieve gain in practice. These include scattering of the beam electrons by the nuclei in the metal and heating of the metal by the electron beam. The effects of scattering are dominated by multiple small-angle scattering events, and the rms change in the electron momentum $\mathbf{p}$ after N scattering events is

$$\langle |\Delta\mathbf{p}_N|^2 \rangle = N\langle |\Delta\mathbf{p}|^2 \rangle \tag{21}$$

where $\langle |\Delta\mathbf{p}|^2 \rangle$ is the rms momentum change in a single scattering event. For small-angle scattering of an electron off a nucleus of atomic number Z this is

$$\langle |\Delta\mathbf{p}|^2 \rangle = \frac{8\gamma_0^2 Z^2}{(\gamma_0^2 - 1)^2}\frac{r_e^2}{b_{max}^2}\ln\left(\frac{b_{max}}{b_{min}}\right) \tag{22}$$

in which $r_e = e^2/4\pi\varepsilon_0 mc^2$ is the classical electron radius, and b_{max} and b_{min} are the maximum and minimum values of the impact parameter [4]. The maximum impact parameter is roughly the radius of an atom in the metal, so that

$$\frac{4}{3}\pi b_{max}^3 n_Z = 1 \tag{23}$$

where n_Z is the number density of nuclei in the metal. The minimum impact parameter may be taken as the point where the small-angle scattering approximation breaks down,

$$\frac{\Delta p}{p_0} \approx \frac{2Zr_e}{\beta_0^2\gamma_0^2\beta_{min}} = 1. \tag{24}$$

Since it enters only logarithmically, the exact value is not critical. Deflection of the electron by scattering becomes important when the electron velocity in the direction of the beam changes enough that the electron is no longer in resonance with the plasma oscillations. This occurs when the delay $\Delta t \approx l_g|\Delta\mathbf{v}|/v_0^2$ in travelling one gain length $(l_g = 1/\kappa'')$ becomes comparable to the time for a plasma oscillation, that is $\omega_p\Delta t = 1$. Since the number of scattering events experienced by an electron traversing a foil of thickness L is $N = \pi b_{max}^2 Ln_Z$, we obtain, after some algebra, the restriction

$$2\kappa''L \leqslant \frac{\gamma_0^2\beta_0^5}{4\pi^2}\frac{\lambda_p}{l_g}\frac{1}{n_Z Z^2 r_e^2 l_g}\frac{1}{\ln(b_{max}/b_{min})} \tag{25}$$

on the overall *intensity* gain, where $\lambda_p = 2\pi c/\omega_p$ is the wavelength corresponding to the plasma frequency.

To illustrate this result with some numbers, we consider a beam of 300-keV electrons with a current density $J_b = 10^{13}$ A/m^2 incident on a lithium foil. We choose lithium because the low atomic number minimises the effects of scattering. In addition, the plasmon damping time is relatively long because the plasmon energy does not overlap inner-shell transition energies. This is related to the fact that lithium, like other alkali metals, exhibits ultraviolet transparency above the plasma frequency. The plasmon energy is 8.05 eV, which corresponds to a wavelength of 175 nm, and the plasmon damping time is 10 fs, so that $\omega_p\tau_p \approx 113 \gg 1$ [5]. The maximum total intensity gain is $2\kappa''L \approx 16$, corresponding to a foil thickness of 20 μm. The rms scattering angle at the exit of the foil is $\theta_{rms} = 0.13$ rad.

Because of the high intensity of the incident electron beam the foil is rapidly heated until it melts. The energy loss by the electron beam is dominated by ionization losses, for which the energy loss per unit length is

$$\frac{dW}{dz} = \frac{n_Z Z e^4}{4\pi \varepsilon_0^2 mc^2 \beta_0^2} \ln\left(\frac{\beta_0^2 \gamma_0 mc^2}{eV_t}\right) \qquad (26)$$

where $V_t \approx 40$ eV is the characteristic energy of the ionization processes in lithium [6,7]. In a 1-ps pulse the foil temperature rises about 600°C.

Finally, at such high electron-beam current density the effects of space charge might be expected to affect the focus of the electron beam on the foil. However, from the envelope equation we find that the length of the focal region is

$$Z_f = \sqrt{\frac{2\beta_0^3 \gamma_0^3 I_A}{J_b}} \qquad (27)$$

where $I_A = \varepsilon_0 mc^3/e =$ Alfven current/4π is a characteristic current [8]. The focal depth for the case we are considering is 20 μm, which is comparable to the foil thickness.

4. Conclusions

As shown above, the coherent amplification of volume plasmons in a metal requires a very intense electron beam, because of the natural damping of the plasmons. The overall gain is constrained by multiple scattering of the electrons in the beam by the nuclei in the metal, and the pulse length is restricted by energy deposition in the foil. Nevertheless, it should be possible to excite coherent plasmons with electron beams that are becoming available now. In particular, beams produced by pulsed-laser irradiated needle cathodes have demonstrated current density in excess of 10^{11} A/m^2 [9]. If this beam is demagnified by a factor of ten the required current density is produced. It remains to be demonstrated that this can be accomplished in picosecond pulses.

The excitation of coherent plasmons is of interest both as a fundamental process in metals and as a source of coherent ultraviolet radiation. The excitation of plasmons can be observed by examining the energy distribution of the electrons emerging from the metal. As coherence develops, the energy lost to the excitation of the plasmons will increase enormously. For 300-keV electrons passing through a 20-μm lithium foil, as discussed above, the mean energy lost to plasmon excitation is about 400 eV. In coherent excitation of plasmons the energy loss at saturation is limited by the width of the excitation resonance. The effect is the same as the resonance error due to scattering of the beam electrons in the metal, as discussed above, and following the same arguments we find that

$$\frac{\Delta\varepsilon}{\varepsilon} = \frac{\beta_0^3 \gamma_0^3}{\gamma_0 - 1} \frac{c}{\omega_p l_g}. \qquad (28)$$

For the conditions discussed above the energy loss corresponds to about 50 keV. At this degree of excitation the linearized equations discussed here are not actually valid, but coherent excitation of plasmons should not be difficult to observe.

In addition, the optical transition radiation produced by the electrons should be strongly enhanced by the coherent bunching of the electrons. Near the plasma frequency and above, the transition radiation is difficult to estimate due to the ultraviolet transparency of lithium, in the case considered. However, the total transition radiation is proportional to e^2, the square of the electronic charge. If we regard a 100% modulated beam as consisting of bunches of N_b electrons radiating as one charge, the enhancement compared with the incoherent beam is N_b. For a total electron-beam current of 100 mA, under the conditions used above, the enhancement is about 10^3. However, the radiation will be compressed into the spectral region around the plasma frequency, enhancing the spectral brilliance even further.

Acknowledgements

The author gratefully acknowledges many helpful discussions with Dr. Leonard Feldman.

References

[1] H. Raether, Excitation of Plasmons and Interband Transitions by Electrons, Springer, Berlin, 1980.
[2] C.A. Brau, unpublished.

[3] J.N. Hodgson, Optical Absorption and Dispersion in Solids, Chapman & Hall, London, 1970, pp. 62–65.

[4] J.D. Jackson, Classical Electrodynamics, 3rd Edition, Wiley, New York, 1998, pp. 640–645.

[5] E.D. Palik (Ed.), Handbook of Optical Constants of Solids, Academic Press, Boston, 1985.

[6] L.D. Landau, E.M. Lifshitz, L.P. Pitaevskii, Electrodynamics of Continuous Media, 2nd Edition, Pergamon, Oxford, 1984, pp. 394–398.

[7] Stopping Powers for Electrons and Positrons, ICRU Report 37, International Commission on Radiation Units and Measurements, Bethesda, MD, 1984.

[8] M. Reiser, Theory and Design of Charged-Particle Beams, Wiley, New York, 1994, pp. 191–202.

[9] C.A. Brau, C. Hernandez Garcia, Nucl. Instr. and Meth. A 475 (2001) 559, these proceedings.

ELSEVIER

Nuclear Instruments and Methods in Physics Research A 475 (2001) 454–457

NUCLEAR
INSTRUMENTS
& METHODS
IN PHYSICS
RESEARCH
Section A

www.elsevier.com/locate/nima

Two-color infrared FEL facility employing a 250-MeV linac injector of Saga synchrotron light source

T. Tomimasu[a,*], M. Yasumoto[b], N. Koga[c], Y. Hashiguchi[c],
Y. Ochiai[c], M. Ishibashi[c]

[a] Free Electron Laser Research Institute Inc. (FELI), c/o Sumitomo Electric Industries Ltd., 1-1-3, Simaya, Konohana,
Osaka 554-0024, Japan
[b] Osaka National Research Institute (ONRI), Osaka 554-0024, Japan
[c] Industry Promotion Division, Saga Prefectural Government, Osaka 554-0024, Japan

Abstract

A two-color infrared free electron laser (FEL) facility is proposed. This FEL facility will employ a new 250-MeV linac injector of the Saga synchrotron light source (SLS). The linac has two operation modes: short macropulse mode of 1 μs at 250 MeV is for injection to a 1.4-GeV storage ring and long macropulse mode of 13 μs at 40 MeV is for the two-color FEL facility. The two-color FEL uses a single electron beam and simultaneously provides both infrared (IR) and far-IR laser pulses for pump-probe studies of quantum-well structures and studying vibrational relaxation of molecules. The Saga SLS will be operated in 2004 to promote material science, bio-medical and industrial applications in Kyushu. © 2001 Published by Elsevier Science B.V.

1. Introduction

In the previous paper, we proposed four free electron laser (FEL) facilities employing a new 150-MeV linac injector of a Saga synchrotron light source (SLS) and discussed electron injectors suitable for FEL generation as well as the beam qualities of the available injectors for FEL and positron generations [1]. However, the plan has been improved by employing a new 250-MeV linac injector and a two-color FEL facility instead of a 150-MeV linac injector and the four FEL facilities.

The higher injection energy facilitates high-current beam storage and the two-color FEL facility enables pump-probe experiments in mid-IR and far-IR range.

We are planning to operate the two-color FEL facility [2] employing the 250-MeV linac injector of the Saga SLS for scientific researches and industrial applications in Tosu City, Saga Prefecture, Kyushu. Tosu City is 25 km north-east of Saga City and 25 km south of Fukuoka City. The project is promoted by the Science and Technology Agency and the Saga Prefectural Government.

The Saga SLS consists of a compact and low emittance 1.4-GeV storage ring with eight 3.05-m long straight sections for six insertion devices and the 250-MeV linac injector. We can

*Corresponding author. Tel./fax: +81-42-563-3826.
E-mail address: tomimasu@sweet.ocn.ne.jp, hcc01112@nif-tyserve.or.jp (T. Tomimasu).

0168-9002/01/$ - see front matter © 2001 Published by Elsevier Science B.V.
PII: S 0 1 6 8 - 9 0 0 2 (0 1) 0 1 6 2 9 - 1

supply high-brilliant photon beams covering wide wavelength range from 0.036 nm (34 keV) to 248 μm (0.005 eV) by using the Saga 1.4-GeV storage ring with a 7.5-T wiggler and the two-color FEL facility. It will be operated and managed by the Saga Prefectural Government before September 2004.

This paper describes the design study of the Saga 1.4-GeV SLS, the 250-MeV linac injector, and the two-color FEL facility.

2. Saga 1.4-GeV storage ring

The main parameters of the Saga 1.4-GeV storage ring are shown in Table 1. The storage ring consists of eight double-bend achromatic cells

Table 1
Main parameters of the Saga storage ring

Electron energy	0.6–1.4 GeV
Beam current and life	300 mA, 5 h
Circumference	75.4 m
Lattice and No. of cells	DBA (double bend achromat) and 8
Straight section for IDS	3.0 m
Emittance (ε)	12 nm rad
Tunes	$v_x = 7.1835$, $v_y = 2.2181$
Momentum compaction	0.008074
Energy spread	0.000672
Radiation loss	105.9 keV
Frequency, power and RF voltage	501.0 MHz, 90 kV & 260 kV
Harmonic number	126
Injection energy	250 MeV
Dipoles	11.25° edge focusing
Number	16
Bending radius and magnetic field	3.2 m & 1.459 T
Quadrupoles	
Number (16 QF1, 16 QD, 8 QF2)	40
Length	0.2 m. 0.2 m, 0.3 m
Maximum gradient	27 T/m
Sextupoles	
Number (16 SF, 16 SD)	32
Length	0.15 m, 0.06 m
Maximum strength	800 T/m^2

and has eight fold symmetry with eight straight sections and is of low-energy injection type. The circumference is 75.4 m and the emittance is 12 nm rad at the nominal beam energy 1.4 GeV. The scale is 84% of the MAX-II ring [3] and eight straight sections are 3.0-m long. These are used for six insertion devices and for a septum magnet and RF cavities. The 7.5-T wiggler is a three pole superconducting planar wiggler to shift the synchrotron radiation spectrum to the hard X-ray region (critical energy $E_c = 9.8$ keV), just like the 5-T wigger of the Electrotechnical Laboratory (ETL 0.8 GeV TERAS ring) [4] and the 7-T wiggler of the Louisiana State University, Center for Advanced Microstructure and Devices (LSU 1.5-GeV CAMD ring) [5].

In the early stage of the operation, we expect the stored beam current and its lifetime will be 300 mA at 1.4 GeV and 5 h, respectively. To this end, the injection energy has been increased from 150 to 250 MeV, considering that the injection energies of TERAS and CAMD are 310 and 200 MeV, respectively. In total, 20 beam ports are constructed and more than 20 beam lines can be installed. An HOM damped cavity with SiC beam-duct [6] will be used for stable storage of high-current beam.

3. Saga 250-MeV linac for electron beam injection and FEL oscillation

The Saga 250-MeV linac is operated in two modes; 1- and 13-μs macropulse operations. The schematic layout of the 250-MeV linac and the two-color FEL facility is shown in Fig. 1. The 250-MeV electron beam with macropulse length of 1 μs is for the storage ring injection, and the 40-MeV electron beam with macropulse length of 13 μs is for FEL oscillation. The 15–40-MeV beam is used at the two-color FEL facility, that is, FEL facilities 1 and 2.

The electron beam consists of a train of several ps, 0.6-nC microbunches repeating at 22.3125 or 89.25 MHz just like the Free Electron Laser Research Institute (FELI) linac [7]. The 1-μs macropulse operation mode at the 250 MeV is for electron injection. An electron charge of about

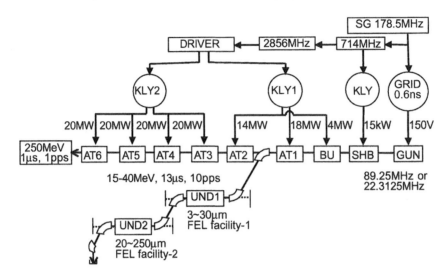

Fig. 1. Schematic layout of the 250-MeV linac and the two-color FEL facility.

12 nC is injected to the storage ring per second at 22.3125 MHz micropulse and 1 Hz macropulse operation. The beam energy will be ramped from 250 MeV to the operation energy after beam storage.

The Saga 250-MeV linac consists of a 120-keV thermionic triode gun, a 714-MHz prebuncher (sub-harmonic buncher, SHB), a 2856-MHz standing-wave type buncher, and six Electrotechnical Laboratory type waveguides. The accelerating waveguides (AT-1~AT-6) with a length of 2.93 m are of linearly narrowed iris type to prevent beam blow up (BBU) effect at high peak current acceleration [8].

An electron gun with a dispenser cathode (EIMAC Y646B) and a grid pulser (Kentech Instruments, Ltd. UK) usually emit 600-ps pulses of 2.3 A at 89.25 MHz. These pulses are compressed to 60 A × 10 ps by the prebuncher and the buncher.

The rf sources are a 714-MHz klystron (1VA88R, 15 kW) for the prebuncher and a 2856-MHz klystron (E3729, 36 MW) for the buncher and the first two accelerating waveguides. At the injection mode, a 2856-MHz klystron (E3712, 80 MW) is used for the following four accelerating waveguides. At this mode, the E3729 type klystron is operated at 36 MW in the 13 μs FEL mode and a 1 μs macropulse electron beam is

Table 2
Beam parameters of the Saga linac at injection and FEL application

At injection	Electron energy	250 MeV
	Energy spread (FWHM)	0.5%
	Peak current	150 A
	Beam radius	0.5 mm
	Normalized emittance	25×10^{-6} m rad
	Micropulse charge	0.6 nC
	Micropulse duration	4 ps
	Micropulse separation	44.8 ns
	Macropulse duration	~1 μs
	Macropulse repetition frequency	1 Hz
At FEL application	Electron energy	40 MeV
	Energy spread (FWHM)	~1%
	Peak current	100 A
	Beam radius	0.5 mm
	Normalized emittance	25×10^{-6} m-rad
	Micropulse charge	0.6 nC
	Micropulse duration	6 ps
	Micropulse separation	11.2 ns
	Macropulse duration	~13 μs
	Macropulse repetition frequency	10 Hz

accelerated to the end of the AT-6 (up to 250 MeV). At the FEL mode, the 13-μs macropulse electron beam is accelerated up to 40 MeV at the end of the AT-1. Table 2 shows the beam parameters of the Saga linac at injection and FEL application.

4. Two-color FEL facility

Simultaneous lasing in the mid-IR range has been achieved at CLIO with two same undulators connected in series by changing their gaps [9] where a single beam and a common optical cavity is used. Recently, their color difference is improved to the order of 5 μm and two-color FELs are delivered to the user simultaneously. Quasi-simultaneous FEL operation in the mid-IR and far-IR range has been achieved at the Stanford FEL Center [10] where interleaved electron beams of different energies are accelerated by a single linac, separated and then used to independently drive the mid-IR facility and the far-IR facility. They lase on alternate macropulses and the optical beams are delivered to two users. FELI has achieved the first simultaneous two-color oscillations in the mid-IR and far-IR ranges by using the FELI linac, beam transport systems, and the two FEL facilities [2].

In the previous experiment at the FELI, the rise-times of the mid-IR FEL and far-IR FEL macropulses are 5 and 9 μs, respectively. Therefore, the 13-μs macropulse beam at 15–40 MeV facilitates mid-IR FEL and far-IR FEL generation and their applications. An expected micropulse FEL energy is 30 μJ and an average power is 2.6 kW per macropulse at the FEL facilities-1 and -2. Simultaneous two-color FELs on the same electron macropulse are used for performing pump-probe studies of quantum-well structures [11] and for studying vibrational relaxation of molecules. A schematic layout of the two-color FEL facility including undulators and optical cavities are shown in Fig. 1. The two-color FEL facility can supply high-power-density photon beams of 10-GW/cm^2 level covering the wave-length range from 3.1 μm (0.4 eV) to 248 μm (0.005 eV).

5. Conclusion

The two-color FEL facility employing the new 250-MeV linac injector of the Saga SLS has been designed. The Saga 250-MeV linac has two operating modes: short macropulse mode of 1 μs at 250 MeV is for injection to a 0.6–1.4-GeV storage ring and long macropulse mode of 13 μs at 15–40 MeV is for the two-color FEL generation to perform pump-probe studies of quantum-well structures and to study vibrational relaxation of molecules.

References

[1] T. Tomimasu, M. Yasumoto, S. Sato,Y. Ochiai, M. Ishibashi, T. Soda, Free Electron Laser in Asia, KAERI, Taejon, June 8–10, 1999, p. 166.

[2] A. Zako, Y. Kanazawa, Y. Konishi, S. Yamaguchi, A. Nagai, T. Tomimasu, Nucl. Instr. and Meth. A 429 (1999) 136.

[3] A. Andersson, M. Eriksson, L.-J. Lindgren, P. Rojsel, S. Werin, Nucl. Instr. and Meth. A 343 (1994) 644.

[4] S. Sugiyama, H. Ohgaki, K. Yamada, T. Mikado, M. Koike, T. Yamazaki, S. Isojima, C. Suzawa, T. Keishi, J. Synchrotron Rad. 5 (1998) 437.

[5] V.M. Borovikov, B. Craft, M.G. Fedurin, V. Jurba, V. Khlestov, G.N. Kulipanov, O. Li, N.A. Mezentsev, V. Saile, V.A. Shkaruba, J. Synchrotron Rad. 5 (1998) 440.

[6] T. Koseki, M. Izawa, Y. Kamiya, Technical Report of ISSP No.2980, May 1995, pp. 1–8.

[7] T. Tomimasu, E. Oshita, Y. Kanazawa, A. Zako, Nucl. Instr. and Meth. A 429 (1999) 141.

[8] T. Tomimasu, IEEE Trans. NS-28 (3) (1981) 3523.

[9] D. Jaroszynski, et al., Phys. Rev. Lett. 74 (1995) 2224.

[10] K.W. Berryman, T.I. Smith, Nucl. Instr. and Meth. A 375 (1996) 6.

[11] F. Glotin, J.M. Ortega, R. Prazares, C. Rippon, Nucl. Instr. and Meth. B 144 (1998) 8.

ELSEVIER

Nuclear Instruments and Methods in Physics Research A 475 (2001) 458–461

NUCLEAR
INSTRUMENTS
& METHODS
IN PHYSICS
RESEARCH
Section A

www.elsevier.com/locate/nima

Field emission from entangled carbon nanotubes coated on/in a hollow metallic tube

Yoshiko Tokura[a], Yoshiaki Tsunawaki[a,*], Nobuhisa Ohigashi[b], Seiji Akita[c],
Yoshikazu Nakayama[c], Kazuo Imasaki[d], Kunioki Mima[e], Sadao Nakai[e]

[a] Department of Electrical Engineering and Electronics, Osaka Sangyo University, 3-1-1 Naka-gaito, Daito, Osaka 574-8530, Japan
[b] Department of Physics, Kansai University, 3-3-35 Yamate-cho, Suita, Osaka 564-8680, Japan
[c] Department of Physics and Electronics, Osaka Prefecture University, 1-1 Gakuen-cho, Sakai, Osaka 599-8531, Japan
[d] Institute for Laser Technology, 2-6 Yamadaoka, Suita, Osaka 565-0871, Japan
[e] Institute of Laser Engineering, Osaka University, 2-6 Yamadaoka, Suita, Osaka 565-0871, Japan

Abstract

Field emission properties of entangled carbon nanotubes were studied for an electron beam source of Cherenkov or Smith–Purcell free electron laser. The cathode was made of carbon nanotubes which were mixed with a very small amount of resin and coated on/in a hollow metallic tube with outer diameter of 0.5 mm. The emission current was as high as 2.2 mA with a fluctuation of <4%. It seems that some entangled nanotubes were frayed under the high electric field and then electrons were emitted mainly from their tips. Reduction of the work function of the carbon nanotubes was observed with the degradation of vacuum pressure in the experimental apparatus. © 2001 Elsevier Science B.V. All rights reserved.

PACS: 41.60.Cr; 41.75.Fr; 52.59.Rz

Keywords: Carbon nanotube; Multi-walled nanotube; Field emission; Field emitter; Electron source

1. Introduction

It is well known that field emitted electrons from a very fine needle have very small energy spread and very high brightness [1]. These properties are very interesting in an electron beam source for a free electron laser (FEL) [2], gyrotron [3], monotron [4] and so on. We have also been studying a field emission cathode of tungsten wire [5] for a

Cherenkov or Smith–Purcell FEL which are expected to be sufficiently small to be used in a typical laboratory unscreened from X-ray radiation. It is, however, very difficult to achieve stable emission with DC current higher than 100 μA from a tungsten tip probably because the vacuum in the chamber degrades at higher current.

After carbon nanotubes were discovered in 1991 [6], much research has been devoted to them as good, small electron beam sources especially for flat panel displays where an array of parallel nanotips is required [7–10]. From the viewpoint of only high current electron source for FEL, it seems that the requirement of parallelism of nanotubes is

---

*Corresponding author. Tel.: +81-72-875-3001; fax: +81-72-870-8189.

E-mail address: ytsuna@elec.osaka-sandai.ac.jp
(Y. Tsunawaki).

0168-9002/01/$ - see front matter © 2001 Elsevier Science B.V. All rights reserved.
PII: S 0 1 6 8 - 9 0 0 2 (0 1) 0 1 6 2 8 - X

not always necessary. In this work, we have studied some elementary properties of electron beam current from entangled carbon nanotubes in/on a hollow metallic tube and then obtained as high as 2.2 mA.

2. Experimental

Carbon nanotubes were synthesized by a usual DC carbon arc in an atmosphere of helium gas. We prepared a cathode containing them mixed with a very small amount of resin in/on a hollow stainless tube (outer- and inner-diameter of 0.5 and 0.25 mm, respectively). Figs. 1 and 2 show the front surface of the cathode and its micrograph, respectively. It is seen in Fig. 2 that carbon nanotubes are entangled.

The field emission experiments were carried out in a vacuum chamber (inner diameter of 146 mm, height of 170 mm) evacuated to $\sim 3 \times 10^{-9}$ Torr. The apparatus consists of only a carbon nanotube cathode and an anode of molybdenum plate with 30×30 mm^2. The anode was grounded. The distance between their electrodes was adjusted to be about 100, 200 and 400 μm.

3. Results and discussion

Both the cathode and anode current were measured. They were almost the same within experimental error. This fact means that the

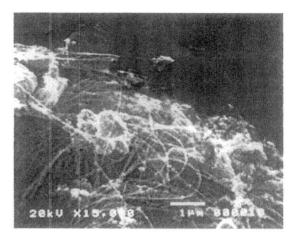

Fig. 2. Micrograph of entangled carbon nanotubes.

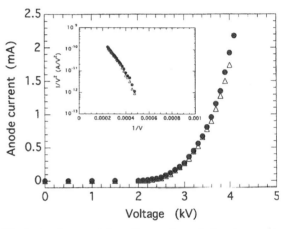

Fig. 3. Anode current depending on the cathode voltage with inset the corresponding Fowler–Nordheim plot.

emitted electrons do not arrive at the wall of the chamber but entirely at the anode. Fig. 3 shows the anode current dependence on the cathode voltage when the distance between the anode and cathode is ~ 200 μm. Solid circle and open triangle show the results obtained on different days. It is seen that the current starts up at the threshold voltage of ~ 2 kV and increases with the cathode voltage. The current of 2.2 mA is significantly higher than that from usual cathode of carbon nanotubes. The data is also shown in Fig. 3 as $\log(I/V^2)$ vs. $1/V$ where I and V are anode current and cathode voltage, respectively. They fall on a

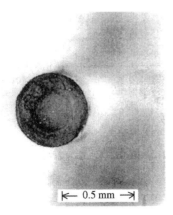

Fig. 1. Carbon nanotubes coated in/on a hollow stainless tube.

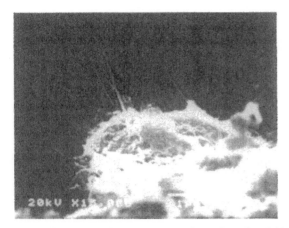

Fig. 4. Micrograph of carbon nanotubes after the field emission.

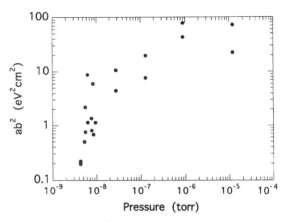

Fig. 5. Values of ab^2 depending on the vacuum pressure.

slightly curved line rather than a straight line. It is probably due to the nonuniform local field depending on the radius of nanotube tip [11]. A micrograph of the cathode tip after the measurements is shown in Fig. 4. It is seen that some needles stand up towards the anode. It seems that entangled tubes were frayed by the high electric field and then electrons were emitted mainly from the tip of each carbon nanotube.

As is well known, Fowler–Nordheim (F–N) equation is written as follows [12]:

$$\frac{I}{V^2} = 1.4 \times 10^{-6} S \frac{\beta^2}{\phi}$$
$$\times \exp\left[9.87\frac{1}{\sqrt{\phi}}\right] \exp\left[-6.53 \times 10^7 \frac{\phi^{3/2}}{\beta}\frac{1}{V}\right]$$
$$= a \exp\left[-b\frac{1}{V}\right] \tag{1}$$

where S, and β and ϕ are the emission area (cm^2), the geometrical factor (1/cm) and the emitter work function (eV), respectively. Eq. (1) leads to

$$ab^2 = 5.97 \times 10^9 \, S\phi^2 \exp\left[9.87\frac{1}{\sqrt{\phi}}\right]. \tag{2}$$

After the emission current vs. cathode voltage were measured in various chamber pressures, the value of ab^2 was estimated from F–N fitting by the method of least squares. Fig. 5 shows its depen-

dence on the vacuum pressure of the chamber. It is observed that ab^2 increases with the vacuum pressure. According to Eq. (2), under the condition that S does not change, the value of ab^2 is minimum at $\phi = 6.09$ eV, i.e. it increases with the decrement or increment of ϕ from 6.09 eV. If ϕ of the carbon nanotube is assumed to be 5 eV (same as pure carbon), it might be said that the work function of carbon nanotube reduces with the vacuum pressure due to the adsorption of atmospheric gas molecules on the cathode surface. This phenomenon seems to be in agreement with the result obtained by Dean et al. [13] using single-walled carbon nanotubes. However, the F–N equation might not be valid directly for carbon nanotubes because the radius of the nanotubes is comparable to the tunneling barrier [14]. Furthermore, it also seems that F–N relation is not strictly applicable for entangled carbon nanotubes which give the ensemble current as sum of the contributions from each nanotube with different properties.

It is expected that the vacuum degradation and the increment of emission current (I) bring fluctuation of the current (δI). It was, however, observed as shown in Fig. 6 that the relative fluctuation ($\delta I/I$) decreased with the product of vacuum pressure (P) and current. $\delta I/I$ was much higher for $PI < 10^{-9}$ Torr mA due to the low S/N ratio based on very small current near the threshold. On the other hand, it was lower than 0.04 for $PI > 10^{-8}$ Torr mA.

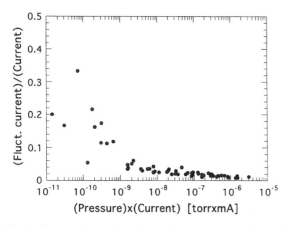

Fig. 6. Current fluctuation depending on the product of pressure and current.

The observation of the electron beam image on a fluorescent screen suggested that electrons were emitted from insular portions of the cathode which were linked by creation/expansion of emission area with the increase of the cathode voltage accompanying the enlargement of brightness at each emission site.

4. Conclusions

We have performed very high field emission current with small fluctuation from entangled carbon nanotubes pasted in/on a hollow stainless tube: 2.2 mA with fluctuation of <4%. It is considered that some nanotubes were frayed from entangled tubes under the high electric field and then electrons were emitted mainly from their tips. Reduction of the work function of the carbon nanotubes was observed with the degradation of vacuum pressure in the experimental apparatus. Further studies are aimed at homogeneous emission with higher DC current for the experiments of a Cherenkov or Smith–Purcell FEL.

Acknowledgements

We would like to thank Dr. T. Taguchi of Setsunan University for his helpful discussion on the field emission experiment. This work was supported in part by a grant-in-aid for Scientific Research from the Ministry of Education, Science, Sports and Culture of Japan.

References

[1] W.P. Dyke, W.W. Dolan, Adv. Electron. Phys. 8 (1956) 90.
[2] C.A. Brau, Nucl. Instr. and Meth. A 393 (1997) 426.
[3] M. Garven, S.N. Spark, A.D.R. Phelps, Digest of 19th International Conference on Infrared Mill, Waves, Sendai, Japan, 1994, p. 472.
[4] K. Yokoo, T. Ishihara, Int. J. Infrared Mill. Waves 18 (1997) 1151.
[5] Y. Tsunawaki, Y. Tokura, M. Kusaba, N. Ohigashi, K. Mima, K. Imasaki, S. Nakai, M. Shiho, Nucl. Instr. and Meth. A 429 (1999) 299.
[6] S. Iijima, Nature 354 (1991) 56.
[7] Y. Saito, S. Uemura, K. Mamaguchi, Jpn. J. Appl. Phys. 37 (1998) L346.
[8] D.N. Davydov, P.A. Sattari, D. AlMawlawi, A. Osika, T.L. Haslett, J. Appl. Phys. 86 (1999) 3983.
[9] J.S. Suhand, J.S. Lee, Appl. Phys. Lett. 75 (1999) 2047.
[10] H. Murakami, M. Hirakawa, C. Tanaka, H. Yamakawa, Appl. Phys. Lett. 76 (2000) 1776.
[11] Y. Nakayama, S. Akita, Synthetic Metal 117 (2001) 207.
[12] C.A. Spindt, I. Brodie, L. Humphrey, E.R. Westerberg, J. Appl. Phys. 47 (1976) 5248.
[13] K.A. Dean, B.R. Chalamada, Appl. Phys. Lett. 76 (2000) 375.
[14] W. Zhu, C. Bower, O. Zhou, G. Kochanski, S. Jin, Appl. Phys. Lett. 75 (1999) 873.

ELSEVIER

Nuclear Instruments and Methods in Physics Research A 475 (2001) 462–469

NUCLEAR
INSTRUMENTS
& METHODS
IN PHYSICS
RESEARCH
Section A

www.elsevier.com/locate/nima

Utilization of CTR to measure the evolution of electron-beam microbunching in a self-amplified spontaneous emission (SASE) free-electron laser (FEL)

A.H. Lumpkin*, B.X. Yang, W.J. Berg, Y.C. Chae, N.S. Sereno, R.J. Dejus, C. Benson, E. Moog

Advanced Photon Source, Argonne National Laboratory, Building 401, 9700 South Cass Avenue, Argonne, IL 60439, USA

Abstract

We report on the first measurements of the z-dependent evolution of electron-beam microbunching as revealed through coherent transition radiation (CTR) measurements in a visible self-amplified spontaneous emission free-electron laser experiment. The increase in microbunching was detected by tracking the growth of the visible CTR signals as generated from insertable metal mirrors/foils after each of the last three undulators. The same optical imaging diagnostics that were used to track the z-dependent intensity of the undulator radiation (UR) were also used to track the electron beam/CTR information. Angular distribution, beam size, and intensity data were obtained after each of the last three undulators in the five-undulator series, and spectral information was obtained after the last undulator. The exponential growth rate of the CTR was found to be very similar to that of the UR and consistent with simulations using the code GENESIS. © 2001 Elsevier Science B.V. All rights reserved.

1. Introduction

The interest in self-amplified spontaneous emission (SASE) free-electron laser (FEL) physics has been growing in recent years. This has been spurred by some early successes in the cm-wave and far-infrared (FIR) regime [1,2] and the speculation that the process should scale to the X-ray regime [3,4]. Fundamental to the process is the longitudinal microbunching of the electron beam (e-beam) as it copropagates with the undulator radiation (UR) along the length, z, of the undulators [5–8]. As the fraction of micro-

bunched beam increases along z, so does the coherent radiation that leads to the exponential gain regime. The UR has been routinely measured in several experiments [2,9], but the *complementary* information in the e-beam has only been measured at the exit of the last (only) undulator in the UCLA/LANL SASE experiment at 13 μm via coherent transition radiation (CTR) [10]. Additionally, a more direct time-domain experiment at Stanford University was able to display the longitudinal striations in the charge distribution at $\lambda = 60$ μm, the wavelength of their oscillator experiment [11]. These experiments were reported in the last two international FEL conferences. The first experiment used the properties of coherent transition radiation generated by the SASE-induced microbunching to test their models [10].

*Corresponding author. Tel.: +1-630-252-4879; fax: +1-630-252-4732.

E-mail address: lumpkin@aps.anl.gov (A.H. Lumpkin).

0168-9002/01/$ - see front matter © 2001 Elsevier Science B.V. All rights reserved.
PII: S 0 1 6 8 - 9 0 0 2 (0 1) 0 1 6 8 8 - 6

We have taken advantage of an operating SASE FEL in the visible wavelength regime (near 530 nm) to extend the CTR-based techniques to z-dependent angular distribution, beam size, and intensity measurements. The z-dependent growth in intensity is particularly of interest since we find the effective gain length for the UR and CTR to be so similar, as expected from the microbunching model. In addition, we have obtained hundreds of CTR angular distribution images that exhibit structures that strongly suggest interference effects in a two-surface geometry, electron-beam/photon-beam overlap effects, and electron-beam pointing information.

These experiments were the culmination of our goals to generate sub-0.5-ps e-beam bunches in a thermionic RF gun resulting in 100-A peak current, to measure such short pulses with a CTR interferometer (and even do this nondestructively with coherent diffraction radiation (CDR)) to characterize the core beam brightness, to demonstrate SASE gain with such a beam, and then to make the microbunching measurements using CTR. The first three goals are addressed in more detail in two complementary papers in these proceedings [12,13]. This paper will focus on the z-dependent microbunching measurements. An estimate of CTR photon intensity has been given by the authors of Ref. [10] (with additional background in Refs. [14,15]). Integration of their equations over angular space results in

$$N_{\mathrm{CTR}} \approx \frac{\alpha (N_{\mathrm{b}} b_{\mathrm{n}})^2}{4\sqrt{\pi} k_{\mathrm{r}} \sigma_z} \left(\frac{\gamma}{n k_{\mathrm{r}}} \right)^4 \left(\frac{\sigma_x^2 + \sigma_y^2}{\sigma_x^3 \sigma_y^3} \right) \qquad (1)$$

where α is the fine structure constant, N_{b} is the number of electrons, b_{n} is the bunching fraction, k_{r} is the radiation wave number (and thus the beam modulation wave number), n is the harmonic number, $\sigma_{x,y}$ are the transverse beam sizes, and σ_z is the longitudinal size. It is noted that higher charge intensities and smaller transverse size would enhance CTR production. It is also noted that since our visible wavelength is 20 times shorter than the 13-µm experiment, the number of electrons in a cubic wavelength volume is relatively reduced. However, the fact that the radiation is enhanced at the visible fundamental wavelength (and harmonics) opens up the experiment to imaging diagnostics.

2. Experimental background

These experiments were performed at the Advanced Photon Source (APS) during operations with a thermionic RF gun beam accelerated to 217 MeV by the S-band linac normally used as part of the injector system for the 7-GeV storage ring [16,17]. This accelerator and diagnostics have been described previously [18]. This beam is transported to the set of five undulators in the Low-Energy Undulator Test Line (LEUTL) tunnel. A schematic of the experiment is given in Fig. 1. It is not to scale, and there is an approximately 40-m transport line between the three-screen emittance station and the entrance of the undulators.

Schematic of APS SASE FEL Experiment

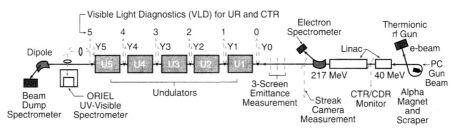

Fig. 1. A schematic of the APS SASE FEL experiment. The e-beam gun sources, the linac, the CTR/CDR interferometer station, the e-beam spectrometers, and the undulators with some diagnostics are indicated.

In this case, the RF gun cavity was operated with a higher gradient than that used for the injection operations, and it was optimized for peak current by using a CTR interferometer located after the first accelerator at ~40 MeV energy [12]. This optimization process was critical to the eventual success of the experiments. The RF gun includes an α magnet to inject the beam into the linac (see Fig. 1). We determined that there is a sensitive balance of RF gun gradient, gun current, α-magnet current, α-magnet scraper setting (low energy), and RF phasing to be investigated. A scan of the α-magnet current was done first to find the peak CTR signal. The CTR signal depends directly on N_b^2 and inversely on the bunch length. The position of the low-energy scraper in the α-magnet chamber was also scanned, and the beam relative intensity from a RF beam position monitor (BPM) sum signal and the CTR signal recorded. A clear N_b^2 dependence of the CTR for the core of the beam was observed. The scraper also rejects the low energy tail of the beam distribution with its inferior emittance. The interferometer scans were performed with several scraper settings to verify whether the core had been selected in the longitudinal profile. A bunch length of ~450 fs (FWHM) was calculated as described in Ref. [12]. An example of the longitudinal profile is shown in Fig. 2. The 450-fs (FWHM) distribution

is almost an order of magnitude shorter than the APS photocathode (PC) gun's micropulse also used in SASE experiments this year [9], but it also has correspondingly less charge. Since the slippage length ($N\lambda$) is about 180 μm in the visible experiment, the RF gun micropulse length is comparable to it. The charge was measured in a downstream Faraday cup, and the peak current determined to be about 100 A.

The beam was then transported to the LEUTL tunnel [19], where a set of five undulators with a total magnetic structure length of 12.0 m is installed. The properties of the undulators and the diagnostics stations have been previously described [20]. Very briefly, the undulator cells have a period of 3.3 cm and a length $L = 2.4$ m. They have a fixed gap with a field parameter $K = 3.1$. There is roughly a 0.38-m space between each undulator. This space is used for diagnostics and focusing and steering elements before the first undulator and after each of the five installed undulator sections. A schematic of these stations is shown in Fig. 3. The screens on the first actuator include positions for a YAG/mirror, a mirror at 45°, and a thin (6-μm) Al foil mounted with its surface normal to the beam direction (in the last three stations). This thin foil serves two functions:

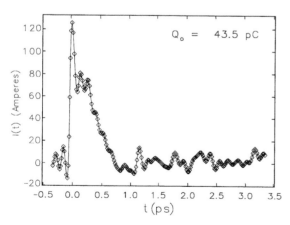

Fig. 2. An example of the electron-beam longitudinal profile as calculated from the CTR autocorrelation. A bunch length of ~450 fs (FWHM) is indicated.

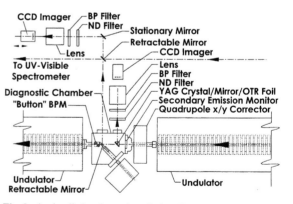

Fig. 3. A detailed schematic of the diagnostics and beam-matching station before and after each undulator. The first actuator includes positions for a YAG:Ce crystal plus 45° mirror, a 45° mirror, and a thin Al foil (after the last three undulators). The second actuator holds a mirror whose surface is at 45° to the beam direction. A digital camera with ND filters and BP filters is used to view each inserted screen.

(1) to block the strong, visible UR, and (2) to generate optical transition radiation (OTR) or CTR as the e-beam transits the foil/vacuum interfaces. A digital camera views the e-beam images from the YAG and the reflected undulator radiation from the mirror. The second actuator, located 63 mm downstream, involves a retractable mirror at 45° to the beam direction and another digital camera and lens for both near-field and far-field (focus at infinity) imaging. This visible light detector (VLD) system provides both beam size and angular distribution data, respectively. Both neutral density (ND) filters and bandpass (BP) filters are selectable by two filter wheels in front of the cameras.

Since the UR is more than 500 times brighter than OTR/CTR after undulator 5, the thin opaque foil was critical for separating the two sources of radiation. Bandpass filters could not be used to see the enhanced CTR with UR present because it is predicted to be at or near the fundamental wavelengths of the SASE [10]. The camera system that views the 45° mirror can thus be used to obtain the UR and the off-axis radiation out to ±10 mrad as well as the OTR/CTR radiation. For OTR, the opening angle is expected at ±1/γ = 2.3 mrad where γ is the Lorentz factor. The estimate of the diffraction-limited CTR cone angle [10] for an axisymmetric beam of size σ (100 μm) is $\theta_d = (\sqrt{2} nk_r\sigma)^{-1} = 0.6$ mrad. Since the spacing between the foil and the second metal mirror is similar to the formation length $\gamma^2\lambda \approx 90$ mm, we actually see interference phenomena in the CTR signals in analogy to OTR interferometers [21,22].

At the end of the last undulator, a remotely controlled pickoff mirror can be used to redirect the UR or CTR to an Oriel UV-visible spectrometer. Spectral effects were also observed in the angular distribution data using bandpass filters. The basic experiments involved the acquisition of digital images of the UR after each of the undulators. After verification, there was noticeable z-dependent intensity growth in the UR, the CTR data were recorded by inserting the thin Al foil at each station. Both near-field and far-field data were taken so that e-beam source size and the angular distribution data were available.

3. Microbunching results and discussion

3.1. Experimental results (z-dependent)

The evolution of the e-beam's microbunching is revealed in the evolution of the angular distribution images at stations VLD 0, 3, 4, 5. The VLD 0 station is unique in that it has no undulator before it and no installed thin, blocking foil in the Y0 location. In Fig. 4a, the dim angular distribution image of the OTR from the single 45° mirror at VLD 0 is shown. The opening cone angle is consistent with the expected 1/γ = 2.3 mrad for the 217-MeV beam. However, at the end of undulator 3, we see striking changes in the Fig. 4b VLD 3 image (also with focus at infinity). The inner lobes are at about ±1.9 mrad, and there are two vertically localized hot spots on the inner lobes. There are weak indications of other vertical fringes at −4.4 mrad. In Fig. 4c, the VLD 5 image indicates an asymmetric evolution since only the lower vertical peak is enhanced by about a factor 40 over the VLD 3 peak. Note the neutral density filter (ND) change from 0.0 to 1.0 and the vertical intensity scale change from 100 to 200 between the VLD 3 (4b) and VLD 5 (4c) data, respectively.

Initial assessments of these and other images support the interpretation that the fringes are indeed due to the interference of the two CTR sources, the forward CTR from the thin foil and the backward CTR from the 45° pickoff mirror. The vertical angle fringe locations can be reproduced within about 10% by our existing OTR interferometer code. The breaking of the azimuthal symmetry of the CTR images is related to the nonaxisymmetric e-beam transverse size (2–3 times larger horizontally) and to the misalignment of the beams. In Fig. 5, the calculated interference pattern is shown for a beam divergence of 0.4 mrad, wavelength interval of 500–580 nm, and foil spacing of 63 mm. Since no known polarized optical component is in front of the camera, the angular distribution would involve the contribution from both parallel (ipar) and perpendicular (iperp) polarization components or the total (w) signal indicated by the thin, solid-line curve. For completeness, the curve denoted tau is the difference over the sum of the two polarization

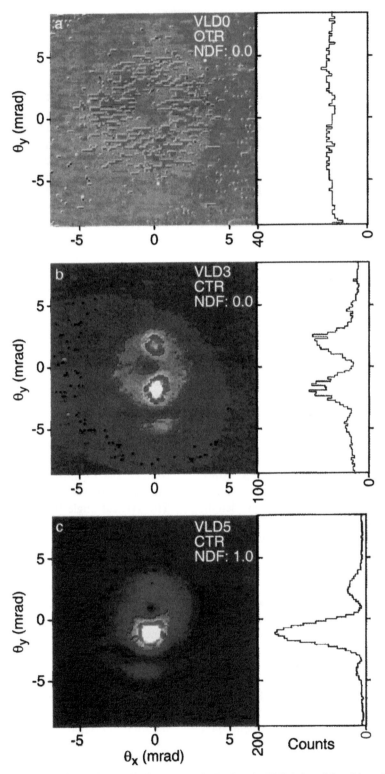

Fig. 4. A composite of OTR and CTR angular distribution images obtained at the VLD 0, 3, and 5 positions. Note the ND filter values and the changes in image intensity along the z-direction of the undulators. The θ_x and θ_y distributions are indicated with a y-intensity profile (sampled through the center of the pattern) displayed on the right hand side.

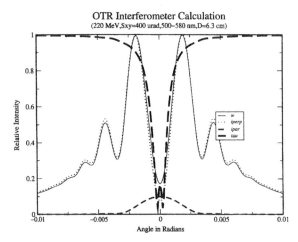

Fig. 5. A calculated OTR interferogram that reproduces the fringe pattern observed in the angular distribution images.

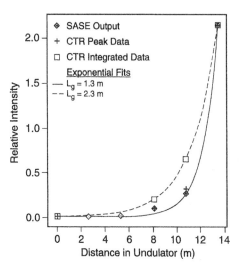

Fig. 6. The z-dependence of the SASE signal, CTR peak signal, and angle-integrated CTR signal with the thermionic RF gun beam. The SASE UR is about 200 times brighter than the CTR, but growth rates are very similar as expected from microbunching models.

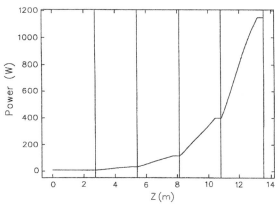

Fig. 7. GENESIS code results for the growth of UR along the undulators using the measured properties of the thermonic gun beam as input.

components. Full symmetry would be expected in the OTR and anticipated in the CTR *if* the longitudinal microbunching occurs uniformly in the transverse slice. This is evidently not the case.

Processing of the 100 images taken at each z location was performed and the averaged intensity growth tracked. In Fig. 6, we see the clear exponential signature corresponding to a gain in length $L_g = 1.3$ m in the UR (SASE) signals represented by the diamond symbol. The ~ 500 times weaker CTR signal (plus symbol) has a very similar gain length fitted at 1.4 m. The CTR angle-integrated (and wavelength integrated) data (square symbol) show the longer gain length of 2.3 m. We attribute the bright peaks in the CTR images to the microbunching on the fundamental at 537 nm.

3.2. GENESIS code results

We also used the code GENESIS [23] to calculate the z-dependence of the bunching fraction with the e-beam parameters ($\varepsilon_{n_x} = 16\pi$ mm mrad, $\varepsilon_{n_y} = 8\pi$ mmmrad, $I_{peak} = 100$ A, and $\sigma_E = 0.1\%$) and an assumed match into the undulators. Fig. 7 shows a plot of the exponential growth of the UR. The approximate gain length is consistent with the data of Fig. 6. In Fig. 8, the calculated bunching fraction of 0.2% at the end of

undulator 5 with an increase by a factor of 8 after undulator 1 is also consistent with the CTR growth using Eq. (1). Additionally, the observed e-beam transverse size of $\sigma_x = 200\,\mu m$, $\sigma_y = 120\,\mu m$ using the CTR data after undulator 5 is smaller horizontally than expected, and this

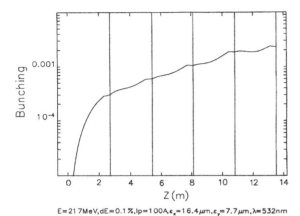

E=217MeV,dE=0.1%,Ip=100A,ε_x=16.4μm,ε_y=7.7μm,λ=532nm

Fig. 8. GENESIS code results for the evolution of the e-beam microbunching fraction along the undulators using the measured properties of the thermionic gun beam as input.

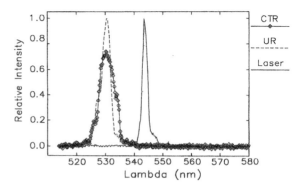

Fig. 9. CTR spectrum (diamonds) and UR spectrum (dashes) obtained after undulator 5 using the microbunched PC gun beam. The narrowband CTR spectrum is at the SASE fundamental wavelength near 530 nm, but it is about 200 times weaker in intensity based on the ND filters employed. The alignment laser spectrum (solid line) centered at 543.5 nm is also displayed in the plot for reference purposes.

may be related to the enhanced emissions of the transverse core of the beam. An average gain length of ~1.6 m is calculated over all five undulators, which is in reasonable agreement with the experiment and the value of 1.3–1.6 m based on Xie's parameterization of SASE gain [24]. The sub-0.5-ps electron bunch also leads to a GENESIS-calculated narrow optical pulse that may have relevance to eventual user experiments.

3.3. UV-visible spectrometer data

On a separate run with PC-gun beam in July 2000, we have obtained the evidence of spectral narrowing. OTR has a broad-band spectrum through the visible regime, but under the SASE gain conditions, we see distinct narrow-band, CTR emissions at the fundamental, consistent with Eq. (1) when the thin foil is inserted after undulator 5. Fig. 9 shows an example of the UR spectrum and the CTR spectrum. The ND filters at the entrance of the spectrometer were used to attenuate, by at least 100, the brighter UR signal. There is an indication that the CTR spectral width is a little larger than that of UR. Due to the shot-to-shot energy jitter, there is no definitive evidence of a small centroid shift as reported by the authors of Ref. [10].

4. Summary

In summary, we have evidence for the direct verification of the bunching fraction evolution using CTR along the undulators in a visible light SASE FEL experiment. The intensity growth rate in the CTR signal is similar to the UR growth rate as predicted. In addition, the angle-resolved data images reveal for the first time interference effects as well as localized, hot spots (peaks) in the distribution. These features may be used to optimize the critical coalignment of the e-beam and photon beam and thus to optimize FEL performance. We look forward to further experiments at higher gain with nine undulators and development of a more complete model of these effects.

Acknowledgements

The authors acknowledge discussions with J. Rosenzweig (UCLA) and A. Tremaine (UCLA) on FIR CTR issues; the assistance of M. Erdmann and T. Powers; the comments of J. Lewellen and S. Milton; and the support provided by G. Decker, P. Den Hartog, K.J. Kim, and J. Galayda (all from the APS). This work is supported by the US

Department of Energy, Office of Basic Energy Sciences, under contract No. W-31-109-ENG-38.

References

[1] T.J. Orzechowski, et al., Nucl. Instr. and Meth. A 250 (1986) 144.

[2] M. Hogan, et al., Phys. Rev. Lett. 81 (1998) 4897.

[3] K.-J. Kim, Nucl. Instr. and Meth. A 393 (1997) 147.

[4] J. Rossbach, Recent SASE free electron laser results, Proceedings of the EPAC 2000, Vienna, June 25–30, 2000, and references therein, in preparation.

[5] Y.S. Derbenev, A.M. Kondratenko, E.L. Saldin, Nucl. Instr. and Meth. A 193 (1982) 452.

[6] R. Bonifacio, C. Pellegrini, L. Narducci, Opt. Commun. 50 (1984) 373.

[7] K.-J. Kim, Nucl. Instr. and Meth. A 250 (1986) 396.

[8] J.-M. Wang, L.H. Yu, Nucl. Instr. and Meth. A 250 (1986) 484.

[9] S.V. Milton, et al., Phys. Rev. Lett. 85 (2000) 988.

[10] A. Tremaine, et al., Phys. Rev. Lett. 81 (1998) 5816.

[11] Kenneth N. Ricci, Todd I. Smith, Phys.Rev.SpecialTopics-Accelerators, Beams, Vol. 3 (2000) 032801-1.

[12] A.H. Lumpkin, et al., Development of a coherent transition radiation-based bunch length monitor with application to the APS rf thermionic gun beam optimiza-tion, Nucl. Instr. and Meth. A 475 (2001) 476, these proceedings.

[13] A.H. Lumpkin et al., First measurements of sub-picose-cond electron-beam structure by autocorrelation of coher-ent diffraction radiation, Nucl. Instr. and Meth. A 475 (2001) 470, these proceedings.

[14] Y. Shibata, et al., Phys. Rev. E 50 (1994) 1479.

[15] J.B. Rosenzweig, G. Travish, A. Tremaine, Nucl. Instr. and Meth. A 365 (1995) 255.

[16] J. W. Lewellen, et al., Proceedings of the 1998 Linac Conference, Chicago, ANL-98/28, Vol. 2, 1999, p. 863.

[17] M. Borland, Proceedings of the 1993 Particle Accel. Conference, New York, 1993, p. 3015.

[18] A.H. Lumpkin, W.J. Berg, B.X. Yang, M. White, Time-resolved imaging for the APS linac beams, Proceedings of the 1998 Linac Conference, Chicago, ANL-98/28, Vol. 1, 1999, p. 529.

[19] S.V. Milton, et al., Nucl. Instr. and Meth. A 407 (1998) 210.

[20] E. Gluskin, et al., Nucl. Instr. and Meth. A 429 (1999) 358.

[21] L. Wartski, et al., J. Appl. Phys. 46 (1975) 3644.

[22] A.H. Lumpkin, et al., Nucl. Instr. and Meth. A 296 (1990) 150;
D.W. Rule, et al., Nucl. Instr. and Meth. A 296 (1990) 739.

[23] S. Reiche, Nucl. Instr. and Meth. A 429 (1999) 243.

[24] M. Xie, Proceedings of the 1995 Particle Accel. Con-ference, Dallas, Texas, 1995, p. 183.

ELSEVIER

Nuclear Instruments and Methods in Physics Research A 475 (2001) 470–475

**NUCLEAR
INSTRUMENTS
& METHODS
IN PHYSICS
RESEARCH**
Section A

www.elsevier.com/locate/nima

First measurements of subpicosecond electron beam structure by autocorrelation of coherent diffraction radiation [☆]

A.H. Lumpkin[a,*], N.S. Sereno[a], D.W. Rule[b]

[a] *Advanced Photon Source, Bldg. 32, Argonne National Laboratory, 9700 South Cass Avenue, Argonne, IL 60439, USA*
[b] *Carderock Division, Naval Surface Warfare Center, West Bethesda, MD, USA*

Abstract

We report the initial measurements of subpicosecond electron beam structure using a nonintercepting technique based on the autocorrelation of coherent diffraction radiation (CDR). A far infrared (FIR) Michelson interferometer with a Golay detector was used to obtain the autocorrelation. The radiation was generated by a thermionic rf gun beam at 40 MeV as it passed through a 5-mm-tall slit/aperture in a metal screen whose surface was at 45° to the beam direction. For the observed bunch lengths of about 450 fs (FWHM) with a shorter time spike on the leading edge, peak currents of about 100 A are indicated. Also a model was developed and used to calculate the CDR from the back of two metal strips separated by a 5-mm vertical gap. The demonstrated nonintercepting aspect of this method could allow on-line bunch length characterizations to be done during free-electron laser experiments. © 2001 Elsevier Science B.V. All rights reserved.

Keywords: Bunchlength; Subpicosecond; Coherent diffraction radiation

1. Introduction

A recurrent theme of several recent conferences and workshops that addressed diagnostics for free-electron lasers (FELs) has been the designation of the nonintercepting feature as critical [1–4]. This feature is particularly of interest in the measurement of longitudinal bunch profile, since this can be used with a charge measurement to determine the e-beam peak current. This latter parameter is one of the most essential to obtaining and under-

standing the FEL's gain. Many of the operating FEL laboratories use techniques that intercept (coherent transition radiation screens) or interrupt (zero-phasing RF) the beam transport to the undulators [5,6]. We report here the first (to our knowledge) measurement of sub-0.5-ps e-beam profiles using a nonintercepting technique based on coherent diffraction radiation (CDR). In a previous measurement by Shibata et al. [7], the forward CDR and the electron beam struck a downstream radiation pick-off mirror after the circular aperture. In our experiment the radiation was generated by a thermionic rf gun beam at 40 MeV as it passed through a 5-mm-tall slit/aperture in a metal screen whose surface was at 45° to the beam direction. The backward CDR was evaluated with a far infrared (FIR) Michelson

---

[☆] Work supported by the US Department of Energy, Office of Basic Energy Sciences, under contract No. W-31-109-ENG-38.

*Corresponding author. Tel.: +1-630-252-4879; fax: +1-630-252-4732.

E-mail address: lumpkin@aps.anl.gov (A.H. Lumpkin).

interferometer that used a Golay cell as the detector. The longitudinal profile was evaluated from the fast Fourier transform (FFT) of the autocorrelation and the use of the minimal phase approximation [8]. The beam was transported at the same time to a downstream electron spectrometer. The results have been compared to complementary, intercepting coherent transition radiation (CTR) and streak camera measurements. Since the micropulse charge was typically 30–50 pC, the macropulse was either 8 or 40 ns long, and the beam energy was only 40 MeV, the CDR-based technique appears to be applicable to extensions of CTR techniques at several FEL laboratories (e.g. Vanderbilt [5], Stanford [9], Jefferson Lab [6], etc.) that have beams with more integrated charge in a macropulse. A model was also developed and used to calculate the CDR from the back of two metal strips separated by a 5-mm vertical gap centered on the beam axis. We also performed complementary measurements using CTR.

2. Experimental background

The Advanced Photon Source (APS) injector linear accelerator (linac) consists of an S-band thermionic RF gun and α magnet that allow a macropulse of 8–40 ns in length to be injected into the accelerator as shown in Fig. 1 [10]. The

micropulses each have about 20–50 pC of charge, so bunch lengths of less than 500 fs (FWHM) are needed in order to reach the target 100-A peak currents. In gun optimization studies [11] we have found specific regimes where the core of the beam has transverse emittances of the order of 8–12 π mm mrad. The nominal beam properties at the station where these measurements were performed are given in Table 1.

The standard beam profile station on the linac uses an intercepting Chromox screen with a video camera. For investigating bright beams, we have modified the station at 40 MeV. In this case one actuator has a Ce-doped YAG single crystal with mirror at 45° to the beam for beam position and profile information with high conversion efficiency. The second actuator with a kinematic positioner has a metal screen with a 5-mm-tall slit aperture machined into it beginning 1.5 mm above the beam centerline when it is inserted. This design

Table 1
APS linac beam properties at Station-1 in the low-emittance mode (RF Thermionic Gun)

RF frequency (MHz)	2856
Beam energy (MeV)	40–50
Micropulse charge (pC)	20–50
Micropulse duration (ps)	0.2–1.0 (FWHM)
Macropulse length (ns)	40, 8
Macropulse repetition Rate (Hz)	6
Normalized emittance (π mm mrad)	~12±4 (1σ)

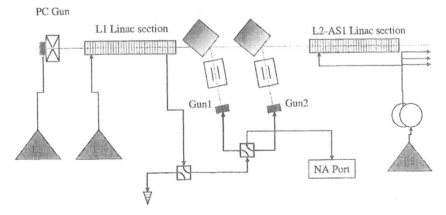

Fig. 1. A schematic of the initial portion of the injection linac for the APS. In these experiments the beam from thermionic RF gun #2 was accelerated to 40 MeV by the L2-AS1 accelerating structure after which the diagnostics station was located (courtesy of J. Lewellen, ANL).

was a compromise dictated by the need to use an alignment laser to position properly the Michelson interferometer and detector below the beamline. The alignment laser was injected into the beamline upstream of the location so that the angle of specular reflection could be determined from the metal screen. By steering the e-beam on the centerline of the beamline, the solid metal is intercepted for optical transition radiation (OTR) or CTR studies. Steering the beam vertically a few mm allowed the beam to pass through the aperture and to be viewed on the downstream electron spectrometer. With this latter steering, the CDR experiments were performed. In addition to the visible beam image downstream, we also used the rf beam position monitor (BPM) sum signal as a relative current monitor before and after the CTR/CDR screen. We tracked these signals during the course of the interferometer mirror position scans.

The CTR and CDR are generated from the metal surface or "metal strips", respectively, seen by the beam. The radiation leaves the beam pipe through a crystalline quartz window and is collimated by a 100-mm focal length crystalline quartz (z-cut) plano-convex lens. A Michelson interferometer is used to analyze the spectrum of the coherent radiation or, equivalently, to perform an autocorrelation of the emitted transition radiation pulse as described in Ref. [12]. An Inconel-coated beam splitter is used to generate the two beams used in the autocorrelation. One beam's path length can be adjusted by a computer-controlled mirror stage. A Golay detector, Model OAD-7, obtained from QMC Instruments Limited, was used as the broadband FIR detector. Signal levels of a few hundred mV were obtained with 200-mA macropulse average current. The analog data were processed and digitized with an APS-built gated integrator module and Hewlett-Packard waveform digitizer. EPICS was the platform used to track the signal-versus-mirror position, α-magnet current, or scraper position [13].

3. Analytical background

A model was developed and used to calculate the CDR spectral emissions from the back of two

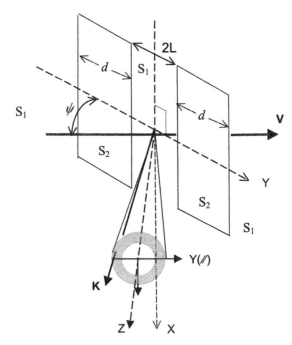

Fig. 2. Schematic of geometry for diffraction radiation production by two finite width, infinitely long, reflecting strips. Here the *y*-axis is in the plane containing the velocity vector *v* and the normal to the plane of the aperture. Radiation is emitted in the direction of *k*.

metal strips separated by a vertical gap centered on the beam axis, as schematically shown in Fig. 2. The spectral emissions were calculated for nominal cases at 50 MeV, 100-pC/μ pulse, 80 pulses. The calculation was compared to measured spectra reported by Shibata et al. [7]. Sensitivities of the CDR spectrum to beam offset from the center of the slot and to slots of various widths were assessed. Model guidance supported the 5-mm vertical gap for our nominal 1-ps case at 40–50 MeV. A brief outline of the approach is given below.

Coherent radiation by a bunch of electrons can be expressed as the product of a term representing the radiation process for a single particle and a term, which takes into account how much of the total charge in the bunch radiates together, constructively in phase. Thus the spectral, angular distribution of the radiation can be expressed as

$$\frac{\mathrm{d}^2 W}{\mathrm{d}\omega\, \mathrm{d}\Omega} = \frac{\mathrm{d}^2 W_1}{\mathrm{d}\omega\, \mathrm{d}\Omega} B_N, \tag{1}$$

where $\mathrm{d}^2 W_1/\mathrm{d}\omega\,\mathrm{d}\Omega$ is the spectral angular distribution for the single particle radiation process, either transition radiation (TR), synchrotron radiation, or diffraction radiation (DR), for example. In the present case, $\mathrm{d}^2 W_1/\mathrm{d}\omega\,\mathrm{d}\Omega$ is taken to be the single-particle diffraction radiation process. B_N is the coherence factor for a bunch of N electrons. The term B_N is related to the square of the Fourier transform of the spatial distribution of the bunch:

$$B_N = N + N(N-1)F(\boldsymbol{k}), \tag{2}$$

with

$$F(\boldsymbol{k}) = [Q^{-1}\mathfrak{I}\{\rho(\boldsymbol{r})\}]^2. \tag{3}$$

Here

$$\mathfrak{I}\{\rho(\boldsymbol{r})\} = \int \rho(\boldsymbol{r})\mathrm{e}^{-\mathrm{i}\boldsymbol{k}\cdot\boldsymbol{r}}\mathrm{d}\boldsymbol{r} \tag{4}$$

is the bunch charge form factor, i.e., the Fourier transform of the charge distribution of total charge $Q = Ne$. The first term in Eq. (2) yields the incoherent radiation produced by N electrons in the bunch, while the second term gives the coherent production, which is proportional to N^2. A simple Gaussian model for the bunch form factor is used to calculate CDR for various cases relevant to the APS linac.

Note that, for $F(\boldsymbol{k})$ to be sizeable, the beam rms radius σ and the wavelength of interest λ must be such that for angles θ of order $1/\gamma$,

$$\sigma \leqslant \sqrt{2}\gamma\bar{\lambda}, \tag{5}$$

where $\bar{\lambda} = \lambda/2\pi$. The wavelengths of interest are determined by the longitudinal part of $F(\boldsymbol{k})$ such that the rms bunch length

$$\sigma_z \leqslant \bar{\lambda}. \tag{6}$$

Clearly a relativistic effect allows the beam radial dimensions to be of the order of γ times $\lambda/2\pi$ and radiate coherently, while the wavelength for coherent radiation must be greater than the longitudinal dimension of the bunch ($\sim 2\sigma_z$). Since the number of electrons N in a bunch is typically quite large, $F(\boldsymbol{k})$ can actually be much less than unity and there will still be coherent radiation.

A computer program to calculate coherent diffraction radiation as a function of wavelength was written and applied to the general cases of

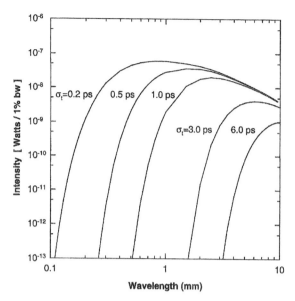

Fig. 3. Calculated spectra of CDR from a slit for a 50-MeV electron beam with 8 nC total charge and bunch lengths of $\sigma_t = 0.2, 0.5, 1.0, 3.0$ and $6.0\,\mathrm{ps}$.

interest here [14]. Fig. 3 shows the calculated spectra of CDR from a slit for a 50-MeV electron beam comprising 8 nC in a macropulse for bunch lengths from $\sigma_t = 0.2$–6 ps. As expected, the form factor results in significant enhancements of radiation at wavelengths about three times longer than the bunch length.

4. Experimental results and analysis procedures

All the data have been obtained since the last FEL conference in 1999. The thermionic RF gun gradient was deliberately set higher than for the standard settings used for generating beam for injection into the storage ring. The higher gradient enhances micropulse charge, and when used with the proper α magnet setting and low-energy scraper setting, results in a brighter core beam.

For simplicity the beam was first evaluated using the CTR signals from Station-1 by steering onto the beam centerline. We then translated the beam upward and sent the beam through the aperture. The CDR FIR signal was about three times lower than the CTR FIR signal but still

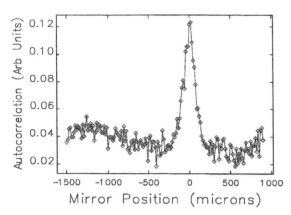

Fig. 4. The CDR autocorrelation from the FIR interferometer. Note the horizontal scale of mirror position in μm would be doubled for optical path difference.

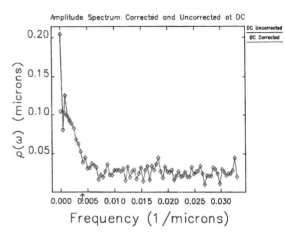

Fig. 5. The amplitude spectrum derived from the autocorrelation and corrected spectrum at long wavelength (low wavenumber).

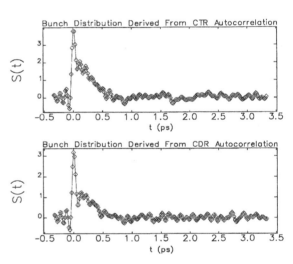

Fig. 6. Longitudinal bunch profiles derived from the corrected autocorrelation amplitude spectrum using CTR (upper) and using CDR (lower). The bunch distribution is sub-0.5 ps (FWHM) for the 40-MeV beam.

sufficient to obtain an autocorrelation as seen in Fig. 4. A fast Fourier transform was performed to assess the spectral content. Using the observed amplitudes from 20 to $40\,cm^{-1}$ as reference, the amplitudes for longer wavelengths were extrapolated [15] (as shown in Fig. 5). The corrected spectrum was used in recalculating the autocorrelation, and then the longitudinal profile was calculated using the minimal phase approximation of Lai and Sievers [8]. In Fig. 6 we show a comparison of the CTR and CDR data, respectively. A similar longitudinal distribution was obtained in both cases with a FWHM distribution of ~450 fs. The bunch distribution shows a spike, which was determined to be on the leading edge by the zero-phasing RF technique, a streak camera measurement, and the effects of inserting the low-energy scraper in the α magnet. The limiting resolution of the streak camera prevented the measurement of the spike width. The streak camera did detect a low-intensity, long time tail which was outside the FIR detector response. Note that the single-particle DR production is explicitly wavelength dependent. This difference between TR and DR is not yet included in the analysis of the data but is small since $\gamma\bar{\lambda} > 2.5\,mm$, the aperture radius. The difference would become even smaller at higher gamma. For the nonintercepting CDR case, the energy distribution was observed downstream of the aperture's location,

and the RF BPM sum signals from locations before and after the aperture were used to confirm the transmission of beam. There was no observable difference in the energy profiles with or without the aperture in place, but the orthogonal y-spatial profile data supported steering to minimize clipping of the beam halo. These particular data are with a 40-ns-long macropulse, but we subsequently obtained an autocorrelation with an 8-ns-long macropulse that was optimized for self-amplified spontaneous emission (SASE) gain experiments [13].

5. Summary

In summary, we have obtained the first, to our knowledge, nonintercepting measurements of sub-0.5-ps electron beam bunch profiles using backward CDR. The measurements were done with $\gamma < 100$ and for quite small charges in a micropulse (20–50 pC). The total macropulse charge of 1–4 nC and these other conditions support the application to several operating FEL experiments, which could benefit from the nonintercepting feature. The simplest extensions would occur at labs that already use a CTR-based monitor, such as Vanderbilt, Stanford, and Jefferson Lab. The scaling also indicates that at higher γ the CDR approaches the CTR signal strengths for an appropriate aperture so even the Linac Coherent Light Source project may benefit.

Acknowledgements

The authors acknowledge W. Berg for engineering assistance, N. Arnold for software support, and J. Galayda for project support.

References

[1] A.H. Lumpkin, Nucl. Instr. and Meth. A 393 (1997) 170.

[2] A.H. Lumpkin, Beam Diagnostics for Future FELs, in: P.G. O'Shea, H.E. Bennett (Eds.), Free-Electron Laser Challenges, SPIE, Vol. 2988, 1997, pp. 70–77.

[3] G. Neil, Photon-and Electron-Beam Characterizations (WGVII), Future Light Source Workshop, April 6–9, 1999, Argonne National Laboratory, ANL-00/8, 2000.

[4] W. Barry, Measurement of Subpicosecond Bunch Profiles Using Coherent Transition Radiation, Proceedings of the 1996 Beam Instrumentation Workshop, Argonne, IL, May 6–9, 1996, AIP Conf. Proc. 390 (1997) 173.

[5] F. Amirmadhi, et al., Nucl. Instr. and Meth. A 375 (1995) 95.

[6] P. Piot, et al., Performance of the Electron Beam Diagnostics at Jefferson Lab's High Power Free Electron Laser, Proceedings of the 1999 Particle Accel. Conference, 1999, pp. 2229–2231.

[7] Y. Shibata, et al., Phys. Rev. E 52 (1995) 6787.

[8] R. Lai, A. J. Sievers, AIP Conf. Proc. 367 (1996) 312 and references therein.

[9] K.N. Ricci, T.I. Smith, Phys. Rev. Special Topics—Accelerators and Beams 3 (2000) 032801.

[10] J. Lewellen, et al., Operation of the APS rf Gun, Proceedings of the 1998 Linac Conference, 1999, pp. 863–865.

[11] N.S. Sereno, M.D. Borland, A.H. Lumpkin, Setup of the APS rf Thermionic Gun Using Coherent Transition Radiation to Produce High Brightness Beams for SASE FEL Experiments, Proc. XXth International Linac Conference, SLAC-R-561, Vol. 1, p. 134, Monterey, CA, August 21–25, 2000.

[12] A.H. Lumpkin, et al., Electron Beam Bunch Length Characterization Using Incoherent and Coherent Transition Radiation on the APS SASE FEL Project, Proceedings of the 1999 International Free-Electron Laser Conference, Nucl. Instr. and Meth. A 445 (2000) 356.

[13] A.H. Lumpkin, et al., Development of a CTR-Based Bunch Length Monitor with Application to the APS Thermionic rf Gun Beam Optimization, Nucl. Instr. and Meth. A 475 (2001) 476, these proceedings.

[14] D. Rule, Argcdr5.sm, August 1999.

[15] N. S. Sereno, OAG Applications Script, Argonne National Lab., January 2000, unpublished.

ELSEVIER

Nuclear Instruments and Methods in Physics Research A 475 (2001) 476–480

NUCLEAR
INSTRUMENTS
& METHODS
IN PHYSICS
RESEARCH
Section A

www.elsevier.com/locate/nima

Development of a coherent transition radiation-based bunch length monitor with application to the APS RF thermionic gun beam optimization ✩

A.H. Lumpkin[a],*, N.D. Arnold[a], W.J. Berg[a], M. Borland[a], U. Happek[b], J.W. Lewellen[a], N.S. Sereno[a]

[a] *Advanced Photon Source, Building 32, Argonne National Laboratory, 9700 South Cass Avenue, Argonne, IL 60439, USA*
[b] *Physics Department, University of Georgia, Athens, GA, USA*

Abstract

We report further development of an EPICS-compatible bunch length monitor based on the autocorrelation of coherent transition radiation (CTR). In this case the monitor was used to optimize the beam from the S-band thermionic RF gun on the Advanced Photon Source (APS) linac. Bunch lengths of 400–500 fs (FWHM) were measured in the core of the beam, which corresponded to about 100-A peak current in each micropulse. The dependence of the CTR signal on the square of the beam charge for the beam core was demonstrated. We also report the first use of the beam accelerated to 217 MeV for successful visible wavelength SASE FEL experiments. © 2001 Elsevier Science B.V. All rights reserved.

Keywords: Bunch length; Subpicosecond; Coherent transition radiation

1. Introduction

An interest in generating beams of sufficient brightness to support self-amplified spontaneous emission (SASE) free-electron laser (FEL) experiments continues to drive the development of beam characterization techniques at the Advanced Photon Source (APS). In this case we have developed an EPICS-compatible version of a

✩ Work supported by the US Department of Energy, Office of Basic Energy Sciences, under contract No. W-31-109-ENG-38.
*Corresponding author. Tel.: +1-630-252-4879; fax: +1-630-252-4732.
E-mail address: lumpkin@aps.anl.gov (A.H. Lumpkin).

bunch length monitor based on the autocorrelation of coherent transition radiation (CTR) [1]. We have taken advantage of the technique to observe sub-0.5-ps micropulse structure in a beam from a thermionic RF gun, and have developed procedures for optimization of the gun's performance by appropriate adjustments of gun current, the RF power to the gun cavity, the α-magnet current, and a scraper installed on the low energy particle trajectory as guided by the CTR signal strength [2]. The processing of the data was done using a series of application scripts developed over the past year [3]. The bunch profile was calculated using a fast Fourier transform (FFT) of the autocorrelation followed by an application of the minimal phase approximation of Lai and Sievers [4]. Bunch lengths of the core of the beam corresponded to peak currents of $\sim 100\,\text{A}$ in each micropulse of the S-band macropulse. The dependence of the CTR signal on the square of the beam charge for the beam core was demonstrated. Although the charge per pulse was lower than initially simulated several years ago [5], the peak current and the beam emittance implied sufficient brightness to support SASE FEL gain experiments. We also report the first successful use of the beam accelerated to 217 MeV for such visible wavelength gain experiments.

2. Experimental background

The APS thermionic RF gun has been described previously as well as its use as the source of electrons for injection into the main storage ring [6]. In this case the gun was operated at a higher gradient in order to increase the charge per micropulse, and hence peak current. For a target of 100 A, the 40–50 pC micropulse charge available implied a bunch length of less than 0.5 ps would be needed. In order to address this regime, we have implemented a far infrared (FIR) Michelson interferometer at the 40-MeV station in the linac. Table 1 lists the basic beam parameters at this location, and Fig. 1 shows a schematic of the experiment. The initial measurements were done with a 40-ns-long macropulse, but subsequently we were able to operate with only the 8-ns macropulse.

Table 1
APS linac beam properties at station-1 in the low-emittance mode (RF thermionic gun)

Rf Frequency (MHz)	2856
Beam Energy (MeV)	40–50
Micropulse (pC)	20–50
Micropulse duration (ps)	0.2–1.0 (FWHM)
Macropulse length (ns)	40.8
Macropulse repetition rate (Hz)	6
Normalized emittance (π mm mrad)	$\sim 12 \pm 4$ (1σ)

2.1. Coherent transition radiation monitor

The bunch length monitor is based on the autocorrelation of FIR CTR generated as the charged-particle beam strikes the metal mirror oriented at 45° to the beam direction. This particular interferometer has a novel, compact design using an Inconel-coated beam splitter [1]. It was constructed at the University of Georgia under an ANL contract. A Golay cell, model OAD-7 from QMC Ltd., is used as the FIR detector. Its entrance window transmits well from $20\,\mu\text{m}$ to 1 mm. The moveable arm of the interferometer and the data acquisition are handled via EPICS. Complementary bunch length measurement information is available at the end of the linac using either an optical transition radiation (OTR) conversion mechanism and a C5680 Hamamatsu streak camera or the RF zero-phasing technique combined with the electron spectrometer. The approximate temporal resolutions of these two systems are 1.5 ps (FWHM) and 0.6 ps (FWHM), respectively. These two techniques can provide supplementary asymmetry information.

2.2. Recent CTR system development

In the earliest configurations of the CTR system we had used a digitizing oscilloscope to process the Golay cell signal. In the last year we have added a transimpedance amplifier, modified the APS-built gated integrator module, and added a Hewlett-Packard waveform digitizer. The latter item has allowed us to evaluate the analog waveforms and to archive them.

The EPICS scripts have been augmented to allow tracking of CTR signal during the α-magnet current, low-energy scraper, or interferometer scans. The RF BPM sum signals just before the intercepting CTR screen location are also monitored during these scans for normalization data if needed. The scripts track the 6-Hz pulse rate.

The data analysis/processing tools have been particularly enhanced. Scripts compatible with the APS self-describing data set (SDDS) format [7] have been written. These scripts apply the FFT routine to the autocorrelation data which gives the square of the bunch spectrum. Using the

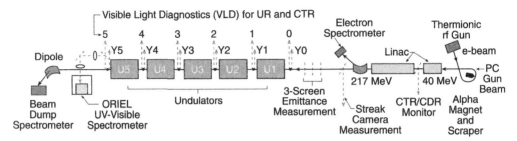

Fig. 1. A schematic of the APS SASE FEL experiment showing the electron gun, accelerator, CTR diagnostics station, and the undulators with their diagnostics.

method of Lai and Sievers [4] involving the minimal phase approximation, the phase spectrum is reconstructed from the amplitude spectrum by computing a principal value integral. Once the phase spectrum is obtained, an inverse FFT is performed to derive the microbunch profile.

3. Thermionic RF gun beam optimization

A sequence of steps has been developed to optimize the RF gun for brighter beams using the high gradient for the gun cavity. We start with a scan of the α-magnet current to find the minimum bunch length. The power to the gun and the cathode heater current are adjusted to produce 1–2 nC in a train of 23–25 S-band bunches. The beam is focused at the 40-MeV station on a YAG:Ce scintillating crystal, and then the CTR converter screen is switched into the beamline. During the scan, the RF gun phase must be adjusted linearly to compensate for path length changes in the α-magnet. Fig. 2 shows a sample α-magnet scan with a CTR signal peak at 175 A. The curve represents the digitized output of the Golay detector.

Once the minimum bunch length has been found, an α-magnet scraper scan is performed. Simulations suggest that the microbunch has a low-emittance, high-energy core beam and a high-emittance, low-energy tail. The scraper scan is used to optimize removal of the low-energy tail. Fig. 3 shows a scraper scan where the CTR signal is plotted versus scraper position. As the scraper position value increases from the "out" position at 9.0 cm, the low-energy tail is first intercepted. The

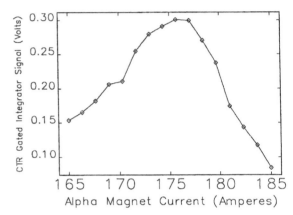

Fig. 2. A plot of the variation of CTR signal versus α-magnet current.

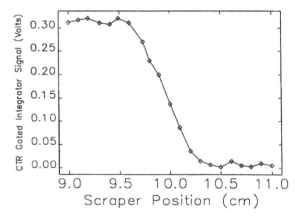

Fig. 3. A plot of the CTR signal versus scraper arm position in the α-magnet.

edge of the core beam is found at ~9.5 cm. As shown in Fig. 4, the CTR signal (plotted versus the nearby RF BPM sum signal) exhibits the expected quadratic dependence on the number of particles

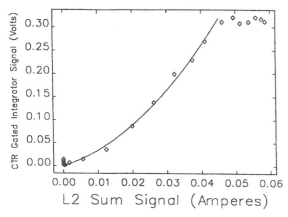

Fig. 4. A plot of the CTR signal versus the RF BPM sum signal. The solid line is a quadratic fit to the data showing the CTR's dependence on the square of the number of particles in the beam core.

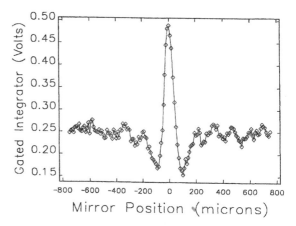

Fig. 5. An autocorrelation of the FIR CTR using the Michelson interferometer.

only in the core beam [8,9]. The low-energy (late time) tail of the beam does not contribute to the CTR signal so we can scrape away the first third of the beam without decreasing the *detected* CTR signal. This is *not* a saturation effect in the CTR mechanism.

After determining the desired scraper position, the FIR interferometer was employed to measure the autocorrelation of the digitized CTR signal. Typically 151 or 201 steps are done with a scan step of $7.5\,\mu m$ and a 10-shot average per point. Fig. 5 shows the result for $\sim 1\,nC$ in 23 S-band micropulses. Note that the width of the peak in microns is doubled for the optical path differences. It is approximately $120\,\mu m$ (FWHM). After the correction is applied at low frequencies, the bunch profile is broadened and the dips on either side of the autocorrelation are flattened (since the autocorrelation should be positive). In Fig. 6 the derived bunch profile from the corrected autocorrelation spectrum is shown. There is a high-current peak greater than $100\,A$ and a lower-current shoulder with an overall width of $\sim 400\,fs$. The streak camera data and the zero phasing technique used in other runs of the gun are consistent with a spike on the leading edge of the bunch.

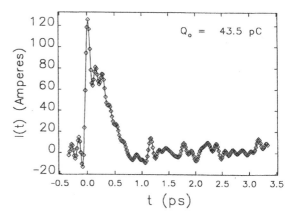

Fig. 6. The derived longitudinal bunch profile for the autocorrelation after a correction is applied at low frequencies.

4. SASE FEL test

The core beam was accelerated to 217 MeV, the charge measured in a Faraday Cup, the energy spread measured in the spectrometer, and the emittance measured using the three-screen technique. The values were found to be $Q = 1\,nC$, $\sigma_\varepsilon \sim 0.1\%$, and $\varepsilon_n \sim 12\,\pi\,mm\,mrad$, respectively. The beam was transported to the low-energy undulator test line (LEUTL) hall and through the five 2.4-m undulators. Diagnostics stations after each undulator [10] were used to measure the growth of the 537-nm fundamental intensity. An

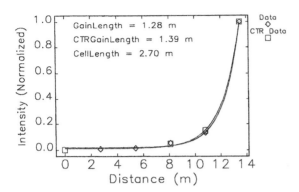

Fig. 7. The intensity of visible undulator radiation (diamonds) versus the undulator length. The exponential growth behavior is clearly seen with a gain length $L_g \sim 1.3$ m.

exponential growth behavior was measured with a SASE gain length of $L_g \sim 1.3$ m as shown in Fig. 7 [2]. These are the first (to our knowledge) such measurements using a thermionic RF gun beam. In addition, the complementary CTR intensities due to z-dependent microbunching evolution also show a similar exponential growth rate (see Ref. [11]). Although the bunch length is comparable to the slippage length, the gain length is in agreement with expectations based on Xie's parameterization of SASE gain [12].

5. Summary

In summary, a CTR-based monitor has been developed into an on-line technique for optimizing the thermionic RF gun beam's bunch length at the sub-0.5-ps and 100-A peak current regime. This optimized beam core was further characterized and had sufficient brightness to support visible light SASE FEL experiments. The studies with this beam will continue, but a recently installed chicane bunch compressor in the APS linac will allow the

relocated CTR monitor to be applied to the PC gun beam as well.

Acknowledgements

The authors acknowledge the support of S.V. Milton, K.-J. Kim, and J.N. Galayda of APS. They also thank Geoff Krafft of Jefferson Lab for discussions on their CTR system.

References

[1] A.H. Lumpkin, et al., Nucl. Instr. and Meth. A 445 (2000) 356.

[2] N.S. Sereno, M.D. Borland, A.H. Lumpkin, Setup of the APS rf Thermionic Gun Using Coherent Transition Radiation to Produce High Brightness Beams for SASE FEL Experiments, Proceedings of the XXth International Linac Conference, SLAC-R-561, Vol. 1, p. 134, Monterey, CA, August 21–25, 2000.

[3] N.S. Sereno, unpublished scripts.

[4] R. Lai, A.J. Sievers, AIP Conf. Proc. 367 (1996) 31, and references therein.

[5] M. Borland, An Improved Thermionic Microwave Gun and Emittance-Preserving Transport Line, Proceedings of the 1993 PAC, IEEE, 1993, pp. 3015–3017.

[6] J. Lewellen, et al., Operation of the APS rf Gun, Proceedings of the 1998 Linac Conference, 1999, pp. 863–865.

[7] M. Borland, et al., Doing Accelerator Physics Using SDDS, UNIX, and EPICS, 1995 ICALEPS Proceedings, Chicago, Fermilab Report CONF-96/069, 1996, pp. 382–391.

[8] U. Happek, E.B. Blum, A.J. Sievers, Phys. Rev. Lett. 67 (1991) 2962.

[9] W. Barry, Measurement of Subpicosecond Bunch Profiles Using Coherent Transition Radiation, Proceedings of the 1996 Beam Instrumentation Workshop, Argonne, IL, May 6–9, 1996, AIP Conf. Proc. 390 (1997) 173.

[10] E. Gluskin, et al., Nucl. Instr. and Meth. A 407 (1998) 210.

[11] A.H. Lumpkin, et al., Utilization of CTR to Measure the Evolution of Electron-Beam Microbunching in SASE FEL, Nucl. Instr. and Meth. A 475 (2001) 462, these proceedings.

[12] M. Xie, Proceedings of the 1995 PAC, IEEE, Dallas, 1995, pp. 183–185.

ELSEVIER

Nuclear Instruments and Methods in Physics Research A 475 (2001) 481–486

NUCLEAR
INSTRUMENTS
& METHODS
IN PHYSICS
RESEARCH
Section A

www.elsevier.com/locate/nima

Photon diagnostics for the study of electron beam properties of a VUV SASE-FEL

Ch. Gerth*, B. Faatz, T. Lokajczyk[1], R. Treusch, J. Feldhaus

Deutsches Elektronen-Synchrotron DESY, Notkestraße 85, D-22603 Hamburg, Germany

Abstract

A single-pass free-electron laser operating in the self-amplified spontaneous-emission (SASE) mode at around 100 nm is currently under test at the TESLA Test Facility at DESY. After first observation of SASE in February 2000, the photon beam has been characterized by different techniques. We present the methods of VUV photon diagnostics that were used to measure the spectral and angular distribution of the photon beam and how these properties are affected by the electron beam energy and orbit in the undulator. © 2001 Elsevier Science Ltd. All rights reserved.

PACS: 41.60.Cr; 42.60.Jf

Keywords: SASE; Free electron laser; X-ray laser; VUV; Photon beam diagnostics; CCD camera

1. Introduction

The TESLA collaboration is currently developing a Free-Electron Laser (FEL) at the TESLA Test Facility (TTF) at DESY [1]. The FEL is based on the principle of self-amplified spontaneous emission (SASE) and operates in the Vacuum Ultraviolet (VUV) energy region. In the SASE process, coherent radiation is emitted by a high-current, low-emittance electron beam during a single pass through a high-precision undulator. The spontaneous radiation emitted in the first part of the undulator overlaps and interacts with the electron bunch and the resulting charge density modulation (micro-bunching) amplifies the power and coherence of the radiation. Since the early theoretical work on the SASE process [2–4], considerable theoretical and experimental efforts have been made to study the physics of SASE. Experimental evidence of SASE in the infrared region 5–16 μm has been reported recently by several groups [5–7] with FEL gains up to 10^5 [8]. Efforts in reducing the wavelength to the visible began with Babzien et al. [9] who observed SASE at 633 nm, and FEL operation in the SASE mode at 530 nm was demonstrated by Milton et al. [10]. A decisive milestone on the way to Å wavelengths has been achieved in February 2000 when the first FEL output in SASE mode at around 100 nm was observed at the TTF [10]. The requirements for SASE at even shorter wavelengths are increasingly demanding in terms of electron beam quality and steering through the undulator. In this paper, we report on the VUV photon diagnostics used at the

---

*Corresponding author. Tel.: +49-40-8998-1841; fax: +49-40-8998-4475.

E-mail address: christopher.gerth@desy.de (C. Gerth).

[1] Present address: debis Systemhaus, D-44227 Dortmund, Germany.

TTF FEL for the determination of the spectral and angular distribution of the photon beam at around 100 nm. We show that these techniques are useful on-line monitors for electron beam properties such as the electron energy, pulse-to-pulse energy fluctuations and the electron orbit in the gain region of the undulator. Photon diagnostics for the characterization of electron beam parameters have already been applied successfully to FELs working in the infrared and visible (e.g. Refs. [12,13]).

2. VUV photon diagnostics

For the complete characterization of the FEL photon beam properties, an experimental station for photon diagnostics, including a grating monochromator and various detectors, provides all the instrumentation necessary to measure the photon pulse intensity and its angular, spectral and temporal distribution. New detection concepts for the (200–40 nm) VUV region have been employed in order to measure all SASE specific properties on a single pulse basis. The detectors and the principle layout of the photon diagnostics unit have been described in detail in Ref. [14].

FEL photon diagnostics is particularly challenging in the VUV region. Due to the high absorbance of radiation below 200 nm by any material, neither window materials to outcouple the radiation nor attenuating filters are available. Thus, in order to cover the full dynamic range of intensity from spontaneous undulator emission to SASE in saturation (about 5 orders of magnitude), different types of detectors, suitable for operation under ultra-high-vacuum (UHV) conditions, are employed. Since the photon diagnostics cannot be separated from the undulator and the accelerator vacuum, all components were assembled under cleanroom conditions. This avoids dust particles which could migrate to the accelerator cavities. In addition, all devices are fully remote controlled because radiation background in the accelerator tunnel prevents access during operation.

3. Spectral distribution

For the determination of the spectral distribution, the photon beam is deflected by a plane mirror onto the entrance slit of a commercial 1 m normal-incidence monochromator. The width of the entrance slit can be varied by a precise piezo actuator from 1 to 195 µm. For the initial FEL commissioning phase, the monochromator has been equipped with a 1200 lines/mm spherical grating. Spectra of the dispersed FEL radiation are recorded by a thinned, back-illuminated UV-sensitive CCD[2] that has been placed in the focal plane of the spherical grating, attached directly to the monochromator vacuum.

The upper part of Fig. 1 presents an image of the dispersed FEL radiation with an acquisition time of 30 s. The FEL was operated in single-bunch mode with 1 Hz repetition rate. Despite a 20 cm lead shielding of the camera, the background radiation in the accelerator tunnel caused a noticeable pixel damage and, therefore, a background image had to be subtracted and a median filter was applied. The horizontal (x-) axis of the image corresponds to the dispersive direction and the vertical (y-) axis is parallel to the entrance slit. The full CCD image covers a wavelength range of 20 nm in the dispersive direction. The intensity distribution in y-direction reflects the vertical beam profile and is mainly defined by a 5-mm-diameter aperture in front of the monochromator.

The spectral distribution, i.e. the integral intensity in vertical direction, is depicted in the lower part of Fig. 1. The wavelength scale has been calibrated with the use of a hollow-cathode lamp [15] and a Hg-lamp. The spectral distribution of the FEL radiation is centered at 92.1 nm with a full-width at half-maximum (FWHM) of 0.74 nm.

When an electron beam passes through a planar undulator, it emits electro-magnetic radiation at the wavelength

$$\lambda_{\mathrm{ph}} = \frac{\lambda_{\mathrm{u}}}{2\gamma^2}\left(1 + \frac{K^2}{2}\right) \qquad (1)$$

---

[2] CCD camera system ATC 300 L from Photometrics, Tucson, Arizona 85706.

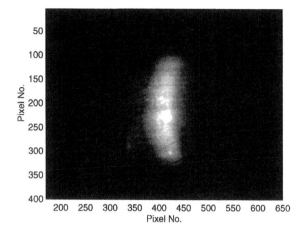

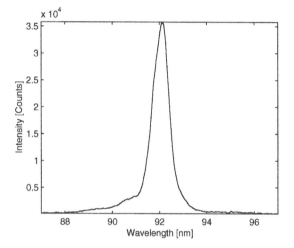

Fig. 1. Upper part: CCD image of the dispersed FEL radiation in SASE mode taken in the focal plane of a 1 m normal-incidence monochromator. Lower part: Spectral distribution obtained by integration of the image above in vertical direction.

Table 1
Measured parameters of the SASE-FEL at the TESLA Test Facility (phase I) at DESY (August 2000)

Parameter	
Photon beam	
Wavelength λ_{ph}	181–80 nm
Energy	7–15 eV
Angular divergence at 91 nm	0.7(3) mrad (FWHM)
Max. FEL gain[a]	3×10^3
Undulator	
Length	13.5 m
Gap	12 mm
Period λ_u	27.3 mm
Peak magnetic field B_u	0.46 T
Electron beam	
Energy	181–272 MeV
No. of bunches per bunchtrain	1–10
Bunch separation	1 μs
Repetition rate	1 Hz

[a] taken from Ref. [11].

where $\gamma = E_e/mc^2$ is the relativistic factor of the electrons and $K = eB_u\lambda_u/2\pi m_e c$ the undulator parameter with the peak magnetic field B_u and the undulator period λ_u. The FEL at the TTF (phase I) consists of three fixed-gap undulator modules [16] with a length of 4.5 m each; the wavelength λ_{ph} of the FEL radiation can be varied by changing the electron beam energy E_e. Since K and λ_u are precisely known (see Table 1), the electron beam energy can be determined from the measured wavelength. The determination of the absolute wavelength can be achieved with an accuracy of 0.5%; this results in an accuracy of 0.25% for the electron energy. For instance, the

spectrum of Fig. 1 centered at 92.1(5) nm corresponds to an electron beam energy of 254(1) MeV. This value is in good agreement with an energy of 250(5) MeV determined from the deflection of the electron beam by a bending magnet. Meanwhile, SASE has been observed between 80 and 181 nm at the TTF FEL. The corresponding parameters are summarized in Table 1.

A series of three single-pulse spectra, centered at around 110 nm, is shown in Fig. 2(a). Each spectrum represents the spectral distribution of the FEL radiation emitted by a single bunch. The spectra were taken subsequently in intervals of several seconds with a slit width of 195 μm which results in an instrumental bandwidth of 0.17 nm. The variation of the centre position of the spectra reflects the energy variation of the electron beam. Here, the shift of 0.2 nm (0.2%) corresponds to an energy jitter of the electron beam of 0.1% (see Eq. (1)). Fig. 2(b) depicts a spectrum which was accumulated during an interval of 30 s, i.e., this spectrum represents the spectral distribution averaged over 30 subsequent electron bunches. As a consequence of the energy jitter, the width of the spectrum of 1.1 nm (FWHM) is larger than the 0.8 nm (FWHM) for each of the single-bunch spectra (Fig. 2(a)). The shoulder at the long

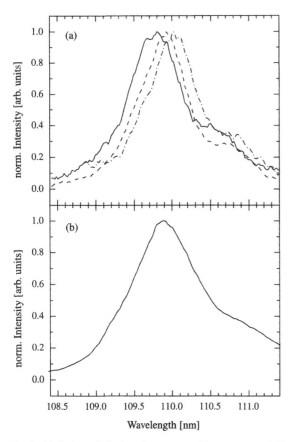

Fig. 2. (a) Series of single-pulse spectra; (b) spectrum of 30 pulses.

Table 2
Parameters of the back-illuminated and the intensified CCD cameras

	Photometrics ATC 300L	Lavision Nano Star 25
Principle	Back-illuminated Thinned CCD	Fluorescence screen, Optics 1:2.17, MCP
No. of pixels	1024 × 1024	1280 × 1024
Pixel size	24 μm	6.7 μm
Shutter	mechanical	MCP
Readout time	8 s	125 ms
Min. exposure time	1 ms	5 ns
Resolution	24 μm	≈12 μm

back-illuminated and the intensified CCD camera are compared in Table 2.

4. Angular distribution

The vertical photon beam profile, recorded at a distance of about 12 m behind the undulator, is shown in the upper part of Fig. 3. A 10×10 mm^2 PtSi-photodiode [18] with a 1-mm-diameter spherical aperture in front was moved in steps of 1 mm through the photon beam, and for each step the signal of 5 subsequent pulses was accumulated;[3] the FEL was operated in single-bunch mode with 1 Hz repetition rate. The solid line represents a fit of a Gaussian profile with a width of 7.7(7) mm (FWHM) and a centre at a vertical position of 3.6(2) mm. The zero position of the detector coordinate system has been aligned to the centre axis of the undulator with an accuracy of ± 0.5 mm with the use of a theodolite. Similar vertical displacements of the photon beam have been observed on the CCD image of the normal-incidence spectrometer, depending on the steering of the electron beam into the undulator. The direction of the electron beam in the gain region can be varied by a few 10^{-4} radians without much change in gain such that the light is emitted in different directions. This has been corroborated by using steerers (corrector coils) along the last

wavelength side is possibly caused by a fraction of electrons within the bunch with slightly smaller energy.

The minimal interval between two single-pulse spectra is limited by the readout time (8 s) of the CCD camera. To overcome this restriction, the back-illuminated CCD camera has been replaced by an intensified CCD camera [17] with a readout time of 125 ms. It utilizes a fluorescent screen (P46; decay time $(1/e)$: 140 ns) in the focal plane of the monochromator which is imaged through a Suprasil viewport. The ICCD camera is equipped with a micro-channel-plate (MCP) as an intensifier which operates as a fast shutter with exposure times down to 5 ns. This enables one to select single pulses from a sequence of pulses and to study, e.g., the variation of electron beam parameters within a bunchtrain. The parameters of the

---

[3] For further details on the signal registration of the photodiode and beam profile recording see Ref. [19].

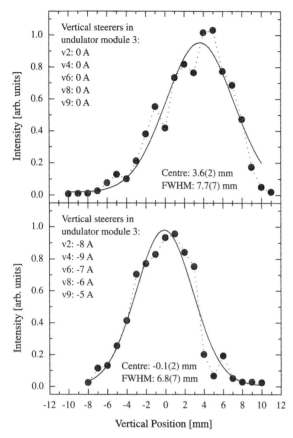

Fig. 3. Vertical beam profile of FEL radiation in SASE mode at 91 nm; (a) without and (b) with the effect of vertical electron beam steerers v2–v9 in the last undulator module.

undulator module to deflect the electron beam only in this area. Starting from the situation of a vertically displaced beam with all steerers turned off (Fig. 3, upper part), it was possible to deflect the electron beam down using five steerers such that the centre of the FEL radiation moved to the nominal zero position (Fig. 3, lower part). Again, in this case the light intensity did not change significantly. These results suggest that the on-line observation of the photon beam position might be very useful as a monitor when the electron beam orbit is adjusted, e.g. using beam based alignment techniques [20].

As in the case of the spectral distribution, the photon beam profile shown in Fig. 3 is most probably broadened by a spatial and angular jitter of the electron beam in the undulator. Hence, single-pulse measurements are required to avoid such a superposition. In a first approach, the fluorescent light of different crystals, such as Ce:YAG and $PbWO_4$, has been imaged with a conventional CCD camera. Using crystals with fast decay channels in conjunction with a gated ICCD camera it will be possible to observe the intensity distribution of single FEL pulses.

5. Summary

The methods used for the determination of the spectral and angular distribution of the VUV radiation generated by the SASE-FEL at the TESLA Test Facility at DESY and a selection of characteristic results have been presented. The angular and spectral distribution of the FEL radiation are particularly sensitive to the electron beam energy and orbit in the undulator. The techniques presented can therefore be used as on-line monitors for electron beam parameters in order to characterize the electron beam quality and stability.

Acknowledgements

The authors are indebted to all members of the TESLA collaboration for their contributions to the TTF at DESY and for successfull operation of the FEL. A list of the participating institutes can be found in Ref. [11].

References

[1] A VUV free electron laser at the TESLA test facility at DESY, Conceptual Design Report, DESY Print, June 1995, TESLA-FEL 95-03; J. Roßbach, Nucl. Instr. and Meth. A 375 (1996) 269.
[2] A.M. Kondratenko, E.L. Saldin, Part. Accl. 10 (1980) 207.
[3] R. Bonifacio, C. Pellegrini, L.M. Narducci, Opt. Commun. 50 (1984) 373.
[4] C. Pellegrini, J. Opt. Soc. Am. 52 (1985) 259.
[5] R. Prazeres, J.M. Ortega, F. Glotin, D.A. Jaroszynski, O. Marcouillé, Phys. Rev. Lett. 78 (1997) 2124.
[6] M. Hogan, C. Pellegrini, J. Rosenzweig, A. Varfolomeev, S. Anderson, K. Bishofberger, P. Frigola, A. Murokh, N.

Osmanov, S. Reiche, A. Tremaine, Phys. Rev. Lett. 81 (1998) 289.

[7] D.C. Nguyen, R.L. Sheffield, C.M. Fortgang, J.C. Goldstein, J.M. Kinross-Wright, N.A. Ebrahim, Phys. Rev. Lett. 81 (1998) 810.

[8] M.J. Hogan, C. Pellegrini, J. Rosenzweig, S. Anderson, P. Frigola, A. Tremaine, C. Fortgang, D.C. Nguyen, R.L. Sheffield, J. Kinross-Wright, A. Varfolomeev, A.A. Varfolomeev, S. Tolmachev, Phys. Rev. Lett. 81 (1998) 4867.

[9] M. Babzien, I. Ben-Zvi, P. Catravas, J.-M. Fang, T.C. Marshall, X.J. Wang, J.S. Wurtele, V. Yakimenko, L.H. Yu, Phys. Rev. E 57 (1998) 6093.

[10] S.V. Milton, et al., Phys. Rev. Lett. 85 (2000) 988.

[11] J. Andruszkow, et al., Phys. Rev. Lett. 85 (2000) 3825.

[12] A.H. Lumpkin, Nucl. Instr. and Meth. A 296 (1990) 134.

[13] T.I. Smith, H.A. Schwettman, Nucl. Instr. and Meth. A 812 (1991) 366. T.I. Smith, H.A. Schwettman, Instr. and Meth. A 429 (1999) 358.

[14] R. Treusch, T. Lokajczyk, W. Xu, U. Jastrow, U. Hahn, L. Bittner, J. Feldhaus, Nucl. Instr. and Meth. A 445 (2000) 456.

[15] K. Danzmann, M. Günther, J. Fischer, M. Kock, M. Kühne, Appl. Opt. 27 (1988) 4947.

[16] J. Pflüger, Nucl. Instr. and Meth. A 445 (2000) 366.

[17] CCD camera Nanostar S25 from LaVision, Göttingen, Germany; see http://www.lavision.de/cameras/index2.htm.

[18] K. Solt, H. Melchior, U. Kroth, P. Kuschnerus, V. Persch, H. Rabus, M. Richter, G. Ulm, Appl. Phys. Lett. 69 (1996) 3662.

[19] R. Treusch, Ch. Gerth, T. Lokajczyk, J. Feldhaus, Nucl. Instr. and Meth. A 467–468 (2001) 30.

[20] P. Castro, B. Faatz, K. Flöttmann, Nucl. Instr. and Meth. A 427 (1999) 12.

ELSEVIER

Nuclear Instruments and Methods in Physics Research A 475 (2001) 487–491

NUCLEAR
INSTRUMENTS
& METHODS
IN PHYSICS
RESEARCH
Section A

www.elsevier.com/locate/nima

Evaluation of electron bunch shapes using the spectra of the coherent radiation

Mitsumi Nakamura[a,*], Makoto Takanaka[a], Shuichi Okuda[a], Takahiro Kozawa[a], Ryukou Kato[a], Toshiharu Takahashi[b], Soon-Kwon Nam[c]

[a] *The Institute of Scientific and Industrial Research, Osaka University, 8-1 Mihogaoka, Ibaraki, Osaka 567-0047, Japan*
[b] *Research Reactor Institute, Kyoto University, Kumatori, Sennan, Osaka 590-0494, Japan*
[c] *Department of Physics, Kangwon National University, Chunchon Kangwon-do 200-701, South Korea*

Abstract

The bunch shape of single-bunch electron beams, generated with a 38 MeV L-band linear accelerator, was evaluated using the longitudinal bunch form factor at the Institute of Scientific and Industrial Research at Osaka University. The single-bunch beams are being used in experiments for generating self-amplified spontaneous emission and coherent radiation. In the present experiments, the energy of the electron beam was 27 MeV, the energy spread 1.1% FWHM, and the electron charge in a bunch 13.5 nC. The form factor was obtained from the spectrum of the coherent transition radiation measured with a Martin–Pupplet interferometer. A streak measurement was also performed in the same configurations and the results were compared to those for the measurements of the coherent radiation. By using these two methods, the performance of a chicane-type bunch compressor was investigated. © 2001 Elsevier Science B.V. All rights reserved.

PACS: 41.85.Qg; 41.75.Ht; 41.60.−m; 42.25.Kb

Keywords: Coherent radiation; Transition radiation; Bunch form factor; Bunch compressor; Electron linear accelerator; Streak camera

1. Introduction

For research applications of short electron bunches generated with an electron linear accelerator (linac), the diagnostics of the bunches are important. In general, a streak camera is used to evaluate a bunch shape. Another method recently developed is to use the coherent radiation emitted from an electron bunch. This method is based on the fact that the spectrum of the coherent radiation is determined by the bunch shape [1–6].

At the Institute of Scientific and Industrial Research (ISIR) at Osaka University, single-bunch electron beams generated with a 38 MeV L-band linac are being used for experiments of pulse radiolysis, self-amplifield spontaneous emission (SASE) [7] and coherent radiation. For advanced studies in these research fields, it is important to measure and control the electron bunch shape.

*Corresponding author. Tel.: +81-6-6879-8511; fax: +81-6-6875-4346.
E-mail address: mitumi03@sanken.osaka-u-ac.jp (M. Nakamura).

In this study, the spectrum of the coherent transition radiation emitted from the single-bunch beam was measured to evaluate the electron bunch shape. The coherent synchrotron radiation is a basic process in the SASE and the radiation intensity strongly depends on the bunch shape. The observation of the bunch shape by using a streak camera was also performed in the same configurations and the results were compared to those for the measurement of the coherent radiation. The performance of a chicane-type bunch compressor (BC) installed in the beam line was investigated by using the two methods.

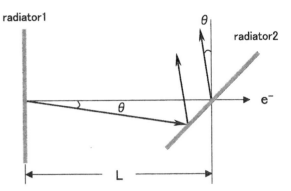

Fig. 1. Schematic diagram for transition radiation emitted from two radiators.

2. The electron bunch shapes and the spectra of the coherent radiation

Radiation, such as transition radiation and synchrotron radiation, which has a broad spectrum is emitted from an energetic electron beam. The intensity of radiation emitted from an electron bunch as generated with a linac is given by

$$P(v) = p(v)N[1 + (N - 1)f(v)] \tag{1}$$

where $p(v)$ is the intensity of radiation emitted from an electron, N the number of electrons in a bunch, v the wavenumber of radiation, and $f(v)$ the bunch form factor. The factor $f(v)$ has a value from 0 to 1. For a relatively large value of $f(v)$ the intensity of the coherent radiation is proportional to the square of N.

The factor $f(v)$ is given by the Fourier transform of a normalized distribution function of electrons in a bunch $S(x)$ as

$$f(v) = \left| \int S(x) \exp(i2\pi v x)\, dx \right|^2. \tag{2}$$

This expression indicates that the distribution function of electrons is given from the bunch form factor which is derived from the spectrum of the coherent radiation by using Eq. (2).

When an electron crosses a metal foil, transition radiation is emitted backward and forward from the surfaces of the foil. Fig. 1 shows the configuration in the present experiments schematically. Transition radiation emitted from the two radiators located on the electron beam trajectory at a

distance L was superposed. The intensity of transition radiation from an electron in this configuration is given by

$$p_e(v) = \frac{2\alpha\beta^2 \sin^2\theta}{\pi^2 v(1 - \beta^2 \cos^2\theta)^2}\left(1 - \cos\frac{L}{Z_f}\right) \tag{3}$$

where α is the fine-structure constant, β the ratio of the velocity of the electron to that of light in vacuum, θ the angle between the electron trajectory and the path of the emitted radiation, and Z_f the formation length [8].

The highest time resolution in the method to obtain a bunch shape by using a streak camera is in a subpicosecond range. In the method using the coherent radiation, the resolution can be made comparatively high because the resolution is determined by the spectral range in the measurement. Hence, this method will be useful for monitoring an ultra-short electron bunch.

3. Experimental method

A single-bunch electron beam generated with the L-band linac at ISIR was used in the present experiments. The beam parameters are as follows: the energy was 27 MeV, the energy spread 1.1% FWHM, and electron charge in a bunch 13.5 nC. The experimental set-up is shown in Fig. 2. The coherent transition radiation measured was the superposition of the forward radiation from an Al foil and the backward radiation from an Al-coated mirror which is oriented at 45° from the beam axis.

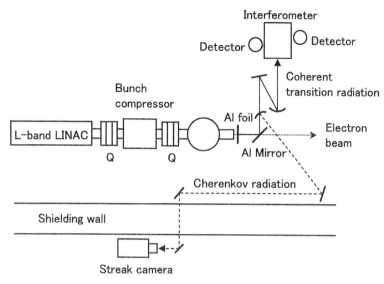

Fig. 2. Experimental set-up.

The distance between the Al foil and the Al-coated mirror was 150 mm. The beam trajectory was found to be relatively stable during the measurement because the accelerated beam was transported straight from the accelerating waveguide and the change in the beam energy didn't cause a change in the trajectory. Radiation was reflected by the Al-coated mirror and introduced into the Martin–Pupplet interferometer, in which wire grid polarizers with a wire spacing of 25 μm were installed as light beam splitters. Calibration of the interferometer was made with a high-pressure mercury lamp. Liquid-helium cooled silicon bolometers were used as light detectors. The resolution in wavenumber Δv was $0.2\,cm^{-1}$ and the range of wavenumber was $5–25\,cm^{-1}$ in the measurement. The intensity fluctuation of the light incident on the interferometer due to the change in the electron beam conditions was monitored by measuring the light from the first splitter.

Streak measurements of the Cherenkov radiation in the visible region were performed by inserting a mirror in the light path as shown in Fig. 2 to observe the electron bunch shapes in the time domain under the same configurations. An interference filter at a wavelength of 478.5 nm with a band width of 11.3 nm was used in the measurements. The time resolution was 0.9 ps.

The measurement of the spectrum of the coherent transition radiation and the measurement with a streak camera were performed, in turn, for the same beam conditions.

Four electro-magnetic bends are installed in the chicane-type BC [9]. For the energy of electrons of 27 MeV the orbit radius and the bending angle in each magnet are 731 mm and 0.27 rad, respectively. The magnetic field strength is 0.105 T.

4. Results and discussion

Typical electron bunch shapes measured with a streak camera are shown in Fig. 3. Small fluctuations which appear on the profiles are due to an insufficient intensity of incident light. As shown in Fig. 3(a), the electron bunch length is 29 ps FWHM in the case without the operation of the BC, and the bunch shape seems to have a triangular shape. The spectrum of the coherent transition radiation measured with the interferometer under the same experimental conditions is shown in Fig. 4. The longitudinal bunch form factor obtained from the measured spectrum is shown in Fig. 5. The effect of the transverse spread of an electron beam in calculating the intensity of the coherent radiation is negligible in the present

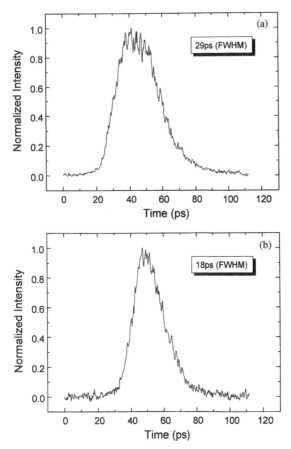

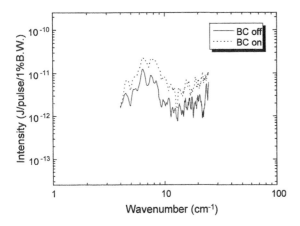

Fig. 3. Typical electron bunch shapes observed with a streak camera without the operation of the BC (a), and with the operation of the BC (b).

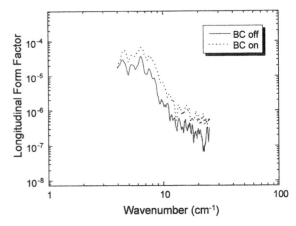

Fig. 5. Longitudinal bunch form factors obtained from the spectra of the coherent transition radiation without the operation of the BC (solid curve), and with the operation of the BC (dotted curve).

experiments. In order to obtain the distribution function of electrons in a bunch from the bunch form factor according to Eq. (2), the wavenumber dependence of the form factor up to near unity is necessary. Because the wavenumber range is limited, as shown in Fig. 5, the longitudinal form factor obtained from the measurement is compared to that obtained by the Fourier transform of the triangular distribution function of electrons with a bunch length of 29 ps FWHM, which is assumed from the results of the streak measurement. As shown in Fig. 6(a), the two form factors thus obtained agree well.

In the case with the operation of the BC, the bunch shape observed with a streak camera and the measured spectrum of the coherent radiation are shown in Figs. 3(b) and Fig. 4, respectively. From the result of the streak measurement the bunch length is found to decrease to 18 ps FWHM, maintaining a triangular shape. The intensity of the coherent radiation increases by about 4 times by operating the BC. These results indicate that the BC worked effectively on compressing the electron bunch even for a relatively small energy spread in the electron beam. The form factor in this case agrees well with that of the triangular shape with a bunch length of 18 ps FWHM, as shown in Fig. 6(b).

The bunch form factor at a relatively high wavenumber was not obtained by a streak

Fig. 4. Spectra of the coherent transition radiation without the operation of the BC (solid curve), and with the operation of the BC (dotted curve).

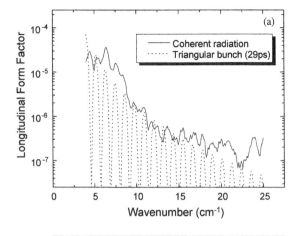

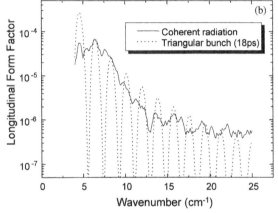

Fig. 6. Comparison of the longitudinal bunch form factors obtained from the experimental results without the operation of the BC (a), and with the operation of the BC (b), to those from the calculation for triangular bunch shapes. The bunch lengths of 29 ps FWHM (a) and of 18 ps FWHM (b) were assumed according to the results of streak measurements, respectively.

measurement. The form factor of the single-bunch beam obtained in the present work is considerably large in a far-infrared region. This result suggests that the coherent synchrotron radiation is important in the SASE processes in the experiments at ISIR.

5. Conclusions

The bunch shape of the single-bunch beam of the L-band linac at ISIR was evaluated by using the spectrum of the coherent transition radiation and by a streak measurement. The results obtained by these two methods agreed well under the assumption of the triangular bunch shape. By using these methods, the effective compression of the bunch with the operation of the BC was observed. The bunch form factor obtained from the spectrum of the coherent transition radiation suggested the importance of the coherent synchrotron radiation in the SASE process.

Acknowledgements

This work was supported in part by a Grant-In-Aid for Scientific Research from the Ministry of Education, Science, Sports and Culture of Japan, and one of the authors (S. Nam) wishes to acknowledge the financial support of Korea Research Foundation Grant (KRF-99-005-D00042).

References

[1] Y. Shibata, K. Ishi, T. Takahashi, T. Kanai, M. Ikezawa, K. Takami, T. Matsuyama, K. Kobayashi, Y. Fujita, Phys. Rev. A 45 (1992) 8340.
[2] Y. Shibata, K. Ishi, T. Takahashi, T. Kanai, F. Arai, S. Kimura, T. Ohsaka, M. Ikezawa, Y. Kando, R. Kato, S. Urasawa, T. Nakazato, S. Niwano, M. Yoshioka, M. Oyamada, Phys. Rev. E 49 (1994) 785.
[3] U. Happek, A.J. Sievers, E.B. Blum, Phys. Rev. Lett. 67 (1991) 2962.
[4] Gi. Schneider, R. Lai, W. Walecki, A.J. Sievers, Nucl. Instr. and Meth. A 396 (1997) 283.
[5] A. Murokh, J.B. Rosenzweig, M. Hogan, H. Suk, G. Travish, U. Happek, Nucl. Instr. and Meth. A 410 (1998) 452.
[6] T. Watanabe, M. Uesaka, J. Sugahara, T. Ueda, K. Yoshii, Y. Shibata, F. Sakai, S. Kondo, M. Kondo, H. Kotaki, K. Nakajima, Nucl. Instr. and Meth. A 437 (1999) 1.
[7] R. Kato, R.A.V. Kumar, T. Okita, S. Kondo, T. Igo, T. Konishi, S. Okuda, S. Suemine, G. Isoyama, Nucl. Instr. and Meth. A 455 (2000) 164.
[8] L. Wartski, J. Marcou, S. Roland, IEEE Trans. Nucl. Sci. NS-20 (1972) 544.
[9] S. Okuda, M. Nakamura, K. Yokoyama, Nucl. Instr. and Meth. A 445 (2000) 351.

ELSEVIER

Nuclear Instruments and Methods in Physics Research A 475 (2001) 492–497

NUCLEAR
INSTRUMENTS
& METHODS
IN PHYSICS
RESEARCH
Section A

www.elsevier.com/locate/nima

Electron bunch shape measurement using coherent diffraction radiation

B. Feng[a],*, M. Oyamada[b], F. Hinode[b], S. Sato[b], Y. Kondo[c], Y. Shibata[c],
M. Ikezawa[d]

[a] *FEL Center, Vanderbilt University, Box 1816 B, Nashville, TN 37212, USA*
[b] *Laboratory for Nuclear Science, Faculty of Science, Tohoku University, Sendai 982, Japan*
[c] *Department of Applied Physics, Faculty of Engineering Tohoku University, Sendai 980, Japan*
[d] *Research Institute for Scientific Measurements, Tohoku University, Sendai 980, Japan*

Abstract

Coherent diffraction radiation, generated by relativistic electron bunches passing through an aperture in a metallic foil, has been studied to diagnose the beam bunch shapes at the Tohoku 300 MeV linear accelerator. Using a polychrometer and fast data conversion method from frequency to time domain, we are able to realize a real-time, non-intercepting beam bunch shape diagnosis. The initial experimental results are presented. © 2001 Published by Elsevier Science B.V.

Keywords: Free electron laser; Electron bunch shape; Polychrometer; Coherent diffraction radiation

1. Introduction

The coherent synchrotron radiation (CSR) effect of electron bunches emitting from a bending magnet was first observed experimentally with a 300 MeV Linac at Tohoku University [1]. Since then the coherent effects in transition, Cherenkov and undulator radiation were established by other experiments [2–5]. The coherent radiation (CR) of a bunched electron beam produces much higher power than that of incoherent radiation. The CR power is proportional to the square of the electron number in a bunch rather than the electron number as in the case of incoherent radiation.

Moreover, the spectral distribution of the CR contains the information about the electron distribution in the bunch. This provides a method of measuring the CR spectrum to derive a bunch form factor and therefore to obtain the shape of a bunch. The measurement of electron bunch shape has been conducted using the CSR, coherent transition radiation (CTR), as well as the coherent diffraction radiation (CDR) [2,5]. As the CDR contributes less effect on the electron beam comparing with that of CTR and CSR, it is a much more suitable radiation source for monitoring the electron beam bunch shapes, especially for a real-time diagnosis. The CR spectrum is usually measured by a Michelson interferometer or by a grating spectrometer, which takes several to 30 min to measure the radiation spectrum. During such a long period, the electron beam might be

---

*Corresponding author. Tel.: +1-615-343-6446; fax: +1-615-343-1103.

E-mail address: bibo.feng@vanderbilt.edu (B. Feng).

0168-9002/01/$ - see front matter © 2001 Published by Elsevier Science B.V.
PII: S 0 1 6 8 - 9 0 0 2 (0 1) 0 1 6 0 5 - 9

changed, and what is measured is the average of the radiation spectrum. Obviously, those measurement methods are not suitable for real-time measurements of the bunch profile.

In this paper, we present our initial results of bunch shape measurement using a polychrometer and CDR. The polychrometer includes a couple of gratings and a ten-channel detector array. It can measure simultaneously ten different wavelengths and another ten points if changing the grating. This method is suitable for a real-time, non-intercepting diagnosis of the bunch profile. Following this introduction, we give a description of the diffraction radiation (DR) from apertures and the CR. We also show the method of extracting the bunch shape information from the CDR spectrum. Then, we briefly describe the polychrometer and the experimental setup, and show the initial experimental results. We present the measured spectrum of CDR from aluminum foils with a circular hole and with a slit, as well as that of CTR from an aluminum foil. As an example, we derive the electron bunch shape from measured spectrum of the CDR.

2. Coherent radiation from an electron bunch

2.1. Diffraction radiation

The DR is generated while a relativistic charge passes through an aperture in a metallic foil. The simplest aperture is a circular hole or a slit. According to the previous work [5], the DR intensity emitted from a relativistic electron passing through a circular aperture in an ideally conducting screen is described by

$$\frac{dI(\omega)}{d\omega\, d\Omega} = \frac{e^2 \beta^2 \sin^2 \theta}{\pi^2 c(1 - \beta^2 \cos^2 \theta)^2} \cdot \left(\frac{\pi a}{\beta\gamma\lambda}\right)^2 J_0^2\left(\frac{\pi a}{\lambda}\sin\theta\right)$$
$$\times K_1^2\left(\frac{\pi a}{\beta\gamma\lambda}\right) \quad (1)$$

where e is electron charge, $\beta = v/c$, γ the energy factor, θ the direction angle to the beam, and a the diameter of the aperture. The J_0 and K_1 are the Bessel function of zeroth order and the modified Bessel function of first order, respectively. The first part in the equation, separated by a dot, represents

the intensity of transition radiation (TR) emitted from the electron passing through the ideally conducting screen in vacuum, and the second part corresponds to the DR effect.

The DR intensity emitted by an electron passing through the center of a slit of width a, which opens along the x-axis, can be described as [6,7]

$$\frac{dI(\omega)}{d\omega\, d\Omega} = \frac{e^2}{8\pi^2 cf^2(f^2 + k_y^2)} \frac{k^2}{}$$
$$\times \left\{ 2e^{-fa}(f^2 + k_x^2) - \frac{2\alpha^2 e^{-fa}}{(f^2 + k_y^2)} \right.$$
$$\left. \times \left[(f^2 - k_y^2)\cos ak_y - 2fk_y \sin ak_y\right] \right\} \quad (2)$$

with $k = 2\pi/\lambda$ and

$$f = \sqrt{k_x^2 + \left(\frac{k}{\beta\gamma}\right)^2}$$

$$k_x = k \sin\theta \cos\varphi$$

$$k_y = k \sin\theta \sin\varphi$$

where λ is the radiation wavelength, γ the electron energy factor, θ and φ the standard polar angles.

A characteristic frequency $\omega_c = c\gamma/a$ can be defined to describe the DR property both from circular hole and from slit as shown in Eqs. (1) and (2). For the radiation frequency $\omega \gg \omega_c$, the radiation intensity is strongly reduced. For $\omega \ll \omega_c$, the DR distribution is similar to that of the TR, but the DR intensity is relatively lower. At the limit of aperture $a \to 0$, the DR becomes coincident with the TR. For $\omega \sim \omega_c$, the DR effects become obvious, even though its intensity is somewhat less than that of TR.

2.2. Coherent radiation

The radiation intensity emitting from N electrons can be described by [8]

$$I_{tot}(\omega) \approx NI_1(\omega) + N^2 F(\omega)I_1(\omega) \quad (3)$$

where $I_1(\omega)$ is the radiation intensity from one electron, $F(\omega)$ is the beam bunch factor, read as,

$$F(\omega) = \left| \int dz\, S(z)\exp\left[i\frac{\omega}{c}z\right] \right|^2 \quad (4)$$

where $S(z)$ is the normalized longitudinal distribution function of electrons in a bunch. In Eq. (3), the first term is the incoherent emission component. The second term, $N^2 F(\omega) I_1(\omega)$, describes the coherent part, which takes into account the phase relation between different electrons. The intensity of CR is proportional to the square of the number of electrons in a bunch.

2.3. Electron bunch shape

As shown in Eq. (3), a measurement of the coherent emission gives the longitudinal bunch form factor $F(\omega)$ and hence provides information about the longitudinal bunch distribution function $S(z)$. Therefore, the electron distribution in a bunch can be obtained from the inverse Fourier transformation of the form factor in Eq. (4), given by [9]

$$S(z) = \frac{1}{\pi c} \int_0^\infty [F(\omega)]^{1/2} \cos[\varphi(\omega) - \omega z/c] \, d\omega \quad (5)$$

where $\varphi(\omega)$ is the phase calculated from the observed form factor by the Kramers–Kronig relation:

$$\varphi(\omega) = -\frac{\omega}{\pi} \int_0^\infty \frac{\ln[F(\kappa)/F(\omega)]}{(\kappa^2 - \omega^2)} \, d\kappa. \quad (6)$$

3. Experimental setup

The experimental schematic diagram is shown in Fig. 1. R represents one of the radiators, which is an aluminum foil or an aluminum foil with a circular hole or a slit. When electron beam passes through the center of the radiator with $45°$ entrance angle, it emits radiation both in the forward direction along the electron beam and in the backward direction of specular reflection from the foil. In the experiments, the backward radiation is reflected by a fold mirror M1 and is collected by a parabolic reflection mirror P, whose focal length is 975 mm. The radiation point is set at the focus point of the parabolic mirror. Therefore, a parallel radiation light from P is reflected by mirror M2 and is finally focused in a polychrometer, a multi-channel detector array with a couple

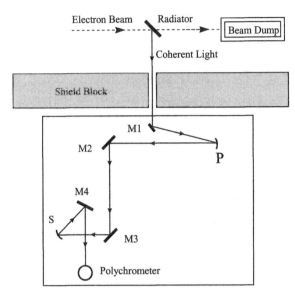

Fig. 1. Setup of the experiment. Radiation is generated from the radiator, collected by the fold mirror M1 and the parabolic mirror P, and finally focused onto a polychrometer.

of gratings. The radiation acceptance angle, limited by the diameter of the reflection mirror and the optical system, is about 31 mrad. There are three types of radiators mounted on a big turntable disk in which one of the radiators can be set on the electron beam trajectory. Three pieces of aluminum foil with circular holes, diameters 5, 10 and 15 mm, respectively, are used as the radiators. The thickness of the foil is about 100 μm. A foil with a 10 mm slit is also used in the DR measurement. In order to compare with the CTR, an aluminum foil of thickness of 15 μm is also used. A BeO plate is also placed on the disk to monitor the electron beam position.

The schematic diagram of the polychrometer [10] is shown in Fig. 2. The CR is reflected by a plane mirror M, dispersed by a spherical grating G, and is focused on the array of ten InSb detectors, which are placed on the Rowland circle. Two gratings with groove width of 1 or 2 mm can be switched by a step motor, resulting in the measured wavenumber range from 6.1 to 13.1 cm^{-1} and from 12.2 to 26.2 cm^{-1}, respectively. The signal from the detectors is processed by amplifier, sample hold circuit and ADC in a

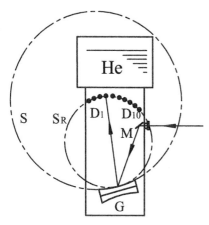

Fig. 2. Schematic diagram of the polychrometer. Radiation is reflected by a plane mirror M, dispersed by a spherical grating G, and focused on the array of ten InSb detectors.

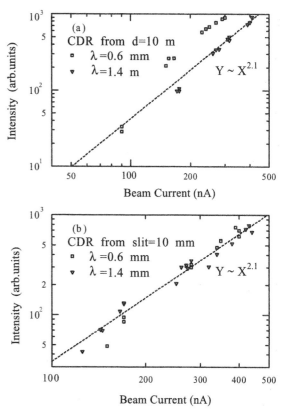

Fig. 3. Dependence of the CDR intensities on the electron beam currents: (a) from a hole with diameter of 10 mm; (b) from a slit with gap of 10 mm.

personal computer. A black body radiator (a high voltage mercury lamp) is used to calibrate the whole measuring system.

4. Experimental results

The experiments are conducted on the Tohoku 300 MeV linac with RF frequency of 2856 MHz at the Laboratory of Nuclear Science, Tohoku University. The electron beam energy of 150 MeV is used for this experiment and its energy spread is estimated about 0.5%. The macropulse duration is 2.5 μs and the reputation is 50 Hz. The beam current is measured by a secondary emission monitor. The beam current can be changed by adjusting a slit far from the radiator position, which do not effect on the electron beam energy. In the experiments, the beam current is changed from 70 to 500 nA. The beam is focused on the radiator with the transverse radius of about 3 mm.

The CR property, i.e., the radiation intensity is proportional to the square of the beam current, is confirmed in the experiment. The dependence of the intensities on the electron beam currents are shown in Fig. 3(a) and (b), where the radiation wavelengths are 0.6 and 1.4 mm. The DR emits from the aluminum foils with a 10 mm hole and a 10 mm slit, respectively. As shown in the figure, the intensity is approximately proportional to the

square of the beam current and therefore is proportional to the square of the electron number N in a bunch. This confirms that the observed radiation is the CDR.

In Fig. 4, the typical observed spectrum of CTR and CDR from different radiators is shown. The aperture diameters of the radiators are 5, 10 and 15 mm, respectively, and the slit gap is 10 mm. Two gratings, corresponding to the measured wavenumber range from 6.1 to 13.1 cm^{-1} and from 12.2 to 26.2 cm^{-1}, respectively, are used to measure the spectra in Fig. 4. The polychrometer has ten channels; however, two of them were broken during the experiments. The beam current is normalized to 1 nA for all of the radiation. As shown in the figure, the CR intensity obviously reduces with increasing the aperture of radiators. The intensity of CTR is relatively higher than that

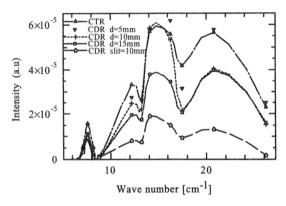

Fig. 4. Spectra from CTR and CDR with different aperture.

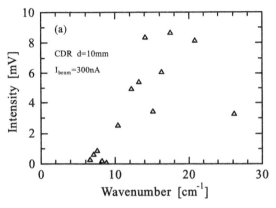

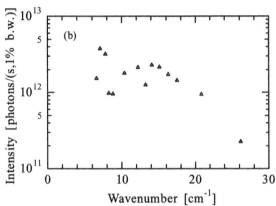

Fig. 5. Spectrum of CDR from a hole with diameter of 10 mm (a) and its calibrated spectrum (b). The average beam current is 300 nA.

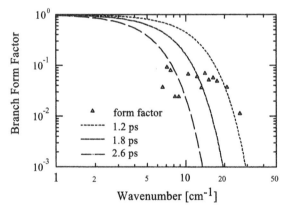

Fig. 6. The bunch form factor derived from the observed spectrum.

from other radiators with larger aperture at the same beam condition.

According to the CR property, the electron bunch shape can be derived from the radiation spectrum. As an example, a typical CDR spectrum from a radiator of 10 mm aperture is shown in Fig. 5(a) and its calibrated intensity in unit of number of photons per second, per 1% bandwidth is shown in Fig. 5(b). The observed wavenumber range is between 6.1 and 26.2 cm^{-1}. The average beam current is 300 nA, corresponding to the electron number of 5.3×10^6 in a bunch. Comparing with the theoretical incoherent DR intensity, the observed CDR intensity is enhanced nearly by a factor of the electron number in a bunch. From Eq. (3), the CDR intensity $I_{\text{tot}}(\omega) \sim N^2 F(\omega) I_1(\omega)$, where the intensity, $I_1(\omega)$, from a single electron can be derived from Eq. (1). According to the measured radiation spectrum in Fig. 5, the measured electron current in the experiment and the calculation CDR spectrum from a single electron, the bunch form factor, shown in Fig. 6, is derived from Eq. (3). For a Gaussian distribution, the bunch form factor can be described as $\exp[-(2\pi\sigma\nu)^2]$, where ν is the wavenumber and σ the rms width of the distribution. The theoretical calculation of bunch form factors is also shown in Fig. 6, where the Gaussian distributions of electron bunch with rms width of 1.2, 1.8 and 2.6 ps are assumed. In order to construct the electron distribution in a bunch, an entire bunch form factor is required. In practice, as the limitation of

the experiments and also of the CR itself, it is impossible to observe the entire CR spectrum. However, in the low wavenumber region the

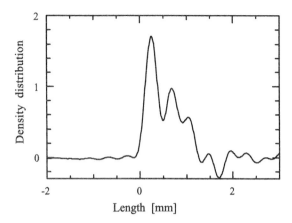

Fig. 7. The electron bunch shape derived from the bunch form factor.

bunch form factor, theoretically, becomes gradually unit and in the high wavenumber region the bunch form factor is relatively small resulting in small effect on the electron bunch shapes. In this example, 100 points are inserted between the measured spectral region with the polynomial interpolation. Outside this region, the bunch form factor with theoretical values of the 1.8 ps rms width Gaussian distribution are used. Total 512 points are used for the inverse Fourier transformation to derive the electron bunch profile from Eq. (5), shown in Fig. 7. The measured bunch length is about 0.7 mm (FMHW), which agrees with that in the early CTR and CDR experiments [11,12].

5. Discussion and conclusion

In summary, we observed the CDR from an aluminum foil with circular hole and slit. The CDR spectra in millimeter and submillimeter range were measured by a polychrometer. Depending on the measured spectrum the electron bunch profile was derived. This experiment demonstrated that a real-time, non-intercepting beam bunch length diagnosis for a 50 Hz bunch repetition may be realized using a polychrometer and fast data conversion method from frequency to time domain. However, there were a couple of disadvantages in the experiment. First, the number of measured data points for the radiation wavenumber was only ten for one shot so the wavelength range covered was insufficient. Even if we can change the grating to obtain another ten data point, the data were not measured from same electron pulse. The electron distribution in bunch should be different from shot to shot. In fact, the bunch length was found to be very unstable even though the beam energy and current was stable. Another disadvantage of this frequency domain method was that as we could not measure the entire spectrum, some artificially inserted data of the bunch form factor are used to derive electron bunch shape. In spite of the above problems, it is possible to use this bunch shape diagnosis to optimize the parameters such as the phase and power of the bunching section at Tohoku 300 MeV linac.

Acknowledgements

One of the authors, B. Feng, is grateful to The Japan Society for the Promotion of Science for financial support (ID No. P96216).

References

[1] T. Nakazato, et al., Phys. Rev. Lett. 63 (1989) 1245.
[2] T. Takahash, et al., Phys. Rev. E 48 (1993) 4674.
[3] T. Takahashi, et al., Phys. Rev. E 50 (1994) 4041.
[4] Y.U. Jeong, et al., Phys. Rev. Lett. 68 (1992) 1140.
[5] Y. Shibata, et al., Phys. Rev. E 52 (1995) 6787.
[6] M.L. Ter-Mikaelian, High-Energy Electromagnetic Processes in Condensed Media, Wiley, New York, 1972, p. 383.
[7] M. Castellano, et al., Nucl. Instr. and Meth. A 394 (1997) 275.
[8] J.S. Nodvick, D.S. Saxon, Phys. Rev. 96 (1954) 180.
[9] R. Lai, A.J. Sievers, Phys. Rev. E 50 (1994) R3342.
[10] Y. Kondo, Res. Rep. Lab. Nucl. Sci. Tohoku Univ. 26 (1993) 323.
[11] Y. Shibata, et al., Phys. Rev. E 49 (1994) 785.
[12] Y. Shibata, et al., Phys. Rev. E 52 (1995) 6787.

ELSEVIER

Nuclear Instruments and Methods in Physics Research A 475 (2001) 498–503

NUCLEAR
INSTRUMENTS
& METHODS
IN PHYSICS
RESEARCH
Section A

www.elsevier.com/locate/nima

Sensitivity of the CSR self-interaction to the local longitudinal charge concentration of an electron bunch ☆

R. Li

Thomas Jefferson National Accelerator facility, 12000 Jefferson Avenue, Newport News, VA 23606, USA

Abstract

Recent measurements of the coherent synchrotron radiation (CSR) effects indicated that the observed emittance growth and energy modulation due to the orbit-curvature-induced bunch self-interaction are sometimes bigger than predictions based on Gaussian longitudinal charge distributions. In this paper, by performing a model study, we show both analytically and numerically that when the longitudinal bunch charge distribution involves concentration of charges in a small fraction of the bunch length, enhancement of the CSR self-interaction beyond the Gaussian prediction may occur. The level of this enhancement is sensitive to the level of the local charge concentration. © 2001 Elsevier Science B.V. All rights reserved.

1. Introduction

In designs of next-generation accelerators, high-energy electron beams of short bunches with high charges are often transported through magnetic bending systems. For example, in order to obtain the high-peak-current beams required by future linear colliders and short-wavelength free-electron laser (FEL) drivers, bunch compression chicanes consisting of several magnetic bends are often used: after an energy correlation along the bunch length is imposed on the bunch by upstream radiofrequency (RF) cavities, high-charge bunches are transported through a chicane where the final bunch length can be manipulated by employing the energy dependence of path length

through the chicane. The Jefferson Lab infrared FEL driver energy recovery recirculator [1] and combiner rings in the compact linear collider (CLIC) design [2] are other examples of high-charge microbunches traversing through magnetic bending systems.

As an electron bunch goes through a bend, each electron emits synchrotron radiation. Coherent synchrotron radiation (CSR) occurs for radiation wavelength longer than the bunch length, as a result of the orbit-curvature-induced electromagnetic self-interactions within the bunch. These self-interactions, which are dominated by the collective synchrotron radiation forces at high energy, will induce energy modulation in the bunch, and may cause degradation of beam quality via the energy dependence of particle orbits in the bend. Understanding this CSR effect is crucial for designs of bending systems in future accelerators so as to meet the stringent design

☆ This work was supported by the U.S. Dept. of Energy under Contract No. DE-AC05-84ER40150.

E-mail address: lir@jlab.org (R. Li).

requirements on the preservation of small transverse and longitudinal emittances.

The analyses of the CSR self-interaction [3,4], based on a rigid-line-charge model, can be applied for *general* longitudinal charge distributions. However, since the analytical results for Gaussian beams or beams with linear density are explicitly given, one usually applies these results to estimate the CSR effects using measured or simulated rms bunch lengths. Recently, simulations [5,6] were developed to study the CSR effect on bunch dynamics for *general* bunch distributions. However, due to the lack of knowledge of the detailed longitudinal phase-space distributions when simulating a CSR experiment, such simulations are sometimes carried out assuming a *Gaussian* phase-space distribution with the rms bunch length derived from bunch length measurements. Recent CSR experiments [7–9] indicated that the measured energy modulation and emittance growth are sometimes bigger than results predicted by these previous analyses and simulations [6] using a Gaussian beam. To understand this observed enhancement of the CSR effect, it is instructive to study the CSR self-interaction for a compressed bunch which may be highly non-Gaussian. In general, the final distribution of a fully compressed bunch is determined by many factors, including the details of the electron source, upstream space-charge interaction, RF structure, wake function and optics — all of which vary with different designs. In this paper, instead of treating the general distribution, we carry out a model study. In this model, an initial Gaussian line bunch, with linear and second order (due to RF curvature) energy correlation, was mapped to the end of a magnetic chicane by the first- and second-order longitudinal optics. Analysis and numerical simulation were then performed within the limit of classical electrodynamics to study the longitudinal CSR self-interaction force on the fully compressed bunch. This study reveals a general feature of the CSR self-interaction: whenever there is longitudinal charge concentration in a small fraction of a bunch length, enhancement of the CSR effect beyond the Gaussian prediction could occur; moreover, the level of this enhancement is sensitive to the level of the local charge concentration within a bunch. This sensitivity should be given serious consideration in designs of future electron accelerators and when one interprets measurement results of CSR experiments, so as to avoid underestimation of the CSR-induced phase-space dilution for high-charge short electron bunches traversing bending systems.

2. Bunch compression optics

In order to study the CSR self-interaction for a compressed bunch, let us first find the longitudinal charge distribution for our model bunch when it is fully compressed by a chicane (here we assume that the perturbation of bunching process by CSR is negligible because most of the CSR perturbations take place at the end of the chicane when the bunch length reaches its minimum value). We will first study the bunch compression for a beam with zero uncorrelated energy spread $\delta_{un} = 0$, and then generalize the results for $\delta_{un} \neq 0$. Consider an electron bunch with N electrons. The longitudinal charge density distribution function of the bunch at time t is

$$\rho(s,t) = N e \, n(s,t), \qquad \int n(s,t)\,\mathrm{d}s = 1 \qquad (1)$$

where s is the longitudinal distance from a reference electron ($s > 0$ for bunch head), and $n(s,t)$ is the longitudinal density distribution of the bunch. As the bunch passes through the chicane, s varies with t for each electron and consequently, the density distribution also depends on t. Let us identify each electron by the parameter μ, which is the initial longitudinal position s_0 of the electron at $t = t_0$:

$$s_0 \equiv s(\mu, t_0) = \mu. \qquad (2)$$

At $t = t_0$, let the bunch be a line aligned on the design orbit at the entrance of a bunch compression chicane, with a Gaussian longitudinal density distribution and the initial rms bunch length σ_{s_0}:

$$n(s_0, t_0) = n_0(\mu) = \frac{1}{\sqrt{2\pi}\sigma_{s_0}} e^{-\mu^2/2\sigma_{s_0}^2}. \qquad (3)$$

To compress the bunch using the chicane, a linear energy correlation is imposed on the bunch by an upstream RF cavity, along with a slight second-order energy correlation due to the curvature of the RF wave form. With δ_1 and δ_2, the linear and second-order correlation coefficients ($\delta_1, \delta_2 > 0$ and $\delta_2 \ll \delta_1$), the relative energy deviation $\delta = \Delta E / E_0$ from the design energy E_0 is then

$$\delta(\mu, t_0) = -\delta_1 \frac{\mu}{\sigma_{s_0}} - \delta_2 \left(\frac{\mu}{\sigma_{s_0}} \right)^2 \qquad (4)$$

where we assume no uncorrelated energy spread. When the beam propagates to the end of the chicane at $t = t_f$, the final longitudinal coordinates of the electrons are

$$s(\mu, t_f) \simeq s(\mu, t_0) + R_{56}\delta(\mu, t_0) + T_{566}[\delta(\mu, t_0)]^2 \qquad (5)$$

$$= s(\mu, t_0)\left(1 - \frac{R_{56}\delta_1}{\sigma_{s_0}}\right) - \alpha[s(\mu, t_0)]^2 \qquad (6)$$

where $R_{56} = dL/d\delta$ and $T_{566} = d^2L/2d\delta^2$, with L denoting the path length of an electron through the chicane, and $\alpha \equiv (R_{56}\delta_2 - T_{566}\delta_1^2)/\sigma_{s_0}^2$. To obtain the maximum compression of the bunch, one designs the chicane by choosing R_{56}, the initial bunch length σ_{s_0} and the initial energy correlation δ_1 to satisfy

$$1 - R_{56}\delta_1/\sigma_{s_0} = 0, \quad s_f \equiv s(\mu, t_f) = -\alpha[s(\mu, t_0)]^2. \qquad (7)$$

For typical bunch-compression chicanes, one has $R_{56} > 0$ and $T_{566} < 0$; therefore $\alpha > 0$, which implies $s_f \leqslant 0$ from Eq. (7). Using Eqs. (7) and (2), we have

$$\mu^2 = -s_f/\alpha \quad (\alpha > 0, \; s_f \leqslant 0). \qquad (8)$$

The final longitudinal density distribution $n(s_f, t_f)$ can then be obtained from charge conservation $n_0(\mu) \, d\mu = n(s_f, t_f) \, ds_f$ and from Eq. (8), which gives

$$n(s_f, t_f) = \frac{1}{\sqrt{2\pi}\sigma_{s_f}} \frac{e^{s_f/\sqrt{2}\sigma_{s_f}}}{\sqrt{-s_f/\sqrt{2}\sigma_{s_f}}} H(-s_f) \qquad (9)$$

where $H(s)$ is the Heaviside step function, and σ_{s_f} denotes the rms of the final longitudinal

distribution

$$\sigma_{s_f} \equiv \sqrt{\langle s_f^2 \rangle - \langle s_f \rangle^2} = \sqrt{2}\alpha\sigma_{s_0}^2. \qquad (10)$$

The first moment of the distribution in Eq. (9) is $\langle s_f \rangle = -\sigma_{s_f}/\sqrt{2}$.

The distribution in Eq. (9) obtained for $\delta_{un} = 0$ is divergent at $s_f = 0$. For a realistic beam, uncorrelated energy spread δ_{un} should be added to Eq. (4) (here we assume δ_{un} has a Gaussian distribution with $\langle \delta_{un} \rangle = 0$, and rms width δ_{un}^{rms}). As a result, the final longitudinal phase-space distribution can be obtained by combining Eqs. (4), (7) and (10), yielding

$$s_f \simeq -(\sigma_{s_f}/\sqrt{2}\delta_1^2)\delta^2 + R_{56}\delta_{un} \qquad (11)$$

and the final effective rms bunch length becomes

$$\sigma_{s_f}^{eff} \equiv \sqrt{\langle s_f^2 \rangle - \langle s_f \rangle^2} = \sigma_{s_f}\sqrt{1 + a^2} \qquad (12)$$

with σ_{s_f} given by Eq. (10), and a the intrinsic bunch width normalized by the rms bunch length (see Fig. 1)

$$a = R_{56}\delta_{un}^{rms}/\sigma_{s_f} \qquad (13)$$

which characterizes the local charge concentration (non-Gaussian feature) of the compressed density. For example, when $\sigma_{s_0} = 1.26$ mm, $R_{56} = 45$ mm, and $\delta_1 = 0.028$, the compression condition in Eq. (7) is satisfied. With $\alpha = 0.08$ mm^{-1}, and $\delta_{un}^{rms} = 0.0012$, Eq. (10) gives the final compressed bunch length $\sigma_{s_f} = 0.18$ mm with normalized width $a = 0.3$. An example of the longitudinal

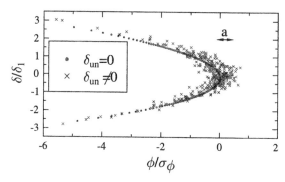

Fig. 1. Example of the longitudinal phase-space distribution for a compressed beam with RF curvature effect.

phase-space distribution described by Eq. (1) is shown in Fig. 1.

3. CSR for a compressed beam

To study the CSR self-interaction for a compressed bunch in steady-state circular motion, we first enlist the formulae established by earlier analyses [3,4] for steady-state CSR self-interaction of a rigid-line bunch. Let the longitudinal density distribution function of the bunch be $\lambda(\phi)$ for $\phi = s/R$, with R the orbit radius. The rms bunch length is σ_s and the rms angular width is $\sigma_\phi = \sigma_s/R$. The longitudinal collective force via space charge and CSR interaction is then derived from the scalar and vector potentials Φ and $\mathbf{A}$:

$$F_\theta(\phi) = \frac{e}{\beta c}\left(\frac{\partial}{\partial t}\Phi - \boldsymbol{\beta}\frac{\partial}{\partial t}\mathbf{A}\right)$$
$$= \frac{-Ne^2}{R^2}\int_0^\infty \frac{1-\beta^2\cos\theta}{2\sin(\theta/2)}\frac{\partial}{\partial\phi}\lambda(\phi-\varphi)\,d\varphi \tag{14}$$

where $\boldsymbol{\beta} = \mathbf{v}/c$, $\beta = |\boldsymbol{\beta}|$, $\gamma = 1/\sqrt{1-\beta^2}$, and θ is an implicit function of φ via the retardation relation $\varphi = \theta - 2\beta\sin(\theta/2)$. In this paper, we treat only the high-energy case when $\gamma \gg \theta^{-1}$ and $\theta \simeq 2(3\varphi)^{1/3}$. In this case, $F_\theta(\phi)$ is dominated by the radiation interaction

$$F_\theta(\phi) \simeq \frac{-Ne^2}{R}\int_0^\infty \frac{\beta^2\sin\theta/2}{1-\beta\cos\theta/2}\frac{\partial}{\partial\phi}\lambda(\phi-\varphi)\,d\varphi$$
$$\simeq \frac{-2Ne^2}{3^{1/3}R^2}\int_0^\infty \varphi^{-1/3}\frac{\partial}{\partial\phi}\lambda(\phi-\varphi)\,d\varphi \tag{15}$$

and the CSR power due to the radiation interaction is

$$P = -N\beta_\theta c\int F_\theta(\phi)\lambda(\phi)\,d\phi. \tag{16}$$

Results for the longitudinal collective force and the CSR power for a rigid-line Gaussian beam are [3,4]:

$$\lambda^{\text{Gauss}}(\phi) = \frac{1}{\sqrt{2\pi}}e^{-\phi^2/2\sigma_\phi^2} \quad \left(\sigma_\phi \gg \frac{1}{\gamma^3}\right) \tag{17}$$

$$F_\theta^{\text{Gauss}}(\phi) \simeq F_g g(\phi), \quad F_g = \frac{2Ne^2}{3^{1/3}\sqrt{2\pi}R^2\sigma_\phi^{4/3}} \tag{18}$$

with

$$g(\phi) = \int_0^\infty \frac{(\phi/\sigma_\phi - \phi_1)}{\phi_1^{1/3}}e^{-(\phi/\sigma_\phi-\phi_1)^2/2}\,d\phi_1 \tag{19}$$

$$P^{\text{Gauss}} \simeq \beta_\theta c\frac{N^2 e^2}{R^2\sigma_\phi^{4/3}}\frac{3^{1/6}\Gamma^2(2/3)}{2\pi} \tag{20}$$

where $\Gamma(x)$ is the Gamma function.

Next, we analyze the CSR self-interaction for a compressed bunch. The longitudinal distribution for a compressed bunch $\lambda^{\text{cmpr}}(\phi)$ with intrinsic width due to $\delta_{\text{un}} \neq 0$ is the convolution of the longitudinal density distribution $\lambda_0^{\text{cmpr}}(\phi)$ with $\delta_{\text{un}} = 0$ and a Gaussian distribution $\lambda_m(\phi)$:

$$\lambda^{\text{cmpr}}(\phi) = \int_{-\infty}^\infty \lambda_0^{\text{cmpr}}(\phi-\varphi)\lambda_m(\varphi)\,d\varphi \tag{21}$$

$$\lambda_0^{\text{cmpr}}(\phi) = \frac{1}{\sqrt{2\pi}\sigma_\phi}\frac{e^{\phi/\sqrt{2}\sigma_\phi}}{\sqrt{-\phi/\sqrt{2}\sigma_\phi}}H(-\phi) \tag{22}$$

$$\lambda_m(\phi) = \frac{1}{\sqrt{2\pi}\sigma_{m\phi}}e^{-\phi^2/2\sigma_{m\phi}^2}, \quad \sigma_{m\phi} = \frac{R_{56}\delta_{\text{un}}^{\text{rms}}}{R} \tag{23}$$

where $\lambda_0^{\text{cmpr}}(\phi)$ is obtained from Eq. (9). Combining Eq. (15) with Eq. (15), and denoting a as the intrinsic width of the bunch relative to the rms bunch length (see Fig. 1)

$$a = \frac{\sigma_w}{\sigma_s} \quad (\sigma_w = R_{56}\delta_{\text{un}}^{\text{rms}}) \tag{24}$$

one finds the steady-state CSR longitudinal force for a compressed bunch (with $0 < a < 1$) under the condition $\sigma_\phi > \sigma_{m\phi} \gg \gamma^{-3}$:

$$F_\theta^{\text{cmpr}}(\phi) = \int_{-\infty}^\infty F_{\theta_0}^{\text{cmpr}}(\varphi)\lambda_m(\phi-\varphi)\,d\varphi. \tag{25}$$

It can be shown that $F_{\theta_0}^{\text{cmpr}}(\varphi)$ in Eq. (25) is

$$F_{\theta_0}^{\text{cmpr}}(\phi) \simeq \frac{-2Ne^2}{3^{1/3}R^2}\int_0^\infty \varphi^{-1/3}\frac{\partial}{\partial\phi}\lambda_0^{\text{cmpr}}(\phi-\varphi)\,d\varphi$$
$$= -2^{1/4}F_g\frac{dG(y)}{dy} \tag{26}$$

with $y = \phi/\sigma_\phi$ and F_g given in Eq. (18), and

$$G(y) = H(-y)e^{-|y|/\sqrt{2}}|y|^{1/6}\Gamma\left(\frac{2}{3}\right)\Psi\left(\frac{2}{3},\frac{7}{6};\frac{|y|}{\sqrt{2}}\right)$$
$$+ H(y)y^{1/6}\Gamma\left(\frac{1}{2}\right)\Psi\left(\frac{1}{2},\frac{7}{6};\frac{y}{\sqrt{2}}\right) \qquad (27)$$

where $\Psi(a,\gamma;z)$ is the degenerate hypergeometric function

$$\Psi(\alpha,\gamma;z) = \frac{1}{\Gamma(\alpha)}\int_0^\infty e^{-zt}t^{\alpha-1}(1+t)^{\gamma-\alpha-1}\,dt.$$
$$(28)$$

As a result, we have

$$F_\theta^{cmpr}(\phi) = \frac{2^{1/4}F_g}{\sqrt{2\pi}a^{5/6}}f\left(\frac{\phi}{a\sigma_\phi};a\right) \qquad (29)$$

$$f(y;a) = \int_{-\infty}^{\infty} G(ax)(y-x)e^{-(y-x)^2/2}\,dx. \qquad (30)$$

Similarly, the CSR power can also be obtained for the compressed bunch using Eq. (16) with $\lambda^{cmpr}(\phi)$ in Eq. (21) and $F_\theta^{cmpr}(\phi)$ in Eq. (29), which gives

$$\frac{P^{cmpr}}{P^{Gauss}} \simeq 0.75\frac{I(a)}{a^{5/6}},$$

$$I(a) = -\int_{-\infty}^{\infty} f\left(\frac{\phi}{a\sigma_\phi};a\right)\lambda^{cmpr}(\phi)\,d\phi. \qquad (31)$$

Numerical integration shows that $|f(y;a)|_{max}$ — the maximum of $|f(y;a)|$ for fixed a — is insensitive to a for $0 < a < 1$, as depicted in Fig. 2. As a result, for a compressed bunch with fixed σ_ϕ, we found from Eq. (29) the amplitude of the CSR force $F_\theta^{cmpr}(\phi)$ varies with $a^{-5/6}$. Therefore, in

contrast to the well-known scaling law $R^{-2/3}\sigma_s^{-4/3}$ for the amplitude of the longitudinal CSR force for a Gaussian bunch, a bunch described by Eq. (21) has $|F_\theta^{cmpr}|_{max} \propto R^{-2/3}\sigma_s^{-1/2}\sigma_w^{-5/6}$ with $\sigma_w = R_{56}\,\delta_{un}^{rms}$ denoting the intrinsic width of the bunch. Likewise, for $a = 0.1$, 0.2, and 0.5, we found from numerical integration that $I(a) \simeq 0.76$, 0.90 and 1.02, respectively, and correspondingly $P^{cmpr}/P^{Gauss} \simeq 3.9$, 2.6 and 1.4. This dependence of the amplitude of the CSR force and power on the intrinsic width of the bunch for a fixed rms bunch length, manifests the sensitivity of the enhancement of the CSR effect on the local charge concentration in a longitudinal charge distribution.

In Figs. 3 and 4, we plot the longitudinal density function for various charge distributions with the same rms bunch lengths (except the $\sqrt{1+a^2}$ factor in Eq. (12)), and the longitudinal CSR collective forces associated with the various distributions. The amplitude of F_θ^{cmpr} in Fig. 4 agrees with the $a^{-5/6}$ dependence in Eq. (29). Good agreement of the analytical result in Eq. (29) with the simulation result [6] for the CSR force along the example distribution in Fig. 1 is shown in Fig. 5. Figs. 3 and 4 show clearly that the more locally concentrated are the charges than the smooth behavior of a Gaussian beam, the more enhancement will result from the CSR self-interaction forces. This general feature of the CSR interaction and emission is not limited to the RF curvature effect described in this paper [10];

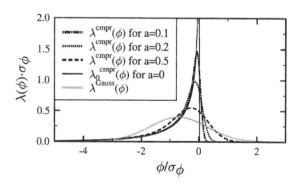

Fig. 3. Longitudinal charge distribution for a compressed beam with relative width described by a, compared with a Gaussian distribution. All the distributions here have the same angular rms size σ_ϕ.

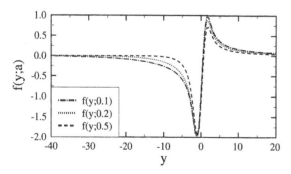

Fig. 2. $f(y;a)$ vs. y for $a = 0.1$, 0.2, and 0.5.

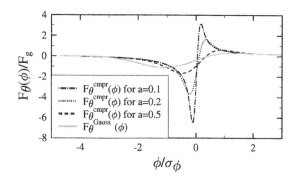

Fig. 4. Longitudinal CSR force along the bunch for various charge distributions illustrated in Fig. 4.

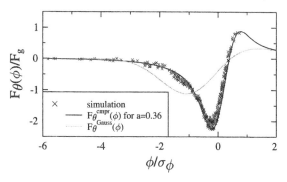

Fig. 5. Comparison of the analytical and numerical results of the longitudinal CSR force along the bunch for the example distributions illustrated in Fig. 1. The transverse rms size σ_x used in the simulation is $\sigma_x \simeq 3\sigma_s$.

furthermore, our simulation shows that the enhancement also occurs in the transient regime. Comparing with a Gaussian beam, the local charge concentration in a small fraction of the bunch length can lead to a bigger emittance growth for a compressed or over-compressed beam in a chicane, and a larger energy spread can be observed if the compressed beam is sent to a spectrometer magnet downstream of the chicane, where the beam can further interact with itself via CSR. The frequency-domain analysis and the effect of interaction with off-axis particles will be discussed elsewhere.

Acknowledgements

This work was inspired by the CSR measurement led by H.H. Braun at CERN. The author is also grateful for valuable discussions with J.J. Bisognano, P. Piot, C. Bohn, D. Douglas, G. Krafft and B. Yunn for the CSR measurement at Jefferson Lab. This work was supported by the U.S. Dept. of Energy under Contract No. DE-AC05-84ER40150.

References

[1] G.R. Neil, et al., Phys. Rev. Lett. 84 (2000) 662.
[2] J.P. Delahaye, et al., Proceedings of the 1999 Particle Accelerator Conference (IEEE No. 99CH36366), Vol. 1, p. 250.
[3] Y.S. Derbenev, et al., DESY Report No. TESLA-FEL-95-05, 1995, unpublished.
[4] J. Murphy, et al., Part. Accel. 57 (1997) 9.
[5] M. Dohlus, T. Limberg, Nucl. Instr. and Meth. A 393 (1996) 494.
[6] R. Li, Nucl. Instr. and Meth. A 429 (1998) 310.
[7] H.H. Braun, et al., Phys. Rev. Lett. 84 (2000) 658.
[8] H.H. Braun, et al., Phys. Rev. ST Accel. Beams 3 (2000) 124,402.
[9] P. Piot, et al., Proceedings of the 2000 European Particle Accelerator Conference, 2000, p. 1546.
[10] G.A. Krafft, DIPAC97, LNF-97/048, Frascati, 45, 1997.

ELSEVIER

Nuclear Instruments and Methods in Physics Research A 475 (2001) 504–508

NUCLEAR
INSTRUMENTS
& METHODS
IN PHYSICS
RESEARCH
Section A

www.elsevier.com/locate/nima

Application of electro-optic sampling in FEL diagnostics

X. Yan[a], A.M. MacLeod[a], W.A. Gillespie[a], G.M.H. Knippels[b,*], D. Oepts[b], A.F.G. van der Meer[b]

[a] School of Science and Engineering, University of Abertay Dundee, Bell Street, Dundee DD1 1HG, UK
[b] FOM-Institute for Plasma Physics "Rijnhuizen", P.O. Box 1207, 3430 BE, Nieuwegein, The Netherlands

Abstract

The electro-optic sampling technique has been used for the full characterization (both amplitude and phase) of freely propagating pulsed electromagnetic radiation (such as FEL pulses, transition radiation) and for the quasistatic electric field of relativistic electron bunches. Measurements of the electron bunch length and shape have been performed with both non-intercepting and intercepting methods. Sub-picosecond time resolution has been obtained with this electro-optic detection technique. © 2001 Elsevier Science B.V. All rights reserved.

PACS: 41.75.Ht; 41.60.Cr; 41.85.Ew

Keywords: Electro-optic sampling; ZnTe; FEL; Electron bunch; Ultrafast diagnostics.

1. Introduction

Over the last decade, there has been an increasing interest in the use of ultra-short relativistic electron bunches in free electron lasers (FELs) and other accelerator-based light sources. A diagnostic for the bunch length and shape will be crucial for the success of these machines. Also, a detailed characterization of the optical output is frequently required for the analysis of the data obtained in the experiments that make use of the FEL as a radiation source.

It was demonstrated that electro-optic sampling is an excellent technique to measure the longitudinal electric field profile of both optical pulses [1] and electron pulses [2]. The electro-optic

sampling technique is based on the linear electro-optic effect (also known as Pockels effect) [3]: when an electric field is applied to an electro-optic crystal, it induces birefringence. The birefringence can then be probed by a synchronized ultra-short Ti:Sapphire (Ti:S) laser pulse. The induced birefringence causes the initially linearly polarized optical probe beam to acquire a phase difference between its polarization components parallel and perpendicular to the induced optical axis of the sensor crystal, and therefore the probe beam will become elliptically polarized. The degree of ellipticity can be measured by using suitable polarization optics. By scanning the delay between the Ti:S pulse train and the FEL (or the electron) pulse train, a cross-correlation of the incident electric field of the FEL (or the relativistic electron) pulse is obtained. Note that with this electro-optic detection technique, an asymmetric pulse shape can be measured with sub-picosecond

*Corresponding author. Tel.: +31-30-6096999; fax: +31-30-6031204.

E-mail address: knippels@rijnh.nl (G.M.H. Knippels).

time resolution and without time-reversal ambiguity. Furthermore, information about the electric field, including its sign, is obtained, in contrast to most other techniques that yield only information on the intensity.

The experiments described in this paper have been carried out at the free-electron laser for infrared experiments (FELIX) facility [4].

2. Characterization of the FEL optical output

Fig. 1 schematically illustrates the experimental setup for the electro-optic measurement of the FEL optical pulse duration and shape. A Ti:S laser (12 fs pulses at 100 MHz) is actively synchronized to the master RF-clock and is used as the probe beam [5]. The Ti:S beam is linearly polarized at 45° with respect to the polarization direction of the FEL radiation. The Ti:S beam and the FEL beam are both focused by a parabolic mirror onto a 500-μm-thick crystal of $\langle 1\,1\,0 \rangle$-oriented ZnTe. The Ti:S laser probes the birefringence of the crystal induced by the incident FEL pulse, and the signal is measured with detector (a) or (b) respectively. For small phase retardation, the intensity of the Ti:S pulse emerging from the second polarizer (in detector (a)) is proportional to the square of the electric field of the FEL pulse, while the response of each diode of the balanced detector (b) is proportional to the electric field strength of the FEL micropulse.

One can measure the intensity profile (the time-averaged value of E^2) of the FEL micropulse with detector (a), by scanning the optical delay between the FEL pulse train and the Ti:S pulse train. The time resolution is determined by the jitter ($\sim 1\,\mathrm{ps}$) that is present during the scan (In case of poor synchronization, adequate time-resolution can be obtained by using single-shot or differential-optical-gating methods) [6]. We measured the intensity profile of the FEL micropulse at a wavelength of 150 μm and at different cavity detuning lengths ΔL. For a large value of ΔL, the FEL pulse develops a leading edge that can be fitted well with an exponential function. The time-constant of the exponential is determined by the cavity loss α and the applied cavity detuning ΔL and is given by $\tau = 2\Delta L/c\alpha$. When the optical pulse has an exponential leading edge, the time-bandwidth product (fwhm value) for different cavity detunings is in the range of 0.2–0.3 [7].

Using detector (b) in Fig. 1, one can sample the electric field of an FEL micropulse. The Ti:S pulse repetition frequency is phase-locked to the electron pulse repetition frequency and therefore also to the FEL micropulse at saturation, when the pulse shape has become stationary. So, at a fixed setting of the optical delay line, the Ti:S pulses sample the FEL output at a fixed position in the micropulse. However, this is only true with regard to the envelope of the FEL micropulse. The phase of the FEL carrier wave shifts in each round-trip with respect to the envelope by an amount $\Delta\Phi = 2\pi(2\Delta L/\lambda)$, due to the applied cavity detuning. This phase shift leads to a beat oscillation in the observed signal with $f_{\mathrm{beat}} = -c\Delta L/\lambda L$, [8] where L is the FEL cavity length (6 m). The phase of the beat signal varies from macropulse to macropulse, because the optical field develops from spontaneous emission and its phase is not fixed to the Ti:S pulses. With the aid of the following rapid-scanning technique, both the phase and the amplitude of the optical micropulse field can be measured in a single macropulse ($\leqslant 10\,\mu\mathrm{s}$).

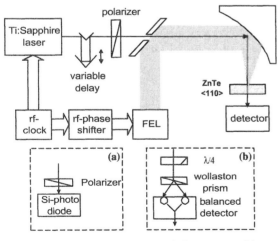

Fig. 1. The electro-optic cross-correlation setup with two different detectors. For detector (a) $I_{\mathrm{EO}} \propto E^2$, for detector (b) $I_{\mathrm{EO}} \propto E$.

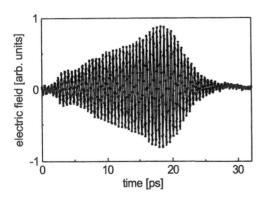

Fig. 2. The full electric-field profile of the shortest measured FEL micropulse at 150 μm wavelength.

The repetition rate of the FEL micropulse can be changed electronically via a voltage controlled RF-phase shifter, which sweeps the electron pulse train over the Ti : S pulse train at a speed of a few ps per μs. The complete electric-field profile is thereby measured within a few microseconds. An additional advantage of this rapid-scanning technique is that on a time scale of a few microseconds there is negligible jitter ($\leqslant 50$ fs) between the Ti : S pulse train and the electron pulse train [8]. Provided that the effective FEL cavity detuning is set equal to $k\lambda/2$ (k is an integer number) when the Ti : S laser pulse train is sampling the FEL micropulse, the beat signal measured with this rapid-scanning technique just corresponds to the electric-field profile of the FEL micropulse (assuming that the micropulse shape is constant during the macropulse). Fig. 2 shows the electric-field profile of the shortest measured FEL micropulse at 150 μm wavelength. The observed number of electric-field cycles of the micropulse is 26 at the fwhm, corresponding to 18 optical field cycles of the intensity profile at the fwhm.

3. Characterization of ultrashort relativistic electron pulses

Fig. 3 shows the experimental setup used for the electro-optic sampling of relativistic electron

pulses ($E = 46$ MeV, $Q = 200$ pC), which are produced by the radio-frequency linear accelerator at FELIX. The electron bunch profile is measured at the entrance of the undulator (position (a) in Fig. 3) with two different methods, both using detector (b) described in Fig. 1.

As a first experiment, the coherent transition radiation (CTR) of the electron bunch at position (a) is measured with the electro-optic detection technique. In our experiment, the CTR is generated by placing an aluminum foil in the path of the electron beam inside the vacuum pipe close to the entrance of the undulator. The CTR is taken out of the beam pipe through a crystalline quartz window. It is then made to propagate collinearly with the Ti : S beam and both beams are focused onto a sensor crystal of ZnTe by a parabolic mirror. By using the rapid-scanning technique, the electron pulse train is swept over the Ti : S pulse train within a macropulse, and the measured field strength of the CTR of the electron bunch is illustrated in Fig. 4. In contrast to the usual interferometric CTR methods [9], our measurement clearly shows an asymmetric profile and allows unambiguous identification of the leading and trailing edge. The effect of absorption and re-emission by ambient water vapor is evident in the dotted curve in Fig. 4(a), and it can be reduced strongly by purging the CTR path with dry nitrogen (see the solid line in Fig. 4(b)). However, the profile of the measured CTR pulse is clearly different from the electron bunch profile, due to the emission and focusing properties of the CTR. It can be seen that the low-frequency terms are absent, as only the high-frequency part of the CTR is effectively focused and sampled.

With the same setup shown in Fig. 3, the electric field is also measured in a second experiment at the entrance of the undulator inside the vacuum pipe, while the FEL is lasing properly. The electron beam has a diameter of approximately 1 mm at this position and the crystal is located at a distance of $R = 6$ mm from the center of the electron beam. The Ti : S laser pulse is linearly polarized and propagates parallel to the electron beam direction. Fig. 5 shows measurements of the observed electric-field profiles. In case of Fig. 5(a), the large

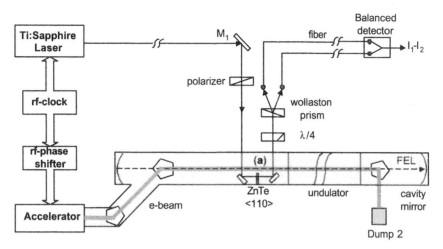

Fig. 3. Experimental setup of the electro-optic measurement of the shape of relativistic electron bunches. The shaded parts indicate the vacuum housing for the electron beam.

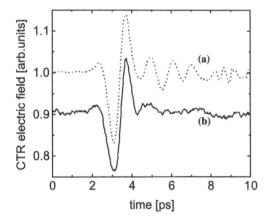

Fig. 4. Electro-optic measurements of the coherent transition radiation (CTR). In (a) the measured CTR pulse profile with strong absorption and re-emission of the water vapor in the ambient air is shown. In (b) the measured CTR pulse profile with the effect of purging with dry nitrogen is shown.

peak (marked by an arrow) represents the direct field of the electron bunch. The direct electric field of the bunch at the crystal will always have the same sign, and the other parts of the signal that have both positive and negative signs are therefore attributed to the wake-fields excited by the electron bunches, due to unavoidable discontinuities in the beam pipe and in the measurement

system itself[1]. The shortest electron bunch duration we have measured is about 1.7 ± 0.2 ps fwhm, as illustrated in Fig 5(b), and the time resolution (estimated as $\tau \approx 2R/\gamma c$) [12] is about 0.44 ps in this configuration. The maximum electric field strength at the ZnTe crystal position can be estimated as $E = Q/2\pi\varepsilon_0 R L_b$, where Q is the total charge of the electron bunch and L_b is its effective length [12]. In our case, the electric field strength is approximately 12 kV/cm at the crystal, indicating a sensitivity of our setup of around 1 kV/cm for a $S/N = 1$. In contrast with the CTR measurement, the DC electric-field component is now effectively sampled, and accurate information on the electron bunch length is obtained. The accelerator settings used to drive the FEL at FELIX result in a clearly asymmetric pulse shape. The steep rising edge of the electron bunch is consistent with the observed enhancement of the coherent spontaneous emission, which can considerably facilitate the startup of a short-pulse FEL at longer wavelengths [13]. We measured the duration of electron bunches as a function of the phase setting of the prebuncher input RF-field. It has been seen that the electron bunch profile is quite sensitive to the setting of prebuncher and the fwhm length was measured to change between 1.7 and 4.1 ps.

---

[1] Similar signals have been observed by Fitch et al. [10,11].

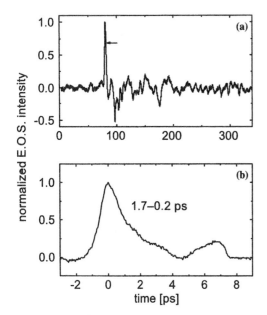

Fig. 5. The electric-field profiles measured at the entrance of the undulator inside the vacuum pipe, while the FEL is running properly. (a) Scanned over several hundred picoseconds, the large positive peak marked by an arrow is the direct field of the passing electron bunch, while the other parts of the signal are attributed to the effects of the wake fields excited by the electron bunches in the beam pipe. (b) The electron bunch profile measured inside the vacuum pipe has a clearly asymmetric pulse shape.

4. Conclusion

In conclusion, the electro-optic sampling technique has broadband detection capability for the full characterization (both amplitude and phase) of freely propagating pulsed electromagnetic radiation (such as FEL pulses, transition radiation) and for the field of relativistic electron bunches. Electro-optic measurements of the length and the shape of relativistic electron bunches have

been performed with both non-intercepting and intercepting methods. Sub-picosecond time resolution has been obtained with this electro-optic detection technique. In addition, it can be used to measure the wake-fields inside the beam pipe.

Acknowledgements

We thank Dr. W. Seidel, Dr. T.I. Smith and Dr. C.W. Rella for valuable discussions. This work is part of the research program of the Stichting FOM, which is financially supported by the 'Nederlandse Organisatie voor Wetenschappelijk Onderzoek (NWO)'. X. Yan, A.M. MacLeod, and W.A. Gillespie are grateful to the UK Engineering and Physical Sciences Research Council for financial support (contract GR/M13602).

References

[1] Q. Wu, X.C. Zhang, Appl. Phys. Lett. 67 (1995) 3523.
[2] X. Yan, et al., Phys. Rev. Lett. 83 (2000) 3404.
[3] C.C. Davis, Lasers and Electro-optics: Fundamentals and Engineering, Cambridge University Press, Cambridge, 1996, p. 472.
[4] D. Oepts, et al., Infrared Phys. Tech. 36 (1995) 297.
[5] G.M.H. Knippels, et al., Opt. Lett. 23 (1998) 1754.
[6] C.W. Rella, et al., Opt. Commun. 157 (1998) 335.
[7] A.M. MacLeod, et al., Phys. Rev. E 62 (2000) 4216.
[8] G.M.H. Knippels, et al., Phys. Rev. Lett. 83 (1999) 1578.
[9] P. Kung, et al., Phys. Rev. Lett. 73 (1994) 967.
[10] M.J. Fitch, et al., Internal report, FERMILAB-TM-2096.
[11] M.J. Fitch, et al., Proceedings of the 1999 Particle Accelerator Conference, New York, 1999, p. 2181–2183.
[12] D. Oepts, et al., Proceedings of the 20th International FEL Conference, Abs-45, Williamsburg, 1998.
[13] D.A. Jaroszynski, et al., Phys. Rev. Lett. 71 (1993) 3798.

ELSEVIER

Nuclear Instruments and Methods in Physics Research A 475 (2001) 509–513

NUCLEAR
INSTRUMENTS
& METHODS
IN PHYSICS
RESEARCH
Section A

www.elsevier.com/locate/nima

Direct observation of beam bunching in BWO experiments

I. Morimoto[a,*], X.D. Zheng[b], S. Maebara[c], J. Kishiro[d], K. Takayama[d],
K. Horioka[a], H. Ishizuka[e], S. Kawasaki[c], M. Shiho[a,c]

[a] Tokyo Institute of Technology, 4259 Nagatsuda, Midori-ku, Yokohama 226-8502, Japan
[b] Beijing Normal University, 19 out-of-Xinjiekou Street, Beijing 100875, China
[c] Japan Atomic Energy Research Institute, Fusion Research Establishment, Naka-machi, Ibaraki 311-0193, Japan
[d] High Energy Accelerator Research Organization, 1-1 Oho, Tsukuba, Ibaraki 305-0801, Japan
[e] Fukuoka Institute of Technology, Wajiro, Higashi-ku, Fukuoka 811-0295, Japan

Abstract

Backward Wave Oscillation (BWO) experiments using a Large current Accelerator-1 (Lax-1) Induction Linac as a
seed power source for an mm-wave FEL are under way. The Lax-1 is typically operated with a 1 MeV electron beam, a
few kA of beam current, and a pulse length of 100 ns. In the BWO experiments, annular and solid beams are injected
into a corrugated wave guide with guiding axial magnetic field of 1 T. In the BWO with annular beam an output power
of 210 MW at 9.8 GHz was obtained. With a solid beam the output power was 130 MW, and an electron beam bunching
with the frequency of 9.6–10.2 GHz was observed by a streak camera. © 2001 Published by Elsevier Science B.V.

PACS: 84.40.Fe

Keywords: BWO; Beam bunching; X-band FEL; Seed power source; Pre-buncher

1. Introduction

At JAERI, we started research on a GW-class
X-band FEL using the induction linac (JLA;
~ 4 MeV, ~ 3 kA, ~ 160 ns) [1]. In this FEL
system, a Backward Wave Oscillator (BWO) is
expected to work as a seed power source for the
FEL and a pre-buncher (Fig. 1) [2,3]. For this
purpose, using the smaller induction linac Lax-1 a
preliminary study is carried out. The output power
of 210 MW was obtained at 9.8 GHz [4]. The RF
power is enough as a seed power of FEL, but beam
bunching characteristics in the BWO are not

clearly observed in previous work. In this paper
the first observation of beam bunching with the
BWO is presented.

2. BWO characteristics

In this BWO, the corrugated wave guide as a
slow wave structure is designed as in Fig. 2 [4].
The corrugated function is

$$R(z) = R_0 + h \cos (2\pi z/15.0 \text{ mm}),$$

$$R_0 = 13.70 \text{ mm} \quad h = 2.8 \text{ mm},$$

pitch number is 10. Using this set up in the BWO,
from the dispersion relation the output RF is only
TM_{01} mode at 9.8 GHz (Figs. 3 and 4). Using the

*Corresponding author. Tel.: +81-29-282-5102.
E-mail address: iwao@felsun1.tokai.jaeri.go.jp
(I. Morimoto).

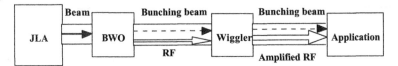

Fig. 1. Schematic view of JLA-X-band FEL.

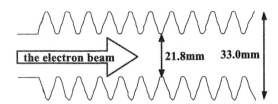

Fig. 2. The corrugated wave guide as a slow wave structure for 9.8 GHz BWO (The mean radius R_0 is 13.7 mm, the depth h is 2.8 mm, and the pitch number is 10).

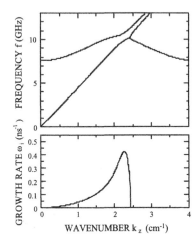

Fig. 4. Wave number vs. frequency and growth rate. The output frequency is 9.8 GHz in this BWO.

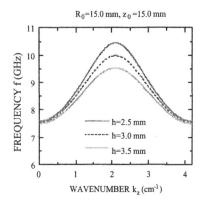

Fig. 3. The corrugated wave guide as a slow wave structure for 9.8 GHz BWO (The mean radius R_0 is 13.7 mm, the depth h is 2.8 mm, and the pitch number is 10).

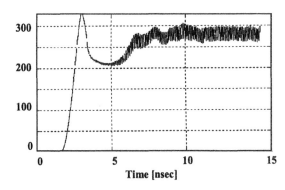

Fig. 5. Microwave power calculation in outlet of slow wave structure using MAGIC code.

MAGIC code, an output power of 270 MW is expected (Fig. 5).

The characteristics of the BWO are shown in Table 1.

3. Beam bunching observation

As schematically shown in Fig. 6, an extracted electron beam (700 keV, $2 \sim 3$ kA, 100 ns) using the induction linac Lax-1 is transported to the corrugated wave guide. The quartz window is set in the outlet of the corrugated wave guide. The Cherenkov light is radiated when the beam hits the quartz window. The radiated Cherenkov light is measured by a streak camera (time resolution is 4.0 ps, sweep speed is 30~670 ps/mm). Simultaneously, the RF output power is measured. The streak data are measured around the peak current of the beam (Fig. 7).

We studied two types of beams in the beam bunching experiment. One was an annular beam, the other was a solid beam.

Using the annular beam (the inner cathode diameter was 12.0 mm, the outer cathode diameter was 16.0 mm, the beam current was 2.7 kA, the accelerated voltage was 710 kV), a beam bunching was measured in the RF output power of 210 MW [4]. The typical waveform of the beam current and the RF output power is shown in Figs. 7 and 8, respectively. The frequency of 9.8 GHz was estimated by FET procedure described elsewhere [3].

The result of a bunching measurement is shown in Fig. 9a. The upper figure of Fig. 9 is the position of the beam in the corrugated wave guide. The position of the camera was adjusted so that the camera did not observe the diode part directly (Fig. 6), this means that only the Cherenkov light

was observed from the quartz window. The lower figure of Fig. 9 is the picture from a streak camera (sweep speed is 67 ps/mm). From that picture a bunching of an annular beam was not clearly observed.

When the solid beam (the cathode diameter is 16 mm, the beam current is 1.9 kA, the accelerated voltage is 710 kV) was injected, the RF output power was 60% of 210 MW. The frequency in this case is the same. The picture from a streak camera is shown in Fig. 9b. In this case, the beam bunching was clearly obtained. Fig. 10 shows a time dependence of the light intensity. The beam bunching frequency was estimated from 9.6 to 10.2 GHz. This corresponded to the RF frequency of 9.8 GHz within the experimental error.

4. Discussion

In the experiments, we always observe the bunching with the solid beam but with the annular

Table 1
BWO parameters

Accelerated voltage	900 kV
Beam current	1.5 kA
Pitch length of slow wave structure	15 mm
Pitch number of slow wave structure	10
Inner radius of wave guide	10.9 mm
Outer radius of wave guide	16.5 mm
Wave guide length	15 cm
Axial magnetic field	1 T
Designed frequency	9.6 GHz
Expected output power	270 MW

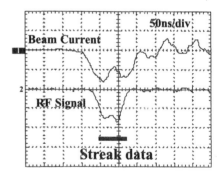

Fig. 7. The upper wave is the beam current 2.7 kA. The down wave is the RF output power of 210 MW. Streak data is measured around the peak current.

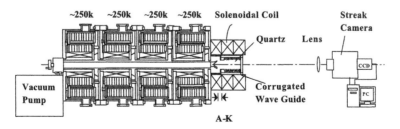

Fig. 6. Schematic view of the BWO.

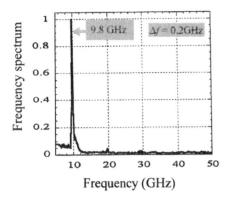

Fig. 8. FFT of oscilloscope data, The output frequency is monochromatic 9.8 GHz.

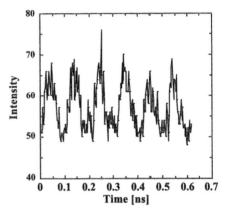

Fig. 10. A time dependence of the intensity of the Cherenkov light.

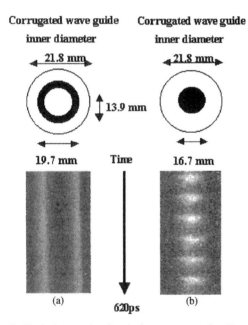

Fig. 9. Typical example of optical measurement for 67 ps/mm at the outlet of the corrugated wave guide: (a) in the annular beam the beam bunching was not clear and (b) in the solid beam the beam bunching was clear.

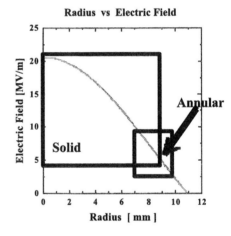

Fig. 11. A longitudinal electric field distribution of 210 MW TM_{01} mode in a cylindrical wave guide (radius 10.9 mm).

the bunching is not clearly seen. The bunching is possibly caused by the coupling of the RF of TM_{01} mode with a beam. When we calculate electric field strength within the corrugated wave guide, it is found that the strength of the electric field in the central part is four times higher than that of the peripheral part (Fig. 11). At present, we think that the bunching is closely related to the strength of the electric field.

Considering the fact that we can always observe higher RF output power with annular beams, why the beam bunching is not clearly observed in the annular beams is open for future study.

From the viewpoint of using BWO as a pre-buncher for FELs, to what extent beam bunching is maintained as another important item to be determined.

5. Conclusion

We always observed the beam bunching with the solid beam. Although the RF output power is not maximum, the BWO with the solid beam always produced a high RF output power of more than 100 MW level, which is enough as a seed power of FEL. Moreover, the BWO with solid beam always produced strong beam bunching.

From the engineering point of view we can reasonably expect that a BWO with the solid beam works as a seed RF power and pre-buncher for X-band FELs.

References

[1] S. Maebara, et al., Nucl. Instr. and Meth., in this conference, WE-4-06.
[2] S. Kawasaki, et al., Nucl. Instr. and Meth. A 341 (1994) 316.
[3] X. Zheng, et al., Nucl. Instr. and Meth. A 407 (1998) 198.
[4] X.D. Zheng, et al., 13th International Conference on High-Power Particle Beams, Nagaoka, Japan, June 25–30, 2000.

ELSEVIER

Nuclear Instruments and Methods in Physics Research A 475 (2001) 514–518

NUCLEAR
INSTRUMENTS
& METHODS
IN PHYSICS
RESEARCH
Section A

www.elsevier.com/locate/nima

Study on interference between far-IR to mm-wave CSR from consecutive electron bunches at BFEL RF-Linac ☆

Zhu Junbiao*, Zhang Guoqing, Li Yonggui, Xie Jialin

Institute of High Energy Physics, Academia Sinica, P.O.Box 2732, Beijing 100080, People's Republic of China

Abstract

Coherent bending magnet or undulator radiation due to a train of electron bunches is treated as radiation from a multi-slit diffraction array. Based on this model, we numerically analyse the interference among coherent synchrotron radiation emitted from consecutive bunches in a train of bunches, which are accelerated by a 30-MeV RF-linac at BFEL. Some interesting results are as follows: (1) Rapidly oscillating radiation enhancement due to interbunch interference is overlapped on the single bunch spectrum. (2) It consists of a series of spectrum lines corresponding to harmonics of the RF fundamental. (3) Main maximum positions are determined by the "diffraction condition". (4) Total intensity is about the square of the number of bunches participating in interference as single bunch intensity. Experimental design to measure interbunch interference at BFEL with the sub-mm and mm-wave Michelson interferometer is presented. © 2001 Elsevier Science B.V. All rights reserved.

PACS: 41.60.Ap; 41.75.Ht; 07.60.Ly; 07.57.Hm

Keywords: Coherent synchrotron radiation; Interbunch interference; Sub- and mm-wave Michelson interferometer

1. Introduction

Study of coherent synchrotron radiation (CSR) for new intense coherent radiation sources such as a broad-band bending magnet, or narrow-band undulator radiation in the far-IR to mm-wave region is of great importance. The characteristics of radiation are closely related to the charge, shape, longitudinal length of an electron bunch,

☆ Work supported by Nature Science Foundation of China Grant No. 19875066, Director Foundation of IHEP, and National High Technology-Laser Technology Field Foundation.

*Corresponding author. Tel.: 8610-62563455; fax: 8610-62554583.

E-mail address: jbzhu@xinmail.com (Z. Junbiao).

and the number of electrons in it. Radiation is coherent if the wavelength is longer than or comparable to the bunch length. The CSR power is nearly the number of bunch electrons times as high as the incoherent SR one. The CSR spectrum was first observed by using the Tohoku 300-MeV RF-linac with an mm-wave grating spectrometer [1]. The CSR intensity is relevant to the phase relationship among consecutive bunches in the bunch train. Wingham discussed the phenomenon of interference theoretically in the far-IR region [2]. Shibata et al. and Ciocci et al. observed mm-wave CSR spectra experimentally in the bending and undulator fields with their polarizing and Fabre–Perot interferometers, thus verifying that there exists coherence among interbunches, respectively [3,4]. The experiment at ENEA [5] showed that this kind of interbunch interference leads to a great enhancement in CSR harmonics exciting and propagating in the waveguide FEL oscillator, which can be explained by Doria's theory [6].

BFEL [7] consists of the thermionic cathode gun, alpha magnet, 30 MeV RF-linac, undulator and optical resonator. Its short bunches are a potential source of strong coherent radiation in the far-IR to mm-wave region. A train of macropulses of 4-μs length at the repetition rate of 3–12 Hz is generated by the injector of the RF-linac. Each macropulse consists of a series of micropulses of 4-ps length and 350-ps period. In this paper, we first discuss the single bunch CSR intensity, and then we present a novel model for the interbunch interference among separate bunches and draw out some interesting characters. Experimental design used to observe interferogram at BFEL is given. In the last section conclusions are given.

2. CSR intensity due to a single bunch

The radiation intensity of a single bunch due to correlation among electrons in it is studied (for more detailed discussion refer to Ref. [8]). When the optical path difference (OPD) between any electron and bunch centre in the bunch to the observation point is equal to or smaller than half of the wavelength, the CSR from the bunch along the tangential direction of its trajectory in the bending magnet is

$$P_{CSR}^{SB}(\lambda, \sigma_L) \approx P_{ISR}(\lambda)[1 + (N_e - 1)f(\lambda, \sigma_L)] \quad (1)$$

$$f(\lambda, \sigma_L) = \left| \int dz S(\sigma_L, z) \exp(i2\pi\sigma_L/\lambda) \right|^2 \quad (2)$$

where $f(\lambda, \sigma_L)$ is called the normalised bunch form factor and is the Fourier transform of the bunch electron distribution $S(\sigma_L, z)$. $P_e(\lambda)$ is the single electron radiation power and $P_{ISR}(\lambda)$ the incoherent SR power from a single bunch. From Eqs. (1) and (2) we observe that (1) as the wavelength λ is comparable to or longer than the longitudinal bunch length σ_L, the CSR power is about N_e times as high as the ISR one. Two kinds of bunches are

produced at BFEL. N_e are 6.3×10^8 and 1.25×10^9, respectively. (2) $f(\lambda, \sigma_L)$ can be measured through the ratio of P_{CSR} to $P_{ISR} \times N_e$, and $S(\sigma_L, z)$ can be derived with Fourier transform. P_{ISR} is given by Schwinger's equation:

$$P_{ISR}(\lambda) = \frac{3^{5/2} e c \gamma^7}{160\pi^2 \rho^2} I y^3 \left[\int_y^\infty K_{5/3}(\eta) \, d\eta \right] \quad (3)$$

where $y = \lambda_c/\lambda$, λ_c is the critical wavelength, $K_{5/3}(\eta)$ the modified Bessel function. In the cases of long $\lambda \gg \lambda_c$ [9] and short wave limits $\lambda \ll \lambda_c$, $P_{ISR}(\lambda)$ is often expanded into an approximate series. It also can be exactly expanded as an infinite series [10]:

$$\int_y^\infty K_v(\eta) \, d\eta = h \left\{ \frac{e^{-y}}{2} + \sum_{m=1}^\infty e^{-y\cosh(mh)} \frac{\cosh(vmh)}{\cosh(mh)} \right\}. \quad (4)$$

With the above formula, $P_{CSR}(\lambda)$ as a function of wavelength from $\lambda_c = 3.1\,\mu m$ to 20 mm for $\sigma_L = 4, 3, 2, 1$ ps and beam current $I = 3 \times 10\,\mu A$ is numerically calculated for Gaussian (a) and rectangular (b) density distributions. The calculated results are shown in Fig. 1. It follows that (1) as $\lambda \geq \sqrt{2\pi}\sigma_L$, $P_{CSR}(\lambda)$ rises rapidly to its maximum; as $\lambda \leq \sqrt{2\pi}\sigma_L$, $P_{CSR}(\lambda)$ decays to P_{ISR}. The enhancement factor is about N_e. (2) The intensity enhances more steeply for shorter bunches. (3) The maximum of $P_{CSR}(\lambda)$ spreads forward from the shorter wave end and into the far-IR region with the bunch length being compressed, so shortening the bunch length is beneficial for the increase in the far-IR P_{CSR}. For the rectangular bunch, the fluctuation in intensity is very large as $\lambda < \sigma_{eff}$ due to the oscillating characteristics of sinc function.

3. CSR intensity for a bunch train

By analogy with the multi-slit diffraction, a simple and novel model to study CSR enhancement due to the interbunch interference in a bending magnet is shown in Fig. 2. The corresponding model in the undulator field is symmetrical about the incident axis of the beam. A train of bunches with the bunch length σ_L, bunch period

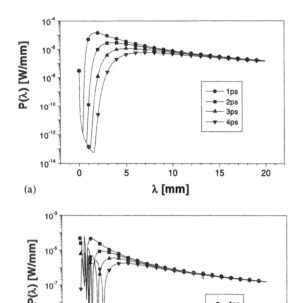

(a)

(b)

Fig. 1. CSR power spectra as a function of wavelength from critical wavelength to 20 mm emitted, respectively, by the single Gaussian bunch (a) and rectangular bunch (b) of the energy 30 MeV, average beam current 3.1 μA, the number of bunch electrons 1.25×10^9 and bunch length 1, 2, 3, 4 ps, respectively.

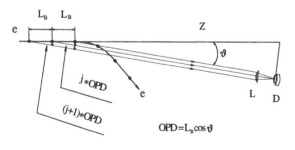

Fig. 2. Schematic of the model for inter-bunch interference. L and D stand for the TPX collecting lens and the detector, respectively.

L_B, and the number of bunches N_B participating in interference is injected into the bending magnet along the Z-axis. CSR is emitted by

electrons of consecutive bunches in the macropulse along the tangent trajectory. Assuming that during the macropulse, the bunch frequency is steady within half of the CSR wavelength generated, there is a definite phase relation among bunches relative to the observation point and therfore their emitted CSR pulses. OPD between adjacent bunches is $L_B \cos \vartheta$. The total intensity due to the bunch train is the product of two factors; one is from the intrabunch interference, the other from the interbunch interference. The total field as a function of the phase difference ζ between the jth and $(j+1)$th bunch in the train relative to the observation point is given by

$$E_{CSR}^{TB}(\lambda) = \left[\sum_{n=1}^{N_e} \exp\left(i\frac{2\pi}{\lambda} x_{jn}\right) \right] \frac{1 - e^{-iN_B\zeta}}{1 - e^{-i\zeta}} \quad (5)$$

$$\zeta(\lambda, L_B, \theta) = \frac{2\pi}{\lambda} n L_B \cos\theta = 2\pi \frac{v}{v_{RF}} \cos\theta \quad (6)$$

where θ is the angle between Z-axis and the emitting point to the observation one. v_{RF} is RF fundamental, and v the harmonics. So, total CSR power emitted by the train of bunches is

$$P_{CSR}^{TB}(\lambda, \sigma_L, L_B) \approx P_{CSR}^{SB}(\lambda, \sigma_L) \left[\frac{\sin(\cos\theta N_B \pi L_B/\lambda)}{\sin(\cos\theta \pi L_B/\lambda)} \right]^2. \quad (7)$$

From Eq. (7), we observe that the interbunch interference intensity is modulated by the single bunch intensity. The interference factor shows rapidly oscillating characteristics. By using the bunch parameters at BFEL, the calculated CSR power due to a train of bunches as a function of wavelength for Gaussian (a) and rectangular (b) bunches is given in Fig. 3, respectively. From Fig. 3 it follows that (1) CSR intensity enhancement due to interbunch interference is superposed over the spectrum due to interference of intrabunch electrons. (2) There are a series of main maxima, and they become regular with wavelength. (3) Main maxima are located at $L_B \cos\theta = m\lambda$, or $v = m v_{RF}/\cos\theta$ (m is the interference order). The intensity is $P_{CSR,MAX}^{TB} = N_B^2 P_{CSR}^{SB}$, and the pulsewidth is about $\Delta\lambda/\lambda \approx 1/(mN_B)$. (4) There are a series of sub-maxima and one minimum between two main maxima: the minimum is P_{CSR}^{SB}.

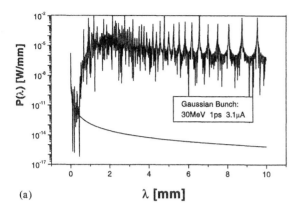

(a)

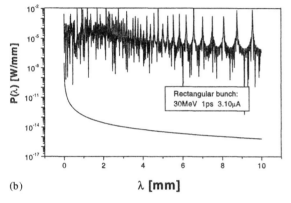

(b)

Fig. 3. Wavelength dependence of CSR spectra from a train of Gaussian (a) and rectangular bunches (b). Calculated parameters of the bunch are given in figures.

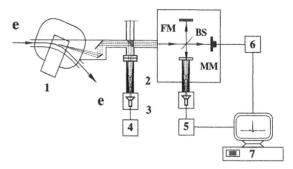

Fig. 4. Experiment used to measure inter-bunch interference. fixed mirror (FM); movable mirror (MM); beam splitter (BS). (1) bend magnet; (2) IR condensing cones; (3) detectors; (4) oscilloscope; (5) data acquisition; (6) step motor; (7) computer.

enhancement is up to two or three orders of magnitude due to the interbunch interference.

4. Experimental design

Experimental design to measure interbunch interference at BFEL is presented in Fig. 4. Relativistic bunches accelerated by the 30-MeV RF-linac are injected into a 45° bending magnet in the vacuum chamber to generate CSR. The chamber gap of 22 mm will not suppress mm-waves lower than 10 mm. The radius of the electron orbit is 150 mm. At the distance of 400 mm from the emitting point, a gold-coated parabolic reflector of 80-mm diameter is used to collect about 100-mrad CSR radiation. The parallel beam of CSR, after passing the parabolic reflector, through a 3-m length low-loss copper pipe and a 6-mm thick TPX window, is guided into a polarising Michelson interferometer operating in the sub- and mm-wave region. The interferometer consists of two hollow retroreflectors of 63-mm diameter as its fixed and movable mirrors and two wire grids of 5-micron polariser and beam splitter. The distance of the movable mirror along the axis direction of a ball screw varies up to 110 mm, which makes the interferometer's resolution higher than the RF fundamental. Acquisition of data and the motion of the movable mirror are automatically controlled by a computer. The experiment shall be performed with two objectives: the first is

(5) The interval between two main maxima expands gradually and coherence enhances with wavelength. (6) The axial radiation frequency is lower than the off-axis one, $\cos\theta > \cos\theta$. (7) Only when the resolution of interferometer is higher than the RF fundamental ν_{RF} is the interbunch interference observable; otherwise, the effect disappears. The other two curves are used to compare CSR power with ISR one.

It must be noted that in the bending magnet, only a few bunches participate in interference and are collected by the detector within the detector response time, so the expected enhancement is only several times. But in the undulator, the number of bunches participating in the interbunch interference due to their radiation coaxiality and the cavity function is greatly increased, and the

to obtain the single-bunch CSR spectrum, and the second to obtain the inter-bunch interferogram.

5. Conclusions

A simple and novel model for the inter-bunch interference in the bending magnet or undulator is suggested. Total CSR power due to both intra-bunch and interbunch interference is numerically calculated and analysed. Results show that in cases of bend and undulator fields, radiation enhancements from interbunch interference are predicted to have several times and more than 2 orders of magnitude, respectively. Main maximum intensities and positions are determined. Several interesting conclusions are given first. The experimental design to measure the inter-bunch interference with the polarising Michelson interferometer is presented.

References

[1] T. Nakazato, M. Oyamada, Phys. Rev. Lett. 3 (1989) 1245.
[2] D. Wingham, J. Phys. Rev. D 35 (1987) 2584.
[3] Y. Shibata, T. Takahashi, K. Ishi, Phys. Rev. A 44 (1991) R3445.
[4] F. Ciccio, R. Bartolini, A. Doria, et al., Phys. Rev. Lett. 70 (1993) 928.
[5] G.P. Gallerano, A. Doria, E. Giovenale, Nucl. Instr. and Meth. A 58 (1995) 78.
[6] A. Doria, G.P. Gallerano, E. Giovenale, Phys. Rev. Lett. 80 (1998) 2841.
[7] J.L. Xie, J.J. Zhuang, Nucl. Instr. and Meth. A 331 (1993) 204.
[8] Junbiao Zhu, Nucl. Instr. and Meth. A 447 (3) (2000) 587.
[9] J.B. Murphy, Nucl. Instr. and Meth. A 346 (1994) 571.
[10] V.O. Kostroun, Nucl. Instr. and Meth. A 172 (1980) 371.

ELSEVIER

Nuclear Instruments and Methods in Physics Research A 475 (2001) 519–523

NUCLEAR
INSTRUMENTS
& METHODS
IN PHYSICS
RESEARCH
Section A

www.elsevier.com/locate/nima

An optical resonator with insertable scraper output coupler for the JAERI far-infrared free-electron laser

R. Nagai*, R. Hajima, N. Nishimori, M. Sawamura, N. Kikuzawa, T. Shizuma, E. Minehara

Free-Electron Laser Laboratory, Advanced Photon Research Center, JAERI 2-4, Shirakata-Shirane, Tokai, Ibaraki 319-11, Japan

Abstract

The performance of an optical resonator featuring an insertable scraper output coupler was evaluated for the JAERI far-infrared free-electron laser. An efficiency factor of the resonator was introduced for evaluation. The efficiency factor was derived from the amount of the output coupling and diffractive loss of the optical resonator, which were calculated by using an optical mode calculation code, using the iterative computation called Fox–Li procedure. As a result of the evaluation, it was found that the insertable scraper coupler was the most suitable for the far-infrared free-electron lasers. Dependencies of insertion direction and scraper radius were also investigated to find out the optimum geometry of the insertable scraper coupler. It was found that the optimum direction of the scraper was parallel to the wiggling plane of the electron beam and the efficiency of the optical resonator increased with the enlargement of the scraper radius. © 2001 Elsevier Science B.V. All rights reserved.

PACS: 41.60.Cr; 95.85.Gn; 42.60.Da

Keywords: Free-electron laser; Far-infrared; Optical resonator; Efficiency

1. Introduction

A free-electron laser based on a superconducting RF linac at Japan Atomic Energy Research Institute (JAERI-FEL) has lased in far-infrared region [1] and achieved quasi-CW lasing with averaged power of over 1 kW, the initial goal of the R&D program [2]. To realize a high-power and high-efficient FEL system, optimization of the high-efficient output coupler of the optical resonator is indispensable and is an important factor of the over 1 kW lasing of the JAERI-FEL. Usually,

the FEL radiation produced in the optical resonator must be output coupled through the coupler of the resonator. This is accomplished in a number of ways, depending on the wavelength, radiation power and geometry of the optical resonator. The output coupling of the optical resonator is accomplished using a partially transmitting dielectric mirror, a Brewster plate beam splitter, or by cutting away part of the FEL radiation by a hole or a small scraper. In far-infrared region, a partially transmitting dielectric mirror and Brewster plate beam splitter are not so suitable because of characteristics of the dielectric material.

An optical resonator system, which consists of metal-coated end mirrors and a partially cut away output coupler, allows a wide broadband

*Corresponding author. Tel.: +81-29-282-6752; fax: +81-29-282-6057.

E-mail address: r_nagai@popsvr.tokai.jaeri.go.jp (R. Nagai).

0168-9002/01/$ - see front matter © 2001 Elsevier Science B.V. All rights reserved.
PII: S 0 1 6 8 - 9 0 0 2 (0 1) 0 1 6 0 0 - X

operation. This output coupler has some disadvantages, however. The output coupling changes with the wavelength, since the mode diameter on the end mirror expands with the square root of the wavelength. The coupler introduces additional diffractive loss due to scattering from the edge of the coupler. For the far-infrared FEL, output coupling and diffractive loss of the optical resonator are important parameters. The performance of the FEL system is degraded by the addition of diffractive losses. Most of these losses are additional diffractive losses due to scattering from the edge of the coupler. Reducing this factor improves the performance of the FEL system.

For the above reasons, the insertable scraper output coupler was chosen for the JAERI-FEL. The coupler allows easy adjustment of the output coupling for efficiency optimization of the optical resonator. The diffraction increases in transport of the FEL radiation to the user room, as the surface area of the output coupler is not so large compared to the wavelength. The insertable scraper coupler has a large output coupling area compared to the center-hole coupler. Hence, the diffraction in transport of the FEL radiation is minimized by use of the insertable scraper coupler.

2. Evaluations and results

Performance of the optical resonator with the output coupler was investigated at optical wavelength of $22\,\mu m$ by using an optical mode calculation code. The output coupling and diffractive losses with the dominant eigenmode of the resonator were calculated using Fox and Li iterative computation procedure [3] and were used to extract the efficiency factor of the optical resonator. The efficiency factor of the optical resonator was introduced for the evaluation of the optical resonator performance. The efficiency factor of the optical resonator was calculated by the amount of output coupling and diffractive loss. The output coupling and diffractive loss were found from the dominant eigenmode of the resonator. The calculation model contains two perfectly reflecting mirrors and four apertures. The effects of the undulator duct and bending magnet

ducts are taken into account by the four apertures at both ends of the undulator and outer side of the bending magnet ducts. The optical guide effect of the electron beam is not taken into account in this calculation. The optical resonator is near-concentric in geometry as shown in Fig. 1. Parameters of the optical resonator are listed in Table 1.

The efficiency factor is introduced as follows. The efficiency of an FEL is defined as

$$\eta_{\text{fel}} = \eta_{\text{out}}\eta_{\text{ext}} = \frac{P_{\text{fel}}}{P_{\text{beam}}} \qquad (1)$$

where η_{out} is the output coupling efficiency of the optical resonator, η_{ext} is the efficiency of FEL energy extraction P_{fel} is the power of the FEL radiation, and P_{beam} is the power of the electron beam. The output coupling efficiency

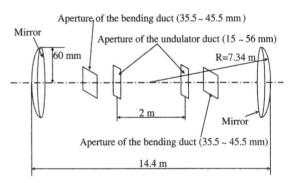

Fig. 1. Nominal configuration of the optical resonator for the JAERI-FEL.

Table 1
Parameters of the optical resonator for the JAERI-FEL

Length of the resonator	14.4 m
End mirror radius	60 mm
Curvature of the end mirror	7.34 m
Undulator duct length	2 m
Scraper mirror radius	10 mm
Reflectivity of the mirrors	99.4%
Aperture of undulator duct	15×56 mm
Aperture of upstream bending duct	45.5×35.5 mm
Aperture of downstream bending duct	35.5×45.5 mm
Undulator to upstream bending aperture	1.11 m
Undulator to downstream bending aperture	1.25 m
Bending aperture to upstream mirror	5.09 m
Bending aperture to downstream mirror	4.95 m

η_{out} is defined as

$$\eta_{\text{out}} = \frac{\alpha_{\text{out}}}{\alpha_{\text{loss}}} \tag{2}$$

where α_{out} is the output coupling and α_{loss} is the total loss of the optical resonator. The total loss contains reflective loss of the end mirrors, diffractive loss and output coupling. The optical resonator of the JAERI-FEL has 1.2% reflective loss. If the FEL is lasing in the spiking-mode and the electron bunch length is longer than slippage length, the efficiency of FEL energy extraction η_{ext} is represented by the following equation [4]:

$$\eta_{\text{ext}} \cong \rho \sqrt{\frac{\varDelta}{\alpha_{\text{loss}} L_{\text{c}}}} = \sqrt{\frac{4\pi N_w \rho^3}{\alpha_{\text{loss}}}} \tag{3}$$

where ρ is the FEL parameter, $\varDelta$ is the slippage length, $L_{\text{c}} = \lambda/4\pi\rho$ is the cooperation length, and N_w is the period number of the undulator. The FEL parameter, slippage length and cooperation length do not depend on the optical resonator. Here, the efficiency factor of the optical resonator η_{opt} is introduced and defined as

$$\eta_{\text{opt}} = \frac{\alpha_{\text{out}}}{\alpha_{\text{loss}}^{3/2}} \tag{4}$$

The FEL radiation power is represented as the optical resonator related term and the electron beam related term,

$$P_{\text{fel}} = \eta_{\text{opt}} \left(\sqrt{4\pi N_w \rho^3} P_{\text{beam}} \right) \tag{5}$$

The performance of the optical resonator is gauged by the efficiency factor of the optical resonator.

Configurations of the output coupler are shown schematically in Figs. 2 and 3. The center-hole coupler cuts away part of the FEL radiation by a hole in the center of the end mirror. The insertable scraper coupler cuts away part of the FEL radiation by an insertable small mirror in front of the end mirror, which is inserted in parallel or perpendicular to the wiggling plane of the electron beam. In the FEL system using a planar undulator, the aperture of the undulator duct is a long and narrow rectangle as shown in Fig. 3. Hence, the performance of the optical resonator depends on the insertion direction of the scraper. The

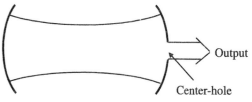

(a) Center-hole output coupler

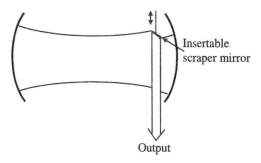

(b) Insertable scraper output coupler

Fig. 2. Configuration of the output couplers.

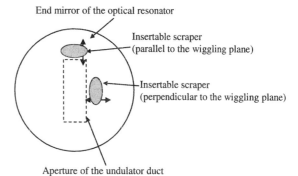

Fig. 3. Insertion direction of the scrapers.

efficiency factor of the optical resonator for the center-hole coupler and insertable scraper couplers are shown as a function of the total loss in Fig. 4, and they have maximum values when the total loss is around 4%. The output coupler degrades the efficiency of the optical resonator because the coupler introduces additional diffractive loss due to scattering from the edge of the output coupler. The insertable scraper coupler is more efficient than the center-hole coupler as shown in Fig. 4. This means that the addition of diffraction due to output coupling by the insertable scraper is less

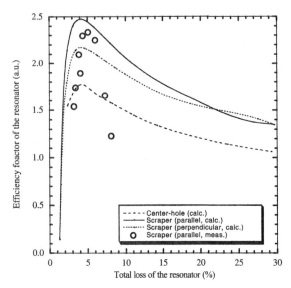

Fig. 4. Efficiency factors of the optical resonator for the JAERI-FEL are shown as a function of total loss. The solid, dotted and dashed lines represent the efficiency factor for the parallel insertion of scraper, perpendicular insertion of scraper and center-hole coupler, respectively. The open circles represent the measured efficiency factor for parallel insertion scraper.

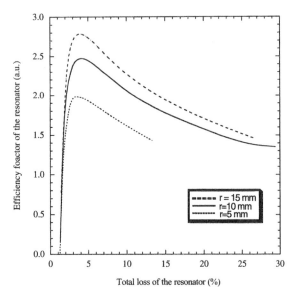

Fig. 5. Efficiency factors of the optical resonator for various scraper radii are shown as a function of the total loss. The dashed, solid and dotted lines represent the efficiency factor for 15 mm radius, 10 mm radius and 5 mm radius, respectively.

than that of the center-hole. The additional diffractive loss is decreased by inserting the scraper parallel to the wiggling plane of the electron beam rather than by the inserting the scraper in perpendicular. Scraper radius dependence on the insertable scraper performance for parallel insertion is shown in Fig. 5. The scraper performance becomes more efficient with increasing scraper radius. This means that the addition of diffraction due to output coupling by the insertable scraper becomes less, the larger the scraper size.

Measured efficiency factors of the scraper coupled optical resonator for the JAERI-FEL are plotted in Fig. 4. Comparison between the measured and calculated factors indicates that there is another loss related to the optical resonator, which is due to its misalignment. Diffractive losses rapidly increase for such misalignment, since the optical resonator has near-concentric geometry and tight apertures. The measured efficiency factor of the resonator is degraded due to the increase of diffractive loss. The measured efficiency factors decrease rapidly

for the large total loss, since lasing mode shifts spiking-mode to normal-mode. The spiking-mode needs very large net gain, so the lasing mode shifts to normal-mode for the large total loss.

3. Conclusion

The performance of an optical resonator with insertable scraper output coupler was evaluated for the JAERI far-infrared free-electron laser. This output coupler offers the following advantages. The coupler allows easy adjustment of the output coupling for optimization of the efficiency of the optical resonator, and enables the wide broadband operation. Diffraction in transport of the FEL radiation is minimized by use of the scraper coupler. As a result of the evaluation, it was found that the insertable scraper coupler was the most suitable for far-infrared free-electron laser.

Dependencies of insertion direction and scraper radius were also investigated to find out optimum geometry of the insertable scraper coupler. It was found that the optimum insertion direction of the scraper was parallel to the wiggling plane of the electron beam and the efficiency of the optical resonator increased with the enlargement of the scraper radius.

References

[1] E. Minehara, et al., Nucl. Instr. and Meth. A 429 (1999) 9.
[2] E. Minehara, et al., Nucl. Instr. and Meth. A 445 (2000) 183.
[3] G. Fox, T. Li. Bell System Tech. J. 40 (1961) 453.
[4] N. Piovella, et al., Phys. Rev. E 52 (1995) 5470.

ELSEVIER

Nuclear Instruments and Methods in Physics Research A 475 (2001) 524–530

NUCLEAR
INSTRUMENTS
& METHODS
IN PHYSICS
RESEARCH
Section A

www.elsevier.com/locate/nima

Calculation of intracavity laser modes in the case of a partial waveguiding in the vacuum chamber of the "CLIO" FEL

Rui Prazeres*, Vincent Serriere[1]

LURE, 209d, Université de Paris Sud, BP 34, 91898 Orsay cedex, France

Abstract

A new method to calculate the propagation of the laser field in a waveguide is presented here. This method uses the "Fast Fourier Transform", without mode decomposition, and it is compatible with the propagation both in a waveguide and in the free space. This is interesting in the case of the FEL, when the optical cavity is not completely guided, and when the guiding occurs softly or only in a limited part of the waveguide; for example, when the input wave is not guided at the entrance and becomes guided at the end. A numerical code, using this method for the wave propagation, is described here. It gives the total losses of the optical cavity, and the transverse profile of the laser at any point of the cavity. This code allowed us to calculate the best parameters of the "CLIO" infrared FEL, corresponding to the minimum of losses. © 2001 Published by Elsevier Science B.V.

PACS: 41.60.Cr; 42.79.Gn; 84.40.Az

Keywords: Free electron laser; Waveguide

1. Introduction

For the Free Electron Lasers (FEL) operating in the far infrared, the diffraction of the light inside the optical cavity, necessitates the use of a waveguide in place of the vacuum chamber of the undulator. In some cases, this waveguide does not match the total length of the optical cavity, because of the free space required for the magnetic dipoles which are guiding the electron beam. As a consequence, the propagation of the laser field in

the optical cavity involves both waveguide propagation and free space propagation. On the other hand, the undulator vacuum chamber (i.e. the waveguide) must have a flat cross-section in order to allow a minimum value for the undulator gap. We can consider, in first approximation, that the waveguide is of horizontal plane type. As a consequence, the guiding of the wave only occurs in the vertical plane. The numerical operator of wave propagation, which is defined here, is a reduction to a plane waveguide of a more general method used for rectangular waveguides [1]. The absorption in the walls of the waveguide is not considered here. The absorption is more important for the high order waveguide modes, but these modes are principally lost by diffraction, which is due to clipping of the laser profile by the mirror or

---

*Corresponding author. Tel.: +33-1-64-46-80-91; fax: +33-1-64-46-41-48.

E-mail address: prazeres@lure.u-psud.fr (R. Prazeres).

[1] Present address: ESRF, BP 220, F-38043 Grenoble Cedex, France.

by the waveguide entrance. The other numerical methods which have been used until now are mostly based on a decomposition in the waveguide proper modes [2,3]; they are more cumbersome and imprecise because they require a large number of modes.

2. Wave propagation

2.1. In free space

The theory of diffraction at short distance gives the evolution of a field distribution $\Psi(x, y, z)$ when propagating along the z-axis [4]. Considering the complex amplitude $\Psi(x, y, 0)$ at $z = 0$, the field $\Psi'(x', y', L)$ at distance $z = L$ is given by a convolution of Ψ with the "Green function" $G(x,y)$:

$$\Psi'(x, y, L) = \Psi(x, y, 0) * G(x, y) \qquad (1)$$

where

$$G(x, y) = -\frac{ik}{2\pi L}\, e^{ikL} e^{ik(x^2+y^2)/2L}.$$

The convolution with the Green function can be calculated numerically as a simple product, in Fourier space, by $\mathrm{TF}[G](u,v)$, where $\mathrm{TF}[G](u,v)$ is the Fourier Transform of $G(x,y)$:

$$\mathrm{TF}[G](u, v) = e^{ikL} e^{-i\pi\lambda L(u^2+v^2)}. \qquad (2)$$

In the design of a numerical code [5], we can define a propagation operator [**P**] which is a combination of three operators:

$$\vec{\Psi}' = [\mathbf{P}]\vec{\Psi} = [\mathrm{FFT}^{-1}][\times\mathrm{TF}(G)][\mathrm{FFT}]\vec{\Psi} \qquad (3)$$

where [FFT] is the 2D "Fast Fourier Transform", $[\times\mathrm{TF}(G)]$ is an operator which makes a product, element by element, with $\mathrm{TF}[G](u,v)$, and $[\mathrm{FFT}^{-1}]$ is the inverse FFT.

2.2. In a waveguide

The proper modes TE_{pq} and TM_{pq} of a rectangular waveguide are well known [6]. As shown in Section 2.1, the convolution $\Psi(x, y) * G(x, y)$ gives the propagation in the free space, i.e. a priori not in a waveguide. Nevertheless, we can verify [1] that this method is still valid for the propagation of the proper modes in a waveguide. The convolution of a proper mode with $G(x,y)$ gives

$$\vec{E}_{pq}(x, y, 0) * G(x, y) =$$
$$\times \vec{E}_{pq}(x, y, 0)\exp(ikL)\exp(i\delta\varphi) \qquad (4)$$

where

$$\delta\varphi = -\frac{\pi\lambda L}{4}\left(\frac{p^2}{a^2} + \frac{q^2}{b^2}\right)$$

corresponds to the phase shift of the proper mode TE_{pq} or TM_{pq} for the waveguide propagation when $f \gg f_c$ where f_c is the cut-off frequency, a and b are the transverse dimensions of the waveguide. As a consequence, this propagation operator can also calculate the longitudinal evolution in a waveguide of a linear combination of proper modes, i.e. to any profile $\Psi(x, y)$. This method can be extended [1] to the general case where f also may be close or smaller than f_c; using in Eq. (4) the exact expression for the phase shift of the modes $\{p,q\}$:

$$\Delta\varphi = \frac{\omega L}{v_\varphi} = \frac{2\pi L}{\lambda}\left[1 + \left(\frac{p\lambda}{2a}\right)^2 + \left(\frac{q\lambda}{2b}\right)^2\right]^{-1/2}.$$

The advantage of this method of calculation is the full compatibility with the case of a guided wave (waveguide propagation), an unguided wave (free space propagation) and any intermediate regime where the guiding occurs softly or only in a limited part of the waveguide.

The numerical operator [**P**] is a numerical approximation of the convolution with $G(x,y)$. The operator [**P**] uses the FFT routine, applied to a wave $\Psi(x, y)$ which is sampled in the domain $[[\Delta x, \Delta y]]$. In normal conditions, as in Section 2.1, the amplitude of Ψ must be equal to zero on the boundary of the domain $[[\Delta x, \Delta y]]$; otherwise, the result of the FFT is false. For the FFT of a proper mode, the field $\vec{E}_{pq}(x, y)$ is extending in the whole domain $[[\Delta x, \Delta y]]$, and the final result of the FFT should be false. Nevertheless, if the size of the FFT domain $[[\Delta x, \Delta y]]$ is equal to the transverse periodicity of the proper modes, then the result of the FFT is correct. This corresponds to the condition $\Delta x = 2a$ and $\Delta y = 2b$, where a and b are the transverse dimensions of the waveguide. When this condition is respected, the transverse periodicity of the proper modes is equal to the periodicity of the FFT, and we can replace the convolution by the operator [P]:

$$\vec{E}'_{pq}(x, y, L) = [P]\vec{E}_{pq}(x, y, 0) =$$
$$\times \vec{E}_{pq}(x, y, 0) * G(x, y). \quad (5)$$

In the following part of the paper, we will always consider that the operator [P] is defined in the area [[2a,2b]].

Let us consider now a given wave front $\vec{\Psi}(x, y)$ at $z=0$, which can be decomposed in a sum $\vec{\Psi}_m(x, y)$ of proper modes TE_{pq} and TM_{pq}

$$\vec{\Psi}_m(x, y) = \Sigma(a_{pq}.\vec{E}_{TEpq} + b_{pq}\vec{E}_{TMpq}) \neq \vec{\Psi}(x, y). \quad (6)$$

The profile Ψ is generally nonperiodic, whereas Ψ_m is periodic. The periodicity of $\vec{\Psi}_m(x, y)$ comes from the periodicity of the proper modes. The wave $\vec{\Psi}_m$ is only equal to Ψ in the waveguide area [[a,b]]. In summary, the propagation in free space can be calculated using the operator [P] applied to the nonperiodic profile $\vec{\Psi}(x, y)$, whereas the propagation in a waveguide must be calculated using the same opertator [P] applied to the periodic profile $\vec{\Psi}_m(x, y)$. This profile $\vec{\Psi}_m$ can be obtained from $\vec{\Psi}$ using an operator of mirror symmetries, called "Mosaic transform", which is described below. The mode decomposition is not required here. This method of calculation of propagation is very rapid because it only uses one FFT and one FFT^{-1}.

2.3. The propagation operator

The periodic profile $\vec{\Psi}_m(x, y)$ is deduced from $\vec{\Psi}(x, y)$ by the application of an operator [M] that we call "Mosaic transform". This operator involves mirror symmetries which are equal to the symmetries of the proper modes TE_{pq} and TM_{pq}. The case of a plane waveguide is more simple because it only uses the symmetry in the vertical axis. The relevant diagram of symmetry for the "Mosaic transform" of the polarization $E_x(x,y)$ is shown in Fig. 1. The left part of the figure displays the input wave $\vec{\Psi}(x, y)$ in the FFT domain [[$\Delta x, \Delta y$]], with $\Delta y = 2b$. There is no condition for Δx because it is a plane waveguide: no symmetry is required in the horizontal axis, and the propagation is free. The profile $\vec{\Psi}$ is only defined in the waveguide area, for $-b/2 < y < b/2$. The white part must be filled with the function $\vec{\Psi}(x, y)$ by mirror symmetries according to Fig. 1. The symmetry properties of TE_p and TM_p are the same. However, the polarizations E_x and E_y of the same proper mode have opposite signs in the symmetries. Therefore, there are two mosaic operators, [Mx] and [My], respectively, for the two polarizations Ψ_x and Ψ_y. Fig. 1 shows the operator [Mx], which corresponds to a mirror symmetry with a sign inversion of Ψ. This inversion can be observed in the representation of the amplitude of the TE_1 mode in Fig. 1. The operator [My] corresponds to the same representation, in Fig. 1, but with a mirror symmetry without sign inversion of Ψ. These two "Mosaic operators"

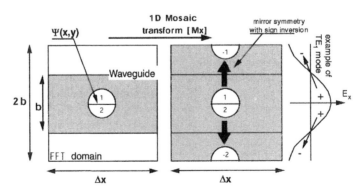

Fig. 1. Mosaic transform [Mx] for the polarization $E_x(x, y)$: the domain on the left is $\Psi(x, y)$ in [[$\Delta x, \Delta y$]], with $\Delta x = 2b$. The domain on the right is $\Psi_m(x, y)$. The curve on the right-hand side represents the TE_1 mode.

can be applied to any input wave $\vec{\Psi}(x, y)$ in order to obtain the mosaic profile $\vec{\Psi}_m(x, y)$ which is used in the waveguide propagation: $\Psi_m^x = [\mathbf{Mx}]\Psi_x$ and $\Psi_m^y = [\mathbf{My}]\,\Psi_y$. The complete operator of waveguide propagation includes some other operators: a diaphragm operator $[\mathbf{D}]$ which sets to zero the field outside of the waveguide, and a couple of resampling operators $[\mathbf{S+}]$ and $[\mathbf{S-}]$. This resampling is used for switching between the FFT domains $[[\Delta x, \Delta y]]$: one domain for the free space areas of the optical cavity, and one restricted domain with the condition $\Delta y = b$ for the waveguide area. In summary, the propagation subroutine will use the following series of operators, applied to the profile $\vec{\Psi}(x, y)$:

$$\vec{\Psi}_x(x, y) = [\mathbf{S-}]\,[\mathbf{D}]\,[\mathbf{P}]\,[\mathbf{Mx}]\,[\mathbf{D}]\,[\mathbf{S+}]\,\Psi_x(x, y)$$

$$\Psi_y'(x, y) = [\mathbf{S-}]\,[\mathbf{D}]\,[\mathbf{P}]\,[\mathbf{My}]\,[\mathbf{D}]\,[\mathbf{S+}]\,\Psi_y(x, y). \quad (7)$$

3. Simulations for "CLIO"

3.1. The numerical code

A numerical simulation, involving the propagation operator described in Section 2.3, has been done in order to calculate the losses of the optical cavity of the FEL when a waveguide is used in place of the undulator vacuum chamber. The numerical code [5] uses several operators: for the reflection on the cavity mirrors, for the transversal gain distribution $G(x,y)$ and for the propagation both in free space or in the waveguide. Fig. 2 shows a layout of the optical cavity. Each mirror is defined by the reflection coefficient for the amplitude, the radius of curvature and the diameter. The gain transverse distribution is a Gaussian function $G(x,y)$, centered on axis, in the center of the undulator. The field of validity of the propagation operator is the same for free space and waveguide propagation: it is limited to the paraxial approximation (small angles). The program calculates the propagation in the cavity of a wave front $\Psi(x,y)$ of complex amplitude, representing the optical field. This wave front is sampled in the transverse plane by 128×128 points, representing a total area of $10 \times 10\,\text{cm}$. The process starts from a Gaussian TEM_{00} field distribution $\Psi_0(x,y)$ in the center of the cavity. This initial field distribution is normalized to $\|\Psi_0\| = 1$. The iterative procedure of propagation in the cavity leads to a convergent solution $\Psi_n(x,y)$, where n is the number of cavity round

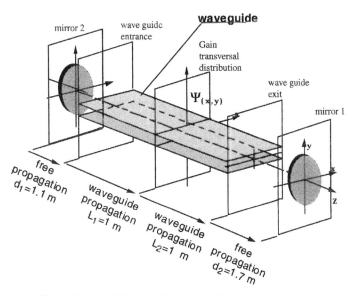

Fig. 2. Layout of the optical cavity used in the numerical code.

trips. Between $n = 20$ and 40 iterations are necessary to obtain a convergent solution, corresponding to the saturation limit. The value of the gain at saturation G_s is not known in advance and depends on the losses of the final mode. This gain value is determined during the iterations, in parallel with the mode, until it reaches the saturation limit. After convergence, the total losses of the cavity can be deduced from the gain value at saturation G_s. The size 10×10 cm of the sampling area is determined in order to fit the maximum width of the laser profile, i.e. on the mirrors. The laser profile cannot be much wider than the mirrors, for which the diameter is $\Phi = 38$ mm in the case of CLIO. On the other hand, the number of points for the sampling depends on the computer capability, but it must be sufficient to allow a reasonable resolution for the sampling in

the middle of the waveguide, where the gain profile fits the electron beam profile: i.e. about $\sigma = 2$ mm for "deep" infrared operation of CLIO ($\lambda > 20 \, \mu$m).

3.2. Numerical results

The aim of these simulations is to obtain the parameters of "CLIO" which are giving a minimum of losses in the optical cavity. In the first set of simulations, the cavity losses have been calculated for various mirror radius of curvature $Ry1$ and $Ry2$ for the vertical axis on both mirrors. The horizontal radius of curvature remains constant with $Rx1 = Rx2 = 3$ m, because the wave is not guided along x-axis. The density plots in Fig. 3 represent the losses versus $Ry1$ and $Ry2$. Four laser wavelengths have been used: $\lambda = 25$, 50, 100

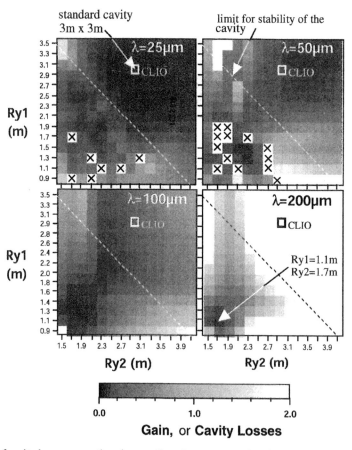

Fig. 3. Density plot of cavity losses versus the mirror radius of curvature $Ry1$ and $Ry2$ on vertical axis for both mirrors.

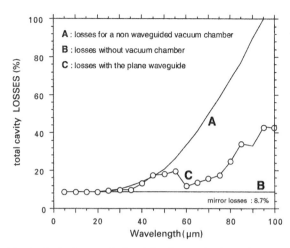

Fig. 4. Cavity losses calculated for the standard set of mirrors $R1 = R2 = 3$ m.

and 200 μm. The small losses correspond to the dark points. The point with the label "CLIO" corresponds to the standard cavity of CLIO with $Rx = Ry = 3$ m on both mirrors. The transverse dotted line corresponds to the theoretical limit of stability of the optical cavity $0 < (1 - L/Ry1)(1 - L/Ry2) < 1$ in vertical plane: only the points above this line are in the stable area. The white points with a cross correspond to a nonconvergent simulation : the laser profile Ψ is not stable along the consecutive cavity round trips. All these points are below the limit of stability of the cavity. At a shorter wavelength, $\lambda = 25$ μm, the minumum of losses corresponds to the standard cavity with symmetric mirrors: $Ry1 = Ry2 = 3$ m. However, for the larger wavelength $\lambda = 200$ μm, the smaller

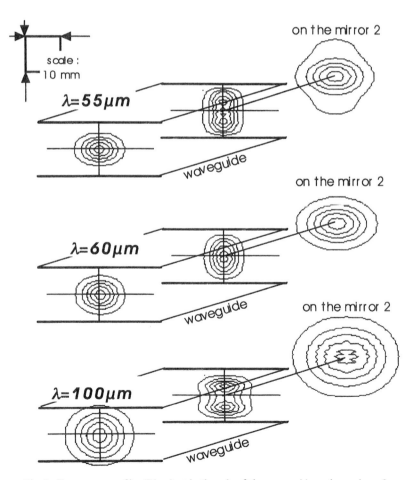

Fig. 5. Transverse profiles $\Psi(x, y)$ at both ends of the waveguide and on mirror 2.

losses are only obtained with toroidal mirrors: with $Ry1 = 1.1$ m and $Ry2 = 1.7$ m. In this configuration, each radius of curvature Ry is equal to the distance between the mirror and the closer entrance of the waveguide (see Fig. 2): $Ry1 = d1$ and $Ry2 = d2$. This point stays below the limit of cavity stability, but this limit only corresponds to a free cavity which is not the case here. Note that for any wavelength larger than 100 μm, the standard cavity of CLIO with $Ry1 = Ry2 = 3$ m gives very large losses which do not allow lasing. This result is in agreement with some calculations already published [3] which lead to the same conclusion by empirical methods. However, this simulation gives the absolute value of the cavity losses, and such an extended view of the FEL losses versus the cavity parameters could not be obtained with empirical methods. We can also observe in Fig. 3 that many points below the limit of stability, at $\lambda = 25$ and 50 μm, are still converging points with low losses (dark points). Only a few of them are marked with a white cross (nonconverging simulation). This means that, in principle, the lasing can be obtained in this area, but the laser may be strongly unstable.

As shown in Fig. 3, working at wavelengths larger than 100 μm requires toroidal mirrors. Therefore, it is interesting for us to know where the wavelength limit of laser operation is with the standard set of mirrors $R1 = R2 = 3$ m. Fig. 4 shows the cavity losses obtained with the simulation for $R1 = R2 = 3$ m. The flat curve B corresponds to a simulation without the vacuum chamber: the losses of 8.7% correspond to the mirror reflectivity $r_1 = 0.99$ and $r_2 = 0.93$. The curve A corresponds to the cavity losses with an unguided vacuum chamber, which is represented

by one iris on both ends with 15 mm in vertical and infinite in horizontal. As expected, the losses are increasing strongly with the wavelength because of the diffraction. The losses with a waveguide are represented by the curve C. They include the mirror losses (8.7%), and the waveguide losses by cutting of the mode at the entrance. Note that there is a minimum of losses for $\lambda = 60$ μm. Fig. 5 displays the transverse profiles at both ends of the waveguide and on the mirror 2. These profiles correspond to the positive direction of propagation of the wave along the z-axis. For $\lambda = 100$ μm, the profile is truncated at the waveguide entrance. At $\lambda = 60$ μm, the laser profiles exhibit a smooth structure, close to a Gaussian profile, which leads to minimum losses. At $\lambda = 55$ μm, the profile exhibits a more complicated structure, which gives larger diffraction losses as for $\lambda = 60$ μm.

References

[1] R. Prazeres, A method of calculating the propagation of electromagnetic fields both in waveguides and in free space, using the Fast Fourier Transform, Eur. Phys. J. Appl. Phys., in press.

[2] K.W. Berrymann, T.I. Smith, Nucl. Instr. and Meth. A 318 (1992) 885; R.L. Elias, et al., Phys. Rev. Lett. 57 (4) (1986) 424.

[3] Li Yi Lin, A.F.G. Van der Meer, Rev. Sci. Instrum. 68 (12) (1997) 4342.

[4] S. Lowenthal, Y. Belvaux, Rev. Opt. 46 (1) (1967) 1.

[5] R. Prazeres, M. Billardon, Nucl. Instr. and Meth. A 318 (1992) 889.

[6] G. Dubost, Propagation libre et guidée des ondes électromagnétiques, Masson, Paris 1995, ISBN:2-225-84792-4.

ELSEVIER

Nuclear Instruments and Methods in Physics Research A 475 (2001) 531–536

NUCLEAR
INSTRUMENTS
& METHODS
IN PHYSICS
RESEARCH
Section A

www.elsevier.com/locate/nima

Transient mirror heating theory and experiment in the Jefferson Lab IR Demo FEL

S. Benson*, Michelle Shinn, G.R. Neil

Jefferson Lab, MS 6A, 12000 Jefferson Ave., TJNAF, Newport News, VA 23606, USA

Abstract

During commissioning of the IR Demo FEL at Jefferson Lab, we noticed that the FEL exhibited a rapid power drop with time when the first set of 3 μm mirrors was used. Though the rate of power drop was unexpected, it was thought that it could be due to a distortion of the mirrors during a time short compared to the thermal diffusion time. This transient distortion might affect the laser more than the steady state distortion. This paper presents some analysis of the transient mirror heating problem and some recent experimental results using different mirror substrates and coatings. It is found that the behavior of the first mirror set cannot be reconciled with the observed power falloff if a linear absorption is assumed. The power drop in more recent experiments is consistent with linear thermal analysis. No anomalous transient effects are seen. © 2001 Elsevier Science B.V. All rights reserved.

PACS: 41.60.Cr; 42.60.Da

Keywords: High power; Resonators; Optics

1. Introduction

In the previous work [1] we described the steady state behavior of a near-concentric resonator with losses only at the output coupler. Subsequent measurement of the distortion induced using a CO_2 laser agreed with the distortion predicted by that paper. It was also found that the power of the IR Demo laser saturated at a value similar to that predicted by that theory when a resonator with CaF_2 mirrors was used [2]. During initial attempts to lase at high power with sapphire mirrors we found that the laser saturated at a very low power

_____
*Corresponding author. Tel.: + 1-757-269-5026; fax: + 1-757-269-5519.
E-mail address: felman@jlab.org (S. Benson).

and that the power dropped extremely rapidly from the kilowatt level down to a few hundred watts (see Fig. 7 from Ref. [2]). This behavior motivated us to study the transient heating effects in a high power FEL. The optical cavity parameters for the IR Demo FEL are given in Table 1. The magnification M, defined as the ratio of the mode size on the mirrors to the mode size at the waist in the cavity, was chosen to be 101 as a compromise between mirror heating and resonator stability issues.

The design details of the accelerator have been reported previously in Ref. [3]. The accelerator can deliver up to 240 kW of continuous electron beam power. The laser efficiency is up to 1.0% at full power and 1.5% at low power. With a 40 cm Rayleigh range and 10% output coupling, the CW

Table 1
Design parameters for the IR Demo optical resonator

Parameter	Value
Length	8.0105 m
Mirror radius of curvature	404.5 cm
Rayleigh range	40 cm
Magnification	101
Mirror diameter	5 cm
Mirror tilt tolerance	1.5 μrad $\sqrt{\lambda\,(\mu m)}$
Typical output coupler reflectivity	90%
HR reflectivity	>99.5%
Coating absorption	<0.1%

intensity at the cavity center could be as high as 2.8 MW/cm^2. With a magnification of 101 the intensity at the mirrors is 28 kW/cm^2. The gain is difficult to measure but is of the order of 100% per pass as judged from the turn-on time and the dependence of the efficiency versus number of round trips per gain pass.

2. Steady state mirror distortion

In Ref. [1] it was shown that the mirror distortion in the case of a Gaussian absorption pattern on a mirror whose edge is held at a constant temperature has the form

$$\delta z(\chi) = \frac{P_\ell}{8\pi F}\left[\gamma + \ln\left(\frac{2a^2}{w_m^2}\right) - 1.17 R_{20}(\chi)\right.$$
$$\left. + 0.44 R_{40}(\chi) - 0.17 R_{60}(\chi)\right] \tag{1}$$

where P_ℓ is the output laser power, a is the mirror radius, w_m is the $1/e^2$ laser mode radius on the mirror, R_{mn} is the Zernike circle polynomial of radial order m and azimuthal order n, and γ is the Euler–Mascheroni constant, equal to 0.57722. The quantities F (the figure of merit for the mirror) and χ are defined by

$$\chi = \frac{r}{2w_m} \quad \text{and} \quad F = \frac{k_{th}}{(h\alpha_B + \alpha_s(1 + 1/t_c))\alpha_e} \tag{2}$$

where k_{th} is the thermal conductivity of the mirror substrate, h is the mirror thickness, t_c is the output coupler transmission, α_B is the bulk absorption coefficient, α_s is the coating absorption, and α_e is

the thermal expansion coefficient of the mirror substrate. The quantity in parentheses in the denominator of the expression for F is the total absorption of the mirror. From this solution, it is possible to calculate the change in the Rayleigh range and the aberration for a given set of mirrors, laser wavelength, and power output. In Ref. [1] we assumed that only one mirror suffered distortion. To better model the IR Demo performance, we assume here that both mirrors have equal distortion. With this assumption, the change in the Rayleigh range is given by

$$\frac{\Delta z_R}{z_R} \cong \frac{1.17 P_\ell}{16 F \lambda}\frac{M}{\sqrt{M-1}}. \tag{3}$$

Note that the ultimate change in the Rayleigh range will be larger than the value calculated by this formula when the change is large due to the positive feedback of the mirror heating. As an example, for the IR Demo a 60% initial change will lead to a 100% change when the change is calculated self-consistently.

The aberration may be calculated from the higher order Zernike polynomial coefficients in Eq. (1). The resulting equation can be rewritten in terms of the change in the Rayleigh range:

$$\frac{\delta z(0)}{\lambda} = 0.332\frac{\sqrt{M-1}}{M}\frac{\Delta z_R}{z_R}. \tag{4}$$

For large magnification and moderate changes in the Rayleigh range, the aberration is much smaller than one wave. For a magnification of 101 and a 100% change in the Rayleigh range, the aberration is only 3% of a wave.

For very large changes in the Rayleigh range, or smaller magnifications, the aberration may limit the laser power.

3. Transient mirror distortion

When the laser first turns on there is a period during which the energy deposited in the mirrors has not had a chance to spread laterally. The mirror distortion should therefore have the shape of the optical mode instead of the shape given in Eq. (1). This transient regime is present for times short compared to the time necessary for the heat to spread laterally by a mode radius, given by

$$t_0 = \frac{w_m^2}{8\kappa} \qquad (5)$$

where κ is the thermal diffusivity of the mirror substrate. Note that, if the mirrors distort in such a way to change w_m, the characteristic time t_0 changes as well. The temperature rise of a half plane on which a Gaussian mode is incident is given by the following equation [4]:

$$T(r,z,t) = \frac{2P}{\rho C (4\pi\kappa)^{3/2}}$$
$$\times \int_0^t \frac{1}{(t'+t_0)\sqrt{t'}} \exp\left[\frac{-z^2}{4\kappa t'} - \frac{r^2}{4\kappa(t'+t_0)}\right] dt'. \qquad (6)$$

The distortion versus time may be calculated by integrating Eq. (6) along the z coordinate. The integral over t may then also be carried out. When the resulting equation is multiplied by the expansion coefficient we get

$$\alpha_e \int_0^\infty T(r,z,t)\, dz$$
$$= \frac{P_{abs}\alpha_e}{4\pi k_{th}}\left[\text{Ei}\left(\frac{r^2}{4\kappa(t_0+t)}\right) - \text{Ei}\left(\frac{r^2}{4\kappa t_0}\right)\right] \qquad (7)$$

where Ei is the tabulated exponential integral [5]. For times small compared with t_0 the expression in brackets can be approximated by the Gaussian mode shape at the mirror. For large times, the term in brackets approaches Eq. (1). The distortion at the mode center versus time is given by

$$\delta z(t) = \frac{P_{abs}\alpha_e}{4\pi k_{th}}\ln\left(1 + \frac{t}{t_0}\right). \qquad (8)$$

For small times, the Zernike coefficients will be different from those in Eq. (1). If the Gaussian mode is expanded in the same Zernike polynomials, the coefficients are

$$c_1 = -0.283, \quad c_2 = 0.275, \quad c_3 = -0.186,$$
$$\text{and} \quad c_4 = 0.110.$$

If these are compared with those of Eq. (2), it is found that the aberration amplitude is about the same as for the steady state case. The curvature coefficient is smaller by a factor of 4. This means that the relative importance of radius of curvature effects and aberrations may be reversed for transient versus steady state operation. It is likely

that aberrations are always the limiting factor for transient distortion while they are only a factor in steady state for cavities with small magnification.

Though we can calculate the change in the aberrations and Rayleigh range from a given power absorption on a mirror, it is difficult to predict how this distortion affects the laser. The most definitive way to determine the effect is to do a full two-dimensional simulation that includes the mirror distortion and cavity phase advances and calculates the saturated power. This can be done by a simulation code such as FELIX, developed at Los Alamos. Modeling a deep UV system using FELIX [6] found that the power drops rapidly with increasing wavefront distortion for a small mirror surface distortion per unit intensity. Let us define the mirror surface distortion per unit incident intensity as Δ. The quantity Δ is a measure of the mirror coating absorption or the expansion coefficient of the mirror substrate. For large values of Δ, the circulating power falls off sufficiently fast to keep the total wavefront distortion of the circulating mode constant with respect to Δ. This is shown in Fig. 1. It can also be a function of the repetition rate if one assumes that the right axis is the power per micropulse instead of the total power. Finally, the power versus time will be similar to the power versus Δ as long as the distortion is linearly dependent on time. The latter assumption is true during the transient regime.

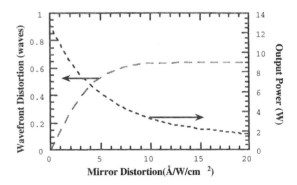

Fig. 1. Wavefront distortion and output power versus the mirror distortion per unit intensity for a UV unstable ring resonator with scraper output coupling. The shape of the distortion was assumed to follow the mode shape.

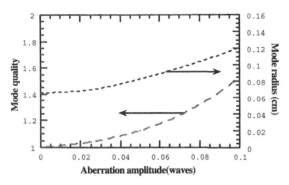

Fig. 2. Mode quality M^2 and waist spot radius as a function of Gaussian aberration added to both mirrors. The total wavefront distortion will be four times this value.

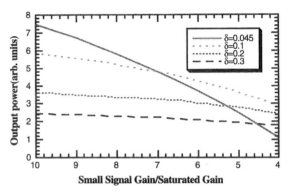

Fig. 3. Output power versus the ratio of the small signal gain to the saturated gain for four different cavity length detunings. A detuning of 0.045 corresponds to the maximum in the detuning curve for a gain ratio of 10. Though the power is much higher for small detuning and high gain, it is not as high when the gain drops.

One can also do a simple Fox and Li type analysis of the with a phase error added to the cavity. This was done using the commercial code PARAXIA for a cavity with parameters of the IR Demo FEL operating at 3 μm. The mode quality factor M^2 and the waist spot size increase as a function of the Gaussian mirror distortion amplitude are shown in Fig. 2. Clearly a distortion in the mirror of 1/10 of a wave is very deleterious to mode quality and should strongly affect the laser gain. On the other hand, a distortion of less than 1/20 wave should have quite a small effect on the laser. Note that the fact that the cavity is nearly concentric makes the total aberration worse. The aberrations from each of the two mirrors add linearly due to the degeneracy of the cavity. For a smaller magnification, the mirror distortion induced aberration will be larger (see Eq. (4)) but the aberration from the two mirrors will not necessarily add linearly because the phase advance may alter or even reverse the wavefront distortion after propagating to the other mirror.

Another question to answer is how much the Rayleigh range can change before the gain is greatly reduced. For a system like the IR demo the answer is that the Rayleigh range must increase by a factor of four to decrease the gain by a factor of two. Note that, for this large a change in the Rayleigh range, the aberration will not necessarily be negligible. For a smaller magnification or in the transient regime the aberration will almost certainly dominate the reduction in laser gain.

Finally, one must remember that it is the saturated gain that is important in determining the output power. The dependence of the saturated gain on the small signal gain is a complicated function of the cavity length detuning and the gain-to-loss ratio. If one assumes the cold cavity parameters of the IR Demo and uses Dattoli's approximation for the saturated gain versus the gain-to-loss ratio [7], one finds the curves in Fig. 3 for the output power versus the ratio of the small signal gain to the saturated gain. One can see that the power is much more sensitive to the ratio near the synchronous point than further out in the detuning curve.

4. Experimental results

In long pulse lasing measurements using a set of rather lossy 3 μm mirrors (absorption for each mirror estimated at 0.4%) we found the power curves shown in Fig. 4. The pulse length in each curve is 200 ms. The laser was operated at the peak of the detuning curve at 3.1 μm. The lower curve is for 18.7 MHz operation and the upper curve is with all the same parameters but with 37.4 MHz operation. In the absence of mirror distortion, the power should have been independent of time for the two curves and equal to the initial power level.

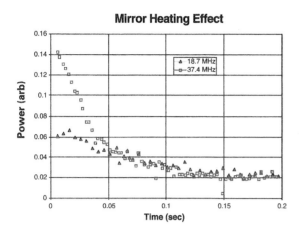

Fig. 4. Power versus time for long single pulse operation. The two curves are for the same laser and accelerator settings except that the repetition rate of the two differs by a factor of two.

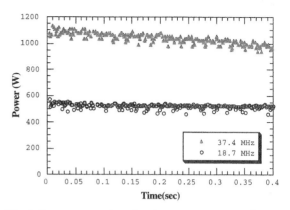

Fig. 5. Power versus time for 400 ms macropulses for two different micropulse repetition rates. The resonator used 3 µm mirrors with low loss mirror coatings.

Instead, the power for both cases falls rapidly and is the same after 50 ms. This implies that the laser efficiency allowed at 37.4 MHz is exactly half that allowed at 18.7 MHz. Note that the characteristic time for these mirrors from Eq. (5) is 0.33 s, so the behavior seen is definitely occurring in the transient regime. Also note that the behavior is qualitatively consistent with the behavior seen in Fig. 1. As noted above, the data in Fig. 1 can be regarded as the power per micropulse train as a function of the repetition rate. One expects the power per micropulse to drop by a factor of two when the repetition rate is doubled. This means that the power should be independent of the repetition rate when the distortion is high enough. Also note that the initial slope of the curve is proportional to the repetition rate. This implies that the power reduction is due to thermal loading, which is linear in the initial power.

When the response seen in Fig. 7 is analyzed using the equations in Section 4, one finds that the absorption required for the aberration amplitude to reach $\frac{1}{10}$ of a wave in 30 ms (the time it takes for the experimental power to drop by a factor of two) is 2.8%. The reflectivity of the high reflector is 99.6%, so linear absorption this high is not possible. This seems to indicate that some non-linear absorption mechanism is present in the coatings of these mirrors.

Power versus time curves for two 400 ms pulses using a different set of 3 µm mirrors is shown in Fig. 5. The behavior is dramatically different from that shown in Fig. 4. The power falloff is linear over the entire macropulse and is quite small. Some of the falloff may be due to mirror steering effects caused by the mirror heating. The coherent harmonics were seen to move sideways during the long pulse. Clearly, no transient behavior is seen since the changes in power are small during the characteristic time for the mirror of 0.3 s.

When similar data are taken for CaF_2 mirrors, whose characteristic time is 2 s, we find the behavior seen in Fig. 6. Here the falloff is more dramatic but the fall-time is comparable to the characteristic time and one expects that the characteristic time will be reduced as the mirrors heat up. Thus, the behavior seen cannot really be classified as occurring in the transient regime but must be simulated using a self-consistent time-dependent theory.

Curves similar to those in Fig. 6 were taken as a function of cavity length detuning. We expected that the falloff in the power, at least on a relative scale, would go down with increasing cavity length detuning. We fit the power versus time at the start of the long pulse to a line of the form $P = P_0(1 - bt)$ and plotted the value of b versus the detuning. This is shown in Fig. 7. The behavior is exactly opposite to what one would

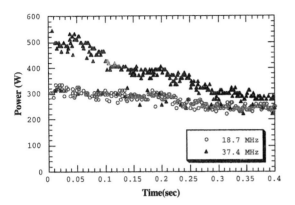

Fig. 6. Power versus time for 400 ms macroupulses. The FEL was operated at 5 μm and used CaF_2 mirrors with coatings similar to those on the sapphire mirrors used in Fig. 5.

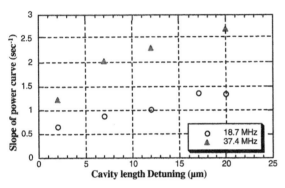

Fig. 7. Relative slope of the power versus time as a function of cavity length detuning for two different micropulse repetition rates using the 5 μm resonator optics.

naively expect. The effect seems to be due only to the saturated gain in the absence of mirror distortion and not at all dependent on the small signal gain.

5. Conclusions

Though the transient response agrees qualitatively with the experimental results, the quantitative agreement is not as good. The only behavior that was clearly in the transient regime was the lossy 3 μm mirrors. The other mirrors showed behavior consistent with adiabatic heating effects. The cavity length detuning dependence is puzzling and needs further study using a more complete four-dimensional FEL model. The transient behavior also needs more study with a self-consistent time-dependent model.

Acknowledgements

Many thanks to all the technical staff at Jefferson Lab who helped make this work possible. This work was supported by the U.S. Department of Energy under contract DE-AC05-84-ER40150, the Office of Naval Research, the Commonwealth of Virginia, and the Laser Processing Consortium.

References

[1] S.V. Benson, et al., Nucl. Instr. and Meth. A 407 (1998) 401.
[2] S.V. Benson, G.R. Neil, M. Shinn, in: S. Basu, S.J. Davis, E.A. Droko (Eds.), Gas, Chemical and Electrical Laser and Intense Beam Control and Applications, SPIE 3931, 2000, p. 243.
[3] C.L. Bohn, et al., Performance of the accelerator driver of Jefferson Laboratory's free-electron laser, Proceedings of the 1999 Particle Accelerator Conference, New York, 1999, p. 2450.
[4] N.N. Rykalin, Y.L. Krasulin, Sov. Phys. Dokl. 163 (1966) 87 (English Transl. 10 (1966) 659).
[5] G. Arfken, Mathematical Methods for Physicists, Section 5.11, Academic Press, New York, NY, 1970.
[6] J.C. Goldstein, B.D. McVey, Thermal distortion limits on the performance of XUV free-electron lasers configured with a multifacet ring resonator, High Heat Flux Engineering, SPIE 1739, 1992.
[7] G. Dattoli, P.L. Ottaviani, Logistic equation, FEL dynamics and self induced harmonic generation, presented at the 20th International Free-electron Laser Conference, Williamsburg, VA, August 1998.

ELSEVIER

Nuclear Instruments and Methods in Physics Research A 475 (2001) 537–544

NUCLEAR
INSTRUMENTS
& METHODS
IN PHYSICS
RESEARCH
Section A

www.elsevier.com/locate/nima

A test of the laser alignment system ALMY at the TTF-FEL

S. Roth[a], S. Schael[b,1], G. Schmidt[a,*]

[a] *Deutsches Elektronen-Synchrotron DESY, Notkestraße 85, D-22603 Hamburg, Germany*
[b] *Max-Planck-Institut für Physik, Föhringer Ring 6, D-80805 München, Germany*

Abstract

The laser alignment system ALMY was tested at the 15 m long undulator section of the TESLA test facility. The positions of the undulator modules relative to each other have been determined with a precision of 0.1 mm, limited by the accuracy of the mechanical support of the sensors. Additionally, ALMY allows one to measure movements or drifts over several days and we found that the undulator components are stable within 10 μm. The resolution of the sensors is better than 2 μm over a distance of 15 m. © 2001 Elsevier Science B.V. All rights reserved.

PACS: 29.17.+w; 29.50.+r; 81.05.Gc

Keywords: Linear accelerators; Alignment; FEL; Undulator

1. Introduction

New alignment techniques have to be established for the construction of a future linear collider or light source. For the proposed TESLA collider [1], the integration of an X-ray free electron laser (FEL) is planned. It will use the effect of Self Amplified Spontaneous Emission (SASE). In such SASE FEL, a tightly focused electron beam of high charge density is sent through a long undulator. The focusing is achieved by separated or integrated quadrupoles. The SASE effect results from the interaction of the electron bunch with its own radiation field created by the motion inside the undulator. This interaction can only take place if the electron and the photon beams overlap. To keep the electron beam inside the undulator on a straight line the precise alignment of the individual undulator modules and the focusing elements with respect to each other is crucial.

A FEL working in the vacuum ultraviolet (VUV) has been constructed at the TESLA Test Facility (TTF). At the TTF-FEL [2] the electron beam position must be straight with transverse deviations of less than 10 μm rms over the entire 15 m long undulator. Therefore, the magnetic axis of the undulator with a superimposed FODO structure must be aligned with about the same accuracy. For the alignment of the three individual undulator segments a commercially available interferometer system has been used which reaches

---

*Corresponding author. Inst. Für Beschleunigerphysik, Univesität Dortmund, Maria Goeppert Mayer Strasse 2, 44221 Dortmund, Germany. Tel.: +49-231-755-5382; fax: +49-231-755-5383.

E-mail address: schmidt@delta.uni-dortmund.de (G. Schmidt).

[1] Now at I. Phys. Inst., RWTH Aachen, D-52056 Aachen, Germany.

a precision of the order of 5 μm. The alignment system uses reference marks on the undulator which have a known offset to the magnetic axis of the undulator. The magnetic axis of the whole undulator is estimated to be straight within 30 μm [3].

As the laser interferometer is a manual system it has the disadvantage that it cannot be used during operation of the machine. Therefore, it cannot deliver a continuous monitoring of the positions of the undulator components. As an alternative we tested the ALMY [4] system. It is a multi-point alignment system that has been developed for the muon spectrometer of the ATLAS detector at the Large Hadron Collider. It uses an infrared laser beam, acting as alignment reference, which traverses several transparent silicon sensors. The sensors measure the laser beam position in both transverse coordinates. Thermal effects like density fluctuations of the air can influence the straightness of the laser beam. Such effects are shielded by means of an aluminium tube around the laser beam.

2. The transparent silicon sensors and the ALMY system

The optical sensors have to combine high position resolution and high light transmission. To optimize the transmission thin films of amorphous silicon (a-Si) are used as photosensitive material. The amorphous silicon strip sensors were produced at Heimann Optoelectronics and provide high precision position measurement at relatively low cost. CVD techniques are used to deposit the 1 μm thick photo-sensitive layers onto a 0.5 mm thick glass substrate. High-quality polished parallel glass wafers minimise uncertainties in the deflection of the traversing laser beam. The a-Si film is sandwiched by two 0.1 μm thick electrodes of indium-tin oxide (ITO) which are segmented into two orthogonal strip rows. The bottom electrode acts as ohmic contact while the top electrode forms a Shottky diode which is operated at about 3 V bias voltage. The strip pitch of about 300 μm has been optimized to the typical laser beam diameters of 3–5 mm. The structure of the sensors is shown in Fig. 1. Position resolutions of 1 μm over the whole sensor surface have been measured and transmission rates above 90% at $\lambda = 790$ nm have been achieved [4].

The readout electronics is integrated inside the sensor module. In Fig. 2, a complete sensor module is shown. The photocurrents of all strips are multiplexed, amplified and digitized. These values are stored into a memory which can be read out by a VME bus system. The system can be read out with a rate which is limited to about one measurement per second at maximum.

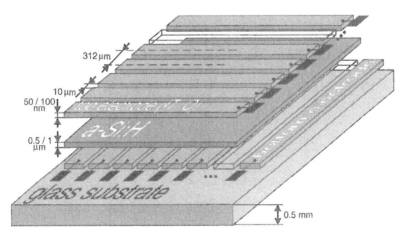

Fig. 1. Cross-section of the photosensitive detector.

Fig. 2. Complete module including sensor and readout electronics is shown.

3. The test setup

The undulator of the TTF-FEL consists of three undulator modules interspersed with four diagnostic modules containing wire scanners and beam position monitors [5]. The magnetic axis of the individual undulator modules itself has been measured using a 12 m long bench [3]. To build the undulator section inside the linac tunnel both ends of each undulator module and the diagnostic modules have to be lined up. As the alignment is done separately for the horizontal and the vertical coordinates, this gives in total 20 reference marks.

In Fig. 3, a view along the TTF undulator section is shown. For a first test of the ALMY system the sensors were placed at the alignment marks which determine the horizontal positions. Due to lack of space the last alignment mark has been used for installation of the laser optics. The laser sends a collimated laser beam with a diameter of 3–5 mm through all nine sensors. The laser beam is shielded against temperature gradients and fluctuations using aluminium tubes.

The readout of the silicon strip detectors is done by each sensor module individually and the digitized signal height of each strip is sent via RS232 connection to a data acquisition program running on a PC. Here a Gauss fit is performed to the shape of the measured beam profiles. The mean value from this fit is taken as the position measurement.

4. Measurement results

The laser alignment system ALMY has shown that it works within the background of radiation and electronic noise of the linac tunnel. It took data without any interruption during 5 days of linac operation with electron beam. Nine detectors were installed. The positions and movements of the sensors could be monitored all the time during this period.

Fig. 3. Test setup of the laser alignment system at the TTF undulator.

A comparison between the measured positions and the design positions of the alignment marks can be seen in Fig. 4. The design position of the reference marks contains the offset of the reference mark to the magnetic axis of the undulator. The difference of measurement and design gives the displacement of the individual components to the magnetic axis of the undulator. It is shown in the lower part of Fig. 4. The measurement error is influenced mainly by the mechanical assembly of the sensors onto the alignment marks which has a precision in the range of 0.1 mm. Within this error

one would conclude from this measurement that the undulator forms a straight line with exception of the components at both ends of the undulator section.

The setup has been operated in the linac for four days. Every 30 s a measurement was performed and the result written to disk. This allowed us to monitor the sensor positions continuously and to look for movements of the individual components, either in the form of oscillations or in the form of drifts. The result is shown in Fig. 5. One observes oscillations of the measured result with amplitudes

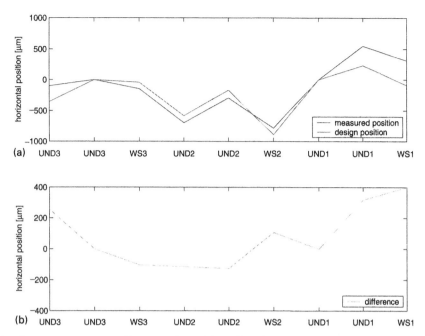

Fig. 4. Result of the alignment measurement. Shown are the positions of the three undulator modules (UND1, UND2, UND3) and of three of the four diagnostic monitors (WS1, WS2, WS3).

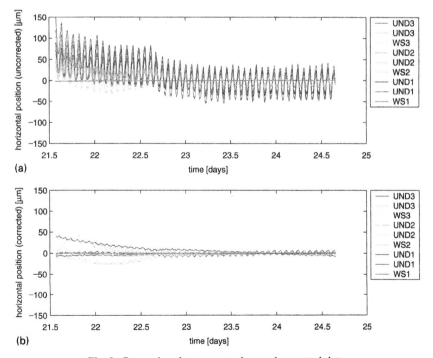

Fig. 5. Comparison between raw data and corrected data.

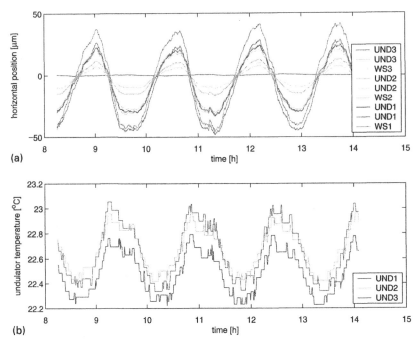

Fig. 6. Correlation between sensor alignment and temperature.

of up to 50 μm and periods of about 40 min. As can be seen in Fig. 6 these oscillations are correlated with the temperature variations in the climatized hut, where the undulator is placed. The amplitude is proportional to the distance of the sensor from the laser.

The observed oscillations are caused by changes of the laser beam direction by 3 nrad due to the temperature change of 0.3°C. As we are not interested in movements of the reference laser beam but in the potential movements of the undulator components with respect to each other, we put again a straight line through two of the components. The difference of the measured position from the straight line is then independent of changes in the laser beam direction. The result is shown in the bottom part of Fig. 5 and shows that the corrected measurements show a reduced dependence on the temperature variations.

During a period of about 1 h we took data every second. These data are analysed in Fig. 7. First all measurements are shown corrected for the changes of the laser beam direction as explained before.

The next plot of Fig. 7 gives the mean value of these single measurements averaged over 5 min. The resulting curve is much smoother than before and movements in the micron range are easily detectable. The resulting curves show the movement of the individual sensors to the reference axis. The spatial resolution of the sensors is calculated out of the position noise and it varies between 0.7 μm near the tail of the laser beam and 2 μm at both ends of the alignment distance.

A comparison with other alignment methods is shown in Fig. 8. In-between the expectable errors the measurements show good agreement. Only at both ends of the undulators some deviations are visible. Further investigations are needed to understand if there are systematic errors explaining this effect.

5. Conclusions

The laser alignment system ALMY was shown to work within the background of radiation and electronic noise inside the linac tunnel. With an

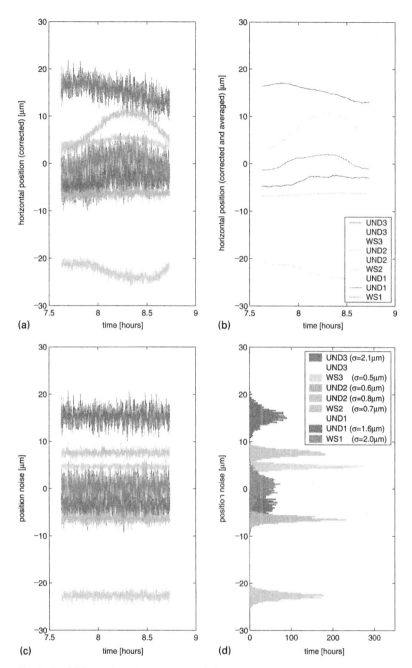

Fig. 7. Sensitivity and spatial resolution of the sensors in the current alignment setup.

improved fixation of the sensors to the undulator and individually calibrated sensors of the ALMY system, it will be possible to measure online the position of the undulator components with an accuracy of better than 0.03 mm. Nevertheless, further test should be done here at DESY to

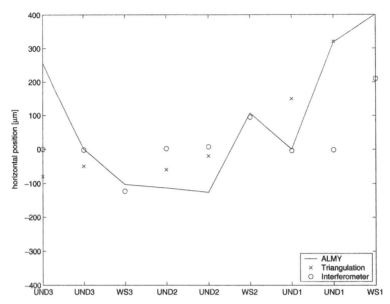

Fig. 8. Comparison of three different alignment measurements. Data for interferometer and triangulation measurement taken from Refs. [6,7].

investigate if the number of sensors can be increased without reducing the accuracy of the measurement and to check the usable distance where the ALMY system works.

However, it should be possible to use ALMY as a fast alignment system for complete beam line sections which could be up to 15 m long.

Acknowledgements

We would like to thank J. Brehling and the HASYLAB workshop for the construction of the detector mounting and T. Vielitz for his help during the installation of the test setup at the TTF undulator section.

References

[1] R. Brinkmann, G. Materlik, J. Rossbach, A. Wagner (Eds.), DESY 1997–048 and ECFA 1997–182, May 1997.
[2] TTF-FEL Conceptual Design Report, TESLA–FEL 95–03, DESY, June 1995;
J. Rossbach, et al., Nucl. Instr. and Meth. A 375 (1996) 269.
[3] J. Pflüger, H. Lu, T. Teichmann, Nucl. Instr. and Meth. A 429 (1999) 386.
[4] W. Blum, H. Kroha, P. Widmann, Nucl. Instr. and Meth. A 377 (1996) 404;
M. Fernandez Garcia, et al., Nucl. Instr. and Meth. A 461 (2001) 213
[5] U. Hahn, J. Pflueger, G. Schmidt, Nucl. Instr. and Meth. A 429 (1999) 276.
[6] J. Pflüger, et al., DESY group HASYLAB, private communication.
[7] J. Prenting, et al., DESY group ZMEA, private communication.

ELSEVIER

Nuclear Instruments and Methods in Physics Research A 475 (2001) 545–548

NUCLEAR
INSTRUMENTS
& METHODS
IN PHYSICS
RESEARCH
Section A

www.elsevier.com/locate/nima

First results of the high resolution wire scanners for beam profile and absolute beam position measurement at the TTF

G. Schmidt, U. Hahn*, M. Meschkat, F. Ridoutt

Deutsches Elektronen-Synchrotron DESY, Notkestr. 85, 22603 Hamburg, Germany

Abstract

In the TESLA Test Facility, wire scanners are used to measure the electron beam profile and position. The intended use of the wire scanners (to center the electron beam in the Free Electron Laser undulator) requires an especially precise alignment of the wire scanners with respect to the undulator axis. The wire scanners should define a reference axis with respect to the external reference system of the undulator with an accuracy better than 30 μm. The wire scanners allow a beam profile measurement, which will be used to optimize and match the beam optics for the undulator. First experimental results of beam position and profile will be presented and discussed. © 2001 Elsevier Science B.V. All rights reserved.

PACS: 41.60.C

Keywords: Free-electron laser; Beam position monitor; Wire scanner

1. Introduction

The undulator at the TESLA Test Facility-Free Electron Laser (TTF-FEL) consists of three undulator modules. The narrow spaces between and at both ends of the undulator modules are used for diagnostics. Two types of monitors are included in a so-called "diagnostic block". RF-type cavity monitors for horizontal and vertical beam position measurement and vertical and horizontal wire scanners allow a measurement of the beam profile and position [1]. The beam position can be measured with respect to external reference marks so that the position of the beam before, after, and between the undulator modules

at several fixed points is known. Fig. 1 shows a photograph of the diagnostic block between two undulator sections. The principle set-up of the wire scanners is sketched in Fig. 2. Each wire scanner fork (wire holder) is equipped with three wires: one tungsten wire of a diameter of 20 μm, and two carbon wires with a diameter of 5 μm. This allows one to select the wire as a function of beam size, beam intensity and required resolution. Electrons hitting a wire in turn produce scattered electrons, which are detected by scintillation counters. Behind each pair of wire scanners, a scintillation counter is installed covering nearly 360° around the beam tube [2].

The main difficulties that had to be overcome were:

- *The cleaning and assembly of the complicated diagnostic block.* Together with the attached

---

*Corresponding author. Tel.: +49-40-8998-3807; fax: +49-40-8994-3807.

E-mail address: ulrich.hahn@desy.de (U. Hahn).

0168-9002/01/$ - see front matter © 2001 Elsevier Science B.V. All rights reserved.
PII: S 0 1 6 8 - 9 0 0 2 (0 1) 0 1 5 9 7 - 2

Fig. 1. Diagnostic block between two undulator sections.

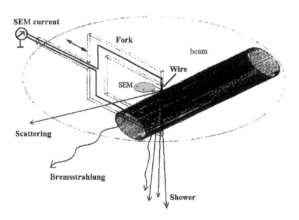

Fig. 2. Possibilities to detect the wire scanner signals.

components (like RF feedthroughs), the assembly had to take place under clean room conditions according to TTF specifications (clean room better than class 100) [3].

- *The alignment and stretching of wires.* The tightening of the wires in the fork is critical. It is difficult to stretch the wires to a straight line and fix them. After installation, the wires have to pass the narrow RF shielding slit (width 1.5 mm) in the diagnostic block during the scan of the beam profile. The alignment is extremely crucial since small deviations from the center of the slit result in the damage of all wires.

- *The calibration of the wire scanner in the diagnostic block.* The goal of this calibration is block with an accuracy of better than 30 μm in both planes. This calibration was done using a vertical coordinate measuring machine (CMM). A local clean room was installed in front of the machine. The diagnostic block was under clean room conditions during the entire calibration measurement. To measure the wire position a microscope was mounted on the CMM. It allowed one to see the thin wires and to measure the distance from the wires to the reference planes. Reproducibility was achieved using a linear encoder connected to the linear driving system of the wire scanner.

2. Results

Fig. 3 presents the results of all eight wire scanner measurements recorded during FEL operation of the TTF linac. The electron beam profile measurements were made with the 20 μm thick tungsten wire, which was chosen to ensure sufficient intensity on the scintillation counter; the signal from the carbon wires, on the other hand, was too small to be detected. By reducing the background (less beam loss before and inside the undulator area) and reducing the electron beam spot size, the signal-to-noise ratio can be improved.

The measured profiles indicate that the beam offset along the entire undulator stays below 300 μm during SASE operation. The spot size is somewhat larger than expected. Measurements have shown that a part of the spot size is generated by dispersion.

Each undulator has a superimposed FODO structure of 10 quadrupoles per undulator [4] In addition to the wire scanners, 10 beam position monitors (BPM), which allow the measurement of the beam position inside the quadrupoles, are installed along the undulator chamber [5].

Fig. 4 gives a comparison between the beam position measured by the wire scanners and the BPMs inside the first two undulators. It is clear

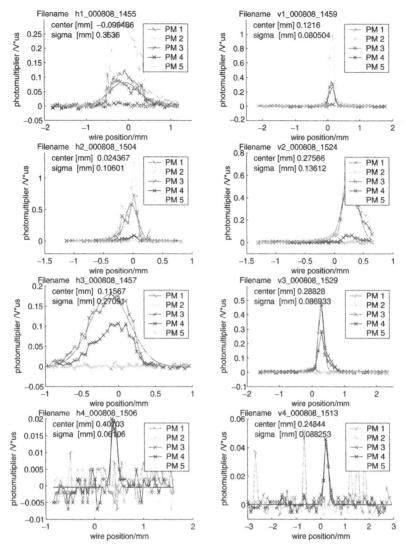

Fig. 3. Measurement of beam profiles using the wire scanner. Each graph shows the reading of five different photomultipliers (PM1 to PM5) located at different places along the undulator. The left column shows the signal of the horizontal wire scanners before (h1), after (h4), and in between the undulator sections (h2, h3). The right column shows the same results for the vertical direction. Only photo multiplier downstream the used wire scanner show any signal. From top graph down to the lower one, the profiles show the development of the spot size along the undulator.

that the BPMs show some random position deviations, due to the fact that they had not been calibrated. The fit of a betatron motion to both the 20 BPMs and the four wire scanner readings results in a similar betatron amplitude and phase of the beam motion.

The measurement of a difference orbit, implemented by a kick before the undulator, allows one to verify the beam optics inside the undulator by a comparison of the expected and measured betatron motion. Fig. 5 shows good agreement between the measured and calculated difference orbits. The betatron motion is only fitted in amplitude and phase to the measured difference orbit. Both measurements (wire scanner and BPMS) result in the same difference orbit. The

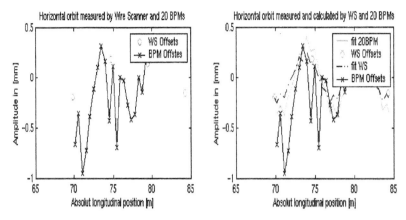

Fig. 4. BPM and wire scanner measurement.

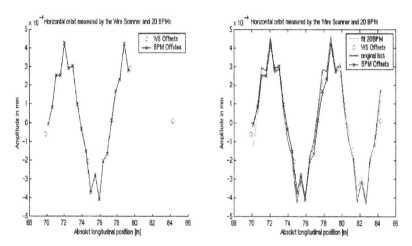

Fig. 5. Difference orbit by implementing a kick to the beam before it enters the undulator.

agreement with the calculated betatron motion is very good ($\sim 50\,\mu$m).

3. Conclusions

The eight wire scanners allow one to measure the beam profile and the beam position with high reproducibility and accuracy (better than $30\,\mu$m). Good agreement between the calculated and measured phase advance inside the undulator was observed. The position measurements of the wire scanners are also in good agreement with the 20 distributed BPMs.

References

[1] U. Hahn, J. Pflüger, G. Schmidt, Nucl. Instr. and Meth. A 429 (1999) 276.

[2] S. Striganov. G. Schmidt, K. Wittenburg, Estimation of the signal from the wire scanner in the TTF, TESLA Report, TESLA 99-08.

[3] D. Edwards, et al., Proposed cleaning procedures for the vacuum components for the TESLA Test Facility, TESLA Report 15, 1997.

[4] Y. Nikitina, J. Pflüger, Nucl. Instr. and Meth. A 375 (1996) 325.

[5] Hahn, et al., Design and performance of the vacuum chambers for the undulator for the VUV FEL at the TESLA test facility at DESY, Proceedings FEL Conference, 2000.

ELSEVIER

Nuclear Instruments and Methods in Physics Research A 475 (2001) 549–553

NUCLEAR
INSTRUMENTS
& METHODS
IN PHYSICS
RESEARCH
Section A

www.elsevier.com/locate/nima

Performance of a DC GaAs photocathode gun for the Jefferson lab FEL

T. Siggins[a],*, C. Sinclair[a], C. Bohn[b], D. Bullard[a], D. Douglas[a], A. Grippo[a], J. Gubeli[a], G.A. Krafft[a], B. Yunn[a]

[a] *Thomas Jefferson National Accelerator Facility, 12000 Jefferson Avenue, Newport News, VA 23606, USA*
[b] *Fermi National Accelerator Laboratory, P.O. Box 500, Batavia, IL 60510, USA*

Abstract

The performance of the 320 kV DC photocathode gun has met the design specifications for the 1 kW IR Demo FEL at Jefferson Lab. This gun has shown the ability to deliver high average current beam with outstanding lifetimes. The GaAs photocathode has delivered 135 pC per bunch, at a bunch repetition rate of 37.425 MHz, corresponding to 5 mA average CW current. In a recent cathode lifetime measurement, 20 h of CW beam was delivered with an average current of 3.1 mA and 211 C of total charge from a $0.283\,cm^2$ illuminated spot. The cathode showed a $1/e$ lifetime of 58 h and a $1/e$ extracted charge lifetime of 618 C. We have achieved quantum efficiencies of 5% from a GaAs wafer that has been in service for 13 months delivering in excess 2400 C with only three activation cycles. © 2001 Elsevier Science B.V. All rights reserved.

Keywords: Photocathode; FEL; Emittance

1. Gun description

The DC photocathode high brightness source for the Jefferson Lab FEL was proposed [1], specified [2] and then developed through a University of Illinois thesis project [3]. The gun design is similar to the 100 kV polarized electron gun for the CEBAF accelerator [4], which uses a strained lattice GaAs wafer and runs an average of 50 μA with 70–75% polarization. The FEL gun design uses a planar stainless-steel cathode geometry (Fig. 1), with a 10.56 cm gap operated at 320 kV,

and incorporates a bulk GaAs wafer photocathode 3.18 cm in diameter. The exposed wafer surface (2.54 cm in diameter) is activated to negative electron affinity with cesium and nitrogen trifluoride. A frequency-doubled, mode-locked Nd:YLF laser at 527 nm is used to drive the gun with a 6 mm spot illuminating the cathode. The design parameters for the FEL require the gun to deliver 5 mA at 37.425 MHz corresponding to 135 pC per bunch for IR lasing [5]. We have achieved a performance increase by adjusting our GaAs cleaning methods and by using high voltage processing techniques for the electrode structure. The limiting factor in cathode lifetime is ion back bombardment [4]. Higher DC voltage operation is presently prevented by field emission. However,

---

*Corresponding author. Tel:. +1-757-269-5019; fax: +1-757-269-5520.

E-mail address: siggins@jlab.org (T. Siggins).

0168-9002/01/$ - see front matter © 2001 Elsevier Science B.V. All rights reserved.
PII: S 0 1 6 8 - 9 0 0 2 (0 1) 0 1 5 9 6 - 0

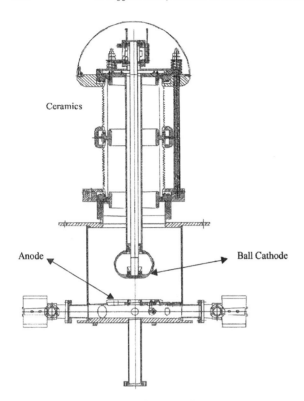

Fig. 1. Mechanical layout of the FEL gun.

simulations indicate that working at higher voltage would result in improved electron beam emittance. We currently have a program underway to study materials for improved high voltage standoff characteristics.

2. Operations

During early operations, the cathode quantum efficiencies (QE) were low—0.1–0.5%. Intermittent arcs occurring on the high voltage electrode structure resulted in permanent cathode degradation and a need to replace the GaAs wafer after a typical 3 month operations period. Several steps were taken to improve the QE and to minimize the occurrence of high voltage arcs. Changes were made in the hydrogen cleaning system [6] used to prepare the GaAs wafer. The original geometry of the hydrogen plasma source in the cleaning system was constricted allowing for recombination in the

plasma and poor cleaning results. The wafer was moved away from the nominal position for cleaning to open up the conductance and enhance the atomic hydrogen flux.

A pre-activation heat clean cycle for the GaAs photocathode was developed for use on the Jefferson Lab polarized source gun [4]. This heat cleaning procedure was transferred to and optimized for use with the FEL gun. The larger mass of the FEL gun required the heat cycle to be run for a longer period to obtain good cleaning results.

From observations made during numerous high voltage conditioning and cesiation cycles, it was determined that a damaged region existed on the stainless-steel ball cathode that could not be removed by our normal mechanical polishing or high voltage processing techniques. The ball cathode was replaced and the electrode structure showed an immediate improvement in performance during and after high voltage conditioning.

We determined that the QE of the photocathode could be damaged during the high voltage conditioning process. Consequently the wafer is now retracted into the ball cathode for protection during high voltage processing, thus preventing damage and minimizing contamination of the photocathode surface.

These changes have resulted in QEs of 5% and a GaAs wafer that has been in service for 13 months delivering in excess 2400 C over three activation cycles. The gun has delivered 5 mA of recirculated CW beam at 37.425 MHz with 135 pC to meet the design parameters for the FEL.

3. Photocathode lifetime

Twenty hours of CW beam at 74.85 MHz were run to measure cathode lifetime. The run started with the cathode at a peak QE of 2.7%. A constant drive laser power was used and the resulting drop in beam current was measured. The FEL driver accelerator was set up to transport 3.60 mA of beam, dropping to 2.56 mA as the QE decayed, delivering a total charge of 211 C. The data in Fig. 2 shows a $1/e$ lifetime of 58 h at an average current of 3.1 mA and a $1/e$ extracted charge lifetime of 618 C.

FEL Cathode Lifetime Measurement

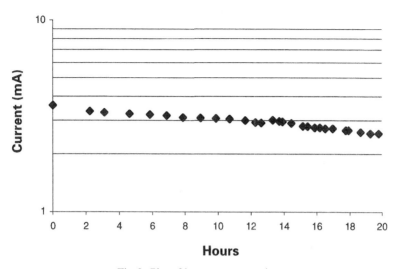

Fig. 2. Plot of beam current vs. time.

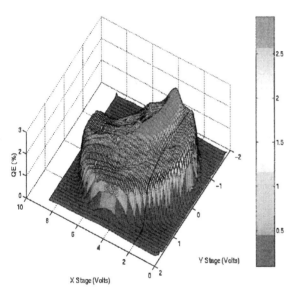

Fig. 3. Cathode QE scan prior to the lifetime run.

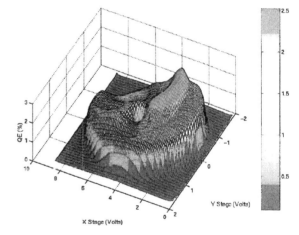

Fig. 4. Cathode QE scan after the lifetime run.

Although we see some signs of chemical poisoning of the cathode, ion back bombardment is the limiting factor in cathode lifetime. Figs. 3 and 4 show quantum efficiency scans of the cathode surface before and after the previously described lifetime run. Ion back bombardment damage to the cathode is seen in Fig. 4 as a hole in the center of the cathode scan following delivery of 20 h of CW beam.

The FEL gun vacuum system currently operates in the mid 10^{-11} Torr range as measured by a RGA. Improving the anode–cathode gap vacuum should increase cathode lifetimes substantially, as indicated by Jefferson Lab's experience with GaAs-based polarized electron sources [4]. These sources operate with at least an order of magnitude lower vacuum ($\lesssim 10^{-12}$ Torr).

4. Emittance measurements

Electron beam emittance data was taken at an energy of 10 MeV using a multi-slit diagnostic located down stream of the FEL gun and a 10 MeV super-conducting RF accelerator module. Measurements were taken from 20 to 135 pC per bunch as shown in Fig 5. These data show that the FEL upgrade design specifications ($\leqslant 30\,\pi$mm mrad at 135 pC) was exceeded [7]. The measured emittance values in Fig. 5 are larger than those quoted in Ref. [5] because the injector set up was changed to support high charge running at 135 pC.

5. Field emission

The FEL gun is currently operated at 320 kV DC with a gradient of 3.9 MV/m at the cathode surface. The stainless-steel electrode structure is routinely conditioned to 420 kV prior to cathode activation. This has provided a very stable environment for the cathode. As the voltage is increased above our operating level, we start to see unacceptable levels of field emission from the cesiated surfaces.

We are currently working on a system that will allow us to activate the cathode without exposing the electrode surface to cesium. One configuration employs a focused cesium source mounted below the anode plate delivering a 18 mm spot of cesium to the photocathode. Another configuration mounts cesium channel sources inside the ball cathode structure and the cesiation of the photocathode is performed there.

Different materials for the electrode structure are currently being studied for their field emission properties in a large area electrode test system. This Field Emission Test System (Fig. 6) uses 15 cm diameter electrodes with Rowgowski profiles and a 125 kV DC power supply. The test cathode is rigidly mounted while the anode is mounted on three micrometer adjustments so the electrode gap and electric field may be varied. Results obtained from this materials survey should allow us to increase the operating voltage of an improved gun design with subsequent improvement of the emittance of the electron beam.

Fig. 6. Field Emission Test System.

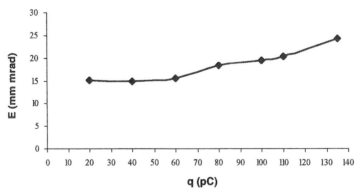

Fig. 5. Plot of emittance vs. charge per bunch.

6. Conclusions

The DC GaAs photocathode gun for the Jefferson Lab FEL is operating at higher voltage and delivering more average current than any bulk GaAs photocathode gun in operation. The good QEs and long cathode lifetimes have been obtained using hydrogen cleaning, an ultra-high vacuum environment and shielding of the wafer during high voltage processing. The gun has met the original design specifications for the Jefferson Lab IR Demo FEL and can routinely deliver 5 mA of CW beam for extended periods. Work being done on improving the gun's vacuum system and raising the operating voltage should allow us to further increase cathode lifetimes and improve the quality of beam delivered.

Acknowledgements

Work supported by U.S. DOE under contract # DE-AC05-84-ER40150, the office of Naval Research and the Commonwealth of Virginia.

References

[1] G.R. Neil, et al., Nucl. Instr. and Meth. A 318 (1992) 212.
[2] C.K. Sinclair, Nucl. Instr. and Meth. A 318 (1992) 410.
[3] D. Engwall, et al., Proceedings of the 1997 Particle Accelerator Conference, Vancouver, BC, IEEE, Piscataway, NJ, 1998, p. 2693.
[4] M. Poelker, et al., Proceedings of the Workshop on Polarized Electron Sources Nagoya, Japan, AIP, New York. to be published.
[5] G.R. Neil, et al., Phys. Rev. Lett. 84 (4) (2000) 662.
[6] C.K. Sinclair, et al., Proceedings of the 1997 Particle Accelerator Conference, Vancouver, BC, IEEE, Piscataway, NJ, 1998, p. 2864.
[7] D. Douglas, et al., Driver accelerator design for the 10 kW upgrade of the Jefferson Lab IR FEL, Proc. of the XXth International Linac Conference, Monterey, CA, August 21–25, 2000.

ELSEVIER

Nuclear Instruments and Methods in Physics Research A 475 (2001) 554–558

NUCLEAR
INSTRUMENTS
& METHODS
IN PHYSICS
RESEARCH
Section A

www.elsevier.com/locate/nima

Jefferson Lab IR demo FEL photocathode quantum efficiency scanner

J. Gubeli*, R. Evans, A. Grippo, K. Jordan, M. Shinn, T. Siggins

Thomas Jefferson National Accelerator Facility, 12000 Jefferson Avenue, MS 6A, Newport News, VA 23606, USA

Abstract

Jefferson Laboratory's Free Electron Laser (FEL) incorporates a cesiated gallium arsenide (GaAs) DC photocathode gun as its electron source. By using a set of scanning mirrors, the surface of the GaAs wafer is illuminated with a 543.5nm helium–neon laser. Measuring the current flow across the biased photocathode generates a quantum efficiency (QE) map of the 1-in. diameter wafer surface. The resulting QE map provides a very detailed picture of the efficiency of the wafer surface. By generating a QE map in a matter of minutes, the photocathode scanner has proven to be an exceptional tool in quickly determining sensitivity and availability of the photocathode for operation. © 2001 Elsevier Science B.V. All rights reserved.

PACS: 41.75.Jv; 41.60.Cr; 07.05.Hd

Keywords: GaAs photocathode; Quantum efficiency; Raster/scanner

1. Introduction

1.1. IR FEL

TJNAF's IR FEL [1] is depicted in Fig. 1. The injector consists of a gun with a cesiated gallium arsenide (GaAs) photocathode, which is illuminated by a frequency doubled, mode-locked Nd:YLF drive laser. The electrons produced by the gun are accelerated by a potential difference of a few hundred kilovolts and enter a 10-MV superconducting radio frequency (SRF) cryomodule. The electron bunches enter the linac in phase with the radio frequency (RF) of an SRF linac and are accelerated to energies in the range of 35–50 MeV. After passing an isochronous and achromatic chicane around the output-coupled laser cavity mirror, the 42-MV electrons enter an IR wiggler. The wiggler consists of a line of magnets with alternating polarity (NSNS, etc.) that causes the electron beam to wiggle (hence the name) and produces the photons that bounce between the mirrors of the optical system. Then the electron bunch leaves the wiggler and passes through another chicane around the highly reflective laser cavity mirror. The electrons proceed around the beam transport and reenter the linac. Unlike the beam's first trip through the linac, it is now out of phase with the RF and is decelerated to 10 MV and sent to a beam dump. The deceleration process allows extensive energy recovery and is responsible for making TJNAF's FEL highly efficient.

---
*Corresponding author. Fax: +1-757-269-5519.
E-mail address: gubeli@jlab.org (J. Gubeli).

0168-9002/01/$ - see front matter © 2001 Elsevier Science B.V. All rights reserved.
PII: S 0 1 6 8 - 9 0 0 2 (0 1) 0 1 6 9 5 - 3

Fig. 1. IR-FEL.

1.2. Motivation

The primary motivation for fabricating the photocathode scanner was to monitor the quantum efficiency (QE) degradation of the photocathode wafer. Experience has shown that photocathode QE falls not only with use, but also as a function of time. With our current periodic operational schedule, there was a need to determine the state of the wafer before each run.

2. Setup

2.1. Optics

A JDS Uniphase helium–neon laser (model 1675) is used to illuminate the photocathode during scans. This laser was chosen mainly because its output wavelength of 543.5 nm closely matches the 527 nm output of the Drive Laser. This single mode 1mW laser also has good pointing stability of $<0.03\,\mu$rad and a small power drift of $\pm 2.5\%$ over an 8-h period [2]. After the laser (Fig. 2), a lens is used to focus the laser spot down to a 0.25 mm radius on the wafer to increase resolution. To reduce space charge effects caused by the focused beam and the low acceleration voltage of 66 V DC, a neutral density filter is used to attenuate the power that is incident on the cathode. A first-surface mirror directs the beam to the two oscillating mirrors of a General Scanning Inc. DX series, closed loop optical scanner. The resulting raster beam travels up through one of the four vacuum viewports in the Light Box assembly to the Light Box Mirror. This mirror reflects the beam to the wafer. The Light Box Mirror is a pyramid-shaped optic with four reflective surfaces and a clear bore through its center for the electron beam to pass. Each of the four surfaces faces a viewport, two of which are used for the Drive

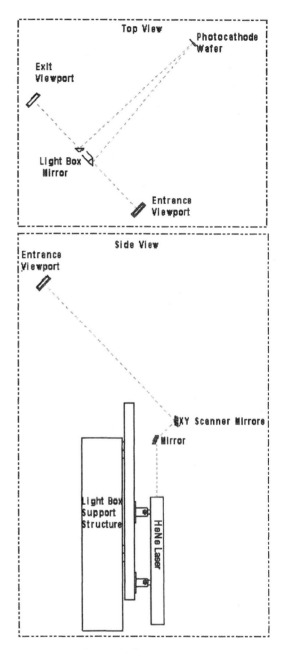

Fig. 2. Photocathode scanner optical layout.

Laser input and output. At the third viewport, opposite the scanner, there is a CCD camera mounted to image the photocathode surface. Because of this arrangement, it is necessary to slightly misalign the scanned beam to the Light Box to avoid saturating the camera.

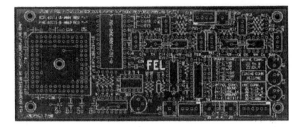

Fig. 3. Photocathode raster board.

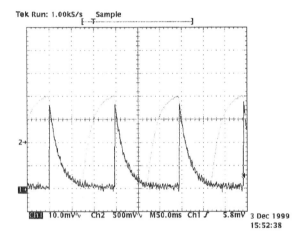

Fig. 4. Photocathode scanner raster output.

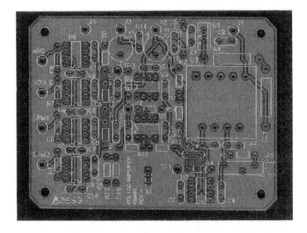

Fig. 5. Scanner logarithmic amplifier board.

2.2. Electronics/data acquisition

A custom-made electronics board (Fig. 3) was designed to both provide the two raster signals to the scanner controller and to output an analog representation of the laser position on the cathode. This raster board uses a PLD chip to output up to 240 discrete steps for the slow X-axis and a continuous ramp for the fast Y-axis. After processing through a digital-to-analog chip, the scope trace of Fig. 4 is the AC output of the raster board with the slow axis on the first channel and the fast axis on the second. With a period of 150 ms, the entire cathode is scanned in 36 s. The second custom board (Fig. 5) is centered around an Analog Devices 759N logarithmic amplifier. The 759N is configured to output 1.5 V for every decade of input current; and with six decades of range, it is sensitive from nanoamps to a milliamp. By biasing the photocathode at 66 V DC, the

current flow resulting from its illumination is measured. The X and Y coordinate positions and the current flow from the cathode are sent to three channels on a VMI VME-3123 digitizer. This 16-channel, 16-bit, analog-to-digital digitizer with a simultaneous "sample and hold" for each channel is able to process 200 k bytes/s/channel.

2.3. Software

EPICS code translates the digitized current into a QE% by using the following relationship. In this equation, I is the beam current, P is the laser power incident on the wafer and λ is the laser wavelength.

$$QE\% = \frac{124I(\mu A)}{P(mW)\lambda(nm)}. \tag{1}$$

An EPICS GUI (Fig. 6) allows a user to input the power and wavelength incident on the wafer as well as select the resolution of the plots. While the default settings should be used, it is necessary to have control over these inputs for setup and testing. The GUI allows users to save comments to both current and old saved scans as well as output any of five different types of plots. The first four plots are 2-D plots showing the wafer as seen by the viewport-mounted camera, the wafer oriented as it is physically, and a difference plot of any two files with either of the above mentioned orientations. With any of the 2-D plots, a color

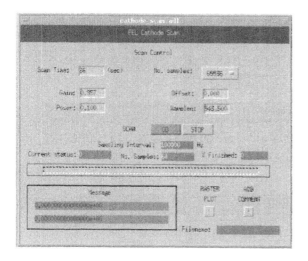

Fig. 6. Photocathode scanner GUI.

scale represents the QE%, and by right-clicking with the mouse on any section of the plot, a specific QE% can be obtained. The last is a 3-D plot that is rendered using MATLAB and can be rotated about either axis. This 3-D plot is also color-scaled to represent the QE%.

3. Testing and results

Several tests and measurements were performed to validate the reliability and accuracy of the photocathode scanner. A measurement of the laser power incident on the wafer was obtained with a sensitive, calibrated power meter placed in front of the input vacuum viewport. The attenuation of the viewport and Light Box Mirror was previously measured by replacing the wafer with a power meter. To confirm the absence of space charge effects, it was simply a matter of inserting neutral density filters in the beam path until the current dropped linearly with laser power. The logarithmic amplifier, digitizer, and software accuracy were verified in one step by substituting the biased wafer with a measured precision resistor. The last test performed was to verify the orientation of the plots. This was accomplished by masking sections of the input viewport and comparing its results with the image from the viewport camera. A sample of a 2-D scan can be seen in

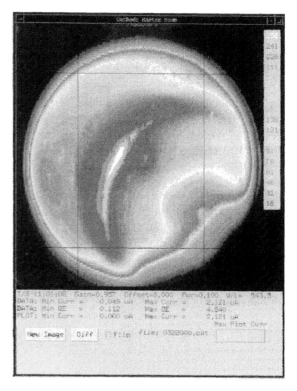

Fig. 7. 2-D photocathode plot.

Fig. 8. 3-D photocathode plot.

Fig. 7. For this scan, the maximum QE was 4.84%, located slightly to the left of center. Fig. 8 is a 3-D plot of the same file, rotated clockwise approximately 50° and tilted to show a near vertical profile.

4. Conclusion

4.1. Summary

The need to quickly determine photocathode operational availability has been met in all respects by the quantum efficiency scanner. Using the raster board to both control the scanning mirrors and to provide a coordinate position of the laser spot on the cathode surface has proven to be a reliable way to scan the wafer. With 240 horizontal and 271 vertical lines of resolution, a scan of more than 65,000 points can be obtained in 36 s. The logarithmic amplifier used to measure the resultant current flow from the biased photocathode is sensitive to changes of a few nanoamps. An EPICS interface to the digitizer card provides a colorful 2-D or 3-D representation of the wafer's quantum efficiency.

Acknowledgements

This work was supported by U.S. DOE Contract No. DE-AC05-84-ER40150, the Office of Naval Research, the Commonwealth of Virginia, and the Laser Processing Consortium

References

[1] M.Shinn, The Jefferson Laboratory IR Demo Project, Proceedings of SPIE, Vol. 2988, 1997, pp. 170–175.
[2] JDS Uniphase, 1600 Series Product Bulletin, 1600 Rev. A 02/00.

ELSEVIER

Nuclear Instruments and Methods in Physics Research A 475 (2001) 559–563

NUCLEAR
INSTRUMENTS
& METHODS
IN PHYSICS
RESEARCH
Section A

www.elsevier.com/locate/nima

Electron beams formed by photoelectric field emission

C. Hernandez Garcia, C.A. Brau*

Department of Physics, Vanderbilt University, Box 1807, Station B, Nashville, TN 37235, USA

Abstract

Previous measurements of electron emission by cw laser-irradiated tungsten needles have been extended to pulsed laser irradiation using a 7-ns long, doubled Nd:YAG laser at intensities up to $10^{11} \, W/m^2$. The results show that the mechanism for emission changes from the tunnelling mechanism observed in the cw case, characterized by exponential voltage dependence, to an ablative mechanism characterized by very weak voltage dependence. The observed emission has a threshold around $10^{11} \, W/m^2$, just below the damage threshold for the needles. Using needles with a 1-μm tip radius at voltages up to 50 kV, current densities as high as $10^{11} \, A/m^2$ were observed. © 2001 Published by Elsevier Science B.V.

PACS: 41.75.Fr; 79.70.+q; 79.60.−i

Keywords: Electron beams; Field emission; Photoemission

1. Introduction

The development of free-electron lasers has always been paced by the development of electron beams of higher normalized brightness. RF photoelectric injectors now produce electron beams with a normalized brightness typically of the order of $10^{11} \, A/m^2$ sr. Pulsed electron beams from laser-irradiated needle cathodes have been reported with peak currents as high as 2 A and estimated brightness as high as $10^{16} \, A/m^2$ sr [1,2]. If these sources can be developed for application to free-electron lasers it will be possible to operate in the ultraviolet at electron energies below 1 MeV [3].

In experiments reported earlier, a 1-W cw argon-ion laser was used to irradiate tungsten and ZrC needle cathodes at an intensity of $10^8 \, W/m^2$ [4,5]. Total current as high as 10 μA was observed, which corresponds to an estimated normalized brightness of $10^{10} \, A/m^2$ sr. The current was observed to be strongly voltage dependent, indicating that under these conditions the current is emitted by a process of photoelectric field emission in which the electrons tunnel out as described by Fowler and Nordheim [6]. The emission is strongly nonlinear, and increases exponentially with laser intensity.

The experiments have now been repeated using a pulsed, doubled Nd:YAG laser to irradiate the needles at intensities up to $10^{11} \, W/m^2$ for times of the order of 7 ns. The results show that the mechanism observed under cw irradiation is replaced by an ablative mechanism under pulsed

*Corresponding author. Tel.: +1-615-322-2559; fax: +1-615-343-1103.

E-mail address: c.a.brau@vanderbilt.edu (C.A. Brau).

irradiation. The effect shows a sharp threshold around $300 \, \text{J/m}^2$, just below the threshold for single-pulse damage to the tip of the needle, and a very weak dependence on voltage, in stark contrast to the cw results.

2. Experimental techniques

The apparatus used in the experiments has been described previously [5]. Polycrystalline tungsten needles are etched to produce a tip radius of the order of $1 \, \mu\text{m}$, and then heated to 2200°C by electron bombardment to produce a smooth, clean surface. To avoid contamination, the background pressure is maintained at $10^{-10} \, \text{Torr}$, and the needles are cleaned periodically by heating them to 1800°C by illuminating them from the side with the 1-W argon-ion laser. The cathode is biased at 10–50 kV relative to the anode at ground potential.

In the present experiments the needles were illuminated end-on with a doubled, Q-switched Nd:YAG laser producing 200 mJ per pulse at 532 nm. The laser beam was attenuated and focused to a diameter of 1.2 mm, and the intensity varied between 10^{10} and $10^{11} \, \text{W/m}^2$ by moving the focal point relative to the tip of the needle. The power was monitored with a Si photodetector with a nominal 1-ns response time, and recorded with a digital oscilloscope with a bandwidth of 500 MHz and a digitizing rate of 2 GHz. The average current was measured in the cathode circuit using a fibre-optic link from the cathode. The peak current was observed by passing the anode current through the 50-Ω input of the oscilloscope. The response time was about 2 ns, and approximately critically damped.

3. Experimental results

A typical current pulse and laser pulse are shown in Fig. 1. The synchronism of the two pulses shown in Fig. 1 is accurate to about a nanosecond. As the laser intensity increases, the onset of the current pulse advances slightly and the pulse becomes shorter, suggesting that the current is saturating as some sort of self-limiting process

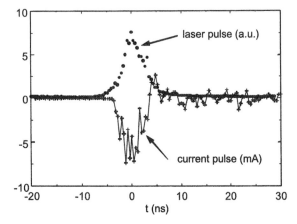

Fig. 1. Typical laser pulse (upper trace, arbitrary units) and current pulse (lower trace, 1 mA/division). The time scale is 10 ns/division.

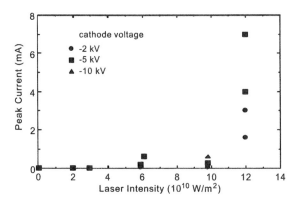

Fig. 2. Dependence of the peak current on the laser intensity.

sets in. A likely possibility is shielding of the surface from the laser by atoms that have already been ablated [7,8].

As shown in Fig. 2, the peak current is a strong function of the laser fluence, and appears to have a threshold near $300 \, \text{J/m}^2$. Above this fluence large current excursions were observed, and single-pulse damage to the tip of the needle was noticeable both in the performance of the needle and in electron micrographs. Below this threshold the needles were observed to accumulate damage over hundreds of shots. The damage could usually be repaired by heating the needle to above 2200°C for a few hours.

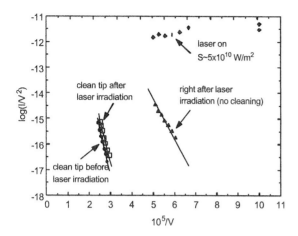

Fig. 3. Fowler–Nordheim plot of the peak current as a function of the voltage.

The effect of voltage is shown in Fig. 3. We see there that ordinary field emission from the needles is well described by the theory of Fowler and Nordheim. On the other hand, the peak current during the laser pulses is almost independent of the voltage. The slight increase in the quantity plotted in Fig. 3 toward lower voltages simply reflects the fact that the ordinate is $\ln(I_{peak})/V^2$.

Also shown in Fig. 3 is the non-laser behaviour of a needle with and without laser damage. After laser irradiation the DC current increases by many orders of magnitude. Electron micrographs show that this is actually due to roughness (damage) on the surface of the needle. After the needle is heated to a high temperature the surface smoothes out and the performance returns to near its original behaviour.

4. Interpretation

Photoelectric field emission has been extensively studied at low intensities [9,10]. It is found that at intensities below about 3.5×10 W/m^2 the emission of electrons near the photoexcitation energy is linear in the laser intensity [11]. It is also found that under these same conditions a substantial fraction of the photoexcited electrons relax to levels just above the Fermi level [12]. However, the high-intensity region has received much less attention [1,2,4,5].

In cw experiments using tungsten needles at intensities of the order of 10^8 W/m^2 it is found that the dependence of the total current on laser intensity and electric field is very strong and can be accurately described by an extension of the Fowler–Nordheim theory of field emission [5]. According to this model the photoexcited electrons rapidly relax to levels just above the Fermi level, where they accumulate [12]. If the electrons relax back into the Fermi sea with a time τ_R, the cw experiments can be interpreted in terms of the ratio τ_R/ρ_E, where ρ_E is the density of electron energy levels in the metal. However, the cw experiments do not permit the identification of the relaxation time and the level density separately.

The response to a pulsed laser depends on the relaxation time and the level density separately, rather than just as their ratio. When the pulse duration is short compared to the relaxation time, the photoemission current is suppressed relative to the cw value since the levels above the Fermi level do not have time to fill up. In the present experiments the peak current is observed to be nearly independent of the voltage, in sharp distinction to the cw results. This implies that the current observed in the pulsed experiments is caused by a different mechanism, and photoemission is suppressed by a relaxation time longer than the laser pulse. Detailed calculations indicate that the relaxation time must actually be longer than 100 ns. However, the slowest relaxation process in a simple free-electron gas is the loss of energy in electron–phonon collisions, and this is expected to occur in 1–10 ns [13]. The long relaxation time may have its origin in the details of the band structure of tungsten. This is supported by the fact that photofield emission could not be detected in cw experiments using ZrC needles [5].

The proximity of the threshold for observing pulsed current to the threshold for damage to the tip of the needle suggests that pulsed emission is related to an ablation process. In fact, it is observed in the present experiments that the surface of the tip of the needle is degraded after typically hundreds or thousands of pulses, as described above. The laser ablation of tungsten has been studied and shows a threshold for melting around $3000 \, \text{J/m}^2$ for nanosecond pulses, which is well above the fluences used in the present experiments [14]. Sub-threshold ablation of tungsten surfaces is less well understood, but it has been observed that the removal of atoms from the surface of tungsten by pulsed laser irradiation begins at fluences around 200–$400 \, \text{J/m}^2$, which is well below the threshold for melting [15,14]. It is likely that the atoms ablated at low fluences are those that are weakly bound to the surface, such as those located at steps in the surface crystal structure.

If ablation occurs, some electrons may be emitted as part of the ablation process. Additional electron emission may be due to field ionization of the ablated tungsten atoms. Just above the surface of the needle the electric field E is of the order of $10^{10} \, \text{V/m}$. In an electric field of this intensity the Coulomb potential of an electron around an atom is strongly modified, and the energy levels of the outermost electrons should be strongly Stark shifted. These two effects should reduce the ionization potential well below the $7.98 \, \text{eV}$ of the isolated atom, and allow the uppermost electron to escape either by passing over the potential barrier or by tunnelling through it. Before the atom is ionized it interacts with the field through its induced dipole moment. The energy of the atom in the electric field is then

$$\mathscr{E}_d = \gamma_{\text{atom}} E^2 \tag{1}$$

where

$$\gamma_{\text{atom}} = \frac{e^2 Z_{\text{eff}}}{m \omega_{\text{eff}}^2} = O(10^{-38} \, \text{C}^2 \, \text{s}^2/\text{kg}) \tag{2}$$

is the atomic polarizability, in which Z_{eff} is the effective number of electrons per atom, and ω_{eff} is the effective frequency of the electrons in the participating levels. At the surface of the cathode tip the energy of a neutral atom is about $-10 \, \text{eV}$ relative to the energy at infinity. An atom created with a kinetic energy of $1 \, \text{eV}$ can therefore travel about $\frac{1}{10}$ of the tip radius from the surface. In such a strong field the atom will probably undergo field ionization and be drawn by the electric field back into the surface. For an electric field of $10^{10} \, \text{V/m}$ and a tip radius of $1 \, \mu\text{m}$, the ion accumulates $1 \, \text{keV}$ of kinetic energy in this process. When the ion

strikes the surface the probability is large (of the order of unity [16]) for sputtering another atom from the surface. In addition, the excess energy is converted into heat at the surface of the tungsten and causes further evaporation of the surface. This runaway process could account for the threshold behaviour of the pulsed current, and for the self-sustained current excursions observed at slightly higher fluences when the electric field is very high.

5. Conclusions

We have explored the emission of electrons from tungsten needles with electric fields of the order of $10^{10} \, \text{V/m}$ under pulsed laser irradiation, up to $10^{11} \, \text{W/m}^2$ for $7 \, \text{ns}$. We find that the emission is related to sub-threshold ablation of tungsten atoms from the surface. This is confirmed by the weak voltage dependence of the emission and the proximity of the threshold for current emission to the threshold for laser damage to the tip. The largest current we have seen is about $100 \, \text{mA}$ from a 1-μm tip, which corresponds to a current density $J = 10^{11} \, \text{A/m}^2$. If the effective electron temperature T_e is not more than about $1 \, \text{eV}$, this corresponds to a normalized brightness $B_{\text{N}} = O(10^{16} \, \text{A/m}^2 \, \text{sr})$ [17]. As discussed elsewhere, this is sufficient to develop tabletop free-electron lasers operating from the far infrared to the ultraviolet portions of the spectrum [3]

References

[1] M. Boussakaya, et al., Nucl. Instr. and Meth. A 279 (1989) 405.
[2] G. Ramian, E. Garate, Presented at the 16th International Free-Electron Laser Conference, Stanford, CA, August 23, 1994.
[3] C.A. Brau, Nucl. Instr. and Meth. A 407 (1998) 1.
[4] C. Hernandez Garcia, C.A. Brau, Nucl. Instr. and Meth. A 429 (1999) 257.
[5] C. Hernandez Garcia, C.A. Brau, in: J. Feldhaus, H. Weise (Eds.), Proc. of the 21st Int. Free Electron Laser Conference, August 23–26, 1999. Elsevier Science B.V., Amsterdam, p. II-71.
[6] R.H. Fowler, L.W. Nordheim, Proc. Roy. Soc. A 119 (1928) 173.

[7] C.P. Grijoropoulos, Lasers, optics, and thermal considerations in ablation experiments, in: J.C. Miller, R.F. Haglund (Eds.), Laser Ablation and Desorption, Academic Press, San Diego, 1998.

[8] O. Bostanjoglo, F. Heinricht, J. Phys. E 20 (1987) 1491.

[9] C.M. Egert, R. Reifenberger, Surf. Sci. 145 (1984) 159.

[10] T. Radon, S. Jaskolka, Surf. Sci. 247 (1991) 106.

[11] D. Venus, M.J.G. Lee, Surf. Sci. 116 (1982) 359.

[12] P.J. Donders, M.J.G. Lee, Phys. Rev. B 41 (1990) 1781.

[13] C. Kittel, Introduction to Solid State Physics, 7th Edition, Wiley, New York, 1996, p. 665.

[14] Z. Toth, et al., Appl. Phys. A 60 (1995) 431.

[15] C. Beleznai, et al., Appl. Surf. Sci. 127 (1998) 88.

[16] E. Hechtl, et al., J. Nucl. Mater. 176 (1990) 874.

[17] J.D. Lawson, The Physics of Charged-Particle Beams, Clarendon Press, Oxford, 1988, p. 160.

ELSEVIER

Nuclear Instruments and Methods in Physics Research A 475 (2001) 564–568

**NUCLEAR
INSTRUMENTS
& METHODS
IN PHYSICS
RESEARCH**
Section A

www.elsevier.com/locate/nima

Research on DC-RF superconducting photocathode injector for high average power FELs ☆

Kui Zhao*, Jiankui Hao, Yanle Hu, Baocheng Zhang, Shengwen Quan, Jiaer Chen, Jiejia Zhuang[1]

Institute of Heavy Ion Physics, Peking University, RFSC Group, Beijing 100871, People's Republic of China

Abstract

To obtain high average current electron beams for a high average power Free Electron Laser (FEL), a DC-RF superconducting injector is designed. It consists of a DC extraction gap, a $1+\frac{1}{2}$ superconducting cavity and a coaxial input system. The DC gap, which takes the form of a Pierce configuration, is connected to the $1+\frac{1}{2}$ superconducting cavity. The photocathode is attached to the negative electrode of the DC gap. The anode forms the bottom of the $\frac{1}{2}$ cavity. Simulations are made to model the beam dynamics of the electron beams extracted by the DC gap and accelerated by the superconducting cavity. High quality electron beams with emittance lower than $3\,\pi$-mm-mrad can be obtained. The optimization of experiments with the DC gap, as well as the design of experiments with the coaxial coupler have all been completed. An optimized $1+\frac{1}{2}$ superconducting cavity is in the process of being studied and manufactured. © 2001 Elsevier Science B.V. All rights reserved.

PACS: 41.60.Cr; 41.75.−i

Keywords: Injector; Superconducting cavity; High average power FEL; Pierce configuration

1. Introduction

Great progress has been achieved in the field of free electron lasers (FELs) since the 1970s. The high average power FEL has great potential applications in the future. Laboratories throughout the world are doing research on this project. In July 1999, an averaged output power of 1.72 kW was obtained at Jefferson Lab using a high brightness photocathode injector and superconducting accelerator [1].

As electron beams with high average current are necessary in order to get high average power FELs [2], high performance injectors are needed. There are three types of traditional electron injectors. The first is the room temperature RF gun. It is only suitable for pulsed mode and low output power levels because the copper cavity wall would induce a great loss of RF power. The second is the superconducting injector. It can work under CW mode and with a high repetition rate. But inserting the photocathode into the Nb cavity could disturb the RF superconductivity, and the connection of the photocathode with the cavity is also complex.

☆ Supported by National Natural Science Foundation of China.

*Corresponding author. Tel.: +86-10-6275-7195; fax: +86-10-6275-1875.

E-mail address: kzhao@pku.edu.cn (K. Zhao).

[1] Guest professor of Peking University.

The third is DC electron gun, which requires very high voltage (about 400–500 kV). If used with a lower-voltage accelerating field, it will induce beam bunch extension and thus diminish the quality of the beam. A DC electron gun was built in Peking University [3] in 1997. Since its voltage was not very high, electron beams were stretched and emittance was increased [4]. To avoid these negative effects, bunchers and pre-accelerating sections are required to work under CW mode.

To solve the problems mentioned above, a DC + RF SC photocathode injector is being studied and developed at Peking University. It is expected to generate mA scaled CW electron beams for high average power FELs.

2. The configuration of DC-RF SC injector

The DC-RF SC injector is designed to work under CW mode and with high average current and high brightness. As shown in Fig. 1, it consists of a DC extraction gap, $1 + \frac{1}{2}$ cell superconducting cavity and coaxial microwave input system. The DC gap has a Pierce extraction structure. The photocathode is fixed on the cathode of the Pierce structure. The anode makes up the bottom of the $\frac{1}{2}$ cell cavity.

The DC-RF SC injector has several advantages [5]:

(1) Compared with a thermal cathode, a photo-cathode injector of the new design can avoid

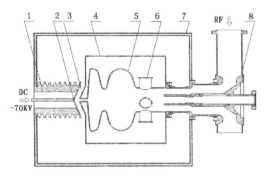

Fig. 1. Configuration of DC-RF SC injector. (1) Ceramics insulation; (2) Photocathode; (3) Pierce gun; (4) LHe tank; (5) SC cavity; (6) HOM coupler; (7) LN shield; (8) Coaxial input coupler.

reverse bombardment of electrons. Photoelectrons can emit only when they are in the accelerating phase of the RF field, which is controlled by the synchronous system between the laser and RF field.

(2) Compared with structures whose photo-cathode is inside the cavity, the RF cavity can work in a very high surface field due to the disappearance of the large dark current on the photocathode.

(3) DC-RF SC injector can be driven by ultra-violet laser. Laser light reflected from the surface of the photocathode cannot hit the inner wall of the cavity because of the narrow beam hole between the DC and RF section. Thus the quenches caused by photoemission from the inner wall are avoided.

(4) The Q degradation caused by the losses of the photocathode material can be avoided because the photocathode is installed outside the superconducting cavity. Thus the accelerating field of the superconducting cavity can be very high and the injector can work under CW mode.

(5) In the new design, the energy of electrons can reach tens, even hundreds of keV through DC accelerating. Further acceleration can be obtained by the accelerating of the $1 + \frac{1}{2}$ superconducting cavity because the velocity β is high at the same time.

(6) The coaxial input coupler can make the radial field symmetric. It helps to reduce the increase in emittance caused by the asymmetric field.

(7) The new installation is a compact structure. The beam transportation distance is short, so that the emittance increase due to space charged effect becomes small. High quality electron beams can be obtained.

3. Main problems of the DC-RF injector

Since the DC-RF photocathode injector is a new concept, there are naturally many problems with it. Every parameter is to be optimized. We will discuss the main problems and difficulties.

3.1. Combination of the DC gap and superconducting cavity

The combination of the DC and RF section is very important to the injector. To solve the compatibility of the photocathode and the superconducting cavity, we move the photocathode out from the superconducting cavity and place it on the cathode of the Pierce gun. We do optimizations on the $\frac{1}{2}$ cell of the cavity in order to increase radial focusing ability. As mentioned above, since the photocathode is outside the cavity, the injector can work under very high surface fields and avoid Q degradation of the cavity.

3.2. Injecting phase

The electrons must be injected into the superconducting cavity at the accelerating phase of the RF field. A phase-lock system is needed to control the injection of the electron beams. To get the best injection phase, we use computer code PARMELA to simulate the injection and transport. The optimized superconducting phase is obtained (see the next section).

3.3. Optimization of the $\frac{1}{2}$ cell

The optimization of the $\frac{1}{2}$ cell is necessary for the DC-RF injector. The peak accelerating field

should be as high as possible in order to accelerate electrons to light velocity as soon as possible. Thus the emittance increase induced by space charged effect could be lowered. E_p/E_{in} is designed to be minimum. E_p is the peak electric field and E_{in} is the field of the center at the bottom of $\frac{1}{2}$ cell cavity. To shorten the distance between the photocathode and superconducting cavity, the end wall of the $\frac{1}{2}$ cell slopes in a little at an angle of 5°. The cone-shaped wall is suitable for the Pierce structure. It also changes the field distribution in the cavity to increase radial focusing effects. The results of the optimization are given in the next section.

4. Computer simulation of the DC-RF structure

We have carried out a lot of optimizations and simulations on the DC-RF structure with codes SUPERFISH, POISSON and PARMELA. Some encouraging results have been obtained.

4.1. Optimization of the superconducting cavity

The optimization of the $\frac{1}{2}$ cell cavity is the most critical step. We use SUPERFISH to optimize the shape of the cavity. The frequency of the superconducting cavity is 1300 MHz. The radius of the cavity is adjusted to get the appropriate frequency, E_p/E_{in} and B_p/E_{in}. Here B_p is the peak magnetic

Table 1
Characteristic parameters of SC cavity changing with the shape of the $\frac{1}{2}$ cell

Length (mm)	Radius of iris (mm)	Maximum radius (mm)	Frequency (MHz)	E_p (MV/m)	B_p (A/m)	E_{in} (MV/m)	E_p/E_{in}	B_p/E_{in}
57.7	35	105.0	1302.76	2.523	4886.86	1.856	1.36	2.63
		106.0	1299.50	2.336	4130.16	1.170	2.00	3.53
57.7	34	104.8	1301.50	2.686	5316.73	2.043	1.31	2.60
		105.0	1300.08	2.489	4795.87	1.911	1.30	2.51
		105.3	1299.26	2.232	4206.93	1.593	1.40	2.64
58.7	35	105.5	1299.97	2.296	4035.07	1.293	1.77	3.12
		105.1	1301.17	2.238	4096.38	1.558	1.44	2.63
59.7	35	105.0	1300.62	2.250	3904.89	1.447	1.55	2.70
		105.2	1300.04	2.287	4009.37	1.318	1.74	3.04
		105.5	1299.34	2.335	4139.50	1.147	2.04	3.61
58.7	34	104.5	1301.60	2.723	5296.09	2.062	1.32	2.57
		104.8	1300.10	2.398	4605.06	1.777	1.35	2.60
		105.0	1299.29	2.198	4179.62	1.604	1.37	2.61

field. By changing the length of the $\frac{1}{2}$ cell, we obtain a different electric distribution at a fixed frequency. We can increase the electric field in the $\frac{1}{2}$ cell by changing the radius of the iris. The relations between these parameters are shown in Table 1.

From Table 1, we see that although a better field distribution can be obtained by changing the length of the $\frac{1}{2}$ cell, the electric field on the center of the end wall is still too low, which will lead to a large E_p/E_{in}. Reducing the radius of the iris can increase the E_{in} field. Considering all the parameters in Table 1, we choose the following. The length of the $\frac{1}{2}$ cell: 57.7 mm; radius of the iris: 34 mm; maximum radius of the cavity: 105 mm; frequency: 1300.08 MHz. The electric field distribution in the cavity and on the axis is shown in Fig. 2.

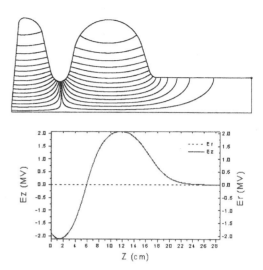

Fig. 2. Electric field distribution in the cavity and on the axis.

4.2. Optimization of the DC pierce structure

An important parameter for optimizing the extraction structure is the slope angle of the anode nose. It affects the emittance and focusing ability. The smaller the slope angle, the stronger the focusing ability. But the decrease of the slope angle will lead to the increase of the distance between photocathode and superconducting cavity, which will increase the emittance. Furthermore, we must consider the matching between the two components. The synchronous phase will be affected by changing the angle of the anode nose. We get the applicable value by computer simulation. Fig. 3 is

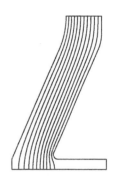

Fig. 3. Pierce gun optimized with Poisson.

Table 2
Simulation results of the DC-RF injector

	Initial conditions		Simulation results	
Electron bunch	Radius	3.0 mm	Anode inclination	65°
	Length	10 ps	Synchronous phase	−50°
	Charge	60 pC	Energy	2.43 MeV
	Emittance	0 mm-rad	Radius	2.8 mm
SC cavity	Average gradient	15 MV/m	E_k/E_k (rms)	2.63%
Pierce gun	Distance between cathode and anode	15 mm	Bunch length	7.8 ps
			ε_x (90%, n)	8.249 mm-mrad
	Cathode voltage	−70 kV	ε_y (90%, n)	8.832 mm-mrad
	Anode voltage	0 kV	ε_z (90%, n)	55.223 keV ps

the optimization of the Pierce gun simulated by POISSON. The angle of the anode nose is 65°. The accelerating gap is 15 mm and the high DC voltage is 70 kV.

4.3. Simulation of beam dynamics

Code PARMELA is used to calculate the performance of the whole injector. The initial conditions and the optimized results at the exit of the injector are listed in Table 2.

5. Conclusion

A DC-RF superconducting injector is designed by the SC laboratory of Peking University. Simulations, optimizations and analyses are carried out to get the appropriate parameters. In order to get about 5 mA average current, the charge of a single bunch should not be lower than 60 pC when the repetition rate is 81.25 MHz ($\frac{1}{16}$ of the cavity frequency). The best Pierce extraction

structure and synchronous accelerating phase of the superconducting cavity are obtained by beam dynamics analysis. High quality electron beams with emittance lower than 3π-mm-mrad can be obtained. We have conducted an analysis on the feasibility of the whole injector. The connection between the Pierce gun and SC cavity is considered carefully. Design calculations and experiments on the coaxial input coupler are complete. Experiments with a prototype injector are in progress.

References

[1] G.R. Neil, et al., Phys. Rev. Lett. 84 (4) (2000) 662.
[2] S. Benson, High Average Power Free-Electron Laser, Proceedings of the PAC 1999, pp. 212.
[3] K. Zhao, et al., Nucl. Instr. and Meth. A 407 (1998) 322.
[4] Yu Chen, Elementary researches on photocathode SC DC-RF gun, Thesis of M.S., Peking University, 1998.
[5] Yang Xi, Design and critical technique researches on high brightness photocathode RF gun, Thesis of M.S., Peking University, 2000.

ELSEVIER

Nuclear Instruments and Methods in Physics Research A 475 (2001) 569–573

NUCLEAR
INSTRUMENTS
& METHODS
IN PHYSICS
RESEARCH
Section A

www.elsevier.com/locate/nima

Simulated performance of the energy-recovery transport system for JAERI-FEL

T. Shizuma*, R. Hajima, E.J. Minehara

Japan Atomic Energy Research Institute, 2-4 Shirakata-Shirane, Tokai, Ibaraki 319-1195, Japan

Abstract

The JAERI superconducting rf linac has been developed to produce a high-power infrared free-electron laser. So far, a stable kW-level laser output has been achieved. In order to increase the average FEL power, an energy-recovery beam line will be installed. In this paper, details of simulated performance of the energy-recovery beam line are described. © 2001 Elsevier Science B.V. All rights reserved.

PACS: 41.60.Cr; 41.85.Ja

Keywords: Free-electron laser; Energy recovery; Beam transport

1. Introduction

At the Japan Atomic Energy Research Institute (JAERI), a free-electron laser (FEL) driven by superconducting linear accelerators has been constructed to produce a high power FEL at far-infrared (20–30 μm) region. Recently, the average output power was increased to 1.7 kW [1]: the goal for the first phase of our project. In order to increase the FEL power, we plan to install an energy-recovery beam line. In this scheme, the electron beam will be re-injected into the same rf modules as used for the acceleration, but at the decelerating phase. The beam power can then be transferred to the rf cavities and recycled for acceleration. Therefore, an electron beam with a higher average current can be accelerated with minimum rf power supplement, which would enable us to increase the average FEL power.

The lattice design of the energy-recovery beam line has been determined [2,3]. The schematic layout is shown in Fig. 1. The first half of the injection beam line consists of an electron gun and an 83.3 MHz sub harmonic buncher (SHB) followed by two units of a 499.8 MHz superconducting rf one-cell cryomodule (PSCA), which is the same as that of the present setup for the nonenergy-recovery mode. In order to merge the injected and re-circulated beams, an achromatic staircase buncher is placed before the main cryomodule. In the re-circulating beam line, major components are two five-cell 499.8 MHz superconducting accelerators (MSCA), two 180° arcs, a half-chicane, an undulator, and chicanes for merging and separating the injected and re-circulated beams.

*Corresponding author. Tel.: +81-29-282-3165; fax: +81-29-282-6057.

E-mail address: shizuma@popsvr.tokai.jaeri.go.jp (T. Shizuma).

0168-9002/01/$ - see front matter © 2001 Elsevier Science B.V. All rights reserved.
PII: S 0 1 6 8 - 9 0 0 2 (0 1) 0 1 5 9 1 - 1

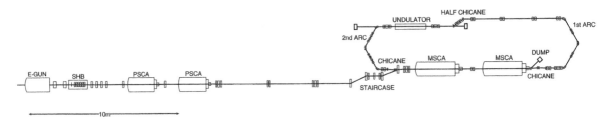

Fig. 1. Schematic layout of the energy-recovery beam line.

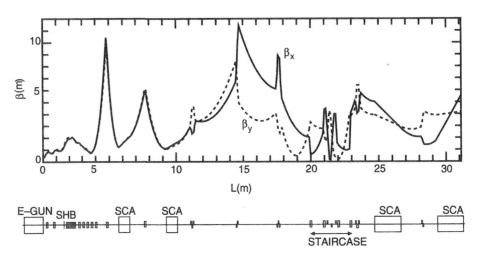

Fig. 2. Horizontal and vertical betatron functions from the electron gun to the main-accelerator exit.

In the following section, the baseline performance of the energy-recovery transport is described. Sensitivity of the baseline design for various parameters is also discussed in Section 3.

2. Baseline performance

2.1. Injection beam line

The transversal beam dynamics along the injection beam line have been simulated with PARMELA [4]. The horizontal and vertical beam envelope functions from the electron gun to the exit of the main-accelerator are shown in Fig. 2. The 2D rf electric fields in the pre- and main-accelerator cavities were calculated with SUPER-FISH. Initial parameters used in the simulation are listed in Table 1. The transversal emittance

increases to 20 πmm-mrad at the pre-accelerator entrance from 14 πmm-mrad at the gun exit due to the linear space charge effect [5]. Since the electrons are accelerated up to 2.6 MeV at the pre-accelerators, the emittance growth by the linear space charge becomes negligible after the pre-accelerators. In the staircase buncher, however, an emittance growth by the space charge is again expected to be large. This is because both the

Table 1
Electron gun parameters

Accelerated voltage	230 kV
Charge per pulse	0.6 nC
RMS pulse width	0.34 ns
RMS beam radius	1.6 mm
Normalized RMS emittance	14 πmm-mrad

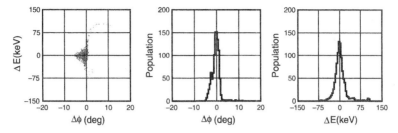

Fig. 3. Various distributions of the longitudinal phase space at the entrance to the main cryomodule.

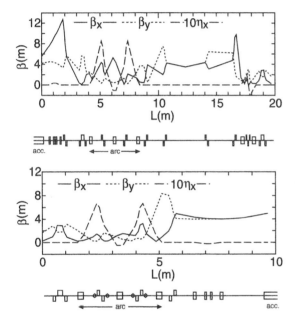

Fig. 4. Horizontal and vertical betatron functions and horizontal dispersion function from the main-accelerator exit to the undulator entrance (a), and from the undulator exit to the re-injection point at the main-accelerator (b).

electron beam energy and radial velocities are modified from magnet to magnet in multiple binding systems, which causes an emittance growth [6]. As a result, the calculated emittance increases to $\varepsilon_x^{n,\mathrm{rms}} = 38\pi$ and $\varepsilon_y^{n,\mathrm{rms}} = 25$ πmm-mrad at the entrance to the main-accelerator from $\varepsilon_x^{n,\mathrm{rms}} = \varepsilon_y^{n,\mathrm{rms}} = 20$ πmm-mrad at the staircase entrance.

The longitudinal motion with space charge included has also been studied with PARMELA. The optimal axial phase distribution at the entrance to the first pre-accelerator was obtained

at the applied voltage of 75 kV and the rf phase of 10° for the SHB. After the 4.5 m drifting space, the electron beam enters the first pre-accelerator. Since the velocity of the electrons is 0.71 with respect to the speed-of-light, both the longitudinal and transversal motions are largely affected by the rf phase of the first pre-accelerator. For optimizing the longitudinal phase and minimizing the transversal emittance increase, the rf phase of the first cavity was chosen as 15° off the maximum beam loading phase toward the debunching direction. Furthermore, the best longitudinal phase space distribution for the entrance to the staircase buncher ($R_{56} = -0.34$ m) was accomplished with 35° off-crest phase toward the bunching direction for the second pre-accelerator. Fig. 3 shows the longitudinal phase space distributions at the entrance to the main cryomodule. The axial beam parameters were obtained as rms bunch length $\sigma_t = 9$ ps and rms energy spread $\Delta E/E = 0.9\%$ at 2.6 MeV.

2.2. Re-circulating beam line

Fig. 4 shows the horizontal and vertical betatron functions and the horizontal dispersion function calculated with DIMAD [7]. The betatron functions are less than about 12 m along the re-circulating beam line. Since the bending angles of the dipoles in the 180° arc and the half-chicane are large (60° and 45°, respectively), the horizontal emittance may increase by coherent synchrotron radiation (CSR) force. To estimate its effect, we have made a numerical calculation with elegant [8]. For a 9 ps(rms) bunched beam with 0.6 nC, the emittance growth by CSR is estimated as $\Delta\varepsilon = 13$

at the first arc and 6 πmm-mrad at the half-chicane [9]. Assuming the growth is independent of the initial emittance, we can estimate a horizontal emittance of 41($= \sqrt{38^2 + 13^2 + 6^2}$) πmm-mrad at the entrance to the undulator. Therefore, it is concluded that the emittance growth by the CSR force is not critical for our FEL commissioning.

In the second half of the re-circulation, the energy acceptance and the second-order chromatic aberration become important issues because of large energy spread of the electron beam after the FEL interaction. In the TDA-3D calculation, the energy spread was estimated to be $\pm 2.4\%$ [3]. At the quadrupoles in the second arc, the horizontal dispersion takes the maximum value of $\eta_x = 0.6$ m. Therefore, the energy acceptance is determined to be $\pm 5\%$ by the quadrupole bore radius (35 mm) which is enough large compared with the value calculated by TDA-3D. Furthermore, the second-order terms of T_{166}, T_{266} and T_{566} can be compensated by sextupoles beside the quadrupoles in the second arc.

3. Sensitivity of the baseline design

3.1. Time jitters

Effects of time jitters at the electron gun have been investigated at the exit of the main-accelerators for the baseline design. The time jitters were set to ± 10, ± 30 and ± 50 ps by changing the initial phase parameters of the electron bunch in the PARMELA calculation. The other parameters such as the rf phase, the rf amplitude and

the magnet strength were the same as those used for the baseline design calculation. The results are summarized in Table 2. The time jitters cause bunch-to-bunch shifts in the mean energy E_m and the bunch centroid ϕ_m. Time jitters of $\Delta T = \pm 23$ ps, which we have achieved so far, would be enough small for the case of the energy-recovery transport. Since the main factor for the emittance growth in the staircase buncher is the space charge discussed in Section 2.1, the final emittance depends on the bunch length, i.e., the smaller bunch length near to the staircase exit results in the larger emittance growth.

3.2. SHB and SCA parameter fluctuation

Sensitivity to the fluctuation of the SHB, PSCA and MSCA parameters was also investigated at the exit of the main-accelerators for the baseline design. In the calculations, the rf phase was varied by $\Delta\phi = \pm 1°$ (5.6 ps), while the rf amplitude was changed by $\Delta E/E = \pm 1\%$. In all the cases, only one parameter was varied with the others being fixed. The results are listed in Table 3. The most sensitive parameters are the rf phase and amplitude of the second PSCA, which cause shifts in the mean values of E_m and ϕ_m. The fluctuation of the first PSCA parameters is less sensitive to the final beam parameters. Since the axial phase space at the entrance to the staircase buncher is optimized by the rf phase of the second PSCA, this parameter is more sensitive to the final performance than those of the SHB and the first PSCA. As discussed in Section 3.1, the horizontal emittance increases with decreasing the bunch length due to the greater

Table 2
Sensitivity for the time jitters (ΔT) at the electron gun

ΔT (ps)	σ_t (ps)	σ_E/E (%)	$\varepsilon_{n,x}/\varepsilon_{n,y}$ (πmm-mrad)	ΔE_m (keV)	$\Delta\phi_m$ (deg.)
Baseline	9.1	0.15	38.1/25.4	0	0
+10	8.7	0.17	38.9/26.1	−3	0.31
−10	9.5	0.14	37.1/24.9	3	−0.40
+30	8.1	0.21	40.9/27.6	−8	0.55
−30	10.4	0.13	35.2/24.0	7	−0.73
+50	7.9	0.27	41.4/28.8	−14	0.68
−50	11.5	0.13	33.6/23.4	11	−0.96

Table 3
Sensitivity for the fluctuation of the rf phase and the rf amplitude in the SHB, the pre-accelerators (PSCA) and the main-accelerators (MSCA)

Element	$\Delta\phi$ (deg.)	$\Delta E/E$ (%)	σ_t (ps)	σ_E/E (%)	$\varepsilon_{n,x}/\varepsilon_{n,y}$ (πmm-mrad)	ΔE_m (keV)	$\Delta\phi_m$ (deg.)
Baseline	0	0	9.1	0.15	38.1/25.4	0	0
SHB	+1	0	8.2	0.19	39.2/27.6	−10	0.74
	−1	0	11.0	0.16	35.5/23.8	8	−0.57
	0	+1	9.6	0.12	35.3/23.7	5	−0.91
	0	−1	9.0	0.20	41.6/28.0	−8	0.52
1st PSCA	+1	0	8.8	0.20	40.0/25.1	−9	0.76
	−1	0	10.0	0.13	35.7/25.9	3	−1.1
	0	+1	8.1	0.17	39.4/26.9	−5	−0.13
	0	−1	10.4	0.14	36.3/24.3	2	−0.16
2nd PSCA	+1	0	11.1	0.17	33.8/24.4	4	−2.8
	−1	0	7.4	0.26	41.0/27.3	−27	2.3
	0	+1	9.3	0.16	38.9/25.5	7	−2.4
	0	−1	8.9	0.20	37.5/25.4	−26	2.1
1st MSCA	+1	0	9.1	0.15	38.1/25.4	−2	−0.01
	−1	0	9.1	0.15	38.1/25.4	0	−0.16
	0	+1	9.1	0.15	38.1/25.4	62	−0.24
	0	−1	9.1	0.15	38.0/25.4	−62	0.07
2nd MSCA	+1	0	9.1	0.15	38.1/25.4	−1	0
	−1	0	9.1	0.15	38.1/25.4	0	0
	0	+1	9.1	0.15	38.4/25.4	62	0
	0	−1	9.1	0.15	38.1/25.4	−62	0

space charge effect at the staircase buncher. Since the MSCA parameters modify only the mean values, the final FEL performance would not be seriously affected by these fluctuations.

4. Summary

The energy-recovery beam line in the second phase project has been designed. Both the transversal and longitudinal electron motions with space charge included were investigated by PARMELA. Fluctuation sensitivity studies of the baseline design for the time jitters, the SHB parameters, and the SCA parameters were also performed. It has been found that the most sensitive parameters were the rf phase and amplitude of the second pre-accelerator. This

sensitivity comes from the change in the axial phase distributions at the entrance to the staircase buncher.

References

[1] N. Nishimori, et al., Nucl. Instr. and Meth. A 475 (2001) 266, these proceedings.
[2] T. Shizuma, et al., Proceedings in the Seventh EPAC, Vienna, 2000.
[3] R. Hajima, et al., Nucl. Instr. and Meth. A 445 (2000) 384.
[4] L.M. Young, LA-UR-96-1835.
[5] B.E. Carlsten, Nucl. Instr. and Meth. A 285 (1989) 313.
[6] B.E. Carlsten, T.O. Raubenheimer, Phys. Rev. E 51 (1995) 1453.
[7] R.V. Servranckx, TRI-DN-93-K233, 1993.
[8] http://www.aps.anl.gov/asd/oag/oagPackages.shtml
[9] R. Hajima, et al., Proceedings in the Seventh EPAC, Vienna, 2000.

ELSEVIER

Nuclear Instruments and Methods in Physics Research A 475 (2001) 574–578

**NUCLEAR
INSTRUMENTS
& METHODS
IN PHYSICS
RESEARCH**
Section A

www.elsevier.com/locate/nima

Design and beam transport simulations of a multistage collector for the Israeli EA-FEM

M. Tecimer*, M. Canter, S. Efimov, A. Gover, J. Sokolowski

Physical Electronics, Tel Aviv University, Ramat Aviv, Tel Aviv 69978, Israel

Abstract

A four stage asymmetric type depressed collector has been designed for the Israeli mm-wave FEM that is driven by a 1.4 MeV, 1.5 A electron beam. After leaving the interaction section the spent beam has an energy spread of 120 keV and 75 π mm mrad normalized beam emittance. Simulations of the beam transport system from the undulator exit through the decelerator tube into the collector have been carried out using EGUN and GPT codes. The latter has also been employed to study trajectories of the primary and scattered particles within the collector, optimizing the asymmetrical collector geometry and the electrode potentials at the presence of a deflecting magnetic field. The estimated overall system and collector efficiencies reach 50% and 70%, respectively, with a beam recovery of 99.6%. The design is aimed to attain millisecond long pulse operation and subsequently 1 kW average power. Simulation results are implemented in a mechanical design that leads to a simple, cost efficient assembly eliminating ceramic insulator rings between collector stages and the associated brazing in the manufacturing process. Instead, each copper plate is supported by insulating posts and freely displaceable within the vacuum chamber. We report on the simulation results of the beam transport and recovery systems and on the mechanical aspects of the multistage collector design. © 2001 Published by Elsevier Science B.V.

PACS: 41.60.Cr

Keywords: Free-electron laser; Depressed collector

1. Introduction

Electron beam energy recovery is essential to the operation of continuous wave, kilowatt level high power free electron lasers. The concept has been utilized before in electrostatic accelerator (EA) driven FELs producing relatively low rep. rate (<10 Hz) far infrared pulses up to 20 μs duration with 5–10 kW output power [1]. It has been only recently that FEL devices based on superconduct-

ing energy recovery RF-linacs have successfully generated (quasi) cw, kilowatt laser beams [2]. There are currently experiments underway to demonstrate the ability of EA-Free Electron Masers (FEM) in providing high power mm-waves from kilowatts [3] to megawatts at pulse lengths up to 100 ms [4]. In Electrostatic Accelerators, sustaining ampere level electron pulses over tens of milliseconds with sufficiently stabilized terminal accelerating voltage requires highly efficient transport and collection (recirculation) of the beam charge with recovery rates R over 99%. The combined charge and energy recovery process also

*Corresponding author.
E-mail address:* tecimer@post.tau.ac.il (M. Tecimer).

0168-9002/01/$ - see front matter © 2001 Published by Elsevier Science B.V.
PII: S 0 1 6 8 - 9 0 0 2 (0 1) 0 1 5 9 0 - X

enhances the overall system efficiency to values exceeding 50% while achieving, as a side benefit, significant reduction in the levels of ionizing radiation and heat dissipation as well.

After lasing has been demonstrated on the Israeli tandem FEM, the current objective is to operate the system at 100 GHz central frequency with milliseconds long pulses in single longitudinal mode, attaining up to 1 kW average output power. The spectral linewidth of the generated radiation is Fourier transform limited, corresponding to 10^{-8}–10^{-9} for the targeted pulse durations; a value that is orders of magnitude smaller than those observed in RF-linac based short pulse FELs. Realization of the desired operation mode, however, imposes stringent requirements on the e-beam transport system and the collector design. The interception of the electron beam current in the terminal due to imperfect beam transport or backstreaming current, leaking out of the collector into the decelerator tube, leads to terminal voltage drop resulting in limited radiation pulse durations and mode hopping [5]. The rate of the voltage drop is determined by the charging current I_c, the recovered current RI_b and the terminal capacitance C. The achievable pulse length t_p is related to the recovery rate R by

$$t_p \leqslant VC/4N(I_b(1-R)-I_c) \qquad (1)$$

N being the number of undulator periods. When charging current I_c and leakage currents out of the terminal are balanced, the FEL operates on a continuous basis. For the Israeli FEM driven by a $V = 1.4$ MV, $I_b = 1.5$ A beam, the condition to

reach millisecond long pulses is $R \geqslant 99.6\%$ (Table 1).

2. Beam transport and multistage collector simulations

The design of a collector, associated system efficiency and the beam transport crucially depend on the energy distribution of the spent beam. This has been determined by the simulations of beam-wave interaction in a waveguide resonator utilizing the FEL3D code [6]. Fig. 1 shows the simulated electron energy distribution after the FEM interaction, under conditions of minimal energy spread and nearly maximal power extraction from the beam (47 kW). At the undulator-exit the beam energy ranges from 1.32 to 1.44 MeV whereby 1.4 MeV is the initial beam energy. Beam parameters resulting from FEL3D code were the input for the beam transport simulations that are carried out using EGUN and GPT codes [7] starting from the undulator exit through decelerator tube into the multistage collector.

2.1. Beam transport from undulator to multistage collector

An 80 kV pierce e-gun delivers 1–1.5 A beam that is injected into a 20 period permanent magnet

Table 1
System Parameters

Current	I_b	1–1.5 A
Voltage	V	1.1–1.5 MV
E-gun voltage		80 kV
Terminal capacit.	C	300 pF
Charg. current.	I_c	225 μA
Energy spread		$<10^{-4}$
Norm. emittance	ε_n	75 π mm mrad
Undulator per.	N	20
Rad. frequency	v	70–110 GHz
Spectral linewidth	$\delta v/v$	10^{-8}–10^{-9}

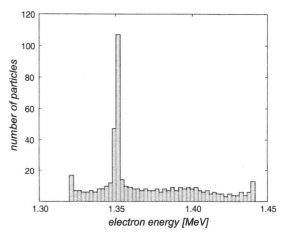

Fig. 1. Simulated electron energy distribution after the FEM interaction [6].

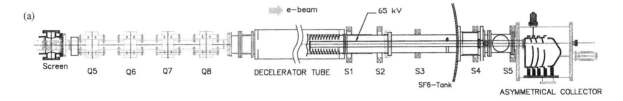

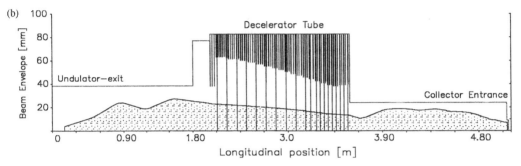

Fig. 2. (a) Layout of the studied beamline section, undulator exit-collector entrance. (b) Particle trajectory simulations defining the beam envelope.

undulator with beam energies ranging from 1.1 to 1.5 MeV corresponding to a laser tuning range of 70–110 GHz. Schematics of the beam transport system following the undulator section and the multistage collector are shown in Figs. 2 and 3, respectively. The present configuration of the collector assembly consists of two 2″ vacuum pipes inserted into the low voltage end of the decelerating tube serving as current collecting electrodes and (by varying their potentials independently) as a simple energy analyzer measuring the energy spectrum of the spent beam as well [8].

Downstream from the interaction section, particularly after the beam deceleration, the beam transmission is complicated by the spent beam's energy spread, amounting to 110–120 keV and a normalized transverse emittance of $75\,\pi$ mm mrad. Behind the undulator, a set of four focusing/defocusing quadrupoles is employed to transport the beam into the decelerator tube. The quadrupole field strengths in conjuction with a relatively large beam radius at the decelerator entrance (~ 40 mm) are set to focus the beam into the beam pipe located at the end of the decelerator column where the potential is chosen to be 65 kV. Five solenoid lenses are required to ensure the beam

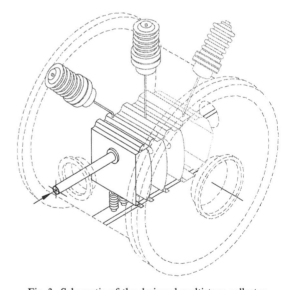

Fig. 3. Schematic of the designed multistage collector.

transport over the drift section between decelerator exit and the depressed collector entrance (Fig. 2). At this part of the beamline, the beam optics design requires an energy acceptance of 55–180 keV according to the beam energy distribution associated with 47 kW power extraction from the

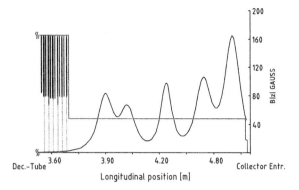

Fig. 4. On axis longitudinal field strengths of the employed solenoid lenses for 1. A beam between the decelerator tube exit and collector entrance.

beam [6]. The solenoid fields are determined such that overfocusing of particles at the low energy range is avoided while providing the necessary guiding for those with high transverse momentum (Fig. 4). Since most of the beam optics components in an internal cavity EA-FEL [3] reside inside a pressurized tank, the electron optics design enables the transport of the 1.1–1.5 MeV beam considering the largest value foreseen for the energy spread and transverse emittance at fixed component positions, by adjusting quadrupole and solenoid current settings.

2.2. Multistage collector

Particle trajectories within the asymmetrical collector are calculated using the GPT code, taking into account 3D space charge and multiscattering of particles from the copper surfaces (Fig. 5). The adopted mechanical and electrical design of the four stage collector has been determined by studying the dependence of the charge and power recovery on the shape, potential and location of the collector electrodes. The collector comprises 5 copper electrodes housed in a grounded vacuum chamber (Fig. 3) that is located outside of the SF_6 pressure tank. The base plate (5th electrode) serves as a reflector for the highest energy electrons. The asymmetrical geometry of the collector causes the beam to enter off center; due to the selected electrode geometry, electric field components are

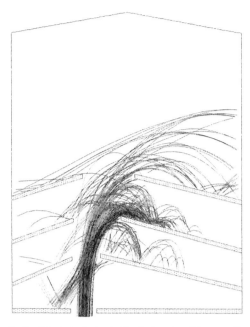

Fig. 5. Particle trajectories in the designed multistage collector.

generated that attract the particles away from the apertures towards the collector-center, reducing thereby the probability of backwards reflection into the decelerator tube [9]. Low energy secondaries with several tens of eV are pushed efficiently back to the plate surface. In addition to the electric fields, a magnetic field is applied from the side of the collector to supress more efficiently the backstreaming of scattered primaries to a level required $(0.2\% \leqslant)$ for the millisecond range FEM operation.

Simulation results are implemented in a mechanical design that leads to a simple, cost efficient assembly, eliminating ceramic insulator rings between collector stages and the associated brazing in the manufacturing process [10]. The copper electrodes are supported by alumina insulator posts. Four ceramic feedthroughs will be used in providing the high voltages to the electrodes. Variations in the order of 5% will be allowed on the potential of the electrode #1 to #3 (Table 2) while operating the FEM. Unlike other collectors employed in EA-FEL's [1,4] the presented one offers the flexibility of variable electrode positioning and voltage optimization for a final improvement on the collection efficiency accounting for real FEL operation conditions. This feature enables the design to be less sensitive for possible erroneous assumptions made on the energy spread and the beam phase-space at collector entrance.

The performance parameters of the four stage collector are listed in Table 2. The electrode voltages are specified with respect to ground recalling that the cathode potential is set to -80 kV. The system efficiency is given by $\eta_s = P_{rf}/P_{in}$, where P_{rf} and P_{in} are the maximum outcoupled mm-wave power and the total electrical power provided to the device by the power supplies, respectively. Corresponding to a resonator design goal with 10% internal cavity losses and 35–40% outcoupling [6], the system efficiency amounts to 50%. The collector efficiency on the other hand is defined by $\eta_c = P_{rec}/P_{ent}$, the ratio of the recovered power by the collector plates and the beam power upon entrance into the collector. The calculated $\eta_c = 70$–75 is based on the beam energy distribution shown in Fig. 1, with a maximum power extraction of 47 kW from the electron pulse at the end of the interaction region. To attain the targeted average mm-wave power $\bar{P}_{rf} = 1$ kW in the continuous pulse train operation, the duty cycle is set to be 2.4%. The resulting thermal load of

nearly 1 kW has to be removed from the collector plates by means of a cooling system.

3. Conclusions

We have discussed a number of critical issues in the beam transport and multistage depressed collector system design of an internal cavity EA-FEM to generate narrow linewidth, tunable mm-wave radiation in continuous millisecond long pulse trains at 1 kW average power. Our estimate of $R = 99.6\%$ for the targeted millisecond pulse duration permits 0.2% (~ 3mA) current loss along the transportline with the predicted 0.2% backstreaming in the presented asymmetrical four stage collector. Further, the collector design concept enables final optimization of the collector structure and the associated collector performance on the basis of "real" FEM operation conditions. The multistage collector is currently in a design and construction stage; it will be incorporated into the FEM structure after experiments with the present two stage collector [8] will be concluded.

Acknowledgements

This work has supported in part by the Israeli Ministeries of Infrastructure and Science.

Table 2
Performance of the four stage collector

Electrode#	Current (mA)	Voltage (kV) (w.r.t.g)
0	18	65
1	375	5
2	590	-20
3	510	-50
Base plate	—	-110
Millisecond pulses:		
Supplied electrical power	P_{in}	85.0 kW
Beam power extraction	ΔP	47.0 kW
mm-wave power	P_{rf}	42.0 kW
Average power:		
mm-wave power	$\bar{P}_{rf}$	1.0 kW
Dissipated power	P_{diss}	1.0 kW
Beam transport loss		~ 3 mA
Backstreaming		0.2%
System efficiency	η_s	50%
Collector efficiency	η_c	70–75%

References

[1] G. Ramian, Nucl. Instr. and Meth. A 318 (1992) 225.
[2] G. Neil, et al., Nucl. Instr. and Meth. A 445 (2000) 192.
[3] A. Gover, et al., Phys. Rev. Lett. 82 (1999) 5257.
[4] W.H. Urbanus, et al., Phys. Rev. E 59 (1999) 6058.
[5] L. Elias, et al., Phys. Rev. Lett 57 (1986) 424.
[6] A. Abramovich, Nucl. Instr. and Meth. A 475 (2001) 579, these proceedings, to be published.
[7] W.B. Herrmannsfeldt, SLAC Rep. 331-1988; Pulsar, GPT Manual 2.51, 1999.
[8] S. Efimov, et al., in: Proceedings of J. Feldhaus, H. Weise (Eds.), Proc. of the 21st Int. Free Electron Laser Conference, August 23–26, 1999, Elsevier Science B.V., Amsterdam, p. 11–89.
[9] R. Hechtel, CREOL-FEL Design Rep., 1991.
[10] L.R. Elias, G. Ramian, Nucl. Instr. and Meth. A 250 (1986) 325.

ELSEVIER

Nuclear Instruments and Methods in Physics Research A 475 (2001) 579–582

NUCLEAR
INSTRUMENTS
& METHODS
IN PHYSICS
RESEARCH
Section A

www.elsevier.com/locate/nima

Optimization of power output and study of electron beam energy spread in a Free Electron Laser oscillator

A. Abramovich[a],*, Y. Pinhasi[a], A. Yahalom[a], D. Bar-Lev[b], S. Efimov[b], A. Gover[b]

[a] Department of Electrical and Electronic Engineering, Faculty of engineering, The College of Judea and Samaria,
P.O. Box 3, Ariel 44837, Israel
[b] Department of Electrical Engineering, Physical Electronics, Ramat-Aviv, Tel-Aviv 69978, Israel

Abstract

Design of a multi-stage depressed collector for efficient operation of a Free Electron Laser (FEL) oscillator requires knowledge of the electron beam energy distribution. This knowledge is necessary to determine the voltages of the depressed collector electrodes that optimize the collection efficiency and overall energy conversion efficiency of the FEL. The energy spread in the electron beam is due to interaction in the wiggler region, as electrons enter the interaction region at different phases relative to the EM wave. This interaction can be simulated well by a three-dimensional simulation code such as FEL3D. The main adjustable parameters that determine the electron beam energy spread after interaction are the e-beam current, the initial beam energy, and the quality factor of the resonator out-coupling coefficient. Using FEL3D, we study the influence of these parameters on the available radiation power and on the electron beam energy distribution at the undulator exit. Simulations performed for $I = 1.5\,\text{A}$, $E = 1.4\,\text{MeV}$, $L = 20\%$ (internal loss factor) showed that the highest radiated output power and smallest energy spread are attained for an output coupler transmission coefficient $T_{\text{m}} \cong 30\%$. © 2001 Published by Elsevier Science B.V.

PACS: 41.60.−m; 41.60.Cr; 41.85.Qg

Keywords: FEL oscillator; Electron beam diagnostic

1. Introduction

Good electron beam transport along the beamline of a Free Electron Laser (FEL) oscillator is essential in order to enable efficient energy exchange between the beam electrons and the electromagnetic wave inside the interaction region. Good transport is particularly important in electron beam energy recovery schemes such as a depressed collector in an Electrostatic Accelerator FEL (EA-FEL) [1–3]. At the entrance to the interaction region all the electrons have very nearly the same energy. In passing through the interaction region electrons may lose or gain a different amount of energy from the electromagnetic wave, depending on their entrance phase; consequently electrons have a large energy spread at the interaction region exit. Since the beam energy spread is generated by the nonlinear interaction process taking place in the resonator, it should be considered in the design of the FEL resonator. The resonator parameters (losses, output coupling coefficient) should be optimized for attaining both maximum output power emission and minimum energy spread [4,5].

*Corresponding author. Fax: +972-3-9066-238.

In this paper, we investigate the electron beam energy spread after the interaction region of the Israeli Tandem FEL [3] using our simulation code FEL3D [6,7]. The results will be used to optimize FEL operation and to determine the most efficient depressed collector voltages [8]. The study led us to the determination of the resonator parameters that provide highest output power and smallest electron energy spread. For the Tandem FEL, the optimal transmission coefficient of the resonator output coupler is $T_m \cong 30\%$.

2. Resonator losses and power out-coupling

Our goal in this study is to find optimal operating conditions for an FEL oscillator. The desired optimization is primarily with regard to the output radiation power of the device. Another parameter of importance is the energy spread of the electron beam after the interaction region. This parameter determines the overall power efficiency that can be obtained in FELs with energy retrieval. In particular it affects the design of a multistage collector in an Electrostatic Accelerator FEL [8].

The model used for this study is depicted in Fig. 1. It is assumed that a single transverse mode $C(z)E(x, y)\exp(jk_z z)$ develops in the interaction region and is amplified along the wiggler. The resonator feedback loop is represented symbolically as a ring cavity, but can be generalized to an arbitrary shape. The output coupling mirror placed at the wiggler exit couples a power fraction T_m externally.

$$P_{out} = T_m P(L_w). \tag{1}$$

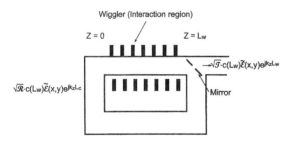

Fig. 1. A schematic of an FEL oscillator.

We lump all the internal losses, including mirror losses, ohmic losses on the waveguide walls, and diffraction losses, into one loss factor in the feedback loop.

Thus

$$P_{loss} = L(1 - T_m)P(L_w) \tag{2}$$

and the power feedback into the wiggler entrance after one round-trip is

$$P(0) = RP(L_w) \tag{3}$$

where R is the round-trip power reflectivity factor

$$R = (1 - L)(1 - T_m). \tag{4}$$

If the gain factor of the FEL, G, satisfies in the small signal (linear) regime the oscillation condition

$$GR > 1 \tag{5}$$

then any input signal $P(0)$ at the wiggler entrance will be amplified in each feedback round trip until the FEL is driven into saturation. At this point, the nonlinear gain G drops till Eq. (5) turns into an equality. The power output at steady state can be written in terms of the radiation power extracted from electron beam $\Delta P = P(L_w) - (P(0))$ using Eqs. (1) and (2)

$$P_{out} = T_m/(1 - R)\Delta P. \tag{6}$$

3. Simulation of electron beam energy spread and output power in our Tandem FEL

We used our simulation codes to determine optimal parameters for the operation of the Tandem FEL [3]. On one hand, we are interested in obtaining maximum output power from the oscillator, and on the other hand, we would like to have minimal electron beam energy spread after the interaction region (for efficient energy retrieval).

The main degree of freedom we have for optimizing the oscillator performances is the mirror reflectivity T_m. Eq. (1) may suggest that increasing T_m increases the output power; however, it would also tend to decrease R and $P(L_w)$, and would eventually break the oscillation condition (5), stopping oscillation altogether. On the other hand, decreasing T_m increases R and $P(L_w)$,

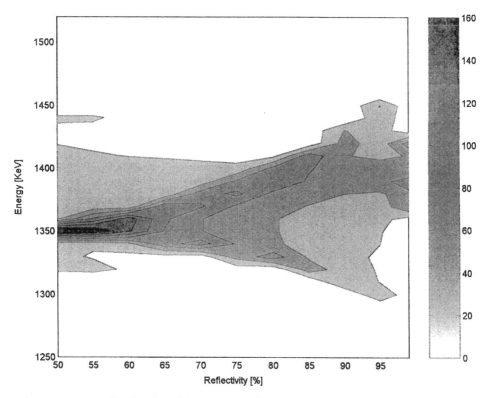

Fig. 2. Electron beam energy spread as function of the resonator reflectivity R (the gray scale indicates the number of electrons per energy interval).

but less power P_{out} couples out (Eq. (1)). Furthermore, considering a finite loss factor L, when T_m grows, the power loss (2) becomes significant relative to the output power (1), and the internal efficiency drops. Clearly, for a given L there must be an optimal value of $0 < T_m < 1$ for which P_{out} is maximal [9]. At the same time, since decreasing T_m drives the FEL deeper into saturation, it also tends to increase the e-beam energy spreads. Therefore, there should be an optimal value of T_m for which the beam energy spread is minimal.

Clearly, the optimization of the FEL oscillator design is a nonlinear problem that requires the use of a nonlinear computer code for simulation of the electron beam interaction with the resonator radiation field at saturation. For this purpose we use our code FEL3D [6,7], which simulates the oscillation buildup process in the FEL by solving exactly the FEL amplifier in each transversal along the wiggler, and, after each round trip, resetting the initial condition according to Eq. (3) (see

Fig. 1). The signal frequency is chosen as the one having maximum small signal gain and minimum threshold for oscillation (5). This process is repeated until steady- state (saturation) is reached. For any given overall round-trip reflectivity value R, the operating point of the oscillator is fully determined. This includes full determination of the radiation power distribution inside the resonator (specifically $P(0)$, $P(L_w)$) and the distribution of the wasted electron beam energies. Note that these parameters are not dependent separately on T_m and L, only implicitly through (4).

Figs. 2–5 display the oscillator optimization study results for the parameters of the Israeli FEL given in Table 1 of Ref. [3]. Fig. 2 presents a map of the e-beam energy distribution after interaction, given as a function of the round trip reflectivity parameter R. The diagram indicates that excessive energy spread happens for large R as the FEL oscillator is driven deep into saturation. To keep the energy spread small, we

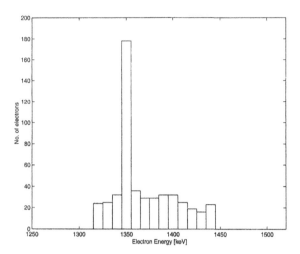

Fig. 3. Electron energy distribution at minimal electron energy spread conditions ($R = 50\%$).

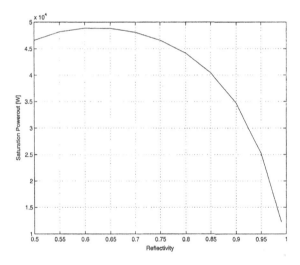

Fig. 4. The total power extracted from the e-beam ΔP and the power at the end of the resonator $P(L_w)$ as a function of the resonator reflectivity R.

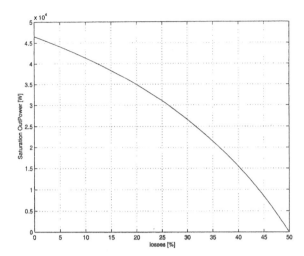

Fig. 5. The FEL output power P_{out} as a function of the out-coupling parameter of the resonator T_m, for different internal loss values L.

an uncontrollable given parameter and only T_m can be varied, either in the design stage, or (if the output coupler is controllable) in real time. For this reason, we provide a set of curves, useful for determining the maximum output power of the FEL oscillator for given loss factor parameter values.

From Fig. 5 we conclude that for an expected internal loss factor $L = 20\%$, maximum output power at the level of 35 kW is expected to be achieved with mirror transmission $T_m = 25-37\%$. Fortunately, the beam energy spread is smallest in the partially overlapping region $R = 55-65\%$ ($T_m = 20-30\%$). This overlapping region would therefore be the optimal operation regime of the FEL oscillator.

will keep $R < 60\%$. Fig. 3 displays the energy distribution for $R = 50\%$ (near oscillation threshold).

Fig. 4 displays the saturation power at the end of the wiggler as a function of the resonator round trip reflectivity R. This curve was obtained by running FEL3D up to the saturation level for the parameters of Table 1 and changing values of R.

The output power of the FEL for a given T_m can be directly calculated from the curve of Fig. 4 using Eq. (1). In practice, usually the loss factor is

References

[1] H.G. Kosmahl, Proc. IEEE 70 (1982) 1325.
[2] L.R. Alias, G. Ramian, R.J. Hu, A. Amir, Phys. Rev. Lett. 57 (1986) 424.
[3] A. Abramovich, et al., Phys. Rev. Lett. 82 (1999) 26.
[4] P. Sprangle, et al., Phys. Rev. A 21 (1980) 302.
[5] T.M. Anthonsen, B. Levush, Phys. Rev. Lett. 62 (1989) 1488.
[6] Y. Pinhasi, et al., Int. J. Electron. 78 (1995) 581.
[7] Y. Pinhasi, et al., Phys. Rev. E 6774 (1996) 54.
[8] M. Tecimer, Nucl. Instr. and Meth. A 475 (2001) 574, these proceedings.
[9] A. Abramovich, et al., Nucl. Instr. and Meth. A 375 (1996) 164.

ELSEVIER

Nuclear Instruments and Methods in Physics Research A 475 (2001) 583–587

NUCLEAR
INSTRUMENTS
& METHODS
IN PHYSICS
RESEARCH
Section A

www.elsevier.com/locate/nima

Performance of the thermionic RF gun injector for the linac-based IR free electron laser at the FEL-SUT

Fumihiko Oda*, Minoru Yokoyama, Masayuki Kawai, Hidehito Koike, Masaaki Sobajima

Kanto technical Institute, Kawasaki Heavy Industries, Ltd., 118 Futatsuzuka, Noda, Chiba 278-8585, Japan

Abstract

Kawasaki Heavy Industries, Ltd. (KHI) has developed a linac-based compact IR free electron laser device and has installed it in the FEL-SUT (IR FEL Research Center of Science University of Tokyo). The FEL device adopts a combination of a multi-cell RF gun with a thermionic cathode and an α-magnet as an injector. The fundamental design of this RF gun is the $\pi/2$ mode standing wave structure. It has two accelerating cells and a coupling cell located on the beam axis, a so-called "on axis coupled structure" (OCS). Characteristics of momentum distribution and micropulse bunch length of the electron beam are compared with beam dynamics simulation results in this paper. We succeeded in obtaining sufficient peak current for FEL lasing with this injector, and the first lasing was achieved on 6 July 2000. © 2001 Elsevier Science B.V. All rights reserved.

PACS: 07.57.Hm; 41.60.C

Keywords: RF gun; Multi-cell; α-magnet; Back bombardment; Thermionic cathode; Microbunch length

1. Introduction

In recent years, thermionic RF guns have been put to use at many facilities as compact and low-cost injectors, as they can generate a beam with sufficiently high brightness. Three FEL facilities using thermionic RF guns with a single cell are in operation, at Duke University [1], Vanderbilt University [2], and Beijing [3]. The beam energy of these RF guns ranges from 0.9–1.2 MeV. To prevent degradation of beam quality through transport to the attached accelerator, some multi-

cell RF guns with higher beam energy were designed at Beijing [4] and Kyoto university [5].

The KHI FEL device is based upon a 32 MeV S-band electron linac [6]. Fig. 1 shows a schematic view of the FEL device. The accelerator section is composed of a multi-cell RF gun, an α-magnet and a 3 m accelerating structure [7].

The cavity shape of the RF gun was designed to minimize the back bombarding power without sacrificing the quality of the output beam and to obtain high shunt impedance to accelerate the electron beam to 2.0 MeV with the field analytic codes EMSYS (2D) and MAFIA (3D) [8]. Fig. 2 shows a cross-sectional view of the newly developed RF gun cavity. It has two accelerating cells and a coupling cell located on the beam axis in an

---

*Corresponding author. Tel.: +81-471-24-0258; fax: +81-471-24-5917.

E-mail address: oda_f@khi.co.jp (F. Oda).

0168-9002/01/$ - see front matter © 2001 Elsevier Science B.V. All rights reserved.
PII: S 0 1 6 8 - 9 0 0 2 (0 1) 0 1 5 8 8 - 1

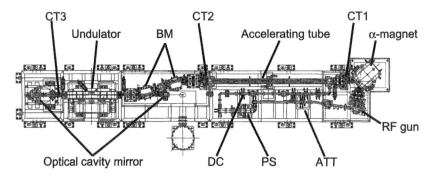

Fig. 1. Schematic view of the FEL device.

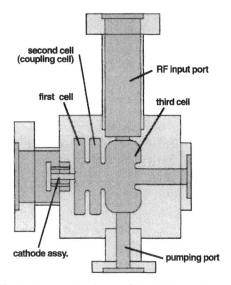

Fig. 2. Cross-sectional view of the OCS type RF gun.

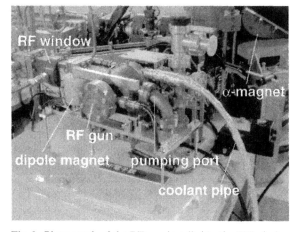

Fig. 3. Photograph of the RF gun installed to the FEL device.

"on-axis coupled structure" (OCS) [9]. The RF gun employs a tungsten based dispenser type cathode whose diameter and maximum current density are 6 mm and more than 16 A/cm², respectively. The resonant frequency and the electric field distribution along the beam axis are measured and the results are in good agreement with the results of the simulations [10].

2. Installation to the FEL device

Fig. 3 is a photograph of the RF gun and the α-magnet installed in the FEL device. RF power of 45 MW, generated by a klystron (Toshiba E-3729),

is divided by a 9 dB directional coupler (DC), while power at 9 MW is transported through a phase shifter (PS) and an attenuator (ATT), and is fed to the gun (see also Fig. 1) [11].

The cavity is pumped by an ion pump (pump capacity is 30 l/s). The vacuum degree is as low as 7×10^{-7} Pa during the operation.

The RF gun and the accelerating tube have individual coolant circuits, whose temperature is controlled individually. Because of the considerable coupling coefficient of the RF input coupler ($\beta = 4.37$), the resonant frequency of the RF gun should shift from the tuned value (2856.0 MHz) under the large beam loading condition. So the temperature of the RF gun cavity can be varied over the range of $40 \pm 6°C$ (temporal fluctuation is less than 0.1°C). A dipole magnet to deflect the back-bombarding electron onto the cathode presents a transverse magnetic field of about 100 G across the cathode. Another dipole magnet at the gun exit compensates the trajectory of the output beam.

3. Characteristics of the output beam

3.1. Temporal current profile

Fig. 4(a)–(c) are the temporal beam current profiles in regions: (a) before the accelerating tube, (b) after the accelerating tube and (c) after the undulator (see also Fig. 1). An average current of more than 200 mA (i.e. more than 60 pC per bunch) for about 1.5 μs flattop was obtained, and almost 100% of the accelerated charge can be transmitted through the undulator, except for the RF rise up time (~0.5 μs).

Macropulse duration and maximum current are limited by the uncontrollable current increase caused by the back bombardment.

3.2. Energy distribution

Fig. 5 shows the relation between input RF power and peak momentum (E_p) of the output

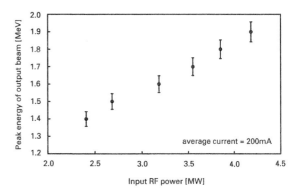

Fig. 5. The relation between input RF power and peak momentum of the output beam.

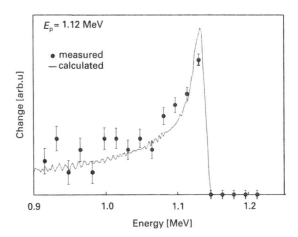

Fig. 6. Measured and calculated momentum distributions of the out put beam for $E_p = 1.12$ MeV case.

beam measured by the energy selection slit at the α-magnet, under the condition of constant beam current. The peak momentum is almost proportional to input power. No significant degradation of the beam quality or transmission through the α-magnet was observed with variations of E_p from 1.2 to 2.1 MeV. Fig. 6 shows the measured and calculated momentum distribution of the output beam for $E_p = 1.12$ MeV case. The measured distribution shows good agreement with the calculation.

3.3. Optimization of bunch compression

Micropulse bunch length was estimated from the temporal structure of the optical transition

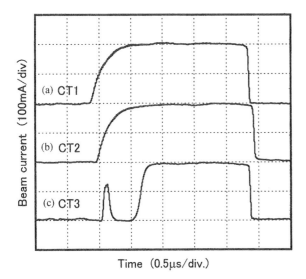

Fig. 4. Temporal beam current profiles at (a) before the accelerating tube (CT1), (b) after the accelerating tube (CT2) and (c) after the undulator (CT3).

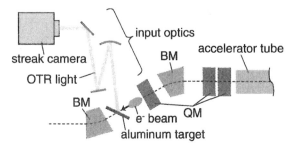

Fig. 7. Experimental set up of the bench length measurement.

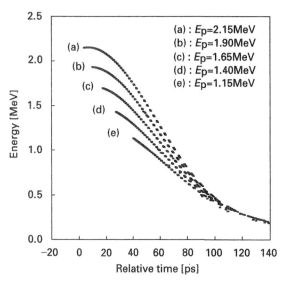

Fig. 8. Simulated energy distributions versus time at the gun exit in a micropulse.

radiation (OTR) light measured by using a streak camera (HAMAMATSU FESCA-200) whose time resolution is better than 200 fs. Fig. 7 illustrates the experimental set up. The OTR light was generated from a mirror polished aluminum target and led by the input optics to the streak camera. Estimated bunch length was less than 2.5 ps (i.e. micropulse peak current is more than 30 A).

Fig. 8 shows electron momentum distributions versus time at the RF gun exit in a micropulse, calculated by beam dynamics simulations. The distribution forms a negative energy chirp. As E_p increases, the distribution becomes curved and unable to be compressed. A beam with energy $1.4\,\mathrm{MeV} \leqslant E_p \leqslant 1.9\,\mathrm{MeV}$ was successfully compressed by the α-magnet.

Now we define a parameter δ as an average energy chirp gradient of the usable beam ($E \geqslant 95\%$ of E_p) as,

$$\delta = (E_p - 0.95E_p)t_d^- \tag{1}$$

where t_d is time delay in a micropulse at the gun exit between the momentum of E_p and $0.95E_p$.

As the peak momentum increases, the δ becomes smaller. The magnetic field gradient of the α-magnet (G) in T/m is expressed as

$$G = (0.2681c^{-1}\beta^{-1}\Delta t^{-1})^2 E \tag{2}$$

where Δt is flight time of the electron through the α-magnet, E is the momentum of the electron in MeV/c, c is the velocity of light in m/s and β is the ratio of light velocity and electron velocity [12]. The optimal field gradient (G_{opt}) in T/m to compress the usable beam to the shortest bunch length is derived from Eqns. (1) and (2) as

$$G_{opt} = \left\{ \frac{0.2681\, C\, (E_p - 0.95\, E_p)\left(\dfrac{\sqrt{E_p}}{\beta_{Ep}} - \dfrac{\sqrt{0.95E_p}}{\beta_{0.95Ep}}\right)}{\delta - L\left(\dfrac{1}{\beta_{E_p}} - \dfrac{1}{\beta_{0.95E_p}}\right)} \right\} \tag{3}$$

where the subscripts of β are the electron momenta in MeV/c and L is the length of the free drift space between the gun and accelerator tube ($= 1.25\,\mathrm{m}$). The δ was derived from an experimentally defined G_{opt}. Comparison between experimental and

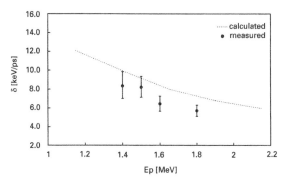

Fig. 9. Experimental and simulated results of the relation between E_p and δ.

calculation results of the relations between E_p and δ are shown in Fig. 9. Measured values have good agreement with the simulations.

4. Discussion

A dispenser cathode 6 mm diameter emits usable charge of more than 60 pC. The surface temperature of the cathode was measured by a comparison type pyrometer to be 980°C. The rather low drive temperature of cathode seems to allow a long cathode life. It is remarkable that no emission drop has occurred for the total drive time of more than 500 h.

The uncontrollable emission increase caused by back bombardment limits the macropulse duration. With higher cathode temperature, more emission can be obtained, but the gradient of the emission increase becomes larger. Since the transverse magnetic field to bend back bombarding electrons is effective in reducing the increase, a cathode of smaller diameter seems to sharply reduce the back bombardment. To obtain the same quantity of usable charge with a cathode 3 mm in diameter, the emission current density should be four times that of the 6 mm diameter cathode, implying a cathode temperature of about 1100°C. The shortening of cathode life does not seem to be significant, because the evaporation of impregnated Ba, the main reason for the emission decrease, occurs at temperatures in excess of 1200°C.

5. Summary

A newly developed thermionic RF gun was installed for the compact FEL device and described. Microbunch peak current of more than 30 A was successfully generated by the combination of the gun and the α-magnet. Measured momentum distributions and the optimized magnetic field of the α-magnet for bunch compression have good agreement with predictions from the calculations. The first lasing was achieved on 6

July, 2000. A small signal gain for the FEL of more than 20% was obtained at a wavelength of 9 μm [13]. This is the first case in which an FEL device with a multi-cell thermionic RF gun achieved lasing. No emission drop has occurred for a drive time of more than 500 h. The long lifetime of the dispenser cathode is advantageous to drive the FEL device constantly as a user's facility.

The emission increase during the macropulse was more rapid than expected. A cathode of smaller diameter seems to be effective to reduce the back bombardment without sacrificing the life of the cathode. We will adopt the smaller (3 mm diameter) cathode and test its performance in the near future.

References

[1] S.V. Benson, W.S. Fann, B.A. Hooper, J.M.J. Madey, E.B. Szarmes, B. Richman, L. Bintro, Nucl. Instr. and Meth. A 296 (1990) 110.
[2] C.A. Brau, Nucl. Instr. and Meth. A 318 (1992) 38.
[3] J.X.J. Zhuang, Y.H.S. Zhong, Y. Li, S. Lin, R. Ying, Y. Zhong, L. Zhang, G. Wu, Y. Zhang, C. Chao, L. Li, Z. Fu, J. Su, Y. Wang, G. Wang, Nucl. Instr. and Meth. A 358 (1990) 256.
[4] Ch, Tang, Y. Lin, J. Xie, D. Tong, Y. Wu, Y. Wang, X. Zhao, Instrum. and Meth. A 421 (1999) 406.
[5] K. Yoshikawa, D. Tsukahara, T. Inamasu, K. Masuda, M. Sobajima, J. Kitagaki, Y. Yamamoto, H. Toku, M. Ohnishi, Instrum. and Meth. A 407 (1998) 364.
[6] H. Kuroda, M. Kawai, A. Iwata, Proceedings of the 12th Russian Synchrotron Radiation Conference, 1998.
[7] M. Yokoyama, F. Oda, A. Nakayama, K. Nomaru, M. Kawai, Nucl. Instr. and Meth. A 429 (1999) 269.
[8] F. Oda, M. Yokoyama, A. Nakayama, E. Tanabe, Nucl. Instr. and Meth. A 429 (1999) 332.
[9] T. Nishikawa, S. Giordano, D. Cartar, Rev. Sci. Instr. 39 (1968) 979.
[10] F. Oda, M. Yokoyama, A. Nakayama, H. Koike, E. Tanabe, Nucl. Instr. and Meth. A 445 (2000) 404.
[11] M. Yokoyama, F. Oda, A. Nakayama, K. Nomaru, H. Koike, M. Kawai, H. Kuroda, Nucl. Instr. and Meth. A 475 (2001) 38, these proceedings.
[12] G. Baria, C. Brassard, P.F. Hinrishsen, J.P. Labrie, IEEE J. Quantum Electron. QE-20 (1984) 637.
[13] M. Yokoyama, F. Oda, H. Koike, A. Sobajima, M. Kawai, K. Nomaru, H. Hattori, H. Kuroda, Nucl. Instr. and Meth. A, to be published.

ELSEVIER

Nuclear Instruments and Methods in Physics Research A 475 (2001) 588–592

NUCLEAR
INSTRUMENTS
& METHODS
IN PHYSICS
RESEARCH
Section A

www.elsevier.com/locate/nima

Experiment and analysis on back-bombardment effect in thermionic RF gun

Toshiteru Kii*, Tomohiko Yamaguchi, Ryuta Ikeda, Zhi-Wei Dong, Kai Masuda, Hisayuki Toku, Kiyoshi Yoshikawa, Tetsuo Yamazaki

Institute of Advanced Energy, Kyoto University, Gokasyo, Uji, Kyoto 611-0011, Japan

Abstract

A serious problem for RF guns with thermionic guns is the back-bombardment effect, which makes the operation of the RF guns unstable owing to the instability of the cathode temperature. To find out how to solve the problem, we tried to evaluate the effect of back-bombardment. We have measured the temperature of cathode surface by an infrared radiation thermometer during the generation of electron beams from our S-band thermionic RF gun with 4.5 cavities. As a consequence, it was found that the temperature of cathode surface increased about 15°C while the beam was on. This implies that current density on cathode surface tends to be unstable because the back-streaming electrons give energy to the cathode. We calculated the amount of the energy transfer by means of an energy-balance equation of thermal transport. Then we compared the result with the total energy of back-streaming electrons calculated by a 2-D particle simulation code KUBLAI. By comparison of two results, we could roughly estimate the energy transfer from the back-streaming electrons within experimental error. © 2001 Published by Elsevier Science B.V.

1. Introduction

RF guns are widely used as injectors of RF linacs for FELs (free electron lasers) because they generate the higher-brightness electron beams with higher current and lower emittance than conventional electrostatic guns. This is due to higher accelerating electric fields, which make it possible to accelerate electrons in a shorter time and reduce emittance growth caused by space charge effect. In general, thermionic cathodes are simpler, easier to treat and more reliable than photocathodes. On the other hand, back-streaming electrons make the surface temperature and current density of the cathode unstable, which means it is difficult to

generate a high-quality beam with pulse width longer than several microseconds. A group at Stanford University [1] proposed and tested a method to avoid this problem by applying a transverse magnetic field, but it is not an ideal solution.

It is necessary to solve the back-bombardment effect for generating high-quality and long-pulse beams, which is in turn important for FELs. A 2-dimensional (2-D) simulation code KUBLAI [2–4] had been developed by our group and electron trajectories were calculated to quantitatively treat the back-bombardment effect. It was numerically found that back-streaming electrons converged and concentrated onto the center of cathode surface.

Recently, we measured the cathode surface temperature during the generation of the electron

---

*Corresponding author. Fax: +81-774-38-3426.

E-mail address: kii@iae.kyoto-u.ac.jp (T. Kii).

0168-9002/01/$ - see front matter © 2001 Published by Elsevier Science B.V.
PII: S 0 1 6 8 - 9 0 0 2 (0 1) 0 1 5 8 7 - X

beam from a 4.5-cavity S-band thermionic RF gun [4–6] and quantitatively evaluated the effect of back-bombardment by comparing the result with that of the particle simulation.

2. Experimental setup

Fig. 1 shows the experimental system. The electron beam generated by the RF gun enters a beam-transport system with focusing magnets, bending magnets, current transformers, beam profile monitors, and Faraday cups. The thermionic tungsten cathode is disk shaped with 3 mm radius and 1 mm thickness. The temperature at the cathode surface was measured by an infrared (IR) fiber thermometer manufactured by CHINO. The thermometer was set 200 mm apart from the center of cathode surface, and the IR light was received through a port of the RF gun. Measured temperature was sampled by digital temperature measurement module (WE-7000 series) with sampling interval of 0.5 s. The beam current was measured by a current transformer at the exit of the RF gun, and incident and reflected RF powers were also measured. The specifications of the RF

gun [5] are shown in Table 1 and those of the infrared thermometer are shown in Table 2.

3. Experimental result

The RF gun was operated with incident RF power of about 3 MW, resonant frequency of 2858.35 MHz and repetition rate of 1 pps. The temperature evolution during the beam generation is shown in Fig. 2. Before the operation of the RF gun, the surface temperature was kept at 1109°C, but due to the back-streaming electrons, the average temperature on the cathode increased to 1120°C. The rapid oscillation in the temperature profile arose from the sampling period of 0.5 s, which is a half of the beam period, and the slow oscillation was caused by a small difference between the beam repetition rate and the sampling frequency. We can get the temperature profile between beam-pulses by rearranging the stored data. The stored profile in 1 s is shown in Fig. 3. The average power given by the back-streaming electrons can be roughly estimated by using the input power dependence of the surface temperature. When the incident RF power is 3 MW, the temperature rise from 1109°C to

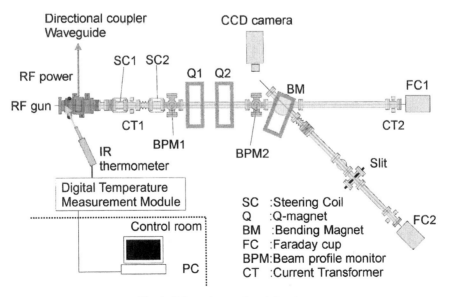

Fig. 1. Schematic experimental system.

Table 1
Specification of the thermionic RF gun

Thermionic RF gun (AET MG-500)	
Resonant frequency	2856 MHz (S-band)
Number of cavity	$4^{1/2}$(side-coupled)
Beam energy	4 MeV
Beam current	500 mA
Pulse width	3 μs @ 10 Hz
Incident RF power	3 MW

Table 2
Specification of an infrared thermometer

Infrared fiber thermometer (CHINO)	
Diameter of the effective area	Ø2 mm
Focus	200 mm
Focusing lens diameter	Ø5 mm
Sampling interval	25 ms
Detector material	Si
Temperature range	600–1800°C

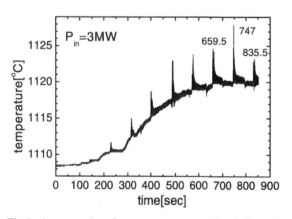

Fig. 2. An example of a temperature profile during the generation of electron beam.

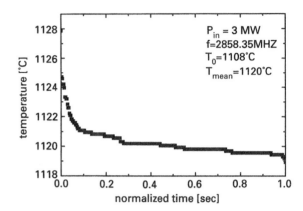

Fig. 3. Sorted temperature profile as the profile for 1 s.

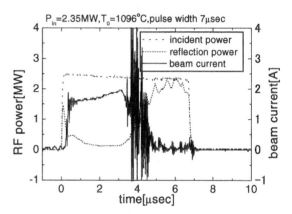

Fig. 4. A waveform of beam current and those of incident and reflection of RF powers.

1120°C corresponds to the power of 0.6 W. When the incident RF power was 2 MW, the average temperature was 1116°C and this temperature rise corresponds to the power of 0.35 W.

Fig. 4 shows the waveform of beam current and those of incident and reflection RF power. The incident RF power was 2.35 MW, resonant frequency was 2858.95 MHz, the pulse width was 7 μs, pulse repetition was 1 pps, and cathode temperature was 1096°C. At around 3 μs of macropulse, the slowly growing beam current suddenly dropped. The sudden changes were also seen in the RF waveform. These phenomena can be described as follows. Due to the backstreaming electrons, the surface temperature and charge density on the cathode increase. As the beam loading increases, the resonant frequency changes and reflected RF power grows; then, almost all RF power is reflected. The large oscillations of the beam current at around 3 μs are due to electric noise detected by the current transformer.

4. Discussion

To evaluate quantitatively the effect of back-bombardment, we need to analyze the above temperature profile. We could obtain the temperature profile after the power was supplied from the back-streaming electrons. But it is more important to obtain the profile while the electron beam hits the cathode surface in a macropulse. This temperature change would change the macropulse structure of electron beam and RF power. An energy balance equation was employed to find the thermal energy supplied by back-streaming electrons.

We could also estimate the energy supply by calculating the total energy of back-streaming electrons [3,4]. First, electron trajectories in RF gun were calculated by the simulation code KUBLAI. Then, the current density was evaluated by comparing the numerical result with the experimental result [6]. Then, the total energy of electrons on the cathode surface was calculated. By comparing these two results, we can evaluate the back-bombardment power of electrons.

For the analysis of the experimental data, we looked at the thermal transport on the cathode surface; thermal conductivity and thermal radiation. The energy balance equation is expressed as

$$C\frac{\mathrm{d}T}{\mathrm{d}\tau} = Q_\mathrm{h} + Q_\mathrm{b} - (AT^4 + BT + X) \qquad (1)$$

where Q_h is the energy supply from the cathode heater, Q_b is the energy supply from the back-streaming electrons. T is the surface temperature in absolute degrees, and τ is the time. We simply assumed A to be a constant for the thermal radiation, B that of thermal conductivity, and X the characteristics of the body of the RF gun. C is a constant for the cathode properties including density, specific heat, and the shape of cathode.

The constants A and B should be determined by comparison with experimental data, since emissivity and conductivity of the cathode are different from those in ideal conditions. We also assumed that Q_h is the power supplied by the cathode heater and the value is about 21 W in the present condition.

As a first step of the analysis, we calculated the average of Q_b, $\bar{Q}_\mathrm{b}$. In order to evaluate the constants, we measured the temperature drop after cutting off the cathode heater. When Q_h and Q_b are zero, it may be rewritten as the following:

$$\frac{\mathrm{d}T}{\mathrm{d}\tau} = a_1 T^4 + a_2 T + a_3. \qquad (2)$$

The constants a_1, a_2 and a_3 were thus evaluated by fitting the equation to the data temperature revolution, as shown in Fig. 5. Next, the remaining constant C was evaluated from the above constants and Eq. (1), with $Q_\mathrm{b} = 0$ and $\mathrm{d}T/\mathrm{d}\tau = 0$. Evaluated values are $a_1 = 3.082$, $a_2 = -4.997 \times \mathrm{e}^{-4}$, $a_3 = 3.98 \times \mathrm{e}^{-12}$, and $C = 1.197$. Then, $\bar{Q}_\mathrm{b}$ was evaluated from the data with the beam on. Finally, the energy given to the cathode by back-streaming electrons was compared with the result of the simulation code KUBLAI, as shown in Fig. 6. As for the experimental data, some causes of error are considered. With an error of 5°C in temperature, for example, the error in the resultant power varies from 40% to 70%. The simulation results are within these errors. Fig. 6 also shows the $\bar{Q}_\mathrm{b}$ values estimated roughly from the average temperature rise described in Section 3. Those are also in good agreement within the above errors. Thus, it is concluded that the energy balance equation gives reasonable estimate of the power given to the cathode by back-streaming

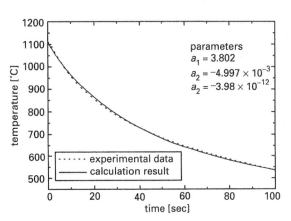

Fig. 5. Comparison of experimental profile with the result of energy balance equation.

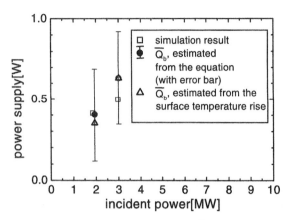

Fig. 6. Comparison of estimated values of $\bar{Q}_b$ from the energy balance equation, simulation, and average temperature rise.

electrons. However, as for the temperature measurement, we need to calibrate the IR thermometer and measure the temperature precisely. Additionally, we also take the actual value of constant C into account for the energy balance equation.

5. Conclusions

We measured the temperature of the cathode surface in a 4.5-cavity S-band thermionic RF gun to make a quantitative evaluation of back-bombardment effect. We could see the temperature rise of approximately 15°C at the maximum during the generation of the electron beam, compared with the stable temperature without the beam. We could estimate the energy supply from the back-streaming electrons by an energy balance equation and compare the results with those from the simulation and experimental results. We could roughly estimate the order of average power supply from back-streaming electrons onto the cathode surface. For the next step, we are planning to calibrate the thermometer, to refine the energy

balance equation and to analyze the whole temperature profile in macropulse timescale. Furthermore, it is expected from the result of 2-D simulation and the damage on the cathode surface that hollow cathodes can reduce the back-bombardment effect with little intensity loss in the high-quality part of the beam [3,4]. We will try the hollow cathode in the near future to see the effect.

Acknowledgements

The authors appreciate Mr. Eiji Tanabe (from AET) for his useful advice on our experiment. We thank Mr. Nishinosono for his technical help in the setups of our experiment. Thanks are also due to Mr. Ohshita from Nissin Electric Inc. for the maintenance of our experimental system and helpful advice. Finally, we appreciate the members of AccLab, NSRF, ICR, Kyoto University for all their help in this study.

References

[1] C.B. McKee, JohnM.J. Maday, Nucl. Instr. and Meth. A 296 (1990) 716.
[2] K. Masuda, Development of numerical simulation codes and application to Klystron efficiency enhancement, Ph.D. Thesis, Kyoto University, 1997.
[3] Takashi Inamasu, Numerical analysis of electron beam in an RF gun, Bachelor Thesis of Kyoto University, 1996.
[4] Yasushi Yamamoto, Takashi Inamasu, Kai Masuda, Masaaki Sobajima, Masami Ohnishi, Kiyoshi Yoshikawa, Hisayuki Toku, Eiji. Tanabe, Nucl. Instr. and Meth. A 393 (1997) 443.
[5] Kai Masuda, Ryuta Ikeda, Tomohiko Yamaguchi, Toshiteru Kii, Kiyoshi Yoshikawa, Tetsuo Yamazaki, The 12th Symposium on Accelerator and Technology, Waco, Japan, 1999.
[6] Ryuta Ikeda, Experiments and analysis on characteristics of 4.5-cavity S-band thermionic RF gun, Master Thesis of Kyoto University, 2000, (in Japanese).

ELSEVIER

Nuclear Instruments and Methods in Physics Research A 475 (2001) 593–598

NUCLEAR
INSTRUMENTS
& METHODS
IN PHYSICS
RESEARCH
Section A

www.elsevier.com/locate/nima

Smith–Purcell experiment utilizing a field-emitter array cathode: measurements of radiation

H. Ishizuka[a,*], Y. Kawamura[a], K. Yokoo[b], H. Shimawaki[b], A. Hosono[c]

[a] Faculty of Engineering, Fukuoka Institute of Technology, Higashi-Ku, Fukuoka 811-0295, Japan
[b] Research Institute of Electrical Communication, Tohoku University, Aoba-ku, Sendai 980-8577, Japan
[c] Advanced Technology R&D Center, Mitsubishi Electric Corporation, 8-1-1 Tsukaguchi-honmachi, Amagasaki, Hyogo 661-8661, Japan

Abstract

Smith–Purcell (SP) radiation at wavelengths of 350–750 nm was produced in a tabletop experiment using a field-emitter array (FEA) cathode. The electron gun was 5 cm long, and a 25 mm × 25 mm holographic replica grating was placed behind the slit provided in the anode. A regulated DC power supply accelerated electron beams in excess of 10 μA up to 45 keV, while a small Van de Graaff generator accelerated smaller currents to higher energies. The grating had a 0.556 μm period, 30° blaze and a 0.2 μm thick aluminum coating. Spectral characteristics of the radiation were measured both manually and automatically; in the latter case, the spectrometer was driven by a stepping motor to scan the wavelength, and AD-converted signals from a photomultiplier tube were processed by a personal computer. The measurement, made at 80° relative to the electron beam, showed good agreement with theoretical wavelengths of the SP radiation. Diffraction orders were −2 and −3 for beam energies higher than 45 keV, −3 to −5 at 15–25 keV, and −2 to −4 in between. The experiment has thus provided evidence for the practical applicability of FEAs to compact radiation sources. © 2001 Elsevier Science B.V. All rights reserved.

MSC: 85.45.Db; 41.75.Fr; 41.60.−m

Keywords: Smith–Purcell radiation; Field-emitter array; Electron source; Electron beam

1. Introduction

Smith–Purcell (SP) radiation has been studied extensively for nearly a half-century [1]. Electron beams with energies between 3 keV [2] and 150 MeV [3] have been used to generate the radiation in a broad band ranging from visible light to millimeter waves. Recent improvements include the production of coherent radiation by short electron bunches, miniaturization of SP devices, and investigation of SP free electron lasers which combine the grating with a resonator. In all cases, a crucial factor is the emittance of an electron beam. Along with photoinjectors, a scanning electron microscope is a favorable beam source in this respect [4], and photoelectric field emission from a needle cathode is also under investigation [5].

In a previous paper we reported the preliminary result of an SP experiment using a field-emitter array (FEA) cathode [6]; the FEA was durable and the electron beams generated therefore were sufficiently stable to be used for systematic measurements. The SP radiation was not identified, however, owing to irrelevant luminescence produced by the electron beam at the grating.

---

*Corresponding author. Fax: +81-92-606-0755.
E-mail address: ishizuka@fit.ac.jp (H. Ishizuka).

0168-9002/01/$ - see front matter © 2001 Elsevier Science B.V. All rights reserved.
PII: S 0 1 6 8 - 9 0 0 2 (0 1) 0 1 5 8 6 - 8

Subsequently, as presented here, the SP light was successfully detected in the whole visible range of the spectrum.

2. Experimental apparatus

2.1. General

Major parts of the experimental setup are described elsewhere [6]. In brief, the vacuum chamber is approximately 20 cm in inner diameter and in height. Electrons emitted from the FEA are accelerated by either a regulated DC power supply or a small Van de Graaff generator to tens of keV in a 5 cm long A–K gap; they then pass through a 1 mm wide slit provided in the anode and graze an optical grating with 1800 rulings/mm and a 30° blaze angle.

Some modifications have been made recently, as follows. First, the fluorescent screen for observing the beam profile was removed from the vacuum chamber, because the fluorescent material P31 released from the screen during electron bombardment formed scattered bright spots over the grating surface when the beam was on and disturbed the measurement of SP light. Second, the grating was reformed for better interception of light emitted inside its plastic body: the thickness of aluminum coating was increased from 0.1 to 0.2 μm, and the periphery of the grating was covered with a 50 μm thick aluminum foil. Third, the optical measurement system was automated to scan the entire region between 300 and 800 nm in a short time and also to process the collected spectral data efficiently. Last, feedback stabilization was applied to the Van de Graaff generator to regulate the output potential.

2.2. Automatic optical measurement system

The block diagram of the automatic optical measurement system is shown on the left side of Fig. 1. The spectrometer was driven by a stepping motor at the command of a personal computer (PC), which controlled the scanning speed, step width and the range of wavelength to be scanned. The scanning speed was set around 1 step/s; if the

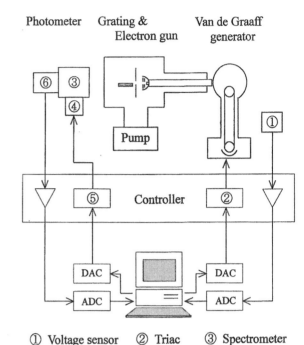

① Voltage sensor   ② Triac   ③ Spectrometer
④ Stepping motor   ⑤ Stepping driver   ⑥ PMT

Fig. 1. Block diagram of operation. For beam energies up to 45 keV, the Van de Graaff generator is replaced by a regulated DC power supply.

spectrometer was driven faster, the spectral peaks observed tended to systematically deviate from those predicted by the theory of SP radiation. After the scan the spectrometer was returned to the initial state, allowing successive measurements for different beam parameters. Usually five signals from the photomultiplier tube (PMT) were transmitted to the PC through an A/D converter at each wavelength. In a similar way, the PC accepted the electron beam current flowing into the grating, though the path is not shown in Fig. 1. The wavelength, the beam current and the PMT output were saved in a file side by side so that the light intensity (normalized by the beam current) was readily plotted against the wavelength.

2.3. Feedback regulation of the Van de Graaff generator

As shown on the right side of Fig. 1, the feedback control system is composed of a voltage sensor, A/D converter, PC, D/A converter and a triac. The voltage sensor was located 50 cm away from the dome of the Van de Graaff generator. It consisted of an electrode exposed to electric fields from the dome and a grounded sector disk, which was rotated to change the mutual capacitance between the electrode and the dome. The induced current was amplified and rectified to generate a DC sensor voltage. The sensor was calibrated for dome potentials up to 40 kV using the regulated DC power supply. The sensor voltage, which showed good agreement with calculation, was then extrapolated for the measurement of higher potentials.

Varying the electric power supplied to the motor that ran the transport belt allowed control of the Van de Graaff generator. For regulation, the sensor voltage was compared with a reference voltage determined by the set point of the dome potential, and the difference fed back negatively to the driving power of the motor via a triac. The PMT output and the beam current signal were passed to the data file when the deviation of the sensor voltage from the reference voltage was less than 1%. After taking five sets of data in this way, the spectrometer was stepped to the next wavelength.

3. Experimental results

3.1. Adjustment of the electron beam

When the electron beam skimmed over the grating, a track was observed visually through a viewing port of the vacuum chamber. Fine control of the beam was carried out, while monitoring the beam track with a CCD camera, by slightly moving the electron gun, changing the focus, and varying the potentials on the upper and lower halves of the anode that shifted the grazing angle. Further adjustment of the beam was practiced to maximize the PMT output. Under the optimum conditions, a narrow track ran uniformly across the grating surface to the end. The current flowing into the anode was negligibly small. The radiation was measured in the plane perpendicular to the

grating surface mainly at 80° with respect to the electron beam.

3.2. Manual measurements of radiation spectra

Our first observation of the SP radiation was made using an optical measurement system having a chopper [6]. Signals from the PMT were averaged over tens of acquisitions and displayed on the screen of Tektronix TDS 540 oscilloscope triggered by the synchronous output of the chopper. The inset of Fig. 2 shows an example of such signals. Here the upper and lower traces are the PMT output and the chopper signal, respectively. The data obtained by this method are shown by closed marks in Fig. 2 for electron beam energies of 35, 40 and 45 keV. (Open circles are data taken with the automated system afterwards.) The beam current was 25 μA and the spectral bandwidth was 40 nm. The central wavelengths of the peaks correspond to $n = -2$ SP radiation.

At this stage the green light due to P31 contaminant was still significantly intense and the SP radiation of higher diffraction orders was not detected clearly. Meanwhile the PMT signal was found to be stable and reproducible, except

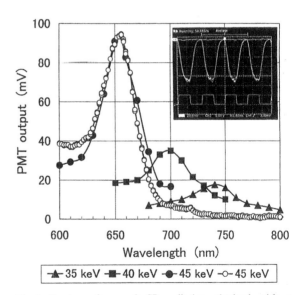

Fig. 2. Spectra of $n = -2$ SP radiation obtained with a chopper-spectrometer system. The inset shows the PMT signal (upper trace) and the synchronous chopper output (lower trace).

for dark current spikes. The measurement of light without the chopper was therefore possible if the PMT output was integrated over a time constant of 0.1 ms or so. The oscilloscope was then replaced by a digital voltmeter, which was used until the introduction of the automatic optical measurement system.

During this period it was also found that electron beams on a 25 μA level caused damage to the grating; the surface was grooved along the beam track and the surrounding portion of the plastic body changed color to dark brown. Because of this, the beam current was reduced to 10 μA or less for most of the following experiment.

3.3. Automatic measurements of radiation spectra

Typical examples of the automatically measured spectra are shown in Fig. 3. Here a 35 keV 10 μA beam and a 60 keV 2 μA beam were produced by the regulated DC power supply and the Van de Graaff generator, respectively. The bandwidth (FWHM) of the spectrometer was set at 20 nm, as seen from the spectrum of He–Ne laser light taken for the calibration of the system.

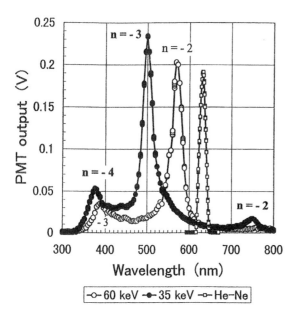

Fig. 3. Spectra obtained with the automatic optical measurement system. The experimental spectrum of He–Ne laser light indicates the spectrometer bandwidth.

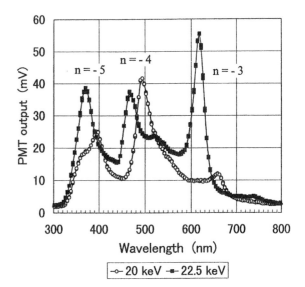

Fig. 4. Radiation spectra at low electron beam energies.

The beam energy was lowered to 15 keV, below which the radiation was not detectable. Fig. 4 shows spectra measured at beam energies of 20 and 22.5 keV. The peaks, from right to left, correspond to the SP radiation of orders $n = -3$, -4 and -5. The bump of the 22.5 keV line around 525 nm is due to the P31 phosphor attached to the grating surface. (The grating was cleaned at intervals to remove the P31 contaminant that reappeared with the passage of time.) The beam current flowing into the grating did not fluctuate significantly during the scan, as shown in Fig. 5. The radiation intensity was nearly proportional to the beam current in consecutive scans.

The dynode photoelectric sensitivity of our PMT has its maximum (70 mA/W) at wavelength $\lambda = 400$ nm. It reduces to 1/3 of the maximum value at 170 and 750 nm, to about 1/5 at $\lambda = 160$ and 800 nm, and drops sharply in the outer regions. The absence of PMT signal for $\lambda < 330$ nm is attributed to ultraviolet absorption in the glass of the viewing window and optical lenses. The height of spectral peaks for $n = -2$ increased with electron beam energy even if the dependence of PMT sensitivity on the wavelength was taken into account. This was not the case for $n = -3$, but the radiation intensity was rather insensitive to the beam energy in successive measurements.

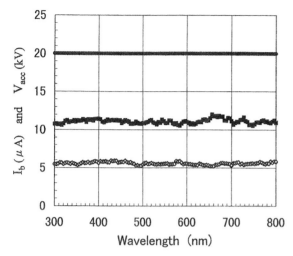

Fig. 5. Stability of electron beam in voltage and current. Top trace: acceleration voltage. Middle and bottom traces: beam current.

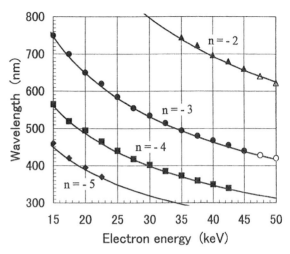

Fig. 6. Comparison of predicted (solid lines) and measured (data points) wavelengths at a radiation angle of 80°.

Radiation spectra were measured for electron beam energies up to 60 keV. Limited by the charging capability of our Van de Graaff generator, higher potentials were not stable enough to allow consistent measurements of spectral characteristics. For beam energies higher than 60 keV, we only mention with certainty that the $n = -2$ SP radiation was observed.

4. Discussion and summary

Measured wavelengths of the radiation are plotted in Fig. 6 against the electron beam energy. Here the closed marks indicate the data obtained with the regulated DC power supply and open marks are those acquired with the Van de Graaff generator. The result is in good agreement with predictions of the SP dispersion relation for $n = -2$ to -5. Widths of the spectra in Figs. 2–4 are reasonably small, considering the bandwidth of the spectrometer and the acceptance angle of the optical collection system. Let us notice that the electron beam energy is significantly lower than in earlier works; Gover et al. [7] and Shih et al. [8] used beams with energies higher than 60 and 80 keV, respectively. Consequently, the range of diffraction order n giving rise to the visible light

extended to -5 in the present experiment, while n was -1 and -2 in these earlier cases.

The measured PMT output reached 0.5 V, indicating that radiation up to 1 pW was received by the PMT. The dependence of radiation intensity on the electron beam parameters has not yet been clarified. Slight differences in beam focusing, grazing angle and the damage to the grating must have affected the radiation intensity. When the beam current was higher than 10 μA, large PMT signals forming irregular spectral peaks were occasionally observed for $\lambda > 500$ nm. Hence many problems still remain in the investigation of the radiation. The chief trouble is not in the FEA, but rather in the damage to the replica grating under the existing circumstances.

Thus far, much experience has been accumulated in the use of FEAs. The electron emission was stabilized by a field-effect transistor [9], and the beam current exceeded 10 mA in pulsed operation [10]. This experiment also allows us to foresee practical applications of FEAs to compact radiation sources.

Acknowledgements

This work was partly supported by a Grant-in-aid for Scientific Research (C) from the Japan Society for the Promotion of Science.

References

[1] J. Walsh, Nucl. Instr. and Meth. A 445 (2000) 214.

[2] J.C. Swartz, et al., IEEE Int. Conf. Plasma Sci. 126 (1998) 343.

[3] Y. Shibata, et al., Phys. Rev. E 57 (1998) 1061.

[4] M. Goldstein, et al., Appl. Phys. Lett. 71 (1997) 452.

[5] C. Hernandez-Garcia, C.A. Brau, Nucl. Instr. and Meth. A 429 (1999) 257.

[6] H. Ishizuka, et al., Nucl. Instr. and Meth. A 445 (2000) 276.

[7] A. Gover, et al., J. Opt. Soc. Am. B 1 (1984) 723.

[8] I. Shih, et al., J. Opt. Soc. Am. B 7 (1990) 345.

[9] H. Ishizuka, et al., Nucl. Instr. and Meth. A 375 (1996) 116.

[10] H. Ishizuka, et al., in: Proceeding of the 1998 International Conference Free Electron Lasers, Elsevier Science Publishers, Amsterdam, 1999, II-47.

ELSEVIER

Nuclear Instruments and Methods in Physics Research A 475 (2001) 599–602

NUCLEAR
INSTRUMENTS
& METHODS
IN PHYSICS
RESEARCH
Section A

www.elsevier.com/locate/nima

Bunching properties of a classical microtron-injector for a far infrared free electron laser

Grigori M. Kazakevitch[a],*, Stanislav S. Serednyakov[a], Nikolai A. Vinokurov[a], Young Uk Jeong[b], Byung Cheol Lee[b], Jongmin Lee[b]

[a] Budker Institute of Nuclear Physics RAS, Academician Lavrentyev 11, Novosibirsk 630090, Russia
[b] Laboratory for Quantum Optics, Korea Atomic Energy Research Institute, P.O. BOX 105, Yusong, Taejon 305-600, South Korea

Abstract

Longitudinal bunching properties of a classical microtron have been investigated by the numerical simulation of the longitudinal motion of accelerated electrons. The simulations were performed for the 12-turn microtron that has been used as an injector for the KAERI far infrared free electron laser. Based on the bunching properties of the electron beam, the temporal distribution of the coherent undulator radiation power during a macro pulse from the free electron laser was calculated. In the calculations, we took into account the dispersion properties of the accelerating cavity and deviations of the bunch repetition rate that were measured by the heterodyne method in real operating conditions of the microtron. The calculation results are compared with the experimental data. © 2001 Elsevier Science B.V. All rights reserved.

Keywords: Microtron; Bunching; Harmonic; Coherent radiation; Far infrared; Undulator

1. Introduction

It is well known that the power of the coherent undulator radiation depends strongly on the longitudinal distribution of the electron charge density in the electron bunches. Analytical calculations of bunching properties of a microtron [1] showed that the decrease of the electron current harmonic amplitudes with the increase of the harmonic numbers is rather slow. This permits us to expect good results when using the classical microtron as an injector for sources of coherent undulator

radiation. In this report, the numerical calculation of the longitudinal charge density distribution in electron bunches accelerated in a microtron was carried out. The calculated values of the current harmonics were used for calculation of the coherent undulator radiation power in the wave-guide chamber of the far infrared (FIR) free electron laser (FEL). Calculated temporal distribution of the coherent radiation power during a macro pulse was compared with the obtained experimental results.

2. Particle motion in microtron

The classical microtron [2] uses a microwave-accelerating cavity situated in a homogeneous magnetic field directed perpendicularly to the

---

*Corresponding author. Current postal address. Laboratory for Quantum Optics, Korea Atomic Energy Research Institute, P.O. Box 105, Yusong, Taejon 305-600, South Korea. Tel.: +82-42-868-8253; fax: +82-42-861-8292.

E-mail address: kaza@kaeri.re.kr (G.M. Kazakevitch).

0168-9002/01/$ - see front matter © 2001 Elsevier Science B.V. All rights reserved.
PII: S 0 1 6 8 - 9 0 0 2 (0 1) 0 1 5 8 5 - 6

cavity axis. Each passage of the electrons through the cavity increases their energy, and correspondingly, their orbit radii also increase. Particles with the largest orbit radii enter the ejection iron tube and go out to the transfer beam line. The internal Melekhin type 1 injection is used in our microtron. This means that the thermionic cathode is placed at the inner flat wall of the cavity and the first turn is disposed fully inside the cavity. In spite of this clever and simple design, the simulation of particle dynamics is rather complicated. It is worth noting that the majority of particles, which leave the cathode, are lost on the cavity walls, but the rest of them are bunched and collimated very tightly. Therefore, the transverse emittances, energy spread, and bunch length for the accelerated particles are very low. Moreover, the longitudinal charge distribution of bunches may have sharp edges, and therefore, the spectrum of the beam current may contain very high harmonics of the accelerating frequency.

The particle dynamics were simulated numerically by using a simplified 2D code. This code does the particle tracking in the medium plane of the microtron from the cathode to the exit.

The simulated particle distributions in the phase space for $\varepsilon = 0.99$, 1.005 and 1.03 (ε is the dimensionless accelerating voltage) are presented in Fig. 1. Using these results, we calculated the longitudinal distribution of electrons in the electron bunches. Histograms of the longitudinal charge distribution in the electron bunch for different values of ε are presented in Fig. 2.

The amplitudes of Fourier-harmonics $A(n)$ of the bunch current versus the harmonics number n are presented in Fig. 3. Relatively high amplitudes of higher harmonics demonstrate the good prospects of the classical microtron as an injector for coherent undulator radiation sources in the FIR range.

3. Coherent radiation power

The power P of coherent undulator radiation to one wave-guide eigenmode can be easily calculated as

$$
P = \frac{4\pi}{c} \frac{K^2}{4 + 2K^2} \left[J_0\left(\frac{K^2}{4 + 2K^2}\right) - J_1\left(\frac{K^2}{4 + 2K^2}\right) \right]^2
$$
$$
\times \frac{\lambda_1 L}{s} \left[\frac{A(\lambda_1/\lambda)}{A(0)} \right]^2 I^2
$$

where c is the velocity of light, K and L are the deflection parameter and the length of the undulator, λ_1 is the wavelength of the accelerating voltage, λ is the wavelength of the undulator radiation, I is the average beam current, $J_0(K^2/4 + 2K^2)$ and $J_1(K^2/4 + 2K^2)$ are Bessel functions of the first kind. The effective cross-section area $s = \int |u|^2 \, dx \, dy / |u(0,0)|^2$ depends on the wave-guide eigenmode $u(x,y)$. For the rectangular wave-guide TE_{01} mode, $s = ab/2$, where a and b are the transverse sizes of the wave-guide.

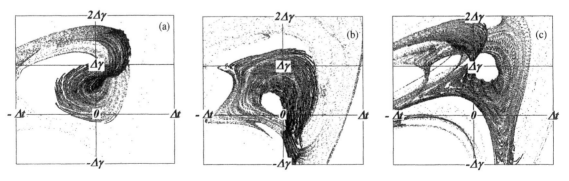

Fig. 1. Particle distributions in the phase plane for different accelerating voltages: (a) $\varepsilon = 0.99$, (b) $\varepsilon = 1.005$ and (c) $\varepsilon = 1.03$; $\Delta t = 15\,\mathrm{ps}$; $\Delta\gamma$ is the energy spread in relative units.

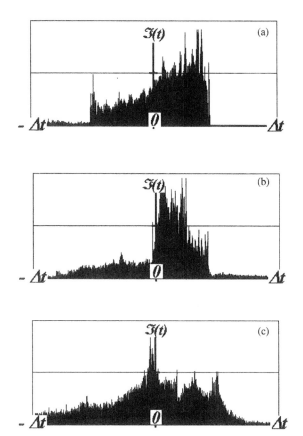

Fig. 2. Longitudinal particle distributions $\Im(t)$ for different accelerating voltages: (a) $\varepsilon = 0.99$, (b) $\varepsilon = 1.005$ and (c) $\varepsilon = 1.03$; $|(t)$ is presented in the relative units, $\Delta t = 15$ ps.

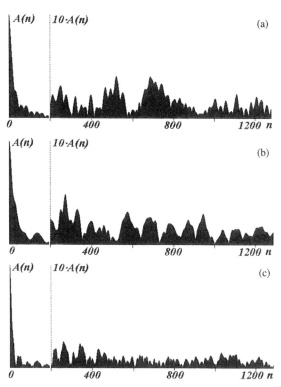

Fig. 3. The electron current spectra for different accelerating voltages: (a) $\varepsilon = 0.99$, (b) $\varepsilon = 1.005$ and (c) $\varepsilon = 1.03$.

4. Time dependence of the accelerating voltage

As it is seen from Figs. 1–3, the longitudinal distribution and consequently, the amplitudes of the high-order harmonics $A(n)$ depend significantly on the accelerating voltage ε. Data for accelerating voltage deviations were obtained by measurements of the bunch repetition rate deviations during a macro pulse. These measurements used a measuring cavity [3,4] and were aimed at optimizing the performance of the microtron for lasing. The simplified modeling of the time dependence of the accelerating voltage based on the bunch repetition rate measurements showed periodical oscillations of the accelerating voltage with period 0.6–0.7 µs during the macro pulse, as is presented in Fig. 4. Due to high-beam loading and coupling of the magnetron oscillator with the accelerating cavity, the time dependence is rather complicated.

5. Measured radiation power

The performed measurements are described in paper [5]. Some parameters of the microtron and undulator are presented in Table 1. A liquid He-cooled Ge–Ga detector that was sensitive to radiation in the wavelength range 30–150 µm was used as the FIR light receiver. FIR radiation was collected in the entrance aperture of the detector (8 mm in diameter) by a quartz lens that served at the same time as a vacuum FIR window and was installed at approximately double focal length distance from the end of the FIR wave-guide and the detector entrance. The detected power was in

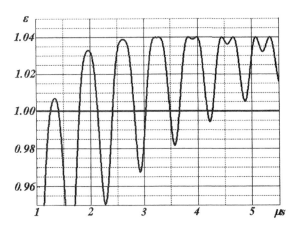

Fig. 4. The time distribution of the accelerating voltage during macro pluse.

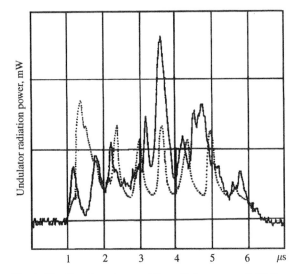

Fig. 5. The time dependence of the radiation power. Solid line-measured, vert. scale is 10 mW/div. Dotted line-calculated, vert. scale is 150 mW/div.

Table 1
Main parameters of electron beam and undulator

Total energy of electrons	7.02 MeV
Macro pulse current	36 mA
Macro pulse duration	5.5 μs
Undulator period	25 mm
Number of undulator periods	80
Undulator K-parameter	1.284
Wavelength of radiation	121 μm
Waveguide cross-section	$2 \times 20 \, mm^2$

the order of 20% of the radiated power due to aperture losses and attenuation in the quartz lens. Detected power always showed a quadratic law dependence on the beam current.

The measured temporal dependence of the coherent undulator radiation power during the macro pulse (solid line) is plotted in Fig. 5. The dotted curve is the result of calculations. In these calculations, we took into account the temporal properties of the detector. Comparison of curves demonstrates a qualitative agreement of the calculations with the experiment. Further improvement of the calculation has to be done to obtain quantitative agreement.

Acknowledgements

We are very appreciative of Prof. G.N. Kulipanov for his permanent interest and useful remarks to the present work.

References

[1] S.P. Kapitza, J. Tech. Phys. V XXIX (6) (1959) 729.
[2] S.P. Kapitza, V.N. Melekhin, The Mictotron, Harwood academic publishers, London-Chur, 1978.
[3] G.M. Kazakevich, Y.U. Jeong, et al., Measurements of intrapulse instabilities in the electron beam of high current microtron, in: Jongmin Lee, Byung Cheol Lee (Eds.), Free Electron Lasers in Asia. Theory, Experiments and Applications, Proceedings of the Fourth Asian Symposium on Free Electron Lasers, 1999, p. 198.
[4] G.M. Kazakevich, Y.U. Jeong, et al., A compact microtron driven by a magnetron as an injector of the far infrared free electron laser, Presented at the 22nd International Free Electron Laser Conference, FEL 2000 Conference, Duke University, 2000.
[5] G.M. Kazakevich, Y.U. Jeong, et al., Start up of lazing in compact far infrared free electron laser driven by 8 MeV microtron, in: G.L. Carr, P. Dumas (Eds.), Accelerator-based Sources of Infrared and Spectroscopic Applications, Proceedings of SPIE, Vol. 3775, 1999, p. 71.

ELSEVIER

Nuclear Instruments and Methods in Physics Research A 475 (2001) 603–607

NUCLEAR
INSTRUMENTS
& METHODS
IN PHYSICS
RESEARCH
Section A

www.elsevier.com/locate/nima

Different focusing solutions for the TTF-FEL undulator

B. Faatz*, J. Pflüger

Hamburger Synchrotronstralungslabor (HASYLAB) at Deutsches Elektronen-Synchrotron (DESY), Notkestr. 85, 22607 Hamburg, Germany

Abstract

While aiming at shorter wavelengths, the SASE-FEL undulators become longer. Therefore, maintaining the small electron beam sizes required, the question of what kind of focusing structures should be used for optimum performance becomes of increasing interest. Present SASE FEL undulators that have been developed and proposed for future FELS both include integrated focusing (TTF-FEL), separate function quadrupoles combined with the natural undulator focusing (LEUTL) and triplet or FODO structures.

In this contribution, the consequence of the different undulator designs in terms of alignment (tolerances) and flexibility will be discussed. © 2001 Elsevier Science B.V. All rights reserved.

PACS: 41.60.Cr

Keywords: Free electron laser

1. Introduction

The first SASE results of the TESLA Test Facility (TTF) FEL have shown that the undulator performs according to predictions [1]. The measured gain is of the order of 3000. This value can certainly increase as soon as the rms tolerance level of 10 μm alignment of the electron beam along the entire undulator is reached, which should be possible with beam-based alignment.

The advantage of the present undulator design with its integrated quadrupoles is that the β-function can be made both small and almost constant [2]. To achieve this, one needs in the order of two quadrupoles per meter for the TTF-

FEL parameters. These have to be aligned with high accuracy. Although the actual alignment of quadrupoles is close to the values needed, one still has to re-steer the beam to compensate for residual quadrupole kicks. In order to know what values of the correctors are needed, beam-based alignment has to be performed. Because the quadrupole strength is fixed, one has to vary the electron beam energy in order to minimize dispersion or determine offsets of the beam position monitors (BPM) [3,4]. For this reason, many steerers and BPMs are needed along the undulator, resulting in a complicated design of the vacuum chamber [5].

As an alternative, one could consider a separate focusing structure. If we do not change the basic design of the undulator, i.e. each segment has a length of approximately 4.5 m, then the quadrupoles, at this moment integrated into the undulator, would have to be taken out. Additional

---

*Corresponding author. Tel.: +49-40-8998-4513; fax: +49-40-8998-2787.

E-mail address: bart.faatz@desy.de (B. Faatz).

0168-9002/01/$ - see front matter © 2001 Elsevier Science B.V. All rights reserved.
PII: S 0 1 6 8 - 9 0 0 2 (0 1) 0 1 5 8 4 - 4

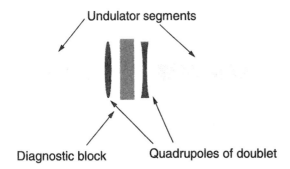

Fig. 1. Layout of the doublet structure with a doublet and a diagnostic block between undulator segments.

space is needed for the quadrupoles between the undulator segments. Due to the length of the segments, a FODO-structure is not possible because of the large phase-advance between quadrupoles at β-functions that are needed. In fact, an average β-function smaller than 9 m is not possible, with $\beta_{max}/\beta_{min} \approx 5$. The only alternatives are doublet or triplet structures. As an example, Fig. 1 shows the doublet structure.

2. Debunching in a separate focusing structure

In any FEL, debunching occurs due to longitudinal velocity spread, i.e. due to energy spread and finite emittance of the electron beam and the betatron oscillation performed by individual electrons. The influence of energy spread effects and emittance has been well recognized and leads to limits on those two parameters for all FELs [6]. Also the influence of focusing has already been described before [7].

The additional problem occurring with the separate focusing is the pathlength difference due to different initial conditions for the individual electrons. An electron with zero transverse position and momentum will enter a next undulator segment at a different time than an electron with an initial angle and offset. Averaged over all electrons, this results in debunching in the space between undulator segments, and therefore a strong decrease in gain. If one looks at the difference in path length of an electron of initial offset and angle (x, x') compared to an on-axis

electron, this can be written as

$$\Delta \ell = \int_0^z (\sqrt{1 + x'^2(\zeta)} - 1)\, d\zeta \approx \frac{1}{2} \int_0^z x'^2(\zeta)\, d\zeta$$

and similar in the y-direction. Averaged over all electrons, this results in

$$\frac{2\delta\ell}{\varepsilon} = \frac{2\langle \Delta\ell \rangle}{\varepsilon} = \beta_0 \int_0^z C'^2(\zeta)\, d\zeta + \gamma_0 \int_0^z S'^2(\zeta)\, d\zeta$$

$$- 2\alpha_0 \int_0^z C'(\zeta)S'(\zeta)\, d\zeta \qquad (1)$$

with α_0, β_0 and γ_0 the Twiss parameters and C' and S' the transfer functions (see for example Ref. [8]). The value for $\delta\ell$ gives an average path length difference, $\sigma_\ell = \langle \Delta\ell^2 \rangle - \langle \Delta\ell \rangle^2$ gives the debunching. From Eq. (1) one can see that the effect becomes smaller for smaller emittance. An alternative would be to reduce the undulator segment length. This would make a smaller and more constant β-function possible, and thus reduce values of the Twiss parameters. This, on the other hand, would result in more quadrupoles to reach the same effective undulator (interaction) length.

In the next section, a numerical example will be worked out for the TTF-FEL parameters.

3. Comparison of integrated with separate function undulator

Because the debunching is most severe at the smallest wavelength, only the 6.4 nm case is studied here.

The following problems will be addressed:

- Increase in saturation length
- Variation in saturation length with effective shotnoise power
- Tolerances on quadrupoles
- Strength of correctors needed to correct quadrupole offsets
- Accuracy of beam position monitors

In Fig. 2, the power growth (a) and bunching (b) along the undulator is shown for an integrated FODO structure (solid line), a separate doublet (dashed line), and triplet structure (dotted line)

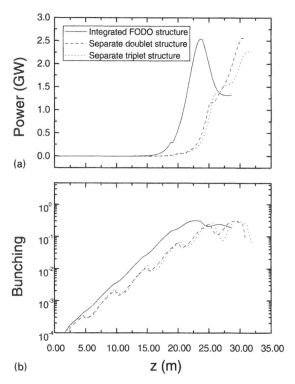

Table 1
Parameters of the TESLA Test Facility FEL with separate doublet focusing

Electron beam	
Energy, $\mathcal{E}_0$	1 GeV
Peak current, I_0	2.5 kA
Rms bunch length, σ_z	50 μm
Rms normalized emittance	2π mm mrad
Rms energy spread	1 MeV
External β-function,	4.5 m
Rms transverse beam size	70 μm
Undulator	
Type	Planar, hybrid
Focusing structure	Separate, doublet
Period, λ_w	27.3 mm
Segment length	4.45 m
Distance between segments	710.0 mm
Peak magnetic field, H_w	0.5 T
Focusing structure	Doublet
Quadrupole optical length	82.0 mm
Quadrupole strength	37.0 T/m
Output radiation	
Wavelength, λ	6–60 nm

Fig. 2. Power (a) and bunching (b) for an integrated FODO lattice (solid line), a separate doublet (dashed line) and a separate triplet (dotted line). The undulator segment length is the same in both cases. In all cases the optics have been adjusted to obtain maximum gain.

with beam parameters as given in Table 1. For the FODO and triplet structure, the average β-function is the same (3 m), for the doublet structure, this value cannot be reached. In this case the minimum (average) value of 4.8 m is used. In case of both doublet and triplet focusing, the power still grows after six undulator segments. With additional segments, the power only increases further by some 25%.

As can be clearly seen in Fig. 2b, at each intersection between undulator segments, the bunching decreases in the case of the doublet and triplet structures. If the intersection length could be reduced, the effect would be less severe. For the integrated FODO structure, it is distributed along the entire undulator and therefore an addition to the 'normal' debunching caused by a warm beam. Because the triplet structure does not seem to give a larger growth rate and the overall length is more

(unless the quadrupole strength is increased), the remainder of this paper is dedicated to the doublet structure.

Because in the doublet structure the power saturates close to the end of the undulator, the sensitivity of saturation length on fluctuation of input power is important. In Fig. 3, the power growth along the undulator is shown for a variation in effective shotnoise power between 8 and 800 W. As can be seen, the power variation at the undulator exit is approximately 20%. This is, for example, comparable to the variation in power after saturation is reached for the integrated FODO-structure shown in Fig. 2.

One of the main advantages of the separate focusing is that the number of quadrupoles can be reduced and the tolerances on undulator alignment becomes more relaxed. The rms value of the electron beam orbit deviation from a straight line should not exceed 20% of the beam size [9], which in this case corresponds to 17 μm. For 50 samples of randomly misaligned quadrupoles, the rms value of this orbit deviation along the undulator has been calculated. The maximum quadrupole

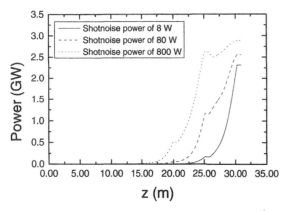

Fig. 3. Power for a separated doublet with an effective shotnoise power of 8 (solid), 80 (dashed) and 800 W (dotted line).

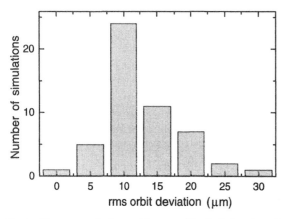

Fig. 4. Histogram of the rms beam orbit deviation for 50 simulations of randomly misaligned quadrupoles within ± 3 μm.

offset for each sample lies within ± 3 μm, a value that can be achieved with beam-based alignment. Fig. 4 shows that in most cases, the rms orbit deviation of the electron beam is within the required 17 μm.

The two questions to be answered are then what strength of correctors are needed to correct the quadrupole center and what BPM resolution is required to detect a quadrupole offset of 3 μm or less using beam-based alignment.

To start with the first, if a corrector would be integrated into the quadrupole it has to be able to shift its center. With a gradient Q T/m, an offset δx results in a dipole field of $Q \times \delta x$. For a

gradient of 37 T/m and a typical offset of 100 μm, the corrector has to have a minimum strength of 3.7 mT. A problem for beam-based alignment in this case could be hysteresis and possibly sextuple terms in the magnetic field. This would be avoided with micro-movers. In this case small steerers could be used to correct for residual first and second magnetic field integrals of the undulator and to do the final alignment. At this moment, both options are under investigation.

For a quadrupole with gradient Q, an offset δx, and length L_Q, the kick is $x' = eQ\delta x L_Q/\gamma mc$. The offset measured at the position of the BPM is then $x'L_u$, with L_u approximately the length of the undulator segment, i.e. 4.5 m (assuming that the undulator can be considered as a drift space). For the values in Table 1, the offset measured due to a kick by a quadrupole with an offset of 3 μm is 10 μm for a 1 GeV beam. This is the resolution that is close to what we have at the moment. For an integrated FODO-structure, simulations have shown that a 1 μm resolution of the BPMs is needed due to the distance of 0.5 m between quadrupoles.

4. Discussion

Under optimum conditions, the integrated FODO-structure used in the TESLA Test facility undulator has a better performance than that of a separate focusing structure. The electron beam bunching does not deteriorate in the gaps between the segments. For the separate focusing doublet structure, one can vary the strength of the quadrupoles, which makes the beam-based alignment procedure easier to perform. In the space between the undulator segments needed for the quadrupoles, conventional technology can be used for diagnostic equipment. Quadrupoles with the specifications needed are available. Matching of betatron phase and matching between electron and radiation phase can be performed independently. Corrector coils integrated into the quadrupoles can be used to align the quadrupoles and at the same time correct residual kicks of the undulator magnetic field. An alternative option would be to use micro-movers to correct quadrupole offsets and use small correctors for final alignment and correction of residual errors of the undulator magnetic field.

Acknowledgements

The authors would like to thank J. Lewellen, S. Reiche, E.L. Saldin, H. Schlarb, G. Schmidt, E.A. Schneidmiller, N.A. Vinokurov and M.V. Yurkov for many useful discussions.

References

[1] J. Andruszkow, et al., Phys. Rev. Lett. 85 (2000) 3825.

[2] J. Pflüger, P. Gippner, A. Swiderski, T. Vielitz, Proceedings of the FEL Conference in Hamburg, Germany, 1999, Nucl. Instr. and Meth. A 445 (2000) 366.

[3] K. Flöttmann, B. Faatz, J. Rossbach, E. Czuchry, Nucl. Instr. and Meth. A 416 (1998) 152.

[4] P. Castro, B. Faatz, K. Flöttmann, Nucl. Instr. and Meth. A 427 (1999) 12.

[5] U. Hahn, P.K. den Hartog, J. Pflüger, M. Rüter, G. Schmidt, E.M. Trakhtenberg, Nucl. Instr. and Meth. A 445 (2000) 442.

[6] W. Brefeld, B. Faatz, Y.M. Nikitina, J. Pflüger, P. Pierini, J. Rossbach, E.L. Saldin, E.A. Schneidmiller, M.V. Yurkov, Nucl. Instr. and Meth. A 375 (1996) 295.

[7] E.T. Scharlemann, J. Appl. Phys. 58 (1985) 2154.

[8] H. Wiedemann, Particle Accelerator Physics, Springer, Berlin, 1995.

[9] B. Faatz, J. Pflüger, Y.M. Nikitina, Nucl. Instr. and Meth. A 393 (1997) 380.

ELSEVIER

Nuclear Instruments and Methods in Physics Research A 475 (2001) 608–612

NUCLEAR
INSTRUMENTS
& METHODS
IN PHYSICS
RESEARCH
Section A

www.elsevier.com/locate/nima

A design study of a staggered array undulator for high longitudinal uniformity of undulator peak fields by use of a 2-D code

K. Masuda*, J. Kitagaki, Z.-W. Dong, T. Kii, T. Yamazaki, K. Yoshikawa

Institute of Advanced Energy, Kyoto University, Gokasho, Uji, Kyoto 611-0011, Japan

Abstract

A staggered array undulator was studied using a 2-D computer code, aiming at high longitudinal uniformity of undulator peak fields. Firstly, insertion of ferromagnetic pieces at both ends of the undulator is found to give higher uniformity of the undulator peak fields, while there are still undesirable leakages of axial magnetic fields in the vicinity of both ends of the undulator. Secondly, to further reduce these leakages, an additional external closed magnetic circuit was designed and found to successfully reduce the leakages as well as giving a much higher uniformity of the undulator peak fields. As a result, it is found in our simulation that, compared with the undulator without the closed magnetic circuit, the transmittance of electrons through the undulator increases by 15%, and consequently the spontaneous emission intensity also increases by 50%. © 2001 Elsevier Science B.V. All rights reserved.

PACS: 41.60.Cr

Keywords: Free electron laser; Staggered array undulator; 2-D unbounded magnetostatic field calculation

1. Introduction

A staggered array undulator, which was originally proposed by a Stanford group [1,2], consists of a solenoid coil and two rows of iron stacks and aluminum half disks as shown in Fig. 1. The advantages of commonly used undulators consisting of permanent magnets include, easy tunability of the free electron laser wavelength by changing the coil current, relatively high undulator fields with a short undulator period, easy fabrication, and low cost.

Previously, by using a 2-D computer code and periodic boundary conditions, we studied dependences of undulator fields on geometrical shapes of ferromagnetic pieces in an infinitely long staggered array undulator, and we obtained an optimal shape of ferromagnetic pieces to give high undulator peak fields [3].

In this paper, for a practical design of a staggered undulator with a finite length, longitudinal uniformity of undulator peak fields is discussed by use of a 2-D numerical code. To achieve highly uniform undulator peak fields, insertion of ferromagnetic pieces at both ends of the undulator and an additional external closed magnetic circuit are considered.

---

*Corresponding author. Tel.: +81-774-38-3442; fax: +81-774-38-3449.

E-mail address: masuda@iae.kyoto-u.ac.jp (K. Masuda).

0168-9002/01/$ - see front matter © 2001 Elsevier Science B.V. All rights reserved.
PII: S 0 1 6 8 - 9 0 0 2 (0 1) 0 1 5 8 3 - 2

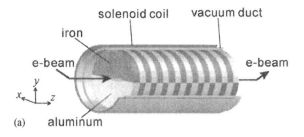

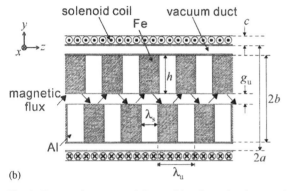

Fig. 1. Staggered array undulator: (a) schematic view, and (b) cross-sectional view.

2. Basic equations and numerical method

The magnetic flux density B can be described by the following Maxwell's equation:

$$\nabla \times \left(\frac{1}{\mu} \nabla \times A\right) = J \tag{1}$$

$$B = \nabla \times A \tag{2}$$

where A is the vector potential, and J is the current density through the solenoid coil. For discussion of the field leakages and peak undulator field uniformity along the axis, a 2-D code is used in this study, which we previously developed for calculating unbounded magnetostatic fields with nonlinearity of ferromagnets' magnetization taken into account [4].

3. Basic design

Fig. 1 shows schematics of the staggered array undulator. As indicated in Fig. 1(a), planer un-dulator fields in the y-direction are produced by

the periodically stacked ferromagnet pieces. Since previous 3-D calculations with a periodic condition in the z-direction predicted a good field uniformity in the x-direction [5], the undulator is assumed to be infinitely long in the x-direction for saving computational efforts. Thus, the aforementioned 2-D code was used in this study. The dimensions of the ferromagnet piece shape used in this study are listed in Table 1, which were previously optimized to give high peak undulator fields [3]. Measured $B-H$ and $\mu-H$ characteristics of an iron piece are taken into account.

First of all, without any closed magnetic circuit (as in the original design by the Stanford group [1,2]), the staggered array undulator of 0.5 m length is set in the middle of a 1.0 m-long solenoid coil. Although the longitudinal uniformity of the solenoid field is good with <2% error, the insertion of the ferromagnet pieces in the solenoid coil is found to result in decreases of the undulator peak fields at both ends as shown in Fig. 2. This poor longitudinal uniformity of the undulator peak fields, of course, would adversely affect electron trajectories and spontaneous emission. In the following section, we discuss design refinements to obtain higher longitudinal uniformity of the undulator peak fields.

4. Design refinements

To improve the longitudinal uniformity of the undulator peak fields, insertion of ferromagnet pieces at both ends of the undulator and addition

Table 1
Dimensions of the staggered array undulator

Undulator period, λ_u	20 mm
Gap width, g_u	5 mm
Ratio of spacer length, $f = \lambda_s/\lambda_u$	0.4
Height of ferromagnet piece, h	27.5 mm
Solenoid coil thickness, c	3 mm
Outer width of vacuum chamber, $2a$	60 mm
Inner width of vacuum chamber, $2b$	50 mm
Number of undulator periods, N_w	25
Undulator length, L_u	0.5 m

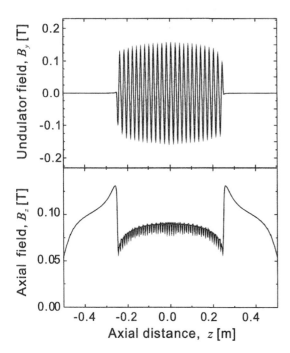

Fig. 2. On-axis magnetic fields by the basic design.

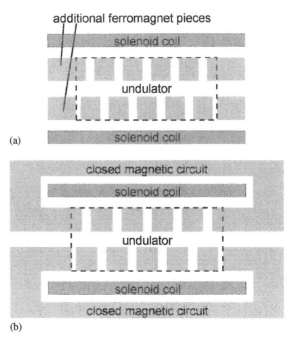

Fig. 3. Schematic views of (a) insertion of additional ferromagnet pieces, and (b) an external closed magnetic circuit.

of an external closed magnetic circuit are studied, which are schematically shown in Figs. 3(a) and (b).

4.1. Insertion of ferromagnet pieces at both ends

To compensate for the decreases of the undulator peak fields at the ends, insertion of ferromagnet pieces was studied. The lengths in the z-direction of the inserted pieces were optimized by trial and error. The optimal shapes of the inserted pieces resulted in higher longitudinal uniformity of undulator peak fields as shown in Fig. 4, while leakages of the axial magnetic field are found outside the undulator ($z = \pm 0.3$ m), which are undesirable for electron beam injection into the undulator gap.

4.2. Addition of a closed magnetic circuit

To reduce the field leakages, the addition of external closed magnetic circuit was then studied. Figs. 5(a) and (b) show magnetic flux lines and on-

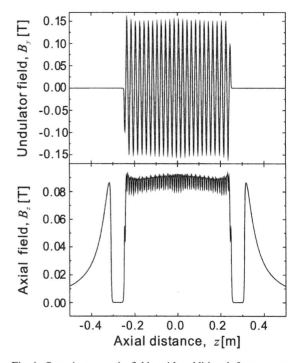

Fig. 4. On-axis magnetic fields with additional ferromagnet pieces inserted.

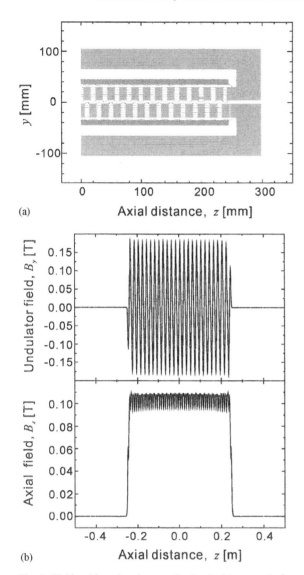

Fig. 5. Fields with a closed magnetic circuit; (a) magnetic flux lines on the $y-z$ plane, and (b) on-axis magnetic fields.

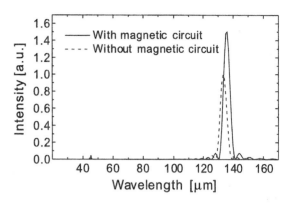

Fig. 6. Spontaneous emission spectra, comparing between the basic undulator design and the refined one with the closed magnetic circuit.

axis fields, respectively, with the optimal closed magnetic circuit for a vacuum solenoid filed of 0.1 T. The field leakages are successfully reduced as expected, and higher longitudinal uniformity of the undulator peak field is achieved as well.

To evaluate performance improvements of the undulator, electron trajectories and spontaneous emissions are calculated for an electron beam. The

initial transverse distribution of the electrons is assumed to be Gaussian. The averaged energy, energy spread, rms diameter, and normalized rms emittance are 4 MeV, 0.1%, 1 mm, and 35π mm mrad, respectively. The transmittance of the electron beam through the 0.5 m-long undulator is found to increase up to 96% from 83% with the basic design described in Section 3 without a closed magnetic circuit. As a consequence of the higher longitudinal uniformity of the undulator peak fields and the higher transmittance of the electron beam, the intensity of the spontaneous emission is found to increase by 50% as shown in Fig. 6.

5. Conclusions

For high longitudinal uniformity of undulator peak fields through a staggered array undulator by compensating for field decreases at the ends, insertion of ferromagnet pieces at the ends, and addition of an external closed magnetic circuit were studied by use of a 2-D numerical code. Both are found efficient, while the former is found insufficient with undesirable field leakages outside the undulator. On the other hand, the optimal closed magnetic circuit is found to result in a perfect longitudinal uniformity of the undulator peak fields without field leakage and an increased electron beam transmittance through the undulator gap as well. Consequently, the spontaneous emission intensity is found to increase by 50%.

References

[1] Y.C. Huang, et al., Nucl. Instr. and Meth. A 318 (1992) 765.
[2] Y.C. Huang, et al., IEEE J. Quantum. Electron. 30 (1994) 1289.
[3] J. Kitagaki, et al., Proceedings of the 21st Interational Free Electron Laser Conference, Hamburg, Germany, August 26–28, 1999, 2000, pp. II–105.
[4] K. Masuda, Development of numerical simulation codes and application to klystron efficiency enhancement, Ph.D Thesis, Kyoto University, 1997.
[5] J. Kitagaki, et al., A design study on electron beam confinement in a staggered array undulator based on a 3-D code, Nucl. Instr. and Meth. A 475 (2001) 613, these proceedings.

ELSEVIER

Nuclear Instruments and Methods in Physics Research A 475 (2001) 613–616

NUCLEAR
INSTRUMENTS
& METHODS
IN PHYSICS
RESEARCH
Section A

www.elsevier.com/locate/nima

A design study on electron beam confinement in a staggered array undulator based on a 3D code

J. Kitagaki, K. Masuda*, Z.-W. Dong, T. Kii, T. Yamazaki, K. Yoshikawa

Institute of Advanced Energy, Kyoto University, Gokasho, Uji, Kyoto 611-0011, Japan

Abstract

A staggered array undulator is implemented to improve the electron beam confinement. A staggered array undulator has an inherent vertical focusing force, but it has no horizontal focusing force, resulting in a beam loss. To improve the electron beam confinement with an additional horizontal focusing force, the geometrical shape of the ferromagnet pieces were refined by use of a newly developed 3D code. As a result, the number of electrons, which can pass through the undulator, was found to increase by 10%, and consequently, the spontaneous emission intensity also increased by 10%. Transmittance of a 4 MeV electron beam is found to be 99.7% for an undulator of 0.5 m length and 5 mm gap width. Also, for a longer undulator of 2.0 m length, the transmittance is found still high up to 97.3%. © 2001 Elsevier Science B.V. All rights reserved.

PACS: 41.60.Cr

Keywords: Free electron laser; Staggered array undulator; 3D magnetostatic field calculation

1. Introduction

A staggered-array undulator was first proposed by the Stanford group [1,2], and it has several advantageous features such as easy tunability of wavelength, relatively high undulator field with short undulator period, easy fabrication, and low cost.

Basic characteristics of a staggered undulator has been studied so far based on a 2D and a 3D code with an assumption of linear magnetization [3,4], and for a practical case, performance characteristics and optimal design parameters have been studied by using a 2D code with nonlinear magnetization taken into account [5,6].

As for the electron beam confinement, a staggered array undulator has an inherent vertical focusing force, while there is no horizontal focusing force, resulting in a beam loss in the undulator. The objective of this study is to improve the electron beam confinement with an additional horizontal focusing force by optimising the geometrical shape of the ferromagnet pieces, by using a newly developed 3D code with nonlinear magnetization taken into account.

2. Basic equations

The following subsets of Maxwell's equations can describe the magnetic field even in the region

*Corresponding author. Tel.: +81-774-38-3442; fax: +81-774-38-3449.

E-mail address: masuda@iae.kyoto-u.ac.jp (K. Masuda).

0168-9002/01/$ - see front matter © 2001 Elsevier Science B.V. All rights reserved.
PII: S 0 1 6 8 - 9 0 0 2 (0 1) 0 1 5 8 2 - 0

where the magnetization has a nonlinearity with respect to field intensity:

$$\nabla \times H = J \tag{1}$$

$$B = \mu H \tag{2}$$

$$\mu = \mu(H) \tag{3}$$

$$\nabla \cdot B = 0 \tag{4}$$

where H, J, B, and μ are, respectively, the field intensity, the source current density in the solenoid coil, the magnetic induction, and the permeability.

The magnetic field intensity H can be expressed as the sum of the source field H_S and the induced field H_M due to ferromagnet magnetization by

$$H = H_S + H_M \tag{5}$$

$$\nabla \times H_S = J \tag{6}$$

$$\nabla \times H_M = 0. \tag{7}$$

From Eq. (7), H_M can be expressed by the reduced scalar potential ϕ [7,8] as

$$H_M = -\nabla \phi. \tag{8}$$

Then, Eqs. (4)–(7) give

$$-\nabla \cdot (\mu \nabla \phi) + \nabla \cdot (H_S) = 0. \tag{9}$$

The magnetic field can then be calculated from Eq. (9) by the Finite Element Method.

3. Basic design of the staggered array undulator

Fig. 1 shows one period of the staggered array undulator that consists of half disks of ferromagnet and aluminum, where design parameters of the undulator are listed in Table 1. Since 2D calculation predicts a high longitudinal uniformity of undulator fields [6], the magnetic field is assumed to be periodic in this study for saving computational efforts in 3D calculations. Measured B–H and μ–H characteristics of an iron piece are taken into account.

Fig. 2(a) shows the calculated undulator field in a half period on the x–z plane ($y = 0$ mm) with a vacuum solenoid field of $B_c = 0.1$ T. The undulator field B_y is seen to decay near the edges of ferromagnet pieces ($x = \pm 30$ mm). For a higher

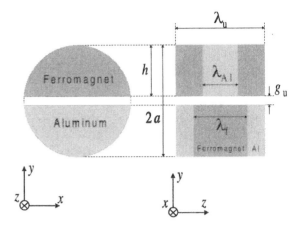

Fig. 1. Schematics of a period of the undulator.

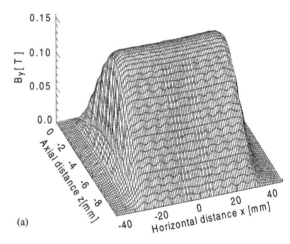

(a)

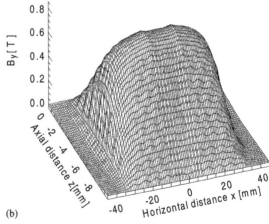

(b)

Fig. 2. Undulator fields B_y's for the basic design on x–z plane ($y = 0$ mm) for vacuum solenoid fields of (a) $B_c = 0.1$ T, and (b) $B_c = 1.0$ T.

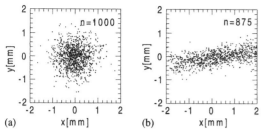

Fig. 3. Beam profiles at (a) the entrance, and (b) the exit of the undulator of the basic design.

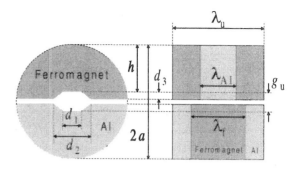

Fig. 4. Schematics of the refined undulator for an enhanced horizontal focusing force.

solenoid field of $B_c = 1.0\,T$, as shown in Fig. 2(b), the undulator field is found to be less uniform in the x-direction due to the nonlinearity of the ferromagnet pieces, while the uniformity is still good enough in the vicinity of the axis ($x = y = 0$) where the electrons pass through.

Electron trajectories are calculated to study basic characteristics, and beam profiles at the entrance and exit of the undulator are shown in Fig. 3. The initial distribution of electrons is assumed to be Gaussian. The averaged energy, energy spread, beam diameter, and normalized emittance are 4 MeV, 0.1%, 1 mm, and 35π mm mrad, respectively. The electrons are found to be focused vertically, and 87.5% of injected electrons can pass through the undulator to the exit. Most of the lost electrons escape from the calculation region through the $x = \pm 2\,mm$ planes, due to the absence of a horizontal focusing force. An additional horizontal focusing force seems to be needed to improve the electron beam confinement. To provide a horizontal focusing force, an x-directionally varying field is to be produced by changing the geometrical shape of the ferromagnet pieces, as described in the Section 4.

4. Refinements of geometrical shape of ferromagnet pieces

Fig. 4 shows a schematic view of one period of the undulator, where the designing parameters λ_u, λ_f, λ_{Al}, g_u, a, and b are the same as those listed in Table 1. For simplicity, parameters d_1 and d_2 are given to be 2 and 5 mm, and, thus, d_3 is the

Table 1
Reference parameters of the undulator

Undulator length (L_u)	0.5 m
Undulator period (λ_u)	20 mm
Gap length (g_u)	5 mm
Ratio of spacer length to λ_u (f)	0.4
Height of ferromagnet piece (h)	27.5 mm
Number of undulator periods (N_w)	25
Solenoid coil length (L_c)	1.0 m
Vacuum solenoid field (B_c)	0.1 T
Beam pipe diameter ($2a$)	60 mm

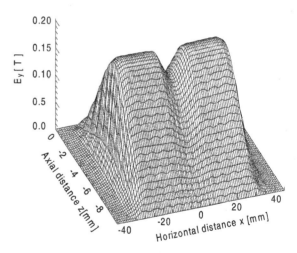

Fig. 5. Undulator field B_y for the refined design on x–z plane ($y = 0$ mm) for a vacuum solenoid field of $B_c = 0.1$ T.

variable parameter. With this tapered shape of the ferromagnet pieces, the undulator field tends to increase with increasing horizontal distance x from the undulator axis as shown in Fig. 5,

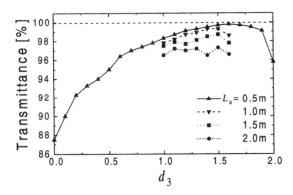

Fig. 6. Electron beam transmittances as functions of d_3 for various undulator lengths L_u's.

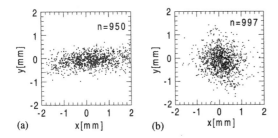

Fig. 7. Beam profiles at the exit of the refined undulator for (a) $d_3 = 0.5$ mm, and (b) $d_3 = 1.5$ mm ($L_u = 0.5$ m).

showing the field on x–z plane ($y = 0$) for $d_3 = 0.5$ mm.

Fig. 6 shows transmittances of injected electrons for various undulator lengths as functions of d_3. We found an optimal $d_3 = 1.5$ mm. Too large d_3's are the results in rather decreased transmittance, due to the horizontal overfocusing force.

Beam profiles for $d_3 = 0.5$, and 1.5 mm are shown in Fig. 7. In Fig. 7(b), for $d_3 = 1.5$ mm, because of an adequate focusing force, the transverse beam profile is found to be kept through the undulator to the exit, and as a consequence, the transmittance is remarkably enhanced up to 99.7%. Accordingly, with this improvement of the electron beam transmittance by 10%, the calculated intensity of spontaneous emission is found to increase by 10%, as well.

Even for longer undulators, the beam confinements are found to be good with the optimal $d_3 = 1.5$, as shown in Fig. 6. Beam transmittances are calculated as 99.3%, 98.7%, and 97.3%, respectively, for undulator lengths of 1.0, 1.5, and 2.0 m.

5. Conclusions

With the basic design without tapering of the ferromagnet pieces, the undulator fields are found to be highly uniform in the x-direction. Since there exist no focusing effects in the x-direction, the transmittance of a 4 MeV electron beam is found to be low, down to 87.5%, for an undulator length of 0.5 m.

Tapered ferromagnet pieces are found effective in inducing focusing forces to the electron beam by providing an x-directional varying field. With an adequate focusing force induced, a high transmittance of up to 99.7% can be achieved for a 4 MeV electron beam, which is very encouraging compared to 87.5% without taper. As a consequence, spontaneous emission intensity is also found to increase by 10%. Further, for a much longer undulator of 2 m, the beam confinement is found still high up to 97.3%.

References

[1] Y.C. Huang, et al., Nucl. Instr. and Meth. A 318 (1992) 765.
[2] Y.C. Huang, et al., IEEE J. Quant. Elect. 30 (1994) 1289.
[3] S. Shimada, et al., Proceedings of the Third Asian Symposium on Free Electron Lasers and Fifth Symposium on FEL Applications, 1997, p. 239.
[4] M. Ohnishi, et al., Nucl. Instr. and Meth. A 407 (1998) 434.
[5] J. Kitagaki, et al., Proceedings of the 21st International Free Electron Laser Conference, Hamburg, Germany, August 26–28, 1999–2000, II–105.
[6] K. Masuda, et al., A design study of a staggered array undulator for high uniformity of the undulator fields based on a 2-D code, Nucl. Instr. and Meth. A 475 (2001) 608, these proceedings.
[7] J. Simkin, et al., J. Num. Meth. Eng. 14 (1979) 423.
[8] T. Fujita, et al., J. Appl. Phys. 59 (1986) 2277.

ELSEVIER

Nuclear Instruments and Methods in Physics Research A 475 (2001) 617–624

NUCLEAR INSTRUMENTS & METHODS IN PHYSICS RESEARCH
Section A

www.elsevier.com/locate/nima

Transient absorption spectroscopy in biology using the Super-ACO storage ring FEL and the synchrotron radiation combination

Eric Renault[a,b,*], Laurent Nahon[a,b], David Garzella[a,b], Daniele Nutarelli[a,b], Giovanni De Ninno[a,b], Matthias Hirsch[a,b], Marie Emmanuelle Couprie[a,b]

[a] *Laboratoire pour l'Utilisation du Rayonnemment Electromagnétique, Bâtiment 209 D, BP 34 Centre Universitaire Paris-Sud, 91 898 Orsay Cedex, France*
[b] *SPAM/DRECAM/DSM Bâtiment 522, CE de Saclay, Gif sur Yvette, France*

Abstract

The Super-ACO storage ring FEL, covering the UV range down to 300 nm with a high average power (300 mW at 350 nm) together with a high stability and long lifetime, is a unique tool for the performance of users applications. We present here the first pump-probe two color experiments on biological species using a storage ring FEL coupled to the synchrotron radiation. The intense UV pulse of the Super-ACO FEL is used to prepare a high initial concentration of chromophores in their first singlet electronic excited state. The nearby bending magnet synchrotron radiation provides, on the other hand a pulsed, white light continuum (UV–IR), naturally synchronized with the FEL pulses and used to probe the photochemical subsequent events and the associated transient species. We have demonstrated the feasibility with a dye molecule (POPOP) observing a two-color effect, signature of excited state absorption and a temporal signature with Acridine. Applications on various chromophores of biological interest are carried out, such as the time-resolved absorption study of the first excited state of Acridine. © 2001 Elsevier Science B.V. All rights reserved.

1. Introduction

Synchrotron Radiation (SR) is an intense and broadly tunable light source covering the entire wavelengths from X-rays to IR. In addition to the broad tunability, bending magnet radiation shows other interesting properties such as its brilliance, its small source dimension, its polarization and its

time structure. It has a very useful time structure for time-resolved spectroscopies, with sub-ns pulses separated by a perfect dark period, repeated with high stabilities at MHz frequencies. The combination of photon sources of different and complementary characteristics represents a very powerful research technique for various scientific areas. A novel approach to two-photon spectroscopy is based on the combined use of the naturally synchronized Storage Ring Free Electron Laser (SRFEL) and synchrotron radiation to study the electronic states of various and biological interest chromophores. A UV-SRFEL, such as the one implemented on the Super-ACO storage

---

*Corresponding author. Present address: Laboratoire de Spectrochimie, Faculte' des Sciences et des Techniques, Universite' de Nantes, 2 Rue de la Houssinière, BP 92208, 44 322 Nantes Cedex 3, France. Tel.: +33-2-51-12-54-28; fax: +33-2-51-12-54-12.

E-mail address: renault@chimie.univ.nantes.fr (E. Renault).

ring, is a unique coherent, tunable and intense UV source, naturally synchronized with the SR, in a one-to-one shot ratio, leading to the possibility of very interesting pump-probe schemes [1,2]. The first experiment of time-resolved fluorescence using a UV-SRFEL was performed with the single-photon technique on a biological chromophore: reduced nicotinamide adenine coenzyme (NADH) [3], while the potential of UV-SRFELs combined with the SR has been reported in surface science at Super-ACO [4,5].

Since the pioneering work in 1980, only a few attempts have been made at pump-probe experiments of molecules using SR and a laser: photoelectron spectroscopy of atomic iodine produced from I_2 [6], photoelectron spectroscopy of N_2 and HCN produced from s-tetrazine [7], and photoionization of atomic iodine produced from CH_3I [8]. These previous experiments were performed using the visible light laser. The shorter wavelength region has recently been approached with the development of solid-state mode-locked lasers, in particular, by Ti : sapphire lasers that oscillate at a high repetition rate compatible with the frequency of an ordinary synchrotron radiation facility as UVSOR [9] and Brookhaven National Laboratory [10]. Presently, the use of a laser combined with the SR in biological sciences is only being developed at the ESRF synchrotron facility for the study of the structure of macromolecule in excited state [11].

Transient absorption spectroscopy is receiving a lot of interest from a wide community of scientists. From the beginning of the development of microsecond flash-photolysis, biologists have investigated the chemistry of exogenous or endogenous molecules of biological interest at different time scales [12–14]. With our SRFEL + SR set-up which presents a natural synchronization, a small jitter (5 ps) and a high stability (around 1%), we can cover the sub-nanosecond range up to 120 ns. This intermediate range appears already quite complementary to the currently used femtosecond set-up involving white continuum generation on the one hand, and to the slower time scales cover by laser-flash photolysis set-ups on the other. The scientific program we want to cover is diverse: first, we will look at antitumor drugs such as a variety

of acridines and porphyrines derivatives [15,16]. They will be studied in solution alone or in interaction with DNA targets. In a second step, we will look at indoles photophysics, in particular, the process of internal proton transfer in the tryptophan zwitterion [17].

Before launching a scientific program, we have demonstrated the feasibility of a transient absorption in terms of flux. The purpose of this paper is to show the time-resolved experiment on an organic molecule: acridine. In the next sections, we will present the UV-SRFEL source, the pump-probe principle and the description of the experimental technique. Then, the differential transient absorption signal, concerning the first single excited state of the acridine will be presented.

2. The Super-ACO storage ring FEL source

The advantage of the FEL over the conventional laser is that it is tunable and powerful over a broader range of wavelengths. The gain medium of an FEL is a high-energy electron beam passing through a transverse sinusoidal permanent magnetic field of an undulator of period λ_0 and peak magnetic field B_0. By ensuring a proper synchronization and transverse overlap between the optical pulses and the electron bunches, the electron/light interaction induces a micro-bunching of the electron bunch. When the radiation is kept in an optical resonator whose length is a submultiple of the recurrence period of the electrons, the stored spontaneous emission becomes more coherent, leading to the laser effect. The first harmonic of the undulator radiation gives the laser wavelength

$$\lambda_{\text{laser}} = \frac{\lambda_0}{2\gamma^2}\left(1 + \frac{\kappa^2}{2}\right) \quad (1)$$

with the deflection parameter: $K = 0.94\ \lambda_0$ (cm) B_0 (T). FELs are easily tuneable by a modification of K, or an energy change (see Eq. (1)). The output power is extracted from the cavity by transmission (about 0.1%) through the dielectric multilayers mirrors and is then carried out towards the application set-up by simple high reflecting plane mirrors.

Table 1
Main characteristics of the Super-ACO FEL

Parameters	Values
Energy range (MeV)	600–800
Lasing current range within two bunches (mA)	100–20
Spectral range (nm)	630–300
Tunability (nm)	10–20
Relative spectral width (%)	10^{-4}
Maximum output power at 350 nm (mW)	300
Pulse width FWHM (ps)	15–60
Repetition rate (MHz)	8.32
Lasing duration (h)	3–9
FEL/SR jitter (ps)	5 ps
Intensity fluctuation (%)	1
Polarization	Linear horizontal
Dominant mode	TEM_{00}

The main Super-ACO FEL characteristics are listed in Table 1.

3. Principle of the pump-probe experiment

3.1. Conceptual background

The principle of the pump-probe experiment resides in a powerful optical excitation to carry a fraction of the molecules in the excited state, which is probed by a second white/tunable light source. The transient absorption detection is a powerful method for the study of relaxation dynamics and the direct identification of the associated excited states and/or transient species via their spectral signature. Different light sources are employed to excite molecules to their excited states. Intense steady-state sources have been a popular excitation source to the triplet state for samples in glasses; long triplet state lifetimes allow large steady-state concentrations of triplets to build up. Conventional flash lamps have provided most of the triplet–triplet absorption spectra recorded up to the 1970s. Their intense and short pulses make them ideal for studying spectra and kinetics on the microsecond timescale, which is in the range of the lifetimes of many triplets in fluid solution. In order to study shorter lifetimes such as those of the

first singlet excited states or those of shorter lived triplets, it is convenient to use Q-switched lasers to get into the nanosecond time range or mode-locked lasers into the picosecond time range.

When an organic molecule, M, is excited it is most likely to go to an excited singlet state,

$$^{1}M + hv \rightarrow {}^{1}M^{*}, \quad k_{ex} \tag{2}$$

where k_{ex} is the intensity dependent excitation rate. The excited singlet state, $^{1}M^{*}$, will then decay back via several independent processes, again both throughout intramolecular and intermolecular pathways. Intramolecularly it can decay by internal conversion, with rate constant k_{ic}, intersystem crossing, k_{isc}, or fluorescence, k_{f}. These decays channels are represented by the following three processes:

$$^{1}M^{*} \rightarrow {}^{1}M + \text{heat}, \quad k_{ic} \tag{3}$$

$$^{1}M^{*} \rightarrow {}^{3}M^{*} + \text{heat} \quad k_{isc} \tag{4}$$

$$^{1}M^{*} \rightarrow {}^{1}M + hv \quad k_{f}. \tag{5}$$

In the usual kinetic scheme these processes are independent, and the singlet state decays exponentially with a rate constant,

$$k_{s} = k_{ic} + k_{isc} + k_{f} \tag{6}$$

which is equal to the inverse of the measured fluorescence lifetime. Triplet states are thus generated directly by the intersystem crossing process, Eq. (4).

Schematically, these elementary processes may be summarized in the energy level diagram displayed in Fig. 1.

Depending on the frequency of the incident radiation, the initial state achieved in the absorption event, Eq. (2), may lie in the vibrational manifold associated with the first electronically excited state (S_1) or that of a higher singlet (S_n). Rapid radiation(t)less deactivation (such as internal conversion and vibrational relaxation) leads to population of the lowest excited singlet state S_1 (Kasha's Rule) in a very short time (rate constant 10^{11}–10^{13} s^{-1}). Nevertheless, in exceptional cases, a radiative emission from S_2 has been observed. These processes are in competition with intersystem crossing from the higher excited singlet states,

Energy

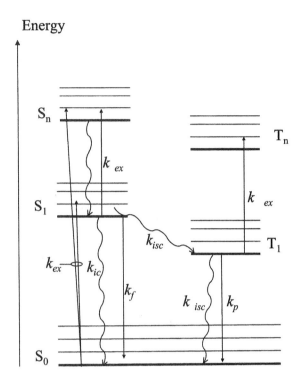

Fig. 1. Electronic transitions between the different electronic energy levels.

but this competition is generally inefficient because of the high rates of internal conversion.

Energy transfer from the singlet manifold (Eq. (4)) is of course, spin-forbidden, but the presence of several small intra- and inter-molecular interactions, serves to make the process weakly allowed. The triplet state, once formed, can undergo similar processes to those of the singlet state such as the reverse intersystem crossing , k'_{isc}, and phosphorescence k_p as shown in Fig. 1.

$$^3M^* \rightarrow {}^1M + \text{heat}, \quad k'_{isc} \tag{7}$$

$$^3M^* \rightarrow {}^1M + hv, \quad k_p. \tag{8}$$

The intramolecular decay is usually observed to be exponential with a constant rate,

$$k_T = k'_{isc} + k_p. \tag{9}$$

Phosphorescence is almost always absent in fluid media because of collision-induced intersystem crossing.

In the presence of radiation of a suitable frequency, the relaxed singlet $^1M^*$ or triplet excited state $^3M^*$, can again absorb photons , thus populating higher excited states with the following characteristics probabilities:

$$^1M^* \rightarrow hv \rightarrow {}^1M^{**}, \quad k'_{ex} \tag{10}$$

$$^3M^* + hv \rightarrow {}^3M^{**} \quad k''_{ex}. \tag{11}$$

These processes, then, are singlet–singlet absorption, $S_n \leftarrow S_1$, or triplet-triplet absorption, $T_n \leftarrow T_1$.

3.2. Experimental technique

Experimental setups for the detection of transient absorption spectra differ greatly, but there are at least two elements common to all of these experiments. First of all, every experiment has a "pump" source to produce excited species. Second, each experiment has a "probe" source to monitor their absorbance.

Fig. 2 shows the technique used to synchronize the FEL to the SR pulses, as well as the method used to diagnose in situ the spatial overlap between pulses. In this system, the Super-ACO storage ring FEL is used as a pump source at an 8.32 MHz repetition rate. A Pockels cell triggered by an electrode pick-up can be employed to reduce the repetition rate if desired. The UV-SRFEL pump was variably delayed by a computer-controlled translation stage, which has a total delay range of 8 ns owing to a reflecting corner-cube moving on a 1.2 m long translation stage (range). In our set-up the accessible delay ranges from $+4.5$ to -3.5 ns with the following convention: a positive delay corresponds to a situation where the UV-SRFEL pump pulse impinges onto the sample before the probe pulse. The white light emitted by a bending magnet, via the SB5 beamline is extracted to generate a white probe pulse in the 400–700 nm range. The pump beam is focused onto the sample with a quartz lens ($f \approx 25$ cm) whereas the probe beam is focused with an achromat lens ($f \approx 10$ cm). Pump and probe beams cross at a 7° angle in the sample which flows through a 1 mm thick quartz cell. The entire white continuum pulse was sent to a $f = 75$ cm imaging spectrometer (Princeton Instruments, Inc.

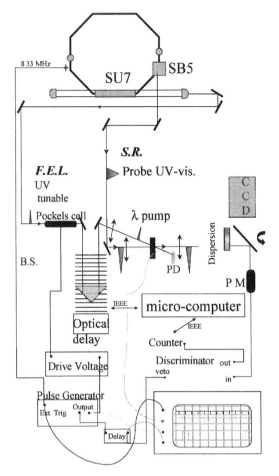

Fig. 2. Experimental arrangement for transient absorption measurements (SU7: undulator; SB5: bending magnet; PD: photodiode; PM: photomultiplier; B.S.: bunch signal).

Model: SpectraPro-750 Monochromator) with three gratings (1200, 300 and 150 lines/mm) and detected by a thermoelectrically cooled CCD camera (TE/CCD-1752-PF/UV, Chip size: 1752×532) in the steady-state operation or by a

photomultiplier (Hamamatsu RS928) in the time-resolved operation.

The control of the spatial overlap between the pulses is a critical part of the setup. To monitor this, we mounted the quartz lens, which is used to focus the pump beam onto the sample, on an X–Y–Z manipulator. The pump beam can then be moved and focused precisely in the region where the probe beam intersects the sample. The interaction region where the two photon beams intersect is optimized with the signal obtained by the CCD camera in the steady-state operation.

The time-resolved experiment is performed at a lower repetition rate, i.e. 83.2 kHz, than the natural one (8.32 MHz). This operation is achieved via the drive voltage of the Pockels cell controlled by a pulse generator. This latter is externally triggered by a signal generated by bunches (8.32 MHz) passing through a pickup electrode. The same pulse generator is used to control the veto of the discriminator of the pump signal obtained by the photomultiplier. (A) Homemade software records the signals of the counter versus the delay time obtained by the controlled translation stage.

3.3. Chemicals

POPOP (1,4-Bis(5-phenyloxazol-2-yl)benzene,2,2′-p-phenylenebis(5-phenyloxazole)) was provided by Exciton Corp, and Acridine was from Prolabo (see Fig. 3). Both are used without further purification. The ethanol used was of spectroscopic grade and obtained from Prolabo.

Absorption spectra were recorded using a Cary 210 (Varian Inc.) spectrophotometer. Typically, a 0.1 mM giving an absorbance of about 0.5 at 350 nm and an absorbance of about 1 at 363 nm in the 0.1 cm path of a quartz cuvette, respectively.

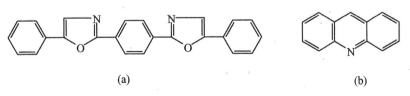

(a)   (b)

Fig. 3. Chemical structure of (a) POPOP and of (b) Acridine.

4. Results and discussion

4.1. Steady-state operation

The absolute intensity of a differential transient absorption spectrum $\Delta A(\lambda, \mathbf{t})$ is directly proportional to the population of the molecule in an excited state. In a first approximation, if we assume that the pump beam interacts only via the transition $S_0 \to S_1$, the saturation fluence is

$$\mathbf{F}_s = 1/\sigma_{a0}(\lambda) \qquad (12)$$

where $\sigma_{a0}(\lambda)$ is the absorption cross-section of the ground state in cm^2 at the wavelength λ. It is necessary to have a pulse fluence of the pump of the same order as the saturation fluence to excite a large fraction of the molecules. The choice of POPOP as a test sample has been determined by the high absorption cross-section at 350 nm ($\sigma_{a0}(350\,nm) = 1.54 \times 10^{-15}\,cm^2$), by the short fluorescence lifetime and by the high quantum yield of fluorescence ($\iota_f = 1.2\,ns$ and $\phi_f = 0.93$). In this case, the saturation fluence equals $\mathbf{F}_s = 6.5 \times 10^{15}$ photons cm^{-2}. The output power of the FEL is limited onto the sample to a maximum of 100 mW, which corresponds to an energy of 12 nJ/pulse (i.e. 2×10^{10} photons/pulse). The pump beam is focused onto the sample within a diameter of $\sim 20\,\mu m$, leading to the possibility of a fluence approximately equal to the saturation fluence.

In Fig. 4(a), are displayed the transmission signals recorded continuously during 5 s with or without the pump (SRFEL) and/or the probe (SR) beams present for a delay time of $+1.5\,ns$ in a quasi-CW operation: the dark current of the CCD camera is acquired without the pump and the probe beams. The emission of fluorescence spectrum of POPOP is obtained with only the pump beam (SRFEL) present. With only the probe beam (SR) present, the resultant signal looks like the transmission through the quartz cell which is similar to a stationary absorption spectrum. When both the pump and the probe beams are present, we observe a decrease of the transmission signal in the red domain (550–600 nm) as a signature of the spatial overlap of the two beams and of the formation of an excited state. This shows the

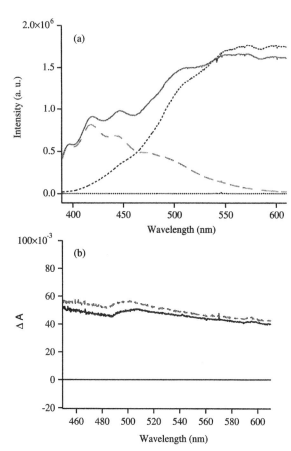

Fig. 4. (a) Transmission signals measured without FEL and SR (dotted line) present, with FEL and without SR (dashed line) present, without FEL and with SR (small dashed line) present and with FEL and SR (solid line) present on 116 μM of POPOP solution in ethanol and a delay time of $+1.5\,ns$ in a quasi-CW operation. (b) Transient differential absorption spectra (delay time: $+1.5\,ns$, solid line), (delay time: $-1.5\,ns$, dashed line) in a quasi-CW operation.

feasibility of the experiment in terms of flux, since it corresponds to a two color effect difference of the sum of the transmission signals for each beam (i.e. bad spatial overlap).

The transient differential absorption spectrum is defined as

$$\Delta A = A_p - A_{nop} \qquad (13)$$

where A_p and A_{nop} are the absorbance in the presence and in the absence of the pump, respectively.

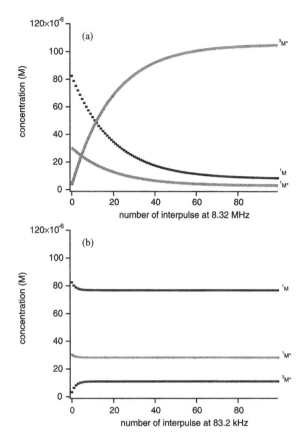

Fig. 5. Evolution of the dependence of the ground state concentration (^1M), of the first singlet excited state concentration ($^1M*$) and of the first triplet excited state ($^3M*$) concentration simulated at +1.5 ns with a pump/probe cycle of (a) 8.32 MHz and (b) 83.2 kHz.

The transient differential absorption spectra determined from signals recorded for two different time delays are shown in Fig. 4(b). It appears clearly that these two very similar curves can be essentially assigned to a long lifetime state from the first triplet state. In order to estimate the contribution of the state distribution of POPOP molecules we have simulated the evolution of the population in a steady-state mode [18] with a three-state model of the photoexcitation–deactivation process developed by Carmichael and Hug [19] (see Fig. 5(a)). This simulation confirms that the transient differential absorption spectra are similar for the two time delays because the long lifetime of the triplet state is in the microsecond range. This long lifetime leads to an accumulation

of molecules in the triplet state via intersystem crossing between two pump/probe cycles.

4.2. Time-resolved operation

It appears clearly in the previous part that the repetition rate of 8.32 MHz is too high and that the transient differential spectra can be essentially assigned to the first triplet state which is built up by an accumulation of molecules in this state via the intersystem crossing between each pump/probe cycle. In Fig. 5(b), is displayed the simulation of the three-state model of the photoexcitation–deactivation process with a repetition rate of the pump/probe cycle reduced by 100. (This is employed for the time-resolved mode.)

In the simplest situation, the pump/probe signal observed is proportional to the population which is not in the ground state. Thus, if the ground state is the only absorbing state, then the pump-probe observable is given by

$$S(t) = A \langle P_{ex}(t) \rangle \tag{14}$$

where A is a constant determined by experimental parameters such as the laser intensity and the probability that the system is found in the excited state $\langle P_{ex} \rangle$. If the probe wavelength is in the spectral region of the existing absorption of the ground state or of the stimulated emission, a bleaching of the solution may be observed. This is the observed situation in our experiment which was performed on the Acridine. The change in transient absorption is shown in Fig. 6, displaying the decay of the transient at 430 nm. The lifetime of the first singlet excited state determined by convolution is $\tau = 1.1(\pm 0.2)$ ns.

Previously, Shapiro and Winn [20] have investigated the fluorescence of Acridine. They reported a dependence of the fluorescence lifetime on several parameters such as the solvent, the excitation and the emission wavelength. Their results for ethanol with a 355 nm excitation and a 396 nm excitation give fluorescence lifetimes of $375(\pm 50)$ and $817(\pm 80)$ ps, respectively. Our results for ethanol are consistent with their data.

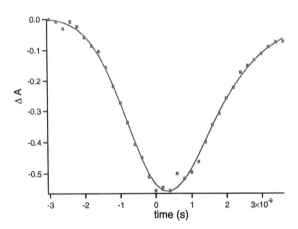

Fig. 6. Pump/probe data for the Acridine in ethanol. The smooth line represents the fit of the data.

5. Conclusion

We have demonstrated the feasibility of doing pump-probe experiment(s) with the flux available during steady-state operation since we have observed a decrease of the transmission signal in the red domain (550–600 nm) as a probe of the spatial overlap of the two beams and the formation of a long lifetime state. This state is the first triplet state which is built up by an energy accumulation of the pump via the intersystem crossing between each pump/probe cycle. The time-resolved operation has been proved here with the study of the lifetime of the first singlet excited state of Acridine. Experiments of this kind whose proof-of-principle has been demonstrated here, should be performed in the near future on different chromophores of biological interest.

Acknowledgements

We gratefully acknowledge the useful discussions with Fabienne Mérola (LURE). We would like to thank the technical staff of LURE for operating the Super-ACO storage ring and especially Thierry Guillou for their technical support. This work was partly supported by EC TMR Network Contract ERBFMRX-C980245. Support for E. Renault under EC RTD Contract No. ERBFMGE-CT98-0102 is acknowledged.

References

[1] L. Nahon, E. Renault, M.E. Couprie, F. Merola, P. Dumas, M. Marsi, I.A. Taleb, D. Nutarelli, R. Roux, M. Billardon, Nucl. Instr. and Meth. Phys. Res. A 429 (1999) 1.

[2] L. Nahon, E. Renault, M.E. Couprie, D. Nutarelli, D. Garzella, M. Billardon, G.L. Carr, G.P. Williams, P. Dumas, Proc. SPIE Int Soc Opt. Eng 3775 (1999) 145.

[3] M.E. Couprie, F. Merola, P. Tauc, D. Garzella, A. Delboulbe, T. Hara, M. Billardon, Rev. Sci. Instr. 65 (1994) 1485.

[4] M. Marsi, M.E. Couprie, L. Nahon, D. Garzella, T. Hara, R. Bakker, M. Billardon, A. Delboulbe, G. Indlekofer, A. Taleb Ibrahimi, Appl. Phys. Lett. 70 (1997) 895.

[5] M. Marsi, R. Belkhou, C. Grupp, G. Panaccione, I.A. Taleb, L. Nahon, D. Garzella, D. Nutarelli, E. Renault, R. Roux, M.E. Couprie, M. Billardon, Phys. Rev. B 61 (2000) R5070.

[6] L. Nahon, L. Duffy, P. Morin, F.F. Combet, J. Tremblay, M. Larzilliere, Phys. Rev. A 41 (1990) 4879.

[7] L. Nahon, P. Morin, M. Larzilliere, I. Nenner, J. Chem. Phys. 96 (1992) 3628.

[8] M. Mizutani, M. Tokeshi, A. Hiraya, K. Mitsuke, J. Synchrotron Radiat. 4 (1997) 6.

[9] M. Mizutani, H. Niikura, A. Hiraya, K. Mitsuke, J. Synchrotron Radiat. 5 (1998) 1069.

[10] G.L. Carr, R. Lobo, C.J. Hirschmugl, J. LaVeigne, D.H. Reitze, D.B. Tanner, Proc. SPIE Int. Soc. Opt. Eng. 3153 (1997) 80.

[11] M. Wulff, F. Schotte, G. Naylor, D. Bourgeois, K. Moffat, G. Mourou, Nucl. Instr. Meth. Phys. Res. A 398 (1997) 69.

[12] G. Porter, Science 160 (1968) 1299.

[13] R.M. Hochstrasser, C.K. Johnson, Laser Focus/Electro Opt. 21 (1985) 100.

[14] J.W. Petrich, J. Breton, J.L. Martin, A. Antonetti, Chem. Phys. Lett. 137 (1987) 369.

[15] W. Sajewicz, A. Dlugosz, J. Appl. Toxicol 20 (2000) 305.

[16] D. Wrobel, J. Goc, R.M. Ion, Photovoltaic and spectral properties of tetraphenyloporphyrin and metallotetraphenyloporphyrin dyes, J. Mol. Struct. 450 (1998) 239.

[17] D. Creed, Photochem. Photobiol. 39 (1984) 537.

[18] E. Renault, L. Nahon, D. Nutarelli, D. Garzella, F. Mérola, M.E. Couprie, Proc. SPIE Int. Soc. Opt. Eng. 3925 (2000) 29.

[19] I. Carmichael, G.L. Hug, J. Phys. Chem. 89 (1985) 4036.

[20] S.L. Shapiro, K.R. Winn, J. Chem. Phys. 73 (1980) 5958.

ELSEVIER

Nuclear Instruments and Methods in Physics Research A 475 (2001) 625–629

NUCLEAR
INSTRUMENTS
& METHODS
IN PHYSICS
RESEARCH
Section A

www.elsevier.com/locate/nima

TJNAF free electron laser damage studies

R.W. Thomson Jr.[a], L.R. Short[a], R.D. McGinnis[a], W.B. Colson[a,*], M.D. Shinn[b], J.F. Gubeli[b], K.C. Jordan[b], R.A. Hill[b], G.H. Biallas[b], R.L. Walker[b], G.R. Neil[b], S.V. Benson[b], B.C. Yunn[b]

[a] Physics Department, Naval Postgraduate School, 833 Dyer Road, Monterey, CA 93940, USA
[b] Free Electron Laser Department, Thomas Jefferson National Accelerator Facility, Newport News, VA 23606, USA

Abstract

Laser material damage experiments were conducted at the Thomas Jefferson National Accelerator Facility (TJNAF) free electron laser (FEL) user laboratory with an average power of 100 W and a power density of 10^4 W/cm^2. The FEL beam bombards the target with a steady stream of tens of millions of pulses per second each containing 50 MW of power in a short burst of ~1 ps. No conventional laser combines these characteristics, and no experiments have previously been done to explore the effects of the FEL pulse. The goal is to develop scaling laws to accurately describe large-scale damage from a MW FEL using small-scale experiments. © 2001 Elsevier Science B.V. All rights reserved.

PACS: 41.60.Cr

Keywords: Free-electron laser; Laser damage

1. Introduction

Laser damage experiments conducted at the Thomas Jefferson National Accelerator Facility (TJNAF) are the first experimental tests that study the damage from a short-pulsed laser at a high repetition rate with a few hundred Watts of average power. Scaling rules can be developed that will allow these small-scale damage experiments to represent the damage from a large, MW-scale laser.

*Corresponding author. Tel.: + 1-831-656-2765; fax: + 1-831-656-2834.
E-mail address: colson@nps.navy.mil (W.B. Colson).

2. TJNAF FEL design and parameters

TJNAF free electron laser (FEL) is the most powerful FEL ever operated. The parameters of the TJNAF FEL are $\bar{P} = 1.7$ kW, $\bar{I} = 5$ mA, $\gamma mc^2 = 48$ MeV, $\hat{I} = 60$ A, $r_b = 100\,\mu$m, $\tau = 0.4$ ps, PRF = 18.7/37.4/74.85 MHz and $\lambda = 3$–6 μm [1]. The requirements for a MW laser are $\bar{I} = 900$ mA, $\gamma mc^2 = 100$ MeV, $I_e/c = 1800$ pC, $\hat{I} = 600$ A, $r_b = 300\,\mu$m, $\tau = 3$ ps, PRF = 500 MHz, and $\lambda = 1\,\mu$m [2]. The significant differences are increases in the peak current by a factor of 10, the repetition rate by a factor of 7, the electron beam energy by a factor of 2, and the pulse length by a factor of 7.

In July 1999, the laser operated continuously at 1.7 kW average power. Since there is no MW-class FEL to perform full-scale experiments, scaling is the only way to determine the effectiveness of a high-power laser. Scaling laws would allow predictions of large area damage from small area experiments. To achieve a power density of $10 \, kW/cm^2$ [3], the 100 W FEL at TJNAF must use a spot size of $1 \, mm^2$, while a 1 kW laser uses a spot size of $10 \, mm^2$.

Scaling of the laser damage will only work, if the heat diffusion is independent of spot size. The characteristic thermal diffusion length represents the distance required for the temperature to drop to $1/e$ times its central value. So, the laser spot size must be much larger than the thermal diffusion length. Then heat will not diffuse outside of the laser spot and the spot will be heated effectively [4].

The pulse train of an FEL is different from other lasers and its interaction with matter at high average power has not been studied. FELs produce short, powerful pulses with a rapid repetition rate. The TJNAF FEL has a pulse length of $\tau = 0.4 \, ps$ and a repetition period of $T = 27 \, ns$. The peak power in each micropulse is 110 MW.

Comparing the TJNAF FEL to another short pulse laser is instructive. The Lawrence Livermore National Labs (LLNL) 1.053-μm Ti:sapphire CPA system [5] has a pulse length $\tau = 0.4 \, ps$, but a pulse repetition rate of only 10 Hz, so the peak power is 25 TW, and the average power is 10 W. Note that the LLNL laser has a much higher peak power than the TJNAF FEL, but the TJNAF FEL has more than one hundred times the average power because of its high duty cycle.

3. FEL experiments

Samples of various materials were irradiated by a laser beam of wavelength $\lambda = 4.825 \, \mu m$ through a calcium fluoride lens with a focal length of 300 mm. Two pulse repetition frequencies (PRF) were used; 74.85 MHz for the phenolic resin sample and 37.425 MHz for all other samples. The average power is 100–103 W. Since a lens focused the beam, the beam area decreased with distance along the direction of propagation to a minimum waist radius of 80 μm at the focal point [6].

An irradiance of $10 \, kW/cm^2$ occurs when the sample is placed 26 mm in front of the focus. As the laser burns into a sample, the intensity actually changes about 10% due to diffraction.

One of the targets irradiated was Slip-cast Fused Silica (SiO_2). A calculation of the thermal diffusion length associated with heating the sample to its melting temperature was performed with a result of $D = 0.02 \, mm$. Therefore, there is no significant thermal diffusion from the $1 \, mm^2$ beam used at TJNAF.

One of the runs using silica punched completely through the material. A digital picture of run two was taken through an optical microscope as shown in Fig. 1.

Although the beam diameter was only 1.1 mm, the melted portion at the surface of the sample measured 5 mm in diameter. The hole is tapered with the melted portion on the back of the sample measuring only 2 mm in diameter.

Examination of the hole from this run through an optical microscope reveals a 1-mm thick layer of melted, and rehardened, SiO_2 filling the hole at the back of the sample. It was clear from the video and the rear power meter that burn-through occurred in run two, but melted material solidified and sealed the hole at the back of the sample. A picture of the back of the target taken through a scanning electron microscope (SEM), that shows

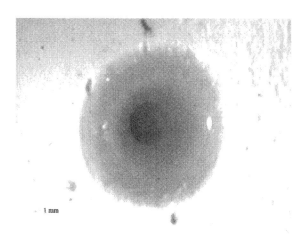

1 mm

Fig. 1. Close-up of damage to slip-cast fused silica in run 2.

the hole from run seven is fairly irregular with a great deal of debris.

The volume of the hole in run two is estimated at 5.6 mm³. The volume of the entire damaged region, including the melted and rehardened portion, is estimated to be 92 mm³. Based on the density of fused silica, $\rho = 2.2\,g/cm^2$, the amount of material removed was 0.012 g, and the amount of material damaged was 0.20 g. The heat energy deposited during run two is 9.7 kJ deposited the 110 second run. The heat of ablation is then 48 kJ/g.

Another material, polyimide fiberglass, measured 11.4 cm × 10.1 cm and 2 mm thick. The damaged area of the sample, after irradiation, is shown in Fig. 2. The sample was irradiated three times from left to right with the sample 26 mm in front of the focus of the beam. Only the first run achieved burn-through of the material, with the entry hole 3 mm in diameter and the exit hole 1.5 mm in diameter. All three holes show significant charring, which adds an additional term to the heat transport equation and impedes ablation. Investigation with an optical microscope reveals a raised lip of material around the face of the hole that does not appear on the fused silica sample and much more roughness. The charred region extends to a diameter of 8 mm for run one, 6.5 mm for run two, and 5.4 mm for run three. The lip height is

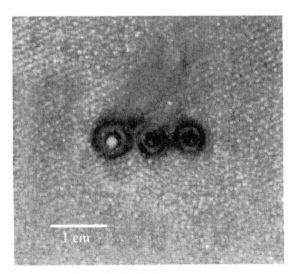

Fig. 2. Polyimide fiberglass target.

0.3 mm for run one, 0.1 mm for run two, and 0.05 mm for run three. These measurements indicate that as the dwell time increases, the radial extent of the damage area increases, and more material is deposited around the edge of the hole. There is no evidence of melted and rehardened material present in the holes as found with the fused silica indicating a different mechanism for ablation in the two samples.

An F2 Epoxy sample measured 10 cm × 11.5 cm and 1.5-mm thick including a 1.6-cm thick polyurethane foam backing. The sample was irradiated three times and, in each case, it appears that the F2 Epoxy was completely penetrated and the ablation of the foam backing had begun, but not completed. The videotape showed flames engulfing the upper portion of the sample and the black charred area extending to the edge of the sample. Significant charring was evident when the sample was viewed with the optical microscope, similar to the polyimide sample. There was also evidence of some melting, but not as much as occurred in the fused silica sample. The holes appear to be filled with the charred debris of the polyurethane backing, making hole depth measurements difficult and rendering penetration depth rates unreliable.

The damaged region extends to a diameter of 11.3 mm for run one, 7.5 mm for run two, and 5.2 mm for run three. There is a lip around each of the holes, but much smaller than the polyimide sample showed. The lip for run one was 0.05 mm and for runs two and three the lip was too small to measure with the optical microscope mechanism. These measurements indicate that as dwell time increases the radial extent of the damage area increases, and more material is deposited around the edge of the hole. When wind is added to the test, debris may be removed from the hole during irradiation.

Two phenolic samples were tested. Phenolic sample 1 is circular with a diameter of 32.5 mm. There is a 7.1 mm hole in the middle of it. It varies in depth from 1.6 mm to 3.2 mm. Sample 1 was irradiated three times with the sample rotated 90° counterclockwise after each run. Run number one did not achieve burn through because of operator intervention. After initial irradiation, the rear camera showed that the sample had ignited. The FEL operators quickly stopped the experiment only to find that the sample had slightly charred on the reverse side and not burned through.

For run numbers two and three, the samples were irradiated for a full 10 s. Though not documented, burn through times for each run were different. Run number two is similar to run number three. The crater is not perfectly cylindrical, since the back face has a slightly smaller area than the front. There is a crater lip that is built up from the damage, approximately 1 mm high. Also, a white crust is present, probably from the separation of the resin into its elements during heating. The white crusts and lips are also evident on the back side for runs two and three.

Phenolic sample 2 is circular with a diameter of 31 mm. There is a 7.1 mm hole in the middle of it. It varies in depth from 1.5 to 3.8 mm. Sample 2 was irradiated seven times and rotated counterclockwise after each.

All runs produced a lip around the entrance of the cavities ranging from 0.1 to 1 mm high. They also produced a white crust that must be some sort of elemental extract of the resin, probably separated during heating. Runs that did not burn through, one and four, created a crater shaped like an inverted cone with a rounded apex, probably due to the Gaussian nature of the FEL's laser beam. Inspection of the back side of run four, showed charring, a sign that the beam almost burned through. Runs that achieved burn through had a cylindrical crater, with the back edge slightly smaller than the front. Lips formed on the reverse side of the sample, just like the front side with the white crust around it.

Run number three is indicative of the runs with irradiances of 12 kW/cm^2, runs one through four. Run six is indicative of the damage produced by the runs with irradiances of 680 kW/cm^2, the other run was number five. Due to the high irradiance, the crater is smaller by a factor of two or three. The beams with more power density were able to punch through the material faster producing a much smaller lip as well. Note the existence of the same white crust but in lesser amounts.

Specific burn-through times were not noted for runs two, three, five and six. Future experiments should measure burn through times more accurately. The recession rate decreases nonlinearly as the exposure time increases. This decline in recession rate could be due to smoke and debris flying out of the crater while the beam is burning through the phenolic material. The smoke and debris impede the laser from doing damage. TJNAF is planning future experiments with wind, which might alleviate this problem.

Two pyroceram samples were irradiated at TJNAF in user laboratory number one. The first sample irradiated is pyroceram sample 1 and the other pyroceram sample 2. Pyroceram sample 1 is an irregular shape with an average depth of 1.4 mm. Fig. 3 is a picture of sample after 3 irradiations numbered from left to right. Pyroceram sample 2 is an irregular shape with an average depth of 1.4 mm.

4. Conclusions

This paper describes the first measurements of laser damage from the newly developed TJNAF FEL and the results could provide the basis for new directions for directed energy research.

The TJNAF FEL, which is capable of several hundred Watts of continuous average power, was used to simulate the damage from a MW-class laser by focusing the beam to a smaller spot size. The eventual goal is to develop

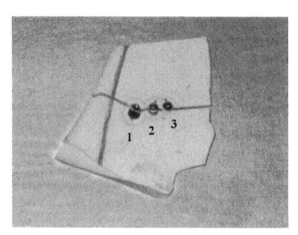

Fig. 3. Pyroceram target.

scaling rules that will reliably predict the damage of a large laser without having to bare the enormous cost of building the large laser first. The experimental data shows that the scaling concept with thermal diffusion calculations is promising. More detailed experiments varying wavelength, power, and spot size may be able to produce scaling laws, which would be invaluable, for future designers.

The extremely short sub-picosecond pulse length of the FEL beam is a result of the electron bunches. The TJNAF FEL has a unique pulse format with a rapid sequence of short, powerful pulses. The peak power in each pulse is about 100 MW lasting for only about one-half picosecond coming at a rate of 37/74 MHz. Other studies have shown that such short pulses may give as much as a factor of ten advantage in reduced fluence required to produce damage [7]. The experiments conducted for this paper began to collect data to show whether this advantage exists, but further experimentation will be required.

The TJNAF FEL is scheduled for an upgrade to 20 kW of power that will allow more flexibility in scaling experiments and further tests of scaling itself. Additional plans are for experiments, which include wind passing over the samples, weighing of the samples before, and after each run, new wavelengths, changing wavelength during irradiation, new pulse formats, and other sample materials. As experimental procedures are refined and the amount of data increases, more thorough analysis of the FEL beam and comparison to other lasers will become possible.

Acknowledgements

The authors are grateful for the support by the Naval Postgraduate School.

References

[1] G.R. Neil, et al., First operation of an FEL in same-cell energy recovery mode, Proceedings of the 21st International FEL Conference, Hamburg, Germany, August 1999, Nucl. Instr. and Meth. A 445 (2000) 192.

[2] D.W. Small, Interaction of laser beams with relativistic electrons, Doctoral Dissertation, Naval Postgraduate School, March 1997.

[3] J.R. Cook, J.R. Albertine, Surf. Warfare, 22 (5) (1997) 17.

[4] J.T. Schriempf, Response of materials to laser radiation: A short course, Naval Research Laboratory, Washington, DC, July 1974, p. 26.

[5] B.C. Stuart, et al., Phys. Rev. B 53 (4) (1996) 1749.

[6] M.D. Shinn, Personal letter, 4 March, 1999.

[7] B.C. Stuart, et al., J. Opt. Soc. Am. B 13 (2) (1996) 459.

ELSEVIER

Nuclear Instruments and Methods in Physics Research A 475 (2001) 630–634

NUCLEAR
INSTRUMENTS
& METHODS
IN PHYSICS
RESEARCH
Section A

www.elsevier.com/locate/nima

Primary experimental studies on mid-infrared FEL irradiation on dental substances at BFEL

Zhu Junbiao[a,*,1], Li Yonggui[a], Liu Nianqing[a], Zhang Guoqing[a],
Wang Minkai[a], Wu Gan[a], Yan Xuepin[a], Huang Yuying[a],
He Wei[a], Dong Yanmei[b], Gao Xuejun[b]

[a] *Institute of High Energy Physics, Academia Sinica, P.O.Box 2732, Beijing 100080, China*
[b] *Stomatological college, Peking University, Beijing 100080, China*

Abstract

A free electron laser (FEL) with its characteristics of wide wavelength tunability, ultrashort pulse time structure, and high peak power density is predominantly superior to all other conventional lasers in applications. Several experimental studies on mid-infrared FEL irradiation on dental enamel and dentine were performed at the Beijing FEL. Experimental aims were to investigate changes in the hardness, ratios of P to Ca and Cs before and after irradiation on samples with a characteristic absorption wavelength of 9.66 µm, in the colors of these sample surfaces after irradiation with different wavelengths around the peak wavelength. The time dependence of temperature of the dentine sample was measured with its ps pulse effects compared to that with a continuous CO_2 laser. FTIR absorption spectra in the range of 2.5–15.4 µm for samples of these hard dental substances and pure hydroxyapatite were first examined to decide their chemical components and absorption maximums. Primary experimental results will be presented. © 2001 Elsevier Science B.V. All rights reserved.

PACS: 41.60.Cr; 42.62.Be; 78.30.−j

Keywords: Free electron laser; Medical application; Infrared absorption spectra; Wavelength selectivity; Short pulse effects

1. Introduction

A conventional laser in dentistry applications possesses many disadvantages difficult to overcome due to its monochrome and continuous or long pulse width. In the early days, Goldman et al. performed the Ruby laser experiment of 694 nm wavelength on dental caries and found that there was seriously thermal side-effect [1]. Stern et al. obtained almost the same conclusion with a 10.6 µm CO_2 laser [2]. Longer laser pulses caused stronger thermal side-effects in samples. In 1989, Keller and Hibst introduced the 2.94 µm Er:YAG laser in dentistry applications due to its wavelength matching with the vibrational absorption peak of H_2O molecules in tooth [3]. Their experiments showed that a great enhancement in photon absorption leads to high efficiency in

---

*Corresponding author. Tel.: +86-1-62563339, fax: +86-1-62561604.

E-mail addresses: zhujb@ihepa.ac.cn, zhjbu@xinmail.com (Z. Junbiao).

[1] Work supported by National High Technology Laser Technology Field Foundation.

ablation. But the rupture of hydroxyapatite structure occurred due to the sudden explosive gasification of water. So, using a long pulse laser whose wavelength matched the characteristic absorption peak of water removal of tooth material was not an ideal choice. Niemz et al. applied a 2.12 μm Ho : YAG laser to irradiate enamel and dentin and found deep slits up to 3 mm and melting phenomena in the sample. Shorter pulse laser therapy for teeth were used, such as 30 ps and 1.053 μm Nd : YLF laser, and 30 ps and 1.064 μm Nd : YAG laser. Microcracks due to these irradiations were examined by dye penetration about 20 μm, in good agreement with what has been done with mechanical tooth drills [4]. Therefore, microcracks due to ultrashort pulses are bearable in dentosurgical applications. Excimer lasers proved to have very small thermal effects, but their ablation rates were very small as well, and more seriously, there was potential danger of violet radiation damage. In addition, a greater increase in temperature (to 42–50°C) in pulp and odontoblast vessel in the pulp cavity caused by either tooth drills or longer pulse laser in surgical operation would strongly excite the patient and cause quite a majority of tissue to die [5].

A LINAC-based free electron laser (FEL) with its characteristics of wide wavelength tunability, ultrashort pulse, and high peak power is predominantly superior to all other conventional lasers in dentistry therapy applications. Comprehensive use of these three parameters must greatly increase its ablation efficiency and decrease patient's pain. Studies on FEL irradiation on tooth dentine by Ogino and Awazu et al. have proven the high efficiency of FEL photo-induced ablation and exact tunability to absorption maximum of tooth material (9.4 μm) in infrared region, and found that 9.4 μm-FEL irradiation caused the selective ablation of PO_4 ions and the reconstruction of disordered atom [6,7]. Ablations in teeth with FEL around the absorption peak of hydroxyapatite (9.5 μm) and between 6.0 and 7.5 μm were experimentally performed in FELIX and Vanderbilt FELC [8]. These investigations showed high ablation efficiency and lower threshold when tuned to the absorption max. Still no thermal cracking in

dentine and small thermal cracking in enamel were observed. In this paper, we report our primary LINAC-based FEL experiments on tooth hard substances in vitro at the Beijing FEL (BFEL).

2. Material and methods

2.1. Tooth sample preparation

A tooth consists of pulp, dentine, and enamel. The enamel is the hardest tissue in the human body. It contains 95% hydroxyapatite (HAP), 4% water and 1% organic materials. Dentine is softer than enamel and consists of 70% HAP, 20% organic materials (mainly elastin and collagen) and 10% water. HAP's chemical formula is $Ca_{10}(PO_4)_6(OH)_2$. Its sub-structure consists of crystalline grains containing some impurities such as Cl^-, F^-, Na^+, K^+, and Mg^{2+} [5]. Experimental samples were from Stomatological College, Peking University.

First, in order to determine the characteristic absorption wavelength of enamel and dentine, samples used to examine FTIR absorption spectra were prepared. The human teeth in vitro were cut by a saw into slices of 1 mm thickness along their longitudinal directions, from which enamel and dentine debris were ground into powder and pressed into films of 0.5 mm thickness, respectively. For the purpose of comparison, the pure HAP (Sigma # H-0252) was pressed into a film of the same thickness.

Second, three facets of complete enamel of 0.5–1 mm thickness cut from the same tooth were prepared to observe changes in P/Ca before and after irradiation with chosen characteristic wavelength, one is for synchrotron radiation X-ray fluorescence (SRXRF) analysis, the other two for scanning electron microscope-X-ray fluorescence (SEM-XRF) analysis.

Third, in order to examine the temperature variation in the pulp cavity when being irradiated with 2–4 ps ultrashort FEL pulses, a complete tooth was horizontally cut off by the waist and the pulp cavity was uncovered, the upper part of the tooth was used for the sample. Its corona shape was kept naturally, the thickness from the tooth surface to the inner wall of pulp cavity was about 3 mm.

2.2. Mid-IR and ultrashort FEL source

The BFEL [9] is an S-band RF LINAC-based FEL with its tunable wavelength region of 7–18 μm whose double pulse time structure is shown in Fig. 1. Within each macropulse of 2–4 μs pulse width and 3 Hz repetition rate are more than 10,000 micropulses of 2–4 ps pulsewidth and 2,856 MHz repetition rate. About 1 μJ laser energy per micropulse corresponds to nearly 5 mJ laser photon energy per macropulse. The FEL beam

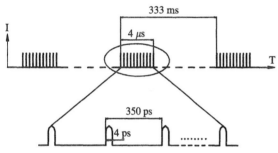

Fig. 1. Time structure of double laser pulses of the S-band RF-LINAC FEL at BFEL.

spot was about $100 \times 300 \, \mu m^2$ depending on slanting degree of beam on the sample surface. A real-time observation setup for FEL irradiation on samples was used. The chosen wavelength laser beam was guided to the surface of a sample that was fixed onto a 3-D moveable pedestal. The experimental status is real-time observed and recorded by a CCD camera mounted onto a stereo microscope and restored in a video tape recorder and computer to treat.

2.3. Mid-IR FT absorption spectra

FTIR absorption spectra of these samples are shown in Fig. 2, from which spectra of enamel and dentine are similar to each other in 8–12 μm with their strongest absorption peak near 9.6 μm due to tooth's main component, HAP, just like the pure HAP. However, there are some evident differences between 6 and 7 μm and in the vicinity of 3 μm, various tissue components and water contained in dentine are much richer than those in enamel. Different absorption peaks located at 6.0, 6.1, 6.4, 6.8 and 7.0 μm correspond, respectively, to amide I, water, amide II, and carbonated apatite. So, the main absorption peak around 9.6 μm is chosen to study mid-FEL-induced ablation and

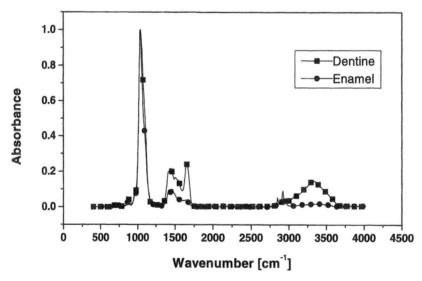

Fig. 2. FTIR absorption spectra of enamel and dentine.

temperature changes of pulp in this experiment. Other absorption peaks are chosen to investigate the ablation of other tooth components in the consecutive experiments. Experiments using the laser wavelength matched with 6.1 μm water absorption peak were not considered to avoid unbearable microcracks due to explosive gasification.

2.4. Irradiation wavelength choice

According to FTIR spectra and experiment on changes in element distribution of enamel before and after irradiation, the irradiation wavelengths and time for three samples were chosen as follows: (1) 6.65 μm and 15 min; (2) 9.64 μm and 10 min; (3) 10.6 μm and 10 min. After irradiation, the element distribution in the irradiated area of sample No.1 was measured and analyzed on the SRXRF station of the Beijing Electron Positron Collider (BEPC), then the peak areas in element spectra of the sample were calculated with the AXIL code. Relative variation in sample elements was obtained with the ratio of the resulting peak area to the control one. Sample Nos. 2 and 3 were scanned and measured with KYKY2800 SEM of 15 kV operating voltage and 75 μA filament current. Scanning areas were chosen to scan over and within the irradiated ones for sample Nos. 2 and 3, respectively. The scanning spot was about 1 μm, the scanning area was about $1.5 \times 0.5 \, mm^2$.

With the view to achieve high ablation efficiency and to minimize patient's pain due to thermal effects, the 9.66 μm FEL wavelength was chosen to ablate and heat the pulp samples. A thermopile was inserted into the pulp cavity and fixed. In the experiment, the defocusing laser beam was perpendicularly shined on the upper surface of the sample and heated it. The temperature rise was read out by the digital thermopile inserted in the pulp cavity.

3. Results and discussion

SRXRF scanning measurement results showed that there exist mainly elements P, Ca, and Sr, as well as sparsely Zn, Fe, etc. in enamel. After irradiation with a wavelength of 9.66 μm, the

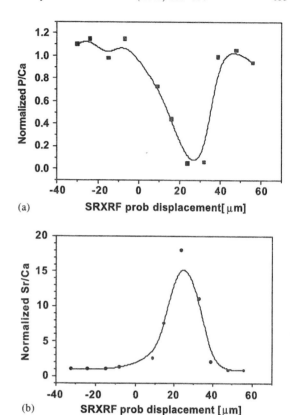

Fig. 3. Relation of normalized P/Ca (a) and Sr/Ca (b) to SRXRF probe displacement for the enamel sample No.1.

elements P, Ca, and Sr in the irradiated area were unevenly lost. Ratios of the peak area ratio P/Ca, Sr/Ca after and before irradiation on the enamel sample No. 1 with the SRXRF beam scanning length along the x-axis are shown in Figs. 3(a) and (b), respectively. From them, it follows that: (1) in the unexposed area, there was no variation in P/Ca or Sr/Ca; (2) P/Ca and Sr/Ca changed greatly as the probe beam scanned into the ablated region; (3) in the core area where the laser energy was centered, the ratios, the former fell whereas the latter rose to its maximum. The later increase in Sr/Ca should not be construed as an increase in Sr, but as a decrease in Ca. Furthermore, it seemed that loss of an element is related to its atomic number Z, i.e. the lighter element in tooth enamel is more easily lost, P < Ca < Sr.

The ratios of P and Ca peak areas in the fluorescent spectra of enamel samples No. 2

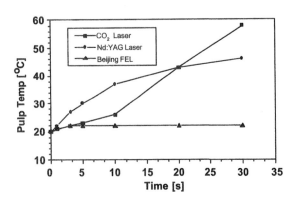

Fig. 4. Comparison of time dependence of temperature in the pulp cavity irradiated with (a) continuous CO_2 laser, (b) continuous Nd : YAG laser [10], and (c) ultrashort (2–4 ps) FEL at BFEL.

(9.64 μm) point-scanned with SEM have been listed: 0.65, 0.612, 0.646, 0.639, 0.663; and for sample No. 3 (10.6 μm), they are: 0.646, 0.656, 0.437, 0.233, 0.477, 0.639, 0.609. It is evident that irradiation of an enamel sample with a non-resonant wavelength does not lead to the loss in the element P, or a decrease in the P/Ca ratio. Conversely, irradiation at the resonant wavelength caused strong loss in the element P, especially in the irradiated area. In this area, a decrease in element P means an enhancement in the hardness of the tooth enamel, which would indicate recrystallization in enamel structure.

The time dependence of temperature in the pulp cavity irradiated with 2–4 ps ultrashort pulse and 9.66 μm characteristic wavelength FEL is shown in Fig. 4. For comparison, curves corresponding to continuous or long pulse CO_2 laser and Nd : YAG laser [10] are listed. From Fig. 4, it follows that temperature in the pulp cavity observed by the thermopile rose to its maximum from the room temperature by 2°C within the previous 3 min. Hereafter, this value is always kept after 30 min. The increment in temperature due to the CO_2 laser was 6°C, that due to the Nd : YAG laser 17°C, within 10 min. When the exposure time reached 30 min, the former climbed almost by 40°C, the latter by 26°C. It was theoretically and experimentally proven that: (1) if the tissue temperature is within 5°C above the normal body temperature of 37°C, there are no clear effects on the body tissue;

(2) if the temperatures is within the hypothermia regime (42–50°C) for several minutes, the majority of tissue appears necrotic; (3) if the temperature is above 50°C, enzyme activity in tissue is greatly decreased, leading to death for the majority of cells. Therefore, an FEL with short (a few ps) pulses seems promising to replace mechanical tooth drills for clinical use because of its high ablation efficiency, and negligible thermal side-effects, and mechanical vibration.

4. Conclusions

Initial experimental results on changes in element distribution in tooth enamel, especially P/Ca and Sr/Ca, after an ultrashort FEL exposure, and on the temperature rise of the pulp cavity were presented. Several conclusions could be derived: (1) loss of elements in the enamel in the exposed area shows P > Ca > Sr; (2) loss of elements irradiated near the resonant wavelengths is stronger than at other wavelengths; (3) a great loss of P indicates an increase in the hardness of the enamel, supporting recrystallization; (4) 2–4 ps ultrashort FEL pulses do not cause a large rise in temperature ($< 5°C$); therefore, it has the potential to replace mechanical tooth drills. Other experiments and measurements are being performed.

References

[1] L. Goldman, P. Hornby, P. Mayer, et al., Nature 203 (1964) 417.
[2] R.H. Stern, J. Vahl, R. Sognnaes, J. Dent. Res. 43 (1972) 873.
[3] U. Keller, R. Hibst, Laser Surg. Med. 9 (1989) 345.
[4] M.H. Niemz, Appl. Phys. B 58 (1994) 273.
[5] M.H. Niemz, Laser–Tissue Interactions, Springer, Berlin, Heidelberg, 1996.
[6] S. Ogino, K. Awazu, T. Tominmasu, SPIE Proc. 2922 (1996) 184.
[7] S. Ogino, K. Awazu, T. Tominmasu, SPIE Proc. 2973 (1997) 29.
[8] M. Ostertag, R. Walker, H. Weber, et al., SPIE Proc. 2672 (1996) 181.
[9] Xie Jialin, Zhuan Jiejia, Li Yonggui, et al., Nuclear Instr. and Meth. A 407 (1998) 356.
[10] M. Frentzen, H. Koort, Int. Dent. J. 40 (1990) 323.

ELSEVIER

Nuclear Instruments and Methods in Physics Research A 475 (2001) 635–639

NUCLEAR
INSTRUMENTS
& METHODS
IN PHYSICS
RESEARCH
Section A

www.elsevier.com/locate/nima

Selective bond breaking in amorphous hydrogenated silicon by using Duke FEL

D. Gracin[a,b], V. Borjanovic[a,c], B. Vlahovic[a,d],*, A. Sunda-Meya[a],
T.M. Patterson[a], J.M. Dutta[a], S. Hauger[e], I. Pinayev[e], M.E. Ware[f], D. Alexson[f],
R.J. Nemanich[f], B. von Roedern[g]

[a] Department of Physics, North Carolina Central University, Durham, NC 27707, USA
[b] Rudjer Boskovic Institute, Zagreb, Croatia
[c] Faculty of Electrical Engineering, Zagreb, Croatia
[d] Jefferson Laboratory, Newport News, VA 23606, USA
[e] Duke University FEL, Durham, NC 27708, USA
[f] North Carolina State University, Raleigh, NC, USA
[g] National Renewable Energy Laboratory, Golden, CO 80401, USA

Abstract

In order to study the possibility of influencing the phase containing predominantly Si–H bonds, while having minimal influence on the surrounding materials, samples of a-Si were exposed to Duke-FEL Mark III radiation. The wavelength of the radiation was selected to fit the absorption maximum of stretching vibrations of Si–H bonds (5 μm). By varying the wavelength in the vicinity of 5 μm, the illumination time and the power density, different types and degrees of structural ordering, of Si–H bonds and Si–Si bonds were obtained, and monitored by Raman spectroscopy. By increasing the energy density, at certain level the crystallization occurs. We were able to demonstrate a direct correlation between short and intermediate range ordering and the wavelength and intensity of the radiation. Using 5 μm at 10 kW/cm² leads to increase in structural disordering. However, increasing power to 60 kW/cm² improves both short and intermediate order in a-Si:H, as demonstrated by Raman spectroscopy. Further increasing power density by an order of magnitude results in crystallization of the sample. © 2001 Elsevier Science B.V. All rights reserved.

PACS: 61.43.Dq; 63.50.+x; 68.55.−a; 81.05.Gc

Keywords: a-Si:H; Structure; Ordering; FEL; Recrystallization; Raman spectroscopy

Hydrogen is crucial for improving the electrical and optical properties of a-Si:H. It not only saturates dangling bonds and decreases density of

defects states in the gap, but it also reduces the structural disorder [1]. However, long-range motion of hydrogen is implicated in light and carrier induced metastable degradation of a-Si:H electrical properties. It has been suggested that the degradation of a-Si:H by illumination, Stabler–Wronski effect (SW), may also be related to the role of hydrogen and its effects on the variations in

---

*Corresponding author. Department of Physics, North Carolina Central University, Durham, NC 27707, USA. Tel.: +1-919-530-7253; fax: +1-919-530-7472.

E-mail address: branko@jlab.org (B. Vlahovic).

0168-9002/01/$ - see front matter © 2001 Elsevier Science B.V. All rights reserved.
PII: S 0 1 6 8 - 9 0 0 2 (0 1) 0 1 5 7 8 - 9

the microscopic structure [2–4]. Experiments suggest that the SW effect is not only related to local breaking of "weak" Si–Si bonds [5], but also to changes beyond the short-range order (SRO, nearest neighbour [6,7]). Thus, the role of hydrogen in the determination of the network order, short range and in particular intermediate-range order (IRO, beyond nearest neighbour), become an important question. Structural changes and defect formations may be caused by photo induced breaking of Si–Si bonds just behind Si–H bonds [8].

In general, the short-range order of the Si network is mainly characterized by the angle and the length distortion between Si–Si bonds [9]. On the other hand, the intermediate order is proportional to the dihedral angle fluctuations in the silicon network. It is shown that distortion of the lengths of Si–Si bonds, the angles between Si–Si bonds as well as the variation in the microscopic structure of the silicon network are affected by the bonding configuration of hydrogen atoms (i.e. monohydride configuration, dihydride configuration, etc. [10]).

In this work, we report on the effect that the irradiation of a-Si : H sample at FEL frequency resonant with the Si–H bonds has on its SRO and IRO. We demonstrate that the direct excitation of the lattice vibration by an intense ultra short laser pulse with an appropriate wavelength can induce improvement of the SRO and IRO at temperatures well below the temperature of deposition, which implies that athermal processes are dominant.

We used samples of a-Si : H, deposited by hot wire deposition at NREL on glass and mono crystalline Si substrate, heated at temperature between 350°C and 420°C. The samples were exposed at Duke-FEL Mark III to radiation in the 4.7–5.2 μm range. Radiation energy per macro pulse was 1–3 mJ, macro pulses are at frequency 10 Hz and of duration 1.3 μs. Each macro pulse incorporates about 3700 micro pulses with duration of 1 ps, separated by about 350 ps.

The laser light was focused on the sample through a calcium fluoride lens on different areas of the sample to vary the power density of illumination. In the following, the power densities are average energies per macro pulse and cm². The

illumination time was between 15 and 20 min in the annealing experiments and several minutes for crystallization. The maximum possible temperature rise was estimated, according to calculations from Ref. [11], to be of the order of 20–100°C for 20–100 W/cm², e.g. 200–300°C below the deposition temperature.

The Raman spectra were taken by double grating spectrometer supplied with water cooled photo multiplier. As a source for the 514.5 nm excitation beam an Ar ion laser, vertically polarized with respect to the scattering plane, was used.

As can be seen from Fig. 1, after irradiation with 10 kW/cm² the intermediate range disorder in the material increases. The ratio of the areas under the TA peak and TO peaks (the I_{TA}/I_{TO} ratio) increase and the width of the TO vibration mode slightly increases. The TA peak (around 160 cm⁻¹) represents the IRO, and involves triads of atoms

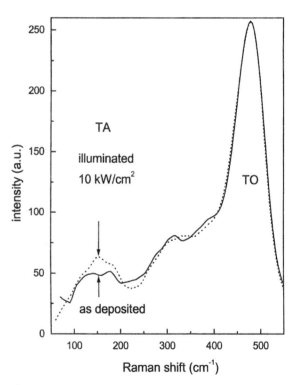

Fig. 1. Raman spectra of a-Si : H sample, as deposited and after 20 min illumination (5 μm and 10 kW/cm²). Spectra are normalised to the same intensity of TO peak.

representing bond bending and the dihedral angle fluctuations [12]. The transverse-optical (TO) vibration mode, peak (around $480\,cm^{-1}$) reflects the degree of the SRO. The lower the position of the TO peak of this spectral band, ω_{TO}, and the larger its width, Γ_{TO}, the lower the SRO [13]. Since the changes in the TO width and peak position are small, it can be concluded that irradiation with low intensity mostly decreased IRO without changing SRO.

After increasing the intensity to about 60 kW/ cm^2, the ratio of the I_{TA}/I_{TO} decreases, suggesting an increase in IRO, see Fig. 2.

Further increase of the irradiation intensity to 90 kW/cm^2 causes additional increase in the IRO. The improvement in IRO is accompanied by slightly narrowing the width of (TO) peak Γ_{TO}, suggesting better SRO. Since the position of the TO peak is not changed one can conclude that the strength, or length, of the Si–Si bonds is the same, so the SRO improvement is related

only to the reduction of the Si–Si bond angle distortion.

The changes in phonon related peaks are closely related to changes in that part of the spectra related to Si–H stretching vibrations. In Fig. 3, there are plots of the spectra of as deposited sample (full line), illuminated with 60 kW/cm^2 (dotted line) and 90 kW/cm^2 (dashed line), while in Fig. 4 are the corresponding spectra related to the Si–Si bonds. Again, the full line represents the as-deposited sample, dotted and dashed lines represent samples illuminated by 60 and 90 kW/cm^2, respectively. As can be seen, the larger the surface under the Si–H stretching peak, the lower the I_{TA}/I_{TO} ratio (the spectra are normalized to the I_{TO} peak and the TA peak in Fig. 4 is lower).

It is worth noting that the increase in the irradiation power intensity leads to an increase in the Si–H stretch mode at $2000\,cm^{-1}$ (see Fig. 3) and at the same time to the decrease in the ratio of the areas under the TA peak and TO peak. This

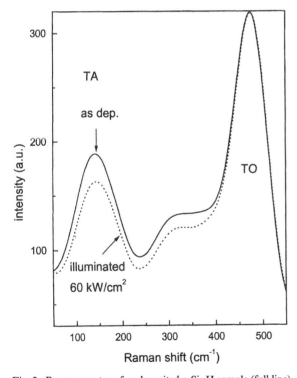

Fig. 2. Raman spectra of as deposited a-Si : H sample (full line) and illuminated for 20 min with 5 μm and 60 kW/cm^2 (dashed line). Spectra are normalised to the same intensity of TO peak.

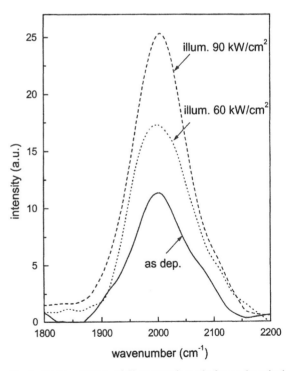

Fig. 3. Raman spectra of Si–H stretch mode for as deposited sample (full line) and the irradiated sample with 60 (dotted line) and 90 kW/cm^2 (dashed line).

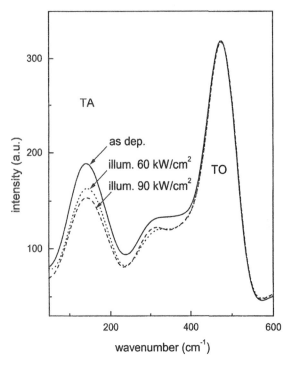

Fig. 4. Raman spectra of Si–Si related part for as deposited sample (full line) and the 20 min irradiated sample with 60 (dotted line) and 90 kW/cm² (dashed line).

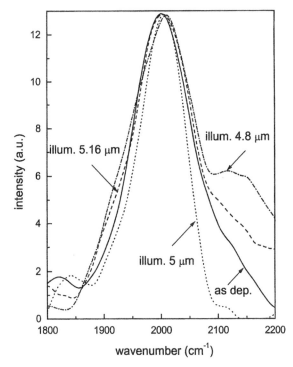

Fig. 5. The spectra of as deposited a-Si : H sample (full line) and illuminated with same energy (80 kW/cm²) and time (15 min) but different wavelengths (denoted on the graph).

suggests that the increase in IRO is connected to an increase in the number of Si–H bonds, which is proportional to the Si–H stretch mode peak. Somehow, irradiation by the Si–H resonant frequency leads to the transformation of the unbound hydrogen, or hydrogen in dihydride configuration, to the monohydride configuration, which leads to the increase in IRO.

We also explored the effect of the wavelength on the IRO and SRO. Fig. 5 shows a spectra of as-deposited sample (full line), and the samples illuminated for the same amount of time by the same power intensity (80 kW/cm²), but with different wavelengths (denoted on the graph). The spectra are normalized to the same peak value in order to illustrate the changes in the peak shape caused by illumination. It appears that only irradiation at the resonant wavelength of the Si–H bonds leads to a narrower peak. Irradiation under the same conditions with wavelengths of 4.8 and 5.16 μm produced wider peaks corresponding

to a Si–H stretch mode in the neighbourhood of 2000 cm⁻¹. This vibration band is a sensitive probe of the local structure, exhibiting a large degree of spectral broadening (~ 100 cm⁻¹) due to the disorder in the material [14].

Further increase of the irradiation intensity to the order of MW/cm² leads to a-Si : H re-crystallization. Fig. 6 shows a sample of a-Si : H before and after irradiation. One can see that irradiation produces the peak, which is characteristic for the poly crystalline Si. The typical peak position of crystallized material was between 516 and 518 cm⁻¹.

In summary, we demonstrated the possibility of changing Si–H bonding configuration at the temperatures well below the deposition temperature by using very low energy of FEL IR radiation. By proper choice of wavelength, in the vicinity of 5 μm, energy density and duration of illumination, it is possible to increase the Si–H bonds ordering by simultaneously increasing the Si–H bonds

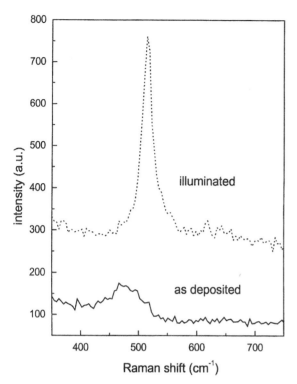

Fig. 6. Raman spectra of as deposited a-Si:H sample and the sample illuminated with 3×10^3 kW/cm^2 for 1 min.

concentration. Thus, establishing a direct correlation between concentration of bonded hydrogen and intermediate and short range structural ordering. Therefore, it can be concluded that irradiation of a-Si samples with low intensity (about 10 kW/cm^2) leads to decrease in the IRO of the material. Irradiation at intermediate intensity (100 kW/cm^2) and with the resonant wavelength of the Si–H bonds increases both IRO, (I_{TA}/I_{TO} lower), and SRO (Γ_{TO} narrower), and also increases the local order in the material (Γ_{2000} narrower).

Acknowledgements

The authors wish to thank A. Soldi for useful suggestions. This work was supported by the NREL subcontract No AAK-9-18675-03.

References

[1] S. Gupta, R.S. Katiyar, G. Morell, S.Z. Weisz, I. Balberg, Appl. Phys. Lett. 75 (1999) 2803.
[2] N. Maley, J.S. Lannin, Phys. Rev. B 36 (1987) 1146.
[3] P.A. Fedders, Y. Fu, D.A. Drabold, Phys. Rev. Lett. 68 (1992) 1888.
[4] P.V. Santos, N.M. Johnson, R.A. Street, Phys. Rev. Lett. 67 (1991) 2686.
[5] H. Fritzsche, Mater. Res. Soc. Symp. Proc. 467 (1997) 19.
[6] J. Fan, J. Kakalios, Phys. Rev. B 47 (1993) 1093.
[7] J.M. Gibson, M.M. Tracy, P.M. Voiles, H.C. Jin, J.R. Abelson, Appl. Phys. Lett. 73 (1998) 3093.
[8] A.H. Mahan, L.M. Gedvilas, J.D. Webb, J. Appl. Phys. 87 (2000) 1650.
[9] Y. Hishikawa, J. Appl. Phys. 62 (1987) 3150.
[10] K. Zellama, L. Chahed, P. Sladek, M.L. Theye, J.H. von Bardeleben, P.R. Cabarrocas, Phys. Rev. B 53 (1996) 3804.
[11] M. Lax, J. Appl. Phys. 8 (1977) 3919.
[12] M. Marinov, N. Zotov, Phys. Rev. B 55 (1997) 2938.
[13] R. Tsu, M.A. Paesler, D. Sayers, J. Non-Cryst. solids 114 (1989) 199.
[14] C.W. Rella, A.V. Akimov, A.F.G. van der Meer, J.I. Dijkhuis, Appl. Phys. Lett. 75 (1999) 2945.

ELSEVIER

Nuclear Instruments and Methods in Physics Research A 475 (2001) 640–644

**NUCLEAR
INSTRUMENTS
& METHODS
IN PHYSICS
RESEARCH**
Section A

www.elsevier.com/locate/nima

Thin-film deposition method assisted by mid-infrared-FEL

M. Yasumoto[a,*], T. Tomimasu[b], A. Ishizu[b], N. Tsubouchi[c],
K. Awazu[c], N. Umesaki[d]

[a] *Photonics Research Institute, AIST, AIST Tsukuba Central 2, Tsukuba, Ibaraki 305-8568, Japan*
[b] *Free Electron Laser Research Institute (FELI), c/o Osaka Works, Sumitomo Electric Industries, Ltd., 1-1-3 Simaya Konohana, Osaka 554-0024, Japan*
[c] *Institute of Free Electron Laser, Osaka University, 2-9-5 Tsuda-yamate, Hirakata, Osaka 573-0128, Japan*
[d] *Japan Synchrotron Radiation Research Institute, 1-1-1 Kouto, Mikazuki, Sayo-gun, Hygo 699-5198, Japan*

Abstract

We propose the novel application of the mid-infrared (MIR) FEL to the thin-film fabrication process. During the application, a substrate on which a thin film is being fabricated by a conventional method is simultaneously irradiated by the MIR FEL. The MIR FEL induces the fabricated molecules into the excited state of the stretching vibration energy, when the photon energy of the MIR FEL corresponds to one of the molecules. Therefore, the method can assist the thin-film fabrication quasi-independent of the substrate temperature. The method has the advantages of application on a temperature sensitive substrate and selective fabrication due to the tunable wavelength of the MIR FEL. In order to realize the method, we developed two thin film fabrication devices; an MIR FEL assisted RF sputtering device and an MIR FEL assisted laser ablation deposition device. For the method, the intensity of the assisted MIR FEL is an important problem. Thus the cross-section of the MIR FEL intensity profile is shown and the propagation discussed. © 2001 Elsevier Science B.V. All rights reserved.

PACS: 41.60.Cr; 42.62.−b; 41.85.Ew; 81.15.Fg

Keywords: Free-electron laser; Mid-infrared; Deposition; Thin-film

1. Introduction

Many applications in material processing using the FEL technique have been done at FELI (iFEL, Osaka Univ.) [1]. The FELI FEL covers the wavelength range of 273 nm to 40 μm using four FEL facilities (FEL-1–FEL-4) [2]. Especially, the mid-infrared (MIR) FEL of $5 \sim 15\,\mu m$ wavelength is most frequently used. The photon-energy of the MIR FEL can excite the stretching vibration energy of a specific molecule due to the energy coincidence. Therefore, the wavelength tunability of the MIR FEL has the possibility of producing a non-thermal process. We have been attempting to apply the mid-infrared (MIR) FEL to the thin-film fabrication process.

For this particular application, the MIR FEL is simultaneously irradiating a substrate on which a thin film is being fabricated by a conventional method. The process is considered a dynamic

*Corresponding author. Tel:. + 81-298-61-3425; Fax: + 81-298-61-5683.
E-mail address: m.yasumoto@aist.go.jp (M. Yasumoto).

material processing with the MIR FEL, while many MIR FEL applications previously performed such as SiC annealing [3] and diamond annealing [4] are classified as static material processing. Dynamic processing has the possibility to realize a new non-thermal process, unlike the thin film coating method using a conventional laser, which is limited by the thermal process [5], namely, that the laser energy is used for the heating of both the thin film and the substrate. Here the MIR FEL can assist the thin-film fabrication quasi-independently of the substrate temperature and assist in the fabrication. In this paper, we discuss the merits, devices, and assignment of the novel thin-film fabrication method.

2. Merits and problems of the method

This novel method has two advantageous compared to other fabrication methods using a conventional laser. First, the method is suitable for thin film fabrication on a temperature sensitive substrate such as a plastic film. The thin-film is generally fabricated on an adequately heated substrate for the film crystallization. Thus the conventional method does not work on a temperature sensitive substrate and temperature resistant materials must be used for the substrate. However, the proposed method has the possibility of fabricating a thin film via a non-thermal process and can be widely applied to many substrate materials including those that are temperature sensitive. This is because the MIR FEL energy can only be transferred to the vibration energy of the specific molecules fabricated on the substrate. The second advantage of this method is the wavelength-selectivity of the MIR FEL. The MIR FEL can possibly be used to control the fabricating of the thin film on the substrate [6]. The advantage is effective in the photo Chemical Vapor Deposition (CVD) process. The specific molecules are only excited and fabricated with specific wavelength tuning. Here the MIR-FEL can fabricate the specific molecules using its wavelength selectivity, although the rate of the photo-CVD generally depends on only the irradiation time.

However, there remains some questions about the method: (1) Heat stacking effect due to the MIR FEL irradiation. The average power of the FELI FEL is low on the tens mW level. Though during the early part of the process the fabricated molecules are only excited, the heat gradually transfers to the substrate under irradiation. (2) Do the MIR FEL and the molecules mainly react in the gas phase or the solid phase on the substrate? If the excitation of the molecules by the MIR FEL is occurred in the gas phase, we can measure the excited state of the molecule by using the laser induced fluorescence (LIF) technique.

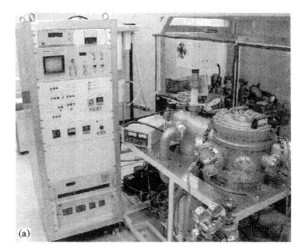

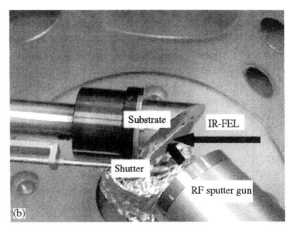

Fig. 1. Photograph of the MIR FEL assisted RF sputtering deposition device: (a) the overview and (b) the substrate in the chamber.

3. Developed devices

We developed two devices specifically for the novel method described above [7]. For both devices, the IR-FEL is simultaneously irradiated on the substrate during the thin-film fabrication process. One device is composed of the RF-sputtering source and the IR-FEL irradiation. The other is composed of the laser ablation deposition and the IR-FEL irradiation. This paper concentrates on the former device. Fig. 1 shows two photographs of (a) the overview and (b) the substrate of the former device. The device has a temperature controlled substrate, an FEL inlet window (ZnSe window) and an RF-sputter gun in a vacuum chamber, which is located under the FEL transport line in the user room at the FELI. Table 1 shows the specifications of the device. The IR-FEL irradiated through the inlet window is directed to the substrate facing the RF sputter gun shown in Fig. 1(b). The substrate can be tunable for the substrate temperature in order to study the MIR FEL assisted effect. Therefore, both the sputtering deposition and the IR FEL irradiation are simultaneously carried out on the substrate.

4. MIR FEL intensity profile propagation for assisted irradiation

The MIR FEL intensity is significantly important for the assisted irradiation without radiation

Table 1
Specifications of the MIR-FEL assisted RF-sputtered thin-film fabrication device

Base pressure	1×10^{-5} Pa
Substrate	Glass, Si (1 1 1), plastic film, etc.
Variable range of substrate temperature	Magnetron RF sputter gun (ULVAC)
RF power	500 W (max)
Sputter gas	Ar or Ar mixed gas
IR-FEL inlet window	ZnSe for $5 \sim 15 \, \mu m$
Distance from the target to the substrate	80 mm

Table 2
Optical parameters of the MIR FEL (FEL-1)

Wavelength (μm)	$4.6 \sim 15$
Micropulse duration (ps)	~ 3
Macropulse duration (μs)	18
Energy of the micropulse (μJ)	~ 30
Energy of the macropulse (mJ)	48

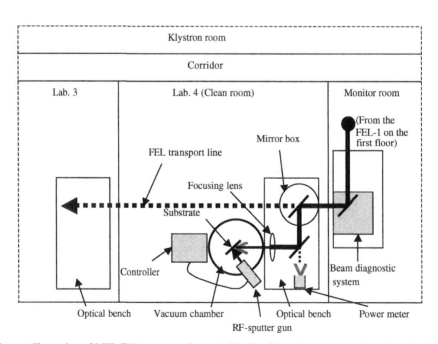

Fig. 2. Layout illustration of MIR FEL transport line and thin film fabrication device in Lab. 4 on the third floor.

damage. This is because at full power, the intense MIR FEL causes ablation on the substrate [8,9]. Conversely, if too weak, one cannot introduce the assisting effect during the processing. Table 2 indicates the optical parameters of the FEL-1. Therefore, one must set the FEL intensity to the correct level.

The MIR FEL is transported from the FEL facility located on the first floor to the user stations on the third floor at the FELI (see Fig. 2). The transport system is composed of a collimator and flat mirrors, and an approximate 15 mm-radius beam is transported to the user station. The irradiated intensity is dependent on the intensity profile of the MIR FEL transported to the

substrate. The intensity profiles were measured with a 2D Pyro-electric detector (Spiricon, Pyrocam 2 & LBA-200). The detector has 32×32 elements in a 2-inch area-sensor. Fig. 3 indicates the intensity profiles of the transported FEL at the wavelengths of (a) $5.0\,\mu m$ and (b) $9.5\,\mu m$ in Lab.4. The intensity profiles are perfectly dependent on the MIR FEL wavelength. Therefore, the assisted intensity also depended on the wavelength and the original beam at the irradiation point has to be estimated or measured.

The laser intensity is generally adjusted by defocusing the laser beam with a lens. Fig. 4 indicates the cross-section of the intensity profile of the $8.5\,\mu m$ FEL focused with the ZnSe lens

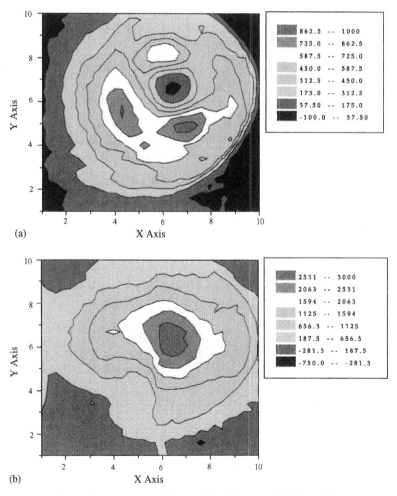

Fig. 3. FEL intensity profiles at the user station in Lab. 4, (a) $5.0\,\mu m$ and (b) $9.5\,\mu m$. The measured area is a 2-in.^{2}

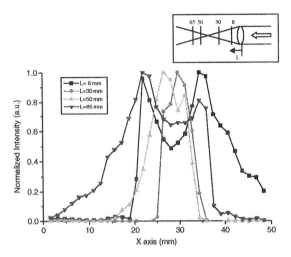

Fig. 4. Cross-section propagation of the focused 8.5 μm FEL intensity profile. The detector is located at a distance *L* from the focusing lens: *L* = 65, 50, 30, and 8 mm.

($f = 351$ mm). The images are also measured with the 2D Pyro-electric detector. The "*L*" in the figure denotes the distance from the focusing lens to the detector. In the range of L = 30–50 mm (near the focal plane), the intensity profiles are roughly estimated to have a Gaussian shape. Thus the radiation intensity is estimated from the Gaussian fitting. On the contrary, outside of the Gaussian range, the intensity profile indicates the reduced shape of the original one (transported shape). Thus the intensity profile of the transported MIR FEL needs to be measured, and the intensity is calculated by the proportional allotment due to the distance between the substrate and the lens. In the former and the latter cases, the intensity of the MIR FEL is desired to be high and low, respectively. In either case, we can usually irradiate the substrate by choosing a suitable focal length for the lens. Furthermore, the top-hat shape modification has been studied for the flat intensity irradiation [10].

5. Conclusions

We propose a novel thin-film fabrication method using the in-situ MIR FEL irradiation. This method has two advantageous points: (1) the wavelength selectivity, and (2) wide application of the substrate. Moreover, we developed two special thin film fabrication devices arranged for the method. The devices have ZnSe windows for the MIR FEL irradiation. The thin-film fabrication using the device is now in progress. Finally, we discuss the irradiated intensity of the MIR FEL, which is the most important problem of the method. The intensity profile of the focused FEL for proper irradiation is measured, and the profile is shown that it totally changed from the transported shape according to the focal point. Therefore, we have to consider the selection of the focal length of the focusing lens for a suitably assisted irradiation.

Acknowledgements

The authors thank Dr. T. Mitsuyu of the FELI for his discussion on the merits of the fabrication method.

References

[1] T. Tomimasu, T. Takii, Y. Kanazawa, A. Zako, A. Nagai, M. Yasumoto, Nucl. Instr. and Meth. B 144 (1998) 1.
[2] T. Tomimasu, T. Takii, T. Suzuki, E. Nishimura, S. Ogino, A. Nagai, M. Yasumoto, Nucl. Instr. and Meth. A 407 (1998) 494.
[3] H. Ohyama, T. Suzuki, K. Nishi, T. Mitsuyu, T. Tomimasu, Appl. Phy. Lett. 71 (1997) 823.
[4] S. Ogino, H. Shiomi, N. Fujimori, K. Awazu, Y. Maeda, Nucl. Instr. and Meth. B 144 (1998) 181.
[5] P. Mei, J.B. Boyce, J.P. Lu, J. Ho, R.T. Fulks, J. Non-Crys. Solids 266–269 (2000) 1252.
[6] T. Mitsuyu, H. Ohyama, T. Suzuki, K. Nishi, T. Tomimasu, Proceedings of the AFEL'97, Osaka, January 21–24, 1997, p. 315.
[7] M. Yasumoto, N. Umesaki, T. Tomimasu, A. Ishizu, K. Awazu, SPIE 3933 (2000) 496.
[8] M. Yasumoto, T. Tomimasu, Y. Kanazawa, A. Zako, N. Umesaki, Proceedings of the AFEL'99, Taejon, Korea, March 8–10, 1999 p. 152.
[9] M. Yasumoto, T. Tomimasu, S. Nishihara, N. Umesaki, Opt. Mat. 15 (2000) 59.
[10] M. Yasumoto, N. Umesaki, T. Tomimasu, Y. Kanazawa, A. Zako, Nucl. Instr. and Meth. A 429 (1999) 65.

ELSEVIER

Nuclear Instruments and Methods in Physics Research A 475 (2001) 645–649

NUCLEAR
INSTRUMENTS
& METHODS
IN PHYSICS
RESEARCH
Section A

www.elsevier.com/locate/nima

Design and construction of an evanescent optical wave device for the recanalization of vessels

Brett A. Hooper[a,b,*], George C. LaVerde[a], Olaf T. Von Ramm[a]

[a] Department of Biomedical Engineering and Center for Emerging Cardiovascular Technologies, Duke University, Box 90319, Durham, NC 27708-0319, USA
[b] Wellman Laboratories of Photomedicine, Massachusetts General Hospital, Harvard Medical School, USA

Abstract

Removing atherosclerotic material without transecting the vessel wall is a common problem for vascular surgeons and interventional cardiologists. The goal of this project is to design and construct a device that uses evanescent optical waves for precise, controlled laser ablation. For laser light incident at an angle to an optic–tissue interface greater than or equal to the critical angle, an evanescent optical wave is launched into the tissue. With evanescent optical waves, there is no free-beam propagation and the laser energy can be confined to a layer less than one wavelength thick at the optic–tissue interface. Several device designs have been proposed and constructed. The Duke University Mark III Infrared Free-Electron Laser is used to study energy deposition and ablation mechanisms at sapphire–tissue and zinc sulfide–tissue interfaces. Ablation experiments on human low-density lipoprotein and aorta tissue are presented. © 2001 Elsevier Science B.V. All rights reserved.

PACS: 41.60.Cr; 42.25; 87.80.−y; 07.60.−j

Keywords: Evanescent waves; Ablation; Atherosclerotic plaque; Catheter; Free-electron laser

1. Introduction

Precise laser surgery external to the body is possible using laser pulses at wavelengths that are strongly absorbed at the surface of tissue. However, these wavelengths (far ultraviolet and far infrared) are not compatible with conventional fiber optics, such as fused silica. The development of an alternative delivery device in the infrared

that confines the laser energy near the tissue surface could lead to a breakthrough for precise laser surgery inside the body. New fibers (hollow glass waveguides and chalcogenide glass) show promise for efficient transport of infrared wavelengths [1,2], and we have built and are testing devices that incorporate evanescent waves at the tip of these delivery devices.

Removing atherosclerotic material without transecting the vessel wall is a common problem for vascular surgeons and interventional cardiologists [3]. The ultimate goal of this research is to develop an evanescent optical wave device that can be used for the safe, controlled laser removal of intravascular material. With evanescent optical

---

*Corresponding author. Department of Biomedical Engineering and Center for Emerging Cardiovascular Technologies, Duke University, Box 90319 Durham, NC 27708-0319, USA. Tel.: +1-919-660-2670; fax: +1-919-660-2671.
E-mail address: hoop@fel.duke.edu (B.A. Hooper).

waves, there is no free-beam propagation and the laser energy can be confined to a layer less than one wavelength thick at the optic–tissue interface [4]. After design and construction, the device is tested in preliminary water and tissue ablation experiments.

2. Evanescent optical waves

Several device designs have been proposed and constructed. All of the device designs incorporate the general evanescent optical wave approach shown in Fig. 1. An evanescent optical wave is launched into the tissue when the refractive index of the optic n_1 is greater than the refractive index of the tissue n_2, and the angle of incidence of the laser light θ_i at the optic–tissue interface is equal to or greater than the critical angle θ_c, defined as the angle for total internal reflection [5]. The two device designs considered in this paper are based on the refractive indices of the optic and the tissue, where the refractive index of water is used to model tissue, since most soft tissue is 70–80% water [6]. Both device designs incorporate a 45° angle of incidence for the laser beam at the optic–tissue interface (θ_i in Fig. 1). This geometry is useful because the light reflected from one side of the tip is still at a 45° angle of incidence on the other side. Fig. 2 plots the critical angle for a

sapphire–water and a ZnS–water interface as a function of wavelength from 2 to 10 μm, and compares the critical angle to the 45° fiber interface angle. ZnS and diamond are good replacements for the sapphire tip for wavelengths longer than about 5 μm, where transmission drops to 40% over an 8 mm pathlength. For sapphire, only wavelengths from 2.70 to 2.86 μm will be evanescent, whereas all wavelengths are evanescent at a ZnS–water and a diamond–water interface with a 45° interface. This is due to the larger refractive index for ZnS (2.24) and diamond (2.38) compared with that of sapphire (1.70) [7]. The important parameters of the evanescent wave are the depth of penetration ∂_{ewave} and the required incident energy for ablation $E_{required}$ [4], and are plotted in Fig. 3 as a function of wavelength from 2 to 10 μm. The short, solid lines are for sapphire and the dashed lines are for ZnS. The evanescent wave penetration depth is between 0.1 and 1 μm over this wavelength range, and the required incident energy for ablation is about 1 mJ for sapphire and varies from 0.2 to 100 mJ for ZnS.

3. Evanescent optical wave device designs

The first design simply consists of a 0.42 mm diameter sapphire fiber with the end polished at

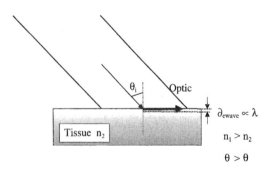

Fig. 1. Diagram of the beamline used to launch evanescent optical waves at a high refractive index optic–tissue interface. The refractive index of the optic n_1 must be greater than the refractive index of the tissue n_2, and the incident angle θ_i must be equal to or greater than the critical angle θ_c for evanescent waves to be generated at the interface.

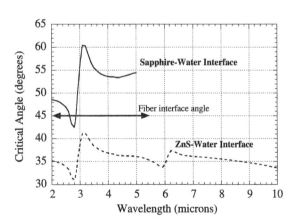

Fig. 2. Critical angle θ_c for evanescent wave generation at a sapphire–water or a ZnS–water interface as a function of wavelength from 2 to 10 μm. Also shown is the 45° fiber interface angle for reference.

45° angles in the shape of a screwdriver tip. This screwdriver-tip sapphire fiber (STSF) "retro-reflector" device design is shown in Fig. 4. A probe laser (HeNe) can also be launched into the device to monitor the dynamics of the ablation at the interface since after interrogation at the optic–tissue interface the probe reflects out the fiber and can be detected using a beamsplitter (see Fig. 4). A catheter device will need to be 1–2 meters

in length. Transmission of the laser light through a 1–2 m solid sapphire fiber is much too low, the fiber will not be flexible enough, and these lengths are currently not available. The solution to this problem comes with the following device design.

The second design consists of a 1 m long, 1.5 mm outside diameter hollow glass waveguide (HGW) with a high refractive index optic (sapphire, ZnS, or diamond) attached to the end. This device design is shown in Fig. 5. The optic tip is 10 mm long and 1.5 mm in diameter with a conical polish at 45°. For wavelengths longer than about 5 μm, ZnS and diamond can be used as a replacement for the sapphire tip. A 10 mm long stainless-steel sleeve is used to hold the optic tip securely onto the end of the HGW. Before placing the tip onto the HGW, transmission through the HGW was measured at 75% for several wavelengths near 3 μm.

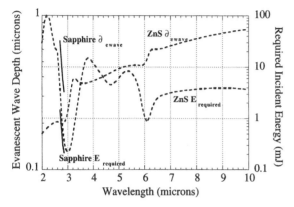

Fig. 3. Evanescent wave depth of penetration ∂_{ewave} and required incident energy for ablation $E_{required}$ [4] at the sapphire–water and ZnS–water interfaces as a function of wavelength from 2 to 10 μm, $E_{required}$ is calculated from the latent heat of vaporization of water: $(2500\,\mathrm{J}/\mathrm{cm}^3 \times \partial_{ewave} \times (\pi\omega_0^2)/2)/(\cos(45°) \times 1-R)$, where $1-R$ is the fraction of incident laser energy absorbed in the tissue.

4. Preliminary ablation studies

Both devices were first tested in water to study the laser ablation mechanisms at the optic–water interface. High-speed imaging of the cavitation bubbles in water was performed using the Mark III IR FEL and an Er:YAG laser (2.94 μm) as the pump laser. Ablation/cavitation of the water is

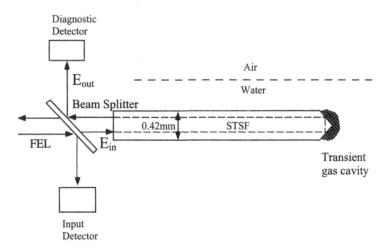

Fig. 4. Screwdriver-tip sapphire fiber (STSF) device with 45° polished tip and the diagnostic beamline for ablation.

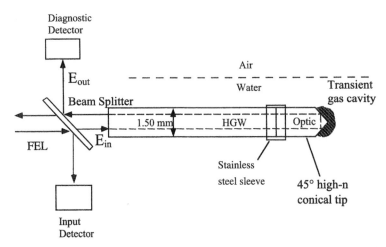

Fig. 5. Hollow glass waveguide with high refractive index conical optic tip (HWG/tip) polished at 45° and the diagnostic beamline.

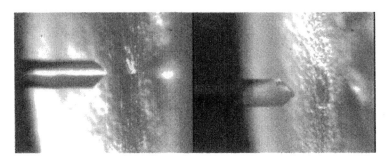

Fig. 6. Porcine aorta endothelium ablation using the STSF with 3.37 μm FEL light (9.3 mJ/1 μs macropulse at 8.6 Hz). Left image: STSF translated into tissue 3 × 100 μm without laser energized. Right image: STSF translated with 10 pulses of FEL light before each 100 μm advancement of the fiber, showing ablation of aorta endothelium.

accompanied by a sharp snapping sound. No damage to the optic surface was noted on either device.

Porcine aorta endothelium was then ablated using the 45° STSF device. Pictures of this experiment are shown in Fig. 6. First, a control was performed by translating the device into the porcine aorta endothelium 3 × 100 μm without the laser energized (the left side of Fig. 6), to be sure that the fiber itself did not cut the aorta. There is a slight indentation caused by the relatively sharp geometry of the device, but no cut. Then, the same translation was performed with 3.37 μm Mark III IR FEL light delivering 10 pulses (9.3 mJ/1 μs macropulse at 8.6 Hz) into the STSF before each

100 μm advance (the right side of Fig. 6). There is very clear ablation that can be seen in the picture on the right side of Fig. 6. This wavelength is not evanescent, though, in the STSF.

Next, an isolation of human low-density lipoprotein (LDL) from whole blood was performed, and the STSF was tested on the LDL. A control was done by translating the device through the LDL without the laser energized, as above. Again, this was done to make sure that the STSF did not cut the LDL. The STSF stretched the LDL, but did not cut it. The same translation was performed with the Er : YAG laser (2.94 μm, 18 mJ/150 μs pulse at 10 Hz) coupled into the STSF. The results of this experiment are shown in Fig. 7. The device

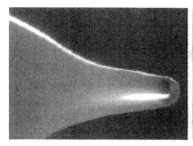

Fig. 7. Human low-density lipoprotein (LDL) ablation using the STSF with Er : YAG laser light. Left image: STSF translated through the LDL without laser energized, to make sure that the STSF itself did not cut the LDL. Right image: STSF translated with Er : YAG laser (18 mJ/150 µs pulse at 10 Hz) 'free-running' during advancement of the STSF. STSF ablates the LDL as fast as it can be advanced through the LDL by hand.

clearly ablated (cut) through the LDL when the laser was energized.

In preliminary experiments with the Mark III FEL and Er : YAG laser, the HGW/sapphire tip device successfully ablated both human LDL and porcine fat inside a glass capillary tube with an inner diameter of about 2.5 mm.

5. Conclusions

Precise, controlled tissue ablation is achieved using these devices. Ablation is controlled because the optic tip must be in physical contact with the tissue to achieve ablation, otherwise the laser energy is retro-reflected from the tip. The STSF and HGW/sapphire tip devices have been successfully tested in the ablation of water, human LDL, and porcine aorta endothelium. The HGW with 45° conical-tip device allows for high-energy transmission over the 1–2 m length necessary for a catheter. The sapphire conical tip can be used for wavelengths from 0.2 to about 5 µm, and ZnS and diamond can be used for wavelengths from 0.4 to about 14 µm and 0.3 to greater than 100 µm, respectively. We plan to construct and test HGW/ZnS tip and HGW/diamond tip devices for their

diagnostic and therapeutic application in the infrared.

Acknowledgements

This work was supported by the U.S. Office of Naval Research (Grants N00014-95-1-1265 and N00014-94-1-0818) and the Duke University Center for Emerging Cardiovascular Technologies.

References

[1] I. Gannot, A. Inberg, M. Oksman, R. Waynant, N. Croitoru, IEEE J. Sel. Top. Quantum Electron. 2 (4) (1996) 880.

[2] L.E. Busse, J.A. Moon, J.S. Sanghera, I.D. Aggarwal, Mid-IR high power transmission through chalcogenide fibers: current results and future challenges, Proceedings of the SPIE, Vol. 2966, 1997, pp. 553–563.

[3] D.R. Holmes, J.F. Bresnahan, Cardio. Clin. 9 (1991) 115.

[4] B.A. Hooper, C. Lin, Y. Domankevitz, R. Rox Anderson, Appl. Opt. 38 (25) (1999) 551.

[5] E. Hecht, A. Zajac, Optics, Addison-Wesley, Reading, MA 1974, p. 81.

[6] G.M. Hale, M.R. Querry, Appl. Opt. 12 (1973) 555.

[7] P. Klocek, (ed.), Handbook of Infrared Optical Materials, Marcel Dekker, New York, 1991.

ELSEVIER

Nuclear Instruments and Methods in Physics Research A 475 (2001) 650–655

NUCLEAR
INSTRUMENTS
& METHODS
IN PHYSICS
RESEARCH
Section A

www.elsevier.com/locate/nima

Selective removal of cholesteryl oleate through collagen films by MIR FEL

Kunio Awazu*, Yuko Fukami

Free Electron Laser Institute, Osaka University, 2-9-5 Tsuda-yamate, Hirakata, Osaka 573-0128, Japan

Abstract

In this paper, two layers tissue model were proposed to evaluate the removal effect. The sample was made of cholesteryl oleate as a model of atherosclerotic legions and collagen as a model of endothelial cells and was exposed to the 5.75 μm FEL. As results, it was found that the amount of the light which transmitted gelatin layer was enough to remove cholesteryl oleate, and that the cholesteryl oleate under the gelatin layer can be removed by the FEL about 15% without structure damage of the gelatin layer as a model of normal endothelial cells. © 2001 Elsevier Science B.V. All rights reserved.

PACS: 42.62.Be; 87.50.Hj; 87.54.Fj

Keywords: FEL; Two layers tissue model; Atherosclerotic lesions; Cholesterol ester; Cholesteryl oleate

1. Introduction

Laser assisted removal of biological tissue is a growing area in laser treatment. Among lasers, Free Electron Lasers (FEL) has advantages for removing tissues selectively because of their wavelength tunability and short pulse structure. Atherosclerotic lesions, called atheromas or fibrofatty plaques, have a lipid core and a fibrous cap. The lesions lie within the vessel wall. Atheromas are pathologic lesions whose growth and development are not as well regulated as that of normal tissue structures. Thus, as with almost all therapies there will be indications and contra-indications [1–4]. Patient selection, or more precisely matching

of the lesion with the treatment, will be the key to successful use of laser-based angioplasty treatments.

The removal of cholesterol esters from the arteriosclerotic region has been particularly difficult because cholesterol bound to fatty acids enters the arterial tissues in a complicated manner. Lasers, however, cannot be tuned to a desired wavelength as the wavelength of most lasers is determined by the type of laser media. Accumulated cholesteryl oleate in the arterial wall are covered with endothelial cells. Therefore, normal endothelial cells should be damaged by the laser irradiation if the cells highly absorb the laser.

An infrared Free Electron Laser (IRFEL) is a tunable pulsed laser in the molecular vibrational region of biomolecules such as water, collagen and fatty acid. At the Osaka University Free Electron Laser Research Institute, we have developed a

*Corresponding author. Tel.: +81-72-896-0411; fax: +81-72-896-0421.

E-mail address: awazu@fel.eng.osaka-u.ac.jp (K. Awazu).

0168-9002/01/$ - see front matter © 2001 Elsevier Science B.V. All rights reserved.
PII: S 0 1 6 8 - 9 0 0 2 (0 1) 0 1 6 9 0 - 4

FEL system in which the wavelength may be adjusted to any desired value ranging from 350 nm to 40 μm [5,6]. Previously, we reported the results of experiments in which the FEL was tuned to 5.75 μm, a wavelength which corresponds to the stretch mode of the ester bonds, and cholesterol ester, albumin and isolated rabbit arterial wall specimens were irradiated. We found that cholesterol ester was selectively ablated without damaging albumin or the normal arterial wall and reported the possibility of selective removal of cholesteryl oleate in the atherosclerotic lesions in vitro using 5.75 μm ($= 1740\,\mathrm{cm}^{-1}$) [7–10].

In this paper, a two layer tissue model is proposed to evaluate the removal effect between cholesteryl oleate, as a model of atherosclerotic lesions and collagen, as a model of endothelial cells, when the tissue model is exposed to the 5.75 μm FEL.

2. Methods

2.1. Sample preparations

Lower layer: Cholesteryl oleate, 0.02 g, (Sigma, #C-9253, $C_{45}H_{78}O_2$, FW = 651.1, St. Louis, MO) was dissolved in 500 μl carbon tetrachloride. An aliquot (10 μl) was dropped on the surface of a BaF_2 crystal plate (12 mm diameter, thickness of 1 mm). After evaporation of the carbon tetrachloride, a thin film of cholesteryl oleate was formed.

Upper layer: Gelatin, 0.1 g, (Sigma:#G-2500) mixed with 10 ml water was dissolved by heating at 50°C. An aliquot (10 μl) was dropped on the cholesteryl oleate layer. After drying, the gelatin layer was formed.

2.2. FEL exposure experiments

The FEL has a complex pulse structure. The structure consists of a train of macropulses, and each macropulse contains a train of 300–400 ultrashort micropulses as shown in Fig. 1. For the present experiment, the width of the macropulse was about 15 μs and the repetition rate was 10 Hz. The separation between micropulses was

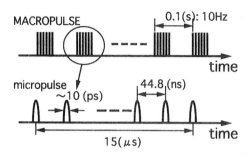

Fig. 1. Pulse structure of FEL. One macropulse is made of 330 micropulses.

Table 1
Conditions of FEL exposure experiments

Wavelength (μm)	7.5
Average power (mW)	5
Power density, P (W cm^{-2})	50
Exposed time, t (s)	3
Energy density, $E = Pt$ (J cm^{-2})	150

45 ns. The width of the micropulse can be estimated to be shorter than the bunch length of the electron beam, which has been measured as 10 ps in the accelerator operation condition for the mid-IR range. The average power used in these experiments was 5 mW with a power density of 50 W cm^{-2} (Table 1).

The FEL beam was directed to an optical microscope through a flexible multi-joint cylindrical tube as shown in Fig. 2. The beam on the object revealed an oval shape. A specimen was placed on an inverted microscope stage and the changes caused by FEL irradiation were monitored continuously by a CCD camera in real time. Images were taken every second by a time-lapse video recorder and stored on an optical magnetic disk.

2.3. Measurements

Infrared absorption of the object before and after FEL radiation was determined by microscopic transmission FTIR (Horiba, Model FT-520). The film thickness was measured by DEKTAK3 (Veeco Instruments).

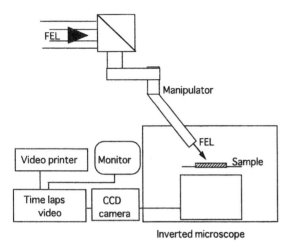

Fig. 2. FEL exposure system.

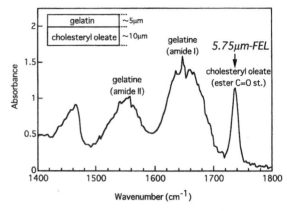

Fig. 3. IR absorption spectrum of gelatin/ cholesteryl oleate. Absorption peak of cholesteryl oleate is at $1740\,\text{cm}^{-1}$ ($5.75\,\mu\text{m}$) and, peaks of protein are at 1550 and $1650\,\text{cm}^{-1}$.

3. Results

3.1. Sample characteristics

The infrared spectra of two layers, gelatin/ cholesteryl oleate, tissue model is shown in Fig. 3. The sharp peak at $1740\,\text{cm}^{-1}$ is the stretching vibration of $C=O$ ester bonding in cholesteryl oleate. The wavelength of $5.75\,\mu\text{m}$ is equal to the wave number of $1740\,\text{cm}^{-1}$. Peaks at 1650 and $1550\,\text{cm}^{-1}$ are due to the amide I mode and II mode of peptide in gelatin, respectively. The characteristics of this two layer model are summarized in (Table 2).

3.2. FEL exposure effect

Microscopic observation after FEL irradiation is shown in Fig. 4, from the left region, cholesteryl oleate layer, gelatin and cholesteryl oleate layers, and gelatin layer, respectively. The FEL was scanned from the left at $3\,\text{mm}\,\text{s}^{-1}$. After the FEL irradiation, the monolayer of cholesteryl oleate was melted, but no change was observed at the monolayer of gelatin. At the two layers of gelatin and cholesteryl oleate, however, the top layer of gelatin was not changed. The bottom layer of cholesteryl oleate was melted, and the irradiation pattern was observed.

Table 2
Properties of 2 layer (gelatin/ cholesteryl oleate) samples

IR absorption peak	
Cholesteryl oleate: A_{ester}	At $1740\,\text{cm}^{-1}$ (ester $C=O$ st.)
Gelatin: A_{II}	At $1550\,\text{cm}^{-1}$ (amide II)
A_{I}	At $1650\,\text{cm}^{-1}$ (amide I)
Ratio of absorbance	$A_{\text{ester}}/A_{\text{I}} \cong 0.8$
	$A_{\text{II}}/A_{\text{I}} \cong 0.7$
Gelatin layer thickness	$\cong 5\,\mu\text{m}$
Cholesteryl oleate layer thickness	$\cong 10\,\mu\text{m}$

3.3. FTIR spectra

The IR spectrum of the monolayer of cholesteryl oleate exposed to FEL is shown in Fig. 5. The peak absorbance at $1740\,\text{cm}^{-1}$, which is derived from the stretching vibration mode of ester, was decreased after FEL irradiation.

Fig. 6 shows IR spectra of the monolayer of gelatin. Gelatin was undamaged after FEL irradiation. No change was observed on the peaks of the stretching vibration mode of amide I bond, $1650\,\text{cm}^{-1}$, and amide II bond, $1550\,\text{cm}^{-1}$. Therefore, the gelatin monolayer suffered no damage due to $5.75\,\mu\text{m}$ FEL irradiation.

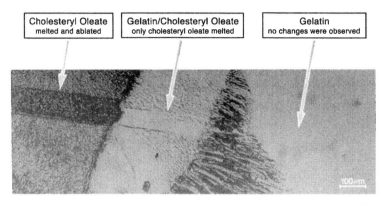

Fig. 4. Microscopic images of gelatin/cholesteryl oleate exposed to 5.75 μm FEL ($50\,W\,cm^{-2} \times 3\,s = 150\,J\,cm^{-2}$)

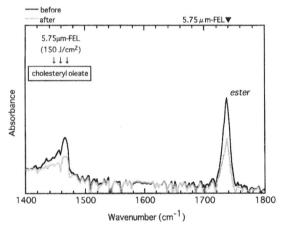

Fig. 5. IR absorption spectra of cholesteryl oleate mono-layer exposed to 5.75 μm-FEL. Absorbance of ester bond of cholesteryl oleate decreased by FEL exposure.

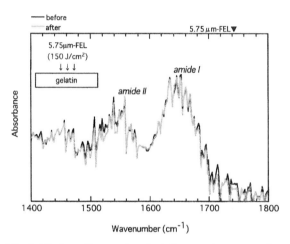

Fig. 6. IR absorption spectra of gelatin mono-layer exposed to 5.75 μm-FELs. No changes were observed.

Fig. 7 shows the IR spectra of two layers of gelatin and cholesteryl oleate. After the FEL irradiation, the absorption peak at $1740\,cm^{-1}$ was reduced. On the other hand, the peaks of the stretching vibration mode of amide I bond, $1650\,cm^{-1}$, and amide II bond, $1550\,cm^{-1}$ were not decreased. As a result, we believe that cholesteryl oleate was removed, but gelatin was not damaged.

4. Discussions

The IR absorption peaks before and after FEL irradiation are summarized in Table 3. From the Table 3 reduction rate of cholesteryl oleate,

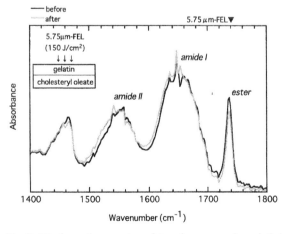

Fig. 7. IR absorption spectra of two layers sample, gelatin/cholesteryl oleate exposed to 5.75 μm FELs. Only cholesteryl oleate decreased by FEL exposures.

Table 3
Discussions for quantitative changes[a]

	C.O.	Gel./C.O.	Gel.
Absorbance at 1740 cm^{-1} (ester)			
$A_{ester} \to A'_{ester}$	$0.599 \to 0.340$	$1.14 \to 0.979$	$(0.020 \to 0.020)$
A'_{ester}/A_{ester}	57%	85%	(100%)
Absorbance at 1550 cm^{-1} (amide II)			
$A_{II} \to A'_{II}$	$(0.050 \to 0.050)$	$1.45 \to 1.45$	$0.312 \to 0.312$
A'_{II}/A_{II}	(100%)	100%	100%
Absorbance at 1650 cm^{-1} (amide I)			
$A_{I} \to A'_{I}$	$(0.070 \to 0.070)$	$0.979 \to 0.979$	$0.212 \to 0.212$
A'_{I}/A_{I}	(100%)	100%	100%

[a] Absorbance before (A) and after (A') exposure, and each ratio of them (A'/A) were summarized. Only cholesteryl ester decreased.

Table 4
Discussions for structural changes of protein[a]

	C.O.	Gel./C.O.	Gel.
Before: A_{II}/A_{I}	—	0.67	0.68
After: A'_{II}/A'_{I}	—	0.67	0.68
Change: $(A'_{II}/A'_{I})/(A_{II}/A_{I})$	—	100%	100%

[a] Ratios of amide II (A_{II}) and amide I (A_{I}); A_{II}/A_{I}, and their changes were summarized. No changes were observed on gelatin.

$(1 - A'_{ester}/A_{ester})$, was calculated to be 0.43 at the monolayer of cholesteryl oleate and 0.15 at the cholesteryl oleate of two layers. This result showed that the amount of the light transmitted by the gelatin layer was enough to remove cholesteryl oleate.

The absorption peaks of the stretching vibration mode of amide I bond, 1650 cm^{-1}, and amide II bond, 1550 cm^{-1} are summarized in Table 4. A_{II}/A_{I} is the ratio between the two vibration modes so that the A_{II}/A_{I} value depends on the structure of gelatin. Therefore, the gelatin structure was not affected by 5.75 μm FEL, because the experimental results showed that the A_{II}/A_{I} value did not change due to FEL irradiation.

From these results, it was found that about 15% of the cholesteryl oleate under the gelatin layer can

be removed by the FEL without damage to the gelatin layer.

5. Conclusions

The two layer tissue model proposed here can be used to evaluate the removal of cholesteryl oleate as a model of atherosclerotic lesions and collagen as a model of normal endothelial cells when the tissue model is exposed to the 5.75 μm FEL. In conclusion, the amount of the light transmitted by the gelatin layer was enough to remove cholesteryl oleate, and about 15% of the cholesteryl oleate under the gelatin layer can be removed by the FEL without structure damage of the gelatin layer.

References

[1] T. Takano, K. Ananuma, J. Kimura, S. Ohkuma, Acta Histochem. Cytochem. 19 (1986) 135.
[2] T. Arai, K. Mizuno, M. Kikuchi, M. Sakurada, A. Miyamoto, A. Kurita, et al., IEEE 3 (1993) 1598.
[3] T. Saito, J. Hayashi, H. Sato, H. Kawabe, K. Aizawa, Med. Electron. Microsc. 29 (1997) 137.
[4] J. Hayashi, T. Saito, K. Aizawa, Lasers Surg. Med. 21 (1997) 287.
[5] T. Tomimasu, K. Saeki, Y. Miyauchi, E. Ohshita, S. Okuma, K. Wakita, et al., Nucl. Instr. Meth. Phys. Res. A 375 (1996) 626.

[6] K. Kobayashi, K. Saeki, E. Ohshita, S. Okuma, K. Wakita, A. Zako, et al., Nucl. Instr. Meth. Phys. Res. A 375 (1996) 317.

[7] K. Awazu, S.L. Jacques, Proceedings of the 12th International Symposium on Analytical Bioscience, 1997, p. 92.

[8] K. Awazu, A. Nagai, K. Aizawa, Lasers Surg. Med. 23 (1998) 233.

[9] K. Awazu, A. Nagai, T. Tomimasu, K. Aizawa, Nucl. Instr. Meth. Phys. Res. B 144 (1998) 225.

[10] Y. Fukami, Y. Maeda, K. Awazu, Nucl. Instr. Meth. Phys. Res. B 144 (1998) 229.

FREE ELECTRON LASERS 2000
V.N. Litvinenko and Y. Wu (Eds.)
2001 Elsevier Science B.V.

RF Photoinjector in CAEP

Li Zhenghong, Hu Kesong, Li Ming, Yang Maorong, Qian Mingquan
Liu Zhiqiang, Du Xingshao, Pan Qing, Deng Renpei

Institute of Applied Electronics, China Academy of Engineering Physics (CAEP)
P. O. Box 919(58), MianYang 621900, P. R. China

Abstract: The RF photoinjector is a source of high-brightness electron beams. Recently a RF photoinjector along with a compact linac has been commissioned in CAEP. The electron beam parameters are: an electron beam energy of 2 MeV with a peak current of 70 A, a micropulse width of 10 ps at 81.25 MHz, and a macropulse duration of 2.5 microseconds with a 3 Hz repetition. This RF photoinjector system consists of a drive laser, a RF cavity, a vacuum chamber for the Cs2Te fabrication, and a high power microwave system. The output power of the drive laser can reach 1 microJ/micropulse at the 4th harmonic with a timing jitter < 2 ps.

Key words: Photocathode, RF cavity, RF photoinjector

Introduction

An intense electron beam is very important for the high-average power and X-ray FELs. The RF photoinjector has been a subject of intense studies in recent years for such purposes. The CAEP RF photoinjector consists of four major components: (1) a Cs2Te fabrication chamber, (2) a RF cavity, (3) a drive laser, (4) a microwave system. The experiments on the RF photoinjector are the subject of this paper. This system has demonstrated a high level of stability and can be extended for FEL and other scientific researches.

1. Arrangement of RF photoinjector

Fig.1 shows a schematic layout of the CAEP RF photoinjector. The 1.3 GHz RF signal is divided into two parts: one for seeding the high-power microwave system, the other is converted down to 81.25 MHz for driving the laser.

2. Drive Laser

The laser oscillator is a diode-pumped self-mode-locked laser and the repetition of the laser is controlled by the driving signal of 81.25 MHz through synchronization. The output of the oscillator is amplified by a lamp-pumped Nd:YAG system, and then pulse-picked at 1~10us durations before the final frequency conversion. The parameters of the drive laser are: a timing jitter below 2ps, a micropulse width of 10ps at 81.25 MHz with 1uJ per micropulse, and a macropulse width of 1-10us with a repetition of 1-10Hz.

3. Cs2Te Cathode

Cs2Te is one of semiconductor photocathode materials. The quantum efficiency (QE) of this material is high at the fourth harmonic of the Nd:YAG laser. The QE reaches 6% immediately after fabrication in the ultra-high vacuum. After some treatments, a quantum efficiency of 1% can remain for more than half a year if the vacuum is kept better than 10^{-6} Pa. In our experiment, Cs2Te cathode is transferred to the RF cavity in the vacuum after fabrication.

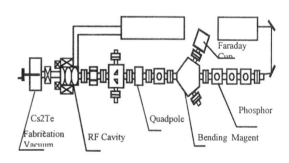

Fig.1 CAEP RF Photoinjector

4. RF Cavity and Microwave System

4.1 Two-and-half Cell RF Cavity

The RF cavity is a L-band structure as shown in Fig.2. The field on the surface of photo-cathode is 37 MV/m according to the design.

4.2. Microwave system

The microwave system is made of a modulator, a KL-28 klystron, and wave-guides. The high-voltage fluctuation in the modulator causes significant phase jitters in the microwave power fed to the RF cavity. By using a bi-line and increasing the number of pulse-forming lines, the high-voltage fluctuation was lowered to: dV_K/V_K=0.2-0.3%, which corresponds to a phase jitter of $d\Phi \approx 2$ degree, or an equivalent timing jitter of 4.3 ps.

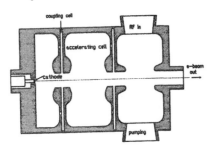

Fig.2 RF Cavity

5. Experimental Results

The measurement setup for the main electron beam parameters such as the beam current, electron energy, and emittance is shown in Fig.3. When the phosphor screen #1 is raised, the electron beam enters the magnet field and some electrons are collected by the Faraday cup.

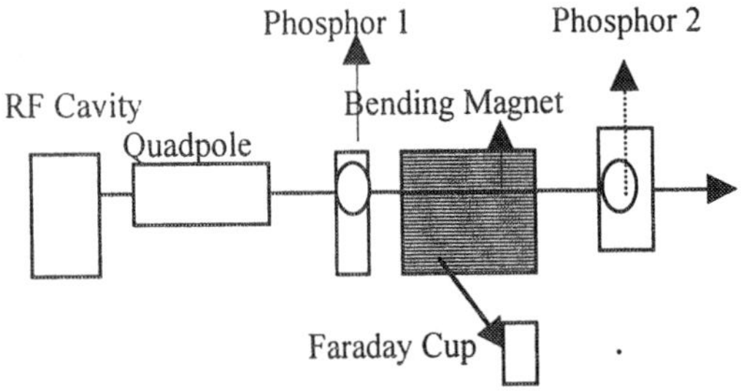

Phosphor 1   Phosphor 2

RF Cavity   Bending Magnet
      Quadpole

Faraday Cup

Fig.3 Schematics for e-beam parameter measurement

Fig.4 is the signal from the Faraday cup when the magnet field is at 0.0416 T. From the measured signal, we calculated that the beam current was 70 A. Fig.5 is the beam image on the second phosphor screen when the beam current is about 40 A.

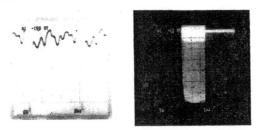

(Micropulse)   (Macropulse)
Fig.4 Electron beam waveform measured by a Faraday cup

Fig.5 Measured spot of electron beam on the 2nd phosphor screen.

6 Summary on Experimental Results

Electron Energy	2MeV
Bean Current	~70A
Micropulse Width	10 ps
Micropulse Repetition	81.25 MHz
Macropulse Width	2.5 μs
Macropulse Repetition	3.125 Hz
Emittance	~4mm·mrad
Q.E.(Cs2Te)	~1%

Table 1 Parameters of the electron beam.

References

1. G. H. Lee, et.al., IEEE Trans. Nucl. Sci. NS-32 (5), 1985.
2. M. Curtin, et.al., Nucl. Instr.& Meth. A 296, 1990.
3. Li Zhenghong et al, "Amplification of ps Laser Pulse" SPIE, vol. 3862, 1999.
4. Yao Chong-guo, Linac for Electron Beam Scientific Press in China, 1980.
5. J. Rosenweig, S.C.Hartman, et al, Nucl. Instru & Meth. A340, p. 219-230. 1994.

FREE ELECTRON LASERS 2000
V.N. Litvinenko and Y. Wu (Eds.)
2001 Elsevier Science B.V.

Photo Injector Test Facility under Construction at DESY Zeuthen*

F. Stephan[a], D. Krämer[b], I. Will[c], and A. Novokhatski[d] for the PITZ Cooperation consisting of

[a]DESY, Notkestraße 85, 22603 Hamburg and Platanenallee 6, 15738 Zeuthen, Germany,

[b]BESSY, Albert-Einstein-Straße 15, 12489 Berlin, Germany,

[c]Max-Born-Institute, Max-Born-Straße 2a, 12489 Berlin, Germany,

[d]TU Darmstadt, Department TEMF, Schloßgartenstraße 8, 64289 Darmstadt, Germany

A Photo Injector Test Facility is under construction at DESY Zeuthen (PITZ) within a cooperation of BESSY, DESY, MBI, and TUD. The aim is to develop and operate an optimized photo injector for future free electron lasers and linear accelerators. First operation of the rf-gun is planned for late autumn 2000. In this paper we want to outline the scientific goals, the planned and existing hardware, the status of the project and new developments.

1. Goals

The scientific goal of the project is to operate a test facility for rf-guns and photo injectors in order to optimize injectors for different applications like free electron lasers, production of flat beams for linear colliders and polarized electron sources. We will make comparisons of detailed experimental results with simulations and theoretical predictions. At the beginning we will concentrate on the development of an optimized photo injector for the subsequent operation at the TESLA Test Facility - Free Electron Laser (TTF-FEL) [1]. This also includes the test of new developed components like the laser, cathodes and beam diagnostics under realistic conditions. After the installation of a booster cavity we will be able to test new concepts for the production of flat beams[2]. On a longer term basis we plan to investigate the design of polarized electron sources.

2. Setup

The experimental setup is shown in figure 1. In the future, the teststand will be complemented by more diagnostics, beam optical components and a booster cavity.

*The project is partially funded by the HGF Vernetzungsfond.

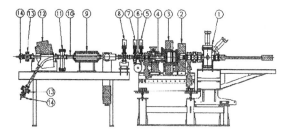

Figure 1. Experimental setup of PITZ in the start-up phase: 1. cathode system, 2. bucking solenoid, 3. main solenoid, 4. coaxial coupler, 5. laser input port, 6. beam position monitor, 7. Faraday cup + view screen, 8. emittance measurement system (slits + pepper pot), 9. quadrupole triplet, 10. wall gap monitor, 11.+13. view screen, 12. dipole, 14. Faraday cup.

3. Schedule and Status

In September 1999 it was decided to built the test facility in DESY Zeuthen. Now the raw construction work is mainly finished and soon we will start the installation of the test stand itself and all the other equipment. We plan to have the first rf inside the gun cavity in November and the first photoelectrons are scheduled for January 2001. A major upgrade of the test stand will take place mid 2002 when a booster cavity will be installed.

4. Laser Development

The Max-Born-Institute develops the photocathode laser for PITZ. Besides the former requirements for the TTF photocathode laser system [3] a new request on the logitudinal shape of the micropulses will be realized by the MBI: the micropulses should have a flat-top profile, 5-20 ps FWHM, with rising and trailing edges shorter than 1 ps. This requires a new laser concept which is shown in figure 2. The key element is the optical-parametric amplifier (OPA). It provides large amplification bandwidth and therefore allows for the amplification of pulses with sharp edges. A grating combination will be used for programming the shape of the micropulses. Wave front deformations will be corrected by computer-controlled optics. An extended version of the field tested TTF photocathode laser will serve as a pump laser for the OPA. In the beginning it will be used to produce the first photoelectrons. Then a continual upgrade to the full laser system follows.

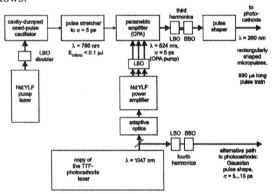

Figure 2. Scheme of the photocathode laser for the generation of micropulses with variable shape.

5. Simulations

One goal of TEMF at TU Darmstadt is the numerical study of the minimum attainable transverse and longitudinal emittance as a function of rf-gun parameters. Figure 3 presents a simulation that takes into account space charge and nonlinear rf forces. It shows that the main part of the transverse emittance is caused by the emission process and that the evolution of the emittance for the first and the second half of the bunch is opposite. The behaviour of the projected emittance seems to be a result of the rf field effects and MAFIA TS2 and ASTRA[4] are shown to be in good agreement. An other topic for TEMF is the development and installation of an on-line simulation program (V-code) [5] that is based on a model of ensembles. It will help to obtain an on-line understanding of the dynamics of the beam. At DESY an ASTRA simulation with a cutted disk structure booster cavity was performed. This cavity provides an average gradient of $\approx$ 12.6 MV/m and boosts the beam up to about 30 MeV. According to that simulation emittances in the sub-mm-mrd regime can be obtained.

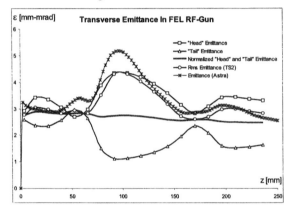

Figure 3. Development of transverse emittance in the rf-gun.

REFERENCES

1. J. Andruszkow et al., First Observation of Self-Amplified Spontaneous Emission in a Free-Electron Laser at 109 nm Wavelength, to be published in Phys. Rev. Lett.
2. R. Brinkmann, Y. Derbenev, K. Flöttmann, A low emittance, flat-beam electron source for linear colliders, EPAC 2000.
3. I. Will, A. Liero, D. Merins, W. Sandner, IEEE Journal of Quantum Electronics, Vol. 34, Oct. 1998, p2020-2028.
4. K. Flöttmann, ASTRA user manual, see www.desy.de/~mpyflo/Astra_dokumentation.
5. A. Novokhatski, T. Weiland, The Model of Ensembles for Beam Dynamics Simulation, PAC99, New York, p.2743.

FREE ELECTRON LASERS 2000
V.N. Litvinenko and Y. Wu (Eds.)
2001 Elsevier Science B.V.

The Pegasus Free Electron Laser Laboratory*

S. Telfer, S. Reiche, J.B. Rosenzweig and P. Frigola

Department of Physics and Astronomy, University of California, Los Angeles,
405 Hilgard Avenue, Los Angeles, California 90095

1. INTRODUCTION

The PEGASUS (Photoelectron Generated Amplified Spontaneous Radiation Source) free electron laser laboratory is currently being commisioned at UCLA. PEGASUS is a linac-based FEL laboratory which was designed with two primary goals in mind. First, the PEGASUS laboratory provides the microwave, laser, and diagnostic infrastructure to test novel integrated RF photocatode electron beam sources, the design of which emphasises application to linac-driven FEL experiments. The first generation of such structures, the Plane Wave Transformer Photoinjector (PWTP) is currently installed, and is in the final stages of commissioning. Second and third generation injectors are currently being designed.

Table 1
PEGASUS Parameters

Value	Parameter
Energy	17 MeV
Energy Spread	1.4 %
Emittance	1 mm-mrad
Charge	1 nC
Pulse width	10 ps
Matched Beam Radius	100 μm
Drive Laser Energy	60 μJ
Undulator Parameter	1.05
Undulator Period	20.5 mm
Number of Periods	98
Beta Function	22 cm

The second design goal of PEGASUS is the study of SASE FEL physics based on a 2m IR FEL. While the beam provided by the PWTP would not be expected to saturate the undulator, extensive GENESIS simulation has shown that the addition of a 1 mm square waveguide will significantly enhance gain and thereby provide a greater chance of saturation [1]. In addition, as part of other ongoing FEL work at UCLA, a two-sage FEL experiment has been proposed which would utilize a short helical undulator. The helical undulator would be placed before the 2m IR undulator in order to reduce the bandwidth of the SASE spectrum.

2. DESCRIPTION OF THE FACILITY

2.1. RF Photoinjector

The PWTP is a novel standing wave S-band photocathode electron source that is designed to provide 1 nC, 10 ps, 1 mm-mrad, 20 MeV electron beams. The PWTP consists of a 60 cm long, 12 cm diameter tank loaded with 11 discs, creating 10 full and two half accelerating cells. The peak gradient is expected to be 60 MV/m. The relatively low gradient improves vacuum running conditions and simplifies RF engineering. The PWTP has a shunt impedance of approximately 50 MΩ/m and a loaded Q of roughly 7000. This makes the PWTP an efficient structure with a moderate fill time of 2-3 μs. The π-mode is a hybrid of the TEM mode supported between the tank and discs, and the TM$_{01}$ which is supported by the disc irises. This hybridization provides excellent cell to cell coupling and a 0-π mode separation of over 500 MHz. Additionally, the RF coupling hole is symmetrized by a vacuum pump port in order to suppress RF multipole fields. The compact design allows for a simple emittance compensation solenoid. Cathode source flexibility is enhanced by a insertable cathode. Sources

*Work supported by DOE grant DE-FG03-98ER45693

at PEGASUS will include OHFC copper, single crystal copper, and Cs_2Te.

2.2. Photocathode Drive Laser

The 60 μJ, 10 ps pulses required to provide the 10 ps, 1 nC electron beam are supplied by a ND:YLF amplifier seeded by a ND:YLF laser oscillator. The oscillator cavity is acousto-optically mode locked at 89.25 MHz, and timing is stabilized by a feedback circuit. The amplifier consists of a doubly folded cavity, diode-pumped ND:YLF rods and a quarter-wave switched Pockel's cell. The pulse repetition rate is 5 Hz, while power output from the amplifier is 2 mJ. The 1053 nm laser is fed into a BBO for doubling and a temperature-controlled LBO for quadrupling. To aid in alignment, the UV light is co-propagated with a HeNe laser which is imaged on the cathode. Additionally, the final turning mirror is mounted on a micrometer which allows the mirror to be positioned as close to the electron beam as possible.

2.3. RF Power System

The RF power system is designed to supply 20 MW of power to the PWT photoinjector. A small amount of the 89.25 MHz mode-locker RF from the laser oscillator is fed into a phase locked oscillator running at 2.856 GHz. This provides a low-level RF signal which is phase-locked to the laser pulse. This CW signal is amplified by a CW amplifier to 30 dBm, and again amplified to 60 dBm by a pulsed triode amplifier. Final amplification of the signal is made by a SLAC XK5 klystron. A SLAC-style transmission line transformer modulator, consisting of a twelve stage pulse forming network, 50 kV power supply, hydrogen thyratron and SCR thyratron trigger system, provides the 4 μs high voltage pulses for the XK5 klystron. The 20 MW of RF power is transported to the PWT by a SF_6 filled Al waveguide system. The XK5 window is protected from standing waves and large reflected voltages by a high power SF_6 filled ferromagnetic isolator. Forward and reflected power are monitored by bi-directional waveguide couplers.

2.4. Electron Beam Diagnostics

Beam charge is measured using impedance matched stripline BPM sum signals, while beam energy is measured using dipole spectrometers. Beam position is monitored using YAG screens in addition to differential BPM signals.

When measuring the emittance of a ultra-high brightness beam like the one expected for the PWTP, space-charge effects must be properly taken into account. A useful figure of merit is the ratio of space-charge to emittance terms taken from the rms envlope equation,

$$R = \frac{2I\sigma_0^2}{I_0\gamma\epsilon_n^2}, \tag{1}$$

where $I_0 = ec/r_e$ is the Alfven current and ϵ_n is the normalized emittance. For the PWTP beam, $R >> 1$ and non-linear space-charge effects clearly dominate. Measurement techniques which rely on linear trensport theory, such as quad scans, will be inaccurate. In such cases it is usually preferable to use a device which separates the beam into emittance-dominated beamlets, such as slits or a pepper pot. However, a slit-based measurement of the PWTP beam would require the slit width to be less than 25 μm. This design requirement presents substantial practical difficulties. As such, emittance measurements at PEGASUS will be made with a quad scans. A theory is being developed at UCLA which will make corrections to the linear transport model.

2.5. Undulator

The PEGASUS undulator is a 2m planar magnet undulator with a 20.5 mmis period and K=1.05. The undulator was constructed in collaboration with the Kurchatov Institute, and SASE was first observed with it at LANL [2]. The undulator enclosed in a vacuum box. A pulsed wire apparatus has been developed at UCLA which was used successfully used in the UCLA/LANL experiment. The apparatus allows for multipole B-field measurement, and thereby facilitates alignment and measurement of the undulator parameter.

REFERENCES

1. S. Reiche, et. al., Presented at the 2000 International FEL Conference.
2. M. Hogan, et. al., Physical Review Letters, 81 22 (1998) 4867.

FREE ELECTRON LASERS 2000
V.N. Litvinenko and Y. Wu (Eds.)
2001 Elsevier Science B.V.

Injector and bunchers of the electron linac for high-intensity single-bunch beam generation at ISIR

Shuichi Okuda*, Tamotsu Yamamoto, Shoji Suemine, Goro Isoyama

The Institute of Scientific and Industrial Research, Osaka University,
8-1 Mihogaoka, Ibaraki, Osaka 567-0047, Japan

1. Introduction

The high-intensity single-bunch electron beams are generated with the 38 MeV L-band (1300 MHz) electron linear accelerator (linac) at the Institute of Scientific and Industrial Research (ISIR) in Osaka University [1, 2]. One of the main application researches of the beams is the development of new light sources in a far-infrared region using the self-amplified spontaneous emission (SASE) [3] and the coherent synchrotron and transition radiation [4].

In order to improve the characteristics of the beam a high-current electron gun [5] was developed. After the installation of the gun in the linac the electron charge in the single-bunch beam increased up to 91 nC/bunch.

The present work was performed to investigate the behaviors of the beams injected from the gun and in the specific buncher system of the linac.

2. Accelerator system

The accelerator system of the ISIR linac is schematically shown in Fig. 1. The linac is equipped with a newly developed 120-keV triode gun, which generates electron beams at a peak beam current of 30 A. The cathode-grid assembly YU-156 (EIMAC) is installed in the gun. The cathode area of the assembly is 3.0 cm^2. The grid pulser of the gun made by using avalanche-type pulsers is installed in a gun tank filled with SF_6 gas. The pulse width is 5 ns FWHM.

The output beam from the gun is focused with two magnetic lenses. The location of the lenses was determined as to minimize the loss of the beam at the wall of the vacuum pipe, according to the beam trajectory which was calculated using the simulation code EGN2e.

The SHPB has a coaxial single-gap cavity at one end of the inner conductor. Pulsed rf of 20 μs duration is supplied to the SHPB by a 20 kW rf amplifier. An electron beam injected at a pulse length shorter than 5 ns from the gun to the first SHPB forms a single-bunch beam.

Two bunchers are driven by a 5 MW L-band klystron. The accelerating waveguide is driven by a 20 MW L-band klystron. The maximum beam energy is 38 MeV for no beam-loading.

3. Behaviors of the output beams from the gun

The characteristics of the single-bunch beams of the ISIR linac were measured and those are listed in Table 1. The maximum charge obtained so far is 91 nC/bunch. Fig. 2 shows the dependence of the charge of the single-bunch beam on the peak current of the beam injected from the gun. This figure shows an approximately linear dependence between the two parameters. Hence, it is expected to obtain the higher charge of the beam by increasing the charge for injection. The rise time of the electron beam pulse at injection in 0.1-0.9 is 1.6 ns. The rise time will be improved to be below 1 ns. By this improvement the charge of electrons in a bunch is expected to exceed 100 nC/bunch.

The normalized rms emittance of the beam was measured by the method using quadrupole magnets and a beam profile monitor, and those are 100-150 πmm mrad at charges below 63 nC/bunch. Such relatively low values seem to indicate that the emittance growth of the beam in the SHPB system is

*Corresponding author. Tel.: +81-6-6879-8511;
fax: +81-6-6875-4346; e-mail: s-okuda@sanken.
osaka-u.ac.jp.

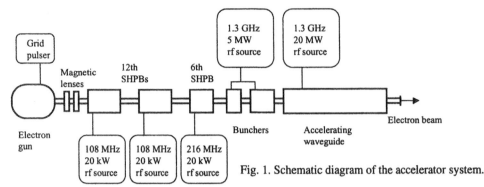

Fig. 1. Schematic diagram of the accelerator system.

not remarkable even at relatively high charges of the beam.

At a charge of the accelerated beam of 91 nC/bunch about 50% of electrons injected from the gun has been lost. The loss can be attributed to the pulse characteristics of the beam from the gun, and that in the beam transport system after the accelerating waveguide.

For the injection from the gun at the longer pulse lengths up to 4 μs under the different operational modes of the SHPBs, multibunch beams are generated in the oscillation experiments of a free-electron laser.

4. Conclusions

In order to increase the intensity of the single-bunch electron beams which are being applied to the SASE and the coherent radiation experiments, the injector system of the L-band linac at ISIR was improved. The charge of the single-bunch beam increased up to 91 nC/bunch. The emittance growth under the transportation of the electron beams in the SHPB system was not remarkable. Higher charges will be

Table 1   Parameters of the single-bunch beam

Energy	27 MeV
Accelerator frequency	1300 MHz
Charge/bunch	91 nC
Peak current/bunch	3-5 kA
Bunch length (FWHM)	20-30 ps
Energy spread (FWHM)	2.5%
Normalized emittance	140 πmm mrad
	(at 63 nC/bunch)

----

obtained by the improvement of the gun pulser.

References

[1] S. Takeda, K. Tsumori, N. Kimura, T. Yamamoto, T. Hori, T. Sawai, J. Ohkuma, S. Takamuku, T. Okada, K. Hayashi, M. Kawanishi, IEEE Trans. Nucl. Sci. NS-32 (1985) 3219.

[2] S. Okuda, Y. Honda, N. Kimura, J. Ohkuma, T. Yamamoto, S. Suemine, T. Okada, S. Takeda, K. Tsumori, T. Hori, Nucl. Instr. and Meth. A358 (1995) 248.

[3] R. Kato, R.A.V. Kumar, T. Okita, S. Kondo, T. Igo, T. Konishi, S. Okuda, S. Suemine, G. Isoyama, Nucl. Instr. and Meth. A445 (2000) 164.

[4] S. Okuda, M. Nakamura, K. Yokoyama, R. Kato, T. Takahashi, Nucl. Instr. and Meth. A417 (1998) 210.

[5] S. Okuda, T. Yamamoto, S. Suemine, S. Tagawa, Nucl. Instr. and Meth. A417 (1998) 210.

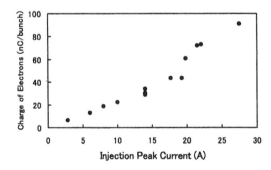

Fig. 3. Relation between the electron charge in the beam and the peak current for injection.

FREE ELECTRON LASERS 2000
V.N. Litvinenko and Y. Wu (Eds.)
2001 Elsevier Science B.V.

Status of the ELBE Accelerator - a Driver for Different Radiation Sources

P. Michel, P. Evtushenko, K. Möller, C. Schneider, J. Teichert, F. Gabriel, E. Grosse
and the ELBE-crew

Forschungszentrum Rossendorf, PSF 510119, D-01314 Dresden, Germany

The ELBE project

At the Forschungszentrum Rossendorf (FZR) a superconducting electron linear accelerator with high brilliance (ELBE) [1] is being built. It will deliver a maximum electron energy of 40 MeV and an average beam current of up to 1 mA. The construction of the ELBE building with caves for housing the accelerator and the experimental equipments has been completed. The accelerator is expected to deliver the first electron beam within the year 2001. By means of the electron beam different kinds of radiation will be produced, such as: Infrared radiatiation from FEL's, MeV-bremsstrahlung, neutrons and X-rays. Different beam requirements resulting from the planned experiments are accomplished by an electronically grid-pulsed thermionic gun, which can be operated in different modes. A further macro pulser allows a very flexible time structure of the beam. The gun is operated at 250 kV. Bunch compression for injection into the first LINAC is done by two bunchers operating at 260 MHz and 1.3 GHz, respectively. For future use with smaller transverse emittance a different type of injector is needed. A new SRF photocathode gun for this injector is being developed. The main accelerator uses standing wave RF-cavities at 1.3 GHz from TTF at DESY. The superconducting Niobium cavities are operated at 2 K in liquid helium. The accelerating gradient is higher than 15 MV/m. Each cavity is driven by a 10 kW klystron amplifier.

Results of 250 keV injector test

The electron beam from the gun and bunching stage (injector) was studied using a diagnostic module. A multislit mask and a beam viewer allow to measure the transverse emittance in the range from 1 to 20 mm mrad. Fig.1 shows the measured transverse geometric emittance of the ELBE injector at 13 and 260 MHz pulse repetition rate. The 260 MHz regime allows operation at very low bunch charges which is substancial to achive low emittance values. For comparison, Fig.1 presents the results of a numerical simulation, too. The influences of electrode geometry, space charge and of the control mesh in front of the cathode are included in the simulation. The results show that the growth of the transverse emittance for higher bunch charges is mainly caused by the control mesh. The strong acceleration field penetrates through the mesh holes, and for the electrons each hole acts as a small lens. Consequently the portion of deflected electrons increases with electron current. In the numerical simulation the field and electron trajectory calculation was performed using the EGUN code [2]. For the electron transmission through the potential barrier of the mesh the free electron gas approximation with a local penetration probability, a Maxwell energy distribution, and an angular

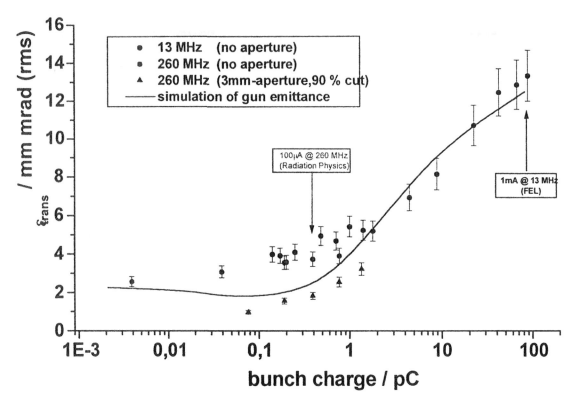

Fig.1: Transverse emittance of the ELBE injector

distribution according to Lambert´s law was applied.

The longitudinal beam parameters were measured by a transverse deflecting 1.3 GHz kicker resonator in front of a vertically bending dipole. The longitudinal phase space distribution can be observed directly on a beam viewer. The micro pulse bunch length can be derived from projection to the time axis. We measured 12 ps in the case of optimal bunching conditions, which are determined by the two buncher phases and the power (see Fig. 2). This value is in good agreement with the results of Parmela [3] calculations.

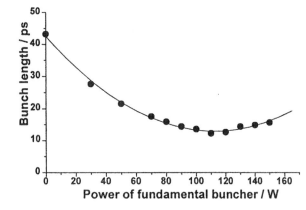

Fig.2: Bunch length of the ELBE injector beam. The power of the pre buncher is fixed at 400 W.

References:
[1] F. Gabriel et al., Nucl. Instr. Meth. **B161-163** (2000) 1143-1147
[2] W.B. Herrmannsfeldt, Report **SLAC-331-UC-28**, Stanford University, 1988
[3] L.M. Young, Los Alamos **LA-UR-96-1835**

FREE ELECTRON LASERS 2000
V.N. Litvinenko and Y. Wu (Eds.)
2001 Elsevier Science B.V.

Design of a Double Focusing Beam Transport System for a 25 MeV Electron Beam

I.V. Volokhine[*], J.W.J. Verschuur, G.J. Ernst, F.F. de Wit, K.J. Boller

University of Twente, P.O. Box 217, 7500 AE Enschede, The Netherlands

J.I.M. Botman

Technical University of Eindhoven, P.O. Box 513, 5600 MB Eindhoven, The Netherlands

A double achromatic and double-focusing electron transport system design is described for the transport of the electron from a racetrack microtron to the undulator. Matching of the 25 MeV electron beam to the undulator appeared to be possible which has been checked by PARMELA simulations. The action of the last bend of the RTM has to be taken into account and was corrected for. The influence of space charge effects in the 100 A electron beam also has been taken into account.

Keywords: *electron beam, transport, double achromatic, dispersion*

1. Introduction

At the University of Twente in co-operation with the Technical University of Einhoven a Free Electron Laser (TEUFEL phase II) [1-2] is being realised with a set-up of two accelerators. The injector is a 6 MeV photo cathode linac producing a high brightness electron beam. The second accelerator is a 25 MeV Racetrack Microtron [3]. The length of the resonator is only 1.846 m. Because most space is occupied by the undulator (1 m length, 40 periods of 25 mm) [4], a compact injection part of the transport system is required. The transport system has to guide the 100 A, 25 ps length bunches from the RTM to the undulator, so as to provide 10 μm radiation. In this paper the design of the transport line, the condition for matching of the electron beam to the undulator, and results of the simulations are described.

2. Design

In the Fig.1 the general layout of the setup and transport line are represented.

physical requirement of matching the electron beam to the undulator as well as obtaining the optimal parameters of the transport system, the second one is the practical design (general layout of the equipment, the demands of the space inside the resonator).

The output electron beam from the RTM has an angle of 12 degrees with the direction of the main axis of the system; the axis of the undulator has to be parallel to it. One of the problems is that the parameters of the output beam from the RTM are not exactly known. Therefore the transport line has to be flexible with respect to the electron beam input parameters. Generally each round trip inside the RTM is double achromatic, exceptions are the last extraction half-turn and the first injection half-turn. The injection half-turn will not have a large dispersion because it has a small radius and the electron energy has the relatively low value of 6 MeV. The last half-turn has a large radius and a high electron energy. Therefore the influence of the last half-turn is important and it is necessary to take the influence of the last bend of the RTM into account. In the fig.2 the results obtained by code

Fig.1 General layout of the setup.

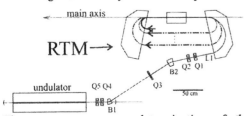

There are two aspects determination of the parameters of the transport line. The first one is the

Fig. 2 Beam radius envelope of the X, Y and dispersion along transport line.

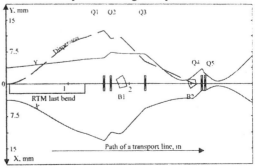

---

[*] Corresponding author. E-mail: I.Volokhine@tn.utwente.nl

TRANSPORT are shown.

It can be seen that the outgoing electron beam from the RTM has a small angle of dispersion and a large shift of it. Thus the action of the RTM's last bend has to be corrected. The transport line includes 2 bends and 5 quadruple lenses. The angle of the RTM bend is 168 degrees and the radius for the electrons with energy 25 MeV is 46 cm. The actions of the quadruples Q1, Q2 and drift L2 provide the matching between the RTM and the transport line. After that first section of a quadruples the reference particles have only a shift of the dispersion in the X direction without a slope. By changing the focus strength of the quadruples Q1 and Q2 and the drift length L2 it is possible to get the same adjustment for different parameters of the output electron beam of the RTM without changing the other elements of the transport line. For the first bend B1 20° and a radius of 28 cm has been chosen in order to avoid strong focusing. The second bend B2 has an angle of 32° in order to get the axis of the undulator and the main axis of the system parallel to each other. The magnetic field strength is 0.7 T, which is almost at the beginning of saturation, in order to get a small size of the injection element. The place of the quadruple lens Q3 between the bends allows adjusting the asymmetric bends in order to get zero dispersion. Two quadruples Q4-Q5 comprise the focusing elements of the transport system. They are placed in the dispersion free region. This allows easy changing of the diameter of the focus spot as well as the position of the waist. This makes the matching between the electron beam and the undulator more flexible.

Thus the transport system has 3 adjusting elements. The first section (Q1, Q2) is for the adjustment of the RTM and the transport line. The second section Q3 is to compensate for the dispersion. The third section (Q4, Q5) provides both the focussing and the matching of the electron beam to the undulator.

An important question is: What is the matching condition of the undulator. From a theoretical analysis [5] one can find the following equations for the waists w_x and w_y of the matched beam in the x and y directions respectively:

$$w_x = \sqrt{\frac{\varepsilon_{RMS}/\sqrt{2}}{2 \cdot \pi^2} \cdot \lambda_x}, \quad w_y = \sqrt{\frac{\varepsilon_{RMS}/\sqrt{2}}{2 \cdot \pi^2} \cdot \lambda_y} \quad (1)$$

in which ε_{RMS} is the RMS-value of the emittance, which can be estimated as 6 π·mm·mrad and λ_x and λ_y are the wiggler focusing lengths, assuming equal focusing in both directions, given by:

$$\lambda_x = \sqrt{2} \cdot \lambda_\beta, \quad \lambda_y = \sqrt{\frac{\lambda_\beta^2 \cdot \lambda_x^2}{\lambda_x^2 - \lambda_\beta^2}} \quad (2)$$

By using the value for the energy 25 MeV, wiggler wavelength 25 mm, and wiggler peak magnetic

field strength 0.67 T we find for the betatron wavelength λ_β a value of 1.596 m. Using equations (3) and (2) we find $w_x = w_y = 1.04$ mm.

In the Fig.3 the results from the simulation by the

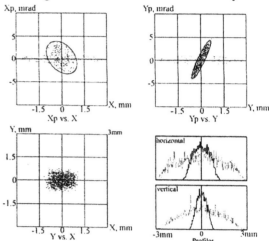

Fig.3 Initial (grey) and final (black) phase-space ellipses (top) and profiles (bottom) obtained by the code PARMELA.

code PARMELA are represented. This code made the dynamic 3-D calculations including space charge effects for a 300 A beam. The calculation with space charge shows that the space charge effect can be neglected. The matrix formalism method and dynamic 3-D calculation give very similar results.

3. Acknowledgement

This joint project has been supported in a part by the Netherlands Technology Foundation (STW).

4. References

[1] G.J. Ernst et al., 13th Int. Free Electron Laser Conf., Sante Fe, USA, 1991,Nucl. Instr. and Meth., A318 (1992), 173.

[2] G.J. Ernst et al., 17th Int. Free Electron Laser Conf., New York, NY, USA, 1995, Nucl. Instr. and Meth., A375 (1996), 26-27.

[3] J.I.M. Botman et al., 12th Int. Free Electron Laser Conf., Paris, France, 1990, Nucl. Instr. and Meth., A304 (1991), 192-196.

[4] J.W.J. Verschuur, et al., 13h Int. Free Electron Laser Conf., Sante Fe, USA, 1991,Nucl. Instr. and Meth., A318 (1992), 847-852.

[5] Charles A. Brau, Free Electron Lasers, Academic Press, INC.

FREE ELECTRON LASERS 2000
V.N. Litvinenko and Y. Wu (Eds.)
2001 Elsevier Science B.V.

Study on the focusing property of a planar wiggler for the infrared FEL

M. Fujimoto, R. Kato, M. Kuwahara, T. Igo, T. Okita, T. Konishi, R. A. V. Kumar, S. Mitani, S. Okuda, S. Suemine and G. Isoyama*

Institute of Scientific and Industrial Research 8-1 Mihogaoka, Ibaraki, Osaka 567-0047, Japan

Abstract

We have studied the focusing property of a planar wiggler for the infrared FEL. The vertical beam size has been measured as a function of the K-value at three positions in the wiggler using screen monitors. The measured beam sizes oscillate as the K-value increases, which indicates that the vertical focusing force is very strong in the wiggler for the electron energy of 11.4 MeV and K-values up to 1.472. This oscillatory behavior of the beam size is explained by a simple model using transfer matrices. A practical method is proposed to realize a small and constant vertical beam size in the wiggler using the strong natural focusing force and the matched beam condition.

1. Introduction

We have been developing the far-infrared free electron laser (FEL) and conducting basic study on self-amplified spontaneous emission (SASE) in the wavelength region longer than 100 μm, using a planar wiggler of the horizontal oscillation type and the L-band electron linac at the Institute of Scientific and Industrial Research, Osaka University [1,2]. A planar wiggler exerts a focusing force in the direction perpendicular to the oscillation plane of the electron beam, which is proportional to the inverse square of the electron energy and to the square of the wiggler magnetic field [3]. The wiggler has a magnetic field up to 0.37 T and the electron energy is lower than those for most of the existing FELs in the Compton regime, so that the focusing force is expected to be stronger. Since no quantitative experimental study has been, to our knowledge, reported on the focusing properties of a planar wiggler for an infrared or far-infrared FEL using a low energy electron beam, we have conducted experiments to study them using the wiggler of the ISIR-FEL at Osaka University.

2. Experiment

A single-bunch electron-beam accelerated using the L-band linac was steered to the FEL through the achromatic beam transport system and the electron beam size was measured in the wiggler.

The characteristics of the electron beam used in the experiment and the main parameters of the wiggler are listed in Table 1. The energy of the electron beam was 11.4 MeV. The period length of the wiggler is 6 cm and the number of periods is 32. The K-value of the wiggler can be varied up to 1.472. The electron beam is focused in the wiggler using a quadrupole doublet. Three screen monitors are installed at the entrance (S1), the centre (S2) and the exit (S3) of the wiggler. The monitors at the entrance and the exit are 85 mm inside from the respective end of the wiggler magnet. A 1 mm thick fluorescent plate, placed at 45° with the beam axis, is used in each screen monitor. The beam profile was measured with a CCD camera and the video signal was processed with an image-processing instrument. The beam size was obtained by the least square fit of a Gaussian function to the measured data. Thus beam sizes have been measured at the three locations as a function of the K-value of the wiggler.

Table 1
Main parameters of the electron beam and the wiggler

Electron beam	
Operation mode	Single bunch
Energy	11.4 MeV
Energy spread (FWHM)	3.8 %
Normalized emittance	~200 π mm mrad
Wiggler	
Total length	1.92 m
Magnetic period	6.0 cm
No. of periods	32
Magnet gap	120-30 mm
Peak field	0.37 T
K-value	0.013-1.472

*Corresponding author.
Tel.: +81-6-6879-8485; fax: +81-6-6879-8489
E-mail address: isoyama@sanken.osaka-u.ac.jp

3. Results and discussions

Fig. 1 shows vertical beam sizes as a function of the K-value. The open squares, the open circles and the filled circles are the measured beam sizes at S1, S2 and S3, respectively. These beam sizes are mean values of five measured values and the error bars are their standard deviations. The variation of the beam size with the K-value is small for S1, but the beam size begins to oscillate for S2 and then the number of oscillations increases for S3. This implies that the focusing force of the wiggler is so strong that vertical beam size oscillates along the z-axis and the number of oscillations increases as the K-value becomes larger.

The vertical beam size calculated using linear beam dynamics theory is given by [4]

$$\sigma_y^2(z) = [1 + \cos(2\sqrt{f}z)]A + [1 - \cos(2\sqrt{f}z)]f^{-1}B$$
$$- \sin(2\sqrt{f}z)f^{-1/2}C , \qquad (1)$$

where y and z are the vertical and the longitudinal coordinates in the wiggler, respectively, A, B and C are constants determined by the beam emittance and Twiss parameters at the entrance of the wiggler, and f is the restoring force given by $f = mK^2/\lambda_w^2\gamma^2$. The solid curve in Fig. 1 shows the fit of Eq. (1) to the measured beam sizes at S3 by the least square method, where as fit parameters we used the beam emittance, Twiss parameters (α_0, β_0) at the entrance of the wiggler, and a coefficient m for f, which are also shown in the figure. The theoretical curve agrees quite well with the experimental data for S3. The normalized emittance of 187 π mm mrad agrees reasonably well with the typical value given in Table 1. The dotted and the dashed curves show beam sizes calculated for S1 and S2, respectively, using Eq. (1) and fit parameters obtained for S3. They agree quite well with the measured beam sizes respectively for S1 and S2. These results show the vertical beam size oscillates several times in the wiggler for a larger K-value.

Here we propose a practical method to make the vertical beam size small and constant along the wiggler using the strong natural focusing force of the wiggler and the matched beam condition. A matched beam is achieved by making $\alpha_0 = 0$ and $\beta_0 = f^{-1/2}$ at the entrance of the wiggler, using the quadrupole magnets in the beam line. In the matched beam condition, the value of the betatron function is constant in the wiggler and equal to 0.156 m for the 11.4 MeV electron beam and a magnet gap of 30 mm. Since the restoring force f is proportional to the inverse square of the electron energy, the matched value of the betatron function is larger as the electron beam energy is higher.

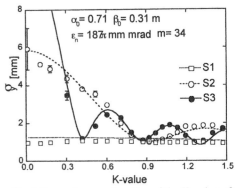

Fig.1 Beam size as a function of the K-value of the wiggler. See text for details.

In case the wiggler has no focusing force, it is well known that the electron beam waist should be made at the centre of the wiggler and the value of the betatron function does not depend on the electron energy [4]. The optimal value of the betatron function is 0.96 m at the waist in a wiggler of length 1.92 m. The matched value of 0.96 m is obtained by the natural focusing force of the magnet gap of 30 mm for the 70 MeV electron beam. This indicates that the matched beam condition using the natural focusing force has an advantage for electron energies up to 70 MeV. In order to realize the matched beam condition, the beam has to be focused at the entrance of the wiggler to make a waist with an appropriate betatron function when the magnet gap of the wiggler is fully opened. In our experimental arrangement, S1 is equipped at a position slightly inside of the wiggler from the entrance and the small difference is not negligible. It is possible to adjust the Twiss parameters at the entrance by making use of the vertical focusing property of the wiggler together with the screen monitor at the center S2. The magnet gap of the wiggler is adjusted to make the vertical betatron wavelength equal to the wiggler length, so that the vertical beam image at the entrance is focused again at the center. Since the betatron wavelength is given by $\lambda_\beta = 2\pi/f^{1/2}$, the condition for the 11.4 MeV electron beam is realized when the magnet gap is 45 mm.

References

[1] R. Kato, S. Kondo, T. Igo, T. Okita, T. Konishi, S. Suemine, S. Okuda, G. Isoyama, Nucl. Instr. And Meth. A445 (2000) 169.
[2] R. Kato, R. A. V. Kumar, T. Okita, S. Kondo, T. Igo, T. Konishi, S. Okuda, S. Suemine, G. Isoyama, Nucl. Instr. And Meth. A445 (2000) 164.
[3] W. B. Colson, Laser Handbook (1990) (North-Holland) vol. 6, p. 122.
[4] H. Wiedemann, Particle Accelerator Physics (1993) (Springer-Verlag) p. 118 (ch.5).

FREE ELECTRON LASERS 2000
V.N. Litvinenko and Y. Wu (Eds.)
2001 Elsevier Science B.V.

Compact Microtron Driven by a Magnetron as an Injector for a Far Infrared Free Electron Laser

Grigori M. Kazakevitch[b,1] , Young Uk Jeong[a], Byung Cheol Lee[a], Sung Oh Cho[a], Sun Kook Kim[a] and Jongmin Lee[a].

[a]Lab. for Quantum Optics, Korea Atomic Energy Research Institute, P.O.Box 105, Yusong, Taejon, 305-600, Rep. Korea.
[b]Budker Institute of Nuclear Physics RAS, Academician Lavrentyev 11, Novosibirsk, 630090, Russia.

Abstract

A 12 turn 8 MeV classical microtron has been improved and is used as an electron beam injector for a compact far infrared (FIR) free electron laser (FEL). The microtron, which is driven by a 2.8 GHz magnetron, provides an accelerated beam current up to 70 mA in the 5.5 µs macro pulse. The frequency of the magnetron is stabilized to $(3-4)\cdot10^{-5}$ with the use of reflected wave from the accelerating cavity. The energy spread and emittance of the electron beam are less than 0.4% and 1-4 mm·mrad, respectively. A measured temporal deviation of the bunch repetition rate of less than 120 kHz could be obtained by optimization of the microtron operating conditions. The compact microtron is successfully employed as a driver for the FIR FEL. With the 7 MeV electrons and macro pulse current of approximately 45 mA, we generated the tunable FIR FEL radiation in the range of wavelengths from 97 to 150 µm. The measured spectrum width was in the range of about 0.5-1.3%. for the deep saturation in the FEL.

A classical microtron has good bunching properties and acceptable values of energy spread and emittances in the electron beam to be used as an injector into coherent FIR radiation sources [1]. Our attempt to use a 12-turn compact microtron as an injector of a bunched electron beam for a compact FIR FEL was successful.

We developed this microtron on the basis of its prototype [2]. During the development significant attention was paid to the stability of the microtron operating parameters and compactness of the facility as a whole. The microtron, including magnetic and vacuum systems, an RF system and a modulator, occupies less than 1.4 m^2.

The microtron RF-system is driven by a 2.8 GHz tunable magnetron, MI-456A, and is designed to provide good frequency stability for the FIR FEL operation. The accelerating frequency was stabilized by the frequency pulling of the magnetron. A cylindrical accelerating cavity of first type [3] with an internal thermionic LaB_6 cathode is used in the microtron. The coupling coefficient of the cavity is about 5.4.

The macro pulse current of the accelerated beam in the microtron is up to 70 mA. The capture coefficient is in the range of 4.5-6.8%. The microtron extracting system efficiency is about 90%.

To provide a flat-top of the accelerating current by the macro pulse duration of about 5.5 µs (Fig. 1, curve-1) we shaped the modulator pulse as shown in curve-2 of Fig. 1.

For choosing the initial magnetron-cavity detuning we took into account estimated values of the accelerating cavity frequency drift by the electron beam loading and the intra pulse frequency instability. They were determined from the temporal measurements of bunch repetition rate variation during the macro pulse [4].

For initial detuning of about 140 kHz, the corresponding maximal value of the accelerated current was ~ 45 mA, and the measured deviations of the intra-pulse bunch repetition rate were in the range of ~100 kHz.

Electron beam emittances were determined from the measured transverse beam sizes via scan of the current of quadruple lenses. The beam profile sizes

---

[1] *Corresponding author: Lab. for Quantum Optics, Korea Atomic Energy Research Institute, P.O. Box 105, Yusong, Taejon, 305-600, Rep. Korea, Tel: +82 42 868 8253, Fax: +82 42 861 8292;E-mail: kaza@kaeri.re.kr

were measured by using a movable Optical Transition Radiation (OTR) screen. The accuracy was ~ 100 μm. These data were used for optimization of the optical parameters of a beam line transport. The beam line is comprised of 3 quadrupole doublets, 6 X-Y steering magnets and 3 bending magnets. It transports the accelerated beam into a FIR FEL resonator. The FEL resonator is composed of a planar wave-guide chamber and two cylindrical curved in the horizontal plane. One of the mirrors has a coupling hole to extract a laser beam from the FIR resonator. The cross section and the length of the wave-guide are 2 mm ×30 mm and 2781 mm, respectively. The coupling coefficient of the FIR resonator is about 1.7%. The wave-guide chamber is placed in the gap of 2-m long, 80-period U-25 undulator with variable magnetic field. Vertical and horizontal sizes of the optimized electron beam at the entrance of the undulator were 0.6 mm and 2.1 mm, correspondingly. Two OTR screens (at the entrance and output of the undulator) and a He-Ne laser system were used for the centering of the electron beam in the wave-guide chamber with accuracy ~ 0.1 mm in the vertical plane and ~ 0.3 mm in the horizontal plane. Under this condition, the loss of the electron beam in the wave-guide was less than 30%. After passing through the wave-guide chamber, the electron beam is directed into a beam dump by a bending magnet. Main parameters of the accelerated electron beam for the FIR FEL are presented in Table 1.

Table 1. Electron beam parameters for the KAERI FIR FEL.

Total energy of electrons:	~7 MeV.
Electron beam energy spread:	<0.4%.
Vertical emittance ε_y:	~ 1.5 mm·mrad.
Radial emittance ε_x:	~ 3.5 mm·mrad.
Macro pulse current:	45-50 mA.
Macro pulse duration:	~5.5 μs.
Intra macro pulse bunch repetition rate deviation:	< 120 kHz.

During the FEL experiments, the FIR radiation passed through a quartz lens which focused it at the entrance of a detector (for power measurements) or made a parallel beam (for the spectrum width measurements). The quartz lens also served as a vacuum window.

By changing the K-parameter of the undulator in the range of 1.05-1.6, we successfully detected a FIR laser beam. The tunability range was 97-150 μm. The measured spectrum width was in the range of 0.5-1.3%. The macro pulse power was ~ 30-50 W. [5]. The oscilloscope trace of the attenuated FIR lasing signal detected by Ge-Ga detector (cooled by liquid He), is shown in Fig. 1, curve-3.

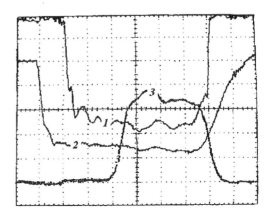

Fig. 1. Oscilloscope traces of signals reletated to the microtron and the FIR FEL. Horizontal scale is 1 μs/div.

1-accelerated current, vert. scale is 10 mA/div.
2-magnetron current, vert. scale is 22.2 A/div.
3- attenuated FIR lasing signal at the output of Ge-Ga detector, vert. scale is 1 V/div.

References

[1] G.M.Kazakevitch et al. "Bunching Properties of a Classical Microtron-Injector for a Far Infrared Free Electron Laser." Presented at the 22nd International Free Electron Laser Conference, FEL 2000 Conference, Duke University, 2000.

[2] G.M.Kazakevich et al. "8 MeV microtron-the injector for an electron synchrotron." Proceed. of second Asian Symposium on Free Electron Lasers, BINP, Novosibirsk, 1995.

[3] S.P.Kapitza and V.N.Melekhin. The Microtron. London. Harwood, 1978.

[4] G.M.Kazakevich et al. "Measurements of Intrapulse Instabilities in the Electron Beam of high Current Microtron." in Free Electron Lasers in Asia. Theory, Experiments and Applications. Jongmin Lee, Byung Cheol Lee, Editors. Proceed. of 4th Asian Symposium on Free Electron Lasers, 198, 1999.

[5] Y.U.Jeong et al. "First Lasing of the KAERI Compact Far-Infrared Free-Electron Laser Driven by a Conventional Microtron." Presented at the 22nd International Free Electron Laser Conference, FEL 2000 Conference, Duke University, 2000.

FREE ELECTRON LASERS 2000
V.N. Litvinenko and Y. Wu (Eds.)
2001 Elsevier Science B.V.

Electron beam energy spectrum measurements at Vanderbilt FEL

B. Feng, R.C Grant, R. Ardrey, J. Kozub and W.E. Gabella

FEL Center, Vanderbilt University, Box 1816 B, Nashville, TN37235

Abstract

The electron beam energy spectrum strongly affects the free electron laser efficiency, especially in the regime of large slippage between the electron bunch and the optical pulse. A fast electron energy spectrometer has been built using a photodiode array measuring the backward optical transition radiation from a thin film of aluminium. We present the measurements of the time resolved electron beam energy spectrum for the Mark III linear accelerator at Vanderbilt University, while lasing in the 2-9 micron range.

1. Introduction

The free electron laser (FEL) at Vanderbilt University is applied in many fields, including materials physics, biophysics and medical sciences, due to its high power, picosecond pulse and widely tuneable wavelength range [1,2]. FEL light is generated when a relativistic electron bunch train passes an N-period wiggler producing spontaneous radiation which is stored in an optical cavity and amplified by succeeding electron bunches. Due to the different velocities of electron bunch and optical micropulse propagating in a wiggler, there is a slippage between them, which is usually described by the longitudinal coupling parameter, $\mu_c = N\lambda / l_b$, where N is the number of periods in the wiggler, λ is the laser wavelength and l_b is the electron bunch longitudinal length. In the case of the Vanderbilt FEL facility, the wiggler period number is 52.5 and the laser wavelength range is 2 to 9.5 microns. The bunch length of less than 0.3 mm is estimated, so the longitudinal coupling parameters are in the range of 0.33 to 1.65. At a wavelength of 6 μm and longer, the longitudinal coupling parameter is greater than one. Therefore the slippage effects become important, and it becomes necessary to measure the time-resolved electron energy spectrum. Several different techniques have been developed to measure the electron energy spectrum, including using transition radiation (TR) [3-5]. We have built a fast electron energy spectrometer using a photodiode array measuring the backward optical transition radiation (OTR) from an aluminized pellicle. Comparing with the earlier work at FELIX [5], we simply use an oscilloscope and a Mac computer with LabVIEW for the data acquisition. There are several ways we can benefit from the electron spectral measurement. Basically, it enhances our understanding of the interaction between the electron micropulse and the optical pulse in the wiggler. It allows us to derive the mean energy extracted from the electron beam. And also it improves the operation and control of the FEL dynamic behavior.

In this paper we describe the fast electron energy spectrometer and the measurement of the time-resolved electron spectrum. We also give a brief description of a high-resolution detector array for OTR and its fast data acquisition electronics. The experimental results are shown in this paper.

2. The Electron Spectrometer

The electron spectrometer consists of a spectrometer magnet, a radiator of optical transition radiation, optical transport lenses, and a fast OTR detector. The magnet is a 90° bend angle dipole with a bend radius of 305 mm and both side pole-edge rotation angles of 15°. The incident electrons are bent by the magnet and dispersed onto an OTR radiator near the focal plane of the magnet. The dispersion is about 3 mm per percent.

The OTR radiator is an 8 μm kapton film with a 40 nm coating of aluminium, which is mounted at an angle of 45° to the electron beam. Electrons emit TR in the forward direction along the electron beam path, and in the backward direction along the direction of specular reflection from the foil. The angle of the peak intensity to the emission direction is $1/\gamma$, where γ is the electron energy factor, and almost all of the TR intensity is distributed around this angle. The OTR is measured by a 32-element photodiode array. The rise time of the photodiode is 50 ns, and it is sensitive to 190 to 1000 nm light, with peak response at 800 nm. The system bandwidth is about 10 MHz, fast enough to measure electron macropulse energy spectrum. The backward TR is focused on the detector through an optical system, which consists of an objective lens of f=200 mm, two mirrors and a zoom lens with f=75~300 mm. The magnification of the system is in the range of 0.5~1.0. The system dispersion can be changed by the zoom lens, and is set to 0.33% per channel, resulting in a full energy spread width of 10%. However the diameter of the radiator is only 25 mm, so only 6% energy dispersion can be observed. A CCD camera is also used to directly observe the electron distribution image on the aluminized film.

Two channels of the detector are selected simultaneously by multiplexers, preamplified and fed into an oscilloscope, which is used for data acquisition through a GPIB interface and LabVIEW. In order to construct an entire electron spectrum by the 32-channel detector, sixteen separate macropulses are measured. Currently the repetition rate for the measurement is 2 Hz, but 30 Hz may be possible. The electron beam is usually stable over such a short time.

3. Experiment results

The Vanderbilt FEL operates in the infrared region from 2 to 9.5 μm using an electron beam energy between 25 to 45 MeV accelerated by a Mark III type linac. The electron beam macropulse width is 8 μs and micropulse width is estimated as 0.7-1.0 ps. The longitudinal coupling parameters are in the range of 0.33 to 1.65. As an example, a three dimension time-resolved energy spectrum is shown in Fig.1 (a), and its contour is in Fig.1 (b), which shows the electron energy spread about 2% (FWHM). For the same beam conditions (beam energy is 33.3 MeV), the spectrum becomes wider while lasing at 5.0 μm, showed in Fig.1 (c), and (d).

In Fig.2, the time-resolved electron energy spectrum is compared with the time-resolved FEL wavelength

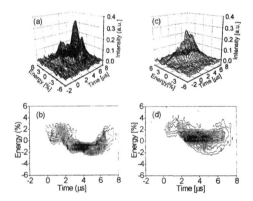

Fig. 1. Typical electron spectrum 3-d and contour plots.

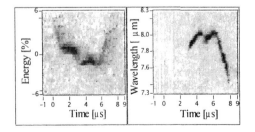

Fig.2 Comparison of time-resolved electron energy spectrum (left) and the FEL wavelenth spectrum (right)

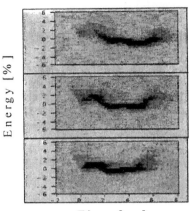

Fig.3 Electron spectrum vs cathode tuning: cold (top), normal (middle) and hot (bottom).

spectrum. The transient effects of beam loading in the electron accelerator shows in the first 2 μs, then lasing begins from 3 μs, resulting in the electron spectrum width increasing. With changing electron beam energy, the laser wavelength varies. After 6 μs, the laser shifts to shorter wavelengths because of the electrons distributed in the higher energy region. The trend of electrons losing energy is observed while lasing. The more energy is extracted by photons, the wider the spectrum becomes; the spectral distribution shifts to low energy. Similar patterns of time-resolved laser spectrum are observed [6], where the change of electron energy results in the change of laser wavelength.

Cathode heating in a RF gun effects the FEL efficiency. It represents not only the change of the electron macropulse current, but also the change of electron beam energy spectrum. Fig.3 shows the electron spectrum while changing the cathode heating. The beam energy is about 33 MeV, lasing at 5 μm. Overheating causes the tail of the electron distribution.

4. Conclusion

We have built a fast electron energy spectrometer with a simple method for data acquisition. We have measured the time-resolved electron beam energy spectrum on the Mark III linear accelerator at Vanderbilt University, while not lasing or lasing at different wavelengths. The cathode heating effects for the electron spectrum are discused. The optical system will be improved including changing to a larger diameter radiator and running at higher repetition.

References

[1] G. Edwards, et al., IEEE J. of Selected Topics in Quantum Electronics, 2 (1996) 810;

[2] W. Gabella, in this conference;

[3] R., Chaput, et al., Mucl. Instr. Meth. A331 (1993) 267;

[4] Lumpkin, et al., Nucl. Instrum. Meth. A231 (1993) 803;

[5] W. A. Gillespie et al., Nucl. Instrum. Meth. A358 (1995) 232;

[6] J. Kozub, et al., in this conference.

FREE ELECTRON LASERS 2000
V.N. Litvinenko and Y. Wu (Eds.)
2001 Elsevier Science B.V.

Dark current at TTFL rf-gun

D. Sertore[a], S. Schreiber[a] K. Zapfe[a] D. Hubert[a] and P. Michelato[b]

[a]Deutsches Elektronen Synchrotron,
Notkestrasse 85,D-22603 Hamburg, Germany

[b]INFN Milano-LASA, Via F.lli Cervi 201, I-20090 Segrate, Italy

During the last run, dark current measurements have been routinely done at TTFL rf-gun. Effects on beam performance and linac operation are presented.

1. Introduction

The TESLA Test Facility Linac (TTFL)[1] uses an RF photoinjector and a superconducting linac to accelerate electrons for the TTF free electron laser (TTF-FEL)[2]. The photoinjector is based on an RF gun operated at 35 MV/m, 100 μs rf flat top pulse length. The electron source is a Cs_2Te photocathodes illuminated by UV laser light. The 4th harmonic (λ= 262 nm) of a Nd:YLF laser[3]is used. Usually a train of bunches, with bunch charge between 1 nC and 8 nC, is generated. The bunch spacing in the train is 1 μs. Dark current is produced along the rf pulse. A possible source of dark current are tips or needles on surfaces exposed to high electric field. The current density due to this field emission is given by the Fowler-Nordheim relation [4]:

$$j \propto (\beta E)^{\frac{5}{2}} exp(-\frac{B}{\beta E}) \qquad (1)$$

where E is the electric field amplitude, β is an enhancement factor due to the geometry of the source and B is a material dependent parameter. The dark current is source of undesired beam halo and radiation along the linac. Therefore is important to understand dark current sources in order to reduce them as much as possible.

2. Measurements

The dark current is measured with a Faraday cup that can be inserted at the gun exit. The

signal from the Faraday cup is detected with an oscilloscope. A typical signal is reported in Fig. 1. The rising and falling edge of the rf pulse are

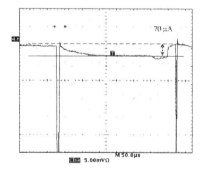

Figure 1. Oscilloscope trace used for dark current measurement. The spikes at the begin and at the end are due to multipactoring during Rf start and stop. The pedestal is due to dark current.

clearly visible as spikes at the beginning and at the end of the pulse. They are due to multipactoring during rise and fall of rf. In between the spikes, a pedestal is visible that grows along the pulse. This pedestal is due to the dark current produced during the Rf pulse. We measure the dark current by measuring the height of the pedestal at the end of the pulse, before the rf stop signal as shown in Fig. 1. From simulations, the source of the dark current is the gun backplane and mainly the region around the cathode. The Rf contact between the cathode and the body of the gun is assured by a CuBe spring. It is possible to image the dark current on a screen downstream from the gun and resolve the spring convolutions. The contribution to the dark current from the photoemissive material is lower than the Faraday Cup sensitivity. The dark current history for the cathodes used is reported in Fig. 2 [5]. The typical initial value of the dark current is below $100\,\mu A$ after the cathode insertion in the gun. The dark current then rises slowly. Often an abrupt increase is observed that can reach also some mA. Reused cathodes have always an initial dark current value higher than their fresh value and the current rise is earlier and faster. On the contrary, degradation of quantum efficiency has not been observed. The experimental evidences collected till now can be summerized as follows:

— Removing the cathode from the gun and pumping the cathode region lowers the dark current but only for short time.
— The dark current value after the cathode insertion depends on the pumping time.
— Reused cathodes have an earlier and faster dark current rise than new cathodes.
— No clear indication of dark current from the photoemissive area has been collected.

The main effect related to dark current during linac operation is radiation losses in the linac components. In addition, beam jumps were observed. These were due to dark current induced charging of the dielectric mirrors used to point the laser to the cathode. The problem has been solved replacing them with mirror having a metallic bulk material.

3. Conclusions

Measurements and induced effects of dark current have been reported. Studies are in progress in order to find and eliminate the dark current sources. New materials and coatings for the

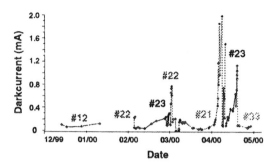

Figure 2. Dark current history for the cathodes used in the gun during the last run. Different cathodes are rappresented by different colors. Note that some cathodes have been used more than once.

spring are investigated. An activity is in progress also for studying new finishing of the Molybdenum cathode plug.

REFERENCES

1. D. A. Edwards (Ed.), TESLA-Collaboration, TESLA Test Facility Linac - Design Report, DESY Print March 1995, TESLA 95-01.
2. J. Andruszkow et al., "First Observation of Self-Amplified Spontaneus Emission in Free-Electron Laser at 109 nm Wavelength", to be published in Phys. Rev. Letters.
3. S. Schreiber et al., "Running experience with the laser system for the RF gun based injector at the TESLA Test Facility linac", NIM **445** (2000) 427-431.
4. J. W. Wang, "Rf properties of periodic accelerating structure for Linear Colliers", SLAC-Report-339.
5. S. Schreiber et al., "Performance status of the rf-gun based injector at the TESLA Test Facility", EPAC 2000 proceedings, Wien.

FREE ELECTRON LASERS 2000
V.N. Litvinenko and Y. Wu (Eds.)
2001 Elsevier Science B.V.

II-21

A diagnostic technique for short electron bunches using coherent off-axis undulator radiation

C.P. Neuman*, W.S. Graves, G.L. Carr

NSLS, Brookhaven National Lab, Upton, NY 11973-5000

P.G. O'Shea

Institute of Plasma Research
Department of Electrical and Computer Engineering,
University of Maryland, College Park, MD 20742

We discuss a technique to measure the relative length of short electron bunches. This technique involves measuring coherent radiation from a short undulator. Wavelengths which are longer than the bunch length are emitted coherently, and the total energy emitted increases with decreasing bunch length. Since only off-axis radiation is measured, the diagnostic is noninterrupting. The diagnostic is designed to be used in conjunction with a single-pass high-gain FEL, which requires short electron bunches. In such an application, a real-time bunch length diagnostic would be useful for optimizing the FEL performance. We review the theory of coherent off-axis undulator radiation and describe plans to install the diagnostic at the Source Development Laboratory (SDL) at Brookhaven National Lab.

PACS Numbers: 29.27.Fh, 29.17.+w, 41.60.Cr
Keywords: beam diagnostic, off-axis undulator radiation, undulator radiation

1. Review of Theory

The theory of coherent off-axis undulator radiation (COUR) has been described [1]. A review follows.

The electric field is derived from the Liénard Wiechert potentials:

$$\vec{E} = \frac{q^2}{4\pi\varepsilon_0}\left\{\frac{1}{\gamma^2 R^2}\frac{\hat{n}-\vec{\beta}}{\left(1-\hat{n}\cdot\vec{\beta}\right)^3}+\frac{1}{cR}\frac{\hat{n}\times\left[\left(\hat{n}-\vec{\beta}\right)\times\dot{\vec{\beta}}\right]}{\left(1-\hat{n}\cdot\vec{\beta}\right)^3}\right\}.$$

(1)

R and $\hat{n}$ are the distance and direction, respectively, from the electron to the observation point. The first term, which depends on the electron's velocity, is neglected. The spectral energy is given by:

$$\frac{d^2 I}{d\omega d\Omega} =$$

$$\frac{\varepsilon_0}{c\pi}\frac{q^2}{(4\pi\varepsilon_0)^2}\left|\int_0^{\frac{L}{\beta_z c}}e^{i\omega\left(t+\frac{R}{c}\right)}\frac{\hat{n}\times\left[\left(\hat{n}-\vec{\beta}\right)\times\dot{\vec{\beta}}\right]}{\left(1-\hat{n}\cdot\vec{\beta}\right)^2}dt\right|^2.\quad (2)$$

A bunch form factor is used to account for the effects of the finite bunch shape. The energy for N_e electrons is given by

$$W_{N_e\text{ electrons}} = W_{1\text{ electron}}[N_e + N_e(N_e - 1)f(\omega)],\quad (3)$$

where $f(\omega)$ is the bunch form factor:

$$f(\omega)=\left|\iint dydz\, S_y(y)S_z(z)e^{-i\frac{\omega}{c}y\sin\theta+i\frac{\omega}{c}z\cos\theta}\right|^2.\quad (4)$$

$S_y(y)$ and $S_z(z)$ are the transverse and longitudinal electron bunch densities, respectively. Eqs. 3 and 4 include the coherence effects of the radiation.

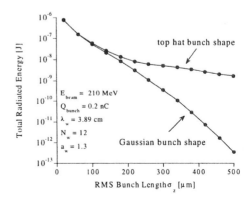

Fig. 1. Effect of bunch length on radiated energy.

* Corresponding author. Tel.: +631-344-7609. Email neuman@bnl.gov.

The effect of changes in bunch length on the total radiated energy is seen in Fig. 1 for both a Gaussian- and a top-hat-shaped bunch.

2. Experiment

An experiment is underway to measure coherent off-axis undulator radiation and to determine its usefulness for measuring electron bunch lengths.

Fig. 2. Undulator magnets (dashed arrows) and vacuum chamber (solid arrow).

2.1 The undulator

The undulator is a 12-pole prototype for the 10m NISUS undulator. The magnets have been mounted in a configuration that will allow the undulator to be removed from the beamline without removing vacuum components. Undulator magnets are usually mounted so that their field is aligned vertically and the electron oscillations take place in a horizontal plane. In this case, the magnets are mounted so that the field is aligned horizontally and the electron oscillations take place in a vertical plane. This allows the top of the undulator to be open so that the undulator can be lowered from the beam line. See Fig. 2.

2.2 Light transport

The layout of the detection components is seen in Fig. 3. The undulator radiation is reflected into the light pipe by a mirror which can be adjusted by an actuator so that it catches off-axis radiation but does not disturb the electron beam. The radiation exits the beamline through a 1" diameter crystal quartz window with a 0° crystal orientation. The 1" diameter light pipe acts as a waveguide for the radiation and allows all of the light entering the light pipe to reach the detector.

2.3 The detector

The detector is a liquid-helium-cooled bolometer which operates at a temperature of 4.2 K. The detector's wavelength response is 14 μm to 1 mm, which is an ideal range for measuring electron bunch lengths from 20 μm to 500 μm. The bolometer element is silicon and is mounted on a diamond substrate. The low temperature operation reduces noise and allows the detector to detect as little as 1 pJ in 1 ms. The detector has a 4-position filter wheel with long wavelength pass filters. The filters have cutoffs at 12.5 μm, 27 μm, 100 μm, and 286 μm. The filters can be used to filter out background radiation. The filters also allow rough spectral measurements to be made.

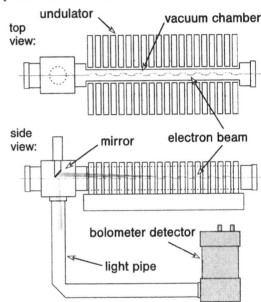

Fig. 3. Experimental setup.

3. Source Development Lab

The experiment will be performed at the Source Development Lab (SDL) at Brookhaven National Laboratory. The SDL is the site of the DUV-FEL, a single-pass FEL which will produce ultraviolet light by passing electrons through the 10 m NISUS undulator.

The 210 MeV linear accelerator, which is now being commissioned, will accelerate up to 1 nC of charge with an emittence of less than 5 μm. The bunch lengths will range from 20 μm to 500 μm. The COUR experiment is installed just after the last section of the linac. If the experiment shows that relative bunch lengths may be measured, the diagnostic may be used while the DUV-FEL is operating.

[1] C.P. Neuman, W.S. Graves, and P.G. O'Shea. Phys. Rev. ST – Accel. Beams 3 (2000) 030701.

FREE ELECTRON LASERS 2000
V.N. Litvinenko and Y. Wu (Eds.)
2001 Elsevier Science B.V.

Recent Developments at the Darmstadt Free Electron Laser

M. Brunken, L. Casper, H. Genz, M. Gopych, H.-D. Gräf, C. Hessler, S. Khodyachykh*, S. Kostial, U. Laier, H. Loos, A. Richter, B. Schweizer, A. Stascheck

Institut für Kernphysik, Technische Universität Darmstadt, D-64289 Darmstadt, Germany

1. Introduction

The Darmstadt Free Electron Laser [1] is driven by the superconducting linear accelerator S–DALINAC providing electron energies of 3–130 MeV for nuclear physics experiments and 25–50 MeV with a repetition frequency of 10 MHz and a macropulse duration of few ms to cw for the FEL. With an undulator parameter K in the range of 0.43 to 1.26 wavelengths between 2.5 to 10 μm are possible. With a maximum peak current of 2.7 A and a resonator Q value of 120, leading to a small signal gain of 3–5 %, lasing was achieved between wavelength of 6.5 and 7.8 μm with a maximum output power of 3 W corresponding to a micropulse energy of 0.3 μJ.

In this paper first investigations regarding the diagnosis of the electron beam by means of electro-optical sampling in a ZnTe crystal as well as by a new tomographic technique are presented. Furthermore, the results of simulations of the spectral intensity distribution of both, spontaneous and stimulated emission with respect to the influence of tapering the undulator magnetic field of the FEL are discussed.

2. Electron-Pulse Length Measurement

A recently developed method for the electron-bunch length measurement based on the electro-optical detection of the bunch's Coulomb field [2] has been applied using an experimental section behind the injector linac. The experimental set-up is shown schematically in figure 1. A ZnTe crystal placed a few mm close to the electron beam becomes temporarily birefringent when

the 2 ps long electron bunches with 6 pC charge pass by. With a synchronised Ti:Sapphire laser of 60 fs pulse duration and suitable polarization optics a change in the laser light polarization with a similar time structure as of the electron bunch distribution should be detectable.

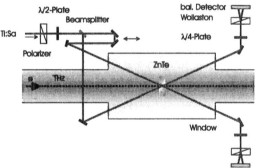

Figure 1. Experimental set-up for the electron-pulse length measurement.

During the last beam time an electro-optical signal could not be detected. The expected change in polarization was of the order of magnitude of 10^{-4} well above the detection limit of 10^{-5}. Also, the remaining phase error of the synchronisation was 4–5 ps, which will be decreased with a new synchronisation set-up. Future experiments will be performed at the TTF accelerator at DESY providing a bunch charge of up to 1 nC with 750 fs duration. In this case the signal is expected to be 3 orders of magnitude larger.

3. Transverse Phase Space Tomography

For diagnosis of the transverse beam parameters a set-up consisting of an optical transition radiation (OTR) target, a CCD camera and a PC with frame-grabber card for imaging the transverse intensity distribution of the electron beam is

*Corresponding author e-mail:
khodyachykh@ikp.tu-darmstadt.de

used. Furthermore, this method was expanded by a computer code written in the Interactive Data Language (IDL) for reconstructing the transverse phase space by tomographic algorithms. The advantage of this method is the capability of reconstructing the beam distribution in phase space without assuming any shape of the electron beam.

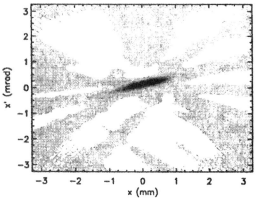

Figure 2. Reconstruction of the horizontal phase-spase distribution.

The accuracy of the filtered backprojection algorithm was properly investigated. In simulations with 18 projections interpolated to 90 projections an asymmetrical phase-space distribution could be reconstructed with an emittance error of less than 15 %. First measurements at the injector linac of the S-DALINAC at an electron energy of 8 MeV showed good agreement of the emittance with respect to the common method. Figure 2 yields the reconstruction of the horizontal phase space distribution with a normalized emittance of $\varepsilon_n = 2.14\ \pi$ mm mrad at 8 MeV.

4. Tapering

In order to increase the efficiency of the energy transfer from electrons to the laser field within the FEL, a detailed study of the tapering processes is in progress. The influence of tapering the undulator magnetic field with a tapering depth α on the parameters of radiation is investigated. Spontaneous emission spectra caused by electrons in a tapered planar undulator have been computed by numerical evaluation of the classical formula.

For the case of stimulated emission the evolution of the efficiency of the electron interaction (fig. 3) was simulated in collaboration with E. L. Saldin, E. A. Schneidmiller and M. V. Yurkov (DESY, Hamburg) by means of the modified version of the simulating code FAST [3].

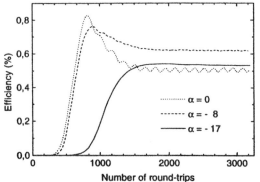

Figure 3. Time evolution of the FEL efficiency for different tapering parameters.

According to these results an increase of the FEL efficiency of approximately 20 % is expected. The region of parameters to observe this effect was defined by this simulation as well. Additionally a regime of very stable operation for our FEL is predicted with the undulator being negatively tapered. The experimental verification will be carried out soon.

5. Outlook

In the next beamtime the optical pulse length will be deduced by means of a background-free autocorrelation set-up based on second harmonic generation in a $AgGaS_2$ crystal. Furthermore experiments concerning the dependence of the spectral intensity distribution of spontaneous and stimulated emission as well as the optical pulse length at various time intervals will be performed.

We gratefully acknowledge the financial support provided by the DFG Graduiertenkolleg "Physik und Technik von Beschleunigern" and by DESY Hamburg.

REFERENCES

1. M. Brunken, S. Döbert, R. Eichhorn et al., NIM A 429 (1999) 21.
2. D. Oepts, G.M.H. Knippels, X. Yan et al., NIM A (in print)
3. E. L. Saldin, E. A. Schneidmiller and M. V. Yurkov, NIM A 429 (1999) 233.

FREE ELECTRON LASERS 2000
V.N. Litvinenko and Y. Wu (Eds.)
2001 Elsevier Science B.V.

Saturated Plasma X-ray Laser at 19 nm

Y. Li[1], J. Dunn[2], J. Nilsen[2], A. Osterheld[2], T. W. Barbee, Jr., and V. N. Shlyaptsev[3]

[1]*Institute of Laser Science and Applications, Lawrence Livermore National Laboratory, Livermore, CA 94550*

[2]*Lawrence Livermore National Laboratory, Livermore, CA 94550*

[3]*Department of Applied Science, University of California at Davis-Livermore, Livermore, CA 94550*

Saturated operation of a tabletop laser plasma X-ray laser at 19 nm is demonstrated with output energy of 2.5 µJ. The narrow beam divergence, high repetition rate, wavelength scalability, short pulse duration, high monochromacity and high brightness make it a potential tool for X-ray laser applications, including seeding a future X-ray free electron laser.

In comparison with other routes towards coherent X-ray sources such as x-ray free electron lasers and high order harmonic generation, plasma X-ray lasers [1] have high output of up to the mJ level and good monochromacity with a bandwidth of the order of 0.01%. This has made them promising tools in many areas of applications such as holography and microscopy of biological structures in the living state, probing of large, dense plasmas relevant to laboratory astrophysics and inertial confinement fusion. They can also be used as the seeds for future accelerator based X-ray free electron lasers operating at fundamental or high harmonics for improving the temporal coherence.

Such plasma X-ray lasers are generated by inducing population inversion in a high temperature high density plasma column with the longitudinal and transverse dimension of ~10 mm and ~0.1×0.1 mm^2. In the case of nickellike Mo studied in this paper, pumping is due to a strong monopole electron collision excitation from the $3d^{10}$ $^1S_0$ ground state to the upper lasing level $3d^9\,4d\,^1S_0$, from which the optical transition back to the ground state is prohibited. A population inversion is generated because the lower lasing state $3d^9\,4p\,^1P_1$ decays rapidly to the ground state via a resonant transition. This then enables lasing at 18.9 nm on the $3d^9\,4d\,^1S_0 \rightarrow 3d^9\,4p\,^1P_1$ transition. Here LS notation is used.

In this paper, we report the demonstration of a laser pumped plasma X-ray laser system using the novel transient collisional excitation (TCE) scheme [2]. The TCE scheme uses a low intensity, long laser pulse (~10^{12}-10^{13} W cm^{-2}, ~ 1 ns) to generate plasma columns from solid targets, which is heated ~1 ns latter by a high intensity, short laser pulse (~10^{15} W cm^{-2}, ~1 ps) to generate the population inversion. The long pulse generates a smooth transverse density profile for the plasma column that allows a better longitudinal propagation of the X-ray laser beam, and the ultrafast heating by the short, intense laser pulse makes it possible to generate unprecedented high gain and enable the X-ray laser to saturate over a small target length.

The saturated operation of the plasma X-ray laser was demonstrated on the compact multipulse terawatt COMET laser system at the Lawrence Livermore National Laboratory [3,4]. The laser occupies two standard optical tables of dimension 1.2 m × 3.6 m. This system is a hybrid CPA laser consisting of a Ti: sapphire oscillator, a regenerative amplifier, and a 4-stage Nd: phosphate glass amplifier. The two final 50-mm diameter amplifiers generate two beams with pulse duration of 600 ps FWHM (full width at half maximum). One of the beams is compressed to generate a short pulse of ~1 ps FWHM duration. The second is sent through a delay line to adjust the delay between the arrivals of the two beams on the target. The two beams were co-aligned to the target chamber where they are focused to form a ~80 µm × 12.5 mm line focus onto solid targets. With the typical energies, the irradiance on the target surface is lower than 10^{12} and 10^{15} W cm^{-2} for the long and the short pulses, respectively. The laser can be fired once every 4 minutes and parameters including energy, pulse shape, pulse separation, near field image and spectrum were monitored.

To make efficient use of the short-lived gain medium, we implemented a travelling wave excitation set-up using a 5-segment stepped mirror which generate group velocity of ~c for the pump laser along the line focus towards the spectrometer, which enhanced the X-ray laser output by up to a factor of 100.

Two diagnostics were used for characterizing the X-ray laser output. One is a time integrating but angularly resolving flat field spectrometer, and the other is a multiplayer mirror imaging system with a magnification of 14.

The x-ray laser output at Ni-like $3d^9\,4d\,^1S_0 \rightarrow 3d^9\,4p\,^1P_1$ 18.9 nm routinely dominates the on-axis spectra [Fig. 1 (a)], and saturation operation of the 18.9 nm X-ray laser was evidenced by its output as a function of the target length [Fig. 1 (b)]. In Fig. 1 (b), the output of the X-ray laser increases nonlinearly with target lengths of up to 8 mm, beyond which the output becomes linear up to target lengths of 10 mm, the maximum target length in this experiment. Fitting to the Linford formula [5], a total gain-length product of ~16.6 is measured, which is a total gain of $1.6×10^6$.

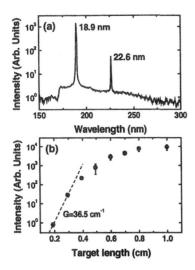

Fig. 1 (a) An on-axis spectrum of the Ni-like Mo X-ray laser from a 10-mm long target, showing the Ni-like 4d $^1S_0 \rightarrow$ 4p $^1P_1$ and 4f $^1P_1 \rightarrow$ 4d $^1P_1$ at 18.9 and 22.6 nm. (b) The Ni-like Mo 18.9 nm X-ray laser output as a function of the target length with an overall gain length product of 16.6. The long pulse and short pulse energy were 1.13 J and 5.02 J, with a delay of 0.7 ns between them. The error bars are due to fluctuation between shots.

We notice that the effective gain decreases continuously with increasing target length before the X-ray laser saturates. The gain starts at 36 cm^{-1} for 3-mm targets and decreases to 5.6 cm^{-1} for 6- to 8- mm targets. We attribute this observation to two facts. The first is the strong spatial variation of the gain coefficients in conjunction with the transverse beam refraction due to the plasma density gradient normal to the target surface. The second is the transient gain lifetime in combination with the non-continuous travelling wave in the experiment. While the refraction deflects the X-ray laser beam from the high-gain, high-density region to the low-gain, low-density region, the segmented travelling wave does not fully utilize the high transient gain with lifetime shorter than the segment length. Both effects deplete the effective gain at long target lengths before the X-ray laser saturates.

The near field beam patterns obtained using the imaging system show typically FWHM of the lasing region of less than 100×100 μm^2. They contains very fine speckle structures, which may be due to the inhomogeneity of the plasma density and gain distribution.

We can also use the imaging system to measure the absolute output of the X-ray laser [6]. The total output energy of the 18.9 nm laser is normally in the range of μJ with the highest being 2.5 μJ. Converting the output energy into intensity using the area of the output aperture and an estimated pulse duration of ~7 ps using previous data with no travelling wave excitation [4], we found the intensity of the X-ray laser output is normally above 1 GW cm^{-2} (see

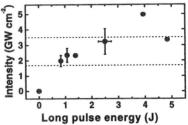

Fig. 2 The 18.9 nm X-ray laser intensity at the output aperture as a function of the long pulse energy at a delay of 0.7 ns and a constant short pulse energy of 5 J. The target length is 1 cm. The dotted lines indicate the range of the theoretical saturation intensities.

Fig. 2), and in most of the cases higher than the theoretically estimated saturation intensities of 1.7-3.5 GW cm^{-2}. This further evidenced the saturated operation and robustness of the system.

We have also demonstrated lasing in materials from Ni-like Zr to Sn [4, 7], and Ne-like Ti to Fe, with wavelengths ranging from 14 nm to 33 nm.

In conclusion, we have demonstrated a saturated table-top plasma X-ray laser at 19 nm with output energy of up to 2.4 μJ with picosecond pulse duration and high brightness. Its robustness and scalability in photon energy, and especially the compactness make it a very promising tool for X-ray applications including seeding an X-ray free electron laser.

The authors would like to thank M. Eckart, R. Ward, L. B. Da Silva, and H. Baldis for their support. We thank J. Hunter, B. Sellick, H. Louis, and T. Demiris for technical support. The work was performed under the auspices of the US Department of Energy by the Lawrence Livermore National Laboratory under Contract No. W-7405-Eng-48.

1. R. Elton, X-Ray lasers (Academic Press, New York, 1990).
2. P. V. Nickles, V.N. Shlyaptsev, M. Kalachnikov, M. Schnürer, I. Will, and W. Sandner, Phys. Rev. Lett. 78, 2748 (1997); J. Dunn, A.L. Osterheld, R. Shepherd, W.E. White, V.N. Shlyaptsev, and R.E. Stewart, Phys. Rev. Lett. 80, 2825 (1998).
3. Y. Li, J. Nilsen, J. Dunn, and A. L. Osterheld, A. Ryabtsev, and S. Churilov, Phys. Rev. A 58, R2668 (1998).
4. J. Dunn, J. Nilsen, A. L. Osterheld, Y. Li, and V. N. Shlyaptsev, Opt. Lett. 24, 101 (1999).
5. G. L. Linford, E. R. Peressini, W. R. Sooy, and M. L. Spaeth, Appl. Opt. 13, 379 (1974).
6. Y. Li, J. Dunn, J. Nilsen, A. L. Osterheld, V. N. Shlyaptsev, J. Opt. Soc. Am. B 17, 1098-1101 (2000).
7. J. Dunn, Y. Li, A. L. Osterheld, J. Nilsen, J. R. Hunter, and V. N. Shlyaptsev, Phys Rev Lett 84, 4834-37 (2000).

FREE ELECTRON LASERS 2000
V.N. Litvinenko and Y. Wu (Eds.)
2001 Elsevier Science B.V.

FEL-beam interaction in a storage-ring: new model versus experimental results

G. De Ninno[1,2], D. Nutarelli[1], D. Garzella[1,2], E. Renault[1,2] et M.E. Couprie[1,2]

[1] LURE, bat 209d, Université Paris-Sud, BP 34, 91898 Orsay Cedex, France[2] SPAM/DRECAM/DSM, bat 522, CE de Saclay, 91191 Gif sur YvetteCedex, France

Abstract

The interaction over many turns between the laser pulse and the electron bunch in a Storage Ring Free Electron Laser induces a saturation mechanism which leads to laser equilibrium. Standard theoretical approaches evaluate the effect of the interaction in terms of a statistical action of the FEL pulse on the whole electron bunch. A new theoretical approach is proposed taking into account that only the electrons which superimpose with the FEL pulse (which is 10 to 20 times thinner than the electron bunch) are perturbed. The main result of this localized interaction is a perturbation of both longitudinal and energy distributions which do not maintain their (laser off) Gaussian profile. Theoretical results are confirmed by experimental observations performed on the ACO and SuperACO FELs at LURE in Orsay. For the latter machine different optics are tested. This allows one to better understand the different role played by the FEL-beam system parameters in the distortion of the longitudinal beam distributions.

1. Introduction

According to the standard theory describing the FEL dynamics, the gain of the amplification process can be evaluated by the Madey theorem [i]. Usually, the gain computation is obtained using the values of the bunch length and the energy spread as rms values of the whole both temporal and energy distributions of the electrons in the bunch [i]. Besides, the Renieri limit allows one to evaluate and laser power and to link it to the induced energy spread [ii]. More complete approches have also been developed [iii]

Recent measurement performed with the Super-ACO FEL operating in the Q-swiched mode have shown some significant discrepancies between the theory and experiments [iv]. In particular, the gain deduced from the laser intensity rise time measuremt is considerably lower than the one predicted by the theory. Direct measurements of the induced energy spread and the bunch lengthening under the FEL interaction show that these values are also considerably lower than those predicted.

We propose here a model of interaction between the laser and the e-bunch, which considers that the laser pulse is considerably shorter than the e-bunch, in order to interpretethe discrepancies observed

2. Description of the model

Only the situation where the laser and the electron bunch are temporally synchronized will be considered here. In this case, the laser pulse develops itself exactly at the center of the electron bunch. In other words, not all the electrons interact with the light pulse in the same way, and the electrons located around the center interact more than the off center ones with the electromagnetic wave (cf. Figure 1).

As the laser temporal width is 10 to 20 times sorter than that of the e-bunch, this localized interaction may lead to an inhomogeneity of the energy exchange and only few electrons are really affected by the laser. In reality, the electrons accomplish the synchrotron oscillations and they are not fixed at one given position in the bunch. The statistical character of the interaction is strongly linked to the difference between the synchrotron oscillation period and the laser rise time and the dynamics can very complicated.

The dynamics of the electrons is described using a system of pass to pass equations for the temporal coordinate τ and the « energy » coordinate ε (the relative energy deviation with respect to the nominal energy of the ring).

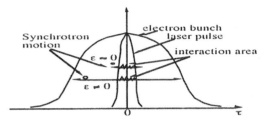

Figure 1. Schematic representation of the localized interaction between the electron bunch and the laser light pulse.

In this system of equations, we take into account the effect of the RF cavity, the random emission of synchrotron radiation, the interaction of the electrons with the vacuum chamber and the laser effect. The latter consists of two contributions: the micro-bunching induced by the laser wave electric field (first order term which is not associated with a variation of the total energy of the bunch) and a term linked to the laser extracted power (second order term which is associated with an effective energy loss by the e-bunch). The laser effect has been considered to take place only for the electrons temporally located at the laser pulse position. In this model, as a first step of analysis, the laser temporal profile is fixed as a Gaussian shape with a constant width, and its integrated intensity evolves according the amplification process. The gain is computed taking into account only the electrons which are really superimposed to the laser pulse.

3. Results of numerical simulations

Several numerical simulations have been performed using this localized interaction model. One example of a result for the Super-ACO case is reported in Figure 2.

The image reported in Figure 2a) shows the temporal evolution of the longitudinal electron density of the bunch. The simulations reported in Figure 2 show that the e-bunch distribution flattens under the effect of the laser action due to the inhomogeneity of the localized interaction. It means that, during the saturation mecanism, the shape of the e-beam distribution changes, not only in term of the rms value of bunch length.

Several more complicated situations are expected for different FEL configurations (the cases of ACO, Super-ACO, Elettra and SOLEIL FELs have been investigated). Very complex evolutions can be observed where a hole may appear at the center of the e-bunch during the laser intensity evolution. One very important parameter seems to be the ratio between the electron synchrotron oscillation half period and the start-up time of the laser. In fact, if this ratio is high, the electrons are almost fixed during the laser intensity evolution and the localized effect is strong (the hole appear). If the ratio is low, the electrons evolve rapidly during the laser intensity evolution and the statistical analysis of the interaction is correct (the effect of the localized interaction is negligible). For the Super-ACO FEL, this ratio is almost 1 and we are in a intermediate situation (only a flattening appears).

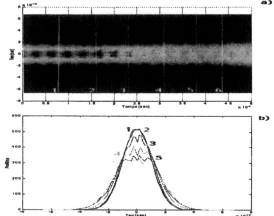

Figure 2 Evolution of the e-beam density longitudinal profile during the laser pulse intensity variation a), longitudinal profiles taken at different steps of the laser intensity evolution b). The vertical direction represents the longitudinal coordinate of the e-bunch density profile (fast scale) and the horizontal direction (slow scale) represent the macro-temporal coordinate.

This approach can explain the discrepancies between the laser gain value estimated according to the « standard » theory and the measurement performed in Q-switched operation at Super-ACO. In fact as only few electrons are really involved in the interaction, the effective laser gain is lower than that estimated taking into account all the electrons in the bunch. The difference between the expected and measured values of bunch lengthening (in term of rms value) and of the induced energy spread (in term of the laser power according to the renieri limit) observed, can also be explained by the localized interaction model.

4. Experimental results

The longitudinal dynamics of the electron bunch during the laser intensity evolution in the Q-switched mode has been experimentally investigated using a double sweep streak camera. This detector allows one to experimentally acquire the numerical image corresponding to the simulation showed in Figure 2 a). The analysis of this numerical image gives the evolution of the e-bunch parameters during the laser intensity variation (cf. Figure 3). Under the laser interaction we observe an increase of the rms bunch length (cf. Figure 3 c) as predicted by the « standard » theory and a displacement of the temporal centroid of the e-bunch due to the laser extracted power, which is not directly related to the local interaction model [v].

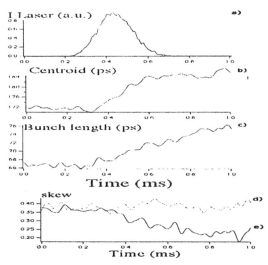

Figure 3 Analysis of the electron beam parameters during the laser intensity evolution in the Q-switched mode: integrated laser intensity a), longitudinal centroid of the electron bunch b), bunch length (rms value) c), symmetry parameter of the e-beam longitudinal distribution (skew) without laser d) and with laser e).

The evolution of the skew parameter of the e-bunch distribution, which is directly related to its symmetry (for a symmetrical shape the skew is zero), is reported in the Figure 3 d) and e). It shows that the symmetry of the e-bunch changes a lot under the laser interaction. This is a signature of the shape change and confirms the hypothesis of the localized interaction model.

5. Conclusion

The combined theoretical and experimental studies of the FEL interaction and saturation mechanisms presented above show the complexity of a storage ring FEL dynamics. In fact the strong coupling between the laser pulse and the electron bunch can lead to a highly nonlinear system. Such a system should be investigated in term of chaos theory and, in principle, stable, pulsed and chaotic solutions can be envisaged.

Besides, as the effect of the localized interaction is more important for powerful FELs, this approach is particularly useful for the study and predictions of the new generation storage ring FEL dynamics and performance.

References
[i] « Stimulated Emission of Bremmstrahlung in a periodic magnetic field », J. M. J. Madey, Jour. Appl. Phys., 42, 5, 1906-1913, (1971).
[ii] « Storage Ring operation of the Free Electron Laser: the amplifier », A. Renieri, Il nouvo Cimento 53B, pp.160 - 178 (1979).
[iii] V. N. Litvinenko et al., NIM A 375, 46 (1996).
[iv] Daniele Nutarelli, PhD Thesis Université Paris XI 20th january 2000.
[v] Roux Raphael, PhD Thesis Université Paris XI 28th january 1999.

FREE ELECTRON LASERS 2000
V.N. Litvinenko and Y. Wu (Eds.)
2001 Elsevier Science B.V.

Design of an insertion device for the FEL-X project

N. Sei[a-], K. Yamada[a], T. Mikado[a], H. Ohgaki[a], T. Yamazaki[b], M. Kawai[c]

[a] *Electrotechnical Laboratory, 1-1-4 Umezono, Tsukuba, Ibaraki 3058568, Japan*
[b] *Institute of Advanced Energy, Kyoto University, Gokasho, Uji, Kyoto 6110011, Japan*
[c] *Kawasaki Heavy Industries, Ltd., 118 Futatsuzuka, Noda, Chiba 2788585, Japan*

Abstract

A plan for hard X-ray generation through FEL-Compton backscattering in the storage ring NIJI-IV, FEL-X project, was proposed at the Electrotechnical Laboratory in 1997. The X-ray energy is about 0.1-1 MeV in the project, so that the electron beam in the storage ring is not lost rapidly through the Compton backscattering process. In order to obtain such hard X-ray beams, it is necessary to generate an infrared FEL at the wavelength of about 10 μm. The infrared FEL experiments will be executed in a long straight section which is located on the opposite side of the ultraviolet FEL line in the NIJI-IV. The maximum length of a new insertion device for the infrared FEL oscillations is about 3.7 m. We will modify a 3-m planar undulator to a 3.6-m optical klystron.

1. Introduction

Several projects aiming to obtain tunable and near-monochromatic high-energy photon beams with a storage ring by FEL-Compton backscattering have been reported in the literature [1]. The energy of photons generated by the FEL-Compton backscattering in these projects was in the γ-ray region, and the electrons in the stored electron beam colliding with the photons were mostly lost. The FEL-X project, however, aimed to generate low-energy photons (0.1 – 1 MeV) and hard X-rays, while avoiding the problem of electron-beam loss [2]. We had already observed FEL-Compton backscattering photons in the gamma ray region with the storage ring NIJI-IV, and from these results roughly designed the FEL-X project [3]. But the practical qualities of the electron beam in the NIJI-IV were worse than the estimated beam qualities in the original design. It was difficult to achieve the FEL oscillation in the far-infrared needed to obtain FEL-Compton backscattering photons in the hard X-ray region.

However, the electron beam qualities were remarkably improved by chromaticity correction, and achieved lasing in the deep ultraviolet region due to this improvement [4]. It was found through this success that a far-infrared FEL could also be obtained by optimizing an insertion device. Though improvements of the electron beam qualities in the NIJI-IV are still being achieved, we describe the design of a new insertion device for the FEL-X project based on the present qualities of the electron beam in this article.

2. Design of an insertion device

The electron-beam in the NIJI-IV is operated at energies ranging from 200 to 450 MeV, and FEL experiments are usually performed at the energy of 310 MeV. Generally, backscattered-photon energy on an axis of the electron beam in the inverse Compton backscattering process, E_γ, is approximately given by

$$E_\gamma = 4\gamma^2 E_l / \left(1 + 4\gamma E_l / m_0 c^2\right), (1)$$

where γ is the electron-beam energy in the unit of its rest mass $m_0 c^2$, and E_l is the energy of the FEL photons. In order to obtain photons with an energy of about 100 keV, therefore, we need an FEL at a wavelength of about 10 μm. There is a conventional laser, the CO_2 laser, in this wavelength region. One can easily obtain optical instruments for CO_2 lasers, such as high-reflection mirrors, so that we aim for the FEL oscillation at 10.6 μm in the first stage of the FEL-X project.

A 3-m planar undulator with bump magnets is modified as a new insertion device for the FEL oscillations in the infrared region. The gap of this undulator can be changed between 35 and 200 mm. It has two bump sections and thirteen magnetic periods, and the length of each period is 20 cm. The magnet material is SEREM-N38H with a remanent field higher than 1.24 T. The new insertion device will be installed in a long straight section which is located on the opposite side of the ultraviolet FEL line in the NIJI-IV. As Fig. 1 shows, available length for the insertion device is about 4.1 m. We plan to install RF contact chambers at both sides of the insertion device so as to suppress the increase of a broad-band impedance on the NIJI-IV. Then the maximum length of the insertion device is about 3.6 m. In the case of a simple planar undulator with 20-cm magnetic period, the periodic number is only 18 and the FEL gain at 10.6 μm would be rather low. We need to improve the 3-m planar undulator to a 3.6-m planar optical klystron. The energy spread of the electron beam in the NIJI-IV depends on the beam current, and it increases up to about 8×10^{-4} at 30 mA per bunch. Therefore, the optimum N_d (number of optical wavelengths passing over the electron in the dispersive section) for the FEL gain is about 90. The length of the dispersive section along the center axis of the electron beam is 72 cm in order to realize such a large N_d. The material of the magnets in the dispersive section is the same as the material of the undulator section, but it is necessary for the thickness of the magnets to be 60 mm. The modified undulator can independently move the undulator section and the dispersive section, and the controllable renge of the gaps is about 75 mm. The deflection parameter K is 8.72 for the FEL oscillations at 10.6 μm.

3. FEL gain

Here we estimate the FEL gain at 10.6 μm by a well-known one-dimensional approximation [5]. The peak electron density reaches up to about 7×10^{16} m^{-3}, and the horizontal and vertical beam size of the electron beam are 0.77 and 0.27 mm, respectively. The wavelength of the FEL oscillation is long, so that diffraction loss should be considered for the estimation of the FEL gain. The narrowest aperture in the optical cavity is the output windows connected with the vacuum chambers of the bending magnets. Their nominal radii are 15 mm, but their effective radii would be down to about 13 mm because of misalignment of the vacuum

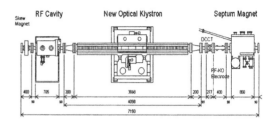

Fig. 1. Layout of the new insertion device and the north-straight section in the NIJI-IV.

chambers. Taking into account the filling factor for the FEL profile, the optimum mirror radius of curvature in a 14.8-m resonator is about 8.5 m. The diffraction loss for a TEM$_{00}$ mode is about 0.2 %, and the maximum FEL gain is estimated to be about 1.1 % by using the above parameters. We can obtain high-reflection mirrors of 99.8 % at 10.6 μm, so that it would be possible to realize FEL oscillations in the far-infrared region.

4. Conclusions

We design a new insertion device for the FEL-X project. A 3-m planar undulator is modified to a 3.6-m optical klystron as the new insertion device. The number of magnetic periods in a single undulator section is 7 with the period of 20 cm. The length and thickness of the dispersive section are 72 cm and 60 mm, respectively. The FEL gain at 10.6 μm would be estimated to be about 1.1 %.

We will improve the vacuum chambers of the bending magnets to expand the aperture and enlarge the FEL gain. A new insertion device (helical undulator or hybrid undulator) would be also examined for the next step in the future.

References

[1] e.g., V. N. Litvinenko et al., Nucl. Instr. and Meth. A407 (1998) 8.
[2] H. Ohgaki et al., Proceedings of the 18th FEL Conference, Nucl. Instr. and Meth. A393 (1997) II-14.
[3] T. Yamazaki et al., Nucl. Instr. and Meth. B144 (1998) 83.
[4] N. Sei et al., Nucl. Instr. and Meth. A445 (2000) 437.
[5] P. Elleaume, J. Phys. (Paris) Colloq. 44 C1 (1983) 353.

FREE ELECTRON LASERS 2000
V.N. Litvinenko and Y. Wu (Eds.)
2001 Elsevier Science B.V.

Chicane Compressor Development for the BNL ATF - Application to SASE FEL

R. Agustsson and J.B. Rosenzweig

UCLA Department of Physics and Astronomy,
405 Hilgard Ave., Los Angeles, CA 90095

A chicane compressor is being designed and constructed at UCLA for implementation at the BNL Accelerator Test Facility. The beam optics, including collective fields, and expected performance of the device has been simulated using TRACE3D and ELEGANT. Based on these studies, as well as constraints due to downstream ATF optics, the chicane magnet specifications were determined. The dipole magnets were designed using AMPERES 3-D magnetostatic modeling, and have been constructed. Implementation of this device at the ATF, as well as initial physics experiments on coherent synchrotron radiation emission (and associated emittance growth) at 70 MeV, and expected performance enhancement of the VISA SASE FEL experiment, are discussed.

1. Introduction

Chicane compressors are now commonly found tools in advanced accelerator[1] and FEL laboratories[2]. Because of an on-going program at UCLA in these devices, their beam physics, and application to FELs, a chicane compressor has been designed and built for the Accelerator Test Facility (ATF) at Brookhaven National Laboratory (BNL). The energy of the beam, and the capability of the ATF to provide a precompressed high brightness electron beam, make this system ideal for studies of compression. The compression of such a beam allows the study of many physical phenomena, including emittance growth due to coherent synchrotron radiation (CSR)[3], which need to be understood in order to proceed with design and construction of X-ray FELs[4]. As the VISA FEL experiment is located at the same facility, the ATF compressor can further test the capability of a compressed beam to drive a SASE FEL.

2. Basic design

The characteristics of the ATF compressor are shown in Table 1, while the layout is shown in Fig. 1. Magnet edge angles were specified at the entrance and exit of the chicane to equalize horizontal and vertical focusing. The bend radius was chosen to allow relatively small bend angles as well as rf phase in the ATF linac.

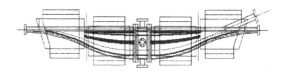

Figure 1. Chicane magnet and vacuum chambers.

Beam Energy	73.5 MeV
Beam Charge	0.20 nC
Energy spread	0.374(relative)
Linac rf phase	17 degrees
Bend angle	20 degrees
Magnetic field	2.1 kG
Initial pulse length	0.24 mm rms
Final pulse length	>0.032 mm rms
TRANSPORT R_{56}	0.606 mm/(δp/p%)
Norm. emit. growth	4.2 mm-mrad

Table 1. ATF compressor characteristics

The beam optics and compression performance were initially simulated using TRACE3D and PARMELA. The bend radius and angle, as well as the simulated beam sizes from these studies then allowed design of the chicane magnets and vacuum vessels. Ports for beam and CSR diagnostics have been included in the vacuum vessel.

3. ELEGANT simulations

Since PARMELA does not incorparate CSR effects, the code ELEGANT was used to model compressor performance as well. The particle input for ELEGANT was obtained from PARMELA simulation of the ATF injector. Emittance growth in the bend plane is approximately 4.2 mm-mrad as shown in Fig. 2. The evolution of rms bunch sizes is displayed in Fig. 3. Note that our vacuum vessel has an aperature of 1cm in the bend region, providing over 30 σ of room for the beam.

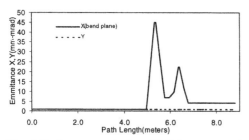

Fig. 2 Emittance evolution (ELEGANT simulation)

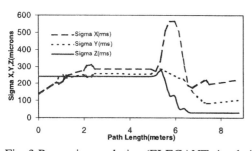

Fig. 3 Beam size evolution (ELEGANT simulation)

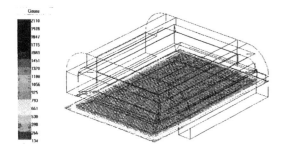

Fig. 5 Central magnet B-field at magnet midplane.

4. Magnet design

Magnet design balanced saturation considerations with implementation constraints. Magnetostatic simulations performed with AMPERES predict good field uniformity with operation of the magnet expected to remain within the unsaturated regime. Field testing is planned for the immediate future with past comparisons yielding higher permeability than simulated with AMPERES.

5. Expected VISA performance

The gain of a SASE FEL is a strong function of the beam current and emittance. As the chicane shortens the bunch by an order of magnitude, it is expected that the gain increases, despite the horizontal emittance increase during compression. This is verified by the GENESIS 1.3 [5] simulation shown in Fig. 5, which shows the expected performance of the beam if it passed through the VISA undulator. The performance is significantly enhanced [6] by the use of compressed beam.

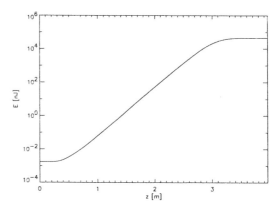

Figure 5. GENESIS simulations of the expected gain due to a 200 pC beam compressed to a 32 micron rms bunch length in the VISA undulator.

6. Acknowledgements

We would like to thank Paul Emma, Ilan Ben-Zvi, John Skiritka and Sven Reiche for their contributions to this work. This work supported by ONR under contract N00014-99-1-0497.

References

1. J.B. Rosenzweig, N. Barov and E. Colby, IEEE Trans. Plasma Sci. 24, 409 (1996).
2. TESLA Test Facility Design Report, DESY (Hamburg, 1995).
3. H. Braun, Phys. Rev. Lett. 84, 658 (2000).
4. R. Tatchyn et al., Nucl. Instr. Meth., A429 (1999).
5. S. Reiche, Nucl. Instr. Meth., A429 (1999) 243.
6. R. Carr et al., The VISA FEL Undulator, Proc. 20th Int. FEL Conference (FEL98).

FREE ELECTRON LASERS 2000
V.N. Litvinenko and Y. Wu (Eds.)
2001 Elsevier Science B.V.

Zero-slippage operation of TEUFEL

Jeroen W.J. Verschuur, Iouri Volokhine, Florens F. de Wit, Jan I.M. Botman*, Klaus-J. Boller

University of Twente; P.O. Box 217; 7500 AE Enschede; the Netherlands.
Technical University of Eindhoven; P.O. Box 513; 5600 MB Eindhoven; the Netherlands

Short pulse operation of a long wavelength FEL can take advantage of a waveguide to slow down the group velocity. In the optimum case the energy travels with the electrons, without slippage. Applying this principle simultaneously with the resonance condition results in a wavelength that is twice the wavelength of the one generated in free space given the energy of the electrons. In TEUFEL these conditions can be met with macroscopic dimensions of the waveguide. In the electron energy range between 3 MeV and 7 MeV and waveguide diameters between 4 and 6 mm various modes can be chosen that show zero slippage. Produced wavelengths are in the range of 0.25 mm to 1.2 mm.

Introduction

When operating a Free Electron Laser in short pulse operation, the slippage becomes significantly large [1]. The main disadvantages of the slippage are that the optical power travels out of the electron pulse and the optical pulse grows larger. Therefore the interaction of the light with the electrons is not optimal anymore. Due to this effect less bunching of the electrons will occur, resulting in a significant reduction in gain [2]. A way to maintain the overlap between the optical pulse and the electron bunch is reducing the group velocity of the light pulse to an extend that it matches the velocity of the electrons. Fortunately application of a waveguide in the FEL can do this. Various transverse modes in a waveguide have different group velocities [3], [4].

The FEL resonance condition and zero-slippage can be fulfilled simultaneously in a waveguide. Advantage of operation in the zero-slippage regime is to allow short pulse operation at high gain. Near zero slippage many interesting effects can show up [5], [6], [2]. In case of full control on the slippage various bunch-length regimes of FEL-working can be exploited [7].

Theory

For operation of a FEL in a waveguide we have to go back to the original form of the resonance condition, which is given by:

$$\frac{\lambda_w}{\beta_z c} = \frac{\lambda_w + \lambda_z}{v_{\varphi z}} \quad (1)$$

with λ_w the undulator wavelength, λ_z the wavelength corresponding to longitudinal wavevector of the generated light, $\beta_z c$ the longitudinal speed of the electrons and $v_{\varphi z}$ the longitudinal phase velocity of the light. The zero-slippage condition is met by setting the longitudinal group velocity v_g of the generated light equal to the longitudinal velocity of the electrons:

$$v_g = \beta_z c \quad (2)$$

The longitudinal velocity of the electrons is given by:

$$\beta_z = \sqrt{1 - \frac{1 + a_w^2}{\gamma^2}} \quad (3)$$

where γ is the relativistic factor of the electrons and a_w is the undulator parameter. Simultaneous fulfilling condition (1) and (2) and substitution of equation (3) gives expressions for the longitudinal and transverse wave vectors as a function of the relativistic factor:

$$k_z = \frac{2\pi}{\lambda_w}\left(\frac{\gamma^2}{1 + a_w^2} - 1\right) \quad (4)$$

and

$$k_\perp = \frac{2\pi}{\lambda_w}\sqrt{\frac{\gamma^2}{1 + a_w^2} - 1} \quad (5)$$

Before the various modes are calculated that fulfil these relations it is worth looking at the generated wavelength in free space. Combining equations (4) and (5) and approximate for $\beta_z \approx 1$ the expression for the vacuum wavelength can be deduced:

$$\lambda_s^{wg} = \frac{\lambda_w}{\gamma^2}\left(1 + a_w^2\right) \quad (6)$$

Note that this wavelength is two times the wavelength generated in vacuum in the absence of a waveguide with the same energy.

To explore the effect of the waveguide it is illustrative to calculate the change in wavelength with respect to the vacuum-generated wavelength as a function of the group velocity change directly from equation (1) expressed in terms of group and phase velocities:

$$\lambda_s = \lambda_w \frac{\frac{v_\varphi}{c} - \beta_z}{\frac{v_g}{c} \cdot \beta_z} \quad (7)$$

The change in wavelength is then given by:

$$d\lambda_s = -\lambda_w \frac{dv_g}{c} \qquad (8)$$

Integrating equation (8) to obtain the deviation from the vacuum generated wavelength gives:

$$\Delta\lambda_s = -\lambda_w \frac{\Delta v_g}{c} \qquad (9)$$

The integration constant is zero to fulfil the condition that the change in wavelength is zero when the change in group velocity with respect to c is zero. In case of zero slippage the group velocity is equal to the longitudinal velocity of the electrons, so the change is group velocity is given by:

$$\Delta v_g = c\beta_z - c \qquad (10)$$

Substitution of equation (10) in equation (9) gives the change in wavelength for the situation where the energy is kept constant but a waveguide is applied that slows down the group velocity to the longitudinal velocity of the electrons:

$$\Delta\lambda_s = -\lambda_w \frac{c\beta_z - c}{c} = \lambda_w (1 - \beta_z) \approx \frac{\lambda_w(1 + a_w^2)}{2\gamma^2}$$

$$(11)$$

where the approximation is valid for relativistic electrons. Further slowing down the group velocity will result in even longer wavelengths ending in cut-off.

Although the longitudinal and transverse wave vectors are smooth functions of the relativistic factor γ (the undulator wavelength λ_w and the undulator constant a_w) not all longitudinal wave vectors are possible given a waveguide. Equation (5) restricts the possible wavelengths for a given waveguide, since the transverse modes have to match the transverse dimensions of the waveguide. This fixes the longitudinal wavelength and the energy of the electrons accordingly. It should be noted that zero-slippage operation in a waveguide is identical to the so-called grazing incidence operation of the FEL. Therefore only modes with a cut-off frequency ω_c that fulfils the condition $\omega_c = ck_\perp > v_z k_w$ $(k_\perp > k_w; \gamma \gg 1)$ can fulfil the zero-slippage condition for a given electron energy. In the next section an overview of the experimental parameters is given to explore (near) zero-slippage operation of TEUFEL.

Experimental parameters

The Free Electron Laser in Twente TEUFEL [8] is very suitable to operate near zero-slippage conditions. The energy of the electrons can roughly be tuned between 3 MeV and 7 MeV. The resonator is a waveguide structure with hole coupling for the electron beam to enter and exit and for the light to exit. The undulator has a period of 25 mm and a length of 1.25 m. Due to the relative low energies, the generated wavelengths are in the several hundred-micron range. Therefore the diameter of the waveguide can have macroscopic dimensions to significantly affect the group

velocity of the generated light. In order to have significant overlap between the electron beam and the waveguide mode, only the lower order modes are suitable. For the TEUFEL parameters, the waveguide diameter should be less than 15 mm for zero-slippage operation on the lowest order (TE$_{11}$) mode.

For three different circular waveguide diameters (4, 5 and 6 mm) the 12 lowest order modes are calculated that fulfil the resonance condition and operate in zero-slippage mode. Results of the calculations in the electron-energy range between 3 MeV and 10 MeV are given in figure 1.

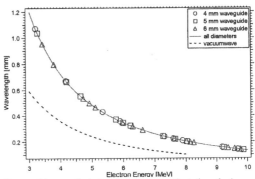

From this graph it is clearly seen that the choice of a waveguide diameter limits the possible electron energy wavelength combinations that are resonant with zero-slippage. The dotted curve shows the energy-wavelength relation for a generation in free space.

Due to the use of hole coupling the losses for transverse modes, that have maximum intensity on axis, are the highest. The diameters of the holes and the mode conversion in the tapered section determine the modes with maximum net gain. No detailed calculations on the net gain are done yet. Various mirrors with different hole sizes are available to perform experiments.

References

1. R. Bonifacio, C. Maroli, N. Piovella; Opt. Comm. 68 (1988) 369.

2. X. Shu; Opt. Comm (1994) 188.

3. A. Doria, G.P. Gallerano, A. Renieri; Opt. Comm. 80 (1991) 417.

4. J.W.J. Verschuur, G.J. Ernst, B.M. van Oerle, D. Bisero, A.F.M. Bouman, W.J. Witteman; Nucl. Instr. Meth. Phys. Res. A393 (1997) 197.

5. Li-Yi Lin, T.C. Marchall; Phys. Rev. Lett 70 (1993) 2403.

6. N. Piovella, V. Petrillo, C. Martoli, R. Bonifacio; Nucl. Instr. Meth. Phys,. Res; A341 (1994) 196.

7. R. Zhang, C. Pellegrini, J. Rosenzweig, G. LeSage, F. Hartemann, D. McDermott, C. Joshi, N. Luhmann, P. Pierini, L. de Salvo, R. Bonifacio; Nucl. Instr. Meth. Phys. Res A341 (1994) 67.

8. G.J. Ernst, W.J. Witteman, J.W.J. Verschuur, R.F.X.A.M. Mols, B.M. van Oerle, A.F.M. Bauman, J.I. M. Botman, H.L. Haagedoorn, J.L. Delhez, W.J.G.M. Kleeven, Infrared Phys. Tech. 36 (1995) 81.

FREE ELECTRON LASERS 2000
V.N. Litvinenko and Y. Wu (Eds.)
2001 Elsevier Science B.V.

Intra-Undulator Measurements at VISA FEL

A. Murokh*, P. Frigola, C. Pellegrini, J. Rosenzweig, A. Tremaine, UCLA
E. Johnson, X. J. Wang, V. Yakimenko, BNL
L. Klaisner, H. D. Nuhn, SLAC
A. Toor, LLNL

UCLA Department of Physics and Astronomy, P.O. Box 951547, Los Angeles, CA 90095-1547

We describe a diagnostics system developed, to measure exponential gain properties and the electron beam dynamics inside the strong focusing 4-m long undulator for the VISA (Visible to Infrared SASE Amplifier) FEL. The technical challenges included working inside the small undulator gap, optimising the electron beam diagnostics in the high background environment of the spontaneous undulator radiation, multiplexing and transporting the photon beam. Initial results are discussed.

1. Introduction

The intra-undulator diagnostics system at the VISA experiment [1] has a dual purpose: (i) align and match the electron beam in the undulator; and (ii) to measure the FEL radiation properties along the length of the undulator [2]. While the second task remains the experimental objective, it is the first one that focused most of the attention of the authors during the initial stage of VISA experiment. For the proper characterisation and optimisation of the FEL process, it is necessary to measure the electron beam trajectory and envelop throughout the length of the undulator, and apply a proper correction (for instance to align the electron beam to the undulator axis with the 20 μm accuracy).

The VISA experiment utilises a strong-focusing undulator [3], with the average electron beam beta-function of about 30 cm (Table 1). Hence, proper diagnostic technique requires sampling period to be 90 cm or less. To accomplish that, the undulator vacuum chamber was equipped with 8 diagnostic ports 50 cm apart. OTR and undulator radiation probes, alignment lasers, and multiplexing optical transport are the major components of the intra-undulator diagnostic system.

2. Diagnostic Probes

One of the most challenging parts of the VISA experiment was the design and fabrication of the diagnostic probes. The purpose of the probes is both to extract an FEL light out of the undulator vacuum chamber, and to enable the electron beam imaging (Fig. 1). When the electron beam path is being intercepted by the outer surface of the probe two-sided mirror, the FEL light is being extracted

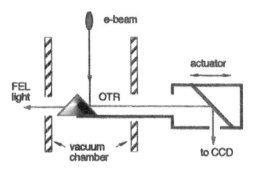

Figure 1: Intra-undulator diagnostic probes diagram.

and directed into the optical transport line. On the opposite side, the inner mirror directs the radiation through the periscope mirror into the imaging system, where the FEL light is being filtered out and the OTR from the mirror surface is used to generate an image of the electron beam.

The miniature openings inside the undulator gap (3.6 mm x 9 mm) put a very stiff limitation on the size of the probe tips. Initially, to maximise the two-sided mirror surface, we built probes with a marginal thickness of 3.3 mm. In addition to the difficulties with the probe insertion in the undulator gap, it turns out that the bellow-coupled vacuum feedthroughs used to actuate the probes tend to tilt after the pump-down, and as a result, the probe tip may not clear the opening, which is not acceptable

Table 1: Relevant parameters for VISA FEL

Nominal Beam Energy	71MeV
Charge	1nC
Undulator Period	1.8 cm
Number of Periods	220
Average β– Function	30 cm
Radiation Wavelength	800nm

* murokh@physics.ucla.edu

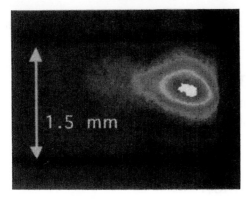

Figure 2: OTR image of the VISA beam in the middle of the undulator

for many reasons. Hence, the latest design utilises 2.3 mm thick probe tips. It is very difficult to polish laser quality mirrors of that size; therefore, the original copper mirrors were replaced with the silicon ones, polished to $\lambda/10$. The mirror polishing has been performed at LLNL, and the probe fabrication was done at SLAC and BNL.

Another important change in the diagnostic system is the use of optical transition radiation (OTR) for beam imaging, instead of the YAG crystals originally proposed. The experimental study [4] at ATF (Accelerator Test Facility) demonstrated, that the YAG crystal scintillators get saturated by the high brightness electron beam (beam of quality identical to the VISA design parameters). As a result, information about the size and shape of electron bunches, which is critical to analyse the FEL performance, can not be obtained from YAG images (Fig. 3), which consistently overestimate the beam radius. To avoid this problem we have to use OTR diagnostics, which provides correct information about the beam transverse profile. However, the OTR has a disadvantage of lower intensity. Indeed, the OTR intensity is few orders of magnitude smaller than the undulator radiation within the bandwidth of the CCD camera; hence we need to filter the undulator radiation in order to obtain a usable OTR image.

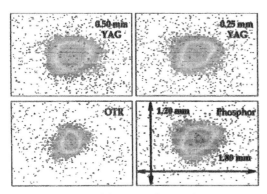

Figure 3: Images of the same electron beam, taken with 4 different methods. Comparison favours OTR as the diagnostics of choice.

Fortunately, linear polarisation and harmonic spectral structure of the undulator emission allows to surpress it with respect to the radially polarised broad band OTR (Fig.2). At the present, the cold mirror periscope in the imaging optics provides 2 orders of magnitude noise reduction. In the future, if necessary, we plan to use polarising cubes to have even smaller fraction of the FEL light to enter the imaging optics. Another "side-effect" of the low intensity OTR implementation is that to get a clear performance the total resolution of the imaging system had to be changed from 4 µm/pixel as designed, up to 10 µm/pixel.

3. Alignment Laser System

Once the imaging system operates, the electron beam can be matched into the undulator. Yet, for the trajectory studies [5], it is necessary to have a reference line. A diode laser system is used to provide a reference line for the electron beam, and also to help with the alignment of optical components.

To this end we use a fiber–coupled diode laser at 632 nm, providing a single mode output, which we focus at the middle of the undulator, located 6 m away. Even though we are diffraction limited, we can achieve sub-millimeter spot sizes throughout the whole length of the undulator. CCD cameras on both sides of the undulator are used to monitor and periodically realign the reference laser beam line. This line is integrated into the interferometric undulator alignment system [6], to overlap the magnetic axis of the undulator. Once aligned, the laser is found to be stable within 20-30 microns from shot-to-shot. That is less than the typical horizontal jitter of the electron beam at the ATF, which is generally small due to the use of the energy collimator after the linac section [7].

The transport line for the FEL light consists of an imaging lens array to multiplex the radiation beam extracted from the different diagnostic ports [2]. As any system with the large number of optical elements, our optical transport line requires a narrow band coating (Ti:Sapphire); hence, an additional diode laser at 780 nm was added for optical alignment purposes. The two laser lines are collinear by virtue of a cold mirror.

This work was performed under the DOE contract No DE-FG03-92ER40693.

REFERENCES

[1] http://www-ssrl.slac.stanford.edu/VISA
[2] A. Murokh *et al.*, Proc. of FEL 99, p. II-119 (2000)
[3] G. Rakowski *et al.*, Proc. of PAC 99 (1999)
[4] A. Murokh *et al.*, Submitted to World Scientific
[5] A. Tremaine *et al.*, Same Proceedings
[6] R. Ruland *et al.*, Proc. of FEL 99 (2000)
[7] X.J. Wang *et al.*, Proc. of PAC 99, p. 3495 (1999)

FREE ELECTRON LASERS 2000
V.N. Litvinenko and Y. Wu (Eds.)
2001 Elsevier Science B.V.

THE PROJECT OF HIGH POWER SUBMILLIMETR-WAVELENGTH FREE ELECTRON LASER

V.P.Bolotin, N.G.Gavrilov, D.A.Kairan*, M.A.Kholopov, E.I.Kolobanov, V.V.Kubarev,

G.N.Kulipanov, S.V.Miginsky, L.A.Mironenko, A.D.Oreshkov, M.V.Popik, T.V.Salikova,

M.A.Sheglov, O.A.Shevchenko, A.N.Skrinsky, N.A.Vinokurov, P.D.Vobly.

Budker Institute of Nuclear Physics, Acad. Lavrentyev Prospect 11, 630090 Novosibirsk, Russia

Abstract A 100-MeV 8-turns accelerator-recuperator intended to drive a high-power infrared FEL is presently under construction in Novosibirsk. The first stage of the machine includes one-turn accelerator-recuperator that contains complete RF-system.

Keywords accelerator-recuperator, free electron laser, magnetic system.

Introduction

The efficiency of the beam power conversion into the FEL radiation power is rather low, being typically not more than a few percent. For high FEL power applications, therefore, it is necessary to recover the beam power after the FEL interaction. The main reason for the energy recovery, in addition to a simple energy saving, is the dramatic reduction of the radiation hazard at the beam dump.

One possible method of the energy recovery is returning the beam energy to the radio-frequency (RF) accelerating structure, which was used to accelerate it [1,2]. If the length of path from the accelerator through the FEL to the accelerator is chosen properly, the deceleration of particles will occur instead of acceleration. Therefore, the energy will be returned to the accelerating RF field (in other words, the beam will excite RF oscillations in the accelerating structure together with the RF transmitter). This mode of an accelerator operation was demonstrated at the Stanford HEPL [3]. Recently, the first high power FEL based on accelerator-recuperator was successfully commissioned [4]. An obvious extention of this approach is the use of multipass recirculator [5,6], instead of a simple linac.

Increasing the number of passes can reduce the cost and the power consumption. However, the threshold current for instabilities also decreases. The "optimal" number of passes is determined by both effects [7]. A possible layout of such FEL is shown in Fig.1

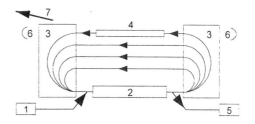

Fig. 1. The layout of an FEL with the accelerator-recuperator: 1-injector; 2-RF accelerating structure 3-180-degree bends; 4-FEL magnetic system; 5-beam dump; 6-mirrors; 7-output light beam.

The high power infrared FEL for the Siberian center of photochemical research (presently under construction) is the implementation of this approach.

First stage of the FEL.

A high-power infrared FEL driven by a 100-MeV, 8-turns accelerator-recuperator is under construction in Novosibirsk [8]

* Corresponding author. Tel. +7 3832 394003, fax +7 3832-342163, e-mail kairan@inp.nsk.su

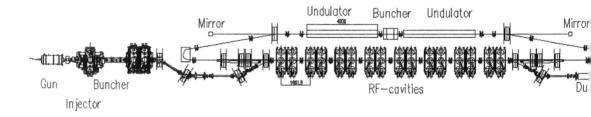

Fig. 2 Layout of the first stage of the high power free electron laser.

The first stage of the machine includes one turn accelerator-recuperator, that contains full-scale RF-system (Fig. 2). Number of turns is reduced to facilitate the commissioning. Main parameters of the accelerator-recuperator are listed in Table 1.

Table 1. Parameters of the accelerator-recuperator.

RF wavelength, m	1.66
Number of RF cavities	16
Amplitude of accelerating voltage per cavity, MV	0.8
Injection energy, MeV	2
Final electron energy, MeV	14
Bunch repetition rate , MHz	2 - 22.5
Average current, mA	4 – 50
Final electron energy dispersion, %	0.2
Final electron bunch length, ns	0.02 - 0.1
Final peak electron current, A	50 – 10

The FEL is installed on the single returning turn of the accelerator-recuperator. The FEL consists of two undulators, a magnetic buncher, two mirrors of optical resonator, and an outcoupling system. Both undulators are planar and electromagnetic. The length of each undulator is 4 m, the period is 120 mm, the gap is 80 mm. The deflection parameter K is controllable from 0 to 1.2. We can operate with one or two undulators, and with or without the magnetic buncher. Both mirrors of the optical are identical. They are spherical, made of polished copper, and are water cooled. The outcoupling system contains two or four adjustable planar 45° copper mirrors (scrapers). These mirrors scrape the radiation inside the optical

resonator and redirect it to user. This scheme preserves the main mode of optical resonator and effectively reduces amplification of higher modes. The buncher is simple three-pole electromagnetic wiggler. It is used to optimize the relative phasing of undulators. The expected FEL parameters are shown Table 2.

Table 2. Expected FEL parameters.

Wavelength, mm	0.1...0.2
Pulse length, ns	0.02...0.1
Peak power, MW	1...7
Average power, kW	0.6...7

Conclusion.

Reliable operation of the 2 MeV injector at average current 10 mA was achieved last year. The measured beam parameters are suitable for the beginning of the commissioning of the main accelerator. The assembly of the RF system is in progress. The commissioning is scheduled for next year.

References
[1] A. N. Skrinsky and N. A. Vinokurov, Proc. 6th Nat. Conf. on Charge Particle Accelerators, JINR, vol. 2, Dubna, p. 233 (1979).
[2] A. N. Skrinsky and N. A. Vinokurov, Preprint INP 78-88, Novosibirsk (1978).
[3] T. I. Smith et al., Nucl. Instr. and Meth. A 259 (1987) 1.
[4] G.R. Neil, et al., Physical Review Letters, Volume 84, Number 4 pp. 662-665 (2000).
[5] R. E. Rand, Rerciculating Electron Accelerators, Harwood Academic Publishers, 1984.
[6] N. G. Gavrilov et al., IEEE J. Quantum Electron., QE-27, pp. 2626-2628, 1991.
[7] N. A. Vinokurov et al., Proc. of SPIE Vol. 2988, p. 221 (1997).
[8] N.G.Gavrilov, et al. Proceedings of SPIE, Vol. 2988 pp. 185-187 (1996).

FREE ELECTRON LASERS 2000
V.N. Litvinenko and Y. Wu (Eds.)
2001 Elsevier Science B.V.

II-39

The University of Maryland Free-Electron Laser

P. O'Shea, C. Davis, E. Elson, D. Feldman, R. Feldman, J. Harris, J. Neumann, A. Shkvarunets,
Department of Electrical and Computer Engineering, University of Maryland, College Park MD
20742
H. Freund , SAIC H. Bluem, A. Todd, Advanced Energy Systems

1. Introduction

The Maryland Infrared Free-Electron Laser (MIRFEL) is under construction at the University of Maryland. MIRFEL, based on the Advanced Energy Systems CIRFEL, is a source of far-infrared (FIR) light for physics, medical, and materials science research. A project to investigate vibrational modes in DNA is currently planned. MIRFEL construction will be complete by early 2001. This paper discusses the design and planned applications for MIRFEL.

2. MIRFEL Design

MIRFEL is being constructed from the Compact Infrared Free-Electron Laser (CIRFEL), which was originally built to develop applications-focused research FELs [1], and which first lased in 1996 [2]. MIRFEL has a number of unique attributes ideally suited for FIR research, including high peak power (> MW), high average power (~2 W), short pulse radiation (5 ps), and rapid tunablility. Another important feature is the 7 ns pulse separation. Unlike FELs with sub-nanosecond pulse spacing, the pulse spacing in MIRFEL is long enough to allow samples to cool or relax between pulses. Furthermore, MIRFEL incorporates a ps-pulsed Nd:YLF photocathode drive laser that is phase locked to the electron beam and FEL optical output. The drive-laser can produce harmonics at 1.047, 0.524, 0.349, and 0.262 μm, and can be used for experiments in conjunction with the primary FEL output.

The main components of the FEL are the electron accelerator, photocathode electron source, UV drive laser, RF power system, and the FEL wiggler and optics. Because MIRFEL was built as an FEL and is not a retrofitted accelerator, it has a very compact footprint. The entire FEL, exclusive of the drive laser and RF system, occupies a space of only 8' x 12'. The other component sizes are modest as well, totaling about 52 square feet. Such compact FELs can be easily and inexpensively sited at places such as Universities and medical institutions.

3. Accelerator

The MIRFEL accelerator is a high-brightness electron source capable of producing electron pulses with energies up to 14 MeV and peak currents in excess of 150 A.

The accelerator consists of two separate units, the gun and the booster. The gun is a 9-MeV, 20 cm long, 3 1/2 cell S-band structure. A 6 mm diameter Mg disk serves as a photocathode[1]. Afterwards, the photocathode will be upgraded to an LaB_6 disk.

Downstream from the gun is a 2-cell booster cavity, identical to the central two cells of the gun. This booster can raise the electron energy to 14 MeV. RF power is provided by a single ITT 2960 klystron system with feed forward control [3]. The macropulse in the gun consists of a train of micropulses, each 5 - 10 ps long, with a micropulse repetition rate of 142.8 MHz. The macropulse is 10 μs long, and has a repetition rate of up to 10 Hz [4]. The expected normalized rms emittance is 5 μm with 1 nC per bunch

4. Wiggler and Optical Cavity

In the original CIRFEL configuration, the accelerator was connected to the wiggler through a 90-degree bend. The optical cavity mirrors were totally reflecting, and light was outcoupled through a hole in the downstream mirror. A smaller hole in the upstream mirror was used for alignment by a HeNe laser [5].

The new MIRFEL configuration will abandon this geometry. Instead, the wiggler and optical cavity will be colinear with the accelerator and beamline. This will provide a more compact and rapidly tunable system. The electron beam will enter and leave the optical cavity through small holes in the cavity mirrors.

The two currently available MIRFEL wigglers are of a well-proven, robust design developed at LANL and used there and at Princeton. The first was used to demonstrate lasing in the 8 - 20 μm region at Princeton. This is a 73 period permanent magnet wiggler, with the length of each period measuring 13.6 mm. The wiggler gap is 6 mm, and the RMS wiggler parameter (a_w) is 0.20 [1]. This wiggler has been successfully tested here at Maryland using the taught wire technique, and will be the first wiggler installed at MIRFEL (Phase I).

A second permanent magnet wiggler, having 50 periods of length 10 mm, will be tested and installed at a later date. It is believed that this wiggler will allow increased output power and shorter wavelength output light [2]. A 3-cm period wiggler for lasing at 50-150 μm will also be constructed (Phase II).

5. Applications

Prohofsky [6], Lilley [7], and Lim [8] claim that two mechanisms control the chemical reactions of large bio-molecules in living organisms: molecular structure and vibrational dynamics. While biomolecular structure (i.e. the

DNA double helix)is relatively well understood, a complete understanding of vibrational dynamics has not been experimentally determined. Grasping the dynamics is thought to be the key to understanding fundamental processes such as DNA replication and protein interaction.

Large biomolecules move primarily in vibrational modes. These "accordion-like" vibrations along the backbone of the DNA involve thousands of base pairs. These vibrations are predicted to be involved with energy transfer through the DNA, including determining points for DNA melting. DNA also undergoes conformational changes where the strands bend into different shapes[9]. It has been shown that some DNA-protein interactions require specific conformations of DNA[7] and that DNA conformations can help modulate chemical reactions as they occur[8].

Laser light is extremely useful for studies related to DNA interactions, because it can serve as both a trigger to initiate reactions and a time resolved probe to monitor the outcome of those reactions. MIRFEL is extremely well suited to perform these tasks as most biologically relevant chemical reactions can be initiated with light between 3 μm and 30 μm; many of the vibrational modes in complex molecules fall in the far infrared spectrum between 30 μm and 200 μm[10].

The laser can be used to generate vibrational modes through interaction with electric dipoles inside the DNA double helix structure. Under normal conditions, a dipole in the presence of an electric field feels a torque, $L = p \; X \; E$, and will rotate accordingly. The dipoles in the DNA will feel the same torques when exposed to laser light. The flexible bonds between these dipoles allows sections of the DNA to rotate and initiate vibrational modes inside the molecule. Therefore, the applied E-M field from MIRFEL will be able to drive the different vibrational modes and conformational changes within the DNA.

While the long term goal of exploring DNA dynamics is to determine exactly how or if certain conformational changes play a role in gene expression, we must first build a larger experimental database of knowledge about DNA dynamics. In the past, some experiments used optical, ultraviolet, or x-ray sources to conduct these types of studies, but optical and UV light is easily scattered by organic material. Additionally, the same material can be easily damaged by x-ray radiation [11]. Some researchers used weak infrared sources, but the high absorption of infrared light in water was difficult to overcome. The radiation from MIRFEL is nondestructive and is not scattered by organic material. Capable of producing megawatts of peak power, MIRFEL would be a prime candidate to gather more data regarding the existence of these modes using Fourier Transform Infrared (FTIR) Spectroscopy. The next step would be to determine if infrared pulses from MIRFEL could produce effects resulting in downstream gene expression. In order to accomplish this task, we plan to bring together four groups at the University of Maryland: the Department of Electrical and Computer Engineering, the University of Maryland Biotechnology Institute, the Institute for Plasma Research, and the Department of Radiation Oncology (at the University of Maryland, Baltimore). In preliminary studies, alterations in DNA as a result of MIRFEL radiation could be calculated using DNA fingerprinting to detect changes in chromatin conformation. Detecting functional alterations will be accomplished by gene expression profiling (Clontech Atlas (r) 1.2).

6. Conclusion

As MIRFEL nears completion in 2001, the University of Maryland will gain a unique light source. MIRFEL's frequency range, optical power, and pulse separation allow it to be used effectively in experimental investigations of DNA dynamics, a field that has never been fully explored. This research, to be conducted in an interdisciplinary setting will make important contributions to genetics while helping to demonstrate the utility of Free Electron Lasers to the scientific community as a whole.

References

[1] I.S. Lehrman, et al. Proc. SPIE 2522 (1995) 451.

[2] I.S. Lehrman, et al. Nucl Instr. and Meth. A. 393 (1997) 178.

[3] R. Hartley, I.S. Lehrman, J. Krishnaswamy. Nucl. Instr. And Meth. A 375 (1996) ABS22

[4] I.S. Lehrman, et al. Nucl. Inst. And Meth. A 358 (1995) ABS5.

[5] I.S. Lehrman, et al. 1995 PAC Conf., May 1-5, Dallas, TX, 1995.

[6] Prohofsky, E. Statistical mechanics and stability of macromolecules : application to bond disruption, base pair separation, melting, and drug dissociation of the DNA double helix. New York, N.Y. : Cambridge University Press, 1995

[7] D. Lilley ed. DNA Protein: Structural Interactions. New York: Oxford University Press, 1995.

[8] M. Lim, T. Jackson, and P. Anfinrud. "Ultrafast rotation and trapping of carbon monoxide dissociated from myoglobin." Nature Structural Biology. 4.3 (1997): 209-214.

[9] L.L. Van Zandt, and VK Saxena. Physical Review A. 39.5 (1989): 2672-2674.

[10] D.D. Dlott, and M.D. Fayer. IEEE Jour. Of Quan. Elec. 27.12 (1991): 2697 – 2713.

[11] GS Edwards. Optical Engineering. 32.2 (1993) : 314-319.

FREE ELECTRON LASERS 2000
V.N. Litvinenko and Y. Wu (Eds.)
2001 Elsevier Science B.V.

Free-electron maser experiment on testing high-gradient accelerating structures

N.S.Ginzburg[b], A.V.Elzhov[a], I.N.Ivanov[a], A.K.Kaminsky[a], V.V.Kosukhin[a],
S.V.Kuzikov[b], N.Yu.Peskov[b*], M.I.Petelin[b], E.A.Perelstein[a], S.N.Sedykh[a],
A.P.Sergeev[a], A.S.Sergeev[b], I.Syratchev[c] and I.Wilson[c]

[a]*Joint Institute for Nuclear Research, Dubna, 141980, Russia*
[b]*Institute of Applied Physics RAS, Nizhny Novgorod, 603600, Russia*
[c]*CERN, Geneva, Switzerland*

Abstract

This paper is devoted to progress in the JINR-IAP FEM experiment based on a 1 MeV / 200 A / 200 ns LINAC. This FEM generates 30.7 GHz / 100 ns / 50 MW RF-pulses with a record efficiency of 35% for FEM-oscillators. The high efficiency together with high stability of single-mode single-frequency operation is provided by the use of a reversed guide magnetic field and a novel Bragg resonator with a step of phase of corrugation.

The project to use this high power FEM source for studying surface heating effects in 30 GHz accelerating structures is discussed. A special cavity will be used to enhance magnetic fields at the cavity surface. The cavity will be studied with regard to 200-500 °C surface pulsed heating stress of the cavity as a consequence of 10^6 RF-pulses, which are planned to be produced at a repetition rate of 0.5-1 Hz.

In recent years a single-mode high-efficiency FEM-oscillator has been investigated in a joint experiment between JINR (Dubna) and IAP RAS (N.Novgorod) [1, 2]. The JINR-IAP FEM generates 30.7 GHz / 100 ns / 50 MW RF-pulses with a record efficiency of 35% for FEM-oscillators. The main features of the FEM-oscillator are the use of a reversed guide magnetic field and a novel Bragg resonator with a step of phase of corrugation.

The Bragg resonator consists of two corrugated waveguide sections with a π step of phase between the corrugations at the point of connection. In this resonator the Q-factor of the fundamental mode at the frequency of precise Bragg resonance is significantly larger than the Q-factors of the other modes.

Consequently, the selective property of a resonator of this type is drastically improved compared to a traditional two-mirror Bragg resonator. Results of simulations and experiments demonstrated that a single-mode single-frequency operation regime can be achieved in a wide region of the FEM parameters (see [2] in details).

The reversed guide magnetic field regime [3, 4] was chosen for FEM operation. Computer simulations show that this regime, which is far from the cyclotron resonance, provides high-quality beam formation in the tapered wiggler section and has a very low sensitivity to the initial spread of the beam parameters [5]. This was corroborated by results of experiments conducted at the LINAC LIU-3000 (0.8 MeV / 200 A / 200 ns / 0.5 Hz) where an efficiency of 35% was achieved, the highest obtained to date for millimeter wavelengths FEM. The measured

---

*Corresponding author. Tel.: +007 8312 384575;
Fax: +007 8312 362061; E-mail: peskov@appl.csi-nnov.ru

spectrum width corresponds to the excitation of the fundamental mode at the frequency of 30.7 GHz and does not exceed 0.1%.

A further development of these experiments is to use this high-power FEM source to study surface heating effects in 30 GHz accelerating structures, which are proposed for the CLIC two-beam accelerator presently under construction at CERN [6, 7]. With the specially designed RF test cavities, the RF-power will be able to create pulsed surface temperature rises from 50 to 500 °C. For the experiments the number of pulses generated by FEM at a repetition rate of 0.5-1 Hz will be limited to the range of 10^4 to 10^6 depending on the maximum surface temperature rise.

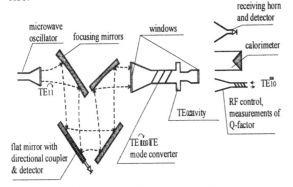

Fig. 1. Schematic of the experimental set-up.

The proposed experimental set-up is shown in Fig.1. The FEM generates the RF-power, which is radiated in the rotating TE_{11} mode. The four-mirror system and the $TE_{11} \rightarrow TE_{01}$ mode converter will be used to transport the RF-power from the FEM to the test cavity. The four-mirror transmission line is an efficient way of transporting the RF Gaussian pulse. One mirror has a built-in directional coupler to control both the incident and reflected powers. After a certain number of pulses the Q-factor of the cavity will be monitored using a network analyzer to detect early signs of surface damage.

The test cavity is a simple pillbox cavity, excited through the small coupling hole in the left end-cap

(see Fig.2). To avoid possible breakdown problems at the cavity surface, the test cavities will operate in the TE_{01} mode. The resonant frequency of the cavities will be mechanically tuned to the frequency of the power source.

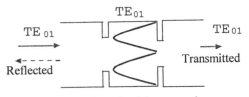

Fig. 2. Schematic of the test cavity

According to the simulations, the maximum magnetic field is expected in the vicinity of the end-caps near the coupling holes; the distribution of the magnetic field is shown in Fig.2. The heating produced is proportional to the square of the magnetic field. The cavity is similar to the well-known Fabri-Perot interferometer, where the absence of reflection is due to the difference in radii of the holes in the left and right end-caps. With a cavity loaded Q-factor ~ 500, the temperature rise in the cavity is expected to be up to 500°C.

The experiments on LIU-3000 are currently in progress.

Frequency, GHz	30 ± 1
RF-power, MW	30
Pulse duration, ns	100
Repetition rate, Hz	$0.5 \div 1$
Total number of pulses (up to)	10^6
Temperature on the cavity wall, °C (up to)	500

Table 1. General parameters of the proposed experiments

References

[1] A.K.Kaminsky, N.S.Ginzburg, A.A.Kaminsky, N.Yu.Peskov, S.N.Sedykh, A.P.Sergeev, A.S.Sergeev, Proc. of FEL'96 Conf., Rome, Italy, 1996, p.II-81.

[2] N.S.Ginzburg, A.K.Kaminsky, A.A.Kaminsky, N.Yu.Peskov, S.N.Sedykh, A.P.Sergeev, A.S.Sergeev, Phys. Rev. Lett. 84 (2000) 3574.

[3] A.A.Kaminsky, A.K.Kaminsky, S.B.Rubin, Particle Accelerators 33 (1990) 189.

[4] M.E.Conde, G.Bekefi, Phys. Rev. Lett. 67 (1991) 3082.

[5] N.Yu.Peskov, S.V.Samsonov, N.S.Ginzburg, V.L.Bratman, Nucl. Instr. and Meth. in Phys. Res. A407 (1998) 107.

[6] J.P.Delahaye e.a., Proc. of PAC'99, New York, p.252.

[7] I.Wilson, CLIC Note 52, Oct. 1987.

FREE ELECTRON LASERS 2000
V.N. Litvinenko and Y. Wu (Eds.)
2001 Elsevier Science B.V.

Design of high power Cerenkov FEL for industrial applications

A. Doria, G.P. Gallerano, L.Giannessi, E.Giovenale

ENEA- Divisione di Fisica Applicata, P.O. Box 65, 00044 Frascati - Italy

Abstract ENEA is involved in the design of an industrial long wavelength Free Electron Laser in the framework of a BRITE project of the European Community. The role of our laboratory in Frascati is to focus our attention to the design of a medium or high average power 100 GHz FEL.

1 Introduction

The European Community has founded a project for the study and design of a microwave FEL for industrial applications. The source to be realised must be suitable to operate in an industrial environment (safety problems must be considered), must have some constraints on costs, both for the realisation and for operation and maintenance, finally the power should be adequate for different kind of applications. Some applications (like material processing) require high CW power (in the range from 10 Watt to 1 kWatt), but other could need high peak power.

In order to satisfy all these requirements we have decided to study at Frascati the project for a Cerenkov FEL (CFEL) capable to operate in three different regimes depending on the electron beam source chosen. The advantage of a Cerekov FEL respect to conventional undulator FEL is due to the fact the the electron beam, travelling above the surface of a thin dielectric film, couples longitudinally with the electromagnetic field of the evanescent wave of the waveguide [1]. This implies that no relativistic beam is needed to generate radiation in the microwave or millimetre wave range. The result is an increased efficiency of the emission process.

2 Description of the project

The source studied is a CFEL operating at a central frequency of 100 GHz; to obtain this goal we can make a combination of the three fundamental parameters that rule the synchronism in a CFEL. This three parameters are the electron energy (or its relativistic factor γ), the dielectric film thickness and the dielectric constant. In order to satisfy the European Community requirements a 100 kV electron beam acceleration has been chosen because this is the limit allowed for the industrial operation. The corresponding relativistic factor is γ=1.2. With a single slab geometry waveguide [1], having a dielectric film of 350 μm of thickness and a dielectric constant ε=7, we obtain a radiation emission at λ=3mm corresponding to a frequency of 100 GHz. Three different cases have been analysed; to each correspond a different e-beam current and thus a different dynamical regime.

For all the three cases we have evaluated the amount of power achievable considering both the losses of the dielectric (including the guiding effect) and the saturation effect. As far as the saturation problem is concerned the theoretical FEL group in Frascati has developed different theories for different dynamical regimes. In the low gain regime the behaviour follows the one of the conventional lasers where the saturation is ruled by the saturation intensity parameter that for a FEL

$$G(In) = g_{0MAX} \left[1 - \exp\left(-\frac{\pi}{2} \frac{In}{I_s} \right) \right] \bigg/ \frac{\pi}{2} \frac{In}{I_s} \qquad (1)$$

can be written as follows [2]:

where g_{0MAX} is the gain coefficient and:

$$I_s = \frac{\gamma I I_{AV}}{g_{0MAX} 4N(4\pi\varepsilon_0)c} \qquad \text{and} \qquad N = \frac{L}{2\gamma^2 \lambda_{emiss}}$$

I_s is the saturation intensity and I_{AV} is the Alfven current.

In the high or medium gain regime, the Ginzburg-Landau equation must be solved leading to a solution that is similar to the logistic function [3]. The saturation is now ruled by the Pierce parameter ρ.

$$I(\tau) = \frac{I_0}{9} \frac{\exp(\sqrt{3}\rho\tau N)}{1+\left(I_0/9\rho I_e\right)\left\{\exp(\sqrt{3}\rho\tau N)-1\right\}} \quad \text{where}$$

$$\rho = \left(\frac{\pi g_{0MAX}}{N^3}\right)^{1/3} \quad (2)$$

Where I_0 is the intensity of the spontaneous emission and I_e is the one of the e-beam.

For the first case we have considered an electron current of 250 mA in order to realise a CW low gain device operating in a oscillatory regime. An output power radiation of 50 W can be obtained from a 25 kW e-beam with coupling losses for the optical cavity optimised at 7% value. Saturation is reached after 200 round trips (the cavity length is L=25cm) as shown in Fig.1:

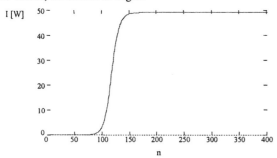

Fig.1 – Power vs. round trip number. Case I

But if we increase the e-beam current up to I=1A we enter in the medium high gain regime. The gain is not high enough to reach saturation in a single pass (the gain coefficient is g0=6.6 and we should need a waveguide slab 5 times longer to have a single pass regime) and thus we need again

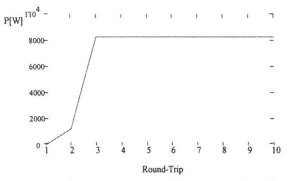

Fig.2 – Power vs. round trip number. Case II

an optical cavity. With an optimised coupling of 10% we can reach saturation after 4-5 round trips with a power of about 8 kW. This case is represented in Fig.2.

The last case considered uses a I=250 A of e-beam with which it is possible to reach saturation during a single passage inside the slab 25 cm long. The gain coefficient is now very high (g_o=821) and eq.(2) must be used to calculate the saturation behaviour. From Fig.3 we can see that saturation is reached at the end of the interaction region with an approximated power of 10 MW. The maximum power achievable with no losses in the waveguide would be I=ρI_e. In this case ρ=0.506 and the e-beam power I_e=25 MW and thus we would have I=12.5 MW.

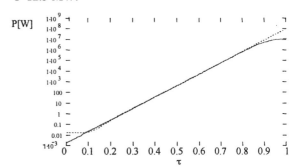

Fig.3 – Power vs. normalised transit time (solid line). The dashed line indicates the start-up of the process, but contains no saturation. This calculation includes the waveguide losses.

3 Conclusions

As a conclusion we can say that the three system analised differ only for the electron beam current (I_1=250 mA, I_2=1 A, I_3=250 A). This means that with a proper e-beam source we can satisfy different application requirements at 100GHz remaining within the EC constraints. The e-beam sources used in commercial high power klystrons are in general suitable for the proposed devices.

References

[1] F.Ciocci, G.Dattoli, A.Doria, G.Schettini, A.Torre, J.E.Walsh, Il Nuovo Cimento, 10D, (1988), 1.
[2] G.Dattoli, S.Cabrini, L.Giannessi, V.Loreto, C.Mari, Nucl. Instr. and Meth., A318, (1992), 495.
[3] G.Dattoli, L.Giannessi, P.L.Ottaviani, M.Carpanese, Nucl. Instr. and Meth., A393, (1997), 133.

FREE ELECTRON LASERS 2000
V.N. Litvinenko and Y. Wu (Eds.)
2001 Elsevier Science B.V.

Design Study of a Waveguide Resonator for an Infrared FEL at ELBE

G. P. Gallerano[a], A. Gover[b], E. Grosse[c], W. Seidel[c], M. Tecimer[b], A. Wolf[c], R. Wünsch[c]

[a]ENEA, Dipartimento Innovazione, Centro Ricerche Frascati, P.O. Box 65-00044 Frascati, Italy

[b]Dept. of Physical Electronics, Tel-Aviv University, Ramat-Aviv 69978, Israel

[c]Forschungszentrum Rossendorf, PSF 510119, D-01314 Dresden, Germany

At the Rossendorf Radiation Source ELBE [1] special emphasis will be put on the production of intense FEL radiation. For the IR wavelength range from 10 to about 150 microns a variable gap Halbach type undulator [2] housed in a big vacuum chamber has been transferred to Rossendorf from ENEA/Frascati.

Kinetic energy	12 - 40 MeV
Charge per bunch	50 pC
Repetition rate	13 MHz
Norm. transverse emittance	15 mm*mrad
Energy spread	< 90 keV
Bunch length (rms)	1 – 10 psec
Undulator period	50 mm
Number of periods	45
Undulator gap	(13) 16 – 30 mm
Undulator parameter K_{rms}	0.65 – 1.6 (1.9)

Table 1
Parameters of the ELBE beam and of the ENEA undulator.

Using the ENEA undulator at the superconducting accelerator of ELBE one needs an additional vacuum chamber for the electron beam ensuring an ultra-high vacuum (10^{-8} mbar). To optimize laser gain and minimize diffraction losses we envisaged a rectangular chamber 10 mm high and 40 mm wide within the undulator and sufficiently wide outside of it. Such a chamber guarantees free propagation at small wavelengths up to roughly $20\,\mu$m. Longer waves are guided in the vertical direction of the undulator and propagate freely horizontally and outside of the undulator in either direction. A trade-off between minimum mode area and resonator stability leads to a

Rayleigh range z_R=1 m in the horizontal direction resulting in a radius of curvature R_h=5.939 m. Within the undulator the optical field is described by a hybrid mode consisting of a waveguide mode in the vertical and a Gaussian one in the horizontal direction. If the dominant electric field component is parallel to the wide dimension (TE_{01} mode) the Ohmic losses in the waveguide walls are negligibly small [3]. At the undulator en-

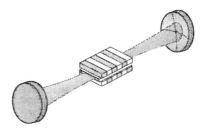

Figure 1. *Schematic of the partially waveguided resonator.*

trance and exit planes the field is converted into a freely propagating wave which is reflected back by the corresponding mirror and then coupled into the waveguide where losses due to mode conversion occur. The coupling losses can be minimized by selecting an appropriate radius of curvature R_v in the vertical plane in dependence on the wavelength of the radiation.

In contrast to the resonator used at FELIX [4] the distance between waveguide exit and the corresponding mirror is rather long, resulting in a small Fresnel number F_n. For the guided radiation wavelengths, the mirror radius of curvature in the vertical plane has to reproduce the fundamental waveguide mode at the waveguide en-

Resonator length L_{R}	11.53 m
Horizontal Rayleigh range z_{R}	1.00 m
Waveguide length L_{wg}	2.36 m
Waveguide height h_{wg}	10 mm
Waveguide width w_{wg}	40 mm

Table 2
Parameters of the optical resonator.

trance as close as possible. For sufficiently small Fresnel numbers, the optimal radius R_{v} is approximately given by the distance between the waveguide exit and the mirror ($R_{\mathrm{v}} = (L_{\mathrm{R}} - L_{\mathrm{wg}})/2 = 4.585$ m). The exact values of the optimal radii of curvature R_{v} will be determined by means of the code GLAD [5]. At the very shortest wavelengths we use spherical mirrors with $R_{\mathrm{v}} = R_{\mathrm{h}}$. Fig. 2 illustrates the transverse dimensions of the optical field.

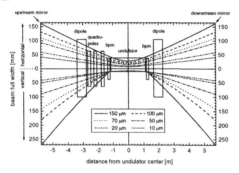

Figure 2. *Horizontal and vertical dimensions of the optical beam in the partially waveguided resonator.*

Detailed numerical simulations have been performed to estimate the expected laser gain and the average power. Fig. 3 shows the single-pass gain as a function of electron energy E_e^{kin} and undulator parameter K_{rms} calculated with the parameters displayed in the tables above. A bunch length of 1.5 psec (rms) has been assumed. The expected average laser power is displayed in Fig. 4. Here optical losses of 10% have been assumed independent of the wavelength. In a first stage the undulator parameter K_{rms} will be restricted to values below 1.3 while at a later stage a smaller gap with $K_{\mathrm{rms}} = 1.6$ will be achieved.

Improving the beam energy spread both single-pass gain and output power can be raised consid-

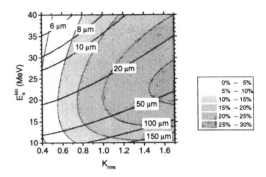

Figure 3. *Single-pass gain calculated for the ENEA undulator with a 10 mm waveguide. A 1.5 psec rms pulse length and 90 keV electron energy spread have been assumed.*

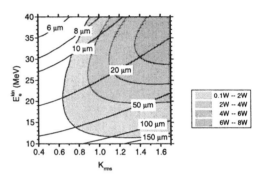

Figure 4. *Average laser power as a function of electron beam energy E_e^{kin} and undulator parameter K_{rms} assuming 10% optical losses and optimal values of cavity detuning and outcoupling fraction.*

erably. In particular an energy spread smaller than 30 keV would increase the gain and the output power by roughly a factor of three.

References:

[1] F. Gabriel et al., *Nucl. Instr. Meth.* **B161-163** (2000) 1143-1147

[2] F. Ciocci et al., *Nucl. Instr. Meth.* **A250** (1986) 134

[3] L. Elias and J. Gallardo, *Appl. Phys.* **B31** (1983)

[4] Li Yi Lin and A. F. G. van der Meer, *Rev. Sci. Instrum.* **68** (1997) 4342

[5] GLAD, G. Lawrence, *Applied Optics Research*, Austin, TX, USA

FREE ELECTRON LASERS 2000
V.N. Litvinenko and Y. Wu (Eds.)
2001 Elsevier Science B.V.

JAERI design options for realizing a compact and stand-alone superconducting rf linac-based high power free-electron laser

E J. Minehara*, T. Yamauchi, M. Sugimoto, M. Sawamura, R. Hajima, R. Nagai, N. Kikuzawa, N. Nishimori, and T. Shizuma

Free-Electron Laser Lab. at Tokai, APRC, Kansai, JAERI,
2-4 Shirakata-Shirane, Tokai, Naka, Ibaraki 319-1195, Japan

Abstract

In the beginning of February 2000, the JAERI free-electron laser (FEL) exceeded our 1-kW design goal and finally reached 2.34kW in quasi-continuous wave (quasi-CW) operation, after about 2 years of improvements since the first lasing of 0.1kW. The quasi-CW operation regime in the JAERI FEL was decided upon and introduced to minimize capital and operational costs in the beginning phase of the feasibility study. We will summarize and discuss briefly the many good features of the JAERI design, options for the existing and future facilities, higher-extraction efficiency, energy-recovery upgrading paths, and a few new design studies for a high-gradient S-band superconducting rf linac driver for both a short-wavelength X-ray FEL and a table-top one.

1. Introduction

A conventional laser has problems of heat loss in the drivers, gain media, and optical resonator mirrors. A free-electron laser (FEL) solves the heat loss problem of the gain media by quick-removal of the heat loss from the system. A superconducting rf linac (SCL) solves the heat loss problem of the laser driver. The SCL based FEL [1] can solve two of the three heat loss problem; a full beam energy recovery [2,3] can simplify the SCL driver, cut a major part of the capital and operational costs for the SCL, and minimize radiation hazards and the need for a radiation shielding wall.

The JAERI FEL [4] exceeded our 1-kW design goal and finally reached 2.34kW in quasi-continuous wave (quasi-CW) operation by increasing extraction efficiency. In the next step, increasing extraction efficiency and electron beam power in the near future will boost FEL laser power.

Here we would like to remove the optical resonator mirror problem from our discussion. In the following, we will discuss the design options available to JAERI in order to realize a compact, possibly movable and stand-alone SCL-based high power FEL.

2. JAERI design options

2.1. Zero-boil-off (ZBO) or cryogen-free cryostat

The easy operation and easy maintenance or maintenance-free feature of the SCL based FEL have been fully or partially demonstrated with a Zero-Boil-Off (ZBO) cryostat and a very long-life refrigerator system inside the cryostat. Several ten thousands of superconducting MRIs, several Maglev trains and several tens of cryogen-free superconducting magnets, which are used everyday and are commercially available, have a design concept similar to the superconducting magnets with the ZBO cryostat of the SCL based FELs.

2.2. Maintenance-free and easy-operation He gas refrigerator

The world's first Zero-boil-off (ZBO) or cryogen-free cryostat is cooled down by a Joule-Thomson (JT) and Gifford-McMahon (GM) composite refrigerator and a GM one. Four of the JAERI ZBO cryostats could be run continuously for one year with no malfunctioning and no shortage of liquid He during the 1995 Japanese fiscal year. This excellent record does not include scheduled and unscheduled power and water failures. Including the frequent malfunctions that occurred in the initial phase of 1993, we have not experienced a shortage of liquid Helium in the cryostats for these 8 years.

Since 1993, four GM refrigerators used to cool 50K and 10K duplex heat-shields have been run for 24 hours, 355 days a year, without any malfunctioning.

2.3.Energy recovery

A 180 degrees reflected or half-turn geometry of the recirculated energy recovery scheme will be used instead of the 360 degrees (full-turn) scheme to improve the recovery efficiency of the superconducting cavities up to 100% in low energy. Time reversal configuration was realized for accelerated and decelerated electron beams inside the cavity. A potential-energy-recovering DC decelerator coupled with the electron gun will be used to maximize the wall-plug efficiency, and to minimize the radiation hazards.

After we can make sense of so-called partial and full energy recovery techniques in superconducting rf cavities, we can cut 95-99% of the rf power supplies and can remove the heavy shielding wall. And we can simplify some components in the SCL FEL driver; for instance, an adjustable coupler, higher mode couplers, an rf amplifier, a water cooling system, a personnel radiation safety system and a power panel. No radiation, including neutrons and gamma-rays, can be expected in the full energy recovered SCL based FELs. Only small amounts of X-ray radiation will be detectable in the beam dump after the beam energy recovery. Energy recovery experiments with a MIT Baytes 360-degree loop have been done at Jefferson lab; demonstrating nearly perfect recycling, beam power after the lasing, except for the injection, remained at 25% total beam power. Finally, we may realize very high wall-plug efficiency using the full energy recovery in the SCL based FELs.

2.4.Quasi- and true- continuous wave operations

A stand-alone, possibly-movable and compact industrial FEL is designed to use a few refrigerators to cool 45W or several tens of watts of rf-heating in CW operation. After minimizing heat invasion and stand-by losses in the cryostat by removing a large number of cables and wiring, we can probably cut 3/4 of the stand-by losses. As we have developed a high cooling power and compact 4K refrigerator recently, 8W, 12W, and 15W ones have been commercially available, with a 25W one under preparation. The single He gas 4K refrigerator has an enough capacity to cool a quarter CW or a quarter duty cavity, and two such refrigerators are able to cool a half CW or a full CW at half duty.

Optimizing the repetition rate and pulse duration—in other words, longer pulse duration than rise-time and a certain duty factor—we can run the SCL based FEL in quarter CW, half-quarter CW, or quarter-quarter CW. To keep the beam power constant with a single refrigerator, we have to increase the repetition rate or peak current. Our current operation is quite equivalent to half-quarter-quarter CW. The quasi-CW operation is

the best way to realize very high wall-plug efficiency, because we can reduce the number of He refrigerators, which consumed the largest part of electricity.

3.Future directions

There are a few future plans for high power tunable Industrial FELs driven by a compact and stand-alone UHF SCL, a high-gradient and low-duty S-band SCL driver for a short-wavelength X-ray FEL, and a table-top S-band SCL. Industrial programs on SCLs and SCL-based high power

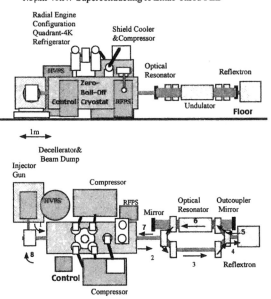

FELs are now under way to build a stand-alone and compact prototype. An example of the industrial SCL-based FEL is illustrated in Fig.1.

Figure 1. Illustration of A 1.3micron 40kW Industrial Super-conducting rf Linac-based Free-Electron Laser for Heavy Metal Processing.

References
[1] E.J.Minehara et al., pp159-161, in the proceedings of Particle Accelerator Conference, 1995, Dallas.
[2] T.Shizuma , et al., "Injector Design for the JAERI-FEL Energy Recovery Transport ", to be published in the EPAC2000 proceedings.
[3] R.Hajima, T.Shizuma, and E.J.Minehara, "Study on Emittance dilution in the JAERI-FEL Energy –Recovery Transport", to be published in the EPAC2000 proceedings.
[4] N.Nishimori, et al., "A Thermionic Electron Gun System for the JAERI Superconducting FEL", to be published in the EPAC2000 proceedings.

FREE ELECTRON LASERS 2000
V.N. Litvinenko and Y. Wu (Eds.)
2001 Elsevier Science B.V.

SASE Experiments In A μFEL

A. Bakhtyari, H. L. Andrews, J. H. Brownell, M. F. Kimmitt, J. E. Walsh

Department of Physics & Astronomy, Dartmouth College, Hanover, NH 03755, USA

A SEM was used to investigate Smith-Purcell Free Electron Laser Radiation. The collected data show pre-threshold beating that suggests the presence of Self-Amplification of Spontaneous Emission (SASE).

Keywords: SEM; Grating; Electrons; Smith-Purcell

1. Introduction

In 1953 S.J. Smith and E.M. Purcell discovered that a charged particle moving over a periodic conductive surface (a grating) produces radiation [1]. Over the intervening decades interest in grating induced radiation has been sustained by the potential for use as an easily tuned source of both coherent and incoherent radiation.

2. Earlier Experiments

Operation of a grating coupled source of coherent radiation in the FIR region of the spectrum has already been described (Urata et. al. [2]). A retired scanning electron microscope was modified to serve as a source of the e-beam. Radiation is produced on a diffraction grating mounted in the electron beam focal region. The apparatus is shown in Figure 1.

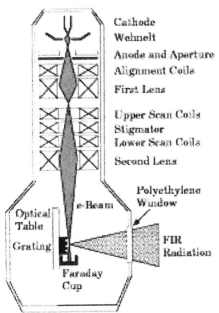

Figure 1 Diagram of the vacuum chamber of the modified SEM.

In 1998 Urata et. al. achieved the first lasing with this basic apparatus [2]. The plots of output radiation power vs. electron beam current in [2] show a spontaneous radiation region, in which the power rises linearly with the beam current, as well as a high gain region with a power vs. current dependence of $P \sim I^n$. Where n was typically of the order of 4. The specific slope of 4 in the high gain region had no apparent basis in theory.

The exact slope and threshold for the onset of the high gain regime depends sensitively on the e-beam focal parameters. In order to investigate this further a number of modifications have been carried out.

3. Modifications

In a second modification step the system was completely automated and the console was replaced by a PC and controlled via the programming environment LabView.

These changes promise the possibility of:
- Rapid data acquisition
- Better control over the system by means of exploitation of the given parameter space and introduction of feedback elements

In order to measure the electron beam current passing over the grating, a carbon Faraday-Cup is mounted underneath the grating. It is connected to an oscilloscope via a BNC feedthrough.

A powerful tool for beam diagnostic in this apparatus is the "profilometer" (see Figure 2). It consists of a set of three grounded wires with known diameter and vertical displacement. As the beam sweeps over one of the wires it casts a shadow, which alters the signal in the Faraday cup collecting the electrons at the bottom of the column. By relating the rise time of that signal to the width of the wire one can deduce the diameter of the electron beam at the position of the wire.

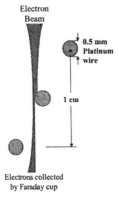

Figure 2 Sketch highlighting the working principle of the Profilometer.

Figure 3 shows the electron beam diameter measured with the profilometer at three different locations within the beam and their evolution with the changing focusing lens current.

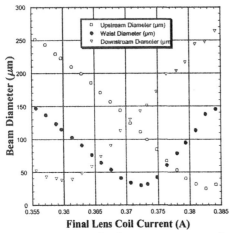

Figure 3 Beam focal parameters vs. lens current.

3.1 Data and Analysis

A scan of output power vs. beam current features the familiar spontaneous radiation region but also a high gain region with a much steeper rise of output power with current (Figure 4). However, the data show also beating-like oscillations which are sometimes very pronounced -- even more so than in this plot. Manipulation of the part of the data which corresponds to the oscillating region (Figure 5) reveals a radiation power vs. beam current dependence of $P \sim I^{4/3}\cos(bI^{1/3})$ suggesting the onset of SASE.

When reviewing the previous data obtained by Urata et. al. [2] one can already see a hint for these oscillations.

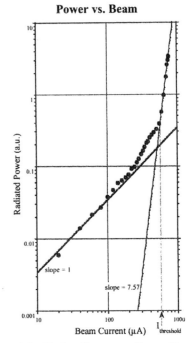

Figure 4 Radiation Power vs. Beam Current

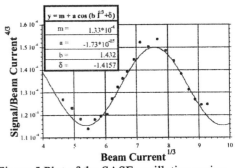

Figure 5 Plot of the SASE oscillation region

The reason the oscillations are more pronounced in the present data is that the electron beam in the automated system is not yet as well collimated as before the modifications. Due to the larger divergence of the beam the lasing threshold is being pushed to higher beam currents and the pre-threshold oscillations are being more emphasized.

Financial supported by USARO Grant DAAD 19-99-1-0067 and NSF Grant ECS 0070491 is gratefully acknowledged.

References

[1] S. J. Smith and E. M. Purcell, "Visible light from localized surface charges moving across a grating", *Phys. Rev.*, 92, p. 1069 (1953)
[2] J. Urata et. al., "Superradiant Smith-Purcell Emission", *Phys. Rev. Lett.*, 80, p.516 (1998)

FREE ELECTRON LASERS 2000
V.N. Litvinenko and Y. Wu (Eds.)
2001 Elsevier Science B.V.

European Thematic Network Project for an Industrial Free Electron Laser at 10-100GHz

A.I. Al-Shamma'a[1], J. Lucas[1], R.A. Stuart[1], P.J.M. Van Der Slot[2]

(1) The University of Liverpool, Dept. of Elec. Eng. and Electronics, Brownlow Hill, Liverpool L69 3GJ, U.K.
(2) Nederlands Centrum Voor Laser Research, P.O. Box 2662, Enschede, 7500 CR, Netherlands.

Abstract

The Industrial Free Electron Laser (IFEL) thematic group has been operating now for 18 months and during this period the group has examined suitable IFEL designs and a range of possible applications. This paper describes the possible designs of an electron beam/em wave interaction region namely conventional, Cerenkov and Gaussian mode operating at 100kV and frequencies from 10 to 100GHz. The programme of work for the next 18 months is reviewed with all the possible industrial applications.

1. Introduction

The IFEL Thematic Group has been operating now for 18 months and a critical examination of its progress has been undertaken in order to set out the plans for the final 18 months of the project. Previously [1] the partners from UK, Netherlands, Italy, Ireland and France have been split in to two groups, A and B. Group A was concerned with IFEL design and group B was concerned with IFEL applications.

It was intended that all the partners would set out initially with the aim of producing a plan to make an IFEL and to identified its associated industrial applications.

The progress achieved so far, and a summary of the future work, will now be outlined for both groups of partners. However, it must be appreciated that the group as a whole needs to be specifically focused on the development of the European IFEL and those industrial applications, which can only be realistically undertaken by having the IFEL.

2. IFEL Operating Targets

These have been given in table 1 and are still the final goals. There are four possible electron beam/EM wave interaction regions, which have been considered so far namely Cerenkov, Ubitron, Conventional and Smith – Purcell. These aspects only occupy 10% of the design considerations for the IFEL, with the remaining 90% being common for all designs. All these four techniques have now been considered in reasonable depth to allow the final IFEL design to be specified and the IFEL to be realised before the end of the Thematic project.

3. The EC IFEL Design Study

The results of the design study has shown that the following IFEL structures should be realised as shown below. The approach in the end has turned out to be very conservative. Three IFEL electron beam/EM wave reactions will be operated within the common IFEL structure. These will be centred around the following IFELs,

1. The 10GHz conventional IFEL will use the design presently being operated with the Liverpool (UK) IFEL.
2. The 50GHz Cerenkov IFEL will use the design presently being operated with the Twente (Netherlands) IFEL
3. The 100GHz design is a conventional design with the rectangular waveguide being replaced by a Gaussian waveguide design.

Operating Frequency Bands	Upper	100GHz
	Middle	50GHz
	Lower	10GHz
Output Power	Continuous	2kW
	Pulsed	100kW Peak
		2kW Average
IFEL Type	Cerenkov	
	Ubitron (Cylindrical waveguide)	
	Standard (Rectangular waveguide)	
	Smith-Purcell	
Beam Energy	100keV	
Beam Current	1A	
Cost	IFEL	50kEuro
	CO_2/Yag	500kEuro

Table 1 The IFEL Operating Targets

4. IFEL Applications

The object was to use existing microwave sources, namely 896MHz and 2.46GHz microwave sources, to undertake a preliminary investigation of the process. This approach has allowed a clear indication of whether microwaves at high frequencies would be more beneficial to carry out the application. It has been decided that only those strong applications will be pursued which fully support the need to produce an IFEL operating in the frequency range 10 to 100GHz.

The following projects have been undertaken so far, Microwave Plasma, Microwave Heating, Medical Applications, Vitrification and Concrete Scabling, Microwave Plasma for Welding, Fashion Cloth Marketing, Microwave Chemistry, Exhaust Gas Pollution Removal, Microwave Thermal Chemistry, Propagation of Microwaves in Sea Water, Hybrid Plasma/CO_2 Laser System, Welding, Ceramic Material Processing, UV Generation, Fibre Optic Sensors for Microwave Detection.

5. Summary

The IFEL during the first 18 months has examined suitable IFEL designs and a range of possible applications. The group size was 26 partners. The main aim was to develop the IFEL to operate with frequencies up to 100GHz which is way beyond the capabilities of the magnetron device (<2.46GHz). In fact the IFEL is targeted to operate between the magnetron and the infra red power laser as given in table 2. As a technology development the IFEL would be able to justify the research effort. However, when coupled with new applications, which have an optimised operating characteristics within the FEL frequency band, then the need to develop the IFEL becomes commercially desirable.

Type	Wavelength	Power
Magnetron (3GHz)	100mm	8kW
IFEL (100GHz)	3mm	1kW
Infra red power laser based upon the CO_2 laser + chemical reaction (1000GHz or 1THz)	300μm	10W
CO_2 Power laser (30THz)	10μm	5kW

Table 2 The IFEL Characteristics

Based upon a critical analysis of the results of the Thematic activities during the first 18 months, a compact plan has now been put forward to realise the IFEL and four major applications. The partner list has now been reduced from 26 to 15, with two partners, namely The University of Liverpool and Twente (NCLR), now undertaking activities under both group A and group B.

The partners who are now leaving the Thematic group are encouraged to apply directly to the EC for RTD grants under the next round of funding as some of the applications using magnetrons are both innovative and commercially viable.

6. Acknowledgments

The authors wish to thank the EC funding for their financial support towards this project, contract no. BRRT-CT98-5052.

7. References

[1] J. Lucas, R.A. Stuart and A.I. Al-Shamma'a, Nucl. Instr. and Meth. A429 (1999) II 95.

[2] A.I. Al-Shamma'a, J. Lucas and R.A. Stuart, Nucl. Instr. and Meth. A445 (2000) II 47.

[3] J. Lucas, EC Thematic Network New Proposal, 1999.

FREE ELECTRON LASERS 2000
V.N. Litvinenko and Y. Wu (Eds.)
2001 Elsevier Science B.V.

Projects for Two-Color Pump-Probe Studies at the Radiation Source ELBE

W. Seidel, A. Büchner, W. Enghardt, P. Evtushenko, F. Gabriel, P. Gippner, E. Grosse, D. Kalionska, U. Lehnert, P. Michel, A. Schamlott, W. Wagner, D. Wohlfarth, A. Wolf and R. Wünsch[a]

[a]Forschungszentrum Rossendorf, Postfach 510119, 01314 Dresden, Germany

The radiation source **ELBE** [1] at Dresden-Rossendorf is centered around a superconducting **EL**ectron accelerator of high **B**rilliance and low **E**mittance, constructed to produce up to 40 MeV electron beams of 1 mA at 100% duty cycle. This new facility (see Fig.1) will deliver secondary beams of different kinds starting in 2001. Special emphasis will be given to the production of intense infrared radiation from Free-Electron Laser (FEL) devices. Different undulators (see table) will be installed in order to cover the IR wavelength range from 5 to about 250 micrometers.

In addition to the wide wavelength range which is essential for investigations in semiconductor research, physical chemistry and biomedicine, plans are pursued to supplement the FEL undulators such that two-color pump-probe studies with possibly large wavelength differences between pump and probe become feasible.

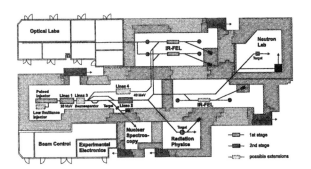

Figure 1. Layout of the radiation source ELBE

The tunable hybrid undulator and two-color operation

For short wavelengths we shall install two undulator units of 34 periods each, which are of the same structure as the DESY-TTF [2] device.

The undulator will be set up such that the gap of the two units can be varied independently [3]. For an optimized phase-matching of the two units the distance in between them is adjustable over a wide range. The predicted laser gain is large enough for lasing with a singel unit (Fig.3).

Figure 2. The DESY undulator (2xU27x34)

New applications for two-color operation [4] are opened by running a mode with different gaps. Additional tapering should result in an increase of the extraction efficiency [5,6]. This new set-up is very promising for pump-probe applications as the two colors are synchronized within 1 ps.

Numerical simulations

Calculations have been performed in order to predict quality and quantity of the produced radiation. Fig.3 shows the calculated [7] single-pass gain of one (left panel) and two (right panel) undulator units as a function of the electron beam energy E and the undulator parameter K_{rms} for pulses with 50 pC charge, 1.5 ps rms length, 90 keV energy spread and 15 mm*mrad normalized transverse emittance.

Table 1

FEL-undulators at ELBE and their radiation. The asterisk (∗) indicates an operation with a waveguide in the undulator (cf. G.P. Gallerano et al., contribution to this conference).

undulator	N	period [mm]	wavelenght [μm]	av.power [W]	pulse energy [μJ]	rate [MHz]	width [ps]
U27x68 (hybrid, DESY)	68	27	5 - 20	20	2	13	0.3 - 3
U50x45 (Halbach, ENEA)	45	50	15 - 150∗	10	1	13	1 - 10
U90x28 (electromagnetic, under discussion)	28	90	20 - 250∗			13	3 - 20

15% single-pass gain assumed in the calculations.

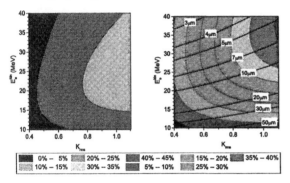

Figure 3. Single-pass gain for the DESY undulator

Two-color experiments with different undulators

The independent tuning of two optical beams over a broad range of wavelengths requires a system with two undulators, each driven by its own electron beam. The energy of both electron beams has to be changed indepently in a brad energy range. Necessarily, the two optical beams must be synchronized precisely in such a way that the pump pulse and the probe pulse time difference is established on the picosecond level. This synchronization can be accomplished using alternate electron bunches arriving at 26 MHz. Following a proposal made by E. Crosson et al. [8] an additional energy modulator cavity provides a longitudinal impulse of alternating sign to subsequent electron bunches. Spatial separation is done by means of a magnetic separation system, which deflects bunches of different momenta into different paths. This will allow pump-probe experiments with independently tunable wavelengths from two parallel FELs running with 13 MHz.

Pump-probe studies with X-rays and IR-radiation

The electron beam switching system described above can also be used (see Fig.1) to direct one of the two beams to a target producing quasi-monochromatic X-rays which are then delivered to the object under investigation. For example, the X-rays may modify the properties of a sample and these modifications will be detected by means of the IR-radiation from the FEL.

The interaction of relativistic electrons with solids (particularly with crystalline structures) and laser radiation can be used for the installation of non-conventional sources of intense and quasi-monochromatic X-rays. A 20 MeV ELBE electron beam can produce:

- Channeling radiation (CR) in the energy range 10 keV ≤ E ≤ 50 keV
- X-rays in the energy range 1 keV ≤ E ≤ 10 keV from Compton back-scattering (CBS).
- Transition radiation (TR) with E ≤ 0.5 keV.

Radiobiological experiments are in preparation to measure the energy-dependent biological efficiencies of quasi -monochromatic X-rays to cells.

REFERENCES

1. F. Gabriel, et al., Nucl. Instr. and Meth., **B161-163** (2000) 1143
2. J. Pflüger, H. Lu, D. Köster and T. Teichmann, Nucl. Instr. and Meth. **A407** (1998) 386
3. A. Büchner et al., FEL confernce 1999, Hamburg
4. R. Prazeres et al., Nucl. Instr. and Meth. **A407** (1998) 464
5. E.L. Saldin, E.A. Schneidmiller, M.V. Yurkov, Phys. Lett. **A185** (1994) 469
6. D. Jaroszynski et al., Phys. Rev. Lett. **72**, 2387 (1994)
7. S.V. Benson, CEBAF TN♯94-065
8. E. Crosson et al., FEL conference 1996, Rome

FREE ELECTRON LASERS 2000
V.N. Litvinenko and Y. Wu (Eds.)
2001 Elsevier Science B.V.

Parametric chirped-pulse amplification of mid-infrared FEL pulses

G.M.H. Knippels, D.Oepts, A.F.G. van der Meer

FOM-Institute for Plasma Physics `Rijnhuizen', P.O. Box 1207, 3430 BE Nieuwegein, The Netherlands

T. Nie, A.M. MacLeod, W.A. Gillespie

School of Science and Engineering, University of Abertay Dundee Bell Street, Dundee, DD1 1HG, United Kingdom

The current generation of mid-infrared FELs is capable of producing tunable picosecond and femtosecond pulses with energies of many microjoules. Highest reported values for the peak power are from CLIO (~1 GW) [1], and record pulse energies as high as 1.9 mJ for 16 ps long pulses have been reported with RAFEL [2]. In this contribution we present a generally applicable amplification scheme that could boost the energy for most operational mid-infrared FELs well into the mJ-range.

The amplification scheme is based on parametric interaction in a non-linear crystal between the mid-IR FEL output and a much more powerful pump laser pulse from a conventional laser running at a shorter wavelength (preferably around 1 μm) [3].

If silver thiogallate ($AgGaS_2$) is used as the non-linear crystal [4], the fundamental (1.064 μm) wavelength of Nd:YAG is suitable as powerful pump beam and IR wavelengths from the FEL can be amplified in the 3-12 μm range. The IR transparency range and phase-matching characteristics of $AgGaS_2$ limit this tuning range. Other candidate materials for 1-μm pumping are $LiInS_2$ and $HgGa_2S_4$ [5]. The parametric amplification scheme has been demonstrated on conventional lasers to give mJ's in 14-ps long pulses, using a parametric generator stage to provide the IR seed pulse [6].

The basic set-up is quite simple and requires propagation of the Nd:YAG pulse together with the FEL pulse (for example at 6 μm wavelength) through a crystal of sufficient length (typically around 10 mm for $AgGaS_2$). Photons at the pump wavelength of 1.064 μm are converted into a 6-μm photon and another photon that takes the remainder of the energy and momentum (in this case at 1.293 μm). This process is therefore also commonly referred to as difference-frequency generation (DFG).

Of course the Nd:YAG laser pulse has to have spatial and temporal overlap with the FEL pulse while it co-propagates through the crystal. This temporal overlap is achieved through active synchronisation of the Nd:YAG laser and spatial overlap is achieved by a high-power beam 'combiner' that is highly reflective for 1.064 μm and has good transmission in the mid-IR range. Note that it is also possible to achieve phase-matching with a small angle between the beams if no suitable optic is available or if spatial separation is preferred.

At FELIX we have a Nd:YAG with 70-100 ps long pulses repeated at 10 Hz. Each pulse has about 100 mJ and the laser is actively synchronised to within 10 ps with the FEL. This laser has been used in first experiments to demonstrate the feasibility of the proposed amplification scheme. Also computer simulations have been performed with a code called SNLO[1]. In first experiments performed at FELIX at a wavelength of 6 μm, pulse energies of 3 μJ in 1 ps have been amplified up to 80 μJ (simulated value 115 μJ). The power density of the pump laser (typically around 1 GW/cm^2 for Ag-GaS$_2$) together with the crystal length determines the maximum amplified energy. Basically, the peak power of the FEL pulse incident on the crystal is amplified to the level of the peak power of the pump laser. If the crystal is longer, back conversion occurs, i.e. photons from the FEL are converted to the pump beam again.

In case of the 100-ps Nd:YAG pulse and a 1-ps FEL pulse, there clearly is a large mismatch in pulse duration. Provided that saturation of the amplification process is achieved in the available crystal length, this mismatch can actually be exploited. When a grating pair stretches the FEL pulse in time, the incident peak power at the crystal will be lowered and amplification to the same level as the pump pulse is again possible. In this way an additional factor of 100 increase in pulse energy could be obtained and simulated pulse energies of several mJ have been observed for realistic parameters. Of course recompression of the pulse is necessary after the amplification process to convert this increase in pulse energy to an increase in peak power. Peak powers on the order of 1 GW are observed in the simulations with a pulse length of 1-2 ps. See Fig. 1 for a proposed experimental layout.

---

[1] SNLO nonlinear optics code available from A.V. Smith, Sandia National Laboratories, Albuquerque, NM 87185-1423, USA.

This stretching of the pulse followed by amplification and recompression is a widely used technique in Ti:Sapphire amplifiers and is commonly called chirped-pulse amplification (CPA). In these amplifiers the stretching is necessary to avoid optical damage caused by the peak intensity of the amplified pulse. In our case the stretching is just a convenient way to extract more energy out of the long Nd:YAG pulse.

When considering amplification of picosecond pulses it is important to take the phase-matching bandwidth into account. Typically, for $AgGaS_2$ this is about 15 cm^{-1} (or 1 ps) for 10-mm crystal length, and the bandwidth is inversely proportional with crystal length. This limited phase-matching bandwidth has consequences for amplifying femtosecond pulses because the maximum allowed crystal length is then only a few mm. Only limited amplification can be achieved since the maximum peak power density of the pump laser is fixed. Because of the fact that the amplification is typically exponential, the scheme appears to be limited to somewhere around 800-fs pulses when $AgGaS_2$ is used. Other crystals like $HgGa_2S_4$ have more attractive properties for femtosecond pulses. A 4-mm long $HgGa_2S_4$ crystal would provide the same amplification factor as a 10-mm long $AgGaS_2$ crystal (at 6 μm wavelength) and would amplify 500 fs short pulses if the optical quality proves to be sufficient.

In addition to a limited phase-matching bandwidth also gain-narrowing during the amplification process reduces the spectral width of the amplified pulse. In the situation where the chirped-pulse configuration is used, typically a reduction of the spectral width by 20 % has been observed.

The proposed amplification scheme seems also very promising for the generation of very narrow bandwidth (< 0.5 cm^{-1}) and still rapidly tunable infrared pulses with several mJ of energy per pulse. In this case it is necessary that the Nd:YAG pump laser has a Fourier-transform limited spectrum (~ 0.15 cm^{-1}), which should be the case for injection-seeded systems like the one we are using. Basically, the wide bandwidth infrared pulse would be transmitted through a monochromator that limits the spectral width to around 0.15 cm^{-1}. The pulse energy is then of course reduced by a factor of 100-1000, depending on the initial spectral width and characteristics of the spectrometer. Furthermore, it is stretched in time to around 100 ps. The narrow band infrared pulse can now be amplified in a longer $AgGaS_2$ crystal (typically around 20 mm long) and should be able to reach pulse energies of several mJ's. This would lead to an increase of the spectral intensity by five orders of magnitude. Rapid tuning is achieved by tuning the monochromator over several wave numbers, until the phase-matching bandwidth of the crystal is reached. At this point a small angular rotation of the crystal is sufficient to tune over the whole FEL spectrum that is incident on the monochromator.

Acknowledgement
This work is part of the research program of the "Stichting voor Fundamenteel onderzoek der Materie (FOM)," which is financially supported by the "Nederlandse Organisatie voor Wetenschappelijk Onderzoek" (NWO). T.N., A.M.M., and W.A.G. are grateful to the UK Engineering and Physical Sciences Research Council for financial support (Grant Code No. GR/M13602).

[1] Private communication with J.M. Ortega
[2] D.C. Nguyen, et al, NIM A **429**, 125-130 (1999).
[3] V.G. Dmitriev, G.G. Gurzadyan, D.N. Nikogosyan, in "Handbook of Nonlinear optical crystals", (2nd Edition, Springer Verlag, Berlin 1997).
[4] See ref [3], page 132.
[5] See ref [3], page 231.
[6] H.J. Krause and W. Daum, Appl. Phys. B 56, 8-13 (1993).

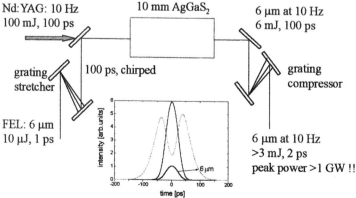

Figure 1. Proposed experimental set-up for the parametric chirped-pulse amplification scheme. A 10 mJ, 1-ps FEL pulse is first stretched to 100 ps, then amplified up to a few mJ's, and recompressed back to 2 ps. The simulated pulse profile is shown in the inset.

FREE ELECTRON LASERS 2000
V.N. Litvinenko and Y. Wu (Eds.)
2001 Elsevier Science B.V.

Status of the Stanford Picosecond FEL Center (1)

R.L. Swent

Stanford Picosecond FEL Center

W. W. Hansen Experimental Physics Laboratory

Stanford University

Stanford, CA 94305-4085 USA

1. Introduction

The Stanford Picosecond FEL Center, established in 1990, supports active experimental programs based on our mid-and far-infrared FELs. Our operating parameters are well-suited to experiments in molecular and carrier dynamics in materials science and biomedicine. Table 1 lists the operating parameters for the mid-IR and far-IR lasers. We also have tunable near-IR conventional lasers synchronized at a picosecond level to both FELs. In the mid-infrared we have two-color pump-probe experiments, cavity ringdown spectroscopy, and a number of scanning near-field microspcopy experiments. In the far-infrared the experiments have concentrated on the spectroscopy of semiconductor quantum structures in ultrahigh magnetic fields. We have also been studying the FEL dynamics of modulated desynchronism to obtain very short optical pulses.

Table 1. Operating Parameters for the FELs

	Mid-IR (STI)		Far-IR (FIREFLY)	
Wavelength	3 – 12	μm	15 – 70	μm
Micropulse Width	0.7 – 3	ps	2 – 10	ps
Micropulse Repetition Rate	84.6	ns	84.6	ns
Macropulse Width	5	ms	5	ms
Macropulse Repetition Rate	20	Hz	20	Hz
Micropulse Energy	1	μJ	1	μJ
Average Power	1.2	W	1.2	W
Spectral Bandwidth	Transform-Limited Gaussian		Transform-Limited Gaussian	
Spectral Stability	0.01% rms		0.01% rms	

2. Scientific Results

Vibrational photon echo techniques as well as pump-probe techniques have been used to study the molecular dynamics of CO bound at the active site of myoglobin (Mb-CO). These experiments exploit the ability of the infrared vibrational echo to eliminate the inhomogeneous broadening contribution to the line width and to provide a direct measure of homogeneous dephasing. Figure 1 shows the pure dephasing rate of Mb-CO as a function of temperature in the solvents trehalose and ethelyne glycol:water (Eg:OH). The pure dephasing rate has a temperature-dependent part and also a dependence on solvent viscosity. Trehalose has essentially infinite viscosity over the temperature range shown, while ethylene glycol

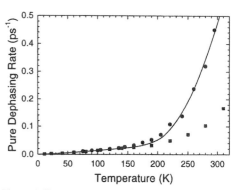

Figure 1. Temperature dependent pure dephasing rates of Mb-CO in the solvents trehalose (■) and in 50:50 ethylene-glycol:water (EgOH; ●). The line through the EgOH data is the fit to the viscoelastic theory added to the temperature dependence in trehalose.

has a glass transition temperature near 130 K and a rapidly decreasing viscosity above this temperature. A viscoelastic theory has been proposed that asserts that the internal protein dynamics are strongly constrained by the motion of the protein surface. As the solvent viscosity decreases and the protein surface becomes free to move, more protein motions are allowed and the dephasing rate rises. The solid line is a fit to this theory that uses the temperature dependence from the trehalose data and the viscosity of EgOH. The theory does a remarkable job of reproducing the data.

Figure 3 shows the results of a nonlinear mixing exeperiment between far-IR and near-IR beams in a bulk GaAs sample. Two sidebands of the near-IR frequency were observed, spaced at the frequency of the far-IR radiation. As shown in the inset, the sidebands are at energies below the band gap where there are no real states and they are therefore not resonantly enhanced. As the temperature varies in the experiment from T= 125 K to 175 K, the sidebands exchange intensity. This can be partially accounted for by the interplay between the absorption of the fundamental and the sideband itself. Another important observation is the appearance of an odd sideband without intentional breaking of the inversion symmetry. The THz power dependence clearly demonstrates entry into the non-perturbative strong field regime and is consistent with recently predicted non-monotonic behavior.

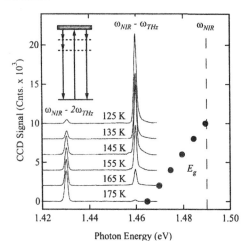

Figure 2. Non-resonant sidebands from virtual states below the energy gap in bulk GaAs.

Experiments on near-field imaging of biological cells were performed in the spectral range from 5 µm to 8.5 µm. Chalcogenide glass fibers with a 100µm core were used to deliver the infrared radiation to a metal-coated pyramid tip with a sub-wavelength opening at the apex, which served as a near-field probe. Sharp pyramid tips with radii of curvature significantly less than 1µm were produced by mechanical grinding. These tips were completely coated with gold (150 nm thickness) by vacuum evaporation and scanned across fine polishing paper (0.05 µm grain size) until a detectable IR signal appeared. Measurements performed with a scanning electron microscope have shown that a typical aperture size was about 1µm. This method enabled us to create high quality optical probes with a light source well localized at the apex of the tip. Images of single sperm cells have been taken in the spectral region containing absorption lines of proteins and DNA (5.5 – 8.5 µm). These images show the tail (~0.5 µm diameter) clearly resolved and have demonstrated a sensitivity high enough to allow spectrally sensitive detection of less than 1 picogram of material.

Cavity ring-down spectroscopy (CRDS) is a general, high sensitivity technique for measuring absorption. In CRDS, light from a laser source is injected into a high-Q optical resonator that encloses the sample of interest. When the light source is interrupted, light trapped inside the optical resonator decays exponentially. An absorption spectrum of the sample species is obtained by measuring the decay rate as a function of wavelength, and subtracting the background decay rate for the empty resonator. Recent measurements of a 4 nm C_{60} film on a 1 mm CaF_2 substrate have demonstrated that the decay time can be measured to an accuracy of 0.2%. This corresponds to a single-shot absorption sensitivity better than 1×10^{-5}. With modest improvements and the use of averaging, it should be possible to achieve a sensitivity of 1×10^{-6}, an order of magnitude better than that of an FTIR.

(1) Work supported in part by the Office of Naval Research, Grant No. N00014-94-1-1024.

FREE ELECTRON LASERS 2000
V.N. Litvinenko and Y. Wu (Eds.)
2001 Elsevier Science B.V.

User Applications for the TAU
Tandem Electrostatic Accelerator FEL at Long Wavelengths

A. Gover, A. L. Eichenbaum, S. Efimov, J. Sokolowski, M. Tecimer, Y. Yakover
- Tel-Aviv University, Tel - Aviv
A. Abramovich, M. Canter, Y. Pinhasi, A. Yahalom - The College of Judea and Samaria, Ariel
I. Schnitzer, Y. Shiloh - Rafael, Haifa

Electrostatic Accelerator (EA) FELs have been used primarily for spectroscopy of semiconductor materials and devices in the Tera Hertz regime [1]. We recently identified user interests in additional applications in the long wavelength regime; primarily material processing, imaging, bio-medical and atmospheric studies.

A new radiation user facility for such long wavelengths application is being built now on the premises of the College of Ariel on the basis of the Tandem EA-FEL.

The concept of the FEL user facility is described schematically in Fig. 1. A layout design of the user rooms is shown in Fig. 2. The FEL radiation will serve three kinds of user stations:
1. Material processing stations.
2. Spectroscopy and imaging experiments
3. Atmospheric experiments

The material processing stations requiring high average power will be located closest to the radiation source. Spectroscopic and imaging experiments as well as biological experiments, which require lower power will be situated at more remote locations. The atmospheric experiments station will be located on the roof of the building, where we intend to install a large mm-wave antenna.

The demonstrated operating parameters of the FEL and the parameters planned for the next stage of development are listed in Table 1. A number of development tasks are now being pursued in order to obtain the goal radiation parameters and to make them available to the users. These include completion of the mm-wave radiation transmission line system transporting the radiation beam to the user rooms, improvement of e-beam optics and design of a multistage collector essential for attaining high average power operation. Installation of an external 5 m diameter antenna dish is planned for a later stage.

A number of user applications are being planned for the initial stages of operation of the user facility as follows:

1. Spectroscopic/Imaging/Biological stations
* Sub-wavelength resolution microscopy /spectroscopy of H.T.S.C. materials.
* Imaging of biological tissues.
* Study of damage to biological tissues.
* Spectroscopic study of excitations and of nanometer structures in semiconductors

2. Material Processing Stations
* Sintering of ceramic layers on ceramic or metal surfaces.
* Melt texturing of H.T.S.C. (YB CO) thin films for improved crystalline quality.
* Surface treatment of metals for improvement of corrosion and wear resistance.

3. Atmospheric Station
* Remote sensing and mapping of clouds, aerosols, dust, gases.
* Study of wide band communication channels.
* Wireless energy transmission.

A number of users have already expressed interest in pursuing research on these applications in the user facility. Realization of these applications awaits completion of the FEL development and funding of the equipment for the user laboratories. Acknowledgment: This research is funded by grants of the Israeli Ministries of Science, Infrastructure and Defence.

Reference

1. L. Elias et al, Phys. Rev. Lett. <u>57</u>, 424 (1986)

Table 1: Parameters of the tandem FEL

Parameter	Present	Next
Tuning range	70–130GHz	50–130GHz
Peak intensity	10 kW	30 kW
Average power	-------	1 kW
Pulse duration	5 - 30 µS	5 – 1000 µS
Beam dimension	5 cm	5 mm
Spatial coherence	Diffraction limited	Diffraction limited
Temporal Coherence $\Delta f/f$	$< 10^{-5}$	$< 10^{-7}$

Fig. 1: User Center for FEL Radiation

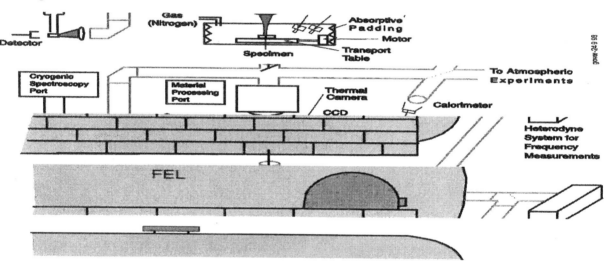

Fig. 2: Radiation User Facility

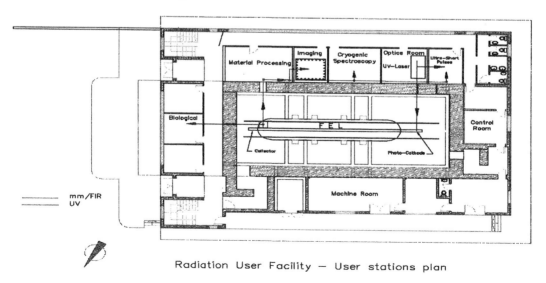

Radiation User Facility — User stations plan

FREE ELECTRON LASERS 2000
V.N. Litvinenko and Y. Wu (Eds.)
2001 Elsevier Science B.V.

Activation of Mg implanted into GaN using a Free Electron Laser

Guoqing Zhang[1,a], Yonggui Li[a], Junbiao Zhu[a], Mingkai Wang[a], Gang Wu[a], Xueping Yang[a], Zhijian Yang[b], Yuzhen Tong[b] and Guoyi Zhang[b]

[a]Free Electron Laser Laboratory, Institute of High Energy Physics,
Chinese Academy of Chinese, P. O. Box 2732-16, Beijing 100080, P. R. China
[b]Department of Physics, Mesoscopic Physics Laboratory,
Peking University, Beijing 100871, P. R. China

Abstract

In this paper, activation of implanted dopants in GaN using a Free Electron Laser(FEL) is reported. The FEL has two main characteristics: wavelength tuneability and ultrashort pulse operation($\sim$4ps) with intense peak power($\sim$MW). The FEL is then used to process the samples which are held at room temperature. The wavelength was selected to be $11.17\mu m$, which corresponds to the absorption peak of Mg doped GaN. Hall effect measurements showed an increase of carrier density for samples. Initial results suggest activation of the implanted dopants. FEL processing of Mg doped GaN appears to be an attractive option compared to furnace processing at high temperatures.

Key Words: Free Electron Laser, Semiconductor Compounds

PACS: 41.60.Cr, 71.20.Nr

1. Introduction

Gallium nitride(GaN) is a promising semiconductor material for high-temperature, high-frequency, and high-power devices due to its electrical, thermal, and mechanical properties. The interest in developing blue light-emitting diodes(LEDs) and laser diodes(LDs) has focused on GaN[1]. Recently, GaN-based blue LEDs and LDs have been fabricated from epilayers using metal-organic vapor phase epitaxy(MOVPE) and molecular beam epitaxy(MBE). Achievement and control of substantial activation of p-type dopants in GaN remains a critical issue vis à vis improved performance of devices fabricated in this materials. The most commonly used p-type dopant is magnesium(Mg) which substitutes on Ga sites. One-to-two orders of magnitude higher atomic concentration of Mg must be incorporated into GaN to achieve the desired hole concentration at room temperature[2]. This incorporation reduces the hole mobility due to the enhanced carrier-impurity scattering processes[3]. Therefore, in order to improve the commercial viability of GaN devices, activation of implanted dopants

in GaN is necessary. In the process technology for GaN devices, it is important to establish the annealing process after implanted dopants for removal of induced damage and activation of dopants. However, extremely high temperatures are required to activate dopants in GaN using conventional thermal annealing techniques. It is desirable to lower the annealing temperature for practical fabrication. There have been several attempts to remove the implantation damage and activate implanted dopants in semiconductor films at low temperature using ultraviolet pulsed lasers such as nitrogen lasers and excimer lasers[4,5]. It was found that molten semiconductor films regrows epitaxially on a substrate once the surface is melted by high-energy laser pulse. These annealing methods were based on the melt-recrystallization cycle using lasers with pulse widths greater than 1ns. The direct excitation of lattice vibration by an intense ultrashort laser pulse may induce the reconstruction of disordered atoms and activation of dopants even at room temperature. However, to our knowledge, there is no attempt to apply a laser tuned to $11.17\mu m$ which corresponds to the strong ab-

---

[1]Corresponding author. e-mail: zhanggq@lnat.ihepa.ac.cn

sorption peak of Mg doped GaN. The main reason is the lack of high-power tunable infrared lasers necessary for this process. The free electron laser (FEL) has two main characteristics: wavelength tuneability and ultrashort pulse operation($\sim$4ps) with intense peak power($\sim$MW). In this paper, we describe the results of applying a FEL to activate Mg dopants in GaN film. The experimental results show that FEL activation is effective for Mg doped GaN films. To our knowledge, this is the first attempt to activate Mg implanted into GaN using FEL.

2. Experimental results

At this writing, the substrates used in this experiment were α-Al$_2$O$_3$ (0001) oriented, double-polished wafers. Mg-containing GaN films were grown by conventional two-flow metal-organic chemical vapor deposition[6,7]. Trimethylgallium(TMG) and ammonia(NH$_3$) were used as reactants and hydrogen and nitrogen as carrier gases,magnesium as dopant. The growth temperatures for the GaN buffer and the single crystal epilayer were 550°C and 1050°C, respectively. During the MOCVD process, flow rates of the TMG, NH$_3$, H$_2$, N$_2$, and Mg were 20μmol/min, 3.5sl/min, 0.5sl/min, 0.5sl/min, and 0.3μmol/min, respectively. The thickness of sample, measurement by a cross-section microscopy, is about 1.5μm. The hole concentration of Mg doped GaN is 2.50×10^{14}. The hole mobility of the sample is 10.6cm^2/Vs.

The Fourier transform infrared(FTIR) grazing incidence reflectivity measurement was carried out at room temperature using a Nicolet infrared spectrometer (Magna-IR 750 Series II) and the grazing incidence angles were fixed at 80°. The resolution of the spectrometer was 4cm^{-1}. The analyzed area was selected to be about 5$\times$5mm^2. The present experiments were performed using Beijing Free Electron Laser(BFEL). Our BFEL delivered macropulses with a duration of 4μs at a repetition rate of 3.125Hz. Each of these macropulses consisted of ultrashort ($\sim$ps) micropulses at 350ps intervals with a maximum power of 9MW[8,9]. The wavelength of BFEL used in the experiment was 11.17μm which corresponds to the strong absorption peak of Mg doped GaN. After the FEL irradiation treatment, the hole concentration and the hole mobility of Mg doped GaN are 2.50×10^{15} and 20.4cm^2/Vs, respectively.

3. Conclusions

We have applied BFEL to activate Mg implanted into GaN at room temperature. We selected the wavelength which corresponds to the strong absorption peak of Mg doped GaN, i.e., 11.17μm. The measurement results indicate that activation of Mg doped GaN is possible using FEL at room temperature. The hole concentration of Mg doped GaN sample was increased ten times more than before irradiation. An investigation of photoluminescence(PL) measurements of Mg doped GaN will be published later. It should be pointed out that the concept of the present study can be adopted to a variety of materials processing by employing the wide range of tuneability in a FEL.

References

[1] S. Nakamura, M Senoh, N. Iwasa, and S. Nagahama, Appl. Phys. Lett. 67(1995)1868.

[2] U. Kaufmann, M. Kunzer, M. Maier, H. Obloh, A. Ramakrishan, B Santic, and P. Schlotter, Appl. Phys. 72(1998)1326.

[3] T. W. Weeks, Jr., M. D. Bremser, K. S. Ailey, E. P. Carlson, W. G. Perry, and R. F. Davis, Appl. Phys. Lett. 67(1995)401.

[4] S. Y. Chou, Y. Chang, K. H. Weiner, T. W. Sigmon, and J. D. Oarsons, Appl. Phys. Lett. 56(1990)530.

[5] S. Ahmed, C. J. Barbero, and T. W. Sigmon, Appl. Phys. Lett. 66(1995)712.

[6] J. Han, T-B Ng, R. M. Biefeld, M. H. Crawford, and D. M. Follstaedt, Appl. Phys. Lett. 71(1997)3114.

[7] H. Amano, N. Sawaki, I. Akasaki, and Y. Toyoda, Appl. Phys. Lett. 48(1986)353.

[8] Jialin Xie, Jiejia Zhuang et al., Nucl. Instr. and Meth. A358(1995)256.

[9] Jialin Xie, Jiejia Zhuang et al., Nucl. Instr. and Meth. A407(1998)146.

FREE ELECTRON LASERS 2000
V.N. Litvinenko and Y. Wu (Eds.)
2001 Elsevier Science B.V.

Thermal Ablation Studies at the S-DALINAC FEL: a 3-dimensional Approach

M. Brunken, H. Genz, H.-D. Gräf, S. Khodyachykh, H. Loos, A. Richter, B. Schweizer *, A. Stascheck

Institut für Kernphysik, Technische Universität Darmstadt, D-64289 Darmstadt, Germany

1. Motivation

Laser ablation of biological tissue has found growing fields of application during the last decade [1]. Among the physically different interaction mechanisms between laser radiation and tissue, the regime of thermal interaction enables efficient removal of tissue. The amount of collateral thermal damage to the surrounding tissue needs to be controlled carefully. FELs in the infrared can be tuned to absorption maxima of water molecules resulting in significantly smaller thermal damage zones in soft tissues. In the present contribution first experimental results observed at the S-DALINAC [2] are compared to a prediction of a simulation code described below.

2. Thermal ablation experiments at the S-DALINAC FEL

Samples of bovine cornea, bovine liver and human cartilage were irradiated at a wavelength of 7.0 μm, different average powers and different macropulse time structures. The FEL beam was focused down to a diameter of 140 μm. We achieved a thermal laser-tissue interaction. To determine the ablation efficiency and the collateral thermal damage the samples were investigated by light microscopy and scanning electron microscopy.

3. Numerical simulation of laser irradiation of tissue

For a precise control of thermal tissue damage, the knowledge of the temperature distribution inside the tissue during and shortly after laser irra-

diation is necessary. Several one-dimensional calculations of this effect have been published [3], but with a focus diameter of the order of the laser penetration depth, the assumptions for these models do not hold for our experiments. Analytical calculations in two or three dimensions can only be performed under very special initial and boundary conditions. To model temperature distributions that are in better agreement to the real physical system, a numerical code for the calculation of the temperature distribution in tissue during laser irradiation was developed. The code takes into account heat generation by laser absorption and heat diffusion inside the tissue. Heat conduction by radiation and convection were neglected. FEL micropulses were not modeled since a single micropulse can't generate an ablation effect. Thus, one has to solve

$$\rho \cdot c \cdot \frac{\partial}{\partial t} T(r, z, t) = \lambda \cdot \Delta T(r, z, t) + S(r, z, t), \quad (1)$$

where T, ρ, c and λ are the temperature, the density, the specific heat capacity and the thermal conductivity of the tissue, respectively (thermal parameters of water were assumed), and S is a heat-source term due to laser absorption inside the tissue. To exploit the symmetry of the problem, a cylindrical coordinate system was used. The temperature was calculated on a rectangular grid in the r-z-plane. The time scale was also divided into steps Δt. This leads to discrete expressions for the differential operators in eq. 1. Rearranging the equations for each knot leads to a set of linear equations describing the temperatures at time step $(n + 1) \cdot \Delta t$ as function of the temperatures at time step $n \cdot \Delta t$. By treating the heat diffusion in the r- and z-direction as

*Corresponding author e-mail:
schweizer@ikp.tu-darmstadt.de

two distinct processes, which take place one after the other in each time step (alternating direction implicit, ADI-method), it is possible to rearrange the matrices in tridiagonal form. The resulting matrix equations can be solved very efficiently.

4. Results

For the simulation, 250 knots in the r- and 250 in the z-direction were chosen, producing a grid of 62500 knots. A time period of 25 ms was simulated in 4000 time steps assuming a 10 ms FEL macropulse duration and a 15 ms cool-down phase. The full spatial and temporal temperature evolution has been calculated.

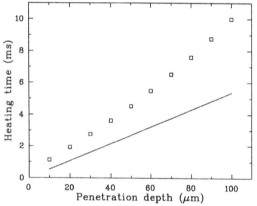

Figure 1. Heating times with (sqares) and without (solid line) thermal diffusion.

The time for a temperature increase of $100°C$ in the center of the laser focus was extracted from the calculation and compared to a tissue heating without diffusion. It turned out that, independent of the penetration depth d, with diffusion about twice the laser energy is needed to start thermal ablation than without diffusion (see fig. 1).

Even though vaporization is not included in the model, an estimate of the size of the thermal damage zones results from the calculation. The simulation run is stopped when the water vaporization temperature of $100°C$ is reached in the center of the laser focus. The tissue region with a temperature higher than $60°C$ in this time step represents the thermal damage zone. Biological cells exposed to this temperature are irreversibly damaged. Figure 2 shows the calculated damage zones for different laser penetration depths. The FEL

beam in our experiments had a penetration depth of approximately 18 μm, so it compares best with the $d = 10$ μm calculation. The strong variation in size of thermal damage zones calculated indicates that a proper choice of wavelength with a small penetration depth is crucial to obtain good control over thermal damage.

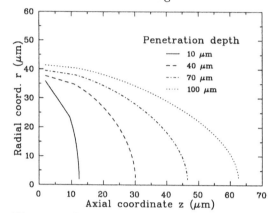

Figure 2. Calculated thermal damage zones.

5. Summary and Outlook

In this work we have presented a numerical model of heat diffusion in tissue under laser ablation. The influence of heat diffusion on the energy necessary to start thermal ablation was calculated. The size of thermal damage zones for different laser-penetration depths was computed and can be compared to the experiments performed at the Darmstadt FEL. Future refinements of the simulation code will include the vaporization process of tissue after it has been heated to $100°C$. Further thermal ablation experiments will be accomplished.

We gratefully acknowledge the financial support provided by the DFG Graduiertenkolleg "Physik und Technik von Beschleunigern".

REFERENCES

1. M. H. Niemz, Laser-Tissue Interaction, Springer-Verlag Berlin Heidelberg, 1996.
2. M. Brunken et. al., Status of the Darmstadt Free-Electron-Laser, NIM A (to be published).
3. A. Olmes et. al., Modelling of infrared soft-tissue photoablation process, Appl. Physics B 65, 659–666 (1997).

FREE ELECTRON LASERS 2000
V.N. Litvinenko and Y. Wu (Eds.)
2001 Elsevier Science B.V.

Second Harmonic Generation in CdTe at 22 μm generated by the JAERI FEL

N. Kikuzawa*, T. Yamauchi, R. Nagai, E.J. Minehara

Free Electron Laser Laboratory, Advanced Photon Research Center, Kansai Research Establishment, Japan Atomic Energy Research Institute (JAERI), Tokai-mura, Naka-gun, Ibaraki 319-1195 Japan

Abstract. The second harmonic generation (SHG) experiment of a sample of CdTe was carried out with the free-electron laser beam at Japan atomic energy research institute (JAERI). The SHG signal converted from the fundamental wavelength 22 μm in the CdTe plate was observed using a liquid-nitrogen cooled MCT detector. The conversion efficiency of SHG was experimentally measured to be 3×10^{-5} %/(MWcm^{-2}), which was almost equal to the theoretical value.

PACS: 41.60.Cr, 42.65.Ky
Key words: free-electron laser, second harmonic generation, CdTe crystal, conversion efficiency

1. Introduction

Techniques to measure ultrashort pulses are necessary to find some critical information to understand the high extraction efficiency of the free electron laser (FEL). A second harmonic generation (SHG) technique is often used to measure the ultrashort pulse width of femtosecond laser, whose instrument is called as an autocorrelator. Frequently-used crystals of the SHG for an autocorrelator are KDP (potassium dihydrogen phosphate) or BBO (beta-barium borate) crystal. However the autocorrelator is not commercially available for the FIR wavelength range, because it is very difficult to find SHG crystals in the far infrared (FIR) wavelength range. Since the study of SHG has been performed using Q-switched CO_2 laser with wavelength 10.6 μm and pulse width ~200 ns, it is understood that the cadmium telluride (CdTe) has high transmittance, high conversion efficiency and easy tunability [1].

As the JAERI FEL usually oscillates in the range of 16 to 28 μm, the development of a SHG autocorrelation system is required to measure the ultrashort pulse width of picosecond FEL in the FIR wavelength. First of all, the SHG experiment using a sample of CdTe crystal was carried out with the JAERI FEL on the condition of the wavelength 22 μm and the macropulse width 400 μs.

2. Experimental Setup

The CdTe crystal used here is the shape of the wedged angle 5°, its size 9 mm x 28 mm, and the maximum thickness 0.98 mm. The coherence length l_c of the CdTe is estimated to be 119 μm at the wavelength of 22 μm. The thickness at the center of the CdTe is 595 μm, which corresponds to the five coherence lengths. The electric field vector of FEL

beam was almost parallel to the [1 1 1] direction inside the CdTe crystal.

The FEL beam was out-coupled through a KRS-5 window by a scraper mirror inside an optical cavity. The wavelength was tuned from 16 μm to 28 μm by changing an undulator gap and the electron beam energy. The fundamental wavelength of 22 μm was fixed for the high power and the stable operation of the JAERI FEL. The output laser power was controlled by the position of the scraper mirror. The extracted FEL beam was transported from the accelerator room to the experimental room by mirrors.

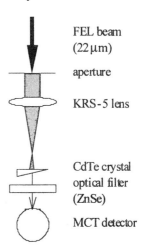

FEL beam
(22 μm)

aperture

KRS-5 lens

CdTe crystal

optical filter
(ZnSe)

MCT detector

Fig. 1. Experimental setup.

An experimental setup used in first experimental demonstration of SHG is shown in fig.1. The input laser power on the CdTe crystal was controlled by an aperture (0.28 cm^2) and by the KRS-5 plate. The macropulse laser power was 12 W, and then the micropulse power was estimated to be 240 kW. The CdTe crystal was placed behind 3 mm from focus point of a KRS-5 lens (f = 35 mm) to avoid the laser induced damage. Therefore, the laser intensity was ~120 MW/cm^2 on the CdTe crystal. The SH signal was measured by a liquid-nitrogen cooled MCT detector. The SHG output was separated from the unconverted fundamental by a few pieces of ZnSe plates which absorbed the fundamental signal. The conversion efficiency was calculated by the ratio of the SHG signal to fundamental signal.

3. Experimental Result

The optical path of the laser beam was set normal to the $(\bar{1}10)$ surface of the crystal and the linear polarization of the FEL beam was almost parallel to the [1 1 1] direction for generating the maximum SHG power. The polarization had the angles of 135° to the [1 1 0] direction.

After adjusting its angle for maximum SHG output, the absolute conversion efficiency was experimentally determined using the measured signal in fig.2. The ratio between the signal with the CdTe crystal and without it is 1.7 as shown in fig.2. The signal from the CdTe includes both signals of second harmonic and fundamental wavelength, which partially passes through the CdTe crystal. Therefore, the ratio of SHG signal to fundamental signal $S(2\omega)/S(\omega)$ is estimated to be 1.4. The conversion efficiency α is approximately given by [2]

$$\frac{S(2\omega)}{S(\omega)} = \frac{P\sqrt{T_1(\omega)T_1(2\omega)}\alpha T_2(2\omega)\eta(2\omega)}{PT_1(\omega)(1-\alpha)T_2(\omega)\eta(\omega)}$$

then $\alpha = 3\times10^{-3}$ %, where P is the input power into the CdTe crystal, T_1 the transmittance of the CdTe crystal, T_2 the transmittance of the ZnSe optical filter and η the detectivity of the MCT detector. As the micropulse power is estimated to be ~120 MW/cm^2, the conversion efficiency is rewritten as $\gamma \approx 3\times10^{-5}$ %/(MWcm^{-2}).

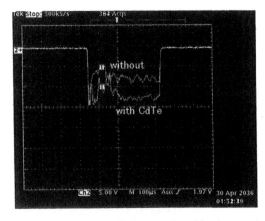

Fig. 2. SH signal with CdTe crystal and fundamental signal without CdTe.

4. Discussion

The second harmonic intensity is given by [3]

$$I(2\omega) = \frac{21.2d_{14}^{2}l_c^{2}I(\omega)^2}{n(\omega)^2 n(2\omega)\lambda(\omega)^2},$$

where $\lambda(\omega)$ is the fundamental wavelength; $n(\omega)$ and $n(2\omega)$ are the refractive index at the frequency ω and 2ω, respectively; l_c is the coherence length $\lambda(\omega)/4[n(2\omega)-n(\omega)]$; $I(\omega)$ is the fundamental laser intensity (W/cm^2) at the frequency ω; and d_{14} is the relevant effective nonlinear coefficient (cm/statV). The specific properties of CdTe are as follows $d_{14}^{2}/n(\omega)^2 n(2\omega)=0.35\times10^{-15}$ (cm^2/statV^2), $l_c = 119$ μm at wavelength 22 μm. First, as the angle between the [1 1 1] direction and the linearly polarized direction is 10°, the useful input power for SHG becomes 0.96. Secondly, the input power component of the [1 1 0] direction is 0.5 which was estimated from the angles 45° between the linear polarization and the [1 1 0] direction. Therefore, the main second harmonic conversion efficiency, which is parallel to the [1 1 1] direction, is calculated to be $\gamma =I(2\omega)/I(\omega)^2 \sim 2\times10^{-}$ %/(MWcm^{-2}), which nearly equals to the experimental value.

The CdTe plate was tested in a FIR autocorrelator with the noncollinear (background-free) SHG method. As the observed SHG signal was very low, the improvement of the system is tried, such as application of a new sample which is a tellurium (Te) crystal. The Te crystal is birefringent and allows phase-matched second harmonic generation with high conversion efficiency [4]. The type of the FIR autocorrelator becomes collinear.

5. Summary

The second harmonic signal generated from input wavelength 22 μm of FEL was observed using CdTe crystal. The conversion efficiency was experimentally obtained to be ~3×10^{-5} %/(MWcm^{-2}) which was almost equal to the theoretical value.

As the JAERI FEL oscillates in the wavelength range of 16 to 28 μm, the CdTe plate will be used for the FIR autocorrelation system.

References

[1] C.K.N.Patel, Phys. Rev. Lett., **16**, 613 (1966).
[2] T. Yamauchi, et al., to be published in Jpn. J. Appl. Phys. (2000).
[3] M.S.Piltch, C.D.Cantrell and R.C.Sze, J.Appl. Phys. **47**,.3514 (1976).
[4] C.K.N.Patel, Phys. Rev. Lett., **15**, 1027 (1965).

Author index

661

665

Printed and bound by CPI Group (UK) Ltd, Croydon, CR0 4YY

08/05/2025

01864931-0003